HOMOGENEOUS CATALYSIS FOR UNREACTIVE BOND ACTIVATION

HOMOGENEOUS CATALYSIS FOR UNREACTIVE BOND ACTIVATION

Edited by

ZHANG-JIE SHI
Professor of Chemistry
College of Chemistry and Molecular Engineering
Peking University
Beijing, China

WILEY

Cover Image: iStockphoto © saicle
Cover Design: Wiley

Published by John Wiley & Sons, Inc., Hoboken, New Jersey.
Published simultaneously in Canada.

Library of Congress Cataloging-in-Publication Data:

Homogeneous catalysis for unreactive bond activation / edited by Dr. Zhang-Jie Shi, professor of chemistry, College of Chemistry and Molecular Engineering, Peking University, Beijing, China.
pages cm
Includes bibliographical references and index.
ISBN 978-1-118-45223-3 (hardback)
1. Homogeneous catalysis. 2. Carbon, Activated. 3. Chemical bonds. I. Shi, Zhangjie, 1974- editor.
QD505.H648 2014
541′.395–dc23

2014008375

Printed in the United States of America.

10 9 8 7 6 5 4 3 2 1

CONTENTS

PREFACE

Transition-metal-catalyzed C–X functionalization is the most important, challenging, and rapidly-evolving research field. With continuous success in developing direct activations and straightforward transformations of inactivated chemical bonds in the past several decades, organic synthesis has been evolving by minimizing the preactivation of substrates, alleviating the waste production, and shortening the synthetic steps. Homogeneous catalysis, especially transition-metal-catalyzed transformations in these fields, exhibits its predominant features, making unpredicted and unprecedented transformations possible, apart from the traditional organic transformations. As a result of the recent achievements in the field of homogeneous catalysis of inactivated chemical bonds, we feel there is great value in summarizing these recent advances in such an important and expanding field so that a wider audience from both laboratory and industry would be able to appreciate the significant contributions made by numerous preeminent chemists. By focusing the spotlights on homogeneous transition-metal catalysis in the activation of inactivated chemical bonds, seven chapters have been gathered together in this book, with regard to the different types of chemical bonds, including C–Cl, C–F, C–N, C–S, C–O, C–H, and C–C bonds. This undertaking was initiated by contacting distinguished Chinese chemists who made significant contributions in related fields and asking them to contribute chapters to this book.

In the first part, homogeneous catalysis in the activation and transformation of C–Cl and C–F is summarized. Apart from C–Br and C–I, organochlorides show their great advantages with easier availability, lower cost, and higher atom economy. Different ligand sets and catalyst systems that have been developed in the past several decades have proved to be the most useful way to promote the efficiency. The progress has shown their importance in both academy and industry. Comparably, the hotspot of C–F activation shows their greater importance in academy than in industry due to the availability of the starting materials. The basic activation modes of C–F bonds *via* transition-metal catalysis are highlighted, mostly focused on various reactions of cross-coupling, nucleophilic substitution, and dehydrofluorination and their intensive mechanisms.

The second part is focused on the activation of carbon–heteroatom bonds, including C–N, C–S, and C–O bonds, which broadly exist in natural and synthetic molecules. Transition-metal-catalyzed C–N bond activation, mainly focused on the oxidative addition pathway, is discussed in detail with the demonstration of its importance and challenge. The understanding of the catalytic pathway is also discussed computationally and experimentally. Carbon–sulfur bonds widely exist in natural products, pesticides, and drugs, and their activation, cleavage, and transformation becomes more and more important in organic chemistry. In Chapter 5, carbon–sulfur bonds linking to both heteroaryl and aryl are activated, covering palladium and/or copper catalysis and even metal-free processes.

In comparison, oxygen-contained molecules are more abundant. After a brief introduction of the importance and challenge in the "inert" C–O bond activation as well as the pioneering stoichiometric reactions, transition-metal-catalyzed C–O activation and following transformations are thoroughly discussed from the aspects of the reaction scopes, experimental and computational efforts to investigate the mechanisms, and the existing applications.

The C–H bonds are ubiquitous and are the most fundamental groups in organic chemistry. The discrimination, activation, and chemical transformation of the particular C–H bond among numerous C–H bonds in the starting materials, products, and even solvents is notorious. Therefore direct selective C–H functionalization presents tremendous intellectual challenge and almost invaluable rewards in synthetic chemistry, thus regarded as the "Holy Grail of Chemistry." In the past half century chemists have never ceased their steps in pursuit of their magnificent goals in this field and have made remarkable progress. With the understanding of feasible pathways to cleave the C–H bonds with transition-metal complexes through different strategies, in Chapter 6 different transformations of various types of C–H bonds are discussed, ranging from fundamental theoretical and experimental studies to the emergence of systems of immediate applicability.

Different from the activation of carbon–heteroatom bonds, carbon–carbon bond cleavage induces the reformation and reorganization of the skeleton of organic molecules, thus stimulating the new tactics of organic synthesis. In Chapter 7 the development of homogeneous transition-metal-catalyzed C–C bond cleavage reactions is outlined, sorted by the types of substrates. Various transformations are consecutively discussed from strained molecules to unstrained molecules, emphasizing the mechanisms, scopes, and limitations.

The editor and contributing authors of this book sincerely hope that it will be a valuable collection of the intellectual discovery in this charming and challenging research field. Due to the work abundance of exciting has been done in this field, we can only cover a fraction of it from the references. We hope that both the contributors and our readers forgive our less informative, but also less unwieldly topics.

I would like to thank the contributors for their chapter submissions done so with great dedication. The support and constructive advice given by the reviewers and Jonathan T. Rose in Wiley's editorial department throughout the publication process were greatly appreciated.

ZHANG-JIE SHI

Department of Chemistry and Molecular Engineering
Peking University
Beijing, China

CONTRIBUTORS

WANG-JUN GUO, Department of Chemistry, University of Science and Technology of China, Hefei, Anhui, People's Republic of China

BI-JIE LI, College of Chemistry and Molecular Engineering, Peking University, Beijing, China

HU LI, College of Chemistry and Molecular Engineering, Peking University, Beijing, China

SHUANG LUO, College of Chemistry and Molecular Engineering, Peking University, Beijing, China

ZHANG-JIE SHI, College of Chemistry and Molecular Engineering, Peking Uniersity, Beijing, China

XIAO-BING WAN, College of Chemistry, Chemical Engineering and Materials Science, Soochow University, Suzhou, China

ZHONG-XIA WANG, Department of Chemistry, University of Science and Technology of China, Hefei, Anhui, People's Republic of China

SHANG-DONG YANG, State Key Laboratory of Applied Organic Chemistry, Lanzhou University, Lanzhou, China

Da-Gang Yu, College of Chemistry and Molecular Engineering, Peking University, Beijing, China

Zhengkun Yu, Dalian Institute of Chemical Physics, Chinese Academy of Sciences, Dalian, Liaoning, People's Republic of China

Fei Zhao, College of Chemistry and Molecular Engineering, Peking University, Beijing, China

CHAPTER 1

CATALYSIS IN C–Cl ACTIVATION

ZHONG-XIA WANG and WANG-JUN GUO
Department of Chemistry, University of Science and Technology of China, Hefei, Anhui, People's Republic of China

Homogeneous Catalysis for Unreactive Bond Activation, First Edition. Edited by Zhang-Jie Shi.

1.1 INTRODUCTION

Over the past decades, transition-metal-catalyzed activation and transformation of carbon–halide bonds have achieved great progress. The cross-coupling reactions based on carbon–halide bond activation play an important role in the synthesis of many drugs, natural products, optical devices, and industrially important starting materials [1–3]. Until the late 1990s, the coupling counterparts were predominantly iodides and bromides. Organic chlorides were rarely used due to their low reactivity [4, 5]. However, organic chlorides are cheaper and more widely available compounds. Their use as electrophilic partners would economically benefit a number of industrial processes [5, 6]. On the other hand, organic chlorides such as polychlorobiphenyls and chlorophenols are environmental pollutants. Dechlorination and detoxification of hazardous organic chlorides are also greatly concerned [7, 8]. In view of the above-mentioned reasons, activation and transformation of C–Cl bonds of organic chlorides attracted extensive attention in the past 10–15 years. A range of effective and efficient catalyst systems have been developed. The construction of new C–C and C–heteroatom bonds through catalytic activation of C–Cl bonds has been carried out [9–17]. In this chapter, we summarize the main advances of transition-metal-catalyzed activation and transformation of C–Cl bonds of organic chlorides.

1.2 REDUCTIVE DECHLORINATION

Organic chlorides are manufactured on a large scale and used widely in a variety of chemical industries. Organic chlorides are also often environmental pollutants, such as chlorofluorocarbons (CFCs), 1,1,1-trichloro-2,2-bis(4-chlorophenyl)ethane (DDT), chlorinated dibenzo-*p*-dioxins (dioxins), polychlorinated biphenyls (PCBs), and chlorophenols. Hence dechlorination of organic chlorides is of great interest in detoxification of hazardous chlorinated organic compounds and in the development of synthetic methodologies.

A range of methods have been developed for dechlorination of organic chlorides, including dechlorinations via photochemical reaction [18], microbial biodegradation [19, 20], oxidative dehalogenation or degradation [21, 22], and reductive dechlorination. In this section, we focus on transition-metal-catalyzed reductive dechlorination of organic chlorides. The cobalamin-mediated reductive dehalogenation is not included. An excellent perspective about this topic has been published recently [23].

1.2.1 H_2 as Reductant

The hydrogenolysis of C–Cl bonds of both alkyl and aryl chlorides can be performed in the presence of transition metal catalysts, mainly Ni, Pd, Rh, and Ru. This topic was reviewed previously [7, 24, 25].

Pd/C is a widely employed catalyst for the reductive dechlorination by H_2. Sajiki et al. [26] reported a mild and efficient one-pot hydrodechlorination of aromatic chlorides using a Pd/C–Et_3N system as the catalyst. When 1 mol% Pd/C (10%) and

1.2 equiv of Et_3N are employed, the dechlorination reaction of 4-chlorobiphenyl in MeOH can proceed at room temperature under 1 atm H_2 and gives 100% conversion of the chloride. The reaction can be applied to wide aromatic substrates and tolerates a range of functional groups except nitro and furyl groups, which are hydrogenated simultaneously. Studies also show that lipophilic bases such as Me_2NH, Me_3N, $i\mathrm{Pr}_2\mathrm{NEt}$, $i\mathrm{Pr}_3\mathrm{N}$, DBU, $PhNH_2$, and $PhNEt_2$ greatly enhanced the efficiency of the reaction compared with the less lipophilic NH_3 or ethylendiamine, whereas the aromatized heterocyclic bases such as pyridine or quinoline strongly suppressed the hydrodechlorination [27]. In the presence of a quaternary onium salt (Aliquat 336) as the phase-transfer agent, 50% aqueous KOH also leads to a good result in a multiphase system consisting of a hydrocarbon solvent (isooctane). A synergistic activation effect was observed for KOH and the phase-transfer catalyst [28]. A similar multiphase system was used for a Pd/C-, Pt/C-, or Raney-Ni-catalyzed dechlorination of γ-hexachlorocyclohexane (lindane) under normal H_2 pressure, generating benzene as the final product. In the presence of KOH and Aliquat 336, the reaction was shown to proceed via the consecutive dehydrochlorination and hydrodechlorination reaction stages, which are also co-promoted by Aliquat 336 and aqueous KOH as mentioned above. In the absence of a base, the reaction proceeds through a removal of a pair of chlorines from lindane at every reaction step by zerovalent metal followed by reduction of metal with hydrogen [29].

H_2O or an 80% H_2O–EtOH mixture is shown to be a suitable solvent for the Pd/C–catalyzed hydrodechlorination of aryl chlorides. The reaction proceeds at room temperature under normal H_2 pressure, leading to complete dechlorination in a short reaction time. The catalyst system tolerates functional groups such as F, CF_3, OH, and C(O)Ph [30]. Nan and co-workers [31] carried out the Pd/C–catalyzed regioselective dechlorination of 2,4-dichloropyrimidines at room temperature and normal H_2 pressure, forming 2-chloropyrimidines in excellent yields [Eq. (1.1)].

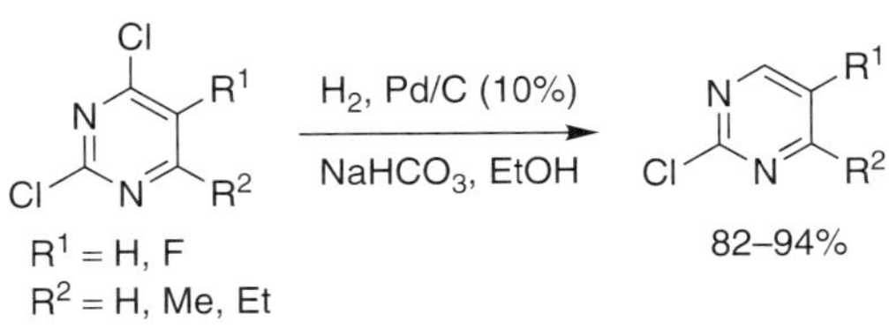

(1.1)

Dehalogenation of aryl or benzyl bromides and chlorides can be performed using Pd/AlO(OH) and a hydrogen balloon under solvent-free conditions. The reaction proceeds at room temperature and gives excellent yields in a short time. The reaction is compatible with a cyano group, but it converts a nitro group into an amino group and converts the carbonyl group of an aldehyde or ketone into a hydroxy group [32]. Complete hydrodechlorination of DDT and its derivatives 1,1-dichloro-2,2-bis(4-chlorophenyl)ethene (DDE) and 1,1-dichloro-2,2-bis(4-chlorophenyl)ethane (DDD) can be achieved using a hydroxyapatite-supported Pd nanoparticle catalyst (Pd^0HAP) using H_2 (10 atm) as the reducing agent [Eq. (1.2)]. The Pd^0HAP catalyst shows good reusability without significant loss of activity [33].

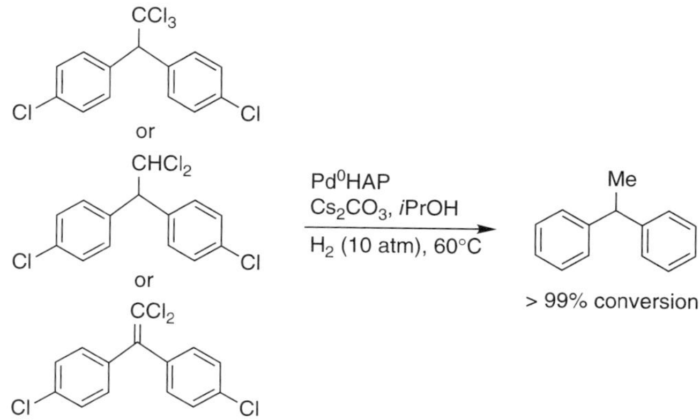

(1.2)

A $Pd(PPh_3)_4$-catalyzed selective dechlorination of 2,3-dichloronitrobenzene can be achieved under normal H_2 pressure, forming 3-chloronitrobenzene in over 90% selectivity. 3-Chloronitrobenzene can be further transformed into nitrobenzene, 3-chloroaniline, or aniline, depending on the catalyst concentration [34].

Esteruelas and co-workers [35] carried out rhodium-nanoparticle-catalyzed hydrodechlorination of aryl chlorides. The rhodiumnanoparticles are formed from bis(imino)pyridinerhodium(I) complexes **1** under a hydrogen atmosphere and in the presence of KO*t*Bu. Under a constant atmospheric pressure of hydrogen, the nanoparticles catalyze the dehalogenation of chlorobenzene, 1,2-, 1,3-, and 1,4-dichlorobenzene, 1,2,4-trichlorobenzene, fluorobenzene, 2-, 3-, and 4-chlorobiphenyl, and 4,4′- and 3,5-dichlorobiphenyl. The best catalyst precursor is the complex with Ar = p-$CF_3C_6H_4$. With this complex, the quantitative dehalogenation of chlorobenzene occurs after 15 min, whereas those of 1,2- and 1,3-dichlorobenzene take place after 30 min. This process is faster than those of Rh-catalyzed hydrodechlorination reported earlier [7]. This catalytic system also leads to the hydrogenation of benzene, toluene, *p*-xylene, styrene, α-methylstyrene, biphenyl, aniline, phenol, and pyridine. A Hg(0) poisoning test reveals that homogeneous and heterogeneous catalysis coexist during the dehalogenation reactions, whereas the hydrogenation processes are heterogeneous [35].

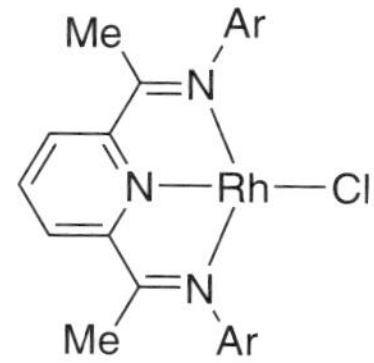

1

Ar = Ph, o-MeC_6H_4, o-EtC_6H_4,o-$i PrC_6H_4$, o-$t BuC_6H_4$, p-MeC_6H_4, p-$MeOC_6H_4$, p-$CF_3C_6H_4$, 2,6-$Me_2C_6H_3$, 2,6-$Et_2C_6H_3$, 2,6-$i Pr_2C_6H_3$

Roucoux and co-workers [36] performed dechlorination–hydrogenation of mono- and multichlorinated anisoles using rhodium nanoparticles in suspension or on silica support as the catalyst and H_2 as a reductant, giving a mixture of methoxycyclohexane and cyclohexanone.

Some supported nickel, palladium, platinum, and ruthenium also catalyze gas-phase dechlorination of chloroarenes or chlorofluorocarbon using H_2 at high temperature. These reactions are beyond the scope of this section and not discussed here.

1.2.2 ROH or ROM as Reductant

Recently a Ni(I)-NHC complex **2** was proved to catalyze dehalogenation of 1-bromo-4-fluorobenzene and 1-chloro-4-fluorobenzene at room temperature. However, the dechlorination reaction is incomplete even after 48 h, giving fluorobenzene in moderate yield [37]. A Ni(0)–NHC system is more effective. The combination of $Ni(acac)_2$, IMes·HCl, and Na*Oi*Pr efficiently catalyzes dehalogenation of functionalized aryl chlorides, bromides, iodides, and polyhalogenated hydrocarbons in refluxed THF. The catalyst system is compatible with functional groups such as OH, CN, NH_2, CF_3, MeS, and MeO groups and nitrogen-containing heterocycles. Deuterium incorporation experiments confirmed that the hydrogen atom is introduced from sodium isopropoxide [38]. This catalytic system is very similar to the $Pd(dba)_2$/SIMes·HCl/KOMe system reported earlier [39]. In both systems, alkoxide attack at the metal center followed by reductive elimination of the arene from the metal(II)–hydride complex is suggested as a likely pathway.

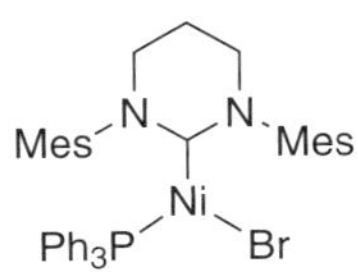

2

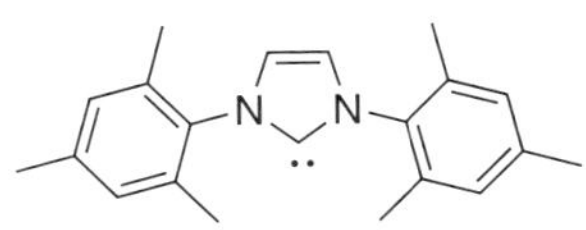

IMes

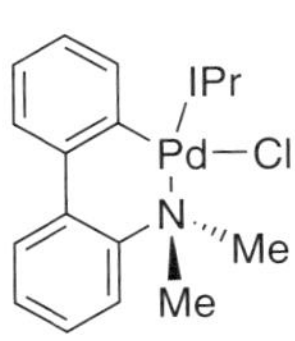

3

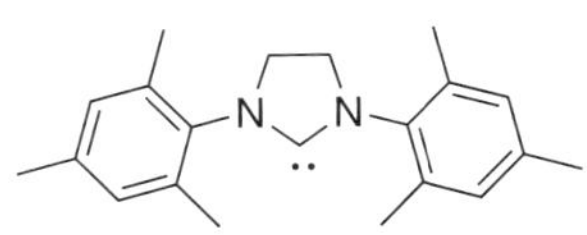

SIMes

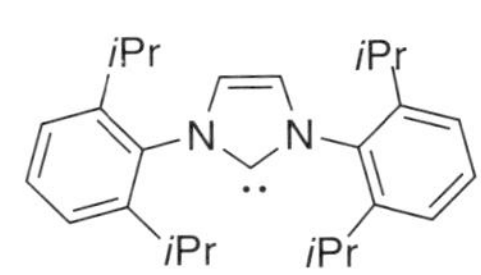

IPr

The NHC-stabilized dinuclear palladium complex $[Pd(\mu\text{-Cl})Cl(IPr)]_2$ is highly active in catalyzing hydrodehalogenation of polychlorinated phenyls and biphenyls. The reaction proceeds at 80°C with isopropanol as the hydrogen source and NaOH as base at very low catalyst loadings, resulting in benzene or biphenyl in excellent yields [40]. Both (IPr)Pd(allyl)Cl and **3** efficiently catalyze hydrodechlorination of aryl chlorides. The former catalyzes dechlorination reaction in *i*PrOH under either thermal heating or microwave radiation conditions. The latter displays higher catalytic activity. It efficiently catalyzes a dechlorination reaction of aryl chlorides in *i*PrOH at room temperature in the presence of NaO*t*Bu [41, 42]. The combination of palladium salts and phosphine or phosphite is also an effective catalyst for the dehalogenation of organic halides. For example, a $Pd(OAc)_2/PPh_3/K_2CO_3$ system catalyzes dehalogenation of aryl halides and α-haloketones with an alcohol as the hydrogen donor [43]. A $Pd_2(dba)_3/(2,4\text{-}t\text{BuC}_6H_3O)_3P/NaOt\text{Bu}$ system catalyzes dechlorination of aryl chlorides in *i*PrOH at 80°C [44]. Guram and co-workers [45] carried out hydrodechlorination of a number of functionalized aryl chlorides using a $Pd(dba)_2/P(Ar)R_2/K_2CO_3$ catalyst system in *i*PrOH at 80°C [$P(Ar)R_2$ = **4–8**]. The catalyst system is compatible with MeC(O), PhC(O), NO_2, NH_2, MeO, and ethenyl groups in the aryl chlorides, but the presence of ester and aldehyde groups leads to side reactions. On the other hand, they performed catalytic oxidation of sterically hindered aliphatic alcohols and benzylic alcohols using a $Pd(dba)_2$/**4**/base system and PhCl as the oxidant in toluene, giving ketones or aldehydes in excellent yields. The bases can be K_2CO_3, K_3PO_4, or NaO*t*Bu, depending on the substrates [45].

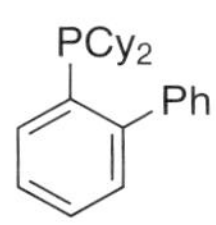

4

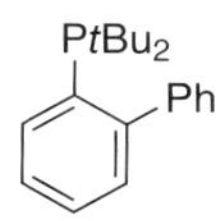

5

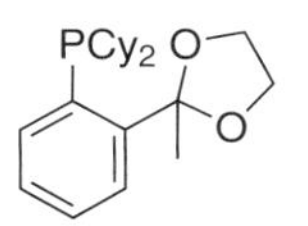

6

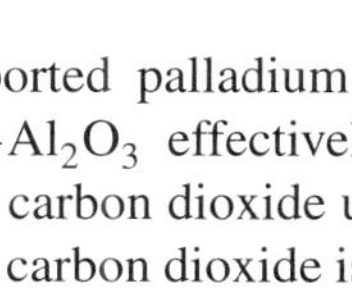

7

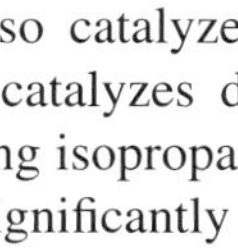

8

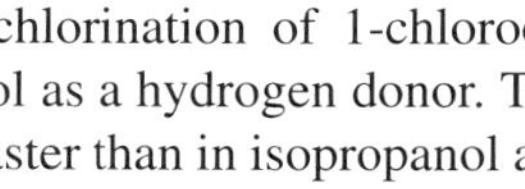

9

The supported palladium also catalyzes dechlorination of alkyl or aryl chlorides. $Pd/\gamma\text{-}Al_2O_3$ effectively catalyzes dechlorination of 1-chlorooctadecane in supercritical carbon dioxide using isopropanol as a hydrogen donor. The reaction in supercritical carbon dioxide is significantly faster than in isopropanol at atmospheric pressure [46]. A Pd–carbon nanotube drives hydrodehalogenation of aryl chlorides and bromides in the isopropanol (for chlorides) or cyclohexanol (for bromides) at a low Pd content (2.3%) and in the absence of any ligands [47].

$RuHCl(H_2)_2(PCy_3)_2$ and $RuH_2(H_2)_2(PCy_3)_2$ catalyze dechlorination of aryl chlorides in alcohols rapidly and completely. The catalytic systems are tolerant of a variety of functional groups and are efficient in dechlorinating multichlorinated arenes. The mechanism involves a transfer hydrogenation step with participation of the alcohol. The catalytic species may be generated *in situ* from the air-stable precursor $[RuCl_2(COD)]_n$ and PCy_3 [48]. Cp*Rh complexes also exhibit remarkable catalytic activity for hydrodechlorination of aryl chlorides in refluxed 2-butanol with high tolerance toward a variety of functional groups. $[Cp^*RhCl_2]_2$ and $Cp^*Rh(OAc)_2{\cdot}H_2O$ show high catalytic activity and KOH, Cs_2CO_3, and Cy_2NMe are highly effective bases [49].

1.2.3 Hydrosilanes as Reductant

Group 8–10 metals can catalyze dechlorination of aryl, alkenyl, or alkyl chlorides using hydrosilanes as the reducing agents. Oliván and co-workers [50] studied groups 8 and 9 metal-catalyzed dechlorination of 1,2,4-trichlorobenzene using $HSiEt_3$ as a reducing agent. A range of metal complexes, including $FeCl_2(PPh_3)_2$, $RuHCl(PPh_3)_3$, $RuH_2(CO)(PPh_3)_3$, $RuHCl(CO)(AsPh_3)_3$, $RuHCl(CO)(P\textit{i}Pr_3)_2$, Os $HCl(CO)(P\textit{i}Pr_3)_2$, $OsH_2Cl_2(P\textit{i}Pr_3)_2$, $CoCl(PPh_3)_3$, $RhH_2Cl(P\textit{i}Pr_3)_2$, $IrH_2Cl(P\textit{i}Pr_3)_2$, $IrCl(PPh_3)_3$, and $IrH_2(SiEt_3)(COD)(PCy_3)$, were screened. The 3*d* metal complexes and $IrH_2(SiEt_3)(COD)(PCy_3)$ were proven to be unactive. The osmium and other iridium derivatives are less effective than the derivatives of ruthenium and rhodium and undergo deactivation. The hydrogenolysis is sequential and selective according to the sequence 1,2,4-trichlorobenzene > dichlorobenzene > chlorobenzene > benzene [50]. Similar results were observed by using a catalytic reductive system consisting of $[Rh(\mu\text{-}Cl)(COE)_2]_2$, PPh_3, and $HSiEt_3$ [51]. The dehalogenation reaction of aryl chlorides, bromides, and iodides can also be carried out in an ionic liquid, IL-$OPPh_2$, using $PdCl_2$ as the catalyst and $HSiEt_3$ as reducing agent. This reaction is competitive with the silylation of aryl halides. In the presence of a base, Cs_2CO_3, the reaction of aryl halides with $HSiEt_3$ under the conditions similar to that of dehalogenation results in aryltriethylsilanes in good yields (Scheme 1.1) [52]. This type of competitive reaction was also observed in a Rh-catalyzed reaction of PhCl with $HSiEt_3$. In the absence of a base, reaction of PhCl with $HSiEt_3$ catalyzed by $(PyInd)Rh(C_2H_4)_2$ [PyInd = 2-(2′-pyridyl)indolide] at 80°C gives benzene and Et_3SiCl. In the presence of $LiN\textit{i}Pr_2$, catalytic C–Si coupling was observed, to produce Et_3SiPh [53].

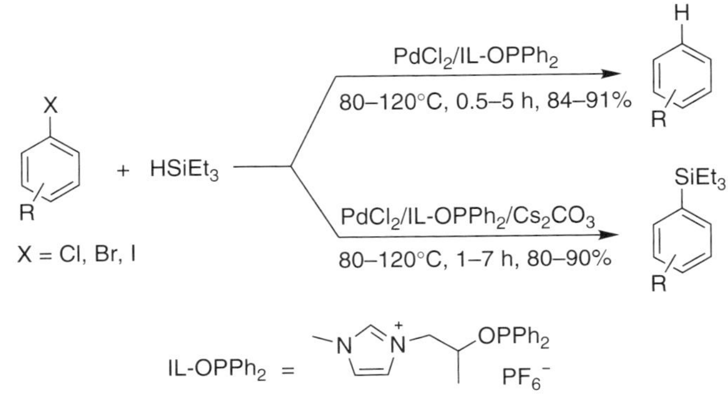

Scheme 1.1. Pd-catalyzed dehalogenation and silylation of aryl halides.

Polymethylhydrosiloxane can be used as a mild reductant in Pd-catalyzed dechlorination of aryl chlorides. Catalytic amounts of $Pd(OAc)_2$ in combination with polymethylhydrosiloxane and aqueous KF lead to rapid dechlorination of electron-neutral, -rich, or -poor chloroarenes at room temperature. Ketones, amides, esters, nitriles, ethers, borate esters, and amines are tolerated by the reaction system, whereas phenols and carboxylic acids are not [54].

Catalytic dehalogenation of fluorinated and chlorinated ethylenes can be performed using $(PPh_3)_3RhCl$ as the catalyst and Et_3SiH as the reducing agent. This reaction has an intermolecular preference for C–F bond activation versus C–Cl bond activation and has an intramolecular preference for C–Cl bond activation versus C–F bond activation. Specifically, dehalogenation of chlorofluorethylenes resulted in the observation of only vinyl fluoride rather than vinyl chloride. A proposed mechanism for the dehalogenation reaction is shown in Figure 1.1 based on the studies of a hydride complex intermediate, the stereochemistry of elimination, and kinetics. In this catalytic process the dominant pathway is proposed to involve rhodium(I), and spectroscopic studies indicate that rhodium hydride species are important intermediates [55, 56].

The reductive dechlorination of alkyl chlorides using hydrosilanes through the catalysis of groups 8–10 metal complexes has been reported. Pincer nickel complex **10** efficiently catalyzes dehalogenation of alkyl chlorides, bromides, and iodides in the presence of Ph_2SiH_2 and Na*Oi*Pr. Two chloride substrates, tetradecyl chloride and 1-chloroadamantane, and a range of bromides and iodides were tested. The yields are between 73% and 99%. $Me(EtO)_2SiH$ is also a good hydride source, but only in conjunction with NaOMe as base and THP as solvent. The reaction appears to proceed through a radical mechanism. The proposed catalytic cycle is shown in Figure 1.2. All the nickel-containing species in the cycle—that is, the nickel chloride, hydride,

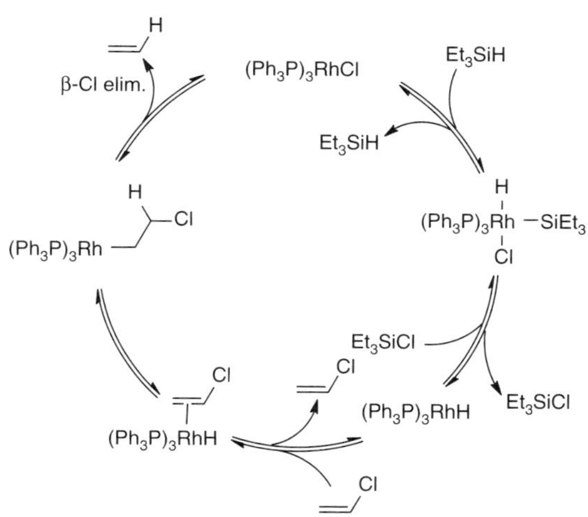

Figure 1.1. Proposed mechanism for the catalytic dehalogenation of chlorinated and fluorinated ethylenes using $(PPh_3)_3RhCl$ and Et_3SiH.

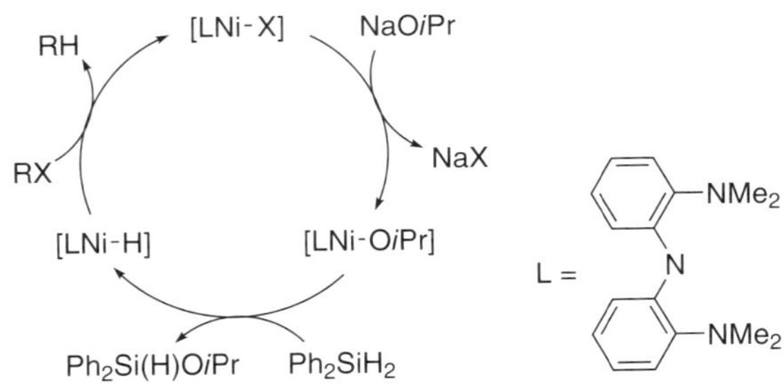

Figure 1.2. Proposed catalytic cycle for hydrodehalogenation reactions of organic halides.

and alkoxide complexes—have been isolated and characterized, and their reactivity is consistent with the proposed catalytic cycle [57].

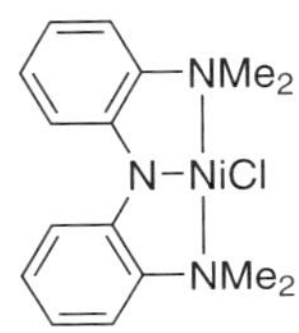

10

A cationic pincer Ir(III) hydride complex, **11**, was found to be a versatile and highly active catalyst for reduction of a broad spectrum of alkyl halides by Et_3SiH. The reaction is carried out either in chlorobenzene or in neat alkyl halide with low catalyst loadings. Primary, secondary, and tertiary alkyl chlorides can be dechlorinated in excellent yields. The mechanistic studies reveal a unique catalytic cycle. The cationic iridium hydride activates the silane through complexation. The resulting complex acts as a silylating reagent to lead to formation of a silyl-substituted halonium ion through transformation of "Et_3Si^+" to the halide. Then the bridged halonium ion is reduced by the nucleophilic iridium dihydride formed following silyl transfer, and the cationic iridium hydride complex is thus regenerated (Figure 1.3) [58, 59].

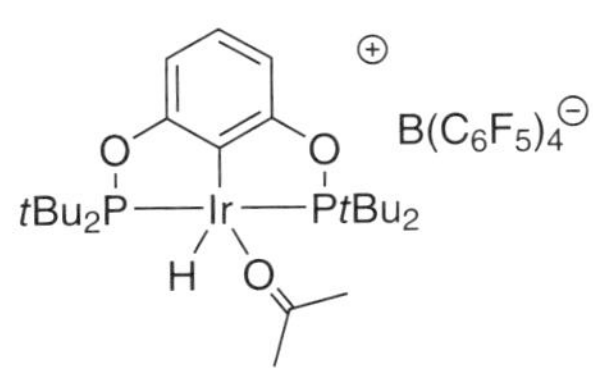

11

Rhodium and ruthenium complexes were found to catalyze dechlorination of hexachlorocyclohexanes. $RhCl(PPh_3)_3$ and $RhH_2Cl(PiPr_3)_2$ catalyze the dechlorination of γ-hexachlorocyclohexane to benzene at 70°C in *p*-xylene.

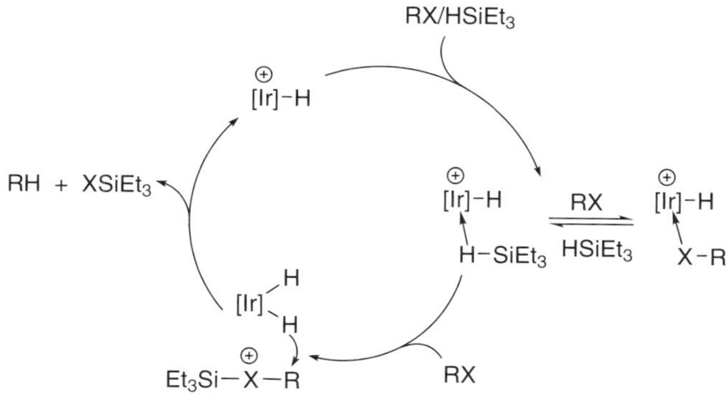

Figure 1.3. Proposed catalytic cycle for iridium-catalyzed reduction of RX by Et_3SiH.

$RhCl(PPh_3)_3$ is more active than $RhH_2Cl(P\textit{i}Pr_3)_2$, the former driving fully dehalogenation of γ-hexachlorocyclohexane to benzene in 31 min. Ruthenium complexes $RuHCl(PPh_3)_3$, $RuH_2Cl_2(P\textit{i}Pr_3)_2$, and $RuHCl(\eta^2\text{-}H_2)(P\textit{i}Pr_3)_2$ also catalyze the dehalogenation of γ-hexachlorocyclohexane, but they show lower catalytic activity than the rhodium complexes and lead to a mixture of cyclohexene and cyclohexane. $RhCl(PPh_3)_3$, $RhH_2Cl(P\textit{i}Pr_3)_2$, and $RuHCl(PPh_3)_3$ also catalyze dehalogenations of α- and δ-hexachlorocyclohexanes. The rate of the dehalogenation decreases in the sequence γ- > α-> δ-hexachlorocyclohexane [60].

In addition, both $PdCl_2/Et_3SiH$ and $[Ph_2PCH_2CH_2NMe_2]Pt(Me)Cl/HSiMe_2Ph$ systems are effective in reductive dechlorination of alkyl chlorides [61, 62].

1.2.4 Formic Acid or Its Salts as Reductant

Palladium catalysts are predominant for the dechlorination of chlorinated organic compounds using formic acid or its salts as the reductants. Palladium on carbon catalyzes dehalogenation of aromatic chlorocarbons using sodium formate as the reducing agent in water at room temperature. At 100°C in the presence of excess HCOONa the catalytic system drives further reduction of the aromatic rings or the functional groups attached on the aromatic rings, depending on the structure of the reactants [63]. A palladized foam nickel (Pd/Ni) catalyst efficiently transfers 4-chlorophenol into phenol in water using formic acid as the reductant [64], whereas the chitosan-supported palladium is effective in catalyzing dechlorination of 2-chlorophenol in the presence of sodium formate [65].

The combination of $Pd(OAc)_2$, 2-(di-*tert*-butylphosphino)biphenyl, and sodium formate in MeOH is an effective homogeneous catalytic system for dechlorination of aryl chlorides. The reaction gives excellent yields and is compatible with functional groups such as MeC(O)NH, COOMe, C(O)Me, and MeO. It was shown that sodium formate, rather than methanol, is the hydride source in hydrodechlorination through control experiments [66]. The combination of $Pd(PhCN)_2Cl_2$, 1,10-bis(diphenylphosphino)ferrocene (dppf), and sodium formate in DMA efficiently catalyzes dechlorination of substituted aryl chlorides at 75°C. It was found that the substrates with electron-donating groups are more reactive than those with electron-withdrawing groups. Based on the *in situ* IR experiments, the reaction is proposed to proceed via decarboxylation and reductive elimination, and the decarboxylation is the key step [67]. A $RhCl(PPh_3)_3$-catalyzed dechlorination of polychloroarenes was also carried out in the presence of sodium formate in *i*PrOH. The hydrogenolysis is sequential and selective according to the sequence 1,2,4-trichlorobenzene > dichlorobenzene > chlorobenzene [68].

1.2.5 Borane or Sodium Borohydride as Reductant

A combination of Ni/C (5 mol%), PPh_3 (20 mol%), and stoichiometric amounts of $Me_2NH{\cdot}BH_3/K_2CO_3$ catalyzes dechlorination of (functionalized) aryl chlorides in refluxing acetonitrile, giving reduced arenes in high yields. The method is highly tolerant of moisture and compatible with functional groups such as CN, CF_3, COOEt, RC(O)NH, C(O)NHR, and OMe. However, ketones and olefins are unfortunately reduced competitively with that occurring at the C–Cl site [69]. A homogeneous catalytic system consisting of $(Ph_3P)_2NiCl_2$ and Ph_3P catalyzes dehalogenation of aryl chlorides, bromides, and iodides using $Me_2NH{\cdot}BH_3/K_2CO_3$ in CH_3CN at room temperature. This method tolerates wider functional groups than the Ni/C system, including CN, CF_3, NHR, OR, OH, COOR, RC(O)NH, C(O)NHR, keto, and vinyl groups. Both bromides and iodides are more readily and selectively reduced over the corresponding aryl chlorides under these conditions [70]. $NaBH_4$ is also an effective reductant for the catalytic dechlorination of aryl chlorides. Schwartz and co-workers [71] carried out complete dechlorination of polychlorinated biphenyls using titanocene dichloride as a pre-catalyst together with $NaBH_4$ and pyridine in glyme solvents at 125°C under a N_2 atmosphere. O'Hare and co-workers [72] carried out dechlorination of polychlorinated biphenyls by using ring-functionalized titanocene dichloride compounds as the pre-catalysts under Schwartz's conditions. This method does not require use of toxic additives such as pyridine, but leads to relatively low product yields. A combination of $PdCl_2$(dppf) and $NaBH_4$ has been proven to be effective in dechlorination of highly chlorinated benzenes and aroclors. The catalytic activity and selectivity are solvent- and base-dependent. The solvents studied can be ranked in order of decreasing performance as follows: DMA > DMF > diglyme > DMSO > CH_3CN > THF. The presence of TMEDA leads to improvement of the yields in all cases, except for DMSO. The selectivity can be tuned by a judicious choice of conditions. For example, the dechlorination of pentachlorobenzene leads to 90% 1,2,3,4-tetrachlorobenzene in DMA/TMEDA, but to 80% 1,2,4,5-tetrachlorobenzene in DMS/TMEDA [73, 74]. In an ionic liquid, $[N\text{-pentylpyridinium}]^+[\text{closo-}CB_{11}H_{12}]^-$, (dppf)$PdCl_2$-catalyzed dechlogenation of polychloroarenes proceeds more rapidly than in a organic solvent. Dehalogenation reaction of 1,2,4-trichlorobenzene gives almost exclusive dechlorination at the 4-position. $(PPh_3)_2PdCl_2$ and (dppe)$PdCl_2$ are also active catalysts for the dechlorination reaction [75]. Cobalt(II) phthalocyanines are also effective toward the reductive dechlorination of atrazine using sodium borohydride as a reducing agent.

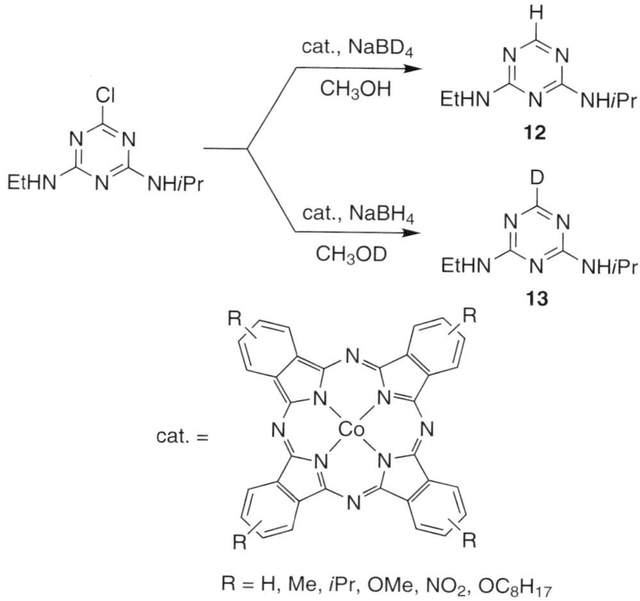

Scheme 1.2.

Among the functional cobalt(II) phthalocyanines the cobalt(II) phthalocyanine bearing nitro groups at the peripheral position is the most efficient. It leads to the reduced product in a 98% yield in 30 min at room temperature. In the reaction, the use of $NaBD_4$ in MeOH results in **12** and the use of $NaBH_4$ in CD_3OD gives **13** (Scheme 1.2). The formation of **12** and **13** in the reactions justifies that the hydrogen that replaces the -Cl group present at the aromatic ring in the dechlorinated product **12** comes from methanol in the reduction reaction [76].

1.2.6 Grignard Reagents as Reductant

In the dehalogenation of organic halides using alkyl Grignard reagents as reductants, the Grignard species delivers a hydride via intramolecular β-hydride elimination. Cp_2TiCl_2, $FeCl_2$, $FeCl_3$, and $Fe(acac)_3$ were found to be effective catalysts for the dechlorination of aryl chlorides using Grignard reagents as the reduing agents under mild conditions. Cp_2TiCl_2-catalyzed dechlorination reactions can proceed well in THF at room temperature in the presence of excess *n*BuMgBr. The chlorides include PhCl, *o*-, *m*-, and *p*-ClC_6H_4OMe, *p*-ClC_6H_4OH, and 1- and 2-chloronaphthalene. The reaction of 2,4-dichloroanisole gives a mixture of PhOMe, *o*-ClC_6H_4OMe, and *p*-ClC_6H_4OMe [77]. Both $FeCl_2$ and $FeCl_3$ catalyze dechlorination of electron-rich aryl chlorides such as chloroanisole, chloroaniline, and chlorotoluene by β-H-containing Grignard reagents. The reaction was carried out in THF at 50°C and gives good to excellent yields. The Grignard reagents without β-H such as

MeMgCl, $PhCH_2MgCl$, and $CH_2{=}CHMgBr$ lead to poor yields. This method is not applicable to the activated aryl chlorides, bromides, and iodides due to dominating cross-coupling [78]. When *t*BuMgCl was employed as the reductant, $Fe(acac)_3$ efficiently catalyzes dechlorination of 4-chloro-2-methylquinoline, *p*-ClC_6H_4OMe, and 1,2- and 1,3-$Cl_2C_6H_4$ in THF at 20–35°C. It is worth noting that reaction of 1,2- and 1,3-$Cl_2C_6H_4$ gives PhCl rather than benzene. This catalytic system is also applicable in dehalogenation of aryl bromides and iodides, and the reactions can be completed at 0°C with lower catalyst loadings [79].

1.2.7 Hydrazine as Reductant

Hydrazine can be employed as a reducing agent in the catalytic dechlorination of aryl chlorides. Palladium on carbon is an effective catalyst for this transformation. The combination of Pd/C, NaOH, or NaO*t*Bu and hydrazine hydrochloride can lead to dehalogenation of aryl chlorides, bromides, or iodides at room temperature [80]. The combination of Pd/C, Na_2CO_3, and hydrazine hydrochloride catalyzes complete dechlorination of trichlorobenzenes and polychlorinated biphenyls under mild conditions. The catalyst system can be reused for several cycles. Ultrasonication of the heterogeneous catalyzed reaction increases the dechlorination rate remarkably [81–83]. A reactive comparison of formic acid, isopropanol, hydrazine, and H_2 as reductants for the Pd-catalyzed hydrodechlorination of chlorobenzene in water at ambient temperature was performed. The results show that hydrazine is effective as a H-donor for the hydrodechlorination under alkaline conditions. However, the reaction is slower than with H_2 by a factor of 30. Formic acid is as reactive as H_2 under acidic and neutral conditions, but less reactive under alkaline conditions. Isopropanol is less reactive by about five orders of magnitude than H_2 [84].

1.3 FORMATION OF C–C BONDS

1.3.1 Suzuki Reaction

The Suzuki coupling is one of the most important transition-metal-catalyzed cross-coupling reactions and may be the most studied cross-coupling reaction because of wide functional group tolerance, use of stable and nontoxic organoboron reagents, mild reaction conditions, and the ease of handling and separating byproducts from its reaction mixtures. Organic bromides and iodides were widely employed as the electrophilic substrates in its early stage. However, cheaper but less reactive organic chlorides have been more and more used in recent years due to development of new catalyst systems.

1.3.1.1 Palladium Catalysts

(i) PHOSPHINE LIGANDS. In 1997 Shen [85] found that $(Cy_3P)_2PdCl_2$ catalyzes coupling of activated aryl chlorides with aryl boronic acids with CsF as a base. Next Fu and co-workers [86, 87] reported that a combination of Pd_2dba_3 and P*t*Bu_3 in the presence of Cs_2CO_3 or KF efficiently catalyzes cross-coupling of arylboronic acids with aryl chlorides, including deactivated aryl chlorides such as 4-chloroanisole and 4-chloroaniline. It was noted that the ratio of P*t*Bu_3 to Pd significantly affects the reaction. The 1:1 ratio furnishes a very active catalyst, and the 2:1 ratio slows the reaction remarkably. KF is superior to Cs_2CO_3 as the base. Under the optimized conditions, the reaction can be performed with low catalyst loadings, achieving almost 10,000 turnovers for the reaction of *o*-chlorobenzonitrile [87]. The combination of $Pd(OAc)_2$, *n*BuPAd_2, and K_3PO_4 exhibits higher turnovers for the Suzuki coupling of activated, unactivated, and deactivated aryl chlorides. For example, the TON for the reaction of *o*-chlorobenzonitrile with phenylboronic acid reaches 69,000 [88]. In addition, in the reaction of chloro-aryl triflate with arylboronic acid the Pd/P*t*Bu_3 system leads to the C–Cl bond activation with excellent selectivity although general reactivity is ArOTf ≫ArCl. By contrast, $Pd(OAc)_2/PCy_3$ exhibits the conventional pattern of reactivity, giving a coupling product through C–OTf bond cleavage (Scheme 1.3) [87].

(*p*-$RC_6H_4PR'_2)_2PdCl_2$ (R = H, CF_3, OMe, NMe_2; R′ = *t*Bu, Cy) is very effective in catalyzing cross-coupling of heteroaryl chlorides with a diverse range of aryl or heteroarylboronic acids or esters. Among the complexes studied, (*p*-$Me_2NC_6H_4$P*t*$Bu_2)_2PdCl_2$ exhibits the highest catalytic activity. The coupling reactions give the high product yields (88–99%) and up to 10,000 turnover numbers [89, 90]. Ma et al. [91] synthesized air-stable 2,4,6-$(MeO)_3C_6H_2PCy_2\cdot HBF_4$ and found that in combination with $Pd(OAc)_2$ it is an efficient catalyst for the coupling of aryl chlorides with arylboronic acids. Deactivated or bulky aryl chlorides as well as functionalized aryl chlorides such as 2-$NO_2C_6H_4Cl$ and 4-$CHOC_6H_4Cl$ can be employed [91].

Besides catalyzing aryl–aryl coupling, Pd–PR_3 systems are also effective in catalyzing Suzuki coupling of alkenyl–aryl, alkyl–aryl, and alkyl–alkyl. Pd_2dba_3/P*t*Bu_3 efficiently catalyzes coupling of alkenyl chlorides with arylboronic acids in THF using KF as a base. It was well known that in Pd-catalyzed cross-coupling

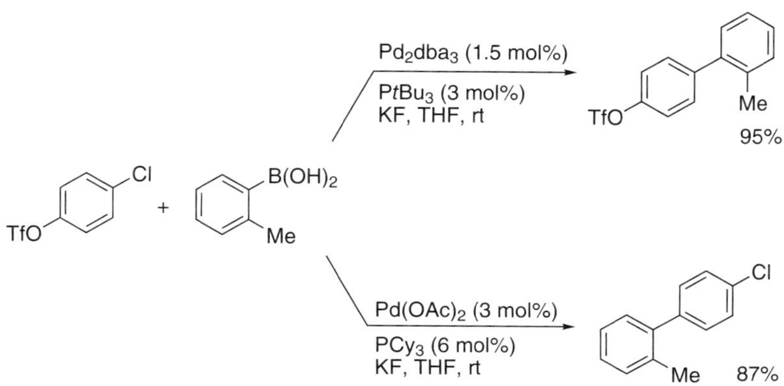

Scheme 1.3. Pd-catalyzed selective cleavage of C–Cl and C–O bonds of chloro-aryl triflate.

reactions, vinyl chlorides are usually significantly more reactive than aryl chlorides. However, a competition experiment revealed that chlorobenzene is more reactive than 1-chlorocyclopentene when $Pd_2(dba)_3/PtBu_3$ is employed as the catalyst [Eq. (1.3)] [87]. $Pd_2(dba)_3/PtBu_3$ also catalyzes cross-coupling of cyclopentylboronic acid and *p*-tolyl chloride in good yield. More examples of cycloalkyl–aryl coupling were given using $Pd(OAc)_2/n BuPAd_2$ as the catalyst and Cs_2CO_3 as a base. Cyclobutyltrifluoroborates, cyclopentyltrifluoroborates, and cyclohexyltrifluoroborates can be coupled with aryl chlorides using the catalyst system. However, reaction of $i PrBF_3K$ leads to a mixture of *i*Pr-Ar and *n*Pr-Ar, and the ratio of two products is ligand-dependent. Compared with $n BuPAd_2$, $PtBu_3$ and $PhPtBu_2$ are more selective but less reactive [92, 93].

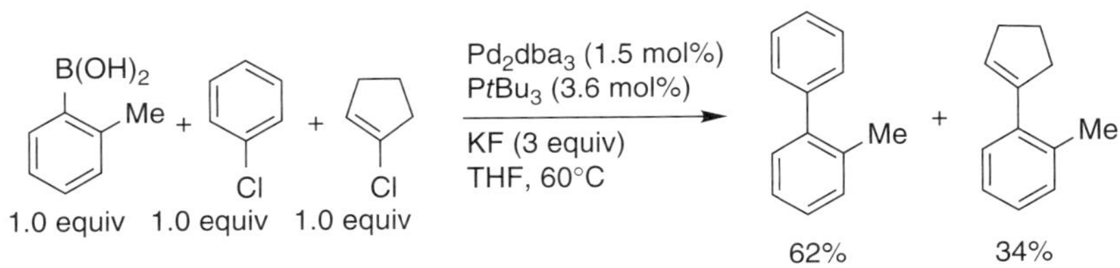

(1.3)

The combination of $2,4,6\text{-}(MeO)_3C_6H_2PCy_2\cdot HBF_4$ and $Pd(OAc)_2$ catalyzes coupling of alkenyl chlorides with arylboronic acids and shows good selectivity in the reaction of β-chlorobutenolides with arylboronic acids. The relatively inert C–Cl bond rather than the more reactive lactonic, allylic C–O bond in β-chlorobutenolides are activated [e.g. Eq. (1.4)] [91, 94].

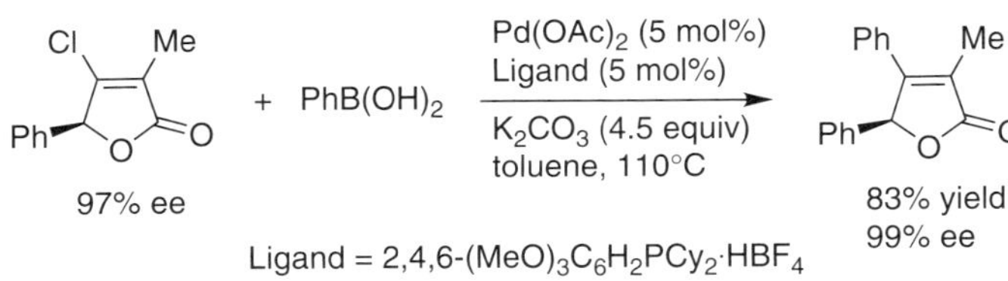

(1.4)

$Pd(PPh_3)_4$ catalyzes coupling of benzyl chlorides with aryl or heteroaryl boronic acids using sodium carbonate as a base in a mixture of DME/water (2:1) [95]. Pd_2dba_3/PCy_3 catalyzes reaction of primary alkyl chlorides that contain β-hydrogen atoms with alkyl-9-BBN derivatives. The process is compatible with a variety of functional groups, including silyl ethers, acetals, olefins, amines, nitriles, and esters [96].

Dialkylphosphinobiphenyls (**4**, **5**, and **14a–18b**) developed by Buchwald and co-workers [16] represent a class of highly efficient ligands for palladium-catalyzed Suzuki coupling. In 1998 Buchwald and co-workers [97] reported that **14a**-$Pd(OAc)_2$ system efficiently catalyzes cross-coupling of aryl chlorides or bromides and arylboronic acids using CsF or K_3PO_4 as a base. The reaction requires 0.5–2 mol% Pd loadings and is applicable to activated and deactivated aryl chlorides. Structures **4**, **5**, and **14b** are excellent in Pd-catalyzed coupling of the sterically hindered substrates, giving three *ortho*-substituted biaryls in excellent yields [98]. Further studies showed that **15a** is a more widely applicable ligand for Pd-catalyzed Suzuki couplings. The substrates include aryl, heteroaryl, and alkenyl chlorides as well as aryl, heteroaryl, or alkyl boron reagents with a wide range of functional groups [99–105]. This catalyst system also catalyzes formation of three *ortho*-substituted biaryls, the activity being comparable to that of Pd–**16** [106]. The Pd–**15b** system is excellent for the coupling of alkylboron reagents with aryl chlorides. Reaction of functionalized primary alkyltrifluoroborates with a variety of aryl chlorides catalyzed by $Pd(OAc)_2$–**15b** in the presence of K_2CO_3 gives corresponding cross-coupling products in good to excellent yields [107, 108]. Modification of **15a** by adding a sulfonate group results in a water-soluble ligand, **17**. The combination of **17** and $Pd(OAc)_2$ gives a highly active catalyst for the Suzuki couplings of highly functionalized aryl or heteroaryl chlorides with aryl or alkyl boronic acids in aqueous media [109]. Both **15a** and **18** are suitable ligands for the Pd-catalyzed coupling of heteroaryl chlorides and heteroaryl boronic acids and esters which results in a wide variety of heterobiaryls [101]. Selected examples of the Suzuki coupling catalyzed by Pd–Buchwald ligands are listed in Table 1.1.

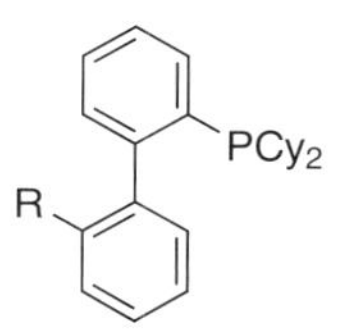

14a R = NMe_2 (Davephos)
14b R = *i*Pr

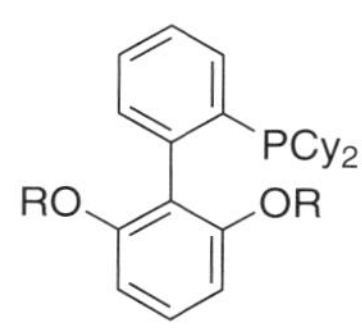

15a R = Me (SPhos)
15b R = *i*Pr (RuPhos)

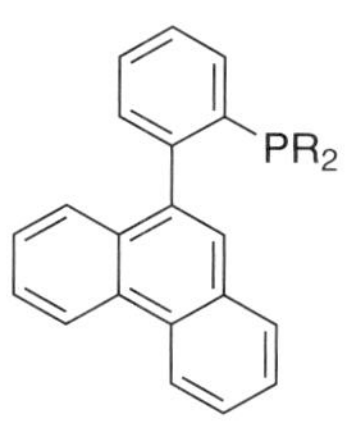

16a R = Cy
16b R = Ph

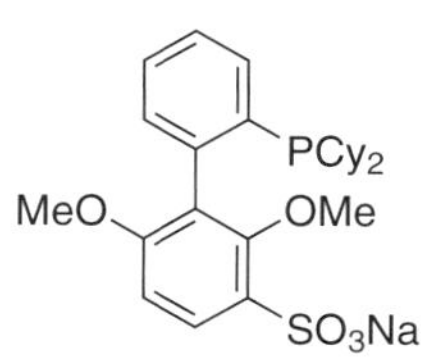

17

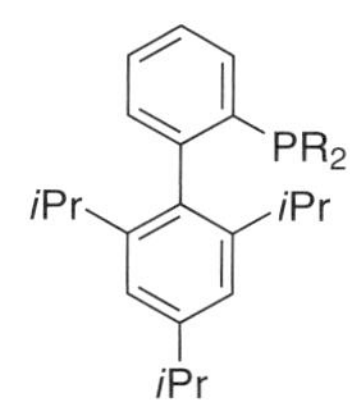

18a R = Cy (XPhos)
18b R = Ph

The high activity and longevity of catalysts based on **15a** are attributed to the ability of this ligand to stabilize and maximize the concentration of the monoligated intermediates, which are particularly reactive in oxidative addition and transmetalation processes. The stabilization of Pd(0) intermediates is believed to be a result of favorable interactions of the aromatic π system with the Pd center which is supported by DFT computational studies and demonstrated by X-ray crystallography of **15a**–Pd–dba complex. The X-ray diffraction analysis reveals a Pd(0) η^1-arene

TABLE 1.1. Suzuki coupling catalyzed by Pd–Buchwald ligands

Entry	Catalyst (mol%)	Chloride	Boron Reagent	Base	Solvent	Temperature (°C)	Time	Yield (%)	Reference
1	$Pd(OAc)_2$–**14a** (2.0)	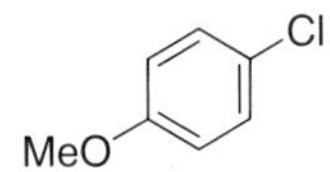	$PhB(OH)_2$	CsF	Dioxane	r.t.	19–30 h	92	97
2	$Pd(OAc)_2$–**14a** (0.5)			K_3PO_4	Dioxane	100	19–30 h	93	97
3	$Pd(OAc)_2$–**14a** (2.0)	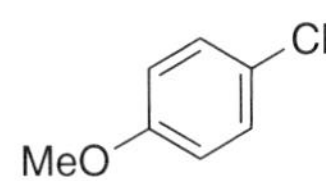	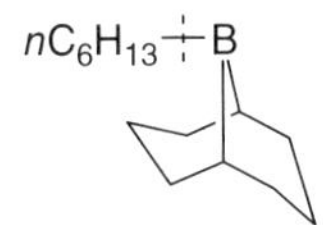	CsF	Dioxane	50	19–30 h	88	97
4	$Pd(OAc)_2$–**5** (0.2)	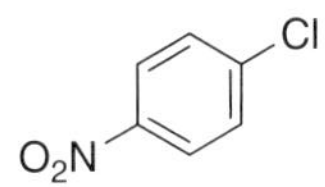	$PhB(OH)_2$	KF	THF	r.t.[a]	8 h	98	98
5	$Pd(OAc)_2$–**5** (1.5)	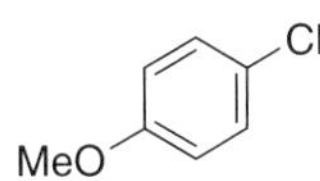	$PhB(OH)_2$	KF	THF	r.t.	21 h	92	98
6	$Pd(OAc)_2$–**5** (1.0)	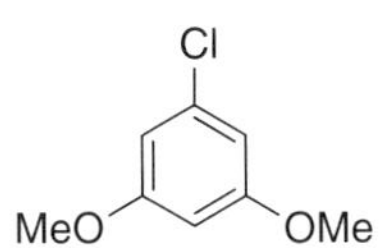	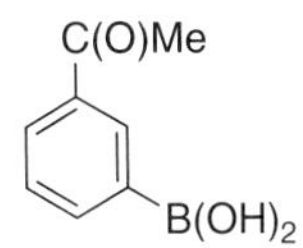	KF	THF	r.t.	7 h	90	98
7	$Pd(OAc)_2$–**4** (1.0)	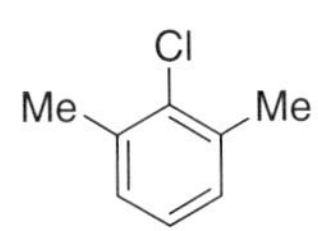	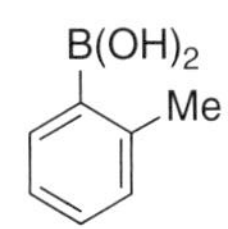	K_3PO_4	Toluene	100	17 h	88	98

8	$Pd(OAc)_2$–**14b** (1.0)			K_3PO_4	Toluene	100	3 h	92	98
9	$Pd(OAc)_2$–**15a** (0.02)	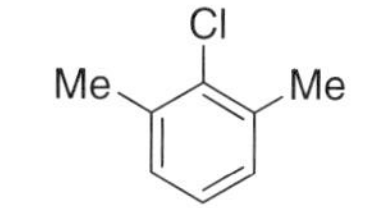	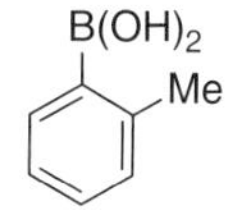	K_3PO_4	Toluene	100	20 h	97	99
10	$Pd(OAc)_2$–**15a** (0.2)			K_3PO_4	Toluene	90	12 min	98	99
11	$Pd(OAc)_2$–**15a** (1)	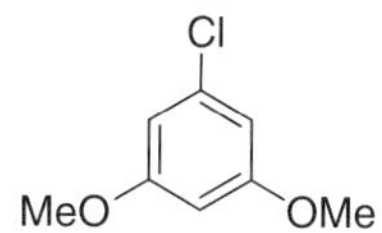	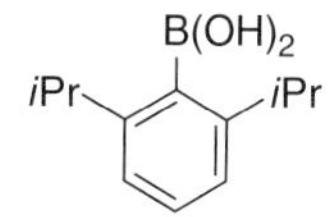	K_3PO_4	Toluene	100	20 h	99	99
12	$Pd(OAc)_2$–**15a** (0.05)	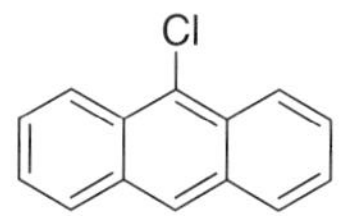	$MeB(OH)_2$	K_3PO_4	Toluene	100	24 h	99	99
13	$Pd(OAc)_2$–**15a** (0.005)	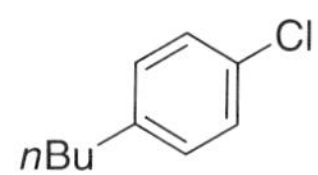	$PhB(OH)_2$	K_3PO_4	Toluene	100	10 h	96	100
14	$Pd(OAc)_2$–**15a** (1)	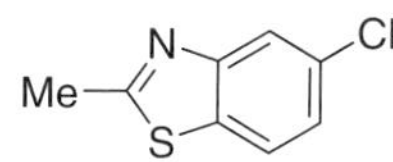	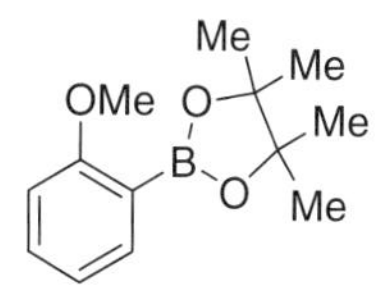	K_3PO_4	$THF–H_2O$ (10:1)	r.t.	24 h	91	100
15	$Pd(OAc)_2$–**15a** (1)	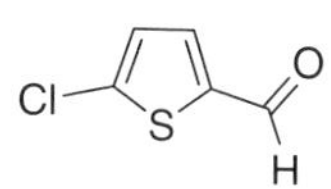	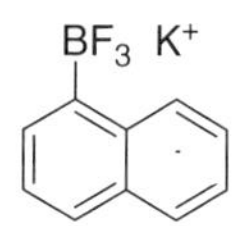	K_2CO_3	MeOH	reflux	16 h	85	102

(*continued*)

TABLE 1.1. (*Continued*)

Entry	Catalyst (mol%)	Chloride	Boron Reagent	Base	Solvent	Tempe-rature (°C)	Time	Yield (%)	Reference
16	Pd_2dba_3–**16a** (1)	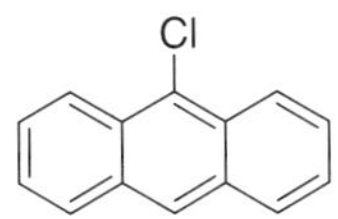	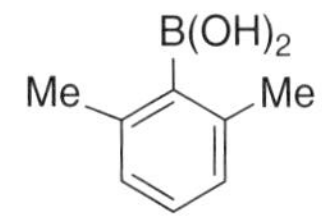	K_3PO_4	*o*-Xylene	110		82	106
17	Pd_2dba_3–**16b** (2)	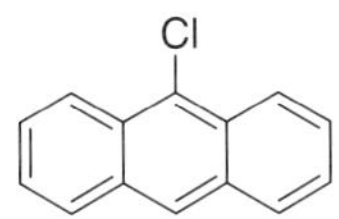	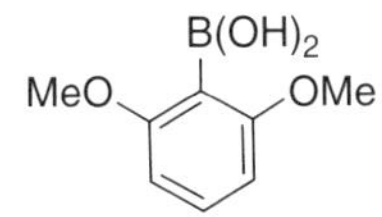	K_3PO_4	*o*-Xylene	110		93	106
18	$Pd(OAc)_2$–**17** (1)	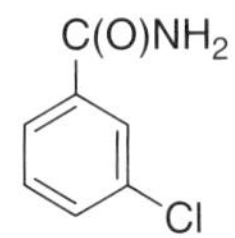	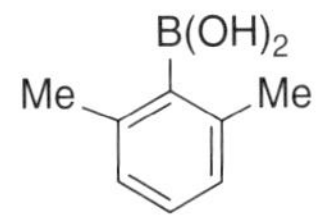	K_2CO_3	H_2O	100	8 h	99	109
19	$Pd(OAc)_2$–**17** (1)	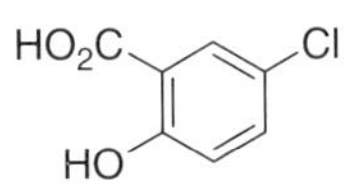	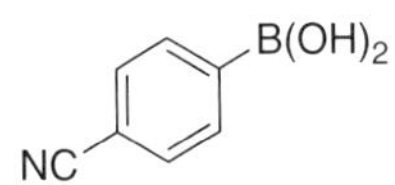	K_2CO_3	H_2O	80	2–8 h	92	109
20	Pd_2dba_3–**18a** (2)	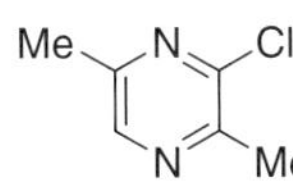	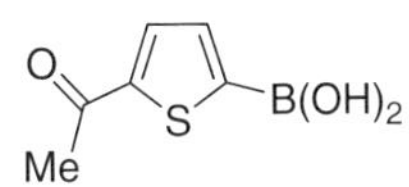	K_3PO_4	*t*-amyl alcohol	100	4–10 h	98	101
21	Pd_2dba_3–**18a** (1)	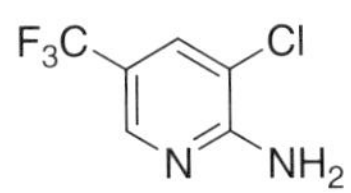	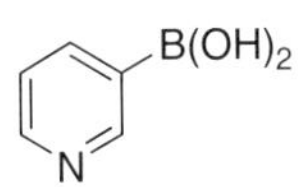	K_3PO_4	*n*BuOH	120	18–24 h	95	101

22	$Pd(OAc)_2$–**18b** (0.5)	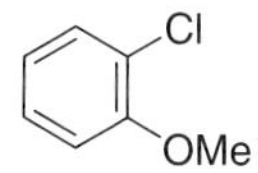	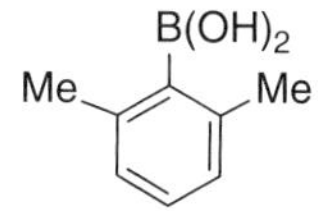	K_3PO_4	THF	40		98	100
23	(**15a**)$_2$$PdCl_2$ (5)	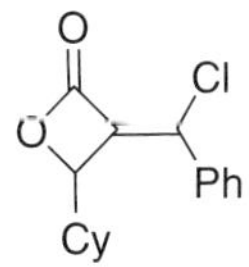	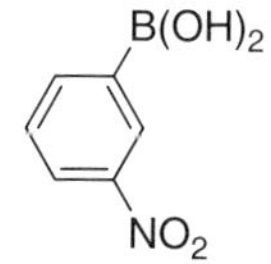	K_3PO_4	Toluene	Reflux	3 h	81	105
24	(**15a**)$_2$$PdCl_2$ (5)	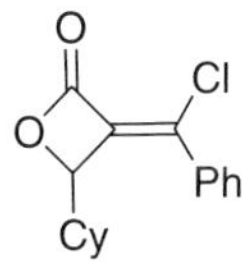	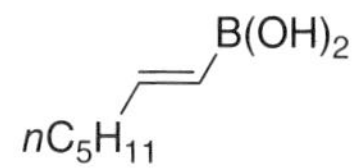	K_3PO_4	Toluene	Reflux	1.5 h	91	105
25	$Pd(OAc)_2$–**15a** (5)	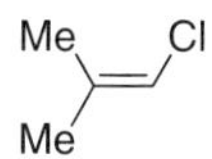	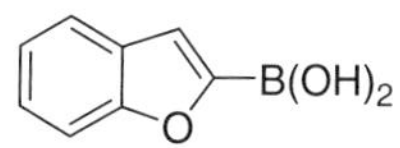	CsF	*i*PrOH	85	6 h	83	104
26	$Pd(OAc)_2$–**15b** (2)	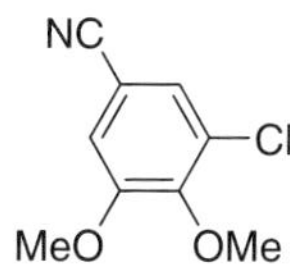	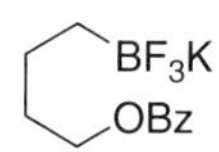	K_2CO_3	Toluene–H_2O (10:1)	80	24 h	80	107
27	$Pd(OAc)_2$–**15b** (2)	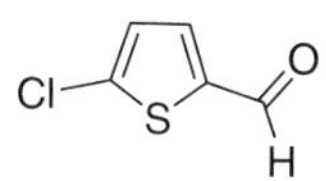	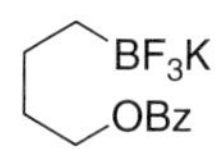	K_2CO_3	Toluene–H_2O (10:1)	80	24 h	83	107

[a]r.t., room temperature.

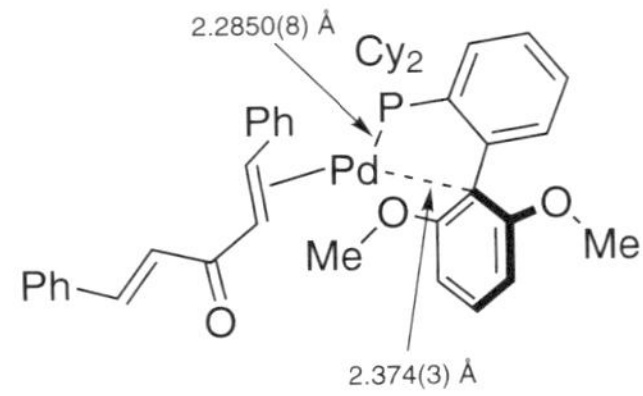

Figure 1.4. Pd(0) η^1-arene interaction with the *ipso* carbon.

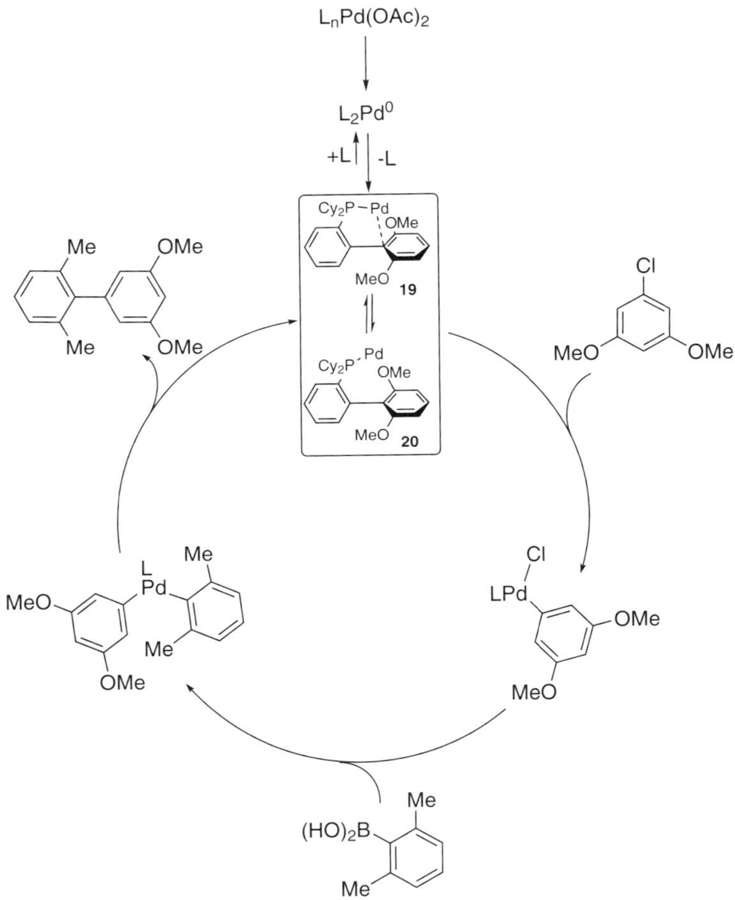

Figure 1.5. Proposed reaction pathway for the Suzuki coupling using **15a**–Pd(0) as a catalyst.

interaction with the *ipso* carbon as shown in Figure 1.4. The DFT computations and NMR studies show that both the Pd–arene and Pd–O interactions contribute to the stability of the Pd complex intermediates [100, 110]. A proposed reaction pathway for the Suzuki coupling using **15a**–Pd(0) is presented in Figure 1.5 [100].

Following the success of Buchwald's biphenyl ligands, some new biaryl ligands were developed in which aromatic heterocycles are used as a replacement for the phosphine-substituted ring of Buchwald biphenyl ligands. Beller and co-workers [111] found that phosphino-substituted *N*-aryl pyrroles are excellent ligands in the palladium-catalyzed coupling of aryl and heteroaryl chlorides with phenylboronic acid. The combination of **21** and Pd_2dba_3 shows the best activity in most cases. The system allows highly efficient couplings of electron-rich and electron-poor aryl chlorides with phenylboronic acid under mild conditions, shows exceedingly high turnover numbers, and tolerates a variety of functional groups such as C(O)Me, CHO, CN, F, CF_3, and OMe and aromatic heterocycles [111]. Kwong and co-workers [112] revealed that **22a**–Pd_2dba_3 is highly effective for the coupling of sterically hindered aryl chlorides and aryl boronic acids, giving tetra-*ortho*-substituted biaryls. Surprisingly, it was found that the diphenylphosphino compound (**22a**) is much more

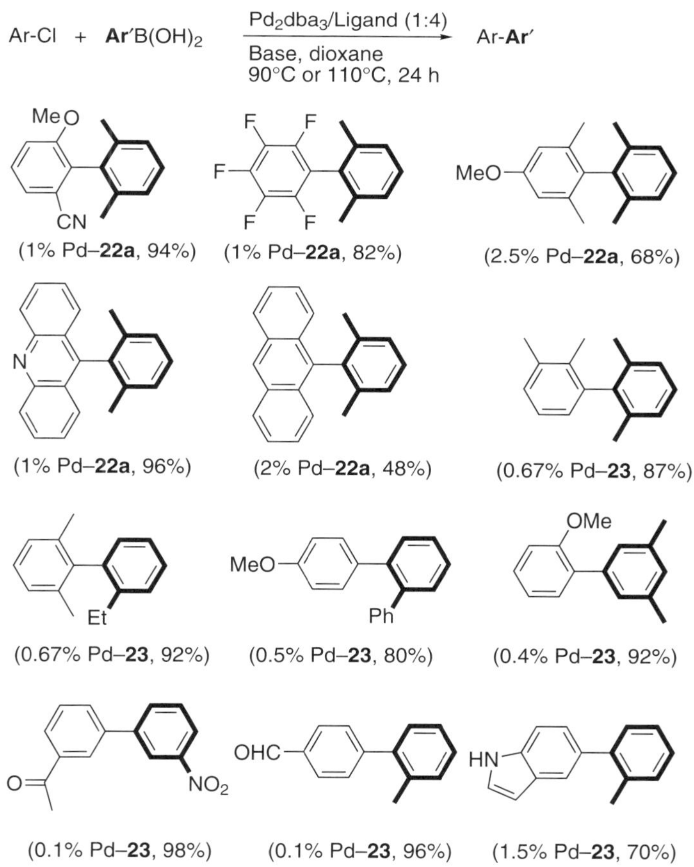

Scheme 1.4. Note that Cs_2CO_3 was employed as a base when **22a** was used as the ligand, and $K_3PO_4{\cdot}H_2O$ was employed as a base when **23** was used as the ligand.

efficient than its dialkylphosphino analogues such as Cy_2PAr and tBu_2PAr in the Pd-catalyzed reactions [Eq. (1.5)]. Usually electron rich and bulky phosphines are more effective ligands for the Pd-catalyzed cross-couplings because electron-rich phosphine ligands promote oxidative addition to palladium, and increasing the steric bulk around the metal accelerates the rate of reductive elimination. The results reported by Kwong and co-workers seem to violate the rationale. On the other hand, triarylphosphanes are air-stable and readily prepared, and hence this ligand is attactive for practical use. The structurally similar ligand **23** is also excellent in Pd-catalyzed Suzuki couplings of aryl, heteroaryl, or alkenyl chlorides with arylboronic acids. Selected examples of Suzuki coupling of aryl chlorides using **22a** or **23** are shown in Scheme 1.4 [112].

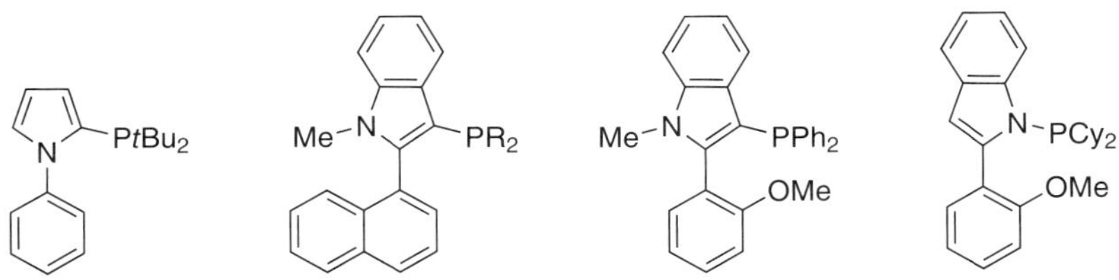

21 **22a** R = Ph **22b** R = Cy **23** **24**

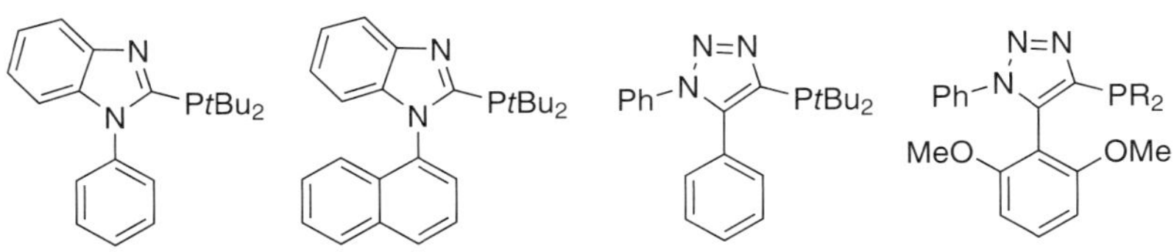

25 **26** **27** **28a** R = *t*Bu **28b** R = Cy

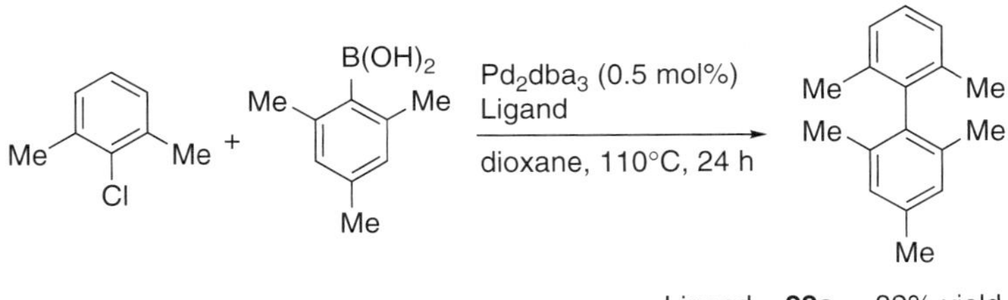

(1.5)

Ligand = **22a** 82% yield
22b 10% yield

The same research group reported a series of 1-phosphino-2-aryl-indoles as ligands in the Pd-catalyzed Suzuki coupling. 1-(Dicyclohexylphosphino)-2-(2-methoxyphenyl)-1*H*-indole, **24**, is structurally related to **23** and exhibits the best catalytic activity in combination with $Pd(OAc)_2$. $Pd(OAc)_2$/**24** drives reactions of a variety of unactivated aryl and heteroaryl chlorides with aryl boronic acids with low catalyst loadings and tolerates a range of functional groups. The catalytic activity is comparable to that of a Pd–**23** system [113]. Both *N*-aryl-2-(dialkylphosphino)-imidazole and -benzimidazole, **25** and **26**, catalyze cross-coupling of unactivated and deactivated aryl chlorides with phenylboronic acid with 0.01–0.05 mol% $Pd(OAc)_2$ loadings, giving excellent product yields. However, the systems are not good for sterically hindered aryl chloride and require excess ligands (Pd:L = 1:10) [114]. When imidazole ring in **25** is replaced with a 1,2,3-triazole ring, a new ligand **27** is formed. $Pd(dba)_2$/**27** displays similar catalytic properties to $Pd(OAc)_2$/**25** in the coupling of unactivated and functionalized aryl chlorides with arylboronic acids, but requires a little higher catalyst loadings (0.1 mol%). $Pd(dba)_2$/**28a** is more effective and is applicable to a wider substrate scope, including aryl and heteroaryl chlorides and sterically hindered aryl chlorides. $Pd(dba)_2$/**28b** also catalyzes formation of tri-*ortho*-substituted biaryls, but displays lower activity than $Pd(dba)_2$/**28a** and leads to moderate product yields [115–117].

Ferrocene-based ligands have also received considerable attention for the Suzuki coupling of organic chlorides due to electron-rich and sterically bulky properties of ferrocene [118]. Both **29** and **30** are similar to Buchwald biphenyl-based ligands and hence were expected to be good ligands for Pd-catalyzed Suzuki coupling. $Pd(OAc)_2$/**29** was shown to be effective for the catalytic coupling of aryl chlorides and arylboronic acids using K_3PO_4 as a base in dioxane at 95°C. Also, 0.1 mol% $Pd(OAc)_2$ loadings can drive the reaction of unactivated and deactivated aryl chlorides in excellent yields [119]. The reaction using $Pd(OAc)_2$/**29** can be carried out in water at room temperature for the activated chlorides or at 60°C for the unactivated chlorides, but requires higher catalyst loadings (2–5 mol%) [120]. $Pd(OAc)_2$/**30a** shows lower catalytic activity than $Pd(OAc)_2$/**4** in catalyzing reaction of aryl chlorides with $PhB(OH)_2$. It was found that the reaction between $Pd(OAc)_2$ and **30a** results in formation of palladacycle [Eq. (1.6)], which exhibits a little higher catalytic activity than the mixture of $Pd(OAc)_2$ and **30a** [121]. Pd_2dba_3–**30b** also shows good catalytic activity in the Suzuki cross-coupling of activated and unactivated aryl chlorides as well as 1-chlorocyclopentene. Arylboronic acids and nBuB$(OH)_2$ can be employed as the nucleophilic substrates [122].

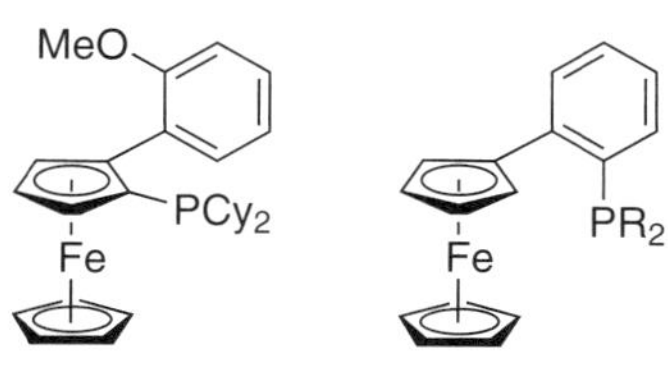

29 **30a** R = Cy **30b** R = Ph

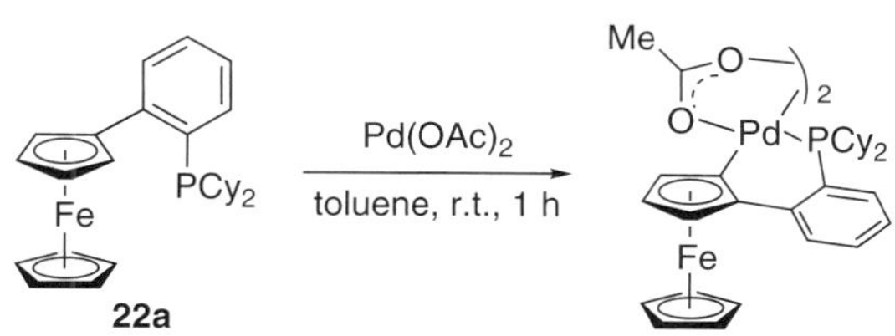

22a

(1.6)

Richards and co-workers [123] revealed that the combination of $Pd_2(dba)_3$ and electron-rich and bulky ligand **31** leads to similar yield to $Pd_2(dba)_3$/P*t*Bu_3 in catalyzing the coupling of 4-chlorotoluene and phenylboronic acid. Further tests showed that **32a** is essentially inactive due to poor electron-donating ability, while **32b** and **32c** provide a moderate yield of cross-coupled product [123]. However, a combination of $Pd_2(dba)_3$ and very bulky ligand **33a** does exhibit good catalytic activity in the coupling of arylboronic acids and aryl chlorides, including unactivated and deactivated ones. By contrast, $Pd_2(dba)_3$/**33b** shows low catalytic activity for the similar coupling [124]. Both **34** and **35** are excellent ligands in Pd-catalyzed coupling of aryl chlorides and arylboronic acids. The reactions are compatible with a range of functional groups such as NO_2, CHO, CN, COOMe, OMe, and aromatic heterocycles. The reaction catalyzed by $Pd_2(dba)_3$/**34** can employ *n*BuB$(OH)_2$ as the nucleophilic reagent, giving excellent product yields [125, 126].

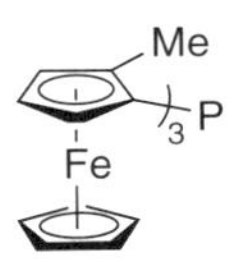

31

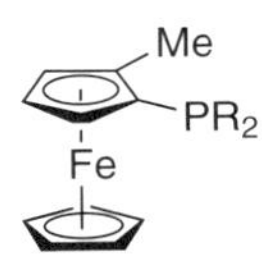

32a R = Ph
32b R= Cy
32c R = *t*Bu

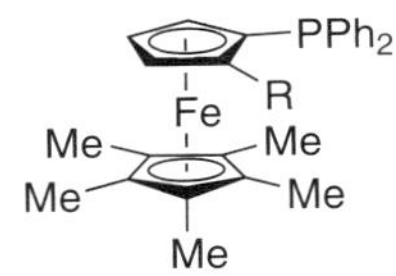

33a R = TMS
33b R = H

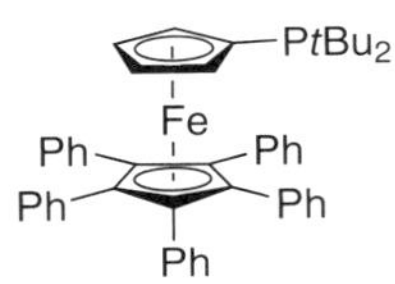

34

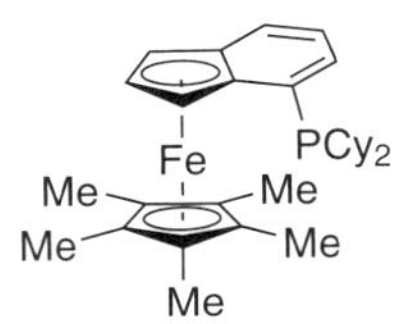

35

Hor and co-workers [127, 128] reported two types of hemilabile ferrocene-based ligands, **36** and **37**, which contain strongly coordinating phosphine and weakly basic pendants. The hemilabile property is expected to promote a facile switchable protection and deprotection mechanism at the active metal center. The results of invesitigation show that the ligands are highly efficient and effective in promoting Suzuki couplings of aryl chlorides and aryl boronic acids. The reaction proceeds through a Pd(0)/Pd(II) process based on isolatation and characterization of reaction intermediates such as Pd(0) complexes and their oxidative addition products with C_6F_5I [127, 128].

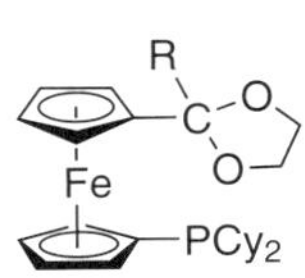

36a R = Ph
36b R = H

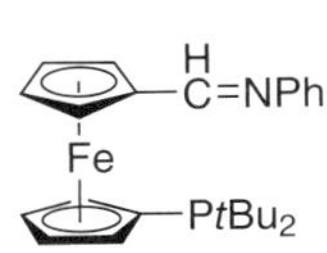

37

1,1-Bis(di-*tert*-butylphosphino)ferrocene/Pd$(OAc)_2$ was found to catalyze cross-coupling of arylboronic acids and heteroaryl chlorides such as 4-amino-2-chloro-5-nitropyrimidine, aminochloropyrimidines and aminochloropyridines with relatively high palladium loadings [129, 130]. Other 1,1′-bisphosphinoferrocenes are less effective for Pd-catalyzed Suzuki coupling of aryl chlorides. The combination of ferrocenyltetraphosphine, **38**, and $[PdCl(\eta^3\text{-}C_3H_5)]_2$ also displays good catalytic efficiency. It catalyzes coupling of a variety of chlorides with aryl boronic acids in good yield in the presence of 1–0.01 mol% palladium at 130°C. The NMR studies show that above 100°C the mononuclear complex **39** is the predominant palladium(II) species. The better results obtained using **38** rather than using PPh_3 or dppe are believed to result from geometric factors more than electronic factors [131].

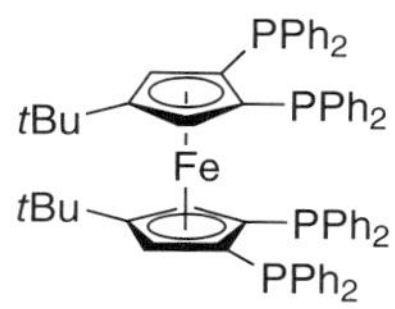

38

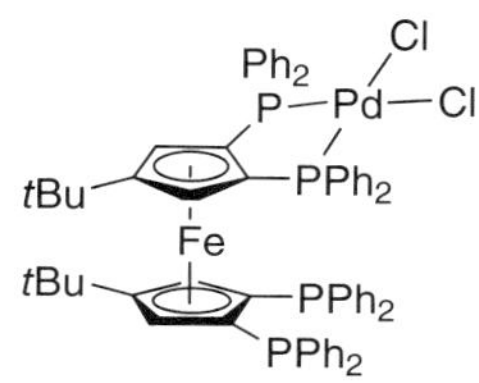

39

The ruthenocene-based phosphine ligand **40** is structurally similar to **34**. The combination of **40** and Pd$(dba)_2$ exhibits excellent catalytic ability in the Suzuki coupling of sterically hindered aryl chlorides, generating tri- or tetra-*ortho*-substituted biaryls in high yields. The system also efficiently catalyzes coupling of activated and deactivated aryl chlorides as well as heteroaryl chlorides with aryl boronic acids (Scheme 1.5) [132].

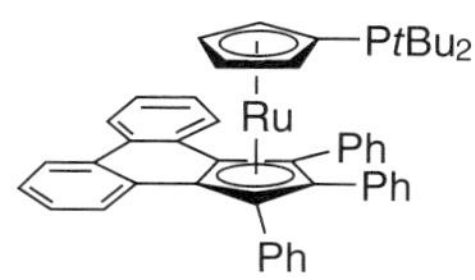

40

Other electron–rich and sterically bulky monophosphine ligands such as **41a–44** are also effective in Pd-catalyzed Suzuki coupling. Pd$(OAc)_2$/**41a–41c** catalyzes the reaction of aryl and heteroaryl chlorides with arylboronic acids in moderare to excellent yields. Both Pd$(OAc)_2$/**41b** and Pd$(OAc)_2$/**41c** show higher activity than Pd$(OAc)_2$/**41a**, and the system is compatible with functional groups including CHO, C(O)Me, and CN groups. Tri-*ortho*-substituted biphenyls can be prepared in moderate yields using the catalysts [133]. Pd$(OAc)_2$/**42** displays higher activity than Pd$(OAc)_2$/**41** in catalyzing coupling of aryl chlorides, the former giving excellent product yields with lower catalyst loadings (0.1–0.6 mol% Pd) [134]. The combination of 9-fluorenyldicyclohexylphosphine **43** and Na_2PdCl_4 catalyzes coupling of aryl or heteroaryl chlorides with arylboronic acids or thiopheneboronic acids in dioxane or butanol (for the heteroaryl couplings). Na_2PdCl_4/**44a** catalyzes coupling

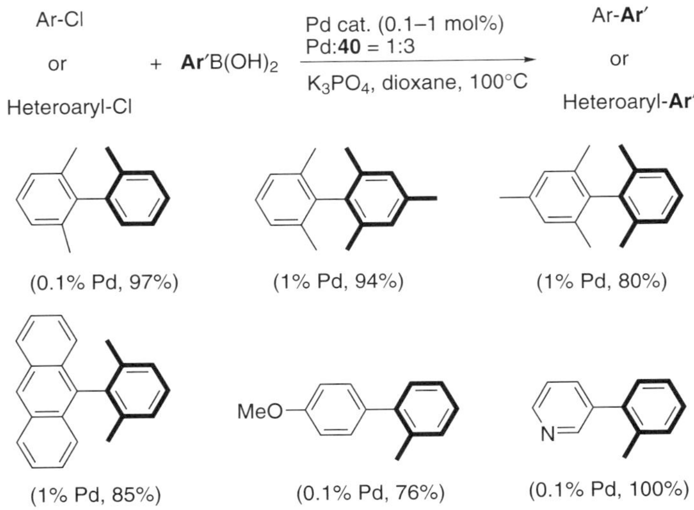

Scheme 1.5. Pd(dba)$_2$/**40**-catalyzed cross-coupling of aryl or heteroaryl chlorides with arylboronic acids.

of aryl and heteroaryl chlorides with aryl or heteroarylboronic acids in water with high product yields and low catalyst loadings. Sterically bucky substrates are also applicable. Na_2PdCl_4/**44b** is effective in catalyzing coupling of thiophene- and furan-boronic acids with heteroaryl chlorides in aqueous butanol [135, 136]. Plenio and co-worker [137] designed a DMSO/n-heptane biphasic system for the Suzuki coupling of aryl chlorides and $PhB(OH)_2$ using polymeric Pd catalysts. The catalysts are generated from Na_2PdCl_4 and modified Ad_2PR (**45**) or Buchwald ligands (**46**) with soluble polyethylene glycol tags. DMSO constitutes the catalyst phase since the Pd catalyst modified with a polar phase tag dissolves in the polar solvent, while n-heptane forms the product phase. Separation of the coupling products from the catalyst can be performed by a simple phase separation of the two room-temperature immiscible solvents. There is no apparent leaching of the catalyst into the heptane solution (<0.05%). The catalysts are highly active and can remain almost constant over more than five reaction cycles [137].

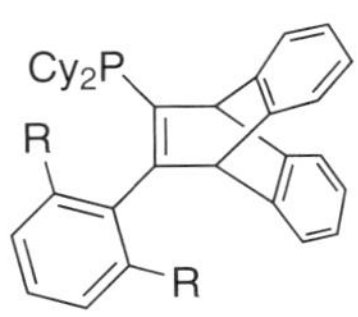

41a R = H
41b R = Me
41c R = OMe

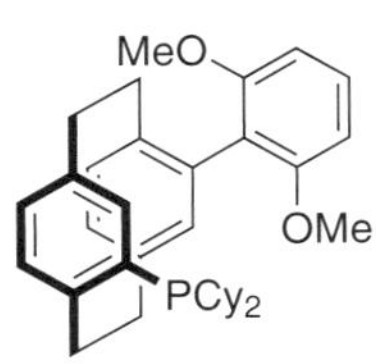

42

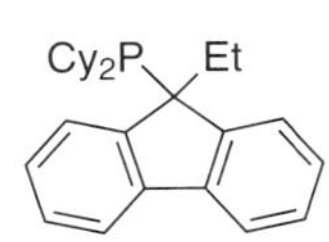

43

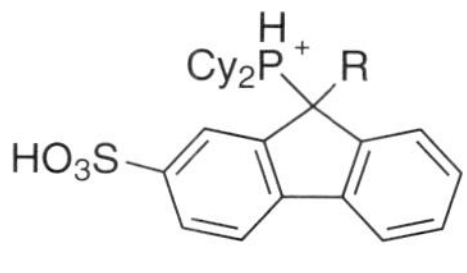

44a R = Et
44b R = *p*-$(CH_2)_3C_6H_4SO_3H$

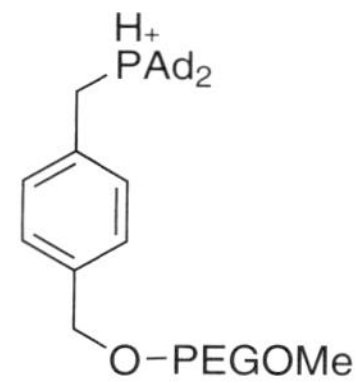

45

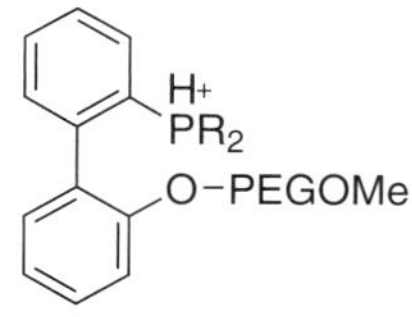

46a R = Cy
46b R = *t*Bu

Several P,O-chelate ligands (**6, 47–52**) were reported to be effective in Pd-catalyzed Suzuki coupling of aryl chlorides. $Pd(dba)_2$/**6** efficiently catalyzes cross-coupling of a wide variety of arylboronic acids with electron-poor, electron-rich, and *ortho*-substituted aryl chlorides. However, $Pd(dba)_2$/**47** is less effective under similar conditions. Spectroscopic and single-crystal X-ray diffraction studies of the reaction intermediates show that **6** functions as a P,O-chelating ligand (**53**), whereas **47** functions as a nonchelating ligand (**54**). It seems that a P,O-chelating system is the most suitable for Suzuki arylation reactions involving certain substrates and conditions [138, 139].

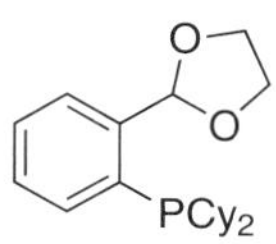

47

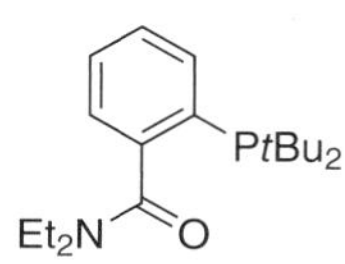

48

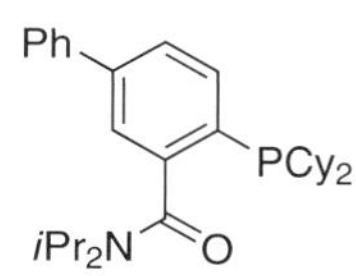

49

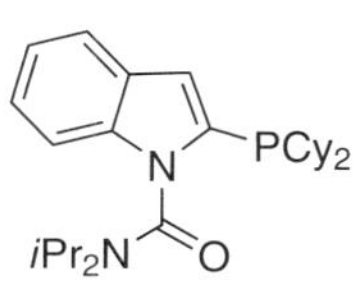

50

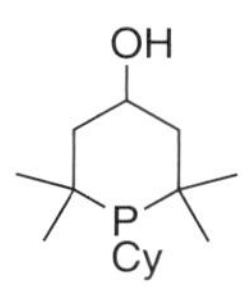

51

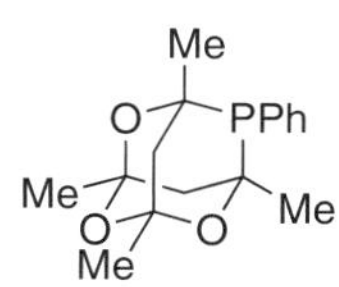

52

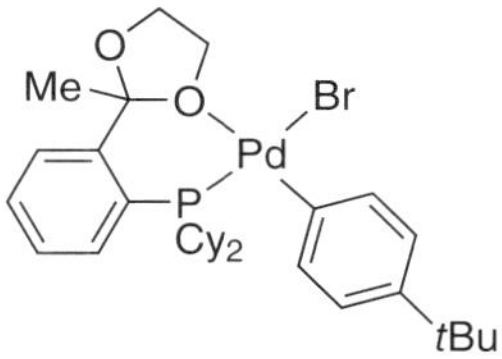

53

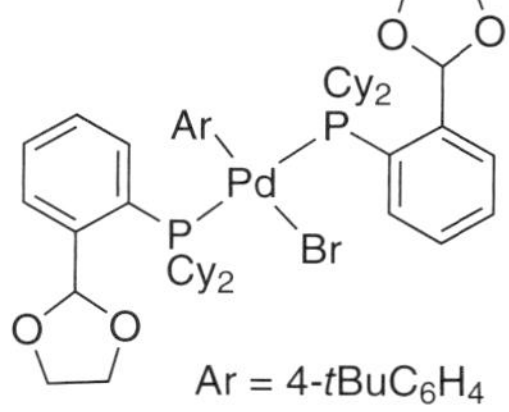

54

Palladium in combination with **48–50** efficiently catalyzes Suzuki coupling of unactivated and sterically bulky aryl chlorides with arylboronic acids in the presence of K_3PO_4 with low palladium loadings (0.01–0.5 mol%). The catalyst systems

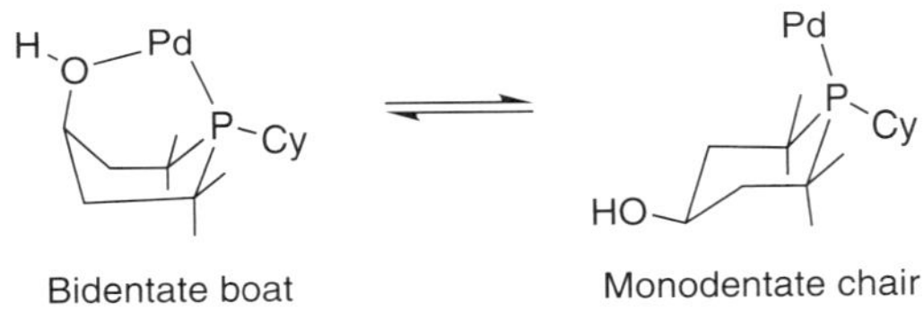

Scheme 1.6. Bi- and monodentate complexes of ligand **51**.

are compatible with a variety of functional groups and nitrogen-containing heterocycles. Pd_2dba_3/**48** also catalyzes the coupling of 1-chlorocyclopent-1-ene with phenylboronic acid or of aryl chlorides with alkylboronic acids. The effectiveness of the ligands may result from their hemilabile property. During the catalytic process the hemilabile group can perform "on and off" dynamic interaction to stabilize the metal center whenever needed [140–142]. Another hemilabile ligand **51** is postulated such that a remote 4-hydroxyl substituent could function to stabilize the catalyst in a boat conformer, but generate the reactive monodentate species through a ring flip to the chair conformer of **51** as shown in Scheme 1.6. However, **51** is less effective than **48–50** in Pd-catalyzed Suzuki coupling, although $Pd(OAc)_2$/**51** catalyzes the coupling of aryl chlorides with arylboronic acids in good yields [143]. Compound **52**, as an electron-rich and sterically bulky ligand, matches P*t*Bu_3 with respect to coupling conditions and yields in the Pd-catalyzed Suzuki reaction. It is not clear if the oxygen atoms play a part in stabilizing the metal center during the catalysis [144].

Structure **55** is a suitable ligand for palladium-catalyzed coupling of 9-alkyl-9-BBN with 1,1-dichloro-1-alkenes [Eq. (1.7)]. The reaction shows good selectivity for preparation of chlorinated internal alkenes of Z-configuration along with suppression of the two-fold substitution side reaction. Structure **55** was proven to be a P,P-chelate ligand in complexation with palladium [145]. The above catalytic selectivity is ascribed to the geometrical constraints of the complex engendered by a large bite angle and rigidness of the ligand which lead to a faster decomposition rate of the unusually persistent palladium(0)–chloro-olefin intermediate [146].

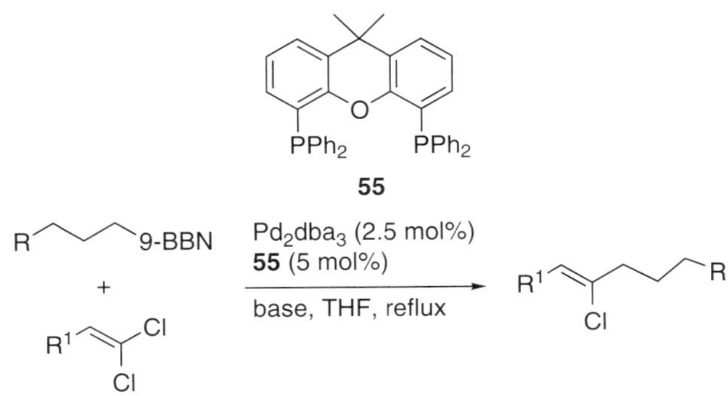

(1.7)

Some P(III) compounds with P-heteroatom bonds such as P–O, P–N, and P–Cl bonds were used as ligands in palladium-catalyzed Suzuki coupling reactions. Palladium complexes processing phosphinous acid ligands **56–58** efficiently catalyze the Suzuki coupling of electron-rich and electron-poor aryl chlorides as well as 3-chloropyridines in THF or dioxane [147–149]. Structure **57**-catalyzed coupling of aryl chlorides can be carried out in aqueous media under microwave irradiation [150]. In addition, **56** also efficiently catalyzes coupling of arylboronic acids with 1-chloro-1-cyclopentene to generate styrene derivatives in good yields, whereas **57** exhibits less activity under similar conditions [148].

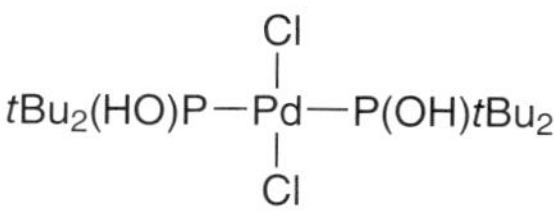

56

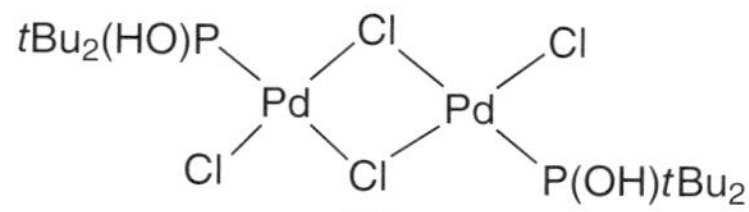

57

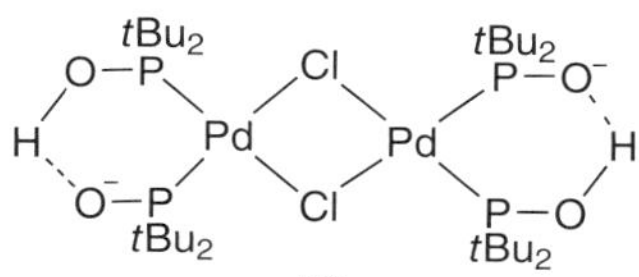

58

Pd/P(OR)$_3$ (R = bulky alkyl or aryl) systems show good activity only for the activated aryl chlorides [151]. Chiral ligands **59a–c** in combination with Pd$_2$dba$_3$ form efficient catalysts that drive asymmetric coupling of very bulky aryl chlorides with arylboronic acids at low palladium loading (0.05–0.5 mol%). Up to 91% yield and 78% ee can be achieved for one set of coupling partners [Eq. (1.8)] [152]. A structurally related phosphine **60** is also very efficient in Pd-catalyzed Suzuki coupling of sterically hindered aryl substrates. Tri-*ortho*-substituted biphenyls can be prepared in excellent yields with extremely low palladium loading (up to TON 5000). But this syatem cannot catalyze formation of tetra-*ortho*-substituted biphenyls such as 2,2′,6,6′-tetramethylbiphenyl [153]. Structure **61** is a phosphine chloride that in combination with Pd$_2$dba$_3$ displays higher catalytic activity than Pd/P(OR)$_3$ and can drive cross-coupling of unactivated and deactivated aryl chlorides with arylboronic acids in good to excellent yields [154].

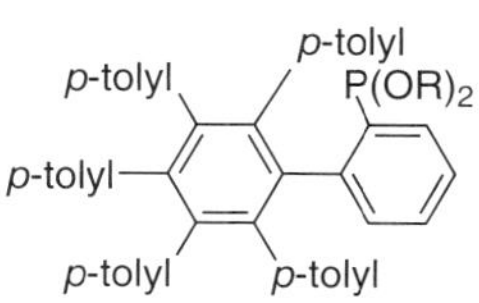

59a-c

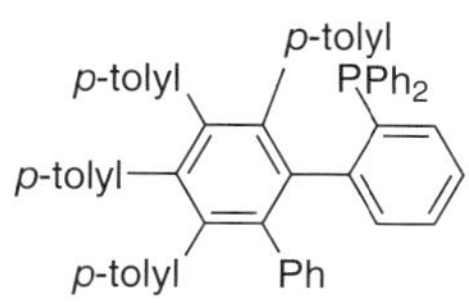

60

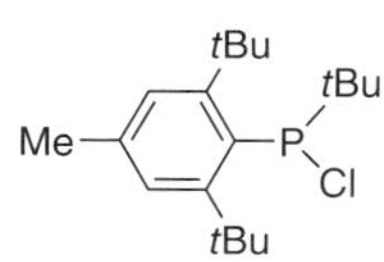

61

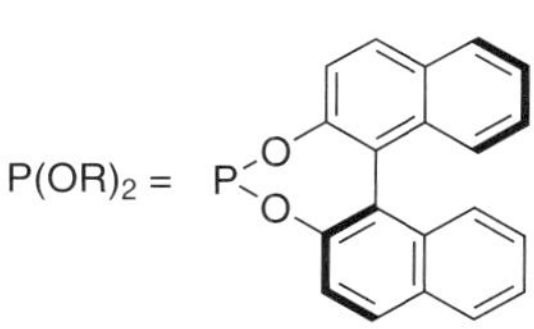

59a

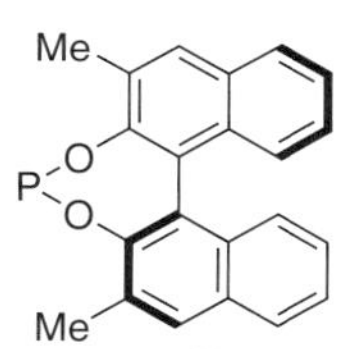

59b

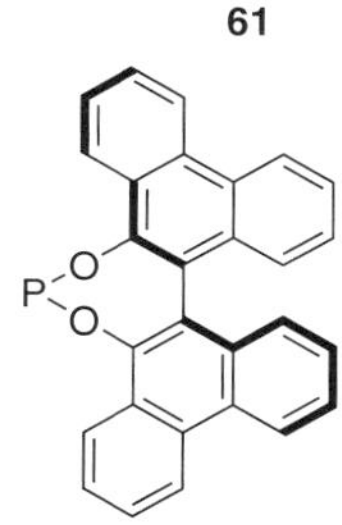

59c

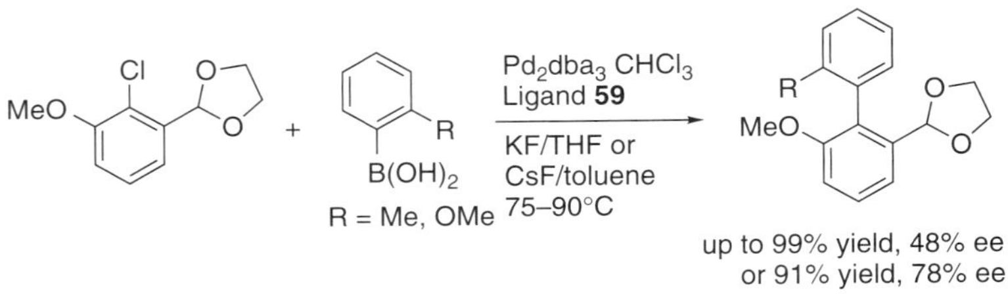

(1.8)

Several aminophosphines were used as ligands in Pd-catalyzed Suzuki coupling of organic chlorides. P(*i*BuNCH$_2$CH$_2$)$_3$N in combination with Pd(OAc)$_2$ catalyzes cross-coupling of a wide variety of aryl bromides and chlorides with arylboronic acids, affording the desired biaryls in excellent yields. However, relatively high Pd(OAc)$_2$ loadings (4 mol%) are required for the chloride substrates [155]. A palladium complex with sterically hindered diaminochlorophosphine ligand **62** catalyzes the synthesis of tetra-*ortho*-substituted biphenyls through the coupling of aryl bromides or chlorides with arylboronic acids. However, for chloride substrates the catalytic system leads to only moderate yields [156]. Both **63** and **64** are electron-rich, bulky, and hemilabile ligands. They in combination with Pd$_2$(dba)$_3$ show good catalytic activities in the Suzuki coupling of aryl chlorides. Pd$_2$(dba)$_3$/**63** exhibits a wide substrate scope, including a variety of electron-rich, electron-poor, and functionalized aryl chlorides as well as heteroaryl chlorides [157, 158].

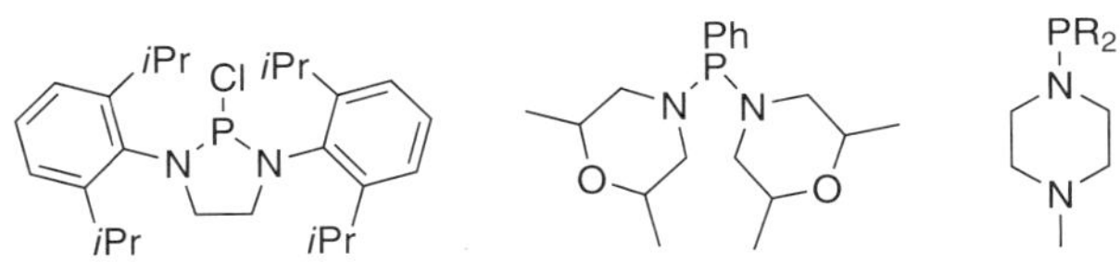

62 **63** **64a** R = *i*Pr
64b R = Cy

(ii) *N*-Heterocyclic Carbene (NHC) Ligands. *N*-Heterocyclic carbenes show high σ-basicity and low π-acidity. Compared with phosphines, NHCs are stronger σ-donor ligands. A lot of palladium–NHC complexes have been prepared and displayed excellent catalytic properties in cross-coupling reactions. Several excellent reviews on this subject have been published recently [17, 159, 160].

Herrmann et al. [161] first reported the use of Pd–NHC complex **65** to catalyze the reaction of 4-chloroacetophenone with arylboronic acid. Further studies found that the combination of Pd(OAc)$_2$ and **66** exhibits very low catalytic activity for the coupling of unactivated aryl chlorides [162]. This can be improved through modification of ligands by changes of substituents, steric hindrance, and linkers. Each of the modified ligand precursors **67–70** in combination of Pd(OAc)$_2$ shows good activity in catalyzing the reaction of unactivated aryl chlorides with PhB(OH)$_2$ [162]. In particular, Pd(OAc)$_2$/**67** catalyzes the coupling of electron-rich and electron-poor aryl chlorides with PhB(OH)$_2$ in DMF-H$_2$O in good to excellent yields [163].

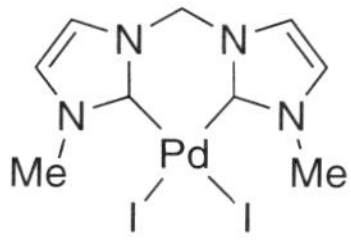

65

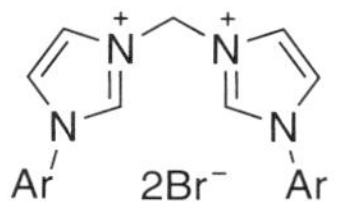

Ar = 4-MeOC$_6$H$_4$,
3,5-Me$_2$C$_6$H$_3$
or mesityl

66

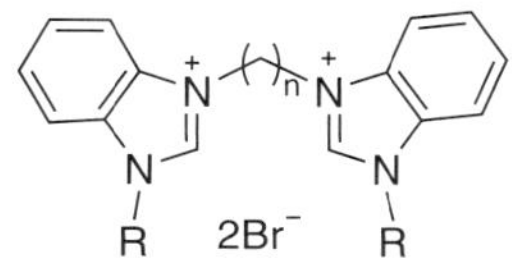

n = 1-4, R = 2,4,6-Me$_3$C$_6$H$_2$CH$_2$
n = 2 or 4, R = 3,4,5-(MeO)$_3$C$_6$H$_2$CH$_2$

67

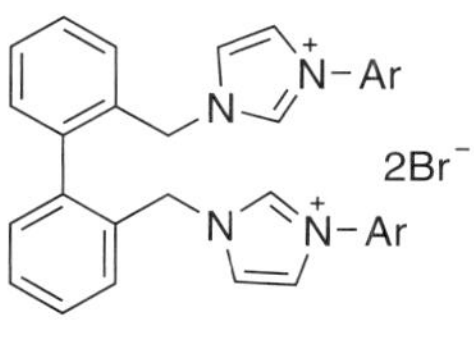

68

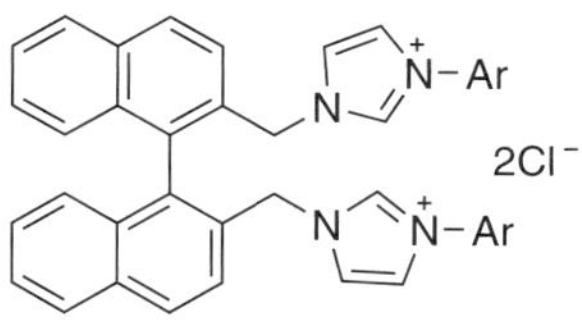

Ar = mesityl

69

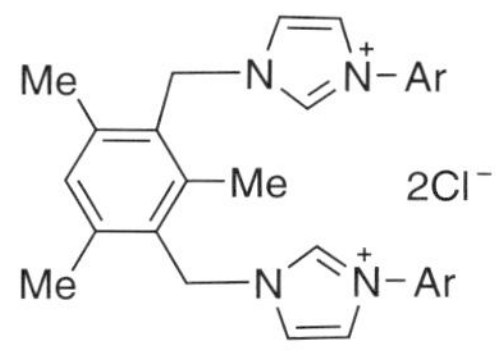

70

The catalytic property of complexes strongly depends on the steric bulk of the NHC ligand. The bis-carbene palladium complex **71** does not show any catalytic activity in the coupling of 4-chloroanisole with phenylboronic acid [164]. Whereas complex **72a** exhibits excellent catalytic activity for the reaction of electron-rich and electron-poor aryl chlorides with arylboronic acids in dioxane at room temperature, affording the desired biphenyls in almost quanitative yields [165]. Both **72b** and **72c** show moderate activity, being comparable with corresponding **73a** and **73b**. It seems that the degree of backbone saturation—and, hence, electron-donating capability—of the complexes plays a relatively small role in the complex's catalytic activity, whereas the steric hindrance of the R groups is a decisive factor. Indeed, **72c** and **73b** show higher activity than **72b** and **73a** [166].

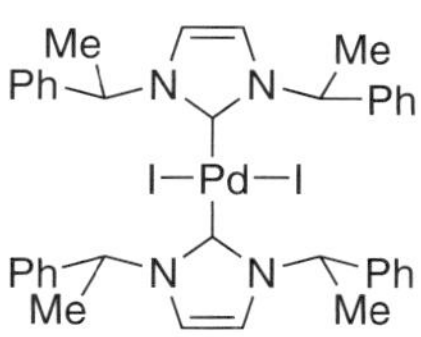

71

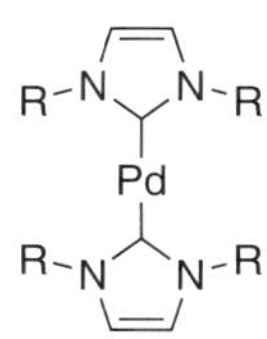

72a R = Ad
72b R = tBu
72c R = 2,6-iPr$_2$C$_6$H$_3$

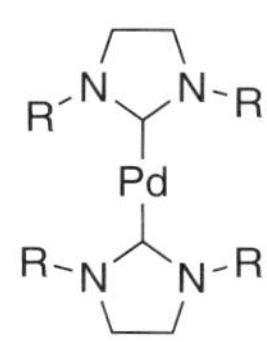

73a R = tBu
73b R = 2,6-iPr$_2$C$_6$H$_3$

Structures **74a–74d** are NHC–palladium polymer self-supported and reusable catalysts. Structures **74a–74c** are not soluble in water, while **74d** is water-soluble. Each of them can catalyze cross-coupling of electron-rich and electron-poor aryl chlorides with arylboronic acids in water at 80–90°C (for **74a–74c**) or at room temperature (for **74d**) with low catalyst loadings (0.05 mol%). Among them, **74c** is more reactive in the Suzuki cross-coupling of chloroarenes, including both deactivated and hindered aryl chlorides with different types of arylboronic acids. Experimental facts showed that the organometallic polymers **74a–74c** could be the source of production of active soluble Pd species in the form of either Pd nanoparticles or fragmented NHC–Pd complexes. Furthermore, the capsular structure of these lipophilic polymers in water provides a means of entrapment of nanoclusters in a hydrophobic region, thus accelerating the reaction in pure water in the absence of any co-organic solvent. In addition, **74d**, as a homogeneous catalyst, can be easily separated from other water-soluble species in the reaction mixture using a simple dialysis technique. This is highly significant for industrial processes [167–169].

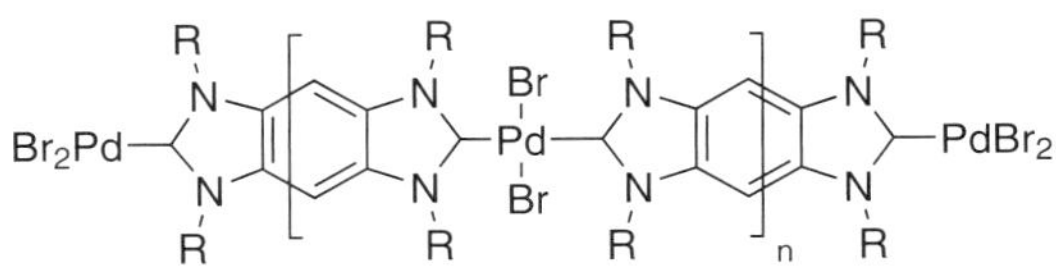

74a R = CH_2Ph **74b** R = *n*-hexyl
74c R = *n*-dodecyl **74d** R = $(CH_2CH_2O)_3Me$

The combination of 2 equiv of NHC precursors **75–79** and appropriate palladium sources such as $Pd(OAc)_2$ and Pd_2dba_3 results in active catalysts for the Suzuki coupling of organic chlorides. In the $Pd(OAc)_2$/**75**- (R = R′ = aryl) or $Pd(OAc)_2$/**76**-catalyzed reactions, KF·18-crown-6 was found to be an effective base and THF is the suitable solvent. Under these conditions, $Pd(OAc)_2$/**76** can catalyze coupling of aryl or vinyl chlorides with aryl, vinyl, and methyl boron reagents in high yields. Tri- and tetra-*ortho*-substituted products are efficiently produced [170]. When R and R′ are alkyl groups in **75**, the reaction of $Pd(OAc)_2$/**75**-catalyzed Suzuki reaction effectively proceeds in DMF/H_2O using Cs_2CO_3 as a base [171]. Pd_2dba_3/**75** (R = R′ = 2,6-$i$$Pr_2C_6H_3$) also catalyzes the reaction of $C_6H_{13}C{\equiv}CB(OMe)_3^-Li^+$ with activated, nonactivated, and deactivated chloroarenes, leading to good to excellent product yields [172]. The combination of palladium salts and **77–79** exhibits catalytic activity comparable to that of a Pd/**75** system in the cross-coupling of aryl chlorides. It seems that the effect of the NHC ring size is very limited [173–175].

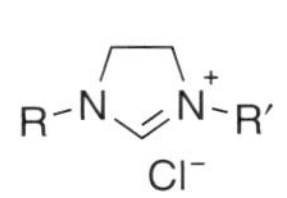

R = 2,4,6-$Me_3C_6H_2CH_2$
$PhCH_2$, $(CH_2)_2OMe$
R′ = 3,4,5$(MeO)_3C_6H_2CH_2$
CH_2Ph, CPh_3, *n*Bu
or R = R′ = 2,6-*i*$Pr_2C_6H_3$

75

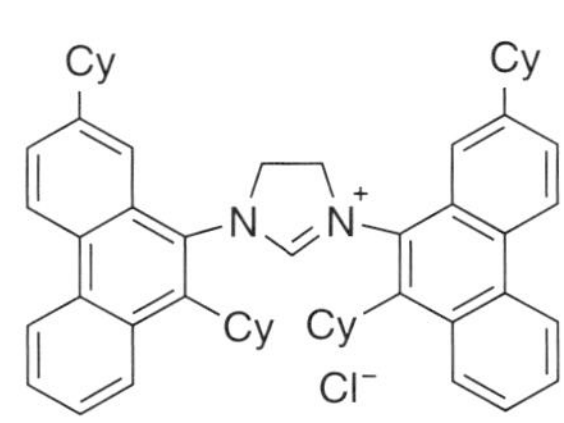

76

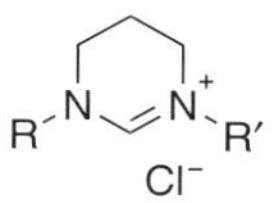

77

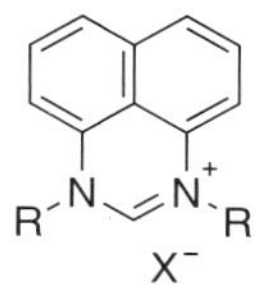

78

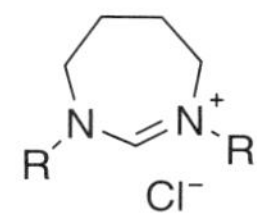

79

It was noted that the catalytic system consisting of mixtures of palladium and ligand precursors gives better yields in the coupling reactions compared to the isolated biscarbene palladium complexes [166, 173]. Glorius and co-workers [176] found that complex **80** is catalytically inactive in the Suzuki coupling at room temperature, whereas a 1:1 mixture of $Pd(OAc)_2$ and **81** prepared *in situ* displays high activity under the same conditions, catalyzing the reaction of hindered and unhindered, activated and deactivated aryl chlorides with aryl boronic acids efficiently. Both **82** and **83** are highly effective pre-catalysts for the Suzuki coupling of sterically hindered aryl chlorides or heteroaryl chlorides with arylboronic acids in alcohol. The former also efficiently catalyzes the coupling of allyl or benzyl chlorides with phenylboronic acid. These experimental findings support the fact that a Pd^0(NHC) species may be the active catalyst [177, 178].

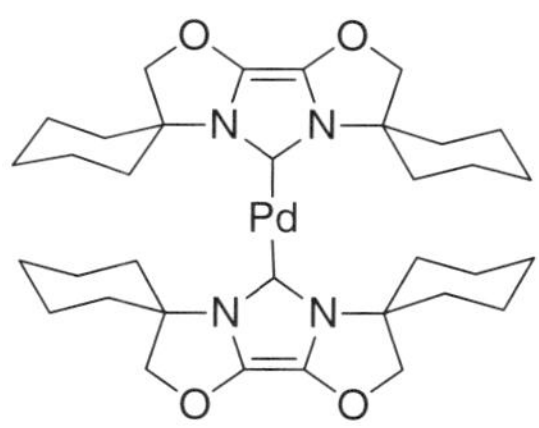

80

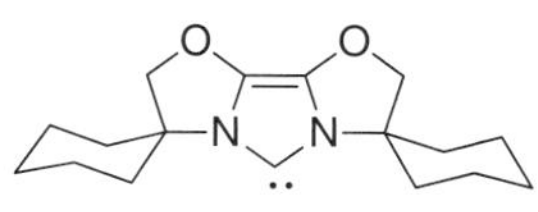

81

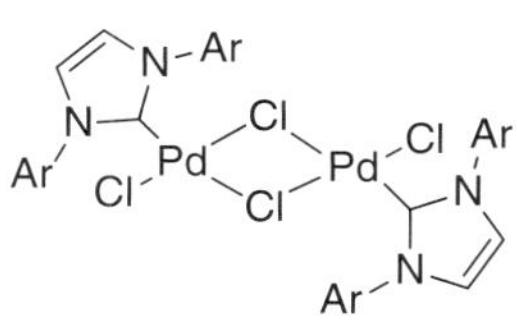

Ar = 2,6-$i$$Pr_2C_6H_3$

82

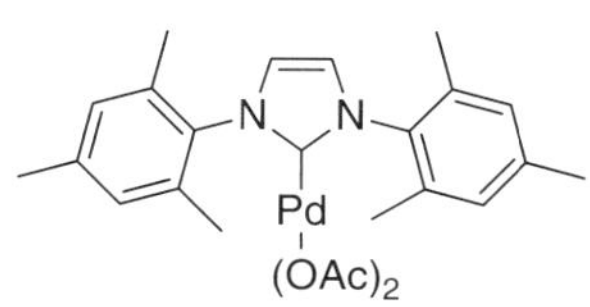

83

The mono-carbene palladium complexes with another ancillary ligand (anionic or neutral) were proven to be excellent pre-catalysts for the Suzuki coupling. These pre-catalysts were believed to liberate an active [(NHC)Pd(0)] species into the catalytic cycle after appropriate activation. However, the ancillary ligand is also crucial to their catalytic performance besides the dominant NHC ligand. In 1999 Herrmann and co-workers [164] reported mixed NHC-phosphine complex of palladium (**84**) for Suzuki coupling reaction of aryl chlorides. In the reactions, PCy_3 is indispensable. Both **85** and **86** show higher catalytic activity than **84** for the Suzuki coupling of aryl chlorides. Electron-rich or -poor and sterically hindered aryl chlorides can be employed, and the reaction proceeds effectively under mild conditions [179, 180]. $ArBF_3K$ and alkyl-9-BBN can also be employed as the nucleophilic reagents in **86**-catalyzed Suzuki coupling of aryl chlorides [179, 181]. Jin et al. [182] evaluated the activity of **87a–88c** in catalyzing the reaction of p-ClC_6H_4OMe with p-$MeC_6H_4B(OH)_2$ in *i*PrOH at room temperature. The allylpalladium complexes **87b** and **88b** exhibit higher activity than the corresponding cyclopentadienylpalladium complexes **87a** and **88a**, respectively, whereas **87c** and **88c** show low activity [182]. Nolan and co-workers [183] found that at elevated temperature, **87c** displays catalytic activity similar to that of **87b**. It was thought that the substitution at the allyl moiety is only useful to activate the catalyst at lower temperatures, forming active catalytic species [(NHC)Pd(0)]. Complex **87b** also permits the coupling of deactivated, functionalized, or bulky aryl chlorides with boronic acids at room temperature at low catalyst loadings (0.05 mol%). At elevated reaction temperature, catalyst loadings as low as 50 ppm effect Suzuki–Miyaura reactions with aryl chlorides [183].

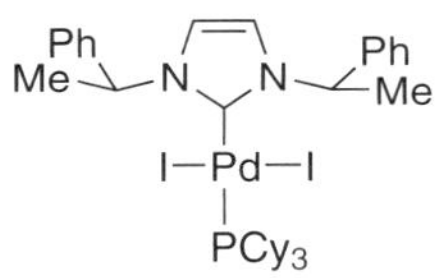

84

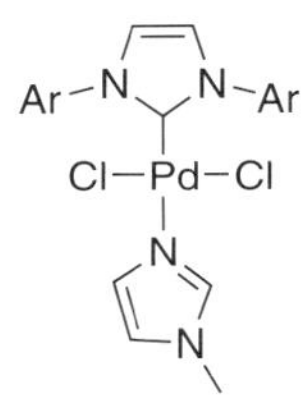

85

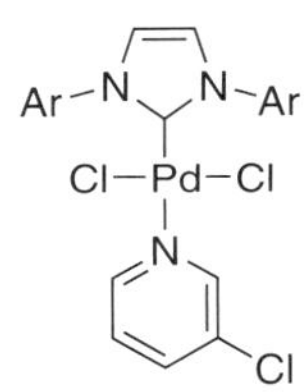

86

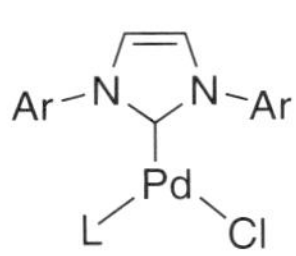

87a L = Cp
87b L = 1-Ph-allyl
87c L = allyl

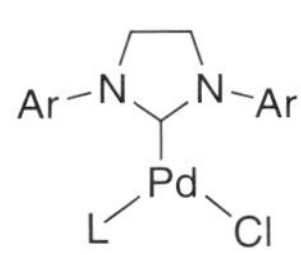

88a L = Cp
88b L = 1-Ph-allyl
88c L = allyl

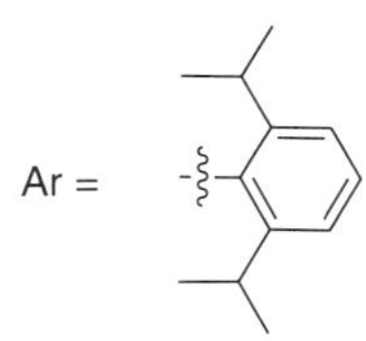

(iii) Palladacycles. Palladacycles are among the most widely used catalyst systems for the C–C bond formation reactions. However, the early studies showed that the P,C-palladacycles **89–91** are low active in catalyzing the Suzuki reaction of unactivated aryl chlorides [184–187]. Use of tricyclohexylphosphine as co-ligand was found to greatly increase the catalytic activity of the palladacycles. It leads to the transformation of a dimeric palladacycle to a monomer—for example, from **90a** to **92**. The **90a**/$2PCy_3$ system is an extremely active catalyst for the reaction of activated, and deactivated aryl chlorides with arylboronic acids. The catalyst loading is as low as 0.005 mol% for the coupling of 4-chloroanisole with phenylboronic acid, leading to 100% conversion of 4-chloroanisole in 17 h. For the activated aryl chlorides such as p-$ClC_6H_4NO_2$, extremely high turnover numbers of up to 1 million are observed. Similar precursors **90b** and **90c** in combination with PCy_3 show slightly lower activity, whereas **93**/PCy_3 is superior to the other systems. The combination of **93** and PCy_3 catalyzes the Suzuki coupling of deactivated, activated and sterically hindered aryl chlorides with TONs of up to 2 million. The high activity obtained with these catalysts is not due to any spectacular increase in rate, but rather to greater catalyst longevity. Catalyst longevity is conferred by the π-acidic nature of the phosphite ligand through coordination stabilization to active intermediate. Hence increasing the π-acidity of the supporting ligand leads to high total turnover numbers. Indeed, the activity order of **93** > **90a** > **90b** > **90c** is in line with decreasing π-acidity of the ancillary ligands. It is also important that the ratio of palladium to PCy_3 is maintained at 1:1. Higher loadings of PCy_3 result in a very rapid diminution in activity. This is ascribed to "overcoordination" of the active catalyst [188–191].

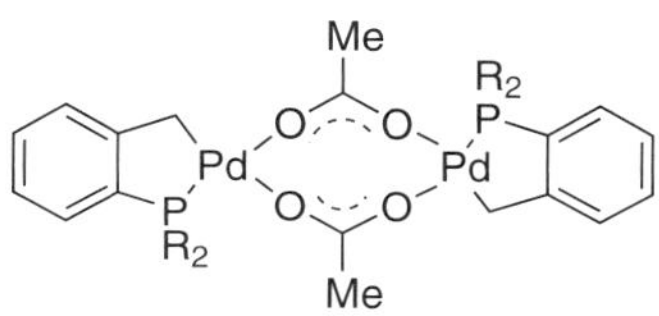

89 R = *o*-MeC_6H_4

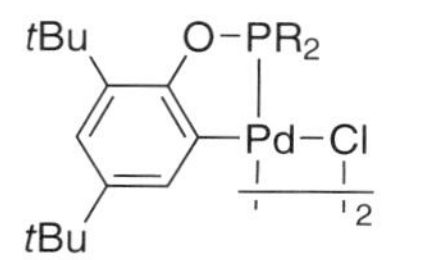

90a R = 2,4-*t*$Bu_2C_6H_3O$
90b R = Ph
90c R = *i*Pr

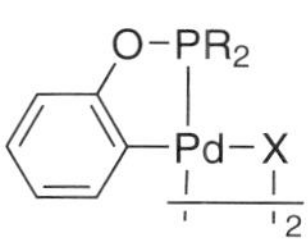

91 R = Ph, *t*Bu
X = Br, Cl

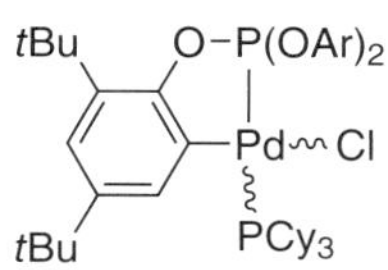

Ar = 2,4-*t*$Bu_2C_6H_3$
92

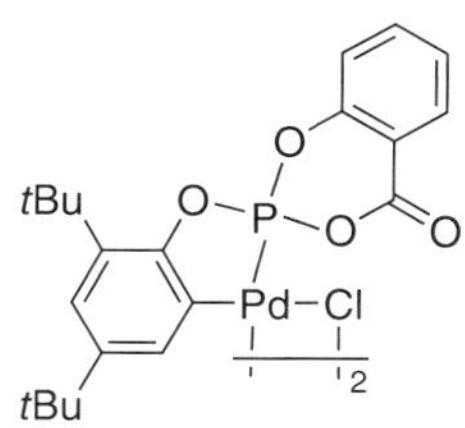

93

Similarly to C,P-palladacycle **91**, C,N-palladacycle **94** shows essentially no activity in the coupling of 4-chloroanisole and phenylboronic acid, whereas in the presence of PCy_3 **94** exhibits excellent activity for above coupling, which shows TONs of up to 8000. In this case a monomeric PCy_3 adduct of palladium **95** is formed as the pre-catalyst [192]. The replacement of PCy_3 in **95** with Cy_2P(*o*-biphenyl) or P*t*Bu_3 gives a good catalyst for the Suzuki coupling of aryl chlorides, but their performances are poorer than that of **95**. The replacement of TFA in **95** with another anionic group also affects the catalytic activity, the approximate order being Cl∼Br > TFA∼ OTf > I [193].

The active catalytic species was assumed to be zerovalent monophosphine palladium "$PdPR_3$" formed as shown in Scheme 1.7. This hypothesis is supported by several experimental facts, including the existence of induction times in the catalytic process, the identification of **99**, and the negative effect on catalytic activity of doubling the concentration of PCy_3 in catalysts formed *in situ*. In this proposed mechanism, the palladacycle complexes seem to only act as a palladium source [192, 193].

However, **95** or **94**/PCy$_3$ shows activity that is far higher than that of preformed palladium(0) complex [Pd(diene)(PCy$_3$)] or complexes formed *in situ* from PCy$_3$ and either [Pd$_2$(dba)$_3$] or Pd(OAc)$_2$ [86, 194]. Complex **97** formed from **96** and PCy$_3$ exhibits similar catalytic property in the coupling of 4-chloroanisole and phenylboronic acid [193].

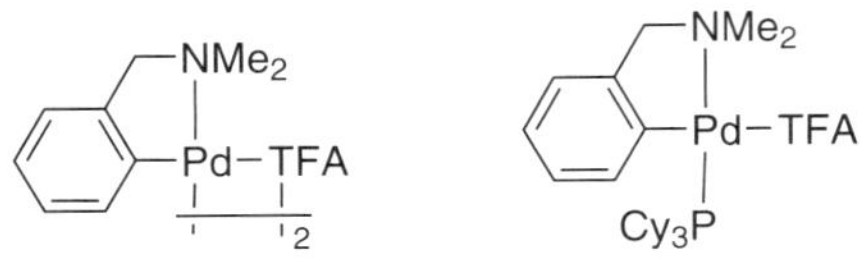

94 **95**

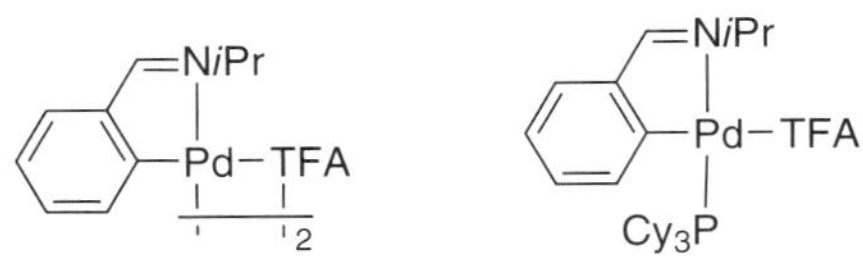

96 **97**

Nájera and co-workers [195–199] found that the oxime-derived palladacycles **100a–d** suit for a wide scope of substrates in the Suzuki coupling. The palladacycles catalyze cross-coupling of aryl, benzyl, and allyl chlorides with arylboronic acids or potassium aryltrifluoroborates in MeOH/H$_2$O, DMF/H$_2$O, acetone/H$_2$O, or H$_2$O without the requirement of phosphine additives. However, the reaction requires the presence of TBAB. The palladacycles **100b** and **100d** show higher catalytic activity than do the other analogues. Activated and deactivated aryl or heteroaromatic chlorides can be employed as the electrophiles. However, the reaction of deactivated aryl chlorides such as 4-chloroanisole with arylboronic acids usually gives relatively low yields and requires higher catalyst loadings compared with **95** and **97**. The reaction of activated aryl, benzyl, and allyl chlorides catalyzed by **100d** leads to good to excellent results. Potassium aryltrifluoroborates shows better reactivity with deactivated aryl chlorides using **100b** under refluxed H$_2$O or microwave radiation conditions [195–197]. Using Binap as a ligand, **100b** efficiently catalyzes vinylation of aryl and heteroaryl chlorides with potassium alkenyltrifluoroborates in DMF.

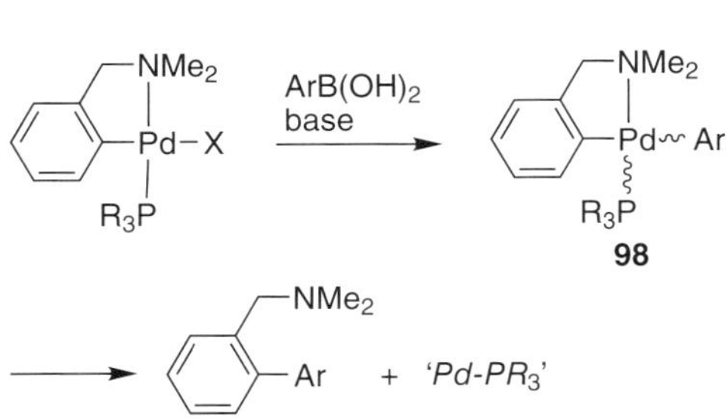

Scheme 1.7. Proposed formation route of the active catalytic species Pd-PR$_3$.

The combination of $Pd(OAc)_2$ and Binap displays the same effect [198]. Under microwave radiation, the vinylation reaction of unactivated and deactivated aryl chlorides can be carried out using **100d** as catalyst, tris(*tert*-butyl)phosphonium tetrafluoroborate as ligand, and tetra-*n*-butylammonium hydroxide as co-catalyst. Both alkenylboronic acids and potassium alkenyltrifluoroborates are suitable nucleophilic reagents. Each of the reactions not only gives normal cross-coupling product, but also small amount of α-isomer [Eq. (1.9)]. The conditions are also suitable for **100d**-catalyzed coupling of allyl or benzyl chlorides with aryl or alkenylboron reagents [199].

Palladacyclic complexes containing either tertiary or secondary alkylphosphine ligands **101** catalyze coupling of 4-chloroanisole and phenylboronic acid in excellent yields [200]. Replacement of the alkylphosphine ligands in **101** with a sterically demanding NHC results in another palladacyclic complex **3**, which have been demonstrated to be excellent in the Suzuki coupling of aryl chlorides with arylboronic acids. The catalyst is applicable for a wide range of substrates, including activated, deactivated, and bulky aryl chlorides. Di- and tri-*ortho*-substituted biaryls are formed in excellent yields at room temperature in short reaction times (50–75 min) [201].

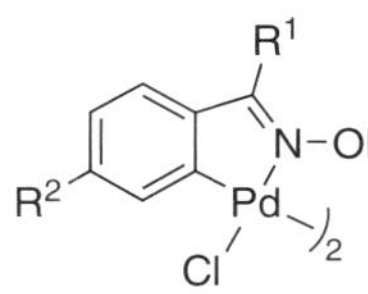

100a R^1 = Me, R^2 = H
100b R^1 = Me, R^2 = OH
100c R^1 = *p*-HOC_6H_4, R^2 = OH
100d R^1 = *p*-ClC_6H_4, R^2 = Cl

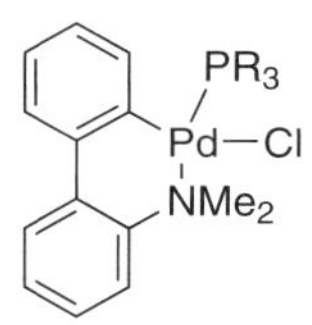

101 PR_3 = $HPNor_2$
or $HPCy_2$
or PCy_3

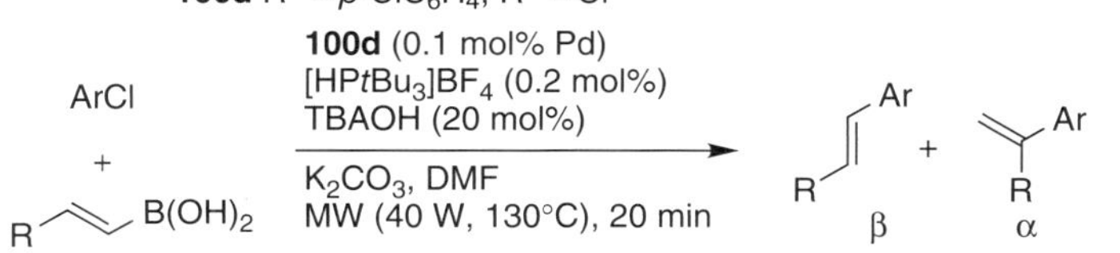

(1.9)

Two ferrocene-based C,N-palladacyclic complexes **102** and **103** are also active catalysts for the coupling of aryl chlorides with arylboronic acids. Both are applicable for the coupling of unactivated, deactivated, and sterically hindered aryl chlorides with 0.5 mol% (for **102**) or 0.05 mol% (for **103**) catalyst loadings and lead to excellent product yields. However, the TONs of both are lower than that using **95** [202, 203].

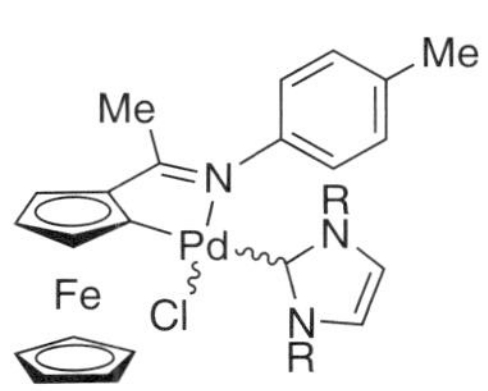

102 R = 2,6-*i*$Pr_2C_6H_3$

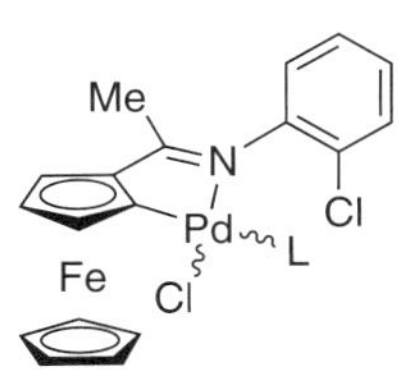

103a (L = 1-Ph-2-(Cy_2P)C_6H_4
103b (L = DavePhos)

C,S-palladacycle **104** cannot be cleanly transferred to mono- or bis-PCy_3 adducts by treating with PCy_3. The combination of **104** and a phosphine such as PCy_3, $PtBu_3$, and PR_2(*o*-biphenyl) (R = Cy, *t*Bu) shows good activity in the Suzuki coupling of chloroarene substrates, but not as high as that shown by either N- or P-based analogues [204]. C,O-palladacycle **105** is applicable for a wide range of substrates, including activated, unactivated, deactivated, and sterically hindered aryl chlorides, heteroaryl chlorides, and alkenyl chlorides as well as aryl, heteroaryl, and alkenylboron reagents (Scheme 1.8). It also catalyzes cross-coupling of alkyl bromides with aryl or alkylboron reagents in good yields [205].

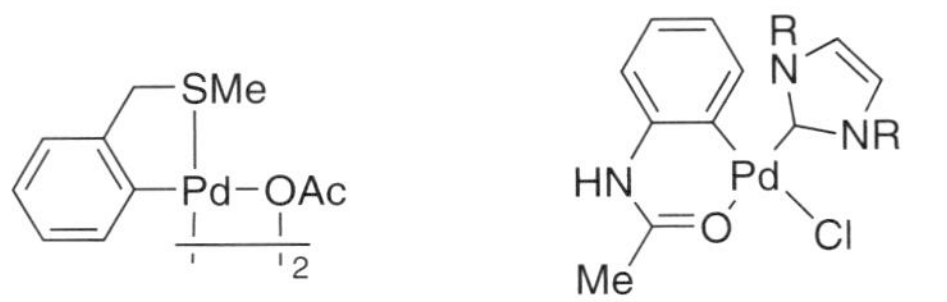

104

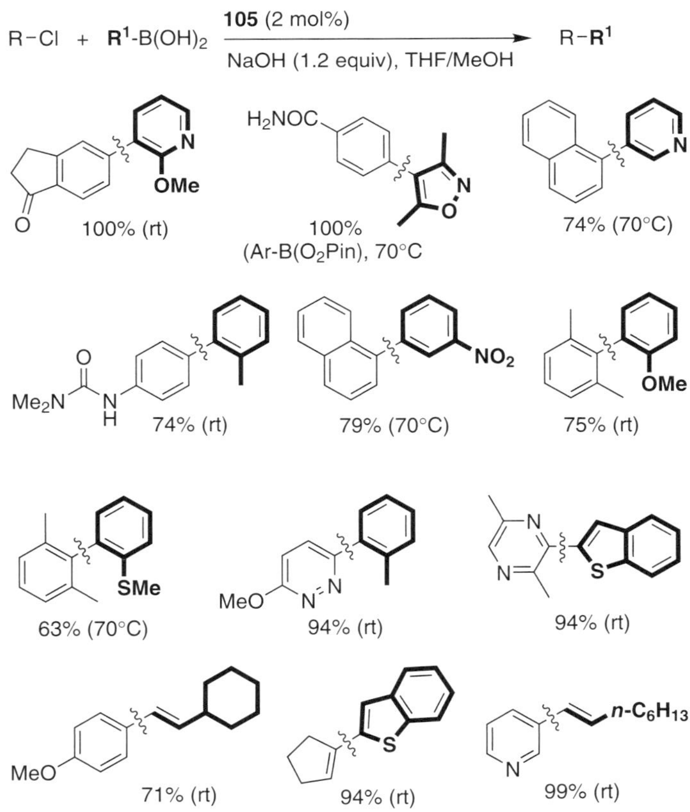

105 R = 2,6-$iPr_2C_6H_3$

Scheme 1.8. C,O-palladacycle **105**-catalyzed cross-coupling of aryl or alkenyl chlorides.

(iv) Nitrogen-Based Ligands. Several monodentate nitrogen ligands in combination with appropriate palladium source show catalytic activity in the cross-coupling of aryl chlorides. 1-Alkylimidazoline or 1-alkylbenzimidazole palladium complexes **106** and **107** are able to catalyze reaction of activated and deactivated aryl chlorides with phenylboronic acid. The reaction proceeds effectively in dioxane at 80°C using Cs_2CO_3 as a base, resulting in the desired biphenyls in 83–97% yields [206]. The combination of $PdCl_2(MeCN)_2$ and 1-alkylbenzimidazole **108** shows similar catalytic activity in toluene at 80°C, but no requirement of a phosphine as co-ligand [207]. $Pd(OAc)_2$/2DABCO system is also effective, but displays only mederate activity [208]. If the reaction is carried out in a continuous capillary microreactor, much higher reaction efficiency can be achieved. In the microreactor, the coupling products were obtained mostly in near quantitative yield within a 4-h residence time at 50°C [209].

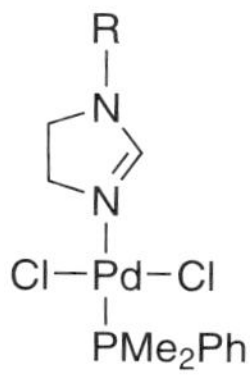

106

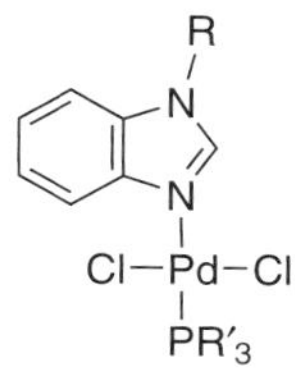

107 PR′3 = PPh3 or PMe2Ph

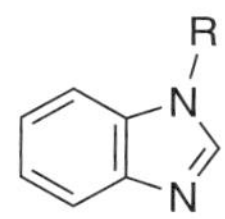

108

Palladium complexes bearing bidentate nitrogen ligands **109** and **110** were prepared from $Pd(cod)Cl_2$ and corresponding ligands. Each of them catalyzes coupling of benzyl or activated aryl chlorides with $PhB(OH)_2$ in excellent yields, whereas for the reaction of unactivated aryl chlorides such as chlorobenzene, low to moderate yields are achieved [210]. Bispyridyl complex **111** catalyzes allyl, benzyl, and activated aryl chlorides to couple with arylboronic acids in water in good to excellent yields. For the electron-rich aryl chloride such as p-$NH_2C_6H_4Cl$ and o-$AcNHC_6H_4Cl$, low to moderate yields were obtained. Complex **111** also catalyzes the coupling of alkylboron reagents such as $n$$BuB(OH)_2$ and $(MeBO)_3$ with activated aryl chlorides, but long reaction times are required [211, 212].

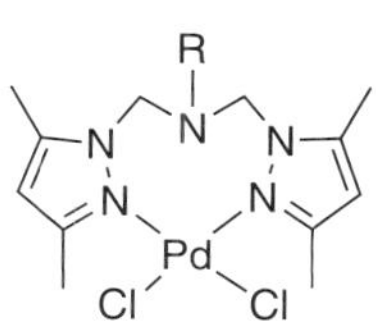

R = 2,6-Me2C6H3
mesityl
benzyl

109

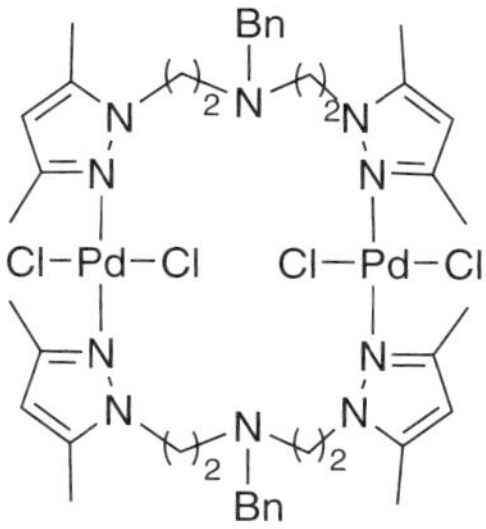

110

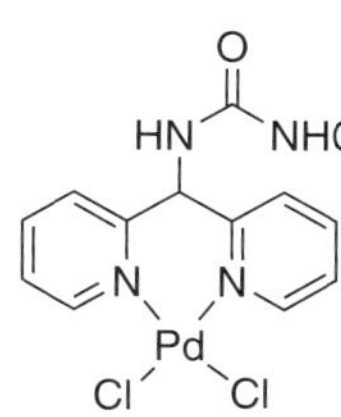

111

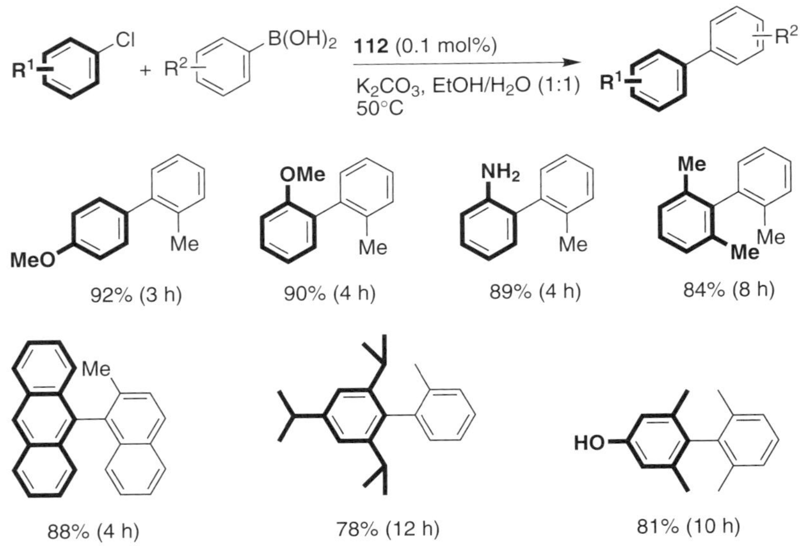

Scheme 1.9. Complex **112**-catalyzed Suzuki coupling of deactivated and bulky chlorobenzenes.

The β-diketiminatophosphane Pd complex **112** was found to be a powerful catalyst for the coupling of deactivated and hindered aryl chlorides under mild conditions with 0.1 mol% Pd loading (Scheme 1.9). This catalytic system also proved to be highly effective in one-pot multiple couplings, including couplings of di-, tri-, and tetra-chlorobenzenes with various arylboronic acids (Scheme 1.10) [213].

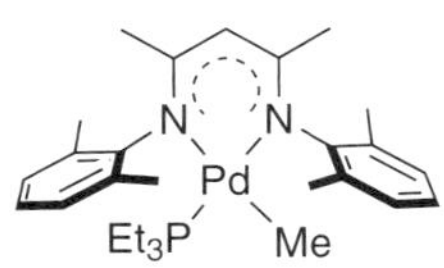

112

1.3.1.2 Nickel Catalysts

(i) Phosphine Ligands. In the 1990s, several phosphine-nickel-catalyzed cross-couplings of aryl chlorides and arylboronic acids in a homogeneous or heterogeneous system were reported. The phosphine ligands used include dppf, dppe, triarylphosphines, and 3,3′,3′′-phosphinidynetris(benzenesulfonic acid)trisodium salt (TPPTS). The reactions gave moderate to excellent yields, depending on catalysts and reactivity of the substrates [214–218].

The Ni-PPh_3 catalytic system mentioned above requires elevated reaction temperature for the Suzuki coupling of aryl chlorides. Hu and co-workers reported that $Ni(PR_3)_n$ formed *in situ* from $Ni(cod)_2$ and PPh_3 or PCy_3 are able to catalyze the

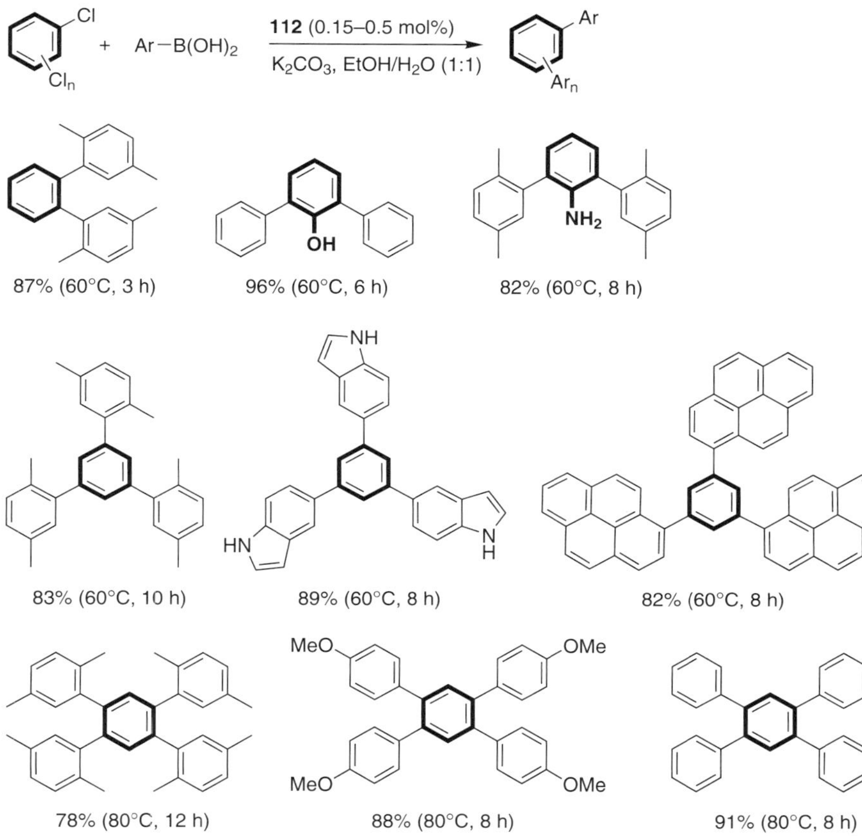

Scheme 1.10. **112**-catalyzed Suzuki coupling of di-, tri-, and tetra-chlorobenzenes.

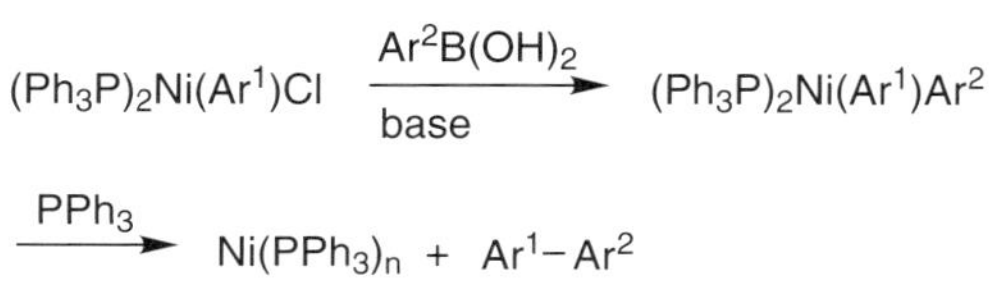

Scheme 1.11. Formation of catalytically active species from $Ni(PPh_3)_2(Ar)Cl$ and PPh_3.

coupling of electron-rich or electron-poor aryl chlorides with arylboronic acids at room temperature. $Ni(PPh_3)_n$ generated from $Ni(PPh_3)_2Cl_2$ and *n*BuLi in the presence of ArCl displays similar catalytic effect. However, in the absence of ArCl the species formed from $Ni(PPh_3)_2Cl_2$ and *n*BuLi is inactive [219]. In the presence of PPh_3 or PCy_3, $Ni(PPh_3)_2(naph)Cl$ shows similar catalytic properties in the Suzuki coupling because the reaction of $Ni(PPh_3)_2(naph)Cl$ with $ArB(OH)_2$ is believed to generate the same catalytically active species (Scheme 1.11) [220, 221].

The combination of ferrocene-based bisphosphine ligand **113** and $Ni(cod)_2$ is also active in catalyzing the cross-coupling of electron-rich and -poor aryl chlorides in refluxed THF, leading to moderate to excellent product yields. The ferrocene-based bisphosphine-containing polymers **114** and **115** behave similarly to **113**, but the polymers could be recovered and reused. The ferrocene-based monophosphine ligand **116** together with $Ni(cod)_2$ exhibits good catalytic property at room temperature. However, its polymeric form and $Ni(cod)_2$ shows much lower activity than does **116**/$Ni(cod)_2$ [221, 222].

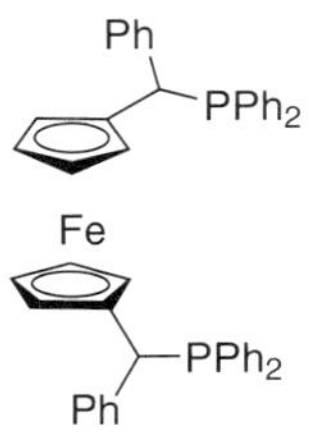

113

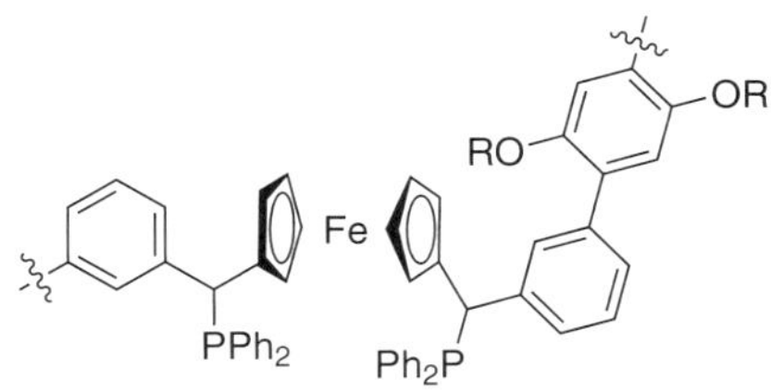

114

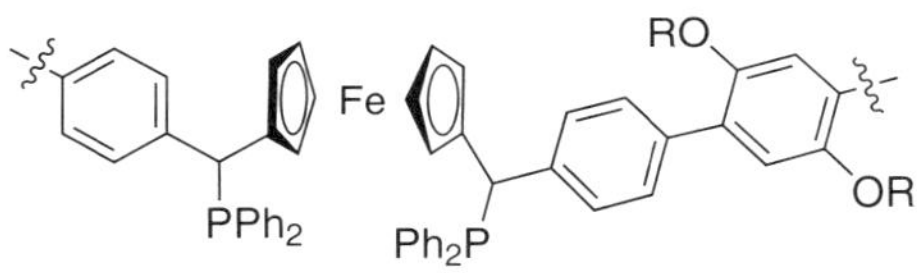

115

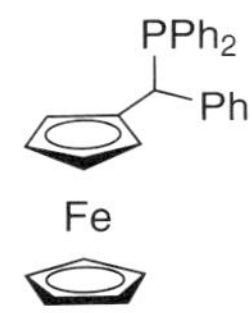

116

(ii) Diamine Ligands. Fu and co-worker [223] reported that diamine ligand **117** in combination with $NiBr_2$·diglyme catalyzes cross-coupling of unactivated primary and secondary alkyl chlorides with primary alkylboron reagents at room temperature (Scheme 1.12).

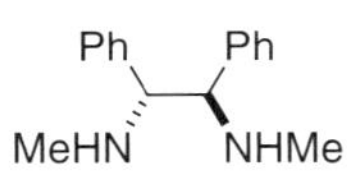

117

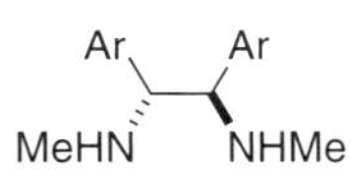

118 Ar = 1-naphthyl

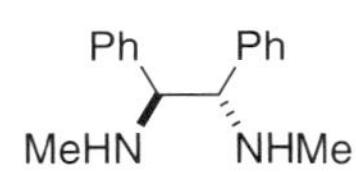

119

The analogue **118**/$NiBr_2$ catalyzes asymmetric coupling of racemic secondary alkyl chlorides that bear proximal arylamines with primary alkylboron reagents [Eq. (1.10)]. This is a nitrogen-directed enantioselective Suzuki reaction. A substrate that lacks an amino substituent leads to poor enantioselectivity of the coupling product. The suitable spatial relationship between the amino group and the chloride and conformation of the chloride substrates are also important for obtaining good enantioselectivity. Isolation of the chloride and the amino group by longer carbon chain or conformational restriction of the chloride substrate result in lower or even no enantioselectivity (Scheme 1.13) [224]. The analogue **119**/$NiBr_2$ catalyzes a series of racemic acylated bromohydrins to couple asymmetrically with primary alkylborane.

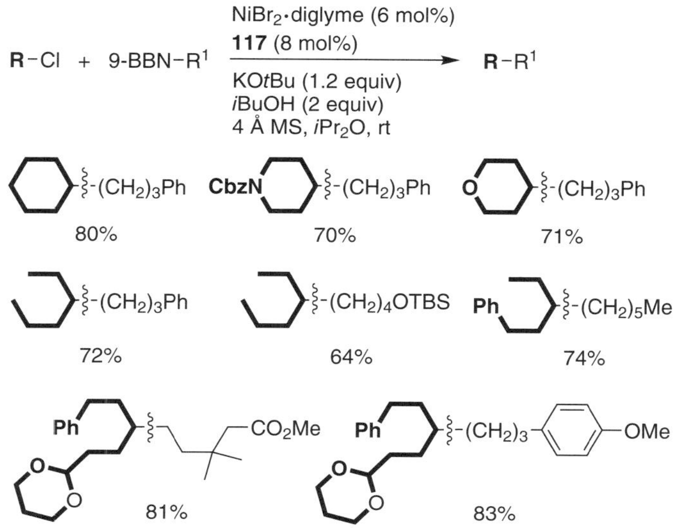

Scheme 1.12.

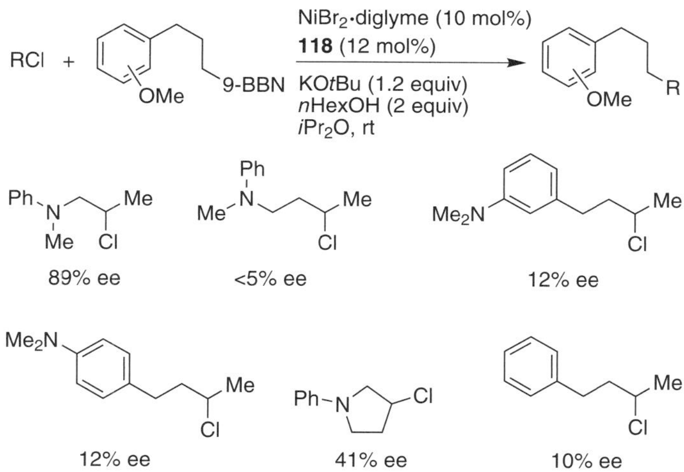

Scheme 1.13.

An example using unactivated secondary alkyl chloride was reported [Eq. (1.11)] [225].

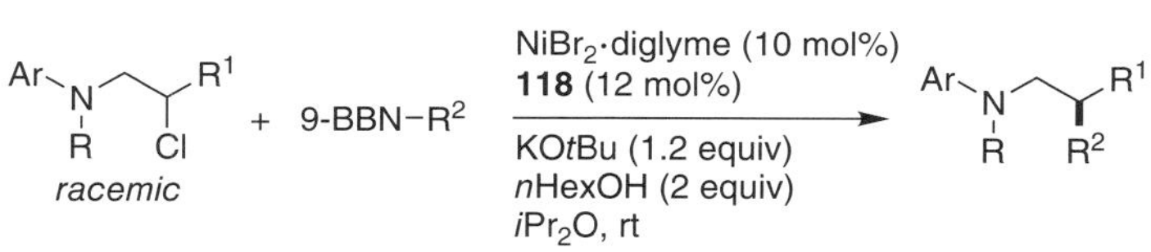 (1.10)

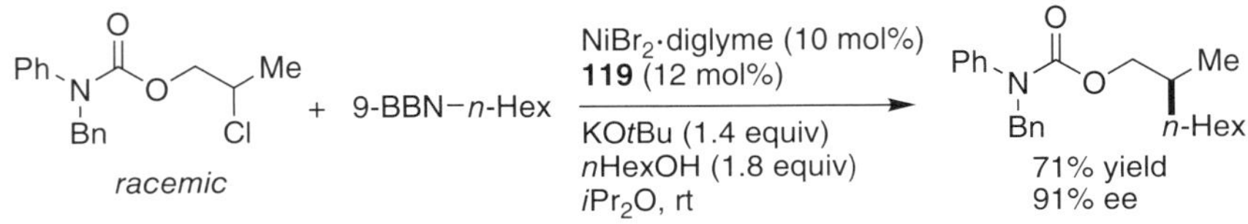

(1.11)

(iii) Amino Alcohol Ligands. The $NiCl_2$·Glyme/prolinol system was found to catalyze the reaction of primary and secondary alkyl chlorides with arylboronic acids or potassium aryltrifluoroborates, forming cross-coupling products in moderate to good yields (Scheme 1.14). When potassium aryltrifluoroborates are employed as the nucleophilic reagents, only 1 equiv of the boron reagents is required. Heteroaryl and *ortho*-substituted organoborons can also be used [226, 227].

(iv) NHC Ligands. NHC–Ni(0) complex **120** catalyzes the coupling of PhCl with $PhB(OH)_2$ in refluxed toluene using KO*t*Bu as a base. The stoichiometric reaction of **120** with electron-rich or electron-poor aryl chlorides generates oxdative addition species **121** [228]. Each of the cationic complexes **122–124** exhibits catalytic activity in the coupling of activated and deactivated aryl chlorides with arylboronic acids. The activity order is approximately **122** > **123** > **124**. In addition, the catalytic reaction using **123** and **124** requires PPh_3 as the co-ligand, whereas **122** can effectively catalyze the cross-coupling without the need for a catalytic amount of PPh_3. However, for unreactive aryl chlorides at low catalyst loading, the presence of PPh_3 accelerates the reaction [229, 230].

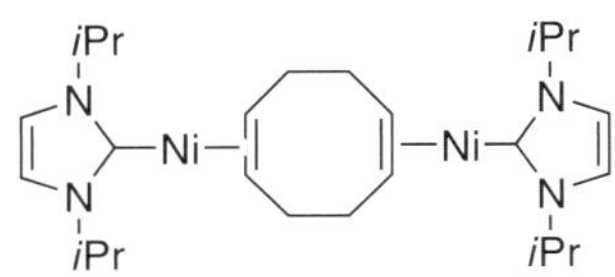

120

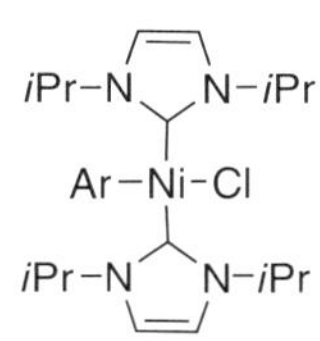

121

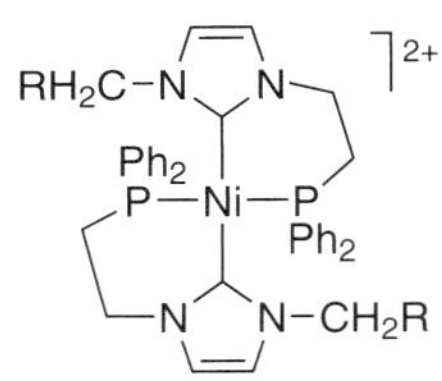

122 R = 1-naphthyl

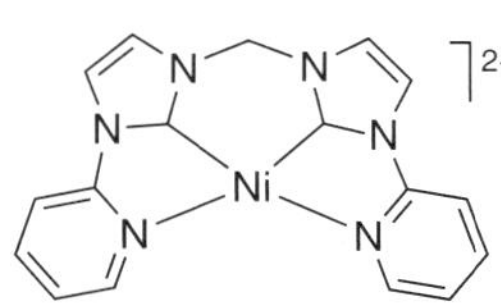

123

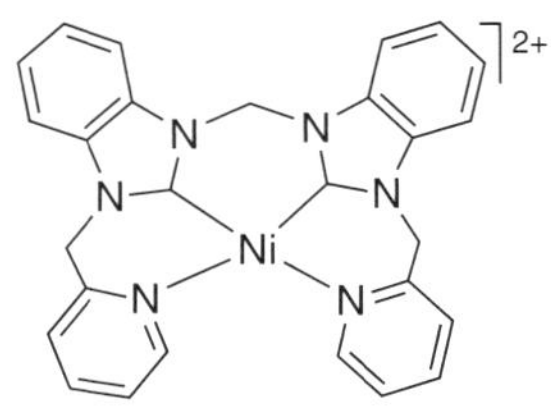

124

1.3.1.3 *Other Metals.* CuI was found to catalyze cross-coupling of primary alkyl chlorides with aryl boronate ester, leading to desired products in moderate yields [Eq. (1.12)] [231]. $[RhCl(cod)]_2$/dppp catalyzes cross-coupling of $NaBPh_4$

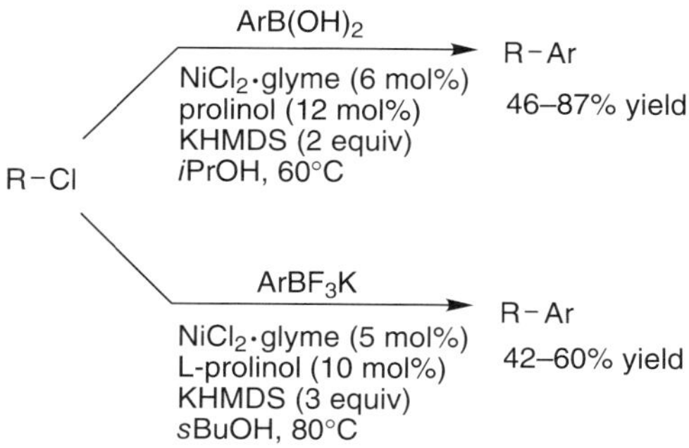

Scheme 1.14.

with activated aryl chlorides such as *p*-ClC$_6$H$_4$COOEt, *p*-ClC$_6$H$_4$CF$_3$, and 2- or 3-chloropyridine in moderate to good yields. However, treatment of 4-chlorobenzonitrile with NaBPh$_4$ under the same conditions results in a complex mixture of mono-, di-, and triphenylated products, in which the chloro moiety remained in each of them [232].

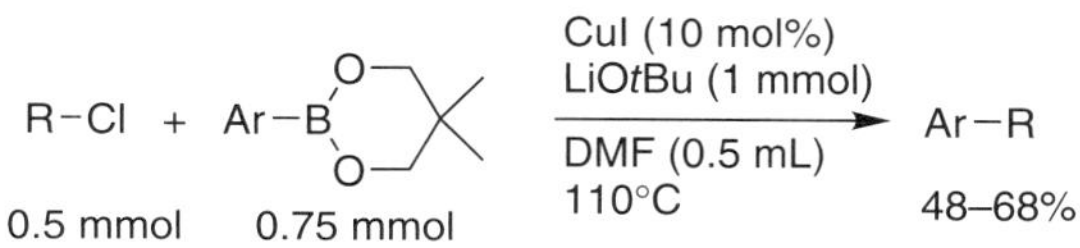

(1.12)

R = *n*Pr, *n*Bu, *n*C$_6$H$_{13}$, *n*C$_{12}$H$_{25}$, Ph(CH$_2$)$_4$
Ar = Ph, 4-ClC$_6$H$_4$, 4-*t*BuC$_6$H$_4$, 2-thienyl

1.3.2 Negishi Reaction

The Negishi coupling was found in 1970s and represents a powerful tool for the construction of carbon–carbon bonds because of the ready availability and high functional-group compatibility of organozinc compounds. Before 2000, organic bromides and iodides were predominantly employed as electrophilic substrates in the reaction. The use of aryl chlorides as electrophiles is scarce, although the first Negishi cross-coupling using aryl chloride as an electrophile was reported in 1980 [233]. However, significant progress has been achieved in recent years. A range of catalysts have been developed, which efficiently catalyze cross-coupling of aryl, alkenyl, and alkyl chlorides with organozinc reagents.

1.3.2.1 Palladium Catalysts. Palladium is the most common metal used in the Negishi cross-coupling. In 2001 Fu and co-worker [234] found Pd(P*t*Bu$_3$)$_2$ to be an effective catalyst for the cross-coupling of aryl and vinyl chlorides with aryl or alkylzinc reagents (Table 1.2). The catalytic system tolerates functional groups such as MeC(O), COOMe, CN, B(OR)$_2$, and even NO$_2$ and is applicable to the sterically hindered substrates. This is the first general method for accomplishing palladium-catalyzed Negishi reactions of aryl and vinyl chlorides. The further

TABLE 1.2. Pd(P*t*Bu$_3$)$_2$-Catalyzed Cross-Coupling of Aryl or Vinyl Chlorides with Aryl or Alkylzinc Chlorides

$$\mathrm{RCl} + \mathrm{R^1ZnCl} \xrightarrow[\text{THF/NMP, 100°C}]{\text{Pd(P}t\text{Bu}_3)_2\text{ (2mol\%)}} \mathrm{R{-}R^1}$$

Entry	RCl	Product	Yield (%)[a]
1	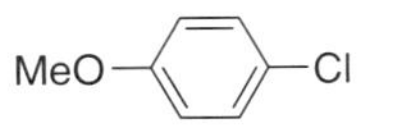	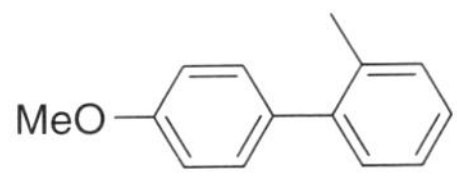	94
2	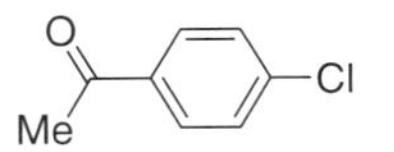	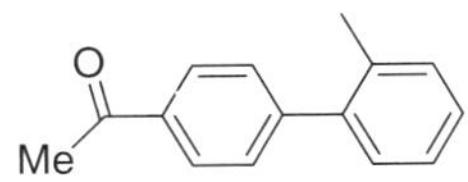	50
3	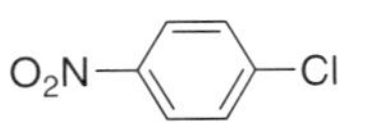	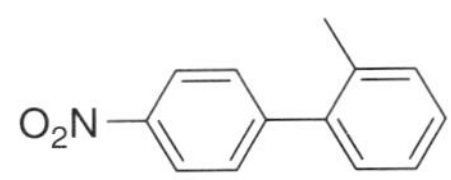	92
4	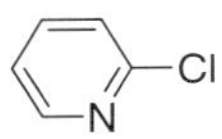	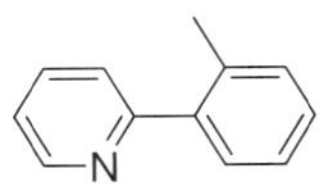	92
5	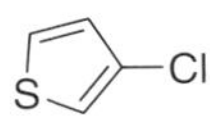	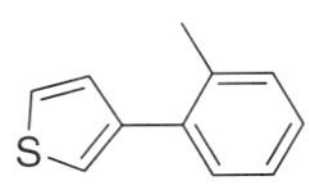	87
6	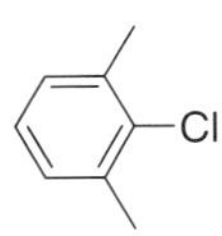	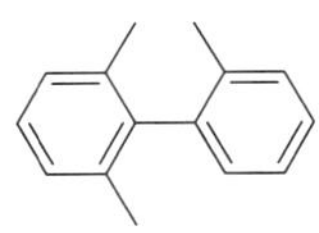	91
7	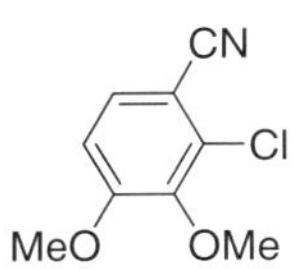	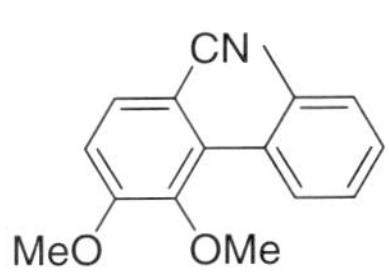	76
8	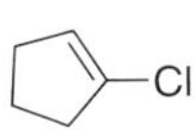	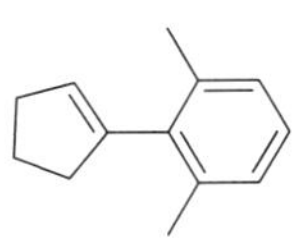	96
9	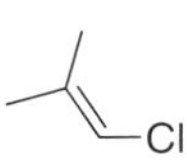	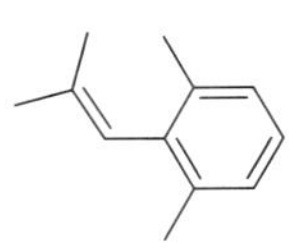	92

TABLE 1.2. (*Continued*)

Entry	RCl	Product	Yield (%)[a]
10	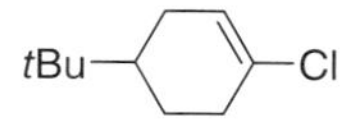		86[b]
11	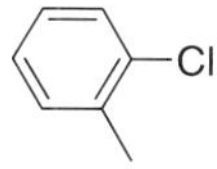	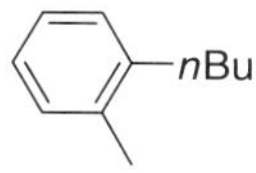	83
12	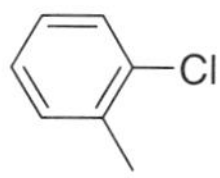	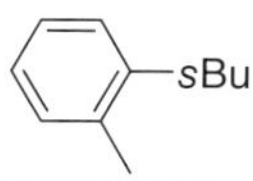	70[c]

[a]Isolated yield, average of two runs.
[b]Includes 2% 4-*tert*-butyl-1-*n*-butylcyclohexene.
[c]Includes 8% 2-*n*-butyltoluene.

studies proved that the $Pd_2(dba)_3$/$t$$Bu_3P{\cdot}HBF_4$-catalyzed Negishi reaction of aryl chlorides was accelerated under microwave radiation [235]. If $PCyp_3$ (Cyp = cyclopentyl) is employed as the ligand and in the presence of *N*-methylimidazole, $Pd_2(dba)_3$ can catalyze the cross-coupling of β-hydrogen-containing primary alkyl iodides, bromides, tosylates, and a few chlorides with a variety of alkyl-, alkenyl-, and arylzinc halides in THF/NMP (2:1) at 80°C. The process tolerates functional groups such as esters, amides, imides, nitriles, and heterocycles [236]. $Pd(PPh_3)_4$ was found to catalyze polyethynylation of polychloroarenes through reaction with ethynylzinc reagents in refluxed THF [237].

Phosphinous acid is also a suitable ligand for Pd-catalyzed Negishi coupling. $t$$Bu_2POH$–Pd complexes, ($t$$Bu_2POH)_2PdCl_2$ (**56**) and [($t$$Bu_2POH)PdCl_2]_2$ (**57**), catalyze coupling of activated, unactivated, and deactivated aryl chlorides with arylzinc chlorides in THF/NMP, giving biaryls in moderate to good yields. ($t$$Bu_2POH)_2PdCl_2$ also catalyzes coupling of acyl chlorides with arylzinc reagents, forming ketones in excellent yields. The Cy_2POH–Pd complex, $[(Cy_2PO{\cdots}H{\cdots}OPCy_2)PdCl]_2$, exhibits higher catalytic activity and wider scope of substrates than do the $t$$Bu_2POH$–Pd complexes. $[(Cy_2PO{\cdots}H{\cdots}OPCy_2)PdCl]_2$ catalyzes cross-coupling of aryl or heteroaryl chlorides with arylzinc reagents in NMP under mild conditions, affording biaryls in 75–93% yields. This catalyst tolerates a range of functional groups such as alkoxy, alkylthio, amino, ketone, cyano, nitro, and ester groups and sterically hindered aryl halides and arylzinc reagents without compromising yields. The coupling of aryl chlorides with vinylzinc chloride or alkylzinc chloride was also performed by the same catalyst [148, 238].

Buchwald ligands apply widely in palladium-catalyzed cross-coupling reactions. A few of the ligands are very effective in Pd-catalyzed cross-coupling of (hetero)aryl chlorides with (hetero)aryl- or alkylzinc reagents. The combination of Pd_2dba_3 and **15b** is an extremely active catalyst for the Negishi cross-coupling of aryl chlorides. This catalytic system can catalyze synthesis of very hindered biaryls and tolerates a variety of functional groups and heterocyclic substrates (Table 1.3) [239].

TABLE 1.3. Pd_2dba_3/15b-Catalyzed Cross-Coupling of (Hetero)aryl Chlorides with Arylzinc Chlorides

$$\text{Ar–ZnCl} + \text{Ar}^1\text{Cl} \xrightarrow[\text{(Pd:RuPhos = 1:2)}]{\text{Pd}_2\text{dba}_3,\ \textbf{15b},\ \text{THF, 70°C, 15 h}} \text{Ar–Ar}^1$$

Entry	Pd Loading (%)	ArZnCl	Product	Yield (%)[a]
1	0.1	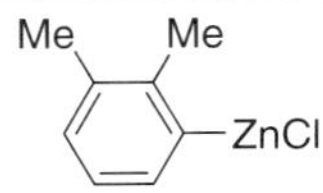	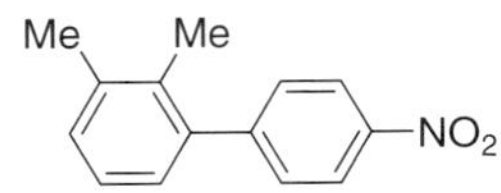	94
2	1	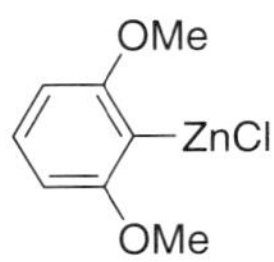	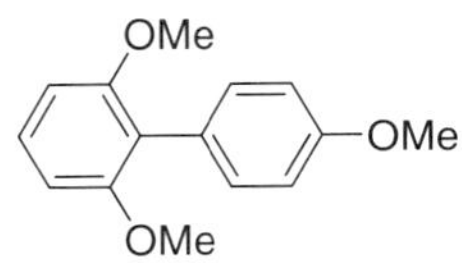	100
3	0.25	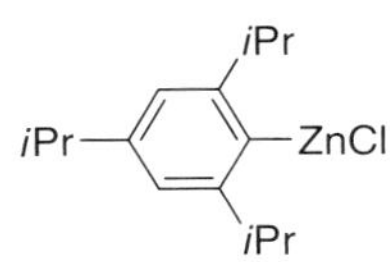	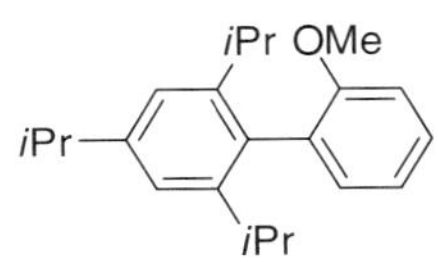	97
4	0.1	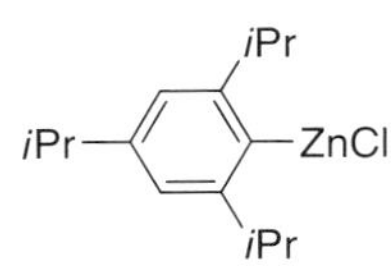	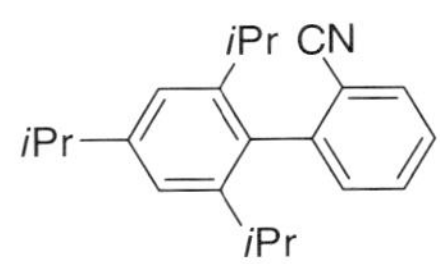	94
5[b]	1	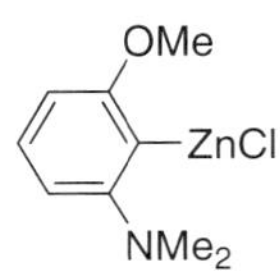	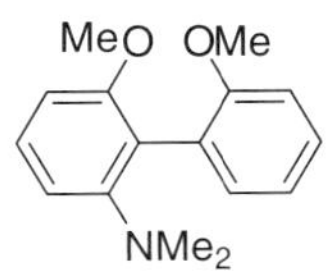	76
6	1	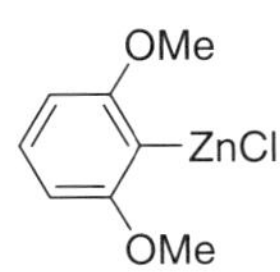	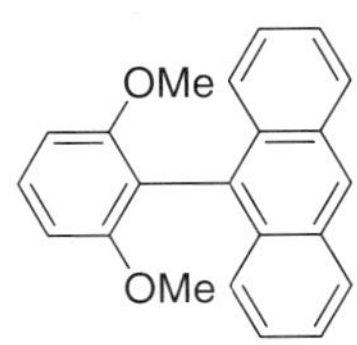	86
7	1	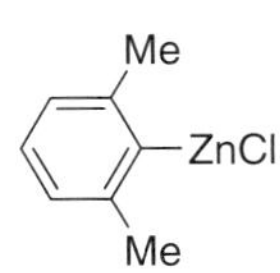	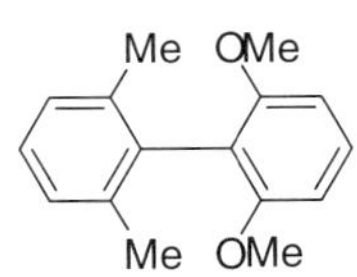	83

TABLE 1.3. **(*Continued*)**

Entry	Pd Loading (%)	ArZnCl	Product	Yield (%)[a]
8[c]	0.25	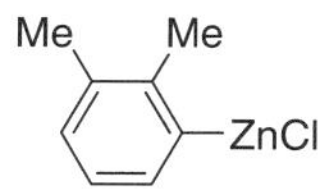	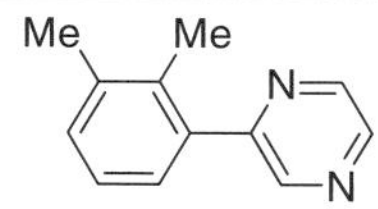	100
9	1	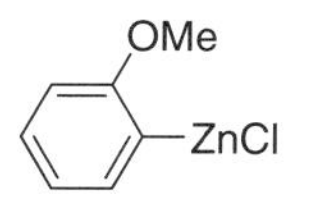	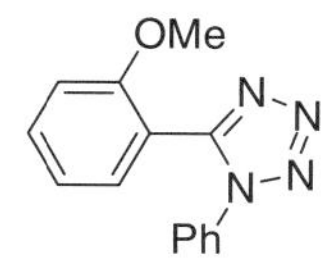	92

[a]Isolated yield.
[b]Reaction ran at 100°C.
[c]Reaction ran at room temperature.

Structure **18a** was proven to be a good ligand for palladium-catalyzed cross-coupling of heteroarylzinc reagents such as 2-pyridylzinc bromides, and 2-thienylzinc bromide with aryl or heteroaryl chlorides [240].

The $Pd(OAc)_2$–**125** system is an effective catalyst for the cross-coupling of aryl bromides and activated aryl chlorides with secondary alkyl halides (Scheme 1.15). The process exhibits mild reaction conditions, excellent yields, and high ratios of secondary to primary coupling products (>20:1).

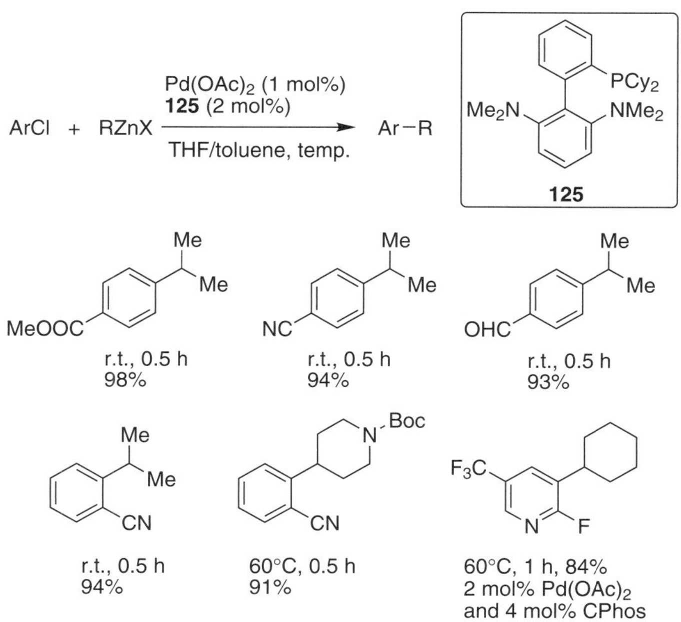

Scheme 1.15.

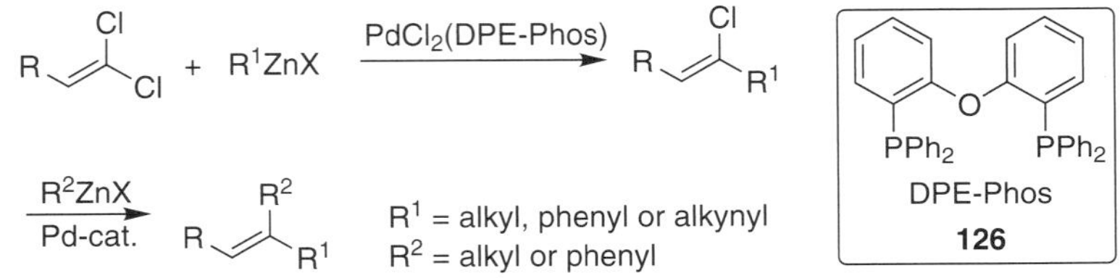

Scheme 1.16.

Negishi and co-workers [241–243] used $PdCl_2$(DPE-Phos) as the catalyst to perform coupling of 1,1-dichloro-1-alkenes with alkylzinc halides, dialkylzincs, phenylzinc bromide or alkynylzinc chlorides. The reaction gives monoalkylated or monoalkynylated products with *trans*-selectivity. The remaining chlorine atom can be further substituted using PhZnBr or R_2Zn catalyzed by $Pd(OAc)_2$-$PCyp_3$ or $Pd(PtBu_3)_2$ (Scheme 1.16).

NHC-Coordinated palladium complexes, **86** and **127**, are effective catalysts for the Negishi coupling of aryl and alkyl chlorides. Complex **127** catalyzes the preparation of sterically hindered biaryls and heterobiaryl through coupling of (hetero)aryl bromides or chlorides with (hetero)arylzinc reagents. The reaction proceeds under mild conditions with low catalyst loadings, giving the desired products in good to excellent yields [244]. The complex also efficiently catalyzes coupling of (hetero)aryl bromides or chlorides with secondary alkylzinc reagents. The reaction proceeds smoothly in THF/toluene at room temperature and leads to the cross-coupling products in high yields. The production of isomeric coupling products are effectively supressed by this catalyst [245]. Complex **86** also catalyzes the Negishi coupling of (hetero)aryl bromides or chlorides with aryl- or alkylzinc reagents. But the catalytic activity seems to be lower than that of **127**, especially for aryl chloride substrates. The reaction of 2-chloropyridine and 2-chloroquinoline with aryl- or alkylzinc reagents was carried out in the presence of **86** [246]. However, it was found that **86** effectively catalyzes alkyl–alkyl Negishi coupling, and solvent polarity has dramatic effects on chemoselectivity. In a less polar solvent the cross-coupling of alkyl bromides with alkylzinc reagents proceeds smoothly, whereas the reaction of alkyl chlorides with alkylzinc reagents requires a more polar solvent. This leads to performance of cross-coupling of bromochloroalkanes with two different alkylzinc reagents in a pot through changing solvent polarity (Scheme 1.17). In addition, excess lithium halide additives are also crucial in the catalyzed coupling reactions [247, 248].

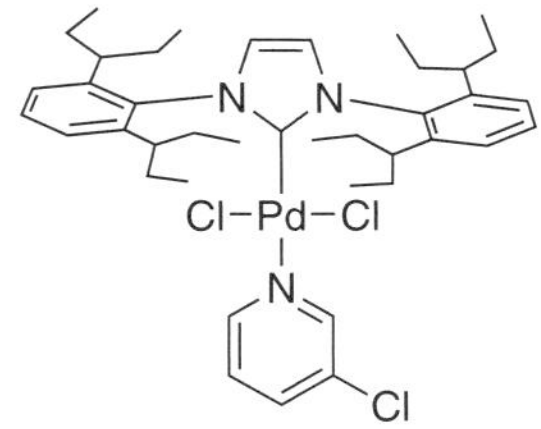

127

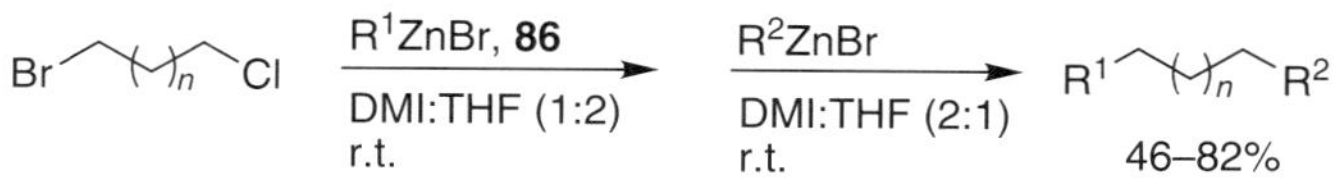

Scheme 1.17.

1.3.2.2 Nickel Catalysts. In 1998 Miller et al. [249] reported a nickel- or palladium-catalyzed cross-coupling of (hetero)aryl chlorides and arylzinc reagents in the presence of phosphine ligands such as PPh_3, $P(OiPr)_3$, and dppf, in which functional groups CN, COOMe, C(O)Et, and NH_2 are tolerated. In 1999 the cross-coupling of aryl chlorides and functionalized alkylzinc iodides was carried out using Ni(0)/PPh_3 as the catalyst. The reaction requires 1.4 equiv of LiCl as an additive and gives good to excellent yields. The catalytic system is compatible with functional groups such as CN, CHO, COOEt, and PhC(O) on the aromatic rings and *t*BuCOO, COOEt, CN, PhCOO, and Cl in the zinc reagents [250]. A very similar result was obtained by the same group using a Ni/C–PPh_3 catalytic system in THF [251]. The combination of $Ni(acac)_2$ and PPh_3 catalyzes cross-coupling between functionalized benzylic zinc reagents and (hetero)aryl chlorides, bromides, and tosylates. Only activated aromatic chlorides were employed as the electrophilic substrates and gave good to excellent yields (Scheme 1.18) [252].

A $NiCl_2$–$P(NC_5H_{10})_3$ system catalyzes the reaction of aryl bromides or aryl chlorides with diarylzinc reagents with low catalyst loadings [Eq. (1.13)]. The scope of substrates is very wide. A large variety of electronically activated, unactivated, deactivated, and *ortho*-substituted, heterocyclic, and functionalized halides were employed. Functional groups tolerated include trifluoromethyl, acetal, ketone, ether, ester, lactone, amide, imine, aniline, and alkene groups. Some heteroaromatic chlorides are also suitable electrophilic substrates, including pyridine, quinoline and pyrimidine derivatives. The experimental facts support a Ni(I)/Ni(III) mechanism (Figure 1.6),

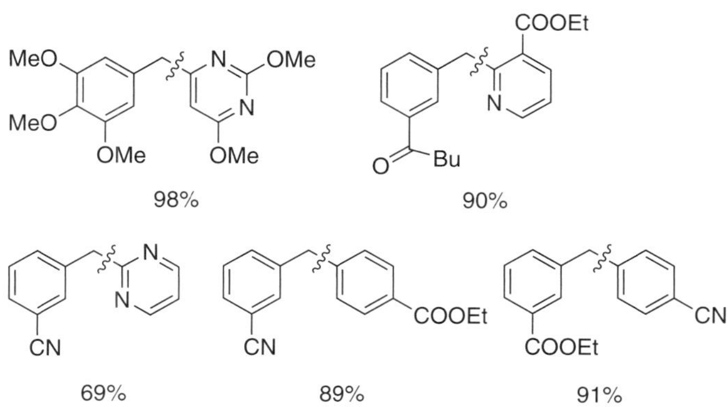

Scheme 1.18. The cross-coupling products of functionalized benzylic zinc reagents and (hetero)aryl chlorides.

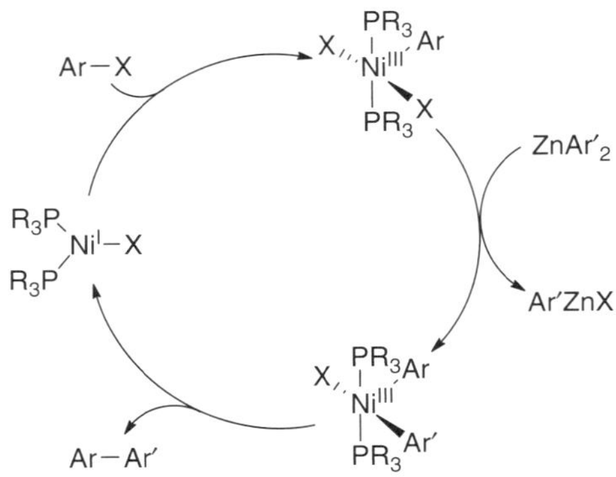

Figure 1.6. Possible mechanism of the Negishi reaction using aminophosphine-based nickel catalytic system.

whereas the involvement of nickel-based nanoparticles in the catalytic cycle could be ruled out [253].

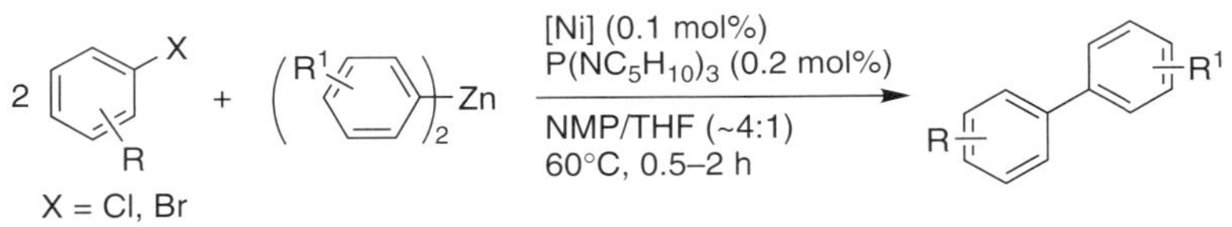

(1.13)

The multidentate ligand-coordinated nickel complexes or the combination of nickel salts and multidentate ligands are also catalytically active for the Nigishi coupling of chloride substrates. A $Ni(acac)_2$–DPEPhos system catalyzes cross-coupling of aryl, heteroaryl, and alkenyl bromides or chlorides with aminoalkylzinc bromides in a mixed solvent of THF and NMP at room temperature. Several examples using chlorides as the electrophilic substrates were included with excellent yields (Scheme 1.19) [254].

Several pincer-type of nickel complexes are active for the coupling of aryl chlorides and arylzinc chlorides. Amido pincer nickel complexes **128** and **129** efficiently catalyze the cross-coupling of aryl chlorides and arylzinc chlorides in a mixture of THF and NMP at 70°C, and 1 mol% **128** or **129** leads to the cross-coupling product of *p*-$ClC_6H_4NMe_2$ with PhZnCl nearly quantitatively. For most aryl chlorides, including those unactivated and deactivated, an extremely low catalyst loading is required; 0.02 mol% **128** or **129** could drive the reaction of *p*-ClC_6H_4OMe with *p*-MeC_6H_4ZnCl to completion. However, the catalytic system is not very effective to the sterically hindered reactants, requiring stronger reaction conditions and often giving low product yields. The reaction tolerates a range of functional groups such as keto, ester, trifluoromethyl, cyano, amide, and pyridyl, but is uncompatible with nitro and formyl groups [255]. P,N,O–Pincer nickel complexes **130a–131** display higher catalytic activity for the coupling of aryl chlorides. They can catalyze reactions

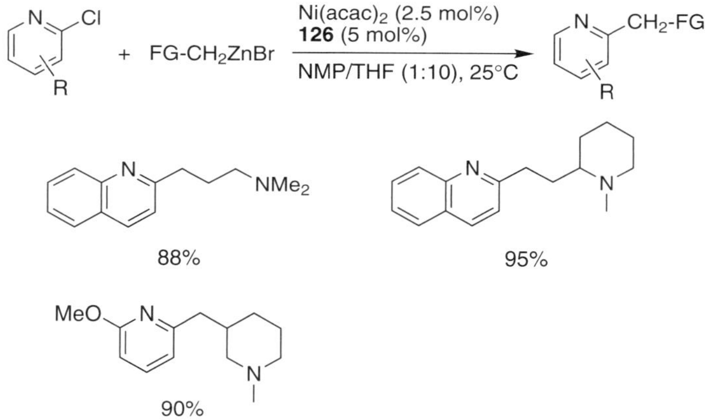

Scheme 1.19.

of aryl chlorides with arylzinc reagents at room temperature, including deactivated aryl chlorides. Complexes **130a** and **131** catalyze room temperature reactions of a broad spectrum of aryl chlorides and tolerate a range of functional groups. A low catalyst loading is required in most cases. *p*-Chlorobenzonitrile and hindered substrates require higher catalyst loadings. In addition, electron-deficient zinc reagents also require higher catalyst loadings and afford relatively poor results in the catalytic coupling [256].

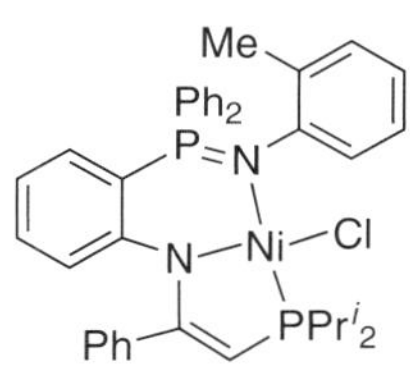

128

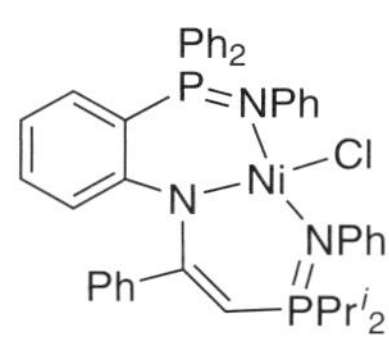

129

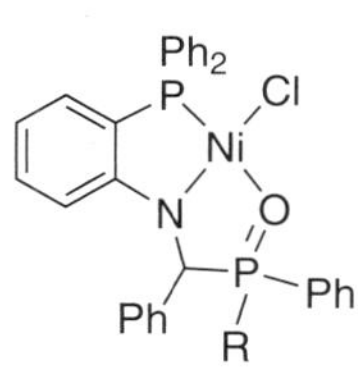

130a R = Ph
130b R = Et

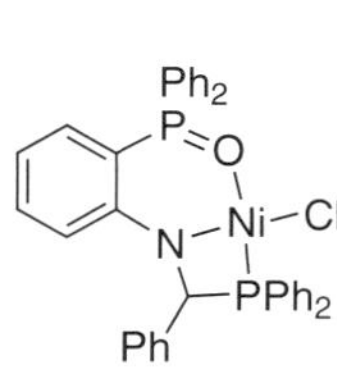

131

NHC-based pincer nickel complex cations, **132–134**, are also active catalysts for the reaction of *p*-MeC_6H_4ZnBr with aryl chlorides. The activity of complex **129** is higher than that of both **132** and **134**, but a little lower than that of **128** and **129**. Complex **133** catalyzes the cross-coupling reactions in a 1:1 mixture of THF and NMP at

70°C, giving 55–99% product yields. The catalyst loadings are low (0.1–2 mol%) and the reactions are compatible with the functional groups such as ester, keto, cyano, and amide [257].

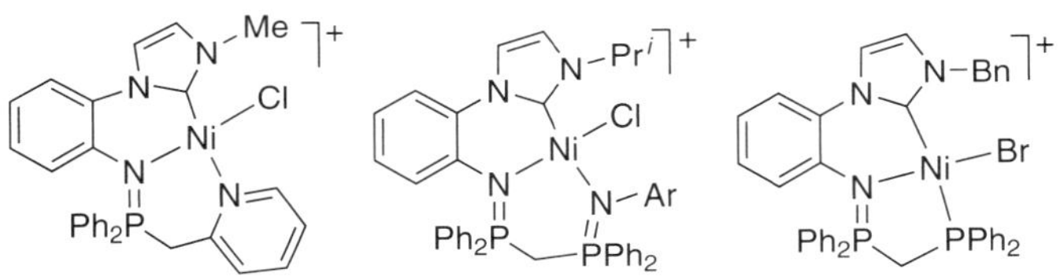

132 133 134

Two binuclear nickel–NHC complex cations, **135** and **136**, also exhibit good catalytic activity for the Negishi cross-coupling of aryl chlorides. The detailed studies of the catalytic properties of **135** show that it can catalyze cross-coupling of a variety of unactivated aryl chlorides, heterocyclic chlorides, aryl dichlorides, and vinyl chloride with arylzinc reagents in THF-NMP (1:1) at 80°C, leading to biaryls and terphenyls in good to excellent yields. Under the same conditions, complex **123** leads to lower product yields than do the binuclear ones. This is attibuted to possible bimetallic cooperativity [258].

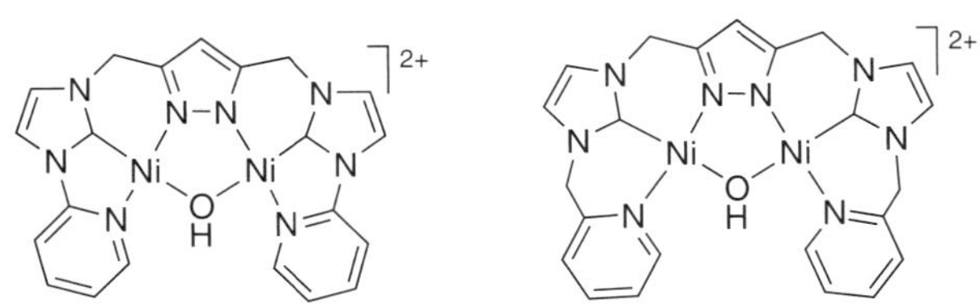

135 136

Nickel complexes supported by pyridine bisoxazoline (Pybox) ligands efficiently catalyze the Negishi coupling of primary and secondary alkyl bromides and iodides with alkylzinc reagents [259]. The combination is also effective with regard to the cross-coupling of secondary benzylic chlorides or secondary allylic chlorides with alkylzinc reagents. The reaction of benzylic chlorides leads to 75–99% *ee* and 39–89% yields of cross-coupling products using $NiBr_2$/(*S*)-*i*Pr-Pybox as a catalyst, and the reaction of allylic chlorides results in 69–98% *ee* and 54–97% yields using $NiBr_2$/(*S*)-$BnCH_2$-Pybox as the catalyst (Scheme 1.20) [260, 261].

1.3.2.3 Other Metals. Cobalt and copper complexes also catalyze the Negishi cross-coupling of organic chlorides. $CoBr_2$ catalyzes reaction of benzyl chlorides with functionalized arylzinc bromides in acetonitrile, giving diarylmethanes in good to excellent yields [262]. $Cu(acac)_2$ catalyzes cross-coupling of α-chloroketones with alkylzinc halides at room temperature [Eq. (1.14)]. Both cyclic and acyclic α-chloroketones as well as primary and secondary alkyl zinc reagents are

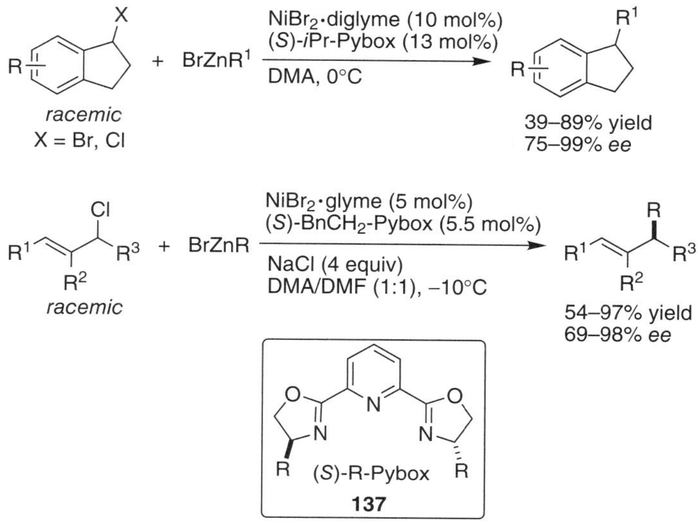

Scheme 1.20.

suitable substrates. This method represents a general strategy for the synthesis of α-branched ketones [263].

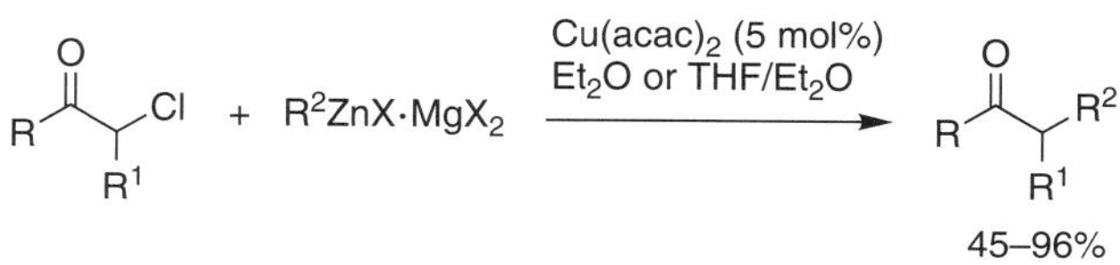

(1.14)

1.3.3 Kumada Reaction

The Kumada reaction is one of the powerful tools to construct new C–C bonds. Compared with the Suzuki, Negishi, and Stille reactions, the Kumada reaction shows poorer functional group compatibility. However, on the other hand, the Kumada reaction offers a more direct approach to the desired products when the substrates tolerate the background reactivity of a Grignard reagent. Hence a great number of studies have been carried out, and wide substrates and catalysts have been explored. Development of new ligands and catalysts also promotes the use of unreactive and less expensive chlorides in the Kumada reaction.

1.3.3.1 Palladium Catalysts

(i) PHOSPHINE LIGANDS. The combination of $Pd(OAc)_2$ and PCy_3, **138** or **139**, can catalyze cross-coupling of aryl Grignard reagents with primary alkyl chlorides in NMP or NMP/THF at room temperature. Complexes **140a** and **140b** prepared from **139a/b** and $PdCl_2(PhCN)_2$ exhibit an improved catalytic performance, which allows a lower catalyst loading. Good to excellent yields of the coupling products were obtained, and active functional groups such as COOMe, PhC(O), and CN are tolerated [264, 265].

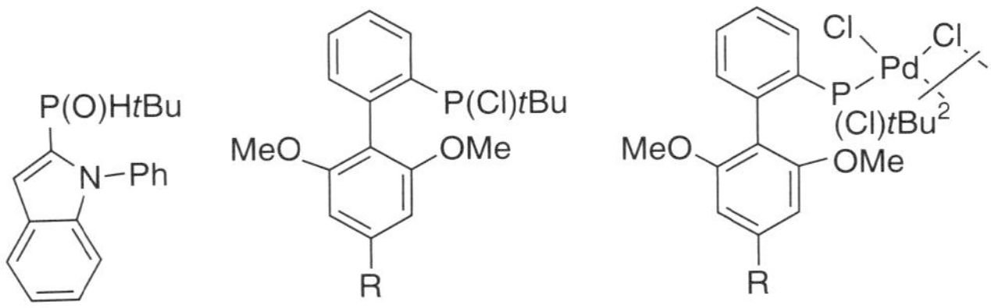

138 **139a** R = H **139b** R = OMe **140a** R = H **140b** R = OMe

The $Pd_2(dba)_3/4tBu_3P{\cdot}HBF_4$ system was demonstrated to catalyze the reaction of *p*-$MeOC_6H_4MgBr$ with *o*-MeC_6H_4Cl in high yield under microwave radiation [235]. Knochel and co-workers [266, 267] reported cross-coupling of heteroaryl chlorides such as chloropyridines, chloroquinolines, and chlorodiazines with aryl Grignard reagents in the presence of $Pd(dba)_2/PtBu_3$ or dppf. The reactions can be run at low temperature and hence functionalized chloroarenes or functionalized Grignard reagents can be employed. $Pd_2(dba)_3/PCy_3$ catalyzes *ortho*-selective cross-coupling of dichlorobenzene derivatives with directing groups such as OH, CH_2OH, NH_2, NHAc, and NHBoc [Eq. (1.15)]. $Pd_2(dba)_3$/**141** is more effective than $Pd_2(dba)_3/PCy_3$ for dichlorophenol and dichloroaniline [268].

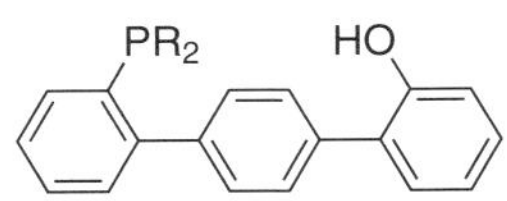

R = Cy or Ph

141

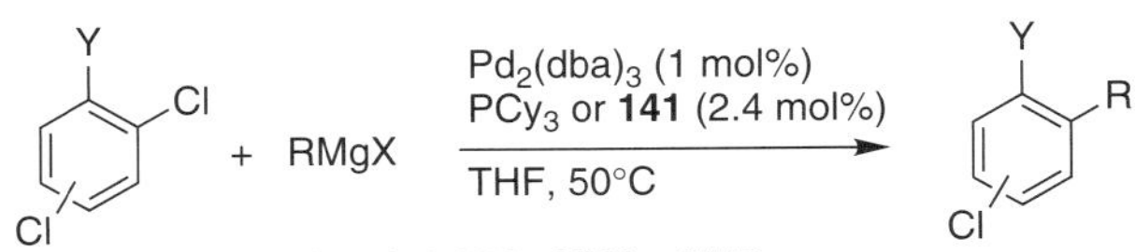

Y = OH, CH_2OH, NH_2, NHAc, NHBoc
R = aryl, $Me_2C{=}CH$ (1.15)

Besides **138** mentioned above, tBu_2POH and **142** are good ligands for Pd-catalyzed cross-coupling of chloride substrates. Complexes **56–58** have proved to be efficient catalysts for cross-coupling of inactivated aryl chlorides with aryl Grignard reagents, resulting in corresponding biaryls in 51–99% yields [269]. Complex **56** is also effective for the formation of di- and tri-*ortho*-substituted biaryls from sterically hindered aryl chlorides and aryl Grignard reagents [270].

The combination of $Pd(OAc)_2$ and **142** in a ratio of 1:2 catalyzes coupling of deactivated aryl chlorides or 1-chlorocyclopentene with aryl Grignard reagents [271].

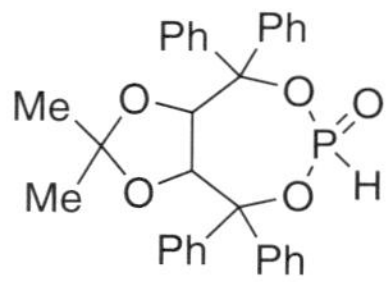

142

(ii) *N*-HETEROCYCLIC CARBENE LIGANDS. Nolan and co-workers [272, 273] were the first to report palladium/imidazolium salt-catalyzed Kumada cross-coupling using a variety of aryl chlorides as electrophilic substrates. The methodology proves effective for unactivated aryl chlorides, and the $Pd_2(dba)_3$/IPrHCl (IPr = 1,3-bis(2,6-diisopropylphenyl)-imidazol-2-ylidene) system is excellent for this coupling [272]. The reaction was improved by employing $[Pd(\mu\text{-Cl})(Cl)(SIPr)]_2$ (**143**) as the pre-catalyst. A variety of chlorides such as functionalized aryl, heteroaryl, and sterically hindered aryl chlorides can be coupled with aryl Grignard reagents under mild conditions with low catalyst loading (0.2–0.45 mol%). The system allows for the synthesis of highly sterically hindered products, including tri- and tetra-*ortho*-substituted biaryls [273]. *N*-Heterocyclic carbenes are also suitable ligands for palladium-catalyzed coupling reactions of alkyl chlorides. The monocarbene–palladium(0) naphthoquinone complex **144** efficiently catalyzes cross-coupling of primary alkyl chlorides and aryl Grignard reagents, showing higher activity than $Pd(OAc)_2/PCy_3$ system. The reaction proceeds at room temperature and is compatible with functional groups such as CN, COOMe, and amide [274].

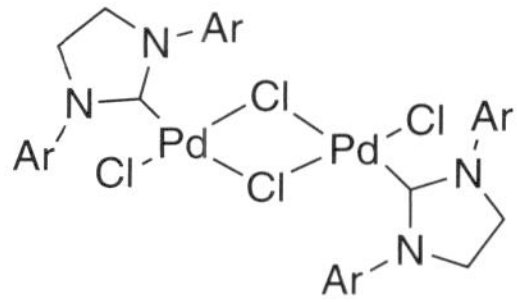

Ar = 2,6-iPr$_2$C$_6$H$_3$

143

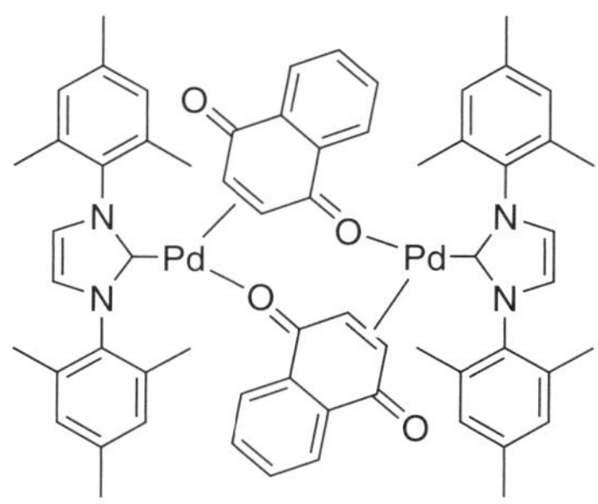

144

Complexes **145**, **146,** and **86** also efficiently catalyze coupling of deactivated and heteroaryl chlorides with arymagnesium reagents at room temperature. Both **86** and **146** are suited for a wider substrate scope, including a variety of heteroaryl and bulky substrates. Complex **86** even leads to the formation of tetra-*ortho*-substituted heterocycle at room temperature [275, 276].

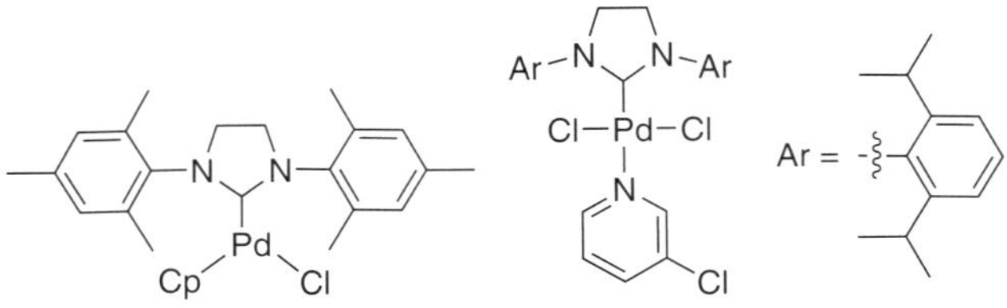

145 **146**

(iii) PALLADACYCLES. 1999 Herrmann et al. [277] showed palladacycle complex **89** can catalyze coupling of PhCl and PhMgBr or MeMgBr in THF. Recently, Wu and co-workers [278] demonstrated that ferrocene-based palladacycle **102** with 2 equiv of LiCl additive is effective for the Kumada coupling of aryl Grignard reagents with electron-rich and electron-poor as well as bulky aryl chlorides [278].

1.3.3.2 Nickel Catalysts

(i) PHOSPHINE LIGANDS. PPh_3 can be used as the ligand for Ni/C–catalyzed cross-coupling of aryl chlorides with aryl or alkyl Grignard reagents, but homocoupling products of aryl chlorides are often formed as the side products [279]. Didentate phosphine ligand **147** in combination with $Ni(MeCN)_2Cl_2$ is effective to catalyze coupling of aryl, heteroaryl, and benzyl chlorides with aryl Grignard reagents. The reaction proceeds smoothly in THF at room temperature with 2 mol% nickel loading [280]. Nickel complexes bearing functionalized phosphine ligands **148a** and **148b** catalyze cross-coupling of aryl Grignard reagents with a variety of aryl chlorides at room temperature in excellent yields. Active functional groups such as COOR, CN, PhC(O), and ArN=CH are tolerated in the presence of additives LiCl (1.2 equiv) and $ZnCl_2$ (10–30 mol%), whereas $C(O)NEt_2$ is tolerated without need of additives [281].

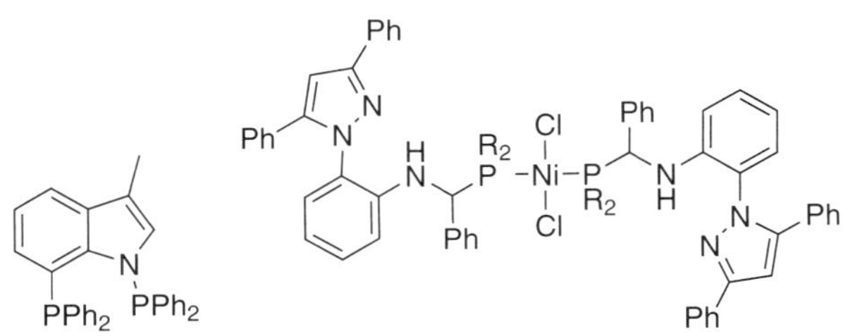

147 **148a** R = Cy
148b R = *i*-Pr

The nickel-complex-bearing hydroxyphosphine ligand **149** is highly effective in catalyzing the reaction of aryl Grignard reagents with alkenyl and aryl halides including chlorides. Experimental and theoretical studies show that the Ni(0) generated from the Ni(II) pre-catalyst, the PO ligand generated from deprotonation of **149** by the Grignard reagent, and the aryl halide form a nickel/magnesium bimetallic intermediate **150**, which leads to activation the aryl–X bond by a cooperative push–pull action of the nucleophilic nickel and Lewis acidic magnesium centers [282, 283].

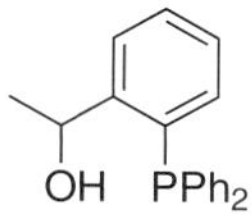

149

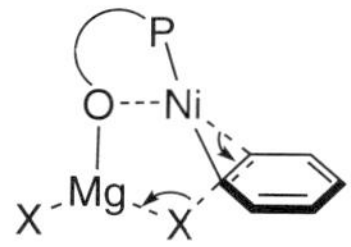

150

Nickel/phosphine oxide systems such as Ni(cod)$_2$/*t*Bu$_2$P(O)H, Ni(acac)$_2$/**151** and NiCl$_2$/**152** are effective for the coupling of arylmeganesium reagents and aryl or heteroaryl chlorides in THF at room temperature. Ni(cod)$_2$/*t*Bu$_2$P(S)H behaves similarly to Ni(cod)$_2$/*t*Bu$_2$P(O)H [147, 284–286].

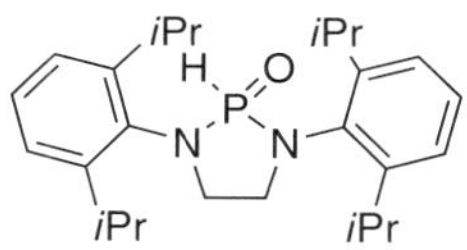

151

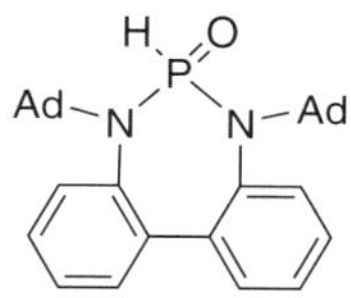

152

(ii) NHC Ligands. Herrmann and co-workers [287] first carried out Ni(acac)$_2$/**153** or **154**-catalyzed coupling of 2-chloropyridine and activated or deactivated aryl chlorides with aryl Grignard reagents at room temperature. The catalytic system shows activity close to that of Ni(acac)$_2$/P*t*Bu$_3$ under the same conditions, whereas Pd$_2$dba$_2$/**154** exhibits much lower activity for the reaction.

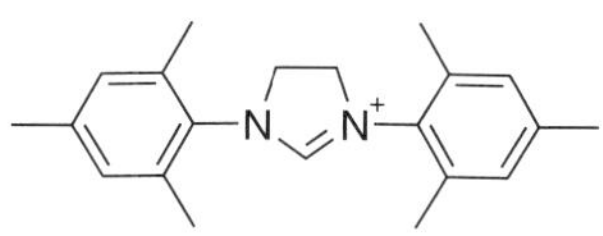

153

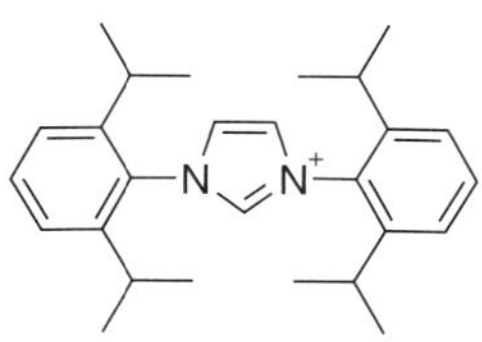

154

Functionalized *N*-heterocyclic carbene nickel(II) complexes **155–159** exhibit good catalytic activity for synthesis of biaryls through activation of aryl chlorides. Among the C,N-chelate nickel complexes **156–159**, **156** shows lowest activity in the reaction of PhMgBr with *p*-ClC$_6$H$_4$OMe, whereas **159** displays highest activity [288, 289].

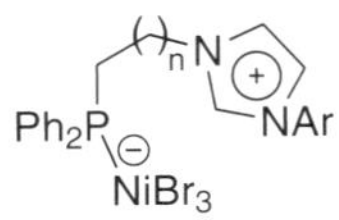

n = 1,2
Ar = 2,4,6-$Me_3C_6H_2$
2,6-$iPr_2C_6H_3$

155

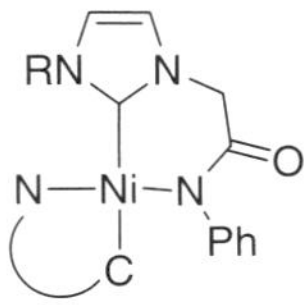

156 R = Me, Bn

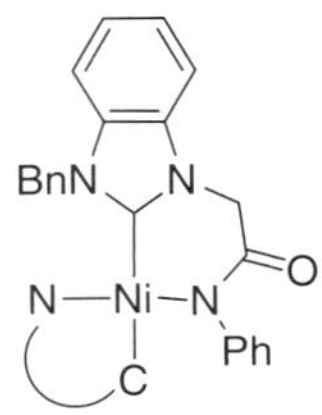

157

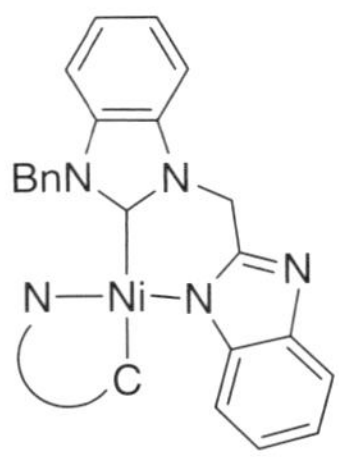

158

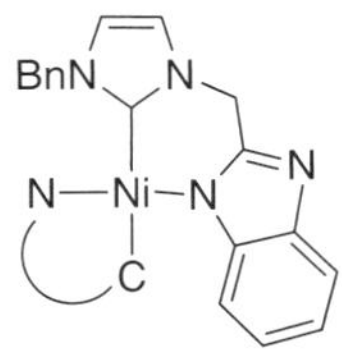

159

Chelated NHC–nickel complexes **123**, **124**, and **160–162** are also active in catalyzing the Kumada coupling of aryl chlorides. Among the derivatives of **160**, the derivatives with R = Me, R′ = 1,4-butyl, and X = Cl or Br show the best activity for the reaction of 4-chloroanisole with phenylmagnesium chloride at room temperature, quantitative conversions being achieved. The derivative with R = Bn, R′ = 1,3-propyl and X = Br give also an excellent catalytic result [290]. Each of **123**, **124**, and **162** is excellent in catalyzing the above reaction, while **161** exhibits lower activity. Complex **123** was demonstrated to be suited for wide chloride substrates, including aryl, heteroaryl, and alkenyl chlorides as well as multichloroarenes [291].

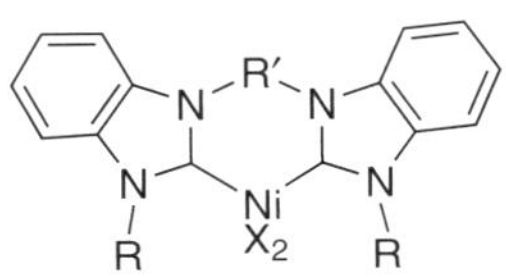

R = Me, Pr, iPr, Bn, Ph
R′ = 1,3-propyl, 1,4-butyl
1,1′-SiMe4, α,α′-xylyl
X = Cl, Br

160

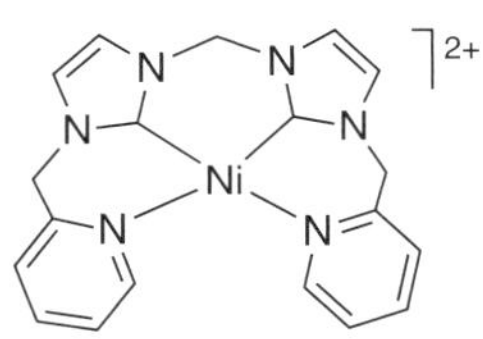

161

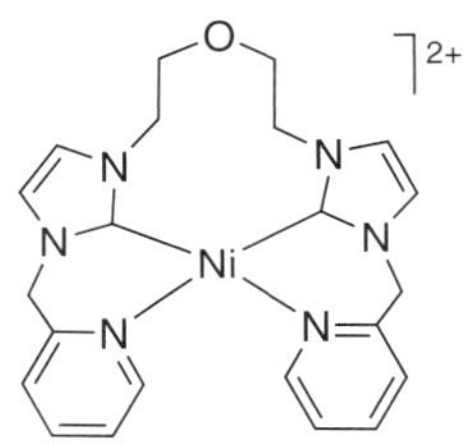

162

(iii) Pincer Nickel Complexes. Amido pincer nickel complex **10** was demonstrated to catalyze double coupling of CH_2Cl_2 with *n*BuMgCl; 25 equiv of CH_2Cl_2 is required and the optimal temperature is −20°C. Unfortunately, this reaction cannot be extended to other multichloroalkanes [292].

Catalysis of unsymmetrical amido pincer complexes of nickel **128**, **129**, and **163a–171b** were evaluated (Table 1.4). Complexes **167** and **169b** were found to be almost inactive. The other complexes catalyze cross-coupling of electron-rich and electron-poor aryl chlorides with aryl Grignard reagents at room temperature or

TABLE 1.4. Cross-Coupling of p-MeC_6H_4MgBr with p-RC_6H_4Cl Catalyzed by Complexes 128–131 and 163a–169b

Entry	Catalyst (mol%)	R	Conditions	Yield (%)
1	**128** (4)	MeO	r.t.[a], 30 h	93
2	**128** (2.5)	MeO	80°C, 6 h	98
3	**129** (2.5)	MeO	80°C, 6 h	98.6
4	**130a** (0.5)	H	r.t., 24 h	95
5	**130a** (2)	MeO	r.t., 24 h	96
6	**130b** (2)	MeO	r.t., 24 h	98
7	**131** (0.5)	H	r.t., 24 h	99
8	**131** (1)	MeO	r.t., 24 h	97
9	**163a** (2.5)	H	r.t., 24 h	80
10	**163b** (2.5)	H	r.t., 24 h	92.8
11	**163b** (4)	MeO	80°C, 12 h	94
12	**164a** (2.5)	H	r.t., 24 h	91
13	**164a** (2.5)	MeO	80°C, 6 h	92
14	**164b** (2.5)	H	r.t., 24 h	>99
15	**165** (2)	H	r.t., 36 h	79
16	**165** (4)	MeO	r.t., 36 h	75
17	**166** (2)	H	r.t., 36 h	95
18	**166** (4)	MeO	r.t., 36 h	94
19	**167** (2)	H	r.t., 36 h	3
20	**168** (4)	MeO	r.t., 36 h	98
21	**169a** (2)	H	r.t., 36 h	94
22	**169a** (4)	MeO	r.t., 36 h	97
23	**169b** (2)	H	r.t., 36 h	9

[a]r.t., room temperature.

at 80°C. Complexes **170** and **171** show the highest activity; they drive reaction of 4-chloroanisole with p-MeC_6H_4MgBr at room temperature with 1–2 mol% catalyst loadings. Complexes **166**, **168**, and **169a** can also efficiently perform the catalytic reaction at room temperature, but require higher catalyst loadings (4 mol%). The activity difference of **165–167** shows the size of the metal chelate rings to be related to the activity. Changes of the coordinate atoms from phosphorus or nitrogen to oxygen significantly enhance the catalytic activity [293–295].

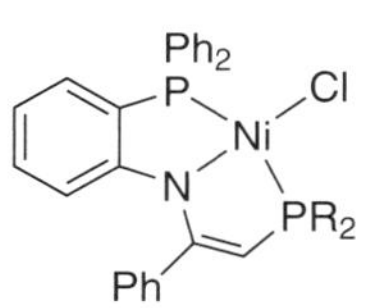

163a R = Ph
163b R = *i*Pr

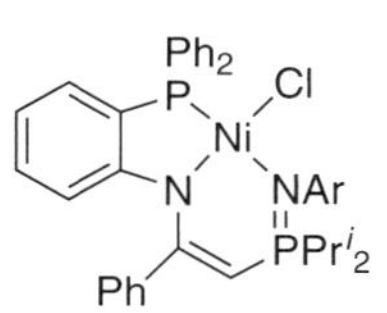

164a Ar = o-MeC_6H_4
164b Ar = 2,6-$iPr_2C_6H_3$

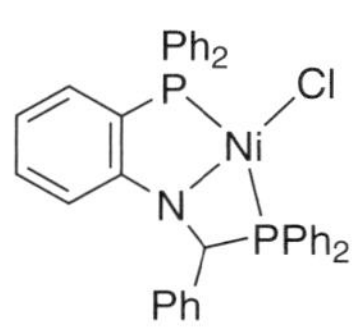

165

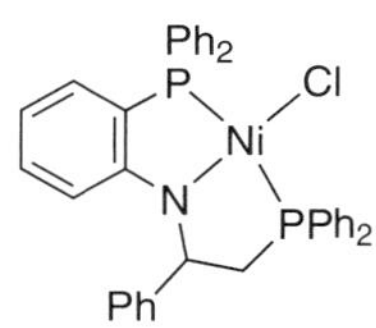

166

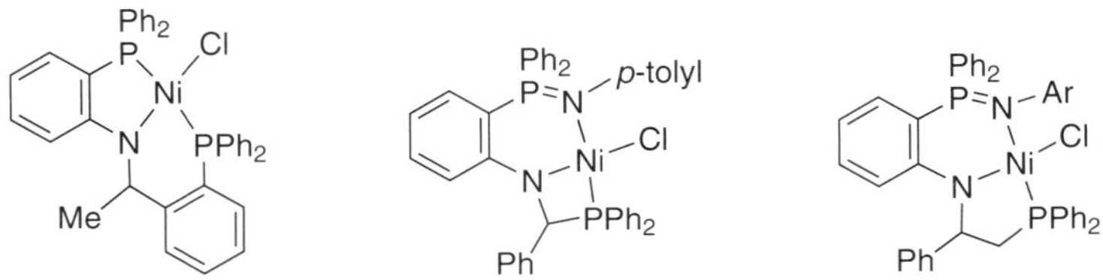

167 **168** **169a** Ar = *p*-MeC$_6$H$_4$ **169b** Ar = 2,6-*i*Pr$_2$C$_6$H$_3$

Each of pincer nickel cations **132–134** are active for the coupling of unactivated aryl chlorides with aryl Grignard reagents. They exhibit very close activity in catalyzing coupling of PhCl with *p*-MeC$_6$H$_4$MgBr, with 1 mol% catalysts at room temperature leading to > 90% biaryl yields. However, for the coupling of deactivated aryl chlorides such as *p*-ClC$_6$H$_4$OMe, both **132** and **133** are more active than **134** [257]. The activity of **132–134** is a little lower than that of the pincer-type complexes of dinuclear nickel **135** and **136**, but much higher than that of **170**. The activity of **135** and **136** is close to that of **130** and **131**. A bimetallic cooperative catalysis by **135** and **136** was proposed, but this needs further evidence to support [296, 297].

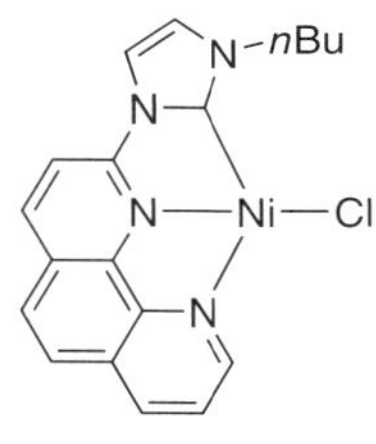

170

(iv) 1,3-Butadiene Additives. In 2002 Kambe and co-workers [298] reported that 1,3-butadiene as an additive greatly promotes Ni-catalyzed cross-coupling of alkyl halides or tosylates and alkyl Grignard reagents [Eq. (1.16)]. This reaction does not require phosphines or other ligands. A plausible reaction pathway is shown in Figure 1.7 [298, 299].

$$n\text{Oct-Cl} + \underset{2\ \text{equiv}}{n\text{BuMgCl}} \xrightarrow[\substack{1,3\text{-butadiene (50 mol\%)} \\ 25^\circ\text{C, 20 h}}]{\text{NiCl}_2\ \text{(3 mol\%)}} \underset{96\%}{n\text{C}_{12}\text{H}_{26}} \qquad (1.16)$$

However, in the presence of (dppf)NiCl$_2$, reaction of 1,3-butadienes, aryl Grignard reagents and alkyl halides result in a three-component coupling (Eq. 1.17). The reaction also proceeds when arylzinc halides are used instead of Grignard reagents [300].

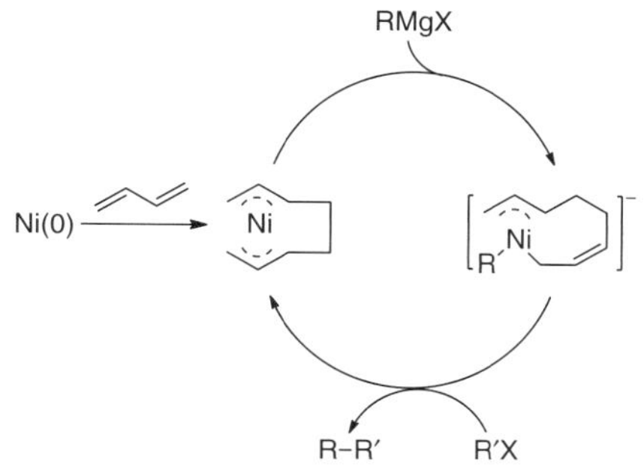

Figure 1.7. A plausible reaction pathway for the Ni-catalyzed cross-coupling in the presence of 1,3-butadienes.

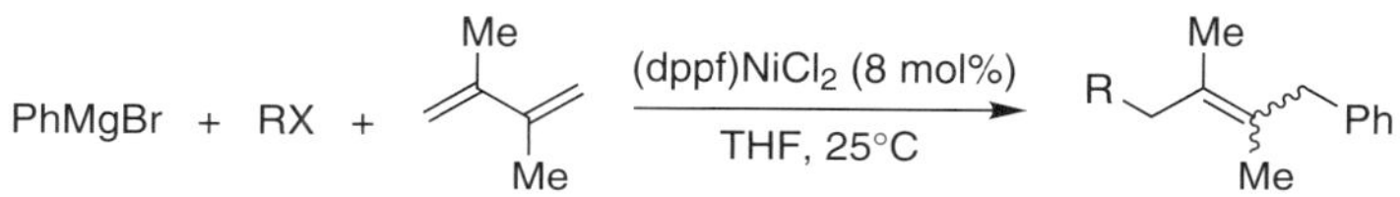

(1.17)

RX = *s*BuCl, yield = 63%, E/Z = 57/43
RX = *t*BuCl, yield = 81%, E/Z = 52/48

1.3.3.3 Iron Catalysts. Iron salts can catalyze cross-coupling of a variety of chorides including activated aryl chlorides, heteroaryl chlorides, alkenyl chlorides, imidoyl chlorides, and even acid chlorides with alkyl or aryl Grignard reagents in THF-NMP or THF (Table 1.5). The reactions proceed rapidly under very mild conditions, giving moderate to excellent product yields depending on substrates [301–307]. Reaction of acid chlorides with Grignard reagents proceeds smoothly at −78°C, affording corresponding ketones in good to excellent yields. Under these conditions, the attack of the Grignard reagent to the resulting ketone is negligible. Selective monoalkylation reactions of dichloroarenes can also be carried out using the $Fe(acac)_3$-catalyzed coupling [305].

In most cases, the nature of the iron salts has little effect on the reaction outcome for the reactions mentioned above. Therefore, the cheap and nonhygroscopic $Fe(acac)_3$ is the most appropriate one from the practical point of view. However, for some substrates $Fe(acac)_3$ is not the best pre-catalyst. For example, for *sec*-alkyl Grignard reagents Fe(salen)Cl (**171**) is better [302]. This complex was also employed in the synthesis of (*R*)-(+)-muscopyridine (**172**) to prepare key intermediate **173** [Eq. (1.18)] [308].

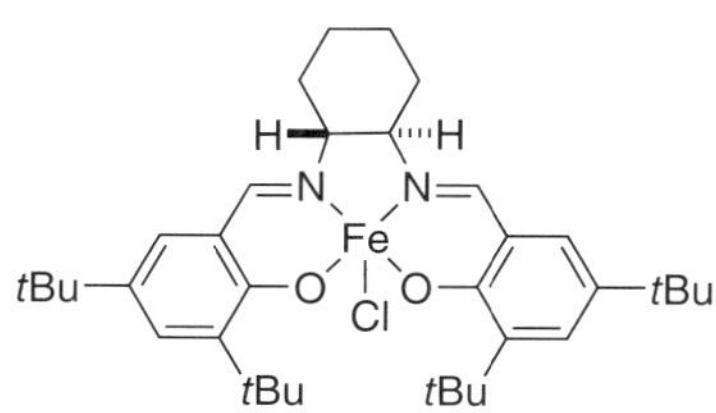

171

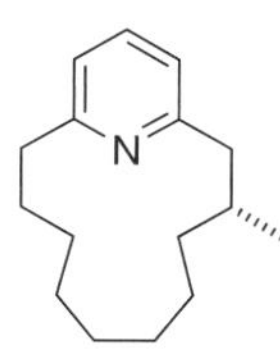

172

TABLE 1.5. Iron-catalyzed Cross-Coupling of RCl and R′MgBr

Entry	R	R′	Fe Catalyst	Conditions	Yield (%)	Reference
1	2-Pyridyl	n-C_6H_{13}	$Fe(acac)_3$ (5 mol%)	THF-NMP, 0°C–r.t., 5 min	91	301
2	2-Pyridyl	n-C_6H_{13}	$Fe(acac)_2$ (5 mol%)	THF-NMP, 0°C–r.t., 5 min	90	301
3	p-$MeO_2CC_6H_4$	n-C_6H_{13}	$Fe(acac)_3$ (5 mol%)	THF-NMP, 0°C–r.t., 5 min	>95	301
4	p-$MeO_2CC_6H_4$	n-C_6H_{13}	$FeCl_2$ (5 mol%)	THF-NMP, 0°C–r.t., 5 min	>95	301
5	p-$F_3CC_6H_4$	n-$C_{14}H_{29}$	$Fe(acac)_3$ (5 mol%)	THF-NMP, 0°C–r.t., 5 min	94	301
6	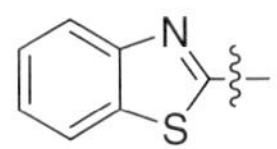	n-$C_{14}H_{29}$	$Fe(acac)_3$ (5 mol%)	THF-NMP, 0°C–r.t., 5 min	68	301
7	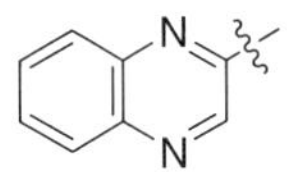	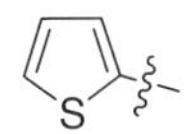	$Fe(acac)_3$ (5 mol%)	THF, −30°C	69	302
8	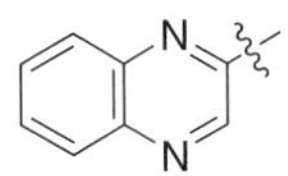	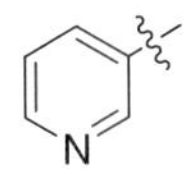	$Fe(acac)_3$ (5 mol%)	THF	82	302
9	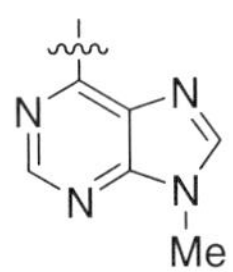	Ph	$Fe(acac)_3$ (5 mol%)	THF	60	302
10	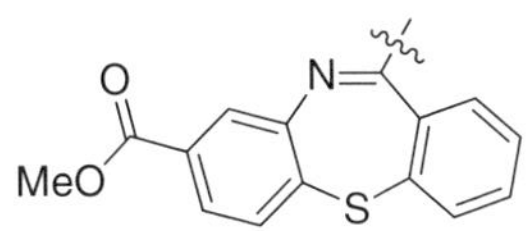	nBu	$Fe(acac)_3$ (5 mol%)	THF-NMP, −78°C, 5 min	89	303

11	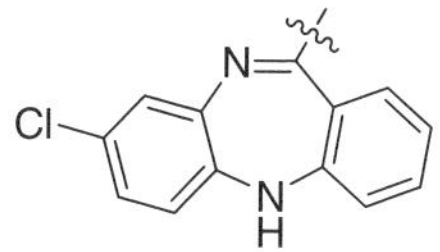		$Fe(acac)_3$ (5 mol%)	THF-NMP, r.t., 5 min	82	303
12	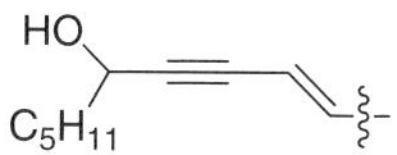	n-$C_{12}H_{25}$	$Fe(acac)_3$ (3 mol%)	THF-NMP, 0°C, 15–60 min	76	304
13	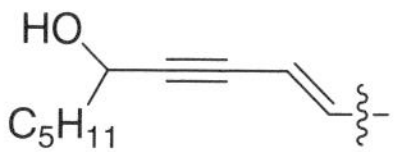	C–C_6H_{11}	$Fe(acac)_3$ (3 mol%)	THF-NMP, 0°C, 15–60 min	72	304
14	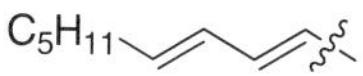	C_2H_5	$Fe(acac)_3$ (3 mol%)	THF-NMP, 0°C, 15–60 min	79	304
15	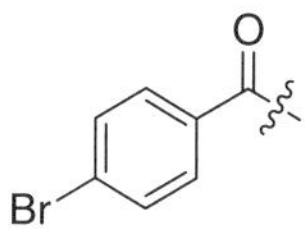	Me	$Fe(acac)_3$ (3 mol%)	THF, −78°C, < 15 min	86	305
16	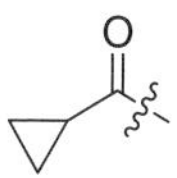	n-C_6H_{13}	$Fe(acac)_3$ (3 mol%)	THF, −78°C, < 15 min	92	305
17	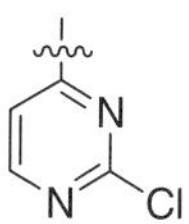	Me	$Fe(acac)_3$	THF, −78°C.	51	305
18	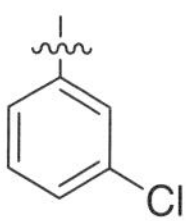	n-C_8H_{17}	$Fe(acac)_3$	THF-NMP, r.t.	77	305

[a]r.t., room temperature.

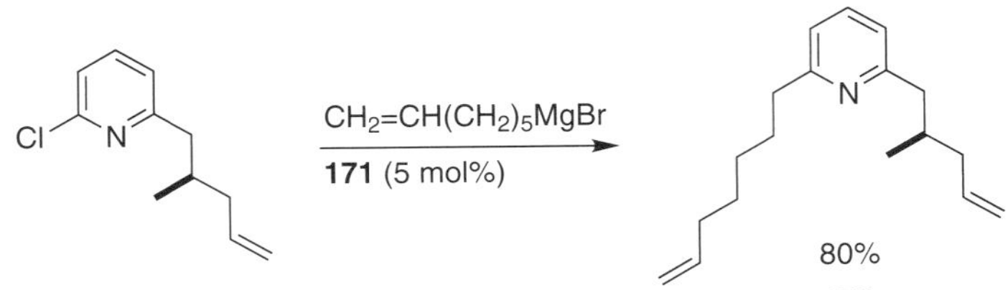

173

(1.18)

Complex **174** or a combination of $FeCl_3$ and appropriate amines such as NEt_3, tmeda, and DABCO shows good catalytic activity for the coupling of aryl Grignard reagents and alkyl bromides or chlorocyclohexane [309, 310]. Complex **175** is suitable for wider substrates. It catalyzes reaction of primary and cyclic or acyclic secondary alkyl chlorides with various aryl Grignard reagents, including sterically demanding ones such as 2,6-dimethylphenylmagnesium bromide [311].

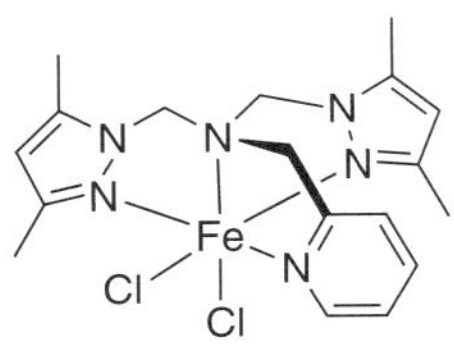

174

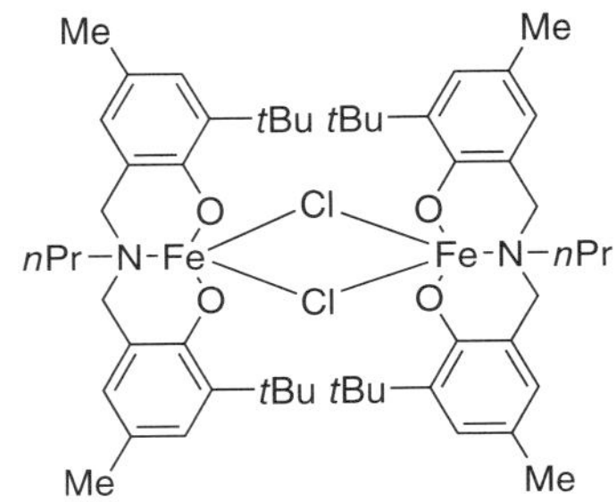

175

Recently, it was reported that NHC–Fe complexes are able to catalyze coupling of aryl or alkyl chlorides with aryl Grignard reagents. Treatment of a mixture of $FeF_3 \cdot 3H_2O$ (3 mol%) and SIPr·HCl (6 mol%) with EtMgBr in THF at 0°C–r.t. for 4 h affords a catalytically active species that efficiently catalyzes cross-coupling of activated, unactivated, and deactivated aryl chlorides with aryl Grignard reagents. The reaction proceeds in THF at 60°C and gives excellent product yields. In this catalytic reaction, the fluoride anion plays a key role in suppressing the undesired homocoupling reaction [312]. The combination of $FeCl_3$ and IPr is very effective in catalyzing reaction of aryl Grignard reagents with various primary, secondary, and tertiary alkyl chlorides as well as polychloroalkanes (Scheme 1.21) [313].

1.3.3.4 Cobalt Catalysts.

Kharasch and Fuchs [314] were the first to report cobalt-catalyzed cross-coupling of vinyl chloride and phenylmagnesium bromide. Cahiez and Avedissian [315] found that the presence of NMP as a co-solvent improves the reaction. Oshima and co-workers [316] demonstrated that $Co(acac)_2$ catalyzes the cross-coupling of pyridyl, quinolyl, and pyrimidyl chlorides with phenyl, benzyl, allyl, and trimethylsilylmethyl Grignard reagents in dioxane at room temperature. Knochel and co-workers [317, 318] reported cross-coupling of aryl Grignard reagents with a variety of heterocyclic chlorides in the presence

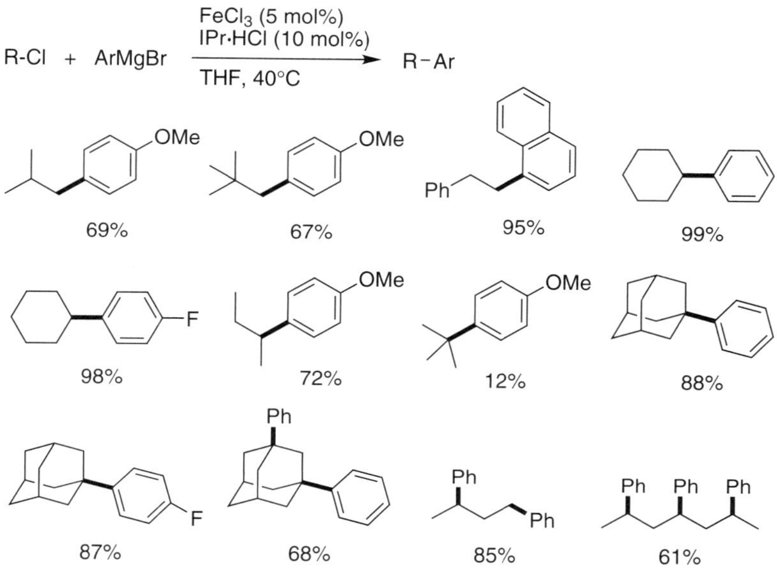

Scheme 1.21. Fe/NHC-catalyzed coupling of alkyl chlorides with aryl Grignard reagents.

of $Co(acac)_2$ or CoX_2 (X = Cl, Br, I). Interestingly, cobalt powder also effectively catalyzes the coupling. Reaction of PhMgCl with 1-chloroisoquinoline or 2-chloroquinoline in the presence of cobalt powder in Et_2O at room temperature results in the desired products in 91% and 88% yield, respectively [318].

Cobalt-catalyzed coupling of alkyl chlorides with Grignard reagents is less successful. The $Co(acac)_2$/tmeda system was found to catalyze coupling of alkyl bromides or iodides with 1-(trimethylsilyl)ethenyl- or 2-trimethylsilylethynylmagnesium reagents at room temperature. However, the reaction of chlorocyclohexane with 1-(trimethylsilyl)ethenylmagnesium bromide under the same conditions gives the cross-coupling product in only 26% yield [319]. Recently, early/late heterobimetallic Zr/Co complexes **176a–176c** were reported to be able to catalyze cross-coupling of *n*OctMgBr with primary and secondary alkyl bromides and chlorides. For the chloride substrates, **176c** exhibits higher catalytic activity, leading to cross-coupling products in 41.7–77.9% yields [Eq. (1.19)] [320].

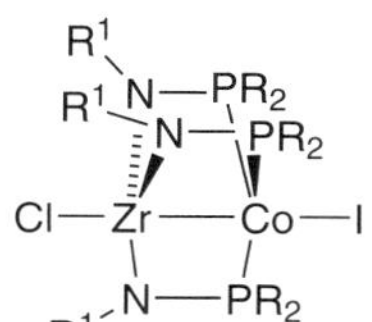

176a R = Ph, R^1 = *i*Pr
176b R = *i*Pr, R^1 = Mes
176c R = R^1 = *i*Pr

$$n\text{OctMgBr} + \text{RCl} \xrightarrow[\text{tmeda (30 mol\%), THF, r.t.}]{\textbf{176}\text{ (5 mol\%)}} n\text{Oct–R} \tag{1.19}$$

1.3.3.5 Copper Catalysts. In 1994 Bäckvall et al. [321] reported that allyl chlorides were regioselectively coupled with arylmagnesium using CuCl·2LiCl as catalyst. Terao et al. [322] found catalytic coupling of alkyl chlorides with Grignard reagents by $CuCl_2$ or CuCl in the presence of 1-phenylpropyne as an additive. In the absence of an alkyne the reaction gives only trace cross-coupling products. Among a series of alkynes, 1-phenylpropyne is the most effective. Primary, secondary, and tertiary alkyl as well as phenyl Grignard reagets can be employed to couple with primary alkyl chlorides [Eq. (1.20)]. The reaction mechanism has not been clarified yet, but a free radical process was ruled out. The role of 1-phenylpropyne is probably to stablize the thermally unstable alkylcopper(I) intermediates through coordination [322].

$$n\text{C}_9\text{H}_{19}\text{–Cl} + \text{RMgCl} \xrightarrow[\text{THF, reflux, 6 h}]{\text{CuCl}_2\text{ (2 mol\%)},\ \text{Ph–}\equiv\text{–Me (10 mol\%)}} n\text{C}_9\text{H}_{19}\text{–R} \tag{1.20}$$

R = *n*Bu >99%; *s*Bu 95%; *t*Bu >98%; Ph 82%

CuI was demonstrated to efficiently catalyze coupling of tertiary Grignard reagents with heteroaryl chlorides in high yields under very mild conditions. This copper-catalyzed reaction selectively converts dichloroazacycles into monosubstituted chloroazacycles and trichloroazacycles into either monosubstituted dichloroazacycles or disubstituted monochloroazacycles (Scheme 1.22) [323].

1.3.4 Stille Reaction

The Stille reaction is transition-metal-catalyzed cross-coupling of organotin compounds and organic halides. Chlorides are less employed as electrophiles in this reaction compared with Suzuki reaction and Kumada reaction. In 1997 Shirakawa et al. [324] reported a Ni-catalyzed Stille reaction of aryl chlorides with vinyl or alkynyltin reagents. Subsequent studies predominantly employed palladium as the catalysts.

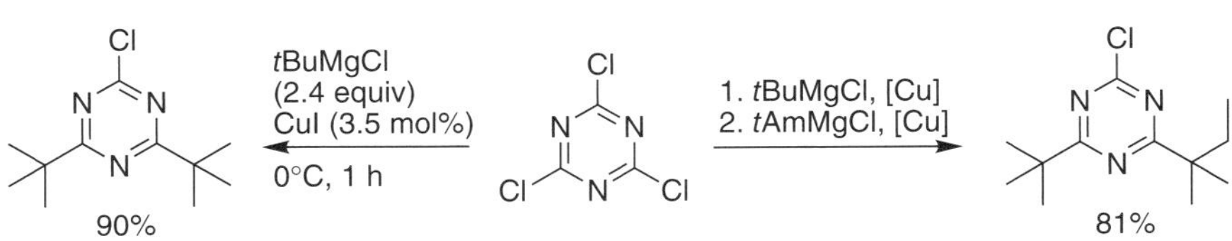

Scheme 1.22. CuI-catalyzed coupling of tertiary Grignard reagents and 2,4,6-trichloro-1,3,5-triazine.

In 2004 the mechanism of palladium-catalyzed Stille reaction was reviewed by Espinet and Echavarren [325].

1.3.4.1 Palladium–Phosphine Catalysts. Fu and co-workers reported general method for Pd-catalyzed Stille reaction of aryl chlorides. P*t*Bu_3 is a highly effective ligand and CsF is found to be a key additive. It is thought that the former enhances the reactivity of the palladium catalyst and the latter enhances the reactivity of the organotin compound. In this reaction unactivated aryl and vinyl chlorides can be coupled with alkyl, allyl, vinyl and aryltin reagents in the presence of CsF in dioxane (Table 1.6). Very hindered biaryls including tri- and tetra-*ortho*-substituted biaryls can be synthesized using this method [326, 327].

The combination of $Pd(OAc)_2$ and Buchwald ligand XPhos (**18a**) catalyzes Stille coupling of aryl chlorides and aryltin reagents in the presence of CsF. Very hindered aryltin such as 2,4,6-trimethylphenyltributyltin and 2,6-dimethoxyphenyltributyltin and deactivated aryl chlorides can be employed as the substrates, producing corresponding biaryl compounds in 61–98% yields [328]. The combination of $Pd_2(dba)_3$ and proazaphosphatrane **177a** or **177b** also exhibits an excellent catalytic effect for the reaction of electron-rich and electron-poor aryl chlorides, heteroaryl chlorides, and vinyl chlorides with aryltin or vinyltin reagents in dioxane. CsF or Me_4NF was employed as the additive. The reaction usually proceeds at 100–110°C with 1.5 mol% palladium loadings. However, the $Pd_2(dba)_3$/**177a**-catalyzed reaction of activated aryl chlorides with phenyltributyltin can go to completion at 60°C. $Pd_2(dba)_3$/**178** is superior to $Pd_2(dba)_3$/**177a** or **b** in catalyzing Stille cross-coupling of sterically hindered substrates. A series of tri- or tetra-*ortho*-substituted biaryls have been synthesized using this catalyst system (Scheme 1.23) [329, 330]. In addition, a palladium–phosphinous acid **56**-catalyzed Stille reaction of aryl chlorides could proceed in water without need of organic cosolvents, forming biaryls in good

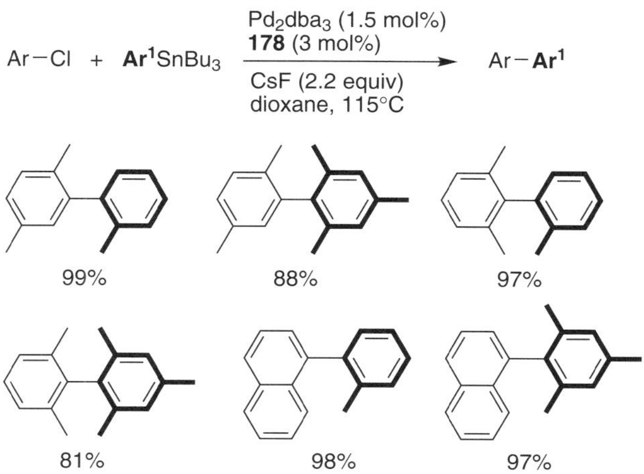

Scheme 1.23. Biaryl coupling catalyzed by $Pd_2(dba)_3$/**178**.

TABLE 1.6. Pd/P*t*Bu$_3$-catalyzed Stille Coupling

$$\text{R–Cl} + \text{R}^1\text{SnBu}_3 \xrightarrow[\text{CsF (2.2equiv), dioxane}]{\text{Pd}_2\text{dba}_3\text{-P}t\text{Bu}_3 \text{ or Pd(P}t\text{Bu}_3)_2} \text{R–R}^1$$

Entry	RCl	R^1	Pd-P Loading	Temperature (°C)	Yield (%)
1	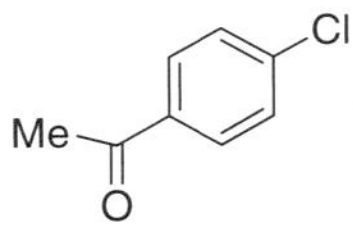	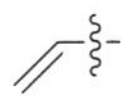	1.5 mol% Pd$_2$dba$_3$ 6 mol% P*t*Bu$_3$	80	87
2	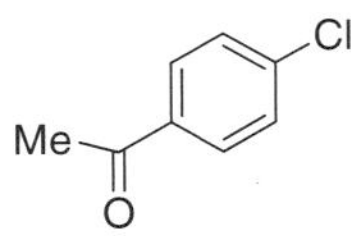	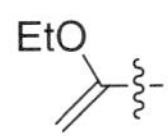	1.5 mol% Pd$_2$dba$_3$ 3 mol% P*t*Bu$_3$	r.t.[a]	86
3	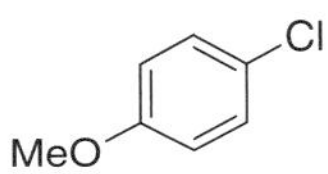	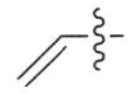	1.5 mol% Pd$_2$dba$_3$ 6 mol% P*t*Bu$_3$	100	82
4	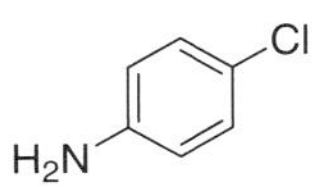	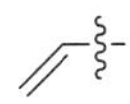	1.5 mol% Pd$_2$dba$_3$ 6 mol% P*t*Bu$_3$	100	61
5	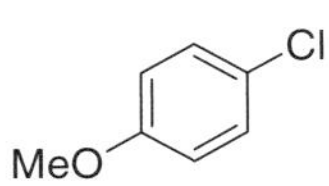	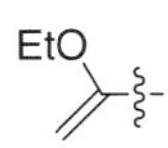	1.5 mol% Pd$_2$dba$_3$ 6 mol% P*t*Bu$_3$	100	98
6	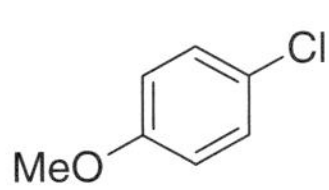	Ph	1.5 mol% Pd$_2$dba$_3$ 6 mol% P*t*Bu$_3$	100	94
7	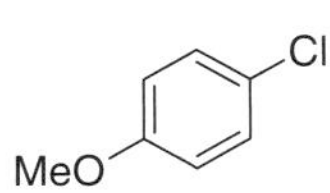	Bu	1.5 mol% Pd$_2$dba$_3$ 6 mol% P*t*Bu$_3$	100	82
8	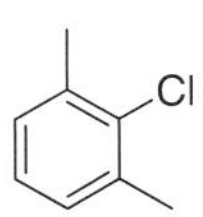	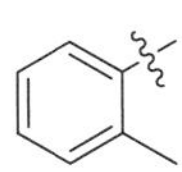	3 mol% Pd(P*t*Bu$_3$)$_2$	100	96
9	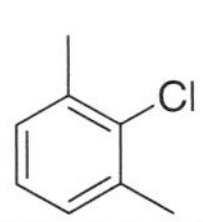	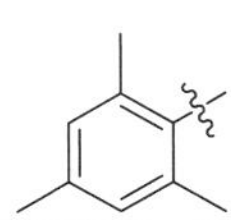	3 mol% Pd(P*t*Bu$_3$)$_2$	100	89

TABLE 1.6. (*Continued*)

Entry	RCl	R^1	Pd-P Loading	Temperature (°C)	Yield (%)
10	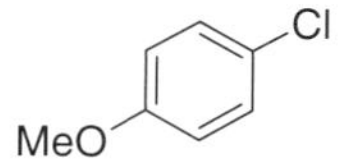	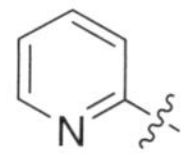	3 mol% $Pd(PtBu_3)_2$	100	47
11	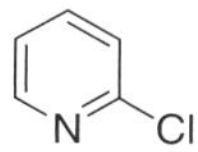	Ph	3 mol% $Pd(PtBu_3)_2$	100	76
12	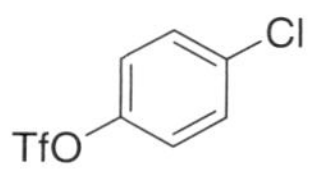	Ph	0.75% Pd_2dba_3 3 mol% $Pd(PtBu_3)_2$	r.t.	93
13	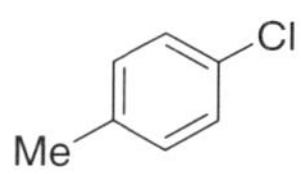	Ph	0.1 mol% $Pd(PtBu_3)_2$	100	92
14	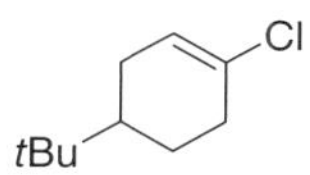	Ph	3 mol% $Pd(PtBu_3)_2$	100	86

[a] r.t., room temperature.

to high yields. The air stability and solubility in water of the palladium complex used in this study greatly facilitate operation of the coupling reaction, product isolation, and catalyst recycling [331].

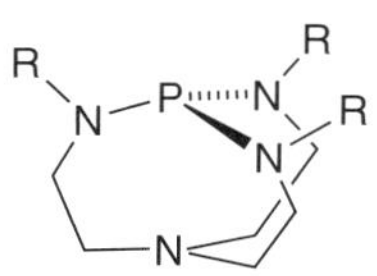

177a R = $PhCH_2$
177b R = *i*Bu

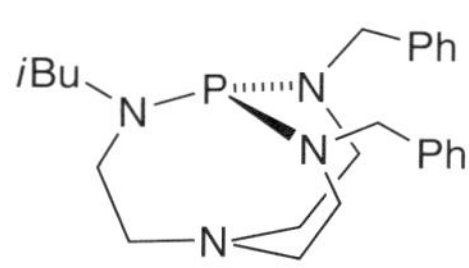

178

1.3.4.2 Palladium–NHC Catalysts. Organ and co-workers [332] reported that **127** catalyzes cross-coupling reaction between heteroaryl stannanes and aryl or heteroaryl chlorides in the presence of CsF and 4-Å molecular sieves. This catalyst system is suitable for thiophene-, furan-, pyrrole-, and thiazole-based organostannanes. The reaction is applicable for the unactivated aryl chlorides under mild conditions (30–80°C) and gives good to excellent yields [332].

1.3.4.3 N,O-Chelate Palladium Catalysts. The catalysts for the Stille reaction of chlorides mentioned above act at elevated temperature, from 30°C to 115°C depending on reactivity of substrates and catalysts used. However, the β-ketiminato Pd complex **179** was found to drive the coupling reaction of aryl chlorides and heteroaryl chlorides with organostannanes in the presence of CsF at room temperature with low catalyst loading (0.5 mol%). The catalyst system is suitable for activated, unactivated, deactivated, and hindered aryl chloride substrates and aryl, vinyl, and allyltin reagents. Electron-rich and electron-poor heteroaryl chlorides can be employed for this coupling (Table 1.7) [333]. The analogue of **179**, triethoxysilyl-functionalized palladium complex **180**, can be immobilized on the surface of silica-coated Fe_3O_4 to form the magnetic–nanoparticle-supported (β-ketiminato)(phosphanyl) palladium complex **181** as a heterogeneous catalyst. Complex **181** exhibits excellent catalytic activity in the presence of CsF in a mixture of EtOH and H_2O at 50°C and the substrate scope is similar to that by using **179** (Table 1.7). An attractive advantage of catalyst **181** is its easy separation using external magnets, which minimizes the loss of catalyst during separation [334].

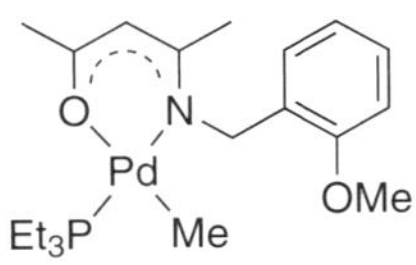

179

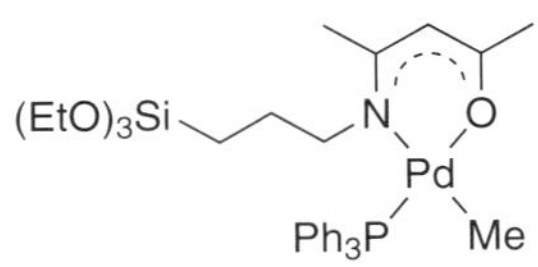

180

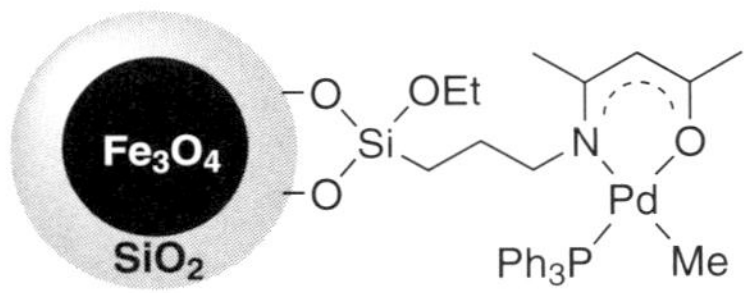

181

1.3.5 Hiyama Reaction

1.3.5.1 Palladium Catalysts

(i) PHOSPHINE LIGANDS. In 1996 Hatanaka and co-workers [335] reported first cross-coupling of aryl chlorides with aryl or alkenylchlorosilanes catalyzed by $(R_3P)_2PdCl_2$ in the presence KF. The reaction leads to arylation or alkenylation of activated aryl chlorides at elevated temperature in good to excellent yields. Hiyama and co-workers [336, 337] improved the reaction by employing the Buchwald ligand **18a** and $[(\eta^3\text{-}C_3H_5)PdCl]_2$ as the catalyst, arylallylsilanes as nucleophiles, and tetra-*n*-butylammonium fluoride (TBAF) as an activator [Eq. (1.21)]. By using this catalyst system, electron-rich and electron-poor aryl chlorides and heteroaryl chlorides were coupled with aryltriallylsilane in good to excellent yields. Tri-*ortho*-substituted biaryls can also be prepared in moderare yields. In addition, $Ph_2Si(allyl)_2$ and $PhSi(allyl)_3$ are also good coupling partners in the biaryl coupling reaction [336, 337]. In these reactions, it is assumed that 3 equiv of TBAF cleaved

TABLE 1.7. Complex 179- or 181-Catalyzed Cross-Coupling of Aryl or Heteroaryl Chlorides with Organostannanes

$$\text{Ar—Cl} + \text{RSnBu}_3 \xrightarrow[\text{CsF (2.0 equiv)}]{\text{cat. (0.5 mol\%)}} \text{Ar—R}$$

Entry	Catalyst	ArCl	R	Conditions	Yield (%)
1	**179**	p-$NO_2C_6H_4Cl$	Ph	THF, r.t.[a], 3 h	92
2	**179**	p-NCC_6H_4Cl	Ph	THF, r.t., 3 h	92
3	**179**	PhCl	Ph	THF, r.t., 4 h	91
4	**179**	p-$MeOC_6H_4Cl$	Ph	THF, r.t., 5h	90
5	**179**	o-MeC_6H_4Cl	o-MeC_6H_4	THF, r.t., 8h	80
6	**179**	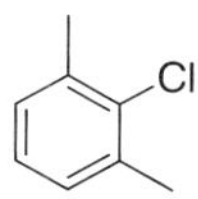	p-MeC_6H_4	THF, 50°C, 15 h	81
7	**179**	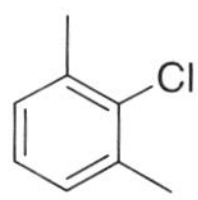	o-MeC_6H_4	THF, 50°C, 15 h	70
8	**179**	PhCl	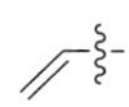	THF, r.t., 6 h	81
9	**179**	PhCl	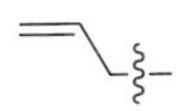	THF, r.t., 6 h	90
10	**179**	2-Chloropyridine	Ph	THF, r.t., 4 h	86
11	**179**	3-Chloropyridine	Ph	THF, r.t., 5 h	89
12	**179**	4-Chloropyridine	Ph	THF, r.t., 4 h	91
13	**179**	4-Chloropyrimidine	Ph	THF, r.t., 8 h	90
14	**179**	2-Chlorothiophene	Ph	THF, r.t., 6 h	88
15	**179**	3-Chlorothiophene	Ph	THF, r.t., 6 h	80
16	**181**	p-$NO_2C_6H_4Cl$	Ph	EtOH-H_2O, 50°C, 2 h	96
17	**181**	PhCl	Ph	EtOH-H_2O, 50°C, 3 h	95
18	**181**	p-$MeOC_6H_4Cl$	Ph	EtOH-H_2O, 50°C, 3 h	92
19	**181**	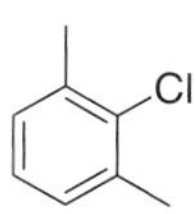	o-MeC_6H_4	EtOH-H_2O, 70°C, 12 h	72
20	**181**	PhCl	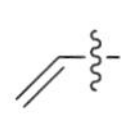	EtOH-H_2O, 50°C, 5 h	87
21	**181**	PhCl	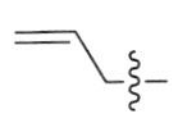	EtOH-H_2O, 50°C, 5 h	90

[a] r.t., room temperature.

the three allyl groups on silicon to form aryltrifluorosilane or its equivalent, and the additional mole of TBAF might have caused the formation of a pentacoordinate silicate species to ease the transmetalation [336]. Indeed, aryltrifluorosilanes have been demonstrated to couple with a wide variety of substituted aryl and heteroaryl chlorides in the presence of 2.5 equiv of TBAF using $Pd(OAc)_2$/**18a** as the catalyst [338].

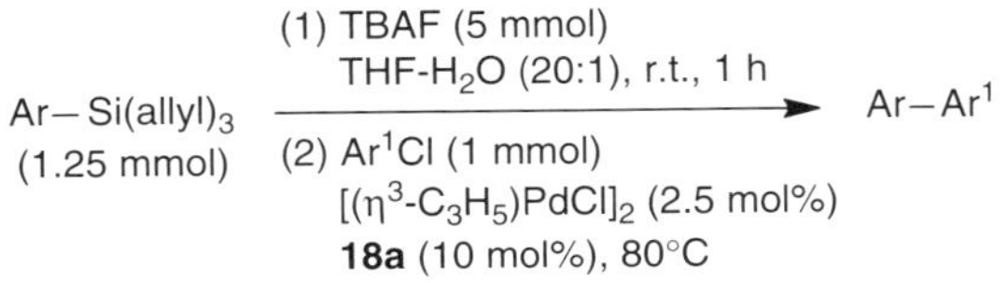

(1.21)

$[(\eta^3\text{-}C_3H_5)PdCl]_2$–**15a** was demonstrated to catalyze stereospecific cross-coupling of (*E*)- and (*Z*)-alkenylsilanolates with aryl chlorides. Di-, tri-, and tetra-substituted alkenyldimethylsilanolates and electron-rich and electron-poor aryl chlorides can be employed as the reaction substrates. The reaction results in high product yields and complete retention of the double bond geometry [339].

The combination of $[(\eta^3\text{-}C_3H_5)PdCl]_2$ or $Pd(dba)_2$ and **15b** or **182** catalyzes trifluoromethylation of aryl chlorides using Et_3SiCF_3 and KF. The reaction proceeds effectively in dioxane at 120–140°C depending on substrates and ligand used, giving corresponding trifluoromethylation products in good to excellent yields (Table 1.8). The catalyst system is compatible with esters, acetals, amides, nitriles, ethers, dialkylamines, and heteroaromatic substituents, but is not compatible with substrates bearing aldehydes, ketones, and OH and NH groups.

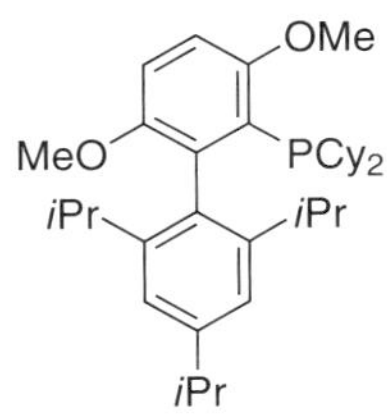

182 (BrettPhos)

This transformation appears to proceed via the generalized catalytic cycle for aryl trifluoromethylation as shown in Figure 1.8. The steps of transmetalation and reductive elemination have been confirmed by formation and characterization of Pd–CF_3 intermediates and NMR spectroscopic studies [340].

Wolf et al. [341] found that the palladium–phosphinous acid complex **58** catalyzes cross-coupling of activated aryl or heteroaryl chlorides with phenyltrimethoxysilane in acetonitrile at 80°C when TBAF is used to activate the arylsiloxanes for transmetalation. NaOH promotes the reaction in H_2O at a higher temperature (135–140°C in a closed vessel) in the absence of commonly used additives such as CsF and TBAF or other surfactants [342]. The Hiyama reaction using phenyltrimethoxysilane and activated aryl chlorides is also catalyzed by $Pd(dba)_2$–**142** in the presence of TBAF in dioxane at 80°C [271].

TABLE 1.8. Pd-Catalyzed Trifluoromethylation of Aryl Chlorides

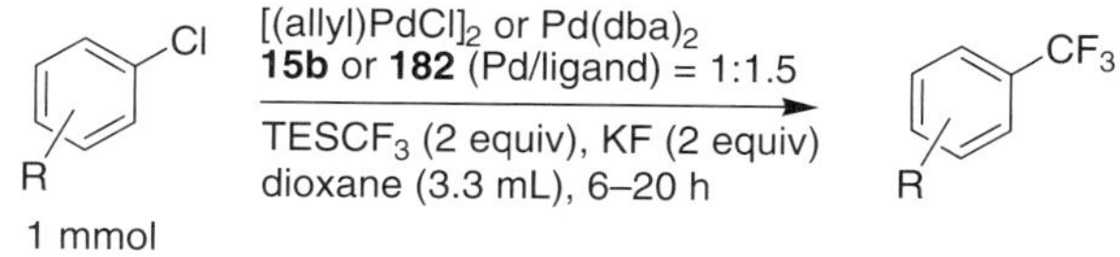

Entry	Aryl Chloride	Pd Catalyst (mol%)	Ligand	Temperature (°C)	Yield (%)
1	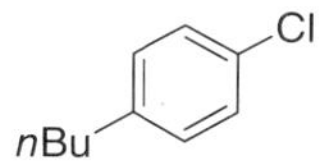	[(allyl)PdCl]$_2$ (3)	**182**	120	80
2	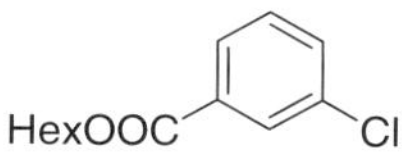	[(allyl)PdCl]$_2$ (3)	**182**	120	83
3	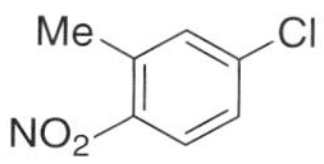	[(allyl)PdCl]$_2$ (3)	**182**	130	70
4	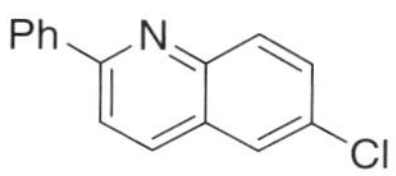	[(allyl)PdCl]$_2$ (3)	**182**	130	82
5		[(allyl)PdCl]$_2$ (3)	**182**	130	76
6	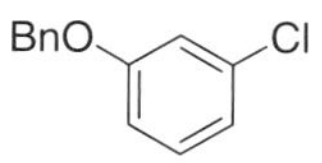	Pb(dba)$_2$ (6)	**182**	130	88
7	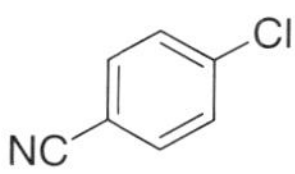	Pb(dba)$_2$ (6)	**182**	130	72
8	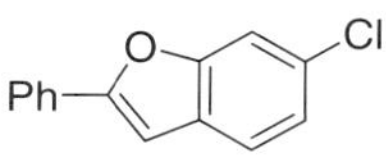	Pb(dba)$_2$ (6)	**182**	130	87
9	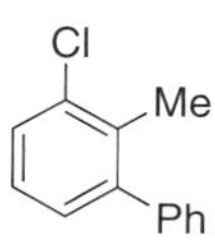	[(allyl)PdCl]$_2$ (4)	**15b**	140	87
10	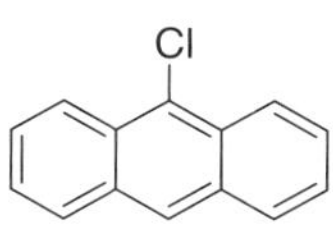	[(allyl)PdCl]$_2$ (4)	**15b**	140	90
11	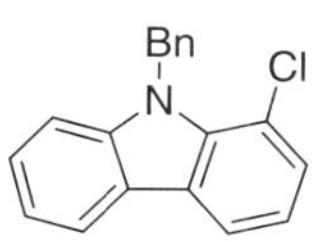	[(allyl)PdCl]$_2$ (4)	**15b**	140	82

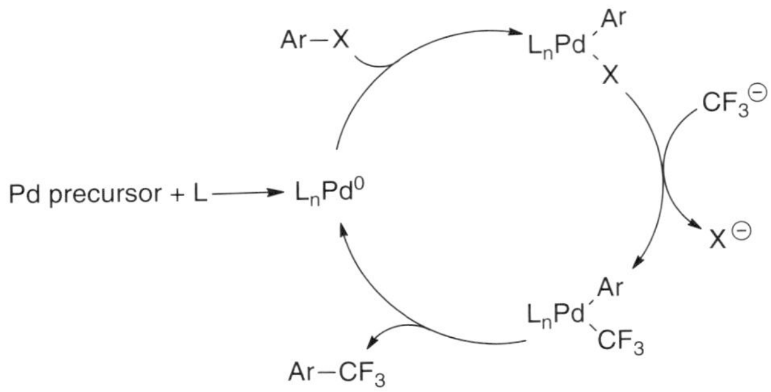

Figure 1.8. Generalized catalytic cycle for aryl trifluoromethylation (L = ligand; Ar = aryl; X = Cl, Br, I, triflate).

(ii) NHC LIGANDS. Both $Pd(OAc)_2$-IPr·HCl–TBAF and **183**–TBAF systems can catalyze the coupling of phenyltrimethoxysilane with activated aryl or heteroaryl chlorides at 80°C. Deactivated aryl chlorides such as 4-chloroanisole and 4-chlorotoluene give poor yields. $Pd(OAc)_2$-IPr·HCl also catalyzes the reaction of $CH_2{=}CHSi(OMe)_3$ with p-$ClC_6H_4C(O)Me$ to give high conversion under the same conditions [343, 344]. Complex **85** catalyzes the cross-coupling of phenyltrimethoxysilane with activated and deactivated aryl chlorides using TBAF as activator in toluene at 120°C, affording biaryls in moderate to excellent yields [345].

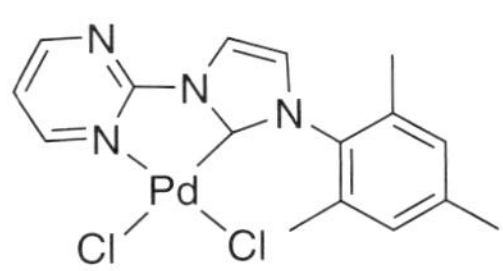

183

(iii) OTHER LIGANDS. $Pd(OAc)_2$/DABCO was found to catalyze the cross-coupling of activated aryl chlorides with $PhSi(OMe)_3$ in the presence of TBAF in dioxane at 80°C. However, this catalyst system is not suitable for unactivated and deactivated aryl chlorides [346]. The combination of $[Pd(allyl)Cl)]_2$ and **64b** catalyzes reaction of $PhSi(OMe)_3$ with 3-chlorofluorobenzene or 4-chlorophenylethanone using TBAF as an activator. Under microwave radiation, the reaction reaches completion within minutes. However, for an unactivated aryl chloride such as 4-chlorotoluene, the reaction affords reduced arene as major product. In this reaction, the catalytically active species is assumed to be **184** [347]. $Pd(OAc)_2$/**185** shows much higher catalytic activity than do $Pd(OAc)_2$/DABCO or the $Pd(allyl)Cl)]_2$/**64b** system. The former can efficiently catalyze the coupling of deactivated aryl chlorides such as 4-chloroanisole and 3,4-$OCH_2OC_6H_3Cl$ with $ArSi(OMe)_3$ in the presence of THAF at 80°C. Functionalized aryl chlorides, heteroaryl chlorides, and alkenyl chlorides can be employed in this cross-coupling [348].

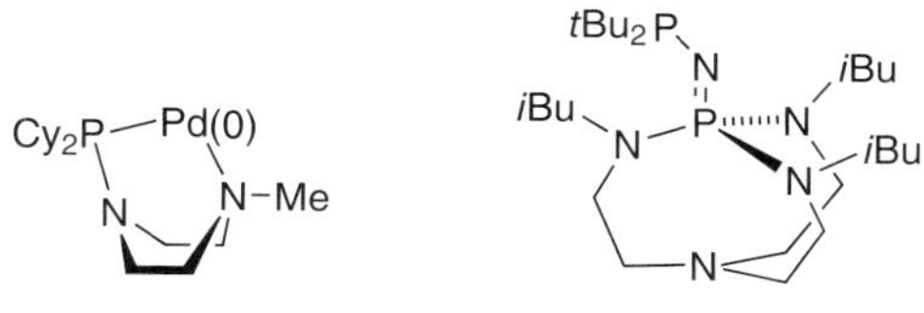

184 185

The β-diketiminatophosphane Pd complex **112** exhibits excellent catalytic activity for the reaction of aryl chlorides with $ArSi(OEt)_3$. The reaction proceeds effectively in H_2O in the presence of NaOH and TBAB at 60°C with 0.5 mol% catalyst loading. The catalyst system is suitable for a wide substrate scope. Electron-rich, electron-poor, and sterically hindered aryl chlorides, heteroaryl chlorides, aryl dichlorides, and 1,2,4,5-tetrachlorobenzene can be employed in this catalytic coupling. Tri-*ortho*-substituted biaryls were synthesized in good yields by this method. The vinylation of 1-chloro-2,4,6-trimethylbenzene was also carried out with triethoxyvinylsilane. In addition, the catalyst promotes a one-pot Hiyama–Heck double coupling of highly hindered substrates (Scheme 1.24) [349].

(iv) Palladacycle Complexes. Alacid and Nájera reported [350–352] oxime-derived palladacycles **100b** and **100d** to catalyze coupling of aryl chlorides with aryl- or alkenyltrialkoxysilanes. Palladacycle **100b** exhibits relatively low activity for

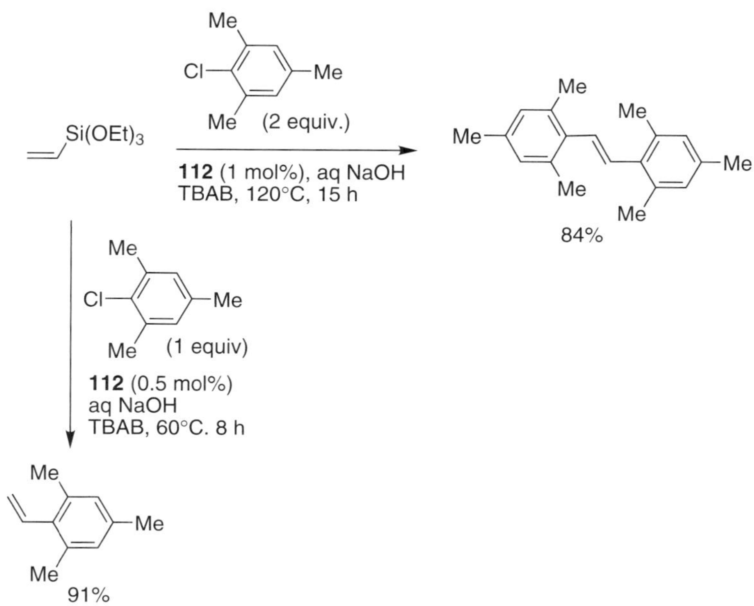

Scheme 1.24. One-pot Hiyama–Heck double coupling.

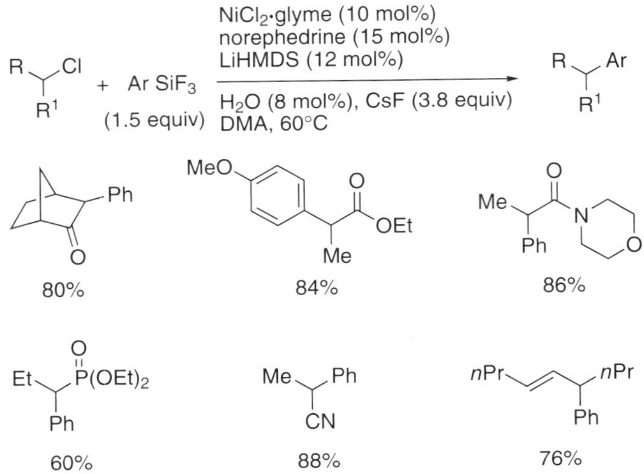

Scheme 1.25. Nickel/norephedrine-catalyzed Hiyama reactions of activated alkyl chlorides.

coupling of aryl chlorides. It catalyzes the reaction of activated aryl chlorides with $RCH{=}CHSi(OR')_3$ (R = H, Ph, n-C_6H_{13}; R′ = Me or Et) in H_2O in the presence of NaOH and TBAB under microwave radiation, giving cross-coupling products in moderate yields [350, 351]. However, under the same conditions, the **100d**-catalyzed cross-coupling of aryl chlorides with aryltrialkoxysilanes affords good results. Electron-rich and electron-poor aryl chlorides and heteroaryl chlorides can be coupled in good to excellent yields within minutes [352].

1.3.5.2 Nickel Catalysts. Nickel-catalyzed Hiyama reaction of organic chlorides is very rarely seen. In 2007 Fu and co-workers [353] reported that activated secondary alkyl chlorides couple with $ArSiF_3$ catalyzed by a nickel/norephedrine complex (Scheme 1.25).

Nakao and co-workers [354] reported a Ni-catalyzed coupling of aryl chlorides with 2-(2-hydroxyprop-2-yl)cyclohexyl-substituted arylsilanes activated by Cs_2CO_3 [Eq. (1.22)]. Using this catalyst system, electron-rich and electron-poor aryl and heteroaryl chlorides were coupled with aryl or heteroarylsilicon derivatives [354].

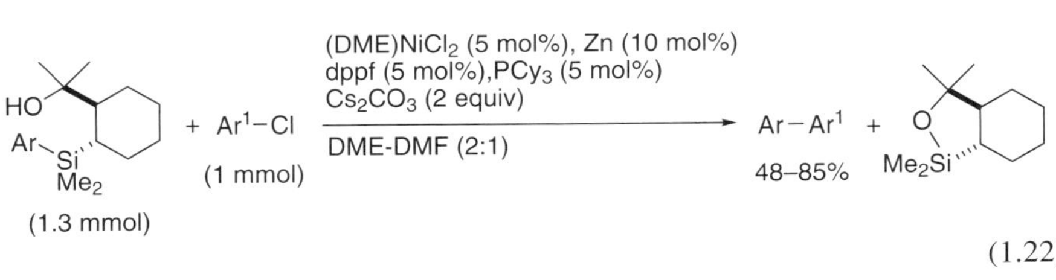

(1.22)

1.3.6 Sonogashira Reaction

Sonogashira reaction is the most important method for the alkynylation of aryl or alkenyl halides. Recently Chinchilla and Nájera [355] reviewed this topic. In this section, we focus on the Sonogashira coupling using chlorides as the coupling partners.

1.3.6.1 Palladium Catalysts

(i) MONOPHOSPHINE LIGANDS. The Pd/C–CuI–PPh_3 catalyst system facilitates Sonogashira coupling of 2-chloropyridine, 2-chloroquinoline, and 2,4-dichloroquinoline with terminal alkynes in water or a mixture of DMA and H_2O [356, 357]. The $Pd(PPh_3)_2Cl_2$–CuI system shows higher catalytic activity than does the Pd/C–CuI–PPh_3 system, the former catalyzing selective coupling of 2,4-dichloropyrimidine with trimethylsilylacetylene in high yield with 0.5 mol% Pd loading. The coupled product is a useful intermediate for the synthesis of medicinally important 4-aryl-5-pyrimidinylimidazoles (Scheme 1.26) [358].

$Pd(PPh_3)_2Cl_2$–CuI also facilitates Sonogashira coupling of benzimidoyl chlorides with 1,6-enynes (Scheme 1.27). When R^3 is an H or alkyl group, normal cross-coupling products are formed. However, when R^3 is an aryl group, quinoline derivatives are obtained in good yields. Formation of the quinoline derivatives should be through further cyclization of the cross-coupling product [359].

The above Pd catalyst systems with PPh_3 as ligand show good activity for only activated aryl chlorides. The combination of $Na_2[PdCl_4]$ (2 mol%), CuI (1.5 mol%), and $(1\text{-Ad})_2P\textit{n}Bn$ or $P\textit{t}Bu_3$ (4 mol%) is applicable for a wider scope of substrates, including electron-rich and electron-poor aryl chlorides and a variety of terminal alkynes. The reaction proceeds in DMSO at 100°C using Na_2CO_3 as base, giving moderate to excellent yields [360]. The catalyst system consisting of $Na_2[PdCl_4]$, CuI and **186** or 9-ethylfluorenyldicyclohexylphosphine (**43**) displays similar catalytic behavior with less Pd loadings (1 mol%) [135, 361].

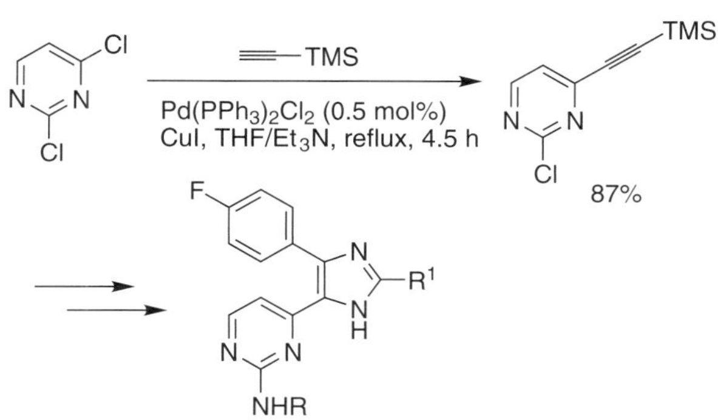

Scheme 1.26. $Pd(PPh_3)_2Cl_2$–CuI-catalyzed selective coupling of 2,4-dichloropyrimidine and trimethylsilylacetylene.

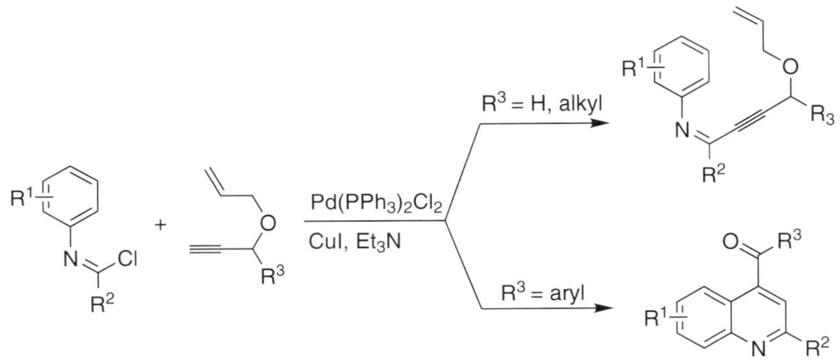

Scheme 1.27. Sonogashira coupling of benzimidoyl chlorides with 1,6-enynes.

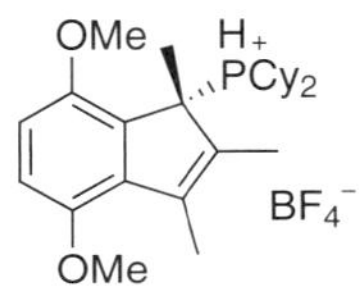

186

Phosphine-free palladium catalyst can also drive Sonogashira coupling of organic chlorides. A palladium complex of azetidine-derived polyamine **187**, [Pd(**187**)Cl]ClO_4, exhibits good catalytic activity in the coupling of electron-rich and electron-poor aryl bromides and chlorides with aryl-, alkyl-, and silylacetylenes using CuI as the co-catalyst and Na_2CO_3 as a base. The reaction proceeds in DMA-H_2O under mild conditions with low Pd loadings. For the deactivated aryl chlorides, the catalytic coupling with phenylacetylene affords good results, while with alkyl or silylacetylenes results in relatively low yields [362].

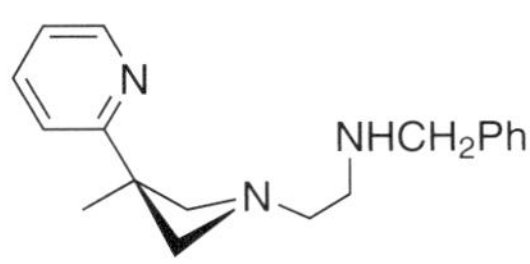

187

The reactions mentioned above require CuI as the co-catalyst. However, this is not essential for a Sonogashira cross-coupling. A series of Cu(I)-free Sonogashira reactions have been reported, including those using chlorides as the electrophilic substrates. An efficient and effective microwave-assisted cross-coupling of terminal alkynes with various aryl chlorides including sterically hindered, electron-rich, electron-neutral, and electron-deficient aryl chloride has been carried out using $Pd(PPh_3)_2Cl_2$–$P\textit{t}Bu_3$–DBU in DMF. Alkenyl and heteroaryl chlorides are also applicable as the electrophilic species. However, when trimethylsilyl acetylene is employed as the nucleophilic partner, the reaction leads to diaryl acetylenes [Eq. (1.23)]. These results indicate that the catalyst system is also effective for the C_{sp}–Si bond activation [363]. Symmetrical diaryl acetylenes can also be synthesized through $Pd(PCy_3)_2Cl_2$-catalyzed Sonogashira reaction between aryl chlorides and 2-methyl-3-butyn-2-ol [Eq. (1.24)]. This reaction is unsatisfactory for the electron-deficient chlorobenzene derivatives such as 1-chloro-4-nitrobenzene and chloropentafluorobenzene, which give low yields and unidentified by products [364].

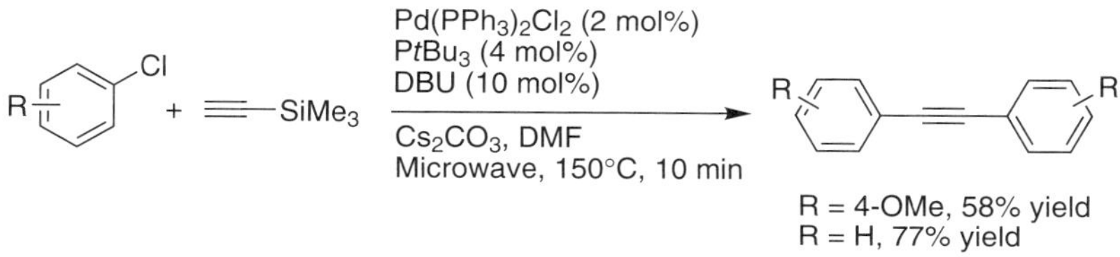

(1.23)

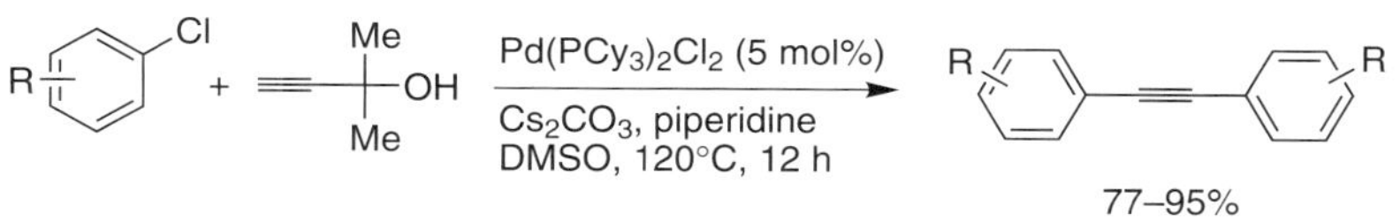

(1.24)

From 2-chloro-3-formylquinoline derivatives and phenylacetylene 1,3-disubstituted pyrano[4,3-*b*]quinolines were synthesized through $Pd(OAc)_2/PPh_3$-catalyzed Sonogashira coupling and then cyclization (Scheme 1.28). PPh_3 and $Pd(OAc)_2$ not only catalyze the cross-coupling, but also facilitate the cyclization reaction by reducing the reaction time and increasing the yield of cyclized products [365]. A similar example is the coupling of 2-chlorophenols with alkynes and following cyclization catalyzed by $Pd(MeCN)_2Cl_2$/**141** (R = Cy) as shown in Eq. (1.25) [366].

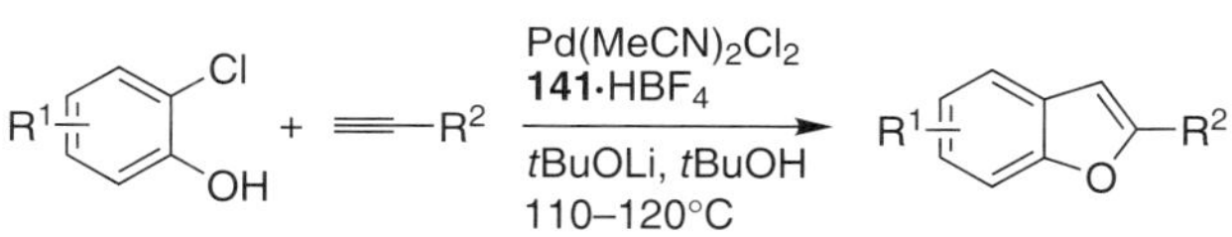

(1.25)

The combination of Buchwald ligands and an appropriate palladium source is also very effective in copper-free Sonogashira coupling of aryl chlorides. Pd/C-XPhos catalyzes coupling of a range of aryl or heteroaryl chlorides with aliphatic or aryl terminal alkynes using K_2CO_3 as base at 110°C, forming corrsesponding products in 72–99% yields [367]. Gelman and Buchwald showed that the $Pd(MeCN)_2Cl_2$–XPhos system give similar results when using Cs_2CO_3 as a base in MeCN at 70–95°C [368]. In this study they found that the addition of a copper

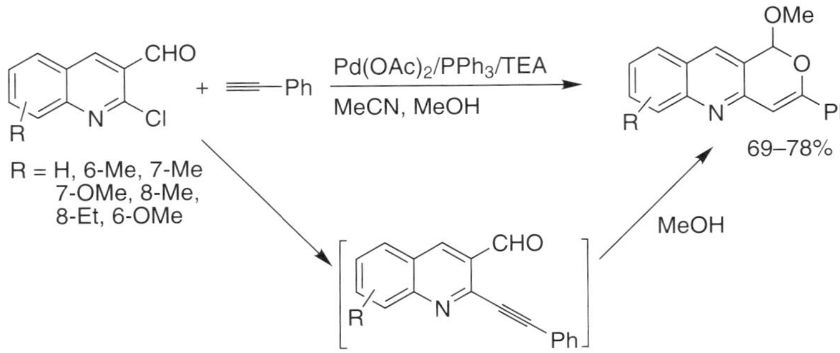

Scheme 1.28. One-pot synthesis of pyrano[4,3-b]quinoline via palladium catayst.

co-catalyst inhibits product formation in the coupling reaction of aryl chlorides with terminal alkynes. Using sulfonated XPhos **188** and $Pd(MeCN)_2Cl_2$, the similar reaction can be carried out in good to excellent yields in a 1:1 mixture of MeCN and H_2O at 60–100°C, but requires higher Pd and ligand loadings (2.5 mol% Pd and 7.5 mol% **188**) [109]. A related ligand **189** is also effective in the $Pd(MeCN)_2Cl_2$-catalyzed reaction of substituted aryl chlorides or 3-chlorothiophene with phenyl, alkyl, and silylacetylenes in the presence of Na_2CO_3 in toluene at 90°C. In most cases the reaction using electron-rich and electron-poor aryl chlorides give good to excellent yields. However, the reaction of 4-chlorobenzonitrile or 1-chloro-2-vinylbenzene results in low yields [369].

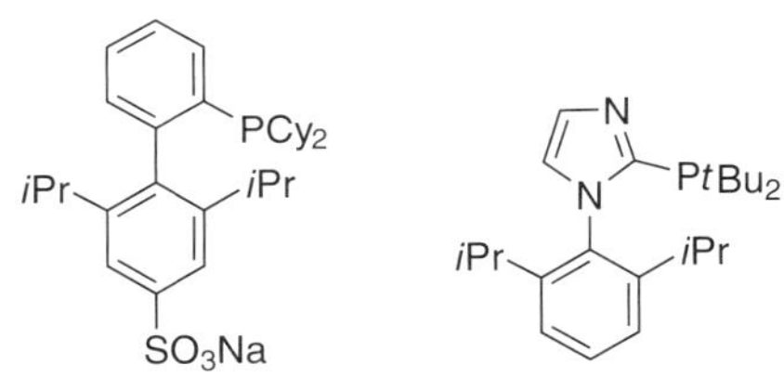

188 **189**

(ii) Multiphosphine Ligands. Multidentate phosphines were also used in palladium-catalyzed Sonogashira coupling of aryl or heteroaryl bromides or chlorides. The tetrapodal phosphine ligand **190** in combination with $[Pd(allyl)Cl]_2$ was found to be effective catalyst for the coupling of aryl and heteroaryl chlorides and bromides with a variety of terminal alkynes using K_2CO_3 as base. For the deactivated aryl chlorides the reaction requires higher temperature (140°C) in DMF without CuI. For the reactive heteroaryl chlorides such as 2-chloropyridine and 2-chloroquinoline the reaction can be carried out at lower temperature (100°C) in the presence of CuI [370–373]. $[Pd(allyl)Cl]_2$–**191** is very effective for coupling of aryl bromides with terminal alkynes, but only catalyzes coupling of activated aryl chlorides such as 4-chloroacetophenone and 4-chlorobenzonitrile with phenylacetylene in good yields [374]. *rac*-BINAP-$Pd(OAc)_2$ can serve as an efficient catalyst for alkynlation of

7-chlorocamptothecins. The reaction does not require Cu(I) co-catalyst and proceeds smoothly in toluene at 100°C using K_2CO_3 as base [Eq. (1.26)] [375].

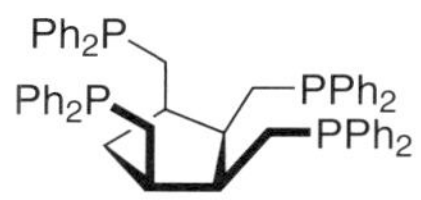

190

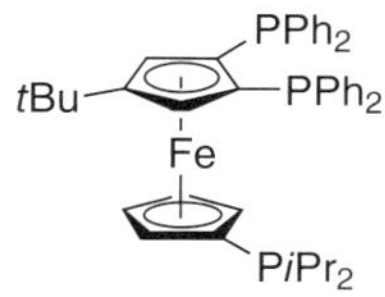

191

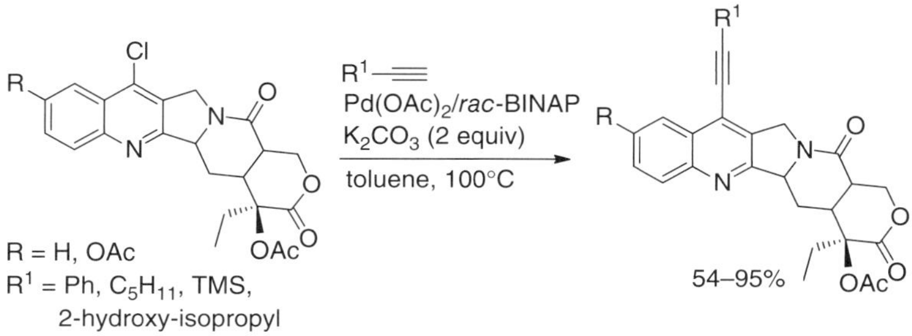

(1.26)

(iii) Palladacycle Complexes. Palladacycles were less successful in catalyzing Sonogashira coupling of organic chlorides. The palladacycle–PTA (PTA = 1,3,5-triaza-7-phosphaadamantane) system (**192**) was found to catalyze coupling of activated and unactivated aryl chlorides with alkyl, alkenyl, and arylacetylenes in the presence of Cs_2CO_3 and TBAB at 80°C. For the activated aryl chlorides, quantitative GC yields could be achieved; however, whereas for the reactions of PhCl, only moderate yields were obtained. This catalyst shows activity very close to that of the Pd(OAc)–PTA (1:3) system [376]. A related complex, pincer-palladium complex **193**, is an active catalyst for cross-coupling of electron-rich and electron-poor aryl chlorides with phenylacetylene using Cs_2CO_3 as a base. This reaction requires high catalyst loading (5 mol%), high temperature (160°C), and $ZnCl_2$ as additive [377].

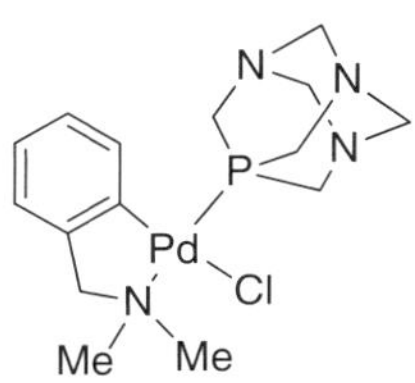

192

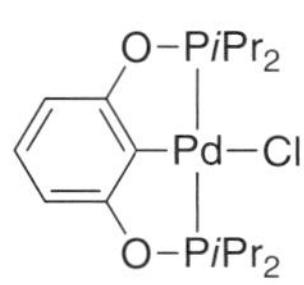

193

(iv) N,O-Chelate Palladium Catalyst. As indicated in Section 1.3.4.3, the magnetic–nanoparticle-supported (β-ketiminato) (phosphanyl)palladium complex **180** is a heterogeneous catalyst and has the advantage of easy separation using external magnets. This complex has been demonstrated to be an efficient catalyst for Stille and Suzuki reactions. It is also active for Sonogashira coupling of aryl chlorides without need of a Cu(I) co-catalyst. The reaction can be performed in H_2O in the presence of 2 equiv of piperidine, 0.5 equiv of TBAB, and 0.5 mol% of **180** at 60°C, resulting in desired products in 74–96% yields. Electron-rich, electron-poor, sterically hindered and functionalized aryl chlorides and aryl, aliphatic, and silylacetylenes are suitable substrates [334].

1.3.6.2 Nickel Catalysts. Nickel-catalyzed Sonogashira cross-coupling is scarce. Alkyl halides are also rarely used as coupling partners in the Sonogashira cross-coupling reaction [355]. Recently, amido pincer–nickel complex **10** was demonstrated to be an active catalyst for the coupling of primary alkyl iodides, bromides, and chlorides with terminal alkynes using Cs_2CO_3 as base and CuI as co-catalyst. For the chloride substrates, $n\text{Bu}_4\text{NI}$ additive and relatively high reaction temperature are required. The reaction tolerates β-H in chloride substrates and a variety of functional groups, giving 57–89% product yields (Table 1.9) [378].

1.3.7 Decarboxylative Cross-Coupling

Since Goossen et al. [379, 380] found that Pd/Cu-catalyzed decarboxylative cross-coupling reactions of aryl carboxylic acids and aryl halides, this method has attracted considerable attention due to widely available, stable, and inexpensive carboxylic acid salts as sources of the carbon nucleophile. The early catalyst systems are suitable for the coupling of aryl bromides. Using a catalyst system consisting of $PdBr_2$, CuBr, 1,10-phenanthroline, and (*o*-biphenyl)P*t*Bu_2 in NMP-quinoline, the coupling can occur between aryl chlorides and aryl carboxylic acids [Eq. (1.27)] [381].

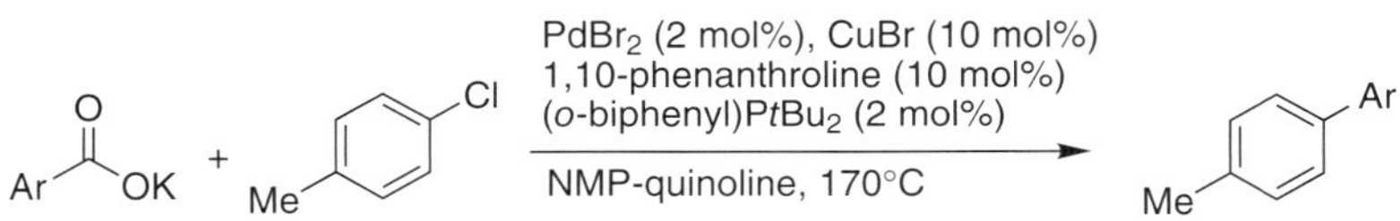

(1.27)

The proposed mechanism for this coupling is shown in Figure 1.9. Reaction of potassium carboxylate and copper salt ***a*** forms copper carboxylate ***b***. CO_2 is released from the copper carboxylate to generate an arylcopper intermediate ***c***. The arylcopper species transfers its aryl group to the arylpalladium(II) complex ***d*** formed by oxidative addition of an aryl halide to a palladium(0) species ***f***, giving a biarylpalladium complex ***e***. Reductive elimination of the biaryl from ***e*** regenerates the active palladium species ***f*** [382].

For the particularly activated carboxylates, palladium alone can catalyze the decarboxylative coupling with aryl halides, including aryl chlorides. Potassium polyfluorobenzoates were demonstrated to couple with aryl bromides, chlorides, and triflates by using $Pd(OAc)_2/PCy_3$ or P(*o*-Tol)$_3$ as the catalyst at 130–160°C. Electron-rich and electron-poor aryl chlorides are suitable substrates, and in most cases the reaction

TABLE 1.9. Sonogashira Coupling of Alkyl Chlorides

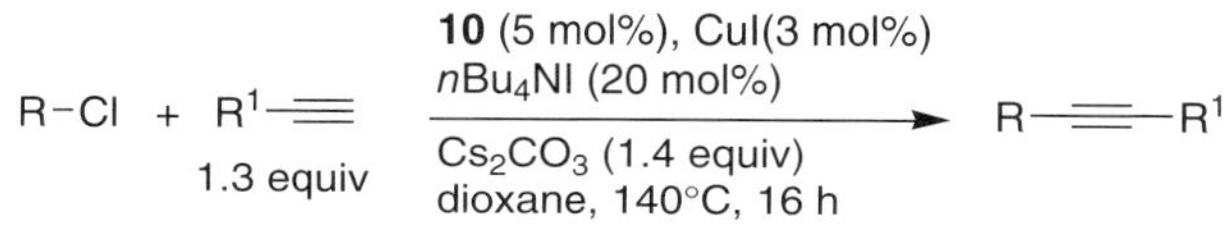

Entry	R-Cl	$R^1C{\equiv}CH$	Yield (%)
1	Octyl-Cl	$nC_6H_{13}C{\equiv}CH$	89
2	$Ph(CH_2)_2Cl$	$AcO(CH_2)_3C{\equiv}CH$	66
3	$EtOC(O)(CH_2)_3Cl$	$iPr_3SiC{\equiv}CH$	62
4	$PhC(O)O(CH_2)_6Cl$	$nC_6H_{13}C{\equiv}CH$	70
5	p-$EtC(O)C_6H_4(CH_2)_3Cl$	$nC_6H_{13}C{\equiv}CH$	57
6	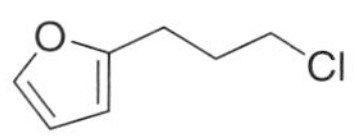	$AcO(CH_2)_3C{\equiv}CH$	64
7	$PhO(CH_2)_2Cl$	$nC_6H_{13}C{\equiv}CH$	88
8	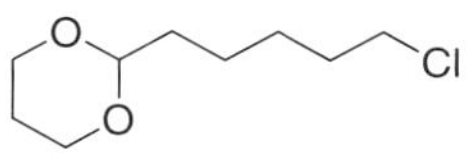	$nC_6H_{13}C{\equiv}CH$	73
9	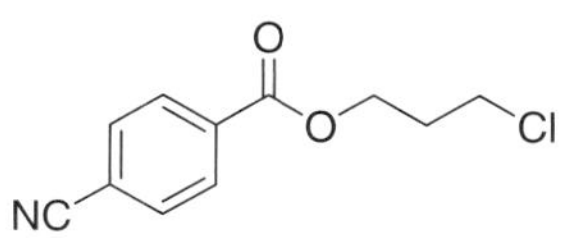	$nC_6H_{13}C{\equiv}CH$	63
10	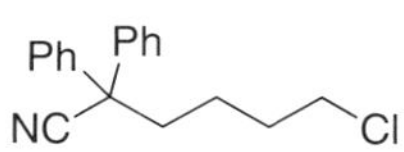	$nC_6H_{13}C{\equiv}CH$	66
11	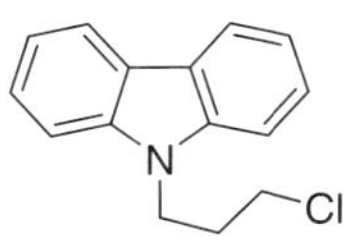	$nC_6H_{13}C{\equiv}CH$	85
12	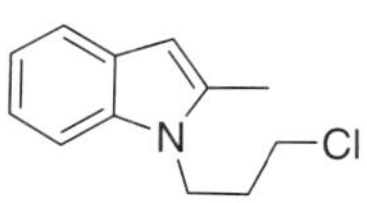	$Ph(CH_2)_2C{\equiv}CH$	75
13	m-$BrC_6H_4(CH_2)_3Cl$	$AcO(CH_2)_3C{\equiv}CH$	58
14	p-$ClC_6H_4(CH_2)_2Cl$	$Ph(CH_2)_2C{\equiv}CH$	72

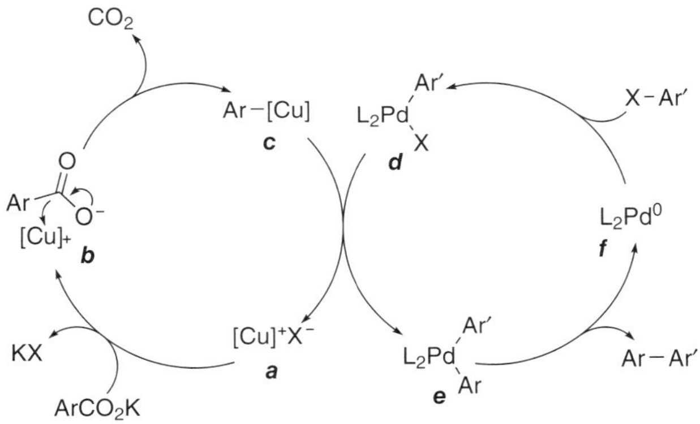

Figure 1.9. The proposed mechanism for Pd/Cu-catalyzed decarboxylative cross-coupling

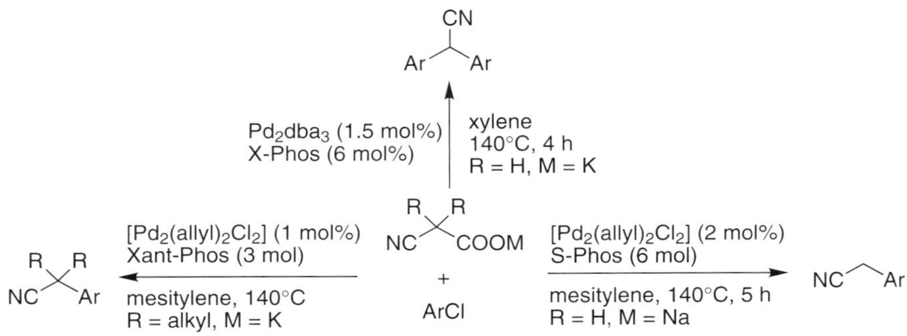

Scheme 1.29. Decarboxylative coupling of cyanoacetates.

affords good to excellent yields [383]. An example of microwave-assisted coupling of 1-methyl-2-pyrrolecarboxylic acid and phenyl chloride using $Pd(PtBu_3)_2$ as catalyst and Cs_2CO_3 as base was also reported. The reaction requires $n Bu_4NCl{\cdot}3H_2O$ as additive and is completed in minutes. This catalytic system is more effective for various aryl bromides [384]. Several Pd-catalyzed decarboxylative coupling using reactive aliphatic carboxylate and aryl chlorides as substrates were reported. The combination of $[Pd_2Cl_2(allyl)_2]$ and SPhos (**15a**) catalyzes decarboxylative coupling of a cyanoacetate salt and its substituted derivatives with aryl chlorides, bromides, and triflates, forming α-aryl nitriles. If XantPhos (**55**) is employed as a ligand, the catalytic reaction can be carried out using quaternary cyanoacetates and aryl bromides or activated aryl chlorides (Scheme 1.29) [385]. The combination of Pd_2dba_3 and XPhos (**18a**) promotes the formation of α-diaryl nitriles from the same starting materials (Scheme 1.35). This process may include α-deprotonation/arylation and subsequent decarboxylative arylation steps [386].

The decarboxylative coupling at the α-carbon atom was also workable for the substrates such as malonate monoester, 2- and 4-nitrophenyl acetates, and

2-(2-azaaryl)acetates. Malonate monoester salts couple with aryl chlorides catalyzed by [$Pd_2Cl_2(allyl)_2$]-RuPhos in mesitylene at 120–140°C to afford α-arylated esters [385]. Potassium 2- and 4-nitrophenyl acetates couple with electron-rich and electron-poor aryl or heteroaryl chlorides using [$Pd_2Cl_2(allyl)_2$] and XPhos (**18a**) as the catalyst at 140°C to form diverse 1,1-diaryl methanes and their derivatives [Eq. (1.28)] [387]. A cross-coupling of 2-$PyCH_2COOK$ and *p*-$CF_3C_6H_4Cl$ was also carried out using Pd_2dba_3-XantPhos as catalyst at 150°C [388].

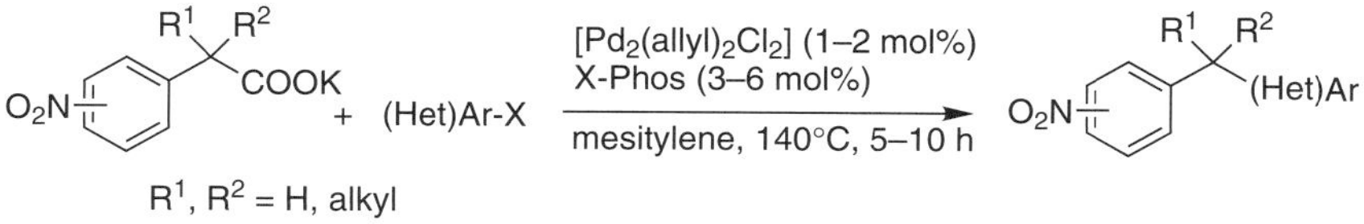

(1.28)

In addition, potassium oxalate monoesters are also good coupling partners in Pd-catalyzed decarboxylative coupling with aryl iodides, bromides, and chlorides. Activated, deactivated, and functionalized aryl chlorides are applicable in this reaction [Eq. (1.29)]. However, sterically hindered *o*-chlorotoluene leads to <5% product yield [389].

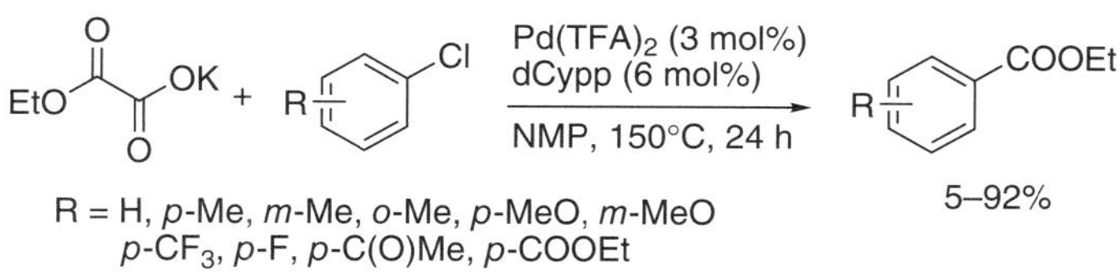

(1.29)

1.3.8 Heck Reaction

Heck reaction of aryl halides is an important method in organic synthesis. Since its discovery in the early 1970s [390, 391], the Heck reaction has been demonstrated to be highly versatile and applicable to a wide range of substrates. In the early stage, aryl chlorides were rarely employed as electrophilic partners due to a strong C–Cl bond. Development of new catalysts, especially the choice of the ligands, make aryl chlorides more and more used in the reaction. In 2002, Littke and Fu [4] reviewed palladium-catalyzed Heck coupling reaction of aryl chlorides before that time.

1.3.8.1 Palladium Catalysts. Palladium is the most widely used catalyst for the Heck reaction of aryl halides. A general mechanism for the Pd-catalyzed Heck reaction is shown in Figure 1.10. It involves oxidative addition, olefin coordination and migratory insertion, β-hydride elimination, and removal of the final product. Palladium remains after this stage in the form of a hydrido complex and is converted back to the initial form by reaction with a base acting as HX acceptor. In this process, the oxidative addition of an aryl halide to a Pd(0) complex is a key step. In contrast to the above-mentioned cross-couplings, the Heck reaction does not

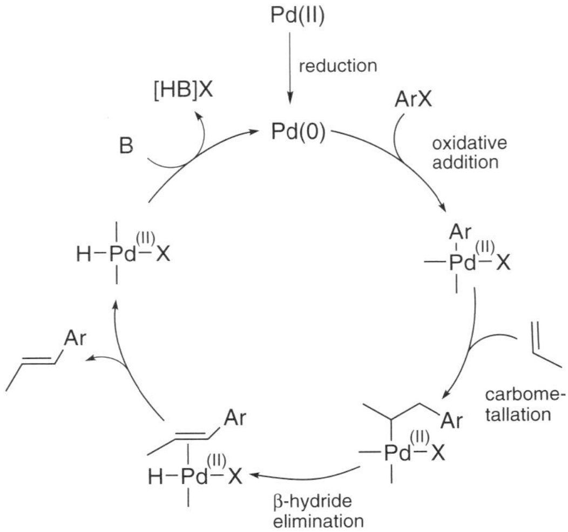

Figure 1.10. Mechanism of the Heck reaction.

include a transmetalation step in its mechanism [392, 393]. There are also alternative mechanisms proposed.

(i) PALLADIUM-PHOSPHINE COMPLEXES. $Pd(OAc)_2/PPh_3$ or P*n*Bu_3 is able to catalyze the reaction of activated aryl chlorides with styrene or electron-deficient alkenes [4]. $PdCl_2(PCy_3)_2/Cs_2CO_3$ can catalyze the reaction of unactivated and activated aryl chlorides with a variety of alkenes in dioxane or toluene, forming *E*-arylated alkenes in good to excellent yields [394, 395]. The $Pd_2(dba)_3$/P*t*Bu_3/Cy_2NMe system is more active than $PdCl_2(PCy_3)_2/Cs_2CO_3$ for the reaction. Activated aryl chlorides can be coupled with an alkene in dioxane at room temperature with 1.5 mol% $Pd_2(dba)_3$ and 3 mol% P*t*Bu_3 loadings. Unactivated and deactivated aryl chlorides can be coupled with an alkene in dioxane at elevated temperature. An unactivated vinyl chloride was also successfully used as the coupling partner in this transformation [396–398]. Herrmann and co-worker [399] demonstrated that the catalytic transformation of unactivated and deactivated aryl chlorides using $Pd_2(dba)_3$/P*t*Bu_3 as the catalyst and NaOAc as the base can be carried out in a nonaqueous ionic liquid, but higher reaction temperature (150°C) is needed. Zapf and Beller [400] found that suitable additive such as *n*Bu_4NBr can improve catalysis of Pd–phosphine systems. Utilizing high catalytic activity of Pd(P*t*Bu_3)$_2$, the stereoselective Heck reaction with tetrasubstituted vinyl ether and organic chlorides was carried out (Scheme 1.30). In the reaction a five-membered chiral, pyrrolidine-based vinyl ether **194**, is β-arylated or β-vinylated with aryl or vinyl chlorides to afford, after hydrolysis, quaternary 2-aryl/vinyl-2-methyl cyclopentanones **197** in moderate to good yields and 89–96% ee. It is propsoed that the reaction proceeds via the neutral π-complex **195** as depicted in Scheme 1.30 [401].

The combination of $Pd(dba)_2$ and *n*BuP(1-Ad)$_2$ exhibits close activity to $Pd_2(dba)_3$/P*t*Bu_3 in catalyzing reaction of electron-rich and electron-poor aryl

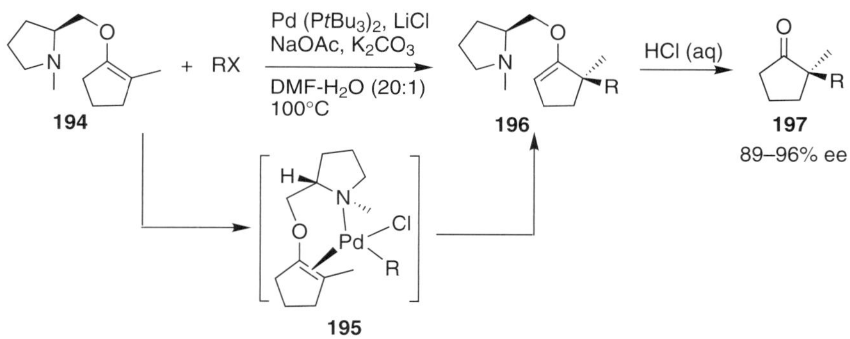

Scheme 1.30. $Pd(PtBu_3)_2$-catalyzed stereoselective Heck reaction of a tetrasubstituted vinyl ether with organic chlorides.

chlorides with alkenes in the presence of K_3PO_4 [402], whereas the $Pd(OAc)_2$/**14a** system shows higher catalytic activity than does the $Pd(dba)_2$/*n*BuPAd$_2$ or $Pd_2(dba)_3$/P*t*Bu$_3$ system, especially for the deactivated aryl chlorides such as 4-chloroanisole. The reaction catalyzed by $Pd(OAc)_2$/**14a** requires *n*Bu$_4$N$^+$OAc$^-$ as the base and can proceed at lower reaction temperature (80°C), giving excellent yields in most cases. However, the reaction using disubstituted alkenes gives only moderate yields under the same conditions [403].

Replacing one of the PR_3 ligands in $PdCl_2(PR_3)_2$ using a nitrogen-containing heterocycle such as 1-alkylimidazoline or 1-alkylbenzimidazole will form highly active catalysts for the Heck reaction. Both **106** and **107** were demonstrated to be highly effective for the coupling of deactivated aryl chlorides with styrene at 80°C [206].

A series of dialkylphosphines such as *t*Bu$_2$PH, Ad$_2$PH, Nor$_2$PH, Cy$_2$PH, **198**, and **199** were also evaluated in the Pd-catalyzed Heck reaction of aryl chlorides with vinyls. The combination of $PdCl_2$ and *t*Bu$_2$PH, Ad$_2$PH, or Nor$_2$PH shows good activity for the reaction of 4-chloroanisole with butyl acrylate using Na_2CO_3 as a base at 140°C, but the activity of each combination is lower than $PdCl_2$/*t*Bu$_3$P system under the same conditions. With lower Pd loadings (0.5 mol%), $Pd(OAc)_2$/*t*Bu$_2$PH or Ad$_2$PH shows close activity to the $Pd(OAc)_2$/*t*Bu$_3$P system [404].

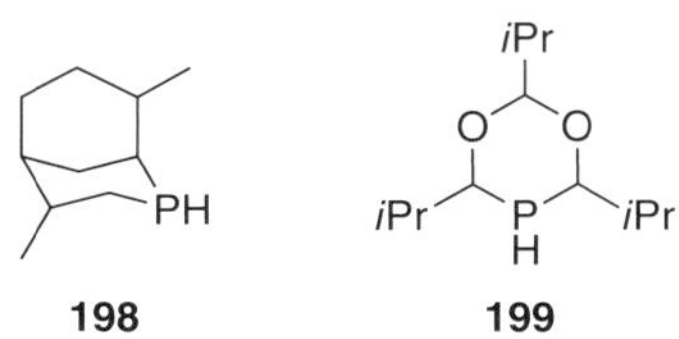

Xiao and co-workers [405] evaluated a series of monodentate and didentate phosphine ligands for Pd-catalyzed reaction of 4-chloroanisole with butyloxyethene and found that $Pd(OAc)_2$/dppp combination in ethylene glycol gives good product yields. Further screening demonstrated combination of $Pd(OAc)_2$ and electron-rich phosphine, *p*-MeO-dppp, to be more effective. Experimental and computational studies show that the arylation is controlled by the oxidative addition step. Electron-rich phosphine ligand promotes oxidative addition of aryl chloride to the metal center. On the other hand, choice of solvent is crucial for this coupling. Ethylene glycol was found to be the best solvent, whereas toluene, DMSO, and DMF are sluggish. DFT calculation supports that a specific hydrogen bonding interaction between a solvent molecule and the chloride of ArCl at the transition state accelerates the oxidative addition. This catalyst system is general and highly efficient. It allows a wide range of activated and unactivated aryl chlorides to couple, α-regioselectively, with electron-rich olefins [405].

(ii) Palladium–Phosphinous Acid or Phosphite Complexes. Li et al. [406] demonstrated that Pd–hosphinous acid complexes **56** and **57** catalyze reaction of an electron-poor aryl chloride with *t*-butyloxyethene in good yields. Beller and Zapf [400, 407] found that the combination of $Pd(OAc)_2$ and various phosphites catalyzes Heck couplings of activated aryl chlorides in the presence of $n\mathrm{Bu_4NBr}$.

(iii) Palladium-NHC Complexes. Herrmann et al. [408] first explored palladium carbenes as catalysts for Heck couplings of activated aryl chlorides in 1995. Subsequent studies have shown that a palladium complex bearing a benzothiazole carbene (**200**) ligand has low activity for the coupling of chlorides, only catalyzing the coupling of 1-chloro-4-nitrobenzene with butyl acrylate or ethyl cinnamate [409, 410]. Both (1,3-dimesitylimidazol-2-ylidene)(naphthoquinone)palladium(0) **144** and (1,3-dimesitylimidazol-2-ylidene)(benzoquinone)palladium(0) **201** show good catalytic activity for the Heck coupling of activated and nonactivated aryl chlorides in $n\mathrm{Bu_4NBr}$ as an ionic liquid at 140–160°C, whereas **202** exhibits low activity under the same conditions [411].

An intramolecular Heck coupling of chlorides was carried out using Pd_2dba_3/**203** as catalyst and Cs_2CO_3 as base at 140°C [Eq. (1.30)]. The reaction of chlorides requires TBAB or TBAN as additive and longer reaction time than corresponding bromides or iodides [412]. A related intramolecular Heck coupling occurs in alphatic

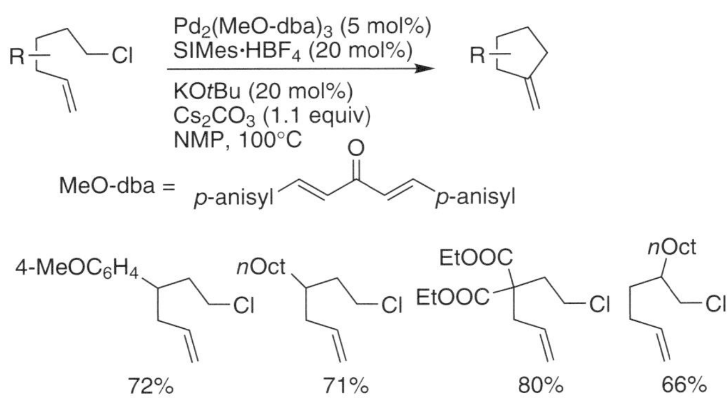

Scheme 1.31. Alkyl–Heck reactions of unactivated alkyl chlorides.

halide systems (Scheme 1.31). In the reaction, $Pd_2(MeO\text{-}dba)_2$ and SIMes·HBF_4 together are employed as a catalyst and KO*t*Bu and Cs_2CO_3 as bases. The cyclization of 6-chlorohex-1-ene derivatives proceeds at 100°C in NMP in moderate to good yields. For corresponding bromide substrates, the reaction requires lower temperature (65°C) in MeCN and gives slightly higher yields [413].

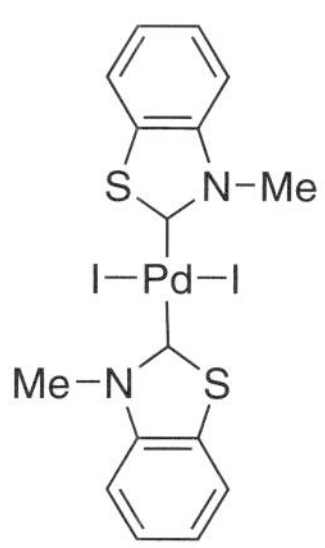

200

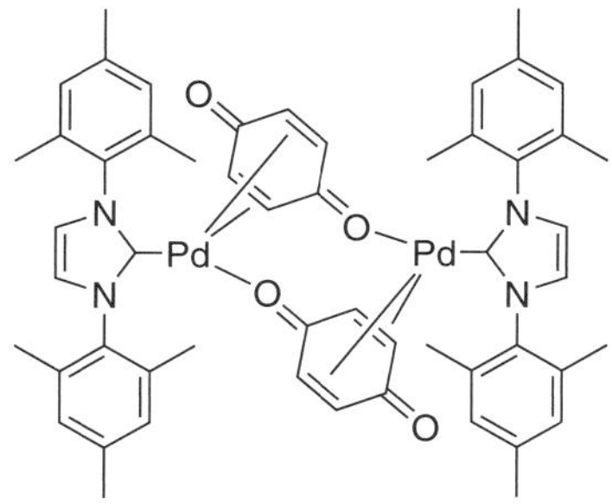

201

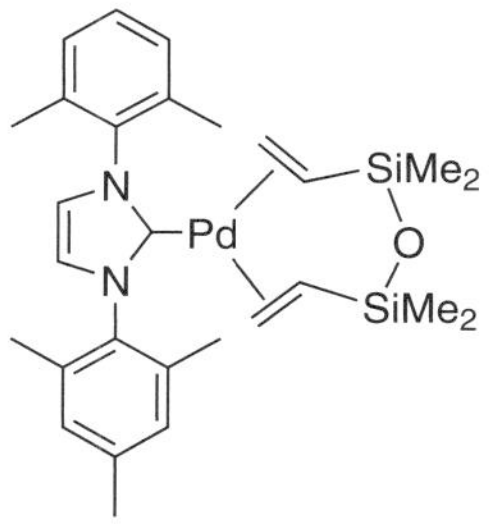

202

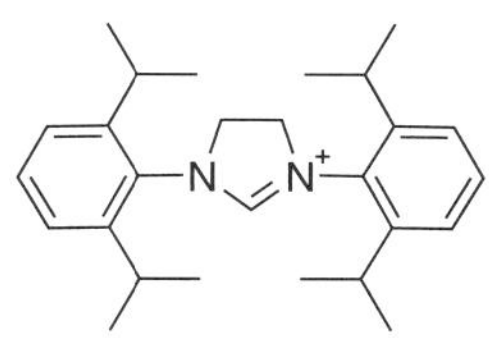

203

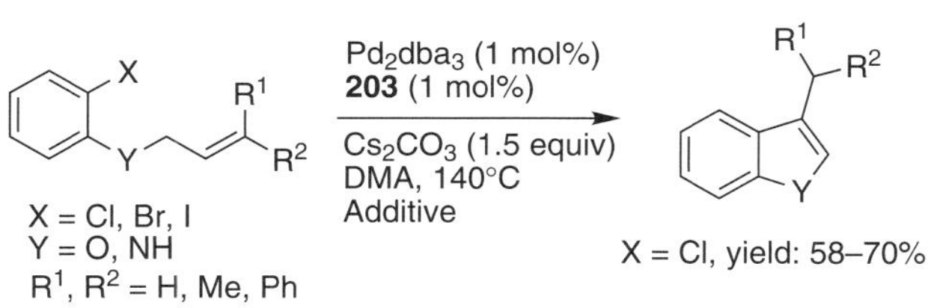

(1.30)

A series of pincer or chelate NHC–palladium complexes **204–208** have been demonstrated to be effective catalysts for the Heck reaction of electron-poor aryl chlorides [414–419]. Both **209** and **210** display good catalytic activity for the coupling of activated, unactivated, and deactivated aryl chlorides with alkenes in TBAB at 140°C. Complex **209** needs a longer catalyst activation time than does **210**. This is ascribed to a slower ligand dissociation of **209** than that of **210** due to strong binding of two bidentate NHC/amido ligands in **209**. In complex **210** a weaker binding ligand has replaced one of the two bidentate carbene/amido ligands in **209**, which leads to a faster catalyst activation. In addition, the two complexes also efficiently catalyze Heck coupling of sterically hindered aryl bromides [420, 421].

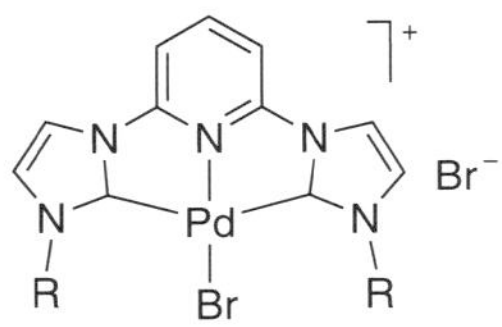

R = alkyl, aryl

204

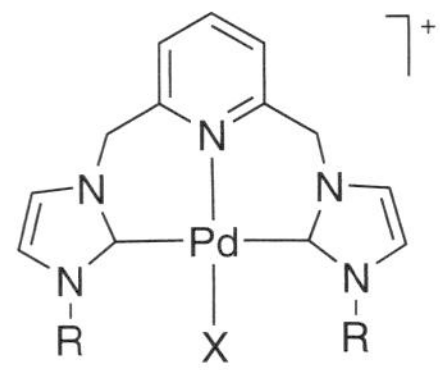

X = Cl, Br

205

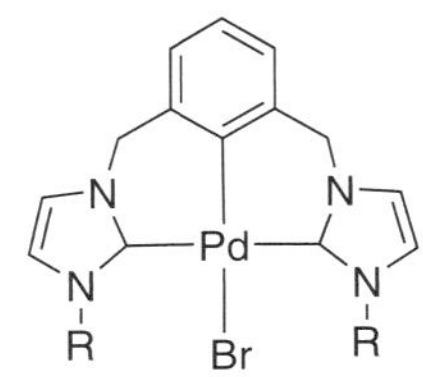

206

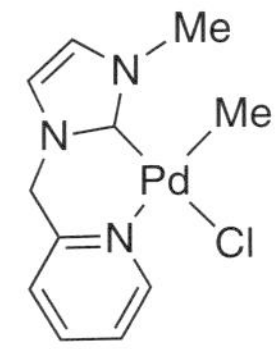

207

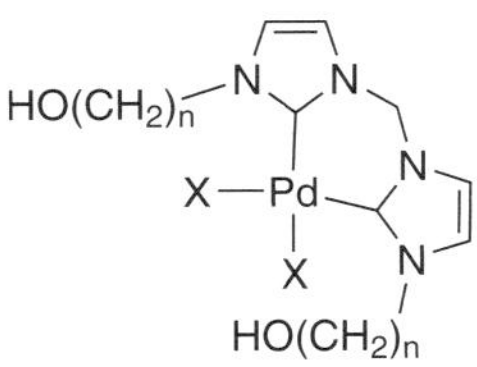

n = 2, X = I
n = 3, X = Br

208

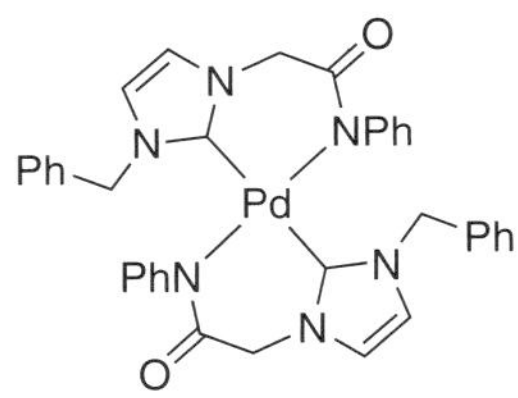

209

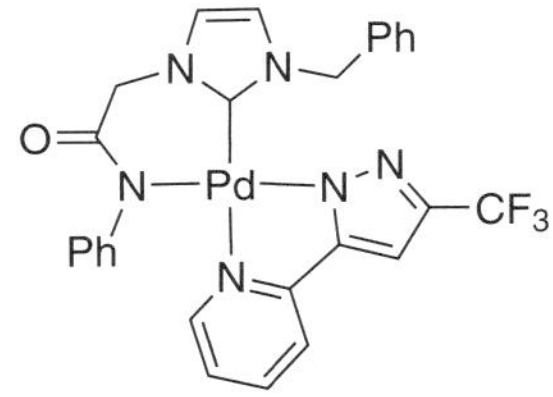

210

(iv) N,N- OR N,O-CHELATE PALLADIUM COMPLEXES. The N,N-chelate Pd complex, *N*-acetyl-*N*,*N*-bis(pyrimid-2-yl)amine palladium dichloride, was proven to catalyze coupling of 4-chloroacetophenone with styrene in the presence of TBAB in moderate yield, but reaction of chlorobenzene results in low yield [422]. The β-diketiminatophosphane Pd complex **112** is an efficient catalyst for coupling of electron-rich, electron-poor, and sterically hindered aryl or heteroaryl chlorides with a variety of alkenes. The reaction requires TBAB as additive and proceeds effectively in DMF-H_2O (Scheme 1.32) [423].

Complex **211** catalyzes the Heck coupling of unactivated or deactivated aryl chlorides with styrenes using LiOH·H_2O as base in DMA at 160°C. It has been shown by thermogravimetric analysis that this complex is stable up to 350°C. Hence it is expected that the metal amidate unit is retained in the catalytic cycle at the reaction temperature used [424]. An N,N,O,O-chelate bimetallic Pd complex **212** was also demonstrated to be active for the Heck coupling of deactivated 4-chloroanisole and styrene at 160°C using $Ca(OH)_2$ as base. It is thought that the macrocyclic ligand can perform the controlled release and recapture of Pd and provides stabilization during the process of catalysis at high temperature. This controlled release of Pd prevents the formation of inactive Pd agglomerates under harsh conditions. Hence the complex shows good catalytic activity and the reaction does not need TBAB as the stabilizer. It was also noted that addition of LiBr improves the product yield of the coupling between styrene and 4-chloroanisole. LiCl as Lewis acid may coordinate either to the chlorine or to the oxygen of the methoxy group of chloroanisole. As a result, the electron density is drawn from the donor substituent and from the aromatic ring. Both situations would weaken the C–Cl bond and make the oxidative addition of the

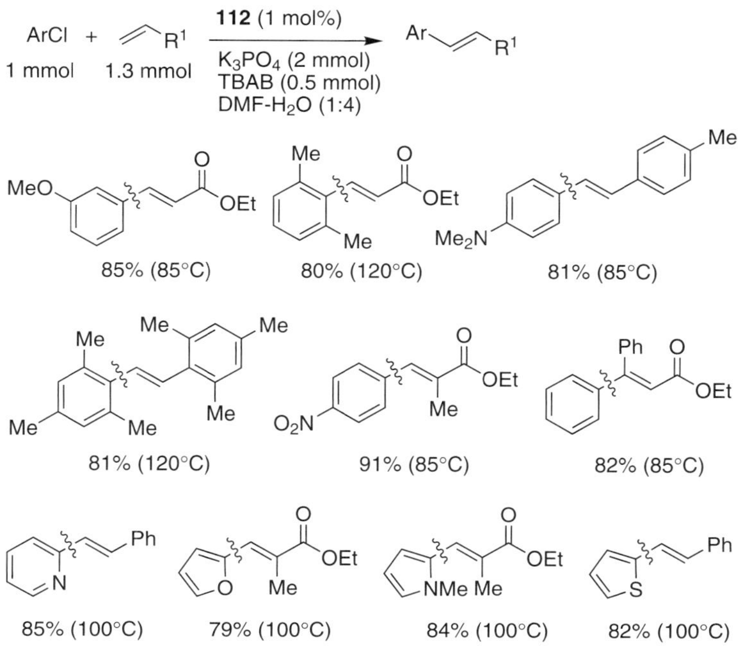

Scheme 1.32. Complex **112**-catalyzed Heck reaction of aryl chlories.

chloroarene to Pd(0) more feasible [425].

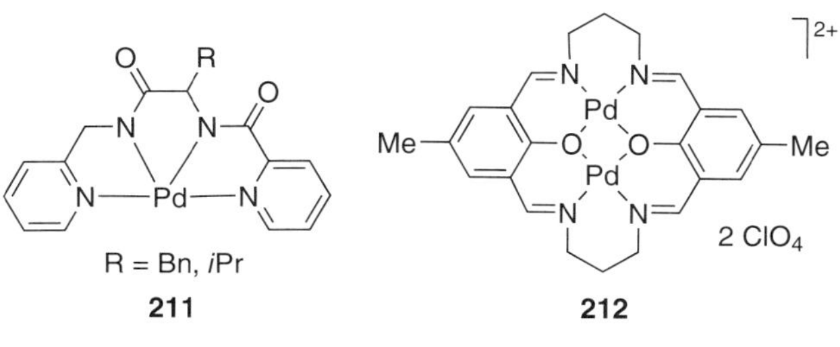

211 **212**

(v) PALLADACYCLES. Herrmann et al. [426, 427] were first to use palladacycle **89** as catalyst for Heck coupling of activated aryl chlorides. This reaction requires $n\text{Bu}_4\text{NBr}$ as a co-catalyst. When PPh_4Cl is used to replace $n\text{Bu}_4\text{NBr}$ as the co-catalyst, **89** can catalyze coupling of PhCl with styrene in excellent yield. However, reaction of 4-chloroanisole under the same conditions leads to low conversion [277]. By using $[t\text{Bu}_3\text{PH}]\text{BF}_4$ as an additive, using aqueous DMF, PEG-200, or a mixture of $[\text{bmim}]\text{PF}_6$ ionic liquid and dioxane as the reaction medium, and using Cy_2NMe as base and under microwave radiation condition, Larhed and co-workers [428, 429] carried out **89**-catalyzed coupling of activated, unactivated, and deactivated aryl chlorides with various alkenes. In this reaction, $[t\text{Bu}_3\text{PH}]\text{BF}_4$ is a key additive. Microwave heating is more effective than thermal heating [428, 429]. The sulfur- and nitrogen-containing palladacycles **213–216** are only effective for Heck reactions of activated aryl chlorides [430–432]. However, the oxime-derived palladacycles **100b** and **100d** can catalyze Heck coupling of unactivated or deactivated aryl chloride in moderate yields [433, 434]. The imine-derived palladacycle **217** catalyzes coupling of 3-chloroanisole with ethyl acrylate in the presence of TBAB in moderate yield [435]. The nitrogen-containing palladacycle **218** can also catalyze coupling of unactivated and activated aryl chlorides with methyl acrylate in good yields [436].

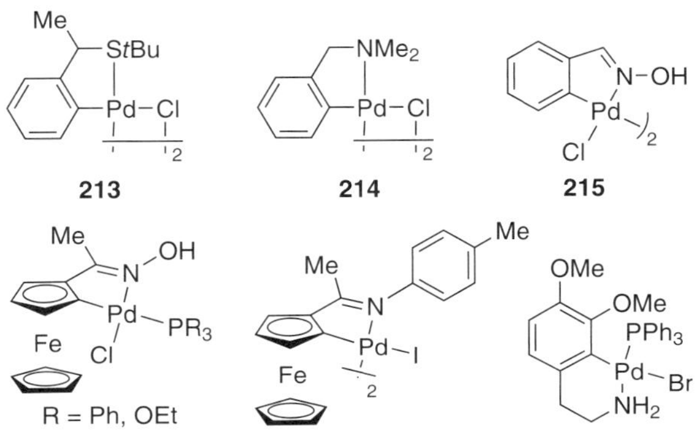

213 **214** **215** **216** **217** **218**

Both pincer–palladium complexes **219** and **220** are able to catalyze coupling of activated aryl chlorides, including cyano, nitro, and acyl group-substituted chlorobenzenes, but they are inactive for the unactivated aryl chlorides [437, 438], whereas P,C,P-pincer–Pd **221** efficiently catalyzes coupling of electron-rich and -poor aryl chlorides with styrene [439].

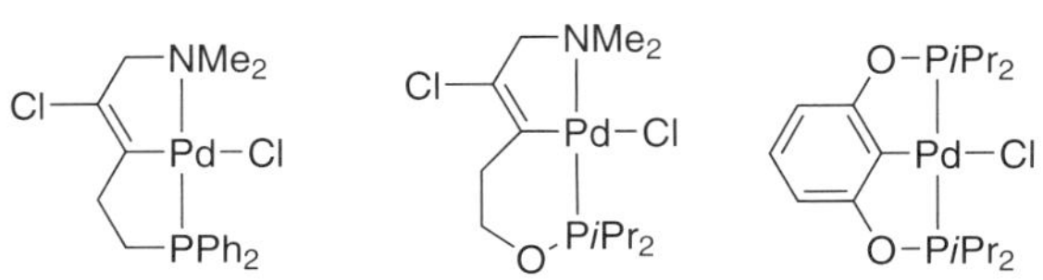

219 **220** **221**

(vi) Other Palladium Catalysts. Köhler and co-workers [440] demonstrated that $Pd(OAc)_2$ catalyzes coupling of activated aryl chlorides with styrene at 160°C in the presence of TBAB without need of organic ligands. At such a temperature, the catalytic active species may be palladium nanoparticles. Li and co-workers [441] reported another Heck coupling of activated aryl chlorides catalyzed by palladium nanoparticles generated from allylpalladium chloride at 120°C. Nacci and co-workers [442] found that in a molten mixture of the solvent tetrabutylammonium bromide (TBAB) and the base tetrabutylammonium acetate (TBAA), palladium nanoparticles generated from $Pd(OAc)_2$ can efficiently catalyze the coupling reaction of deactivated chloroarenes with various substituted olefins, including 1,1- and 1,2-disubstituted olefins. Layered double hydroxide and Merrifield resin-supported nanopalladium(0) catalysts developed by the simple ion-exchange technique followed by reduction also shows excellent catalytic behavior for the coupling of activated, unactivated, and deactivated aryl chlorides in the presence of TBAB by either microwave or thermal heating [443]. Other supported palladium catalysts such as palladium/montmorillonite, Pd/Al_2O_3, Pd/SAPO-31, and Pd/C usually catalyze the coupling of activated aryl chlorides in good yields and have low activity or inactivity for unactivated and deactivated aryl chlorides [444–447].

1.3.8.2 Nickel and Cobalt Catalysts. Nickel is rarely used in catalyzing the Heck reaction, especially for chloride substrates. Recently, Jamison et al. reported a Heck-type coupling of benzyl chloride derivatives with ethylene or α-olefins using the combination of $Ni(cod)_2$ and $PCyPh_2$ or PCy_2Ph as the catalyst [Eq. (1.31)]. The reaction requires excess Et_3SiOTf and proceeds effectively at room temperature. The catalyst system is compatible with various substituted aromatic groups except nitro-substituted ones. The catalyst system is also sensitive to the substituents at the benzylic position. For example, α-methyl benzyl chloride cannot be catalytically coupled with an alkene. An important feature of this reaction is its selectivity of forming 1,1-disubstituted olefins from α-olefins [448].

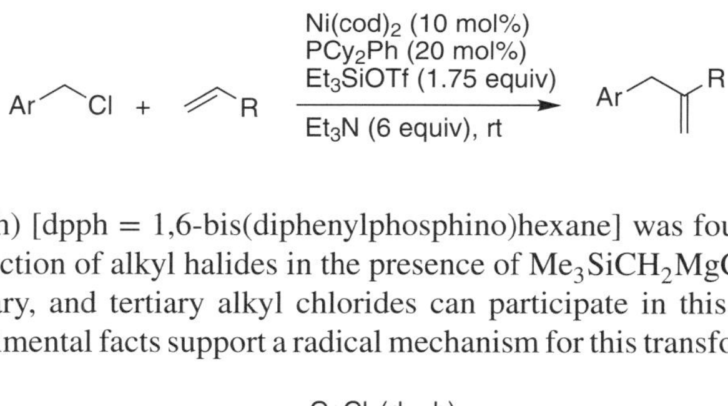

(1.31)

$CoCl_2$(dpph) [dpph = 1,6-bis(diphenylphosphino)hexane] was found to catalyze styrylation reaction of alkyl halides in the presence of Me_3SiCH_2MgCl in ether. Primary, secondary, and tertiary alkyl chlorides can participate in this coupling [Eq. (1.32)]. Experimental facts support a radical mechanism for this transformation [449].

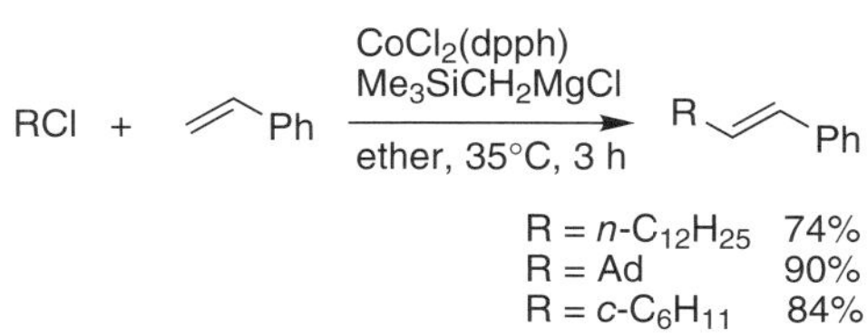

(1.32)

1.3.9 C–H Functionalization with Organic Chlorides

Direct C–H functionalization with organic halides is an attractive strategy for the construction of C–C bonds. This method is more straightforward and, in some cases, more atom economical compared with the cross-couplings using organometallic nucleophiles such as Grignard reagents, organozinc reagents, organotin reagents, and organoboron reagents. The use of aryl chlorides makes this methodology more attractive due to their lower cost and the diversity of available compounds.

1.3.9.1 α-Arylation of Carbonyl and Related Compounds. Transition-metal-catalyzed α-arylation of carbonyl and related compounds using aryl halides has attracted considerable attention. α-Arylation of ketones, aldehydes, esters, amides, nitriles, imines, and nitro compounds have been carried out. More and more aryl chlorides are being employed as the electrophilic substrates in these reactions [450, 451].

(i) α-ARYLATION OF KETONES. Kawatsura and Hartwig [452] were first to use aryl chlorides as electrophiles in the α-arylation of ketones. They carried out α-arylation of a variety of ketones with electron-rich and electron-poor aryl chlorides using $Pd(dba)_2$-**222** or $Pd(OAc)_2$-PR_3 (R = Cy or *t*Bu) as catalyst and NaO*t*Bu as base. Subsequently, Beller and co-workers [453] showed that $Pd(OAc)_2$-*n*BuPAd$_2$ give better catalytic results compared with $Pd(OAc)_2$-PR_3 (R = Cy or *t*Bu) for the arylation of ketone enolates using aryl chlorides. Grasa and Colacot [454] synthesized and characterized **222**–Pd complexes, **223a–f**, and demonstrated that both **223d** and **223e** are very active catalysts for the reaction of ketones with aryl chlorides (Scheme 1.33) in the presence of NaO*t*Bu at 100°C.

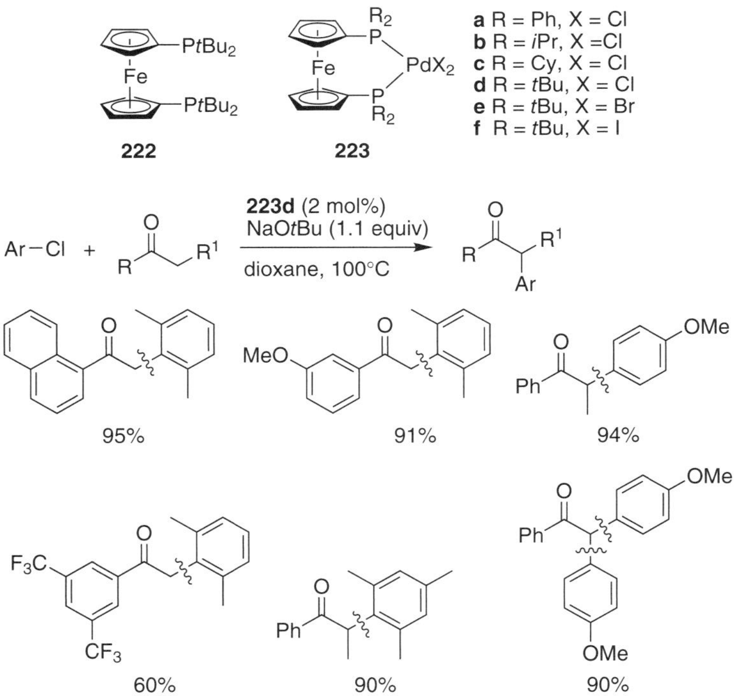

Scheme 1.33. **223d**-Catalyzed α-atylation of ketones.

The combination of Buchwald ligands such as **5**, **14a**, **14b**, **18a**, and **224** and an appropriate Pd source is also active for the α-arylation of various ketones with aryl chlorides [455]. For example, $Pd(OAc)_2$/XPhos catalyzes the synthesis of tricyclo-[7.3.1.$0^{2,7}$]-trideca-2,4,6-trien-13-ones through intramolecular α-arylation of 2-(2-halo-benzyl)-cyclohexanones (Scheme 1.34) [456]. The palladacycle complex of XPhos (**225**) efficiently catalyzes α-arylation of (hetero)aryl methyl ketones with aryl chlorides [457] or an intramolecular α-arylation of ketones **226** derived from Suzuki reaction, forming tetracyclic indole dervatives (Scheme 1.35) [458].

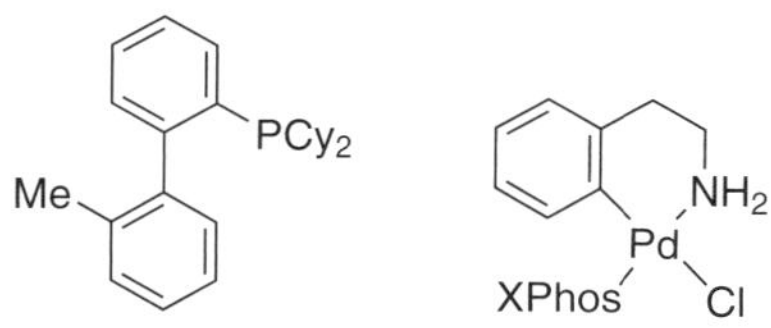

224 (MePhos) **225**

1,3,5,7-Tetramethyl-2,4,8-trioxa-6-phenyl-6-phosphaadamantane (**52**) and Pd_2 dba_3 forms $Pd(\mathbf{52})_2$·dba. This complex catalyzes α-arylation of propiophenone with various aryl bromides and chlorides. Both chlorobenzene and 4-chlorotoluene can be catalytically coupled using NaO*t*Bu as base at 70°C, giving corresponding products in 75% and 78% yields, respectively [459]. Sterically congested diaminochlorophosphine **62** is a good ligand for Pd-catalyzed α-arylation of ketones. $Pd(dba)_2$–**62** catalyzes the α-arylation reaction of various aryl ethyl ketones with electron-rich aryl chlorides such as 3- or 4-chloroanisole, 3- or 4-chlorotoluene, and 1-chloro-3,5-dimethoxybenzene in the presence of NaO*t*Bu [460]. The combination of P,N-ligand **227** and $[Pd(cinnamyl)Cl]_2$ catalyzes the α-arylation reaction of acetone with aryl chlorides, bromides, and iodides using Cs_2CO_3 as base. It was noted that the use of electron-poor aryl halides leads to lower product yields than does the use of electron-rich or neutral substrates. This substrate scope limitation

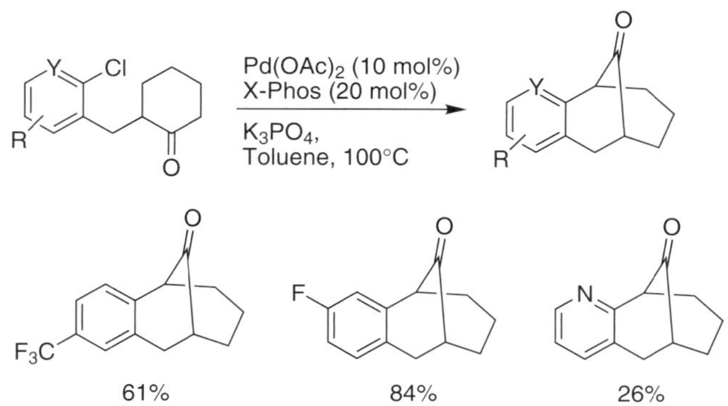

Scheme 1.34.

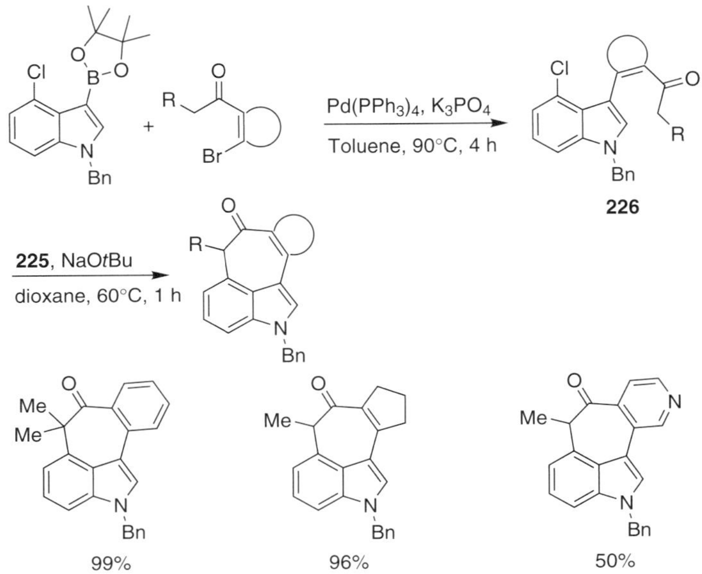

Scheme 1.35.

can be improved for electron-deficient aryl bromides by using **228** to replace **227** as the ligand [461].

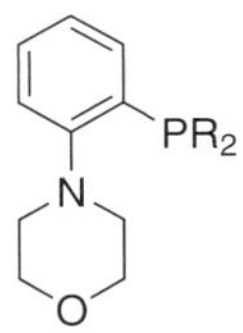

227 R = 1-Ad
228 R = 2-tol

NHCs are another class of important ligands for Pd-catalyzed α-arylation of ketones with aryl chlorides. Nolan and co-workers [462–465] reported that NHC–Pd complexes such as (SIPr)Pd(allyl)Cl (**88c**) and (IPr)Pd(acac)Cl are efficient catalysts for the α-arylation reaction of aryl or heteroaryl alkyl ketones with electron-rich and electron-poor aryl/heteroaryl chlorides under mild conditions. (IPr)Pd(OAc)$_2$ could catalyze a polycondensation reaction through α-arylation of 4-haloaryl alkyl ketones [Eq. (1.33)] [466].

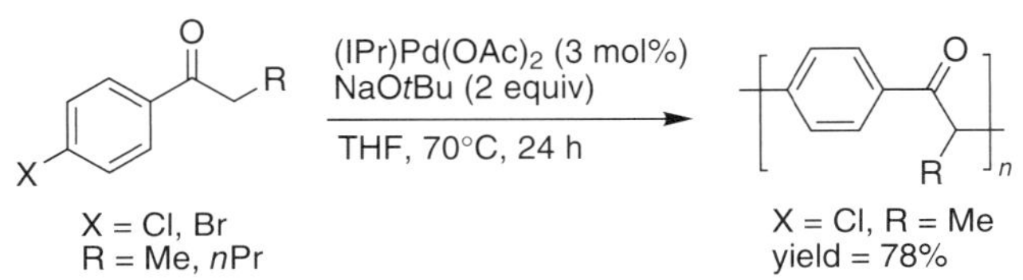

(1.33)

Bertrand and co-workers [467] designed new types of carbene ligands **229–231**. In these cyclic alkylaminocarbenes, one of the electronegative amino susbtituents in NHCs has been replaced by an alkyl group that makes cyclic alkylaminocarbenes more electron-rich than NHCs. Their palladium complexes can catalyze α-arylation of propiophenone with unactivated or sterically hindered aryl chlorides in different catalytic activity. With nonhindered aryl chlorides, [Pd(allyl)(**231**)Cl] displays higher activity compared to that of [Pd(allyl)(**229**)Cl] and [Pd(allyl)(**230**)Cl]. When sterically hindered 2,6-dimethylchlorobenzene is employed as the coupling partner, no catalytic activity has been observed with the palladium complexes of **229** and **231**. By contrast, [Pd(allyl)(**230**)Cl] is active. The activity difference is ascribed to differences of the steric effects of the ligands [467].

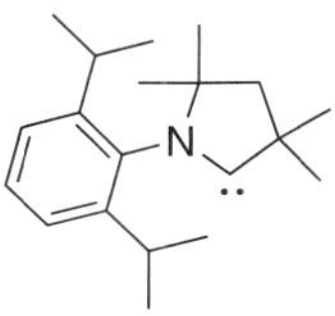

229

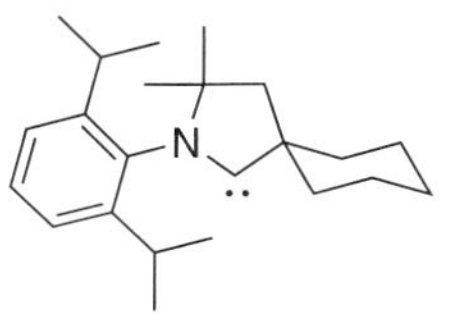

230

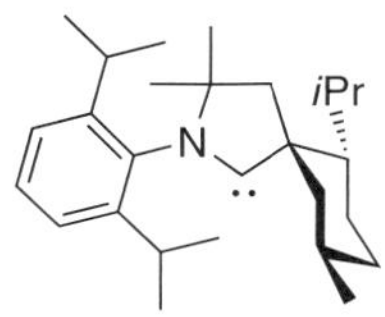

231

Palladacycles are often used as palladium sources for Pd-catalyzed reactions. They combine with other ligands such as phosphines and NHCs to perform catalysis as complex **225**. IPr-palladacycles **3** and **102** catalyze α-arylation of aliphatic or aromatic ketones with aryl chlorides [468, 469]. Complex **102** is suitable for a wide scope of substrates, including ketones with different substituents and various electron-rich, electron-poor, and sterically hindered aryl chlorides. However, reaction of 1-chloro-4-nitrobenzene and 3-chloropyridine with acetophenone results in trace products [469]. Adduct of palladacycle and $HPNor_2$, **101**, was also demonstrated to be highly effective for the α-arylation of propiophenone with 4-chlorotoluene or 4-chloroanisole, and the activity is comparable to that of palladacycle-PCy_3 adduct [200].

Palladium complexes are predominant in catalyzing α-arylation of ketones with aryl chlorides. Quite rare nickel complexes were reported for the reaction. The $NiCl_2$ complex bearing mixed PPh_3/*N*-heterocyclic carbene ligands, (IPr)Ni(PPh_3)Cl_2, can catalyze reaction of propiophenone with unactivated aryl chlorides in low to moderate yields [470]. The combination of Ni(cod)$_2$ and P-Phos (**232**) catalyzes asymmetric α-arylation of 2-methyl-1-tetralone with various aryl halides. However, for PhCl substrate, the reaction gives only 59% yield with 91% ee [471].

Didentate phosphines BINAP and DIFLUORPHOS were demonstrated to be good ligands for the Ni-catalyzed coupling. (*R*)-BINAP-Ni(cod)$_2$ shows excellent catalytic activity and steric selectivity for asymmetric α-arylation of indanones and tetralones with chloroarenes. The reaction proceeds in toluene at 60–80°C using NaO*t*Bu as base, giving moderate to excellent yields and 90–99% ee. Using a combination of Ni(cod)$_2$ and (*R*)-DIFLUORPHOS (**233**) as the catalyst, the coupling of ketones with a range of heteroaryl chlorides has been carried out with enantioselectivities up to

99% ee. By employing [(*R*)-BINAP]Ni(η^2-N≡CPh)] as the catalyst, the coupling of ketones with substituted chlorobenzene can even be carried out at room temperature with good yields and excellent selectivity [472].

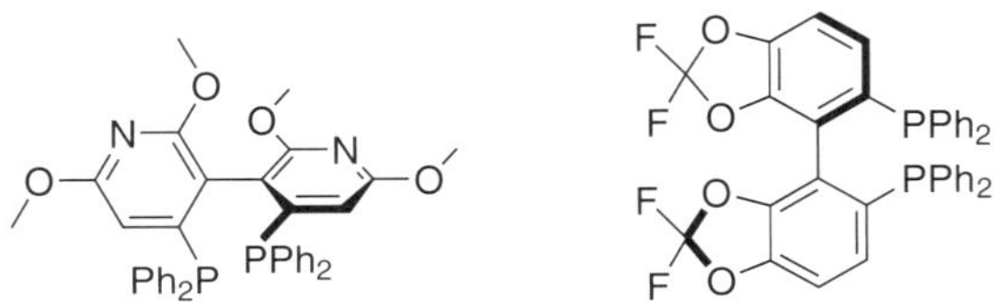

(*R*)-P-Phos **232** (R)-DIFLUORPHOS **233**

(ii) α-ARYLATION OF TRIMETHYLSILYL ENOL ETHERS.. The catalytic coupling of trimethylsilyl enol ethers with aryl halides results in α-arylated ketones. Compared with the coupling of alkali-metal enolates of ketones, the coupling of silyl enol ethers does not require a strong base and hence should improve functional group compatibility and allow the development of enantioselective processes to form acidic tertiary stereocenters in the α-position to a carbonyl group. Coupling of silyl enol ethers also enables more effective regiocontrol of the arylation and allows the insertion of an aryl group at the more substituted α-position of a ketone with two enolizable α-positions. However, this methodology is mainly used for the coupling of aryl bromides and iodides. Relatively rare examples are reported for the coupling of aryl chlorides [450, 473].

In 2006 two groups reported that the combination of P*t*Bu_3, $Pd(OAc)_2$, or $Pd_2(dba)_3$ and fluoride additives such as *n*Bu_3SnF, CsF, ZnF_2, and MnF_2 catalyzes coupling of trimethylsilyl enol ethers with aryl halides. Iwama and Rawal [474] employed *n*Bu_3SnF alone as the additive and the system works effectively for the coupling reaction of electron-rich and electron-poor aryl iodides, bromides, and chlorides with trimethylsilyl enol ether of cyclohexanone in refluxed benzene or toluene [474]. Hartwig and co-workers [473] used a combination of fluorides as additives including *n*Bu_3SnF-CsF, ZnF_2-MnF_2, and ZnF_2-CsF. The combination of $Pd(OAc)_2$, P*t*Bu_3, and *n*Bu_3SnF-CsF efficiently catalyzes coupling of acyclic and cyclic silyl enol ethers with aryl bromides and electron-deficient aryl chlorides in toluene at 85–90°C. Chobanian et al. [475] carried out α-arylation of acetone through Pd-catalyzed coupling of 2-trimethylsilyloxypropene with aryl bromides or chlorides under microwave radiation. The reaction proceeds rapidly in DMF using Pd_2dba_3/SPhos as catalyst and ZnF_2 as additive. Electron-rich and electron-poor aryl and heteroaryl chlorides are applicable, and reaction gives moderate to good yields [475].

(iii) α-ARYLATION OF ALDEHYDES. Both **225** and $Pd(OAc)_2$–**234** catalyze α-arylation of linear and α-branched aldehydes with aryl chlorides. The reaction catalyzed by **226** is suitable for a variety of electron-rich aryl chlorides and gives moderate yields in most cases [476], whereas the $Pd(OAc)_2$–**234** system is suitable for electron-rich and electron-poor aryl chlorides and heteroaryl chlorides and leads to good product yields [477]. The combination of [{Pd(allyl)Cl}$_2$] and QPhos (**34**) is a suitable catalyst for α-arylation of linear and α-branched aldehydes with aryl bromides or chlorides. But examples for aryl chlorides are very limited, including the coupling of 2-methylbutanal with ethyl 4-chlorobenzoate, 4-chlorobenzonitrile, and 3-chloroanisole [478].

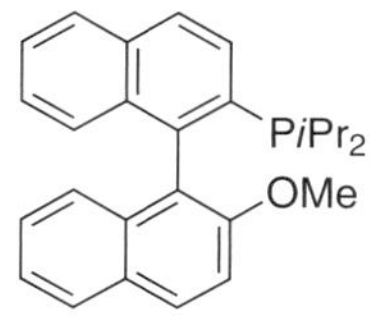

234

(iv) α-Arylation of Carboxylic Acid Esters. Hama and Hartwig [479] found that [(tBu$_3$P)PdBr]$_2$ or the Pd(dba)$_2$–tBu$_3$P combination catalyzes α-arylations of *tert*-butyl propionate or methyl isobutyrate with chloroarenes using NaHMDS as base. Electron-rich and electron-poor chloroarenes are applicable for this transformation. *tert*-Butyl propionate also couples with bulky 2,6-dimethylchlorobenzene under the conditions mentioned above. [(tBu$_3$P)PdBr]$_2$ leads to better yield for this coupling. 2-Chloropyridine was also coupled with sodium enolate of methyl isobutyrate using either [(tBu$_3$P)PdBr]$_2$ or a combination of Pd(dba)$_2$ and tBu$_3$P as the catalyst. Coupling of chloroarenes with the alkali metal enolates of *tert*-butyl acetate cannot be carried out using current conditions [479]. This limitation was overcome by Biscoe and Buchwald [457] through developing new catalyst system. They found that the *t*BuXPhos–palladacycle complex **235** is an effective catalyst for the coupling of chloroarenes with *tert*-butyl acetate using LHDMS as a base. Electron-rich and electron-poor aryl chlorides and heteroaryl chlorides have been coupled with *tert*-butyl acetate with 1 mol% catalyst loading. The reaction proceeds effectively at room temperature and gives excellent yields [457]. In addition, zinc enolates of acetates can also be coupled with chloroarenes using a combination of Pd(dba)$_2$ and QPhos as the catalyst (Scheme 1.36). [(tBu$_3$P)PdBr]$_2$ is inactive for this reaction [479].

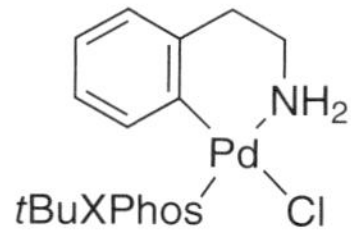

235

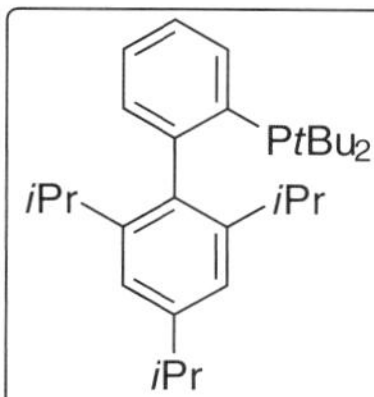

*t*BuXPhos

Protected amino acids were also catalytically α-arylated using aryl bromides or chlorides. Pd(dba)$_2$-tBu$_3$P is an effective catalyst and K_3PO_4 is a suitable base for

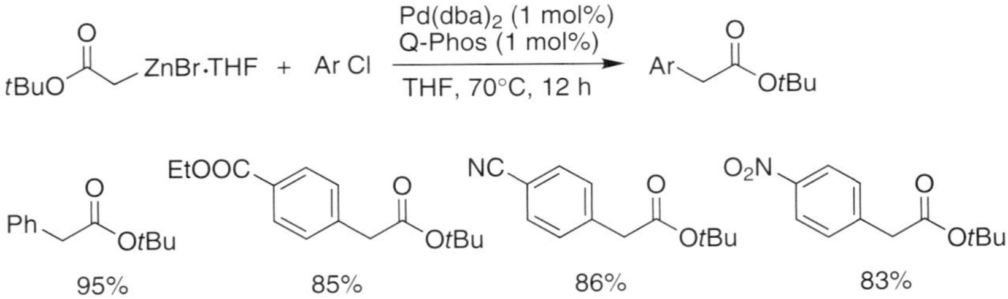

Scheme 1.36.

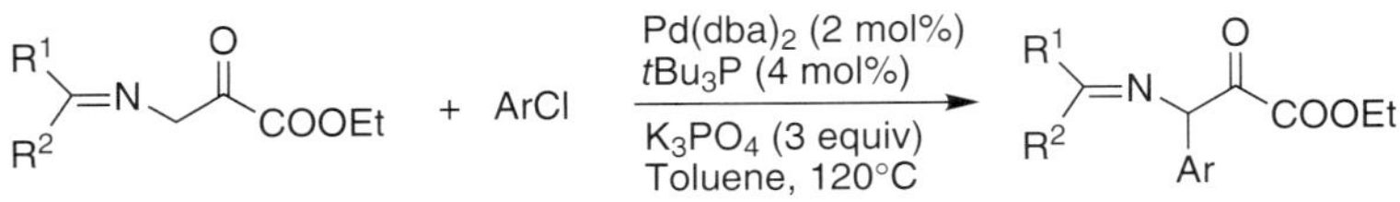

$R^1 = R^2 = Ph$

PhCl, 82%; 2-MeC_6H_4Cl, 81%; 4-$MeOC_6H_4Cl$, 83%; 4-FC_6H_4Cl, 83%; 4-NCC_6H_4Cl, 85%; 4-$CF_3C_6H_4Cl$, 84%; 3-chloropyridine, 80%

R^1 = *p*-$MeOC_6H_4$, R^2 = H

PhCl, 67%; 4-$CF_3C_6H_4Cl$, 67%

Scheme 1.37.

this reaction. Electron-rich, electron-neutral, and electron-poor aryl chlorides can be coupled in moderate to good yields (Scheme 1.37) [480].

Nickel-(*S*)-BINAP was demonstrated to be an effective catalyst for asymmetric α-arylation of α-substituted γ-butyrolactones; 15 mol% $ZnBr_2$ is added as the additive for improving reaction rates and yields. Electron-rich, electron-neutral, and electron-poor aryl chlorides are applicable in this transformation. The reaction gives moderate to excellent yields in most cases and achieves high enantioselectivities [Eq. (1.34)] [481].

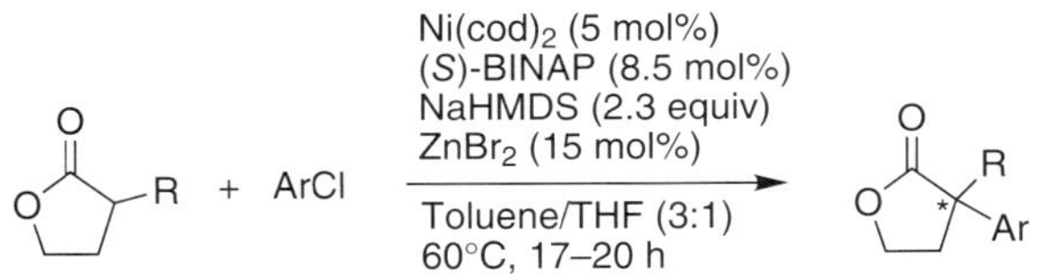

R = Me, Bn, allyl, *n*Pr

25–95% yields
83–99% ee

(1.34)

(v) α-ARYLATION OF AMIDES. The C-3 arylation of *N*-benzyl oxindole has been carried out with aryl bromides or chlorides catalyzed by $Pd(dba)_2$–XPhos. The reaction proceeds in a mixture of THF and toluene using KHDMA as base, giving moderate yields [Eq. (1.35)] [482]. For the highly reactive aryl chlorides, the C-3 arylation can be performed via nucleophilic aromatic substitution without need of a palladium catalyst [483].

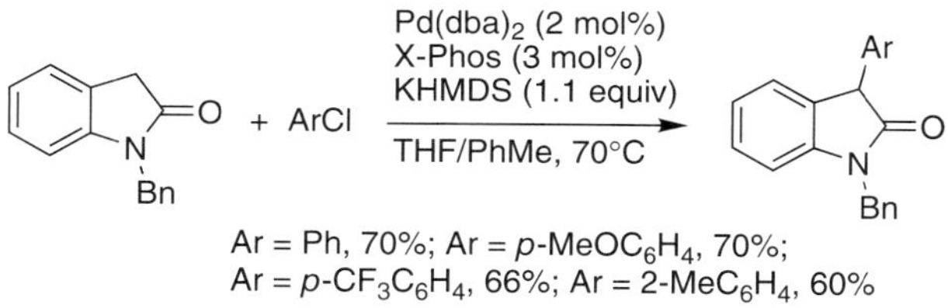

(1.35)

Unprotected oxindoles may lead to C-3 arylation and *N*-arylation with aryl halides under catalytic conditions. Buchwald and co-workers [484] found that $Pd_2(dba)_3$–XPhos catalyzes C-3 arylation of oxindoles with aryl chlorides in the presence of K_2CO_3 [Eq. (1.36)], while the Cu-diamine-based catalyst system leads to *N*-arylation exclusively with aryl bromides or iodides.

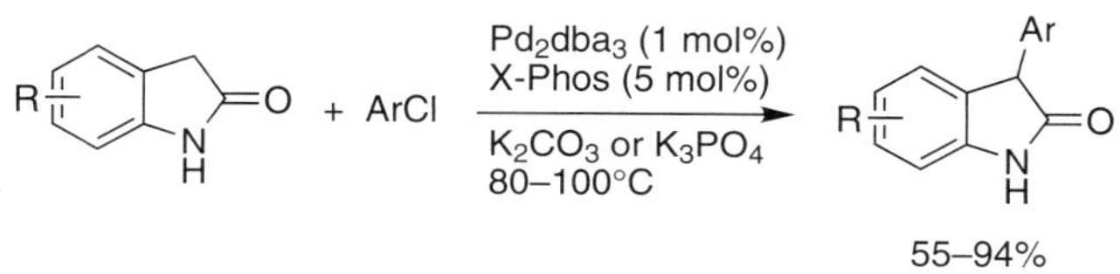

(1.36)

$Pd(OAc)_2$–PCy_3 is effective for intramolecular α-arylation of amides, which generates oxindoles (Scheme 1.38). This catalyst system also catalyzes tandem intra- and

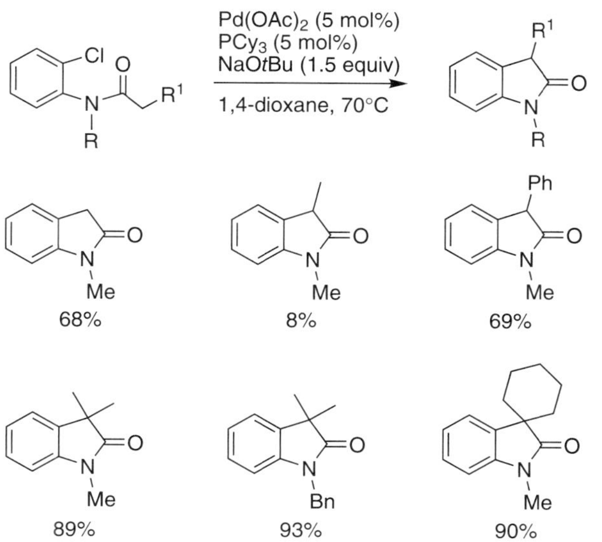

Scheme 1.38. $Pd(OAc)_2/PCy_3$-catalyzed intramolecular α-arylation of amides.

intermolecular arylation processes with higher catalyst loading [Eq. (1.37)] [485]. The $Pd(OAc)_2$–$(1\text{-}Ad)_2P(O)H$ system behaves similarly to $Pd(OAc)_2$–PCy_3 in catalyzing intramolecular α-arylation of amides with aryl chlorides [486]. The chiral Pd–NHC complex **236** catalyzes asymmetric intramolecular α-arylation of amides with aryl chlorides or bromides, but only one example of chloride substrate was presented [Eq. (1.38)] [487].

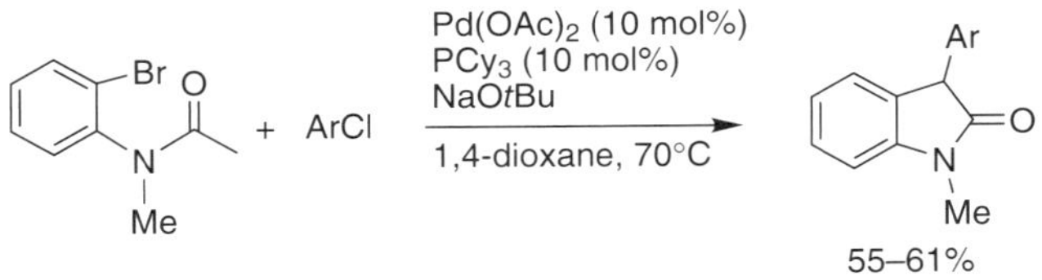

(1.37)

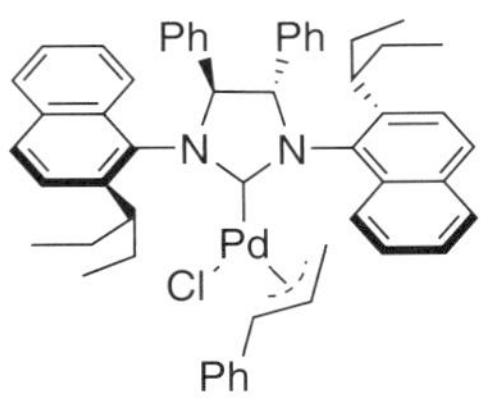

236

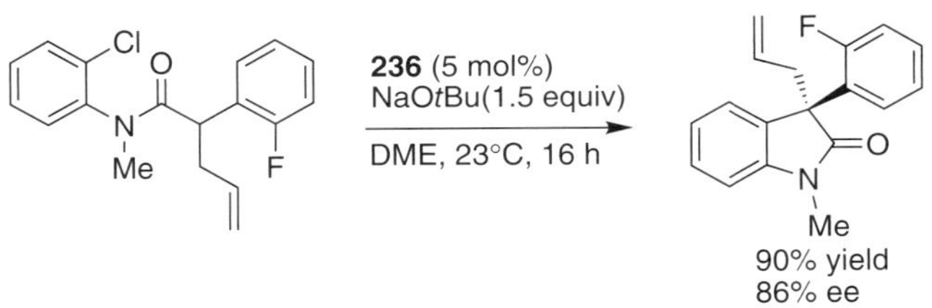

(1.38)

(vi) α-Arylation of Ketimines, Nitriles, and Nitroalkanes. Treatment of an imine with a strong base results in an azaallyl anion. The anion might undergo *C*-arylation to give an arylated imine or undergo *N*-arylation to form an enamine. Barluenga et al. [488] have demonstrated that the reaction occurs exclusively at the C-position when Pd_2dba_3–XPhos or Pd_2dba_3–DavePhos is employed as the catalyst. Pd_2dba_3–XPhos also catalyzes a domino reaction of an imine and *o*-dihalobenzene in which an imine is α-arylated with *o*-dihalobenzene and followed an intramolecular C–N coupling to afford indole derivatives (Scheme 1.39) [488].

Verkade reported that $Pd(OAc)_2$–**177b** is an excellent catalyst for the α-arylation of linear or branch nitriles with electron-rich and electron-poor aryl chlorides using $NaN(SiMe_3)_2$ as base. $Pd_2(dba)_3$–**177b** efficiently catalyzes α-arylation of ethyl cyanoacetate with aryl chlorides using KO*t*Bu as a base. The reaction yields α-aryl cyanoacetates, which are useful intermediates in the preparation of amino alcohols, β-amino acids, and arylacetic acids [489]. Malononitrile can also be α-arylated with

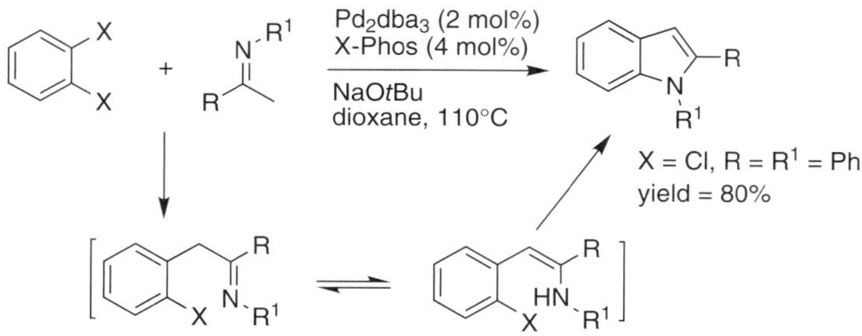

Scheme 1.39.

aryl chlorides employing combination of $Pd(dba)_2$ and imidazolium salt **237a** or **237b** as a catalyst and NaH as a base [490].

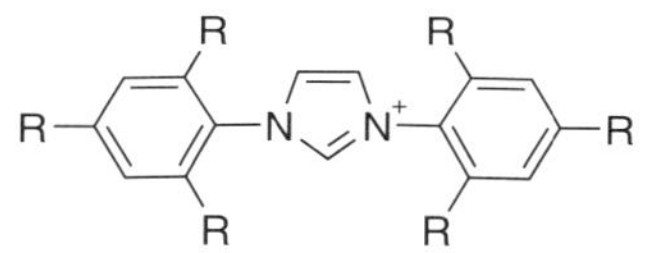

237a R = Me
237b R = Et

Buchwald and co-workers [455, 491] reported that the combination of 2-di-*tert*-butylphosphino-2′-methylbiphenyl and $Pd_2(dba)_3$ is effective for the α-arylation of linear notroalkanes with aryl bromides and chlorides. The reaction uses Cs_2CO_3 as base and proceeds under mild conditions. For chloride substrates, moderate to good yields were obtained. In the presence of enolizable ketone and ester groups the nitroalkane can be chemoselectively arylated.

1.3.9.2 C–H Functionalization of (Hetero)arenes with Organic Chlorides. In 1985 Ohta and co-workers [492] reported arylation of indoles with aryl chlorides catalyzed by $Pd(PPh_3)_4$. Since then a series of examples of C–H functionalization of (hetero)arenes with aryl or alkyl chlorides have been reported. Several excellent reviews cover this topic [493–497].

(i) Palladium Catalysts

Arylation or Alkylation of Arenes via C–H Bond Cleavages with Aryl or Alkyl Chlorides. The combination of $Pd(OAc)_2$ and a series of imidazolinium salts involving **153** and **238a–d** is an effective catalyst for *ortho*-arylation of aromatic aldehydes with aryl chlorides or bromides. Both **153** and **238a** are the best ligand precursors for the reaction. Either $Pd(OAc)_2$–**153** or $Pd(OAc)_2$–**238a** efficiently catalyzes the arylation reactions under mild conditions and leads to good to excellent product yields [Eq. (1.39)]. Reaction of aryl chlorides gives 2-arylated aromatic aldehydes, while reaction of aryl bromides results in 2,6-diarylated aromatic aldehydes [498].

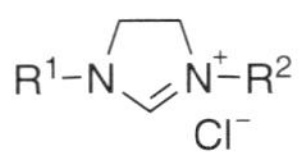

a $R^1 = R^2 = 2,4,6\text{-}(MeO)_3C_6H_2$
b $R^1 = R^2 = 3,4,5\text{-}(MeO)_3C_6H_2$
c $R^1 = 2,4,6\text{-}(MeO)_3C_6H_2$
$R^2 = CH_2CH_2OMe$
d $R^1 = 2,4,6\text{-}(MeO)_3C_6H_2$
$R^2 = CH_2CH_2OEt$

238

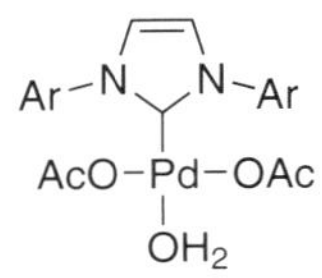

Ar = 2,6-iPr$_2$C$_6$H$_3$

239

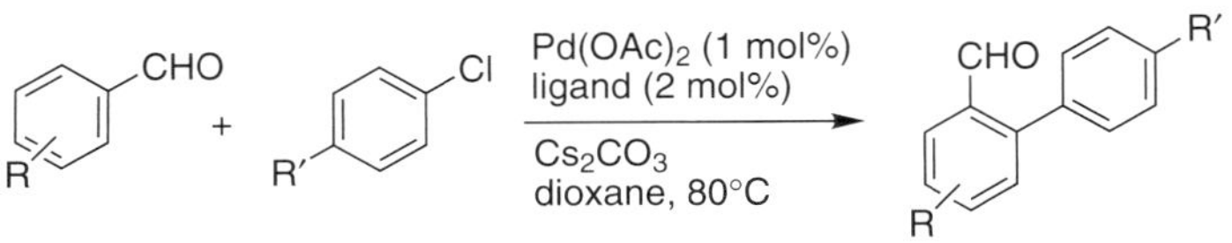

(1.39)

R = H, 4-CHO, 4-NMe$_2$, 4-OMe, 4-tBu, 3,4,5-(OMe)$_3$
R′ = H, Me, OMe, C(O)Me

The NHC–Pd complex **239** catalyzes an intramolecular direct arylation of aryl chlorides, which yields a variety of six- and five-membered ring biaryls (Scheme 1.40). The use of IPr·HCl additive significantly improves the TON of the palladium–NHC catalyst [499]. Similar intramolecular arylation reaction was also carried out using $Pd(OAc)_2$–$PCy_3\cdot HBF_4$ as catalyst and K_2CO_3 as base. This catalyst system is also applicable for the intermolecular direct arylation of 1,3-benzodioxle with aryl chlorides or bromides [500].

A catalytic domino N–H/C–H bond activation process was designed to propare carbazole derivatives as shown in Eq. (1.40). This approach involves an amination and a direct arylation by using anilines and 1,2-dihaloarenes and shows high regioselectivity (Table 1.10) [501].

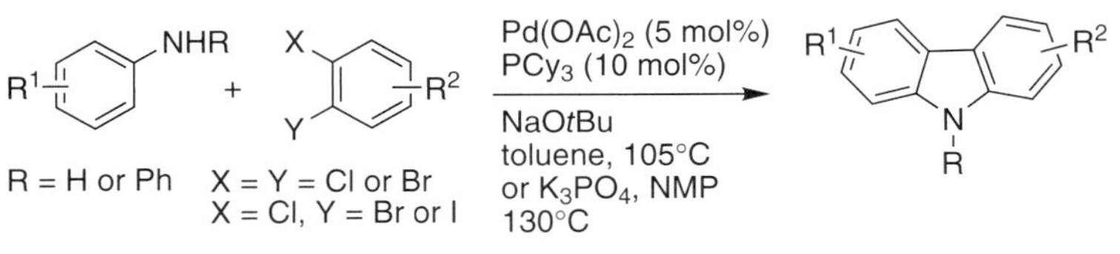

(1.40)

A related consecutive amination and C–H activation reaction between 2-chloranilines and aryl bromides was reported by Bedford and Cazin [502] using $Pd(OAc)_2/PtBu_3$ as catalyst and NaOtBu as base (Scheme 1.41). 4-Chloroquinoline can replace aryl bromides in this reaction if $Pd_2dba_3/PtBu_3$ is used as catalyst and K_3PO_4 as base [503].

Pentafluorobenzene can be arylated with aryl chlorides or aryl bromides using $Pd(OAc)_2$–$PtBu_2Me\cdot HBF_4$ or $Pd(OAc)_2$–SPhos as catalyst. For the aryl chloride or sterically hindered aryl bromide substrates, the $Pd(OAc)_2$–SPhos system is much

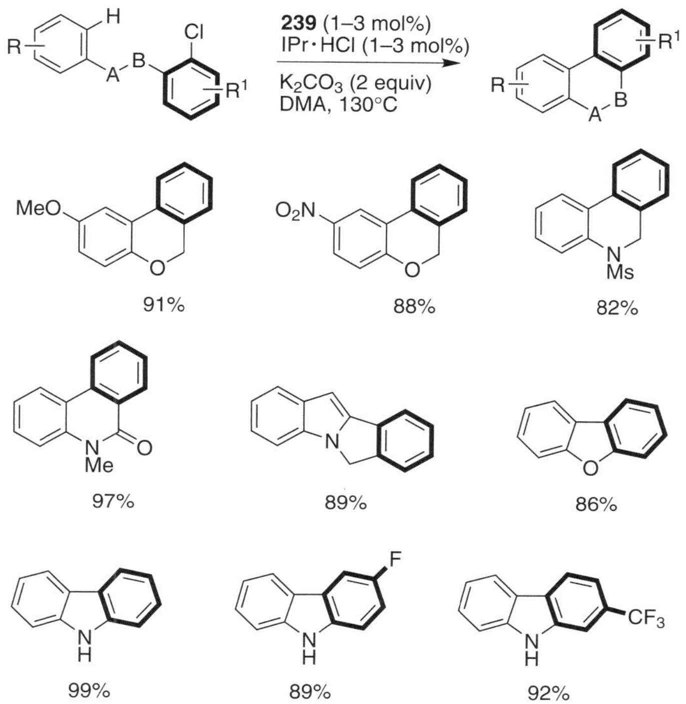

Scheme 1.40. **239**/IPr- catalyzed intramolecular arylation with aryl chlorides.

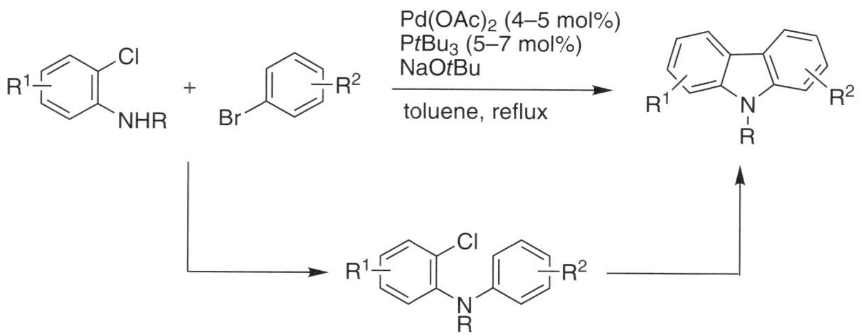

Scheme 1.41.

more effective. Using $Pd(OAc)_2$–SPhos as catalyst, electron-rich, electron-poor, and bulky aryl chlorides can be coupled in excellent yields. Other perfluoroaromatic compounds involving 2,3,5,6-tetrafluorotoluene, 2,3,5,6-tetrafluoroanisole, 1,3,5-trifluorobenzene, and 2,3,5,6-tetrafluoropyridine have also been arylated in high yields [504, 505].

TABLE 1.10. Direct Arylation-Based Domino Reaction[a]

Entry	R	R[1]	Chloride	Product	Yield (%)
1	Ph	H	1,2-Dichlorobenzene	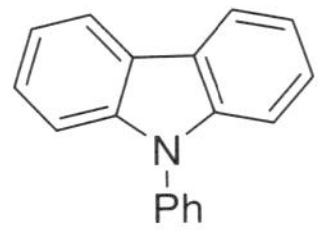	85
2	Ph	H	1-Bromo-2-chloro-benzene	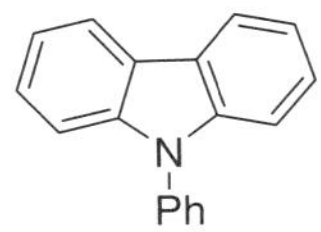	88
3	H	2,4-Me_2	1,2-Dichlorobenzene	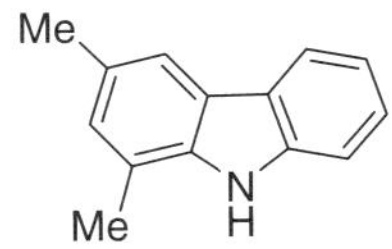	63
4	H	2-MeO	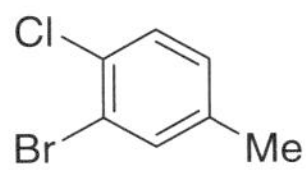	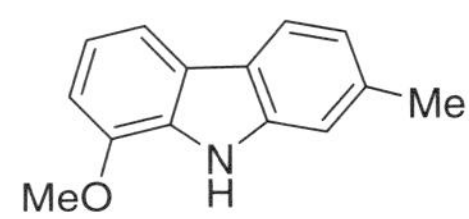	80
5	H	H	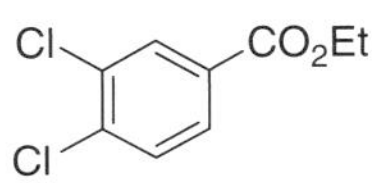	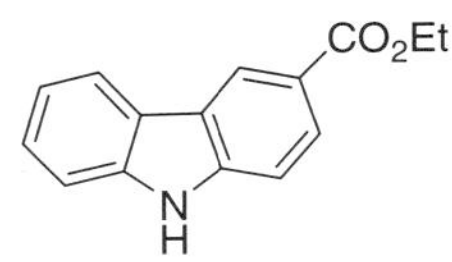	62
6	H	4-Me	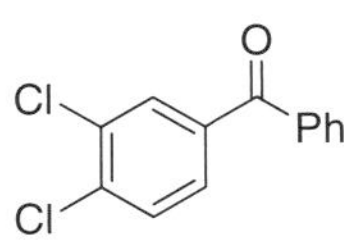	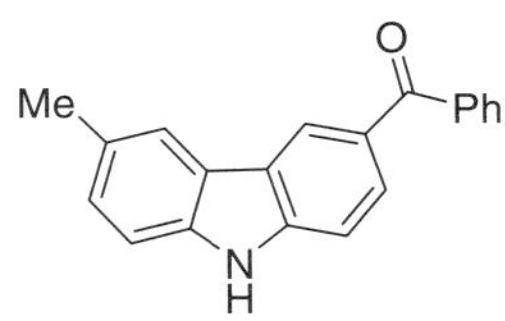	76
7	H	4-Me	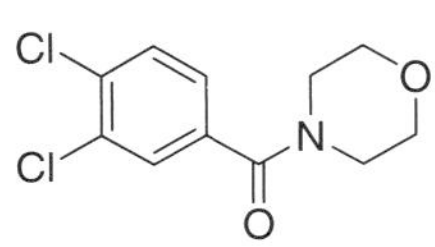	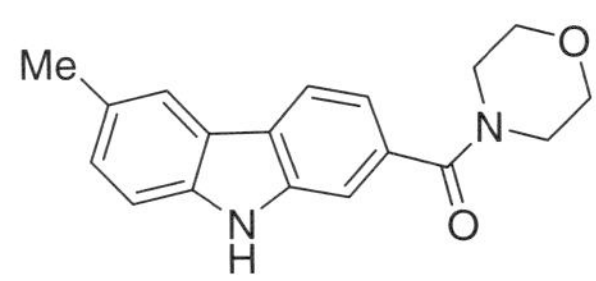	71

[a]Reaction conditions are the same as shown in Eq. (1.40).

The Pd-catalyzed alkylation of arenes with alkyl chlorides has also been carried out. The intramolecular C–H functionalization of α-chloroacetanilides proceeds effectively in toluene at 80°C using $Pd(OAc)_2$/2-(di-*tert*-butylphosphino)biphenyl as catalyst and Et_3N as base, forming substituted oxindoles in 76–99% yields [506]. The *ortho* alkylation of benzoic acids with alkyl chlorides or bromides was carried out using $Pd(OAc)_2$ at 115–140°C (Scheme 1.42). It was demonstrated

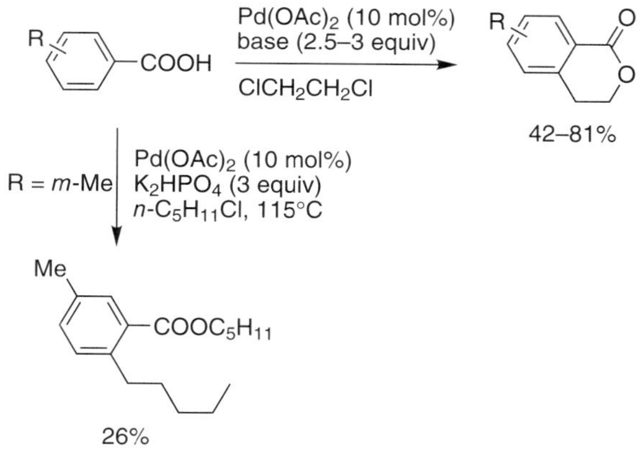

Scheme 1.42.

that **240** does not undergo an *ortho* alkylation under the same conditions as used above. This experimental fact does not support a mechanism in which nucleophilic substitution occurs first, followed by C–H activation initiated by oxidative addition to the intramolecular halide. This finding combined with data from kinetic isotope effect studies also do not support a Friedel–Crafts-type reaction mediated by the Pd^{II} catalyst or by trace quantities of HCl generated from Pd^{II} salts. The reaction may proceed through the intermediate **241** generated by *ortho* C–H insertion with benzoic acid [507].

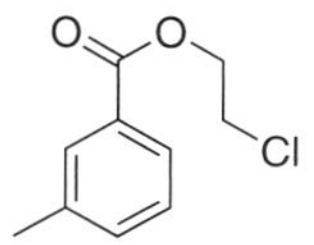

240

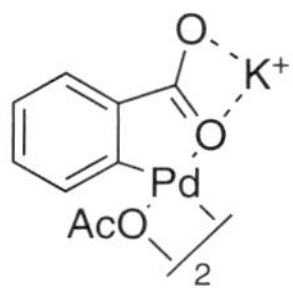

241

Arylation or Alkylation of Heteroarenes via C–H Bond Cleavages with Aryl or Alkyl Chlorides. A series of heterocycles have been arylated or alkylated with aryl or alkyl chlorides. The combination of $Pd(OAc)_2$ and ferrocenyl triphosphane **242** is an effective catalyst for α-arylation of electron-rich, electron-poor, and poly-substituted furans, thiophenes, and pyrroles with electron-deficient aryl chlorides using KOAc as base and TBAB as additive. The reaction was carried out in DMAc at 150°C and requires low catalyst loading (0.1–0.5 mol%). 2-Propylthiazole was arylated with electron-deficient aryl chlorides under the same conditions, forming 5-aryl-2-propylthiazoles [508]. NHC–Pd complexes also catalyze the same reactions as above, but with lower catalytic activity [509]. $Pd(tBu)_2$ behaves excellently in catalyzing α-arylation of thiophene or benzothiophene with electron-rich or electron-poor aryl chlorides in the presence of LiO*t*Bu in DMF at 100°C [510]. The system composed of $Pd(OAc)_2$ and 2-(dicyclohexylphosphino)biphenyl is effective for the arylation of indoles, pyrroles, and furans by electron-rich and electron-poor aryl chlorides using Na_2CO_3 (for indoles) or K_3PO_4 (for pyrroles and furans) as the base [511]. A combination of $Pd(OAc)_2$ and 2-(di-*tert*-butylphosphino)biphenyl (**5**) results in formation of condensed pyrroloindoles via intramolecular C–H bond functionalization of pyrroles [Eq. (1.41)] [512].

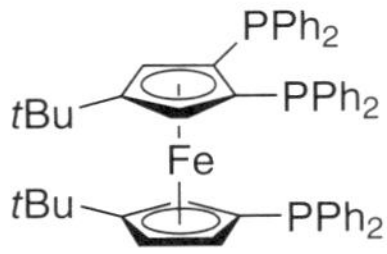

242

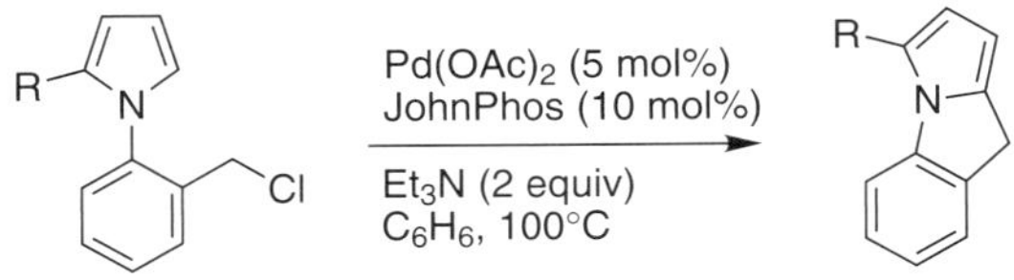

(1.41)

PdCl(dppb)(allyl) was demonstrated to be effective for the reaction of 2-substituted thiazoles with electron-deficient aryl or heteroaryl chlorides [513]. It also catalyzes 2-arylation of benzothiazole or benzoxazole by electron-deficient aryl or heteroaryl chlorides in the presence of Cs_2CO_3 [514, 515]. Strotman et al. [516] developed a method for Pd-catalyzed arylation of oxazole with high regioselectivity at both C-5 and C-2 positions. Thus, the combination of $Pd(OAc)_2$ and Ad_2PnBu or **243** in polar solvents such as DMA leads to arylation of oxazole at C-5 position, whereas $Pd(OAc)_2$–**15b** in nonpolar solvents such as toluene results in arylation of oxazole at C-2 position (Scheme 1.43). This method is applicable to a variety of electron-rich or electron-poor aryl and heteroaryl electrophiles, including bromides, chlorides, and triflates. Hoarau and co-workers [517, 518] observed similar condition-dependent regioselectivity in Pd-catalyzed arylation of oxazole and thiazole-4-carboxylate esters with aryl chlorides or bromides. They demonstrated both base-assisted nonconcerted metalation–deprotonation (nCMD) and concerted metalation–deprotonation (CMD) as two potent operating mechanisms in the reactions.

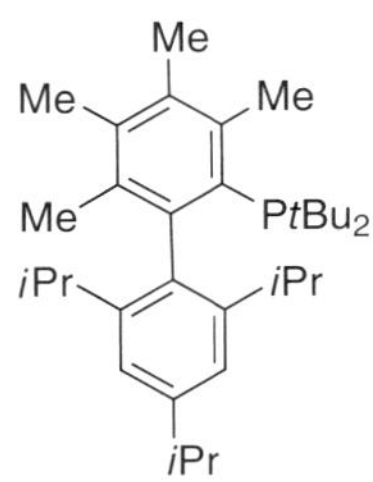

243

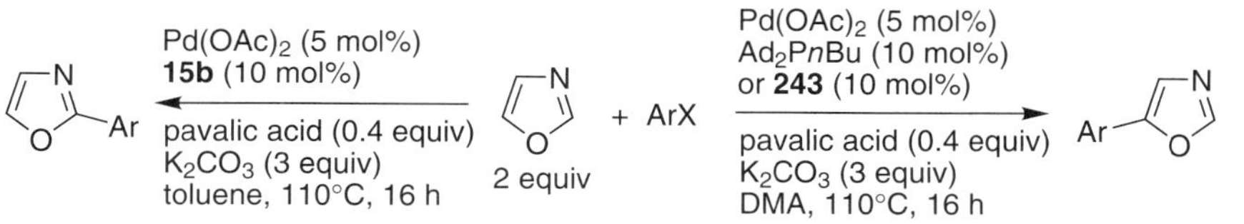

Scheme 1.43.

Pd-catalyzed arylation of imidazoles with aryl chlorides or bromides also shows positional selectivety. C-5 arylated imidazole can be obtained by the reaction in DMA using NHC–Pd complex **244** as the catalyst and KOAc as base. C-2 arylated imidazole is formed through reaction in toluene using $Pd(OAc)_2$–Ad_2PnBu as the catalyst and NaO*t*Bu as the base (Scheme 1.44). In an *N*-alkyl imidazole molecule, the C-2 position bears the most acidic C–H bond, the C-5 position shows high reactivity toward electrophilic substitution, and the C-4 position is relatively less reactive (Figure 1.11). In the presence of a strong base NaO*t*Bu, deprotonation of the C-2 position becomes efficient. This process is also facilitated by complexation of the palladium complex to the N-3 nitrogen (Scheme 1.45). The C-5 selectivity can be explained by electrophilic metalation deprotonation (EMD) because the 5-position is

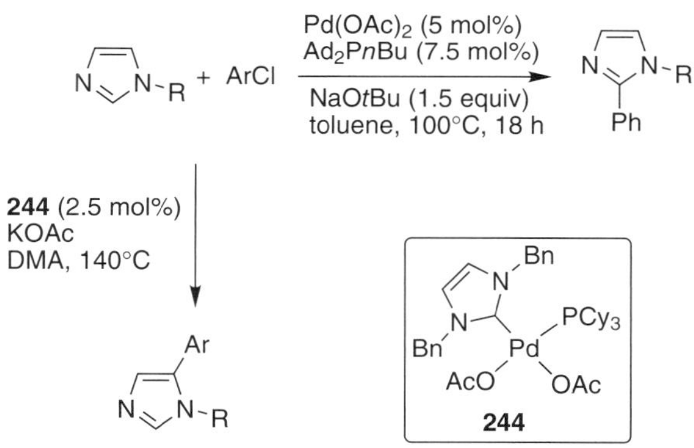

Scheme 1.44. C-2 and C-5 arylation of imidazoles.

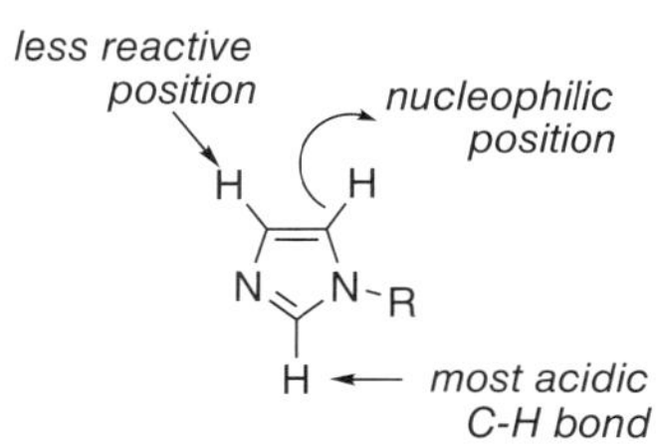

Figure 1.11. General reactivity trends of imidazole.

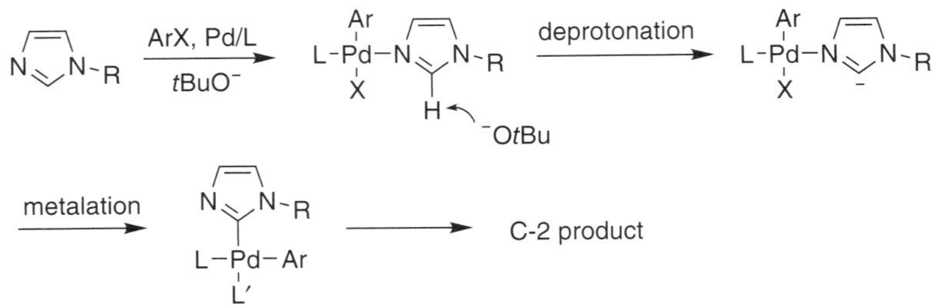

Scheme 1.45. A proposed mechanism for the formation of a C-2 product.

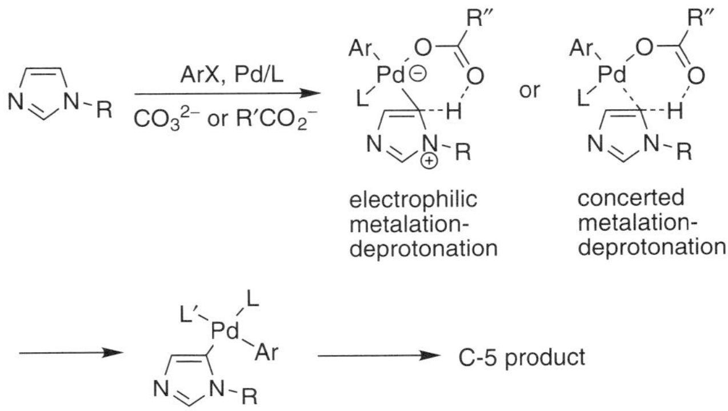

Scheme 1.46. A proposed mechanism for the formation of a C-5 product.

more nucleophilic than the 2- and 4-positions; standard electrophilic aromatic substitutions occur preferentially at the C-5 position (Scheme 1.46) [519, 520].

Pd-catalyzed arylation of adenine derivatives by aryl chlorides occurs at the C-8 position [Eq. (1.42)]. Both electron-rich and electron-poor aryl chlorides are suitable electrophilic substrates. CuI is used as an additive [521, 522]. The $Pd(OAc)_2/Ad_2P\mathit{n}Bu/K_3PO_4$ system catalyzes C-8 arylation of caffeine by electron-rich aryl chlorides in NMP without need of CuI. This catalyst system is also effective for the arylations of other heterocycles involving thiophene, benzothiophene, benzofuran, oxazole derivatives, thiazole, benzothiazole, 1-butylimidazole, and 1-methyl-1,2,4-triazole [523].

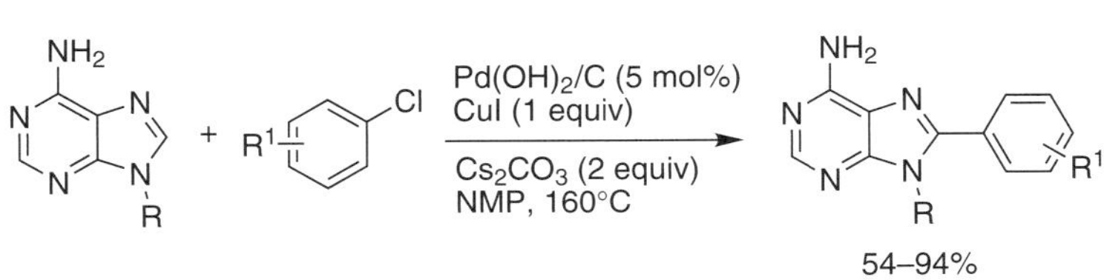

(1.42)

$Pd(OAc)_2$–PCy_3 catalyzes arylation of 1,2,3-triazoles with electron-rich or electron-poor aryl chlorides using K_2CO_3 as base under either conventional or microwave heating. Reaction of 1,4-disubstituted 1,2,3-triazoles with aryl chlorides in the presence of $Pd(OAc)_2/PCy_3/K_2CO_3$ forms C-5 arylated trizoles. Under the microwave radiation, bulky aryl chlorides such as 2,6-dimethylchlorobenzene and 2-chlorobiphenyl can be catalytically coupled with 1,2,3-triazoles. The C-4 arylation of 1,5-disubstituted 1,2,3-triazoles with aryl chlorides has also been carried out using the same catalyst and conditions. However, the product yields are low under either thermal heating or microwave radiation. Reaction of 1-butyl-1,2,3-triazole with aryl chlorides using the same catalyst under microwave radiation or thermal heating affords 1-butyl-5-aryl-1,2,3-triazole (major) and 1-butyl-4,5-diaryl-1,2,3-triazole (minor) [524, 525].

Palladium-catalyzed alkylation of heterocycles by alkyl halides has also been achieved. Reaction of ethyl oxazole-4-carboxylate with allyl chlorides or benzyl chlorides using $Pd(OAc)_2$/P(biphenyl-2-yl)Cy_2 as catalyst and Cs_2CO_3 as base affords 2-alkylated oxazoles via C–H activation. This reaction condition is also suitable for the coupling of ethyl oxazole-4-carboxylate with vinyl chloride [526]. $Pd(OPiv)_2$/2-Ph_2P-2′-(Me_2N)biphenyl catalyzes benzylation of a variety of heterocyclic aromatic compounds by benzyl chlorides. The reaction was performed in toluene at 110°C using PivOCs and Cs_2CO_3 as the bases. The heterocycles tested involve thiazoles, thiophenes, indolizines, imidazopyrimidine, 1,2,3-triazole, oxazole, and furans. The regioselectivity of the reactions is consistent with that of arylation reactions [527]. $[PdCl(\eta^3\text{-}C_3H_5)]_2$-$n$$Bu_3P$ catalyzes C-2 alkylation of benzoxazole and 5-substituted oxazoles by primary alkyl chlorides using LiO*t*Bu as the base [528].

Arylation via sp^3 C–H Bond Cleavages with Aryl Chlorides. Pd-catalyzed arylations of sp^3 carbons by aryl chlorides have been carried out. Benzylic position arylation of aryl(azaaryl)methanes with aryl chlorides or bromides is catalyzed by $PdCl_2(MeCN)_2$–Cy_3P in the presence of $CsOH{\cdot}H_2O$, forming triarylmethanes. Both electron-rich and electron-poor aryl chlorides are applicable in this transformation [529]. Rousseaux et al. [530] developed a method of Pd-catalyzed intramolecular alkane C–H arylation from aryl chlorides. A series of cyclobutarenes, indanes, indolines, dihydrobenzofurans, and indanones have been synthesized by this protocol. $Pd(OAc)_2$–$t$$Bu_3P$-$K_2CO_3$ is a suitable catalytic system for synthesis of cyclobutarenes (Scheme 1.47), while combination of $Pd(OAc)_2$, Cy_3P, Cs_2CO_3, and PivOH is effective for the synthesis of indolines, dihydrobenzofurans, and indanones (Scheme 1.48). Synthesis of indanes employs $Pd(OAc)_2$-Cyp_3P as catalyst and K_2CO_3 as base. Good to excellent yields were obtained in DMF at 140°C [e.g., Eq. (1.43)] [530].

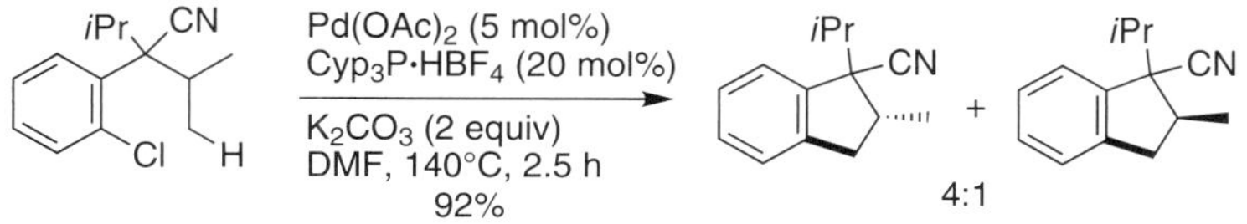

(1.43)

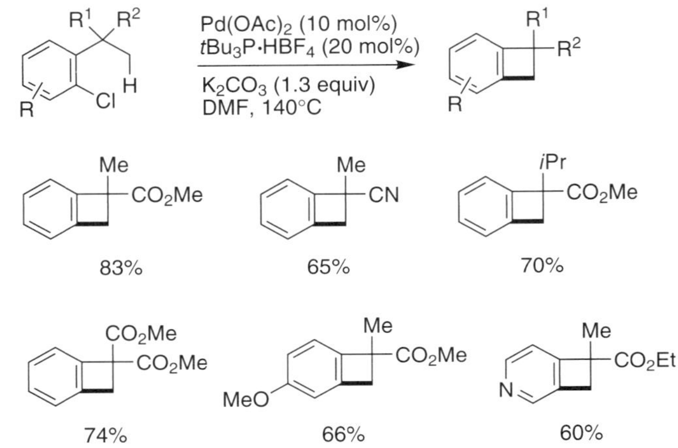

Scheme 1.47.

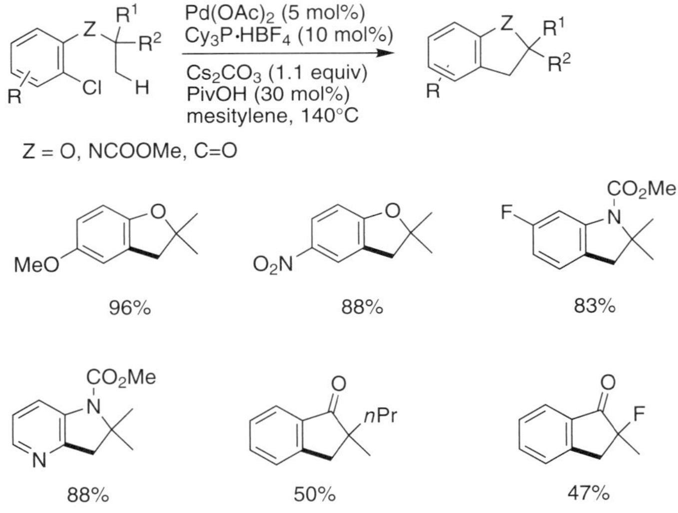

Scheme 1.48.

(ii) Ruthenium Catalysts. $[RuCl_2(p\text{-cymene})]_2$ was demonstrated to be a good catalyst for the *ortho* C–H bond functionalization of 2-phenylpyridine with aryl or heteroaryl chlorides [Eq. (1.44)]. K_2CO_3, KOAc, and KOPiv are often used as bases. NMP, diethyl carbonate, and H_2O are frequently used solvents. The catalytic activity is higher in H_2O than in an organic solvent. Electron-rich and electron-poor aryl chlorides and heteroaryl chlorides (such as 2-chloropyridine, 2-chloro-6-methylpyridine, 2-chloroquinoline, and 2- or 3-chlorothiophene) are applicable as the electrophilic substrates. When excess aryl chlorides (2.5 equiv) are employed, diarylated products are predominant. 1-Pyrazolyl and 2-oxazolyl were also used as directing groups in the Ru-catalyzed arylation and behave similar to the pyridyl group [531–533]. The addition of co-ligand $Ad_2P(O)H$ in the Ru catalyst system appears to give results similar to that of $[RuCl_2(p\text{-cymene})]_2$ alone for the 2-arylpyridines. However, the catalyst system consisting of $[RuCl_2(p\text{-cymene})]_2$, $Ad_2P(O)H$, and K_2CO_3 in NMP efficiently catalyzes functionalization of differently substituted ketimines with aryl chlorides. The reaction gives monoarylated products in moderate to good yields [534].

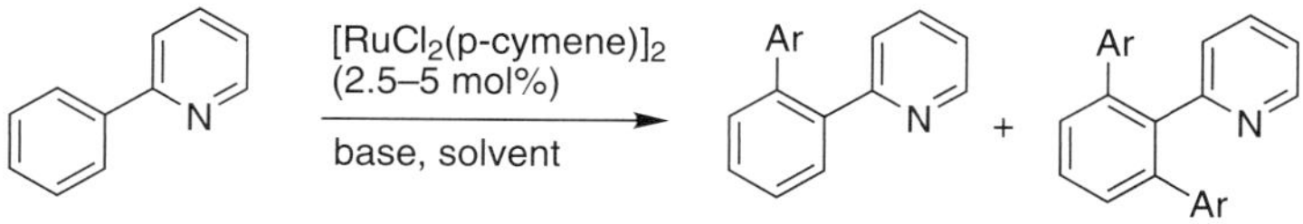

(1.44)

$[RuCl_2(p\text{-cymene})]_2$ also catalyzes CH arylation of pyridyl aryl ethers in the presence of $MesCO_2H$ [Eq. (1.45)]. Importantly, the directing group, 2-pyridyls, can be removed after the arylation [535].

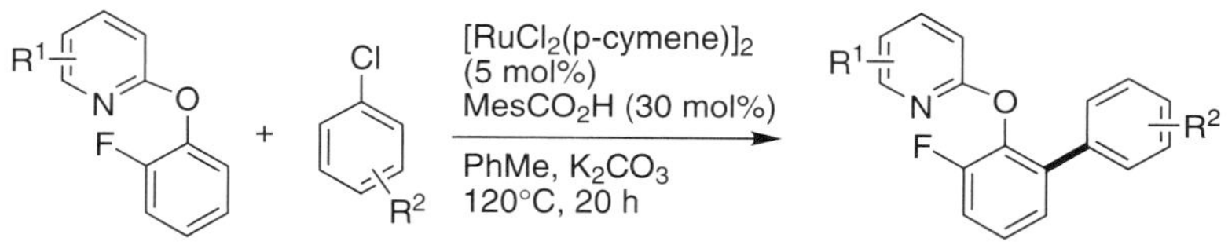

(1.45)

Ru complex **245** was prepared from the reaction of $[RuCl_2(p\text{-cymene})]_2$ with 2-phosphinoethene. It is an excellent catalyst for the C–H functionalization of 2-arylpyridines with electron-donating and electron-withdrawing aromatic chlorides in NMP using K_2CO_3 as a base. The monoarylation and diarylation of 2-arylpyridine through functionalization of C–H bonds are controllable. When an equimolar amount of aryl chlorides are employed, the monoarylated products are obtained. If excess aryl chlorides (2.5 equiv) are employed, the diarylated products are predominant [536].

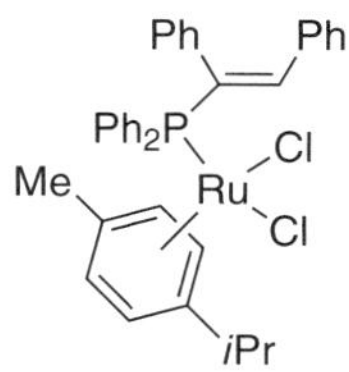

245

$[RuCl_2(p\text{-cymene})]_2$ in water, $[RuCl_2(p\text{-cymene})]_2\text{-}PPh_3$, $Ru_3(CO)_{12}$, $RuCl_3\cdot nH_2O\text{-}PPh_3$, and $Ru/CeO_2\text{-}PPh_3$ show good catalytic activity for the C–H functionalization of benzoquinoline with aryl chlorides, giving 10-arylbenzoquinolines [533, 537, 538]. Both $RuCl_3\cdot nH_2O\text{-}PPh_3$ and $Ru/CeO_2\text{-}PPh_3$ catalyst systems were examined with various chloride substrates. $Ru/CeO_2\text{-}PPh_3$ shows higher catalytic activity for the electron-rich aryl chlorides and are applicable for heteroaryl chlorides such as 2-chloropyridine and 2-chlorothiophene [537, 538].

1,2,3-Triazolyl is also an effective directing group for the Ru-catalyzed C–H arylation of an phenyl ring with electron-rich and electron-poor aryl chlorides as arylating reagents [Eq. (1.46)] [539].

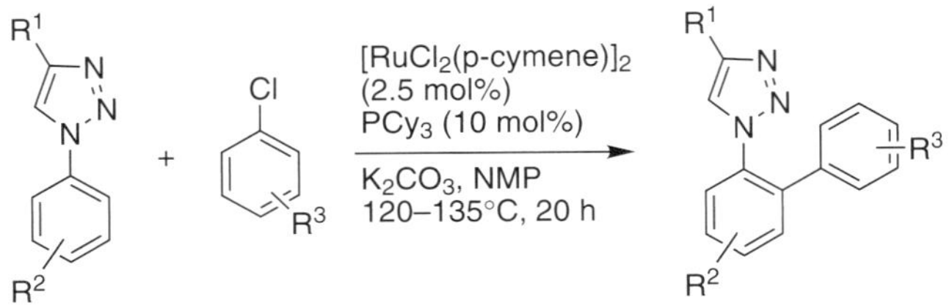

(1.46)

In addition, direct C–H alkylation with alkyl chlorides can be carried out using $[RuCl_2(p\text{-cymene})]_2$ as catalyst [Eq. (1.47)]. Benzyl chlorides and n-$C_6H_{13}Cl$ are suitable electrophilic reagents [540, 541].

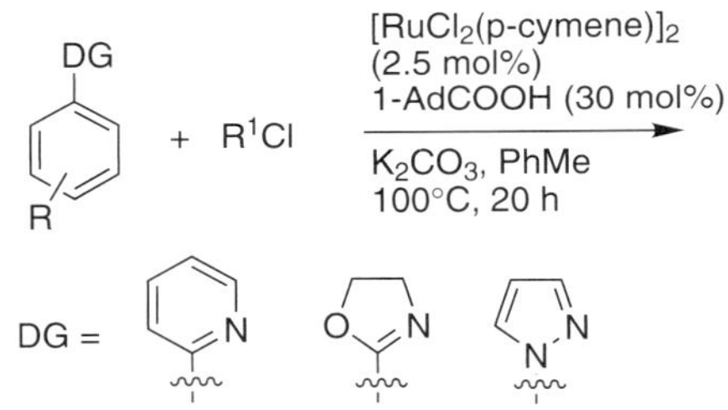

(1.47)

(iii) Iron and Cobalt Catalysts. $FeCl_3$ was demonstrated to catalyze the coupling of benzene with substituted bromo- or chloro-benzene in the presence of LiHMDS and *N,N*-dimethylethylenediamine (DMEDA). This catalyst system is effective for bromides. For the chloride substrates, low yields are achieved [Eq. (1.48)] [542]. The combination of $CoBr_2$, DMEDA, and LiHMDS catalyzes the same reaction under the same conditions and produces slightly higher yields [543].

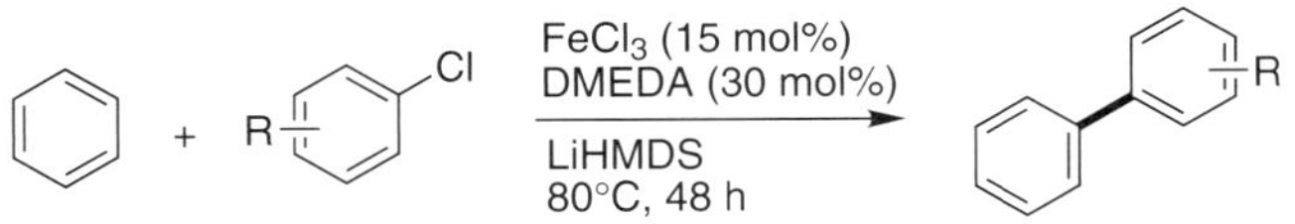

(1.48)

R = 4-OMe 44%
R = H, 52%
R = 4-Me 34%
R = 4-Ph <5%

Nakamura and co-workers [544] found cobalt-catalyzed *ortho*-alkylation of benzamide with alkyl chloride through directed C–H bond activation. $Co(acac)_2$,

$Co(acac)_3$, and $CoBr_2$ are effective catalysts, suggesting that they are converted into the same catalytic species under the reaction conditions. CyMgCl and DMPU are indispensable, otherwise the reaction will not proceed. Primary and secondary alkyl chlorides are applicable for this transformation. However, reaction of *t*-BuCl results in an isobutylated product that suggests β-eliminative rearrangement of a σ-cobalt intermediate or intermediacy of a π-complex. MeO-, F-, and Ph-substituted benzamide, as well as 1-naphthylamide, can be *ortho*-alkylated using this catalyst system. However, 4-fluorobenzamide and *N*-methylisonicotinamide tend to give the dialkylated products [544].

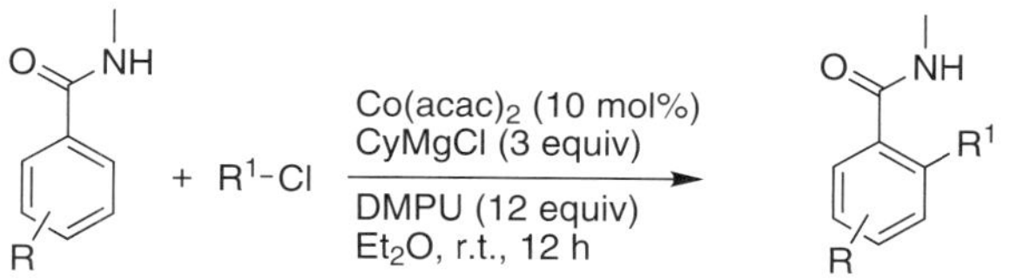

(1.49)

1.4 FORMATION OF C–N BONDS

Transition-metal-catalyzed C–N bond formation has played an important role in the preparation of important compounds in biological, pharmaceutical, and materials science [545, 546]. Since Ullmann and Goldberg developed the initial method of metal-catalyzed amination on the basis of the organocuprate chemistry more than a century ago, transition-metal-catalyzed cross-coupling of aryl halides with nitrogen-containing compounds has achieved great progress [547]. Aryl bromides and iodides were widely used as the electrophilic partners for the coupling in the past decades, whereas in the last decade aryl chlorides attracted a lot of attention. A range of new ligands and catalysts were developed for the coupling of aryl chlorides with various amines or related compounds. Several recent reviews cover this topic [13, 15, 545, 548, 549].

1.4.1 Copper Catalysts

It has been known for a long time that aryl or vinyl halides react with nitrogen-containing compounds such as amines, amides, imides, and carbamates in the presence of copper to form C–N bonds [13, 550]. Aryl or vinyl iodides and bromides are often employed and chlorides are only limitedly used in this transformation.

(i) N,N-LIGANDS. Buchwald and co-workers [551, 552] reported an efficient catalyst system consisting of CuI and racemic *trans*-1,2-cyclohexanediamine (**246a** or **246b**) for the *N*-amidation of aryl halides or the *N*-arylation of heterocycles in the presence of a base such as K_3PO_4, K_2CO_3, Cs_2CO_3, or NaO*t*Bu. For aryl chlorides the *N*-amidation was carried out using CuI–**246b** as catalyst and K_2CO_3 as base

(Scheme 1.49). DMEDA can be used to facilitate intramolecular amidation reactions [Eq. (1.50)] [551, 552].

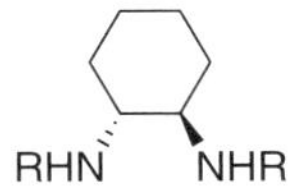

246a R = H
246b R = Me

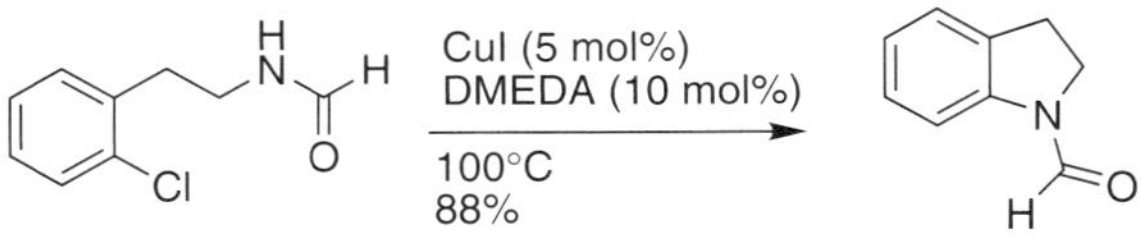

(1.50)

CuOAc/DMEDA-catalyzed amination allows the efficient preparation of a range of *N*-functionalized indoles [Eq. (1.51)]. The reaction employs K_3PO_4 or K_2CO_3 as base and proceeds in toluene at 110°C. When X = Cl, the cyclization reaction cannot proceed under the reaction conditions [553]. This shows that the vinyl chloride is more reactive than the aryl chloride in this reaction. An intramolecular amination of vinyl chlorides has been carried out using CuI/DMEDA as catalyst. The reaction is very efficient, giving β-lactams in almost quantitative yields [Eq. (1.52)] [554].

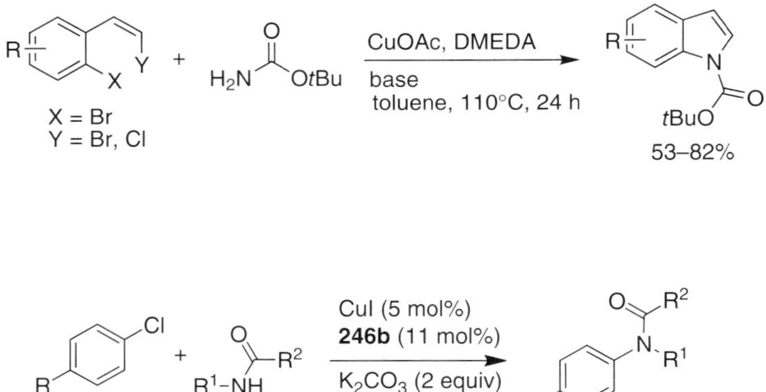

(1.51)

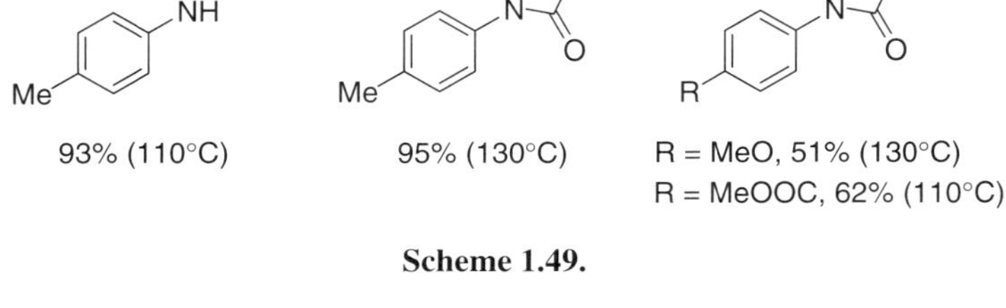

Scheme 1.49.

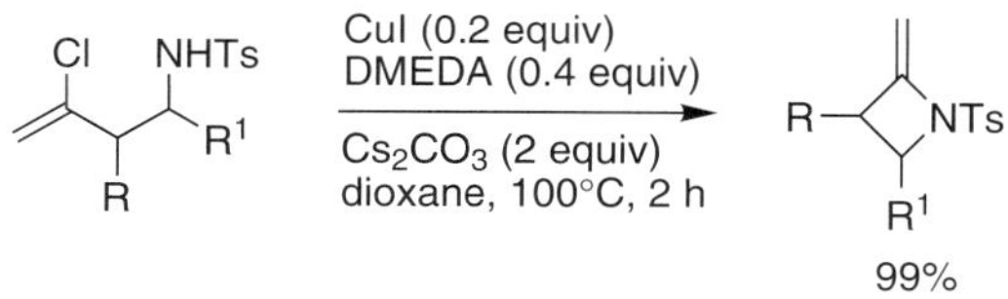

(1.52)

Fu and co-workers [555] developed a CuCl-catalyzed *N*-arylation method of nitrogen-containing heterocycles and aliphatic amines by aryl halides in water using (1*E*, 2*E*)-oxalaldehyde dioxime as the ligand and TBAB as the phase transfer catalyst. This method is suitable for aryl iodides and bromides. Aryl chloride exhibits low reactivity in this catalytic transformation. Only one example was presented in which KI is added to promote the reaction [Eq. (1.53)].

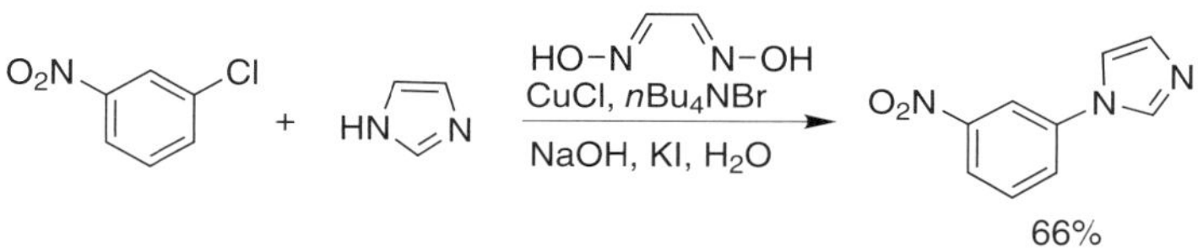

(1.53)

Through experimental and DFT calculation studies, the mechanism of Cu-diamine-catalyzed aryl aminiation is proposed as shown in Figure 1.12. The neutral Cu^{I}(amidate)(diamine) species is the most active copper species. Oxidative addition of this species to aryl halides is the rate- and turnover-determining step in the catalytic cycle [556].

(ii) Proline Derivatives. L-Proline and related compounds are excellent ligands for the Cu-catalyzed C–N bond formation between nitrogen-containing compounds and

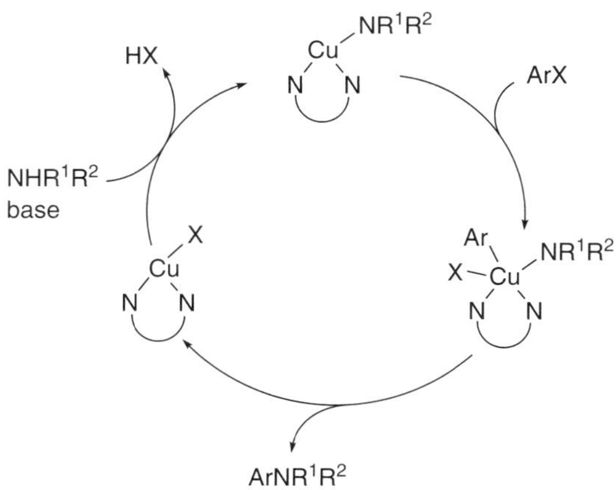

Figure 1.12. Proposed catalytic cycle for copper-catalyzed amidation reactions.

aryl iodides or electron-deficient aryl bromides [557, 558]. A CuI/L-proline-catalyzed intramolecular C–N coupling from aryl chloride has been reported [Eq. (1.54)] [559]. L-Proline has also been used as the ligand for an intermolecular condensation followed by a copper-catalyzed intramolecular C–N coupling reaction to form polycyclic quinolines [Eq. (1.55)]. Iodides, bromides, and chlorides can be used in this transformation. However, for the chloride substrates, only moderate yields were achieved [560].

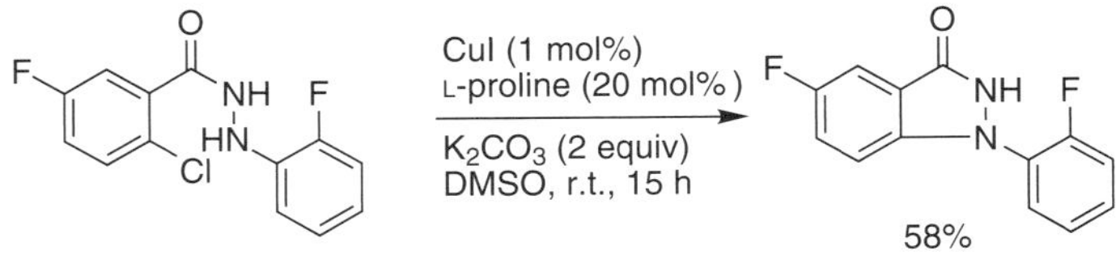

(1.54)

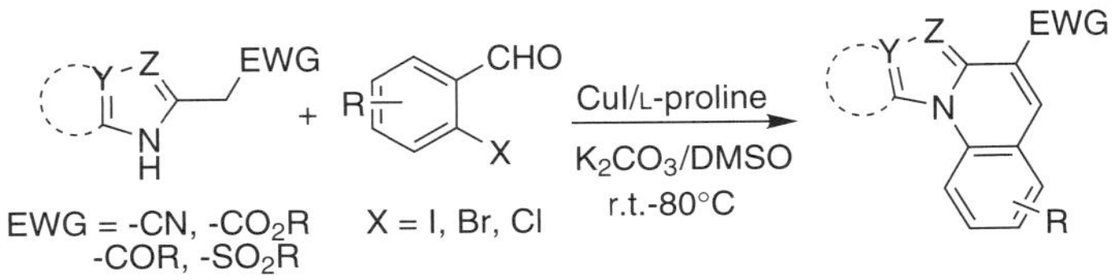

(1.55)

You and co-workers [561] found **247** to be an efficient ligand for the Cu-catalyzed *N*-arylation of imidazoles with aryl or heteroaryl bromides and activated aryl chlorides. Zhao and co-workers [562] carried out CuI/**248**-catalyzed cross-coupling of PhCl with indole, acetamide, or 2-nitrobenzenamine in DMF in low yields.

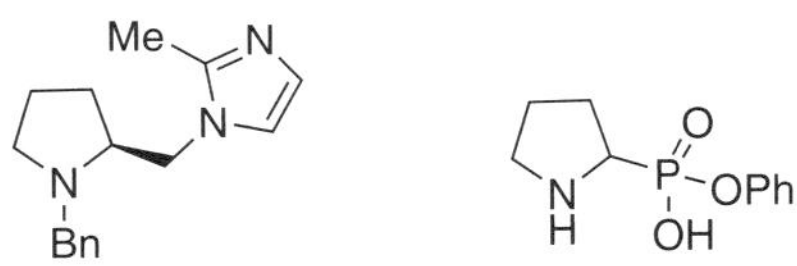

247 **248**

(ii) O,O- AND O,N-LIGANDS. It has been reported that *N*-hydroxyimides (**17a–c**) are efficient ligands for Cu-catalyzed *N*-arylation of pyrrole, imidazole, and indole. The protocols could tolerate an array of functional groups such as nitrile, nitro, and free primary amine on the aryl iodides, bromides, and activated aryl chlorides. The reaction is proposed to proceed through the O,O-chelate copper intermediates [563].

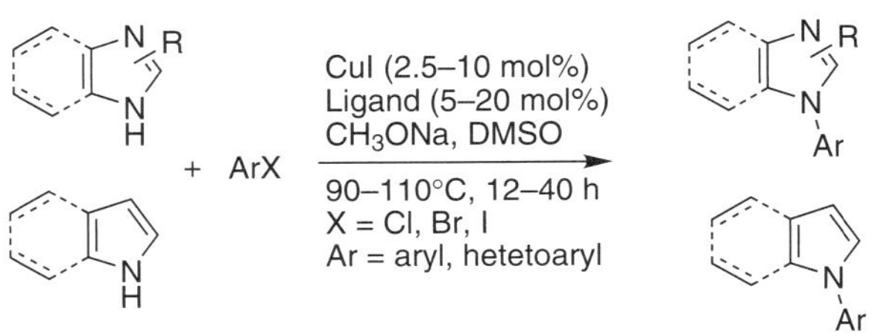

(1.56)

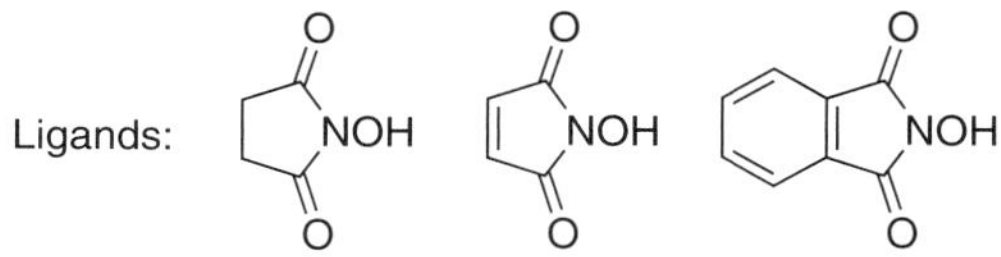

249a 249b 249c

Apatite supported Cu, CuFAP, is believed to be a O,O-chelate heterogeneous Cu catalyst (Figure 1.13). It can catalyze *N*-arylation of aromatic heterocycles involving imidazole, pyrrole, pyrazole, benzimidazole, and piperidine with chloro- and fluoroarenes in DMF at 120°C using K_2CO_3 as base. Activated chloro- and fluoroarenes give excellent coupling results, while unactivated chloro- and fluoroarenes give moderate yields [564].

Water-soluble substituted sulfonato-Cu(salen) complex **250** was demonstrated to catalyze the cross-coupling of aryl halides and alkyl amines [Eq. (1.57)]. The temperature and base are the key parameters in the process. The system is efficient for aryl iodides and bromides as well as for activated chloroarenes [565].

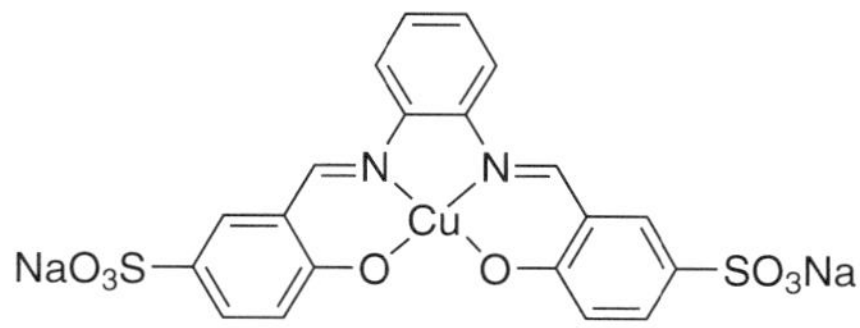

250

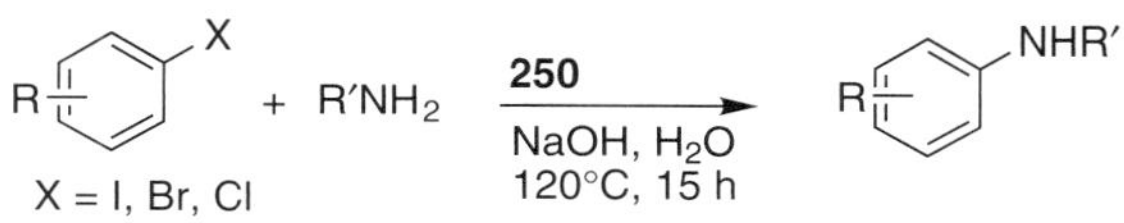

(1.57)

(iv) Other Cu Catalysts. In recent years, Cu-catalyzed *N*-arylation of aryl halides with aqueous ammonia, amines, amides, amidines, benzazoles, phenylurea, and *N*-heterocycles were improved with ligand-free copper catalysts or nanocopper particles [566–573]. Aryl iodides and bromides often display high reactivity in these catalytic transformation. Cu_2O-catalyzed amination of aryl chlorides and CuI nanoparticles-catalyzed arylation of heterocycles by aryl chlorides also give excellent results. Other couplings using aryl chloride substrates often afford low to moderate yields.

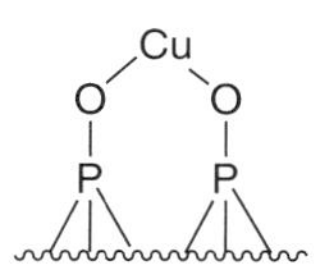

Figure 1.13.

1.4.2 Palladium Catalysts

The direct Pd-catalyzed C–N bond formation between aryl halides and amines (1° and 2° aliphatic or aromatic amines, imides, amides, sulfonamides, sulfoximines) in the presence of a stoichiometric amount of base is known as the Buchwald–Hartwig cross-coupling. Four generations of Pd catalysts have been developed so far for the amination process according to Hartwig [15, 574]. The first-generation Pd catalysts employ P(*o*-tolyl)$_3$ as the ligand, and this catalyst system is not suitable for the amination of aryl chlorides. The second-generation catalysts use chelating aromatic phosphines as the ligand. Palladium combined with carbenes, hindered alkylphosphines such as P*t*Bu$_3$, Ph$_5$FcP*t*Bu$_2$, (biaryl)PR$_2$, (heterobiaryl)PR$_2$, and caged P(NRR′)$_3$ are the third-generation catalysts. Hindered ferrocenyl alkyl bisphosphines in combination with palladium are the fourth-generation catalysts. The mechanism of Pd-catalyzed amination has been studied with numerous ligand systems, and the proposed catalytic cycle is shown in Figure 1.14 [574, 575].

1.4.2.1 The Second-Generation Catalysts. In the past years, few of the second-generation catalysts can be used for amination of chloroarenes [575]. Xantphos (**55**) has been used as the ligand for the synthesis of aryl carbamic acid 2-trimethylsilylethyl esters [576] and diverse 2-(*N*-acylamino)pyrimidines [577] from aryl chlorides. In addition, it can be used for the amination of 4-chloroquinazolines with the cyclic secondary amino groups of 3-amino-1*H*-pyrazoles in combination of Pd$_2$dba$_3$/Na$_2$CO$_3$, while amination of 4-chloroquinazolines with the primary amino groups of 3-amino-1*H*-pyrazoles proceeds exclusively via S$_N$Ar substitution in the presence of HCl [578]. Other bisphosphines such as air-stable dipyridylphosphine ligand *rac*-P-Phos **232** [579] and GemPhos **251** [580] have also been synthesized and applied to the palladium-catalyzed aromatic amination of haloarenes with variable amines.

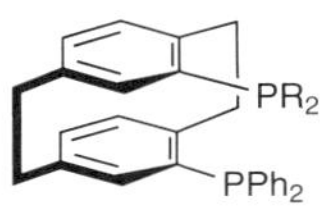

R = Cy or Ph

251

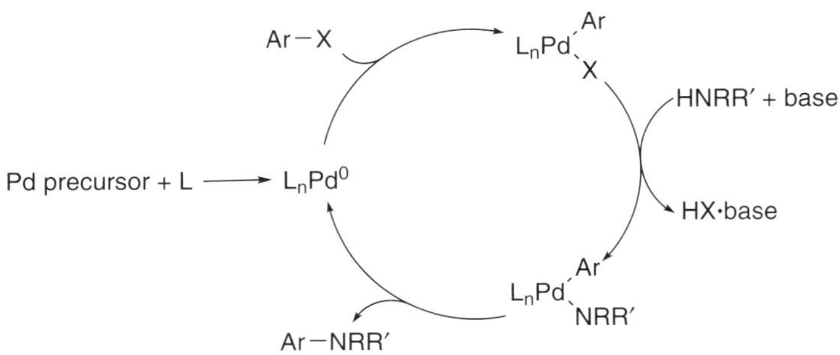

Figure 1.14. Generalized catalytic cycle for aromatic amination.

1.4.2.2 The Third- and Fourth-Generation Catalysts

(i) PHOSPHINE LIGANDS. Pd(0)–P*t*Bu_3 system was proven to be effective for the aminations of chloroarenes under mild conditions. Electron-deficient and electron-donating aryl chlorides can react with aliphatic and aromatic amines, indole as well as amides using this catalyst system [581]. When $LiN(SiMe_3)_2$ is employed as nucleophilic reagent, the same catalyst system leads to formation of anilines from aryl chlorides or bromides after aqueous workup [Eq. (1.58)] [582]. Pd(0)–PCy_3 system is also active for the coupling of activated and unactivated aryl chlorides with aliphatic or aromatic amines using NaO*t*Bu as the base [583].

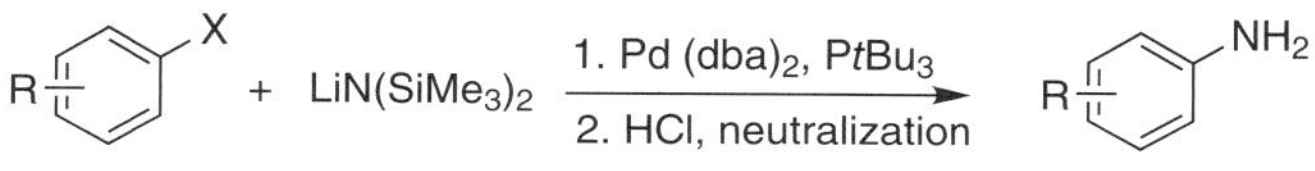

(1.58)

X = Cl, Br

Palladacycles can be used as palladium source for the amination reactions. Palladacycle **90a** or **94** in combination with P*t*Bu_3 is an effective catalyst for amination of aryl chlorides. For example, 0.1 mol% **90a** in combination with 0.2 mol% P*t*Bu_3 drive reaction of 4-chloroanisole with morpholine to completion in 17 h at 100°C in toluene [584, 585].

tBu_2PCH_2*t*Bu, **252**, and **253** are also good ligands for Pd-catalyzed amination of chloroarenes. *t*Bu_2PCH_2*t*Bu in combination with Pd_2dba_3 or $Pd(OAc)_2$ efficiently catalyzes amination of electron-deficient and electron-donating aryl chlorides using aromatic amines or alkyl amines in the presence of NaO*t*Bu [586]. Combination of **252** and $[(allyl)PdCl]_2$ is highly effective for reaction of primary and secondary aryl amines as well as indole with various chloroarenes, including electron-deficient and electron-donating aryl chlorides and multichloroarenes [587]. The activity of **253**–$Pd(OAc)_2$ for the amination of aryl chlorides is comparable to those of Pd_2dba_3/**41a–41c** systems, but the latter is suitable for wider scope of substrates [133, 588, 589].

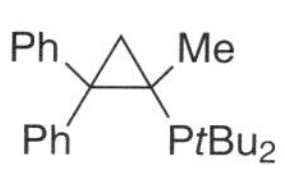

252

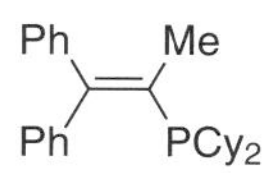

253

Both cyclopentadiene-based ligands **43** and **254** are effective for the amination of aryl chlorides or bromides when combined with a palladium compound such as Na_2PdCl_4 (for the former) or $Pd(dba)_2$ (for the latter). Na_2PdCl_4–**43** is suitable for the activated and unactivated aryl chlorides. $Pd(dba)_2$–**254** is suitable for a wider range of chloride substrates, but reaction of primary alkyl amines using this catalyst gives only low to moderate yields [135, 590].

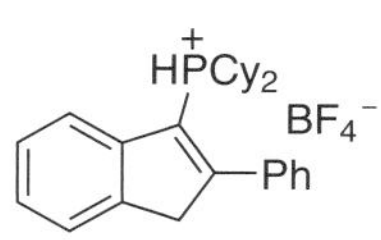

254

Buchwald's biarylphosphine ligands are highly effective in Pd-catalyzed amination of aryl or vinyl halides, including chlorides [575]. Among the series of ligands, **14a**, **18a**, **243**, **255**, **256**, and *t*BuXPhos in combination with appropriate palladium sources are effective for the amination of aryl or vinyl chlorides. Polymer-supported dialkylphosphinobiphenyls are also effective ligands in the catalytic amination reactions of aryl chlorides [591]. Palladacycles **257** with Buchwald ligands such as SPhos (**15a**), RuPhos (**15b**), XPhos (**18a**), and BrettPhos (**182**) are also highly effective for the catalytic aminations. Some representative results are listed in Table 1.11. In these reactions, the catalytically active species is believed to be the monoligated Pd(0) complex [L_1Pd], which exists in equilibrium with the [L_2Pd] species. Both the steric hindrance of the ligands and the metal/ligand ratio were demonstrated to markedly affect the reaction rate. A more sterically hindered ligand may be easier to form the active [L_1Pd] species [575, 592]. Buchwald and co-workers also proved that in the case of 2′-monosubstituted biaryl phosphines, palladacycles **258** can form during the course of a Pd-catalyzed C–N bond-forming reaction. The transformation of the palladacycles into the catalytically active species is slow. Hence the rate of catalyst activation is low. In the case of 2′,6′-disubstituted biaryl ligands, palladacycle formation is avoided and hence optimal catalyst activity is achieved [593, 594]. In addition, the interactions of the aromatic π system with the Pd center as shown by Figure 1.4 (Section 1.3.1.1) is believed to promote catalyst stability and increase electron density at the Pd center.

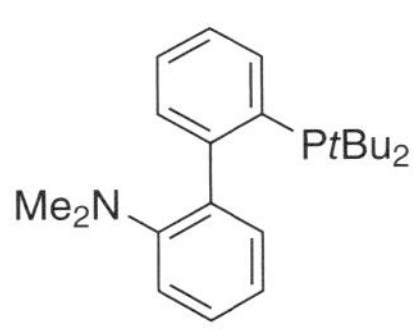

255

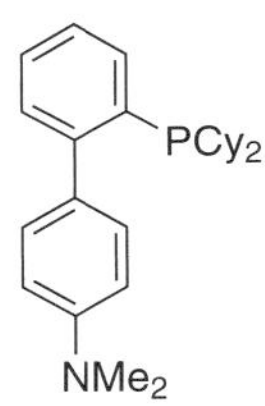

256

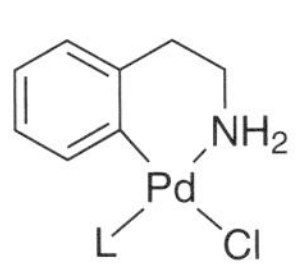

L = SPhos (**a**)
RuPhos (**b**)
XPhos (**c**)
BrettPhos (**d**)

257

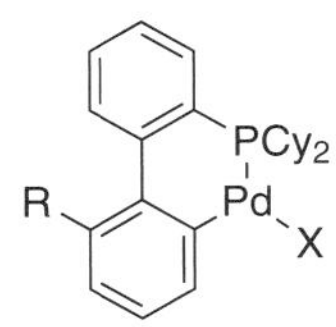

258

The combination of XPhos and Pd_2dba_3 catalyzes the coupling of aryl chlorides with alkyl or aryl amines as well as indole in alcohol or water [595]. The combination of P(*t*-Bu)$_2$-*o*-biphenyl (JohnPhos) and $Pd_2(dba)_3$ or $Pd(OAc)_2$ is a widely applicable catalyst that facilitates amination reactions of electron-neutral, -rich, and -poor aryl chlorides and heteroaryl chlorides with a variety of aliphatic and aromatic amines as well as benzophenone imine and benzophenone hydrazone. The coupling of bulky aryl chlorides such as 2,6-dimethylchlorobenzene with bulky arylamines such as 2,6-diisopropylaniline can also be performed using this catalyst

TABLE 1.11. Amination of Aryl Chlorides Catalyzed by Palladium with Buchwald Ligands

Entry	Pd Source (mol%)	Ligand (mol%)	Condition	Product	Yield (%)	Reference
1	Pd_2dba_3 (0.5)	XPhos (2)	K_2CO_3, *t*BuOH 100–110°C	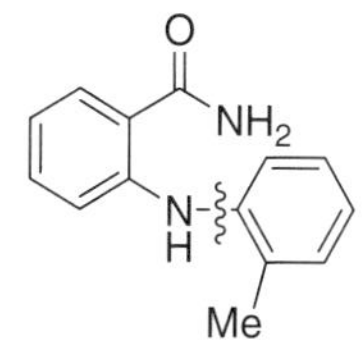	87	595
2	$Pd(OAc)_2$ (2)	XPhos	Cs_2CO_3, toluene 100–110°C	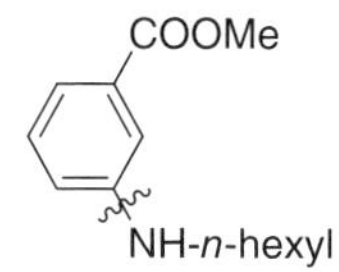	92	595
3	Pd_2dba_3 (0.5)	XPhos (2)	KOH, H_2O 110°C	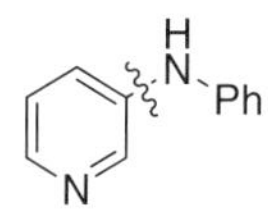	93	595
4	$Pd(OAc)_2$ (1)	JohnPhos (2)	NaO*t*Bu, toluene, 80°C, 23 h		92	596
5	Pd_2dba_3 (4)	JohnPhos (16)	NaO*t*Bu, toluene, 80°C, 20 h	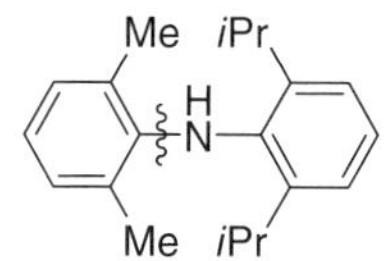	73	596

(*continued*)

TABLE 1.11. (*Continued*)

Entry	Pd Source (mol%)	Ligand (mol%)	Condition	Product	Yield (%)	Reference
6	Pd_2dba_3 (0.5)	JohnPhos (2)	NaO*t*Bu, toluene, 80°C, 2.5 h	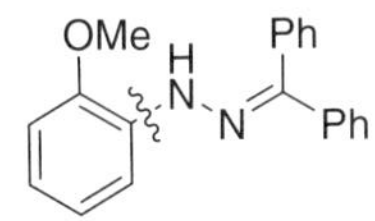	91	596
7	Pd_2dba_3 (1–2)	XPhos (4–8)	$LiN(TMS)_2$ THF, 65°C,	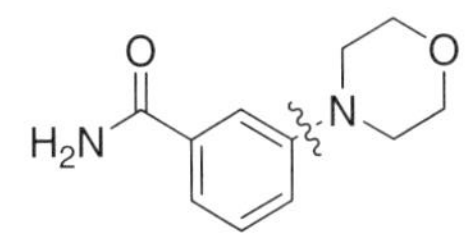	81	597
8	Pd_2dba_3 (2–2.5)	*t*BuXPhos (8–10)	NaO*t*Bu, toluene 80–100°C	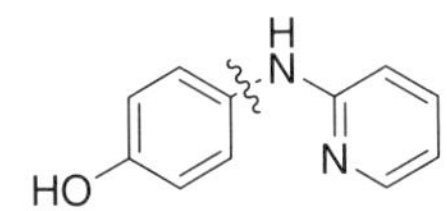	92	597
9	Pd_2dba_3 (2.5)	**243** (10)	K_3PO_4, toluene 100°C	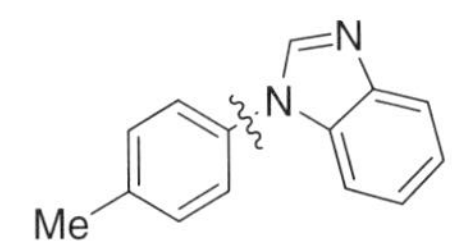	97	597
10	Pd_2dba_3 (0.5)	**14a** (1.5)	NaO*t*Bu, toluene 80°C	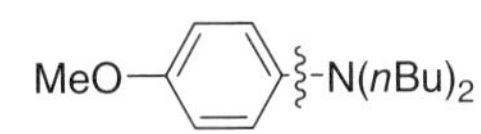	90	97

11	Pd_2dba_3 (0.5)	**14a** (1.5)	NaO*t*Bu, toluene 80°C	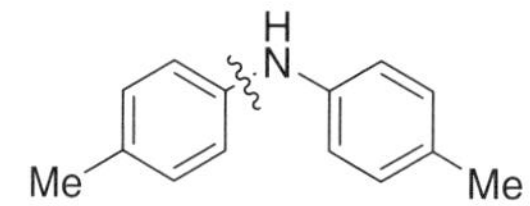	93	97
12	Pd_2dba_3 (0.5)	**256** (1.5)	NaO*t*Bu, toluene 100°C	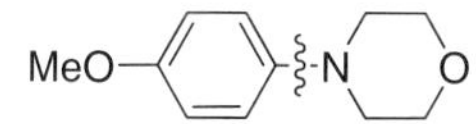	91	594
13	Pd_2dba_3 (0.5)	**256** (1.5)	NaO*t*Bu, toluene r.t.[a]	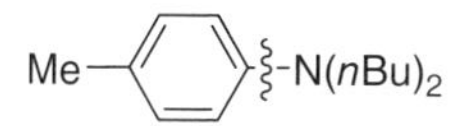	66	594
14	**225** (0.1)	XPhos (0.1)	NaO*t*Bu, dioxane 100°C, 10 min	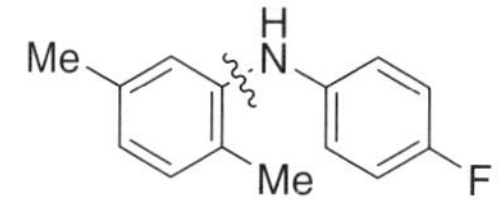	93	598
15	**225** (0.1)	XPhos (0.1)	NaO*t*Bu, dioxane, 100°C, 10 min	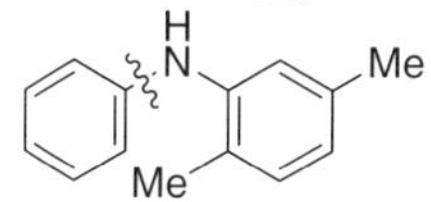	96	598
16	**257a**		NaO*t*Am, dioxane, r.t., 2 h	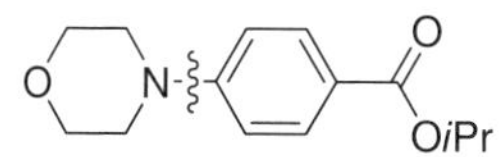	82	598
17	**257b**		NaO*t*Am, dioxane, r.t., 20 min	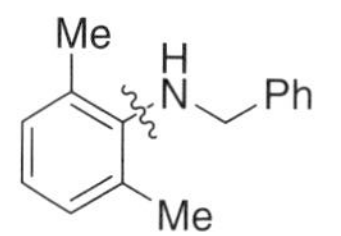	99	598

(*continued*)

TABLE 1.11. (*Continued*)

Entry	Pd Source (mol%)	Ligand (mol%)	Condition	Product	Yield (%)	Reference
18	**257d** (0.05)	**182** (0.05)	NaO*t*Bu, *n*Bu_2O, 80–100°C	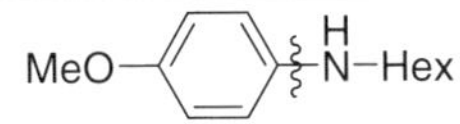	88	599
19	**257d** (0.01)	**182** (0.01)	NaO*t*Bu, *n*Bu_2O, 80–100°C	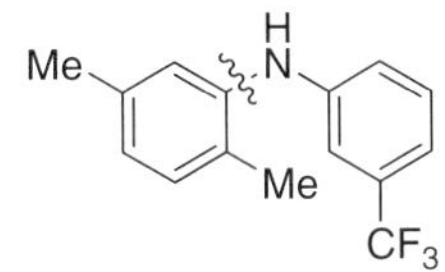	97	599
20	**257d** (0.01)	**15b** (0.01)	NaO*t*Bu, dioxane, 110°C	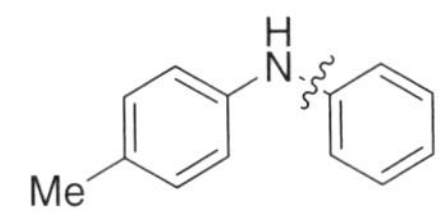	99	600
21	**257d** (0.05)	**15b** (0.05)	NaO*t*Bu, dioxane, 110°C	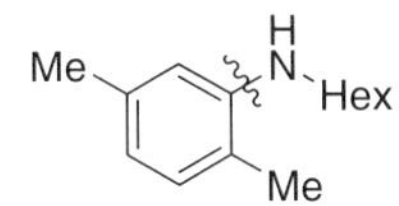	95	600

[a] r.t., room temperature.

system. The above reactions used NaO*t*Bu as the base. For the substrates bearing base-sensitive functional groups, the $Pd_2(dba)_3/PCy_2$-*o*-biphenyl/K_3PO_4 system is more suitable. The reaction conditions tolerate the presence of enolizable ketones, methyl esters, nitriles, and nitro groups [596, 601]. Sulfonamides and sulfamides were also *N*-arylated with aryl chlorides catalyzed by the combination of $Pd_2(dba)_3$ and XPhos or **14a** [602, 603]. Palladacycles **225** and **257** exhibit excellent catalytic properties. The coupling of aryl chlorides with aryl amines catalyzed by 0.1 mol% **225** can reach completion in 10 min in the presence of 0.1 mol% XPhos in dioxane at 100°C. Both **257a** and **257b** can drive the couplings at or below room temperature. This allows the presence of active functional groups in the reactants [598]. **257d** is highly active in catalyzing the coupling of primary aliphatic amines and anilines with aryl chlorides at low catalyst loadings [599]. Buchwald and co-workers also disclosed a catalyst incorporating effect. The mixed catalyst not only encompasses the substrate scope and activity that each catalyst exhibits alone, but also manifests its own reactivity. For example, in the synthesis of triarylamines the multiligand system was more efficient than a system based on either ligand independently [600].

Surry and Buchwald [604] developed new efficient systems for the palladium-catalyzed coupling of ammonia with aryl bromides or chlorides based on ligands Xphos and **255**, which allow the selective synthesis of either anilines or di- or triarylamines as desired. Triarylamines were also prepared from an aniline and two different aryl halides in "one-pot" catalyzed by the mixture of $Pd_2(dba)_3$ and JohnPhos [605]. The use of Pd_2dba_3 with bulky, electron-rich ligands **103a** or **103b** and $LiN(TMS)_2$ as the base can perform the coupling of amines with aryl bromides or chlorides containing alcohol, phenol, ketone, or amide groups [606]. A coupling of vinyl chlorides with amines was also carried out using Pd_2dba_3–**103b** as catalyst and NaO*t*Bu as base. Reaction of primary amines results in imines and secondary amines gives enamines (Scheme 1.50) [607].

This methodology was also used to construct nitrogen-containing heterocycles through intramolecular or intermolecular coupling of amines with aryl/vinyl chlorides, including oxindoles, thienopyrroles, azaindoles, indoles, carbazoles, indolo[3,2-*b*]carbazoles, and cyclic ureas [608–614]. For example, reaction of 2-halo-arylacetic acids and various alkylamines and anilines, followed by a $Pd(OAc)_2$/XPhos-catalyzed intramolecular amidation under aqueous conditions, yields *N*-substituted oxindoles [608]. A palladium-catalyzed reaction of *gem*-dichlororovinyl-thiophene or -pyridine and a boronic acid via a tandem intramolecular C–N and intermolecular Suzuki coupling process gives corresponding substituted thienopyrroles or azaindoles [Eq. (1.59)]. SPhos, XPhos, and DavePhos are effective ligands for this transformation [609–611]. A Rh-catalyzed

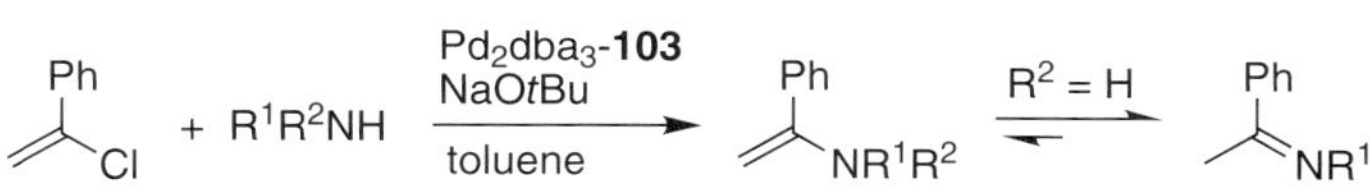

Scheme 1.50. $Pd_2(dba)_3$/**103**-catalyzed amination of vinyl chlorides.

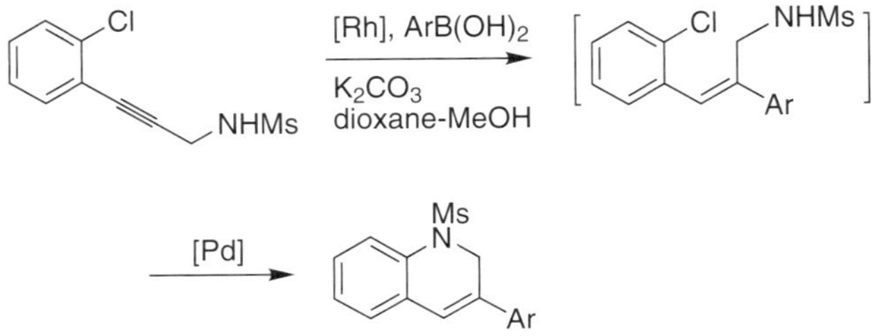

Scheme 1.51.

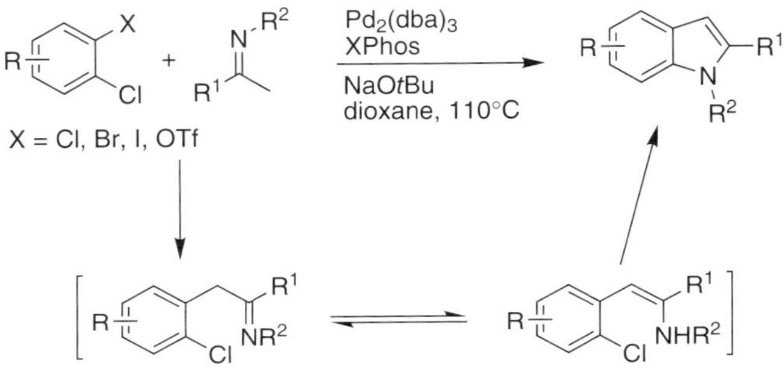

Scheme 1.52.

arylation of alkyne with $ArB(OH)_2$ followed by an intramolecular amination of aryl chlorides forms dihydroquinolines (Scheme 1.51). In this domino reaction, the arylated intermediates formed by a Rh-catalyzed process do not need to isolate. The intramolecular aryl amination can be carried out using $Pd(OAc)_2$–XPhos as catalyst and K_2CO_3 as base [612]. 2-(2-Haloalkenyl)aryl bromides or chlorides can be similar when coupling with primary amines catalyzed by the combination of Pd and Buchwald ligands, affording 1-substituted indoles. For the coupling of 2-(2-haloalkenyl)aryl chlorides with $PhNH_2$, Pd_2dba_3–**103b** as catalyst will produce the best results [Eq. (1.60)] [615]. The same strategy was used to construct 9-phenyl-9*H*-carbazole and indolo[3,2-*b*]carbazoles through palladium-catalyzed double *N*-arylation of aniline as shown in Eqs. (1.61) and (1.62), respectively [616]. The combination of palladium and Buchwald ligands also catalyzes a cascade process as shown in Scheme 1.52. The first step of the reactions is an intermolecular α-arylation of an imine, and the second step is an intramolecular aromatic amination, forming indoles as the final products. Pd_2dba_3 is a suitable palladium source, and XPhos was demonstrated to be the best ligand for this transformation [613].

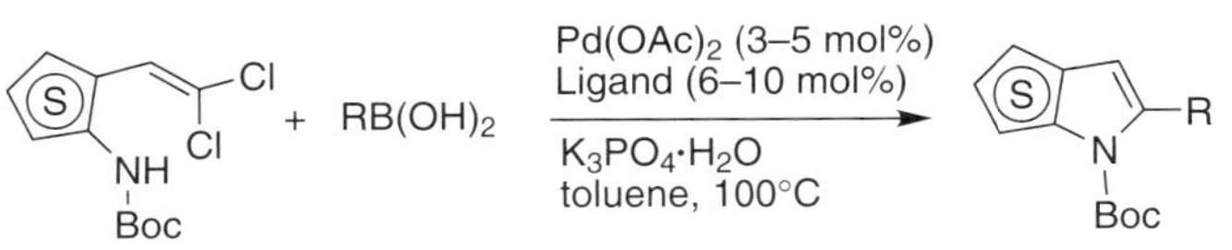

(1.59)

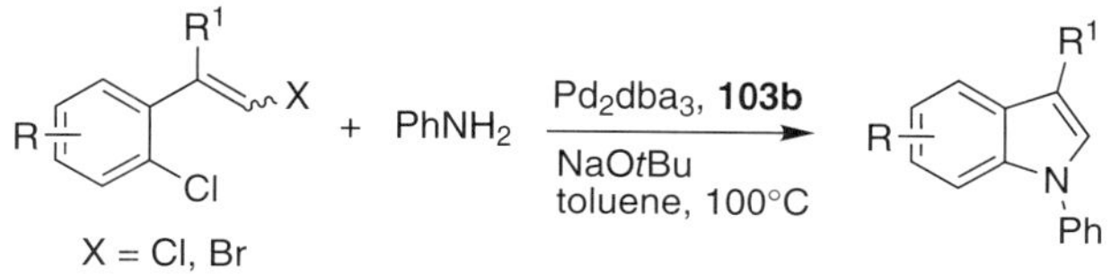
(1.60)

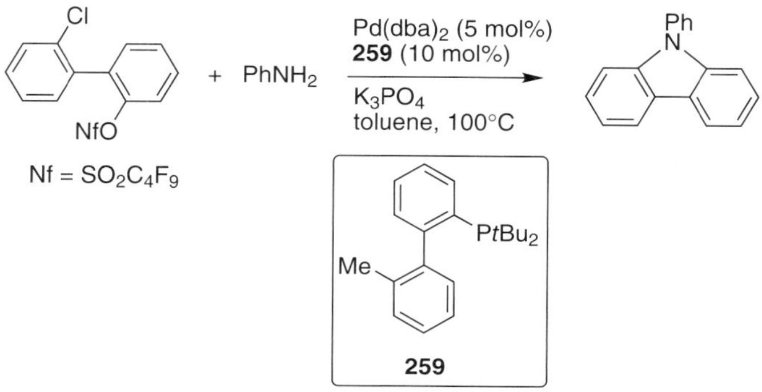
(1.61)

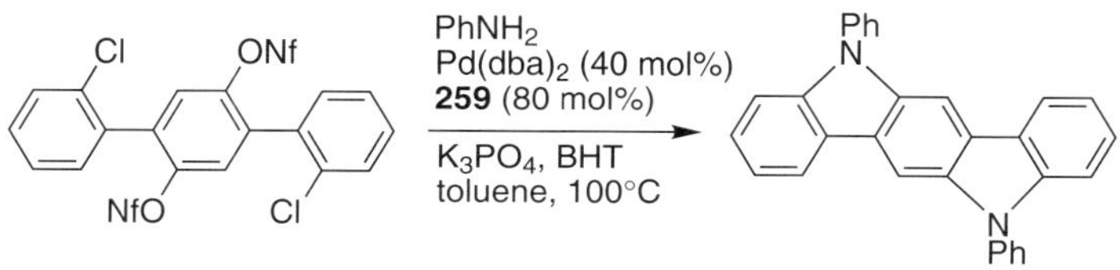
(1.62)

The catalyst systems composed of palladium and Buchwald ligands have been applied in pharmaceutical synthesis and modification of biomolecules through catalytic amination of aryl chlorides or bromides. Several amination processes with aryl chlorides have been developed on a larger scale [575].

Several dinuclear palladium complexes derived from Buchwald ligands and appropriate palladium compounds were also synthesized (Scheme 1.53). The first step of the reactions forms a palladacycle **260** as indicated by Buchwald. The palladacycle further reacts with Pd(cod)X_2 to yield dinuclear palladium complexes **261**. Complex **261a** (R = Me) was demonstrated to be an active catalyst for the amination of electron-rich aryl chlorides with aliphatic or aryl amines using NaO*t*Bu as base in THF at room temperature [617].

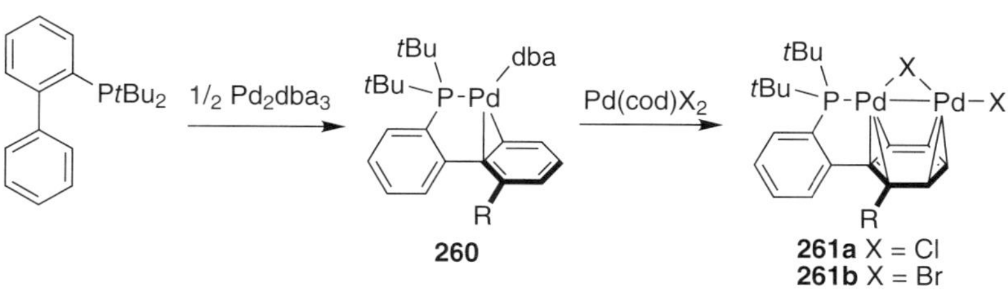

Scheme 1.53.

Beller and co-workers [114, 618] developed a series of heteroaryl-containing biaryl phosphine ligands such as **21**, **25**, **26**, and **262a–265**, which are excellent in palladium-catalyzed amination of chloroarenes. The combination of $Pd(OAc)_2$ and any of **262a–264** catalyzes amination of electron-rich and -poor aryl chlorides as well as heteroaryl chlorides with aliphatic and aromatic primary or secondary amines in the presence of NaO*t*Bu. Ligands **21** and **264–265b** in combination with $Pd(OAc)_2$ can catalyze the arylation of ammonia with aryl bromides in the presence of NaO*t*Bu. The unactivated and bulky aryl chlorides and heteroaryl chlorides are applicable as electrophilic substrates when **265a** is employed as the ligand, resulting in primary aromatic amines in good to excellent yields. The reactions are highly selective to form monoarylated products of ammonia. A subsequent reaction of the resulting aniline derivatives to the corresponding diaryl amines was not observed in the reactions using either bromides or chlorides [619]. Tomkinson and co-workers [620] carried out Pd-catalyzed coupling of hydroxylamines with aryl chlorides, bromides, and iodides using **266** as the ligand and Cs_2CO_3 as the base. In the catalytic reactions, only one example using aryl chloride, PhCl, was exhibited in moderate product yield.

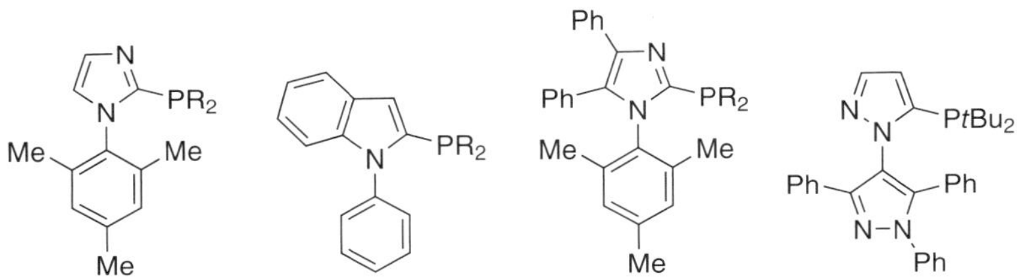

262a R = Cy
262b R = *t*Bu
263 R = Cy
264 R = *t*Bu
265a R = *t*Bu
265b R = Ad
266

Ferrocene-based phosphines are also highly effective in Pd-catalyzed amination reactions of chloroarenens. $Pd(dba)_2$/**34** system is an excellent catalyst for the amination of activated and deactivated aryl chlorides or bromides with aliphatic and aromatic amines in the presence of NaO*t*Bu or K_3PO_4 as base. Sterically hindered aryl chlorides such as 2,5-dimethylchlorobenzene and 2,6-dimethylchlorobenzene are applicable in this transformation. The system is compatible with various reactive functional groups on aryl halides, including CN, COOMe, PhC(O), and NO_2 [126]. Either the combination of didentate phosphine ligand **267** and $Pd(OAc)_2$ or the palladium complex **268** is a highly reactive and long-lived catalyst for coupling heteroaryl and aryl chlorides, bromides, and iodides with amines, especially primary amines. Most reactions take place in high yield with 0.001–0.1 mol% catalyst loadings. Electron-rich and sterically congested aryl chlorides can be employed as the electrophilic substrates [621–623]. Complex **268** also catalyzes coupling of aryl halides with ammonia or lithium amide, forming primary aromatic amines as the major products [624]. Hartwig and co-workers [622] indicated that the high efficiency of this catalyst system for the amination process should be ascribed to (1) a rigid backbone, (2) strong electron donation, and (3) steric bulk of the ligand. The rigidity of the backbone causes the ligand to bind to palladium tightly enough to prevent displacement by amines and basic heterocycles. Strong electron donation of the ligand promotes the oxidative addition of less reactive haloarenes. The steric hindrance of the ligand disfavors diarylation, facilitates the generation of the (chelate)Pd(0) intermediate, and promotes reductive elimination from the arylpalladium amido complexes [622].

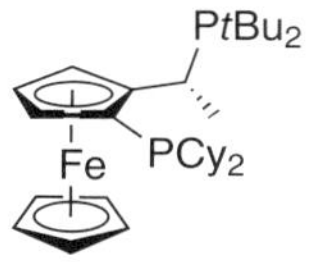

267

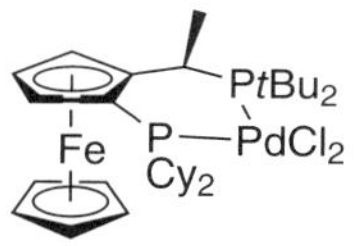

268

P,O-chelate ligand **6** in combination with $Pd(dba)_2$ shows good catalytic activity in the amination of aryl chlorides with primary and secondary aliphatic or aromatic amines in the presence of NaO*t*Bu [625]. Benzamide-derived phosphines **48** and **269** in combination with $Pd(OAc)_2$ are very effective catalysts for the amination of electron-rich, electron-poor, and sterically hindered aryl or heteroaryl chlorides with various amines such as primary and secondary aliphatic and aromatic amines and secondary cyclic amines. An alkenyl chloride was also coupled with PhNHMe using $Pd(OAc)_2$/**48** as catalyst. The catalytic system tolerates a variety of functional groups on chlorides including CN, COOMe, OH, MeC(O), NO_2, and $P(O)Ph_2$ as well as nitrogen-containing heterocycles. The catalyst loadings can be as low as 0.01 mol% [626]. Ligand **270** is similar to Buchwald ligands and shows good catalytic property in combination with $Pd(dba)_2$ or $Pd(OAc)_2$, which drives coupling of electron-rich and -poor aryl chlorides with a variety of aliphatic or aromatic amines with 1–2 mol% Pd loadings [627]. $Pd_2(dba)_3$–**42** is also an excellent catalyst for the aromatic amination of chloroarenes with aliphatic and aromatic amines [134]. P,N-chelate ligands **271a** and **271b** in combination with $[Pd(allyl)Cl]_2$ or $[Pd(cinnamyl)Cl]_2$ are good catalysts for the coupling of aryl and heteroaryl chlorides with various amines and related substrates such as primary alkyl- and arylamines, cyclic and acyclic secondary amines, N–H imines, hydrazones, lithium amide, and ammonia. Electron-rich and -poor aryl chlorides with functional groups such as RO, CF_3, PhC(O), COOH, COOMe, Ac, MeC(O)NH, and vinyl groups were employed in this transformation. Chemoselectivity of $[Pd(cinnamyl)Cl]_2$/**271b** in catalyzing amination of aryl chlorides was also studied. Primary amines exhibit higher reactivity than secondary amines in the same reaction system. Reaction of 1,4-bromochlorobenzene with 4-anisidine catalyzed by $[Pd(cinnamyl)Cl]_2$/**271b** is also chemoselective. The amination at the bromide position of 1,4-bromochlorobenzene was achieved in 76% isolated yield [628]. A study of chemoselectivity of amination of aryl chlorides using $[Pd(cinnamyl)Cl]_2$/**272** was also carried out. Through competition reactions the relative preference of this catalyst system for amine coupling partners is: linear primary alkylamines and imines > unhindered electron-rich primary anilines, primary hydrazones, *N*,*N*-dialkylhydrazines, and cyclic primary alkylamines > unhindered electron-deficient primary anilines, α-branched acyclic primary alkylamines, hindered electron-rich primary anilines ≫ cyclic and acyclic secondary dialkylamines, secondary alkyl/aryl and diarylamines, α,α-branched primary alkylamines, and primary amides. Utilizing the selectivity of $[Pd(cinnamyl)Cl]_2$/**272** system, a series of di-, tri-, and tetraamines have been synthesized (Table 1.12) [629].

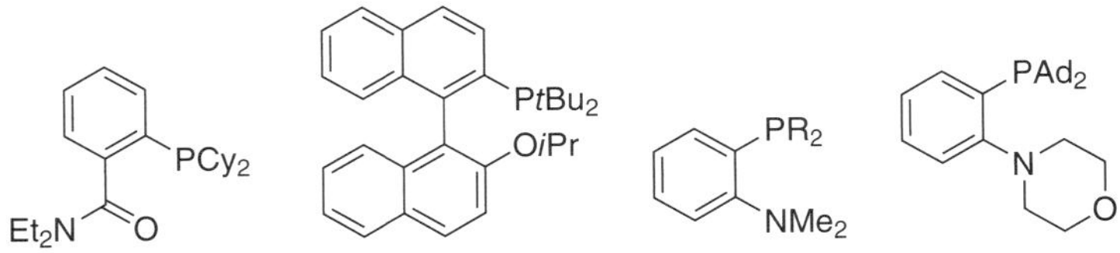

269 **270** **271a** R = *t*Bu **271b** R = 1-Ad **272**

Diaminochlorophosphines **62** or **273** in combination with $Pd(dba)_2$ are good catalysts for aromatic amination using chloroarenes. $Pd(dba)_2$–**62** exhibits higher catalytic activity and is suitable for wider range of chloride and amine substrates, especially for the very bulky substrates [e.g., Eq. (1.63)] [460, 630]. The combination of $Pd_2(dba)_3$ and di(dicyclohexylamino)phenylphosphine catalyzes amination of aryl or heteroaryl chlorides with primary and secondary aliphatic or aromatic amines in the presence of KO*t*Bu [631]. The $Pd_2(dba)_3$/**177b** system is also effective for the amination reactions. Studies show that much better results can be achieved at 100°C than at 80°C using the catalyst system [632, 633]. $Pd_2(dba)_3$/**177b** also catalyzes amination of vinyl bromides and chlorides. Although mostly bromides have been employed in this amination reaction, the exploration for chloride substrates is only preliminary. 1-Chlorocyclopentene was coupled with morpholine or *N*-methylaniline at 115°C using realtively high catalyst loading (2.5 mol% $Pd_2(dba)_3$ and 5 mol% **177b**), affording corresponding aminated products in excellent yields [634]. Phosphine ligand **185** derived from **177b** is also effective in Pd-catalyzed amination of aryl chlorides and bromides. The catalytic system composed of Pd and **185** is very effective for a variety of aryl chlorides and amines, including electron-rich, electron-poor, and sterically hindered aryl chlorides and primary and secondary aliphatic or aromatic amines. Very congested diaryl amines can be synthesized using this catalyst system with low Pd loadings. The system can tolerate a series of reactive functional groups. Several vinyl chlorides and bromides were also coupled with aliphatic or aromatic amines in excellent yields using the catalyst system [635].

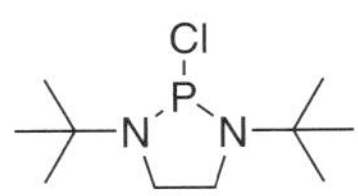

273

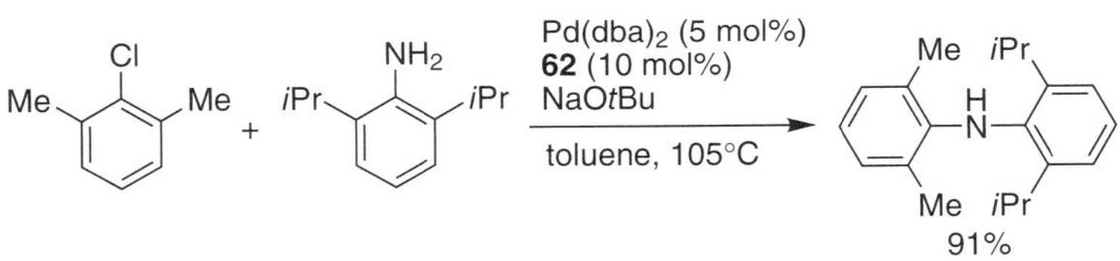

(1.63)

TABLE 1.12. Chemoselective Amination of aryl Chlorides

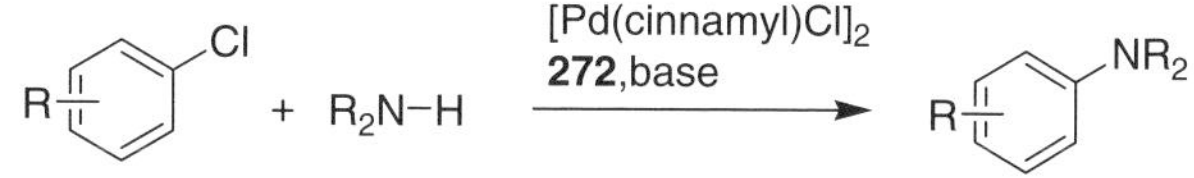

Entry	Aryl Chloride	Product	Conditions	Yield (%)
1	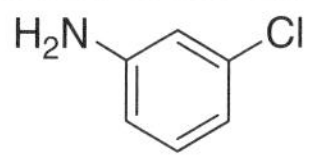	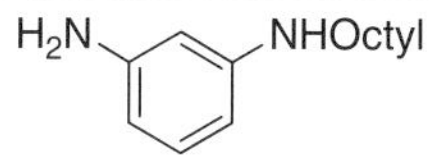	1 mol% Pd, 2 mol% **272**, NaO*t*Bu, toluene, 110°C	74
2	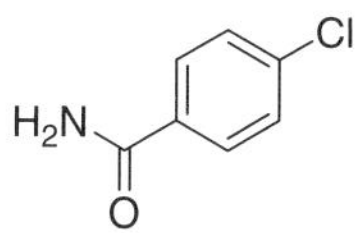	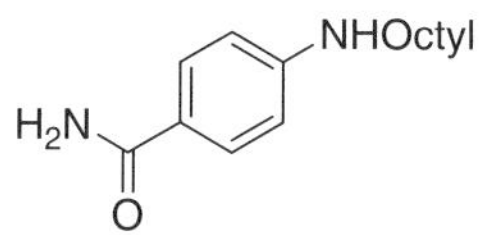	2 mol% Pd, 4 mol% **272**, LiHMDS, NaO*t*Bu (8%), dioxane, 65°C	74
3	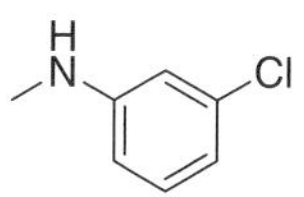	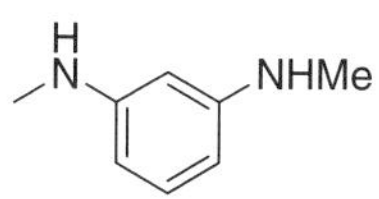	1 mol% Pd, 2 mol% **272**, NaO*t*Bu, toluene, 110°C	83
4	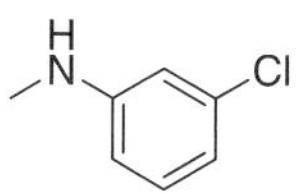	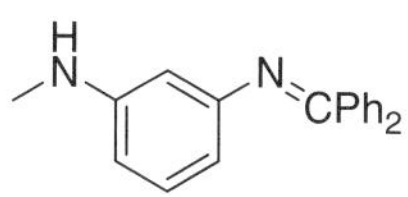	1 mol% Pd, 2 mol% **272**, NaO*t*Bu, toluene, 110°C	92

(*continued*)

TABLE 1.12. (*Continued*)

Entry	Aryl Chloride	Product	Conditions	Yield (%)
5	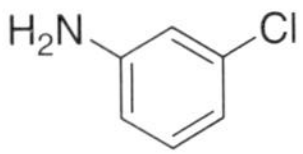	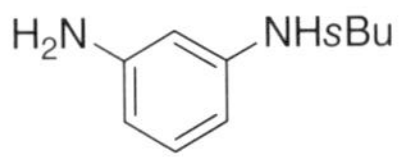	1 mol% Pd, 2 mol% **272**, NaO*t*Bu, toluene, 110°C	81
6	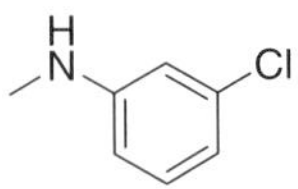	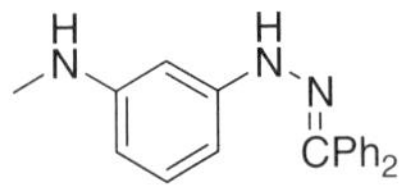	1 mol% Pd, 2 mol% **272**, NaO*t*Bu, toluene, 110°C	80
7	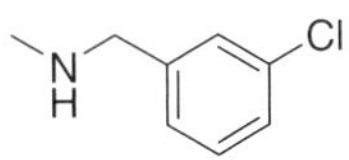	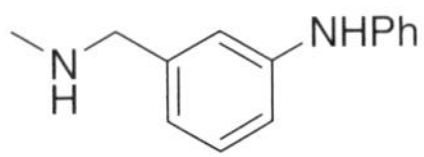	1 mol% Pd, 2 mol% **272**, NaO*t*Bu, toluene, 110°C	82
8	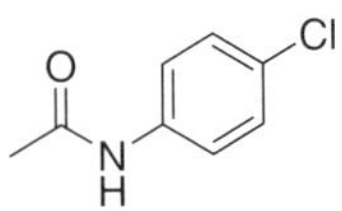	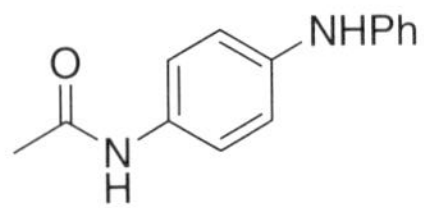	2 mol% Pd, 4 mol% **272**, LiHMDS, NaO*t*Bu (8%), dioxane, 65°C	94
9	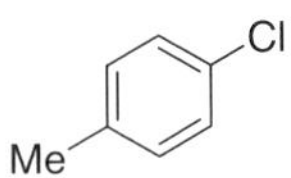	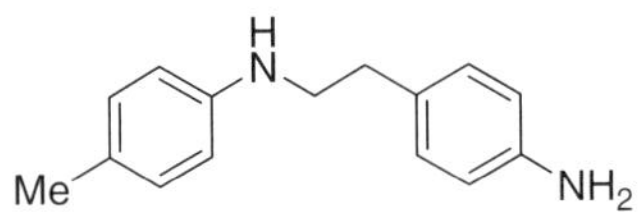	1 mol% Pd, 2 mol% **272**, NaO*t*Bu, toluene, 110°C	69

10	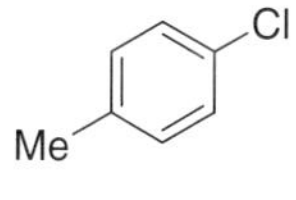	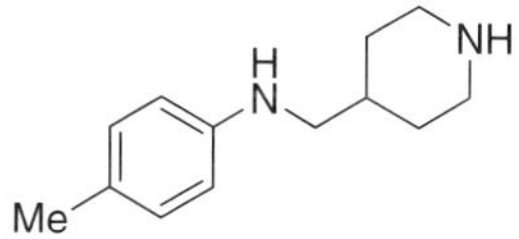	1 mol% Pd, 2 mol% **272**, NaO*t*Bu, toluene, 110°C	81
11	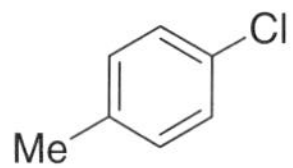	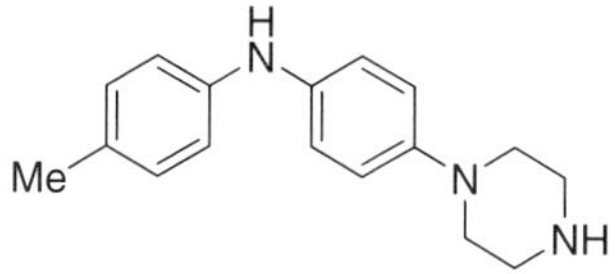	1 mol% Pd, 2 mol% **272**, NaO*t*Bu, toluene, 110°C	79
12	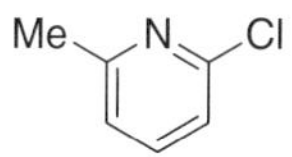	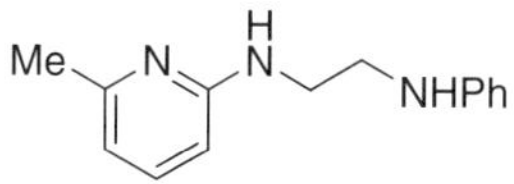	1 mol% Pd, 2 mol% **272**, NaO*t*Bu, toluene, 110°C	94
13	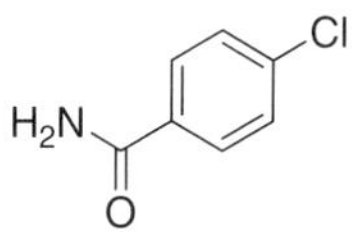	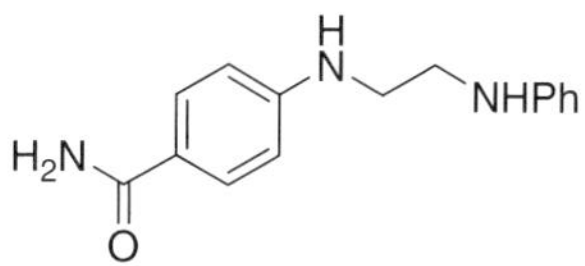	5 mol% Pd, 10 mol% **272**, LiHMDS, NaO*t*Bu (8%), dioxane, 65°C	78

(ii) NHC LIGANDS. Pd–NHC–Catalyzed C–N bond formation has been reported in recent years [17]. The combination of IPr·HCl and $Pd_2(dba)_3$ or NHC–Pd complex $[Pd(IPr)Cl_2]_2$ is an effective catalyst for the amination of aryl chlorides or bromides in the presence of an appropriate base such as KO*t*Bu and KO*t*Am. The catalysts are suitable for electron-rich and electron-poor aryl chlorides and a wide variety of amines such as cyclic and acyclic primary and secondary alkyl or arylamines. Triarylamines can also be synthesized through coupling of primary arylamines with aryl chlorides catalyzed by using $Pd_2(dba)_3$/IPr·HCl [636–638]. The $Pd(OAc)_2$/SIPr·HCl system was found to catalyze an intermolecular *N*-arylation of bulky alkyl or aryl amines with *ortho*-alkynylchloroarenes and then an intramolecular hydroamination to form indoles [Eq. (1.64)] [639]. The combination of SIPr·HCl and $Pd(dba)_2$ exhibits higher catalytic activity than those containing IPr ligand. It can catalyze amination of aryl and heteroaryl chlorides at room temperature with low Pd loading, including deactivated aryl chlorides such as 4-chloroanisole. Primary and secondary aliphatic and aromatic amines as well as imine are applicable for this coupling [640].

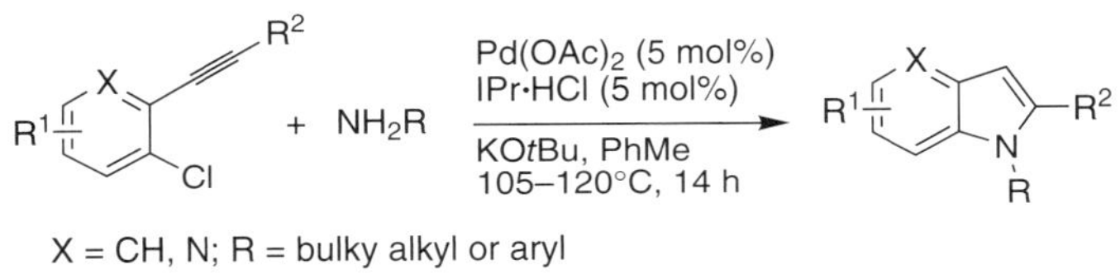

(1.64)

NHC–Pd complexes with anionic acac, allyl, or Cp auxiliary ligands such as [(IPr)(acac)PdCl], **87b**, **87c**, and **88a–88c** are also excellent catalysts for the amination of aryl chlorides. They catalyze coupling of activated and deactivated aryl chlorides with various amines [182, 464, 641, 642]. (NHC)Pd(allyl)Cl, especially **88b**, exhibits extremely high activity. Complex **88b** can catalyze the coupling reaction of primary, secondary, alkyl, or aryl amines with unactivated, neutral, and activated aryl chlorides and bromides at room temperature in minutes. When the reaction temperature was raised to 80°C, this system could be effective with as low as 10 ppm of catalyst. This catalyst is also excellent for extremely hindered substrates; tetra-*ortho*-substituted diarylamines can be produced without loss of activity [641].

NHC–Pd complexes with neutral auxiliary ligands such as 1-methylimidazole and 3-chloropyridine, **85** and **86**, have been reported to be active catalysts in the amination of aryl chlorides. Both complexes effectively catalyze coupling of electron-rich and -poor aryl chlorides with primary, secondary, alkyl, or aryl amines using KO*t*Bu as base [643, 644]. Complex **86** is more active. It can drive the coupling at room temperature or 50°C and leads to formation of very congested arylamines such as tetra-*ortho*-substituted diarylamines (Scheme 1.54) [644].

Recently, Tu et al. [645] synthesized and characterized complex **274**, which is structurally similar to **86**. This complex exhibits higher catalytic activity than does **86** in amination of aryl chlorides. It drives coupling of electron-rich and -poor aryl chlorides with primary and secondary amines with lower catalyst loadings (0.5 mol%) and results in excellent yields. It is also extremely effective in catalyzing

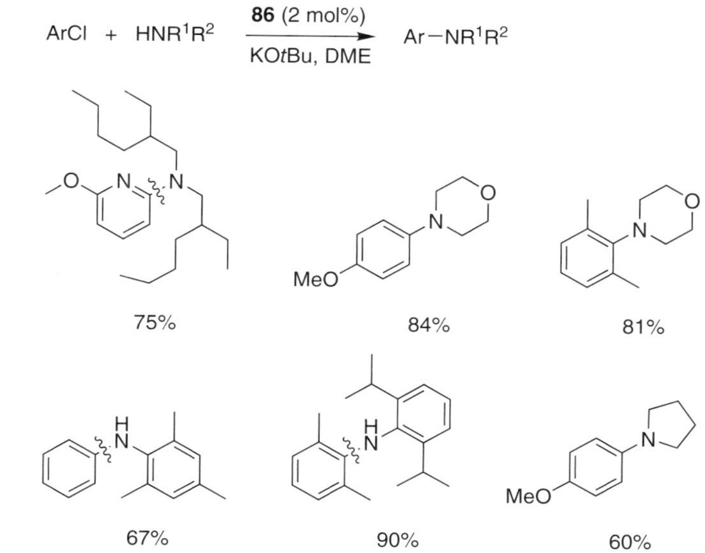

Scheme 1.54. **86**-catalyzed amination of aryl chlorides.

the formation of sterically congested aryl amines such as tetra-*ortho*-substituted diarylamines [645]. Both palladacycle complexes **102** and **275** were demonstrated to be very effective in catalyzing amination of aryl chlorides. The ferrocene-based palladacycle **102** also efficiently catalyzes coupling of bulky aryl chlorides with bulky amines, affording tri- or tetra-*ortho*-substituted diarylamines in excellent yields [646, 647]. A ferrocene-based C,P-didentate ligand precursor **276** in combination with $Pd(OAc)_2$ can catalyze amination of aryl bromides or chlorides. But for the deactivated aryl chlorides, this catalyst system displays only moderate activity [648].

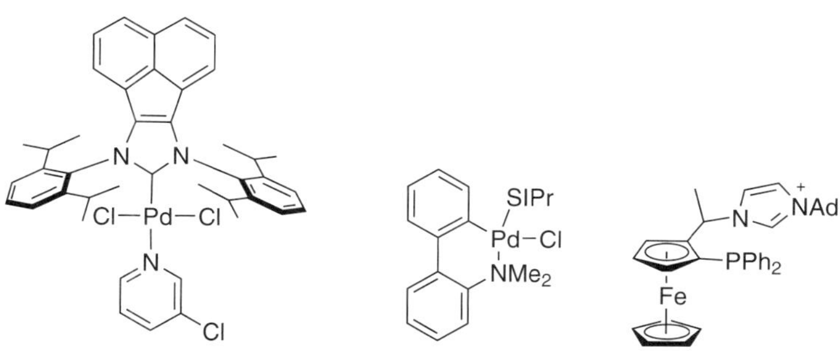

274 **275** **276**

1.4.3 Nickel Catalysts

The initial report of nickel-catalyzed *N*-arylation of chloroarenes with secondary aromatic and aliphatic amines appeared in 1997. The reaction employed dppf or

1,10-phenanthroline as ligand and NaO*t*Bu as base [649]. Since then, a series of studies on nickel-catalyzed-amination of aryl chlorides have been reported [13]. Combination of Ni(0) generated *in situ* and IPr·HCl or SIPr·HCl effectively catalyzes amination of aryl chlorides in the presence of NaO*t*Bu. Cyclic and acyclic alkyl amines and arylamines can be employed in this reaction. Electron-donating, electron-withdrawing, and bulky aryl chlorides are applicable for this transformation [650, 651]. An intramolecular amination of aryl chlorides under similar conditions was also carried out in good yields [Eq. (1.65)]. Combination of Ni(0) and bpy also catalyzes this transformation and leads to comparable yields in most cases [652].

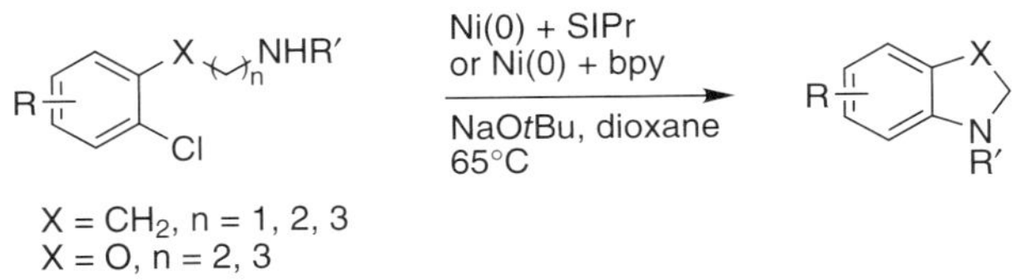

(1.65)

$(Ph_3P)_2Ni(Cl)Ar$ can be employed as a nickel source [653]. In the catalytic process, it could be transferred into a NHC–Ni–phosphine complex that catalyzes the amination of aryl chlorides. Indeed, $(IPr)Ni(PPh_3)$ catalyzes coupling of alkyl or arylamines with aryl bromides or chlorides, although it leads to relatively low yield when aryl chlorides are used as the electrophiles [470].

2,2-Bipyridine is also a suitable ligand for Ni-catalyzed amination of aryl chlorides [654–656]. Recently, 3,5,6,8-tetrabromo-1,10-phenanthroline was used as the ligand for the Ni-catalyzed amination of aryl chlorides. The reaction employs $Ni(acac)_2$ as the nickel source and polymethylhydrosiloxane as a reducing agent, suitable for a wide range of substrates with low catalyst loading [657].

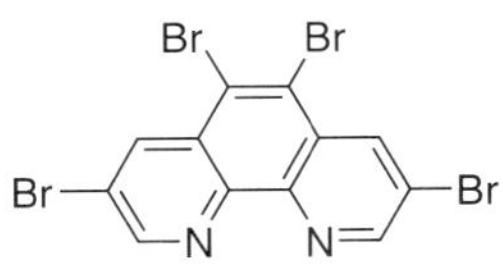

277

Lipshutz and co-workers [279, 658, 659] demonstrated that a nickel-on-charcoal/dppf system efficiently catalyzes amination of chloroarenes using LiO*t*Bu as base in dioxane at 130°C. The amination and Kumada coupling results and spectroscopic analyses including TEM, EDX, ICP-AES, and React-IR reveal that catalysis using nickel-on-charcoal is most likely of a homogeneous rather than heterogeneous nature. A bimetallic system consisting of copper and nickel oxide particles supported within charcoal is also active for the amination of aryl chlorides in the presence of dppf. The catalytic reaction employs LiO*t*Bu as base and uses microwave heating in dioxane, giving excellent yields for several activated and unactivated aryl chloride substrates [660].

1.4.4 Iron and Cobalt Catalysts

A combination of Fe_2O_3, L-proline, and NaO*t*Bu catalyzes coupling of aryl halides with amines in DMSO at 135°C. However, using this catalyst system, only the reaction of PhCl with morpholine was carried out in 51% yield [661]. $CoCl_2$/dppp catalyzes coupling of secondary aliphatic amines with 2-chloropyridine derivatives, 2-chloroquinoline and 2-chloroquinoxaline, using K_2CO_3 as base [662]. From the results reported so far, iron and cobalt catalysts for the amination of aryl chlorides are much less effective than palladium and nickel systems.

1.5 FORMATION OF C–O BONDS

The formation of diaryl ethers and phenols via a C–O cross-coupling reaction represents a powerful and straightforward method in organic synthesis [663]. For roughly the past decade, considerable efforts toward developing methodologies for constructing diaryl ethers have been devoted by synthetic chemists. Metal-catalyzed cross-coupling reactions between aryl halides and alcohols in the presence of a base are of tremendous interest for products relevant to both academic research and industry. The transition-metal-catalyzed C–O bond formation has been discussed in the reviews [545, 550, 664, 665].

1.5.1 Copper Catalysts

Copper-catalyzed Ullmann ether synthesis has been studied for many years because aryl ethers comprise important classes of medicinally active compounds and agrochemicals. The traditional Ullmann-type reactions required high temperatures, use of polar and high-boiling solvents, and stoichiometric quantities of the copper salt [666]. Development of new ligands leads to Cu-catalyzed C–O bond formation more efficiently under milder conditions.

(i) N,N-LIGANDS. 1,10-Phenanthroline has been demonstrated to be an effective ligand for the Cu-catalyzed C–O coupling. Cu/C/1,10-phen catalyzes coupling of aryl bromides and chlorides with phenol derivatives in the presence of Cs_2CO_3 under microwave radiation, but only one example for the coupling of activated aryl chloride was reported [667]. CuI/1,10-phen catalyzes intramolecular C–O bond formation of *ortho*-haloanilides in refluxed DME in the presence of Cs_2CO_3 [Eq. (1.66)]. The chloride shows lower reactivity than the bromides, the former requiring longer reaction time for the transformation [668]. This intramolecular C–O coupling is also catalyzed by CuI/DMEDA. The reaction can start from *o*-dihalobenzenes and primary amides as shown in Eq. (1.67). In this domino process, the Cu-catalyzed C–N coupling occurs at first, and then further cyclization takes place through C–O coupling [669].

The catalyst system also catalyzes an intramolecular C–O coupling of vinyl bromides or chlorides [Eq. (1.68)] [670].

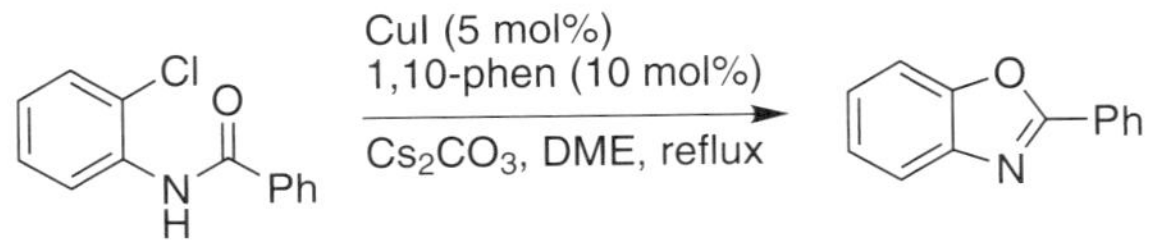

(1.66)

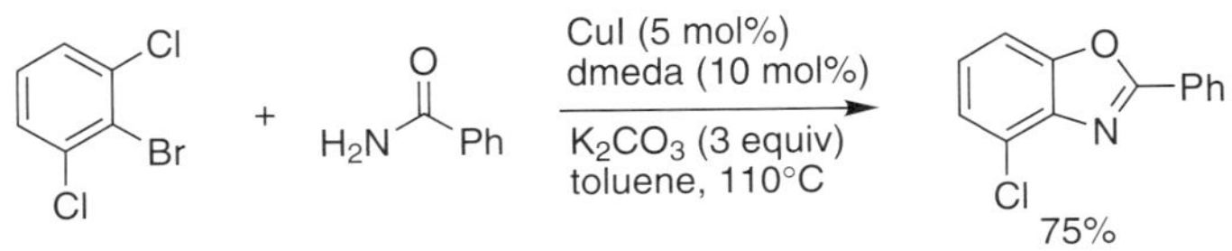

(1.67)

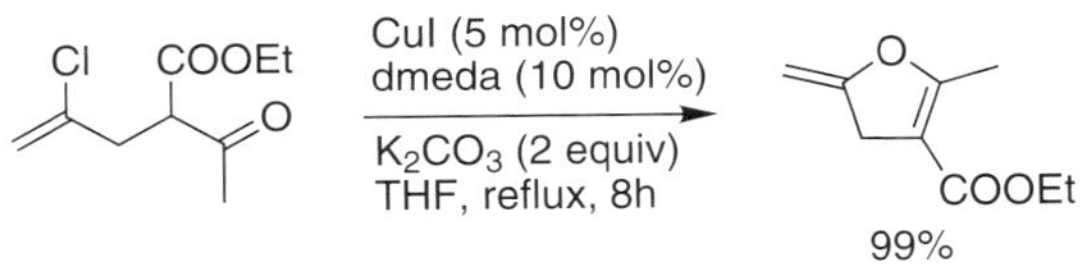

(1.68)

Silica-supported copper complex **278** is also effective in catalyzing coupling of aryl bromides, iodides, or activated aryl chlorides with phenols in the presence of KF. The silica-supported copper could be recovered and recycled by a simple filtration of the reaction solution and used for 10 consecutive trials without loss of its reactivity [663].

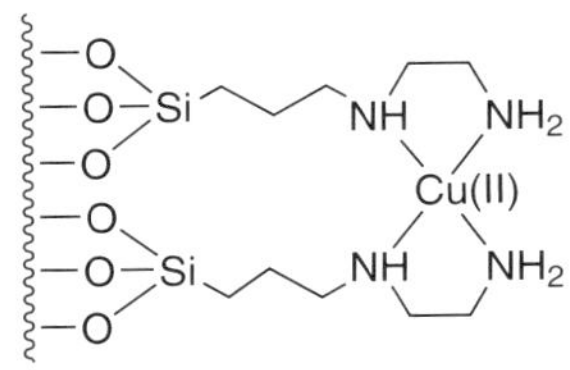

278

(ii) O,O-LIGANDS. Xia and Taillefer [671] proved that the combination of $Cu(acac)_2$ and 2,2,6,6-tetramethyl-3,5-heptanedione is active for the coupling of electron-rich and -poor aryl chlorides with phenols using Cs_2CO_3 as base. In most cases, good to excellent yields can be obtained. Jiang and co-workers [672] employed **279** as the ligand and carried out CuI-catalyzed hydroxylation of aryl iodides, bromides, and chlorides using CsOH as base. The catalyst system is effective for activated and deactivated aryl chlorides and tolerates reactive functional groups such as CHO, CN, COOH, and NO_2. This protocol also provides a one-pot procedure for the direct formation alkyl aryl ethers from aryl halides and alkyl bromides or iodides [672].

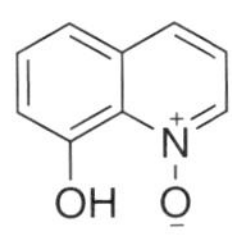

279

(iii) O,N- OR O,S-LIGANDS. (2-Pyridyl)acetone/CuBr catalyzes coupling of aryl chlorides, bromides, and iodides with phenols using Cs_2CO_3 as base in DMSO. For bromides and iodides the reaction results in good to excellent yields in most cases. However, for chloride substrates, only activated ones give good yields. The unactivated and deactivated chlorides give only low to moderate yields. 2-Chloropyridine and 2-chloroquiniline exhibit good reactivity. Their coupling with phenols affords corresponding products in 72–88% yields [673]. Copper(I) thiophene carboxylate (CuTC) complex **280** can catalyze intramolecular C–O coupling of aryl chlorides and bromides under microwave heating, affording cyclization products in 81–99% yields [Eq. (1.69)] [674].

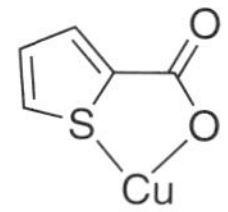

280

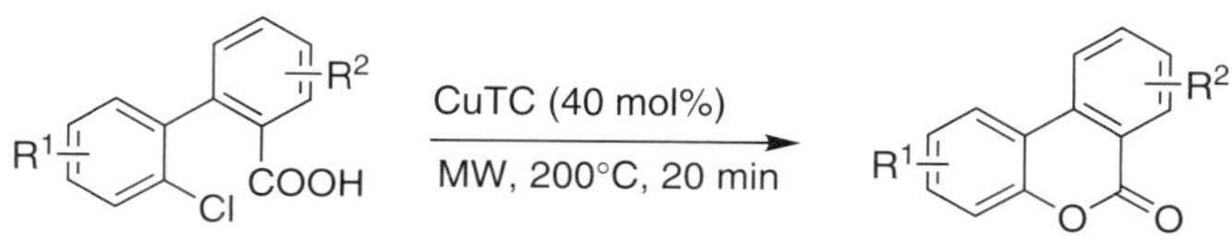

(1.69)

Nano CuO and Cu_2O-nanotube were reported to catalyze the Ullmann-type *O*-arylation of aryl halides with phenols. This methodology offers the competitive advantages of recyclability of the catalyst without loss of catalytic activity. However, nano CuO shows low activity for aryl chlorides [675, 676].

1.5.2 Palladium Catalysts

Buchwald biaryl phosphine ligands are very effective in Pd-catalyzed coupling of aryl chlorides with phenols or alcohols. Ligands **243**, *t*BuXPhos, and **281–283** in combination with an appropriate palladium source effectively catalyze cross-coupling of electron-rich and electron-poor aryl chlorides with various phenol derivatives to form biaryl ethers. A range of functional groups and heteroaromatic substrates can be tolerated using these catalyst systems. A structure–activity relationship study suggests that the ligand bulk plays an important role in C–O bond-forming reactions [677, 678].

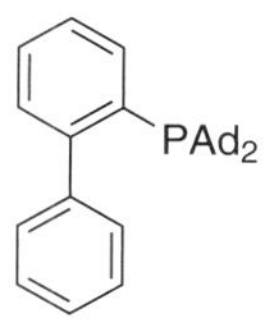

281

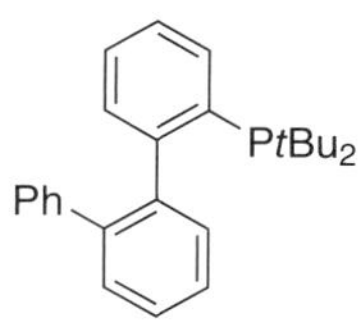

282

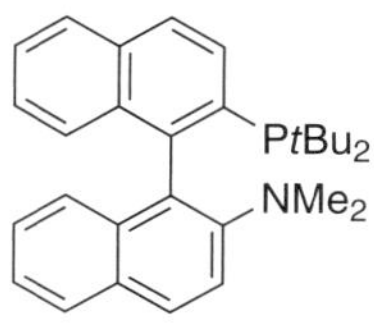

283

The combination of palladium and Buchwald ligands is also effective in catalyzing coupling of aryl chlorides with alcohols. $Pd(OAc)_2$/**259** catalyzes the reaction of deactivated aryl chlorides or bromides with KO*t*Bu to afford aryl *tert*-butyl ether in good to excellent yields [679]. The combination of $Pd(OAc)_2$ and **283** or **284** drives the C–O coupling of aryl chlorides or bromides with primary alcohols such as MeOH, EtOH, BnOH, *n*BuOH, $CH_2{=}CH(CH_2)_2OH$, and $CH_2{=}CH(CH_2)_3OH$ [680]. Recently, Buchwald and co-workers [681] revealed $[(allyl)PdCl]_2$/**285** to be excellent for the coupling of aryl chlorides or bromides with primary and secondary alcohols. This catalyst system is suitable for wide substrate scope of both the aryl halide and alcohol coupling partners. Catalytic studies using this and related ligands show that the substituent in the 6-position of the biarylphosphine ligand scaffold has a profound effect on the catalytic activity of these systems [681]. $Pd(OAc)_2$/**284** was proven to catalyze an intramolecular C–O coupling of aromatic halo alcohols in the presence of Cs_2CO_3. Aryl chlorides and primary and secondary alcohol substrates can be employed in this transformation. The reactions involving aryl bromides bearing stereogenic centers at the carbinol carbon can be effected without racemization via reversible β-hydride elimination. The analogous aryl chloride substrates can also be efficiently cyclized with this catalyst; however, small amounts of epimerization may be observed [682, 683].

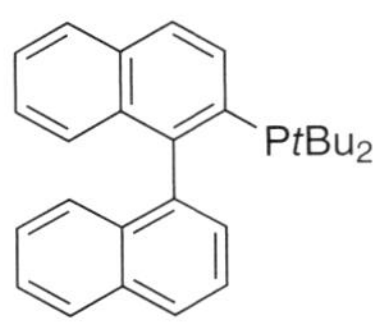

284

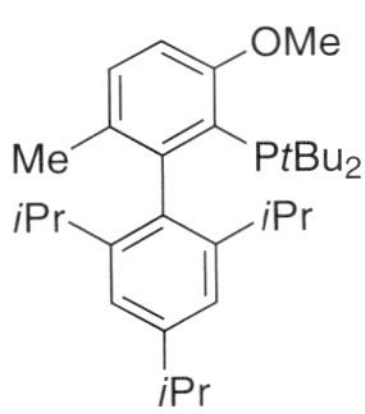

285 (RockPhos)

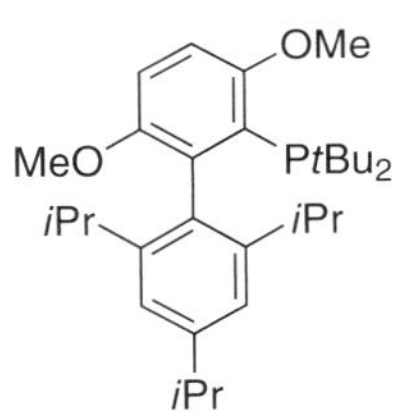

286 (*t*BuBrettPhos)

BrettPhos (**182**) or *t*BuBrettPhos (**286**) in combination with $[(allyl)PdCl]_2$ catalyzes *O*-arylation of ethyl acetohydroximate. $[(allyl)PdCl]_2$/*t*BuBrettPhos is effective for the reactions using aryl chlorides and iodides. The *O*-arylated ethyl acetohydroximates are readily hydrolyzed to give *O*-arylhydroxylamines [684]. The combination of $Pd_2(dba)_3$ and *t*BuXPhos or **243** effectively catalyzes the reaction of aryl chlorides or bromides with NaOH to prepare phenols after being neutralized with HCl. If the initially formed phenoxide is treated with alkyl halides in the presence of the phase-transfer catalyst, alkyl aryl ethers can be achieved in good yields (Scheme 1.55). Treatment of 2-chloroaryl alkynes with KOH under the same conditions affords benzofurans due to further catalytic cyclization of alkynyl phenols [Eq. (1.70)] [685].

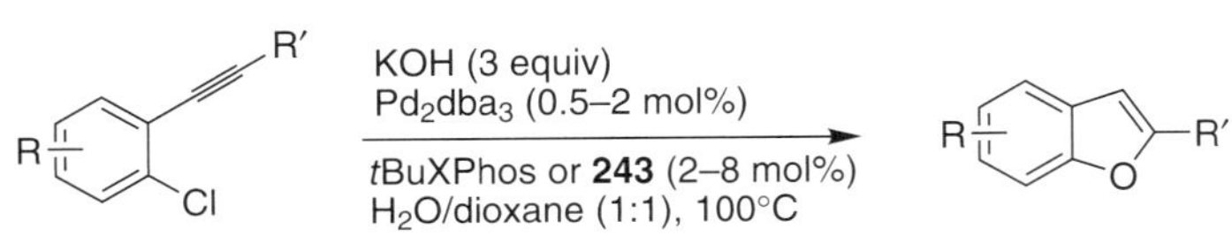

(1.70)

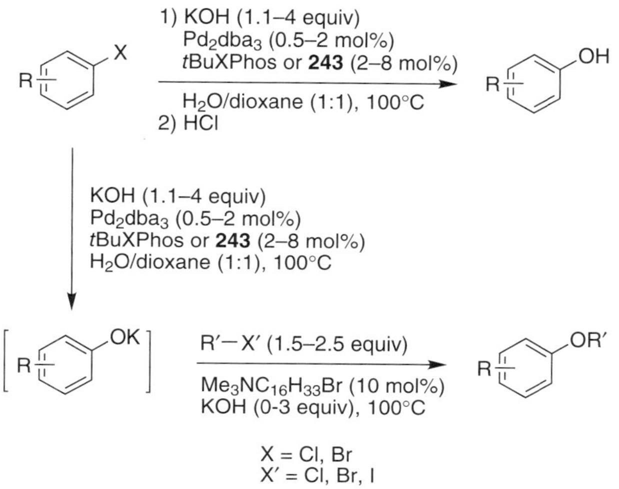

Scheme 1.55. Conversion of aryl halides to phenols or alkyl aryl ethers.

Beller and co-workers [686–689] examined a series of heteroaryl-containing biaryl phosphine ligands. Some of them in combination with an appropriate palladium source display excellent catalytic properties in the C–O bond formation from aryl chlorides or bromides. Ligands **21**, **264**, and **287–289** are effective for the Pd-catalyzed C–O coupling of aryl chlorides. The combination of **289** and $Pd(OAc)_2$ catalyzes coupling of aryl bromides or chlorides with phenols using K_3PO_4 as base [687]. The combination of **289** and $Pd_2(dba)_3$ drives the reaction of aryl bromides or chlorides with KOH in H_2O-dioxane at 100–120°C to form phenols [688], whereas the combination of **289** and $[Pd(cod)(CH_2SiMe_3)_2]$ can perform a room temperature hydroxylation of aryl bromides or chlorides using $CsOH{\cdot}H_2O$ as base. The reaction mechanism was studied. The reaction product **290** of 2,4,6-trimethylbromobenzene with $[Pd(cod)(CH_2SiMe_3)_2]$ in the presence of **289** was isolated and characterized. Treatment of **290** with $n$$Bu_4NOH$ results in 2,4,6-trimethylphenol. Attempts to isolate the key intermediate **291** were unsuccessful. It transforms to final product rapidly at room temperature (Scheme 1.56) [689].

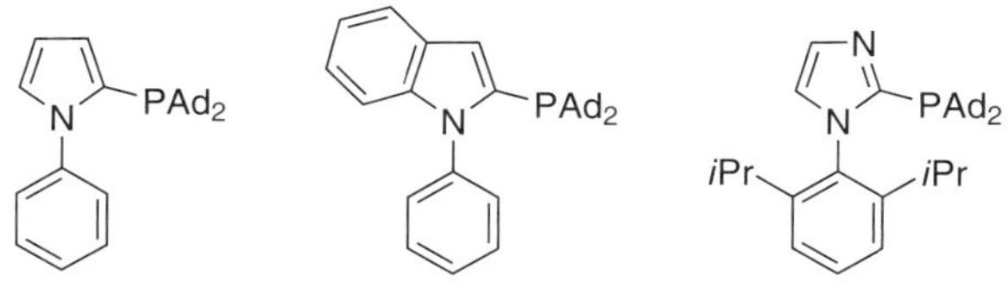

287 **288** **289**

A combination of $Pd(OAc)_2$ and **266** or **292** catalyzes C–O coupling of aryl chlorides or bromides with primary alcohols using Cs_2CO_3 as base in toluene at 80°C.

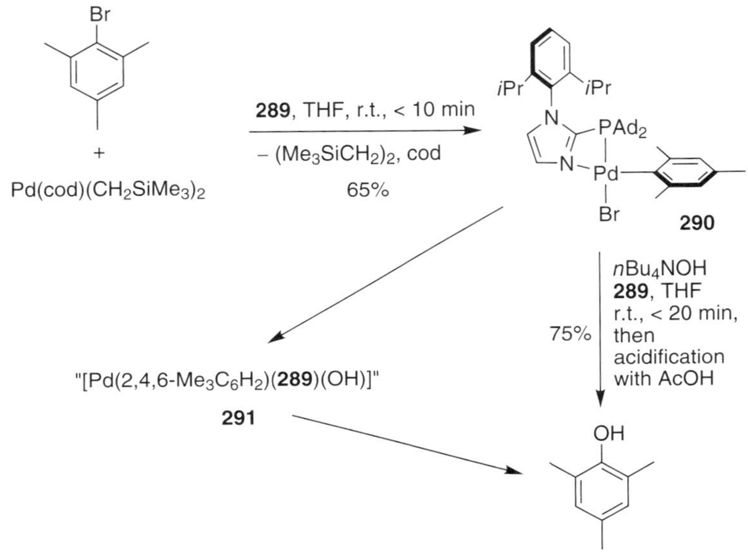

Scheme 1.56. Formation process of phenol catalyzed by [Pd(cod)(CH_2SiMe_3)$_2$]/**289**.

The system tolerates a variety of functional groups. Unactivated aryl chlorides and heteroaryl chlorides were used in this transformation [690].

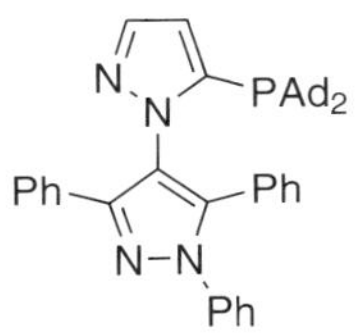

292

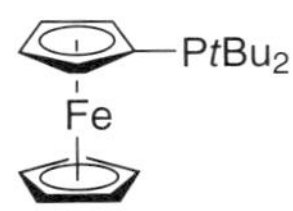

293

Ferrocene-based phosphine ligands such as **34** and **293** were also reported to be effective in Pd-catalyzed C–O bond formation reaction from aryl chlorides or bromides and NaOAr, NaO*t*Bu, or NaOTBS. The Pd(dba)$_2$/**293** system shows catalytic activity similar to that of the Pd(dba)$_2$/P(*t*Bu)$_3$ system for the reaction of hindered aryl halides. However, for the unhindered aryl halides, Pd(dba)$_2$/**293** leads to better results. Pd(dba)$_2$/**34** shows activity comparable to that of Pd(dba)$_2$/**293**. These catalytic systems give relatively low yields for the couplings of deactivated aryl chlorides [126, 691, 692].

1.6 FORMATION OF C–S BONDS

Although transition-metal-catalyzed formation of carbon–heteroatom bonds has achieved great progress, the development of effective C–S bond formation reactions is underdeveloped because of the poisoning to many catalysts by thiols. Compared with other transition metals, palladium and copper in combination with appropriate ligands are often more effective for the C–S bond formation from aryl or vinyl halides. Recently, several reviews discussing the transition-metal-catalyzed the C–S bond formation reactions have been published [693, 694].

1.6.1 Copper Catalysts

CuI was demonstrated to catalyze the coupling of aryl halides with aryl mercaptans in water in the presence of TBAB. However, for the aryl chlorides, relatively low yields were achieved [695]. CuCl–EDA catalyzes coupling of aryl mercaptans with activated aryl chlorides in water at 120°C and gives diaryl sulfides in 23–97% yields, depending on the substrates [696]. In the presence of CuI/bipyridyl and magnesium, PhCl was coupled with Ph_2S_2 in DMF in 58% yield under microwave radiation [697]. A combination of CuBr and 1,2,3,4-tetrahydro-8-hydroxyquinoline catalyzes coupling 1-chloro-4-nitrobenzene with benzenethiol in the presence of K_2CO_3. However, in the presence of KO*t*Bu, coupling of 4-chlorotoluene with benzenethiol proceeds smoothly in DMSO at 110°C without need of catalysts or ligands [698]. Basu et al. demonstrated that CuI–CV (**294**) can catalyze coupling of aryl iodides, bromides, or chlorides with thiols. In the presence of 10 mol% CuI–CV, deactivated aryl chlorides were coupled with PhSH using K_2CO_3 as base to afford diaryl sulfides in good yields [699].

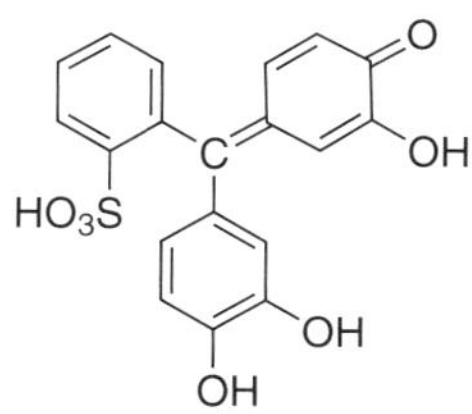

catechol violet (CV)

294

1.6.2 Palladium Catalysts

Lautens et al. [700] carried out a Pd-catalyzed tandem coupling reactions of *gem*-dihalovinyl thiophenols in which an *S*-vinylation reaction was combined with Suzuki carbon–carbon bond formation in a single synthetic operation [Eq. (1.71)]. A combination of $PdCl_2$, SPhos, K_3PO_4, and Et_3N is effective system for the catalytic reactions. Chlorides are applicable for this transformation [700]. The combination of an intramolecular *S*-vinylation reaction of *gem*-dihalovinyl thiophenols with an intermolecular carbonylation reaction results in the synthesis of 2-carbonylbenzo[*b*]thiophene [Eq. (1.72)]. $Pd(OAc)_2$/RuPhos effectively catalyzes this reaction [701]. In the presence of $Pd(OAc)_2$/XPhos, Cs_2CO_3, and H_2O, the reaction of aryl bromides or chlorides with $Na_2S_2O_3$ affords, after treatment with Zn/HCl, benzenethiols [702].

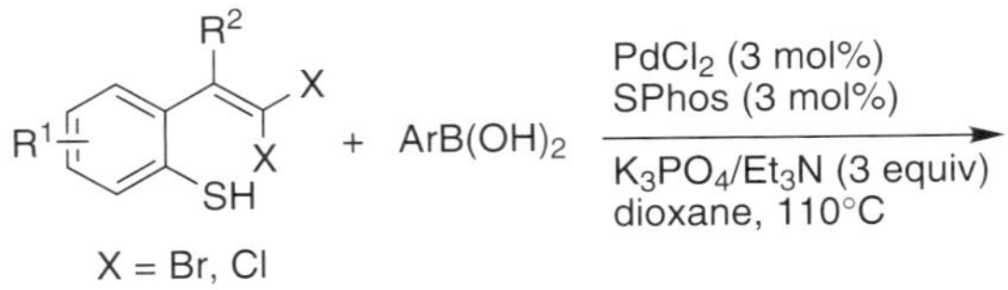

(1.71)

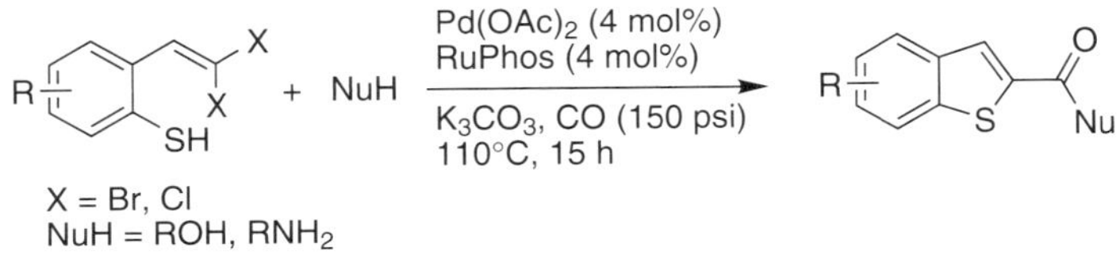

(1.72)

Ferrocene-based ligands are also effective for the Pd-catalyzed C–S couplings. Murata and Buchwald [703] demonstrated that the $Pd(OAc)_2$/DiPPF (**295**) system is efficient for the coupling of aryl bromides or chlorides with primary, secondary, and tertiary thiols and benzenethiols. The combination of CyPF-*t*Bu (**267**) and $Pd(OAc)_2$ is an extremely effective catalyst for the C–S coupling of electron-rich and electron-poor aryl chlorides or bromides with primary, secondary, and tertiary thiols, *i*Pr_3Si-SH, as well as benzenethiols in the presence of a base such as NaO*t*Bu or LiHMDS at 110°C. The reaction requires low catalyst loadings and gives high yields in most cases [704–707]. The reaction mechanism was studied by Alvaro and Hartwig [708] using $Pd(dba)_2$/CyPF-*t*Bu and the catalytic cycle is showed in Figure 1.15.

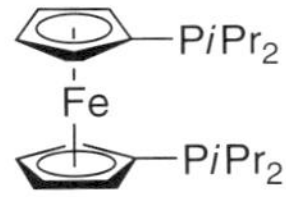

295 (DiPPF)

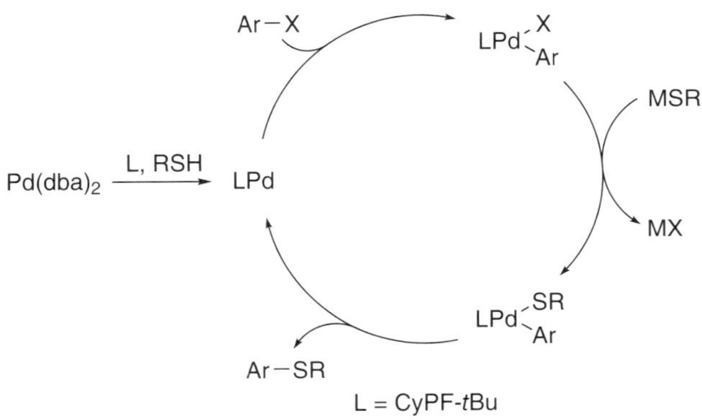

Figure 1.15. Proposed mechanism for the palladium-catalyzed thiation of aryl halides.

The palladium complex possessing phosphinous acid ligand, **56**, is also active for the catalytic coupling of aryl chlorides or bromides with benzenethiol or primary and secondary thiols using Na*Ot*Bu as a base. The coupling of secondary thiols with unactivated aryl chlorides gives moderate yields, whereas deactivated 4-chloroanisole gives only 30% yield [406].

N-Heterocyclic carbene complex of palladium Pd–PEPPSI–Ipent (**127**) also exhibits good catalytic activity for the C–S coupling of aryl chlorides or bromides with aliphatic and aromatic thiols, $i\text{Pr}_3$Si-SH, as well as benzenethiols. Deactivated and bulky aryl chlorides as well as heteroaryl chlorides can be used as the electrophiles. In addition, the reaction can proceed smoothly at 40°C and even at room temperature [709].

1.7 FORMATION OF C–B BONDS

Due to their unique reactivity in cross-coupling reactions, aryl boronic acids and esters remain important reagents widely used in organic synthesis [710, 711]. However, the traditional method for the synthesis of aryl boronic acids and esters is mainly through reaction of aryl lithium compounds or Grignard reagents, which is not compatible with numerous functional groups. In recent years, the development of transition-metal-catalyzed C–B bond formation from halides attracted considerable attention, but the use of chlorides is still limited [712, 713].

1.7.1 Palladium Catalysts

$Pd(dba)_2$–PCy_3 in dioxane was demonstrated to catalyze formation of pinacol arylboronates from activated or unactivated aryl chlorides and bis(pinacolato)diboron at 80°C using KOAc as a base [714]. $Pd(PPh_3)_2Cl_2$–NaOAc catalyzes the same reaction in polyethylene glycol 600 (PEG600) at 90°C, but only activated aryl chlorides can be coupled [715]. $Pd(PPh_3)_2Cl_2$ also catalyzes a coupling of benzyl chlorides with pinacolborane using $i\text{Pr}_2$NEt as a base [716]. The combination of Pd_2dba_3, $Ad_2P\textit{n}Bu$, and KOAc catalyzes borylation of 8-chloro- or bromo-quinilines with bis(pinacolato)diboron and following arylation with aryl chlorides or bromides in one pot, affording 8-arylquinolines in excellent yields [717].

Buchwald ligands in combination with an appropriate palladium source show highly catalytic activity for borylation of aryl or ethenyl chlorides. Pd_2dba_3–**18a** or $Pd(OAc)_2$–**15a** efficiently catalyzes the cross-coupling of electron-rich, electron-poor, sterically hindered aryl chlorides or heteroaryl chlorides with bis(pinacolato)diboron. The former requires an elevated reaction temperature, while the latter drives the reactions at room temperature [Eq. (1.73)] [711]. $PdCl_2(CH_3CN)_2$–**15a** is also effective for arylborylation of aryl chlorides or bromides with pinacol borane at 110°C using Et_3N as base. This catalyst system is also effective for the reaction of ethenyl or heteroaryl chlorides with pinacol borane [718].

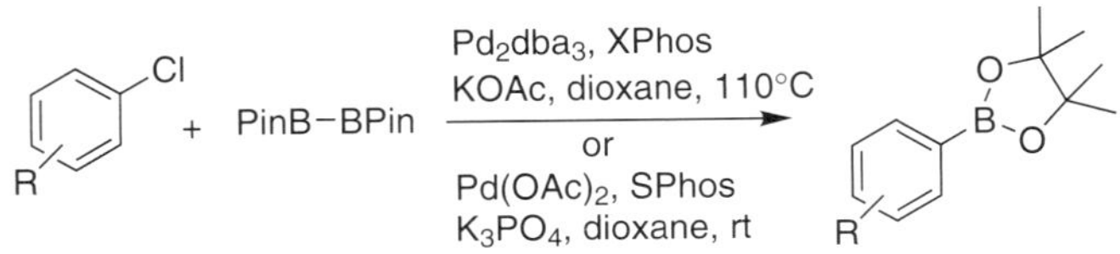

(1.73)

Palladacycle **226** in the presence of Xphos (**18a**) efficiently catalyzes direct synthesis of aryl boronic acids [Eq. (1.74)] [719, 720]. The phosphine ligand **296** in combination with palladacycle **90a** effeciently catalyzes borylation of sterically hindered aryl bromides or chlorides with bis(pinacolato)diboron at low catalytic loading [721].

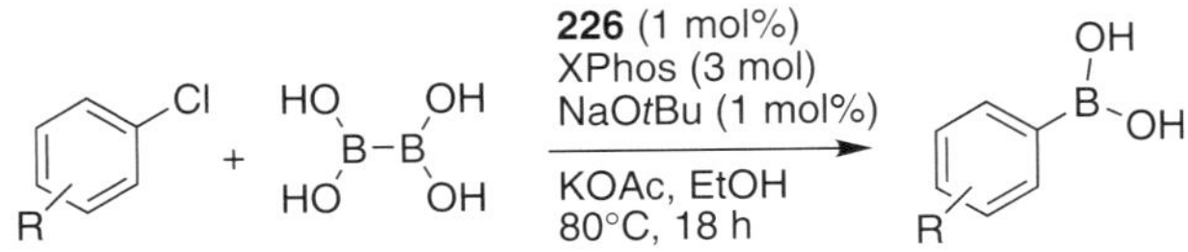

(1.74)

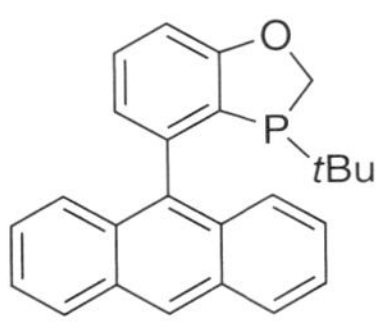

296

A silica-supported phosphane, silica-SMAP, in combination with $PdCl_2$(cod), $Pd(OAc)_2$ or $PdCl_2(pyridine)_2$ was demonstrated to be active in catalyzing the borylation of aryl chlorides and bromides with bis(pinacolato)diboron. The reaction was carried out in benzene at 60°C using KOAc as base. Electron-rich, electron-poor, and very bulky aryl chlorides can be employed as the substrates (Scheme 1.57) [722]. In addition, $PdCl_2$-catalyzed synthesis of substituted and functionalized allylic boronates from allylic chlorides and bis(pinacolato)diboron has also been performed [723].

1.7.2 Nickel Catalysts

Percec and co-workers [724–726] developed a nickel catalytic system for the neopentylglycolborylation of aryl chlorides in toluene at 100°C using Et_3N as base (Scheme 1.58). A combination of 5 mol% $NiCl_2$(dppp) and 10 mol% dppf was found to be the most effective. A diversity of electron-rich and electron-deficient aryl chlorides, including *ortho*-substituted ones, have been employed in this transformation [724, 725]. Further studies demonstrated that the presence of 2 equiv of metal zinc accelerate the transformation and improve the yields. Other metals such as Mn, Mg, and Al can be employed to replace Zn [726]. This catalyst system also displays excellent activity for the borylation of aryl bromides and iodides. $NiCl_2(PMe_3)_2$ has

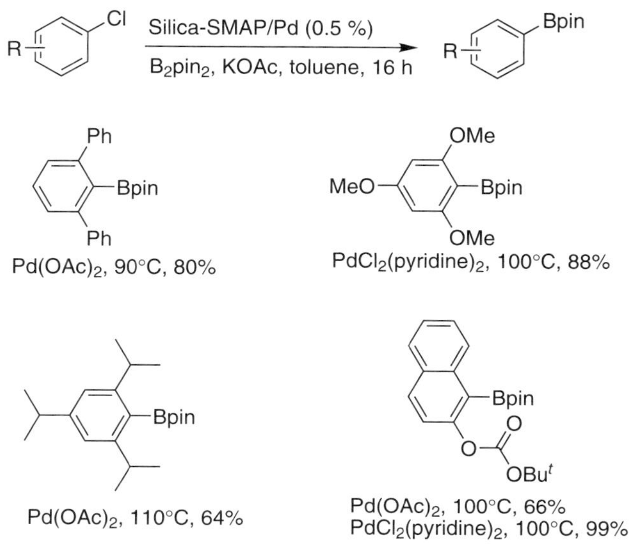

Scheme 1.57. Silica-SMAP/Pd-catalyzed borylation of aryl chlorides.

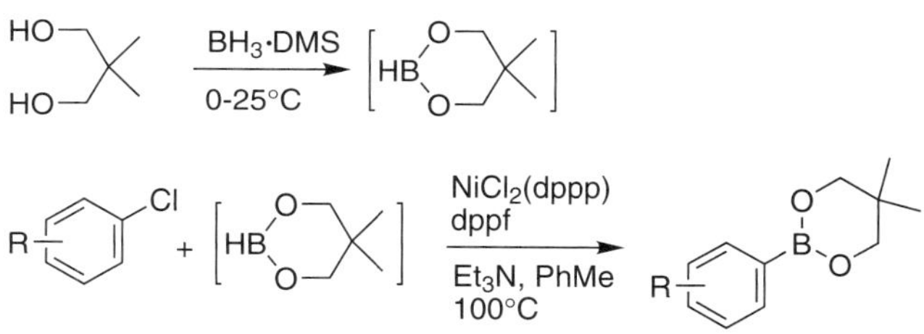

Scheme 1.58. $NiCl_2$(dppp)-catalyzed borylation of aryl chlorides in the presence of dppf.

been used as catalyst for the borylation of aryl chlorides with bis(pinacolato)diboron. CsF is employed as base and $TMSOCH_2CF_3$ is used as additive. Electron-rich, electron-poor, and sterically hindered aryl chlorides are applicable for this borylation reaction, and good to excellent yields can be achieved [Eq. (1.75)] [727].

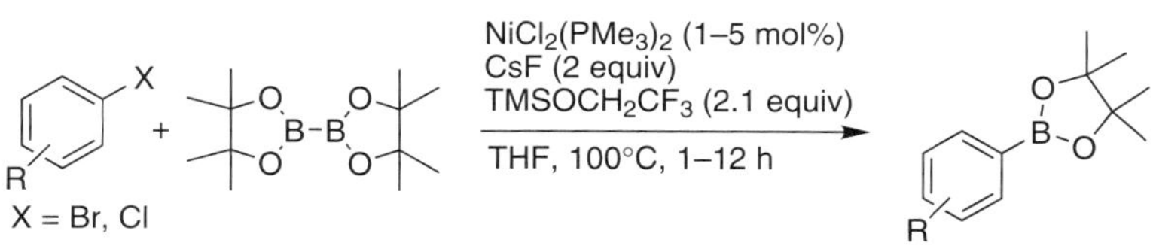

(1.75)

The combination of $NiBr_2$·diglyme and **297** was found to catalyze borylation of alkyl halides by pinB–Bpin (pin = pinacolato = $OCMe_2Me_2CO$) at room temperature. Primary, secondary, and tertiary alkyl bromides and iodides can be employed

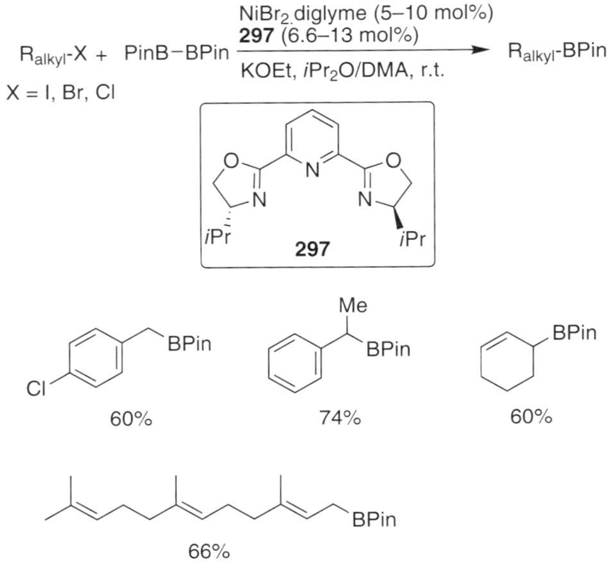

Scheme 1.59. Nickel/**297**-catalyzed borylation of allyl or benzyl chlorides.

in this transformation. As for chloride substrates, only allyl and benzyl chlorides are suitable for this coupling (Scheme 1.59) [728].

1.8 CONCLUSION AND OUTLOOK

In this chapter we summerize transition-metal-catalyzed activation and transformation of (hetero)aryl, alkenyl, and alkyl chlorides. The remarkable advances occurred in the past 10–15 years. These achievements benefit from development of new catalysts and ligands as shown in this review. Function and importance of sterically demanding and electron-rich ligands in transition-metal-catalyzed C–Cl bond activation were realized and utilized. The representative ligands include bulky trialkylphosphines [729], Buchwald biaryldialkyl phosphines [16, 575], Hartwig's ferrocene-based phosphines [574, 713], and *N*-heterocyclic carbenes developed by Herrmann, Nolan, Organ, and other groups [159, 730]. Palladium and nickel are predominantly used in the catalytic transformations. However, other metals such as copper, iron, and cobalt are also effective in some reactions. Through the catalytic process, C–C, C–N, C–O, and C–S bonds have been effectively constructed from activated, unactivated, and deactivated as well as sterically hindered organic chlorides. Transition-metal-catalyzed reductive dechlorination was also carried out for the purpose of detoxification of hazardous chlorinated organic compounds and the development of synthetic methodologies.

Although significant progress has been achieved, there are still quite a few issues that need to be resolved. These issues include (1) further development of the cross-couplings of alkyl chlorides to cover a broader scope of substrates and improve enantioselectivity; (2) raising activity and efficiency of catalysts, thereby allowing catalysts to work in mild temperature conditions and with high turnovers and turnover frequencies; and (3) understanding reaction mechanisms, which will be extremely helpful for the development of more efficient catalytic processes. In addition, functionalization of C–H bonds with organic chlorides should attract more attention in the future.

REFERENCES

1. Diederich, F., Stang, P. J., (1998). *Metal-Catalyzed Cross-Coupling Reactions*, VCH, Weinheim.
2. de Meijere, A., Diederich, F., (2004). *Metal-Catalyzed Cross-Coupling Reactions*, 2nd ed., Wiley-VCH, Weinheim.
3. Hassan, J., Sévignon, M., Gozzi, C., Schulz, E., Lemaire, M. (2002). Aryl–aryl bond formation one century after the discovery of the Ullmann reaction. *Chemical Reviews* **102**, 1359–1469.
4. Littke, A. F., Fu, G. C. (2002). Palladium-catalyzed coupling reactions of aryl chlorides. *Angewandte Chemie International Edition* **41**, 4176–4211.
5. Grushin, V. V., Alper, H. (1994). Transformations of chloroarenes, catalyzed by transition-metal complexes. *Chemical Reviews* **94**, 1047–1062.
6. Murai, S., (1999). *Activation of Unreactive Bonds and Organic Synthesis*, Springer, Berlin.
7. Alonso, F., Beletskaya, I. P., Yus, M. (2002). Metal-mediated reductive hydrodehalogenation of organic halides. *Chemical Reviews* **102**, 4009–4091.
8. Scheutz, C., Durant, N. D., Hansen, M. H., Bjerg, P. L. (2011). Natural and enhanced anaerobic degradation of 1,1,1-trichloroethane and its degradation products in the subsurface—A critical review. *Water Research* **45**, 2701–2723.
9. Magano, J., Dunetz, J. R. (2011). Large-scale applications of transition metal-catalyzed couplings for the synthesis of pharmaceuticals. *Chemical Reviews* **111**, 2177–2250.
10. Bedford, R. B., Cazin, C. S. J., Holder, D. (2004). The development of palladium catalysts for C–C and C–heteroatom bond forming reactions of aryl chloride substrates. *Coordination Chemistry Reviews* **248**, 2283–2321.
11. Jana, R., Pathak, T. P., Sigman, M. S. (2011). Advances in transition metal (Pd,Ni,Fe)-catalyzed cross-coupling reactions using alkyl-organometallics as reaction partners. *Chemical Reviews* **111**, 1417–1492.
12. Burukin, A. S., Vasilév, A. A., Chizhov, A. O., Zlotin, S. G. (2005). Cross-coupling of polychloroarenes with phenylboronic acid and organozinc compounds catalyzed by palladium complexes. *Russian Chemical Bulletin International Edition* **54**, 970–974.
13. Serdyuk, O. V., Abaev, V. T. (2009). Transition metals in the amination of chloroarenes. *Russian Chemical Reviews* **78**, 1031–1045.
14. Fu, G. C. (2008). The development of versatile methods for palladium-catalyzed coupling reactions of aryl electrophiles through the use of $P(t\text{-}Bu)_3$ and PCy_3 as ligands. *Accounts of Chemical Research* **41**, 1555–1564.

15. Hartwig, J. F. (2008). Evolution of a fourth generation catalyst for the amination and thioetherification of aryl halides. *Accounts of Chemical Research* **41**, 1534–1544.
16. Martin, R., Buchwald, S. L. (2008). Palladium-catalyzed Suzuki–Miyaura cross-coupling reactions employing dialkylbiaryl phosphine ligands. *Accounts of Chemical Research* **41**, 1461–1473.
17. Marion, N., Nolan, S. P. (2008). Well-defined *N*-heterocyclic carbenes–palladium(II) precatalysts for cross-coupling reactions. *Accounts of Chemical Research* **41**, 1440–1449.
18. Dichiarante, V., Fagnoni, M., Albini, A. (2009). Eco-friendly hydrodehalogenation of electron-rich aryl chlorides and fluorides by photochemical reaction. *Green Chemistry*, **11**, 942–945.
19. Olaniran, A. O., Igbinosa, E. O. (2011). Chlorophenols and other related derivatives of environmental concern: Properties, distribution and microbial degradation processes. *Chemosphere* **83**, 1297–1306.
20. Field, J. A., Sierra-Alvarez, R. (2008). Microbial transformation and degradation of polychlorinated biphenyls. *Environmental Pollution* **155**, 1–12.
21. Pera-Titus, M., García-Molina, V., Baños, M. A., Giménez, J., Esplugas, S. (2004). Degradation of chlorophenols by means of advanced oxidation processes: A general review. *Applied Catalysis B: Environmental* **47**, 219–256.
22. Bressan, M., d'Alessandro, N., Liberatore, L., Morvillo, A. (1999). Ruthenium-catalyzed oxidative dehalogenation of organics. *Coordination Chemistry Reviews* **185–186**, 385–402.
23. Kliegman, S., McNeill, K. (2008). Dechlorination of chloroethylenes by cob(I)alamin and cobalamin model complexes. *Dalton Transactions*, 4191–4201.
24. Pinder, A. R. (1980). Hydrogenolysis of organic halides. *Synthesis*, 425–452.
25. Urbano, F. J., Marinas, J. M. (2001). Hydrogenolysis of organohalogen compounds over palladium supported catalysts. *Journal of Molecular Catalysis A: Chemical* **173**, 329–345.
26. Sajiki, H., Kume, A., Hattori, K., Hirota, K. (2002). Mild and general procedure for Pd/C–Catalyzed hydrodechlorination of aromatic chlorides. *Tetrahedron Letters* **43**, 7247–7250.
27. Monguchi, Y., Kume, A., Hattori, K., Maegawa, T., Sajiki, H. (2006). Pd/C-Et_3N-mediated catalytic hydrodechlorination of aromatic chlorides under mild conditions. *Tetrahedron* **62**, 7926–7933.
28. Marques, C. A., Selva, M., Tundo, P. (1994). Facile hydrodehalogenation with H2 and PdC catalyst under multiphase conditions. 2. Selectivity and kinetics. *The Journal of Organic Chemistry* **59**, 3830–3837.
29. Zinovyev, S. S., Shinkova, N. A., Perosa, A., Tundo, P. (2004). Dechlorination of lindane in the multiphase catalytic reduction system with Pd/C, Pt/C and Raney-Ni. *Applied Catalysis B: Environmental* **47**, 27–36.
30. Xia, C., Xu, J., Wu, W., Liang, X. (2004). Pd/C–Catalyzed hydrodehalogenation of aromatic halides in aqueous solutions at room temperature under normal pressure. *Catalysis Communications* **5**, 383–386.
31. Zhang, Y.-M., Gu, M., Ma, H., Tang, J., Lu, W., Nan, F.-J. (2008). An efficient synthesis of 2-chloropyrimidines via Pd-catalyzed regioselective dechlorination of 2,4-dichloropyrimidines in the presence of $NaHCO_3$. *Chinese Journal of Chemistry* **26**, 962–964.

32. Chang, F., Kim, H., Lee, B., Park, H. G., Park, J. (2011). Pd-catalyzed dehalogenation of aromatic halides under solvent-free conditions using hydrogen balloon. *Bulletin of the Korean Chemical Society* **32**, 1074–1076.

33. Hashimoto, N., Hara, T., Shimazu, S., Takahashi, Y., Mitsudome, T., Mizugaki, T., Jitsukawa, K., Kaneda, K. (2010). Complete hydrodechlorination of DDT and its derivatives using a hydroxyapatite-supported Pd nanoparticle catalyst. *Chemistry Letters* **39**, 49–51.

34. Angeloff, A., Brunet, J.-J., Legars, P., Neibeckera, D., Souyri, D. (2001). Regioselective dechlorination of 2,3-dichloronitrobenzene into 3-chloronitrobenzene and regioselective dechlorination–hydrogenation into 3-chloroaniline. *Tetrahedron Letters* **42**, 2301–2303.

35. Buil, M. L., Esteruelas, M. A., Niembro, S., Oliván, M., Orzechowski, L., Pelayo, C., Vallribera, A. (2010). Dehalogenation and hydrogenation of aromatic compounds catalyzed by nanoparticles generated from rhodium Bis(imino)pyridine complexes. *Organometallics* **29**, 4375–4383.

36. Huberta, C., Bilé, E. G., Denicourt-Nowicki, A., Roucoux, A. (2011). Tandem dehalogenation–hydrogenation reaction of halogenoarenes as model substrates of endocrine disruptors in water: Rhodium nanoparticles in suspension vs. on silica support. *Applied Catalysis A: General* **394**, 215–219.

37. Davies, C. J. E., Page, M. J., Ellul, C. E., Mahon, M. F., Whittlesey, M. K. (2010). Ni(I) and Ni(II) ring-expanded *N*-heterocyclic carbene complexes: C–H activation, indole elimination and catalytic hydrodehalogenation. *Chemical Communications* **46**, 5151–5153.

38. Desmarets, C., Kuhl, S., Schneider, R., Fort, Y. (2002). Nickel(0)/imidazolium chloride catalyzed reduction of aryl halides. *Organometallics* **21**, 1554–1559.

39. Viciu, M. S., Grasa, G. A., Nolan, S. P. (2001). Catalytic dehalogenation of aryl halides mediated by a palladium/imidazolium salt system. *Organometallics* **20**, 3607–3612.

40. Akzinnay, S., Bisarob, F., Cazin, C. S. J. (2009). Highly efficient catalytic hydrodehalogenation of polychlorinated biphenyls (PCBs). *Chemical Communications*, 5752–5753.

41. Navarro, O., Kaur, H., Mahjoor, P., Nolan, S. P. (2004). Cross-coupling and dehalogenation reactions catalyzed by (*N*-heterocyclic carbene)Pd(allyl)Cl complexes. *The Journal of Organic Chemistry* **69**, 3173–3180.

42. Navarro, O., Marion, N., Oonishi, Y., Kelly, R. A., III, Nolan, S. P. (2006). Suzuki–Miyaura, γ-ketone arylation and dehalogenation reactions catalyzed by a versatile *N*-heterocyclic carbene–palladacycle complex. *The Journal of Organic Chemistry* **71**, 685–692.

43. Chen, J., Zhang, Y., Yang, L., Zhang, X., Liu, J., Li, L., Zhang, H. (2007). A practical palladium catalyzed dehalogenation of aryl halides and α-haloketones. *Tetrahedron*, **63**, 4266–4270.

44. Moon, J., Lee, S. (2009). Palladium catalyzed-dehalogenation of aryl chlorides and bromides using phosphite ligands. *Journal of Organometallic Chemistry* **694**, 473–477.

45. Bei, X., Hagemeyer, A., Volpe, A., Saxton, R., Turner, H., Guram, A. S. (2004). Productive chloroarene C–Cl bond activation: Palladium/phosphine-catalyzed methods for oxidation of alcohols and hydrodechlorination of chloroarenes. *The Journal of Organic Chemistry* **69**, 8626–8633.

46. Aikawa, B., Burk, R. C., Sitholé, B. B. (2001). Catalytic dechlorination of 1-chlorooctadecane in supercritical carbon dioxide. *Applied Catalysis B: Environmental* **32**, 269–280.

47. Kim, J. Y., Jo, Y., Lee, S., Choi, H. C. (2009). Synthesis of Pd-CNT nanocomposites and investigation of their catalytic behavior in the hydrodehalogenation of aryl halides. *Tetrahedron Letters* **50**, 6290–6292.
48. Cucullu, M. E., Nolan, S. P., Belderrain, T. R., Grubbs, R. H. (1999). Catalytic dehalogenation of aryl chlorides mediated by ruthenium(II) phosphine complexes. *Organometallics* **18**, 1299–1304.
49. Fujita, K.-i., Owaki, M., Yamaguchi, R. (2002). Chemoselective transfer hydrodechlorination of aryl chlorides catalyzed by Cp*Rh complexes. *Chemical Communications*, 2964–2965.
50. Díaz, J., Esteruelas, M. A., Herrero, J., Moralejo, L., Oliván, M. (2000). Simultaneous dehalogenation of polychloroarenes and chlorination of $HSiEt_3$ catalyzed by complexes of the groups 8 and 9. *Journal of Catalysis* **195**, 187–192.
51. Esteruelas, M. A., Herrero, J., López, F. M., Martín, M., Oro, L. A. (1999). Dehalogenation of polychloroarenes with $HSiEt_3$ catalyzed by an homogeneous rhodium-triphenylphosphine system. *Organometallics* **18**, 1110–1112.
52. Iranpoor, N., Firouzabadi, H., Azadi, R. (2010). Diphenylphosphinite ionic liquid (IL-$OPPh_2$): A solvent and ligand for palladium-catalyzed silylation and dehalogenation reaction of aryl halides with triethylsilane. *Journal of Organometallic Chemistry* **695**, 887–890.
53. Karshtedt, D., Bell, A. T., Tilley, T. D. (2006). Stoichiometric and catalytic reactions involving Si–H bond activations by Rh and Ir complexes containing a pyridylindolide ligand. *Organometallics*, **25**, 4471–4482.
54. Rahaim, R. J., Jr., Maleczka, R. E., Jr., (2002). Room temperature dehalogenation of chloroarenes by polymethylhydrosiloxane (PMHS) under palladium catalysis. *Tetrahedron Letters* **43**, 8823–8826.
55. Peterson, A. A., McNeill, K. (2006). Catalytic dehalogenation of sp^2 C–F and C–Cl bonds in fluoroand chloroalkenes. *Organometallics* **25**, 4938–4940.
56. Peterson, A. A., Thoreson, K. A., McNeill, K. (2009). Phosphinorhodium-catalyzed dehalogenation of chlorinated and fluorinated ethylenes: Distinct mechanisms with triethylsilane and dihydrogen. *Organometallics* **28**, 5982–5991.
57. Breitenfeld, J., Scopelliti, R., Hu, X. (2012). Synthesis, reactivity, and catalytic application of a nickel pincer hydride complex. *Organometallics* **31**, 2128–2136.
58. Yang, J., Brookhart, M. (2007). Iridium-catalyzed reduction of alkyl halides by triethylsilane. *Journal of the American Chemical Society* **129**, 12656–12657.
59. Yang, J., Brookhart, M. (2009). Reduction of alkyl halides by triethylsilane based on a cationic iridium bis(phosphinite) pincer catalyst: Scope, selectivity and mechanism. *Advanced Synthesis & Catalysis* **351**, 175–187.
60. Esteruelas, M. A., Herrero, J., Oliván, M. (2004). Dehalogenation of Hexachlorocyclohexanes and simultaneous chlorination of triethylsilane catalyzed by rhodium and ruthenium complexes. *Organometallics* **23**, 3891–3897.
61. Boukherroub, R., Chatgilialoglu, C., Manuel, G. (1996). $PdCl_2$-catalyzed reduction of organic halides by triethylsilane. *Organometallics* **15**, 1508–1510.
62. Stöhr, F., Sturmayr, D., Schubert, U. (2002). C–Cl/Si-H exchange catalysed by P,N-chelated Pt(II) complexes. *Chemical Communications*, 2222–2223.
63. Arcadi, A., Cerichelli, G., Chiarini, M., Vico, R., Zorzan, D. (2004). Pd/C–catalyzed transfer reduction of aryl chlorides with sodium formate in water. *European Journal of Organic Chemistry*, 3404–3407.

64. Wang, S., Yang, B., Zhang, T., Yu, G., Deng, S., Huang, J. (2010). Catalytic hydrodechlorination of 4-chlorophenol in an aqueous solution with Pd/Ni catalyst and formic acid. *Industrial & Engineering Chemistry Research* **49**, 4561–4565.

65. Vincent, T., Spinelli, S., Guibal, E. (2003). Chitosan-supported palladium catalyst. II. Chlorophenol dehalogenation. *Industrial & Engineering Chemistry Research* **42**, 5968–5976.

66. Logan, M. E., Oinen, M. E. (2006). Dechlorination of aryl chlorides with sodium formate using a homogeneous palladium catalyst. *Organometallics* **25**, 1052–1054.

67. Zhang, C., Li, X., Sun, H. (2011). Palladium-catalyzed hydrodechlorination of ary chlorides and its mechanism. *Inorganica Chimica Acta* **365**, 133–136.

68. Atienza, M. A., Esteruelas, M. A., Fernández, M., Herrero, J., Oliván, M. (2001). Dehalogenation of polychloroarenes with sodium formate in propan-2-ol catalyzed by $RhCl(PPh_3)_3$. *New Journal of Chemistry* **25**, 775–776.

69. Lipshutz, B. H., Tomioka, T., Sato, K. (2001). Nickel-on-charcoal-catalyzed reductions of aryl chlorides. *Synlett*, SI, 970–973.

70. Lipshutz, B. H., Tomioka, T., Pfeiffer, S. S. (2001). Mild and selective reductions of aryl halides catalyzed by low-valent nickel complexes. *Tetrahedron Letters* **42**, 7737–7740.

71. Liu, Y., Schwartz, J., Cavallaro, C. L. (1995). Catalytic dechlorination of polychlorinated biphenyls. *Environmental Science & Technology* **29**, 836–840.

72. Fletcher, A. E. D., Hyatt, J. D., Ok, K. M., O'Hare, D. (2009). Catalytic dechlorination of polychlorinated biphenyls (PCBs) using amine functionalised titanocenes. *Green Chemistry* **11**, 1343–1348.

73. Lassová, L., Lee, H. K., Hor, T. S. A. (1998). Catalytic dehalogenation of highly chlorinated benzenes and aroclors using $PdCl_2$(dppf) and $NaBH_4$: Efficiency, selectivity, and base support. *The Journal of Organic Chemistry* **63**, 3538–3543.

74. Lassová, L., Lee, H. K., Hor, T. S. A. (1999). Catalytic dechlorination of chlorobenzenes: Effect of solvent on efficiency and selectivity. *Journal of Molecular Catalysis A: Chemical* **144**, 397–403.

75. Zhu, Y., Ching, C., Carpenter, K., Xu, R., Selvaratnam, S., Hosmane, N. S., Maguire, J. A. (2003). Synthesis of the novel ionic liquid [*N*-pentylpyridinium]$^+$ [closo-$CB_{11}H_{12}$]$^-$ and its usage as a reaction medium in catalytic dehalogenation of aromatic halides. *Applied Organometallic Chemistry* **17**, 346–350.

76. Kumari, P. P., Ahmad, S., Chauhan, S. M. S. (2011). Reductive dechlorination of atrazine using sodium-borohydride catalysed by cobalt(II) phthalocyanines. *Tetrahedron Letters* **52**, 7083–7086.

77. Hara, R., Sato, K., Sun, W.-H., Takahashi, T. (1999). Catalytic dechlorination of aromatic chlorides using Grignard reagents in the presence of $(C_5H_5)_2TiCl_2$. *Chemical Communications*, 845–846.

78. Guo, H., Kanno, K.-i., Takahashi, T. (2004). Iron-catalyzed dechlorination of aryl chlorides. *Chemistry Letters* **33**, 1356–1357.

79. Czaplik, W. M., Grupe, S., Mayer, M., von Wangelin, A. J. (2010). Practical iron-catalyzed dehalogenation of aryl halides. *Chemical Communications* **46**, 6350–6352.

80. Cellier, P. P., Spindler, J.-F., Taillefera, M., Cristau, H.-J. (2003). Pd/C–catalyzed room-temperature hydrodehalogenation of aryl halides with hydrazine hydrochloride. *Tetrahedron Letters* **44**, 7191–7195.

81. Rodríguez, J. G., Lafuente, A. (2002). A new advanced method for heterogeneous catalysed dechlorination of 1,2,3-, 1,2,4-, and 1,3,5-trichlorobenzenes in hydrocarbon solvent. *Tetrahedron Letters* **43**, 9645–9647.

82. Rodríguez, J. G., Lafuente, A. (2002). A new advanced method for heterogeneous catalysed dechlorination of polychlorinated biphenyls (PCBs) in hydrocarbon solvent. *Tetrahedron Letters* **43**, 9581–9583.

83. Rodríguez, J. G., Lafuente, A. (2008). Effective elimination of PCBs catalyzed by the palladium/hydrazine system as an ecological sustainable process. *Industrial & Engineering Chemistry Research* **47**, 7993–7996.

84. Kopinke, F.-D., Mackenzie, K., Koehler, R., Georgi, A. (2004). Alternative sources of hydrogen for hydrodechlorination of chlorinated organic compounds in water on Pd catalysts. *Applied Catalysis A: General* **271**, 119–128.

85. Shen, W. (1997). Palladium catalyzed coupling of aryl chlorides with arylboronic acids. *Tetrahedron Letters* **38**, 5575–5578.

86. Littke, A. F., Fu, G. C. (1998). A convenient and general method for Pd-catalyzed Suzuki cross-couplings of aryl chlorides and arylboronic acids. *Angewandte Chemie International Edition* **37**, 3387–3388.

87. Littke, A. F., Dai, C., Fu, G. C. (2000). Versatile catalysts for the Suzuki cross-coupling of arylboronic acids with aryl and vinyl halides and triflates under mild conditions. *Journal of the American Chemical Society* **122**, 4020–4028.

88. Zapf, A., Ehrentraut, A., Beller, M. (2000). A new highly efficient catalyst system for the coupling of nonactivated and deactivated aryl chlorides with arylboronic acids. *Angewandte Chemie International Edition* **39**, 4153–4155.

89. Guram, A. S., King, A. O., Allen, J. G., Wang, X., Schenkel, L. B., Chan, J., Bunel, E. E., Faul, M. M., Larsen, R. D., Martinelli, M. J., Reider, P. J. (2006). New air-stable catalysts for general and efficient Suzuki–Miyaura cross-coupling reactions of heteroaryl chlorides. *Organic Letters* **8**, 1787–1789.

90. Guram, A. S., Wang, X., Bunel, E. E., Faul, M. M., Larsen, R. D., Martinelli, M. J. (2007). New catalysts for Suzuki–Miyaura coupling reactions of heteroatom-substituted heteroaryl chlorides. *The Journal of Organic Chemistry* **72**, 5104–5112.

91. Lü, B., Fu, C., Ma, S. (2010). Application of a readily available and air stable monophosphine HBF4 salt for the Suzuki coupling reaction of aryl or 1-alkenyl chlorides. *Tetrahedron Letters* **51** 1284–1286.

92. Dreher, S. D., Dormer, P. G., Sandrock, D. L., Molander, G. A. (2008). Efficient cross-coupling of secondary alkyltrifluoroborates with aryl chlorides Reaction discovery using parallel microscale experimentation. *Journal of the American Chemical Society* **130**, 9257–9259.

93. Molander, G. A., Gormisky, P. E. (2008). Cross-coupling of cyclopropyl- and cyclobutyltrifluoroborates with aryl and heteroaryl chlorides. *The Journal of Organic Chemistry* **73**, 7481–7485.

94. Lü, B., Fu, C., Ma, S. (2010). Application of dicyclohexyl-(*S*)-trimethoxyphenyl phosphine·HBF4 salt for the highly selective Suzuki coupling of the C–Cl bond in *b*-chlorobutenolides over the more reactive allylic C–O bond. *Chemistry—A European Journal* **16**, 6434–6437.

95. Henry, N., Enguehard-Gueiffier, C., Thery, I., Gueiffier, A. (2008). One-pot dual substitutions of bromobenzyl chloride, 2-chloromethyl-6-halogenoimidazo[1,2-a]pyridine

and -[1,2-*b*]pyridazine by Suzuki–Miyaura cross-coupling reactions. *European Journal of Organic Chemistry*, 4824–4827.

96. Kirchhoff, J. H., Dai, C., Fu, G. C. (2002). A method for palladium-catalyzed cross-couplings of simple alkyl chlorides: Suzuki reactions catalyzed by $[Pd_2(dba)_3]/PCy_3$. *Angewandte Chemie International Edition* **41**, 1945–1947.
97. Old, D. W., Wolfe, J. P., Buchwald, S. L. (1998). A highly active catalyst for palladium-catalyzed cross-coupling reactions: Room-temperature Suzuki couplings and amination of unactivated aryl chlorides. *Journal of the American Chemical Society* **120**, 9722–9723.
98. Wolfe, J. P., Singer, R. A., Yang, B. H., Buchwald, S. L. (1999). Highly active palladium catalysts for Suzuki coupling reactions. *Journal of the American Chemical Society* **121**, 9550–9561.
99. Walker, S. D., Barder, T. E., Martinelli, J. R., Buchwald, S. L. (2004). A rationally designed universal catalyst for Suzuki–Miyaura coupling processes. *Angewandte Chemie International Edition* **43**, 1871–1876.
100. Barder, T. E., Walker, S. D., Martinelli, J. R., Buchwald, S. L. (2005). Catalysts for Suzuki–Miyaura coupling processes: Scope and studies of the effect of ligand structure. *Journal of the American Chemical Society* **127**, 4685–4696.
101. Billingsley, K., Buchwald, S. L. (2007). Highly efficient monophosphine-based catalyst for the palladium-catalyzed Suzuki–Miyaura reaction of heteroaryl halides and heteroaryl boronic acids and esters. *Journal of the American Chemical Society* **129**, 3358–3366.
102. Barder, T. E., Buchwald, S. L. (2004). Efficient catalyst for the Suzuki–Miyaura coupling of potassium aryl trifluoroborates with aryl chlorides. *Organic Letters* **6**, 2649–2652.
103. Molander, G. A., Shin, I. (2011). Synthesis and Suzuki–Miyaura cross-coupling reactions of potassium Boc-protected aminomethyltrifluoroborate with aryl and hetaryl halides. *Organic Letters* **13**, 3956–3959.
104. Thakur, A., Zhang, K., Louie, J. (2012). Suzuki–Miyaura coupling of heteroaryl boronic acids and vinyl chlorides. *Chemical Communications*. **48**, 203–205.
105. Ma, S., Jiang, X., Cheng, X., Hou, H. (2006). Highly efficient Suzuki coupling reaction of α-chloroalkylidene-β-lactones and β-lactams with organoboronic acids. *Advanced Synthesis & Catalysis* **348**, 2114–2124.
106. Yin, J., Rainka, M. P., Zhang, X.-X., Buchwald, S. L. (2002). A highly active Suzuki catalyst for the synthesis of sterically hindered biaryls: Novel ligand coordination. *Journal of the American Chemical Society* **124**, 1162–1163.
107. Dreher, S. D., Lim, S.-E., Sandrock, D. L., Molander, G. A. (2009). Suzuki–Miyaura cross–coupling reactions of primary alkyltrifluoroborates with aryl chlorides. *The Journal of Organic Chemistry* **74**, 3626–3631.
108. Molander, G. A., Shin, I., Jean-Geíard, L. (2010). Palladium-catalyzed Suzuki–Miyaura cross-coupling reactions of enantiomerically enriched potassium β-trifluoroboratoamides with various aryl- and hetaryl chlorides. *Organic Letters* **12**, 4384–4387.
109. Anderson, K. W., Buchwald, S. L. (2005). General catalysts for the Suzuki–Miyaura and Sonogashira coupling reactions of aryl chlorides and for the coupling of challenging substrate combinations in water. *Angewandte Chemie International Edition* **44**, 6173–6177.
110. Barder, T. E., Biscoe, M. R., Buchwald, S. L. (2007). Structural insights into active catalyst structures and oxidative addition to (biaryl)phosphine–palladium complexes via density functional theory and experimental studies. *Organometallics* **26**, 2183–2192.

111. Zapf, A., Jackstell, R., Rataboul, F., Riermeier, T., Monsees, A., Fuhrmann, C., Shaikh, N., Dingerdissen, U., Beller, M. (2004). Practical synthesis of new and highly efficient ligands for the Suzuki reaction of aryl chlorides. *Chemical Communications*, 38–39.

112. So, C. M., Chow, W. K., Choy, P. Y., Lau, C. P., Kwong, F. Y. (2010). Remarkably effective phosphanes simply with a PPh_2 moiety: Application to Pd-catalysed cross-coupling reactions for tetra-ortho-substituted biaryl syntheses. *Chemistry—A European Journal* **16**, 7996–8001.

113. So, C. M., Lau, C. P., Kwong, F. Y. (2007). Easily accessible and highly tunable indolyl phosphine Ligands for Suzuki–Miyaura coupling of aryl chlorides. *Organic Letters* **9**, 2795–2798.

114. Harkal, S., Rataboul, F., Zapf, A., Fuhrmann, C., Riermeier, T., Monsees, A., Beller, M. (2004). Dialkylphosphinoimidazoles as new ligands for palladium-catalyzed coupling reactions of aryl chlorides. *Advanced Synthesis & Catalysis* **346**, 1742–1748.

115. Dai, Q., Gao, W., Liu, D., Kapes, L. M., Zhang, X. (2006). Triazole-based monophosphine ligands for palladium-catalyzed cross-coupling reactions of aryl chlorides. *The Journal of Organic Chemistry* **71**, 3928–3934.

116. Liu, D., Gao, W., Dai, Q., Zhang, X. (2005). Triazole-based monophosphines for Suzuki–Miyaura coupling and amination reactions of aryl chlorides. *Organic Letters* **7**, 4907–4910.

117. Spinella, S. M., Guan, Z.-H., Chen, J., Zhang, X. (2009). Suzuki coupling of heteroaromatic chlorides using highly electron-donating ClickPhos ligands. *Synthesis*, 3094–3098.

118. Fihri, A., Meunier, P., Hierso, J.-C. (2007). Performances of symmetrical achiral ferrocenylphosphine ligands in palladium-catalyzed cross-coupling reactions: A review of syntheses, catalytic applications and structural properties. *Coordination Chemistry Reviews* **251**, 2017–2055.

119. Baillie, C., Zhang, L., Xiao, J. (2004). Ferrocenyl monophosphine ligands: Synthesis and applications in the Suzuki–Miyaura coupling of aryl chlorides. *The Journal of Organic Chemistry* **69**, 7779–7782.

120. Jensen, J. F., Johannsen, M. (2003). New air-stable planar chiral ferrocenyl monophosphine ligands: Suzuki cross-coupling of aryl chlorides and bromides. *Organic Letters* **5**, 3025–3028.

121. Roca, F. X., Richards, C. J. (2003). A ferrocene based palladacyclic precatalyst for the Suzuki cross-coupling of aryl chlorides. *Chemical Communications*, 3002–3003.

122. Kwong, F. Y., Chan, K. S., Yeung, C. H., Chan, A. S. C. (2004). An active ferrocenyl triarylphosphine for palladium-catalyzed Suzuki–Miyaura cross-coupling of aryl halides. *Chemical Communications*, 2336–2337.

123. Pickett, T. E., Roca, F. X., Richards, C. J. (2003). Synthesis of monodentate ferrocenylphosphines and their application to the palladium-catalyzed Suzuki reaction of aryl chlorides. *The Journal of Organic Chemistry* **68**, 2592–2599.

124. Liu, S.-Y., Choi, M. J., Fu, G. C. (2001). A surprisingly mild and versatile method for palladium-catalyzed Suzuki cross-couplings of aryl chlorides in the presence of a triarylphosphine. *Chemical Communications*, 2408–2409.

125. Thimmaiah, M., Fang, S. (2007). Efficient palladium-catalyzed Suzuki–Miyaura coupling of aryl chlorides with arylboronic acids using benzoferrocenyl phosphines as supporting ligands. *Tetrahedron* **63**, 6879–6886.

126. Kataoka, N., Shelby, Q., Stambuli, J. P., Hartwig, J. F. (2002). Air Stable, Sterically hindered ferrocenyl dialkylphosphines for palladium-catalyzed C–C, C–N, and C–O bond-forming cross-couplings. *The Journal of Organic Chemistry* **67**, 5553–5566.

127. Weng, Z., Teo, S., Koh, L. L., Hor, T. S. A. (2004). Efficient Suzuki coupling of aryl chlorides catalyzed by palladium(0) with a P,N heteroligand and isolation of unsaturated intermediates. *Organometallics* **23**, 4342–4345.

128. Teo, S., Weng, Z., Hor, T. S. A. (2006). 1,1′-P/O-ferrocenyl ligands in palladium-catalyzed Suzuki coupling of aryl chlorides. *Organometallics* **25**, 1199–1205.

129. Itoh, T., Sato, K., Mase, T. (2004). A novel practical synthesis of C-2-arylpurines. *Advanced Synthesis & Catalysis* **346**, 1859–1867.

130. Itoh, T., Mase, T. (2005). Direct synthesis of hetero-biaryl compounds containing an unprotected NH_2 group via Suzuki-Miyaura reaction. *Tetrahedron Letters* **46**, 3573–3577.

131. Hierso, J.-C., Fihri, A., Amardeil, R., Meunier, P. (2003). A palladium-ferrocenyl Tetraphosphine system as catalyst for Suzuki cross-coupling and Heck vinylation of aryl halides: Dynamic behavior of the palladium/phosphine species. *Organometallics* **22**, 4490–4499.

132. Hoshi, T., Nakazawa, T., Saitoh, I., Mori, A., Suzuki, T., Sakai, J.-i., Hagiwara, H. (2008). Biphenylene-substituted ruthenocenylphosphine for Suzuki–Miyaura coupling of aryl chlorides. *Organic Letters* **10**, 2063–2066.

133. Doherty, S., Knight, J. G., McGrady, J. P., Ferguson, A. M., Ward, N. A. B., Harrington, R. W., Clegg, W. (2010). *ortho,ortho′*-Substituted KITPHOS monophosphines: Highly efficient ligands for palladium-catalyzed C–C and C–N bond formation. *Advanced Synthesis & Catalysis* **352**, 201–211.

134. Ruan, J., Shearer, L., Mo, J., Bacsa, J., Zanotti-Gerosa, A., Hancock, F., Wu, X., Xiao, J. (2009). [2.2]Paracyclophane-based monophosphine ligand for palladium-catalyzed cross-coupling reactions of aryl chlorides. *Organic & Biomolecular Chemistry* **7**, 3236–3242.

135. Fleckenstein, C. A., Plenio, H. (2007). 9-Fluorenylphosphines for the Pd-catalyzed Sonogashira, Suzuki, and Buchwald–Hartwig coupling reactions in organic solvents and water. *Chemistry—A European Journal* **13**, 2701–2716.

136. Fleckenstein, C. A., Plenio, H. (2008). Efficient Suzuki–Miyaura coupling of (hetero)aryl chlorides with thiophene- and furanboronic acids in aqueous *n*-butanol. *The Journal of Organic Chemistry* **73**, 3236–3244.

137. der Heiden, M., Plenio, H. (2004). Homogeneous catalysts supported on soluble polymers: Biphasic Suzuki–Miyaura coupling of aryl chlorides using phase-tagged palladium–phosphine catalysts. *Chemistry—A European Journal* **10**, 1789–1797.

138. Bei, X., Crevier, T., Guram, A. S., Jandeleit, B., Powers, T. S., Turner, H. W., Uno, T., Weinberg, W. H. (1999). A convenient palladium/ligand catalyst for Suzuki cross-coupling reactions of arylboronic acids and aryl chlorides. *Tetrahedron Letters* **40**, 3855–3858.

139. Bei, X., Turner, H. W., Weinberg, W. H., Guram, A. S. (1999). Palladium/P,O-ligand-catalyzed Suzuki cross-coupling reactions of arylboronic acids and aryl chlorides. Isolation and structural characterization of (P,O)-Pd(dba) complex. *The Journal of Organic Chemistry* **64**, 6797–6803.

140. Kwong, F. Y., Lam, W. H., Yeung, C. H., Chan, K. S., Chan, A. S. C. (2004). A simple and highly efficient P,O-type ligand for Suzuki–Miyaura cross-coupling of aryl halides. *Chemical Communications*, 1922–1923.

141. Dai, W.-M., Zhang, Y. (2005). A family of simple amide-derived air-stable P,O-ligands for Suzuki cross-coupling of unactivated aryl chlorides. *Tetrahedron Letters* **46**, 1377–1381.

142. So, C. M., Yeung, C. C., Lau, C. P., Kwong, F. Y. (2008). A new family of tunable indolylphosphine ligands by one-pot assembly and their applications in Suzuki–Miyaura coupling of aryl chlorides. *The Journal of Organic Chemistry* **73**, 7803–7806.

143. Ullah, E., McNulty, J., Robertson, A. (2009). A novel P,O-type phosphorinane ligand for the Suzuki–Miyaura cross-coupling of aryl chlorides. *Tetrahedron Letters* **50**, 5599–5601.

144. Adjabeng, G., Brenstrum, T., Wilson, J., Frampton, C., Robertson, A., Hillhouse, J., McNulty, J., Capretta, A. (2003). Novel class of tertiary phosphine ligands based on a phospha-adamantane framework and use in the Suzuki cross-coupling reactions of aryl halides under mild conditions. *Organic Letters* **5**, 953–955.

145. Kamer, P. C. J., van Leeuwen, P. W. N. M., Reek, J. N. H. (2001). Wide bite angle diphosphines: Xantphos ligands in transition metal complexes and catalysis. *Accounts of Chemical Research* **34**, 895–904.

146. Liron, F., Fosse, C., Pernolet, A., Roulland, E. (2007). Suzuki–Miyaura cross-coupling of 1,1-dichloro-1-alkenes with 9-alkyl-9-BBN. *The Journal of Organic Chemistry* **72**, 2220–2223.

147. Li, G. Y. (2001). The first phosphine oxide ligand precursors for transition metal catalyzed cross-coupling reactions: C–C, C–N, and C–S bond formation on unactivated aryl chlorides. *Angewandte Chemie International Edition* **40**, 1513–1516.

148. Li, G. Y. (2002). Highly active, air-stable palladium catalysts for the C–C and C–S bond-forming reactions of vinyl and aryl chlorides: Use of commercially available $[(t\text{-Bu})_2P(OH)]_2PdCl_2$, $[(t\text{-Bu})_2P(OH)PdCl_2]_2$, and $[[(t\text{-Bu})_2PO\cdots H\cdots OP(t\text{-Bu})_2]PdCl]_2$ as catalysts. *The Journal of Organic Chemistry* **67**, 3643–3650.

149. Khanapure, S. P., Garvey, D. S. (2004). Use of highly reactive, versatile and air-stable palladium–phosphinous acid complex $[(t\text{-Bu})_2P(OH)]_2PdCl_2$ (POPd) as a catalyst for the optimized Suzuki–Miyaura cross-coupling of less reactive heteroaryl chlorides and arylboronic acids. *Tetrahedron Letters* **45**, 5283–5286.

150. Miao, G., Ye, P., Yu, L., Baldino, C. M. (2005). Microwave-promoted Suzuki reactions of aryl chlorides in aqueous media. *The Journal of Organic Chemistry* **70**, 2332–2334.

151. Zapf, A., Beller, M. (2000). Palladium/phosphite catalyst systems for efficient cross coupling of aryl bromides and chlorides with phenylboronic acid. *Chemistry—A European Journal* **6**, 1830–1833.

152. Kamei, T., Sato, A. H., Iwasawa, T. (2011). Asymmetric Suzuki–Miyaura cross-coupling of aryl chlorides with enhancement of reaction time and catalyst turnover. *Tetrahedron Letters* **52**, 2638–2641.

153. Iwasawa, T., Kamei, T., Watanabe, S., Nishiuchi, M., Kawamura, Y. (2008). A functionalized phosphine ligand with a pentaarylbenzene moiety in palladium-catalyzed Suzuki-Miyaura coupling of aryl chlorides. *Tetrahedron Letters* **49**, 7430–7433.

154. Mai, W., Lv, G., Gao, L. (2007). Monooxychlorophosphine as a novel and efficient ligand for palladium-catalyzed Suzuki–Miyaura cross-coupling of aryl chlorides. *Synlett*, 2247–2251.

155. Urgaonkar, S., Nagarajan, M., Verkade, J. G. (2002). Pd/P(i-$BuNCH_2CH_2)_3N$: An efficient catalyst for Suzuki cross-coupling of aryl bromides and chlorides with arylboronic acids. *Tetrahedron Letters* **43**, 8921–8924.

156. Ackermann, L., Potukuchi, H. K., Althammer, A., Born, R., Mayer, P. (2010). Tetra-*ortho*-substituted biaryls through palladium-catalyzed Suzuki–Miyaura couplings with a diaminochlorophosphine ligand. *Organic Letters* **12**, 1004–1007.

157. Cho, S.-D., Kim, H.-K., Yim, H.-s., Kim, M.-R., Lee, J.-K., Kim, J.-J., Yoon, Y.-J. (2007). Suzuki–Miyaura coupling reaction of aryl chlorides using di(2,6-dimethylmorpholino)phenylphosphine as ligand. *Tetrahedron* **63**, 1345–1352.

158. Clarke, M. L., Cole-Hamilton, D. J., Woollins, J. D. (2001). Synthesis of bulky, electron rich hemilabile phosphines and their application in the Suzuki coupling reaction of aryl chlorides. *Journal of the Chemical Society, Dalton Transactions*, 2721–2723.

159. Fortman, G. C., Nolan, S. P. (2011). *N*-Heterocyclic carbene (NHC) ligands and palladium in homogeneous cross-coupling catalysis: A perfect union. *Chemical Society Reviews* **40**, 5151–5169.

160. Kantchev, E. A. B., O'Brien, C. J., Organ, M. G. (2007). Palladium complexes of *N*-Heterocyclic carbenes as catalysts for cross-coupling reactions—A synthetic chemist's perspective. *Angewandte Chemie International Edition* **46**, 2768–2813.

161. Herrmann, W. A., Reisinger, C.-P., Spiegler, M. (1998). Chelating *N*-heterocyclic carbene ligands in palladium-catalyzed heck-type reactions. *Journal of Organometallic Chemistry* **557**, 93–96.

162. Zhang, C., Trudell, M. L. (2000). Palladium-bisimidazol-2-ylidene complexes as catalysts for general and efficient Suzuki cross-coupling reactions of aryl chlorides with arylboronic acids. *Tetrahedron Letters* **41**, 595–598.

163. Demir, S., Özdemir, I., Çetinkaya, B. (2006). Use of bis(benzimidazolium)–palladium system as a convenient catalyst for Heck and Suzuki coupling reactions of aryl bromides and chlorides. *Applied Organometallic Chemistry* **20**, 254–259.

164. Weskamp, T., Böhm, V. P. W., Herrmann, W. A. (1999). Combining *N*-heterocyclic carbenes and phosphines: Improved palladium(II) catalysts for aryl coupling reactions. *Journal of Organometallic Chemistry* **585**, 348–352.

165. Gstöttmayr, C. W. K., Böhm, V. P. W., Herdtweck, E., Grosche, M., Herrmann, W. A. (2002). A defined *N*-heterocyclic carbene complex for the palladium-catalyzed Suzuki cross-coupling of aryl chlorides at ambient temperatures. *Angewandte Chemie International Edition* **41**, 1363–1365.

166. Arentsen, K., Caddick, S., Cloke, F. G. N. (2005). On the efficiency of two-coordinate palladium(0) *N*-heterocyclic carbene complexes in amination and Suzuki–Miyaura reactions of aryl chlorides. *Tetrahedron* **61**, 9710–9715.

167. Karimi, B., Akhavan, P. F. (2009). Main-chain NHC–palladium polymer as a recyclable self-supported catalyst in the Suzuki–Miyaura coupling of aryl chlorides in water. *Chemical Communications*, 3750–3752.

168. Karimi, B., Akhavan, P. F. (2011). A study on applications of *N*-substituted main-chain NHC–palladium polymers as recyclable self-supported catalysts for the Suzuki–Miyaura coupling of aryl chlorides in water. *Inorganic Chemistry* **50**, 6063–6072.

169. Karimi, B., Akhavan, P. F. (2011). *Chemical Communications* **47**, 7686–7688.

170. Song, C., Ma, Y., Chai, Q., Ma, C., Jiang, W., Andrus, M. B. (2005). Palladium catalyzed Suzuki–Miyaura coupling with aryl chlorides using a bulky phenanthryl N-heterocyclic carbene ligand. *Tetrahedron* **61**, 7438–7446.

171. Özdemir, I., Demir, S., Yasar, S., Çetinkaya, B. (2005). Palladium-catalysed Suzuki reaction of aryl chlorides in aqueous media using 1,3-dialkylimidazolidin-2-ylidene ligands. *Applied Organometallic Chemistry* **19**, 55–58.

172. Torres, G. H., Choppin, S., Colobert, F. (2006). Efficient Suzuki–Miyaura coupling reactions between lithium alkynyltrimethylborates and aryl chlorides. *European Journal of Organic Chemistry*, 1450–1454.

173. Özdemir, I., Demir, S., Çetinkaya, B. (2005). Use of tetrahydropyrimidinium salts for highly efficient palladium-catalyzed cross-coupling reactions of aryl bromides and chlorides. *Tetrahedron* **61**, 9791–9798.

174. Özdemir, I., Gürbüz, N., Gök, Y., Çetinkaya, E., Çetinkaya, B. (2005). Palladium-catalyzed Suzuki–Miyaura reaction of aryl chlorides in aqueous media using tetrahydrodiazepinium salts as carbene ligands. *Synlett*, 2394–2396.

175. Özdemir, I., Alici, B., Gürbüz, N., Çetinkaya, E., Çetinkaya, B. (2004). *In situ* generated palladium catalysts bearing 1,3-dialkylperimidin-2-yline ligands for Suzuki reactions of aryl chlorides. *Journal of Molecular Catalysis A: Chemical* **217**, 37–40.

176. Altenhoff, G., Goddard, R., Lehmann, C. W., Glorius, F. (2003). An *N*–heterocyclic carbene ligand with flexible steric bulk allows Suzuki cross-coupling of sterically hindered aryl chlorides at room temperature. *Angewandte Chemie International Edition* **42**, 3690–3693.

177. Singh, R., Viciu, M. S., Kramareva, N., Navarro, O., Nolan, S. P. (2005). Simple (imidazol-2-ylidene)–Pd–Acetate complexes as effective precatalysts for sterically hindered Suzuki–miyaura couplings. *Organic Letters* **7**, 1829–1832.

178. Diebolt, O., Braunstein, P., Nolan, S. P., Cazin, C. S. J. (2008). Room-temperature activation of aryl chlorides in Suzuki–Miyaura coupling using a $[Pd(\mu\text{-}Cl)Cl(NHC)]_2$ complex. *Chemical Communications*, 3190–3192.

179. O'Brien, C. J., Assen, E., Kantchev, B., Valente, C., Hadei, N., Chass, G. A., Lough, A., Hopkinson, A. C., Organ, M. G. (2006). Easily prepared air- and moisture-stable Pd–NHC (NHC = *N*-heterocyclic carbene) complexes: A reliable, user-friendly, highly active palladium precatalyst for the Suzuki–Miyaura reaction. *Chemistry—A European Journal* **12**, 4743–4748.

180. Tang, Y.-Q., Lu, J.-M., Shao, L.-X. (2011). NHC-Pd(II)-Im (NHC = *N*-heterocyclic carbene; Im = 1-methylimidazole) complexes as efficient catalysts for Suzuki–Miyaura coupling reactions of aryl chlorides. *Journal of Organometallic Chemistry* **696**, 3741–3744.

181. Valente, C., Baglione, S., Candito, D., O'Brien, C. J., Organ, M. G. (2008). *Chemical Communications*, 735–737.

182. Jin, Z., Guo, S.-X., Gu, X.-P., Qiu, L.-L., Song, H.-B., Fang, J.-X. (2009). Highly active, well-defined (cyclopentadiene)(*N*-heterocyclic carbene)palladium chloride complexes for room-temperature Suzuki–Miyaura and Buchwald–Hartwig cross-coupling reactions of aryl chlorides and deboronation homocoupling of arylboronic Acids. *Advanced Synthesis & Catalysis* **351**, 1575–1585.

183. Marion, N., Navarro, O., Mei, J., Stevens, E. D., Scott, N. M., Nolan, S. P. (2006). Modified (NHC)Pd(allyl)Cl (NHC = *N*-heterocyclic carbene) complexes for room-temperature Suzuki–Miyaura and Buchwald–Hartwig reactions. *Journal of the American Chemical Society* **128**, 4101–4111.

184. Beller, M., Fischer, H., Herrmann, W. A., Öfele, K., Brossmer, C. (1995). Palladacycles as efficient catalysts for aryl coupling reactions. *Angewandte Chemie International Edition* **34**, 1848–1849.

185. Bedford, R. B., Welch, S. L. (2001). Palladacyclic phosphinite complexes as extremely high activity catalysts in the Suzuki reaction. *Chemical Communications*, 129–130.

186. Gibson, S., Foster, D. F., Eastham, G. R., Toozeb, R. P., Cole-Hamilton, D. J. (2001). Highly efficient C–C coupling reactions using metallated benzylphosphine complexes of palladium. *Chemical Communications*, 779–780.

187. Bedford, R. B., Hazelwood (née Welch), S. L., Horton, P. N., Hursthouse, M. B. (2003). Orthopalladated phosphinite complexes as high-activity catalysts for the Suzuki reaction. *Dalton Transactions*, 4164–4174.

188. Bedford, R. B., Cazin, C. S. J., Hazelwood (née Welch), S. L. (2002). Simple mixed tricyclohexylphosphane-triarylphosphite complexes as extremely high-activity catalysts for the Suzuki coupling of aryl chlorides. *Angewandte Chemie International Edition* **41**, 4120–4122.

189. Bedford, R. B., Hazelwood (née Welch), S. L., Limmert, M. E., Albisson, D. A., Draper, S. M., Scully, P. N., Coles, S. J., Hursthouse, M. B. (2003). Orthopalladated and -platinated bulky triarylphosphite complexes: Synthesis, reactivity and application as high-activity catalysts for Suzuki and Stille coupling reactions. *Chemistry–A European Journal* **9**, 3216–3227.

190. Bedford, R. B., Hazelwood (née Welch), S. L., Limmert, M. E. (2002). Extremely high activity catalysts for the Suzuki coupling of aryl chlorides: The importance of catalyst longevity. *Chemical Communications*, 2610–2611.

191. Bedford, R. B. (2003). Palladacyclic catalysts in C–C and C–heteroatom bond-forming reactions. *Chemical Communications*, 1787–1796.

192. Bedford, R. B., Cazin, C. S. J. (2001). Highly active catalysts for the Suzuki coupling of aryl chlorides. *Chemical Communications*, 1540–1541.

193. Bedford, R. B., Cazin, C. S. J. (2003). High-activity catalysts for Suzuki coupling and amination reactions with deactivated aryl chloride substrates: Importance of the palladium source. *Organometallics* **22**, 987–999.

194. Andreu, M. G., Zapf, A., Beller, M. (2000). Molecularly defined palladium(0) monophosphine complexes as catalysts for efficient cross-coupling of aryl chlorides and phenylboronic acid. *Chemical Communications*, 2475–2476.

195. Botella, L., Nájera, C. (2002). A convenient oxime-carbapalladacycle-catalyzed Suzuki cross-coupling of aryl chlorides in water. *Angewandte Chemie International Edition* **41**, 179–181.

196. Alonso, D. A., Nájera, C., Pacheco, M. C. (2002). Highly active oxime-derived palladacycle complexes for Suzuki–Miyaura and Ullmann-type coupling reactions. *The Journal of Organic Chemistry* **67**, 5588–5594.

197. Alacid, E., Nájera, C. (2008). First cross-coupling reaction of potassium aryltrifluoroborates with organic chlorides in aqueous media catalyzed by an oxime-derived palladacycle. *Organic Letters* **10**, 5011–5014.

198. Alacid, E., Nájera, C. (2009). General reaction conditions for the palladium-catalyzed vinylation of aryl chlorides with potassium alkenyltrifluoroborates. *The Journal of Organic Chemistry* **74**, 8191–8195.

199. Cívicos, J. F., Alonso, D. A., Nájera, C. (2011). Oxime palladacycle-catalyzed Suzuki–Miyaura alkenylation of aryl, heteroaryl, benzyl, and allyl chlorides under microwave irradiation conditions. *Advanced Synthesis & Catalysis* **353**, 1683–1687.

200. Schnyder, A., Indolese, A. F., Studer, M., Blaser, H.-U. (2002). A new generation of air stable, highly active Pd complexes for C–C and C–N coupling reactions with aryl chlorides. *Angewandte Chemie International Edition* **41**, 3668–3671.

201. Navarro, O., Kelly, R. A., III, Nolan, S. P. (2003). A general method for the Suzuki–Miyaura cross-coupling of sterically hindered aryl chlorides: Synthesis of di- and tri-ortho-substituted biaryls in 2-propanol at room temperature. *Journal of the American Chemical Society* **125**, 16194–16195.

202. Xu, C., Gong, J.-F., Guo, T., Zhang, Y.-H., Wu, Y.-J. (2008). Cyclopalladated ferrocenylimine complexes with dicyclohexylphosphinobiphenyl ligands: Synthesis, crystal structures and their use as highly efficient catalysts for Suzuki reaction of aryl chlorides. *Journal of Molecular Catalysis A: Chemical* **279**, 69–76.

203. Cui, M., Li, J., Yu, A., Zhang, J., Wu, Y.-J. (2008). Carbene adduct of cyclopalladated ferrocenylimine: Efficient catalyst for the Suzuki coupling of sterically hindered aryl chlorides with a weaker base and low catalyst loading. *Journal of Molecular Catalysis A: Chemical* **290**, 67–71.

204. Bedford, R. B., Cazin, C. S. J., Hursthouse, M. B., Light, M. E., Scordia, V. J. M. (2004). Di- and tri-alkylphosphine adducts of S-donor palladacycles as catalysts in the Suzuki coupling of aryl chlorides. *Dalton Transactions*, 3864–3868.

205. Peh, G.-R., Kantchev, E. A. B., Er, J.-C., Ying, J. Y. (2010). Rational exploration of *N*-heterocyclic carbene (NHC) palladacycle diversity: A highly active and versatile precatalyst for Suzuki–Miyaura coupling reactions of deactivated aryl and alkyl substrates. *Chemistry—A European Journal* **16**, 4010–4017.

206. Özdemir, İ., Çetinkay, B., Demir, S. (2004). Synthesis of novel 1-alkylimidazoline and 1-alkylbenzimidazole palladium(II) complexes as efficient catalysts for Heck and Suzuki reactions involving arylchlorides. *Journal of Molecular Catalysis A: Chemical*, **208**, 109–114.

207. Özdemir, İ., Şahin, N., Gök, Y., Demir, S., Çetinkay, B. (2005). *In situ* generated 1-alkylbenzimidazole–palladium catalyst for the Suzuki coupling of aryl chlorides. *Journal of Molecular Catalysis A: Chemical* **234**, 181–185.

208. Li, J.-H., Zhu, Q.-M., Xie, Y.-X. (2006). $Pd(OAc)_2$/DABCO-catalyzed Suzuki–Miyaura cross-coupling reaction in DMF. *Tetrahedron* **62**, 10888–10895.

209. Jin, J., Cai, M.-M., Li, J.-X. (2009). Highly efficient Suzuki coupling of aryl chlorides in a continuous flow capillary microreactor. *Synlett*, 2534–2538.

210. John, A., Shaikh, M. M., Butcherc, R. J., Ghosh, P. (2010). Highly efficient palladium precatalysts of homoscorpionate bispyrazolyl ligands for the more challenging Suzuki-Miyaura cross-coupling of aryl chlorides. *Dalton Transactions* **39**, 7353–7363.

211. Nájera, C., Gil-Moltó, J., Karlström, S. (2004). Suzuki–Miyaura and related cross-couplings in aqueous solvents catalyzed by di(2-pyridyl)methylamine-palladium dichloride complexes. *Advanced Synthesis & Catalysis* **346**, 1798–1811.

212. Nájera, C., Gil-Moltó, J., Karlström, S., Falvello, L. R. (2003). Di-2-pyridylmethylamine-based palladium complexes as new catalysts for Heck, Suzuki, and Sonogashira reactions in organic and aqueous solvents. *Organic Letters* **5**, 1451–1454.

213. Lee, D.-H., Jin, M.-J. (2011). An extremely active and general catalyst for Suzuki coupling reaction of unreactive aryl chlorides. *Organic Letters* **13**, 252–255.

214. Saito, S., Sakai, M., Miyaura, N. (1996). A synthesis of biaryls via nickel(0)-Catalyzed cross-coupling reaction of chloroarenes with phenylboronic acids. *Tetrahedron Letters* **37**, 2993–2996.

215. Indolese, A. F. (1997). Suzuki-type coupling of chloroarenes with arylboronic acids catalysed by nickel complexes. *Tetrahedron Letters* **38**, 3513–3516.

216. Galland, J.-C., Savignac, M., Gentt, J.-P. (1999). Cross-coupling of chloroarenes with boronic acids using a water-soluble nickel catalyst. *Tetrahedron Letters* **40**, 2323–2326.

217. Saito, S., Oh-tani, S., Miyaura, N. (1997). Synthesis of biaryls via a nickel(0)-catalyzed cross-coupling reaction of chloroarenes with arylboronic acids. *The Journal of Organic Chemistry* **62**, 8024–8030.

218. Lipshutz, B. H., Sclafani, J. A., Blomgren, P. A. (2000). Biaryls via Suzuki cross-couplings catalyzed by nickel on charcoal. *Tetrahedron* **56**, 2139–2144.

219. Tang, Z.-Y., Hu, Q.-S. (2006). Triphenylphosphine as a ligand for room-temperature Ni(0)-catalyzed cross-coupling reactions of aryl chlorides with arylboronic acids. *The Journal of Organic Chemistry* **71**, 2167–2169.

220. Fan, X.-H., Yang, L.-M. (2011). Room-temperature nickel-catalysed Suzuki–Miyaura reactions of aryl sulfonates/halides with arylboronic acids. *European Journal of Organic Chemistry*, 1467–1471.

221. Chen, C., Yang, L.-M. (2007). Nickel(II)–aryl complexes as catalysts for the Suzuki cross-coupling reaction of chloroarenes and arylboronic acids. *Tetrahedron Letters* **48**, 2427–2430.

222. Lu, Y., Plocher, E., Hu, Q.-S. (2006). Synthesis of novel bisphosphine-containing polymers and their applications as bidentate ligands for nickel(0)-catalyzed cross-coupling reactions. *Advanced Synthesis & Catalysis* **348**, 841–845.

223. Lu, Z., Fu, G. C. (2010). Alkyl–Alkyl Suzuki cross-coupling of unactivated secondary alkyl chlorides. *Angewandte Chemie International Edition* **49**, 6676–6678.

224. Lu, Z., Wilsily, A., Fu, G. C. (2011). Stereoconvergent amine-directed alkyl–alkyl Suzuki reactions of unactivated secondary alkyl chlorides. *Journal of the American Chemical Society* **133**, 8154–8157.

225. Owston, N. A., Fu, G. C. (2010). Asymmetric alkyl–alkyl cross-couplings of unactivated secondary alkyl electrophiles: Stereoconvergent Suzuki reactions of racemic acylated halohydrins. *Journal of the American Chemical Society* **132**, 11908–11909.

226. González-Bobes, F., Fu, G. C. (2006). Amino alcohols as ligands for nickel-catalyzed Suzuki reactions of unactivated alkyl halides, including secondary alkyl chlorides, with arylboronic acids. *Journal of the American Chemical Society* **128**, 5360–5361.

227. Molander, G. A., Argintaru, O. A., Aron, I., Dreher, S. D. (2010). Nickel-catalyzed cross-coupling of potassium aryl- and heteroaryltrifluoroborates with unactivated alkyl halides. *Organic Letters* **12**, 5783–5785.

228. Zell, T., Feierabend, M., Halfter, B., Radius, U. (2011). Stoichiometric and catalytic CeCl activation of aryl chlorides using an NHC-stabilized nickel(0) complex. *Journal of Organometallic Chemistry* **696**, 1380–1387.

229. Lee, C.-C., Ke, W.-C., Chan, K.-T., Lai, C.-L., Hu, C.-H., Lee, H. M. (2007). Nickel(II) complexes of bidentate *N*-heterocyclic carbene/phosphine ligands: Efficient catalysts for Suzuki coupling of aryl chlorides. *Chemistry–A European Journal* **13**, 582–591.

230. Xi, Z., Zhang, X., Chen, W., Fu, S., Wang, D. (2007). Synthesis and structural characterization of nickel(II) complexes supported by pyridine-functionalized *N*-heterocyclic carbene ligands and their catalytic acitivities for Suzuki coupling. *Organometallics* **26**, 6636–6642.

231. Yang, C.-T., Zhang, Z.-Q., Liu, Y.-C., Liu, L. (2011). Copper-catalyzed cross-coupling reaction of organoboron compounds with primary alkyl halides and pseudohalides. *Angewandte Chemie International Edition* **50**, 3904–3907.

232. Ueura, K., Satoh, T., Miura, M. (2005). Rhodium-catalyzed arylation using arylboron compounds: Efficient coupling with aryl halides and unexpected multiple arylation of benzonitrile. *Organic Letters* **7**, 2229–2231.

233. House, H. O., Ghali, N. I., Haack, J. L., Derveer, D. (1980). Reactions of the l,8-diphenylanthracene system. *The Journal of Organic Chemistry* **45**, 1807–1817.

234. Dai, C., Fu, G. C. (2001). The first general method for palladium-catalyzed Negishi cross-coupling of aryl and vinyl chlorides: Use of commercially available $Pd(P(t\text{-}Bu)_3)_2$ as a catalyst. *Journal of the American Chemical Society* **123**, 2719–2724.

235. Walla, P., Kappe, C. O. (2004). Microwave-assisted Negishi and Kumada cross-coupling reactions of aryl chlorides. *Chemical Communications*, 564–565.

236. Zhou, J., Fu, G. C. (2003). Palladium-catalyzed Negishi cross-coupling reactions of unactivated alkyl iodides, bromides, chlorides, and tosylates. *Journal of the American Chemical Society* **125**, 12527–12530.

237. Sonoda, M., Inaba, A., Itahashi, K., Tobe, Y. (2001). Synthesis of differentially substituted hexaethynylbenzenes based on tandem Sonogashira and Negishi cross-coupling reactions. *Organic Letters* **3**, 2419–2421.

238. Xu, H., Ekoue-Kovi, K., Wolf, C. (2008). Palladium–Phosphinous acid-catalyzed cross-coupling of aryl and acyl halides with aryl-, alkyl-, and vinylzinc reagents. *The Journal of Organic Chemistry* **73**, 7638–7650.

239. Milne, J. E., Buchwald, S. L. (2004). An extremely active catalyst for the Negishi cross-coupling reaction. *Journal of the American Chemical Society* **126**, 13028–13032.

240. Luzung, M. R., Patel, J. S., Yin, J. (2010). A mild Negishi cross-coupling of 2-heterocyclic organozinc reagents and aryl chlorides. *The Journal of Organic Chemistry* **75**, 8330–8332.

241. Tan, Z., Negishi, E.-I. (2006). Widely applicable Pd-catalyzed *trans*-selective monoalkylation of unactivated 1,1-dichloro-1-alkenes and Pd-catalyzed second substitution for the selective synthesis of E or Z trisubstituted alkenes. *Angewandte Chemie International Edition* **45**, 762–765.

242. Shi, J.-C., Zeng, X., Negishi, E.-I. (2003). Highly selective synthesis of (*E*)-3-methyl-1-trialkylsilyl-3-en-1-ynes via *trans*-selective alkynylation catalyzed by Cl2Pd(DPEphos) and stereospecific methylation with methylzincs catalyzed by $Pd(tBu_3P)_2$. *Organic Letters* **5**, 1825–1828.

243. Shi, J.-C., Negishi, E.-I. (2003). Pd-catalyzed selective tandem arylation–alkylation of 1,1-dihalo-1-alkenes with aryl- and alkylzinc derivatives to produce α-alkyl-substituted styrene derivatives. *Journal of Organometallic Chemistry* **687**, 518–524.

244. Çalimsiz, S., Sayah, M., Mallik, D., Organ, M. G. (2010). Pd-PEPPSI-IPent: Low-temperature Negishi cross-coupling for the preparation of highly functionalized, tetra-*ortho*-substituted biaryls. *Angewandte Chemie International Edition* **49**, 2014–2017.

245. Çalimsiz, S., Organ, M. G. (2011). Negishi cross-coupling of secondary alkylzinc halides with aryl/heteroaryl halides using Pd-PEPPSI-IPent. *Chemical Communications* **47**, 5181–5183.

246. Sase, S., Jaric, M., Metzger, A., Malakhov, V., Knochel, P. (2008). One-pot Negishi cross-coupling reactions of *in situ* generated zinc reagents with aryl chlorides, bromides, and triflates. *The Journal of Organic Chemistry* **73**, 7380–7382.

247. Valente, C., Belowich, M. E., Hadei, N., Organ, M. G. (2010). Pd-PEPPSI complexes and the Negishi reaction. *European Journal of Organic Chemistry*, 4343–4354.

248. Hadei, N., Achonduh, G. T., Valente, C., O'Brien, C. J., Organ, M. G. (2011). Differentiating C–Br and C–Cl bond activation by using solvent polarity: Applications to orthogonal alkyl–alkyl Negishi reactions. *Angewandte Chemie International Edition* **50**, 3896–3899.

249. Miller, J. A., Farrell, R. P. (1998). Preparation of unsymmetrical biaryls via Ni- or Pd-catalyzed coupling of aryl chlorides with arylzincs. *Tetrahedron Letters* **39**, 6441–6444.

250. Lipshutz, B. H., Blomgren, P. A., Kim, S.-K. (1999). Substitution reactions of aryl chlorides with organozinc reagents catalyzed by Ni(0). *Tetrahedron Letters* **40**, 197–200.

251. Lipshutz, B. H., Blomgren, P. A. (1999). Nickel on charcoal ("Ni/C"): An expedient and inexpensive heterogeneous catalyst for cross-couplings between aryl chlorides and organometallics. I. Functionalized organozinc reagents. *Journal of the American Chemical Society* **121**, 5819–5820.

252. Schade, M. A., Metzger, A., Hug, S., Knochel, P. (2008). Nickel-catalyzed cross-coupling reactions of benzylic zinc reagents with aromatic bromides, chlorides and tosylates. *Chemical Communications*, 3046–3048.

253. Gerber, R., Frech, C. M. (2011). Negishi cross-coupling reactions catalyzed by an aminophosphine-based nickel system: A reliable and general applicable reaction protocol for the high-yielding synthesis of biaryls. *Chemistry–A European Journal* **17**, 11893–11904.

254. Melzig, L., Dennenwaldt, T., Gavryushin, A., Knochel, P. (2011). Direct Aminoalkylation of arenes, heteroarenes, and alkenes via Ni-catalyzed Negishi cross-coupling reactions. *The Journal of Organic Chemistry* **76**, 8891–8906.

255. Wang, L., Wang, Z.-X. (2007). Efficient cross-coupling of aryl chlorides with arylzinc reagents catalyzed by amido pincer complexes of nickel. *Organic Letters* **9**, 4335–4338.

256. Liu, N., Wang, L., Wang, Z.-X. (2011). Room-temperature nickel-catalysed cross-couplings of aryl chlorides with arylzincs. *Chemical Communications* **47**, 1598–1600.

257. Zhang, C., Wang, Z.-X. (2009). *N*-Heterocyclic carbene-based nickel complexes: synthesis and catalysis in cross-couplings of aryl chlorides with ArMX (M = Mg or Zn). *Organometallics* **28**, 6507–6514.

258. Xi, Z., Zhou, Y., Chen, W. (2008). Efficient Negishi coupling reactions of aryl chlorides catalyzed by binuclear and mononuclear nickel-*N*-heterocyclic carbene complexes. *The Journal of Organic Chemistry* **73**, 8497–8501.

259. Wang, Z.-X., Liu, N. (2012). Nickel-catalyzed cross-coupling with pincer ligands. *European Journal of Inorganic Chemistry*, 901–911.

260. Arp, F. O., Fu, G. C. (2005). Catalytic enantioselective Negishi reactions of racemic secondary benzylic halides. *Journal of the American Chemical Society* **127**, 10482–10483.

261. Son, S., Fu, G. C. (2008). Nickel-catalyzed asymmetric Negishi cross-couplings of secondary allylic chlorides with alkylzincs. *Journal of the American Chemical Society* **130**, 2756–2757.
262. Amatorea, M., Gosmini, C. (2008). Synthesis of functionalised diarylmethanes via a cobalt-catalysed cross-coupling of arylzinc species with benzyl chlorides. *Chemical Communications*, 5019–5021.
263. Malosh, C. F., Ready, J. M. (2004). Catalytic cross-coupling of alkylzinc halides with α-chloroketones. *Journal of the American Chemical Society* **126**, 10240–10241.
264. Frisch, A. C., Shaikh, N., Zapf, A., Beller, M. (2002). Palladium-catalyzed coupling of alkyl chlorides and Grignard reagents. *Angewandte Chemie International Edition* **41**, 4056–4059.
265. Ackermann, L., Kapdi, A. R., Schulzke, C. (2010). Air-stable secondary phosphine oxide or chloride (pre)ligands for cross-couplings of unactivated alkyl chlorides. *Organic Letters* **12**, 2298–2301.
266. Bonnet, V., Mongin, F., Trécourt, F., Quéguiner, G., Knochel, P. (2001). Preparation of polyfunctional pyridines by a palladium(0)-catalyzed cross-coupling of functionalized aryl Grignard reagents. *Tetrahedron Letters* **42**, 5717–5719.
267. Bonnet, V., Mongin, F., Trécourt, F., Quéguiner, G., Knochel, P. (2002). Syntheses of substituted pyridines, quinolines and diazines via palladium-catalyzed cross-coupling of aryl Grignard reagents. *Tetrahedron*, **58**, 4429–4438.
268. Ishikawa, S., Manabe, K. (2007). Highly ortho-selective cross-coupling of dichlorobenzene derivatives with Grignard reagents. *Organic Letters* **9**, 5593–5595.
269. Li, G. Y. (2002). Highly active, air-stable palladium catalysts for Kumada–Tamao–Corriu cross-coupling reaction of inactivated aryl chlorides with aryl Grignard reagents. *Journal of Organometallic Chemistry* **653**, 63–68.
270. Wolf, C., Xu, H. (2008). Efficient synthesis of sterically crowded biaryls by palladium–phosphinous acid-catalyzed cross-coupling of aryl halides and aryl grignards. *The Journal of Organic Chemistry* **73**, 162–167.
271. Ackermann, L., Gschrei, C. J., Althammer, A., Riederer, M. (2006). Cross-coupling reactions of aryl and vinyl chlorides catalyzed by a palladium complex derived from an air-stable H-phosphonate. *Chemical Communications*, 1419–1421.
272. Huang, J., Nolan, S. P. (1999). Efficient cross-coupling of aryl chlorides with aryl Grignard reagents (Kumada reaction) mediated by a palladium/imidazolium chloride system. *Journal of the American Chemical Society* **121**, 9889–9890.
273. Hartmann, C. E., Nolan, S. P., Cazin, C. S. J. (2009). Highly active $[Pd(\mu\text{-}Cl)(Cl)(NHC)]_2$ (NHC = *N*-heterocyclic carbene) in the cross-coupling of Grignard reagents with aryl chlorides. *Organometallics* **28**, 2915–2919.
274. Frisch, A. C., Rataboul, F., Zapf, A., Beller, M. (2003). First Kumada reaction of alkyl chlorides using *N*-heterocyclic carbene/palladium catalyst systems. *Journal of Organometallic Chemistry* **687**, 403–409.
275. Jin, Z., Gu, X.-P., Qiu, L.-L., Wu, G.-P., Song, H.-B., Fang, J.-X. (2011). Air-stable CpPd(NHC)Cl (NHC *N*-heterocyclic carbene) complexes as highly active precatalysts for Kumada–Tamao–Corriu coupling of aryl and heteroaryl chlorides. *Journal of Organometallic Chemistry* **696**, 859–863.
276. Organ, M. G., Abdel-Hadi, M., Avola, S., Hadei, N., Nasielski, J., O'Brien, C. J., Valente, C. (2007). Biaryls made easy: PEPPSI and the Kumada–Tamao–Corriu reaction. *Chemistry—A European Journal* **13**, 150–157.

277. Herrmann, W. A., Böhm, V. P. W., Reisinger, C.-P. (1999). Application of palladacycles in Heck type reactions. *Journal of Organometallic Chemistry* **576**, 23–41.

278. Ren, G., Cui, X., Wu, Y. (2010). Efficient synthesis of biaryls through the Kumada reaction catalyzed by carbene adducts of cyclopalladated ferrocenylimine. *European Journal of Organic Chemistry*, 2372–2378.

279. Tasler, S., Lipshutz, B. H. (2003). Nickel-on-charcoal-catalyzed aromatic aminations and Kumada couplings: Mechanistic and synthetic aspects. *The Journal of Organic Chemistry* **68**, 1190–1199.

280. Ghosh, R., Sarkar, A. (2010). Bidentate P, N-P ligand for nickel-catalyzed cross-coupling of Aryl or benzyl chlorides with ArMgX. *The Journal of Organic Chemistry* **75**, 8283–8286.

281. Xie, L.-G., Wang, Z.-X. (2010). Nickel-catalyzed cross-coupling of non-activated or functionalized aryl halides with aryl Grignard reagents. *Chemistry–A European Journal* **16**, 10332–10336.

282. Yoshikai, N., Mashima, H., Nakamura, E. (2005). Nickel-catalyzed cross-coupling reaction of aryl fluorides and chlorides with Grignard reagents under nickel/magnesium bimetallic cooperation. *Journal of the American Chemical Society* **127**, 17978–17979.

283. Yoshikai, N., Mashima, H., Nakamura, E. (2009). Hydroxyphosphine ligand for nickel-catalyzed cross-coupling through nickel/magnesium bimetallic cooperation. *Journal of the American Chemical Society* **131**, 9590–9599.

284. Li, G. Y., Marshall, W. J. (2002). Air-stable phosphine sulfide ligand precursors for nickel-catalyzed cross-coupling reactions of unactivated aryl chlorides with aryl Grignard reagents. *Organometallics* **21**, 590–591.

285. Ackermann, L., Born, R., Spatz, J. H., Meyer, D. (2005). Efficient aryl–(hetero)aryl coupling by activation of C–Cl and C–F bonds using nickel complexes of air-stable phosphine oxides. *Angewandte Chemie International Edition* **44**, 7216–7219.

286. Jin, Z., Li, Y.-J., Ma, Y.-Q., Qiu, L.-L., Fang, J.-X. (2012). Biphenyl-based diaminophosphine oxides as air-stable preligands for the nickel-catalyzed Kumada–Tamao–Corriu coupling of deactivated aryl chlorides, fluorides, and tosylates. *Chemistry—A European Journal* **18**, 446–450.

287. Böhm, V. P. W., Veskamp, T., Gstöttmayr, C. W. K., Herrmann, W. A. (2000). Nickel-catalyzed cross-coupling of aryl chlorides with aryl Grignard reagents. *Angewandte Chemie International Edition* **39**, 1602–1604.

288. Wolf, J., Labande, A., Natella, M., Daran, J.-C., Poli, R. (2006). Aryl Grignard cross-coupling of aryl chlorides catalysed by new, highly active phosphine/imidazolium nickel(II) complexes. *Journal of Molecular Catalysis A: Chemical* **259**, 205–212.

289. Berding, J., van Dijkman, T. F., Lutz, M., Spek, A. L., Bouwman, E. (2009). N-donor functionalized *N*-heterocyclic carbene nickel(II) complexes in the Kumada coupling. *Dalton Transactions*, 6948–6955.

290. Berding, J., Lutz, M., Spek, A. L., Bouwman, E. (2009). Synthesis of novel chelating benzimidazole-based carbenes and their nickel(II) complexes: Activity in the Kumada coupling reaction. *Organometallics* **28**, 1845–1854.

291. Xi, Z., Liu, B., Chen, W. (2008). Room-temperature Kumada cross-coupling of unactivated aryl chlorides catalyzed by *N*-heterocylic carbene-based nickel(II) complexes. *The Journal of Organic Chemistry* **73**, 3954–3957.

292. Csok, Z., Vechorkin, O., Harkins, S. B., Scopelliti, R., Hu, X. (2008). Nickel complexes of a pincer NN2 ligand: Multiple carbon-chloride activation of CH_2Cl_2 and $CHCl_3$ leads to selective carbon–carbon bond formation. *Journal of the American Chemical Society* **130**, 8156–8157.

293. Wang, Z.-X., Wang, L. (2007). Amido pincer complex of nickel-catalysed Kumada cross-coupling reactions. *Chemical Communications*, 2423–2425.

294. Sun, K., Wang, L., Wang, Z.-X. (2008). Synthesis and characterization of amido pincer complexes of lithium and nickel and catalysis of the nickel complexes in the Kumada cross-coupling. *Organometallics* **27**, 5649–5656.

295. Liu, N., Wang, Z.-X. (2011). Kumada coupling of aryl, heteroaryl, and vinyl chlorides catalyzed by amido pincer nickel complexes. *The Journal of Organic Chemistry* **76**, 10031–10038.

296. Zhou, Y., Xi, Z., Chen, W., Wang, D. (2008). Dinickel(II) complexes of bis(*N*-heterocyclic carbene) ligands containing [Ni_2(μ-OH)] cores as highly efficient catalysts for the coupling of aryl chlorides. *Organometallics*, **27**, 5911–5920.

297. Gu, S., Chen, W. (2009). Pincer complexes of palladium- and nickel-containing 3-butyl-1-(1,10-phenanthrolin-2-yl)imidazolylidene as efficient aqueous Sonogashira and Kumada coupling reactions. *Organometallics* **28**, 909–914.

298. Terao, J., Watanabe, H., Ikumi, A., Kuniyasu, H., Kambe, N. (2002). Nickel-catalyzed cross-coupling reaction of Grignard reagents with alkyl halides and tosylates: Remarkable effect of 1,3-butadienes. *Journal of the American Chemical Society* **124**, 4222–4223.

299. Iwasaki, T., Tsumura, A., Omori, T., Kuniyasu, H., Terao, J., Kambe, N. (2011). Kinetic studies of the Ni-catalyzed cross-coupling of alkyl halides and a tosylate with butyl Grignard reagent in the presence of 1,3-butadiene. *Chemistry Letters* **40**, 1024–1026.

300. Terao, J., Nii, S., Chowdhury, F. A., Nakamura, A., Kambe, N. (2004). Nickel-catalyzed regioselective three component coupling reaction of alkyl halides, Butadienes, and Ar-M (M = MgX, ZnX). *Advanced Synthesis & Catalysis* **346**, 905–908.

301. Fürstner, A., Leitner, A. (2002). Iron-catalyzed cross-coupling reactions of alkyl-Grignard reagents with aryl chlorides, tosylates, and triflates. *Angewandte Chemie International Edition* **41**, 609–612.

302. Fürstner, A., Leitner, A., Méndez, M., Krause, H. (2002). Iron-catalyzed cross-coupling reactions. *Journal of the American Chemical Society* **124**, 13856–13863.

303. Ottesen, L. K., Ek, F., Olsson, R. (2006). Iron-catalyzed cross-coupling of imidoyl chlorides with Grignard reagents. *Organic Letters* **8**, 1771–1773.

304. Seck, M., Franck, X., Hocquemiller, R., Figadère, B., Peyrat, J.-F., Provot, O., Brion, J.-D., Alami, M. (2004). Synthesis of substituted quinolines by iron-catalyzed coupling reactions between chloroenynes and Grignard reagents. *Tetrahedron Letters* **45**, 1881–1884.

305. Scheiper, B., Bonnekessel, M., Krause, H., Fürstner, A. (2004). Selective iron-catalyzed cross-coupling reactions of Grignard reagents with enol triflates, acid chlorides, and dichloroarenes. *The Journal of Organic Chemistry* **69**, 3943–3949.

306. Cahiez, G., Avedissian, H. (1998). Highly stereo- and chemoselective iron-catalyzed alkenylation of organomagnesium compounds. *Synthesis*, 1199–1205.

307. Hocek, M., Dvořáková, H. (2003). An efficient synthesis of 2-substituted 6-methylpurine bases and nucleosides by Fe- or Pd-catalyzed cross-coupling reactions of 2,6-dichloropurines. *The Journal of Organic Chemistry* **68**, 5773–5776.

308. Fürstner, A., Leitner, A. (2003). A catalytic approach to (R)-(+)-muscopyridine with integrated "self-clearance". *Angewandte Chemie International Edition* **42**, 308–311.

309. Xue, F., Zhao, J., Hor, T. S. A. (2011). Iron(II) complexes with functionalized amine-pyrazolyl tripodal ligands in the cross-coupling of aryl Grignard with alkyl halides. *Dalton Transactions* **40**, 8935–8940.

310. Bedford, R. B., Bruce, D. W., Frost, R. M., Hird, M. (2005). Simple iron-amine catalysts for the cross-coupling of aryl Grignards with alkyl halides bearing β-hydrogens. *Chemical Communications*, 4161–4163.

311. Qian, X., Dawe, L. N., Kozak, C. M. (2011). Catalytic alkylation of aryl Grignard reagents by iron(III) amine-bis(phenolate) complexes. *Dalton Transactions* **40**, 933–943.

312. Hatakeyama, T., Nakamura, M. (2007). Iron-catalyzed selective biaryl coupling: Remarkable suppression of homocoupling by the fluoride anion. *Journal of the American Chemical Society* **129**, 9844–9845.

313. Ghorai, S. K., Jin, M., Hatakeyama, T., Nakamura, M. (2012). Cross-coupling of non-activated chloroalkanes with aryl Grignard reagents in the presence of iron/N-heterocyclic carbene catalysts. *Organic Letters* **14**, 1066–1069.

314. Kharasch, M. S., Fuchs, C. F. (1943). Factors influencing the course and mechanisms of Grignard reactions. XI. The effect of metallic halides on the reaction of Grignard reagents with vinyl halides and substituted vinyl halides. *Journal of the American Chemical Society* **65**, 504–507.

315. Cahiez, G., Avedissian, H. (1998). Cobalt-catalyzed alkenylation of organomagnesium reagents. *Tetrahedron Letters* **39**, 6159–6162.

316. Ohmiya, H., Yorimitsu, H., Oshima, K. (2004). Cobalt-catalyzed cross-coupling reaction of chloropyridines with Grignard reagents. *Chemistry Letters* **33**, 1240–1241.

317. Korn, T. J., Schade, M. A., Cheemala, M. N., Wirth, S., Guevara, S. A., Cahiez, G., Knochel, P. (2006). Cobalt-catalyzed cross-coupling reactions of heterocyclic chlorides with arylmagnesium halides and of polyfunctionalized arylcopper reagents with aryl bromides, chlorides, fluorides and tosylates. *Synthesis*, 3547–3574.

318. Korn, T. J., Cahiez, G., Knochel, P. (2003). New cobalt-catalyzed cross-coupling reactions of heterocyclic chlorides with aryl and heteroaryl magnesium halides. *Synlett*, 1892–1894.

319. Ohmiya, H., Yorimitsu, H., Oshima, K. (2006). Cobalt-mediated cross-coupling reactions of primary and secondary alkyl halides with 1-(trimethylsilyl)ethenyl- and 2-trimethylsilylethynylmagnesium reagents. *Organic Letters* **8**, 3093–3096.

320. Zhou, W., Napoline, J. W., Thomas, C. M. (2011). A catalytic application of Co/Zr heterobimetallic complexes: Kumada coupling of unactivated alkyl halides with alkyl Grignard reagents. *European Journal of Inorganic Chemistry*, 2029–2033.

321. Bäckvall, J.-E., Persson, E. S. M., Bombrun, A. (1994). Regiocontrol in copper-catalyzed cross coupling of allylic chlorides with aryl Grignard reagents. *The Journal of Organic Chemistry* **59**, 4126–4130.

322. Terao, J., Todo, H., Begum, S. A., Kuniyasu, H., Kambe, N. (2007). Copper-catalyzed cross-coupling reaction of Grignard reagents with primary-alkyl halides: Remarkable effect of 1-phenylpropyne. *Angewandte Chemie International Edition* **46**, 2086–2089.

323. Hintermann, L., Xiao, L., Labonne, A. (2008). A general and selective copper-catalyzed cross-coupling of tertiary Grignard reagents with azacyclic electrophiles. *Angewandte Chemie International Edition* **47**, 8246–8250.

324. Shirakawa, E., Yamasaki, K., Hiyama, T. (1997). Nickel-catalysed cross-coupling reactions of aryl halides with organostannanes. *Journal of the Chemical Society, Perkin Transactions* **1**, 2449–2450.

325. Espinet, P., Echavarren, A. M. (2004). The mechanisms of the Stille reaction. *Angewandte Chemie International Edition* **43**, 4704–4734.

326. Littke, A. F., Fu, G. C. (1999). The first general method for Stille cross-couplings of aryl chlorides. *Angewandte Chemie International Edition* **38**, 2411–2413.

327. Littke, A. F., Schwarz, L., Fu, G. C. (2002). Pd/P(t-Bu)$_3$: A mild and general catalyst for Stille reactions of aryl chlorides and aryl bromides. *Journal of the American Chemical Society* **124**, 6343–6348.

328. Nabera, J. R., Buchwald, S. L. (2008). Palladium-catalyzed Stille cross-coupling reaction of aryl chlorides using a pre-milled palladium acetate and XPhos catalyst system. *Advanced Synthesis & Catalysis* **350**, 957–961.

329. Su, W., Urgaonkar, S., Verkade, J. G. (2004). $Pd_2(dba)_3$/P(i-$BuNCH_2CH_2)_3$N-catalyzed Stille cross-coupling of aryl chlorides. *Organic Letters* **6**, 1421–1424.

330. Su, W., Urgaonkar, S., McLaughlin, P. A., Verkade, J. G. (2004). Highly active palladium catalysts supported by bulky proazaphosphatrane ligands for Stille cross-coupling: Coupling of aryl and vinyl chlorides, room temperature coupling of aryl bromides, coupling of aryl triflates, and synthesis of sterically hindered biaryls. *Journal of the American Chemical Society* **126**, 16433–16439.

331. Wolf, C., Lerebours, R. (2003). Efficient Stille cross-coupling reaction using aryl chlorides or bromides in water. *The Journal of Organic Chemistry* **68**, 7551–7554.

332. Dowlut, M., Mallik, D., Organ, M. G. (2010). An efficient low-temperature Stille–Migita cross-coupling reaction for heteroaromatic compounds by Pd-PEPPSI-IPent. *Chemistry–A European Journal* **16**, 4279–4283.

333. Lee, D.-H., Taher, A., Ahn, W.-S., Jin, M.-J. (2010). Room temperature Stille cross-coupling reaction of unreactive aryl chlorides and heteroaryl chlorides. *Chemical Communications* **46**, 478–480.

334. Jin, M.-J., Lee, D.-H. (2010). A practical heterogeneous catalyst for the Suzuki, Sonogashira, and Stille coupling reactions of unreactive aryl chlorides. *Angewandte Chemie International Edition* **49**, 1119–1122.

335. Gouda, K.-i., Hagiwara, E., Hatanaka, Y., Hiyama, T. (1996). Cross-coupling reactions of aryl chlorides with organochlorosilanes: Highly effective methods for arylation or alkenylation of aryl chlorides. *The Journal of Organic Chemistry* **61**, 7232–7233.

336. Sahoo, A. K., Oda, T., Nakao, Y., Hiyama, T. (2004). Cross-coupling of triallyl(aryl)silanes with aryl bromides and chlorides: An alternative convenient biaryl synthesis. *Advanced Synthesis & Catalysis* **346**, 1715–1727.

337. Sahoo, A. K., Nakao, Y., Hiyama, T. (2004). A highly effective and practical biaryl synthesis with triallyl(aryl)silanes and aryl chlorides. *Chemistry Letters* **33**, 632–633.

338. Molander, G. A., Iannazzo, L. (2011). Palladium-catalyzed Hiyama cross-coupling of aryltrifluorosilanes with aryl and heteroaryl chlorides. *The Journal of Organic Chemistry* **76**, 9182–9187.

339. Denmark, S. E., Kallemeyn, J. M. (2006). Stereospecific palladium-catalyzed cross-coupling of (E)- and (Z)-alkenylsilanolates with aryl chlorides. *Journal of the American Chemical Society* **128**, 15958–15959.

340. Cho, E. J., Senecal, T. D., Kinzel, T., Zhang, Y., Watson, D. A., Buchwald, S. L. (2010). The palladium-catalyzed trifluoromethylation of aryl chlorides. *Science* **328**, 1679–1681.

341. Wolf, C., Lerebours, R., Tanzini, E. H. (2003). Cross-coupling reactions of aryl chlorides and bromides with phenyltrimthoxysilane catalyzed by palladium–Phosphinous acid complexes. *Synthesis*, 2069–2073.

342. Wolf, C., Lerebours, R. (2004). Palladium–phosphinous acid-catalyzed NaOH-promoted cross-coupling reactions of arylsiloxanes with aryl chlorides and bromides in water. *Organic Letters* **6**, 1147–1150.

343. Lee, H. M., Nolan, S. P. (2000). Efficient cross-coupling reactions of aryl chlorides and bromides with phenyl- or vinyltrimethoxysilane mediated by a palladium/imidazolium chloride system. *Organic Letters* **2**, 2053–2055.

344. Zhang, X., Xia, Q., Chen, W. (2009). Synthesis of tetra- and pentacoordinate palladium complexes of functionalized *N*-heterocyclic carbenes and a comparative study of their catalytic activities. *Dalton Transactions*, 7045–7054.

345. Gu, Z.-S., Shao, L.-X., Lu, J.-M. (2012). NHC-Pd(II)-Im (NHC = *N*-heterocyclic carbene; Im = 1-methylimidazole) complex catalyzed Hiyama reaction of aryl chlorides with aryltrimethoxysilanes. *Journal of Organometallic Chemistry* **700**, 132–134.

346. Li, J.-H., Deng, C.-L., Liu, W.-J., Xie, Y.-X. (2005). $Pd(OAc)_2$/DABCO as an inexpensive and efficient catalytic system for Hiyama Cross-coupling reactions of aryl halides with aryltrimethoxysilanes. *Synthesis*, 3039–3044.

347. Clarke, M. L (2005). First microwave-accelerated Hiyama coupling of aryl- and vinylsiloxane derivatives: Clean cross-coupling of aryl chlorides within minutes. *Advanced Synthesis & Catalysis* **347**, 303–307.

348. Raders, S. M., Kingston, J. V., Verkade, J. G. (2010). Advantageous use of $t\text{Bu}_2\text{P-N=P}(i\text{BuNCH}_2\text{CH}_2)_3\text{N}$ in the Hiyama coupling of aryl bromides and chlorides. *The Journal of Organic Chemistry* **75**, 1744–1747.

349. Lee, D.-H., Jung, J.-Y., Jin, M.-J. (2010). General and highly active catalyst for mono and double Hiyama coupling reactions of unreactive aryl chlorides in water. *Chemical Communications* **46**, 9046–9048.

350. Alacid, E., Nájera, C. (2008). Aqueous sodium hydroxide promoted cross-coupling reactions of akenyltrialkoxysilanes under ligand-free conditions. *The Journal of Organic Chemistry* **73**, 2315–2322.

351. Alacid, E., Nájera, C. (2006). The first fluoride-free Hiyama reaction of vinylsiloxanes promoted by sodium hydroxide in water. *Advanced Synthesis & Catalysis* **348**, 2085–2091.

352. Alacid, E., Nájera, C. (2006). Solvent-less and fluoride-free Hiyama reaction of arylsiloxanes with aryl bromides and chlorides promoted by sodium hydroxide: A useful protocol for palladium recycling and product isolation. *Advanced Synthesis & Catalysis* **348**, 945–952.

353. Strotman, N. A., Sommer, S., Fu, G. C. (2007). Hiyama reactions of activated and unactivated secondary alkyl halides catalyzed by a nickel/norephedrine complex. *Angewandte Chemie International Edition* **46**, 3556–3558.

354. Tang, S., Takeda, M., Nakao, Y., Hiyama, T. (2011). Nickel-catalysed cross-coupling reaction of aryl(trialkyl)silanes with aryl chlorides and tosylates. *Chemical Communications* **47**, 307–309.

355. Chinchilla, R., Nájera, C. (2007). The Sonogashira reaction: A booming methodology in synthetic organic chemistry. *Chemical Reviews* **107**, 874–922.

356. Reddy, E. A., Barange, D. K., Islam, A., Mukkanti, K., Pal, M. (2008). Synthesis of 2-alkynylquinolines from 2-chloro and 2,4-dichloroquinoline via Pd/C–Catalyzed coupling reaction in water. *Tetrahedron* **64**, 7143–7150.

357. Novák, Z., Szabó, A., Répási, J., Kotschy, A. (2003). Sonogashira coupling of aryl halides catalyzed by palladium on charcoal. *The Journal of Organic Chemistry* **68**, 3327–3329.

358. Deng, X., Mani, N. S. (2006). An efficient route to 4-aryl-5-pyrimidinylimidazoles via sequential functionalization of 2,4-dichloropyrimidine. *Organic Letters* **8**, 269–272.

359. Gao, G.-L., Niu, Y.-N., Yan, Z.-Y., Wang, H.-L., Wang, G.-W., Shaukat, A., Liang, Y.-M. (2010). Unexpected domino reaction via Pd-catalyzed Sonogashira coupling of benzimidoyl chlorides with 1,6-enynes and cyclization to synthesize quinoline derivatives. *The Journal of Organic Chemistry* **75**, 1305–1308.

360. Köllhofer, A., Pullmann, T., Plenio, H. (2003). A versatile catalyst for the Sonogashira coupling of aryl chlorides. *Angewandte Chemie International Edition* **42**, 1056–1058.

361. Fleckenstein, C. A., Plenio, H. (2007). 1-Indenyldialkylphosphines and cyclopentadienyldialkylphosphines as ligands for high-activity palladium-catalyzed cross-coupling reactions with aryl chlorides. *Organometallics* **26**, 2758–2767.

362. Lee, D.-H., Lee, Y. H., Harrowfield, J. M., Lee, I.-M., Lee, H. I., Lim, W. T., Kim, Y., Jin, M.-J. (2009). Phosphine-free Sonogashira coupling: Reactions of aryl halides catalysed by palladium(II) complexes of azetidine-derived polyamines under mild conditions. *Tetrahedron* **65**, 1630–1634.

363. Huang, H., Liu, H., Jiang, H., Chen, K. (2008). Rapid and efficient Pd-catalyzed Sonogashira coupling of aryl chlorides. *The Journal of Organic Chemistry* **73**, 6037–6040.

364. Yi, C., Hua, R., Zeng, H., Huang, Q. (2007). Palladium-catalyzed efficient and one-pot synthesis of diarylacetylenes from the reaction of aryl chlorides with 2-methyl-3-butyn-2-ol. *Advanced Synthesis & Catalysis* **349**, 1738–1742.

365. Chandra, A., Singh, B., Khanna, R. S., Singh, R. M. (2009). Copper-free palladium-catalyzed Sonogashira coupling-annulation: Efficient one-pot synthesis of functionalized pyrano[4,3-b]quinolines from 2-chloro-3-formylquinolines. *The Journal of Organic Chemistry* **74**, 5664–5666.

366. Wang, J.-R., Manabe, K. (2010). Hydroxyterphenylphoshine-palladium catalyst for benzo[*b*]furan synthesis from 2-chlorophenols. Bifunctional ligand strategy for cross-coupling of chloroarenes. *The Journal of Organic Chemistry* **75**, 5340–5342.

367. Komáromi, A., Novák, Z. (2008). Efficient copper-free Sonogashira coupling of aryl chlorides with palladium on charcoal. *Chemical Communications*, 4968–4970.

368. Gelman, D., Buchwald, S. L. (2003). Efficient palladium-catalyzed coupling of aryl chlorides and tosylates with terminal alkynes: Use of a copper cocatalyst inhibits the reaction. *Angewandte Chemie International Edition* **42**, 5993–5996.

369. Torborg, C., Huang, J., Schulz, T., Schäffner, B., Zapf, A., Spannenberg, A., Börner, A., Beller, M. (2009). Improved palladium-catalyzed Sonogashira coupling reactions of aryl chlorides. *Chemistry—A European Journal* **15**, 1329–1336.

370. Lemhadri, M., Doucet, H., Santelli, M. (2005). Sonogashira reaction of aryl halides with propiolaldehyde diethyl acetal catalyzed by a tetraphosphine/palladium complex. *Tetrahedron* **61**, 9839–9847.

371. Feuerstein, M., Doucet, H., Santelli, M. (2004). Sonogashira cross-coupling reactions of aryl chlorides with alkynes catalysed by a tetraphosphine–palladium catalyst. *Tetrahedron Letters* **45**, 8443–8446.

372. Feuerstein, M., Doucet, H., Santelli, M. (2005). Sonogashira cross-coupling reactions with heteroaryl halides in the presence of a tetraphosphine-palladium catalyst. *Tetrahedron Letters* **46**, 1717–1720.

373. Feuerstein, M., Doucet, H., Santelli, M. (2006). Sonogashira reaction of heteroaryl halides with alkynes catalysed by a palladium-tetraphosphine complex. *Journal of Molecular Catalysis A: Chemical* **256**, 75–84.

374. Hierso, J.-C., Fihri, A., Amardeil, R., Meunier, P., Doucet, H., Santelli, M., Ivanov, V. V. (2004). Catalytic efficiency of a new tridentate ferrocenyl phosphine auxiliary: Sonogashira cross-coupling reactions of alkynes with aryl bromides and chlorides at low catalyst loadings of 10^{-1} to 10^{-4} Mol%. *Organic Letters* **6**, 3473–3476.

375. Luo, Y., Gao, H., Li, Y., Huang, W., Lu, W., Zhang, Z. (2006). Copper-free Sonogashira reaction using 7-chloro camptothecins. *Tetrahedron* **62**, 2465–2470.

376. Ruiz, J., Cutillas, N., López, F., López, G., Bautista, D. (2006). A copper- and amine-free Sonogashira reaction of aryl halides catalyzed by 1,3,5-triaza-7-phosphaadamantane palladium systems. *Organometallics* **25**, 5768–5773.

377. Eberhard, M. R., Wang, Z., Jensen, C. M. (2002). Investigations into the Pd-catalysed cross-coupling of phenylacetylene with aryl chlorides: Simple one-pot procedure and the effect of $ZnCl_2$ co-catalysis. *Chemical Communications*, 818–819.

378. Vechorkin, O., Barmaz, D., Proust, V., Hu, X. (2009). Ni-catalyzed Sonogashira coupling of nonactivated alkyl halides: Orthogonal functionalization of alkyl iodides, bromides, and chlorides. *Journal of the American Chemical Society* **131**, 12078–12079.

379. Goossen, L. J., Deng, G., Levy, L. M. (2006). Synthesis of biaryls via catalytic decarboxylative coupling. *Science* **313**, 662–664.

380. Goossen, L. J., Rodríguez, N., Melzer, B., Linder, C., Deng, G., Levy, L. M. (2007). Biaryl synthesis via Pd-catalyzed decarboxylative coupling of aromatic carboxylates with aryl halides. *Journal of the American Chemical Society* **129**, 4824–4833.

381. Goossen, L. J., Zimmermann, B., Knauber, T. (2008). Palladium/copper-catalyzed decarboxylative cross-coupling of aryl chlorides with potassium carboxylates. *Angewandte Chemie International Edition* **47**, 7103–7106.

382. Rodríguez, N., Goossen, L. J. (2011). Decarboxylative coupling reactions: A modern strategy for C–C-bond formation. *Chemical Society Reviews* **40**, 5030–5048.

383. Shang, R., Xu, Q., Jiang, Y.-Y., Wang, Y., Liu, L. (2010). Pd-catalyzed decarboxylative cross coupling of potassium polyfluorobenzoates with aryl bromides, chlorides, and triflates. *Organic Letters* **12**, 1000–1003.

384. Bilodeau, F., Brochu, M.-C., Guimond, N., Thesen, K. H., Forgione, P. (2010). Palladium-catalyzed decarboxylative cross-coupling reaction between heteroaromatic carboxylic acids and aryl halides. *The Journal of Organic Chemistry* **75**, 1550–1560.

385. Shang, R., Ji, D.-S., Chu, L., Fu, Y., Liu, L. (2011). Synthesis of a-aryl nitriles through palladium-catalyzed decarboxylative coupling of cyanoacetate salts with aryl halides and triflates. *Angewandte Chemie International Edition* **50**, 4470–4474.

386. Yeung, P. Y., Chung, K. H., Kwong, F. Y. (2011). Palladium-catalyzed decarboxylative arylation of potassium cyanoacetate: Synthesis of α-diaryl nitriles from aryl halides. *Organic Letters* **13**, 2912–2915.

387. Shang, R., Huang, Z., Chu, L., Fu, Y., Liu, L. (2011). Palladium-catalyzed decarboxylative coupling of potassium nitrophenyl acetates with aryl halides. *Organic Letters* **13**, 4240–4243.

388. Shang, R., Yang, Z.-W., Wang, Y., Zhang, S.-L., Liu, L. (2010). Palladium-catalyzed decarboxylative couplings of 2-(2-azaaryl)acetates with aryl halides and triflates. *Journal of the American Chemical Society* **132**, 14391–14393.

389. Shang, R., Fu, Y., Li, J.-B., Zhang, S.-L., Guo, Q.-X., Liu, L. (2009). Synthesis of aromatic esters via Pd-catalyzed decarboxylative coupling of potassium oxalate monoesters with aryl bromides and chlorides. *Journal of the American Chemical Society* **131**, 5738–5739.

390. Mizoroki, T., Mori, K., Ozaki, A. (1971). Arylation of olefin with aryl iodide catalyzed by palladium. *Bulletin of The Chemical Society of Japan* **44**, 581–581.

391. Heck, R. F., Nolley, J. P. (1972). Palladium-catalyzed vinylic hydrogen substitution reactions with aryl, benzyl, and styryl halides. *The Journal of Organic Chemistry* **37**, 2320–2322.

392. Trzeciak, A. M., Ziółkowski, J. J. (2007). Monomolecular, nanosized and heterogenized palladium catalysts for the Heck reaction. *Coordination Chemistry Reviews* **251**, 1281–1293.

393. Knowles, J. P., Whiting, A. (2007). The Heck–Mizoroki cross-coupling reaction: A mechanistic perspective. *Organic & Biomolecular Chemistry* **5**, 31–44.

394. Yi, C., Hua, R. (2006). An efficient palladium-catalyzed Heck coupling of aryl chlorides with alkenes. *Tetrahedron Letters* **47**, 2573–2576.

395. Li, M., Hua, R. (2008). Palladium-catalyzed Heck coupling of 2-vinylpyridine with aryl chlorides. *Applied Organometallic Chemistry* **22**, 397–401.

396. Littke, A. F., Fu, G. C. (1999). Heck reactions in the presence of $P(t\text{-Bu})_3$: Expanded scope and milder reaction conditions for the coupling of aryl chlorides. *The Journal of Organic Chemistry* **64**, 10–11.

397. Littke, A. F., Fu, G. C. (2001). A versatile catalyst for Heck reactions of aryl chlorides and aryl bromides under mild conditions. *Journal of the American Chemical Society* **123**, 6989–7000.

398. Shaughnessy, K. H., Kim, P., Hartwig, J. F. (1999). A fluorescence-based assay for high-throughput screening of coupling reactions. Application to Heck chemistry. *Journal of the American Chemical Society* **121**, 2123–2132.

399. Böhm, V. P. W., Herrmann, W. A. (2000). Nonaqueous ionic liquids: Superior reaction media for the catalytic Heck-vinylation of chloroarenes. *Chemistry—A European Journal* **6**, 1017–1025.

400. Zapf, A., Beller, M. (2001). Palladium catalyst systems for cross-coupling reactions of aryl chlorides and olefins. *Chemistry—A European Journal* **7**, 2908–2915.

401. Datta, G. K., Larhed, M. (2008). High stereoselectivity in chelation-controlled intermolecular Heck reactions with aryl chlorides, vinyl chlorides and vinyl triflates. *Organic & Biomolecular Chemistry*, **6**, 674–676.

402. Ehrentraut, A., Zapf, A., Beller, M. (2000). A new efficient palladium catalyst for Heck reactions of deactivated aryl chlorides. *Synlett*, 1589–1592.

403. Xu, H.-J., Zhao, Y.-Q., Zhou, X.-F. (2011). Palladium-catalyzed Heck reaction of aryl chlorides under mild conditions promoted by organic ionic bases. *The Journal of Organic Chemistry* **76**, 8036–8041.

404. Schnyder, A., Aemmer, T., Indolese, A. F., Pittelkow, U., Studer, M. (2002). First application of secondary phosphines as supporting ligands for the palladium-catalyzed Heck reaction: Efficient activation of aryl chlorides. *Advanced Synthesis & Catalysis* **344**, 495–498.

405. Ruan, J., Iggo, J. A., Berry, N. G., Xiao, J. (2010). Hydrogen-bonding-promoted oxidative addition and regioselective arylation of olefins with aryl chlorides. *Journal of the American Chemical Society* **132**, 16689–16699.

406. Li, G. Y., Zheng, G., Noonan, A. F. (2001). Highly active, air-stable versatile palladium catalysts for the C–C, C–N, and C–S bond formations via cross-coupling reactions of aryl chlorides. *The Journal of Organic Chemistry* **66**, 8677–8681.

407. Beller, M., Zapf, A. (1998). Phosphites as ligands for efficient catalysis of Heck reactions. *Synlett*, 792–793.

408. Herrmann, W. A., Elison, M., Fischer, J., Kocher, C., Artus, G. R. J. (1995). Metal complexes of *N*-heterocyclic carbenes—A new structural principle for catalysts in homogeneous catalysis. *Angewandte Chemie International Edition* **34**, 2371–2373.

409. Calò, V., Nacci, A., Lopez, L., Mannarini, N. (2000). Heck reaction in ionic liquids catalyzed by a Pd–benzothiazole carbene complex. *Tetrahedron Letters* **41**, 8973–8976.

410. Calò, V., Nacci, A., Monopoli, A., Lopez, L., di Cosmo, A. (2001). Heck reaction of β-substituted acrylates in ionic liquids catalyzed by a Pd–benzothiazole carbene complex. *Tetrahedron* **57**, 6071–6077.

411. Selvakumar, K., Zapf, A., Beller, M. (2002). New palladium carbene catalysts for the Heck reaction of aryl chlorides in ionic liquids. *Organic Letters* **4**, 3031–3033.

412. Caddick, S., Kofie, W. (2002). Observations on the intramolecular Heck reactions of aromatic chlorides using palladium/imidazolium salts. *Tetrahedron Letters* **43**, 9347–9350.

413. Firmansjah, L., Fu, G. C. (2007). Intramolecular Heck reactions of unactivated alkyl halides. *Journal of the American Chemical Society* **129**, 11340–11341.

414. Tulloch, A. A. D., Danopoulos, A. A., Tizzard, G. J., Coles, S. J., Hursthouse, M. B., Hay-Motherwell, R. S., Motherwell, W. B. (2001). Chiral 2,6-lutidinyl-biscarbene complexes of palladium. *Chemical Communications*, 1270–1271.

415. Peris, E., Loch, J. A., Mata, J., Crabtree, R. H. (2001). A Pd complex of a tridentate pincer CNC bis-carbene ligand as a robust homogenous Heck catalyst. *Chemical Communications*, 201–202.

416. Schwarz, J., Böhm, V. P. W., Gardiner, M. G., Grosche, M., Herrmann, W. A., Hieringer, W., Raudaschl-Sieber, G. (2000). Polymer-supported carbene complexes of palladium: Well-defined, air-stable, recyclable catalysts for the Heck reaction. *Chemistry—A European Journal* **6**, 1773–1780.

417. McGuinness, D. S., Cavell, K. J. (2000). Donor-functionalized heterocyclic carbene complexes of palladium(II): Efficient catalysts for C–C coupling reactions. *Organometallics* **19**, 741–748.

418. Gründemann, S., Albrecht, M., Loch, J. A., Faller, J. W., Crabtree, R. H. (2001). Tridentate carbene CCC and CNC pincer palladium(II) complexes: Structure, fluxionality, and catalytic activity. *Organometallics* **20**, 5485–5488.

419. Loch, J. A., Albrecht, M., Peris, E., Mata, J., Faller, J. W., Crabtree, R. H. (2002). Palladium complexes with tridentate pincer bis-carbene ligands as efficient catalysts for C–C coupling. *Organometallics* **21**, 700–706.

420. Sie, M.-H., Hsieh, Y.-H., Tsai, Y.-H., Wu, J.-R., Chen, S.-J., Kumar, P. V., Lii, J.-H., Lee, H. M. (2010). A heteroleptic palladium(II) complex containing a bidentate carbene/amido ligand and 3-(trifluoromethyl)-5-(2-pyridyl)pyrazolate: Fast catalyst activation in the Heck coupling reaction. *Organometallics* **29**, 6473–6481.
421. Lee, J.-Y., Cheng, P.-Y., Tsai, Y.-H., Lin, G.-R., Liu, S.-P., Sie, M.-H., Lee, H. M. (2010). Efficient Heck reactions catalyzed by palladium(0) and -(II) complexes bearing *N*-heterocyclic carbene and amide functionalities. *Organometallics* **29**, 3901–3911.
422. Buchmeiser, M. R., Schareina, T., Kempe, R., Wurst, K. (2001). Bis(pyrimidine)-based palladium catalysts: Synthesis, X-ray structure and applications in Heck–, Suzuki–, Sonogashira–Hagihara couplings and amination reactions. *Journal of Organometallic Chemistry* **634**, 39–46.
423. Lee, D.-H., Taher, A., Hossain, S., Jin, M.-J. (2011). An efficient and general method for the Heck and Buchwald–Hartwig coupling reactions of aryl chlorides. *Organic Letters* **13**, 5540–5543.
424. Srinivas, P., Likhar, P. R., Maheswaran, H., Sridhar, B., Ravikumar, K., Kantam, M. L. (2009). N4-tetradentate dicarboxyamidate/dipyridyl palladium complexes as robust catalysts for the Heck reaction of deactivated aryl chlorides. *Chemistry—A European Journal* **15**, 1578–1581.
425. Röhlich, C., Köhler, K. (2010). Tetraalkylammonium-free Heck olefination of deactivated chloroarenes by using a macrocyclic catalyst precursor *Chemistry—A European Journal* **16**, 2363–2365.
426. Herrmann, W. A., Brossmer, C., Ofele, K., Reisinger, C.-P., Priermeier, T., Beller, M., Fischer, H. (1995). Palladacycles as structurally defined catalysts for the Heck olefination of chloro- and bromoarenes. *Angewandte Chemie International Edition* **34**, 1844–1847.
427. Herrmann, W. A., Brossmer, C., Reisinger, C.-P., Riermeier, T. H., Ofele, K., Beller, M. (1997). Palladacycles: Efficient new catalysts for the Heck vinylation of aryl halides. *Chemistry—A European Journal* **3**, 1357–1364.
428. Datta, G. K., Vallin, K. S. A., Larhed, M. (2003). A rapid microwave protocol for Heck vinylation of aryl chlorides under air. *Molecular Diversity* **7**, 107–114.
429. Datta, G. K., von Schenck, H., Hallberg, A., Larhed, M. (2006). Selective terminal Heck arylation of vinyl ethers with aryl chlorides: A combined experimental–computational approach including synthesis of Betaxolol. *The Journal of Organic Chemistry* **71**, 3896–3903.
430. Gruber, A. S., Zim, D., Ebeling, G., Monteiro, A. L., Dupont, J. (2000). Sulfur-containing palladacycles as catalyst precursors for the Heck reaction. *Organic Letters* **2**, 1287–1290.
431. Iyer, S., Ramesh, C. (2000). Aryl–Pd covalently bonded palladacycles, novel amino and oxime catalysts {di-*m*-chlorobis(benzaldehydeoxime-6-C,N)dipalladium(II), di-m-chlorobis(dimethylbenzylamine-6-C,N)dipalladium(II)} for the Heck reaction. *Tetrahedron Letters* **41**, 8981–8984.
432. Iyer, S., Jayanthi, A. (2001). Acetylferrocenyloxime palladacycle-catalyzed Heck reactions. *Tetrahedron Letters* **42**, 7877–7878.
433. Alonso, D. A., Nájera, C., Pacheco, M. C. (2002). Oxime-derived palladium complexes as very efficient catalysts for the Heck–Mizoroki reaction. *Advanced Synthesis & Catalysis* **344**, 172–183.
434. Alacida, E., Nájera, C. (2008). Regioselective Heck reaction of *N*-vinylphthalimide: A general strategy for the synthesis of (*E*)-*N*-styrylphthalimides and phenethylamines. *Advanced Synthesis & Catalysis* **350**, 1316–1322.

435. Wu, Y.-J., Hou, J.-J., Liao, X.-C. (2001). Studies on the catalytic activity of Cyclopalladated ferrocenylimines in Heck reaction. *Acta Chimica Sinica* **59**, 1937–1941.

436. Hajipour, A. R., Rafiee, F., Rouho, A. E. (2011). Applications of a monomeric orthopalladate complex containing mixed phosphorus–nitrogen donors in the Heck reaction. *Tetrahedron Letters* **52**, 4782–4787.

437. Rosa, G. R., Rosa, D. S. (2012). NCP pincer palladacycle as a phosphine-free catalyst precursor for the Heck–Mizoroki coupling of aryl halides. *RSC Advances* **2**, 5080–5083.

438. Consorti, C. S., Ebeling, G., Flores, F. R., Rominger, F., Dupont, J. (2004). On the use of non-symmetrical mixed PCN and SCN pincer palladacycles as catalyst precursors for the Heck reaction. *Advanced Synthesis & Catalysis* **346**, 617–624.

439. Morales-Morales, D., Redón, R., Yung, C., Jensen, C. M. (2000). High yield olefination of a wide scope of aryl chlorides catalyzed by the phosphinito palladium PCP pincer complex: $[PdCl\{C_6H_3(OPPr^i{}_2)_2\text{-}2,6\}]$. *Chemical Communications*, 1619–1620.

440. Kleist, W., Pröckl, S. S., Köhler, K. (2008). Heck reactions of aryl chlorides catalyzed by ligand free palladium salts. *Catalysis Letters* **125**, 197–200.

441. Wang, W., Yang, Q., Zhou, R., Fu, H.-Y., Li, R.-X., Chen, H., Li, X.-J. (2012). Palladium nanoparticles generated from allylpalladium chloride *in situ*: A simple and highly efficient catalytic system for MizorokieHeck reactions. *Journal of Organometallic Chemistry* **697**, 1–5.

442. Calò, V., Nacci, A., Monopoli, A., Cotugno, P. (2009). Heck reactions with palladium nanoparticles in ionic liquids: Coupling of aryl chlorides with deactivated olefins. *Angewandte Chemie International Edition* **48**, 6101–6103.

443. Choudary, B. M., Madhi, S., Chowdari, N. S., Kantam, M. L., Sreedhar, B. (2002). Layered double hydroxide supported nanopalladium catalyst for Heck-, Suzuki-, Sonogashira-, and Stille-type coupling reactions of chloroarenes. *Journal of the American Chemical Society* **124**, 14127–14136.

444. Molnár, Á., Papp, A. (2006). Efficient heterogeneous palladium–montmorillonite catalysts for Heck coupling of aryl bromides and chloride. *Synlett*, 3130–3134.

445. Pröckl, S. S., Kleist, W. (2005). Design of highly active heterogeneous palladium catalysts for the activation of aryl chlorides in Heck reactions. *Tetrahedron* **61**, 9855–9859.

446. Srivastava, R., Venkatathri, N., Srinivas, D., Ratnasamy, P. (2003). Pd-SAPO-31, an efficient, heterogeneous catalyst for Heck reactions of aryl chloride. *Tetrahedron Letters* **44**, 3649–3651.

447. Mukhopadhyay, S., Rothenberg, G., Joshi, A., Baidossi, M., Sasson, Y. (2002). Heterogeneous palladium-catalysed Heck reaction of aryl chlorides and styrene in water under mild conditions. *Advanced Synthesis & Catalysis* **344**, 348–354.

448. Matsubara, R., Gutierrez, A. C., Jamison, T. F. (2011). Nickel-catalyzed Heck-type reactions of benzyl chlorides and simple olefins. *Journal of the American Chemical Society* **133**, 19020–19023.

449. Affo, W., Ohmiya, H., Fujioka, T., Ikeda, Y., Nakamura, T., Yorimitsu, H., Oshima, K., Imamura, Y., Mizuta, T., Miyoshi, K. (2006). Cobalt-catalyzed trimethylsilylmethylmagnesium-promoted radical alkenylation of alkyl halides: A complement to the Heck reaction. *Journal of the American Chemical Society* **128**, 8068–8077.

450. Bellina, F., Rossi, R. (2010). Transition metal-catalyzed direct arylation of substrates with activated sp^3-hybridized C–H bonds and some of their synthetic equivalents with aryl halides and pseudohalides. *Chemical Reviews* **110**, 1082–1146.

451. Johansson, C. C. C., Colacot, T. J. (2010). Metal-catalyzed α-arylation of carbonyl and related molecules: Novel trends in C–C bond formation by C–H bond functionalization. *Angewandte Chemie International Edition* **49**, 676–707.

452. Kawatsura, M., Hartwig, J. F. (1999). Simple, highly active palladium catalysts for ketone and malonate arylation: Dissecting the importance of chelation and steric hindrance. *Journal of the American Chemical Society* **121**, 1473–1478.

453. Ehrentraut, A., Zapf, A., Beller, M. (2002). Progress in the palladium-catalyzed α-Arylation of ketones with chloroarenes. *Advanced Synthesis & Catalysis* **344**, 209–217.

454. Grasa, G. A., Colacot, T. J. (2007). α-Arylation of ketones using Highly active, air-stable (DtBPF)PdX2 (X = Cl, Br) catalysts. *Organic Letters* **9**, 5489–5492.

455. Fox, J. M., Huang, X., Chieffi, A., Buchwald, S. L. (2000). Highly active and selective catalysts for the formation of R-aryl ketones. *Journal of the American Chemical Society* **122**, 1360–1370.

456. Powell, N. A., Hagen, T. J., Ciske, F. L., Cai, C., Duran, J. E., Holsworth, D. D., Leonard, D., Kennedy, R. M., Edmunds, J. J. (2010). Optimization of a Pd-catalyzed intramolecular α-arylation synthesis of tricyclo-[7.3.1.0^{2,7}]-trideca-2,4,6-trien-13-ones. *Tetrahedron Letters* **51**, 4441–4444.

457. Biscoe, M. R., Buchwald, S. L. (2009). Selective monoarylation of acetate esters and aryl methyl ketones using aryl chlorides. *Organic Letters* **11**, 1773–1775.

458. Hellal, M., Singh, S., Cuny, G. D. (2012). Synthesis of tetracyclic indoles via intramolecular α-arylation of ketones. *The Journal of Organic Chemistry* **77**, 4123–4130.

459. Adjabeng, G., Brenstrum, T., Frampton, C. S., Robertson, A. J., Hillhouse, J., McNulty, J., Capretta, A. (2004). Palladium complexes of 1,3,5,7-tetramethyl-2,4,8-trioxa-6-phenyl-6-phosphaadamantane: Synthesis, crystal structure and use in the Suzuki and Sonogashira reactions and the α-arylation of ketones. *The Journal of Organic Chemistry* **69**, 5082–5086.

460. Ackermann, L., Spatz, J. H., Gschrei, C. J., Born, R., Althammer, A. (2006). A diaminochlorophosphine for palladium-catalyzed arylations of amines and ketones. *Angewandte Chemie International Edition* **45**, 7627–7630.

461. Hesp, K. D., Lundgren, R. J., Stradiotto, M. (2011). Palladium-catalyzed arylation of acetone with aryl halides and tosylates. *Journal of the American Chemical Society* **133**, 5194–5197.

462. Viciu, M. S., Germaneau, R. F., Navarro-Fernandez, O., Stevens, E. D., Nolan, S. P. (2002). Activation and reactivity of (NHC)Pd(allyl)Cl (NHC = *N*-heterocyclic carbene) complexes in cross-coupling reactions. *Organometallics* **21**, 5470–5472.

463. Viciu, M. S., Germaneau, R. F., Nolan, S. P. (2002). Well-defined, air-stable (NHC)Pd(Allyl)Cl (NHC = *N*-heterocyclic carbene) catalysts for the arylation of ketones. *Organic Letters* **4**, 4053–4056.

464. Marion, N., Ecarnot, E. C., Navarro, O., Amoroso, D., Bell, A., Nolan, S. P. (2006). (IPr)Pd(acac)Cl: An Easily Synthesized, Efficient, and Versatile Precatalyst for C–N and C–C bond formation. *The Journal of Organic Chemistry* **71**, 3816–3821.

465. Navarro, O., Marion, N., Scott, N. M., González, J., Amoroso, D., Bell, A., Nolan, S. P. (2005). Synthesis of novel (NHC)Pd(acac)Cl complexes (acac = acetylacetonate) and their activity in cross-coupling reactions. *Tetrahedron* **61**, 9716–9722.

466. Matsubara, K., Okazaki, H., Senju, M. (2006). Polycondensation of haloarylketones catalyzed by palladium compounds bearing *N*-heterocyclic carbene (NHC) ligands. *Journal of Organometallic Chemistry* **691**, 3693–3699.

467. Lavallo, V., Canac, Y., Präsang, C., Donnadieu, B., Bertrand, G. (2005). Stable cyclic (alkyl)(amino)carbenes as rigid or flexible, bulky, electron-rich ligands for transition-metal catalysts: A quaternary carbon atom makes the difference. *Angewandte Chemie International Edition* **44**, 5705–5709.

468. Viciu, M. S., Kelly III, R. A., Stevens, E. D., Naud, F., Studer, M., Nolan, S. P. (2003). Synthesis, characterization, and catalytic activity of *N*-heterocyclic carbene (NHC) palladacycle complexes. *Organic Letters* **5**, 1479–1482.

469. Zhang, J., Yang, X., Cui, X., Wu, Y. (2011). Carbene adduct of cyclopalladated ferrocenylimine catalyzed a-arylation of ketones with aryl chlorides or bromides. *Tetrahedron* **67**, 8800–8807.

470. Matsubara, K., Ueno, K., Koga, Y., Hara, K. (2007). Nickel-NHC–catalyzed α-Arylation of acyclic ketones and amination of haloarenes and unexpected preferential *N*-Arylation of 4-Aminopropiophenone. *The Journal of Organic Chemistry* **72**, 5069–5076.

471. Chen, G., Kwong, F. Y., Chan, H. O., Yub, W.-Y., Chan, A. S. C. (2006). Nickel-catalyzed asymmetric α-arylation of ketone enolates. *Chemical Communications*, 1413–1415.

472. Ge, S., Hartwig, J. F. (2011). Nickel-catalyzed asymmetric α-arylation and heteroarylation of ketones with chloroarenes: Effect of halide on selectivity, oxidation state, and room-temperature reactions. *Journal of the American Chemical Society* **133**, 16330–16333.

473. Su, W., Raders, S., Verkade, J. G., Liao, X., Hartwig, J. F. (2006). Pd-catalyzed α-arylation of trimethylsilyl enol ethers with aryl bromides and chlorides: A synergistic effect of two metal fluorides as additives. *Angewandte Chemie International Edition* **45**, 5852–5855.

474. Iwama, T., Rawal, V. H. (2006). Palladium-catalyzed regiocontrolled α-arylation of trimethylsilyl enol ethers with aryl halides. *Organic Letters* **8**, 5725–5728.

475. Chobanian, H. R., Liu, P., Chioda, M. D., Guo, Y., Lin, L. S. (2007). A facile, microwave-assisted, palladium-catalyzed arylation of acetone. *Tetrahedron Letters* **48**, 1213–1216.

476. Martín, R., Buchwald, S. L. (2008). An improved protocol for the Pd-catalyzed α-Arylation of aldehydes with aryl halides. *Organic Letters* **10**, 4561–4564.

477. Martín, R., Buchwald, S. L. (2007). A general method for the direct α-arylation of aldehydes with aryl bromides and chlorides. *Angewandte Chemie International Edition* **46**, 7236–7239.

478. Vo, G. D., Hartwig, J. F. (2008). Palladium-catalyzed α-arylation of aldehydes with bromo- and chloroarenes catalyzed by [{Pd(allyl)Cl}2] and dppf or Q-phos. *Angewandte Chemie International Edition* **47**, 2127–2130.

479. Hama, T., Hartwig, J. F. (2008). Palladium-catalyzed α-arylation of esters with chloroarenes. *Organic Letters* **10**, 1549–1552.

480. Lee, S., Beare, N. A., Hartwig, J. F. (2001). Palladium-catalyzed α-arylation of esters and protected amino acids. *Journal of the American Chemical Society* **123**, 8410–8411.

481. Spielvogel, D. J., Buchwald, S. L. (2002). Nickel-BINAP Catalyzed Enantioselective α-arylation of α-substituted γ-butyrolactones. *Journal of the American Chemical Society* **124**, 3500–3501.

482. Durbin, M. J., Willis, M. C. (2008). Palladium-catalyzed α-arylation of oxindoles. *Organic Letters* **10**, 1413–1425.

483. Mai, C.-K., Sammons, M. F., Sammakia, T. (2010). α-Arylation of 3-aryloxindoles. *Organic Letters* **12**, 2306–2309.

484. Altman, R. A., Hyde, A. M., Huang, X., Buchwald, S. L. (2008). Orthogonal Pd- and Cu- based catalyst systems for *C*- and *N*-arylation of oxindoles. *Journal of the American Chemical Society* **130**, 9613–9620.

485. Lee, S., Hartwig, J. F. (2001). Improved catalysts for the palladium-catalyzed synthesis of oxindoles by amide α-arylation. Rate acceleration, use of aryl chloride substrates, and a new carbene ligand for asymmetric transformations. *The Journal of Organic Chemistry* **66**, 3402–3415.

486. Ackermann, L., Vicente, R., Hofmann, N. (2009). Air-stable secondary phosphine oxide as preligand for palladium-catalyzed intramolecular α-arylations with chloroarenes. *Organic Letters* **11**, 4274–4276.

487. Luan, X., Wu, L., Drinkel, E., Mariz, R., Gatti, M., Dorta, R. (2010). Highly chemo- and enantioselective synthesis of 3-allyl-3-aryl oxindoles via the direct palladium-catalyzed α-arylation of amides. *Organic Letters* **12**, 1912–1915.

488. Barluenga, J., Jiménez-Aquino, A., Valdés, C., Aznar, F. (2007). *Angewandte Chemie International Edition* **46**, 1529–1532.

489. You, J., Verkade, J. G. (2003). A general method for the direct α-arylation of nitriles with aryl chlorides. *Angewandte Chemie International Edition* **42**, 5051–5053.

490. Gao, C., Tao, X., Qian, Y., Huang, J. (2003). A highly efficient catalytic system for cross-coupling of aryl chlorides and bromides with malononitrile anion by palladium carbene complexes. *Chemical Communications*, 1444–1445.

491. Vogl, E. M., Buchwald, S. L. (2002). Palladium-catalyzed monoarylation of nitroalkanes. *The Journal of Organic Chemistry* **67**, 106–111.

492. Akita, Y., Inoue, A., Yamamoto, K., Ohta, A., Kurihara, T., Shimizu, M. (1985). Palladium-catalyzed coupling reaction of chloropyrazines with indole. *Heterocycles* **23**, 2327–2333.

493. Ackermann, L. (2010). Metal-catalyzed direct alkylations of (hetero)arenes via C–H bond cleavages with unactivated alkyl halides. *Chemical Communications* **46**, 4866–4877.

494. Daugulis, O. (2010). Palladium and copper catalysis in regioselective, intermolecular coupling of C–H and C–Hal bonds. *Topics in Current Chemistry* **292**, 57–84.

495. Roger, J., Gottumukkala, A. L., Doucet, H. (2010). Palladium-catalyzed C3 or C4 direct arylation of heteroaromatic compounds with aryl halides by C–H bond activation. *ChemCatChem* **2**, 20–40.

496. Ornellas, S. D., Storr, T. E., Williams, T. J., Baumann, C. G., Fairlamb, I. J. S. (2011). Direct C–H/C–X coupling methodologies mediated by Pd/Cu or Cu: An examination of the synthetic applications and mechanistic findings. *Current Organic Synthesis* **8**, 79–101.

497. Bellina, F., Rossi, R. (2009). Recent advances in the synthesis of (hetero)aryl-substituted heteroarenes via transition metal-catalysed direct (hetero)arylation of heteroarene C–H bonds with aryl halides or pseudohalides, diaryliodonium salts, and potassium aryltrifluoroborates. *Tetrahedron* **65**, 10269–10310.

498. Gürbüz, N., Özdemir, I., Çetinkaya, B. (2005). Selective palladium-catalyzed arylation(s) of benzaldehyde derivatives by *N*-heterocarbene ligands. *Tetrahedron Letters* **46**, 2273–2277.

499. Campeau, L.-C., Thansandote, P., Fagnou, K. (2005). High-yielding intramolecular direct arylation reactions with aryl chlorides. *Organic Letters* **7**, 1857–1860.

500. Campeau, L.-C., Parisien, M., Jean, A., Fagnou, K. (2006). Catalytic direct arylation with aryl chlorides, bromides, and iodides: Intramolecular studies leading to new intermolecular reactions. *Journal of the American Chemical Society* **128**, 581–590.

501. Ackermann, L., Althammer, A. (2007). Domino N–H/C–H bond activation: Palladium-catalyzed synthesis of annulated heterocycles using dichloro(hetero)arenes. *Angewandte Chemie International Edition* **46**, 1627–1629.

502. Bedford, R. B., Cazin, C. S. J. (2002). A novel catalytic one-pot synthesis of carbazoles via consecutive amination and C–H activation. *Chemical Communications*, 2310–2311.

503. Meyers, C., Rombouts, G., Loones, K. T. J., Coelho, A., Maesa, B. U. W. (2008). Auto-tandem catalysis: Synthesis of substituted 11*H*-indolo [3,2-*c*]quinolines via palladium-catalyzed intermolecular C–N and intramolecular C–C bond formation. *Advanced Synthesis & Catalysis* **350**, 465–470.

504. Lafrance, M., Rowley, C. N., Woo, T. K., Fagnou, K. (2006). Catalytic intermolecular direct arylation of perfluorobenzenes. *Journal of the American Chemical Society* **128**, 8754–8756.

505. Lafrance, M., Shore, D., Fagnou, K. (2006). Mild and general conditions for the cross-coupling of aryl halides with pentafluorobenzene and other perfluoroaromatics. *Organic Letters* **8**, 5097–5100.

506. Hennessy, E. J., Buchwald, S. L. (2003). Synthesis of substituted oxindoles from α-Chloroacetanilides via palladium–catalyzed C–H functionalization. *Journal of the American Chemical Society* **125**, 12084–12085.

507. Zhang, Y.-H., Shi, B.-F., Yu, J.-Q. (2009). Palladium(II)-catalyzed ortho alkylation of benzoic acids with alkyl halides. *Angewandte Chemie International Edition* **48**, 6097–6100.

508. Roy, D., Mom, S., Beaupérin, M., Doucet, H., Hierso, J.-C. (2010). A versatile palladium/triphosphane system for direct arylation of heteroarenes with chloroarenes at low catalyst loading. *Angewandte Chemie International Edition* **49**, 6650–6654.

509. Ozdemir, I., Gök, Y., Özeroğlu, Ö., Kaloğlu, M., Doucet, H., Bruneau, C. (2010). *N*-Heterocyclic carbenes: Useful ligands for the palladium-catalysed direct C5 arylation of heteroaromatics with aryl bromides or electron-deficient aryl chlorides. *European Journal of Inorganic Chemistry*, 1798–1805.

510. Tamba, S., Okubo, Y., Tanaka, S., Monguchi, D., Mori, A. (2010). Palladium-catalyzed C–H functionalization of heteroarenes with aryl bromides and chlorides. *The Journal of Organic Chemistry* **75**, 6998–7001.

511. Nadres, E. T., Lazareva, A., Daugulis, O. (2011). Palladium-catalyzed indole, pyrrole, and furan arylation by aryl chlorides. *The Journal of Organic Chemistry* **76**, 471–483.

512. Hwang, S. J., Cho, S. H., Chang, S. (2008). Synthesis of condensed pyrroloindoles via Pd-catalyzed intramolecular C–H bond functionalization of pyrroles. *Journal of the American Chemical Society* **130**, 16158–16159.

513. Gottumukkala, A. L., Doucet, H. (2007). Activated aryl chlorides: Useful partners for the coupling with 2-substituted thiazoles in the palladium-catalysed C–H activation/functionalisation reaction. *European Journal of Inorganic Chemistry*, 3629–3632.

514. Derridj, F., Roger, J., Geneste, F., Djebbar, S., Doucet, H. (2009). Palladium-catalyzed direct heteroarylation of chloropyridines and chloroquinolines. *Journal of Organometallic Chemistry* **694**, 455–465.

515. Derridj, F., Djebbar, S., Benali-Baitich, O., Doucet, H. (2008). Direct arylation of oxazole and benzoxazole with aryl or heteroaryl halides using a palladium–diphosphine catalyst. *Journal of Organometallic Chemistry* **693**, 135–144.

516. Strotman, N. A., Chobanian, H. R., Guo, Y., He, J., Wilson, J. E. (2010). Highly regioselective palladium-catalyzed direct arylation of oxazole at C-2 or C-5 with aryl bromides, chlorides, and triflates. *Organic Letters* **12**, 3578–3581.

517. Verrier, C., Martin, T., Hoarau, C., Marsais, F. (2008). Ethyl oxazole-4-carboxylate: An efficient access to (hetero)aryloxazoles. *The Journal of Organic Chemistry* **73**, 7383–7386.

518. Théveau, L., Verrier, C., Lassalas, P., Martin, T., Dupas, G., Querolle, O., Hijfte, L. V., Marsais, F., Hoarau, C. (2011). Mechanism selection for regiocontrol in base-assisted, palladium-catalysed direct C–H coupling with halides: First approach for oxazole- and thiazole-4-carboxylates. *Chemistry—A European Journal* **17**, 14450–14463.

519. Kumar, P. V., Lin, W.-S., Shen, J.-S., Nandi, D., Lee, H. M. (2011). Direct C5-arylation reaction between imidazoles and aryl chlorides catalyzed by palladium complexes with phosphines and *N*-heterocyclic carbenes. *Organometallics* **30**, 5160–5169.

520. Joo, J. M., Touré, B. B., Sames, D. (2010). C–H bonds as ubiquitous functionality: A general approach to complex arylated imidazoles via regioselective sequential arylation of all three C–H bonds and regioselective *N*-alkylation enabled by SEM-group transposition. *The Journal of Organic Chemistry* **75**, 4911–4920.

521. Sahnoun, S., Messaoudi, S., Brion, J.-D., Alami, M. (2009). A site selective C–H arylation of free-(NH2) adenines with aryl chlorides: Application to the synthesis of 6,8-disubstituted adenines. *Organic & Biomolecular Chemistry* **7**, 4271–4278.

522. Sahnoun, S., Messaoudi, S., Peyrat, J.-F., Brion, J.-D., Alami, M. (2008). Microwave-assisted $Pd(OH)_2$-catalyzed direct C-H arylation of free-(NH_2) adenines with aryl halides. *Tetrahedron Letters* **49**, 7279–7283.

523. Chiong, H. A., Daugulis, O. (2007). Palladium-catalyzed arylation of electron-rich heterocycles with aryl chlorides. *Organic Letters* **9**, 1449–1451.

524. Ackermann, L., Vicente, R., Born, R. (2008). Palladium-catalyzed direct arylations of 1,2,3-triazoles with aryl chlorides using conventional heating. *Advanced Synthesis & Catalysis* **350**, 741–748.

525. Iwasaki, M., Yorimitsu, H., Oshima, K. (2007). Microwave-assisted palladium-catalyzed direct arylation of 1,4-disubstituted 1,2,3-triazoles with aryl chlorides. *Chemistry—An Asian Journal* **2**, 1430–1435.

526. Verrier, C., Hoarau, C., Marsais, F. (2009). Direct palladium-catalyzed alkenylation, benzylation and alkylation of ethyl oxazole-4-carboxylate with alkenyl-, benzyl- and alkyl halides. *Organic & Biomolecular Chemistry* **7**, 647–650.

527. Lapointe, D., Fagnou, K. (2009). Palladium-catalyzed benzylation of heterocyclic aromatic compounds. *Organic Letters* **11**, 4160–4163.

528. Yao, T., Hirano, K., Satoh, T., Miura, M. (2010). Palladium- and nickel-catalyzed direct alkylation of azoles with unactivated alkyl bromides and chlorides *Chemistry—A European Journal* **16**, 12307–12311.

529. Niwa, T., Yorimitsu, H., Oshima, K. (2007). Palladium-catalyzed direct arylation of aryl(azaaryl)methanes with aryl halides providing triarylmethanes. *Organic Letters* **9**, 2373–2375.

530. Rousseaux, S., Davi, M., Sofack-Kreutzer, J., Pierre, C., Kefalidis, C. E., Clot, E., Fagnou, K., Baudoin, O. (2010). Intramolecular palladium-catalyzed alkane C–H arylation from aryl chlorides. *Journal of the American Chemical Society* **132**, 10706–10716.

531. Požgana, F., Dixneuf, P. H. (2009). Ruthenium(II) acetate catalyst for direct functionalisation of sp^2-C–H bonds with aryl chlorides and access to tris-heterocyclic Molecules. *Advanced Synthesis & Catalysis* **351**, 1737–1743.

532. Arockiam, P., Poirier, V., Fischmeister, C., Bruneau, C., Dixneuf, P. H. (2009). Diethyl carbonate as a solvent for ruthenium catalysed C–H bond functionalisation. *Green Chemistry* **11**, 1871–1875.

533. Arockiam, P. B., Fischmeister, C., Bruneau, C., Dixneuf, P. H. (2010). C–H bond functionalization in water catalyzed by carboxylato ruthenium(II) systems. *Angewandte Chemie International Edition* **49**, 6629–6632.

534. Ackermann, L. (2005). Phosphine oxides as preligands in ruthenium-catalyzed arylations via C–H bond Functionalization using aryl chlorides. *Organic Letters* **7**, 3123–3125.

535. Ackermann, L., Diers, E., Manvar, A. (2012). Ruthenium-catalyzed C–H bond arylations of arenes bearing removable directing groups via six-membered ruthenacycles. *Organic Letters* **14**, 1154–1157.

536. Yu, B., Yan, X., Wang, S., Tang, N., Xi, C. (2010). A highly efficient ruthenium(II) catalyst with (1,2-diarylvinyl)phosphine ligands for direct ortho arylation of 2-arylpyridine with aryl chlorides. *Organometallics* **29**, 3222–3226.

537. Miura, H., Wada, K., Hosokawa, S., Inoue, M. (2010). Recyclable solid ruthenium catalysts for the direct arylation of aromatic C–H bonds. *Chemistry—A European Journal* **16**, 4186–4189.

538. Luo, N., Yu, Z. (2010). $RuCl_3 \cdot xH_2O$-catalyzed direct arylation of arenes with aryl chlorides in the presence of triphenylphosphine. *Chemistry—A European Journal* **16**, 787–791.

539. Ackermann, L., Born, R., Vicente, R. (2009). Ruthenium-catalyzed direct arylations of *N*-aryl 1,2,3-triazoles with aryl chlorides as electrophiles. *ChemSusChem* **2**, 546–549.

540. Ackermann, L., Novák, P. (2009). Regioselective ruthenium-catalyzed direct benzylations of arenes through C–H bond cleavages. *Organic Letters* **11**, 4966–4969.

541. Ackermann, L., Novák, P., Vicente, R., Hofmann, N. (2009). Ruthenium-catalyzed regioselective direct alkylation of arenes with unactivated alkyl halides through C–H bond cleavage. *Angewandte Chemie International Edition* **48**, 6045–6048.

542. Liu, W., Cao, H., Lei, A. (2010). Iron-catalyzed direct arylation of unactivated Arenes with aryl halides. *Angewandte Chemie International Edition* **49**, 2004–2008.

543. Liu, W., Cao, H., Xin, J., Jin, L., Lei, A. (2011). Cobalt-catalyzed direct arylation of unactivated arenes with aryl halides *Chemistry—A European Journal* **17**, 3588–3592.

544. Chen, Q., Ilies, L., Nakamura, E. (2011). Cobalt-catalyzed *ortho*-alkylation of secondary benzamide with alkyl chloride through directed C–H bond activation. *Journal of the American Chemical Society* **133**, 428–429.

545. Muci, A. R., Buchwald, S. L. (2002). Practical Palladium catalysts for C–N and C–O bond formation. *Topics in Current Chemistry* **219**, 131–209.

546. Hartwig, J. F. (2006). Discovery and understanding of transition-metal-catalyzed aromatic substitution reactions. *Synlett*, 1283–1294.

547. Senra, J. D., Aguiar, L. C. S., Simas, A. B. C. (2011). Recent progress in transition-metal-catalyzed C–N cross-couplings: Emerging approaches towards sustainability. *Current Organic Synthesis* **8**, 53–78.

548. Schlummer, B., Scholza, U. (2004). Palladium-catalyzed C–N and C–O Coupling—A practical guide from an industrial vantage point. *Advanced Synthesis & Catalysis* **346**, 1599–1626.

549. Bellina, F., Rossi, R. (2010). Regioselective functionalization of the imidazole ring via transition metal-catalyzed C–N and C–C bond forming reactions. *Advanced Synthesis & Catalysis* **352**, 1223–1276.

550. Ma, D.-W., Cai, Q. (2008). Copper/amino acid catalyzed cross-couplings of aryl and vinyl halides with nucleophiles. *Accounts of Chemical Research* **41**, 1450–1460.

551. Klapars, A., Antilla, J. C., Huang, X. H., Buchwald, S. L. (2001). A general and efficient copper catalyst for the amidation of aryl halides and the *N*-arylation of nitrogen heterocycles. *Journal of the American Chemical Society* **123**, 7727–7729.

552. Klapars, A., Huang, X., Buchwald, S. L. (2002). A general and efficient copper catalyst for the amidation of aryl halides. *Journal of the American Chemical Society* **124**, 7421–7428.

553. Hodgkinson, R. C., Schulz, J., Willis, M. C. (2009). Tandem copper-catalysed aryl and alkenyl amination reactions: The synthesis of *N*-functionalised indoles. *Organic & Biomolecular Chemistry* **7**, 432–434.

554. Lu, H. J., Li, C. J. (2006). General and highly efficient synthesis of 2-alkylideneazetidines and β-Lactams via copper-catalyzed intramolecular *N*-vinylation. *Organic Letters* **8**, 5365–5367.

555. Li, X. F., Yang, D. S., Jiang, Y. Y., Fu, H. (2010). Efficient copper-catalyzed *N*-arylations of nitrogen-containing heterocycles and aliphatic amines in water. *Green Chemistry* **12**, 1097–1105.

556. Zhang, S., Ding, Y. (2011). Theoretical study on mechanism of copper(I)-catalyzed cross-coupling between aryl halides and alkylamines. *Organometallics* **30**, 633–641.

557. Ma, D. W., Cai, Q. (2004). L-Proline promoted Ullmann-type coupling reactions of aryl iodides with indoles, pyrroles, imidazoles or pyrazoles. *Synlett*, 128–130.

558. Zhang, H., Cai, Q., Ma, D. W. (2005). Amino acid promoted CuI-catalyzed CN bond formation between aryl halides and amines or N-containing heterocycles. *The Journal of Organic Chemistry* **70**, 5164–5173.

559. Tanimori, S., Kobayashi, Y., Iesaki, Y., Ozaki, Y., Kirihata, M. (2012). Copper-catalyzed synthesis of substituted indazoles from 2-chloroarenes at low catalyst-loading. *Organic & Biomolecular Chemistry* **10**, 1381–1387.

560. Cai, Q., Li, Z. Q., Wei, J. J., Fu, L. B., Ha, C. Y., Pei, D. Q., Ding, K. (2010). *Organic Letters* **12**, 1500–1503.

561. Zhu, L. B., Cheng, L., Zhang, Y. X., Xie, R. G., You, J. S. (2007). Highly efficient copper-catalyzed *N*-arylation of nitrogen-containing heterocycles with aryl and heteroaryl halides. *The Journal of Organic Chemistry* **72**, 2737–2743.

562. Rao, H., Jin, Y., Fu, H., Jiang, Y., Zhao, Y. (2006). A versatile and efficient ligand for copper-catalyzed formation of C–N, C–O, and P–C bonds: Pyrrolidine-2-phosphonic acid phenyl monoester *Chemistry—A European Journal* **12**, 3636–3646.

563. Ma, H. C., Jiang, X. Z. (2007). *N*-Hydroxyimides as efficient ligands for the copper-catalyzed *N*-arylation of pyrrole, imidazole, and indole. *The Journal of Organic Chemistry* **72**, 8943–8946.

564. Choudary, B. M., Sridhar, C., Kantam, M. L., Venkanna, G. T., Sreedhar, B. (2005). Design and evolution of copper apatite catalysts for *N*-arylation of heterocycles with chloro- and fluoroarenes. *Journal of the American Chemical Society* **127**, 9948–9949.

565. Wu, Z. Q., Zhou, L., Jiang, Z. Q., Wu, D., Li, Z. K., Zhou, X. G. (2010). Sulfonato–Cu(salen) complex catalyzed *N*-arylation of aliphatic amines with aryl halides in water. *European Journal of Organic Chemistry*, 4971–4975.

566. Xu, H. H., Wolf, C. (2009). Efficient copper-catalyzed coupling of aryl chlorides, bromides and iodides with aqueous ammonia. *Chemical Communications*, 3035–3037.

567. Xu, H. H., Wolf, C. (2009). Copper catalyzed coupling of aryl chlorides, bromides and iodides with amines and amides. *Chemical Communications*, 1715–1717.

568. Yong, F. F., Teo, Y. C., Chua, G. L., Lim, G. S., Lin, Y. Z. (2011). Efficient ligand-free copper-catalyzed *N*-arylation of amides with aryl halides in water. *Tetrahedron Letters* **52**, 1169–1172.

569. Cortes-Salva, M., Garvin, C., Antilla, J. C. (2011). Ligand-free copper-catalyzed arylation of amidines. *The Journal of Organic Chemistry* **76**, 1456–1459.

570. Lee, H. G., Won, J. E., Kim, M. J., Park, S. E., Jung, K. J., Kim, B. R., Lee, S. G., Yoon, Y. J. (2009). TBAF-assisted copper-catalyzed *N*-arylation and benzylation of benzazoles with aryl and benzyl halides under the ligand/base/ solvent-free conditions. *The Journal of Organic Chemistry* **74**, 5675–5678.

571. Gavade, S., Shingare, M., Mane, D. (2011). Microwave assisted, solvent- and ligand-free copper catalyzed *N*-arylation of phenylurea with aryl halides. *Bulletin of the Korean Chemical Society* **32**, 4167–4170.

572. Verma, S. K., Acharya, B. N., Kaushik, M. P. (2011). Chemospecific and ligand free CuI catalysed heterogeneous *N*-arylation of amines with diheteroaryl halides at room temperature. *Organic & Biomolecular Chemistry* **9**, 1324–1327.

573. Huang, Y. Z., Miao, H., Zhang, Q. H., Chen, C., Xu, J. (2008). Cu_2O: A simple and efficient reusable catalyst for *N*-arylation of nitrogen-containing heterocycles with aryl halides. *Catalysis Letters* **122**, 344–348.

574. Hartwig, J. F. (2008). Carbon-heteroatom bond formation catalysed by organometallic complexes. *Nature* **455**, 314–322.

575. Surry, D. S., Buchwald, S. L. (2008). Biaryl phosphane ligands in palladium-catalyzed amination. *Angewandte Chemie International Edition* **47**, 6338–6361.

576. Mullick, D., Anjanappa, P., Selvakumar, K., Sivakumar, K. R. M. (2010). Carbamic acid 2-trimethylsilylethyl ester as a new ammonia equivalent for palladium-catalyzed amination of aryl halides. *Tetrahedron Letters* **51**, 5984–5987.

577. Vimolratana, M., Simard, J. L., Brown, S. P. (2011). Palladium-catalyzed amidation of 2-chloropyrimidines. *Tetrahedron Letters* **52**, 1020–1022.

578. Shen, Z. L., Hong, Y. M., He, X. F., Mo, W. M., Hu, B. X., Sun, N., Hu, X. Q. (2010). Switching the chemoselectivity in the amination of 4-chloroquinazolines with aminopyrazoles. *Organic Letters* **12**, 552–555.

579. Chen, H. S., Wang, Q. R., Tao, F. G. (2009). A novel palladium-catalyzed amination of aryl halides with amines using rac-P-Phos as the ligand. *Chinese Journal of Chemistry* **27**, 1382–1386.

580. Falk, F. C., Fröhlich, R., Paradies, J. (2011). Coupling of ortho-substituted aryl chlorides with bulky amides. *Chemical Communications* **47**, 11095–11097.

581. Hartwig, J. F., Kawatsura, M., Hauck, S. I., Shaughnessy, K. H., Alcazar-Roman, L. M. (1999). Room-temperature palladium-catalyzed amination of aryl bromides and chlorides and extended scope of aromatic C–N bond formation with a commercial ligand. *The Journal of Organic Chemistry* **64**, 5575–5580.

582. Lee, S., Jørgensen, M., Hartwig, J. F. (2001). Palladium-catalyzed synthesis of arylamines from aryl halides and lithium bis(trimethylsilyl)amide as an ammonia Equivalent. *Organic Letters* **3**, 2729–2732.

583. Reddy, N. P., Tanaka, M. (1997). Palladium-catalyzed amination of aryl chlorides. *Tetrahedron Letters* **38**, 4807–4810.

584. Bedford, R. B., Blake, M. E. (2003). Mixed phosphite-phosphine and phosphinite–phosphine palladacyclic complexes as highly active catalysts for the amination of aryl chlorides. *Advanced Synthesis & Catalysis* **345**, 1107–1110.

585. Bedford, R. B., Cazin, C. S. J. (2003). High-activity catalysts for Suzuki coupling and amination reactions with deactivated aryl chloride substrates: Importance of the palladium source. *Organometallics* **22**, 987–999.

586. Hill, L. L., Moore, L. R., Huang, R., Craciun, R., Vincent, A. J., Dixon, D. A., Chou, J., Woltermann, C. J., Shaughnessy, K. H. (2006). Bulky alkylphosphines with neopentyl substituents as ligands in the amination of aryl bromides and chlorides. *The Journal of Organic Chemistry* **71**, 5117–5125.

587. Suzuki, K., Hori, Y., Kobayashi, T. (2008). A new hybrid phosphine ligand for palladium-catalyzed amination of aryl halides. *Advanced Synthesis & Catalysis* **350**, 652–656.

588. Suzuki, K., Hori, Y., Nishikawa, T., Kobayashi, T. Suzuki, K. Hori, Y., Kobayashi, T. (2007). A novel (2,2-diarylvinyl)phosphine/palladium catalyst for effective aromatic amination. *Advanced Synthesis & Catalysis* **349**, 2089–2091.

589. Doherty, S., Knight, J. G., Smyth, C. H., Jorgenson, G. A. (2008). Electron-rich, bicyclic biaryl-like KITPHOS monophosphines via [4 + 2] cycloaddition between 1-Alkynylphosphine oxides and anthracene: Highly efficient ligands for palladium-catalysed C–N and C–C Bond Formation. *Advanced Synthesis & Catalysis* **350**, 1801–1806.

590. Chen, L., Yu, G.-A., Li, F., Zhu, X., Zhang, B., Guo, R., Li, X., Yang, Q., Jin, S., Liu, C., Liu, S.-H. (2010). Preparation of anionic phosphine ligands in situ for the palladium-catalyzed Buchwald/Hartwig amination reactions of aryl halides. *Journal of Organometallic Chemistry* **695**, 1768–1775.

591. Parrish, C. A., Buchwald, S. L. (2001). Use of polymer-supported dialkylphosphinobiphenyl ligands for palladium-catalyzed amination and Suzuki reactions. *The Journal of Organic Chemistry* **66**, 3820–3827.

592. Christmann, U., Vilar, R. (2005). Monoligated palladium species as catalysts in cross-coupling reactions. *Angewandte Chemie International Edition* **44**, 366–374.

593. Strieter, E. R., Buchwald, S. L. (2006). Evidence for the formation and structure of palladacycles during Pd-catalyzed C–N bond Formation with catalysts derived from bulky monophosphinobiaryl ligands. *Angewandte Chemie International Edition* **45**, 925–928.

594. Pratap, R., Parrish, D., Gunda, P., Venkataraman, D., Lakshman, M. K. (2009). Influence of biaryl phosphine structure on C–N and C–C bond formation. *Journal of the American Chemical Society* **131**, 12240–12249.

595. Huang, X., Anderson, K. W., Zim, D., Jiang, L., Klapars, A., Buchwald, S. L. (2003). Expanding Pd-catalyzed C–N bond-forming processes: The first amidation of aryl sulfonates, aqueous amination, and complementarity with Cu-catalyzed reactions. *Journal of the American Chemical Society* **125**, 6653–6655.

596. Wolfe, J. P., Tomori, H., Sadighi, J. P., Yin, J., Buchwald, S. L. (2000). Simple, efficient catalyst system for the palladium-catalyzed amination of aryl chlorides, bromides, and triflates. *The Journal of Organic Chemistry*, **65**, 1158–1174.

597. Anderson, K. W., Tundel, R. E., Ikawa, T., Altman, R. A., Buchwald, S. L. (2006). Monodentate phosphines provide highly active catalysts for Pd-catalyzed C–N bond-Forming reactions of heteroaromatic halides/amines and (H)N-heterocycles. *Angewandte Chemie International Edition* **45**, 6523–6527.

598. Biscoe, M. R., Fors, B. P., Buchwald, S. L. (2008). A new class of easily activated palladium precatalysts for facile C–N cross-coupling reactions and the low temperature oxidative addition of aryl chlorides. *Journal of the American Chemical Society* **130**, 6687–6688.

599. Fors, B. P., Watson, D. A., Biscoe, M. R., Buchwald, S. L. (2008). A highly active catalyst for Pd-catalyzed amination reactions: Cross-coupling reactions using aryl mesylates and the highly selective monoarylation of primary amines using aryl chlorides. *Journal of the American Chemical Society* **130**, 13552–13554.

600. Fors, B. P., Buchwald, S. L. (2010). A multiligand based Pd catalyst for C–N cross-coupling reactions. *Journal of the American Chemical Society* **132**, 15914–15917.

601. Haddad, N., Salvagno, A., Busacca, C. (2004). Application of the palladium-catalyzed *N*-arylation of hydrazones to deactivated heteroaryl halides in the synthesis of pyrazoles. *Tetrahedron Letters* **45**, 5935–5937.

602. Alcaraz, L., Bennion, C., Morris, J., Meghani, P., Thom, S. M. (2004). Novel *N*-aryl and *N*-heteroaryl sulfamide synthesis via palladium cross coupling. *Organic Letters* **6**, 2705–2708.

603. Burton, G., Cao, P., Li, G., Rivero, R. (2003). Palladium-catalyzed intermolecular coupling of aryl chlorides and sulfonamides under microwave irradiation. *Organic Letters* **5**, 4373–4376.

604. Surry, D. S., Buchwald, S. L. (2007). Selective palladium-catalyzed arylation of ammonia: Synthesis of anilines as well as symmetrical and unsymmetrical di- and triarylamines. *Journal of the American Chemical Society* **129**, 10354–10355.

605. Harris, M. C., Buchwald, S. L. (2000). One-pot synthesis of unsymmetrical triarylamines from aniline precursors. *The Journal of Organic Chemistry* **65**, 5327–5333.

606. Harris, M. C., Huang, X., Buchwald, S. L. (2002). Improved functional group compatibility in the palladium-catalyzed synthesis of aryl amines. *Organic Letters* **4**, 2885–2888.

607. Barluenga, J., Fernández, M. A., Aznar, F., Valdés, C. (2004). Palladium catalyzed amination of vinyl chlorides: A new entry to imines, enamines and 2-amino-1,3-butadienes. *Chemical Communications*, 1400–1401.

608. Poondra, R. R., Turner, N. J. (2005). Microwave-assisted sequential amide bond formation and intramolecular amidation: A rapid entry to functionalized oxindoles. *Organic Letters* **7**, 863–866.

609. Fang, Y.-Q., Yuen, J., Lautens, M. (2007). A general modular method of azaindole and thienopyrrole synthesis via Pd-catalyzed tandem couplings of *gem*-dichloroolefins. *The Journal of Organic Chemistry* **72**, 5152–5160.

610. Fang, Y.-Q., Lautens, M. (2005). Pd-catalyzed tandem C–N/C–C Coupling of gem-dihalovinyl systems: A modular synthesis of 2-substituted indoles. *Organic Letters* **7**, 3549–3552.

611. Fang, Y.-Q., Lautens, M. (2008). A highly selective tandem cross-coupling of gem-dihaloolefins for a modular, efficient synthesis of highly functionalized indoles. *The Journal of Organic Chemistry* **73**, 538–549.

612. Panteleev, J., Zhang, L., Lautens, M. (2011). Domino rhodium-catalyzed alkyne arylation/palladium-catalyzed *N* arylation: A mechanistic investigation. *Angewandte Chemie International Edition* **50**, 9089–9092.

613. Barluenga, J., Jiménez-Aquino, A., Aznar, F., Valdés, C. (2009). Modular synthesis of indoles from imines and *o*-dihaloarenes or *o*-chlorosulfonates by a Pd-catalyzed cascade process. *Journal of the American Chemical Society* **131**, 4031–4041.

614. McLaughlin, M., Palucki, M., Davies, I. W. (2006). Efficient access to cyclic ureas via Pd-catalyzed cyclization. *Organic Letters* **8**, 3311–3314.

615. Willis, M. C., Brace, G. N., Findlay, T. J. K., Holmes, I. P. (2006). 2-(2-Haloalkenyl)-aryl halides as substrates for palladium-catalysed tandem C–N bond formation: Efficient synthesis of 1-substituted indoles. *Advanced Synthesis & Catalysis* **348**, 851–856.

616. Kawaguchi, K., Nakano, K., Nozaki, K. (2007). Synthesis of ladder-type π-conjugated heteroacenes via palladium-catalyzed double *N*-arylation and intramolecular *o*-arylation. *The Journal of Organic Chemistry* **72**, 5119–5128.

617. Christmann, U., Pantazis, D. A., Benet-Buchholz, J., McGrady, J. E., Maseras, F., Vilar, R. (2006). Experimental and theoretical investigations of new dinuclear palladium complexes as precatalysts for the amination of aryl chlorides. *Journal of the American Chemical Society* **128**, 6376–6390.

618. Rataboul, F., Zapf, A., Jackstell, R., Harkal, S., Riermeier, T., Monsees, A., Dingerdissen, U., Beller, M. (2004). New ligands for a general palladium-catalyzed amination of aryl and heteroaryl chlorides. *Chemistry—A European Journal* **10**, 2983–2990.

619. Schulz, T., Torborg, C., Enthaler, S., Schäffner, B., Dumrath, A., Spannenberg, A., Neumann, H., Börner, A., Beller, M. (2009). A general palladium-catalyzed amination of aryl halides with ammonia. *Chemistry—A European Journal* **15**, 4528–4533.

620. Porzelle, A., Woodrow, M. D., Tomkinson, N. C. O. (2009). Palladium-catalyzed coupling of hydroxylamines with aryl bromides, chlorides, and iodides. *Organic Letters* **11**, 233–236.

621. Shen, Q., Shekhar, S., Stambuli, J. P., Hartwig, J. F. (2005). Highly reactive, general, and long-lived catalysts for coupling heteroaryl and aryl chlorides with primary nitrogen nucleophiles. *Angewandte Chemie International Edition* **44**, 1371–1375.

622. Shen, Q., Ogata, T., Hartwig, J. F. (2008). Highly reactive, general and long-lived catalysts for palladium-catalyzed amination of heteroaryl and aryl chlorides, bromides and iodides: Scope and structure–activity relationships. *Journal of the American Chemical Society* **130**, 6586–6596.

623. Shen, Q., Hartwig, J. F. (2008). An air-stable, one-component, highly efficient catalyst for amination of heteroaryl and aryl halides. *Organic Letters* **10**, 4109–4112.

624. Shen, Q., Hartwig, J. F. (2006). Palladium-catalyzed coupling of ammonia and lithium amide with aryl halides. *Journal of the American Chemical Society* **128**, 10028–10029.

625. Bei, X., Guram, A. S., Turner, H. W., Weinberg, W. H. (1999). General and efficient palladium-catalyzed aminations of aryl chlorides. *Tetrahedron Letters* **40**, 1237–1240.

626. Chen, G., Lam, W. H., Fok, W. S., Lee, H. W., Kwong, F. Y. (2007). Easily accessible benzamide-derived P,O ligands (Bphos) for palladium-catalyzed carbon–nitrogen bond-forming reactions. *Chemistry—An Asian Journal* **2**, 306–313.

627. Xie, X., Zhang, T. Y., Zhang, Z. (2006). Synthesis of bulky and electron-rich MOP-type ligands and their applications in palladium-catalyzed C–N bond formation. *The Journal of Organic Chemistry* **71**, 6522–6529.

628. Lundgren, R. J., Sappong-Kumankumah, A., Stradiotto, M. (2010). A highly versatile catalyst system for the cross-coupling of aryl chlorides and amines. *Chemistry—A European Journal* **16**, 1983–1991.

629. Tardiff, B. J., McDonald, R., Ferguson, M. J., Stradiotto, M. (2012). Rational and predictable chemoselective synthesis of oligoamines via Buchwald–Hartwig amination of (hetero)aryl chlorides employing Mor-DalPhos. *The Journal of Organic Chemistry* **77**, 1056–1071.

630. Ackermann, L., Born, R. (2005). Modular diamino- and dioxophosphine oxides and chlorides as ligands for transition-metal-catalyzed C–C and C–N couplings with aryl chlorides. *Angewandte Chemie International Edition* **44**, 2444–2447.

631. Kim, B. R., Cho, S.-D., Kim, E. J., Lee, I.-H., Sung, G. H., Kim, J.-J., Lee, S.-G., Yoon, Y.-J. (2012). Efficient palladium-catalyzed amination of aryl chlorides using di(dicyclohexylamino)phenylphosphine as a PN2 ligand. *Tetrahedron* **68**, 287–293.

632. Urgaonkar, S., Nagarajan, M., Verkade, J. G. (2003). $P[N(i\text{-Bu})CH_2CH_2]_3N$: A versatile ligand for the Pd-catalyzed amination of aryl chlorides. *Organic Letters* **5**, 815–818.

633. Urgaonkar, S., Verkade, J. G. (2004). Scope and Limitations of $Pd_2(dba)_3$/$P(i\text{-BuNCH}_2CH_2)_3N$-catalyzed Buchwald–Hartwig amination reactions of aryl chlorides. *The Journal of Organic Chemistry* **69**, 9135–9142.

634. Reddy, C. R. V., Urgaonkar, S., Verkade, J. G. (2005). A highly effective catalyst system for the Pd-catalyzed amination of vinyl bromides and chlorides. *Organic Letters* **7**, 4427–4430.

635. Reddy, C. R. V., Kingston, J. V., Verkade, J. G. (2008). $(t\text{-Bu})_2PNdP(i\text{-BuNCH}_2CH_2)_3N$: New efficient ligand for palladium-catalyzed C–N couplings of aryl and heteroaryl bromides and chlorides and for vinyl bromides at room temperature. *The Journal of Organic Chemistry* **73**, 3047–3062.

636. Huang, J., Grasa, G., Nolan, S. P. (1999). General and efficient catalytic amination of aryl chlorides using a palladium/bulky nucleophilic carbene system. *Organic Letters* **1**, 1307–1309.

637. Viciu, M. S., Kissling, R. M., Stevens, E. D., Nolan, S. P. (2002). An air-stable palladium/*N*-heterocyclic carbene complex and its reactivity in aryl amination. *Organic Letters* **4**, 2229–2231.

638. Chen, C., Li, Y. F., Yang, L. M. (2007). Efficient synthesis of triarylamines catalyzed by palladium/*N*-heterocyclic carbene. *Journal of Molecular Catalysis A: Chemical* **269**, 158–162.

639. Ackermann, L., Sandmann, R., Schinkel, M., Kondrashov, M. V. (2009). Palladium-catalyzed sequential indole synthesis using sterically hindered amines. *Tetrahedron* **65**, 8930–8939.

640. Stauffer, S. R., Lee, S., Stambuli, J. P., Hauck, S. I., Hartwig, J. F. (2000). High turnover number and rapid, room-temperature amination of chloroarenes using saturated carbene Ligands. *Organic Letters* **2**, 1423–1426.

641. Marion, N., Navarro, O., Mei, J., Stevens, E. D., Scott, N. M., Nolan, S. P. (2006). Modified (NHC)Pd(allyl)Cl (NHC = *N*-heterocyclic carbene) complexes for room-temperature Suzuki–Miyaura and Buchwald–Hartwig reactions. *Journal of the American Chemical Society* **128**, 4101–4111.

642. Schoeps, D., Sashuk, V., Ebert, K., Plenio, H. (2009). Solvent-resistant nanofiltration of enlarged (NHC)Pd(allyl)Cl complexes for cross-coupling reactions. *Organometallics* **28**, 3922–3927.

643. Zhu, L., Gao, T.-T., Shao, L.-X. (2011). Well-defined NHCePd(II)eIm (NHC = *N*-heterocyclic carbene; Im = 1-methylimidazole) complexes catalyzed amination of aryl chlorides. *Tetrahedron* **67**, 5150–5155.

644. Organ, M. G., Abdel-Hadi, M., Avola, S., Dubovyk, I., Hadei, N., Kantchev, E. A. B., O'Brien, C. J., Sayah, M., Valente, C. (2008). Pd-catalyzed aryl amination mediated by well defined, *N*-heterocyclic carbene (NHC)–Pd precatalysts, PEPPSI. *Chemistry—A European Journal* **14**, 2443–2452.

645. Tu, T., Fang, W., Jiang, J. (2011). A highly efficient precatalyst for amination of aryl chlorides: synthesis, structure and application of a robust acenaphthoimidazolylidene palladium complex. *Chemical Communications* **47**, 12358–12360.

646. Broggi, J., Clavier, H., Nolan, S. P. (2008). *N*-Heterocyclic carbenes (NHCs) containing N–C–palladacycle complexes: Synthesis and reactivity in aryl amination reactions. *Organometallics* **27**, 5525–5531.

647. Li, J., Cui, M., Yu, A., Wu, Y. (2007). Carbene adduct of cyclopalladated ferrocenylimine as an efficient catalyst for the amination of aryl chlorides. *Journal of Organometallic Chemistry* **692**, 3732–3742.

648. Shi, J.-C., Yang, P., Tong, Q., Jia, L. (2008). Palladium-catalyzed aminations of aryl halides with phosphine-functionalized imidazolium ligands. *Dalton Transactions*, 938–945.

649. Wolfe, J. P., Buchwald, S. L. (1997). Nickel-catalyzed amination of aryl chlorides. *Journal of the American Chemical Society* **119**, 6054–6058.

650. Gradel, B., Brenner, E., Schneider, R., Fort, Y. (2001). Nickel-catalysed amination of aryl chlorides using a dihydroimidazoline carbene ligand. *Tetrahedron Letters* **42**, 5689–5692.

651. Kuhl, S., Fort, Y., Schneider, R. (2005). Nickel(0)/*N*-heterocyclic carbene complexes catalysed arylation of aromatic diamines. *Journal of Organometallic Chemistry* **690**, 6169–6177.

652. Omar-Amrani, R., Thomas, A., Brenner, E., Schneider, R., Fort, Y. (2003). Efficient nickel-mediated intramolecular amination of aryl chlorides. *Organic Letters* **5**, 2311–1314.

653. Chen, C., Yang, L. M. (2007). Ni(II)–(α-aryl) Complex: A facile, efficient catalyst for nickel-catalyzed carbon-nitrogen coupling reactions. *The Journal of Organic Chemistry* **72**, 6324–6327.

654. Brenner, E., Schneider, R., Fort, Y. (1999). Nickel-catalysed couplings of aryl chlorides with secondary amines and piperazines. *Tetrahedron* **55**, 12829–12842.

655. Desmarets, C., Schneider, R., Fort, Y. (2000). Nickel-mediated amination chemistry. Part 1: Efficient aminations of (het)aryl 1,3-di and 1,3,5-trichlorides. *Tetrahedron Letters* **41**, 2875–2879.

656. Desmarets, C., Schneider, R., Fort, Y. (2001). Nickel-catalysed sequential amination of aryl- or heteroaryl di- and 1,3,5-trichlorides. *Tetrahedron* **57**, 7657–7664.

657. Manolikakes, G., Gavryushin, A., Knochel, P. (2008). An efficient silane-promoted nickel-catalyzed amination of aryl and heteroaryl chlorides. *The Journal of Organic Chemistry* **73**, 1429–1434.

658. Lipshutz, B. H., Ueda, H. (2000). Aromatic amination by heterogeneous Ni^0/C catalysis. *Angewandte Chemie International Edition* **39**, 4492–4494.

659. Lipshutz, B. H., Tasler, S., Chrisman, W., Spliethoff, B., Tesche, B. (2003). On the nature of the 'heterogeneous' catalyst: Nickel-on-charcoal. *The Journal of Organic Chemistry* **68**, 1177–1189.

660. Lipshutz, B. H., Nihan, D. M., Vinogradova, E., Taft, B. R., Boškovic, Z. V. (2008). Copper + nickel-in-charcoal (Cu-Ni/C): A bimetallic, heterogeneous catalyst for cross-couplings. *Organic Letters* **10**, 4279–4282.

661. Guo, D., Huang, H., Xu, J., Jiang, H., Liu, H. (2008). Efficient iron-catalyzed *N*-arylation of aryl halides with amines. *Organic Letters* **10**, 4513–4516.

662. Toma, G., Fujita, K.-i., Yamaguchi, R. (2009). Cobalt-catalyzed C–N bond-forming reaction between *N*-aromatic 2-chlorides and Secondary Amines. *European Journal of Organic Chemistry*, 4586–4588.

663. Miao, T., Wang, L. (2007). Immobilization of copper in organic–inorganic hybrid materials: a highly efficient and reusable catalyst for the Ullmann diaryl etherification. *Tetrahedron Letters* **48**, 95–99.

664. Evano, G., Blanchard, N., Toumi, M. (2008). Copper-mediated coupling reactions and their applications in natural products and designed biomolecules synthesis. *Chemical Reviews* **108**, 3054–3131.

665. Ley, S. V., Thomas, A. W. (2003). Modern synthetic methods for copper-mediated C(aryl)–O, C(aryl)–N, and C(aryl)–S bond formation. *Angewandte Chemie International Edition* **42**, 5400–5449.

666. Tye, J. W., Weng, Z. Q., Giri, R., Hartwig, J. F. (2010). Copper(I) phenoxide complexes in the etherification of aryl halides. *Angewandte Chemie International Edition* **49**, 2185–2189.

667. Lipshutz, B. H., Unger, J. B., Taft, B. R. (2007). Copper-in-charcoal (Cu/C) promoted diaryl ether formation. *Organic Letters* **9**, 1089–1092.

668. Evindar, G., Batey, R. A. (2006). Parallel synthesis of a library of benzoxazoles and benzothiazoles using ligand-accelerated copper-catalyzed cyclizations of *ortho*-halobenzanilides. *The Journal of Organic Chemistry* **71**, 1802–1808.

669. Altenhoff, G., Glorius, F. (2004). A domino copper-catalyzed C–N and C–O cross-coupling for the conversion of primary amides into benzoxazoles. *Advanced Synthesis & Catalysis* **346**, 1661–1664.

670. Fang, Y., Li, C. (2005). CuI-catalyzed intramolecular *O*-vinylation of carbonyl compounds. *Chemical Communications*, 3574–3576.

671. Xia, N., Taillefer, M. (2008). Copper- or iron-catalyzed arylation of phenols from respectively aryl chlorides and aryl iodides. *Chemistry—A European Journal* **14**, 6037–6039.

672. Yang, K., Li, Z., Wang, Z. Y., Yao, Z. Y., Jiang, S. (2011). Highly efficient synthesis of phenols by copper-catalyzed hydroxylation of aryl iodides, bromides, and chlorides. *Organic Letters* **13**, 4340–4343.

673. Zhang, Q., Wang, D., Wang, X., Ding, K. (2009). (2-Pyridyl)acetone-promoted Cu-catalyzed *O*-arylation of phenols with aryl iodides, bromides, and chlorides. *The Journal of Organic Chemistry* **74**, 7187–7190.

674. Thasana, N., Worayuthakarn, R., Kradanrat, P., Hohn, E., Young, L., Ruchirawat, S. (2007). Copper(I)-mediated and microwave-assisted C_{aryl}-$O_{carboxylic}$ coupling: Synthesis of benzopyranones and isolamellarin Alkaloids. *The Journal of Organic Chemistry* **72**, 9379–9382.

675. Zhang, J., Zhang, Z., Wang, Y., Zheng, X., Wang, Z. (2008). Nano-CuO-catalyzed Ullmann coupling of phenols with aryl halides under ligand-free conditions. *European Journal of Organic Chemistry*, 5112–5116.

676. Kim, J. Y., Park, J. C., Kim, A., Kim, A. Y., Lee, H. J., Song, H., Park, K. H. (2009). Cu_2O nanocube-catalyzed cross-coupling of aryl halides with phenols via Ullmann coupling. *European Journal of Inorganic Chemistry*, 4219–4223.

677. Aranyos, A., Old, D. W., Kiyomori, A., Wolfe, J. P., Sadighi, J. P., Buchwald, S. L. (1999). Novel electron-rich bulky phosphine ligands facilitate the palladium-catalyzed preparation of diaryl ethers. *Journal of the American Chemical Society* **121**, 4369–4378.

678. Burgos, C. H., Barder, T. E., Huang, X., Buchwald, S. L. (2006). Significantly improved method for the Pd-catalyzed coupling of phenols with aryl halides: Understanding ligand effects. *Angewandte Chemie International Edition* **45**, 4321–4326.

679. Parrish, C. A., Buchwald, S. L. (2001). Palladium-catalyzed formation of aryl *tert*-butyl ethers from unactivated aryl halides. *The Journal of Organic Chemistry* **66**, 2498–2500.

680. Torraca, K. E., Huang, X., Parrish, C. A., Buchwald, S. L. (2001). An efficient intermolecular palladium-catalyzed synthesis of aryl ethers. *Journal of the American Chemical Society* **123**, 10770–10771.

681. Wu, X., Fors, B. P., Buchwald, S. L. (2011). A single phosphine ligand allows palladium-catalyzed intermolecular C–O bond formation with secondary and primary alcohols. *Angewandte Chemie International Edition* **50**, 9943–9947.

682. Torraca, K. E., Kuwabe, S.-I., Buchwald, S. L. (2000). A high-yield, general method for the catalytic formation of oxygen heterocycles. *Journal of the American Chemical Society* **122**, 12907–12908.

683. Kuwabe, S.-i., Torraca, K. E., Buchwald, S. L. (2001). Palladium-catalyzed intramolecular C–O bond formation. *Journal of the American Chemical Society* **123**, 12202–12206.

684. Maimone, T. J., Buchwald, S. L. (2010). Pd-catalyzed *O*-arylation of ethyl acetohydroximate: Synthesis of *O*-arylhydroxylamines and substituted benzofurans. *Journal of the American Chemical Society* **132**, 9990–9991.

685. Anderson, K. W., Ikawa, T., Tundel, R. E., Buchwald, S. L. (2006). The Selective Reaction of Aryl Halides with KOH: Synthesis of Phenols, Aromatic Ethers, and Benzofurans. *Journal of the American Chemical Society* **128**, 10694–10695.

686. Harkal, S., Kumar, K., Michalik, D., Zapf, A., Jackstell, R., Rataboul, F., Riermeier, T., Monsees, A., Beller, M. (2005). An efficient catalyst system for diaryl ether synthesis from aryl chlorides. *Tetrahedron Letters* **46**, 3237–3240.

687. Hu, T. J., Schulz, T., Torborg, C., Chen, X. R., Wang, J., Beller, M., Huang, J. (2009). Efficient palladium-catalyzed coupling reactions of aryl bromides and chlorides with phenols. *Chemical Communications*, 7330–7332.

688. Schulz, T., Torborg, C., Schäffner, B., Huang, J., Zapf, A., Kadyrov, R., Börner, A., Beller, M. (2009). Practical imidazole-based phosphine ligands for selective palladium-catalyzed hydroxylation of Aryl Halides. *Angewandte Chemie International Edition*, **48**, 918–921.

689. Sergeev, A. G., Schulz, T., Torborg, C., Spannenberg, A., Neumann, H., Beller, M. (2009). Palladium-catalyzed hydroxylation of aryl halides under ambient conditions. *Angewandte Chemie International Edition* **48**, 7595–7599.

690. Gowrisankar, S., Sergeev, A. G., Anbarasan, P., Spannenberg, A., Neumann, H., Beller, M. (2010). A general and efficient catalyst for palladium-catalyzed C–O coupling reactions of aryl halides with primary alcohols. *Journal of the American Chemical Society* **132**, 11592–11598.

691. Mann, G., Incarvito, C., Rheingold, A. L., Hartwig, J. F. (1999). Palladium-catalyzed C–O coupling involving unactivated aryl halides. Sterically induced reductive elimination to form the C–O bond in diaryl ethers. *Journal of the American Chemical Society* **121**, 3224–3225.

692. Shelby, Q., Kataoka, N., Mann, G., Hartwig, J. F. (2000). Unusual *in situ* ligand modification to generate a catalyst for room temperature aromatic C–O bond formation. *Journal of the American Chemical Society* **122**, 10718–10719.

693. Beletskaya, I. P., Ananikov, V. P. (2011). Transition-metal-catalyzed C–S, C–Se, and C–Te bond Formation via cross-coupling and atom-economic addition reactions. *Chemical Reviews* **111**, 1596–1636.

694. Eichman, C. C., Stambuli, J. P. (2011). Transition metal catalyzed synthesis of aryl sulfides. *Molecules* **16**, 590–608.

695. Rout, L., Saha, P., Jammi, S., Unniyamurthy, T. (2008). Efficient copper(I)-catalyzed C–S cross coupling of thiols with aryl halides in water. *European Journal of Organic Chemistry*, 640–643.

696. Herrero, M. T., San Martin, R., Domínguez, E. (2009). Copper(I)-catalyzed *S*-arylation of thiols with activated aryl chlorides on water. *Tetrahedron* **65**, 1500–1503.

697. Kumar, S., Engman, L. (2006). Microwave-assisted copper-catalyzed preparation of diaryl chalcogenides. *The Journal of Organic Chemistry* **71**, 5400–5403.

698. Feng, Y., Wang, H., Sun, F., Li, Y., Fu, X., Jin, K. (2009). A highly efficient and widely functional-group-tolerant catalyst system for copper(I)-catalyzed *S*-arylation of thiols with aryl halides. *Tetrahedron* **65**, 9737–9741.

699. Basu, B., Mandal, B., Das, S., Kundu, S. (2009). Catechol violet as new, efficient, and versatile ligand for Cu(I)-catalyzed C-S coupling reactions. *Tetrahedron Letters* **50**, 5523–5528.

700. Bryan, C. S., Braunger, J. A., Lautens, M. (2009). Efficient synthesis of benzothiophenes by an unusual palladium-catalyzed vinylic C–S coupling. *Angewandte Chemie International Edition* **48**, 7064–7068.

701. Zeng, F. L., Alper, H. (2011). Palladium-catalyzed domino C–S coupling/carbonylation reactions: An efficient synthesis of 2-carbonylbenzo[*b*]thiophene derivatives. *Organic Letters* **13**, 2868–2871.

702. Yi, J., Fu, Y., Xiao, B., Cui, W. C., Guo, Q. X. (2011). Palladium catalyzed synthesis of aryl thiols: Sodium thiosulfate as a cheap and nontoxic mercapto surrogate. *Tetrahedron Letters* **52**, 205–208.

703. Murata, M., Buchwald, S. L. (2004). A general and efficient method for the palladium-catalyzed cross-coupling of thiols and secondary phosphines. *Tetrahedron* **60**, 7397–7403.

704. Fernández-Rodríguez, M. A., Shen, Q. L., Hartwig, J. F. (2006). Highly efficient and functional-group-tolerant catalysts for the palladium-catalyzed coupling of aryl chlorides with thiols. *Chemistry—A European Journal* **12**, 7782–7796.

705. Fernández-Rodríguez, M. A., Shen, Q. L., Hartwig, J. F. (2006). A general and long-lived catalyst for the palladium-catalyzed coupling of aryl halides with thiols. *Journal of the American Chemical Society* **128**, 2180–2181.

706. Fernández-Rodríguez, M. A., Hartwig, J. F. (2010). One-pot synthesis of unsymmetrical diaryl thioethers by palladium-catalyzed coupling of two aryl bromides and a thiol surrogate. *Chemistry—A European Journal* **16**, 2355–2359.

707. Fernández-Rodríguez, M. A., Hartwig, J. F. (2009). A general, efficient, and functional-group-tolerant catalyst system for the palladium-catalyzed thioetherification of aryl bromides and iodides. *The Journal of Organic Chemistry* **74**, 1663–1672.

708. Alvaro, E., Hartwig, J. F. (2009). Resting state and elementary steps of the coupling of aryl halides with thiols catalyzed by alkylbisphosphine complexes of palladium. *Journal of the American Chemical Society* **131**, 7858–7868.

709. Sayah, M., Organ, M. G. (2011). Carbon–Sulfur bond formation of challenging substrates at low temperature by using Pd-PEPPSI-Ipent. *Chemistry—A European Journal* **17**, 11719–11722.

710. Haddenham, D., Bailey, C. L., Vu, C., Nepomuceno, G., Eagon, S., Pasumansky, L., Singaram, B. (2011). Lithium aminoborohydrides 17. Palladium catalyzed borylation of aryl iodides, bromides, and triflates with diisopropylaminoborane prepared from lithium diisopropylaminoborohydride. *Tetrahedron* **67**, 576–583.

711. Billingsley, K. L., Barder, T. E., Buchwald, S. L. (2007). Palladium-catalyzed borylation of aryl chlorides: Scope, applications, and computational studies. *Angewandte Chemie International Edition* **46**, 5359–5363.

712. Ishiyama, T., Miyaura, N. (2004). Metal-catalyzed reactions of diborons for synthesis of organoboron compounds. *The Chemical Record* **3**, 271–280.

713. Hartwig, J. F. (2011). Regioselectivity of the borylation of alkanes and arenes. *Chemical Society Reviews* **40**, 1992–2002.

714. Ishiyama, T., Ishida, K., Miyaura, N. (2001). Synthesis of pinacol arylboronates via cross-coupling reaction of bis(pinacolato)diboron with chloroarenes catalyzed by palladium(0)–tricyclohexylphosphine complexes. *Tetrahedron* **57**, 9813–9816.

715. Lu, J., Guan, Z.-Z., Gao, J.-W., Zhang, Z.-H. (2011). An improved procedure for the synthesis of arylboronates by palladium-catalyzed coupling reaction of aryl halides and bis(pinacolato)diboron in polyethylene glycol. *Applied Organometallic Chemistry* **25**, 537–541.

716. Murata, M., Oyama, T., Watanabe, S., Masuda, Y. (2002). Synthesis of benzylboronates via palladium-catalyzed borylation of benzyl halides with pinacolborane. *Synthetic Communications* **32**, 2513–2517.

717. Zhang, Y., Gao, J., Li, W., Lee, H., Lu, B. Z., Senanayake, C. H. (2011). Synthesis of 8-arylquinolines via one-pot Pd-catalyzed borylation of quinoline-8-yl halides and subsequent Suzuki–Miyaura coupling. *The Journal of Organic Chemistry* **76**, 6394–6400.

718. Billingsley, K. L., Buchwald, S. L. (2008). An improved system for the palladium-catalyzed borylation of aryl halides with pinacol borane. *The Journal of Organic Chemistry* **73**, 5589–5591.

719. Pilarski, L. T., Szabó, K. J. (2011). Palladium-catalyzed direct synthesis of organoboronic acids. *Angewandte Chemie International Edition* **50**, 8230–8232.

720. Molander, G. A., Trice, S. L., Dreher, S. D. (2010). Palladium-catalyzed, direct boronic acid synthesis from aryl chlorides: A simplified route to diverse boronate ester derivatives. *Journal of the American Chemical Society* **132**, 17701–17703.

721. Tang, W. J., Keshipeddy, S., Zhang, Y. D., Wei, X. D., Savoie, J., Patel, N. D., Yee, N. K., Senanayake, C. H. (2011). Efficient monophosphorus ligands for palladium-catalyzed miyaura borylation. *Organic Letters* **13**, 1366–1369.

722. Kawamorita, S., Ohmiya, H., Iwai, T., Sawamura, M. (2011). Palladium-catalyzed borylation of sterically demanding aryl halides with a silica-supported compact phosphane ligand. *Angewandte Chemie International Edition* **50**, 8363–8366.

723. Zhang, P., Roundtree, I. A., Morken, J. P. (2012). Ni- and Pd-catalyzed synthesis of substituted and functionalized allylic boronates. *Organic Letters* **14**, 1416–1419.

724. Moldoveanu, C., Wilson, D. A., Wilson, C. J., Corcoran, P., Rosen, B. M., Percec, V. (2009). Neopentylglycolborylation of aryl chlorides catalyzed by the mixed ligand system NiCl2(dppp)/dppf. *Organic Letters* **11**, 4974–4977.

725. Moldoveanu, C., Wilson, D. A., Wilson, C. J., Leowanawat, P., Resmerita, A. M., Liu, C., Rosen, B. M., Percec, V. (2010). Neopentylglycolborylation of *ortho*-substituted aryl halides catalyzed by $NiCl_2$-based mixed-ligand systems. *The Journal of Organic Chemistry* **75**, 5438–5452.

726. Leowanawat, P., Resmerita, A. M., Moldoveanu, C., Liu, C., Zhang, N., Wilson., D. A., Hoang, L. M., Rosen, B. M., Percec, V. (2010). Zero-valent metals accelerate the neopentylglycolborylation of aryl halides catalyzed by NiCl2-based mixed-ligand systems. *The Journal of Organic Chemistry* **75**, 7822–7828.

727. Yamamoto, T., Morita, T., Takagi, J., Yamakawa, T. (2011). $NiCl_2(PMe_3)_2$-catalyzed borylation of aryl chlorides. *Organic Letters* **13**, 5766–5769.

728. Dudnik, A. S., Fu, G. C. (2012). Nickel-catalyzed coupling reactions of alkyl electrophiles, including unactivated tertiary halides, to generate carbon–boron bonds. *Journal of the American Chemical Society* **134**, 10693–10697.

729. Fleckenstein, C. A., Plenio, H. (2010). Sterically demanding trialkylphosphines for palladium-catalyzed cross coupling reactions—alternatives to $PtBu_3$. *Chemical Society Reviews* **39**, 694–711.

730. Kantchev, E. A. B., O'Brien, C. J., Organ, M. G. (2006). Pd-*N*-heterocyclic carbene (NHC) catalysts for cross-coupling reactions. *Aldrichimica Acta* **39**, 97–111.

CHAPTER 2

HOMOGENEOUS TRANSITION-METAL-CATALYZED C–F ACTIVATION

SHANG-DONG YANG

State Key Laboratory of Applied Organic Chemistry, Lanzhou University, Lanzhou, People's Republic of China

2.1 BACKGROUND

In recent years, C–F bond activation by transition metal centers has attracted a great deal of interest not only because in basic studies the C–F bond is the strongest and most inert single bond in an organic compound, but also because the extraordinary

Homogeneous Catalysis for Unreactive Bond Activation, First Edition. Edited by Zhang-Jie Shi.

properties of fluorocarbons are usually used in pharmaceuticals, agrochemicals, imaging agents, and novel materials [1]. Approximately 30% of all agrochemicals and 20% of all pharmaceuticals contain fluorine, including drugs such as Lipitor, Lexapro, and Prozac. In the past several decades, with the remarkable progress in C–F bond activation by the use of stoichiometric amounts of transition metal complexes, the number of reports on catalytic transformations has been increasing [2]. From the viewpoint of organic synthesis, Various factors such as atom economy, chemo-, regio-, and stereoselective transformations, and catalytic processes have become primary and most essential requirements. Therefore, development of catalytic processes of C–F bond activation, especially selective C–F bond activation, is of practical significance both for laboratory synthesis and industrial production. Although several excellent reviews [3] on mechanistic, structural, and, in some cases, synthetic aspects of C–F bond activation by metal complexes have been published. This chapter will focus on various reactions of cross-coupling, nucleophilic substitution, and selectivity dehydrofluorination by different metals through different activation strategies with an emphasis on the mechanisms, scopes, and their applications in synthetic chemistry. Finally, a summary and outlook of C–F bond activation in organic synthesis is provided.

At present, transition-metal-catalyzed C–F bond activation models have four different ways (Figure 2.1) [4]; the most frequent pathway is oxidative addition, which mainly involves the lower oxidation state complexes of nickel, palladium, and platinum. The second pathway is electronic reagent assistance C–F bond elimination, the third pathway is involved in the hydrodefluorination process, and the last pathway is the nucleophilic attack and involves some strong nucleophilies such as lithium and potassium reagents.

According to the different course of reactions, the account will be summarized in three parts of cross-coupling, nucleophilic substitution, and dehydrofluorination by different metals through different activation strategies. Finally, applications in the organic synthesis of C–F bond activation are also being related.

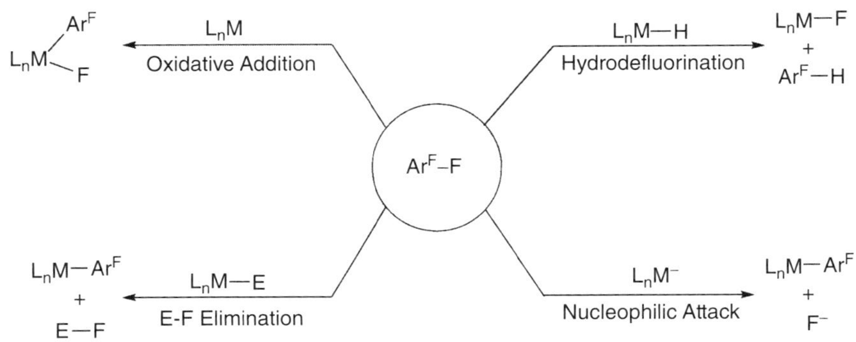

Figure 2.1. Various strategies of C–F bond activation.

2.2 TRANSITION-METAL-MEDIATED CROSS-COUPLING REACTIONS BY C–F BOND ACTIVATION

The transition-metal-catalyzed cross-coupling reaction of organometallic reagents with organic halides or pseudo halides has been extensively studied and employed in many areas of organic synthesis for drug discovery, natural product synthesis, and material science (Figure 2.2) [5]. Therefore the 2010 Nobel Prize was awarded for contributions specifically to the field of palladium-catalyzed cross-coupling reactions. Most of these reactions are initiated generally by the oxidative addition of organic halides to M(0) complexes, so that the scope of organic halides has been limited to aryl, alkenyl, allyl, and benzyl halides. This is probably due to the slow oxidative addition of alkyl halides to low valent transition metals and to the rapid β-hydrogen elimination from the β-alkyl metal intermediates. Although aryl iodides and bromides are most commonly employed, the use of aryl chlorides as cross-coupling participants has recently received increasing attention from the scientific and industrial viewpoints [6]. Actually, there have been excellent reports on Stille and Suzuki–Miyaura reactions of aromatic chlorides by the use of bulky and electron-donating phosphine ligands on the palladium. However, the progress in cross-coupling of aromatic and vinylic fluorides has been much slower than that of chlorides.

The challenges in the activation of C–F bonds rival those of C–H activation. Activation of the C–F bond is of importance due to the environmental hazards associated with the use of fluorocarbons [7]. Fluorocarbons are highly resistant to oxidative degradation, which makes them useful for many applications. The inertness of C–F bonds is a consequence of the strength of the C–F bond and the high electronegativity of fluorine. The energy of the sp^3 C–F bond is typically 120–125 kcal mol^{-1} and the low σ-basicity of the fluorine lone pairs makes fluorocarbons very poor ligands. Compared to their saturated counterparts, fluorinated alkenes and arenes are much more reactive due to the presence of the π-electron system, which is susceptible to nucleophilic attack, and fluoride is a good leaving group [8]. Previously, aryl fluorides were considered to be uncommon coupling partners for the cross-coupling reactions; furthermore, its mechanism, which is the same as that of chloride, bromide, and iodides, is involved in a sequence of oxidative addition, transmetalation,

Figure 2.2. Transition metal-catalyzed cross-coupling reations.

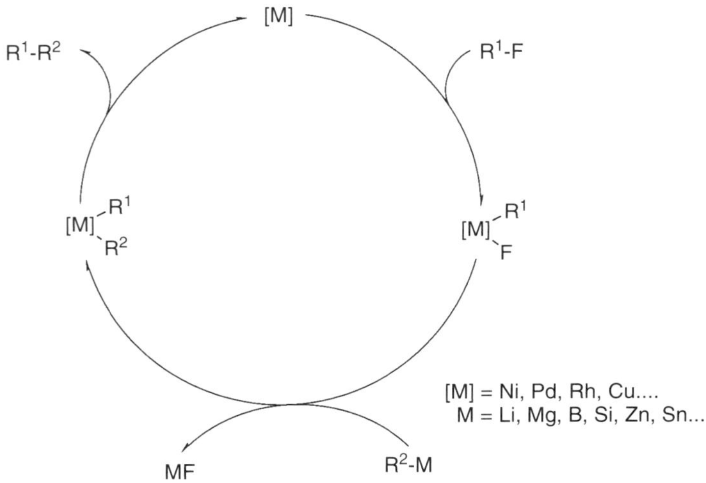

Figure 2.3. Catalytic cycle of transition-metal catalyzed C–F bond acivation.

and reductive elimination (Figure 2.3). The general catalysts include group 9 and 10 transition metals, particularly nickel and palladium, which are highly effective in catalyzing cross-coupling reactions of C–F bond activation. In this part of the chapter, the advancement will be reviewed with different metals.

2.2.1 Nickel-Mediated C–F Bond Activation

Transition-metal-catalyzed cross-coupling reactions between aryl metals and aryl halides are an actively investigated field in organometallic chemistry. Palladium and nickel complexes are very active catalysts and are extensively used to catalyze C–F bond activation. In 1973, Tamao and co-workers [9] reported the first nickel-catalyzed cross-coupling reactions of Grignard reagents with aryl fluorides by sp^2 carbon–fluorine bond activation (Scheme 2.1). Although the regioselectivity and yield were not satisfied, this groundbreaking work provided the possibility of C–F bond activation. Encouraged by this significant advance, many excellent examples of Ni-catalyzed cross-coupling reactions involving C–F bond activation have been published successively.

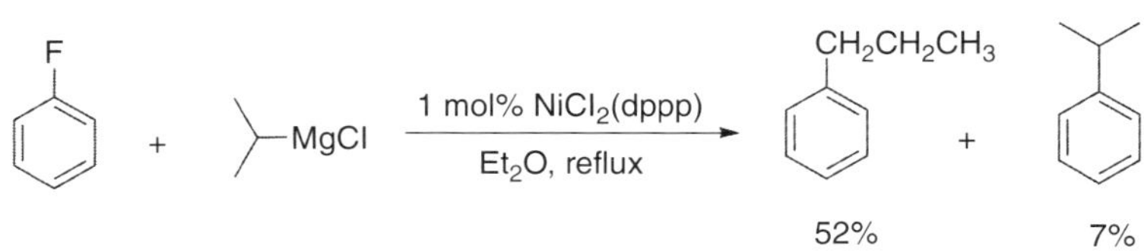

Scheme 2.1.

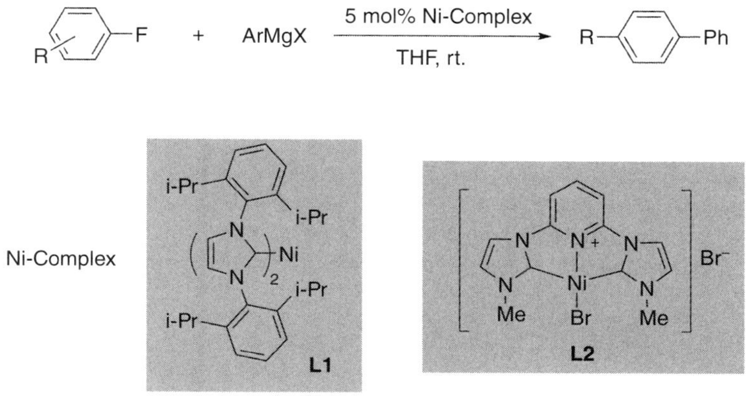

Scheme 2.2.

In 2001, Herrmann and co-workers [10] reported the nickel–NHC complex **L1**-catalyzed cross-coupling reaction of aryl fluorides and aryl Grignard reagents at room temperature (Scheme 2.2). By the use of nickel–NHC complex **L1**, electronically nonactivated aryl fluorides undergo cross-coupling reactions affording a variety of unsymmetrical biaryls. In this reaction, the catalyst ligand plays a pivotal role. Recently, Inamoto et al. [11] developed an NHC-derived pincer-type bis-(imidazolin-2-ylidene) nickel(II) complex **L2** and successfully applied it to Suzuki–Miyaura couplings of aryl fluorides under mild conditions and displayed good functional group tolerance. Different aryl Grignard reagents also were investigated in the reactions.

In many cases, catalytic cross-coupling with palladium and nickel catalysts is facilitated by sterically demanding ligands. For the activation of C–F bonds, a highly electron-rich metal center is required. This criterion can be met by the use of strongly Lewis basic, electron-donating ligands combined with low-valent metals. For example, complexes of nickel(0) with triethylphosphane ligands [12] are known to stoichiometrically activate aromatic C–F bonds. In 2002, Mongin et al. [13] reported that nickel-catalyzed cross-coupling reactions between aryl Grignard reagents and fluoroazines and fluorodiazines occurred in THF at room temperature using commercially available 1,2-bis(diphenylphosphino) ethane, 1,3-bis(diphenylphosphino) propane, or 1,1-bis(diphenylphosphino) ferrocene as ligand. Various fluoro heterocycles such as pyridines, diazines (pyrazine, pyridazine), benzodiazines (quinoxaline), and quinolines were successfully involved in the reaction with phenylmagnesium halides. In 2005, Tamao and co-workers [14] reexamined the cross-coupling reaction of aryl fluorides with Grignard reagents in the presence of nickel and palladium catalysts and revealed that the bidentate phosphine ligands, such as dppe and dppp, are the keys of the Kumada–Tamao cross-coupling reactions with *N*-heterocyclic fluorides (Scheme 2.3). Although different nickel complexes could prompt sp^2 C–F bond activation smoothly to provide the corresponding unsymmetrical biaryls, the yield will be affected by the ligand.

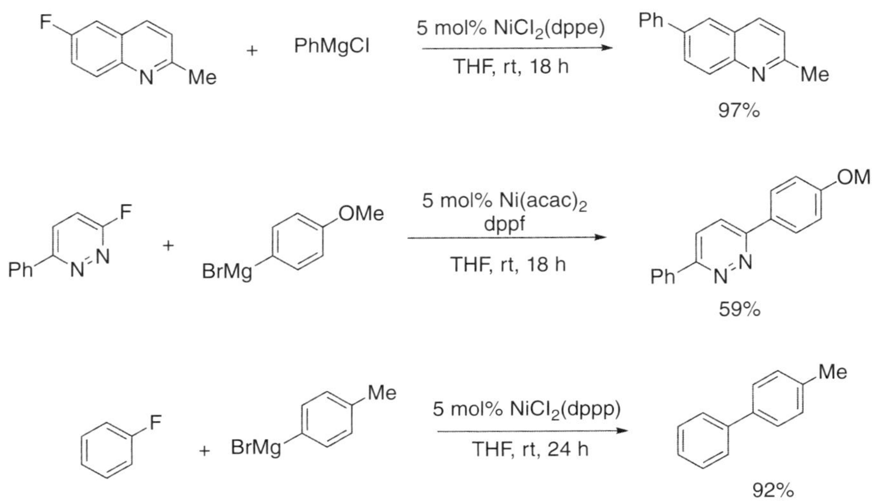

Scheme 2.3.

The design of phosphine ligands is quite important for improvement of catalytic activity. Nakamura designed a useful hydroxy phosphine ligand **L4** that facililated C–F bond activation of fluoroaromatics via nickel/magnesium bimetallic cooperation (Scheme 2.4) [15]. On the basis of an assumption that C–F bond breaking is a turnover-limiting step in the catalytic cycle, the bidentate ligand **L4** can hold nickel and magnesium atoms together to accelerate the Ni-catalyzed cross-coupling

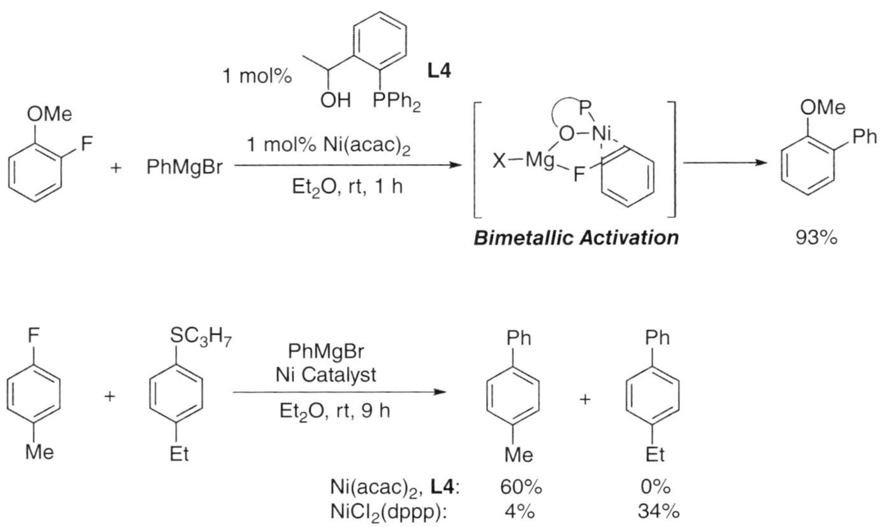

Scheme 2.4.

reactions of aromatic fluorides. Although the reported synthetic results do not necessarily prove this hypothesis, the high reactivity and the chemoselectivity, in particular, are consistent with the assumption. Furthermore, the present catalysis shows intriguing chemoselectivity in which aryl fluorides react faster than sulfides and triflates. Competition experiments of an equimolar mixture of 4-fluorotoluene, 1-ethyl-4-propylthiobenzene, and PhMgBr in the presence of **L4** were carried out and afforded 4-methylbiphenyl (60%) and none of 4-ethylbiphenyl (the aryl sulfide was entirely recovered). If the catalyst is changed to $NiCl_2$ (dppp), the C–F bond activation will be inhibited. This result suggested that the sulfide is a preferred ligand to the catalyst but does not take part in the coupling reaction.

Besides aryl and alkyl phosphines, bulky phosphites **L5** and aminophosphine oxides **L6** are effective ligands for Ni-catalyzed cross-coupling reactions (Scheme 2.5) [16]. Utilization of microwave screening allows for a rational high-throughput screening protocol to quickly identify nickel- or palladium-based catalyst systems for the cross-coupling of aryl fluorides with aryl Grignard reagents. The more sterically hindered of tris(2,4-di-*t*-butylphenyl) phosphite **L5** proved to be a outstanding ligand in this cross-coupling. Moreover, the Ackermann group found that the air-stable secondary phosphine oxides ligand **L6** and **L7** exhibited

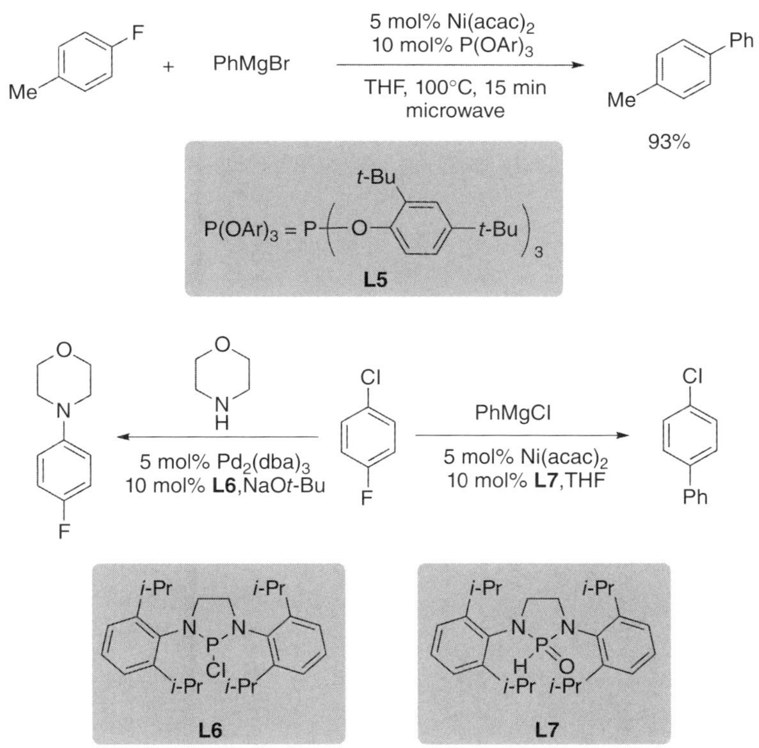

Scheme 2.5.

superior catalytic activity in the activation of C–F bonds [17]. Highly efficient catalysis was achieved with this preligand even at ambient temperature with different Grignard reagents. Finally, a combination of a palladium-catalyzed amination and a nickel-catalyzed Kumada cross-coupling of the resulting electron-rich aryl fluoride were used for the regioselective functionalization of fluorobenzene (Scheme 2.5).

Compared to monofluoroarenes, the selective activation sp^2 C–F bond of polyfluoro-substituted arenes is still very difficult and presents challenges. The replacement of fluorine atoms on an electron-deficient aromatic ring with different nucleophiles by aromatic nucleophilic substitution (S_NAr) has received increasing attention for the synthesis of functionalized aromatic compounds [18]. However, transition-metal-catalyzed sp^2 C–F bond selective activation is only a new area of study. In 2006, Radius and co-workers [19] reported the first examples of Ni-catalyzed Suzuki–Miyaura coupling involving C–F bond activation of fluoroarenes (Scheme 2.6). In the presence of the NHC (*N*-heterocyclic carbene)-stabilized nickel complex **L8** ($[Ni_2(i\text{-}Pr_2Im)_4(cod)]$ ($i\text{-}Pr_2Im$: 1,3-di(isopropyl)imidazol-2-ylidene)), the cross-coupling reactions of octafluorotoluene and perfluorinated biphenyl with phenylboronic acid proceed smoothly. In all cases, the fluoro substituents in the *para* position of the CF_3 or C_6F_5 group are substituted with aryl groups.

Transition-metal-catalyzed cross-coupling reactions of organometallic reagents with carboxylic acid derivatives present a powerful and convergent approach to ketone synthesis [20]. These new approaches have the advantage of avoiding strongly basic and nucleophilic reaction conditions typically associated with protocols such as the Weinreb amide approach. Acid fluorides as potentially versatile acid derivatives were widely used in peptide coupling under conditions that do not lead to epimerization. Zhang and Rovis [21] reported the rapid Ni-catalyzed cross-coupling of acid fluorides with organozinc reagents tolerating epimerizable stereocenters and transferring both groups from zinc in the presence of a versatile

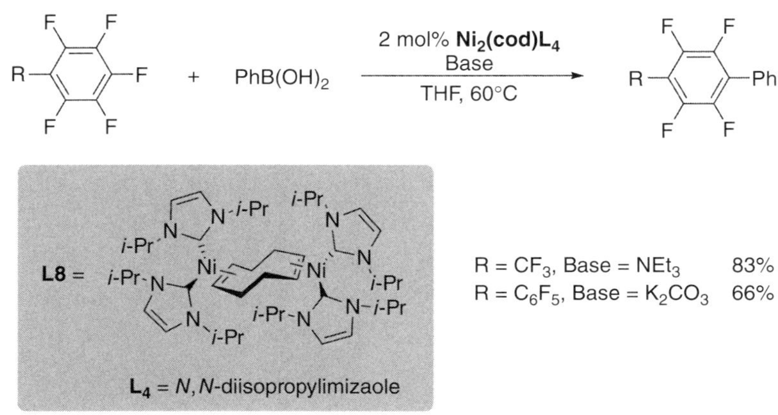

Scheme 2.6.

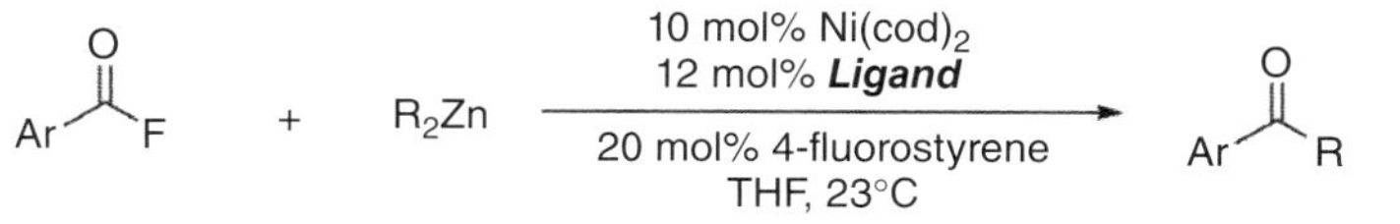

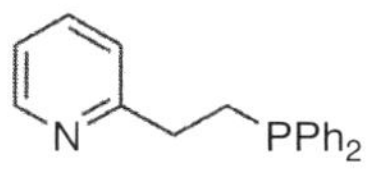

pyphos

R = Ph:

Ligand	***Yield (%)***
pyphos	97
bipy	70
1, 10-phenanthroline	57
dppe	trace

Entry	R_2Zn	R	Yield (%)
1	Me_2Zn	Me	92
2	Et_2Zn	Et	90
3	$i\text{-}Pr_2Zn$	i-Pr	91
4	$(EtO_2CCH_2CH_2)_2Zn$	$EtO_2CCH_2CH_2$	95

Scheme 2.7.

catalyst (Scheme 2.7). [21] The procedure can be applied to a variety of acid derivatives, including pyridyl esters, and tolerates epimerizable stereocenters. A range of functionality is tolerated, including thiophene, alkenes, and acetals. Significantly, α-alkoxides are well tolerated. Impressively, acid chlorides, anhydrides, acyl cyanides, thioesters, and even pyridyl and aryl esters are good reaction partners with this catalyst, providing the coupling product in excellent yields with slightly longer reaction times. In addition, different organozinc reagents exhibited good reaction activity and afforded highest yields.

In order to render this reaction more practical, further optimized reaction conditions were carried out and they found that air-stable $Ni(acac)_2$ may be used as a pre-catalyst in place of the air-sensitive $Ni(COD)_2$. As a testament to the practicality of this protocol, this reaction was conducted using 1.0 mol% $Ni(acac)_2$, 1.2 mol% pyphos, and 2.0 mol% 4-fluorostyrene on a 40 mmol scale to afford ketone in 93% yield after 5 min at 23°C (Scheme 2.8).

One of the current shortcomings of metal-catalyzed ketone synthesis is that epimerizable functionality has not been extensively illustrated to participate. The use of acyl fluorides allows this to occur. Lactate-derived acyl fluoride **1** undergoes

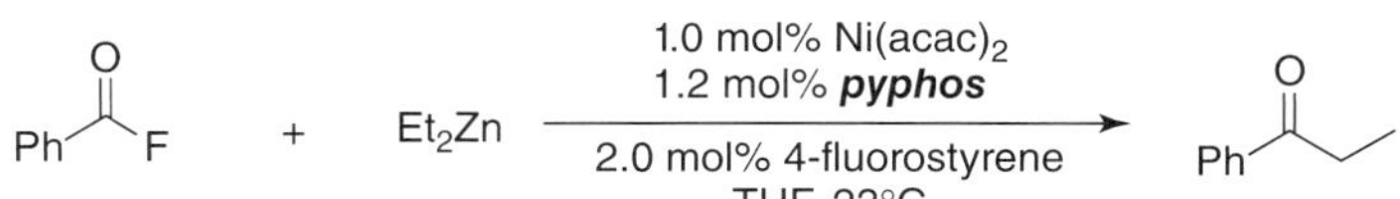

93% yield

Scheme 2.8.

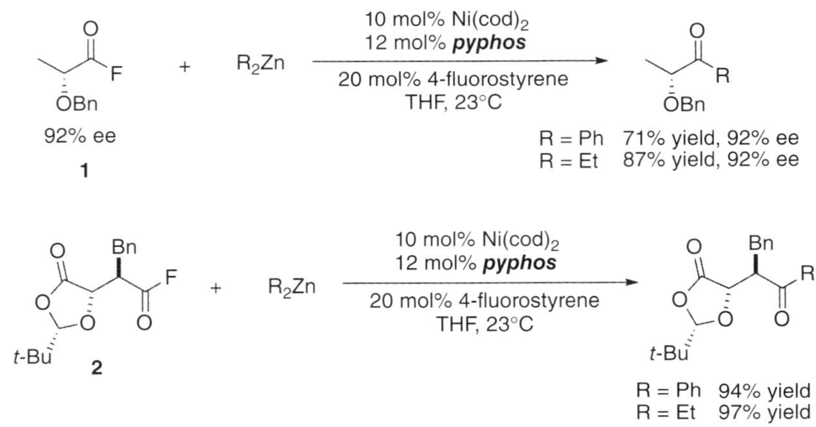

Scheme 2.9.

the reaction in good yield without loss of stereochemical integrity, while **2** also participates well with no elimination or epimerization (Scheme 2.9).

Nowadays the perfluoro- or polyfluoro-diaryl ethers have gained much attention, especially in the materials science, pharmaceutical, and agricultural areas due to their unique chemical and physiological properties [22]. Recently, Cao's group described a novel and efficient method for the synthesis of polyfluoro-substituted unsymmetrical biaryl ethers via oxygen-promoted Ni-catalyzed coupling of arylboronic acids with polyfluoroarenes (Scheme 2.10) [23]. The polyfluorinated arenes presumably captured the phenoxide intermediate efficiently, which made the oxygen insertion proceed smoothly via the S_NAr protocol. The ^{18}O-labeling experiment clearly indicated that the oxygen introduced into the biaryl ether completely originated from very trace amounts of oxygen and acts an efficient Ni(II) oxidant simultaneously in the reaction system. Based on these results, a plausible mechanism for oxygen-promoted Ni-catalyzed coupling of arylboronic acids with fluoroarenes is outlined in Scheme 2.10. Generally, nickel complexes are inert to dioxygen, but the suitable metal's supporting ligand can engender such reactivity. Nickel ions might produce $Ni{-}O_2$ species in the oxidation state of I supported by the acac ligand. Thus, in the first step, molecular oxygen could oxidize $Ni(acac)_2$ to peroxo complex $O_2Ni(acac)_2$ I, which can react with arylboronic acids to generate the plausible intermediate $ArB(OH)_2OONi(acac)_2$ II. The formation of II is rapidly displaced by the formation of III. Hydrolysis of III furnishes two intermediates, $ArNi(acac)_2$ IV and $HOOB(OH)_2$ V. If the reaction is performed in the presence of highly electron-deficient fluorinated aromatic compounds, V can further hydrolyze to the key intermediate, hydrogen peroxide VI, which reacts easily with arylboronic acids to afford phenol VII. Finally, nucleophilic attack of fluoroarenes by phenoxide could generate the biaryl ether VIII, whereas IX was detected as a minor product.

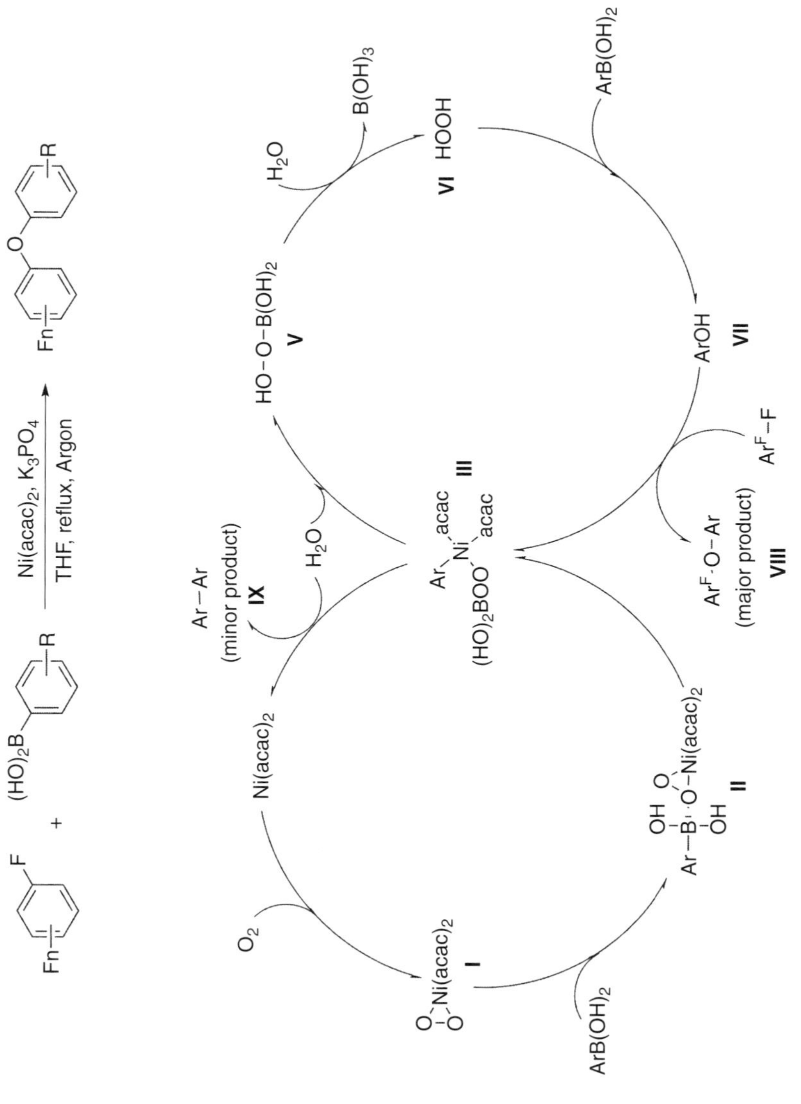
Ni(acac)2, K3PO4
THF, reflux, Argon
Ni(acac)2
I
II
III
V
VI HOOH
VII
VIII
IX
ArB(OH)2
B(OH)3
H2O
O2
ArOH
HO–O–B(OH)2
Ar–Ar
(minor product)
(major product)

Scheme 2.10.

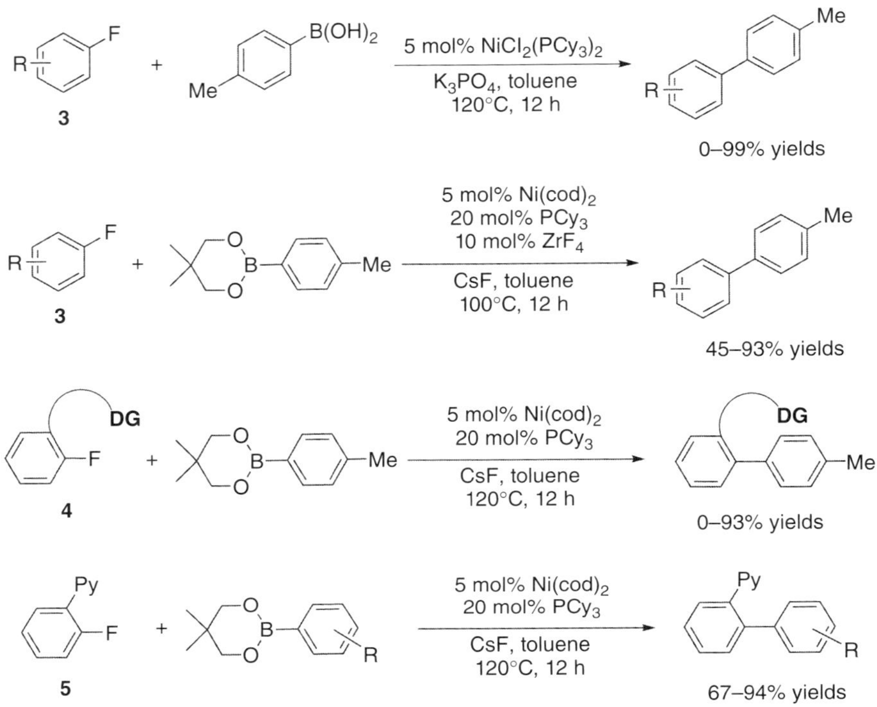

Scheme 2.11.

Very recently, Chatani and co-workers [24]. developed two protocols for the nickel-catalyzed Suzuki-Miyaura cross-coupling of aryl fluorides (Scheme 2.11). The first uses metal fluoride co-catalysts, including ZrF_4 and TiF_4, to significantly accelerate the reaction and thus expand the substrate scope. A range of electron-deficient aryl fluorides **3** and those bearing extended π-systems, such as biaryl-, stilbene-, and naphthalene-based fluorides, can be arylated using Ni/Zr bimetallic catalysis. The second protocol utilizes aryl fluorides bearing a directing group **4**, which facilitate the difficult C–F bond oxidative addition by forming stable cyclometalated intermediates. Directing groups that contain an sp^2-hybridized nitrogen atom **5**, including pyridine, pyrazole, and oxazoline, can successfully promote the cross-coupling reaction with an array of aryl and alkenylboronic esters. The authors propose that the turnover-limiting step changes from being the oxidative addition to the transmetalation, through the introduction of directing groups. The electronic characteristcs of the boronic esters have little impact in the nondirected cross-coupling of aryl fluorides, whereas electron-deficient boronic esters exhibited a superior reactivity in the chelation-assisted reactions.

Utilization of tricarbonylchromium complexes of fluoroarenes **(6)** permits Suzuki and Stille cross-coupling reactions via nucleophilic aromatic substitution to form the corresponding biaryl and styrene Cr complexes (Scheme 2.12) [25].

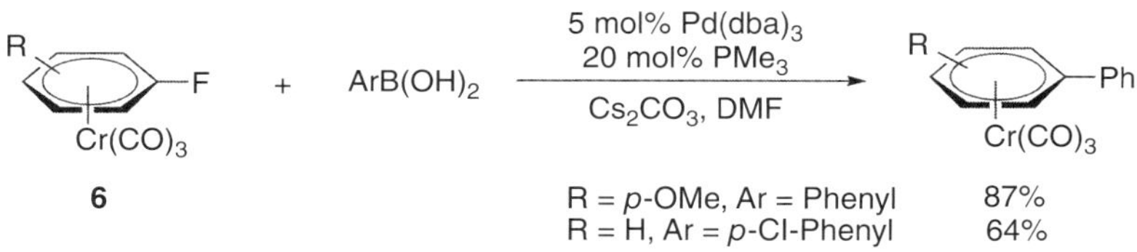

Scheme 2.12.

When *p*-chlorosubstituted phenylboronic acid is used, the cross-coupling product is isolated in 64% yield, which suggests that the reactivity of the C–F bond in a Cr(0)–arene complex (**6**) is greater than that of the aromatic C–Cl bond under the present conditions.

Comparison to the cross-coupling of monofluoroarenes dominating the field of C–F bond activation, reports of cross-coupling of polyfluoroarenes to form new C–C bonds by selective, mono-functionalization of polyfluoroarenes remain scarce [26]. Perutz and co-workers [27] described a carbon–fluorine activation of hexafluorobenzene and pentafluoropyridine using a Ni(0) complex. Their studies confirmed an earlier discovery that *trans*-$Ni(PEt_3)_2(C_6F_5)F$ can be generated from the reaction between C_6F_6, $Ni(COD)_2$ and PEt_3, even though the reaction rate was unusually slow and the isolated yield was only 48%. In contrast, the reaction between $Ni(COD)_2$, PEt_3, and pentafluoropyridine was very rapid and led to *trans*-$Ni(PEt_3)_2(C_5F_4N)F$ (**7**) as the major product (85% by ^{19}F NMR). Activation at the 3- and 4-position was also detected by ^{19}F NMR but the products were not isolated (Scheme 2.13). It is important to note that although pentafluoropyridine itself is prone to nucleophilic substitution at the 4-position of the tetrafluoropyridyl ring, these reactions proceed principally with activation at the 2-position. DFT calculations revealed a novel phosphine-assisted C–F activation. This neighboring group effect involves phosphine ligand and the pyridyl nitrogen and is relevant only for activation at the 2-position [28]. Such a mechanism is operative for nickel phosphine species, but not nickel carbene species. In a subsequent report, the Perutz group described the reactivity of Ni–F complexes with carbon and oxygen nucleophiles. For example, when complex **7** was reacted with Me_3SiOTf, fluoride abstraction took place, generating *trans*-$[Ni(OTf)(C_5F_4N)(PEt_3)_2]$ (**8**). Reaction of **7** with dimethyl zinc and phenyl lithium readily generated complexes **9** and **10**, respectively. Based on these initial studies, Perutz and co-workers [28] reported that **11** could also react with $Bu_3SnCH{=}CH_2$ at room temperature to generate the products **12** in the presence of 10 mol% *trans*-$Ni(PEt_3)_2(C_5F_4N)F$ (**7**) as catalyst, which were identified by ^1H and ^{19}F NMR spectroscopy. Subsequent stoichiometric studies indicated the feasibility of selective C–F activation of 2,4,6-trifluoropyrimidine and octafluoronaphthalene by Ni(0). Fluoropyridyl and pyrimidyl Ni complexes possessing monodentate phosphine ligands such as PEt_3 promote cross-coupling reactions.

Significant advances have been achieved in *sp*^{2} C–F bond activation over the last decade that have enabled cross-coupling reactions to be effected between fluoroarenes and aryl Grignard reagents by using either a nickel or palladium

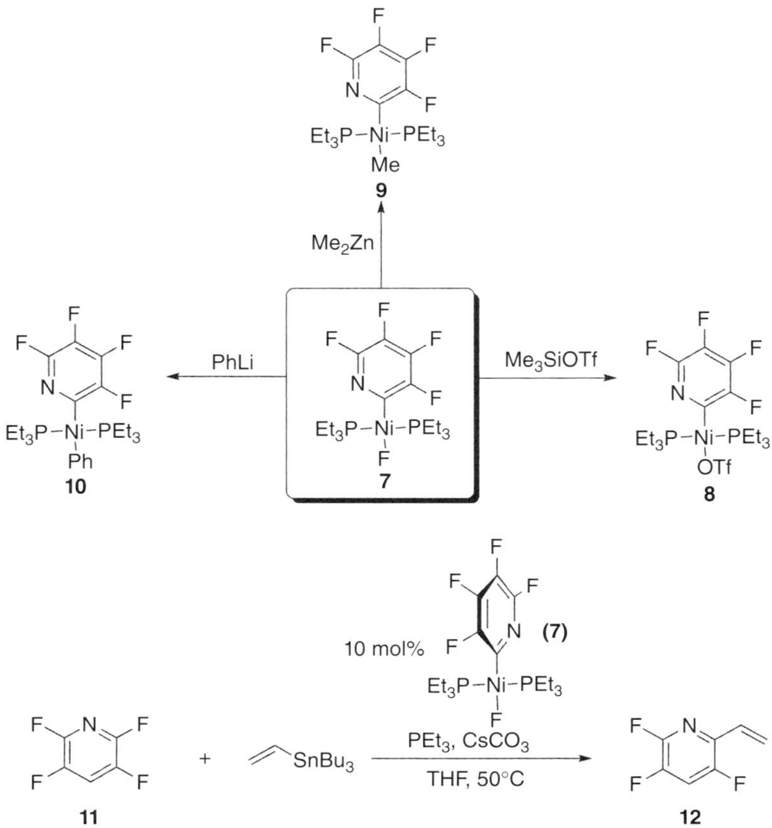

Scheme 2.13.

catalyst. However, examples of aliphatic sp^3 C–F bond fission by use of transition metal complexes are very rare. In 2003, Terao et al. [29] have investigated systematically the transition-metal-catalyzed alkyl–alkyl cross-coupling reactions of alkyl chlorides, bromides, and tosylates with Grignard reagents in the presence of a 1,3-butadiene derivative under mild conditions and extended them to C–F bond cleavage (Scheme 2.14). They disclosed that the first example of C–C bond formation using n-C_8H_{17}-F (**13**) with n-propylmagnesium bromide proceeds smoothly at 25°C in the presence of $CuCl_2$, with 3-butadiene as an additive. However, when no additive is employed, n-C_8H_{17}-F does not react at all, even in the presence of Ni catalyst. These results may imply that 1,3-butadiene plays an important role in stabilizing an active species in the Cu-catalyzed cross-coupling reaction. It is noted that n-octyl fluoride is much more reactive than the corresponding chloride under the present reaction conditions. A plausible reaction pathway involves bis-π-allyl nickel complex **14**, which is generated by the reaction of Ni(0) with 2 equiv of 1,3-butadiene. Grignard reagents (R^1MgX) react with (η3-allyl)$_2$ nickel (**14**) to

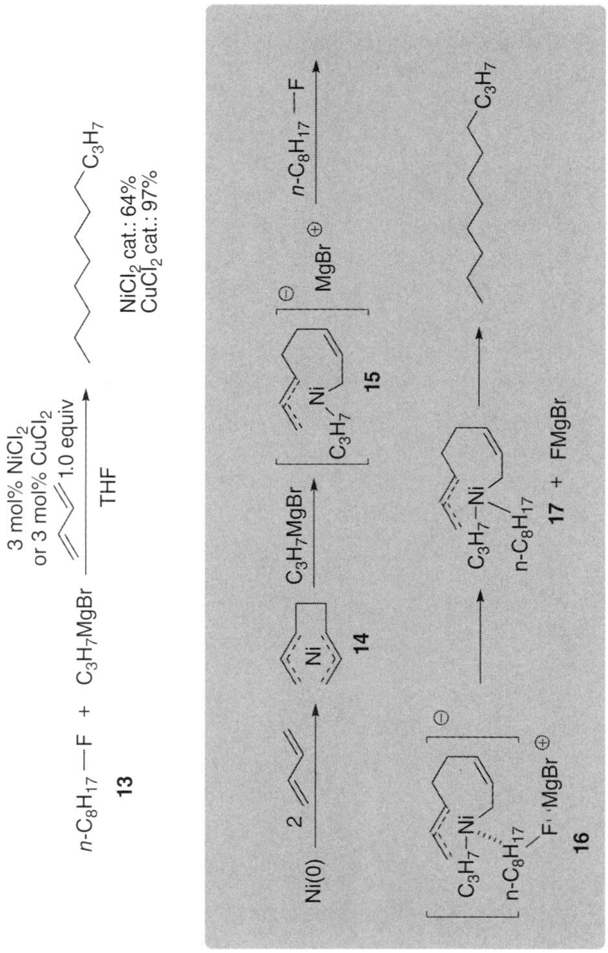

n-C8H17 — F + C3H7MgBr
13
3 mol% NiCl2
or 3 mol% CuCl2
1.0 equiv
THF
C3H7
NiCl2 cat.: 64%
CuCl2 cat.: 97%
Ni(0)
2
Ni
14
C3H7MgBr
C3H7
15
MgBr ⊕
n-C8H17 —F
C3H7—Ni
n-C8H17
F
MgBr ⊕
16
17 + FMgBr

Scheme 2.14.

form $\eta 1,\eta 3$-octadienediylnickelate (**15**). The cross-coupling products would be formed via oxidative addition of alkyl fluorides (R^2-F) to anionic Ni-complex **16**, providing di(alkyl)nickel complex **17**, followed by reductive elimination. In the oxidative addition step, magnesium (the countercation of anionic complex **15**) might play an important role (as a Lewis acid) to activate C–F bonds in fluoroalkanes by complexation.

Based on the former methodology, they developed a fascinatingly Ni complexes catalyze the alkylative dimerization reaction of vinyl Grignard reagents using alkyl fluorides to give the 2-alkyl-3-butenyl Grignard reagent (Scheme 2.15) [30]. When a reaction of *n*-octyl fluoride (**13**) with vinyl Grignard reagent was quenched with D_2O, deuterated compound **18** (*d*-content > 93%) was formed in 92% yield. When CO_2 was introduced after the reaction, carboxylic acid **19** was obtained in 85% yield. These results imply that the butenyl Grignard reagent **20** is formed in the present reaction. As a synthetic application of the thus-formed **20**, we undertook further alkylation by Ni-catalyzed cross-coupling reaction with alkyl halides. Into a solution of **20** prepared *in situ* under conditions similar to those of eq. 2 were added stearyl bromide (3.0 mmol) and 1,3-butadiene (1 mmol). After stirring at 25°C for 14 h, the desired coupling product **21** was obtained in 80% yield, where two different alkyl groups are introduced regioselectively to the butenyl skeleton derived from the vinyl Grignard reagent. 1,3-Butadiene plays an important role in the formation of an active species **L9** in the reactions.

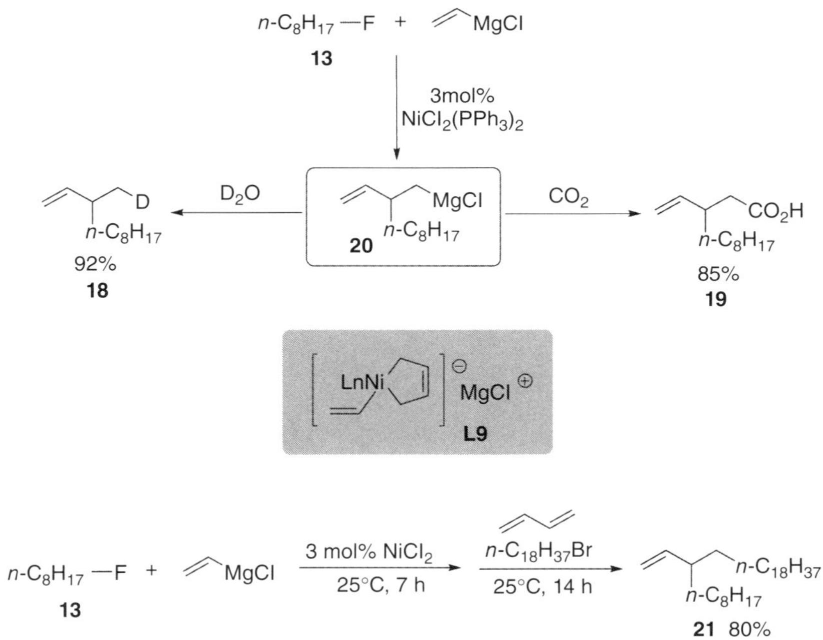

Scheme 2.15.

2.2.2 Palladium-Mediated C–F Bond Activation

In the previous section, we introduced Ni-catalyzed cross-coupling reactions of organic fluorides with Mg/Li reagents (RMgX or RLi). However, the use of such highly reactive reagents is unfortunately complicated and limited by a lack of functional group compatibility or by competitive halogen-metal exchange leading to homocoupling products. The nickel-catalyzed reactions also suffered from stereochemical scrambling of the starting alkenyl halides and metals. Thus, the palladium-catalyzed coupling of aryl and alkenyl halides and triflates with main group organometallics has been very broadly developed [31]. These reactions also involve oxidative addition/transmetalation/reductive elimination sequences. A very wide range of main group organometallics undergo transmetalation to Pd(II); that is, they transfer their R group to palladium in exchange for metal halides or triflates. Transmetalation from Li, Mg, Zn, Zr, B, Al, Sn, Si, and others have been reported. Therefore, a key feature of Pd-catalyzed cross-coupling is functional group tolerance. Organometallic reagents such as tin and boron compounds have proved their usefulness in the synthesis of compounds containing functional groups (COOR, CN, etc.) that are not compatible with organomagnesium and -lithium compounds. Cross-couplings with boron compounds (the Suzuki–Miyaura reactions) and tin compounds (the Stille reactions) are especially useful in syntheses of structurally complicated molecules. In the oxidative addition of organic halides to a Pd center, the order of reactivity is I > OTf, Br, Cl, F. Actually, in the old days, aryl fluorides were considered to be inert for Pd-catalyzed cross-coupling reactions. How to activate aryl and alkenyl fluorides is strategically important. The discovery that organopentafluorosilicates could participate in the cross-coupling process by Tamao and Kumada was greatly improved by Hiyama and Hatanaka to a practical C–C bond-forming reaction using organosilicon compounds [32]. Cross-coupling procedures that utilize such coupling partners with highly covalent C–M bonds have found many applications in the syntheses of natural products and fine chemicals. This is due to a large extent to the simplicity of their synthesis, isolation, and handling, as well as their compatibility with a wide range of functional groups.

In 2005, Saeki et al. [33] reported the nickel- and palladium-catalyzed cross-coupling reaction of polyfluorinated arenes and alkenes with Grignard reagents (Schemes 2.16). They found that the use of nickel catalysts and Grignard reagent is essential for the activation of the carbon–fluorine bond. The palladium catalyst is also effective for the 1,2-difluorobenzene (**22**) and trifluorobenzenes (**23**) to selectively produce the corresponding mono-coupled products while the nickel-based catalyst system affords a mixture of the mono-coupled product and di- or tri-coupled product. From the finding that 1,3- and 1, 4-difluorobenzenes give the cross-coupling products in poor yields, a chelating effect of the adjacent atom is expected to play an important role in promoting oxidative addition of the C–F bond. The reaction of 1,2,3-trifluorobenzene (**23**) using $NiCl_2$(dppp) provides di-coupled product **24** exclusively, whereas the use of the Pd catalyst furnishes mono-coupled product **25**. In sharp contrast to the reactions of fluroarenes, arylzinc reagent **27** was found to be more effective than the Grignard reagent to produce the

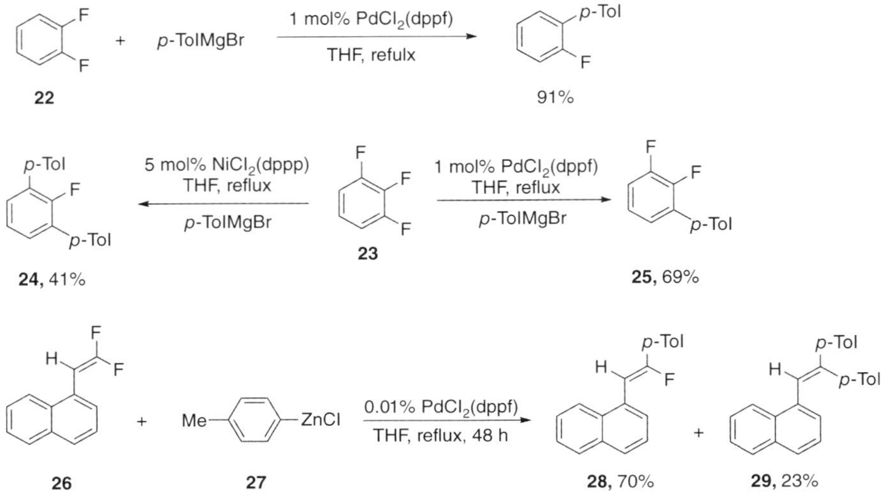

Scheme 2.16.

mono-coupled product (*Z*)-**28** in 70% yield after 48 h of stirring under reflux, along with di-coupled product **29** in 23% yield; the formation of an isomer (*E*)- **28** was not observed. Alkenyl fluorides are applicable to Pd-catalyzed cross-coupling reactions. Selective cleavage of a C–F bond is observed in the cross-coupling reaction of *gem*-difluoroalkene to form (*Z*)-fluoroalkene as a major product; the fluorine atom *trans* to the naphthyl group is selectively replaced because of the steric hindrance of the naphthyl group at the vicinal *cis*-position.

Dankwardt [34] also discovered that phosphine free $Pd(dba)_2$ was able to catalyze the biaryl cross-coupling of aryl fluorides under both microwave and thermal conditions (Scheme 2.17). The author declared that this is the first example of a Kumada–Corriu coupling of an unactivated aryl fluoride utilizing a palladium catalyst under ligandless conditions. In addition, microwave technology allowed rapid optimization of catalyst systems, which identified several ligands for this cross-coupling reaction.

In 2008, Manabe and Ishikawa [35] demonstrated interesting examples of *ortho*-selective cross-coupling of fluorobenzene derivatives with Grignard reagents in the presence of $PdCl_2(PCy_3)_2$ (Scheme 2.18) Directing groups such as hydroxyl (**30, 31**), hydroxymethyl (**32**), and amino (**33**) in aromatics accelerate the palladium-catalyzed cross-coupling reactions at the fluoro group *ortho* to the

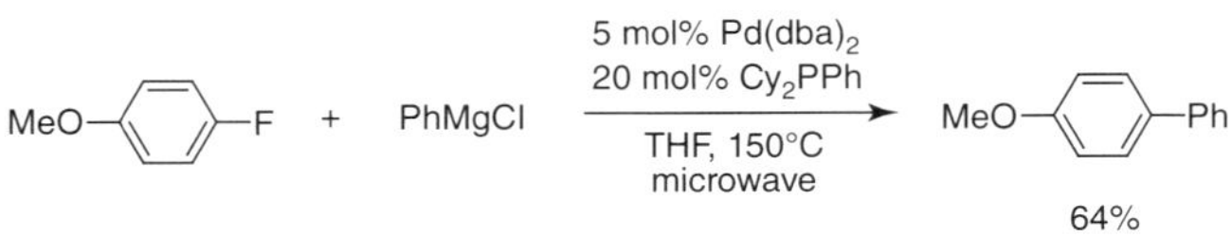

Scheme 2.17.

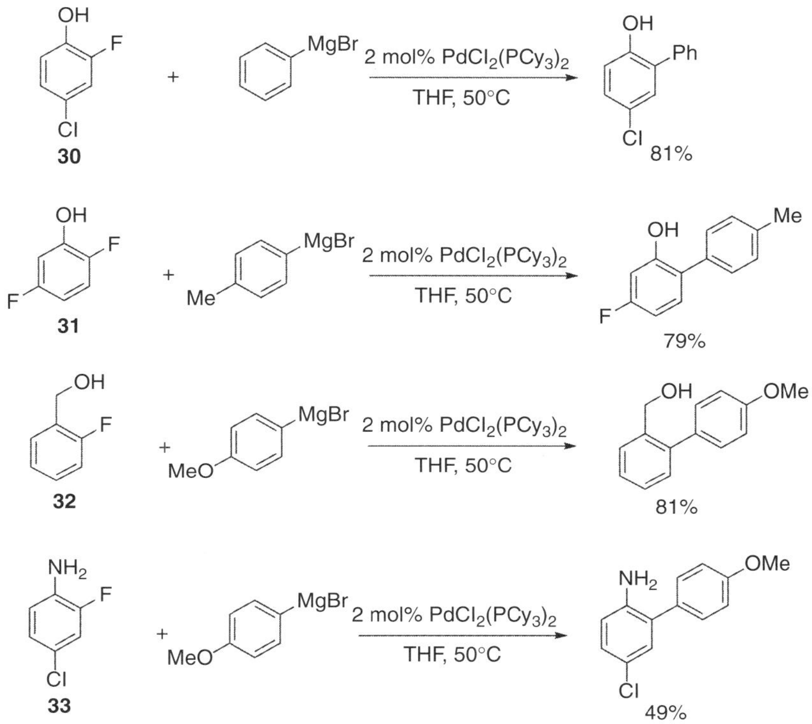

Scheme 2.18.

directing group. As a noteworthy event, fluoro and chloro groups at positions other than *ortho* to the directing groups survive under the reaction conditions. Although the mechanism of activation of the *ortho*-fluoro group is unclear at this moment, this site-selective cross-coupling of fluorobenzene derivatives should expand the usefulness of cross-coupling.

Introduction of electron-withdrawing groups such as nitro and carboxyl groups, which are sensitive to organomagnesium and -lithium reagents, into aryl groups of fluoroarenes is effective for Suzuki–Miyaura and Stille cross-coupling reactions (Scheme 2.19) [36]. As obligate features of the coupling process, the *o*-nitro groups of fluoroarenes not only can function as electron-withdrawing groups but also can coordinate the incoming Pd atom to form the complex **L10** and facilitate C–F bond activation. The reaction tolerates a wide range of boronic acids; however, a nitro group in the 2-position of the fluoroarene is essential for reaction. The method that was developed gives ready access to 2,4-dinitrobiphenyls (**34**) in good yields from Sanger's reagent. Although the C–F bond is much less reactive than C–Cl or C–Br bond, for nitrophenyl fluoride (**34**), Pd^{2+}/BINAP catalyst could facilitate the coupling reaction with $PhB(OH)_2$ to give **35** in an excellent (91%) yield within 4 h. To fully understand the role that the *o*-nitro group played in the successful

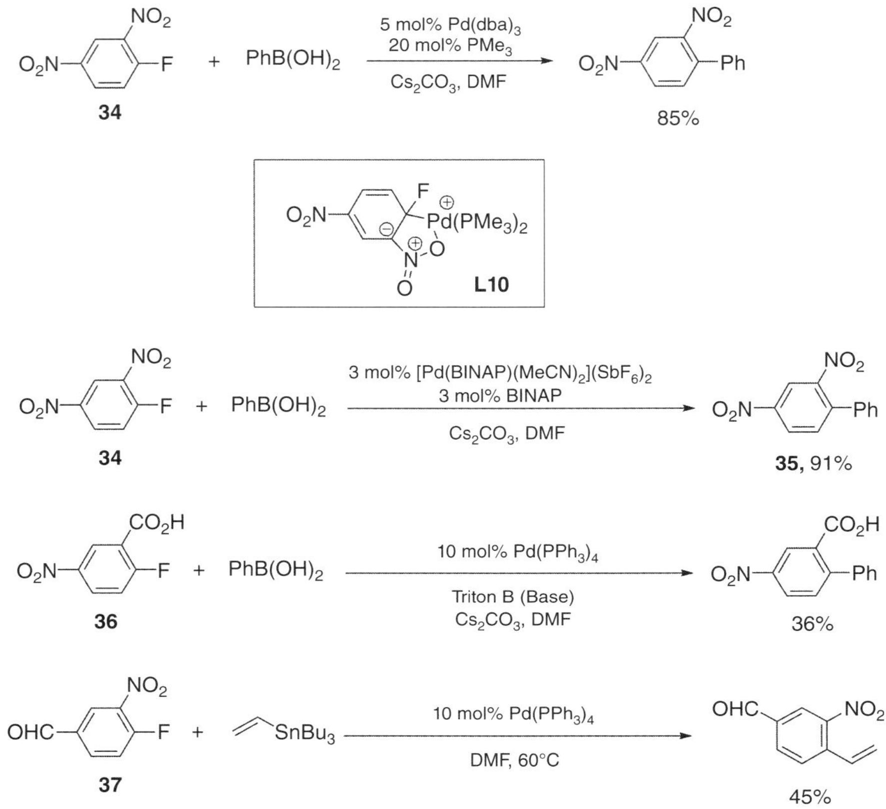

Scheme 2.19.

Pd-catalyzed coupling reactions of aryl fluorides, a computational study using hybrid density functional theory was conducted. On the basis of this computational study and subsequent experimental verifications of its predictions, the authors reported that such reactions are facilitated by stabilization of the transition state by proximal oxyanions. This stabilization allowed the prediction of a nonobvious substrate (the carboxylate anion [36]) that could successfully undergo a Pd-catalyzed coupling reaction. The experimental verification confirms that fluorobenzenes with an *o*-carboxyl (**36**) and *para* electron-withdrawing substituents (**37**), similar to 2-fluoronitrobenzene, are activated toward Stille and Suzuki couplings. This is the first report on Pd-catalyzed coupling reactions of fluorobenzenes that do not have an *o*-nitro group.

Clearly, a 2-nitro group, which not only can function as an electron-withdrawing group, but also can coordinate an incoming palladium atom, is an obligate feature for the coupling process. *Ab initio* calculations of plausible mechanisms (Scheme 2.20) have shown that such a coordination of palladium has a significant influence on the activation energy of the insertion step. The mechanism could be either a concerted

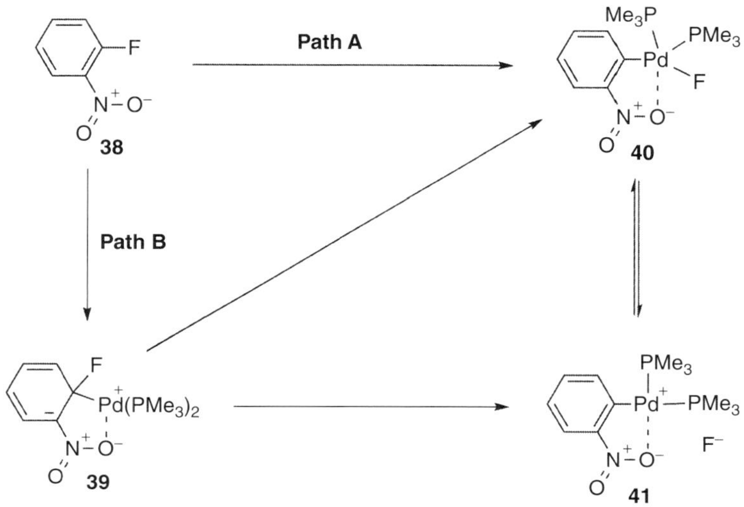

Scheme 2.20.

insertion (Scheme 2.20: path A) or an addition–elimination sequence (Scheme 2.20: path B) via adduct **38** followed by fluoride loss to produce cationic palladium species **39**. Distinction between these cannot be made at this point, and both are facilitated by *ortho*-nitro coordination. The insertion product could be active in the catalytic cycle in either neutral **40** or cationic **39**. The calculations show that either pathway is energetically feasible under the conditions.

In 2010, Paquin's group developed a palladium-catalyzed allylic amination reaction of 3,3-difluoropropenes (Scheme 2.21) [37]. Notably, the key fluorinated

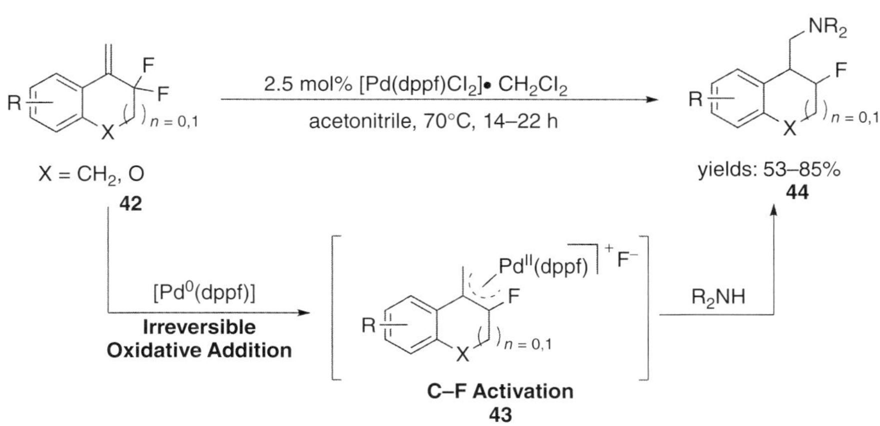

Scheme 2.21.

palladium *p*-allyl intermediate is generated using an allylic C–F bond activation. This reaction allows the efficient synthesis of β-aminofluoroalkenes from readily available 3,3-difluoropropenes (**42**) and amines. Initial oxidative addition of an allylic C–F bond to the palladium(0) catalyst would generate fluorinated palladium *p*-allyl complex **43**. Interestingly, the formation of **44** is thought to be irreversible (i.e., the reductive elimination of a $F^{-}PdII^{+}$allyl complex is kinetically disfavored), which would make this system analogous to the established decarboxylative allylation system developed by Tsuji and co-workers [38].

Shen and Lu developed an efficiently palladium-catalyzed Suzuki–Miyaura coupling reaction of polyfluorophenyl oxazoline through *ortho* C–F activation (Scheme 2.22) [39]. The reaction was compatible with a variety of functional groups. A selective palladium-catalyzed Suzuki–Miyaura coupling reaction of polyfluorophenyl oxazolines (**45**) through ortho C–F activation is described. It was found that reactions with DPPF as the ligand occurred much faster than those with other ligands. A variety of arylboronic acids including challenging functionalized arylboronic acids such as enolizable ketones, aldehyde, cyano, ester, and trifluoromethyl groups were tolerated with the reaction conditions.

Very recently, studies have been presented on the synthesis and reactivity of the palladium fluoro complex *trans*-$[Pd(F)(4\text{-}C_5NF_4)(^iPr_2PCH_2CH_2OCH_3)_2]$ (***L-11***). Complex ***L-11*** can be synthesized by C–F bond activation of pentafluoropyridine. It catalyzes cross-coupling as well as hydrodefluorination reactions of pentafluoropyridine to yield 4-phenyltetrafluoropyridine, 4-tolyltetrafluoropyridine, or 2,3,5,6-tetrafluoropyridine by C–F activation (Scheme 2.23) [40]. It should be mentioned that the bromo complex or chloro complex cannot be applied as catalysts in any of the described catalytic conversions. This observation reflects the fact that fluorocomplexes often exhibit a higher reactivity than their chloro or bromo counterparts. In addition, it was shown that the phosphine ligands $^iPr_2PCH_2CH_2OCH_3$ have a superior influence on the C–F bond cleavage and catalytic C–F activation reactions of pentafluoropyridine when compared to the corresponding P^iPr_3 complexes. However, the role of the methoxy function in these conversions is, so far, unclear. In catalytic experiments, pentafluoropyridine (**47**) could be converted into the 4-aryl-tetrafluoropyridines and into 2,3,5,6-tetrafluoropyridine in the presence of the boronic acids $PhB(OH)_2$, $TolB(OH)_2$, or HBpin when 5 mol% of ***L-11*** is employed as catalyst.

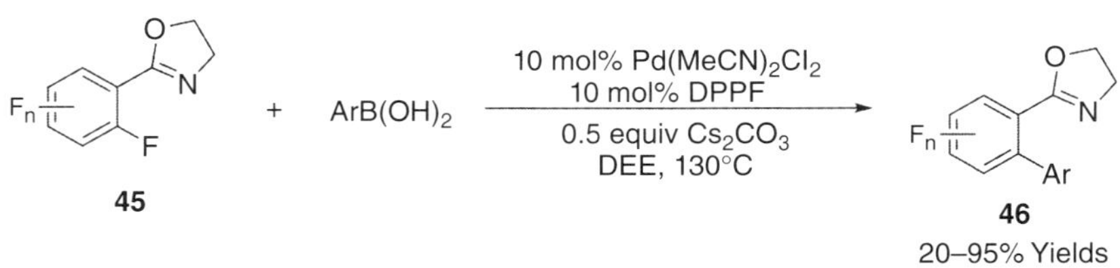

Scheme 2.22.

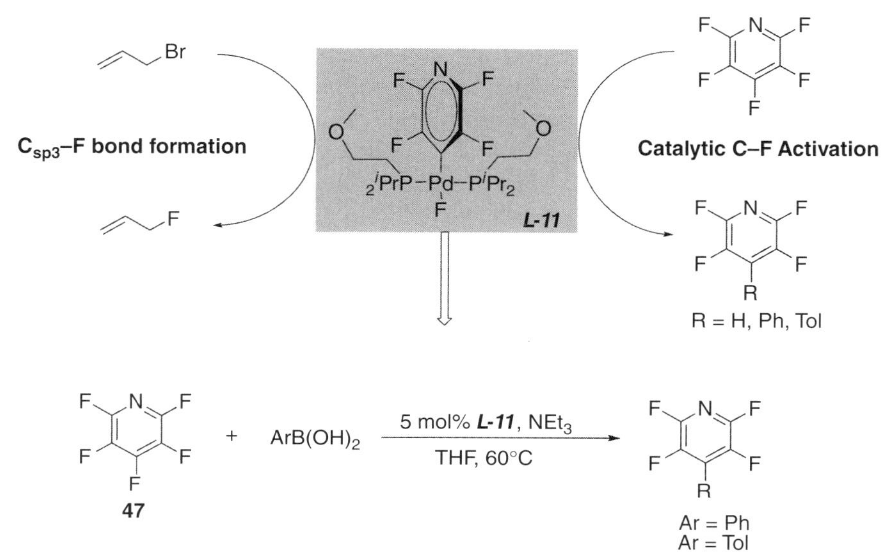

Scheme 2.23.

2.2.3 Platinum-Mediated C–F Bond Activation

Catalytic cross-coupling of aryl fluorides has been a longstanding goal in organometallic chemistry, in part because of the challenge of activating the strong C–F bond. The catalytic C–C bond-forming reactions using aryl fluorides by using the Ni and Pd catalysts have been the most extensively explored.[41] Inspired by these reports, Crespo et al. disclose a [Me_2-Pt(*í*-SMe_2)]$_2$ promoted stoichiometric C–F activation of a series of aryl imines (Scheme 2.24, eq. 1). The

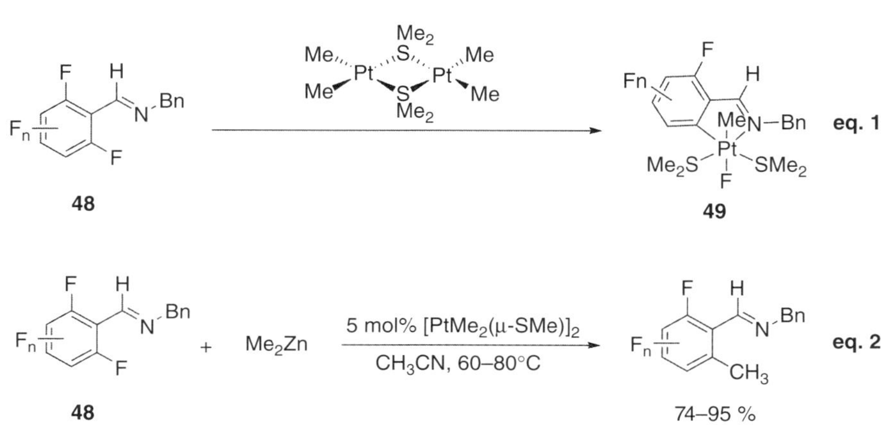

Scheme 2.24.

authors decided to explore the potential for this reaction to be part of a catalytic process. They reasoned that the Pt–F species could undergo transmetalation with an appropriate organometallic reagent. Subsequent reductive elimination would provide the functionalized product and would regenerate the Pt(II) catalyst. The first report on platinum-catalyzed cross-coupling of aryl fluorides was provided by Love and co-workers [42] (Scheme 2.24, eq. 2). Inspired by the reports on $[Me_2Pt(\mu\text{-}SMe_2)]_2$-mediated stoichiometric C–F activation of *ortho*-fluoroaryl imines, they explored the catalytic version of these *ortho*-selective C–F transformations. A series of polyfluoroaryl imines **48** react with Me_2Zn to generate functionalized fluoroarenes **49**. The reaction is selective for *ortho* C–F activation in the presence of weaker aryl C–Br bonds. Outstanding selectivity is also achieved because the substrates are more reactive toward methylation than the products. Extensive mechanistic studies revealed that C–F bond cleavage occurred by oxidative addition. It is noteworthy that reductive elimination of ethane a common occurrence from Pt(IV)–methyl complexes, was not observed.

On the basis of the former work using $[(CH_3)_2Pt(\mu\text{-}SMe_2)]_2$ for catalytic C–F activation, Love and co-workers [43] have discovered that this complex also catalyzes aryl methyl ether formation (Scheme 2.25). A range of silanes and fluorinated aryl imines were tested, Reactions of several other imines with $Si(OMe)_4$ generated moderate to excellent yields of the corresponding aryl methyl ether products.

In addition, Cahoez et al. [44] reported the manganese-catalyzed cross-coupling between activated aryl halides and Grignard reagents in 1999 (Scheme 2.26). In the presence of $MnCl_2$ (10 mol%), BuMgCl readily reacts in THF with aryl fluoride **49** bearing an electron-withdrawing activating group (CH=NR) in the *ortho* position. The manganese-catalyzed procedure compares favorably to the aromatic nucleophilic substitution (S_NAr). From aldimine **49**, in the absence of $MnCl_2$, the substitution takes place at 20°C for 24 h to give **50** in 40% yield. On the contrary, the beneficial

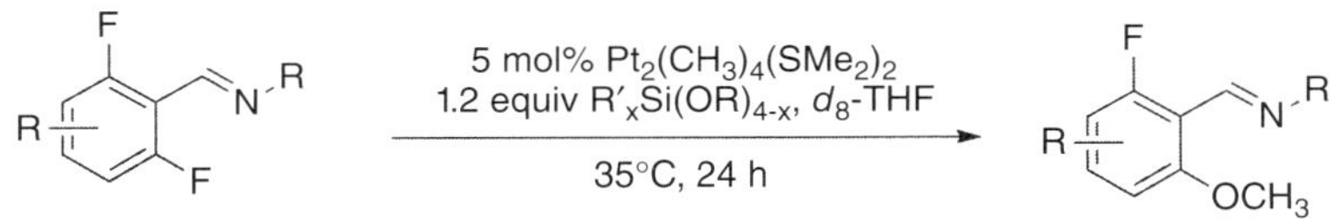

Scheme 2.25.

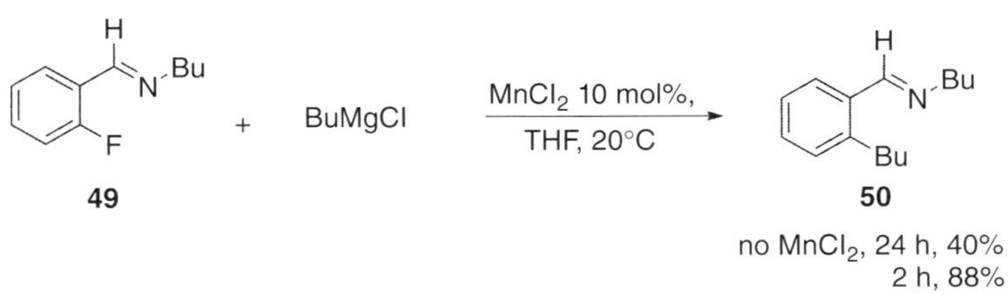

Scheme 2.26.

influence of manganese chloride is obvious; the reaction proceeds faster (2 h instead 24 h) and leads to a much better yield (88% instead 40%), and no adduct of the Grignard reagent to the imino carbon of **49** is observed.

2.2.4 Cobalt- and Rhodium-Mediated C–F Bond Activation

Besides the Ni, Pd, and Pt, other transition metals such as cobalt and rhodium also are used to catalyze C–F bond activation. Knochel and co-workers [45] demonstrated cobalt(II)-catalyzed cross-coupling between aryl cuprates and aryl fluorides (Scheme 2.27). In the presence of $Co(acac)_2$ (7.5 mol%), Bu_4NI (1 equiv), and 4-fluorostyrene (20 mol%) as promoters, organocopper compounds prepared by the transmetalation of functionalized arylmagnesium halides with CuCN·2LiCl undergo the cross-coupling reactions with aryl fluorides **51** and **53** bearing a carbonyl functionality in the *ortho* position, leading to polyfunctional aromatics **52** and **54**, respectively. The presence of the promoters' 4-fluorostyrene and Bu_4NI is crucial for the success of the reaction.

Besides strong nucleophilic reagents, organoboron compounds react with trifluoromethylated alkenes (**55–57**) by C–F bond elimination in the presence of Rh complexes to afford the corresponding coupling product. In 2008, Miura et al. [46] reported the synthesis of *gem*-difluoroalkenes from the reactions of R-(trifluoromethyl)styrenes with arylboronic esters and MeMgCl (Schemes 2.28). With a high level of functional group tolerance, the reaction proceeds through the addition of arylrhodium(I) species across the electron-deficient carbon–carbon double bonds and the subsequent fluoride elimination. Treatment of

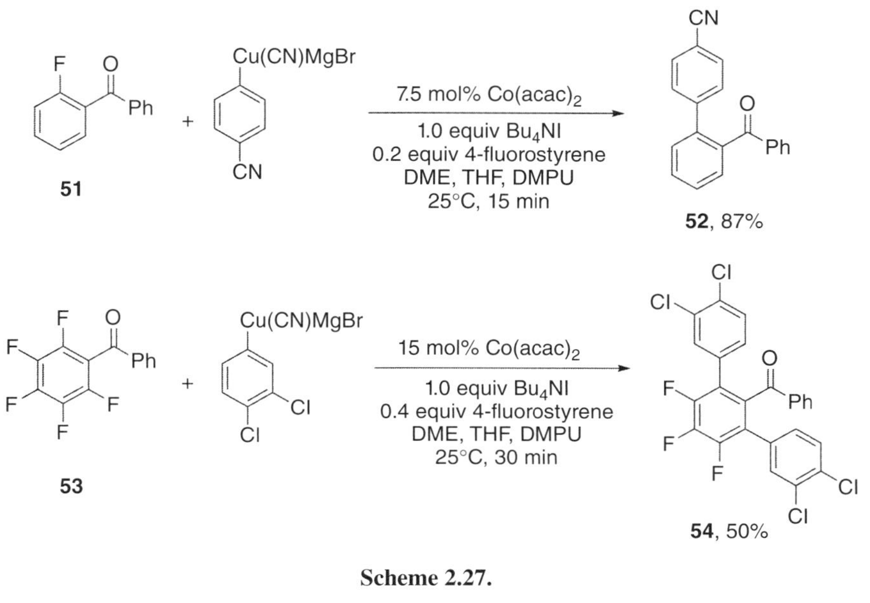

Scheme 2.27.

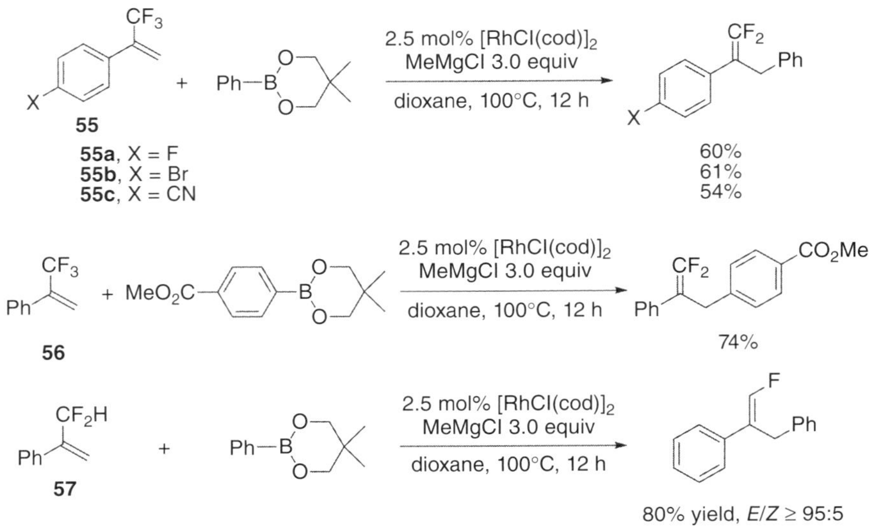

Scheme 2.28.

α-(trifluoromethyl)styrenes with arylboronic esters and MeMgCl in the presence of a rhodium(I) catalyst affords *gem*-difluoroalkenes. The reaction proceeds through the addition of arylrhodium(I) species across the electron-deficient carbon–carbon double bond and the subsequent β-fluoride elimination.

Rhodium-catalyzed single bond metathesis reactions for synthesis of organosulfur compounds have become an important methodology. Rhodium complexes can cleave the SS bond and exchange bonds with SS, HH, CH, PS, PP, and CS bonds. In 2008, Arisawa et al. [47] developed Rh-catalyzed single-bond metathesis of C–F and S–S bonds (Table 2.1). Upon exposure to a catalytic amount of $RhH(PPh_3)_4$ and 1,2-bis(diphenylphosphino)benzene (dppBz) **L12**, aromatic fluorides **58**, disulfides (0.5 equiv), and triphenylphosphine (0.5 equiv) react in refluxing chlorobenzene to give aryl sulfides **59** in high yields. Triphenylphosphine traps fluoride atoms to form phosphine difluoride; both organothio groups of the disulfides react effectively. Interestingly, the fluorine substituent reacts more readily than the chlorine and bromine atoms. Mechanistically, the rhodium complex should be involved in the cleavage of the C–F bond in aryl fluorides and the SS bond in disulfides, and the mode of C–F cleavage is a subject of interest. Nucleophilic aromatic substitution of the aryl fluoride with rhodium thiolate is a possible pathway. However, oxidative addition to the C–F bond or electron transfer of a low-valent rhodium complex can also be conceivable at present.

In 2012, Yu et al. [48] demonstrated a Cu-catalyzed cross-coupling reaction of diaryl disulfides or aryl thiols with pentafluorobenzene using CuBr as the catalyst and utilizing *t*-BuOLi or *t*-BuOK as the base in DMSO at 60°C under an O_2 atmosphere. The corresponding bisarylthiolation products were obtained in moderate to good yields by C–H bond and C–F bond activation (Scheme 2.29).

TABLE 2.1. Rhodium-Catalyzed Single-Bond Metathesis Reactions for Synthesis of Organosulfur Compounds

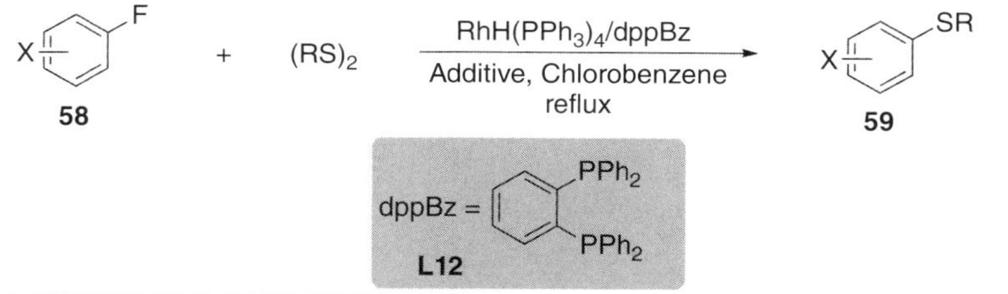

Substrate	Conditions	X	R	Yield
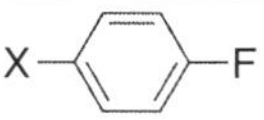	0.25 mol% $RhH(PPh_3)_4$	x = COPh	R = m-$MeOC_6H_4$	89%
	0.5 mol% dppBz	COPh	p-ClC_6H_4	90%
	0.5 equiv PPh_3	COPh	p-$MeOC_6H_4$	87%
		CN	p-$MeOC_6H_4$	91%
		NO_2	p-$MeOC_6H_4$	86%
		$COCH_3$	p-$MeOC_6H_4$	91%
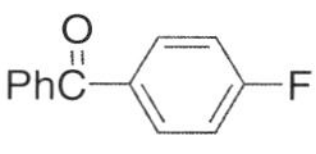	2 mol% $RhH(PPh_3)_4$		R = $MeO_2C(CH_2)_3$	92%
	4 mol% dppBz		$BocNH(CH_2)_2$	92%
	0.5 equiv (p-$MeOC_6H_4)_3P$		n-Bu	81%
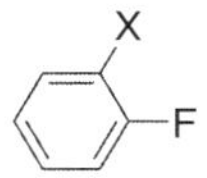	1 mol% $RhH(PPh_3)_4$	x = CN	p-$MeOC_6H_4$	100%
	2 mol% dppBz	NO_2	p-$MeOC_6H_4$	93%
	0.5 equiv PPh_3	$COCH_3$	p-$MeOC_6H_4$	80%
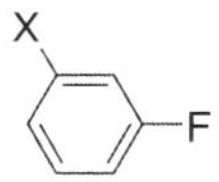	5 mol% $RhH(PPh_3)_4$	x = CN	p-$MeOC_6H_4$	82%
	10 mol% dppBz	NO_2	p-$MeOC_6H_4$	86%
	0.5 equiv PPh_3			
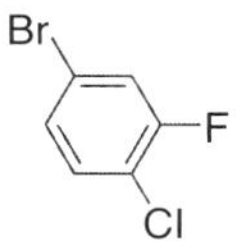	5 mol% $RhH(PPh_3)_4$		R = p-$MeOC_6H_4$	72%
	10 mol% dppBz			
	0.5 equiv PPh_3			

When 1,10-phenanthroline·H_2O and DDQ were added to the above system, monoarylthiolation products, with or without bisarylthiolation products, could be isolated in moderate yields. More experiments were carried out to gain preliminary insight into the mechanism of the catalytic reaction. These results indicated that bisarylthiolation products could be formed from monoarylthiolation products, DDQ may serve an important role in the reaction, diaryl disulfides were very stable in the presence of 1 equiv of DDQ, and the monoarylthiolation products could be obtained with an appropriate amount of DDQ. A plausible mechanism for these transformations was given. Thus, it represents a method for the synthesis of polyfluorinated diaryl sulfides and polyfluorinated triarylbisulfides through C–H bond and C–F bond activation.

Once the feasibility of catalytic cross-coupling of aryl fluorides was demonstrated, efforts began to explore the utility of other metal complexes and to expand the scope

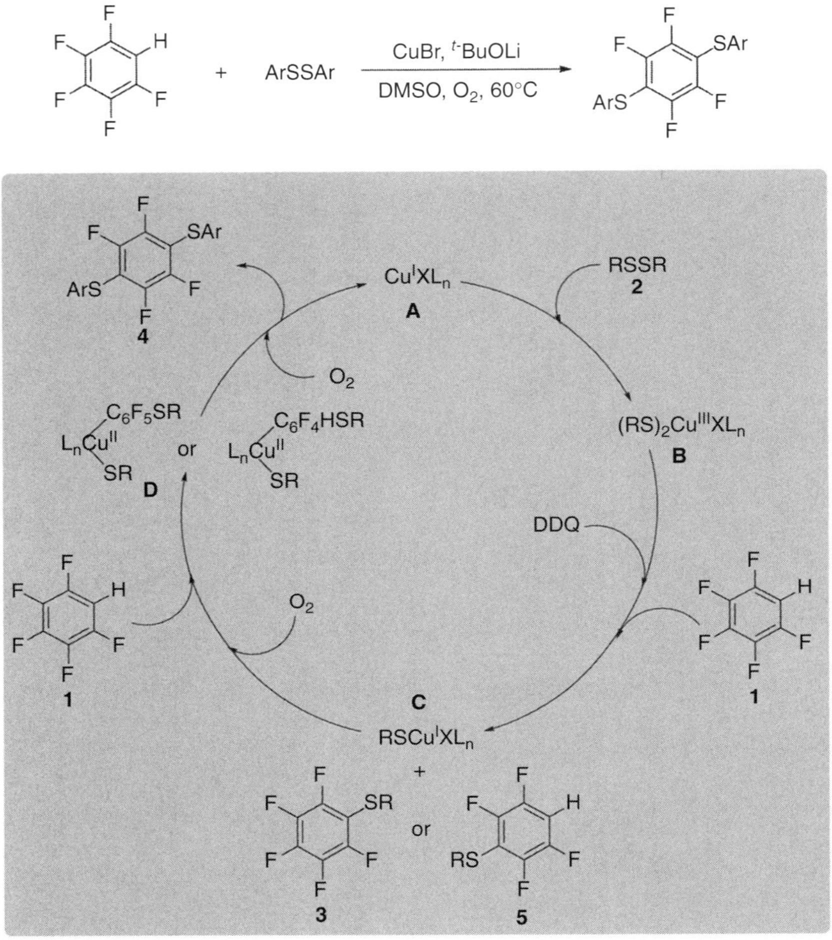

Scheme 2.29.

of fluoroaromatic substrates. Likewise, the use of different nucleophiles was pursued. While most of these accounts dealt with monofluoroaromatics, the use of polyfluoroarenes occasionally was reported. For example, the Murai group discovered a Si–F exchange reaction catalyzed by Rh(I), demonstrating that a polyfluoroarene could be functionalized while still retaining one or more fluorine atoms (Scheme 2.30) [49] to be effective for catalyzing Si–F exchange between aromatic C–F bonds adjacent to acyl or oxazoline functionality. Reaction of pentafluoroacetophenone **61** with hexamethyldisilane for 20 h in toluene in a 130°C oil bath in the presence of 10 mol% $Rh(COD)_2BF_4$ gave a 79–88% yield of 2,3,4,5-tetrafluoro-6-trimethylsilyl acetophenone **62**. Not surprisingly, 2,6-difluoroacetophenone affords the mono-Me_3Si–F exchange product in somewhat lower yields (33–48%). In the case of a related oxazoline derivative, some disilylation accompanied the mono-substituted product.

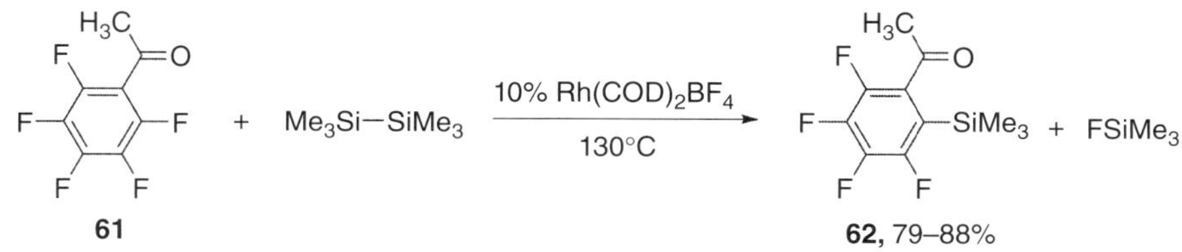

Scheme 2.30.

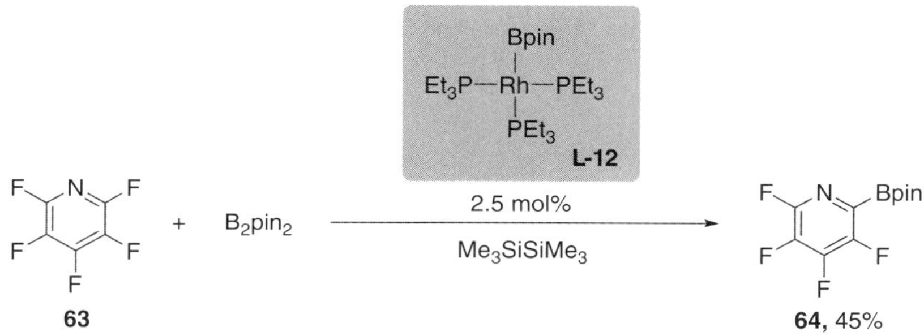

Scheme 2.31.

The authors propose a chelate-assisted mechanism for the initial C–F activation step. It is interesting to note that these catalytic reactions all involve the later transition metal rhodium with a relatively labile Rh–F bond removed as R_3Si–F or HF. However, related catalytic reactions of aromatic C–F bonds have also been discovered for early transition metals and even in the very electropositive lanthanide series.

Recently, Macgregor and co-workers [50] reported a 16-electron rhodium(I)–boryl complex (**L-12**)–catalyzed borylation reaction of pentafluoropyridine (Scheme 2.31). Investigation of the catalytic borylation reaction led to the formation of tetrafluoropyridyl boronate esters, which can provide new fluorinated building blocks. They also experimentally investigated the reactivity of **L-12** in catalytic borylation reactions of pentafluoropyridine **63**. Indeed, pentafluorpyridine was catalytically converted into 2-boryl derivative **64** with 45% yield in the presence of 2.5 mol% **L-12** (based on equimolar amounts of C_5NF_5 and B_2pin_2) in $Me_3SiSiMe_3$ solvent. Borylation of the C–F bond has also been reported by Marder, Perutz, and co-workers.

2.2.5 Other-Metals-Mediated C–F Bond Activation

Several transition metal reagents including both early transition metal salts and late transition metal complexes are quite useful as catalysts for cross-coupling reactions of aromatic and vinylic fluorides. Notwithstanding this significant achievement, cross-coupling reactions of polyfluoroarenes by early transition metal complexes

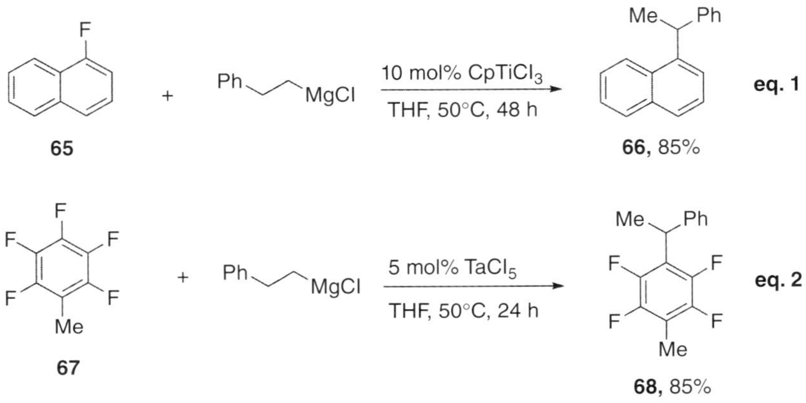

Scheme 2.32.

continue to be rare. The reaction of 1-fluoronaphthalene (**65**) with 3 equiv of 2-phenylethylmagnesium chloride in the presence of a catalytic amount of $CpTiCl_3$ affords 1-(1-phenylethyl)naphthalene (**66**) in good yield via isomerization of the 1-phenylethyl group (Scheme 2.32, eq. 1) [51]. Similarly, one example of the Ta-catalyzed cross-coupling of hexafluorobenzene was reported by the Takahashi group in 2006. The $TaCl_5$-catalyzed cross coupling of hexafluorobenzene with 2-phenethylmagnesium chloride in the presence of 10 mol% catalyst afforded the mono-coupled product in 75% yield. When an excess amount of the Grignard reagent was used, the di-substituted product was obtained in 57% yield, along with 14% of the mono-substituted product. Only the 1-phenethyl products were observed, indicating complete isomerization of the cross-coupling reagent. This was also found by Kumada in Ni(II)-catalyzed cross-coupling with Grignard reagents. It is also worth noting that only the para-disubstituted isomer was detected. Similarly, the reaction of 2,3,4,5,6-pentafluorotoluene (**67**) with phenethylmagnesium chloride generated exclusively 4-(1-phenethyl)-2,3,5,6-tetrafluorotoluene (**68**) (Scheme 2.32, eq. 2).

In 2005, Terao et al. [52] investigated systematically the transition metal-catalyzed alkyl–alkyl cross-coupling reactions of alkyl chlorides, bromides, and tosylates with Grignard reagents in the presence of a 1,3-butadiene derivative under mild conditions and extended it to C–F bond cleavage. In the same year, they disclosed a new and different catalytic system. In the presence of a catalytic amount of Cp_2ZrCl_2, n-octyl fluoride (**69**) reacts with phenethylmagnesium chlorides to afford 2-phenyldecane (**71**) via alkylation of the styrene–zirconate intermediate (**70**) at the benzylic position (Scheme 2.33) [53].

In addition, Ishihara and co-workers [54] reported that the reaction of α,β,β-trifluoroacrylate (**72**) with various types of Grignard reagents in the presence of a catalytic amount of CuBr gave the corresponding addition–elimination products (**73**) in a highly Z-selective manner (Scheme 2.34). During the course of our continuous studies on the reactions of fluorine-containing alkenes having

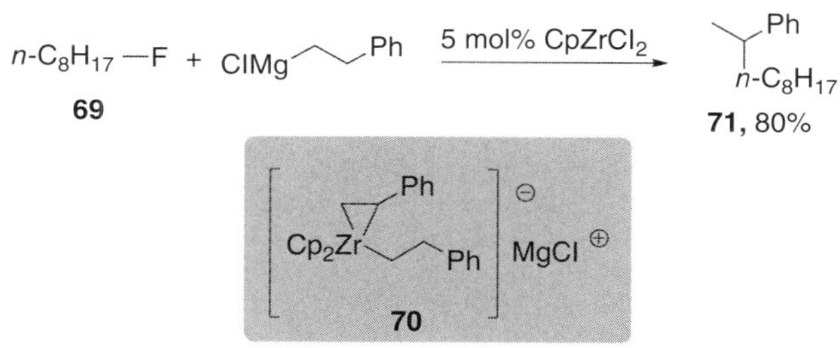

Scheme 2.33.

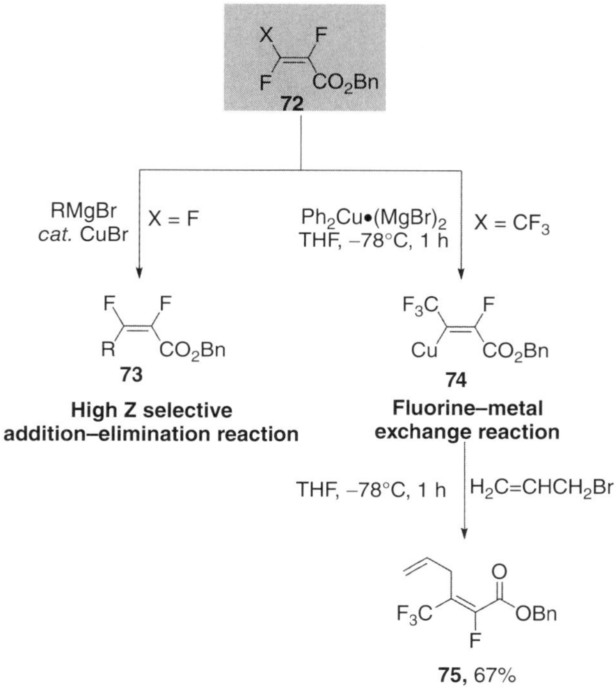

Scheme 2.34.

an electron-withdrawing group with organocopper reagents, they found that the treatment of benzylpentafluorocrotonate (**74**) with Grignard reagents in the presence of copper(I) salt caused the vinylic fluorine–metal exchange reaction, leading to the corresponding β-metallated product (**75**) [55].

A possible reaction mechanism is described in Scheme 2.35. The starting ester **76** undergoes the vinylic fluorine–copper exchange reaction with organocuprate to form the corresponding vinyl copper species **77**, which is stabilized due to the strong electron-withdrawing effect of a CF_3 group and is not susceptible to the reductive elimination leading to **78** at −78°C. By subsequent treatment with

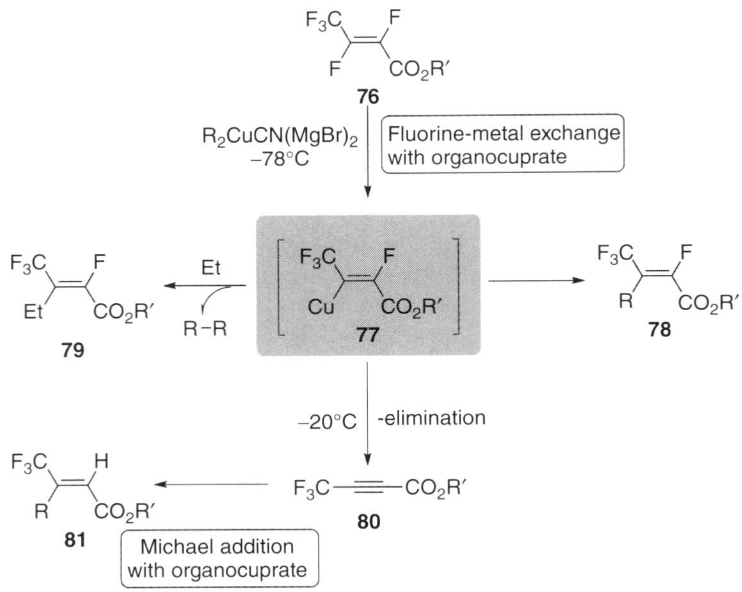

Scheme 2.35.

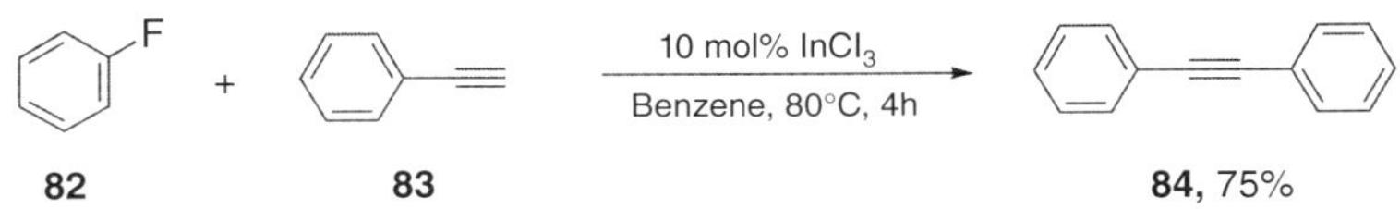

Scheme 2.36.

various electrophiles including H_2O, the vinyl copper species **77** is converted to the corresponding tetrafluorocrotonates **79**, along with the formation of homo-coupling product (R–R). On the other hand, raising the reaction temperature from −78°C to −20°C promotes the elimination of $Cu–F_{14}$ in 2, giving rise to the corresponding alkynoate **80**, which may undergo Michael addition of an excess amount of cuprate to afford 3-substituted -4,4,4-trifluorocrotonate **81**.

Indium salts engender copper-, phosphine ligand-, and palladium-free Sonogashira coupling reactions (Scheme 2.36). In the presence of 10 mol% indium trichloride, the coupling of fluorobenzene (**82**) with phenylacetylene (**83**) affords diphenylacetylene (**84**) in 75% yield [56].

2.3 TRANSITION-METAL-CATALYZED SUBSTITUTION BY C–F BOND ACTIVATION

Electrophilic reagents are typically employed for substitution chemistry on aromatic rings. However, replacement of fluorine on an aromatic ring with nucleophiles by

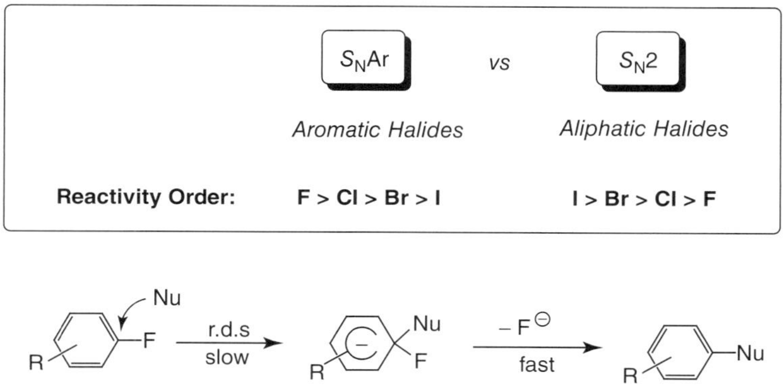

Figure 2.4.

either nucleophilic substitution (S_NAr) or transition-metal-catalyzed coupling with organometallics has seen increasing attention for the synthesis of functionalized aromatic compounds. The strong electron-withdrawing inductive effect exerted by fluorine makes nucleophilic substitutions on an aromatic ring possible. The reactivity of the aryl halides decreases in the order of fluoride > chloride > bromide > iodide, which is entirely opposite to that observed in the S_N2 reactions of aliphatic halides. The big difference arises from the extent of bond breaking between carbon and halogen atoms at the transition state. Addition and elimination pathways are involved in the mechanism of aromatic nucleophilic substitution, in which addition is rate-determining (Figure 2.4) [57]. Therefore, the sterically small and strongly electron-withdrawing fluorine atom activates the addition step. Meanwhile, bond breaking between carbon and halogen atoms plays an important role in the transition state for the S_N2 reaction so that aliphatic iodides react fastest among the aliphatic halides.

In 2007, Jean's group developed convenient access to a variety of carbazole- containing molecules by the tandem Suzuki cross-coupling/S_NAr protocol (Scheme 2.37) [58]. This process was found to be compatible with a variety of electron-withdrawing groups including aldehydes, esters, and sulfones. Furthermore, using this method an efficient total synthesis of the carbazole alkaloid glycosinine was achieved. For example, cross-coupling of an aniline-derived boronic ester **86** with an appropriately substituted dihalide **85** using microwave assisted palladium catalysis in a one-pot procedure provided diaryl intermediate **87**, which was transferred to carbazole by intramolecular fluorine makes nucleophilic substitutions in near quantitative yield. Although the aniline boronic ester failed to undergo the desired CC-S_NA_r coupling, the same compound could be accessed by removal of the sulfonamide functionality, such as exposing tosylated carbazole under Cs_2CO_3 conditions.

The Mizoroki–Heck reaction is a general and convenient method for arylation of olefinic compounds which involved oxidative addition of aromatic halides to Pd(0), carbopalladation to alkenes, followed by β-H elimination to complete the circle [59].

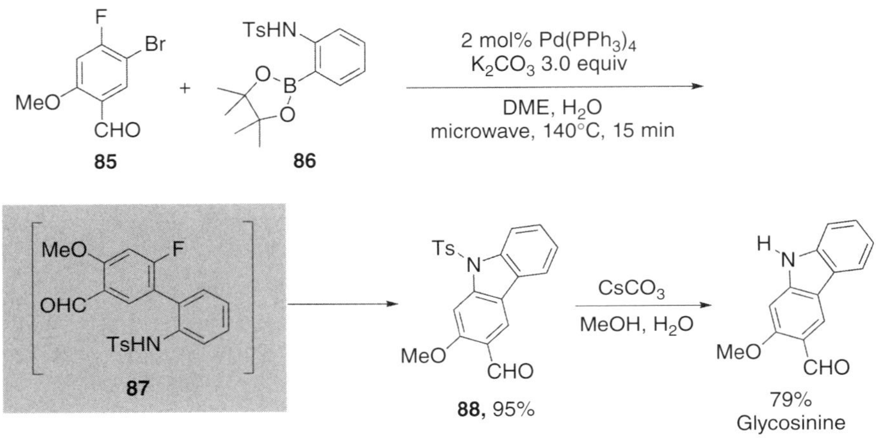

Scheme 2.37.

Under the Mizoroki–Heck reaction conditions, vinylidene fluoride **89** undergoes carbopalladation to afford α-fluorostyrene **90** as a main product that could be obtained for the first time in a one-step reaction instead of difluorostyrene which was formed as the Heck reaction product (Scheme 2.38) [60]. This is the first transformation, in which fluorine elimination occurred, although there is a great possibility of elimination of a competing β-hydride. Furthermore, an α-fluorovinyl substituent is introduced into

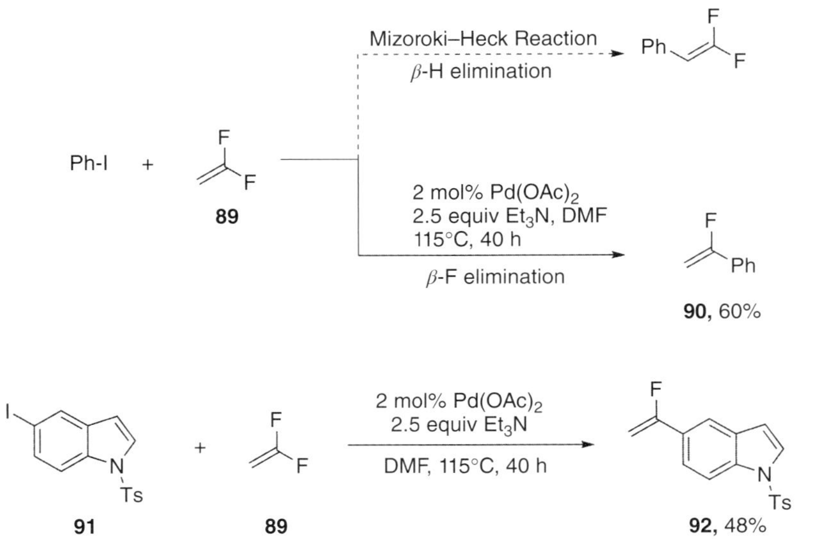

Scheme 2.38.

TABLE 2.2. Rhodium-Catalyzed Fluoride Elimination to Synthesis of *Gem*-Difluoroalkenes

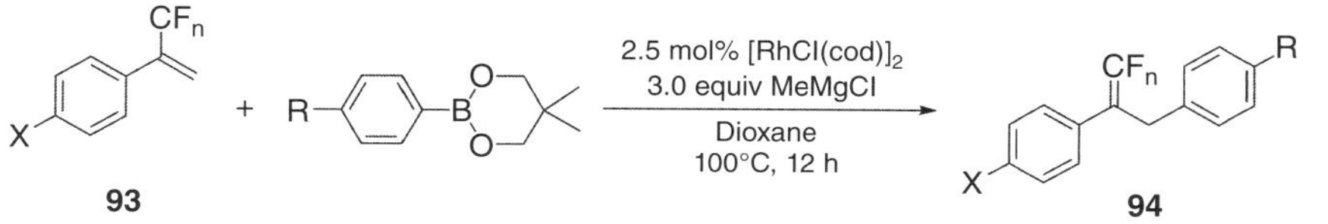

X	R	CF_n	Yield [%]
H	CO_2Me	CF_3	74
F	H	CF_3	60
Br	H	CF_3	61
CN	H	CF_3	54
H	H	CF_2	80 (*E*/*Z* ≥ 95:5)

the 5-position of indole **91** using a catalytic amount of $Pd(OAc)_2$ to obtain **92** in 48% yield.

Besides strong nucleophilic reagents, organoboron compounds react with trifluoromethylated alkenes in the presence of Rh complexes. In 2008, Miura and et al. [61] reported the synthesis of *gem*-difluoroalkenes from the reactions of a mixture of α-(trifluoromethyl)styrene **93**, phenylboronic acid, and MeMgCl in 1,4-dioxane with the presence of $[Rh(OH)(cod)]_2$ at 100°C (Table 2.2). With a high level of functional group tolerance, the reaction proceeds through the addition of arylrhodium(I) species across the electron-deficient carbon–carbon double bonds and the subsequent fluoride elimination to provide another synthetic route to *gem*-difluoroalkenes **94**, which involved a rare example of β-fluoride elimination of an organorhodium(I) complex.

In 2002, Ichikawa's group developed a method of nucleophilic 5-*endo*-tricyclization of 1,1-difluoro-1-alkenes. The disfavored 5-endo-trig cyclizations have been accomplished for 1,1-difluoro-1-alkenes with nitrogen, oxygen, sulfur, and carbon nucleophiles by taking advantage of the properties of fluorine. β,β-Difluorostyrenes bearing a tosylamido-, hydroxy-, or methylsulfinyl group at the α-position undergo intramolecular nucleophilic substitution with a loss of the vinylic fluorine, leading to 2-fluorinated indole, benzo[*b*]furan, and benzo[*b*]thiophene in high yields (Figure 2.5) [62]. 1,1-Difluoro-1-butenes bearing homoallylic tosylamido, hydroxy, mercapto, or iodomethyl group also successfully cyclize via a 5-*endo*-trig process with the *in situ*-generated intramolecular nucleophiles to afford 2-fluoro-2-pyrroline, 5-fluoro-2,3-dihydrofuran, 5-fluoro-2,3-dihydrothiophene, and 1-fluorocyclopentene. The two vinylic fluorines proved to be essential and play a critical role in these "anti-Baldwin" cyclizations. The striking feature in this cyclization is the fact that two fluorine substituents are essential; the corresponding monofluoro compound (R_1 = Bu) reacts very slowly to furnish the desired benzofuran in only 17% yield (at 60°C, 2 h for the difluoro compound, and at 80°C, 43 h for the monofluoro compound). A large amount (80%) of the starting material is

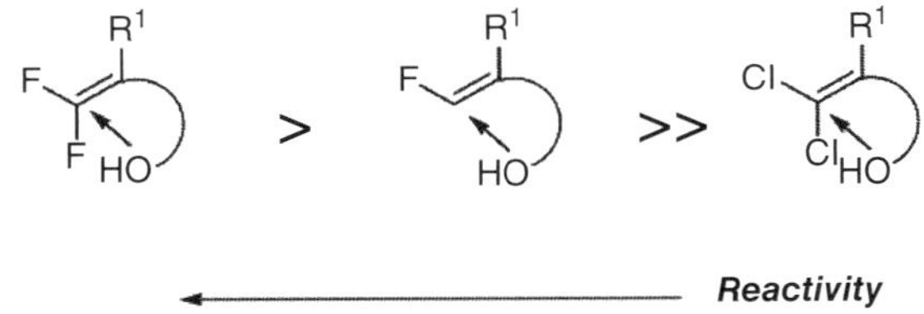

Figure 2.5.

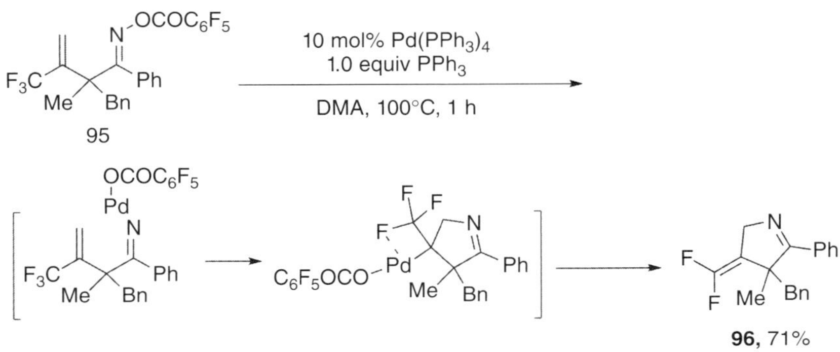

Scheme 2.39.

recovered without any desired cyclization product at 60°C for 8 h in the case of chloro compound.

In 2005, Ichikawa's group developed a method of intramolecular version of "atypical" Mizoroki–Heck reactions involving fluorine elimination to construct fluorine-containing five-membered rings (Scheme 2.39) [63]. With the condition of a catalytic amount of $Pd(PPh_3)_4$, 2-(trifluoromethyl)allyl ketone *O*-pentafluorobenzoyloxime **95** undergoes oxidative addition of the N–O bond, followed by a 5-*endo* mode of alkene insertion to produce 4-difluoromethylene-1-pyrroline **96** in 71% yield via β-fluorine elimination. It is important to show that vinylic fluorines play a crucial role in this Hect-type *5-endo-trig* cyclization and the advantage of the polarized double bond of 1,1-difluoro-1-alkenes.

Very interestingly, unusual reaction selectivities are observed in the nucleophilic transformations of fluorinated iminoesters and dithioesters as compared with that of the corresponding carbonyl compounds. The carbon–heteroatom double bonds bearing a CF_3 group accept nucleophiles at the heteroatom site, leading to fluoroalkenes via formal S_N2' reactions (Scheme 2.40) [64]. The reaction seems to proceed via 1,4-alkylation on imino nitrogen (**97**), followed by defluorination of the trifluoromethyl group. In this procedure, diethylzinc was found to be an efficient reagent for the difluoropropenoates **98** in excellent yield. On the other hand, thiocarbonyl derivatives **99** are able to react with aliphatic organometallic reagents via a thiophilic attack, giving rise to sulfur-stabilized carbanions. It was showed that these

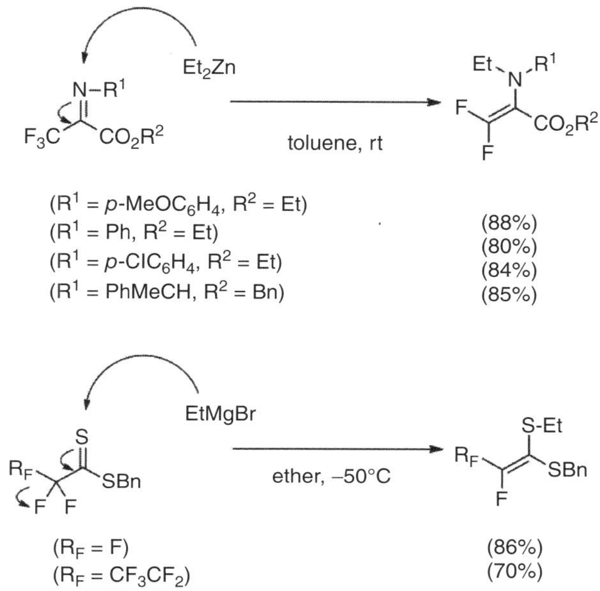

Scheme 2.40.

dithioesters would be good candidates for thiophilic organometallic addition, owing to the possibility of a subsequent fluoride elimination that would be a further driving force for this procedure.

Fluoride-ion-promoted reactions have been widely investigated in organic synthesis. A general, powerful strategy for activation of organosilicon compounds is the use of fluoride (Bu_4NF, TASF, CsF, KF/18-crown-6, etc.) endowed with the strong affinity to silicon. Generally, these reactions are in need of a stoichiometric amount of fluoride sources. From the practical viewpoint, the strong basic conditions occasionally cause side reactions, such as decomposition of the sensitive functional groups and decomposition of the desired products by side reactions involving further nucleophilic attack of fluoride ion. Thus, reaction conditions in which smaller amounts of fluoride can be employed are desirable for the development of more practical and useful protocols. In 1994, Portella's group developed a method of fluoride-ion-catalyzed reaction sequences (including fluoride elimination under the Brook rearrangement condition) that are very efficient and versatile for the synthesis of difluoroenol silyl ethers (Scheme 2.41) [65]. It emphasizes the importance of the choice of the activating fluoride and reveals the electrophilic character of the difluoroenoxysilane, which could be used *in situ* to prepare different difluoromethylene compounds. Under the standard condition, fluoride-promoted trifluoromethylation of acylsilane **100** with trifluoromethylsilane **101** followed by Brook-type rearrangement of the intermediate alkoxide provided a carbanion, which readily eliminates a (reusable) fluoride ion to give difluoroenol silyl ether **102**.

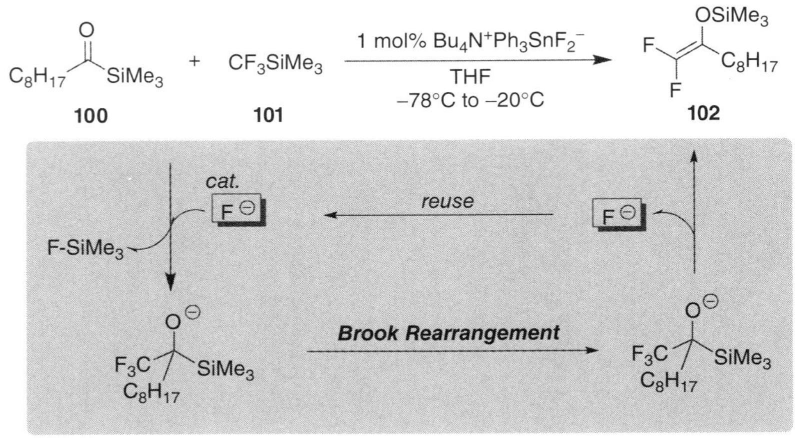

Scheme 2.41.

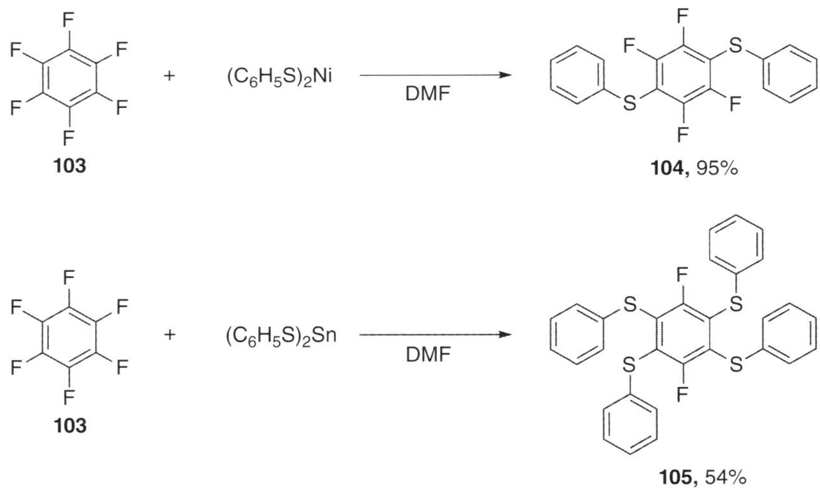

Scheme 2.42.

The reactions of tin(II) arenethiolates or nickel(II) arenethiolates in DMF with hexafluorobenzene had also been explored in depth. In 1988, Peach's group developed a method of replacement of fluorine by the phenylthio group, the reactions of hexafluorobenzene **103** with 2 equiv of nucleophiles, that afforded 1,2,4,5-tetrasubstituted tetrafluorobenzenes **104** regioselectively in near-quantitative yields (Scheme 2.42). [66] Moreover, the 1,4-disubstituted difluorobenzenes were obtained in reasonable yields with bidentate nucleophiles. In the research, DMF was a better solvent than either ethylene glycol/pyridine mixture (1:2) or diglyme for the reaction.

2.4 TRANSITION-METAL-PROMOTED DEHYDROFLUORINATION BY C–F BOND ACTIVATION

Hydrodefluorination of fluoroarenes is also attained via aromatic nucleophilic substitution with hydride. In 1985, Imamoto et al. [67] demonstrated the facile hydrodefluorination of monofluoroarenes upon a new reagent system, $LiAlH_4$–$CeCl_3$ (3:1) in 1,2-dimethoxyethane (DME) or TFH at reflux, which exhibited powerful and characteristic reducing ability (Scheme 2.43). In addition, various organic halides including fluorine compounds were smoothly dehalogenated with this reagent, thereby producing phosphine oxides in excellent yields.

Because of the practical significance both for laboratory synthesis and for industrial production, it is desirable to reduce the quantity of transition metals employed in these transformations. A number of approaches have been reported, especially with metal hydrides activated by transition metal salts however, most of the activating reagents for this process were in stoichiometric amount. Furthermore, the hydrodehalogenation of chloroarenes and fluoroarenes is a difficult task because of their high bond dissociation energies. In 1996, Liao's group demonstrated several splendid examples of reductive catalytic hydrodefluorination of monofluoroarenes **110** by the action of a small amount of transition metal salts (Ni, La, Nb, etc.) and a stoichiometric amount of hydride sources such as NaH and $LiAlH_4$ under mild conditions (Table 2.3) [68].

Palladium-catalyzed C–F bond activation via reductive defluorination of allylic *gem*-difluorides is also useful for the chemo- and stereoselective synthesis of fluoroalkenes (Scheme 2.44) [69]. In the presence of a catalytic amount of $[\eta 3\text{-}C_3H_5PdCl]_2$, dppe, and Et_3N, allylic *gem*-difluorides were treated with 2 equiv of Ph_3SiH at 50°C, affording monofluoroalkenes in high yields with a variety of substrates tolerance. In this transformation, Et_3N played a very important role in the generation of these active hydride species such as $Ph(EtO)SiH_2$ and $Ph(EtO)_2SiH$ by promoting dehydrogenative coupling, which on the top of that could work as reducing agents to generate Pd(0) complexes.

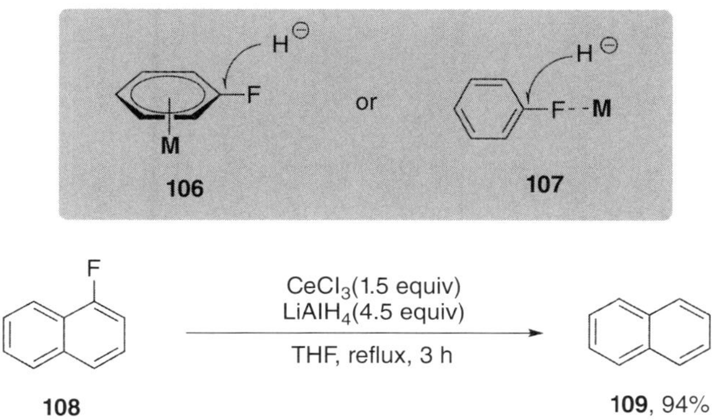

Scheme 2.43.

TABLE 2.3. Transition-Metal-catalyzed Hydrodefluorination

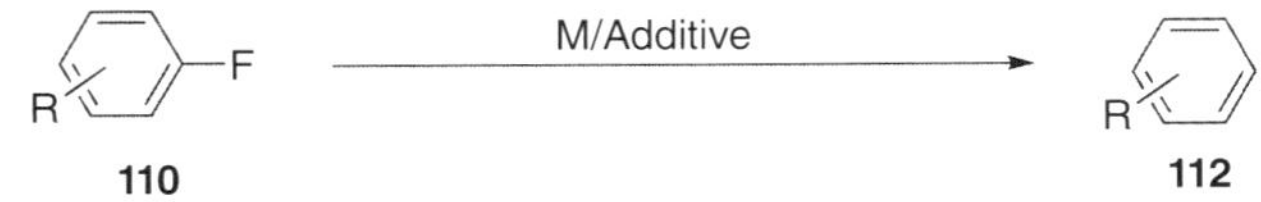

R	M	Additive	Product	Yield [%]
o-Me	3 mol% $Ni(OAc)_2$ 3 mol% $TiCl_4$	2.0 equiv NaH 0.15 equiv *i*-PrONa	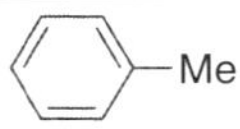	99
H	3 mol% $LaCl_3$	3.0 equiv NaH	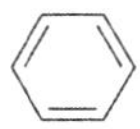	41
H	1.5 mol% $YbCl_3$ 0.15% $NiCl_2$	3.0 equiv NaH		50
H		3.0 equiv NaH	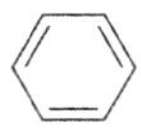	29
p-Ph	3 mol% $NbCl_5$	2.0 equiv LiAlH4	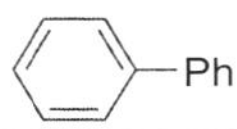	91

In 2006, low-valent niobium-mediated generated *in situ* double activation of C–F/C–H bonds was reported (Scheme 2.45) [70] *o*-Phenyl trifluorotoluene undergoes low-valent niobium-mediated dehydrodefluorination from both benzylic C–F bond and aromatic ortho C–H bond to form 9,9-difluorofluorene, which is readily reduced under the conditions to produce parent fluorene. Although other metal salts,

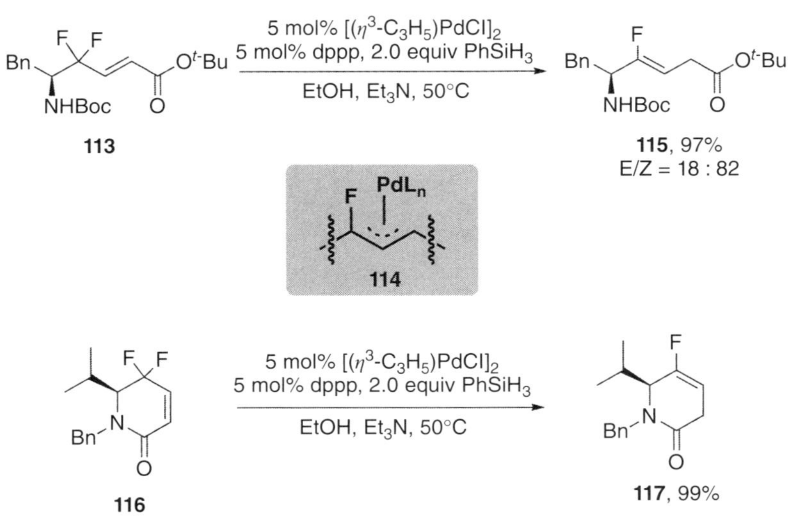

Scheme 2.44.

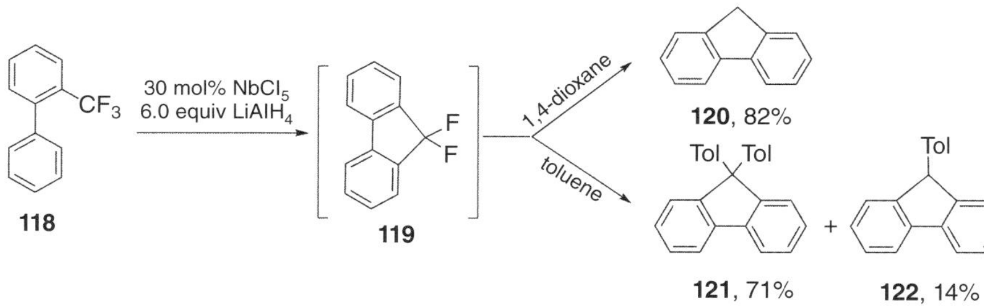

Scheme 2.45.

such as $TaCl_5$ or VCl_3, also took part in this transformation, the results were not satisfied at all. The yield of fluorine was dramatically improved when a DME suspension of $LiAlH_4$ was added to the medium, especially with the use of 1,4-dioxane. Interestingly, when the reaction is carried out in toluene,9,9-ditolylfluorene and 9-tolylfluorene were obtained in 71% and 14% yields respectively. One possible mechanism for the formation of difluorofluorene is that difluorobenzylic niobium species would be generated from the substrate and the low-valent niobium. The CF_3 group was activated not only prior to the aromatic C–F bond, but also prior to C–O, C–S and C–N bonds to give the heteroatom-substituted fluorenes in good yields.

In the presence of a catalytic amount of niobium(V) chloride, trifluoromethylated aromatics are reduced with lithium aluminum hydride to provide toluene derivatives in good yields (Scheme 2.46) [71]. Phenyl trifluorotoluenes bearing substituents such as methyl and fluorine gave the corresponding products in high yields. In particular, it is noteworthy that the aromatic C–F bonds remained intact under the reaction conditions. Stepwise, partial reduction of bis(trifluoromethyl)benzene derivatives proceeds by tuning of the amount of $LiAlH_4$. With large excess amounts of $LiAlH_4$, bis(trifluoromethyl)-benzene gave completely reduced product in 78% yield. In contrast, the partial reduction product was given in 77% yield with 3 molar amounts of the reducing agent. Deuterium labeling experiments suggest that the Nb-catalyzed hydrodefluorination of trifluoromethyl arenes does not proceed via a simple S_N2 mechanism; two of the fluorines in 4-(trifluoromethyl)-1,1′-biphenyl are replaced with deuteriums by $LiAlD_4$, and a (dideuterio)benzylic anion equivalent is formed to give the product after hydrolysis.

In 1994, Aizenberg and Milstein [72] demonstrated that the silyl complex $(Me_3P)_3RhSiMe_2Ph$ reacts quantatively with hexafluorobenzene at room temperature; that is, homogeneous catalytic activation of the strong C–F bonds was achieved with the use of rhodium complexes as catalysts (Scheme 2.47). The pentafluorophenyl rhodium(I) complex with square-planar geometry is obtained [71] chemo- and regioselectively with loss of F-$SiMe_2$Ph. The silyl ligand would facilitate aromatic C–F bond cleavage by electron transfer from the electron-rich Rh complex to C_6F_6 as well as release of fluorosilane, which arises from the great affinity between silicon and fluorine atoms.

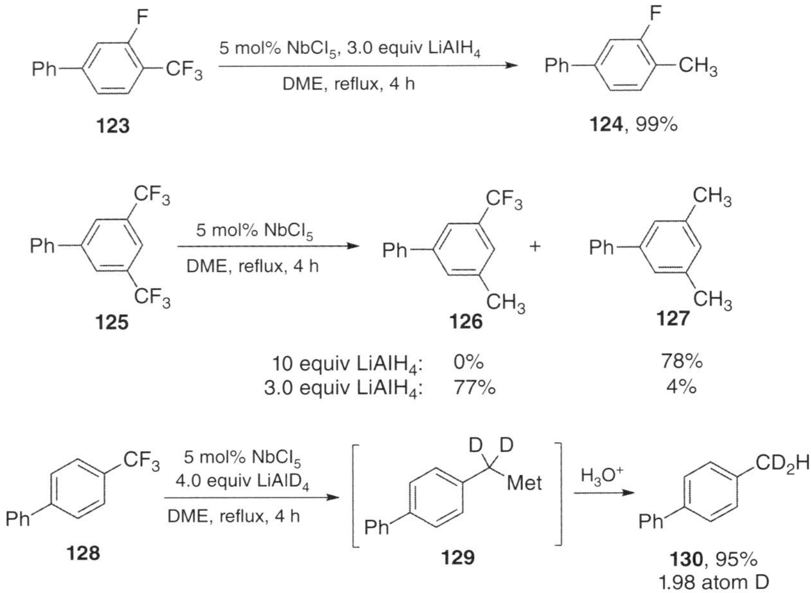

Scheme 2.46.

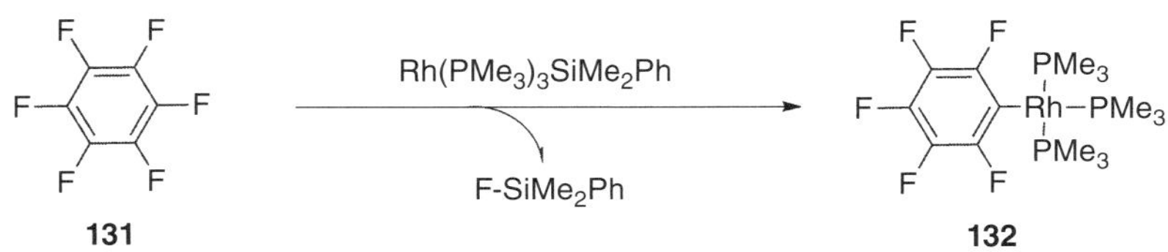

Scheme 2.47.

In 1994, Aizenberg and Milstein [72] reported catalytic hydrodefluorination of aromatic C–F bonds. Rh(I)–silyl complexes can cleave C–F bonds of C_6F_6 and C_6F_5H. Highly regioselective catalysis of C_6F_5H proceeds upon the treatment of $(EtO)_3SiH$ and $C_6F_5Rh(PMe_3)_3$ to yield 1, 4-$C_6F_4H_2$ exclusively (Scheme 2.48) [73]. In combination with Si–H oxidative addition to arylrhodium(I) complex and C–H reductive elimination, a catalytic cycle of Rh-mediated activation of aromatic C–F bonds is completed.

Peterson and McNeill [74] investigated the catalytic dehalogenation of fluoro- and chloroethylenes (Scheme 2.49). Treatment of vinyl fluoride with H-$SiEt_3$ and a catalytic amount of $RhCl(PPh_3)_3$ at 35°C leads to a complete defluorination product. The system has an intramolecular preference for Cl over F removal, an intermolecular preference for F-over Cl-containing alkenes, and a strong preference for sp^2 over sp^3 carbon–halogen bonds. The rate of the dehalogenation was determined by both the substitution pattern and the substituent identity.

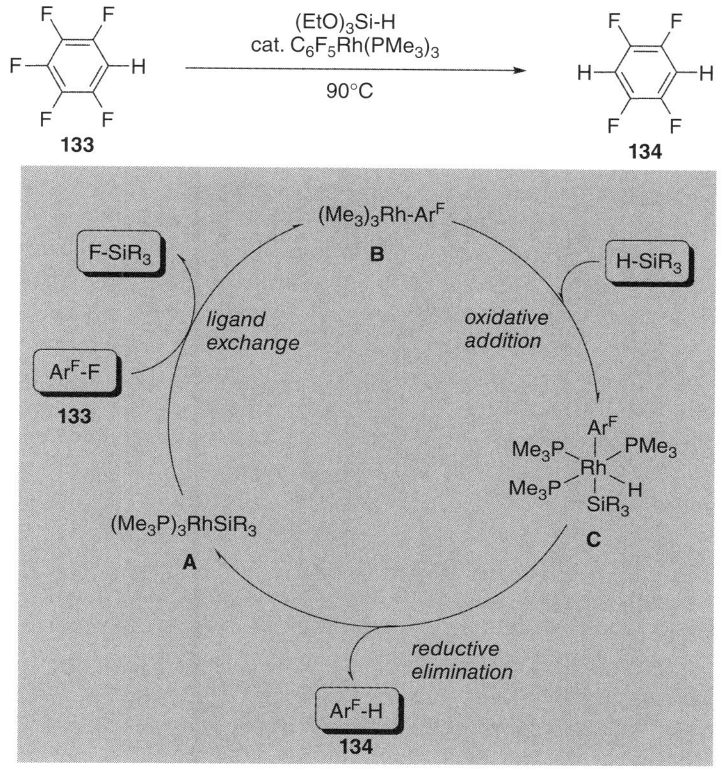

Scheme 2.48.

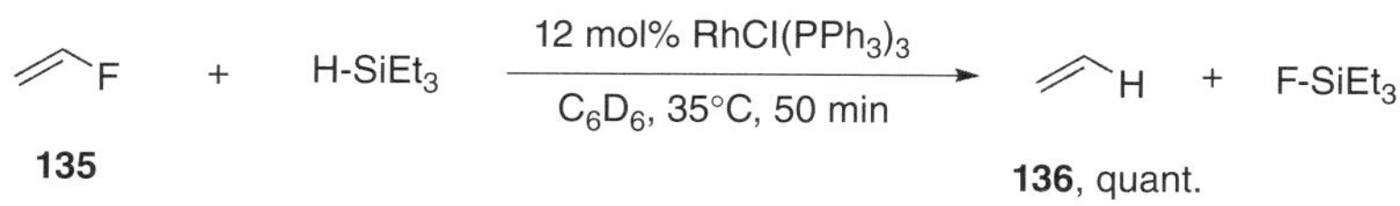

Scheme 2.49.

In 1999, Grushin's group reported selective hydrogenolysis of C–F bonds in most unactivated monofluoroarenes in the presence of homogeneous rhodium catalyst, which is generated from $[(Cy_3P)_2Rh(H)Cl_2]$, H_2, and aqueous alkali. Under the conditions, it promotes the hydrogenolysis of 1-fluoronaphthalene to yield naphthalene in an acceptable yield (Scheme 2.50) [75]. Oxygen plays a crucial role in this catalyzed circle because this homogeneous rhodium catalyst is incapable of hydrogenolysis of fluorobenzene under oxygen-free conditions. On the contrary, even the presence of a trace amount of air demonstrates a high activity for selective hydrogenolysis of fluorobenzene, 4-fluorotoluene, 3-fluoroanisole, and 4-fluoroaniline with the formation of the heterogeneous rhodium catalyst. In fact, the phosphine hydrido Rh species are

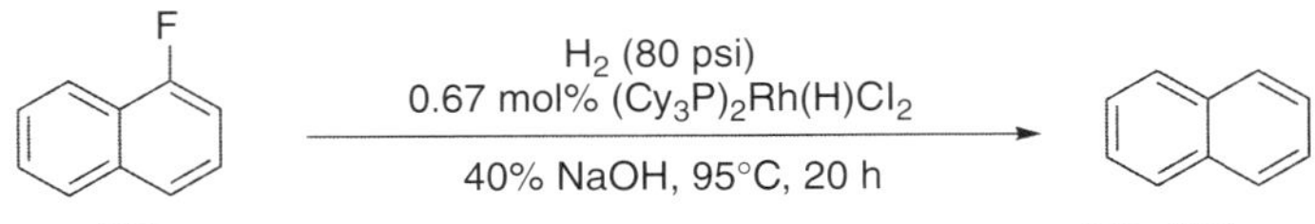

137 **138**, 45% conv.

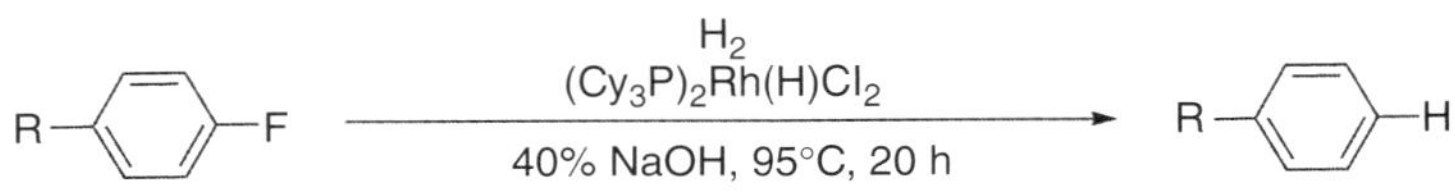

139 **140**

R	$P(H_2)$ (psi)	O_2	Conversion (%)	TON
H	40	+	55	88
NH_2	130	+	95	150
NH_2	130	–	<3	—

Scheme 2.50.

easily oxidized. Once oxidized, the phosphine oxide remains unchanged under the reaction conditions, whereas the Rh^{3+} is reduced with the hydrogen to a low-valent form.

In conventional procedures, H_2 has been widely employed for the catalytic hydrodehalogenation of various organic halides, including organic fluoroaromatics, with the requirement of hydrogen pressure. But it was not necessary. Next, we discuss examples of catalytic hydrogenolysis of fluoroaromatics without using hydrogen gas as the hydrogen atom donor. Ukisu, Urbano, and Dong developed convenient methods respectively including hydrogen transfer from hydrogen donors such as formic acid, secondary alcohols, and their salts. It is very important to note that the catalytic hydrogen-transfer reactions usually proceed under mild conditions without using special apparatus for high pressure. For example, Palladium on carbon (Pd/C) is applicable to hydrogen-transfer defluorination of aromatic fluorides with exposure to aqueous NaOH. Benzene was afforded quantitatively via hydrogen transfer reactions with fluorobenzene at 82°C from 2-propanol, which has the same nature of the hydrogen donor as HCOOK (Scheme 2.51) [76].

Plenty of research proves that not only heterogeneous but also homogeneous systems are able to catalyze C–F bond activation of fluoroarenes via hydrogen-transfer reaction. Aromatic C–F bonds were activated by catalytic Ni(0)/*N*-heterocyclic carbene (NHC) complexes. Using hydrogen-transfer reactions from Et_2CHONa as a

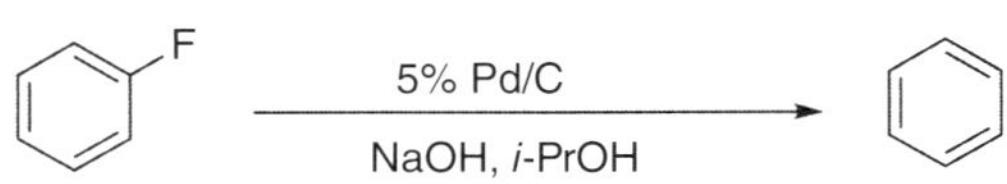

141 **142**

Scheme 2.51.

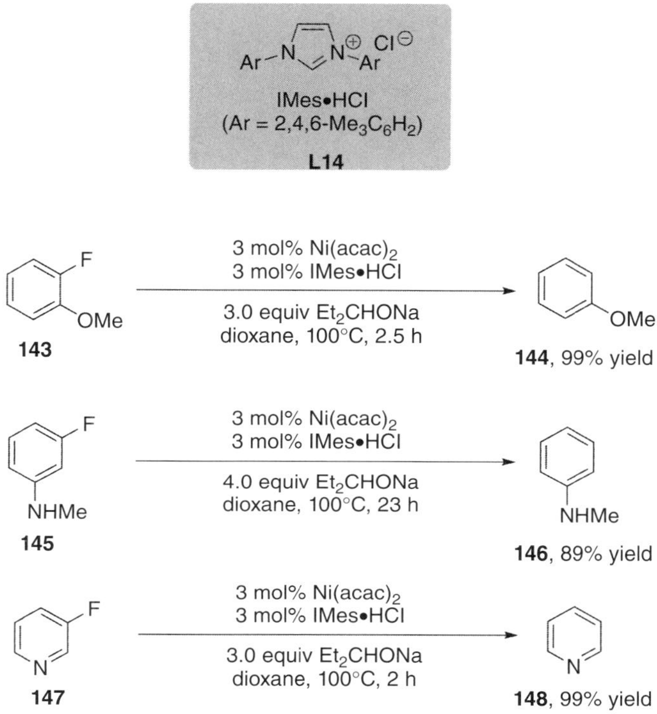

Scheme 2.52.

hydrogen source, various aryl fluorides are converted to the corresponding reduction products in the presence of a catalytic amount of the monocoordinate nickel–carbene complex, which is prepared *in situ* from $Ni(acac)_2$ and IMes·HCl (Scheme 2.52) [77]. The whole procedure involves the formation of Ni(0) by reduction of $Ni(acac)_2$ with Et_2CHONa-activated sodium hydride followed by coordination of the carbine ligand to the metal center. Then oxidative addition of the aryl fluoride to the Ni(0) complex is followed by Et_2CHONa attack at nickel and displacement of the fluorine atom. Finally, reductive elimination of the arene from an intermediate nickel–hydride complex obtained by β-elimination to regenerate the catalyst.

In 2000, Adonin and Starichenko [78] pointed that low-valent Ni complex worked well for *ortho*-hydrodefluorination of pentafluorobenzoic acid. In the presence of 0.01 equiv of $NiCl_2$-2,2′-bipyridine complex and 10 equiv of Zn, *ortho*-monodefluorination of $C_6F_5CO_2H$ proceeds selectively to give 2,3,4,5-tetrafluorobenzoic acid in 93% yield (Scheme 2.53). It is very interesting to note that the use of an increased amount of Ni catalyst (5 mol% Ni complex) and the increased reaction time led to the further removal of *ortho*-fluorine atoms to afford 3,4,5-trifluorobenzoic acid. The replacement of water by ammonium chloride as proton donor led to acceleration of hydrodefluorination.

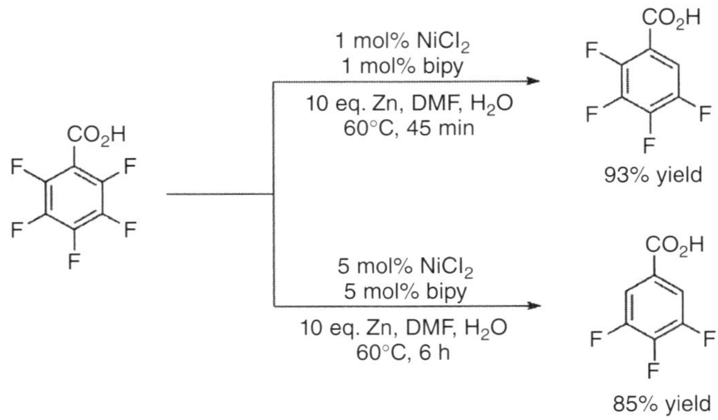

Scheme 2.53.

Lv et al. [79] demonstrated the first example of unusual π–π interaction-assisted C–F bond activation by gold(I) hydrides (Scheme 2.54). Although direct HDF of fluoroaromatics could not be achieved by [(IMes)AuH] due to its low reactivity, the addition of DMAP can effectively promote the reaction process. Through mechanistic studies, it is observed that the weak interaction between [(NHC)AuH] and perfluoroarenes by ^1H NMR and UV–vis spectroscopies; however, this interaction

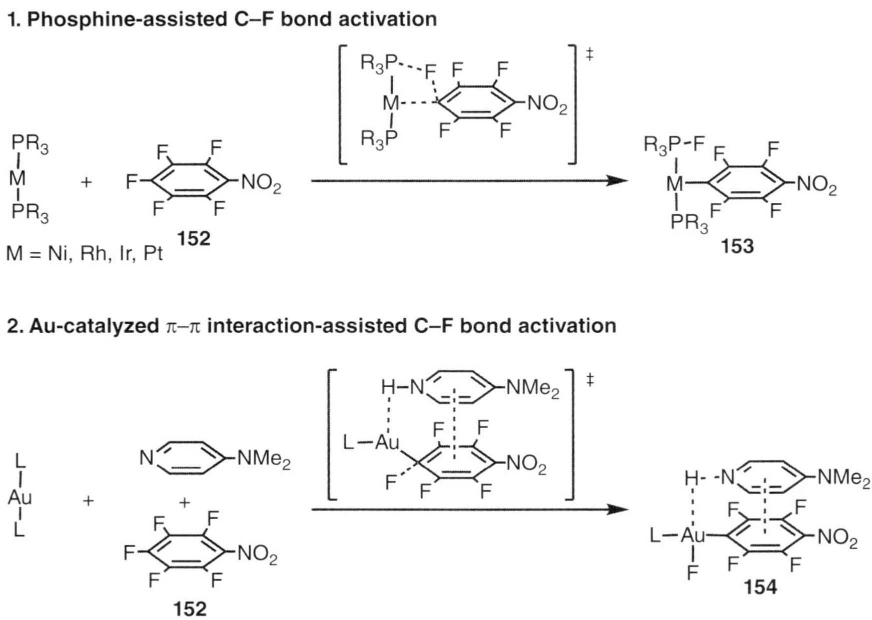

Scheme 2.54.

is not efficient to perform C–F activation through oxidative addition or nucleophilic attack of the C–F bond. More importantly, it is important that the formation of an intermediate between PFNB and DMAP though $\pi-\pi$ interaction could promote the transfer of proton in [(NHC)-AuH], accompanied by C–F bond activation. The"synergistic effect" is prevalent in natural metalloenzymes in activating small molecules, and the success has inspired the development of artificial catalysts capable of unprecedented organic transformations. In this work, the attractive $\pi-\pi$ interaction between organic additives (as electron donors) and the perfluorinated arenes (as electron acceptors) is effective in gold hydride-catalyzed activation of C–F bonds, specifically hydrodefluorination (HDF) of perfluoroarenes catalyzed by Sadighi's gold hydrides [(NHC)AuH] (NHC = *N*-heterocyclic carbene). Although a weak interaction between [(NHC)AuH] and perfluoroarenes was observed from 1H NMR and UV–vis spectroscopies, low reactivity of [(NHC)AuH] toward HDF was found. In contrast, in the presence of *p*-*N,N*-dimethylaminopyridine (DMAP), the HDF of perfluoroarenes with silanes can be efficiently catalyzed by [(NHC)AuH], resulting in mainly the *para*-hydrodefluorinated products with up to 90% yield and 9 turnovers. The yield of the reaction increases with the more electron-withdrawing groups and degree of fluorination on the arenes, and the HDF reaction also tolerates different function groups (such as formyl, alkynyl, ketone, ester, and carboxylate groups), without reduction or hydrogenation of these function groups (Table 2.4). To reveal the role of DMAP in the reactions, the possible $\pi-\pi$ interaction between DMAP and perfluoroarenes was suggested by UV–vis spectral

TABLE 2.4. Au-Catalyzed $\pi-\pi$ Interaction-Assisted C–F Bond Activation

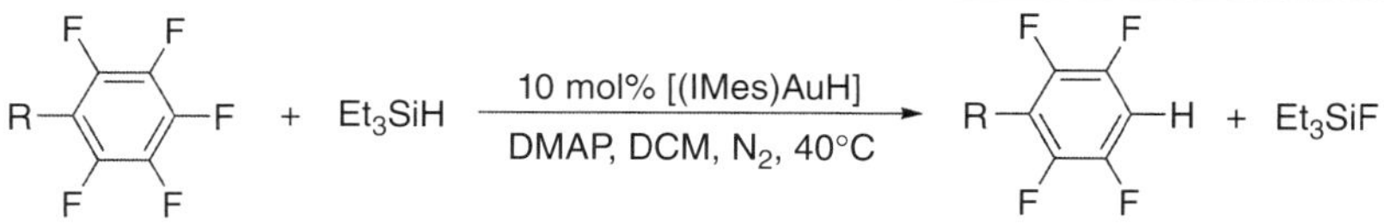

155 156

R	Yield (%)	R	Yield (%)
NO_2	90	OMe	0
CHO	86	≡—TMS	75
CN	90	COMe	65
CF	85	COOEt	55
H	0	COOH	22
F	0		

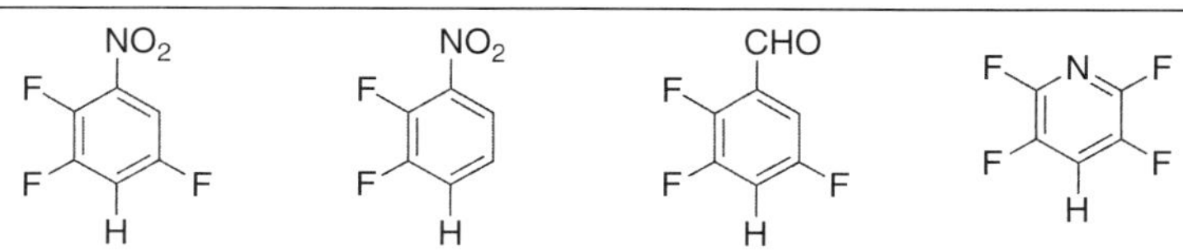

Yields: 67% 53% 42% 90%

titrations, 1H NMR spectroscopic studies, and DFT calculations. Moreover, 1H and 19F-NMR studies show that this π–π interaction promotes hydrogen transfer from [(NHC)AuH] to pyridyl N atom, resulting in C–F bond cleavage. The interpretation of π–π interaction-assisted C–F activation is supported by fact that there are fewer activation barriers in the presence of DMAP (31.6 kcal/mol) than in the absence of DMAP (40.8 kcal/mol) for this reaction. An analysis of the charge distribution and transition state geometries indicates that this HDF process is controlled by the π–π interaction between DMAP and perfluoroarenes, accompanied by the changes of partial atomic charges.

Based on the above experimental and theoretical studies for C–F bond activation mediated by [(IMes)AuH] (a), the following proposed mechanism was depicted in image below (Scheme 2.55). In the absence of DMAP (path I), [(IMes)AuH] interacts with PFNB first, forming complex b, which is supported by 1H NMR experiments, UV–vis spectra, and theoretical studies. Then, the C–F bond is activated via the oxidative addition step to the gold(I) complex, resulting in the formation of gold(III)–fluoride complex f. Subsequent reduction elimination leads to the product and gold(I)–fluoride complex e, which reacts with silane to reform [(IMes)AuH] via H/F exchange. In the presence of DMAP (path II), it was assumed that the interaction among [(IMes)AuH], DMAP, and PFNB would occur via two possible pathways: Either complex b interacts with DMAP or [(IMes)AuH] interacts with the π–π stacking complex of DMAP and PFNB (complex b′). In either case, the generation of protonated DMAP complex c is postulated to be the key step. Then the oxidative addition reaction of the C–F bond to gold(I) occurs in the complex c, affording the DMAP involved gold(III) complex d. Subsequent reductive elimination results in formation of the product and the complex.

2.5 THE APPLICATIONS OF C–F ACTIVATION IN ORGANIC SYNTHESIS

Nowadays, the chemistry of natural products attracts a very lively interest. In the course of the investigation of a complicated substance, the investigator is sooner or later confronted by the problem of synthesis—that is, the preparation of the substance by chemical methods. The workers can have various motives, especially the new developed methods. So next, it is shown that some examples of substances were synthesized via C–F bonds activation. First, 4-Bromo-1,3-difluorobenzene underwent regiospecific nitration to provide the 5-nitro derivative in 99% yield. Then, a palladium-catalyzed Stille coupling with vinyltributyltin provided the styrene in 90% yield. Conjugate addition of triphenylsilanethiol under radical conditions gave the triphenylsilyl-protected thiol in 52% yield (Scheme 2.56) [80]. Nucleophilic cyclization of silyl-protected thiol to the benzothiophene could be accomplished using KOH

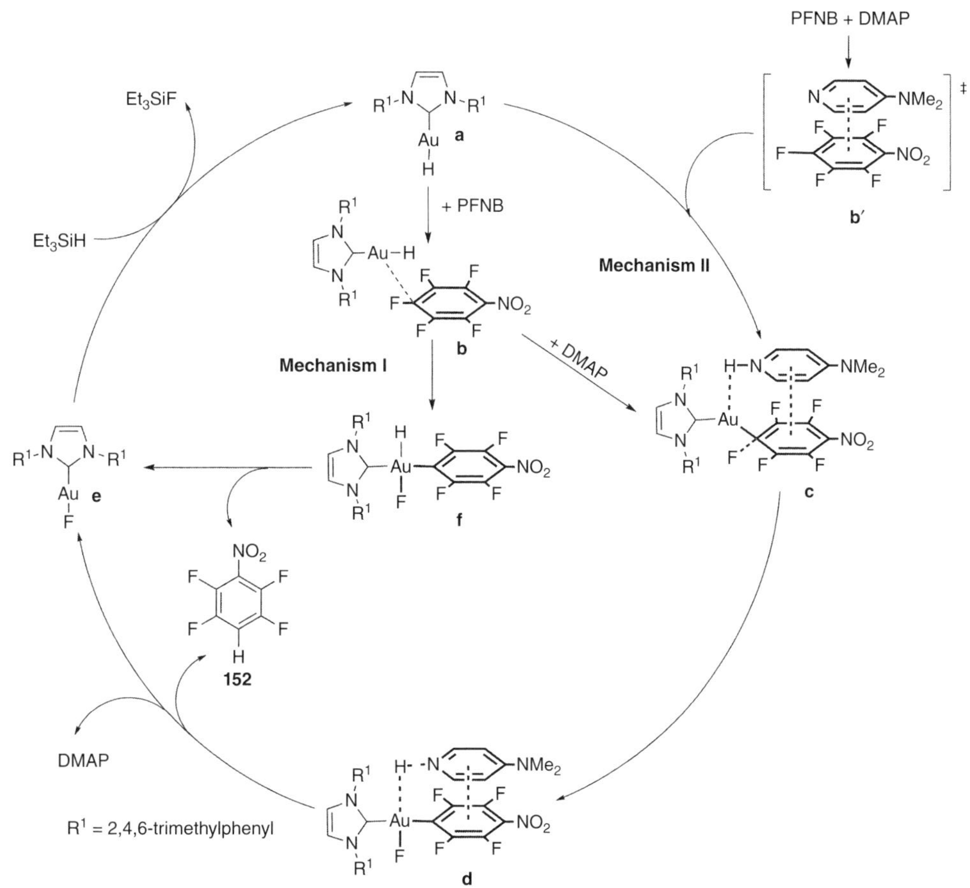

Scheme 2.55.

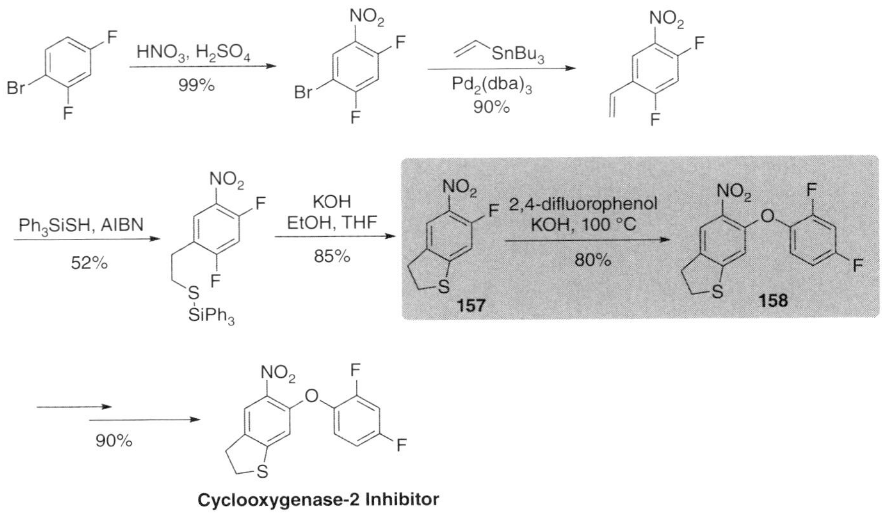

Scheme 2.56.

to remove the triphenylsilyl protecting group while simultaneously generating the reactive potassium thiolate. With the key functionalized heterocyclic ring system in hand, standard chemical transformations then led to the target molecule. Thus nucleophilic aromatic substitution with 2,4-difluorophenol

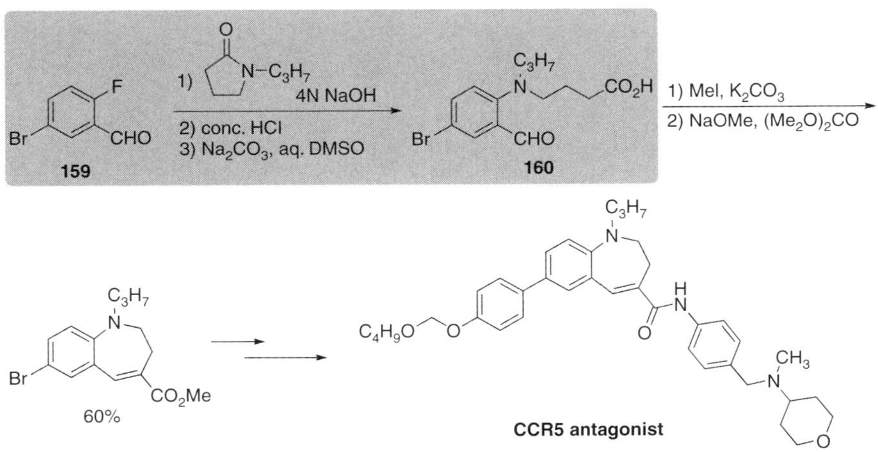

Scheme 2.57.

provided the tricyclic framework. Reduction of the nitro group to the aniline followed by mesylation (MsCl, pyridine) gave sulfonamide as a cyclooxygenase-2-inhibitor.

Takeda Pharmaceutical Company developed the practical method of producing **CCR5**, an orally active CC chemokine receptor 5 (CCR5) antagonist (Scheme 2.57) [81]. 5-Bromo-2-fluorobenzaldehyde in DMSO was added and the whole mixture was refluxed to afford the corresponding acid, which did not need to be purified. The esterification of crude acid with MeI in the presence of K_2CO_3 in DMF followed by an intramolecular Claisen-type reaction triggered by the addition of dimethyl carbonate and 28% sodium methoxide in one pot gave high-quality methyl formate without a byproduct.

A number of quinolone carboxylic acid derivatives, useful antibacterial agents, have been prepared by nucleophilic substitution of suitably designed aryl fluorides as a key reaction. Some of the related synthetic examples are shown below (Scheme 2.58) [82]. Fluorine on C-7 is an important substituent to promote a smooth introduction of an amino group under the conditions of *N*-methylpiperidine in DMSO at 75°C without impurities. Two of the three fluorine atoms in trifluorobenzoylacetic acid ethyl ester can be employed in a stepwise manner to introduce amino functionalities. Several examples of the F-activated S_NAr protocol for the intramolecular ring closures between N-1 and C-8 related to quinolone derivatives have been reported. For example, cyclization of resin-bound enamide in a solution of tetramethylquanidine and dichloromethane at 55°C provided the corresponding resin-bound quinolone, which was reacted with a solution of piperazine in NMP at 110°C, followed by TFA, to form the ciprofloxacin.

The reaction sequences involving regioselective nucleophilic displacement of polyfluorinated arenes are brilliant methodologies to produce quinolone carboxylic acid derivatives. Many quinolone antibacterial agents were synthesized via C–F bond activation, which is nucleophilic displacement of fluorinated arenes, such as levofloxacin, pazufloxacin, and sparfloxacin (Scheme 2.59) [83].

The F-activated S_NAr protocol has been used for the syntheses of a variety of heterocycles and macrolides. By the intramolecular diarylether construction protocol using fluoroarene, the preparation of cycloisodityrocine precursor is achieved in a good yield. For the synthesis of the D-O-E segment of vancomycin, the substrate undergoes S_NAr macrocyclization to afford the cyclization product as a mixture of atropisomers (Scheme 2.60) [84].

Furthermore, in 2008, Barrett and co-workers [85] succeeded in the first total synthesis of dehydroaltenuene B by means of a four-component coupling reaction of 1-fluoro-3,5-dimethoxybenzene with 2-methyl-2-cyclohexenylmagnesium chloride, carbon dioxide, and iodine (Scheme 2.61). Benzyne, obtained from aryl fluoride (170) by lithiation with *n*-BuLi, reacted with the Grignard reagent regioselectively to afford the arylmagnesium chloride intermediate, which then trapped with carbon dioxide and followed by iodolactonization to provide the

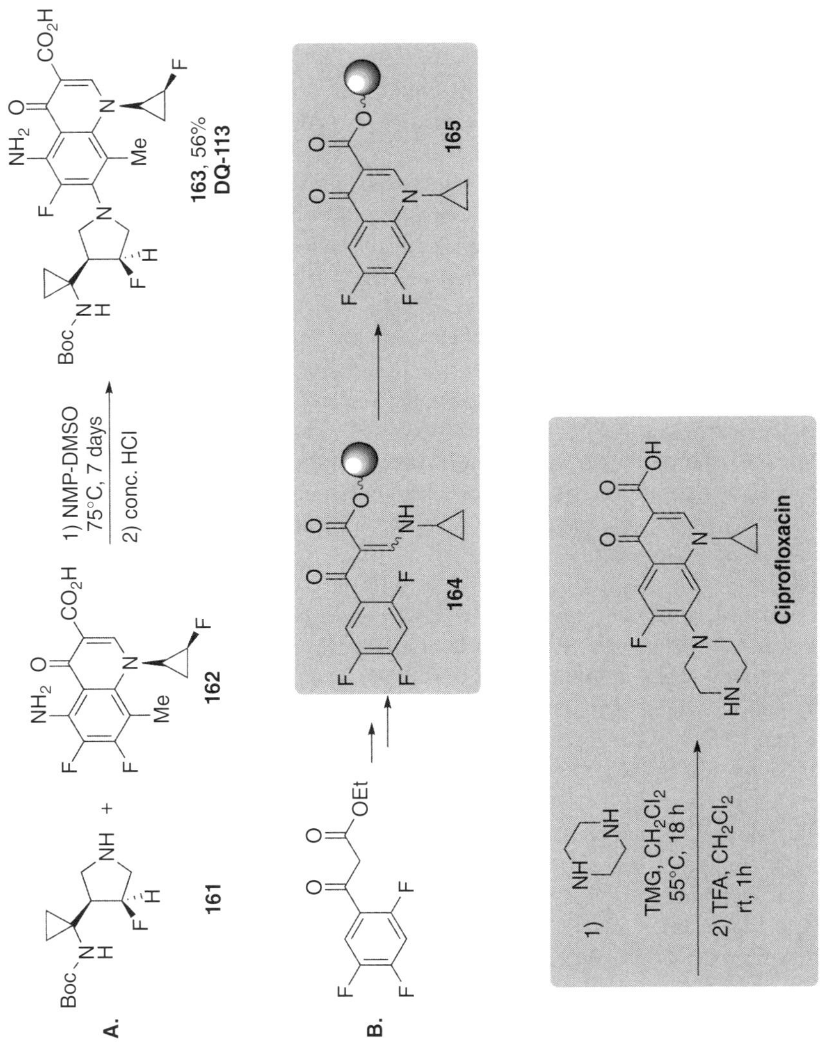

Scheme 2.58.

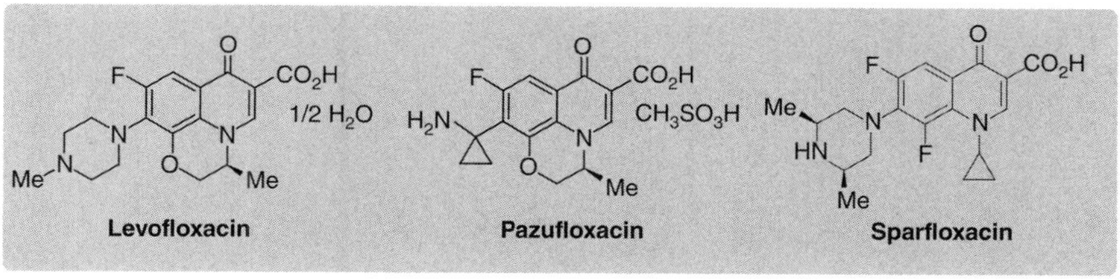

Scheme 2.59.

corresponding iodide in 56% yield. Finally, the dehydroaltenuene B was synthesized from the iodide.

Benzynes from aryl fluorides have been employed for Diels–Alder cycloaddition with furans and thiophenes and coupling with aryl lithiums. The treatment of 1,4-difluoro-2,5-dimethoxybenzene as a novel precursor with a range of 2- and 3-substituted furans, via two-directional benzyne-furan Diels–Alder reactions, was reported by Morton and Barrett in 2005 [86]. Highly substituted anthracenols were provided in the second benzyne–furan Diels–Alder reaction following acid-catalyzed isomerization of the cycloadducts (Scheme 2.62).

In 2006, Barrett and co-workers [87] developed the first asymmetric synthesis of (+)-clavilactone B, a potent antifungal agent via defluorinative benzyne formation. The strategy includes a powerful and convergent sophisticated three-component benzyne coupling with a methylallyl Grignard reagent followed by chiral epoxy-aldehyde to generate two C–C bonds and install the carbon skeleton of clavilactone in an RCM reaction (Scheme 2.63).

In 2008, Senboku' group reported the selective electrochemical carboxylation of α,α-difluorotoluene derivatives, providing the corresponding α-fluorophenylacetic acids in good yields (Scheme 2.64) [88]. Interestingly enough, this reaction, in the presence of CO_2 using a Pt cathode and a Mg anode, was successfully applied to the synthesis of α-fluorinated nonsteroidal anti-inflammatory drgus, namely NSAIDs.

During the research of C–F bond activation, the choice of a countercation of nucleophile is quite important for nucleophilic transformation of organic fluorides. In general, Lewis acids are able to catalyze C–C bond formation reactions of sugar fluorides that act as versatile sugar donors. But in the absence of external Lewis acids, perbenzylated *D*-ribofuranosyl fluorides can also react with diverse Grignard reagents to provide the corresponding β-*C*-glycosides in moderate yields, which was reported by Yokoyama et al. [89] in 1998 (Scheme 2.65). It is implied that magnesium atoms of Grignard reagents or the resultant Mg(II) salts play the role of Lewis acid. Also,

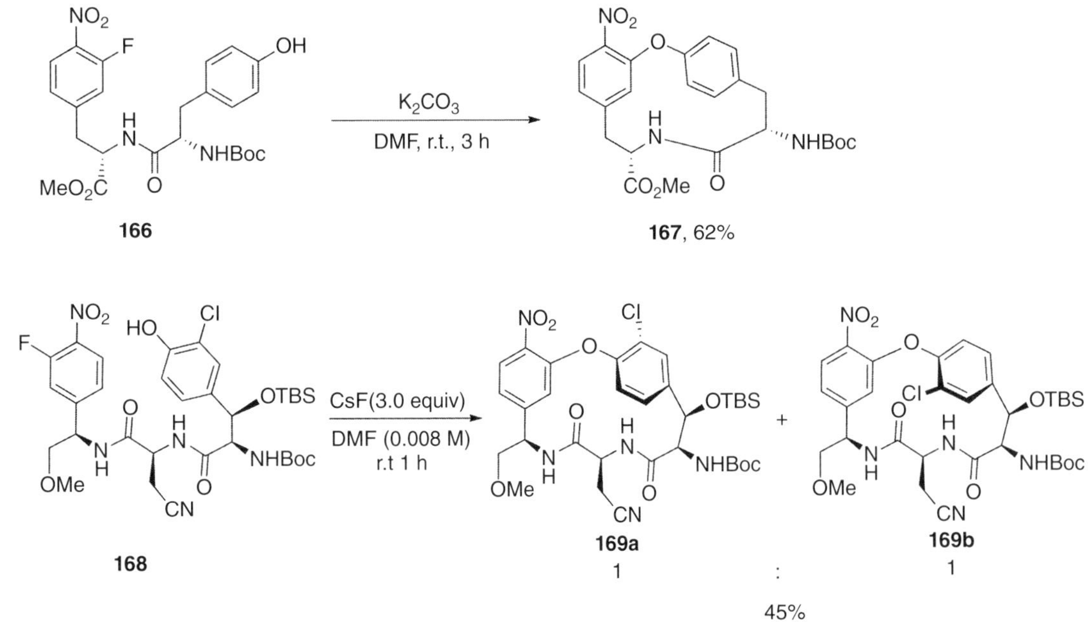
NO2
F
OH
NHBoc
MeO2C
O
166
K2CO3
DMF, r.t., 3 h
CO2Me
167, 62%
HO
Cl
OTBS
OMe
CN
168
CsF(3.0 equiv)
DMF (0.008 M)
r.t 1 h
169a
1
:
169b
1
45%

Scheme 2.60.

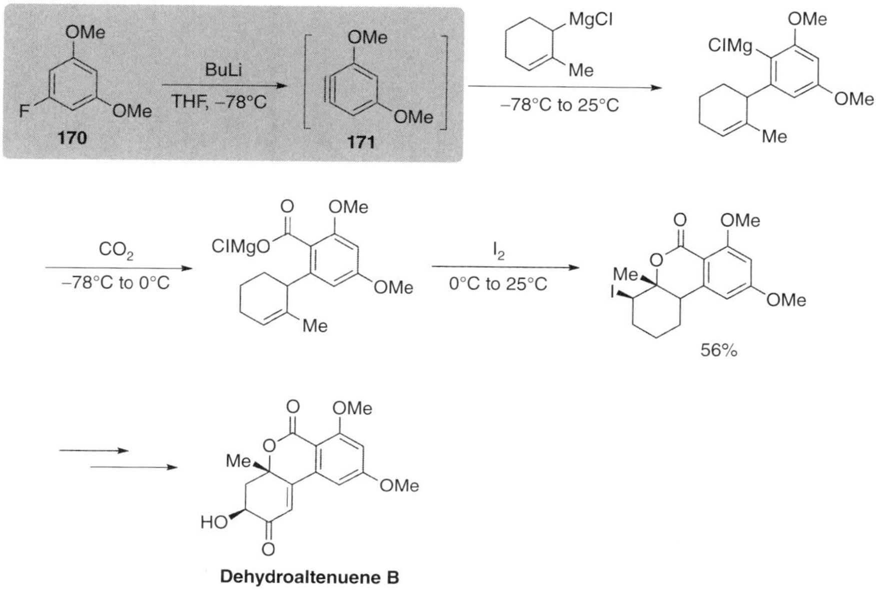

Scheme 2.61.

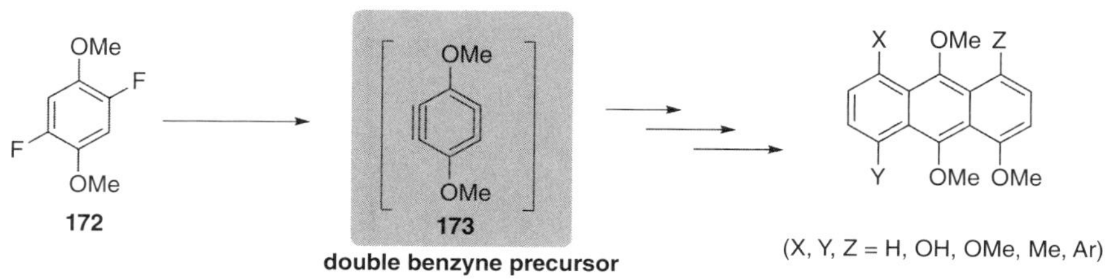

Scheme 2.62.

because of the steric hindrance of 2-alkoxy group, the reaction is considered to precede β-selectivtiy via an oxocarbenium ion.

2.6 SUMMARY AND OUTLOOK

In the past decade, a number of new transformations for carbon-fluorine bond activation have been developed. Impressive advances in the fields of transition metal-mediated selective carbon–fluorine bond activation and applications for

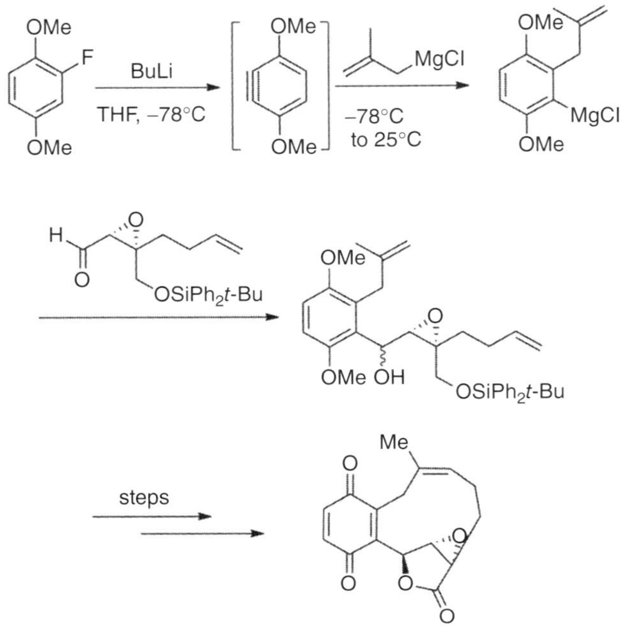

Scheme 2.63.

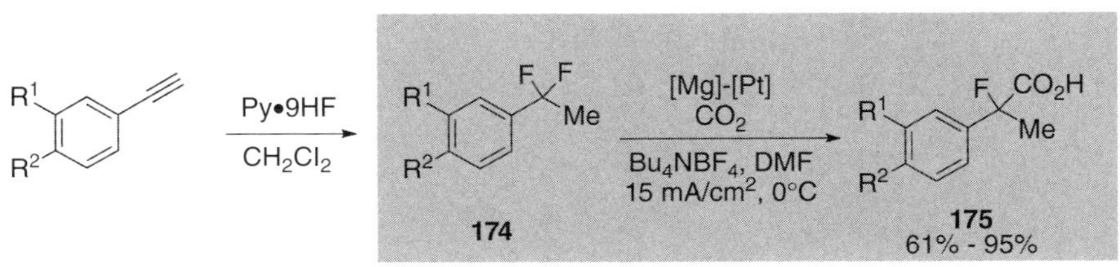

Scheme 2.64.

pharmaceuticals and natural products synthesis have provided a useful methodology for various carbon–fluorine compounds. Despite recent progress, controlled, general and selective carbon–fluorine bond activation remains a major challenge in synthetic organic chemistry; however, because of the importance of fluorine in pharmaceuticals, agrochemicals, materials and PET, there should be a rapid development of new carbon–fluorine bond activation reactions in the coming years.

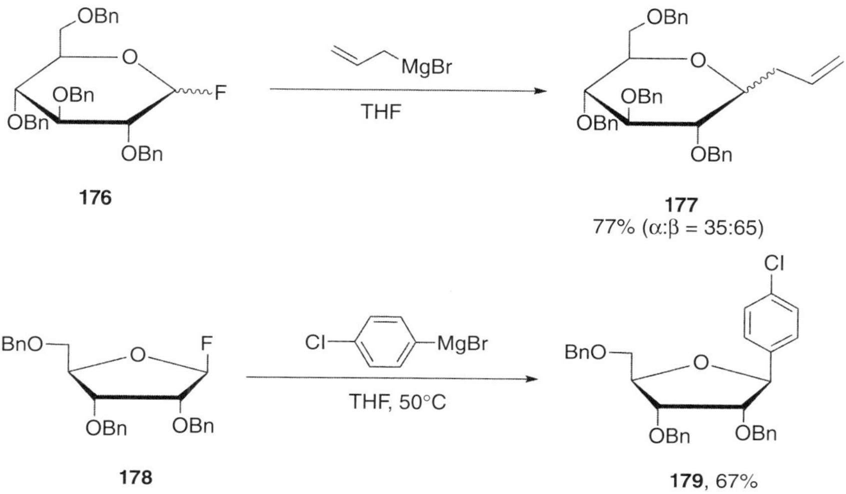

Scheme 2.65.

REFERENCES

1. (a) Blanksby, S. J., Ellison, G. B. (2003). Bond dissociation energies of organic molecules. *Acc. Chem. Res.* **36**, 255. (b) Müller, K., Faeh, C., Diederich, F. (2007). Fluorine in pharmaceuticals: looking beyond intuition. *Science* **317**, 1881. (c) Purser, S., Moore, P. R., Swallow, S., Gouverneur, V. (2008). Fluorine in medicinal chemistry. *Chem. Soc. Rev.* **37**, 320. (d) Jeschke, P. (2004). The unique role of fluorine in the design of active ingredients for modern crop protection. *Chem Bio Chem* **5**, 570. (e) Phelps, M. E. (2000). Positron emission tomography provides molecular imaging of biological processes. *Proc. Natl. Acad. Sci. USA* **97**, 9226. (f) Lasne, M.-C., Perrio, C., Rouden, J., Barré, L., Roeda, D., Dolle, F., Crouzel, C. (2002) Chemistry of $\beta+-$emitting compounds based on fluorine-18. In *Topics in Current Chemistry*, Vol.**222**, Springer, Berlin, p. 201. (g) Ametamey, S. M., Honer, M., Schubiger, P. A. (2008). Molecular imaging with PET. *Chem. Rev.* **108**, 1501. (h) Banks, R. E. (2000). *Fluorine Chemistry at the Millenium. Fascinated by Fluorine*, Elsevier, Amsterdam. (i) Hiyama, T., Kusumoto, T., Morizawa, Y., Shimizu, M., (2000). *Organofluorine Compounds: Chemistry and Applications*, Springer, Berlin. (j) Gladysz, J. A., Curra, D. P., Horvath, I. T. (2004). *Handbook of Fluorous Chemistry*, Wiley-VCH, Weinheim. (k) Tressaud, A., Haufe, G. (2008). *Fluorine and Health*, Elsevier, Amsterdam. (l) Bégué, J.-P., Bonnet-Delpon, D. (2008). *Bioorganic and Medicinal Chemistry of Fluorine*, John Wiley & Sons, New York.

2. (a) Perutz, R. N., Braun, T. (2007). Transition metal-mediated C–F bond activation. In *Comprehensive Organometallic Chemistry III*, Elsevier, Oxford, England, p. 725. (b) Torrens, H. (2005). Carbon fluorine bond activation by platinum group metal complexes. *Coord. Chem. Rev.* **249**, 1957. (c) Crespo, M., Martinez, M., Sales, J. (1992). Intramolecular activation of a C–F bond at platinum(II) in the presence of weaker C–X

bonds (X = H, Cl and Br). *J. Chem. Soc., Chem. Commun.*, 822. (d) Bosque, R., Clot, E., Fantacci, S., Maseras, F., Eisenstein, O., Perutz, R. N., Renkema, K. B., Caulton, K. G. (1998). Inertness of the aryl–F bond toward oxidative addition to osmium and rhodium complexes: Thermodynamic or kinetic origin? *J. Am. Chem. Soc.* **120**, 12634. (e) Erhardt, S., Macgregor, S. A. (2008). Computational study of the reaction of C_6F_6 with $[IrMe(PEt_3)_3]$: Identification of a phosphine-assisted C–F activation pathway via a metallophosphorane intermediate. *J. Am. Chem. Soc.* **130**, 15490. (f) Murai, S. (1999). *Activation of Unreactive Bonds and Organic Synthesis,* Springer, Berlin. (g) Kiplinger, J. L., Richmond, T. G., Osterberg, C. E. (1994). Activation of carbon–fluorine bonds by metal complexes. *Chem. Rev.* **94**, 373. (h) Burdeniuc, J., Jedlicka, B., Crabtree, R. H. (1997). Recent advances in C–F Bond Activation. *Chem. Ber.* **130**, 145. (i) Braun, T.; Perutz, R. N. (2002). Routes to fluorinated organic derivatives by nickel mediated C–F activation of heteroaromatics. *Chem. Commun.*, 2749. (j) Jones, W. D. (2003). Activation of C–F bonds using $Cp^*{}_2ZrH_2$: a diversity of mechanisms. *Dalton Trans.,* 3991.

3. Recent reviews on C–F bonds activation: (a) Clot, E., Eisenstein, O., Jasim, N., MacGregor, S. A., McGrady, J. E., Perutz, R. N. (2011). C–F and C–H bond activation of fluorobenzenes and fluoropyridines at transition metal centers: How fluorine tips the scales. *Acc. Chem. Res.* **44**, 333. (b) Braun, T., Wehmeier, F. (2011). C–F bond activation of highly fluorinated molecules at rhodium: From model reactions to catalysis. *Eur. J. Inorg. Chemi.* **5**, 613. (c) Sun, A. D., Love, J. A. (2010). Cross coupling reactions of polyfluoroarenes via C–F activation. *Dalton Trans.* **39**, 10362. (d) Meier, G., Braun, T. (2009). Catalytic C–F activation and hydrodefluorination of fluoroalkyl groups. *Angew. Chem., Int. Ed.* **48**, 1546. (e) Braun, T., Perutz, R. N. (2002). Routes to fluorinated organic derivatives by nickel mediated C–F activation of heteroaromatics. *Chem. Commun.* **23**, 2749. (f) Osterberg, C. E.; Richmond, T. G. (1994). Activation of carbon–fluorine bonds by oxidative addition to low-valent transition metals. *ACS Symp. Ser.* **555**, 392. (g) Grushin, V. V. (2002). Palladium fluoride complexes: One more step toward metal-mediated C–F bond formation. *Chem. Eur. J.* **5**, 1006. (h) Frisch, A. C., Beller, M. (2005). Catalysts for cross-coupling reactions with non-activated alkyl halides. *Angew. Chem., Int. Ed.* **44**, 674. (i) Albini, A., Monti, S. (2003). Photophysics and photochemistry of fluoroquinolones. *Chem. Soc. Rev.* **32**, 238. (j) Amii, H., Uneyama, K. (2009). C–F bond activation in organic synthesis. *Chem. Rev.* **109**, 2119.

4. Braun, T., Perutz, R. N. (2006). Transition metal-mediated C–F bond activation. In *Comprehensive Organometallic Chemistry III*, Elsevier, Oxford, U.K., Chapter 1, p. 26.

5. (a) Meijere, A., Diederich, F. (2004). *Metal-Catalyzed Cross-Coupling Reactions*, Wiley-VCH, Weinheim. (b) Johansson, S. C., Kitching, M. O., Colacot, T. J., Snieckus, V. (2012). Palladium-catalyzed cross-coupling: A historical contextual perspective to the 2010 Nobel Prize. *Angew. Chem. Int. Ed.* **51**, 5062.

6. (a) Littke, A. F., Fu, G. C. (2002). Palladium-catalyzed coupling reactions of aryl chlorides. *Angew. Chem. Int. Ed.* **41**, 4176. (b) Brennführer, A., Neumann, H., Beller, M. (2009). Palladium-catalyzed carbonylation reactions of aryl halides and related compounds. *Angew. Chem. Int. Ed.* **48**, 4114. (c) Buchwald, S. L. (2008). Cross coupling. *Acc. Chem. Res.* **41**, 1439. (d) Corbet, J.-P., Mignani, G. (2006). Selected patented cross-coupling reaction technologies. *Chem. Rev.* **106**, 2651.

7. Steenland, K., Fletcher, T., Savitz, D. A. (2010). *Environmental Health Perspectives* **118**, 1100.

8. (a) O'Hagan, D. (2008). Understanding organofluorine chemistry. An introduction to the C–F bond. *Chem. Soc. Rev.* **37**, 308. (b) Lemal, D. M. (2004). Perspective on fluorocarbon chemistry. *J. Org. Chem.* **69**, 1.

9. (a) Kiso, Y., Tamao, K., Kumada, M. (1973). Effects of the nature of halides on the alkyl group isomerization in the nickel-catalyzed cross-coupling of secondary alkyl Grignard reagents with organic halides. *J. Organomet. Chem.* **50**, C12. (b) Tamao, K., Sumitani, K., Kiso, Y., Zembayashi, M., Fujioka, A., Kodama, S., Nakajima, I., Minato, A., Kumada, M. (1976) Nickel–Phosphine complex-catalyzed Grignard coupling. I. Cross-coupling of alkyl, aryl, and alkenyl Grignard reagents with aryl and alkenyl halides: General scope and limitations. *Bull. Chem. Soc. Jpn.* **49**, 1958.

10. Böhm, V. P. W., Gstttmayr, C. W. K., Weskamp, T., Herrmann, W. A. (2001). Catalytic C–C bond formation through selective activation of C–F bonds. *Angew. Chem., Int. Ed.* **40**, 3387.

11. Inamoto, K., Kuroda, J.-I., Sakamoto, T., Hiroya, K. (2007). Catalytic activities of a bis(carbene)-derived nickel(II)-Pincer complex in Kumada–Tamao–Corriu and Suzuki–Miyaura coupling reactions for the synthesis of biaryl compounds. *Synthesis,* 2853.

12. (a) Braun, T., Perutz, R. N., Sladek, M. (2001). Catalytic C–F activation of polyfluorinated pyridines by nickel-mediated cross-coupling reactions. *Chem. Commun.,* 2254. (b) Steffen, A., Sladek, M. I., Braun, T., Neumann, B., Stammler, H.-G. (2005). Catalytic C–C coupling reactions at nickel by C–F activation of a pyrimidine in the presence of a C–Cl bond: The crucial role of highly reactive fluoro complexes. *Organometallics* **24**, 4057.

13. Mongin, F., Mojovic, L., Guillamet, B., Trécourt, F., Quéguiner, G. (2002). Cross-coupling reactions of phenylmagnesium halides with fluoroazines and fluorodiazines. *J. Org. Chem.* **67**, 8991.

14. Saeki, T., Takashima, Y., Tamao, K. (2005). Nickel- and palladium-catalyzed cross-coupling reaction of polyfluorinated arenes and alkenes with Grignard reagents. *Synlett,* 1771.

15. Yoshikai, N., Mashima, H., Nakamura, E. (2005). Nickel-catalyzed cross-coupling reaction of aryl fluorides and chlorides with Grignard reagents under nickel/magnesium bimetallic cooperation. *J. Am. Chem. Soc.* **127**, 17978.

16. Dankwardt, J. W. (2005). Transition metal catalyzed cross-coupling of aryl Grignard reagents with aryl fluorides via Pd- or Ni-activation of the C–F bond: An efficient synthesis of unsymmetrical biaryls—Application of microwave technology in ligand and catalyst screening. *J. Organomet. Chem.* **690**, 932.

17. (a) Ackermann, L., Born, R., Spatz, J. H., Meyer, D. (2005). Efficient Aryl–(Hetero)aryl coupling by activation of C–Cl and C–F bonds using nickel complexes of air-stable phosphine oxides. *Angew. Chem., Int. Ed.* **44**, 7216. (b) Ackermann, L., Born, R., Spatz, J. H., Althammer, A., Gschrei, C. J. (2006). Air-stable phosphine oxides as preligands for catalytic activation reactions of C–Cl, C–F, and C–H bonds. *Pure Appl. Chem.* **78**, 209.

18. (a) Inada, K., Miyaura, N. S (2000). Synthesis of biaryls via cross-coupling reaction of arylboronic acids with aryl chlorides catalyzed by $NiCl_2$/triphenylphosphine complexes. *Tetrahedron* **56**, 8657. (b) Zhou, J., Fu, G. C. (2004). Suzuki cross-couplings of unactivated secondary alkyl bromides and iodides. *J. Am. Chem. Soc.* **126**, 1340.

19. Schaub, T., Backes, M., Radius, U. (2006). Catalytic C–C bond formation accomplished by selective C–F activation of perfluorinated arenes. *J. Am. Chem. Soc.* **128**, 15964.

20. (a) Baudoin, O. (2007). New approaches for decarboxylative biaryl coupling. *Angew. Chem. Int. Ed.* **46**, 1373. (b) Gooßen, L. J., Rodríguez, N., Gooßen, K. (2008). Carboxylic acids as substrates in homogeneous catalysis. *Angew. Chem. Int. Ed.* **47**, 3100.

21. Zhang, Y. D., Rovis, T. (2004). A unique catalyst effects the rapid room-temperature cross-coupling of organozinc reagents with carboxylic acid fluorides, chlorides, anhydrides, and thioesters. *J. Am. Chem. Soc.* **126**, 15964.

22. (a) Yeom, C.-E., Kim, H. W., Lee, S. Y., Kim, B. M. (2007). DBU-mediated mild and chemoselective deprotection of aryl silyl ethers and tandem biaryl ether formation. *Synlett*, 146. (b) Jones, L. H.; Randall, A., Barba, O., Selby, M. D. (2007). Synthetic chemistry-led creation of a difluorinated biaryl ether non-nucleoside reverse transcriptase inhibitor. *Org. Biomol. Chem.* **5**, 3431. (c) Behniafar, H., Sedaghatdoost, M. (2011). New fluorinated aromatic poly(ether-amide)s derived from 2,2′-bis(3,4,5-trifluorophenyl)-4,4′-diaminodiphenyl ether and various dicarboxylic acids. *J. Fluorine Chem.* **132**, 276.

23. Zhang, J., Wu, J. J., Xiong, Y., Cao, S. (2012). Synthesis of unsymmetrical biaryl ethers through nickel-promoted coupling of polyfluoroarenes with arylboronic acids and oxygen. *Chem. Commun.* **48**, 8553.

24. Tobisu, M., Xu, T., Shimasaki, T., Chatani, N. (2011). Nickel-catalyzed Suzuki–Miyaura reaction of aryl fluorides. *J. Am. Chem. Soc.* **133**, 19505.

25. (a) Widdowson, D. A., Wilhelm, R. (1999). Palladium catalysed cross-coupling of (fluoroarene)tricarbonylchromium(0) complexes. *Chem. Commun.* 2211. (b) Wilhelm, R., Widdowson, D. A. (2000). Palladium catalysed cross-coupling of (fluoroarene)tricarbonylchromium(0) complexes. *J. Chem. Soc. Perkin Trans.* **1**, 3808.

26. Sun A. D., Love, J. A. (2010). Cross coupling reactions of polyfluoroarenes via C–F activation. *Dalton Trans.* **39**, 10362.

27. Clot, E., Eisenstein, O., Jasim, N., MacGregor, S. A., McGrady, J. E., Perutz, R. N. (2011). C–F and C–H bond activation of fluorobenzenes and fluoropyridines at transition metal centers: How fluorine tips the scales. *Acc. Chem. Res.,* **44**, 333.

28. (a) Braun, T., Perutz, R. N., Sladek, M. I. (2001). Catalytic C–F activation of polyfluorinated pyridines by nickel-mediated cross-coupling reactions. *Chem. Commun.* **2254**. (b) Guihaume, J., Clot, E., Eisenstein, O., Perutz, R. N. (2010). Importance of palladium–carbon bond energies in direct arylation of polyfluorinated benzenes. *Dalton Trans.* **39**, 10510.

29. Terao, J., Lkumi, A., Kuniyasu, H., Kambe, N. (2003). Ni- or Cu-catalyzed cross-coupling reaction of alkyl fluorides with Grignard reagents. *J. Am. Chem. Soc.* **125**, 5646.

30. Terao, J., Watabe, H., Kambe, N. (2005). Ni-catalyzed alkylative dimerization of vinyl Grignard reagents using alkyl fluorides. *J. Am. Chem. Soc.* **127**, 3656.

31. Tsuji, J. (2004). *Palladium Reagents and Catalysts: New Perspectives for the 21st Century,* Wiley, Chichester, U.K..

32. Kiso, Y., Tamao, K., Kumada, M. (1973). Effects of the nature of halides on the alkyl group isomerization in the nickel-catalyzed cross-coupling of secondary alkyl Grignard reagents with organic halides. *J. Organomet. Chem.* **50**, C12–C14.

33. Saeki, T., Takashima, Y., Tamao, K. (2005). Nickel- and Palladium-catalyzed cross-coupling reaction of polyfluorinated arenes and alkenes with Grignard reagents. *Synlett*, 1771.

34. Dankwardt, J. W. (2005). Transition metal catalyzed cross-coupling of aryl Grignard reagents with aryl fluorides via Pd- or Ni-activation of the C–F bond: An efficient synthesis of unsymmetrical biaryls—Application of microwave technology in ligand and catalyst screening. *J. Organomet. Chem.* **690**, 932.

35. Manabe, K., Ishikawa, S. (2008). Ortho-selective Cross-coupling of fluorobenzenes with Grignard reagents: Acceleration by electron-donating ortho-directing groups. *Synthesis*, 2645.

36. (a) Widdowson, D. A., Wilhelm, R. (2003). Palladium catalysed Suzuki reactions of fluoroarenes. *Chem. Commun.*, 578. (b) Mikami, K., Miyamoto, T., Hatano, M. (2004). A highly efficient asymmetric Suzuki–Miyaura coupling reaction catalyzed by cationic chiral palladium(II) complexes. *Chem. Commun.*, 2082. (c) Kim, Y. M., Yu, S. (2003). Palladium(0)-catalyzed amination, Stille coupling, and Suzuki coupling of electron-deficient aryl fluorides. *J. Am. Chem. Soc.* **125**, 1696. (d) Bahmanyar, S., Borer, B. C., Kim, Y. M., Kurtz, D. M., Yu, S. (2005). Proximity effects in the palladium-catalyzed substitution of aryl fluorides. *Org. Lett.* **7**, 1011.

37. Pigeon, X., Bergeron, M., Barabe, F., Dube, P., Frost, H. N., Paquin, J. F. (2010). Activation of allylic C–F bonds: Palladium-catalyzed allylic amination of 3,3-difluoropropenes. *Angew. Chem., Int. Ed.* **49**, 1123.

38. Weaver, J. D., Recio, III, A., Grenning, A. J., Tunge, J. A. (2011). Transition Metal-Catalyzed Decarboxylative Allylation and Benzylation Reactions. *Chem. Rev.* **111**, 1846.

39. Yu, D. H., Shen, Q. L., Lu, L. (2012). Selective palladium-catalyzed C–F activation/carbon–carbon bond formation of polyfluoroaryl oxazolines. *J. Org. Chem.* **77**, 1798.

40. (a) Breyer, D., Braun, T., Kläring, P. (2012). Synthesis and reactivity of the fluoro complex *trans*-[Pd(F)(4-C_5NF_4)($iPr_2PCH_2CH_2OCH_3)_2$]: C–F bond formation and catalytic C–F bond activation reactions. *Organometallics* **31**, 1417. (b) Anderson, C. M., Puddephatt, R. J., Ferguson, G., Lough, A. J. (1989). Oxidative addition of aryl–halogen bonds to platinum(II) and the structure of a complex formed by aryl–fluoride oxidative addition. *J. Chem. Soc., Chem. Commun.*, 1297. (c) Anderson, C. M., Crespo, M., Ferguson, G., Lough, A. J., Puddephatt, R. J. (1992). Activation of aromatic carbon-fluorine bonds by organoplatinum complexes. *Organometallics* **11**, 1177.

41. (a) Crespo, M., Martinez, M., Sales, J. (1992). Intramolecular activation of a C–F bond at platinum(II) in the presence of weaker C–X bonds (X = H, Cl and Br). *J. Chem. Soc., Chem. Commun.*, 822. (b) Crespo, M., Martinez, M., Sales, J. (1993). Effect of fluorine substituents in intramolecular activation of carbon–fluorine and carbon–hydrogen bonds by platinum(II). *Organometallics* **12**, 4297. (c) Ceder, R. M., Granell, J., Muller, G., Font-Barda, M., Solans, X. J. (1993). Effect of fluorine substituents in intramolecular activation of carbon–fluorine and carbon–hydrogen bonds by platinum(II). *Organometallics* **12**, 4297.

42. Wang, T., Alfonso, B. J., Love, J. A. (2007). Platinum(II)-catalyzed cross-coupling of polyfluoroaryl imines. *Org. Lett.* **9**, 5629.

43. Buckley, H. L., Wang, T. G., Tran, O., and Love, J. A. (2009). Selective platinum-catalyzed C–F bond activation as a route to fluorinated aryl methyl ethers. *Organometallics* **28**, 2356.

44. Cahoez, G., Lepifre, F., Ramiandrasoa, P. (1999) Manganese-catalyzed substitution of activated aryl halides (X=Cl, Br and F) and aryl ethers by organomagnesium reagents. *Synthesis*, 2138.

45. (a) Korn, T. J., Schade, M. A., Wirth, S., Knochel, P. (2006). Cobalt(II)-catalyzed cross-coupling between polyfunctional arylcopper reagents and aryl fluorides or tosylates. *Org. Lett.* **8**, 725. (b) Korn, T. J., Schade, M. A., Cheemala, M. N., Wirth, S., Guevara, S. A., Cahiez, G., Knochel, P. (2006). Cobalt-catalyzed cross-coupling reactions of heterocyclic chlorides with arylmagnesium halides and of polyfunctionalized arylcopper reagents with aryl bromides, chlorides, fluorides and tosylates. *Synthesis*, 3547.
46. Miura, T., Ito, Y., Murakami, M. (2008). Synthesis of *gem*-difluoroalkenes via β-fluoride elimination of organorhodium(I). *Chem. Lett.* **37**, 1006.
47. Arisawa, M.; Suzuki, T.; Ishikawa, T. Yamaguchi, M. (2008). Rhodium-catalyzed substitution reaction of aryl fluorides with disulfides: *p*-Orientation in the polyarylthiolation of polyfluorobenzenes. *J. Am. Chem. Soc.* **130**, 12214.
48. Yu, C. M., Zhang, C. L., Shi, X. J. (2012). Copper-catalyzed direct thiolation of pentafluorobenzene with diaryl disulfides or aryl thiols by C–H and C–F bond activation. *Eur. J. Org. Chem.,* 1953.
49. Ishii, Y., Chatani, N., Yorimitsu, S., Murai, S. (1998). Rhodium-catalyzed Si–F exchange reaction between fluorobenzenes and a disilane. Catalytic reaction involving cleavage of C–F bonds. *Chem. Lett.* **27**, 157.
50. Teltewskoi, M., Panetier, J. A., Macgregor, S. A., Braun, T. (2010). A highly reactive rhodium(I)–boryl complex as a useful tool for C–H bond activation and catalytic C–F bond borylation. *Angew. Chem., Int. Ed.* **49**, 3947.
51. Guo, H., Kong, F., Kanno, K., He, J., Nakajima, K., Takahashi, T. Early (2006). Transition metal-catalyzed cross-coupling reaction of aryl fluorides with a phenethyl Grignard reagent accompanied by rearrangement of the phenethyl group. *Organometallics* **25**, 2045.
52. Terao, J., Begum, S. A., Oda, A., Kambe, N. (2005). Zr-catalyzed coupling reaction of alkyl halides, tosylates, and sulfates with β-phenethyl Grignard reagents via styrene–zirconate intermediates. *Synlett*, 1783.
53. Terao, J., Watabe, H., Kambe, N. (2005). Ni-catalyzed alkylative dimerization of vinyl Grignard reagents using alkyl fluorides. *J. Am. Chem. Soc.* **127**, 3656.
54. Yamada, S., Takahashi, T., Konno, T., Ishihara, T. (2007). A novel fluorine–metal exchange reaction of pentafluorocrotonate with organocuprate. Generation of β-metallated tetrafluorocrotonate and its cross-coupling reaction. *Chem. Commun.*, 3679.
55. Knebelkamp, A., Heitz, W. (1991). Synthesis of fluorostyrenes via palladium-catalyzed reactions of aromatic halides with fluoroolefins. *Makromol. Rapid Commun.* **12**, 69.
56. Borah, H. N., Prajapati, D., Boruah, R. C. (2005). A novel indium-catalyzed Sonogashira coupling reaction, effected in the absence of a copper salt, phosphine ligand and palladium. *Synlett*, 2823.
57. (a) Grecian, A. A., Hadida, S., Warren, S. D. (2005). Direct synthesis of trimethyl(2-phenoxyethyl)silanes from aromatic fluorides. *Tetrahedron Lett.* **46**, 4683. (b) Cogolli, P., Maiolo, F., Testaferri, L., Tingoli, M., Tiecco, M. (1979). Nucleophilic aromatic substitution reactions of unactivated aryl halides with thiolate ions in hexamethylphosphoramide. *J. Org. Chem.* **44**, 2642. (c) Katagiri, K., Danjo, H., Yamaguchi, K. (2005). Nucleophilic aromatic substitution reactions of fluorobenzenechromium complexes with P-chiral secondary phosphine–boranes: Synthesis of optically pure P-chiral (dialkyl) arylphosphine–boranes. *Tetrahedron* **61**, 4701. (d) Thomas, S., Collins, C. J., Cuzens, J. R., Spiciarich, D., Goralski, C. T., Singaram, B. (2001). Novel tandem S_NAr amination–reduction reactions of 2-halobenzonitriles with lithium *N,N*-dialkylaminoborohydrides. *J. Org. Chem.* **66**, 1999.

58. St. Jean, D. J., Jr., Poon, S. F., Schwarzbach, J. L. (2007). A tandem cross-coupling/S_NAr approach to functionalized carbazoles. *Org. Lett.* **9**, 4893.

59. Martin, O., (2009). *The Mizoroki–Heck Reaction*, John Wiley & Sons, Ltd., Chichester.

60. (a) Gharat, L. A., Martin, A. R. (1996). 5-(α-Fluorovinyl)-*N*-tosylindole. *Heterocycles* **43**, 185; (b) Gharat, L. A., Martin, A. R. (1996). 5-(α-Fluorovinyl) tryptamines and 5-(α-fluorovinyl)-3-(*N*-methyl-1′,2′,5′,6′,-tetrahydropyridin-3′-and -4′-yl)indoles–5-(α-fluorovinyl)tryptamines. *J. Heterocycl. Chem.* **33**, 197.

61. Miura, T., Ito, Y., Murakami, M. (2008). Synthesis of *gem*-difluoroalkenes via β-fluoride elimination of organorhodium(I). *Chem. Lett.* **37**, 1006;

62. Ichikawa, J., Wada, Y., Fujisawa, M., Sakoda, K. (2002). The nucleophilic 5-*endo-trig* cyclization of 1, 1-difluoro-1-alkenes: Ring-fluorinated hetero- and carbocycle synthesis and remarkable effect of the vinylic fluorines on the disfavored process. *Synthesis*, 1917.

63. (a) Sakoda, K., Mihara, J., Ichikawa, J. (2005). Heck-type 5-*endo-trig* cyclization promoted by vinylic fluorines: Synthesis of 5-fluoro-3*H*-pyrroles. *Chem. Commun.*, 4684. (b) Ichikawa, J., Sakoda, K., Mihara, J., Ito, N. (2006). Heck-type 5-*endo-trig* cyclizations promoted by vinylic fluorines: Ring-fluorinated indene and 3H-pyrrole syntheses from 1,1-difluoro-1-alkenes. *J. Fluorine Chem.* **127**, 489. (c) Ichikawa, J., Nadano, R., Ito, N. (2006). 5-*endo* Heck-type cyclization of 2-(trifluoromethyl)allyl ketone oximes: Synthesis of 4-difluoromethylene-substituted 1-pyrrolines. *Chem. Commun.*, 4425.

64. (a) Uneyama, K., Yang, F.-Y., Hirama, S., Katagiri, T. (1996). Tandem alkylation-defluorination reaction: Synthesis of 2-(*N*-alkyl–*N*-aryl) amino-3,3-difluoropropenoates from 2-(*N*-aryl) imino-3,3,3-trifluoropropanoates. *Tetrahedron Lett.* **37**, 2045. (b) Portella, C., Shermolovich, Y. (1997). Fluorinated ketenedithioacetals. 5. A new synthesis of perfluoroketenedithioacetals from alkyl perfluoroalkanedithiocarboxylates. *Tetrahedron Lett.* **38**, 4063. (c) Mae, M., Amii, H., Uneyama, K. (2000). First synthesis of 3,3-difluoroserine and cysteine derivatives via Mg(0)-promoted selective C–F bond cleavage of trifluoromethylimines. *Tetrahedron Lett.* **41**, 7893.

65. (a) Brigaud, T., Doussot, P., Portella, C. (1994). Synthesis of difluoroenoxysilanes from acylsilanes and trifluoromethyltrimethylsilane (TFMTMS). Dramatic effect of the catalytic fluoride source. *J. Chem. Soc., Chem. Commun.*, 2117. (b) Arisawa, M., Suzuki, T., Ishikawa, T., Yamaguchi, M. (2008). Rhodium-catalyzed substitution reaction of aryl fluorides with disulfides: *p*-Orientation in the polyarylthiolation of polyfluorobenzenes. *J. Am. Chem. Soc.* **130**, 12214

66. (a) MacDougall, C. T., Peach, M. E. (1987). *Sulfur Lett.* **7**, 15. (b) Jesudason, J. J., Peach, M. E. (1988). The thiolate anion as a nucleophile Part XIV. further reactions of tin(II) arenethiolates. *Journal of Fluorine Chemistry* **41**, 357.

67. Imamoto, T., Takeyama, T., Kusumoto, T. (1985). Facile reduction of organic halides and phosphine oxides with $LiAlH_4$–$CeCl_3$. *Chem. Lett.* **14**, 1491.

68. (a) Li, H., Liao, S., Xu, Y. (1996). Catalytic hydrogenolysis of aromatic halides with sodium hydride of nanometric size. *Chem. Lett.* **25**, 1059. (b) Zhang, Y., Liao, S., Xu, Y., Yu, D., Shen, Q. (1997). Reductive dehalogenation of aryl halides by the nanometric sodium hydride using lanthanide chloride as catalyst. *Synth. Commun.* **27**, 4327. (c) Fuchibe, K., Akiyama, T. (2004). Niobium-catalyzed reduction of monofluoroarenes with $LiAlH_4$. *Synlett*, 1282.

69. Narumi, T., Tomita, K., Inokuchi, E., Kobayashi, K., Oishi, S., Ohno, H., Fujii, N. (2007). Facile synthesis of fluoroalkenes by palladium-catalyzed reductive defluorination of allylic *gem*-difluorides. *Org. Lett.* **9**, 3465.

70. (a) Fuchibe, K., Akiyama, T. (2006). Low-valent niobium-mediated double activation of C–F/C–H bonds: Fluorene synthesis from *o*-arylated α,α,α-trifluorotoluene derivatives. *J. Am. Chem. Soc.* **128**, 1434. (b) Fuchibe, K., Mitomi, K., Akiyama, T. (2007). Low-valent niobium-mediated synthesis of indenes: Intramolecular coupling reaction of CF_3 group with alkene C–H bond. *Chem. Lett.* **36**, 24. (c) Fuchibe, K., Mitomi, K., Suzuki, R., Akiyama, T. (2008). C–C coupling reactions of superstrong CF^3 groups with $C(sp^2)$–H bonds: Reactivity and synthetic utility of zero-valent niobium catalyst. *Chem. Asian J.* **3**, 261.

71. (a) Fuchibe, K., Ohshima, Y., Mitomi, K., Akiyama, T. (2007). Low-valent niobium-catalyzed reduction of α,α,α-trifluorotoluenes. *Org. Lett.* **9**, 1497. (b) Fuchibe, K., Ohshima, Y., Mitomi, K., Akiyama, T. (2007). Hydrodefluorinations by low-valent niobium catalyst. *J. Fluorine Chem.* **128**, 1158.

72. Aizenberg, M., Milstein, D. (1994). Catalytic activation of carbon–fluorine bonds by a soluble transition metal complex. *Science* **265**, 359.

73. Tamejiro, H., (2000). *Organofluorine Compounds: Chemistry and Applications*, Springer, Berlin.

74. Peterson, A. A., McNeill, K. (2006). Catalytic dehalogenation of sp^2 C–F and C–Cl bonds in fluoro- and chloroalkenes. *Organometallics* **25**, 4938.

75. Young, R. J., Jr., Grushin, V. V. (1999). Catalytic C–F bond activation of nonactivated monofluoroarenes. *Organometallics* **18**, 294.

76. (a) Ukisu, Y., Miyadera, T. (1997). Hydrogen-transfer hydrodehalogenation of aromatic halides with alcohols in the presence of noble metal catalysts. *J. Mol. Catal. A: Chem.* **125**, 135.

77. (a) Kuhl, S. Schneider, R. Fort, Y. (2003). Catalytic carbon–fluorine bond activation with monocoordinated nickel–carbene complexes: Reduction of fluoroarenes. *Adv. Synth. Catal.* **345**, 341; (b) Desmarets, C., Kuhl, S., Schneider, R., Fort, Y. (2002). Nickel(0)/imidazolium chloride catalyzed reduction of aryl halides. *Organometallics* **21**, 1554.

78. (a) Adonin, N. Y., Starichenko, V. F. (2000). Regioselective *ortho*-hydrodefluorination of pentafluorobenzoic acid by low-valent nickel complexes. *J. Fluorine Chem.* **101**, 65. (b) Adonin, N. Y., Starichenko, V. F. (2000). Hydrogenolysis of C–F bonds in fluorinated aromatic hydrocarbons catalysed by nickel complexes. *Mendeleev Commun.*, 60.

79. Lv, H., Zhan, J.-H., Cai, Y.-B., Yu, Y., Wang, B., Zhang, J.-L. (2012). π–π Interaction assisted hydrodefluorination of perfluoroarenes by gold hydride: A case of synergistic effect on C–F bond activation. *J. Am. Chem. Soc.* **134**, 16216.

80. Black, W. C., Guay, B. (1998). Synthesis of sulfonyl flosulide. *Synthesis*, 1101.

81. Ikemoto, T., Ito, T., Nishiguchi, A., Miura, S., Tomimatsu, K. (2005). Practical synthesis of an orally active CCR5 antagonist, 7-{4-[2-(butoxy)-ethoxy]phenyl}-*N*-(4-{[methyl(tetrahydro-2*H*-pyran-4-yl)amino]methyl}phenyl)-1-propyl-2, 3-dihydro-1 *H*-1-benzazepine-4-carboxamide. *Org. Process Res. Dev.* **9**, 168.

82. (a) Inagaki, H., Takeda, T., Miyauchi, R. N., Kawakami, K., Takahashi, H., Takemura, M. (2004). Practical synthesis of DQ-113, a new quinolone antibacterial agent, by using the intramolecular Horner–Wadsworth–Emmons reaction. *Hetetrocycles* **63**, 699. (b) MacDonald, A., DeWitt, S., Hogan, E. M., Ramage, R. (1996). A solid phase approach to quinolones using the DIVERSOMER® technology. *Tetrahedron Lett.* **37**, 4815. (c) Barrett, D., Tsutsumi, H., Kinoshita, T., Murata, M., Sakane, K. (1995). Synthesis of new 1,8-bridged tricyclic quinolones by a novel intramolecular arylation of

N-1 tethered malonamides. *Tetrahedron* **51**, 11125. (d) Barrett, D., Sasaki, H., Tsutsumi, H., Murata, M., Terasawa, T., Sakene, K. (1995). A concise, practical synthesis of the pyrido[3, 2, 1-i, j]cinnoline ring system of potent DNA gyrase inhibitors. *J. Org. Chem.* **60**, 3928. (e) Zhang, Z.; Zhou, W.; Yu, A. (2004). Synthesis and antibacterial activity of 7-(substituted)aminomethyl quinolones. *Bioorg. Med. Chem. Lett.* **14**, 393 (f) Zhang, Z., Zhou, W. (2005). Arylation of nitromethane: Masked nucleophilic formylation of fluoroquinolones. *Tetrahedron Lett.* **46**, 3855. (g) Choi, D. R., Shin, J. H., Yang, J., Yoon, S. H., Jung, H. (2004). Syntheses and biological evaluation of new fluoroquinolone antibacterials containing chiral oxiimino pyrrolidine. *Bioorg. Med. Chem. Lett.* **14**, 1273.

83. (a) Hayakawa, I., Hiramitsu, T., Tanaka, Y. (1984). Synthesis and antibacterial activities of substituted 7-oxo-2, 3-dihydro-7*H*-pyrido[1, 2, 3-*de*]-1,4-benzoxazine-6-carboxylic acids. *Chem. Pharm. Bull.* **32**, 4907. (b) Atarashi, S., Yokohama, S., Yamazaki, K., Sakano, K., Imamura, M., Hayakawa, I. (1987). Synthesis and antibacterial activities of optically active ofloxacin and its fluoromethyl derivative. *Chem. Pharm. Bull.* **35**, 1896. (c) Todo, Y., Takagi, H., Iino, F., Hayashi, K., Takata, M., Kuroda, H.; Momonoi, K.; Narita, H. (1994). Practical synthesis of T-3761, (*S*)-10-(1-Aminocyclopropyl)-9-fluoro-3-methyl-7-oxo-2,3-dihydro-7*H*-pyrido[1,2,3-*de*]-1,4-benzoxazine-6-carboxylic acid. *Chem. Pharm. Bull.* **42**, 2629. (d) Miyamoto, T., Matsumoto, J., Chiba, K., Egawa, H., Shibamori, K.; Minamida, A., Nishimura, Y., Okada, H., Kataoka, M., Fujita, M., Hirose, T., Nakano, J. (1990). Pyridonecarboxylic acids as antibacterial agents. Part 14. Synthesis and structure-activity relationships of 5-substituted 6,8–difluoroquinolones, including sparfloxacin, a new quinolone antibacterial agent with improved potency. *J. Med. Chem.* **33**, 1645. (e) Shibamori, K., Egawa, H., Miyamoto, T., Nishimura, Y., Itokawa, A., Nakano, J., Matsumoto, R. (1990). Regioselective displacement reactions of 1-cyclopropyl-5,6,7,8-tetrafluoro-4(1*H*)-oxoquinoline-3-carboxylic acid with amine nucleophiles. *J. Chem. Pharm. Bull.* **38**, 2390.

84. (a) Rama Rao, A. V., Gurjar, M. K., Reddy, K. L., Rao, A. S. (1995). Studies directed toward the synthesis of vancomycin and related cyclic peptides. *Chem. Rev.* **95**, 2135. (b) Beugelmans, R., Bigot, A., Bois-Choussy, M., Zhu, J. (1996). A new approach to the synthesis of piperazinomycin and bouvardin: Facile access to cycloisodityrosine via an intramolecular S_NAr reaction. *J. Org. Chem.* **61**, 771 (c) Rama Rao, A. V., Gurjar, M. K., Lakshmipathi, P., Reddy, M. M., Nagarajan, M., Pal, S., Sarma, B. V. N. B. S., Tripathy, N. K. (1997). S_NAr macrocyclisation: A new approach towards the synthesis of D-O-E-segment of vancomycin. *Tetrahedron Lett.* **38**, 7433.

85. Soorukram, D., Qu, T., Barrett, A. G. M. (2008). Four-component benzyne coupling reactions: A concise total synthesis of dehydroaltenuene B. *Org. Lett.* **10**, 3833.

86. (a) Morton, G. E., Barrett, A. G. M. (2005). 1,4-Difluoro-2,5-dimethoxybenzene as a precursor for iterative double benzyne–furan Diels–Alder reactions. *J. Org. Chem.* **70**, 3525. (b) Hart, H., Lai, C.-Y., Nwokogu, G., Shamouilian, S., Teuerstein, A., Zlotogorski, C. (1980). Bisannelation of arenes with bis-aryne equivalents. *J. Am. Chem. Soc.* **102**, 6649. (c) Hart, H., Nwokogu, G. C. (1983). The dilithiation mechanism in the reactions of polyhaloarenes as di-aryne equivalents. *Tetrahedron Lett.* **24**, 5721. (d) Hart, H., Raju, N., Meador, M. A., Ward, D. L. (1983). Synthesis of heptiptycenes with face-to-face arene rings via a 2,3:6,7-anthradiyne equivalent. *J. Org. Chem.* **48**, 4357.

87. Larrosa, I., Da Silva, M. I., Gomez, P. M., Hannen, P., Ko, E., Lenger, S. R., Linke, S. R., White, A. J. P., Wilton, D., Barrett, A. G. M. (2006). Highly convergent three component benzyne coupling: The total synthesis of *ent*-clavilactone B *J. Am. Chem. Soc.* **128**, 14042.
88. Yamauchi, Y., Fukuhara, T., Hara, S., Senboku, H. (2008). Electrochemical carboxylation of α,α-difluorotoluene derivatives and its application to the synthesis of α-fluorinated nonsteroidal anti-inflammatory drugs. *Synlett*, 438.
89. Yokoyama, M., Toyoshima, H., Shimizu, M., Mito, J., Togo, H. (1998). Simple synthesis of aromatic β-C-nucleosides via coupling of aryl Grignard reagents with sugar fluorides. *Synthesis*, 409.

CHAPTER 3

HOMOGENEOUS TRANSITION-METAL CATALYZED C–N ACTIVATION

XIAO-BING WAN
College of Chemistry, Chemical Engineering and Materials Science, Soochow University, Suzhou, People's Republic of China

3.1 BACKGROUND

Recently, transition-metal-catalyzed C–N bond cleavage has become an interesting research subject in organometallic chemistry. For the functionalization of C–N bonds, it was mainly restricted to diazonium salts ($Ar{-}N_2X$), ammonium salts, triazenes, and imidazoles. A review of transition-metal-catalyzed C–N bond cleavage reactions is presented here, with particular emphasis on the reaction conditions, including transition-metal-catalysts, bases, and side reactions for achieving selective coupling. As previous reviews have included studies on palladium-catalyzed cross-coupling reactions using diazonium salts as substrates [1, 2], new developments

Homogeneous Catalysis for Unreactive Bond Activation, First Edition. Edited by Zhang-Jie Shi.

during the period from 2000 to the end of 2012 are mainly discussed herein, and this will, in part, overlap previously published, related articles.

3.2 PALLADIUM-CATALYZED C–N ACTIVATION

The group of Ozerov has prepared several new *N*-methylated diarylamine-based PNP pincer ligands [3]. The reactions of PN(Me)P ligands with PdX_2 (X = OAc, Cl) produced (PN(Me)P)–PdX_2, which may exist in either the neutral or the ionic forms (Scheme 3.1). These complexes are unreactive in the case of PPh_2-bearing ligands, but with the more donating $PPr^i{}_2$-bearing ligands, they dissociate into square planar (PNP)PdX with irreversible loss of MeX under room temperature. Notably, replacement of one or both chlorides by a weakly coordinating triflate stops the C–N cleavage. After extensive mechanistic investigations, the oxidative addition process is ruled out and the association of X^- is involved in the C–N cleavage. The overall ease of the C–N cleavage can be partly ascribed to the rigidity of the pincer backbone favoring a meridionally bound anionic PNP ligand.

Saeki et al. [4] found that aryltriazenes could be directly coupled with areneboronic acids in the presence of $Pd_2(dba)_3$, $P(tBu)_3$, and $BF_3{\cdot}OEt_2$ in DME,

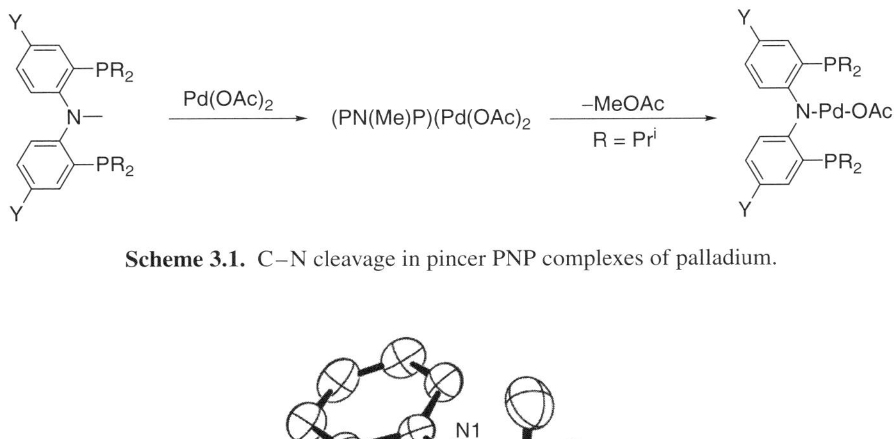

Scheme 3.1. C–N cleavage in pincer PNP complexes of palladium.

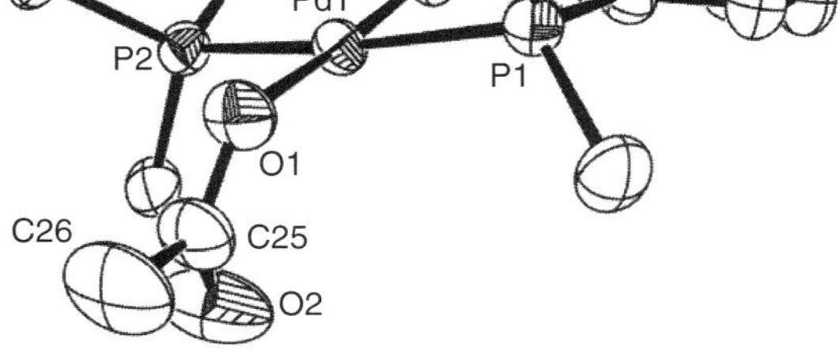

Scheme 3.2. ORTEP drawing (50% probability ellipsoids) of (FPNPiPr)PdOAc.

$$\text{Ar–N=N–NR}_2 + \text{Ar}^1\text{–B(OH)}_2 \xrightarrow[\text{DME, rt}]{\text{BF}_3\cdot\text{OEt}_2\text{ (1.0 equiv)},\ \text{cat.Pd(0)/P(}t\text{Bu)}_3} \text{Ar–Ar}^1$$

up to 98% yield

Scheme 3.3. BF_3 induced Pd-catalyzed cross-coupling reaction of 1-aryltriazenes with boronic acids.

resulting in biaryl products in up to 98% yield (Scheme 3.3). At present, the actual reaction mechanism is not yet clear. The method with a nitrogen-based leaving group has a potential utility in synthetic chemistry, providing a simple, halogen-free, and practical cross-coupling process under mild conditions.

The scope of the transformation with respect to the substituent on the aryl ring of 1-aryltriazenes and boronic acids was shown in Table 3.1. Aryltriazenes bearing electron-donating groups at the *ortho*, *meta*, and *para* positions and a sterically bulky mesityl and naphthyl groups smoothly underwent the desired coupling reaction (entries 1–5 and 13). On the other hand, electron-withdrawing groups including halogens, triflate, and acetyl substituents decreased the yields (entries 6–11). It is

TABLE 3.1. Cross-Coupling Reaction of 1-Aryltriazines with Boronic Acids[a]

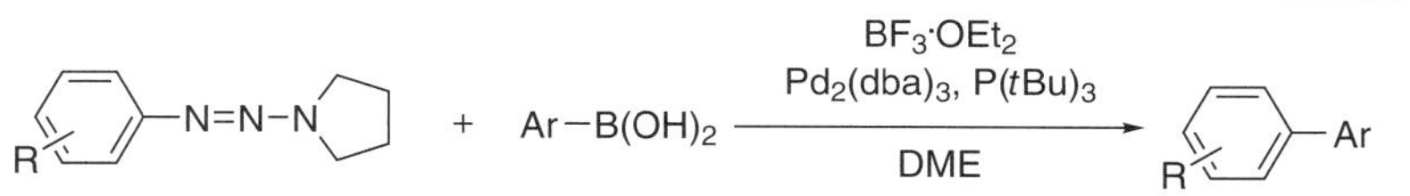

Entry	R	Ar	Yield (%)[b] 10 min, 1 h
1	*o*-Me	*p*-MeO-C_6H_4 (**2**)	81, 91(72)[d]
2	*m*-Me	*p*-MeO-C_6H_4 (**2**)	88, 92(80)[d]
3	*p*-Me (**1**)	*p*-MeO-C_6H_4 (**2**)	98, 98(91)[d]
4	*p*-MeO	*p*-Me-C_6H_4	94, 94
5	2, 4, 6-Me_3	*p*-MeO-C_6H_4 (**2**)	85, 90
6	*p*-F	*p*-MeO-C_6H_4 (**2**)	73, 81
7	*p*-Cl	*p*-MeO-C_6H_4 (**2**)	64, 74(68)[d]
8[c]	*p*-Br	*p*-MeO-C_6H_4 (**2**)	30, 40
9[c]	*p*-l	*p*-MeO-C_6H_4 (**2**)	trace, trace
10	*p*-TfO	*p*-MeO-C_6H_4 (**2**)	43, 52(41)[d]
11	*p*-MeCO	*p*-MeO-C_6H_4 (**2**)	31, 58
12	*p*-Et_2N	*p*-MeO-C_6H_4 (**2**)	97, 97
13	*p*-Me (**1**)	1-naphthyl	88, 90(78)[d]
14	*p*-Me (**1**)	2-thienyl	94, 94(84)[d]
15	*p*-Me (**1**)	2-furyl	0, 0

[a]Reaction conditions: 1-aryltriazene (0.50 mmol), boronic acid (1.0 mmol), $Pd_2(dba)_3$ (2 mol%), $P(^tBu)_3$ (8 mol%), $BF_3\cdot Et_2O$ (0.50 mmol), and DME (5.0 mL) under room temperature.

[b]Yields are determined by GC with eicosane as an internal standard.

[c]5 mol% of $Pd_2(dba)_3$ was used without phosphine ligand.

[d]Isolated yields.

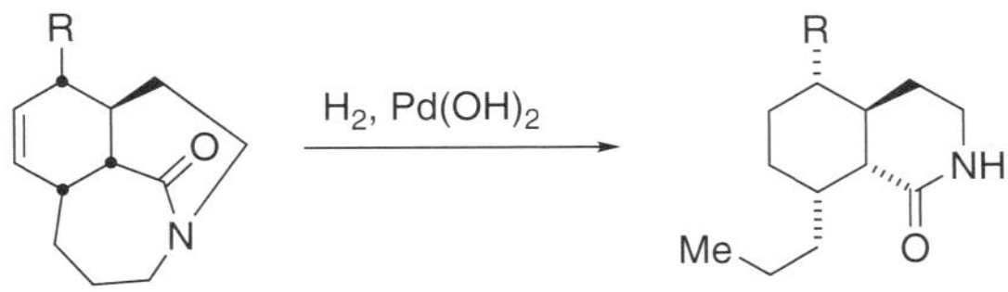

Scheme 3.4. C–N cleavage in bridged lactams.

noteworthy that the triazene moiety exhibited a higher chemoselectivity over the triflate group and the bromine atom as a leaving group under the optimized reaction conditions, leading to the corresponding products in moderate yields (entries 8 and 10). The inertness of the iodophenyltriazene remained was obscure (entry 9). Notably, the electron-rich substituents such as the methoxy, diethylamino, and thienyl groups were also unaffected under the standard conditions (entries 4, 12, and 14), whereas the furyl group displayed a lack of tolerance (entry 15).

In 2005, Aubé and co-workers [5] developed interesting C–N bond cleavage of a variety of strained bi- and tricyclic amides. This bond undergoes convenient breaking in thr presence of $H_2/Pd(OH)_2$, MeI, and DDQ. Notably, the transformation is highly regioselective and lead to breaking the C–N bond that deviate the farthest from its natural planar state (Scheme 3.4). In addition, preliminary experiments that elucidate the mechanisms of these reactions are also described.

Recently, Chen and co-workers [6] reported cleavage of a C–N bond of 3,6-(*N*-*n*-butylimidazolidenyl)pyridazine salt in the presence of Ag_2O and $Pd(OAc)_2$ (Scheme 3.5). Palladation of the resultant *N*-*n*-butylimidazole led to the isolation of two novel dinuclear palladacyclic complexes. Notably, both complexes consist of a $Pd_2C_2N_2$ core, with two palladium atoms doubly bridged by anionic imidazoles in a N,C_5- or N,C_2-coordination fashion. Accompanied by C–N bond cleavage and palladation processes, the addition of an imidazole C–H bond to the C≡N bond of acetonitrile took place in the reaction of 3,6-(*N*-*n*-butylimidazolidenyl)pyridazine and $Pd(OAc)_2$. All compounds have been characterized by spectroscopic techniques, and X-ray molecular structures are also described. Finally, both palladacyclic compounds show good catalytic activities in Heck and Suzuki coupling reactions of activated aryl bromides.

In 2009, Labinger and Bercaw developed a reaction of glyoxal-derived *a*-diimines with palladium acetates, leading to cleavage of the C–N bond in nitromethane (Scheme 3.6) [7]. Palladium nitro complexes were generated in this transformation in which the *a*-diimine ligand was methylated. Notably, the proposed mechanism remains speculative due to the absence of any characterizable intermediates. Alternative mechanisms, such as oxidative addition of the C–N bond to give a Pd(IV) intermediate, cannot be excluded.

Recently, Loh and co-workers [8] in their pioneering work, developed the first palladium catalyzed C–N bond cleavage of arylhydrazines to construct a C–C bond with olefins (Scheme 3.7). Notably, this transformation proceeded smoothly with various aryl hydrazines and acrylates as well as unactivated olefins under mild reaction conditions. With commercially available aryl hydrazines and alkenes as coupling partners

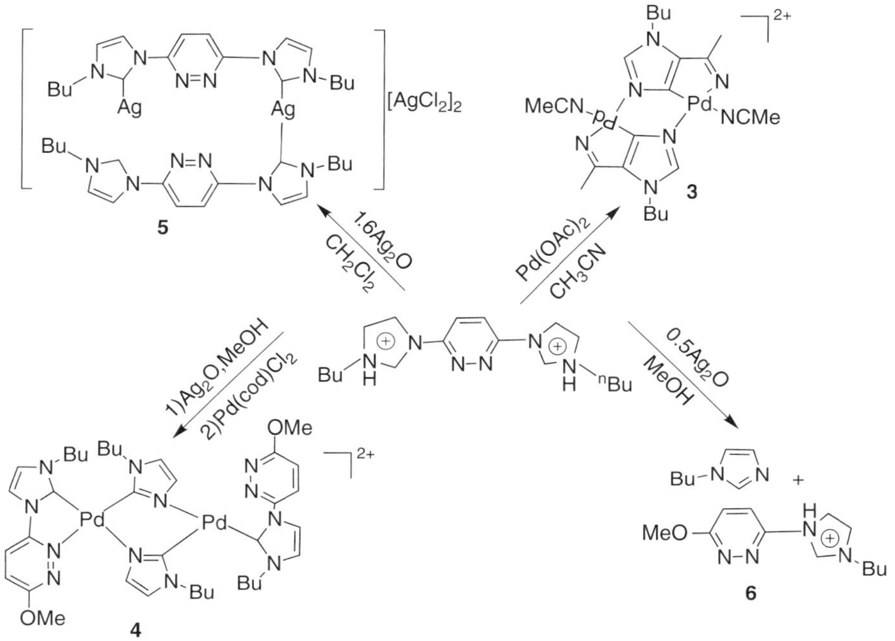

Scheme 3.5. Dinuclear palladacyclic complexes from C–N activation of an imidazolium salt.

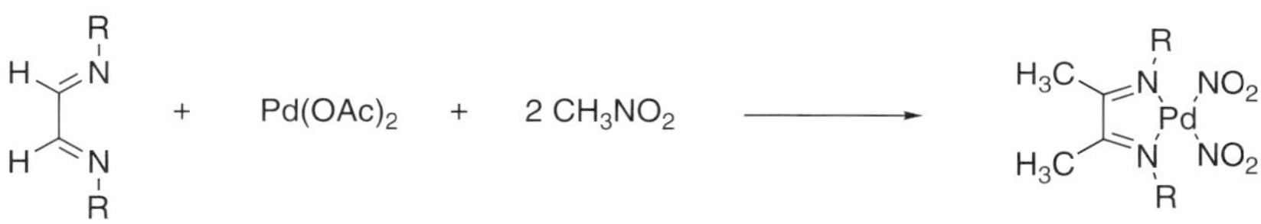

Scheme 3.6. Cleavage of the C–N bond in nitromethane by Pd–α-diimine complexes.

and 1 atm of air as the environmentally benign oxidant, this practical methodology will be very attractive for both academia and industry. This study offered a new mode of Pd–C bond formation, which *in situ* generated an aryl palladium complex that readily participates in the C–C bond formation with olefins.

A plausible catalytic cycle was proposed to elucidate the C–N bond cleavage process (Scheme 3.8). Both the C–N and N–N bonds in palladiaziridine complex **I** were activated by palladium (**II**). The oxidative addition of Pd(0) to the initially formed palladiaziridine complex I cleaved the C–N bond to give the two palladium(II) centered complexes **II**. Protonolysis of complex **II** formed the aryl palladium complex **IV** and the palladiaziridine complex **III**, which decomposed to give palladium(0), nitrogen gas, and water in the presence of air. The complex **IV** stabilized by ligand is the major peak in the absence of olefin. However, it could be further transformed

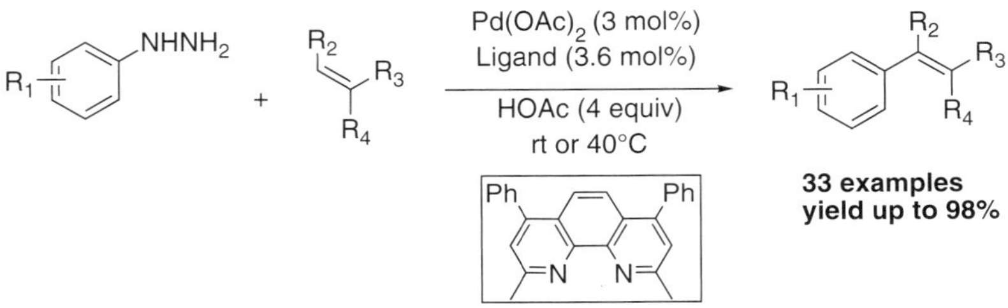

Scheme 3.7. Pd-catalyzed C–C bond formation of arylhydrazines with olefins via C–N bond cleavage.

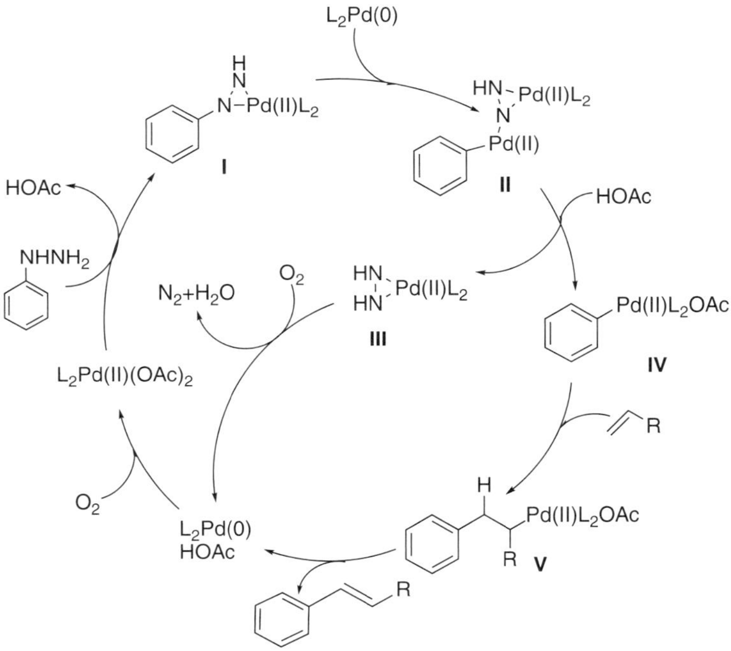

Scheme 3.8. Proposed catalytic cycle.

to intermediate **V** in the presence of olefin. Degeneration of intermediate **V** formed a coupling product and regenerated [$L_2Pd(0)$]. Finally, Pd(0) was oxidized to Pd(II) in air to complete the catalytic cycle with the aid of a ligand.

The generality of this new process was examined with a variety of arylhydrazines under the standard reaction conditions as shown in Table 3.2. A range of arylhydrazines formed the desired products in good yields, including *ortho*-, *para*-, and *meta*-substituted arylhydrazines. Both the electron-poor and -rich functional groups on the *para*-substituted arylhydrazines are tolerated (Table 3.2, entries 2–11). Generally, electron-withdrawing arylhydrazines are more reactive than electron-donating

TABLE 3.2. Pd-Catalyzed Cross-Coupling of Arylhydrazines with *tert*-Butyl Acrylate (8a)[a]

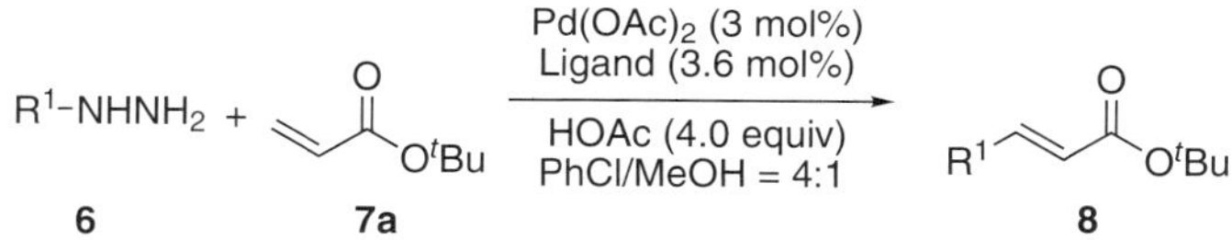

Entry	R^1	T (°C)	Time (h)	Yield (%)[b]
1	Ph(**6a**)	40	2	91% (**8a**)
2	4-MeO-C_6H_4 (**6b**)	40	6	90% (**8b**)
3	4-Me-C_6H_4 (**6c**)	40	2	92% (**8c**)
4	4-F-C_6H_4 (**6d**)	40	2	85% (**8d**)
5	4-Cl-C_6H_4 (**6e**)	40	2	85% (**8e**)
6	4-Br-C_6H_4 (**6f**)	40	2	81% (**8f**)
7	4-I-C_6H_4 (**6g**)	25	10	80% (**8g**)
8	4-CN-C_6H_4 (**6h**)	40	3	93% (**8h**)
9	4-CO_2H-C_6H_4 (**6i**)	25	8	98% (**8i**)
10	4-CF_3-C_6H_4 (**6j**)	25	8	85% (**8j**)
11	4-SO_2NH_2-C_6H_4 (**6k**)	25	8	85% (**8k**)
12	3-Me-C_6H_4 (**6l**)	40	3	93% (**8l**)
13	3-Br-C_6H_4 (**6m**)	40	4	93% (**8m**)
14	3-CO_2H-C_6H_4 (**6n**)	25	8	94% (**8n**)
15	2-Me-C_6H_4 (**6o**)	40	3	73% (**8o**)
16	2-NO_2-C_6H_4 (**6p**)	40	4	75% (**8p**)
17	3,4-$(MeO)_2$-C_6H_3 (**6q**)	40	3	88% (**8q**)
18	3-Cl-4-F-C_6H_3 (**1r**)	40	2	82% (**8r**)
19	3,5-Cl_2-C_6H_3 (**6s**)	40	3	85% (**8s**)
20	2,5-F_2-C_6H_3 (**6t**)	40	4	73% (**8t**)

[a]Unless noted otherwise, the reactions were carried out on arylhydrazine (0.3 mmol), *tert*-butyl acrylate (0.6 mmol) with 3.6 mol% of ligand (0.0108 mmol), and 3 mol% of $Pd(OAc)_2$ (0.009 mmol) in PhCl/MeOH (0.4 mL/0.1 mL).

[b]Isolated yield.

ones. A longer reaction time was required to achieve the corresponding products in good yields when an electron-rich group was present in the arylhydrazines (Table 3.2, entries 2 and 17). The electron-withdrawing 4-hydrazinylbenzoic acid (**6i**) gave an almost quantitative yield under room temperature (Table 3.2, entry 9). Notably, the *ortho*-substituted arylhydrazines also proceeded smoothly to generate the desired product in good yields (Table 3.2, entries 15, 16, and 20). The functional groups tolerance of *ortho*-substituted arylyhydrazines is an advantage over Pd-catalyzed diazonium salts chemistry, which is usually sensitive to steric effect. It is noteworthy that good chemoselectivities were observed when halophenylhydrazine were involved in this new coupling reaction. This feature offers an opportunity for further functionalization at the halide group. The carboxylic acid and sulfonamide functional groups on the aryl ring are also suitable, and these substrates were even more reactive due to their electron deficiency (Table 3.2, entries 9, 11, and 14).

TABLE 3.3. Pd-Catalyzed Cross-Coupling of Arylhydrazine (6) with Various Olefins[a]

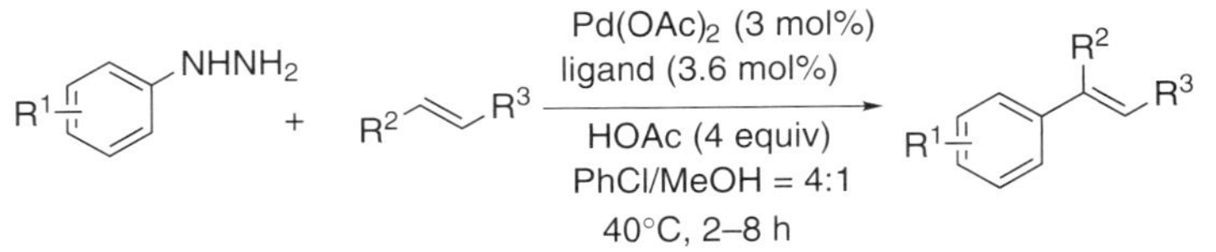

Entry	R^1	R^2	R^3	Yield (%)[b]
1	R^1 = COOH	R^2 = H	R^3 = COO^{n}Bu	**97% (9u)**
2	R^1 = COOH	R^2 = H	R^3 = COOMe	**96% (9v)**
3	R^1 = COOH	R^2 = H	R^3 = COOMe	**98% (9w)**
4	R^1 = CH_3	R^2 = H	R^3 = COO^{n}Bu	**89% (9x)**
5	R^1 = CH_3	R^2 = H	R^3 = COOEt	**86% (9y)**
6	R^1 = CH_3	R^2 = H	R^3 = COOMe	**84% (9z)**
7	R^1 = CH_3	R^2 = CH_3	R^3 = COOEt	**72% (10)**
8	R^1 = CN	R^2 = $PhCH_2CH_2$	R^3 = COOEt	**45% (11)**
9	R^1 = H	R^2 = H	R^3 = CONHPh	**78% (12)**
10	R^1 = H	R^2 = H	R^3 = COOH	**73% (13)**
11	R^1 = H	R^2 = H	R^3 = Ph	**93% (14)**
12[c]	12[c] 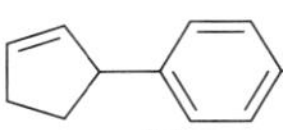 **15**	= H	R = Ph 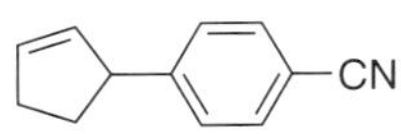 **16**	**80% yield[d]** **15 : 16 = 83 : 17**
13	13 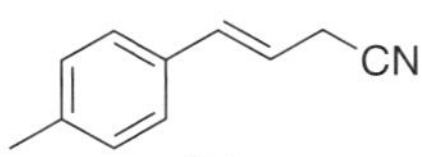 **17**		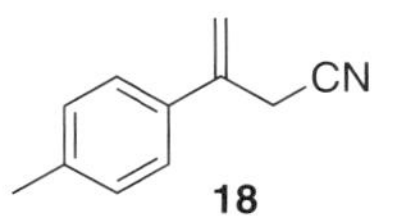 **18**	**85% yield[d]** **17 : 18 = 84 : 16**

[a]Unless noted otherwise, the reactions were carried out on arylhydrazine (0.3 mmol), alkene (0.6 mmol) with 3.6 mol% of ligand (0.0108 mmol), and 3 mol% of $Pd(OAc)_2$ (0.009 mmol) in PhCl/MeOH (0.4 mL/0.1 mL).

[b]Isolated yield.

[c]Used cyclopentene.

[d]Total yield.

Next, a range of acrylates and olefins were evaluated under the standard reaction conditions (Table 3.3). As shown in Table 3.3, the ester moiety of acrylate has little effect on the reaction efficiency. The β-substituent can be present and give the desired coupling product with good stereoselectivity (Table 3.3, entries 7 and 8). It is noteworthy that both acrylic acid and acrylamide were suitable for this reaction to generate cross-coupling products in good yields (Table 3.3, entries 9 and 10). Unactivated olefins such as 3-butenenitrile and cyclopentene are also viable reaction partners, indicating that this reaction is not limited to the activated double bond (Table 3.3, entries 12 and 13).

Zhang and his coworkers have developed a new method for C–N bond cleavage of allylic amines via hydrogen-bond activation, leading to allylic alkylation of

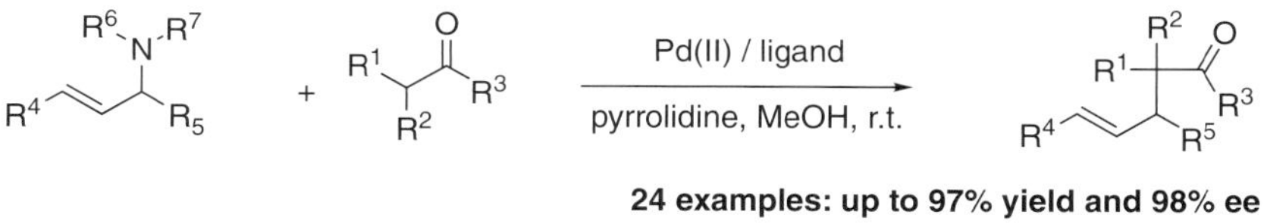

Scheme 3.9. Pd-catalyzed allylic alkylation of carbonyl compounds via C–N bond cleavage.

carbonyl compounds to form C–C bond products catalyzed by Palladium complex (Scheme 3.9) [9]. A series of allylic amines including primary, secondary, and tertiary species and ketones, including those with acid-labile groups, were suitable reaction partners with excellent results. Preliminary attempt at asymmetric allylic alkylation of *N*-(1,3-diphenyl-2-propenyl)pyrrolidine with ketone was performed and gave good yield and excellent enantioselectivity.

They proposed a plausible reaction mechanism (Scheme 3.10), beginning with coordination of the palladium to the double bond to form a Pd–olefin complex. Meanwhile, a hydrogen bond forms between the nitrogen of *N*-allylpyrrolidine and the hydrogen of the alcohol solvent. Thus, the C–N bond of the *N*-allylpyrrolidine is activated by this hydrogen bond action, leading to cleavage of the C–N bond and formation of a stable Pd–allylic complex. Subsequently, the complex is attacked by the enamine generated *in situ*, followed by removal of the palladium to generate the C–C bond product.

This new Pd-catalyzed allylic alkylation could be applied to a range of ketones. As shown in Table 3.4, aliphatic ketones proceeded smoothly in high yields (entries 1–6).

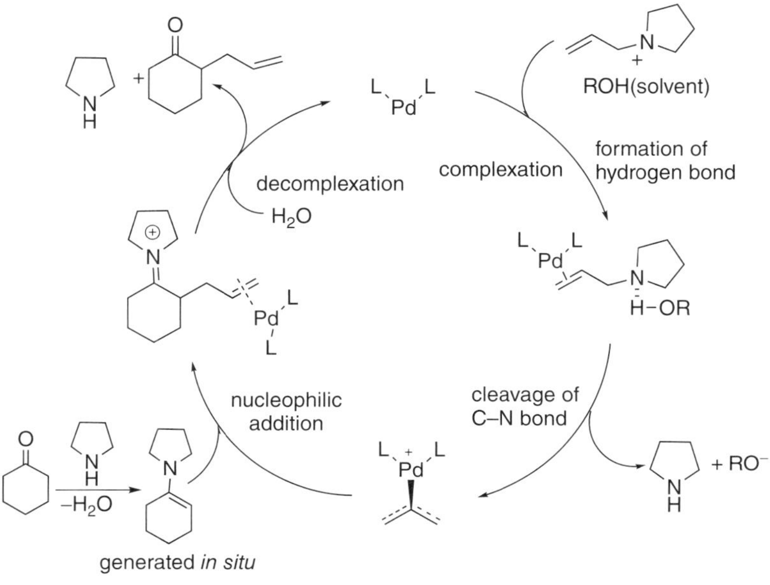

Scheme 3.10. A plausible reaction mechanism.

TABLE 3.4. Pd-Catalyzed Allylic Alkylation with Different Ketones and Aldehydes[a]

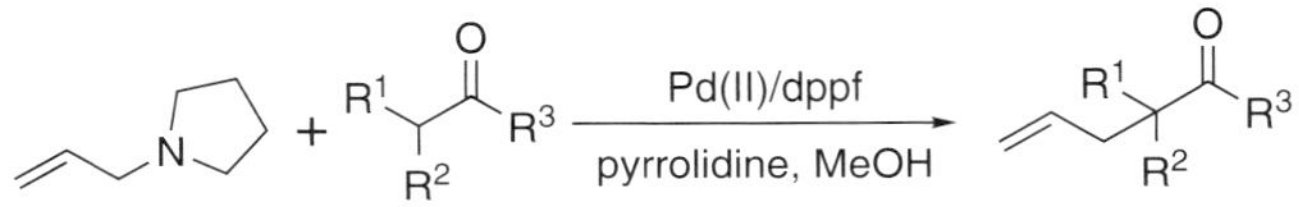

Entry	Ketone/Aldehyde	Time (h)	Yield (%)[b]
1	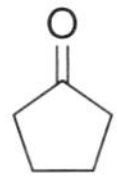	3	95
2	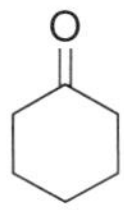	3	95
3	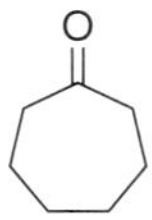	8	93
4	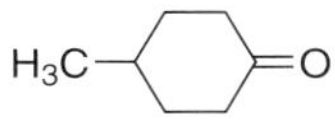 H3C, =O	4	90
5	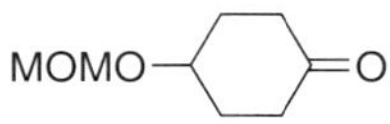 MOMO, =O	4	93
6	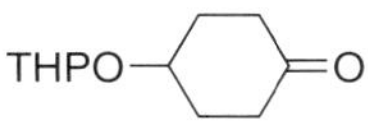 THPO, =O	4	88
7	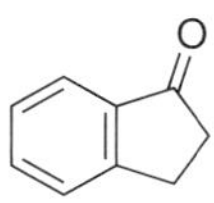	12	92
8	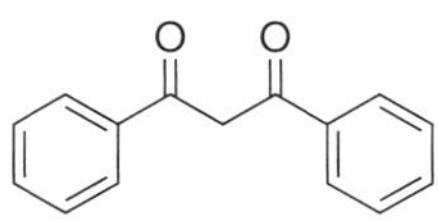	18	97
9[c]	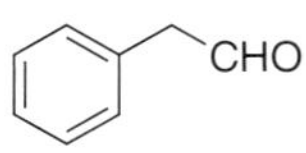 CHO	18	79
10[c]	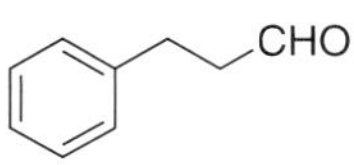 CHO	18	73

[a]Reactions of *N*-allylpyrrolidine (0.50 mmol) with ketones (1.50 mmol) were performed in the presence of pyrrolidine (0.50 mmol) using 6.0 mol% dppf and 2.5 mol% $[Pd(\eta^3\text{-}C_3H_5)Cl]_2$ in MeOH (2 mL) at 20°C.
[b]Isolated yields.
[c]5.0 mmol of aldehyde was used.

The size of cyclic ketone had an obvious influence on the reaction rate, as evaluated by cycloheptanone, which required a longer time to complete the reaction (entries 1–3). It is noteworthy that ketones with acid-labile groups could work well in this catalytic system (entries 5 and 6). For aromatic ketones, both monoketone and diketone afforded high yields (entries 7 and 8). In order to expand the scope of nucleophiles, phenyl acetaldehyde and phenylpropyl aldehyde were also added to the reaction and provided the desired products in good yields (entries 9 and 10). The above results showed that both ketones and aldehydes could serve as efficient nucleophiles, while aliphatic ketones showed higher reactivity than aromatic ketones and aldehydes.

Several kinds of allylic amines were also examined as shown in Table 3.5. The reactions of *N*-allylic cyclic amines with cyclohexanone work well and gave the desired products in high yields (entries 1–3). Both aliphatic and aromatic *N*-allylic acyclic amines afforded excellent results with slightly lower reaction activities (entries 4–8). Furthermore, when primary and secondary allylic amines were used as substrates, the corresponding products were obtained in excellent yields (entries 9 and 10). This showed that this catalytic system is suitable for primary, secondary, and tertiary allylic amines. In addition, excellent yields were also achieved for *N*-(cyclohex-2-enyl)pyrrolidine, (4-phenylbut-3-en-2-yl)pyrrolidine, and *N*-(1,3-diphenyl-2-propenyl)pyrrolidine with substituted allyl groups (entries and 11–13). Notably, the *N*-allylic cyclicamines with higher steric hindrance showed lower reaction activities than those with less steric hindrance (entries 1 and 11–13).

Recently, Tian and co-workers [10] reported a palladium-catalyzed C–N bond cleavage, leading to regioselective and stereospecific cross-coupling of primary allylic amines with boronic acids and boronates (Scheme 3.11). With this methodology, alkenes were achieved in good to excellent yields and with excellent E selectivity. The NH_2 group serves as an effective leaving group in this transformation. Notably, the reaction works well with a variety of substrates, including aryl- and alkenylboronic acids and aryl-, alkenyl-, allyl-, and benzylboronates, and

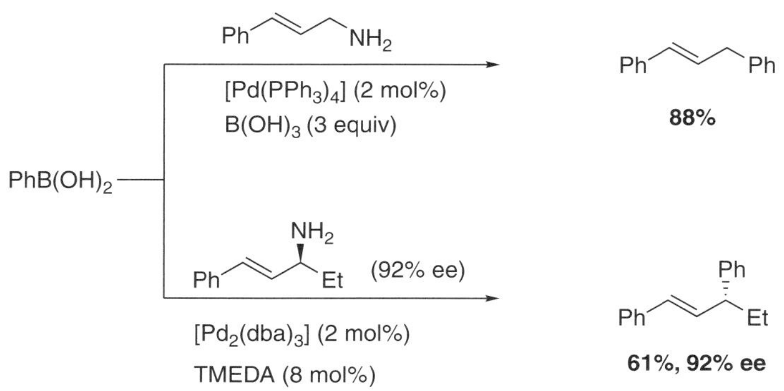

Scheme 3.11. Pd-catalyzed C–N bond cleavage between primary allylic amines and boronic acids.

TABLE 3.5. Pd-Catalyzed Allylic Alkylation with Different Allylic Amines[a]

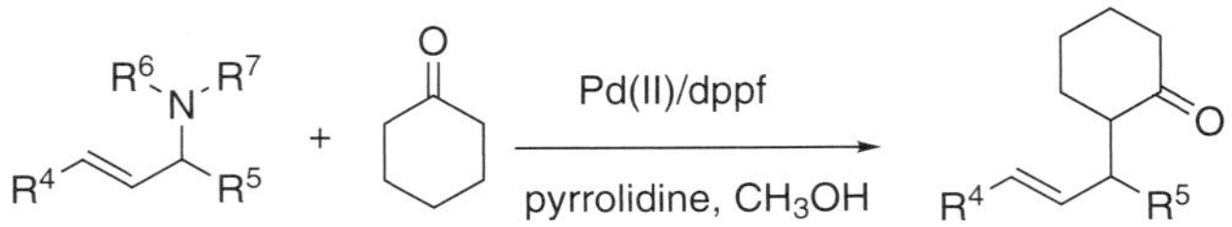

Entry	Allylamine	Time (h)	Yield (%)[b]
1	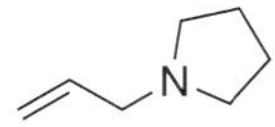	3	95
2	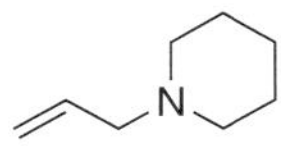	8	92
3	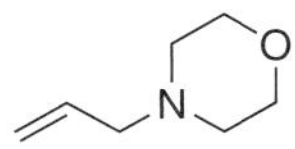	8	91
4	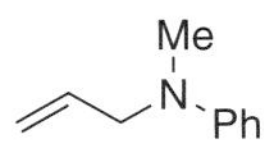	12	88
5	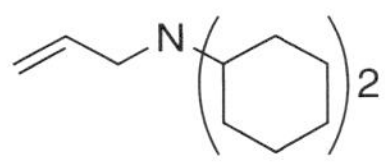	12	78
6	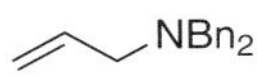	12	90
7	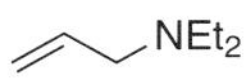	12	93
8	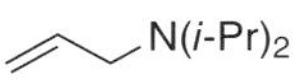	12	83
9	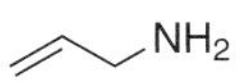	4	94
10	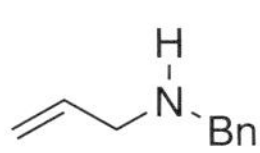	4	92
11	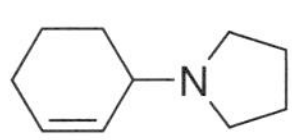	12	96
12	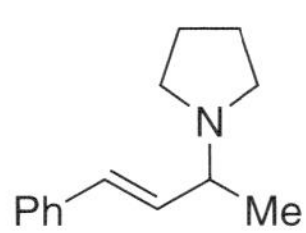	12	95
13	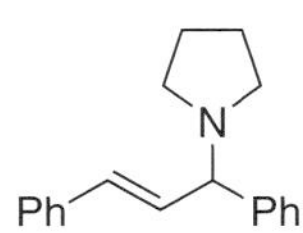	12	95

[a]Allylic amines (0.50 mmol), cyclohexanone (1.50 mmol), pyrrolidine (0.50 mmol), 6.0 mol% dppf, 2.5 mol% $[Pd(\eta^3\text{-}C_3H_5)Cl]_2$, MeOH (2 mL), 20°C.

[b]Isolated yields.

complete transfer of chirality has been achieved when using achiral primary allylic amines as the allylic electrophiles.

Under the standard reaction conditions, a series of primary *E*-allylic amines bearing different β and γ substitutents smoothly undertook reaction with boronic acid **20a** or **20b** in an excellent α-selective fashion to lead to the desired alkenes **22** in good to high yields and with excellent retention of the alkene geometry (Table 3.6, entries 1–7 and 9–11). In sharp contrast, the reaction with Z-allylic amine **19h** generated only the *E* alkene **22a** in 86% yield (Table 3.6, entry 8). On the other hand, the reaction worked well with a variety of aryl-, heteroaryl-, and alkenylboronic acids (Table 3.6, entries 12–22). Moreover, the scope for the reaction was significantly extended by employing aryl-, alkenyl-, allyl-, and benzylboronates as the carbon nucleophiles (Table 3.6, entries 23–29). As demonstrated by the results summarized in Table 3.6, this reaction well tolerated a variety of functional groups, such as alkoxy, aromatic nitro, halo, ester, vinyl, and ketone.

Based upon experimental results and the general mechanism for palladium catalysis, they proposed a plausible reaction pathway depicted in Scheme 3.12 for the cross-coupling of boronic acids with *a*-chiral primary allylic amines. First, the NH_2 group of amine **24** is activated by boronic acid **20**, leading to allylic C–N bond cleavage catalyzed by the palladium(0) catalyst with inversion of configuration to generate π-allylpalladium **27a**. When TMEDA is used as the ligand, complex **27a** undergoes transmetalation followed by reductive elimination to produce alkene **25** and regenerate the palladium(0) catalyst. However, racemization of complex **27a** takes place when a phosphine is used as the ligand, and the reaction loses optical purity by transforming a portion of amine **24** into alkene ent-**25**.

Recently, the Tian's group developed a novel substitution of primary allylic amines with sulfinate salts [11].

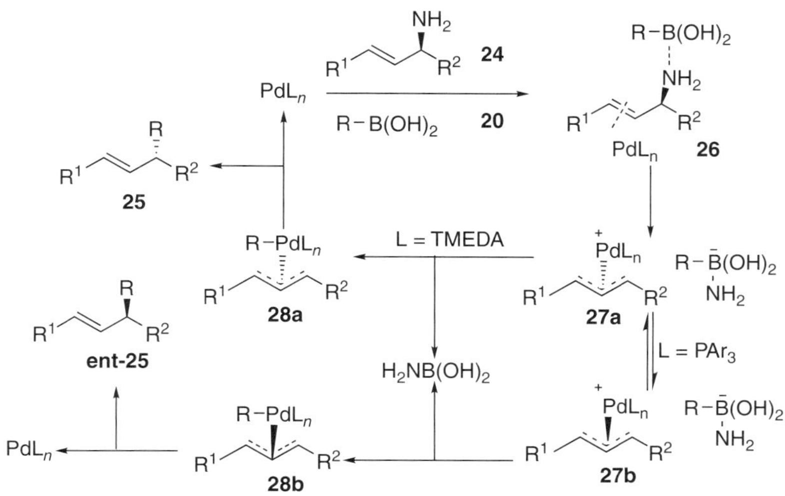

Scheme 3.12. A plausible reaction mechanism.

TABLE 3.6. Pd-Catalyzed Cross-Coupling of Primary Allylic Amines with Boronic Acids and Boronates[a,b]

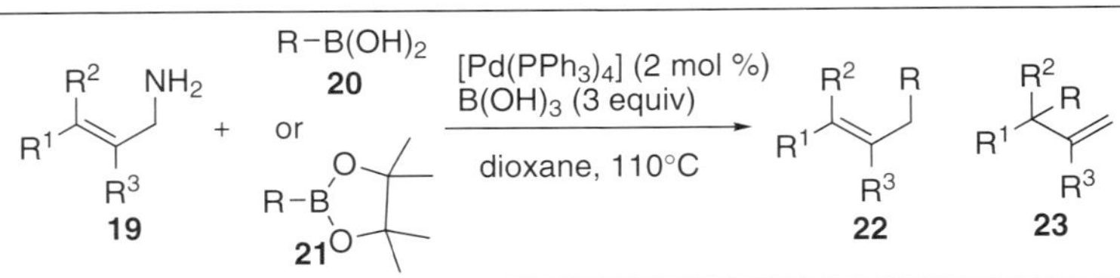

Entry	**19**, R^1, R^2, R^3	**20** or **21**, R	Product (**22/23**)[c]	Yield (%)[d]
1	**19a**, Ph, H, H	**20a**, Ph	**22a** (>99:1)	88
2	**19b**, 4-MeOC$_6$H$_4$, H, H	**20a**, Ph	**22b** (>99:1)	88
3	**19c**, 2-MeOC$_6$H$_4$, H, H	**20a**, Ph	**22c** (>99:1)	87
4	**19d**, 2-NO$_2$C$_6$H$_4$, H, H	**20a**, Ph	**22d** (>99:1)	84
5	**19e**, 3-pyridinyl, H, H	**20a**, Ph	**22e** (>99:1)	80
6	**19f**, 3-thienyl, H, H	**20a**, Ph	**22f** (>99:1)	72
7	**19g**, cyclohexyl, H, H	**20a**, Ph	**22g** (94:6)[e]	76
8	**19h**, H, Ph, H	**20a**, Ph	**22a** (>99:1)	86
9	**19i**, Ph, H, Me	**20a**, Ph	**22h** (>99:1)	92
10	**19j**, Ph, Ph, H	**20a**, Ph	**22i** (>99:1)	93
11[f]	**19k**, H H, H	**20b**, 9-phenanthrenyl	**22j**	84
12	**19a**, Ph, H, H	**20c**, 4-FC$_6$H$_4$	**22k** (>99:1)	80
13	**19a**, Ph, H, H	**20d**, 4-ClC$_6$H$_4$	**22l** (>99:1)	66
14	**19a**, Ph, H, H	**20e**, 4-(MeO$_2$C)C$_6$H$_4$	**22m** (>99:1)	84
15	**19a**, Ph, H, H	**20f**, 4-PhC$_6$H$_4$	**22n** (>99:1)	89
16	**19a**, Ph, H, H	**20g**, 3-(MeO$_2$C)C$_6$H$_4$	**22o** (>99:1)	86
17	**19a**, Ph, H, H	**20h**, 3,5-(CF$_3$)$_2$C$_6$H$_3$	**22p** (>99:1)	85
18[g]	**19a**, Ph, H, H	**20i**, 2,6-Me$_2$C$_6$H$_3$	**22q** (>99:1)	68
19[g]	**19a**, Ph, H, H	**20j**, 2-furyl	**22r** (>99:1)	60
20	**19a**, Ph, H, H	**20k**, 2-naphthyl	**22s** (>99:1)	88
21	**19a**, Ph, H, H	**20l**, 1-naphthyl	**22t** (>99:1)	91
22	**19a**, Ph, H, H	**20m**, (*E*)-PhCH=CH	**22u** (>99:1)	88
23	**19a**, Ph, H, H	**21a**, Ph	**22a** (>99:1)	66
24	**19a**, Ph, H, H	**21b**, 4-MeOC$_6$H$_4$	**22v** (>99:1)	71
25	**19a**, Ph, H, H	**21c**, 4-vinylphenyl	**22w** (>99:1)	70
26	**19a**, Ph, H, H	**21d**, 4-PhCOC$_6$H$_4$	**22x** (>99:1)	60
27	**19a**, Ph, H, H	**21e**, (*E*)-PhCH=CH	**22u** (>99:1)	76
28	**19a**, Ph, H, H	**21f**, CH$_2$=CHCH$_2$	**22y** (>99:1)	71
29	**19a**, Ph, H, H	**21g**, PhCH$_2$	**22z** (>99:1)	83

[a] Amine 19 (0.50 mmol), boronic and 20 (0.60 mmol) or boronate 21 (0.75 mmol), [Pd(PPh$_3$)$_4$] (2 mol%), B(OH)$_3$ (3 equiv), dioxane (5.0 mL), 110°C, 12 h.

[b] Unless otherwise stated, only the *E*-alkene product was obtained.

[c] Determined by ^{1}H NMR spectroscopic analysis.

[d] Yield of the isolated product.

[e] 96:4 *E*/*Z*.

[f] The reaction was run in a sealed tube.

[g] 24 h.

TABLE 3.7. Pd-Catalyzed Substitution of Unbranched Allylic Amines with Sodium Sufinates[a]

29 30 31

Entry	**29**, R^1, R^2, R^3	**30**, R	**31**	Yield (%)[b]
1	**29a**, Ph, H, H	**30a**, Ph	**31a**	92
2	**29b**, 4-$MeOC_6H_4$, H, H	**30a**, Ph	**31b**	90
3	**29c**, 4-ClC_6H_4, H, H	**30a**, Ph	**31c**	82
4	**29d**, 2-$MeOC_6H_4$, H, H	**30a**, Ph	**31d**	67
5[c]	**29e**, 2-$O_2NC_6H_4$, H, H	**30a**, Ph	**31e**	79
7	**29g**, 2-furyl, H, H	**30a**, Ph	**31f**	84
8	**29h**, (*E*)-MeCH=CH, H, H	**30a**, Ph	**31g**	70
9	**29i**, cyclohexyl, H, H	**30a**, Ph	**31h**	80
10[d]	**29j**, H, H, H	**30a**, Ph	**31j**	76
11	**29k**, Ph, Me, H	**30a**, Ph	**31k**	83
12	**29l**, H, H, Ph	**30a**, Ph	**31a**	84
13	**29a**, Ph, H, H	**30b**, 4-$MeOC_6H_4$	**31l**	94
14	**29a**, Ph, H, H	**30c**, 4-MeC_6H_4	**31m**	94
15	**29a**, Ph, H, H	**30d**, 4-ClC_6H_4	**31n**	86
16[c]	**29a**, Ph, H, H	**30f**, 1-naphthyl	**31o**	64
17	**29a**, Ph, H, H	**30g**, 2-naphthyl	**31p**	80
18	**29a**, Ph, H, H	**30h**, Me	**31q**	75
19	**29a**, Ph, H, H	**30i**, n-$C_{16}H_{33}$	**31r**	92
20	**29a**, Ph, H, H	**30j**, $PhCH_2$	**31s**	87

[a]Reaction conditions: amine **29** (0.50 mmol), sulfinate **30** (0.60 mmol), $[Pd(allyl)Cl]_2$ (0.1 mol%), dppb (0.4 mol%), $B(OH)_3$ (2.0 mmol), dioxane (0.5 mL), 100°C, 4 h.
[b]Isolated yields.
[c]The reaction was run for 12 h.
[d]The reaction was run in a sealed tube.

In the presence of catalytic $[Pd(allyl)Cl]_2$, dppb, and excess boric acid, a broad range of α-unbranched primary allylic amines were substituted with a variety of sodium sulfinates in α-selective fashion to afford the desired allylic sulfones in moderate to excellent yields (Table 3.7). It is noteworthy that the β- and γ-positions of the allylic amines could tolerate various groups including aryl, heteroaryl, alkenyl, or alkyl groups. The alkene geometry of (*E*)-allylic amines was retained (entries 1–9, 11, and 13–20); however, it was completely inverted using a (*Z*)-allylic amine as substrate (entry 12). The entire *Z*-to-*E* isomerization could be attributted to the *in situ* generation of a π-allylpalladium intermediate from a palladium catalyst and an allylic amine (see below), leading to the reaction to give an (*E*)-allylic sulfone under the thermodynamic conditions.

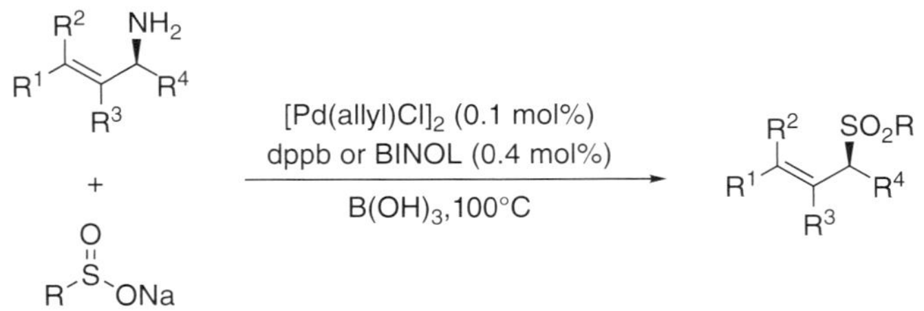

Scheme 3.13. Pd-catalyzed substitution of primary allylic amines with sulfinate salts.

Based upon experimental results and the general mechanism of the Tsuji–Trost reaction, a plausible mechanism was proposed as shown in Scheme 3.14. First, nucleophilic attack of sulfinate **30** on $[Pd(allyl)Cl]_2$ generates the allyl sulfone **34** and the Pd(0) catalyst (PdL_n). The amino group in chiral amine **32** is activated by boric acid, and the cleavage of an allylic C–N bond by the Pd(0) catalyst gives π-allylpalladium **36a** with inversion of configuration. When BINOL is used as a

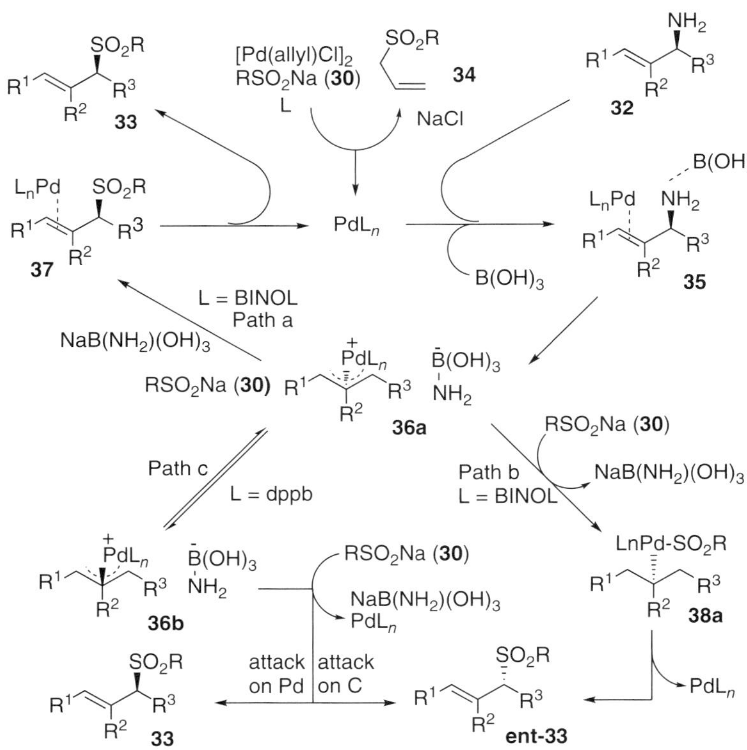

Scheme 3.14. Proposed reaction pathways.

ligand, the allylic carbon of intermediate **36a** is attacked by sulfinate **30** along with inversion of configuration to give the chiral sulfone **33** and regenerate the Pd(0) catalyst (path a). Alternatively, the Pd atom of intermediate **36a** is attacked by sulfinate **30** to generate palladium-*S*-sulfinate **38a**, which then undergoes reductive elimination to form the minor enantiomer ent-**33** (path b). The regioselectivity is dominated by the steric and electronic natures of the R^1 and R^3 groups in intermediates **36a** and **38a**. It is noteworthy that racemization of intermediate **36a** proceeds when a phosphine is used as the ligand, leading to chiral sulfone ent-**33** from chiral amine **32** (path c).

3.3 RUTHENIUM-CATALYZED C–N ACTIVATION

In 2007, Kakiuchi and co-workers [12], in their pioneering works, reported that the reaction of 2-amino-6-methylacetophenone with phenylboronic acid ester catalyzed by $RuH_2(CO)(PPh_3)_3$ generated the phenylation product in good yield via aryl C–N bond cleavage (Scheme 3.15). This reaction is distinguished by two notable features: (1) The coupling reaction proceeds via the oxidative addition of an aryl C–N bond in anilines to the ruthenium complex, and (2) C–C bond formation occurs via transmetalation between the $Ru–NR_2$ species and organoboronates.

Several organoboronates involving aryl, alkenyl, heteroaryl, and alkylboronates can be used as partners in this reaction. Representative results are listed in Table 3.8. In the case of 4-styrylboronate, the yield was decreased slightly due to the undesired polymerization of 4-styrylboronate during the transformation. The reaction using sterically hindered arylboronates including 2-tolyl- and 1-naphthylboronates afforded the corresponding products in 87% and 99% isolated yields, respectively. When the coupling reactions with heteroarylboronates were carried out in the presence of 8 mol% of catalyst, the corresponding products were obtained in 45–77% yields. The reaction with 1-propenylboronate (E:Z = 5:95) formed 2′-(1-*E*-propenyl)pivalophenone as a major product (E:Z = 94:6). In the presence of $RuH_2(CO)(PPh_3)_3$ catalyst, heating of (*Z*)-1-propenylboronate at 60°C for 30 min resulted in isomerization of (*Z*)-1-propenylboronate (E:Z = 5:95) to the corresponding *E*-isomer (E:Z = 95:5). This indicates that the *E*-propenylation product was

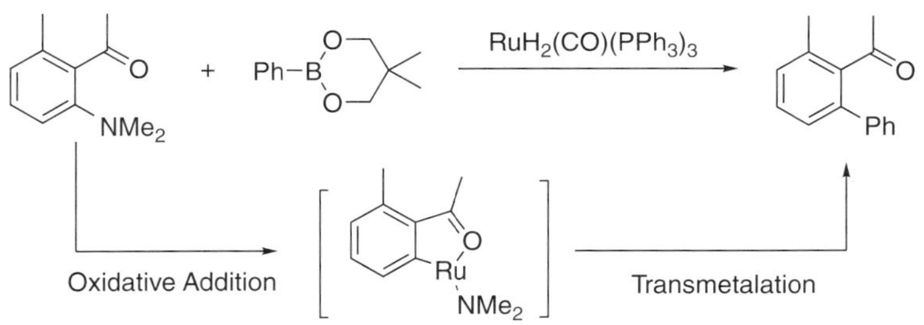

Scheme 3.15. Ru-catalyzed C–C bond formation via the C–N bond cleavage of aniline derivatives.

TABLE 3.8. The Coupling Reaction of *N,N*-Dimethylaminopivalophenone (40) with Various Organoboronates[a]

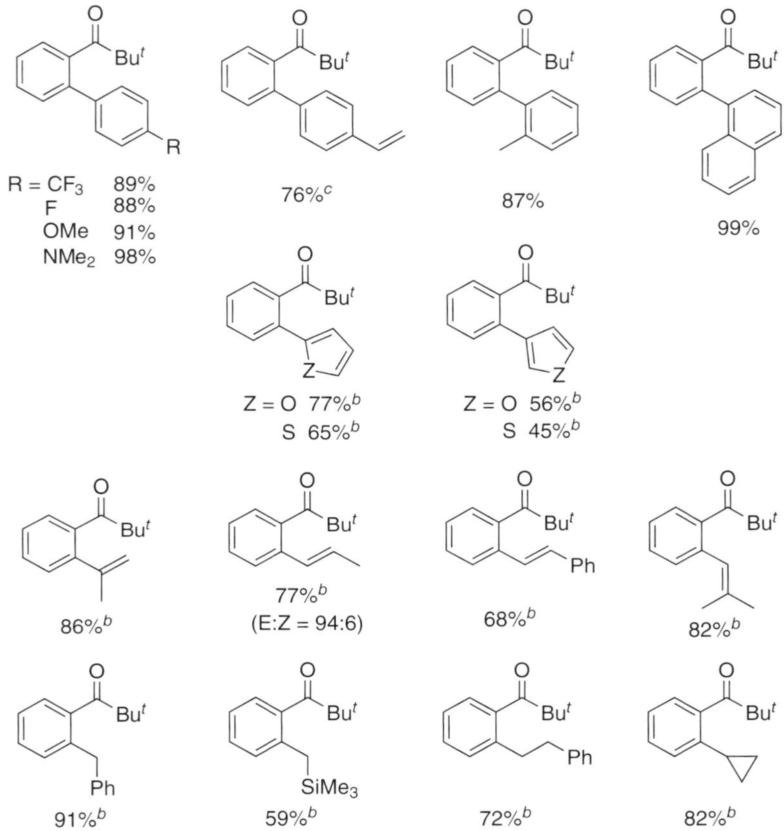

[a]Conditions: $RuH_2(CO)(PPh_3)_3$ **39** (0.02 mmol), toulene (0.5 mL), arylamine **40** (0.5 mmol), organoboronate (0.6 mmol), reflux, 20 h.
[b]Conditions: **39** (0.04 mmol), organoboronate (1.0 mmol).
[c]Conditions: **39** (0.06 mmol), organoboronate (1.5 mmol).

generated via isomerization of the propenylboronate. Alkylboronates, including benzyl-β-phenethyl-trimethylsilylmethyl- and cyclopropylboronates, can be used as substrates in this reaction. In the case of β-phenethylboronate, the α-phenethyl product, generated via β-hydride elimination from the β-phenethylruthenium intermediate followed by the reinsertion of the C=C bond in styrene into the Ru–H bond, was not achieved. Notably, the cyclopropyl group can also be introduced on the aromatic ring with this method without ring opening of the cyclopropyl ring, which is generally reactive toward low-valent transition metals.

In 2009, Kakiuchi and co-workers [13]. developed the first cleavage of unactivated aromatic C–N bond on a late-transition-metal center using *o*-acylanilines and a ruthenium complex (Scheme 3.16). Dramatic improvement of the rate of C–N bond

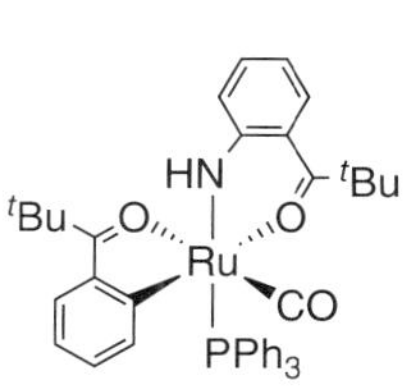

41

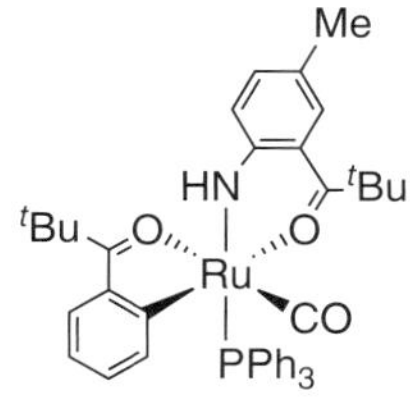

42

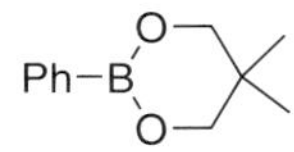

43

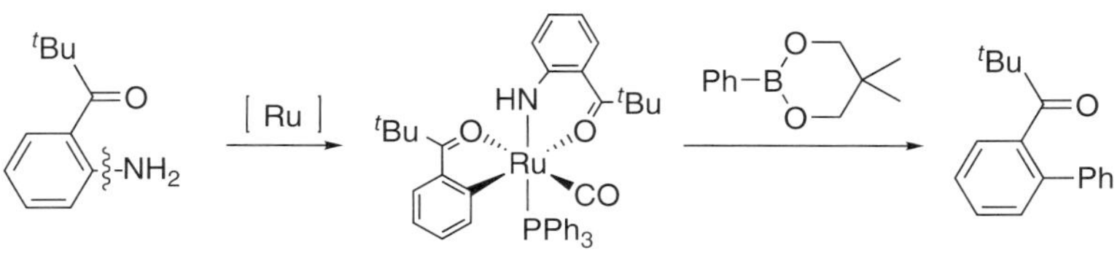

Scheme 3.16. Ru-catalyzed C–C bond formation via the C–N bond cleavage of aniline derivatives.

cleavage was observed with olefins as additives. The couplings of boronate **43** with the carbon fragments of **41** and **42** generated *in situ* via C–N bond cleavage resulted in high yields. Based upon these results, the authors suggested a proposed catalytic cycle, including an oxidative addition, transmetalation, and a final reductive elimination pathway.

Whittlesey's group described a rare example of C–N bond cleavage of an NHC and the unprecedented formation the N-1 bound product **46** from C-2 bound carbene complex **45** [14]. Upon reaction of $Ru(PPh_3)_3(CO)HCl$ with $I^iPr_2Me_2$ (1,3-diisopropyl-4,5-dimethylimidazol-2-ylidene, Scheme 3.17) in THF at 70°C for 16 h, a mixture of products was formed including the C–H activated carbene complex $Ru(I^iPr_2Me_2)'(PPh_3)_2(CO)H$ **44**, $Ru(PPh_3)_3(CO)H_2$, **45**, and **46** in a ratio of approximately 5:3.5:2:1, respectively. Continued heating of the reaction mixture for a total of 5 days (70°C) led to complete conversion to **46**, which was characterized by X-ray crystallography to be the tautomeric imidazole compound $Ru(N\text{-}I^iPrHMe_2)(PPh_3)_2(CO)HCl$ (Figure 3.1). Compound **45** was subsequently

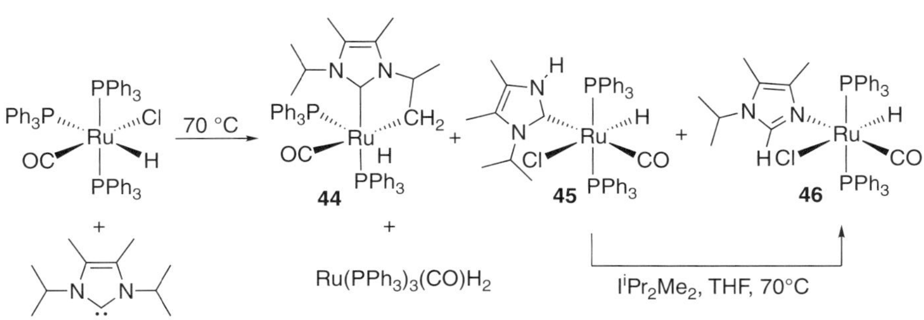

Scheme 3.17. Ruthenium-induced C–N bond cleavage.

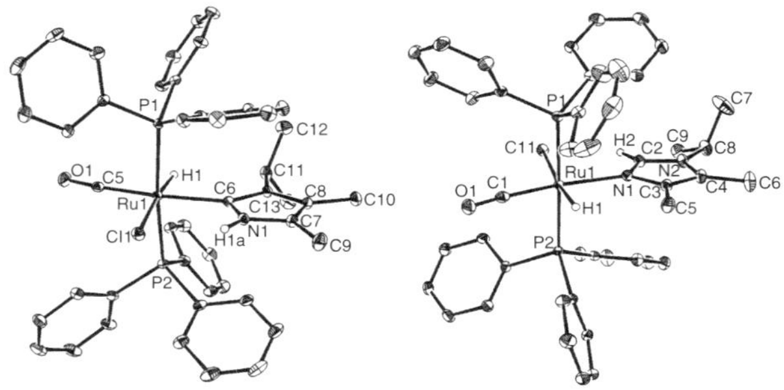

Figure 3.1. Molecular structures together with selected bond lengths (Å) and angles (deg) for (*left*) **47** (Ru(1)–C(6) 2.1282(18), Ru(1)–C(5) 1.884(2), N(1)–C(6) 1.346(2), N(2)–C(6) 1.366(2), N(1)–C(6)–N(2) 103.91(15)) and (*right*) **5** (Ru(1)–N(1) 2.1816(18), Ru(1)–C(1) 1.844(3), N(1)–C(2) 1.323(3), N(2)–C(2) 1.351(3), N(1)–C(2)–N(2) 111.3(2)). Ellipsoids are shown at 30% probability level.

achieved by reaction of complex $Ru(PPh_3)_3(CO)HCl$ with ligand $I^iPr_2Me_2$ under slower, lower temperature conditions (50°C, 6 days) and similarly structurally characterized (Figure 3.1). The need for free carbine ligand in converting **45** to **46** is consistent with a base-catalyzed process.

In 2010, Whittlesey et al. [15] elucidated the C–N bond activation in a series of ruthenium NHC complexes through combination of experimental studies and density functional theory calculations (Scheme 3.18). These results show that prior C–H cleavage of the NHC ligand led to irreversible C–N activation. In the presence

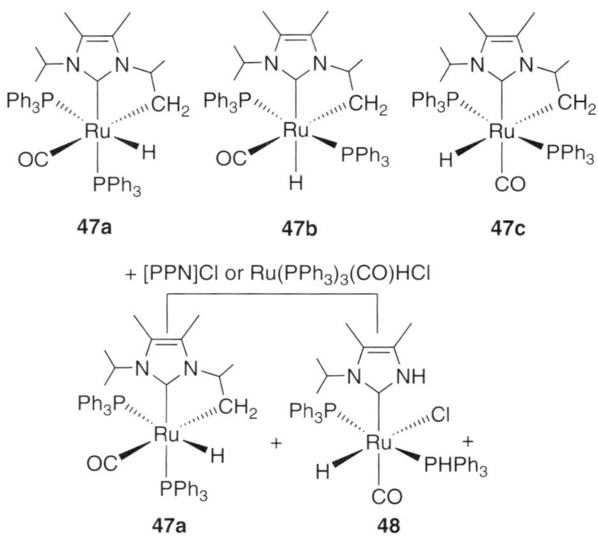

Scheme 3.18. Experimental and computational investigation of C–N bond cleavage in Ru–NHC complexes.

of a proper source of HCl, C–H-activated Ru(I^{i}Pr$_2$Me$_2$)′(PPh$_3$)$_2$(CO)H (**47**) reacts to form Ru(I^{i}PrHMe$_2$)(PPh$_3$)$_2$(CO)HCl (**48**) and propene. The proposed mechanism involves (1) initial isomerization to a *trans*-phosphine isomer **47c**, in which hydride locates *trans* to the metalated alkyl arm, (2) C–N cleavage to generate a ruthenium propene complex with a C-2-metalated imidazole, and (3) final *N*-protonation and propene/Cl$^-$ substitution to give complex **48**. The activation barrier ($\Delta E^{\ddagger}{}_{calcd}$) corresponds to the isomerization and C–N activation process and has a computed value of +24.4 kcal/mol. The C–N cleavage in complex **47c** is promoted by relief of the electronic strain due to the *trans* disposition of the high-*trans* influence between hydride and alkyl ligands. Experimental studies on analogues of complex **47** with different C-4/C-5 NHC backbone substituents such as (Ru(I^{i}Pr$_2$Ph$_2$)′(PPh$_3$)$_2$(CO)H, Ru(I^{i}Pr$_2$)′(PPh$_3$)$_2$(CO)H), or other *N*-substituents (Ru(IEt$_2$Me$_2$)′(PPh$_3$)$_2$(CO)H) indicate that Ph substituents promote C–N cleavage. Density functional theory calculations confirm that Ru(I^{i}Pr$_2$Ph$_2$)′(PPh$_3$)$_2$(CO)H undergoes isomerization/C–N bond activation with a low barrier of only +21.4 kcal/mol. In addition, sterically hindered *N*-alkyl groups also promote C–N bond cleavage (Ru(I^{t}Bu$_2$Me$_2$)′(PPh$_3$)$_2$(CO)H, $\Delta E^{\ddagger}{}_{calcd} = +21.3$ kcal/mol), and in this case the transformation is facilitated by the generation of the highly substituted 2-methylpropene.

3.4 NICKEL-CATALYZED C–N ACTIVATION

MacMillan and co-workers developed the first Suzuki reaction of aryltrimethylammonium salts with boronic acid. A novel catalyst system, capable of activating aryl-ammonium C–N bonds, was reported [16].

The scope of this novel Suzuki coupling with respect to the boronic acid has been evaluated (Table 3.9). Gratifyingly, a wide range of boronic acids bearing both electron-donating groups and electron-withdrawing groups at the *ortho*, *meta*, and *para* positions are tolerated in this transformation. It is noteworthy, that both boronate esters and alkenylboranes could be cross-coupled with aryltrimethylammonium salts under the optimized standard reaction conditions.

The scope of the aryltrimethylammonium triflate was also investigated as shown in Table 3.10. It was found that electron-rich aromatic substrates are less susceptible to oxidative addition. Consequently, a range of electron-rich aryltrimethylammonium salts coupled smoothly with phenylboronic acid (entries 1–5). In addition, aryl systems bearing electron-withdrawing substituents also worked well under the standard reaction conditions (entries 6–8).

Modified purines and its derivatives play an important role in biology, biochemistry, and pharmaceutics. Recently, 6-arylpurine ribonucleosides have been reported to possess cytostatic activity. Liu and Robins [17] developed nickel-based systems with imidazolium carbene as ligands, which catalyze highly efficient Suzuki reaction of arylboronic acids and 6-(benzimidazol-1-yl)-, 6-(imidazol-1-yl)-, and 6-(1,2,4-triazol-4-yl)purine (2′-deoxy)nucleoside derivatives to generate the corresponding 6-arylpurine (2′-deoxy)nucleosides products (Scheme 3.19).

TABLE 3.9. Ammonium Suzuki Couplings of Arylboronic Acids[a]

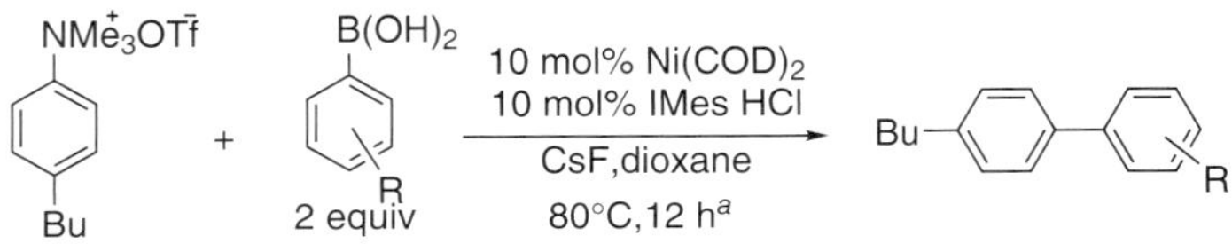

Entry	$ArB(OR)_2$	Product	Yield (%)
1	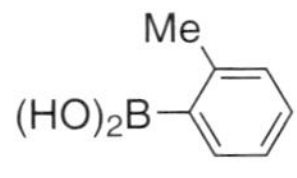	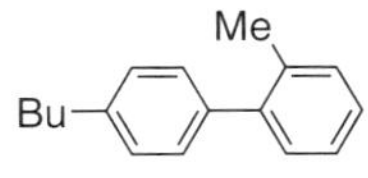	83
2	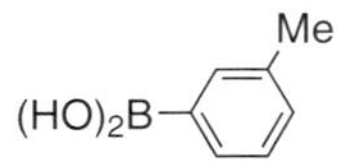	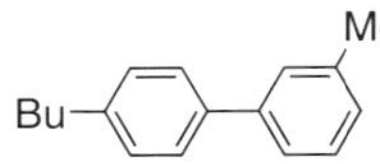	89
3[b]	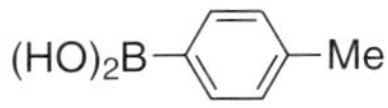	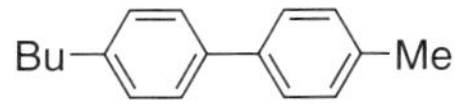	95
4	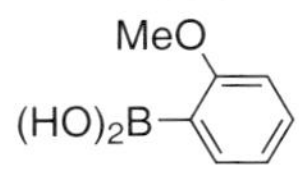	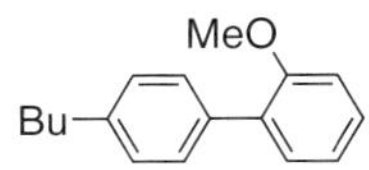	94
5[c]	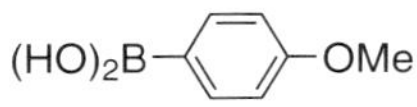	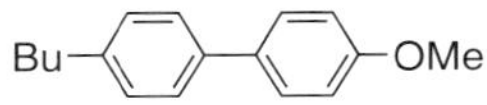	83
6	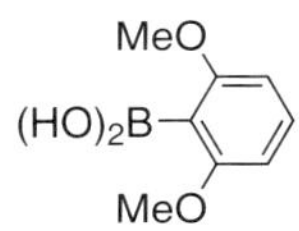	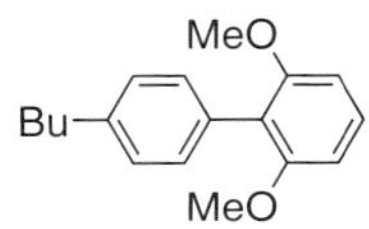	79
7	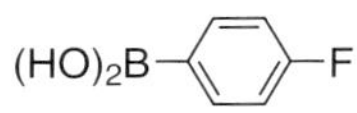	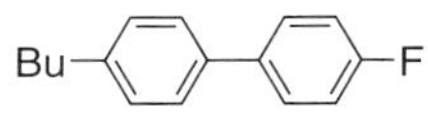	82
8	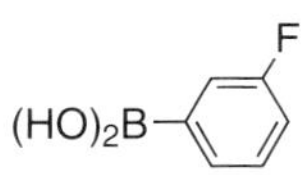	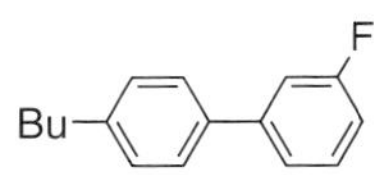	82
9	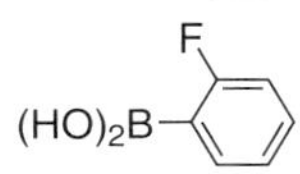	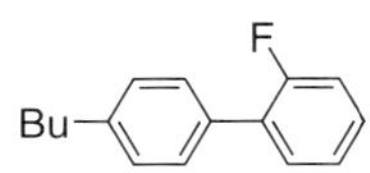	98
10	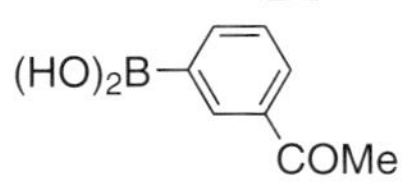	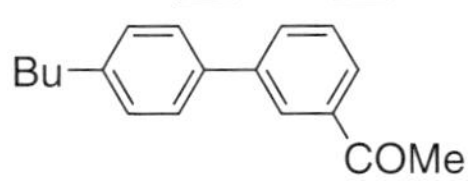	87
11	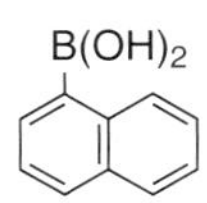	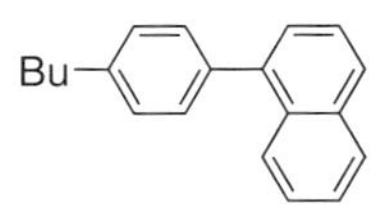	84
12	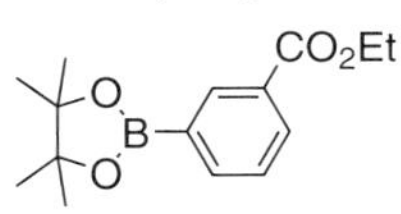	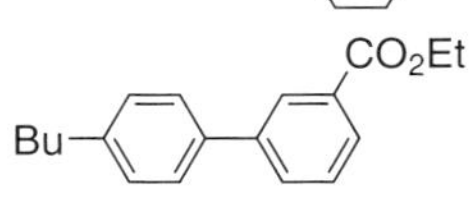	89
13[d]	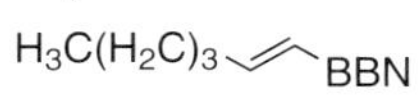	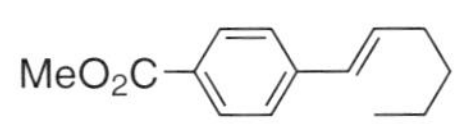	92

[a]Reaction times not optimized.
[b]1.1 equiv of boronic acid was used.
[c]K_3PO_4 was used in place of CsF.
[d]Reaction performed with 4-(carboxylic acid methylester)-*N*,*N*,*N*-trimethylanilinium triflate.

TABLE 3.10. Suzuki Couplings of Aryltrimethylammonium Triflates[a]

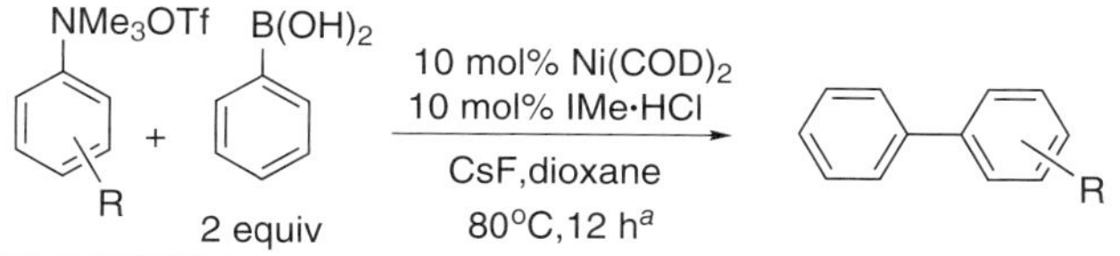

Entry	$ArNMe_3OTf$	Product	Yield (%)
1	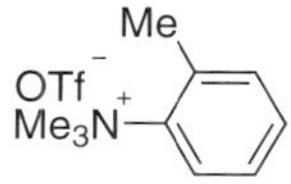	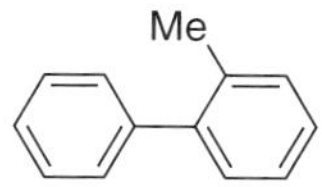	89
2	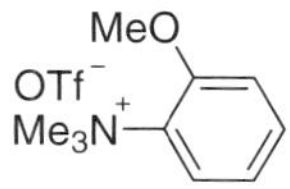	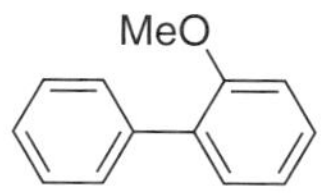	84
3	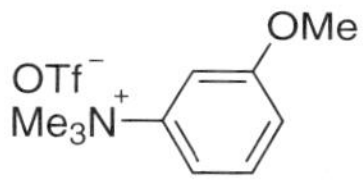	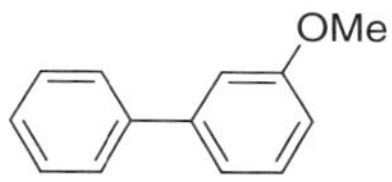	96
4	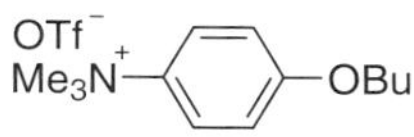	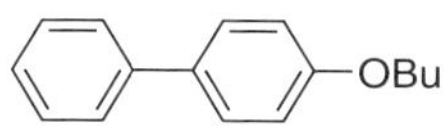	85
5	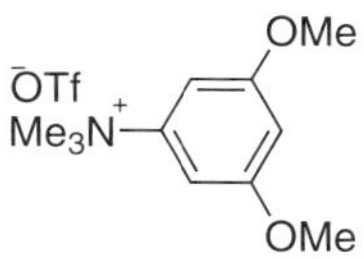	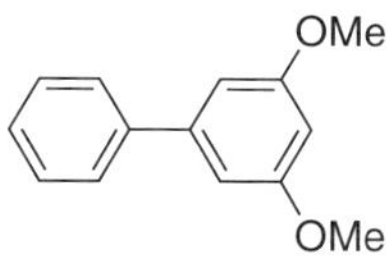	92
6	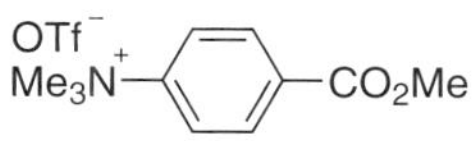		93
7	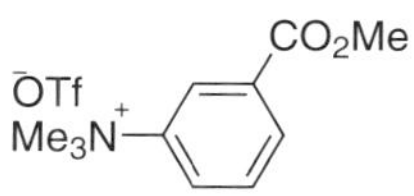	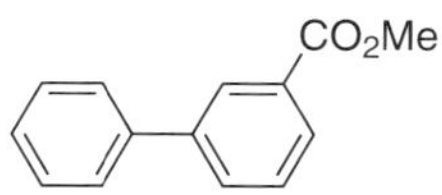	94
8	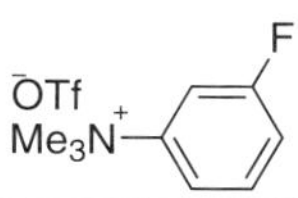	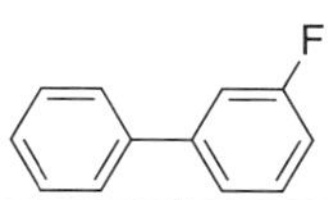	88

[a]Reaction times were not optimized.

Different base/ligand combinations give better results with the imidazole versus triazole substrates. Noteworthy, novel 6-(aryloxy)purine 2′-deoxynucleosides, oxygen-insertion products, were observed as the byproducts with the 6-(1,2,4-triazol-1-yl)purine derivatives.

Both electron-donating and electron-withdrawing arylboronic acids underwent coupling reaction with purines in good yields (Table 3.11, entries 1–3). Notably,

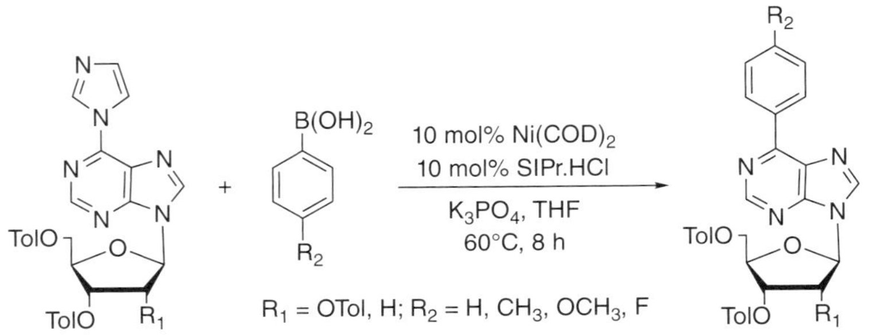

Scheme 3.19. Azoles as Suzuki cross-coupling leaving groups.

TABLE 3.11. Yields of Coupling Products with Varied Substrates

Entry	R_1	R_2	Yield (%)
1	OTol	H	73
2	OTol	CH_3	81
3	OTol	F	78
4	H	H	68
5	H	CH_3	61
6	H	OCH_3	75
7	H	F	65

application of the present methodology to such protected 2′-deoxynucleosides formed the corresponding 6-arylpurine products in good yields (Table 3.11, entries 4–7).

3.5 COPPER-CATALYZED C–N ACTIVATION

Tian and co-workers [18] reported that sulfonyl-activated sp^3 C–N bonds could be cleaved by Grignard reagents in the presence of CuI. A broad range of benzylic, allylic, and propargylic amines smoothly underwent a coupling reaction with Grignard reagents to provide structurally diverse products in moderate to excellent yields and with satisfactory chemo-, regio-, and stereoselectivity (Scheme 3.20). An S_N2 mechanism in the cross-coupling reaction was proposed based upon the synthesis of chiral hydrocarbons from optically pure *a*-branched amine derivatives.

As summarized in Table 3.12, a wide range of double *p*-toluenesulfonyl-activated primary alkylamines (sulfonimides **1**) smoothly underwent cross-coupling reactions with various Grignard reagents in the presence of CuI at 70°C to afford desired products **3** in good to excellent yields and with excellent regioselectivity. The R^1 groups in the sulfonimide **1** could be aryl, heteroaryl, alkenyl, or alkynyl group, and the R^2 groups in the Grignard reagent could be aryl or alkyl group with different

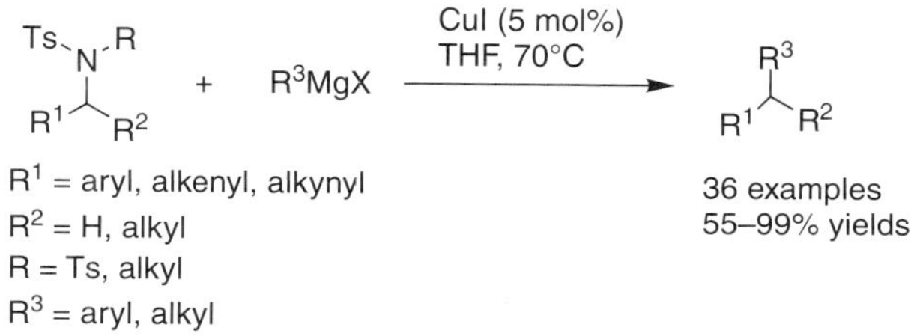

Scheme 3.20. Cross-coupling of Grignard reagents with sulfonyl-activated sp^3 C–N bonds.

TABLE 3.12. Cross-Coupling of Grignard Reagents with Sulonaimides[a]

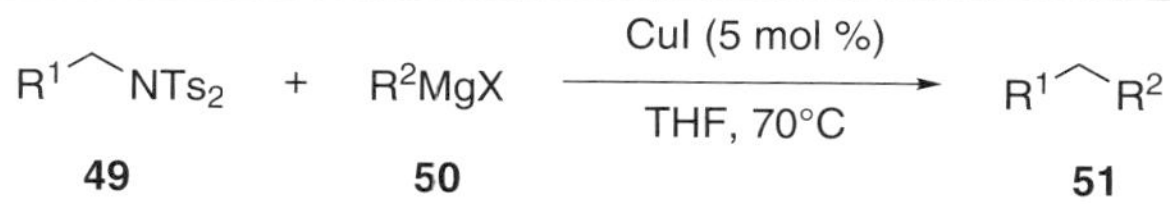

Entry	49	R¹	R²	X	51	Yield (%)[b]
1	**49a**	PMP	Ph	Br	**51a**	93
2	**49a**	PMP	2-MeC_6H_4	Cl	**51b**	92
3	**49a**	PMP	$TMSCH_2$	Cl	**51c**	89
4	**49a**	PMP	cyclopropyl	Br	**51d**	80
5	**49a**	PMP	Me_3C	Cl	**51e**	55
6	**49b**	2-$MeOC_6H_4$	Ph	Br	**51f**	88
7	**49c**	Ph	PMP	Br	**51a**	75
8	**49d**	4-BrC_6H_4	PMP	Br	**51g**	72
9	**49e**	4-ClC_6H_4	PMP	Br	**51h**	65
10	**49f**	4-FC_6H_4	PMP	Br	**51i**	74
11	**49g**	1-napththyl	PMP	Br	**51j**	74
12	**49h**	2-furyl	PMP	Br	**51k**	80
13	**49i**	2-thienyl	PMP	Br	**51l**	86
14	**49j**	CH_2=CH	PMP	Br	**51m**	97
15	**49j**	CH_2=CH	Bn	Cl	**51n**	83
16	**49k**	(*E*)-PhCH=CH	PMP	Br	**51o**	96
17	**49k**	(*E*)-PhCH=CH	$TMSCH_2$	Br	**51p**	93
18	**49k**	(*E*)-PhCH=CH	Me_2CH	Cl	**51q**	92
19	**49k**	(*E*)-PhCH=CH	Me_3C	Cl	**51r**	90
20	**49l**[c]	(*E*)-$BnOCH_2$CH=CH	2-MeC_6H_4	Cl	**51s**[d]	83
21	**49m**[e]	(*Z*)-$BnOCH_2$CH=CH	2-MeC_6H_4	Cl	**51t**[f]	85
22	**49n**	Me_2C=CH	PMP	Br	**51u**	99
23	**49o**	CH_2=C(Me)	PMP	Br	**51v**	96
24	**49p**	phenylethynyl	2-MeC_6H_4	Cl	**51w**	90
25	**49q**	1-heptynyl	Ph	Br	**51x**	88

[a]Reaction conditions: sulfonimide **49** (0.50 mmol), Grignard reagent (0.60 mmol), CuI (5 mol%), THF (1.0 mL), 70°C, 2 h (for entries 14–23, 1 h.

[b]Isolated yield.

[c]98:2 *E*/*Z*.

[d]98:2 *E*/*Z*.

[e]5:95 *E*/*Z*.

[f]5:95 *E*/*Z*.

electronic properties and steric effects. Furthermore, the cross-coupling reaction tolerated a variety of functional groups such as halides, alkenes, benzyl ethers, alkynes, and allyl ethers (Table 3.12, entries 8–10 and 14–25). It is noteworthy that the geometric configuration of the C–C double bond in a sulfonimide was tolerated in the novel cross-coupling reaction (Table 3.12, entries 16–21). Significantly, no rearrangement took place for the C–C multiple bonds in the transformation with double *p*-toluenesulfonylactivated propargylic or allylic amines (Table 3.12, entries 16–22, 24, and 25).

The Cu-catalyzed cross-coupling reaction underwent smoothly between a Grignard reagent and a single *p*-toluenesulfonyl-activated secondary amine (sulfonamide **52**, Table 3.13). In sharp contrast, single sulfonyl-activated benzylic C–N bond could not be cleaved effectively by Grignard reagent under the same conditions. For example, compound **51a** could not be obtained through the cleavage of the benzylic C–N bond in the reaction of sulfonamide **52a** with 4-methoxyphenylmagnesium bromide (Table 3.13, entry 1). Instead, the allylic C–N bond of sulfonamide **52a** was smoothly cleaved to form compound **51o** in good yield.

Further investigation shows that *p*-toluenesulfonyl-activated *a*-branched amines could also serve as suitable electrophiles to react with Grignard reagents in the presence of catalytic CuI (Table 3.14). To achieve chiral hydrocarbons and obtain insights into the reaction mechanism, the author evaluated optically active *a*-branched amine in the new cross-coupling reaction. While enantioenriched sulfonimides were submitted to epimerization under the standard reaction conditions, no loss of optical purity

TABLE 3.13. Cross-Coupling of Grignard Reagents with Sulfonamides[a]

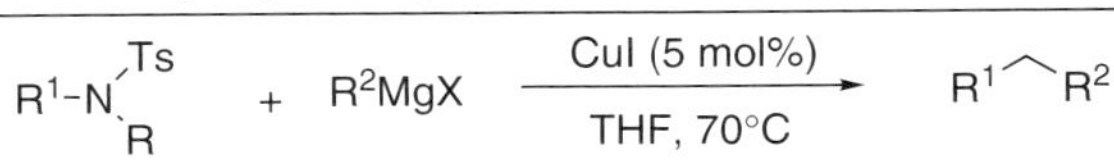

Entry	Sulfonamide		Product		Yield (%)[b]
1[c]	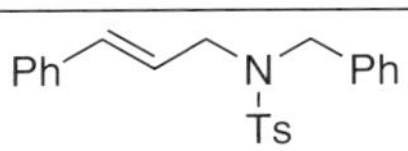	**52a**	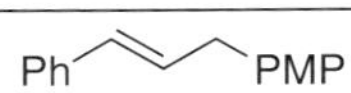	**51o**	63
2[c]		**52b**		**51o**	86
3[d]	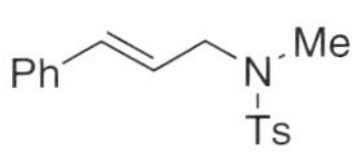	**52b**	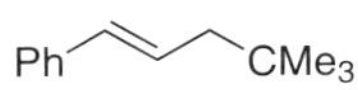	**51r**	86
4[e]	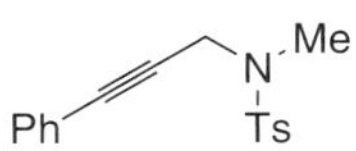	**52c**	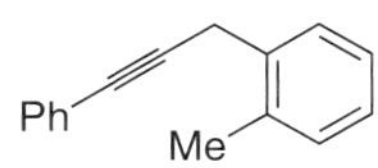	**51w**	77

[a]Reaction conditions: sulfonamide **52** (0.50 mmol), Grignard reagent (0.60 mmol), CuI (5 mol%), THF (1.0 mL), 70°C, 2 h.
[b]Isolated yield.
[c]PMPMgBr was used.
[d]Me_3CMgCl was used.
[e]2-MeC_6H_4MgCl was used.

TABLE 3.14. Cross-coupling of Grignard reagents with *p*-toulene-sulfonyl-activated *α*-branched amines[a]

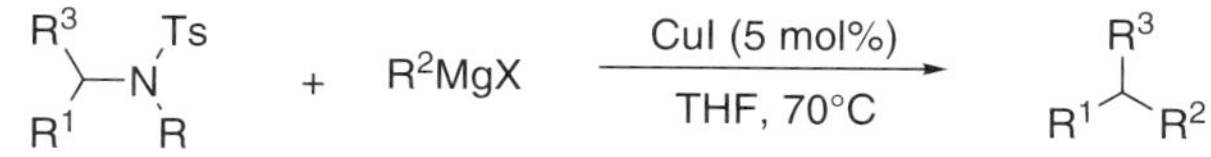

Entry	Amine Derivative		Product		Yield (%)[b]
1[c]	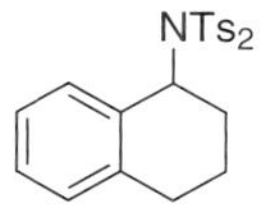	**53a**	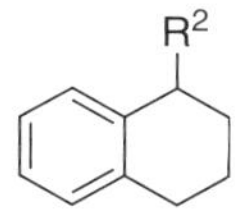	**54a**, R^2=allyl	64
2[d]		**53a**		**54b**, R^2=PMP	75
3[d]	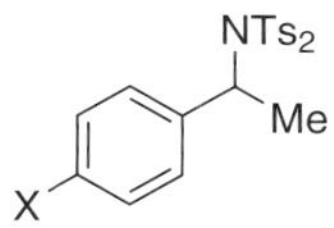	**53b**	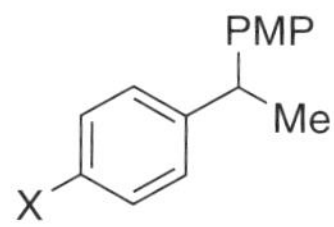	**54c**, X=Cl	58
4[d]		**53c**		**51d**, X=F	59
5[d]	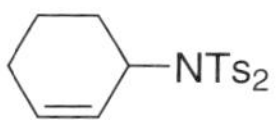	**53d**	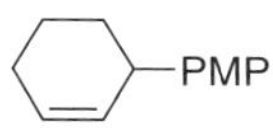	**54e**	72
6[d]	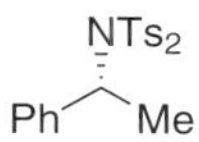	**53e** 99% ee	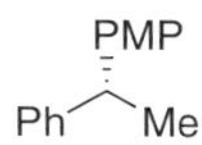	**54f** 44% ee	71
7[e]	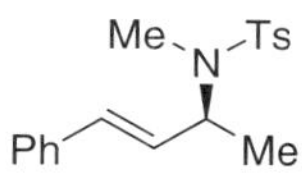	**53f** 95% ee	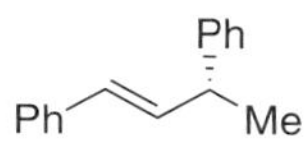	**54g** 95% ee	84

[a]Reaction conditions: amine derivative (0.50 mmol), Grignard reagent (0.60 mmol), CuI (5 mol%), THF (1.0 mL), 70°C, 2 h.
[b]Isolated yield.
[c]Allylmagnesium chloride was used.
[d]PMPMgBr was used.
[e]PhMgBr was used.

was found for the sulfonamides during the transformation. The reaction of sulfonimide **53e** (99% ee) with 4-methoxyphenylmagnesium bromide in the presence of catalytic CuI generated the corresponding product **54f** in only 44% ee (Table 3.14, entry 6). In sharp contrast, the use of sulfonamide **53f** in 95% ee led to the formation of product **54g** as a pure regioisomer without any loss of optical purity (Table 3.14, entry 7). Importantly, configurational inversion was detected for the chiral centers of both **53e** and **53f**. These results indicate that an S_N2 mechanism is involved in the Cu-catalyzed cross-coupling reaction of Grignard reagents with sulfonamides or sulfonimides.

Recently, Chen and Wu, in their pioneering work, developed the first example of cross-coupling of nitroarenes with arylboronic acids catalyzed by copper (Scheme 3.21) [19]. A variety of diaryl ethers bearing different functional groups were obtained in moderate to excellent yields. Importantly, the transformation was

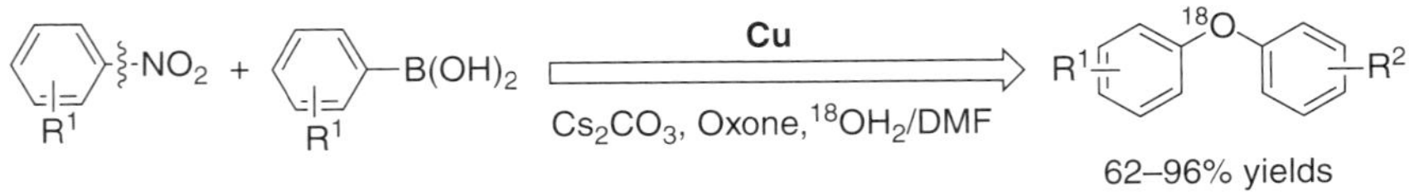

Scheme 3.21. Copper-catalyzed coupling of nitroarenes with arylboronic acids.

not sensitive to air and moisture. Thus, the method provides a simple, practical, and facile procedure to achieve diaryl ethers. Based upon mechanistic experiments using deuterium labeling, the author suggested that the oxygen atom of diaryl ether was derived from water.

3.6 IRON-CATALYZED C–N ACTIVATION

In 2010, Tian and co-workers [20] developed $FeCl_3$-catalyzed cleavage of sp^3 C–N bonds to construct a variety of indene derivatives from *N*-benzylic sulfonamides and disubstituted alkynes (Scheme 3.22) [20].

A number of Brønsted and Lewis acids were investigated in the reaction of *N*-(*p*-toluenesulfonyl)-benzhydrylamine (**55a**) with diphenylacetylene (**56a**) in CH_3NO_2 under room temperature for 24 h. While almost no desired reaction occurred in the presence of catalytic TsOH, H_2SO_4, HCl, $CuCl_2$, $ZnCl_2$, $Pd(OAc)_2$, $AlCl_3$, or $Bi_2(SO_4)_3$, the use of catalytic $FeCl_3$ led to the formation of 1,2,3-triphenyl-1*H*-indene (**57a**) in low 20% yield. Notably, the yield was increased to 66% when the temperature was elevated to 80°C, although the product **57a** was partially consumed via alkylation with sulfonamide **55a** (Table 3.15, entry 1). It's noteworthy that a large-scale synthesis of indene **57a** was smoothly performed based upon this protocol. In the presence of $FeCl_3$, the reaction of sulfonamide **55a** with a diphenylacetylene bearing either an electron-rich or an electron-poor functional group occurred smoothly to give the desired indene derivative with bigger than 99:1 regioselectivity (Table 3.15, entries 2 and 3). By employing functionalized disubstituted alkynes, a variety of groups including alkyl, alkoxycarbonyl, carboxyl,

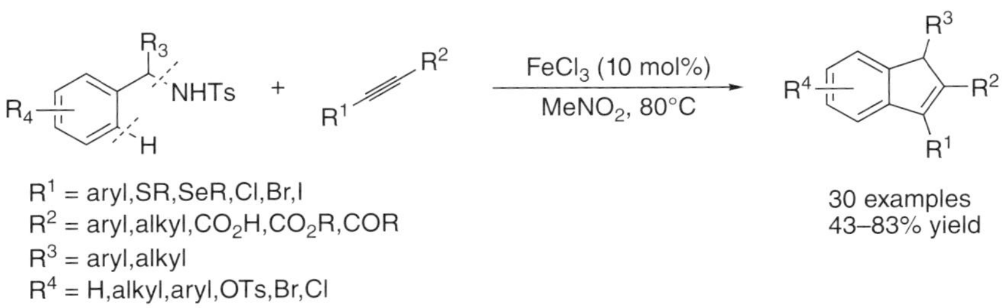

Scheme 3.22. Fe-catalyzed synthesis of indene derivatives from *N*-benzylic sulfonamides and disubstituted alkynes.

TABLE 3.15. $FeCl_3$-Catalyzed Synthesis of Indene Derivatives from Sulfonamide 55a and Disubstituted Alkynes[a]

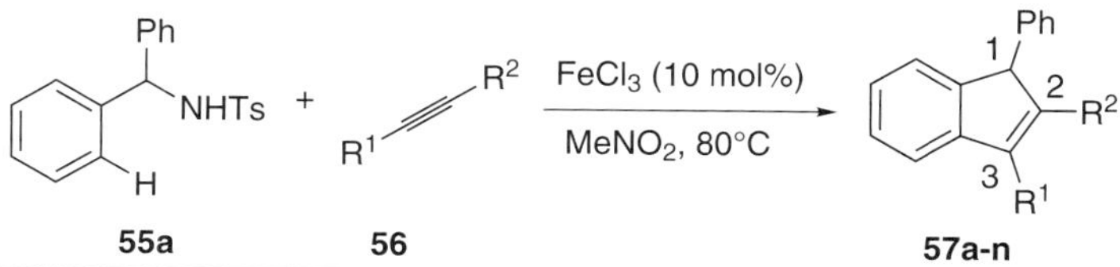

Entry	55	R^1	R^2	Product	Time (h)	Yield[b] (%)
1	**55a**	Ph	Ph	**57a**	12	66
2	**55b**	4-$MeOC_6H_4$	Ph	**57b**	5	75
3	**55c**	Ph	4-$O_2NC_6H_4$	**57c**	24	53
4	**55d**	Ph	*n*-Pr	**57d**	10	60
5	**55e**	Ph	COOEt	**57e**	24	74
6	**55f**	Ph	COOH	**57f**	24	83
7	**55g**	Ph	COPh	**57g**	24	66
8	**55h**	Ph	COMe	**57h**	24	69
9	**55i**	SPh	Ph	**57i**	3	43
10	**55j**	SePh	Ph	**57j**	6	58
11	**55k**	Br	Ph	**57k**	4	72
12	**55l**	Br	*n*-Bu	**57l**	24	61
13	**55m**	Cl	Ph	**57m**	12	66
14	**55n**	I	Ph	**57n**	6	72

[a]Reaction conditions: sulfonamide **55a** (0.20 mmol), alkyne **56** (0.24 mmol), $FeCl_3$ (10 mol%), nitromethane (2.0 mL), 80°C.
[b]Isolated yield.

or acyl group were introduced exclusively into C-2 positions of the desired indene derivatives (Table 3.15, entries 4–8). Moreover, a broad range of alkynyl chalcogenides and alkynyl halides could serve as suitable partners to react with sulfonamide **55a** and, consequently, afforded a convenient access to the corresponding indene derivatives bearing a variety of heretoatoms including sulfur, selenium, chlorine, bromine, and iodine at the C-3 positions (Table 3.15, entries 9–14). Notably, no rearrangement was observed with the C–C double bonds under the standard reaction conditions, and the high regioselectivity promotes the synthetic elaboration of indene derivatives.

A broad range of *N*-benzylic sulfonamides were evaluated to react with substituted alkynes in the presence of catalytic $FeCl_3$ to give structurally diverse indene derivatives (Table 3.16). Notably, the phenyl group instead of 4-chlorophenyl group in the unsymmetric *N*-bisbenzylic sulfonamide **55b** was found to construct a new C–C bond with a disubstituted alkyne in the reaction, although a similar C–C bond-forming reaction took place with one of the two 4-chlorophenyl groups in the symmetric *N*-bisbenzylic sulfonamide **55c** (Table 3.16, entries 1–3). The indene containing polycarbocycles **57r–t** were successfully achieved through this

TABLE 3.16. $FeCl_3$-Catalyzed Synthesis of Indene Derivatives from *N*-Benzylic Sulfonamides and Disubstituted Alkynes[a]

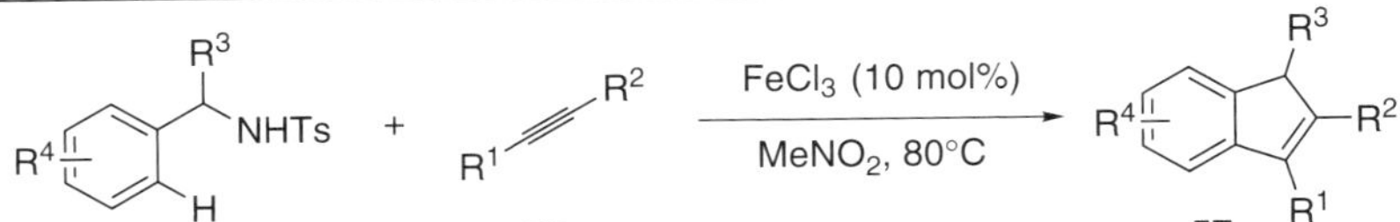

55 **56** **57**

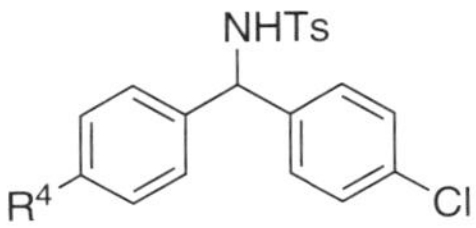

55b, R^4 = H
55c, R^4 = Cl

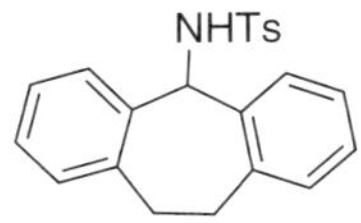

55d

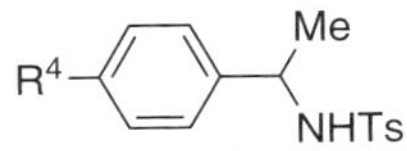

55e, R^4 = H
55f, R^4 = OTs
55g, R^4 = Br
55h, R^4 = Cl

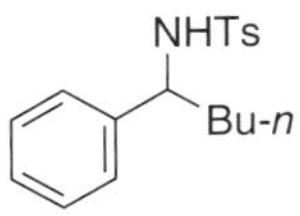

55i

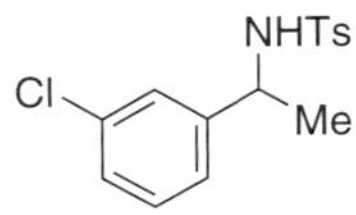

55j

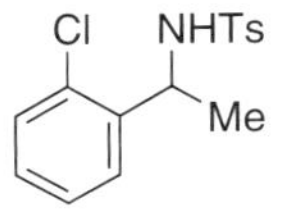

55k

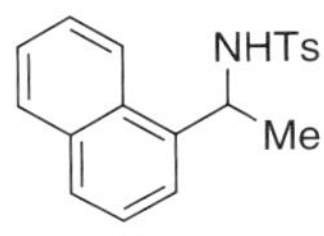

55l

Entry	1	2	Product		Time (h)	Yield (%)[b]
1	**55b**	**56a**	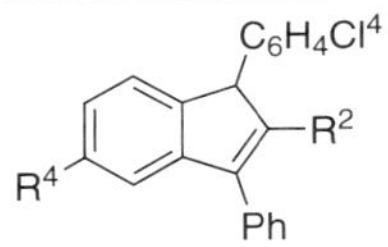	**57o**, R^2 = Ph, R^4 = H	13	61
2	**55b**	**56d**		**57p**, R^2 = *n*-Pr, R^4 = H	18	65
3	**55c**	**56a**		**57q**, R^2 = Ph, R^4 = Cl	24	66
4	**55d**	**56a**	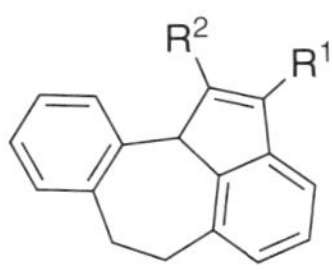	**57r**, R^1 = R^2 = Ph	6	61
5	**55d**	**56d**		**57s**, R^1 = Ph, R^2 = *n*-Pr	24	71
6	**55d**	**56k**		**57t**, R^1 = Br, R^2 = Ph	16	61
7	**55e**	**56a**	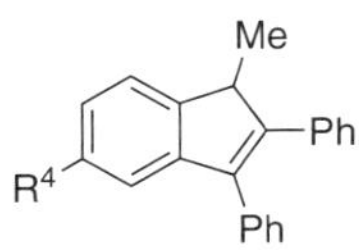	**57u**, R^4 = H	12	56
8	**55f**	**56a**		**57v**, R^4 = OTs	24	70
9	**55g**	**56a**		**57w**, R^4 = Br	24	80
10	**55h**	**56a**	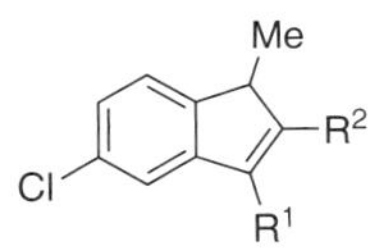	**57x**, R^1 = R^2 = Ph	24	63
11	**55h**	**56d**		**57y**, R^1 = Ph, R^2 = *n*-Pr	24	61
12	**55h**	**56j**		**57z**, R^1 = SePh, R^2 = Ph	24	61

TABLE 3.16. (*Continued*)

Entry	1	2	Product		Time (h)	Yield (%)[b]
13	**55i**	**56a**	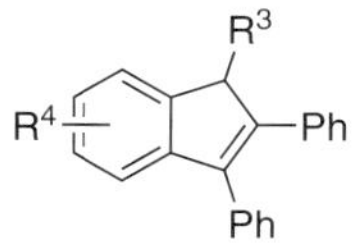	**57aa**, $R^3 = n$-Bu, R^4 = 5-Cl	24	56
14[c]	**55j**	**56a**		**57ab,** R^3 = Me, R^4 = 6-Cl	24	61
15	**55k**	**56a**		**57ac**, R^3 = Me, R^4 = 7-Cl	24	65
16	**55l**	**56h**	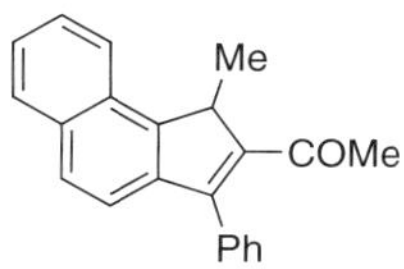	**57ad**	24	61

[a]Reaction conditions: sulfonamide **55** (0.20 mmol), alkyne **56** (0.24 mmol), $FeCl_3$ (10 mol%), nitromethane (2.0 mL), 80°C.
[b]Isolated yield.
[c]4-Chloro-1-methyl-2,3-diphenyl-1*H*-indene (57ab) was obtained in 26% yield as a minor regioisomer.

protocol (Table 3.16, entries 4–6). Moreover, the $FeCl_3$-catalyzed indene underwent synthesis smoothly with a range of *N*-monobenzylic sulfonamides and, importantly, heteroatoms including chlorine, bromine, and oxygen were smoothly introduced into the C-5, C-6, and C-7 positions of indene derivatives, respectively (Table 3.16, entries 7–16).

The same group also developed $FeCl_3$-catalyzed decarboxylative coupling of β-keto acids with sulfonamides via the combination of C–N and C–C bonds cleavage (Scheme 3.23) [21].

A broad range of β-keto acids smoothly underwent decarboxylative alkylation with sulfonamide **59a** to give structurally diverse unsymmetric ketones in moderate to excellent yields (Table 3.17, entries 1–7). This reaction tolerated electron-donating aromatic moieties (Table 3.17, entries 2 and 3); moreover, no regioisomeric alkylation byproduct was observed from the reaction with β-keto

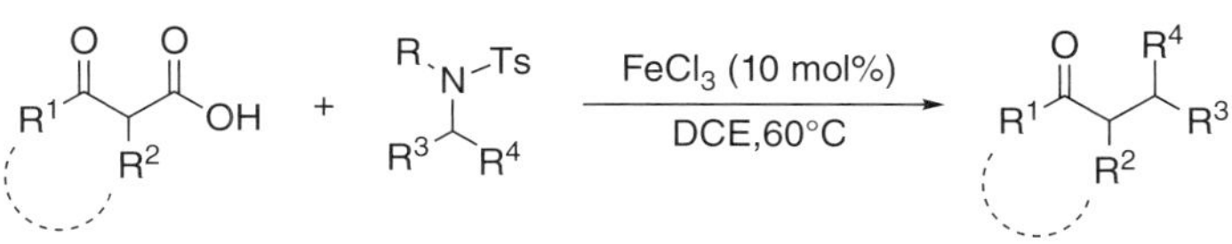

R_1 = aryl,alkyl
R_2 = H,alkyl
R_3 = aryl,alkenyl
R_4 = aryl,alkenyl,alkynyl,alkyl
R = H,alkyl

25 examples
39–98% yield

Scheme 3.23. Catalytic decarboxylative alkylation of β-keto acids with sulfonamides via the cleavage of C–N and C–C bonds.

TABLE 3.17. Decarboxylative Alkylation of β-Keto Acid with Sulfonamide 59a[a]

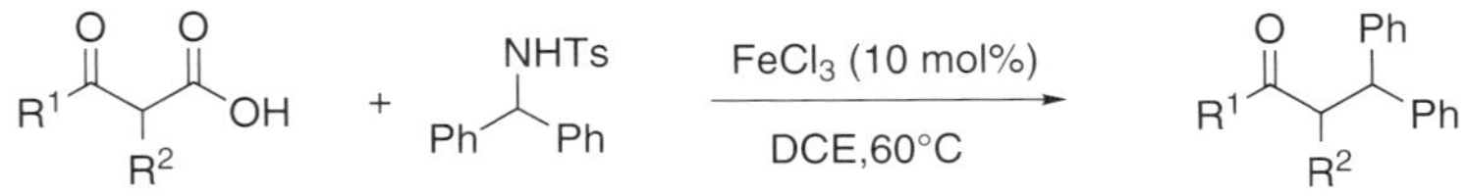

58 **59a** **60a–i**

Entry	β-keto Acid	Product	Time (h)	Yield[b] (%)
	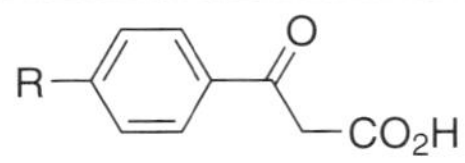	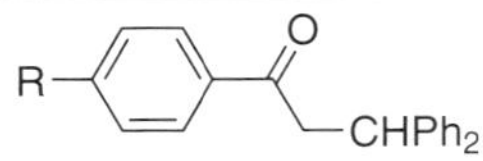		
1	**58a, R = H**	**60a, R = H**	1.5	98
2	**58b, R = OMe**	**60b, R = OMe**	2	92
	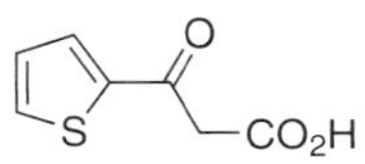	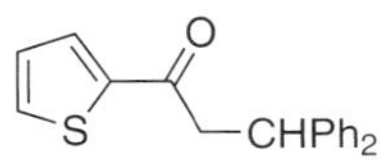		
3	**58c**	**60c**	2	74
	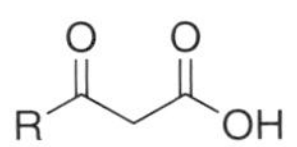	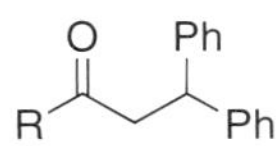		
4	**58d, R = Me**	**60d, R = Me**	2	77
5	**58e, R = $(CH_2)_2Me$**	**60e, R = $(CH_2)_2Me$**	2	92
6	**58f, R = $CHMe_2$**	**60f, R = $CHMe_2$**	2	90
7	**58g, R = CMe_3**	**60g, R = CMe_3**	1.5	79
	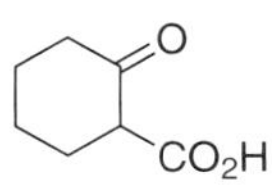	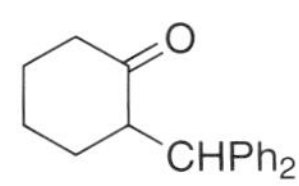		
8	**58h**	**60h**	2	56
	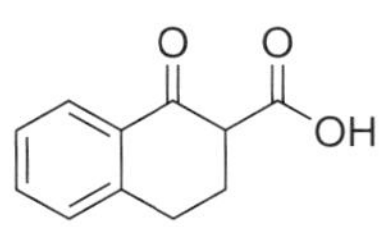	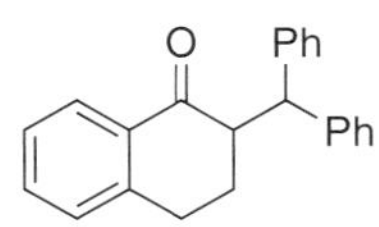		
9	**58i**	**60i**	2	39

[a]Reaction conditions: β-keto acid **58** (0.24 mmol), sulfonamide **59a** (0.20 mmol), $FeCl_3$ (10 mol%), DCE (1.0 mL), 60°C.
[b]Isolated yield.

acids **58e** and **58f** (Table 3.17, entries 5 and 6). α-Alkyl β-keto acids were also evaluated in the $FeCl_3$-catalyzed decarboxylative alkylation reaction, and the desired unsymmetric ketones were achieved in moderate yields (Table 3.17, entries 8 and 9).

A variety of *N*-bisbenzylic sulfonamides proceeded well with β-keto acid **58a** in the presence of catalytic $FeCl_3$ to give the desired 2,2-diarylethyl phenyl ketones in good yields (Table 3.18, entries 1–5). It is noteworthy that both electron-poor and electron-rich groups could be successfully introduced into the corresponding

TABLE 3.18. Decarboxylative alkylation of β-keto acid 58a with sulfonamides[a]

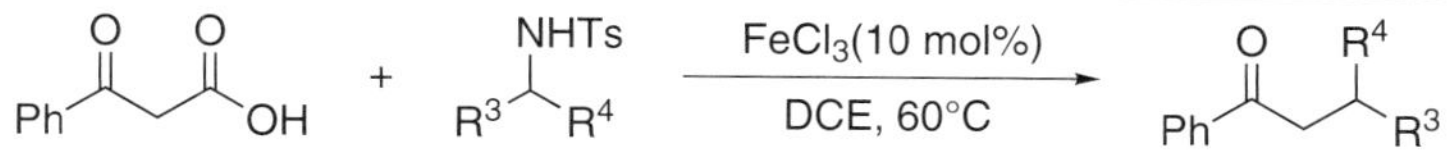

Entry	Sulfonamides	Product	Time (h)	Yield[b] (%)
	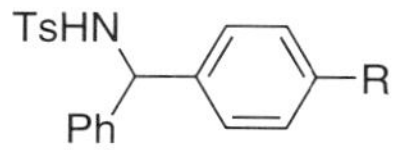	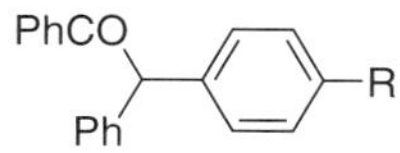		
1[c]	**59b**, R = OMe	**60j**, R = OMe	6	87
2	**59c**, R = Cl	**60k**, R = Cl	1.5	83
	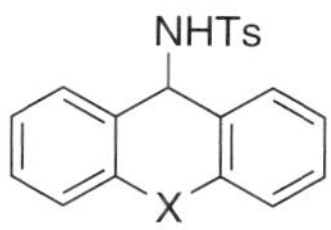	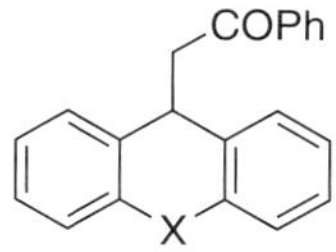		
3[d]	**59g**, R = CH_2CH_2	**60g**, R = CH_2CH_2	1	83
4	**59e**, X = S	**60e**, X = S	1.5	65
5	**59f**, X = O	**60f**, X = O	2	72
	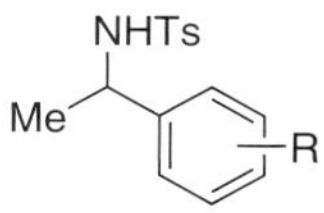	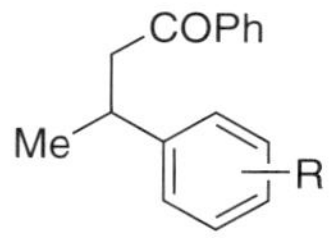		
6	**59g**, R = H	**60o**, R = H	3	52
7	**59h**, R = 4-OMe	**60p**, R = 4-OMe	1.5	59
8	**59i**, R = 2-Me	**60q**, R = 2-Me	1.5	65
	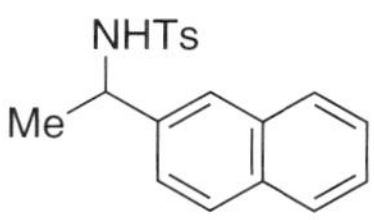	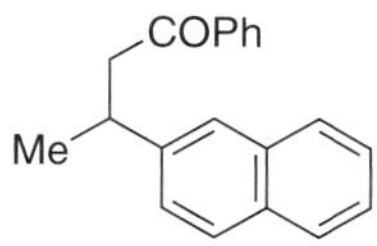		
9	**59j**	**60r**	2	58
	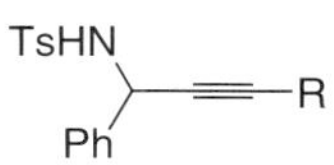	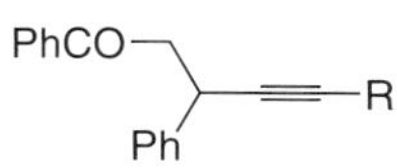		
10	**59k**, R = Ph	**60s**, R = Ph	2	83
11	**59l**, R = *n*-Bu	**60t**, R = *n*-Bu	1	71
	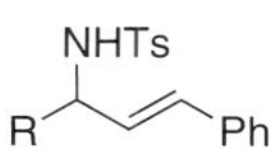	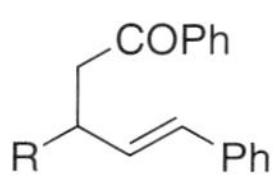		

(*continued*)

TABLE 3.18. ***(Continued)***

Entry	Sulfonamides	Product	Time (h)	Yield[b] (%)
12	**59m**, R = Ph	**60u**, R = Ph	1	73
13	**59n**, R = Me	**60v**, R = Me	1.5	54
	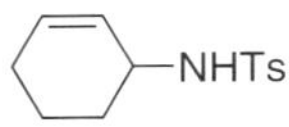	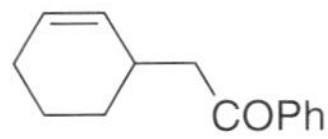		
14	**59o**	**60w**	3	45

[a]Reaction conditions: β-keto acid **58a** (0.24 mmol), sulfonamide **59** (0.20 mmol), $FeCl_3$ (10 mol%), 1,2-dichloroethane (1.0 mL), 60°C.
[b]Isolated yield.
[c]The reaction was run at room temperature.
[d]0.30 mmol of β-keto acid **58a** was used.

ketone products with the sulfonamides as substrates. The carbon–nitrogen bonds of *N*-monobenzylic sulfonamides were also cleaved under the optimized reaction conditions and reacted with *β*-keto acid **58a** to give the ketones in moderate to good yields (Table 3.18, entries 6–9). Notably, no rearrangement occurred for the C–C multiple bonds in the coupling reaction with *N*-(α-alkynylbenzyl)- or *N*-(α-alkenylbenzyl)-*p*-toluenesulfonamides (Table 3.18, entries 10–12). Moreover, *N*-allylic sulfonamides could serve as suitable substrates to react with *β*-keto acid **58a** (Table 3.18, entries 13 and 14). When less reactive sulfonamides such as *N*-benzyl *p*-toluenesulfonamide were submitted to this reaction, no desired product was detected.

Tian's group developed $FeCl_3$-catalyzed cyclization of arylallenes to construct polysubstituted indenes via C–N bond cleavage (Table 3.19) [22]. In the presence of 10 mol% ferric chloride, a wide range of 1,3-di- and 1,1,3-trisubstituted arylallenes were coupled with *N*-benzylic sulfonamide **62a** to give a variety of polysubstituted indenes in good yields with high regioselectivity (Table 3.19, entries 1–9). Moreover, *N*-bisbenzylic, *N*-monobenzylic, and *N*-allylic sulfonamides also served as suitable reaction substrates (Table 3.19, entries 10–20). Notably, both electron-poor and electron-rich groups were successfully introduced into the corresponding indene products by using the sulfonamides with such groups on the aromatic rings, and it is noteworthy that no isomerization was detected for the C–C double bonds. However, the reaction was not suitable to less reactive *N*-alkyl sulfonamides such as *N*-benzyl-*p*-toluenesulfonamide, *N*-allyl *p*-toluenesulfonamide, *N*-cyclohexyl-*p*-toluenesulfonamide, and *N*-(1-adamantyl)-*p*-toluenesulfonamide. It could be attributed to the low stability of the corresponding carbocations generated *in situ* via C–N bond cleavage.

3.7 OTHER-TRANSITION–METAL-CATALYZED C–N ACTIVATION

Gambarotta and co-workers [23] reported the reaction of the diamagnetic cluster complex (TMEDA)2Nb2Cl5Li(TMEDA) with the (3,5-Me2Ph)-(Ad)NLi

TABLE 3.19. Synthesis of Polysubstituted Indenes from Arylallenes and *N*-Alkyl sulfonamides[a]

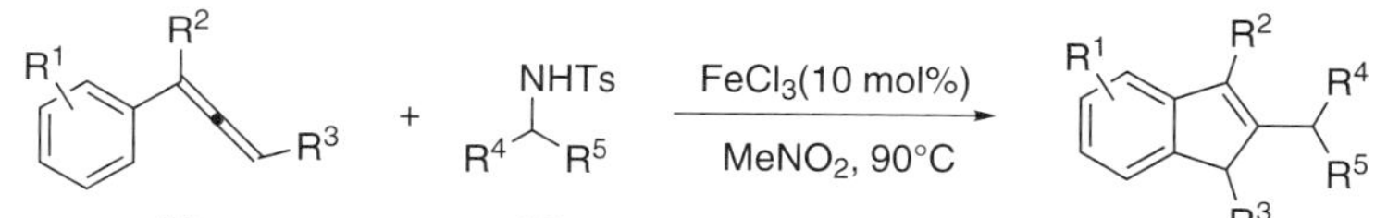

61 62 63

61a, R^1=OTs **61c**, R^1=Cl

61b, R^1=H **61d**, R^1=Br

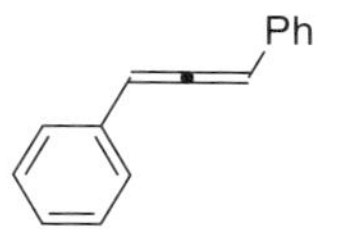

61e

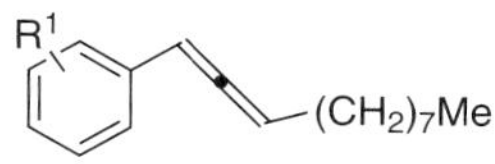

61f, R^1 = H
61g, R^1 = 4-Cl
61h, R^1 = 3-Me
61i, R^1 = 2-cl

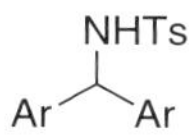

62a, Ar = Ph
62b, Ar = PMP
62c, Ar = 4-ClC_6H_4

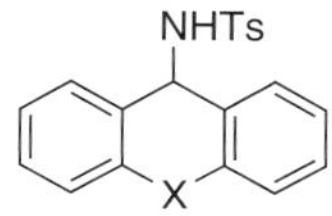

62d, X = $(CH_2)_2$
62e, X = O
62f, X = S

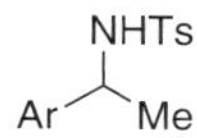

62g, Ar = Ph
62h, Ar = 4-PMP
62i, Ar = 4-ClC_6H_4
62j, Ar = 2-ClC_6H_4

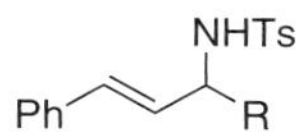

62k, R = Ph
62l, R = Me

Entry	1	2	Product		Time (h)	Yield (%)[b]
1	**61a**	**62a**	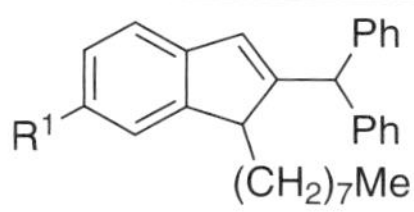	**63a**, R^1 = OTs	12	84
2	**61b**	**62a**		**63b**, R^1 = H	24	65
3[c]	**61c**	**62a**		**63c**, R^1 = Cl	12	72
4	**61d**	**62a**		**63d**, R^1 = Br	20	88
5	**61e**	**62a**	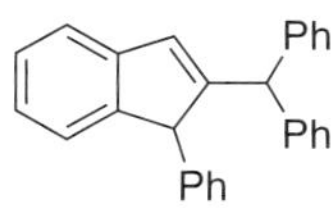	**63e**	24	62
6	**61f**	**62a**	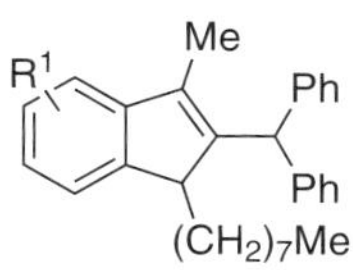	**63f**, R^1 = H	24	60
7	**61g**	**62a**		**63g**, R^1 = 6-Cl	24	72
8	**61h**	**62a**		**63h**, R^1 = 5-Me	24	71
9	**61i**	**62a**		**63i**, R^1 = 4-Cl	24	75
			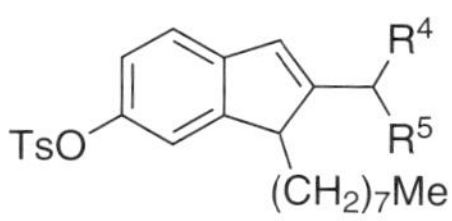			
10	**61a**	**62b**	**63j**, $R^4 = R^5$ = 4-$MeOOC_6H_4$		6	71
11	**61a**	**62c**	**63k**, $R^4 = R^5$ = 4-ClC_6H_4		24	73
12	**61a**	**62d**	**63l**, $R^4 = R^5$ = 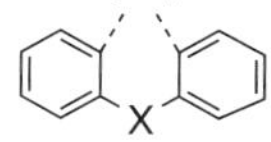	X = $(CH_2)_2$	12	75
13	**61a**	**62e**	**63m**, $R^4 = R^5$ =	X = O	12	59
14	**61a**	**62f**	**63n**, $R^4 = R^5$ =	X = S	12	89

(*continued*)

TABLE 3.19. (*Continued*)

Entry	1	2	Product	Time (h)	Yield (%)[b]
15[d]	**61a**	**62g**	**63o**, R^4 = Ph, R^5 =Me	30	87
16[e]	**61a**	**62h**	**63p**, R^4 = 4-MeoOC_6H_4, R^5 = Me	15	73
17[f]	**61a**	**62i**	**63q**, R^4 = 4-ClC_6H_4, R^5 = Me	24	85
18[g]	**61a**	**62j**	**63r**, **3q**, R^4 = 2-ClC_6H_4, R^5 = Me	36	81
19[h]	**61a**	**62k**	**63s**, R^4 = (*E*)-PhCH = CH, R^5 = Ph	12	66
20[i]	**61a**	**62l**	**63t**, R^4 = (*E*)-PhCH = CH, R^5 = Me	24	70

[a]Reaction conditions: arylallene **61** (0.3 mmol), *N*-alkyl sulfonamide **62** (0.20 mmol), $FeCl_3$ (10 mol%), nitromethane (2.0 mL), 90°C.
[b]Isolated yield.
[c]The structure of compound **63c** was confirmed by single-crystal X-ray analysis (CCDC 889428).
[d]54:46 dr.
[e]55:45 dr.
[f]93:7 dr.
[g]65:35 dr.
[h]50:50 dr.
[i]55:45 dr.

(Ad = adamantyl) amide, resulting in a rare case of C–N bond cleavage and formation of a tetravalent, dinuclear, and diamagnetic complex {(3,5-Me2Ph)(Ad)N]Nb(3,5-Me2Ph)}2(μ-NAd)2+-ether **(64)** containing a Nb–Nb single bond (Scheme 3.24) [23]. In contrast, a similar reaction with [Cy2NLi]4 (Cy = cyclohexyl) yielded a dinuclear complex {[(Cy2N)2Nb]2(*í*-N)}(*í*3-N)Li(TMEDA).

In 1999, Jun's group developed a C–N bond cleavage process catalyzed by Rh(I) complex **65**. The alkylamine system such as compound **66** was applied to the cleavage reaction of the C–N bond [24]. As expected, the electron-rich substituent such as *N,N*-dimethylaminophenyl group underwent the C–N bond cleavage to generate **67a** in excellent 94% yield, exclusively (Table 3.20, entry 1). In addition, the electron-donating ferrocenyl group also gave the similar result (entry 2). However, the electron-poor substituent, such as a 4-trifluoromethylphenyl group, did not form any product of C–N bond cleavage, but instead a small amount of dehydrogenation-hydrolysis product **68d** (entry 4). Even compound **66c** bearing no substituent on the aryl group gave a similar result (entry 3).

The plausible mechanism of these two competitive reactions is shown in Scheme 3.25. The electron-donating group facilitates the C–N bond cleavage to form the intermediate **70**, followed by β-elimination to generate intermediate **71**. Compound **67** could be liberated from intermediate **71** with the formation of intermediate **72**, which then undergoes reductive elimination to form 2-amino-3-picoline and the starting catalyst **65**. Since the electron-poor group weakens the benzylic C–H bond, it is easier for the cleavage of the C–H bond to give intermediate **73**. β-Elimination in intermediate **73** gives ketimine **74**, which is hydrolyzed to give compound **68** along with the hydride complex **75**. Complex **75** hydrogenates alkene to alkane along with the regeneration of catalyst **65**.

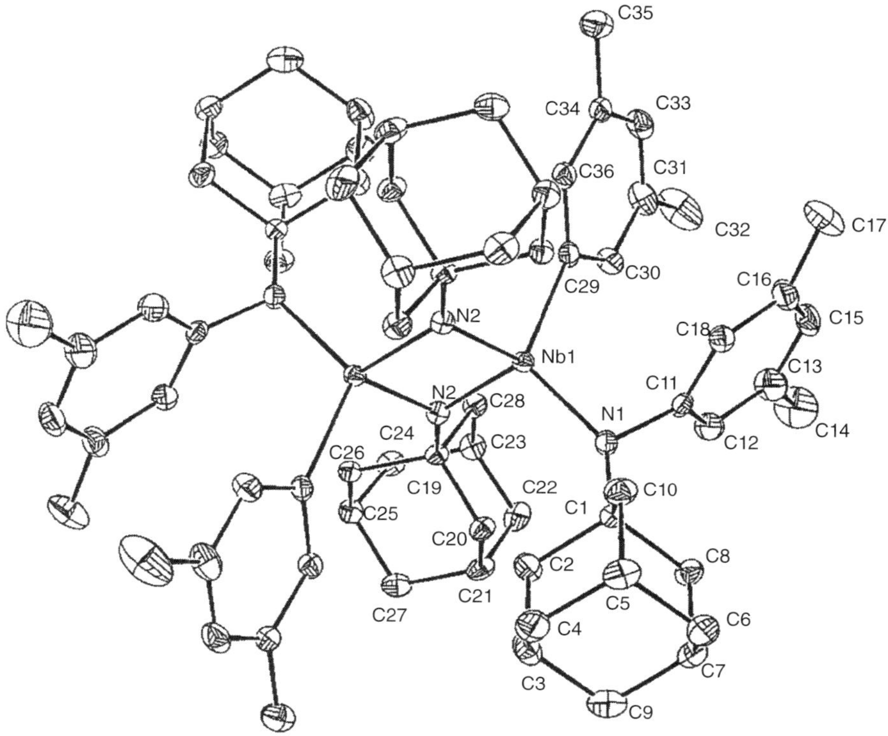

Scheme 3.24. ORTEP drawing of **64**. Thermal ellipsoids are drawn at the 30% probability level.

TABLE 3.20. Catalytic Reaction of 65 and H_2O by 10 mol% of Complex 65

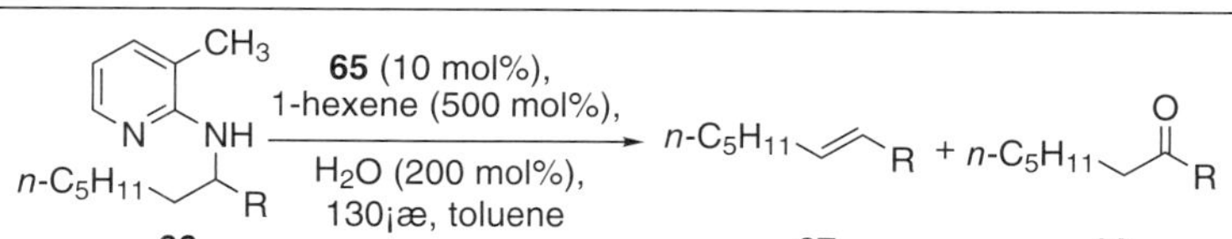

66 **67** **68**

Entry	R(1)	Reactant	Product	Isolated Yield (%)
1	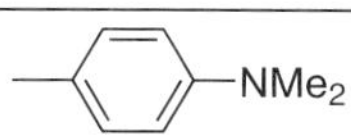	**66a**	**67a**	94
2	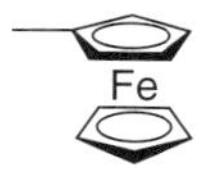	**66b**	**67b**	95
3	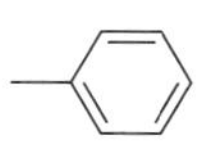	**66c**	**68c**	9
4	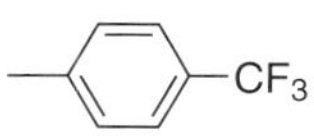	**66d**	**68d**	8

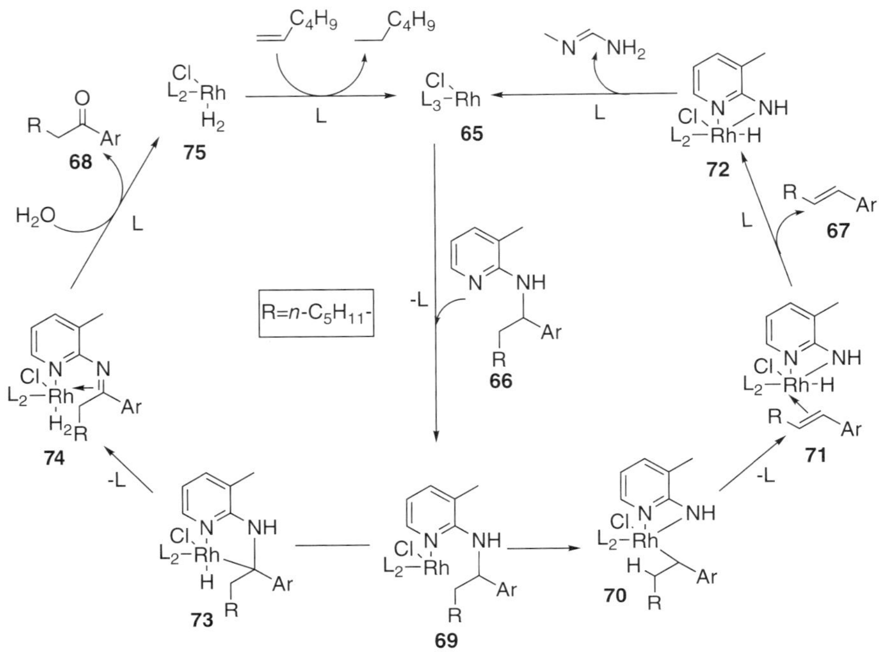

Scheme 3.25. Catalytic cycle for the C–N bond cleavage and dehydrogenation (L = PPh_3).

Wagenknecht and co-workers [25] reported that the reaction of $Co(acac)_3$ with H_2dc3 led to C–N cleavage of the ligand and the construction of a complex of dioxocyclam, Co(dc)(acac), **76** (Scheme 3.26). This complex could be prepared in high yield by the reaction of $Co(acac)_3$ with dioxocyclam. An X-ray crystallographic study shows that the dioxocyclam provides the heretofore unreported *cis* configuration, which has folded along a N–Co–N axis perpendicular to the Co–acac plane.

Ozerov et al. [26] reported an interesting investigation of oxidative addition of an sp^3 C–N bond to a transition metal center (Scheme 3.27). Notably, related N–H oxidative addition is also reported. These N–H and N–C bonds are weaker than those in aliphatic or monoaromatic amines. Solution kinetic studies demonstrate a simple migration of the Me group from a coordinated amine to Rh. It is noteworthy that in the solid state the oxidative addition of N–C bond to Rh takes place in a crystal-to-crystal fashion, introducing one of the two independent molecules into the crystal.

Recently, Tian and co-workers [27] developed a highly efficient benzylic and allylic alkylation of protic nucleophiles with sulfonamides via double Lewis acid-catalyzed cleavage of sp^3 C–N bonds under room temperature.

A variety of tosyl-activated benzylic and allylic amines were evaluated with acetyl acetone (**81a**) at room temperature in the presence of $ZnCl_2$ and TMSCl (Table 3.21). The coupling reaction of tosyl-activated benzylic amines, bearing either electron-rich groups or electron-poor groups on their aryl rings, with acetyl

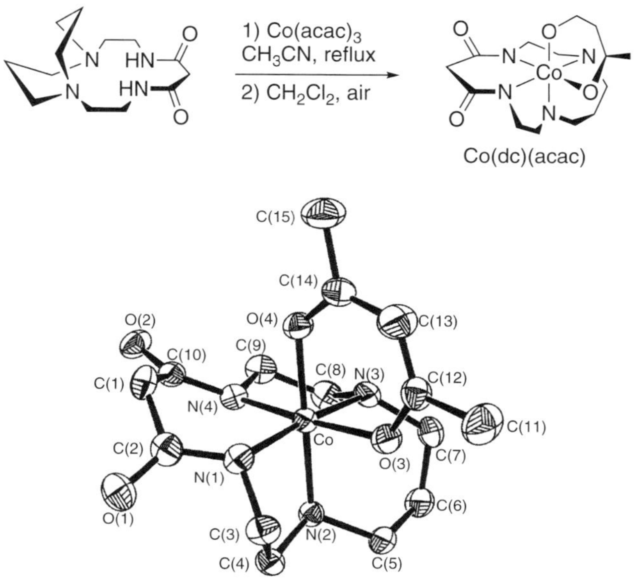

Scheme 3.26. Thermal ellipsoid plot for Co(dc)(acac), **76**, at the 50% probability level. The hydrogen atoms have been omitted for clarity.

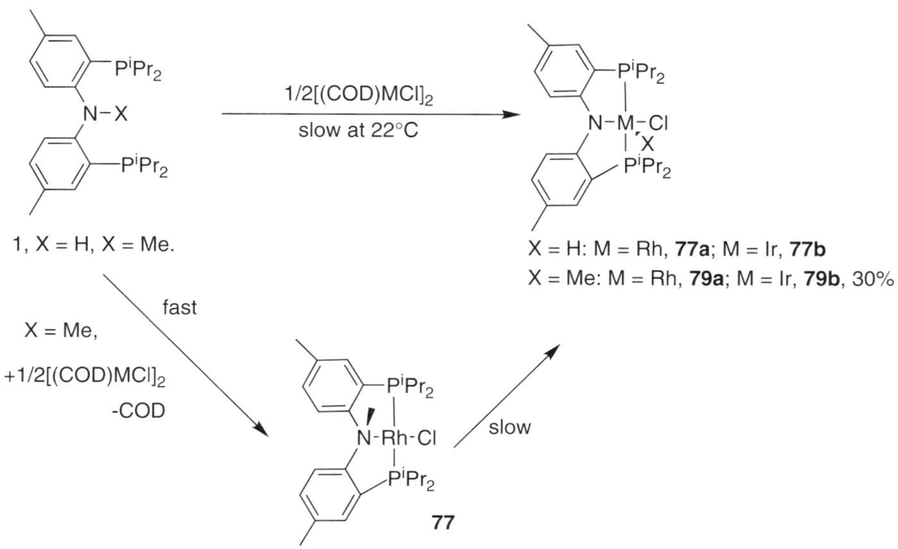

Scheme 3.27. Oxidative addition of N–C and N–H bonds to rhodium and iridium.

acetone (**81a**) occurred smoothly to form the desired 1,3-diketones in good to excellent yields (Table 3.21, entries 1–8). With respect to the allylic alkylation of acetyl acetone (**81a**), tosyl-activated allylic amines also served as suitable reaction partners (Table 3.21, entries 9–13). The attack of acetyl acetone (**81a**) to unsymmetric allylic amine derivatives **80j–80l** took place only at the allylic positions to maintain a maximum degree of conjugation, which was demonstrated by the lack of isomeric products (Table 3.21, entries 10–12). Furthermore, excellent selectivity was observed when tosyl-activated secondary amines served as benzylic and allylic alkylating agents. For example, secondary amine derivative **80af** could

TABLE 3.21. Catalytic Alkylation of Acetyl Acetone with Sulfonamides[a]

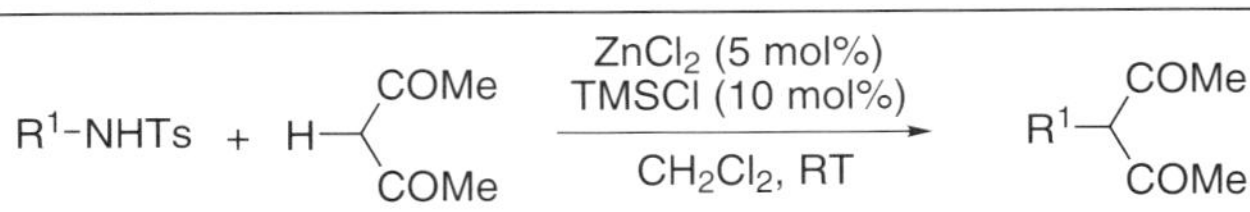

80 **81a** **82aa–82ma**

		80, R^1	Product	Time (h)	Yield (%)[b]
1	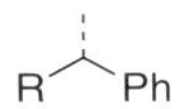	**80a**, R = Ph	**82aa**	4.5	99
2		**80b**, R = PMP	**82ba**	1	98
3		**80c**, R = 4-ClC$_6$H$_4$	**82ca**	6	89
4	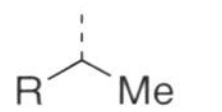	**80d**, R = Ph	**82da**	9	95
5		**80e**, R = PMP	**82ea**	8	96
6[c]		**80f**, R = 4-ClC$_6$H$_4$	**82fa**	30	85
7	**80g**	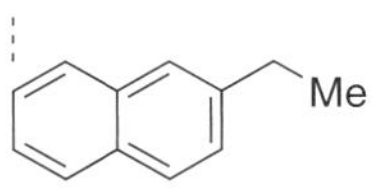	**82ga**	7	90
8	**80h**		**82ha**	24	73
9	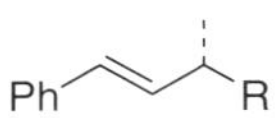	**80i**, R = Ph	**82ia**	5	90
10		**80j**, R = Mc	**82ja**	4	86
11		**80k**, R = *n*Pr	**82ka**	4	70
12	**80l**	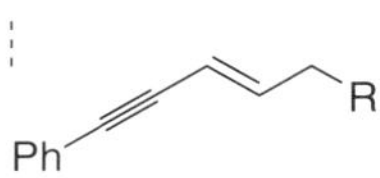	**82la**	25	77
13	**80m**		**82ma**	7	91
14	**80af**	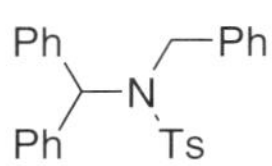	**82aa**	3	98

[a]Reaction conditions: **80** (0.50 mmol), **81a** (1.2 equiv), ZnCl$_2$ (5 mol%), TMSCl (10 mol%), CH$_2$Cl$_2$ (0.50 mL), RT.
[b]Isolated yield.
[c]10 mol% of ZnCl$_2$ and 100 mol% of TMSCl were used. PMP = 4-methoxyphenyl.

donate its benzhydryl group selectively to acetyl acetone (**81a**) in the presence of the $ZnCl_2$ and TMSCl (Table 3.21, entry 14). Notably, the coupling reaction of acetyl acetone (**81a**) with optically active sulfonamide (*R*)-1e (95% ee) was performed to give product **82ea** in nearly racemic form (Table 3.21, entry 5). This result indicates that a carbocation intermediate is generated *in situ* from the sulfonamide and then cross-couples with the nucleophile. Nevertheless, the observed racemization can also be attributed, in part, to the reversible cleavage of sp^3 C–N bonds under the standard reaction conditions. The sulfonamide (*R*)-**80e** underwent racemization smoothly under room temperature, its optical purity decreasing from 95% to 11% in the presence of 5 mol% of $ZnCl_2$ and 10 mol% of TMSCl.

As shown in Table 3.22, the scope of protic carbon nucleophiles is broad in the coupling reaction with tosyl-activated amines. Notably, β-diketones and β-keto esters showed better activity than β-keto amides and malononitrile with respect to reaction time, yield, and the amount of catalyst in the benzylic alkylation of methylene compounds with tosyl-activated amines (Table 3.22, entries 1–6). Activated arenes could undergo a Friedel–Crafts alkylation reaction with tosyl-activated allylic and benzylic amines to form the desired products in good to excellent yields and with bigger than 20:1 regioselectivity (Table 3.22, entries 7–14). Owing to their electron-rich nature, hydroxy, methoxy, and methylmercapto activate arenes in the Friedel–Crafts reaction and dominate the regioselectivity thereof. As expected, the *para* position of aryl ring in the nucleophile, if available, was alkylated by a tosyl-activated amine. In contrast, the *ortho* position was the predominate choice if the *para* position was occupied by another functional group.

Notably, mercapto-substituted arenes were also able to cross-couple with tosyl-activated benzylic and allylic amines, however, the reaction generated the corresponding thioethers instead of C-alkylation products. The coupling reaction of *N*-tosyl benzhydrylamine (**80a**) with 2-methylthiophenol (**83a**) in the presence of $ZnCl_2$ and TMSCl occurred smoothly under room temperature to form thioether **84a** in excellent yield (Table 3.23, entry 1). Likewise, thioether **84b** was achieved from the coupling reaction of 4-chlorothiophenol (**83b**) with *N*-tosyl benzhydrylamine (**80a**) in excellent yield (Table 3.23, entry 2). Next, more protic sulfur nucleophiles were applied to synthesize diverse sulfurcontaining compounds. As expected, aliphatic thiols could serve as excellent substrates in the coupling reaction with tosyl-activated amines (Table 3.23, entries 3–6). Further investigation indicated that readily available thioacetic acid proceeded well with a variety of tosyl-activated allylic and benzylic amines to give the desired thioesters in good to excellent yields (Table 3.23, entries 7–12).

Recently, Ong and co-workers [28] developed C–N bond cleavage of amino-linked NHC (*N*-heterocyclic carbene) mediated by zirconium. Treatment of $Zr(CH_2Ph)_4$ with *t*Bu-amino-linked NHC (**85**-tBu) leads to complexation of the NHC to form complex **86**-tBu as an orange crystalline solid at room temperature (Scheme 3.28). An X-ray crystallographic study of single crystals of **86**-*t*Bu shows the presence of a distorted five-coordinated complex, with the NHC binding via σ-donation to the Zr center with a bond length of 2.393(2) Å. One of the benzyl ligands in **86**-tBu exhibits an acute angle of 98.8(1)° for Zr–C(28)–C(34)

TABLE 3.22. Catalytic Alkylation of Protic Carbon Nuclephiles with Sulfonamides

$$\underset{\mathbf{80}}{R^1\text{-NHTs}} + \underset{\mathbf{81}}{\text{H-}R^2} \xrightarrow[\text{CH}_2\text{Cl}_2,\ \text{RT}]{\text{ZnCl}_2\ (5\ \text{mol\%}),\ \text{TMSCl}\ (10\ \text{mol\%})} \underset{\mathbf{82}}{R^1\text{-}R^3}$$

	80	**81**	Product, **82**	Time (h)	Yield (%)[b]
1	**80a 81b,**	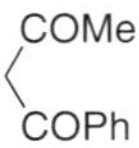	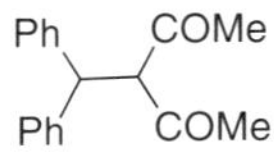	**82ab** 1	96
2	**80a** 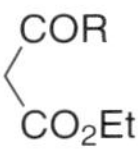	**81c**, R = Me	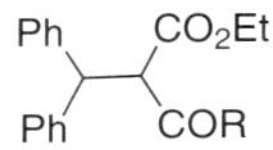	**82ac** 2	94
3	**80a**	**81d**, R = Ph		**82ad** 5	91
4	**80h 81d**		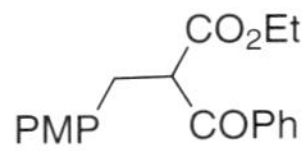	**82hd** 9	88
5	**80a 81e**	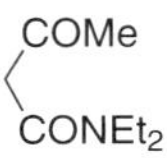	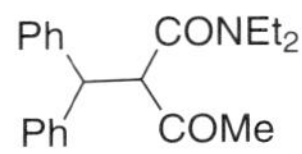	**82ae** 48	62
6	**80a 81f**	$(NC)_2CH$	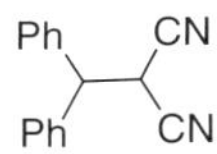	**82af** 24	80
7	**80a** 	**81g**, R = H	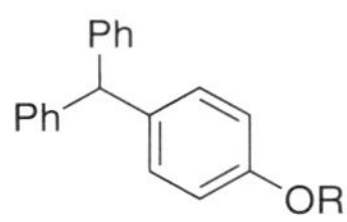	**82ag** 0.5	80
8	**80a**	**81h**, R = Me		**82ah** 1	92
9	**80h 81h**			**82hh** 3	89
10	**80j 81h**		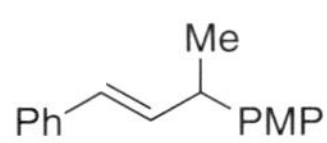	**82jh** 1.5	91
11	**80af 81h**			**82ah** 1	93
12	**80a 81i**, PhSMe		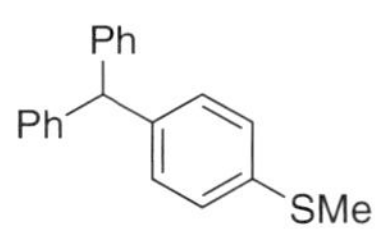	**82ai** 20	82
13	**80a** 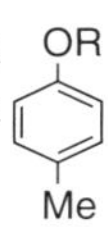	**81j**, R = H	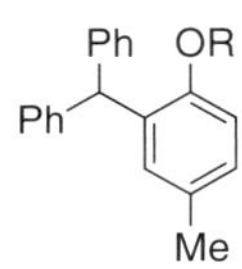	**82ai** 0.5	82
14	**80a**	**81k**, R = Me		**82ak** 1	90

[a]Reaction conditions: **80** (0.50 mmol), **81** (1.2 equiv), $ZnCl_2$ (5 mol%), TMSCl (10 mol%), CH_2Cl_2 (0.50 mL), RT.
[b]Isolated yield.
[c]10 mol% of $ZnCl_2$ and 50 mol% of TMSCl were used.

TABLE 3.23. Catalytic Alkylation of Protic Sulfur Nucleophiles with Sulfonamides

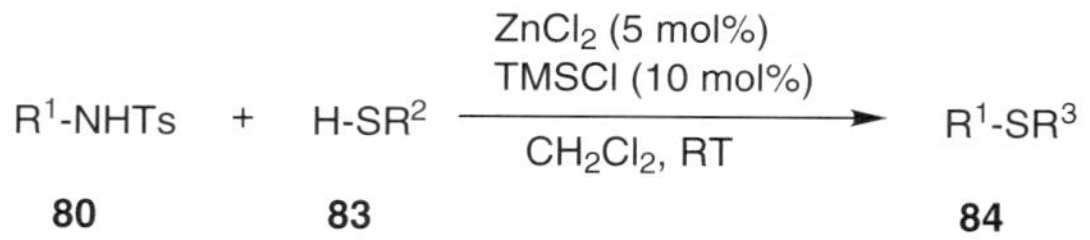

	80	83	84	Time (h)	Yield (%)[b]
1	**80a 83a**	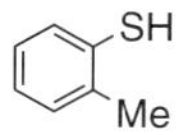	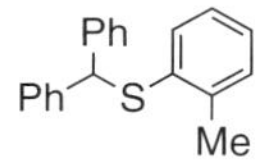	**84a** 1	99
2	**80a 83b**	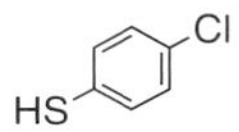	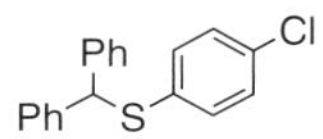	**84b** 1	98
3	**80a 83c**	$PhCH_2SH$	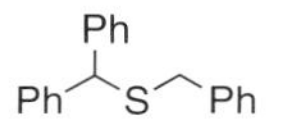	**84c** 1	92
4	**80a 83d**	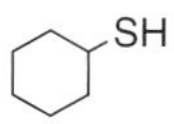	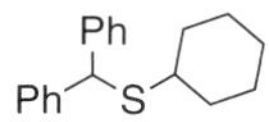	**84d** 1.5	99
5	**80h 83c**			**84e** 2	93
6	**80j 83c**		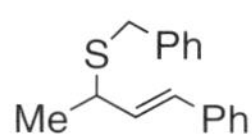	**84f** 2	94
7	**80a 83e**	AcSH	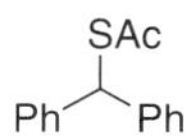	**84g** 0.5	91
8	**80e 83e**		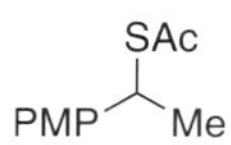	**84h** 2	93
9	**80g 83e**		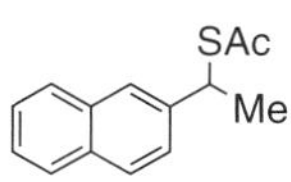	**84i** 1	83
10	**80h 83e**		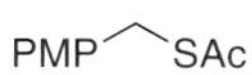	**84j** 1	56
11	**80j 83e**		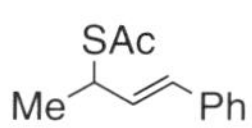	**84k** 0.5	96
12	**80n 83e**		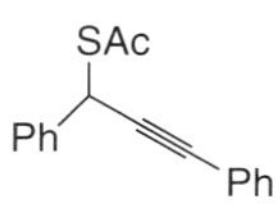	**84l** 1	81

[a]Reaction conditions: **80** (0.50 mmol), **83** (1.2 equiv), $ZnCl_2$ (5 mol%), TMSCl (10 mol%), CH_2Cl_2 (0.50 mL), RT.
[b]Isolated yield.

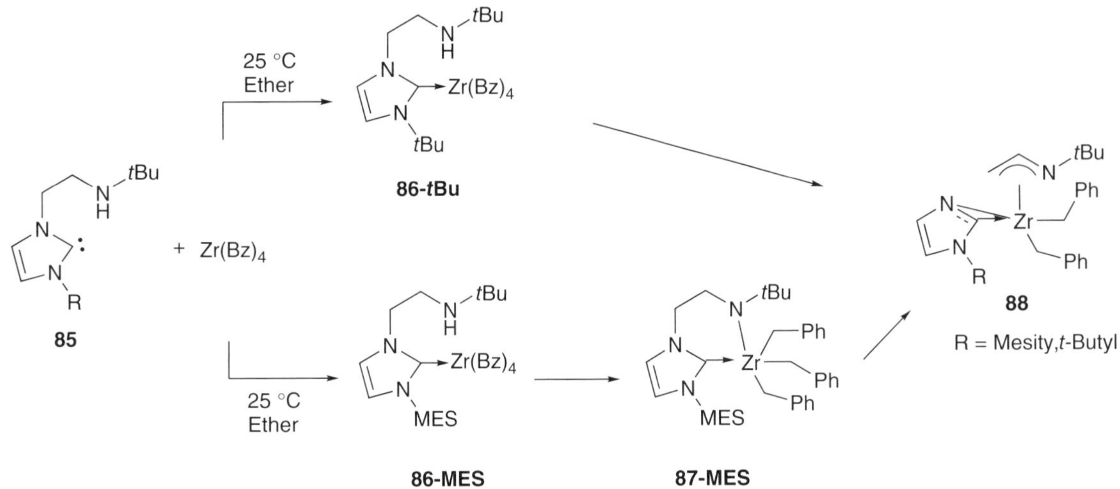

Scheme 3.28. The zirconium benzyl-mediated C–N bond cleavage.

with a nonbonded Zr–C(34) distance of 2.952(2) Å, indicating the presence of a weak η^2-benzyl coordination, as usually observed for electron-poor metal benzyl complexes.

3.8 COMPUTATIONALLY AND EXPERIMENTALLY MECHANISTIC STUDIES

Bertrand and co-workers [29] reported thiyl radical-mediated cleavage of allylic C–N bonds to construct enamines (Scheme 3.29). The catalytic cycle involves abstraction of an allylic C–Hα to a nitrogen atom by a thiyl radical and a subsequent hydrogen transfer from thiol to carbon atom γ to the nitrogen in the allylic radical intermediate. The scope and limitations of the reaction were extensively investigated. The experimental results were interpreted based upon the DFT calculations of the C–Hα BDE in the allylic amines and of the C–Hγ BDE in the product enamines. The efficiency of initial hydrogen transfer is the first requirement for the reaction to occur. A good balance should be demonsrated between S–H BDE and two above-mentioned C–H BDEs. The incidence of stereoelectronic factors was investigated via the NBO calculations carried out on the optimized geometries of the allylic amines.

Using the DFT-GGA method, of Wang's group reported the study of C–N bond cleavage of methylamine on transition metal surfaces such as Cu(1111), Ag(111), Au(111), Ni(111), Rh(111), Pd(111), Pt(111), and Mo(100) (Scheme 3.30) [30]. The calculated results demonstrate that the activation energy of C–N bond cleavage drops as the metal element rises and to the left of the periodic table, generally, which is in agreement with the experimental results. Notably, the steric effects should be considered for the metals displaying high activity and small radius such as Ni, which is different from the small molecules decomposition in which the steric effects could be ignored. The linear relationships between electronic properties (d-band center) and the activation energies are provided. It is expected that the rule can be applied to forecast the activity of metal for other dissociative adsorption systems.

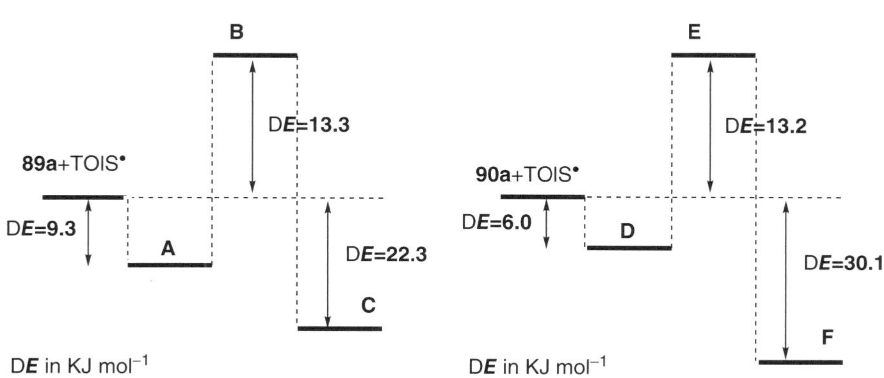

Scheme 3.29. Reaction profiles for TolS•-mediated hydrogen abstractions from **89a** and **90a**.

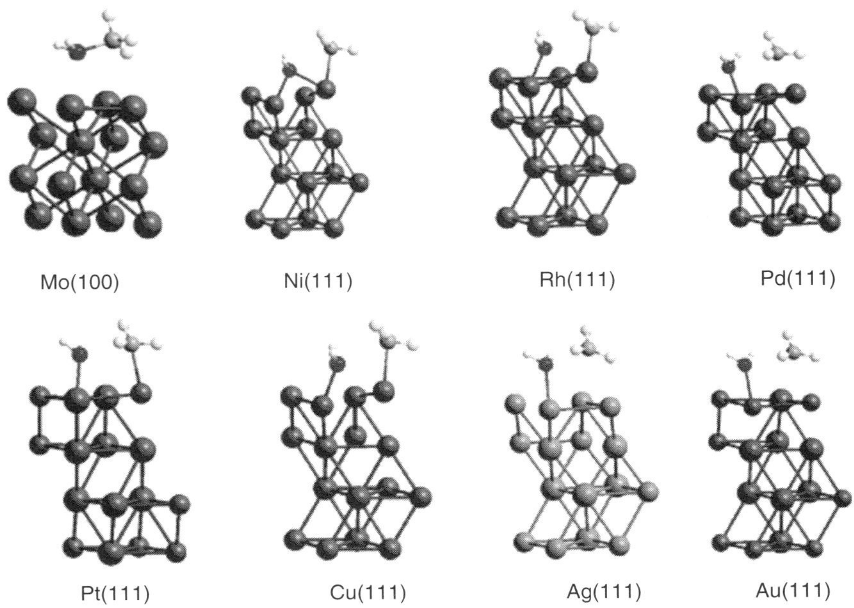

Scheme 3.30. Possible TS structures of C–N bond cleavage during methylamine adsorption on the investigated transition metal surfaces.

Recently, Chen and Li [31] carried out CASPT2//CASSCF investigation on the mechanism of the photolysis of 2,3-diazabicyclo[2.1.1]hex-2-ene (Scheme 3.31) [31]. A tentative description on this photolysis processes was provided via exploring the potential energy surfaces and the surface crossing points for some low-lying excited states. Based upon calculations, on the S1 surface one C–N bond is cleaved first to form the diazenyl biradicals, leading to the photoproduct bicyclobutane via a concerted C–N cleavage and C–C coupling process. The intersystem crossing (ISC) from the S1 state to triplet state was demonstrated to be very efficient, thus accounting for two other photoproducts (butadiene and 1,2-diazabicyclo[3.1.0]hex-2-ene) on the direct irradiation. On the T1 surface, the C–C bond cleavage occurred smoothly to give the triplet hydrazonyl biradical. One deactivation path of this active intermediate is to convert into a singlet biradical through an ISC process. Next, this biradical undergoes a barrierless C–N bond formation to provide the photoproduct 1,2-diazabicyclo[3.1.0]hex-2-ene. Another destination of this triplet hydrazonyl biradical is to generate the intermediate 2-allyl-diazomethane via a C–N cleavage. This intermediate will cleave another C–N bond to give a terminal carbene species, which is readily converted into butadiene and bicyclobutane via 1,2-hydrogen shift and addition reaction to C=C double bond, respectively.

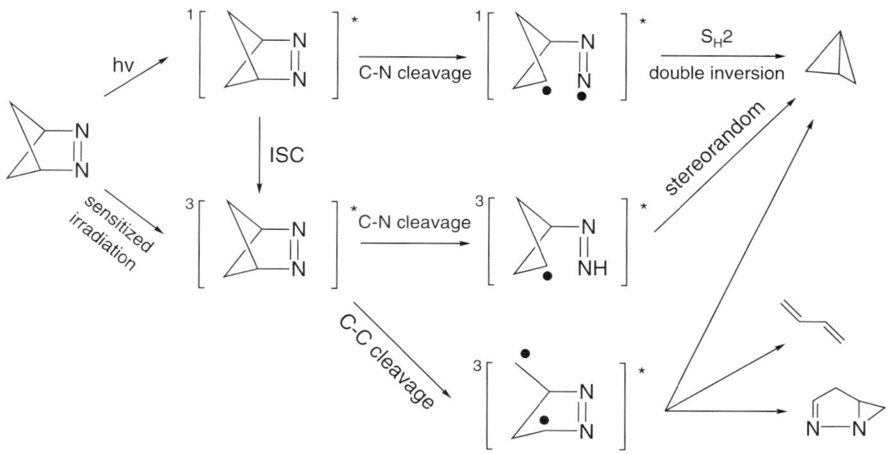

Scheme 3.31.

3.9 SUMMARY AND OUTLOOK

This review summarizes the impressive development of the transition-metal-catalyzed C–N activation in synthetic methodology. Although important achievements have been obtained, this concept is still in its infancy. A survey through the literature indicates that the number of publications from 2001 to 2011 has increased year by year. We hope that this short review will inspire the synthetic community to continue the search for new C–N reactions. Further investigations will likely find application in total synthesis and medicinal chemistry.

REFERENCES

1. Roglans, A., Pla-Quintana, A., Moreno-Mañas, M. (2006). *Chem. Rev.* **106**, 4622.
2. Taylor, J. G., Moro, A. V., Correia, C. R. D. (2011). *Eur. J. Org. Chem.* 1403.
3. Fan, L., Yang, L., Guo, C., Foxman, B. M., Ozerov, O. V. (2004). *Organometallics* **23**, 4778.
4. Saeki, T., Son, E.-C., Tamao, K. (2004). *Org. Lett.* **6**, 617.
5. Lei, Y., Wrobleski, W. G., Golden, J. E., Powell, D. R., Aubé, J. (2005). *J. Am. Chem. Soc.* **127**, 4552.
6. Ye, J., Zhang, X., Chen, W. and Shimada, S. (2008). *Organometallics* **27**, 4166.
7. Golisz, S. R., Hazari, N., Labinger, J. A., Bercaw, J. E. (2009). *J. Org. Chem.* **74**, 8441.
8. Zhu, M-K., Zhao, J-F., Loh,T.-P. (2011). *Org. Lett.* **13**, 6308.
9. Zhao, X., Liu, D., Guo, H., Liu, Y., Zhang, W. J. *Am. Chem. Soc.* **133**, 19354.

10. Li, M.-B., Wang, Y., Tian, S-K. *Angew. Chem. Int. Ed.* **51**, 2968.
11. Wu, X.-S., Chen, Y., Li, M-B., Zhou, M-G., Tian, S.-K. (2012). *J. Am. Chem. Soc.* **134**, 14694.
12. Ueno, S., Chatani, N., Kakiuchi, F. (2007). *J. Am. Chem. Soc.* **129**, 6098.
13. Koreeda, T., Kochi, T., Kakiuchi, F. (2009). *J. Am. Chem. Soc.* **131**, 7238.
14. Burling, S., Mahon, M. F., Powell, R. E., Whittlesey, M. K., Williams, J. M. J. (2006). *J. Am. Chem. Soc.* **128**, 13702.
15. Häller, L., J. L., Page, M. J., Erhardt, S., Macgregor, S. A., Mahon, M. F., M. Naser, M. A., Vélez, A., Whittlesey, M. K. (2010). *J. Am. Chem. Soc.* **132**, 18408.
16. Blakey, S. B., MacMillan, D. W. C. (2003). *J. Am. Chem. Soc.* **125**, 6046.
17. Liu, J., Robins, M. J. (2004). *Org. Lett.* **6**, 3421.
18. Li, M.-B., Tang, X-L., Tian, S.-K. (2011). *Adv. Synth. Catal.* **353**, 1980.
19. Zhang, J., Chen, J., Liu, M., Zheng, X., Ding, J., Wu, H. (2012). *Green Chem.* **14**, 912.
20. Liu, C.-R., Yang, F-L., Jin, Y-Z., Ma, X.-T., Cheng, D.-J., Li, N., Tian, S.-K. (2010). *Org. Lett.* **12**, 3832.
21. Yang, C.-F., Wang, J.-Y., Tian, S.-K. (2011). *Chem. Commun.*, **47**, 8343.
22. Liu, C.-R., Wang, T.-T., Qi Q.-B., Tian, S.-K. (2012). *Chem. Comm.* **48**, 10913.
23. Tayebani, M., Feghali, K., Gambarotta, S., Bensimon, C. (1997). *Organometallics* **16**, 5084.
24. Jang, S.-H., Jun, C.-H. (1999). *Bull. Korean Chem. Soc.* **20**, 30.
25. Hu, C., Chin, M. R., Nguyen, T. D., Nguyen, K. T., Wagenknecht, P. S. (2003). *Inorg. Chem.* **42**, 7602.
26. Ozerov, O. V., Guo, V., Papkov, V. A., Foxman, B. M. (2004). *J. Am. Chem. Soc.* **126**, 4792.
27. Liu, C.-R., Li, M.-B., Yang, C.-F., Tian, S.-K. (2009). *Chem. Eur. J.* **15**, 793.
28. Hu, Y.-C., Tsai, C.-C., Shih, W.-C., Yap, G. P. A., Ong, T.-G. (2010). *Organometallics* **29**, 516.
29. Escoubet, S., Gastaldi, S., Timokhin, V. I., Bertrand, M. P., Siri, D. (2004). *J. Am. Chem. Soc.* **126**, 12343.
30. Li, J., Li, R.-F., Wang, G.-C. (2006). *J. Phys. Chem. B* **110**, 14300.
31. Chen, H., Li, S. (2006). *J. Org. Chem.* **71**, 9013.

CHAPTER 4

CATALYTIC CARBON–SULFUR BOND ACTIVATION AND TRANSFORMATIONS

ZHENGKUN YU
Dalian Institute of Chemical Physics, Chinese Academy of Sciences, Dalian, Liaoning, People's Republic of China

4.1 BACKGROUND

Chemical and catalytic C–S bond activation by transition metals has been extensively studied due to its importance in petroleum industry [1–3]. Hydrodesulfurization (HDS) is among one of the two most important steps in the processing of petroleum, by which sulfur is removed from hydrocarbons during the refinement of petroleum. Inefficient removal of sulfur during this process usually results in noxious sulfur oxides upon fuel combustion, which has been considered as the major cause to air pollution. Ni/Mo and Mo/Co sulfide-based heterogeneous catalysts are effective for the removal of most organosulfur compounds, but removing substituted thiophenes, dibenzothiophenes (DBT), and so on, still remains a challenge. To improve catalyst efficiency for removing these organosulfur compounds is strongly desired [4–8]. In this aspect, homogeneous transition metal compounds have been explored with the goal of providing the fundamental insight into C–S bond cleavage in these organosulfur compounds.

Homogeneous Catalysis for Unreactive Bond Activation, First Edition. Edited by Zhang-Jie Shi.

Transition-metal-promoted C–S activation or cleavage has been applied in organic transformations for diverse synthesis [9, 10]. Recently, cross-coupling has been paid much attention in the construction of new chemical bonds, and organosulfur compounds have exhibited increasing importance as promising coupling partners. This chapter summarizes the recent advances in transition-metal-mediated C–S bond activation and transformations and also gives a brief overview of C–S bond activation by stoichiometric transition metal compounds.

4.2 C–S BOND ACTIVATION BY TRANSITION METAL COMPOUNDS

C–S bonds can be activated by transition metals to form novel sulfur compounds or sulfide complexes. This transformation allows insight into the mechanistic aspects of two related heterogeneous catalytic HDS processes [11]. Transition-metal-mediated bond cleavage reactions have been well investigated and established for various C–S bonds. Herein, an overview of C–S bond cleavage by stoichiometric transition metal compounds will be presented, although a variety of reports have been documented in this area [1–3, 12–26]. The insertion of metal atoms to C–S bonds in thiophenes and other organosulfur compounds have been extensively studied to get a better understanding of the heterogeneous HDS mechanisms. Dibenzothiophene, benzothiophene (BT), thiophene, and their methyl-substituted derivatives as well as heteroatom-containing functional group-substituted thiophene derivatives can be used as the substrates. Platinum(0) complexes are usually utilized for this purpose.

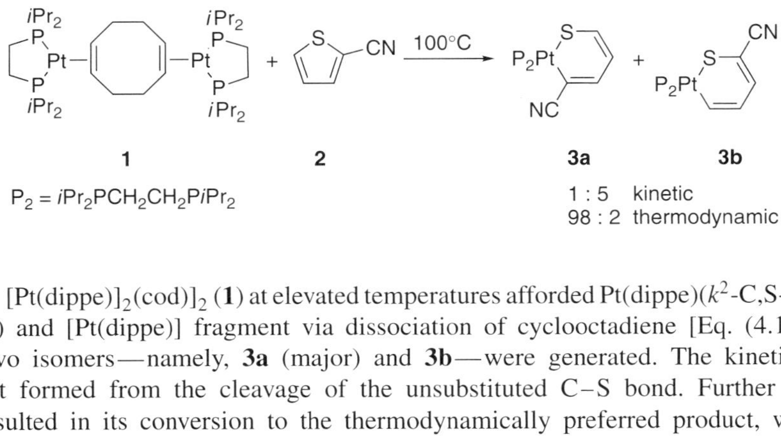

(4.1)

$[Pt(dippe)]_2(cod)]_2$ (**1**) at elevated temperatures afforded Pt(dippe)(k^2-C,S-C_4H_3S) (**3**) and [Pt(dippe)] fragment via dissociation of cyclooctadiene [Eq. (4.1)] [12]. Two isomers—namely, **3a** (major) and **3b**—were generated. The kinetic product formed from the cleavage of the unsubstituted C–S bond. Further heating resulted in its conversion to the thermodynamically preferred product, which is presumably generated from the cleavage of the substituted C–S bond. Dimeric rhodium(I) complex $[Rh(dippe)(\mu\text{-}H)]_2$ (**4**) reacted with dibenzothiophene (**5**) to form the C–S cleavage product $[Rh_2(dippe)_2(\mu\text{-}SC_{12}H_9)(\mu\text{-}H)]$ (**6**) [Eq. (4.2)] [13]. By increasing the temperature to 135°C, **4** also efficiently reacted with 4-methyldibenzothiophene and 4,6-dimethylthiophene to produce products of type **6**. Treatment of 4-(2′-pyridyl)dibenzothiophene (**7**) with ruthenium carbonyl cluster $Ru_3(CO)_{12}$ formed dimeric Ru(II) complex **8** [Eq. (4.3)], where the ligand

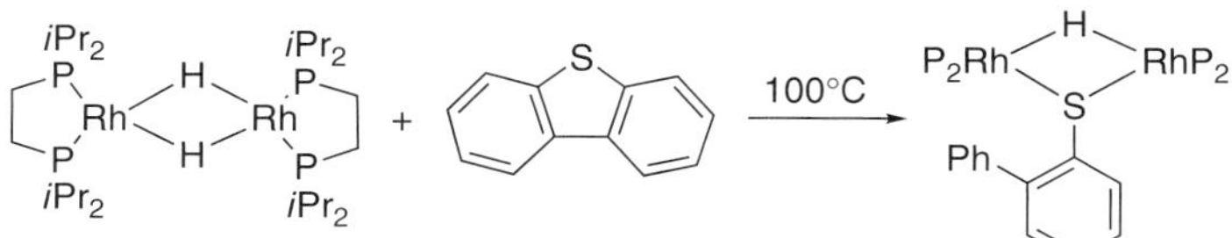

(4.2)

4 5 6

$P_2 = iPr_2PCH_2CH_2PiPr_2$

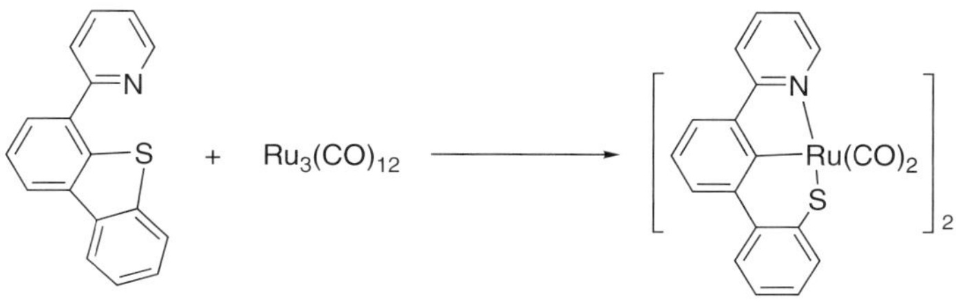

(4.3)

7 8

donates a dianion of 3′-(2″-pyridyl)-1,1′-biphenyl-2-thiol [14]. The tridentate-N,C,S ligand provides a pincer structure consisting of a six-membered thiaruthenacycle and a five-membered azaruthenacycle. Photolysis of a hexane solution containing $Fe(CO)_5$ and CS_2 led to desulfurization and formation of a novel cluster $[\{Fe_2(CO)_6\}_2(\mu\text{-}C_2S_3)]$ (**9**) [Eq. (4.4)] [15]. Activation of the C=S bond is selective

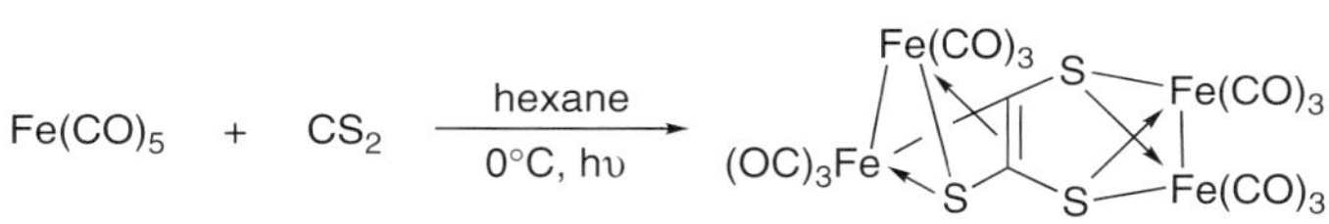

(4.4)

9

for the reaction with $Fe(CO)_5$, and the unique C_2S_3 ligand acts as a bridge between two $Fe_2(CO)_6$ units and forms an unusual ferrathiacyclobutene ring. C–S bond cleavage has also been achieved by means of the reaction of pyridylmethylthioether (**10**) with the oxorhenium(V) ion, that is, $Bu_4N[ReOCl_4]$ (**11**) [Eq. (4.5)] [16].

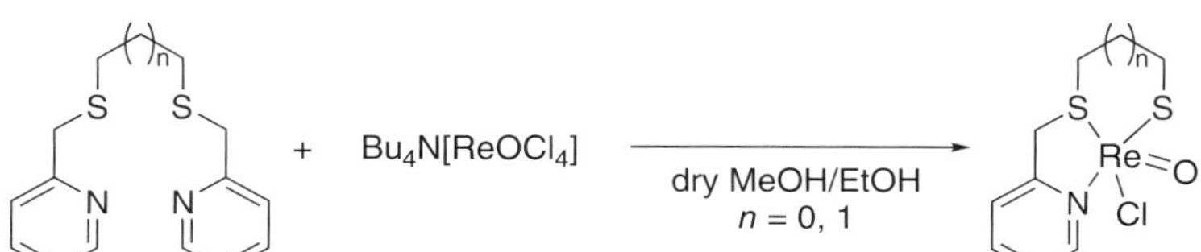

(4.5)

10 11 12

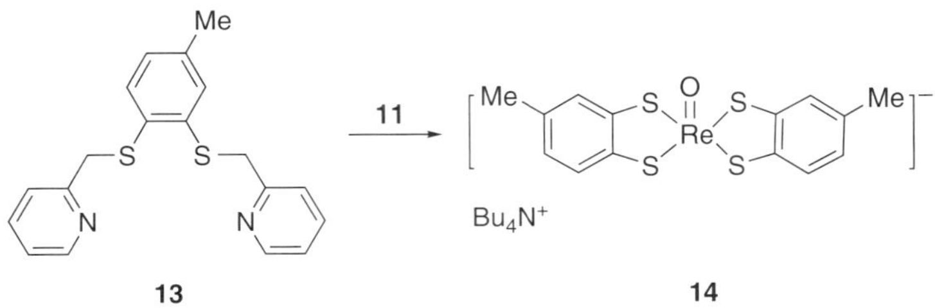

(4.6)

The structural results for **12** evidenced the unexpected cleavage of one of the C–S bonds within **11** during metal complexation. However, treatment of 3,4-bis(2-pyridylmethylthio)-5-methyltoluene (**13**) with **11** under the same conditions led to the formation of a binary oxorhenium(V) complex of 3,4-dimercapto-toluene (**14**) [Eq. (4.6)]. Aerobic reaction of $Co(O_2CMe)_2{\cdot}4H_2O$ with thioether-containing acyclic pyrazine amide hexadentate ligand 1,4-bis[*o*-(pyrazine-2-carboxamidophenyl)]-1,4-dithiobutane (**15**) furnished complex **16**, resulting from a C–S bond cleavage reaction triggered by an acetate ion as a base, having CoN_2(pyrazine)N'_2(amide)S(thioether)S(thiolate) coordination [Eq. (4.7)] [17].

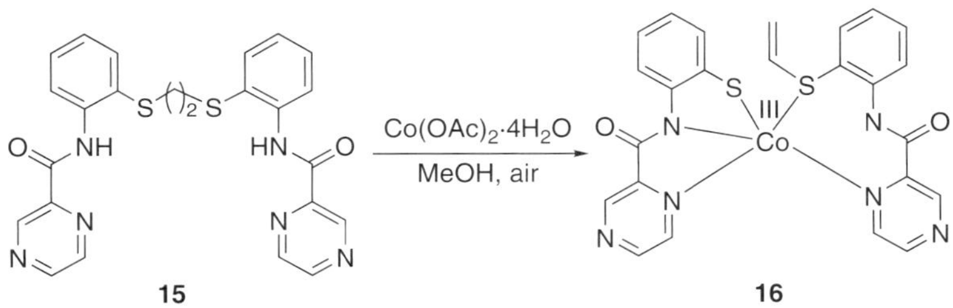

(4.7)

Nickel-mediated C–S cleavage of sulfoxides containing sp^2-and sp^3-hybridized carbon bonds attached to the sulfur atom was reported [18]. The reaction of dinuclear complex $[Ni(NHC)]_2(COD)$ (**17**) with two equivalents of dimethyl sulfoxides (DMSO) or phenyl methyl sulfoxide (PMSO) in toluene smoothly yielded products **18** by means of an sp^3 C–S(O) bond cleavage of DMSO and sp^2 C–S(O) bond cleavage of PMSO [Eq. (4.8)].

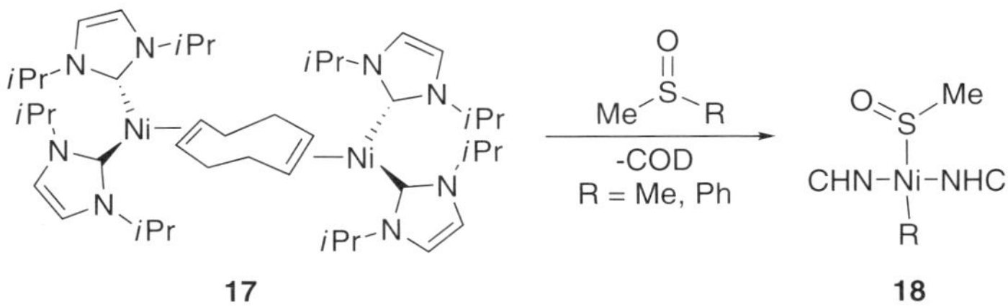

(4.8)

Reaction of trithiolato-bridged complex $[Mo_2Cp_2(\mu\text{-SMe})_3(MeCN)_2](BF_4)$ with acyclic (e.g., Et_2S) or cyclic (e.g., thiirane, thietane, tetrahydrothiophene, 1,4-dithiane, and 1,4-thioxane) thioethers and chalcogenophenes (benzothiophene and dibenzothiophene) in dihaloalkanes led to either (di)thioether- and halide-bridged compounds or μ-sulfido complexes [9]. The complexes of other transition metals such as tungsten [19], tantalum [20], iridium [21], copper [22], zirconium [23], manganese [24], chromium [25], and uranium [26] have also been applied to activate or cleave C–S bonds. All these organometallic systems can be chosen as the models for homogeneous HDS of thiophene and related organosulfur compounds in order to get a deeper insight into the heterogeneous HDS of petroleum feedstocks.

4.3 CATALYTIC C–S CLEAVAGE IN THIOESTERS

By means of transition metals as catalysts, catalytic carbon–carbon cross-coupling reactions can be employed as the most powerful and flexible transformations in organic synthesis [27]. The cross-coupling reactions usually involve the interaction of an organometallic reagent with an electrophilic organohalide or related analogue. Organosulfur compounds have also been documented to act as electrophilic coupling partners [28–30]. Organosulfur compounds can be readily added to low-valent transition metals [12–26], but the key to turn the reaction catalytic is activation of the very stable bond formed between the catalytically active metal and the soft sulfur atom. The nucleophilic organometallic reagent for the following transmetalation step is crucial for the reaction to smoothly proceed. In 1997, Liebeskind and co-authors [31] reported palladium- and nickel-catalyzed cross-coupling reactions of tetramethylene sulfonium salts with organoboron, -tin, and -zinc reagents under mild conditions, and then documented the first examples of palladium-catalyzed, copper-mediated cross-coupling between thiol esters (thioesters) and boronic acids under base-free conditions [32]. The so-called "Liebeskind–Srogl cross-coupling" features the usage of stoichiometric amount of a sacrificial copper(I) carboxylate compound. Copper(I) thiophene-2-carboxylate (CuTC) [33] was found to act as an effective thiophilic metal reagent because boron atoms only demonstrate low thiophilicity and organoboron compounds exhibit relatively low nucleophilic reactivity.

Thioesters and boronic acids did not undergo the cross-coupling reactions in the presence of a palladium catalyst [34]. However, the palladium-catalyzed thioester-boronic acid cross-coupling occurred when simple alkylating agents were present [Eq. (4.9)] [35]. Alkylative conversion of the very stable palladium-thiolate bond in intermediate **21** to a labile palladium–thioether bond is presumed to be crucial to the catalytic cycle. Such a cross-coupling efficiently afforded ketones **20** as the products. As a new synthetic methodology, substituted alkynes were also prepared by means of palladium-catalyzed, copper(I) carboxylate-mediated thioalkyne-boronic acid cross-coupling [36]. Using heteroaromatic thioethers as the substrates, the pseudo-Liebeskind–Srogl reactions occurred efficiently to form

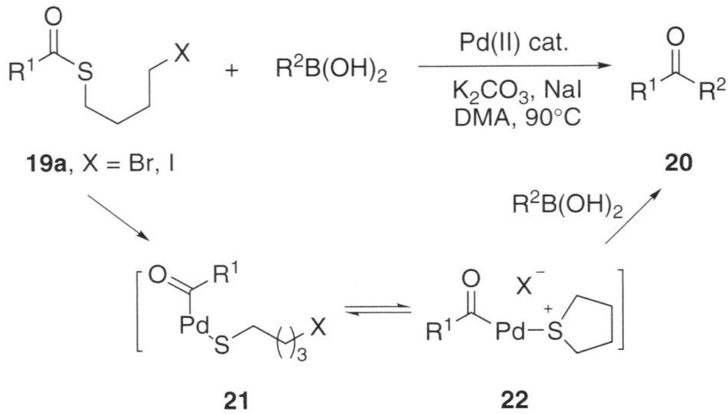

(4.9)

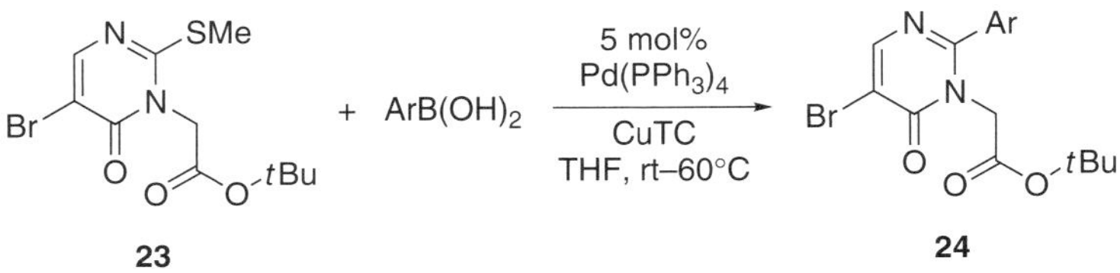

(4.10)

functionalized heteroarenes [Eq. (4.10)] [37, 38]. With B-alkyl-9-BBN (**26**) as the nucleophilic reagent, the cross-coupling of thioesters (**25**) proceeded in the presence of a base such as Cs_2CO_3 [Eq. (4.11)] [39], revealing an additive effect (base effect) different from the Liebeskind–Srogl cross-coupling reactions of thioorganics with other boronic acids. This method was successfully applied for the synthesis of protected benzamidines (**28**) [40]. The SEM-protected thiopseudourea **27** functioned as the amidine-forming cross-coupling partner under Liebeskind–Srogl reaction conditions, and the fully protected benzamidines **28** were obtained in good to excellent yields [Eq. (4.12)].

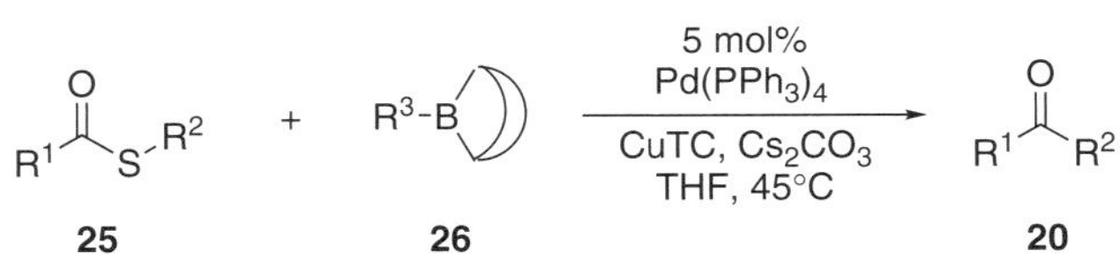

(4.11)

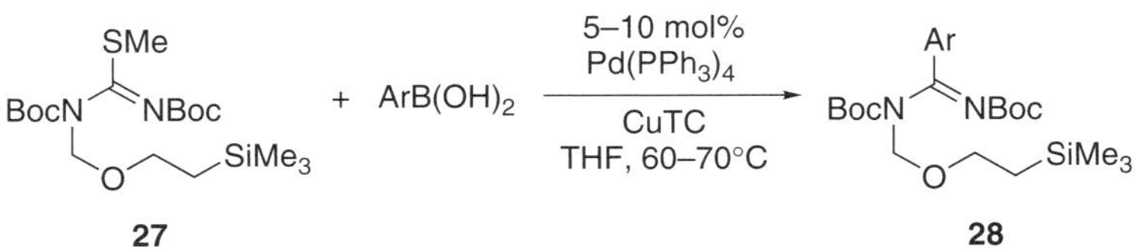

(4.12)

By means of other reagents, Liebeskind–Srogl cross-coupling has also exhibited its versatility in carbon–carbon bond formation [29]. Organoindiums were reported as the coupling partners with thioesters. With $Pd(MeCN)_2Cl_2$ as catalyst, the reaction of thioesters with aryl, primary, and secondary alkyl organoindiums proceeded to efficiently afford aryl alkyl ketones **30** [Eq. (4.13)] [41]. This protocol has shown two advantages over the cross-coupling of thioesters with organoboron and -tin reagents: (a) No copper(I) reagent was required, and (b) no base was necessary to activate the

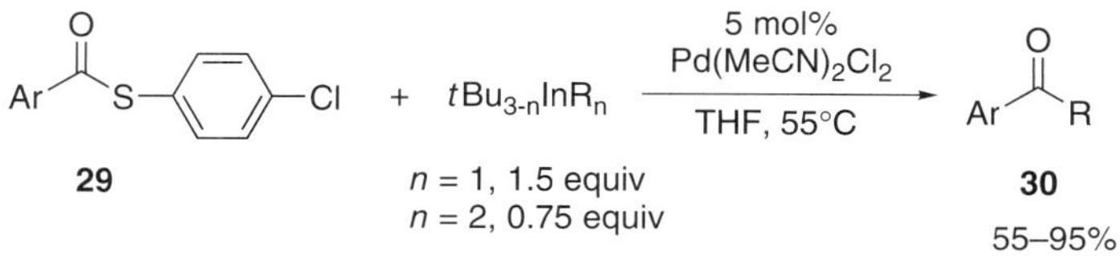

(4.13)

organoindium reagents [39]. $Pd(PPh_3)_4$ also catalyzed the cross-couplings of thioesters with alkyl- and arylzinc reagents [Eq. (4.14)] [42]. In a similar fashion, nitriles were efficiently synthesized from the reactions of alkyl-, benzyl- or arylthiocyanates, which is considered as a complement to the classic cyanation of aryl halides by using a cyanide source and a transition metal catalyst [Eq. (4.15)] [43].

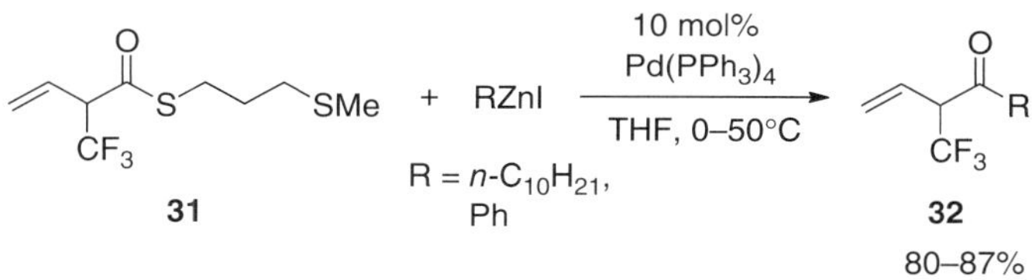

(4.14)

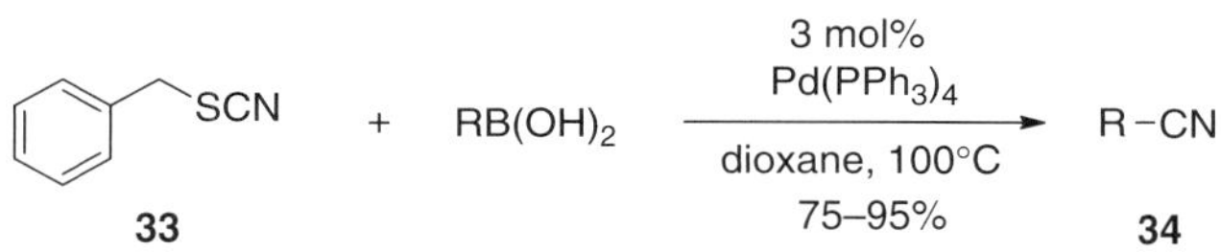

(4.15)

Functionalized α-amino acid thioesters **35a** derived from *N*-protected peptides were reacted with aryl, heteroaryl, or alkenyl boronic acids in the presence of stoichiometric CuTC and $Pd_2(dba)_3$ catalyst to efficiently form *N*-protected peptidyl ketones **36a** [Eq. (4.16)] [44]. The triethylphosphite ligand played a key role in inhibiting undesired palladium-catalyzed decarbonylative β-elimination of the amino thioester substrates. The reaction proceeded at room temperature in the absence of a base, demonstrating a high tolerance to different functionalities. This synthetic methodology provides a potential route to the C-terminal or side-chain modification of proteins.

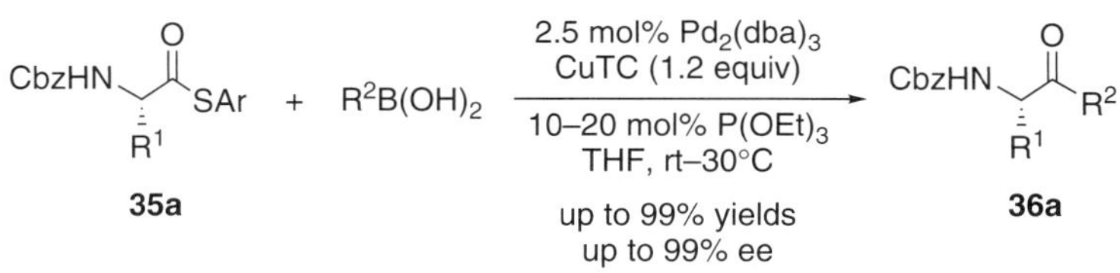

(4.16)

The anaerobic reactions of thioorganics with boronic acids usually require the presence of a catalytic palladium source and stoichiometric amount of Cu(I) carboxylate. Unexpectedly, Liebeskind and co-workers [45] reported a mechanistically unprecedented system for the construction of carbon–carbon bonds: the copper-catalyzed cross-coupling of a thioorganic with a boronic acid under aerobic conditions [Eq. (4.17)]. In the reaction sequence, the Cu(I) ion pairs with the thiolate in a thermodynamically strong Cu–SR bond, while a full equivalent of the borophilic carboxylate counterion drives the $B(OH)_2$ moiety to become $R'C(CO)OB(OH)_2$, which implies that the reaction can be rendered catalytic in Cu(I) if a Cu(I) oxygenate

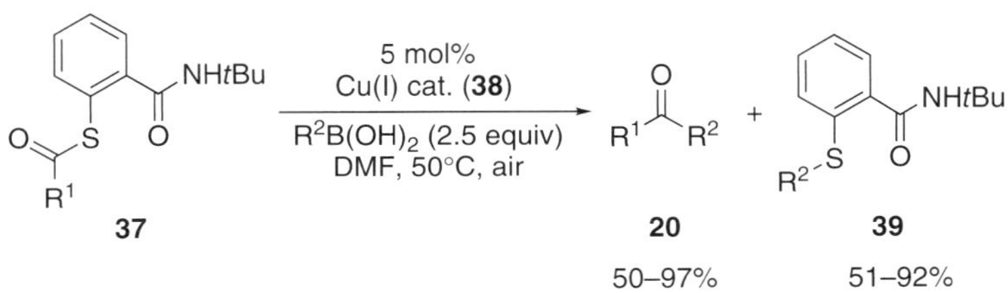

(4.17)

could be regenerated *in situ* from Cu–SR. With thioesters **37** bearing the S-pendant -(2-C_6H_4CONH*t*-Bu) as the substrates and 5 mol% Cu(I) 3-methylsalicylate (**38**) as catalyst, the cross-coupling successfully afforded ketones **20** and thioether **39** in the presence of equivalent of the boronic acid (total 2.5 equiv). With a similar strategy, highly enantiopure peptidyl ketones **36b** were synthesized from peptidic S-acylthiosalicylamide (**35b**) and boronic acids [Eq. (4.18)] [46]. Neither metal-binding nor oxidation-sensitive peptide residues interfered with the reaction. Liebeskind–Srogl cross-coupling was successfully used for the construction of

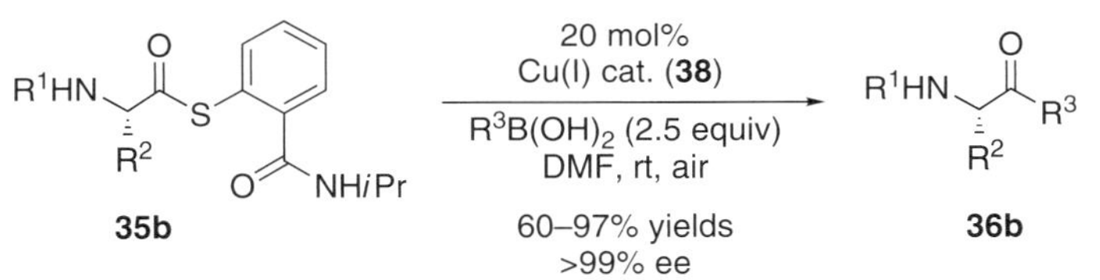

(4.18)

some medicinally important natural products such as verbenachalcone [47]. In order to design a small-molecule chemical analogue at the metallothionein system, an N–O reactant acts the same way as the S–S reactant of a biological system, and copper(I) can also be rendered catalytically viable in the presence of thiolate [Eq. (4.19)] [48]. Both the Cu(I) and internal *O*-methyl oxime moiety are essential for the catalytic turnover. The S–N trap **41** was also obtained during the work-up procedure, suggesting that large excess of a nucleophilic reagent (Ar′M) and Cu(I) reagent, as well as a palladium pre-catalyst, was not necessary. This procedure may be classified as the third generation of Liebeskind–Srogl cross-coupling.

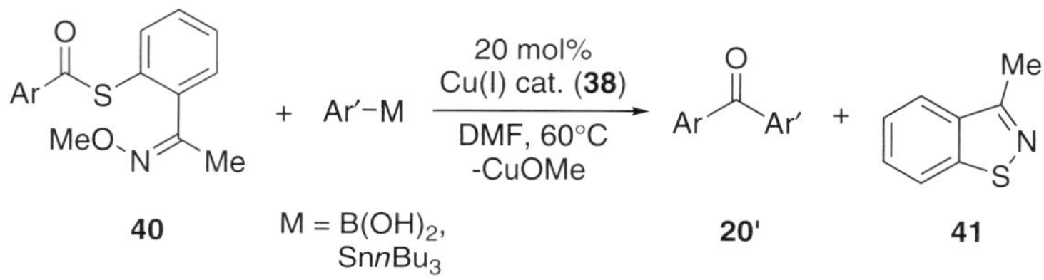

(4.19)

4.4 CATALYTIC C–S CLEAVAGE IN DITHIOACETALS

In organic synthesis, dithioacetal functionalities are very useful protecting, stabilizing, and promoting groups due to their easy deprotection under acidic conditions [49–64]. Because a strong bond can be readily formed between a catalytic metal and the soft sulfur atom and dithioacetals are sensitive to acids, organic transformations of dithioacetals are usually carried out under basic conditions.

With the organic base DBU as catalyst, α-alkenoyl ketene dithioacetals **42**—that is, the divinyl ketones [65] with terminal *gem*-dialkylthio substituents—reacted with ethyl isocyanoacetate (**43**) to afford C_2-tethered pyrrole/oxazole pairs **44** under mild conditions [Eq. (4.20)] [66]. This reaction proceeds via [5 + 1] annulation of **42** and **43**, ring-opening of the intermediate **45**, and then two consecutive [3 + 2] cycloadditions. The chemoselective fragmentation of the cyclohexanone intermediate **45** is crucial for the formation of the target product **44**. In a similar fashion, base-catalyzed

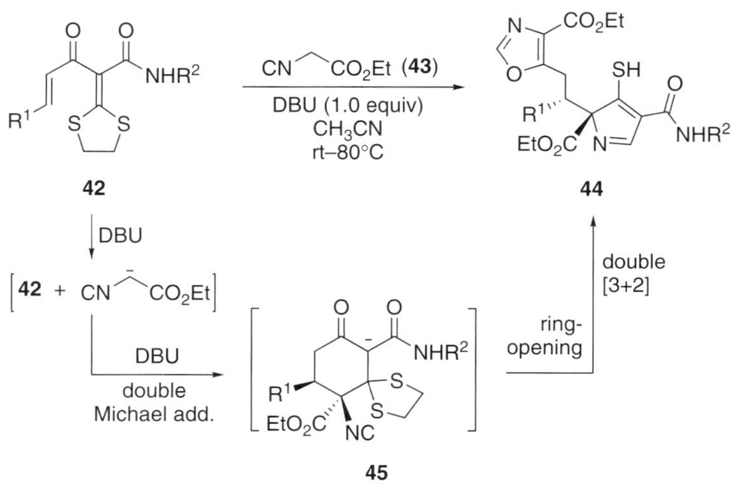

(4.20)

1,3-dipolar cycloaddition of acyclic ketene dithioacetals **46** with carbanions derived from activated methylene isocyanides efficiently afforded substituted 1H-pyrroles **47** [Eq. (4.21)] [67]. Introduction of different substituents such as tosyl, carbalkoxy, aryl, cyano, nitro, acetyl, benzoyl, and cyclic amino, etc. to the three

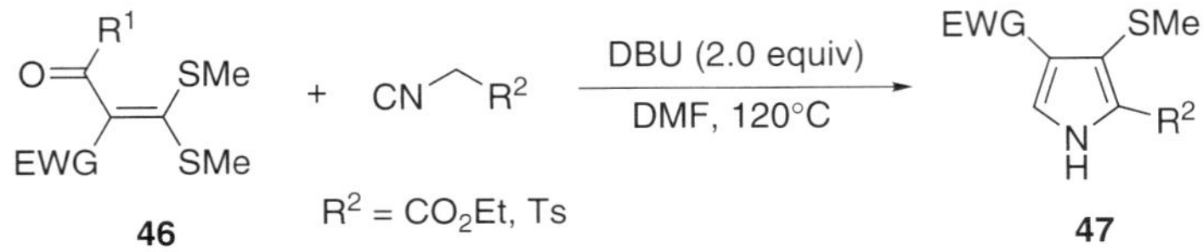

(4.21)

positions of the present pyrrole ring could be effectively controlled. With acyl ketene dithioacetals **48** as the substrates, *t*BuOK-catalyzed one-pot synthesis of 3-amino/alkylthio-cyclobut-2-en-1-ones was successfully achieved [Eq. (4.22)] [68]. The reaction was initiated by deprotecting **48** with *t*BuOK base, and the subsequent

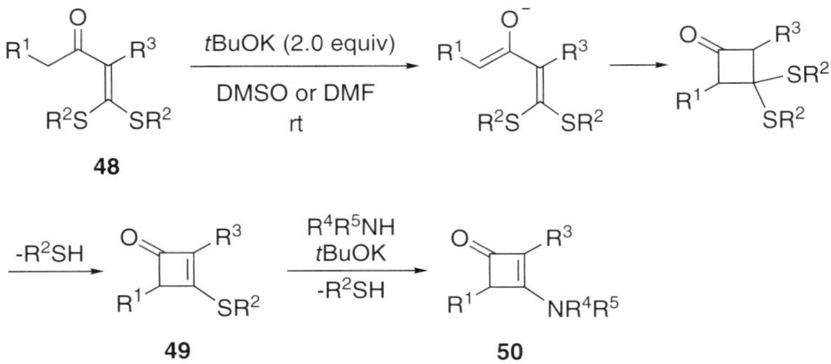

(4.22)

ring-closing, elimination, and substitution produced the desired products cyclic enones **49** and **50**. The three-component reaction of acetyl ketene dithioacetal **51**, an aldehyde, and a ketone with *t*BuOK base (4 equiv) formed polysubstituted phenols **52** via a [4 + 1 + 1] annulation and subsequential metal-free oxidative aromatization in air at room temperature [Eq. (4.23)] [69]. The highly chemo- and regioselective sequence consists of an aldol condensation, intermolecular Michael addition, intramolecular Michael addition, and elimination of ethanethiol. However, in the absence of an aldehyde the two-component reaction of alkenyl ketene dithioacetal **54** with an aryl methyl ketone under similar conditions formed 2-cyclohexanones of type **53a** (i.e., **53b**) as the product through a [5 + 1] annulation [Eq. (4.24)] [70]. In the presence of *t*BuOK and iodine, compounds **53b** were easily transformed to iodo-diarylketones of type **52** in 76–82% yields. At an elevated temperature such as 120°C, the reaction of dithioacetals **55** with nitroethane gave *p*-terphenyls **56** in the presence of DBU (4 equiv) in 71–88% yields [Eq. (4.25)] [71]. Such a [5 + 1] annulation strategy may be applied in the synthesis of highly substituted phenols and cyclohexanones [69–72]. By varying reaction temperatures, [5C + 1N] annulation of dithioacetals formed pyridine derivatives [71, 73]. In the presence of NH_4OAc, 1,1-bisalkylthio-1,4-pentanedienes **57** were conveniently transformed to pyridines **58** or **59** [Eq. (4.26)] [73]. Using a relatively weak inorganic base such as K_2CO_3, the reactions of α-alkenoyl ketene diethylthioacetals **60** reacted with diethyl malonate (**61**) to efficiently produce highly functionalized vinylogous thiol esters **62**

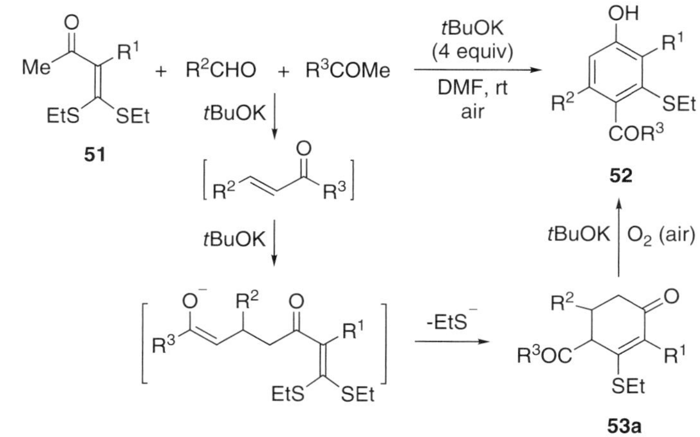

(4.23)

(4.24)

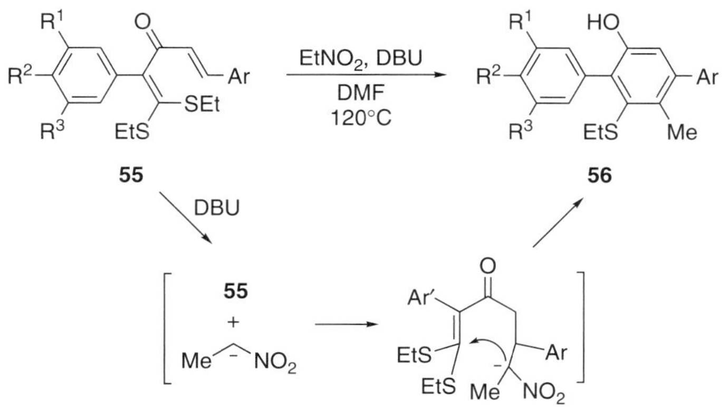

(4.25)

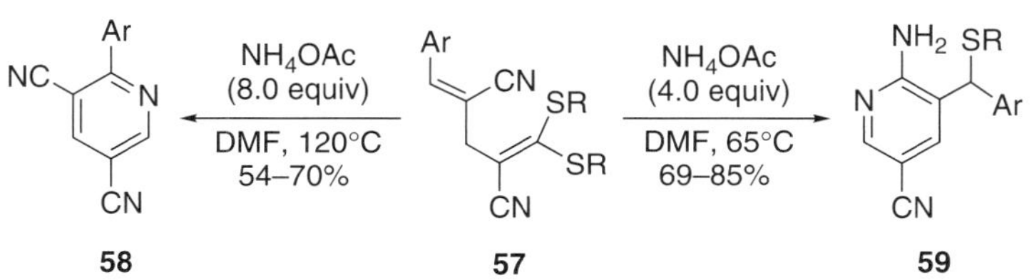

(4.26)

[Eq. (4.27)] [74]. The reaction involves a base-catalyzed *retro*-Michael ring-opening of the cyclohexenone intermediate to generate **62**, which could be easily cyclized under Brønsted acid catalysis.

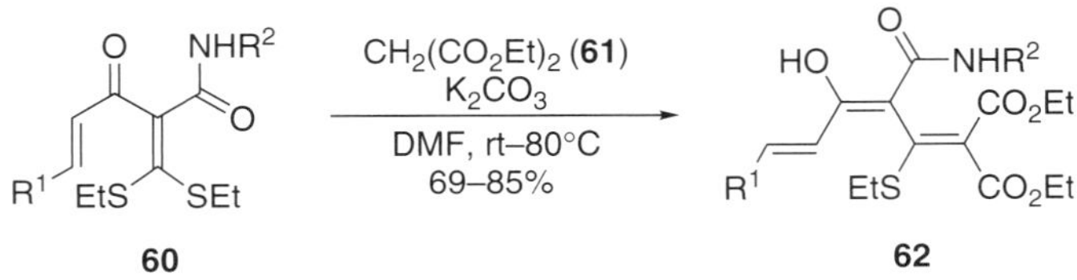

(4.27)

Recently, Yu and Yu [75] reported Brønsted acid-promoted direct alkenylation of indoles with α-oxo ketene dithioacetals **63** [Eq. (4.28)]. In the presence of trifluoroacetic acid (TFA) in CH_2Cl_2, the reaction proceeded smoothly, resulting in the mono- or disubstituted products **64** and **65** under the controlled conditions. The monoindole products **64** were found to be the important precursors to marine indole alkaloids meridianin derivatives. Compounds **64** were condensed with guanidines to produce meridianin derivatives. Hydrobromic acid could act as a unique catalyst for

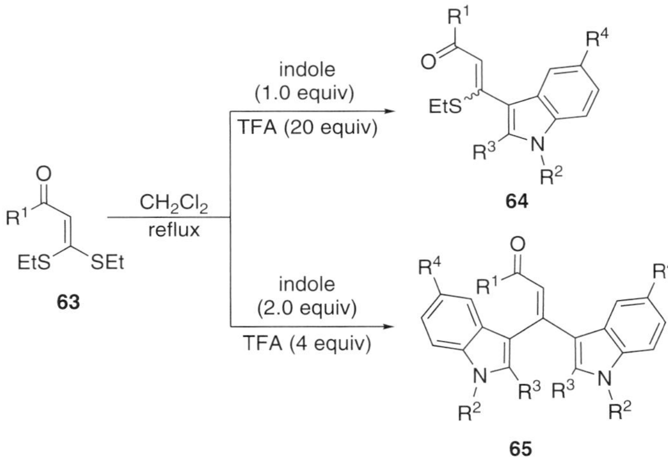

(4.28)

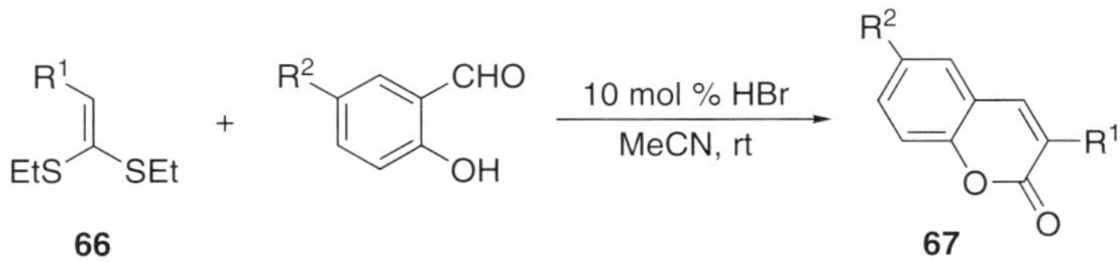

(4.29)

the carbon–carbon bond formation with ketene dithioacetals through C–S cleavage [Eq. (4.29)] [76], affording coumarins (**67**). By reacting **66** with benzoquinones under similar conditions, benzofurans were also obtained. Lewis acid $BF_3{\cdot}OEt_2$ catalyzed the domino carbocationic rearrangement of 2-indolylcyclopropyl ketene dithioacetals **68a** in CH_3NO_2 to generate β-ketocarbothioate **69** [Eq. (4.30)], while the 3-indolylcyclopropyl analogues **68b** underwent similar cyclization in the presence of TFA/CH_2Cl_2 to furnish pentaleno fused indolodiketones **70** [Eq. (4.31)] [77].

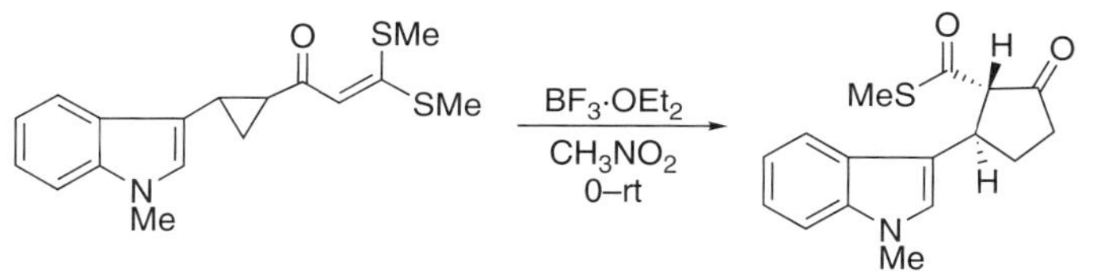

68a **69**, 61% (4.30)

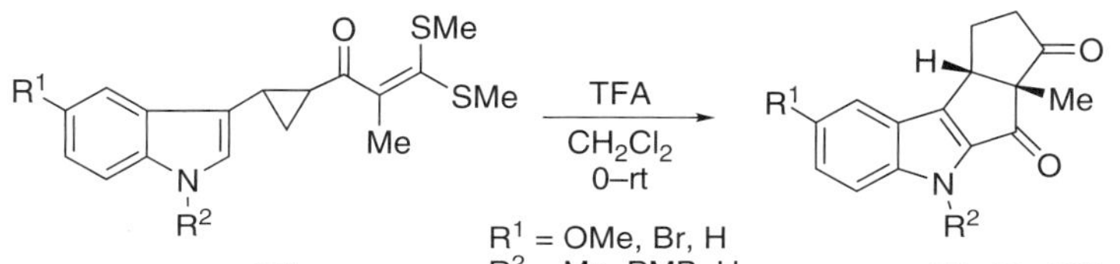

68b **70**, 62–65% (4.31)

Base- and acid-catalyzed transformations of dithioacetals have been well investigated, but transition-metal-promoted reactions of their carbon–sulfur bonds have seldom been realized until recently. Liu and co-workers reported $CuBr_2$-catalyzed synthesis of coumarins from the same reaction as shown in Eq. (4.29) [78], while the reactions of α-electron-withdrawing group-substituted ketene dithioacetals **71** with quinones **72** in acetonitrile at room temperature produced benzofurans (**73**) in the presence of $CuBr_2$ (2 mol%) as catalyst and $BF_3{\cdot}OEt_2$ (10 mol%) as co-catalyst [Eq. (4.32)] [79]. In a similar fashion, polyfunctionalized 4H-chromenes **75** and dihydrocoumarins **76** were prepared by $CuBr_2$-catalyzed cross-coupling of benzyl acohols with ketene dithioacetals **74** [Eq. (4.33)] [80].

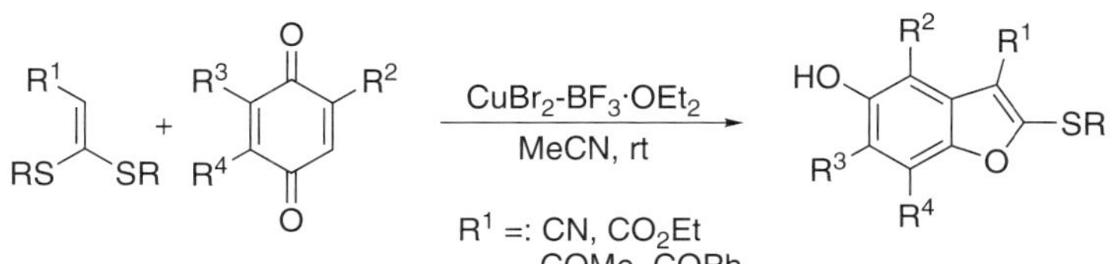

71 **72** **73** (4.32)

Rare examples of transition–metal-catalyzed C–S cleavage in α-oxo ketene dithioacetals **63** were also documented by using 20 mol% $Pd(PPh_3)_4$ as catalyst, CuTC as mediator, and boronic acids as the nucleophiles [Eq. (4.34)] [81]. Trisubstituted alkenes **77** and polyfunctionalized conjugate dienes **78** were efficiently obtained. This synthetic protocol has provided a potential route to highly functionalized conjugate dienes and polysubstituted alkenes. With ketene dibenzylthioesters (**79**) as substrates, aerobic $Cu(OAc)_2$-catalyzed desulfitative cross-coupling with

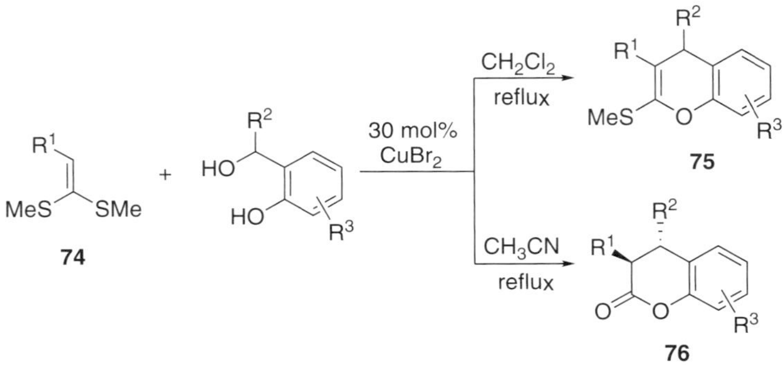

(4.33)

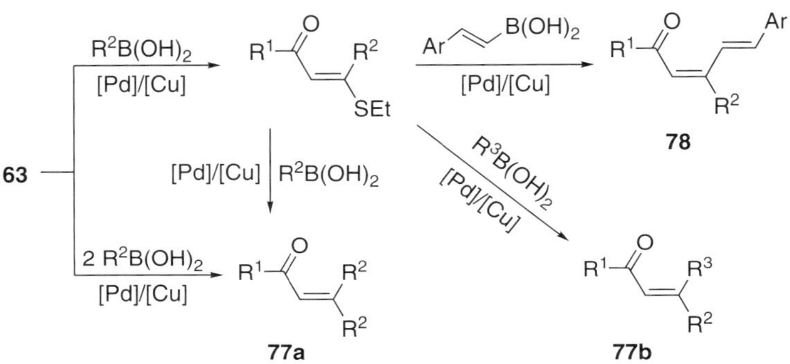

(4.34)

boronic acids produced tetrasubstituted alkenes **80** under relatively harsh conditions [Eq. (4.35)] [82]. Iron salt $FeCl_3$ was also used to promote the transformation of the C–S bonds in 1,3-dithanes **81** with ammonia and secondary amines [Eq. (4.36)] [83]. In a similar fashion and by using hydrazine hydrate, substituted pyrazoles were synthesized [83, 84].

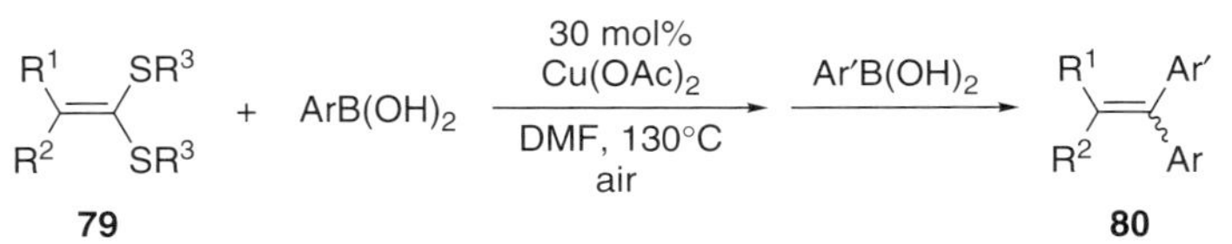

(4.35)

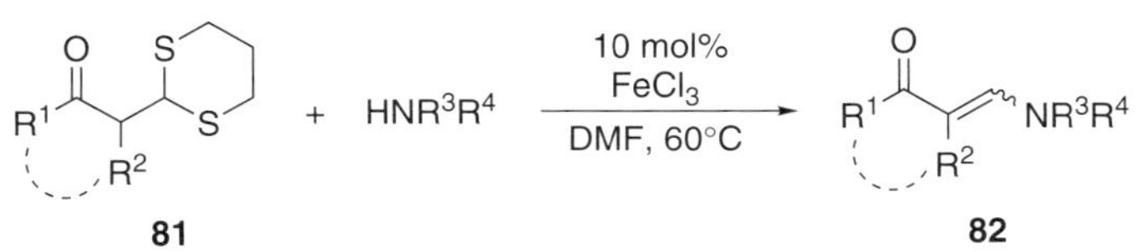

(4.36)

The organic transformations of oxidized dithioacetals were also documented. Using 2-(2,2,2-trifluoroethylidene)-1,3-dithiane-1-oxide (**83**) as the trifluoromethyl-ketene equivalent, triflic anhydride mediated the extended Pummerer annulation reactions with phenols, forming trifluoromethylbenzo[*b*]furans **84** [Eq. (4.37)] [85]. The methylthio group in the products can be further converted to form highly functionalized trifluoromethylbenzo[*b*]furans. α-Alkenoylketene dithioacetals **85** were reacted with $Na_2S{\cdot}9H_2O$ in DMF at 80°C to form 2,3-dihydro-thiopyran-4-ones (**86**) via a formal [5C + 1S] annulation [Eq. (4.38)] [86]. The methodology is featured simplicity of manipulation, ready availability of substrates, and broad scope of the products, rendering the present [5 + 1] annulation a promising route to *S*-heterocycles. Radical conditions were also reported to promote the cyclization of dithioacetals **87**, leading to 2,3-disubstituted benzo[*b*]thiophenes **88** [Eq. (4.39)] [87].

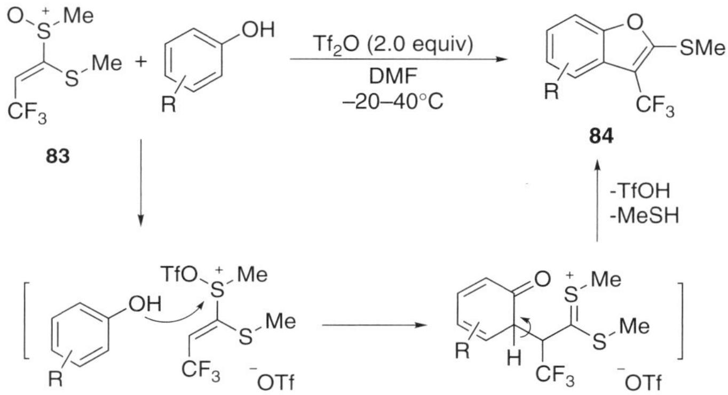

(4.37)

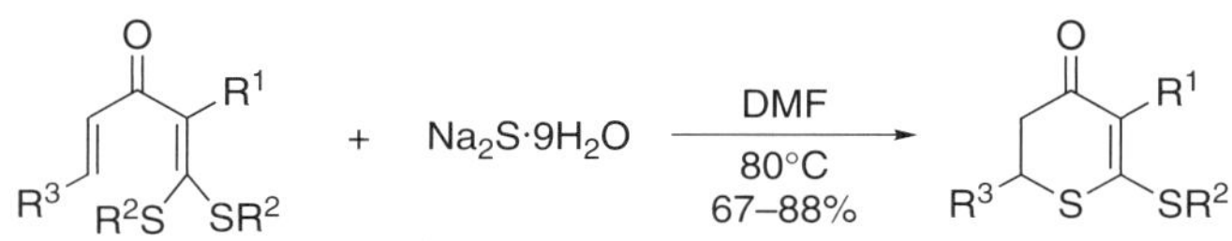

85 86

(4.38)

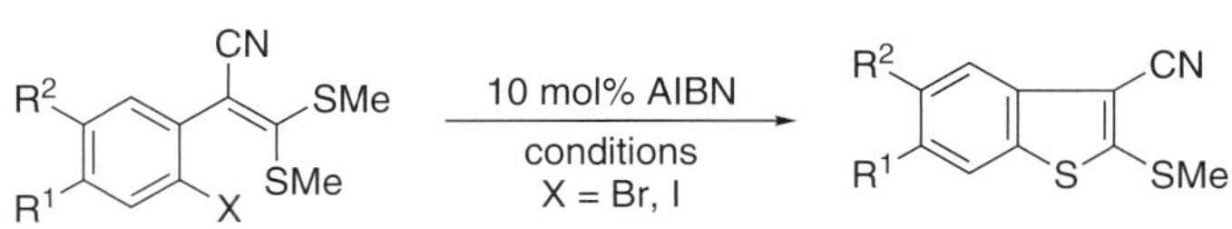

87 88

(4.39)

4.5 DIVERSE CATALYTIC C–S CLEAVAGE

Although organosulfur compounds are usually considered to be catalyst poisons [88], they have been proved to be versatile reagents for many transition-metal-catalyzed reactions as discussed above. In the early literature, palladium catalysts were explored for promoting C–S bond cleavage. Tanaka and co-workers reported palladium-catalyzed thioesterification of terminal alkynes with *O*-methyl *S*-phenyl thiocarbonate (**89**) under very mild conditions [Eq. (4.40)] [89]. In the catalytic reaction sequence, the oxidative addition step proceeded at room temperature upon mixing $Pd(PCy_3)_2$ with **89** in hexane, generating species *trans*-$Pd(SPh)(COOMe)(PCy_3)_2$ (**91**). Under similar conditions, the reactions of thiol esters **19b** promoted by $Pt(PPh_3)_4$ produced the decarbonylative addition products **90b** [Eq. (4.41)] [90]. Platinum-catalyzed regio- and stereoselective

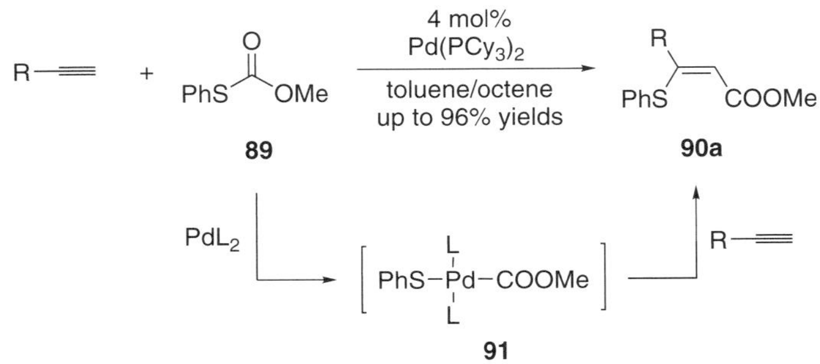

(4.40)

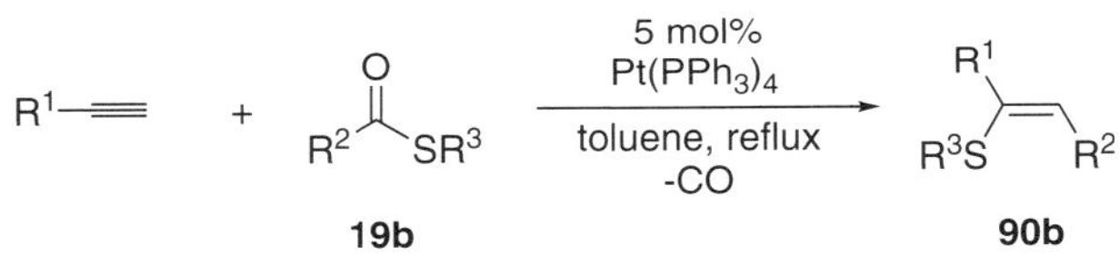

(4.41)

arylthiolation of internal alkynes was also realized by decarbonylative addition of arylthioesters [91]. 5-Methylpyridine-2-thiolate-bridged dinuclear Pt(III), Pt(II), or Pd(II) complexes (1 mol%) were successfully applied as catalysts for the desulfitative reduction of pyridine-2-thiol by H_2 (60 atm) in DMF at 150°C [92]. Rhodium-catalyzed alkyne carbothiolation was realized by using aryl methyl sulfides **91** as the substrates [93]. Such a reaction resulted in reincorporation of the original arene functional group and a methyl sulfide moiety into the products of type **92** [Eq. (4.42)].

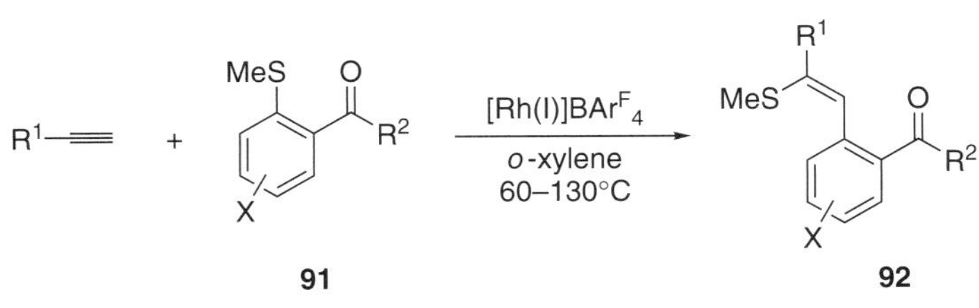

(4.42)

Tetrasubstituted alkenes are usually bestowed with potential bioactivity. Vinyl 2-pyrimidyl sulfide was reported as a platform for tetrasubstituted alkene (**94**) synthesis. The last step of the synthetic procedure is a palladium-catalyzed C–C cross-coupling via C–S cleavage of the intermediate species sulfide **93** [Eq. (4.43)] [94]. In a similar fashion by using boronic acids instead of Grignard reagents as the nucleophiles in the presence of CuTC, trisubstituted alkenes were prepared [95]. Palladium nanoparticles generated *in situ* from the *N,N*-dimethylacetamide (DMA) solutions of PdX_2 ($X = Cl^-$, OAc^-, CF_3COO^-) or $Pd_2(dba)_3$ by reduction with alkyl silanes selectively catalyzed the cross-coupling of silanes R_3SiH with phenyl and vinyl thioethers **95**, forming the corresponding thiosilanes **96**, and so on [Eq. (4.44)] [96]. Under the standard Liebeskind–Srogl cross-coupling conditions, bisarylthio-cyclobutenedione **97** reacted with both organostannanes and (hetero)arylboronic acids to give symmetrical disubstituted cyclobutenediones **98** [Eq. (4.45)] [97]. Even though Knochel's organocuprate method allows the

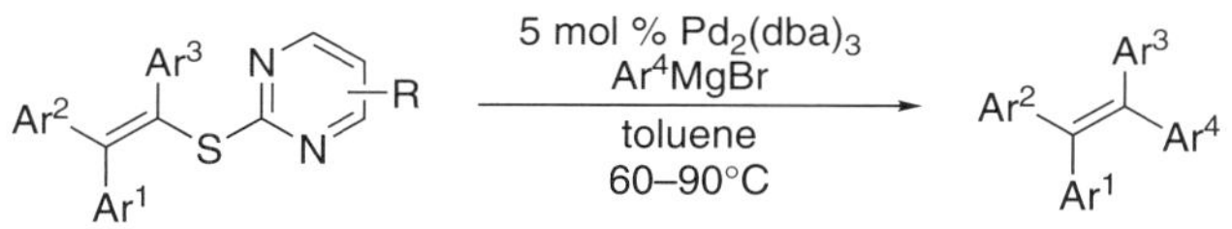

93 **94**

(4.43)

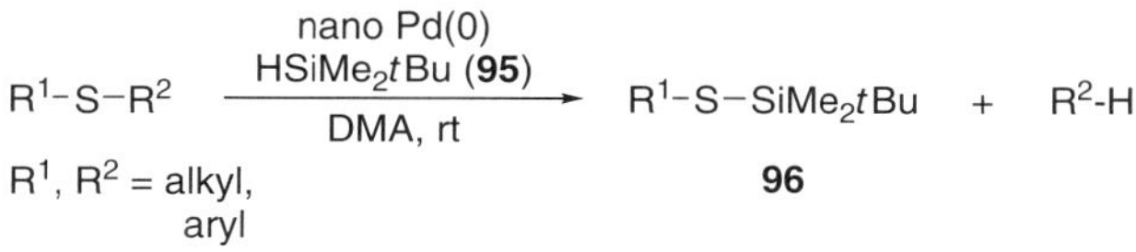

96

(4.44)

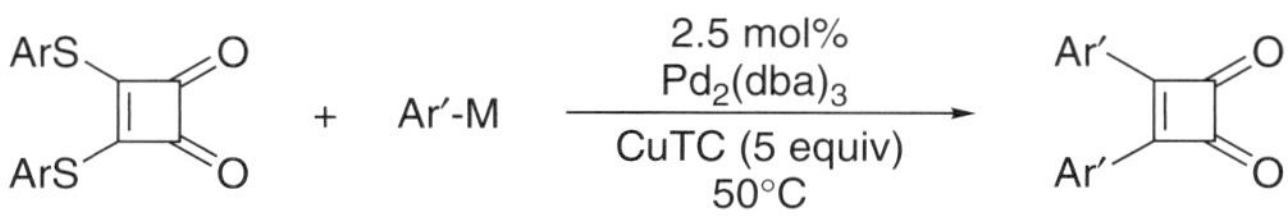

97 **98**

(4.45)

introduction of two different groups at the vinyl positions, the present methodology uses readily available reagents and catalysts that are air stable. Double cross-couplings also occurred in 2,5-dimethylthio-oxazole (**99**) with organozinc reagents by one-pot Pd/Ni-catalysis [Eq. (4.46)] [98], and the selectivities and yields were very good. This protocol has been applied to functionalize dye molecules [99].

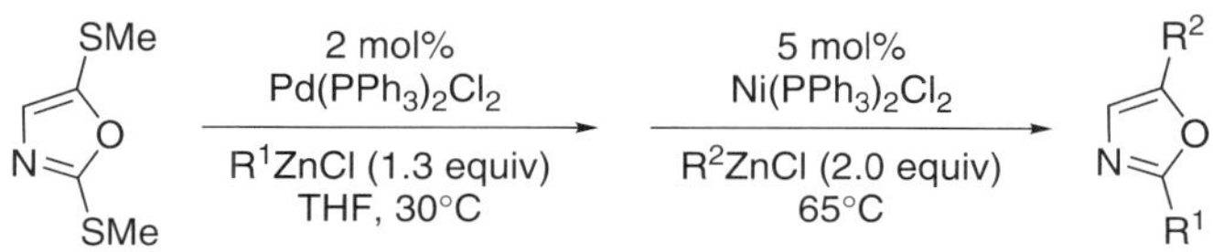

99 **100**

(4.46)

The Liebeskind–Srogl cross-coupling of cyclic thioamides or thioureas (**101**) with alkenylboronic acids, vinyl- and (hetero)arylstannanes, and arylsiloxanes resulted in the corresponding desulfitative products **102** [Eq. (4.47)] [100]. (*o*-Bromoaryl/heteroaryl)acrylonitrile (**103**) underwent intramolecular heteroarylation to form polycyclic heterocycles **104** by palladium-catalyzed C–Br and C–S bond activation [Eq. (4.48)] [101]. Under very similar conditions, palladium-catalyzed reactions of bromothiophenes **105** with alkynes afforded sulfur-based heterocycles and fulvenes **106** [Eq. (4.49)] [102]. For the formation of **106**, both C–S cleavage and formation occurred during the reaction.

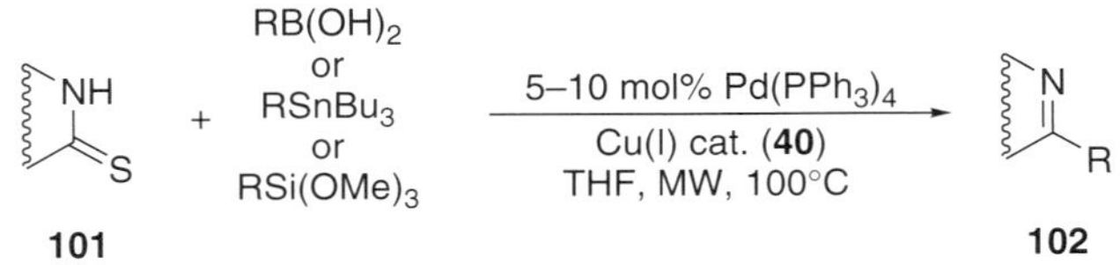

(4.47)

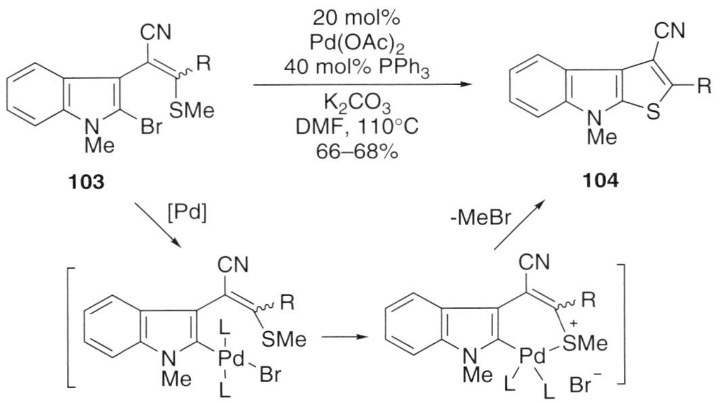

(4.48)

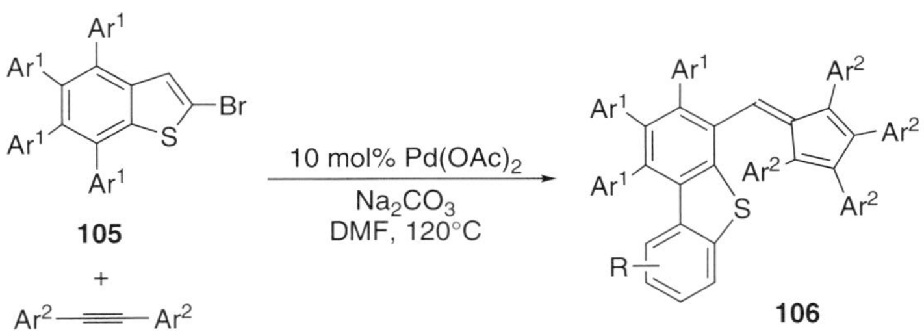

(4.49)

Nickel has also been known to activate C–S bonds [98]. With $NiCl_2(dppp)$ as catalyst, functionalized pyrimidyl methyl sulfide **107** underwent cross-coupling with Grignard reagents, while self-coupling occurred in the presence of $NiCl_2/PPh_3/Zn$, affording the desulfitative products **108a** and **108b**, respectively [Eq. (4.50)] [103]. Similar results were obtained in the synthesis of 1,10- bis(methyl)fluoranthene [104] and alkenes [105]. In the presence of an NHC ligand, namely, SIPr (ligand precursor: 1,3-bis(2,6-diisopropylphenyl)imidazolinium chloride, SIPrHCl),

$Ni(COD)_2$-catalyzed alkenylative cross-coupling of alkylsulfides **109** with arylmagnesium bromides produced alkenyl-aryl coupling products **110** by cleaving both a sp^3 C–S and a sp^2 C–S bonds in **109** [Eq. (4.51)] [106]. The C–S bond of an alkenyl sulfide can also be cleaved by sequential Ag(I)-mediated alkylation of the C–S bond and Ni(II)-catalyzed cross-coupling of the activated C–S bond with Et_2Zn [Eq. (4.52)] [93].

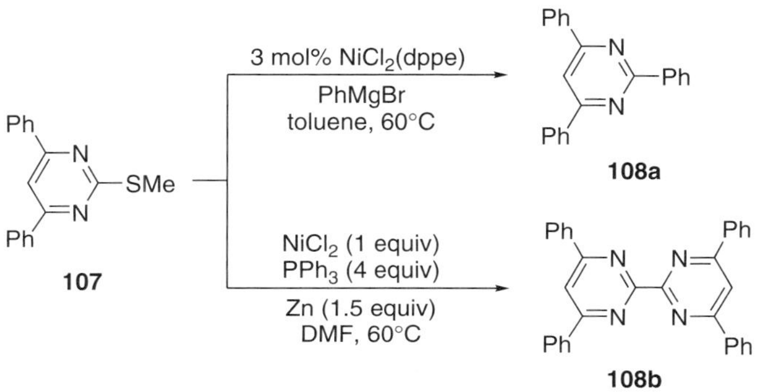

(4.50)

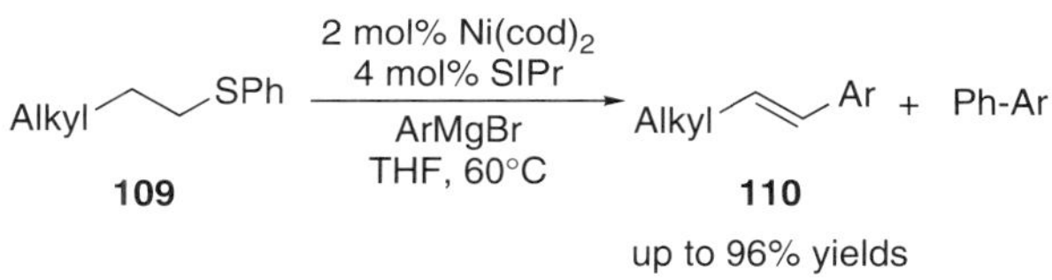

(4.51)

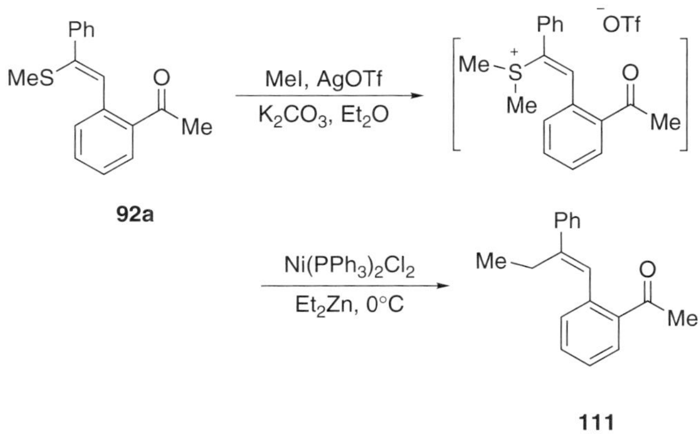

(4.52)

As demonstrated in Liebeskind–Srogl cross-coupling, copper has proved to be an effective promoter for C–S cleavage. Copper(I)-catalyzed, α-selective, allylic alkylation reactions between phosphorothioate esters **112** and organomagnesium reagents were documented [107], selectively furnishing the challenging alkylation, vinylation, and arylation at the α-position of **112** [Eq. (4.53)]. This novel synthetic protocol features a wide scope of substrates. Starting from allylic chlorides by using sodium diethylphosphorothioate as a stoichiometric additive, high α-selectivity was obtained. Copper(II) salt $CuSO_4$ was reported to mediate the intra- and intermolecular C–S cross-coupling sequence in **114**, producing 2-(arylthio)arylcyanamides (**115**) through *in situ* generation of the catalytically active Cu(I) species [Eq. (4.54)] [108]. Using a Rh(I) complex catalyst, direct functionalization of heteroarenes **116** such as benzothiazoles, benzoxazoles, and benzothiophene was performed by

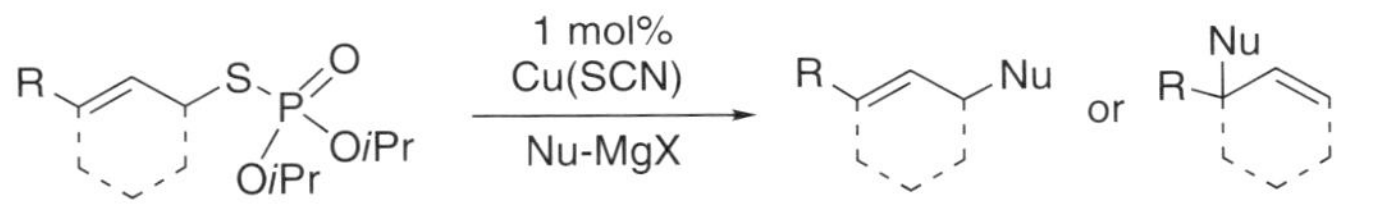

112 **113a** (major) **113b** (minor) (4.53)

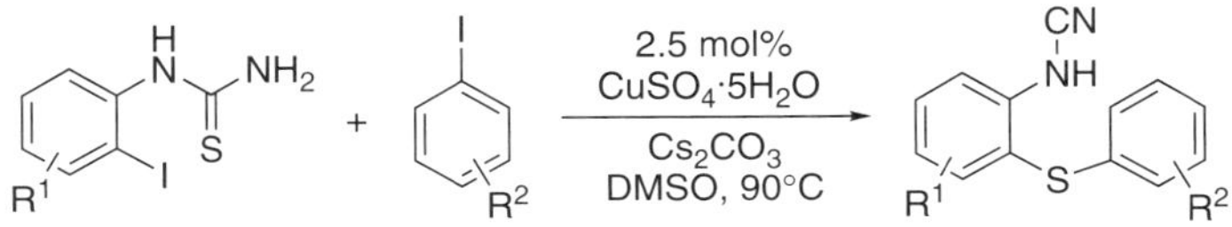

114 **115** (4.54)

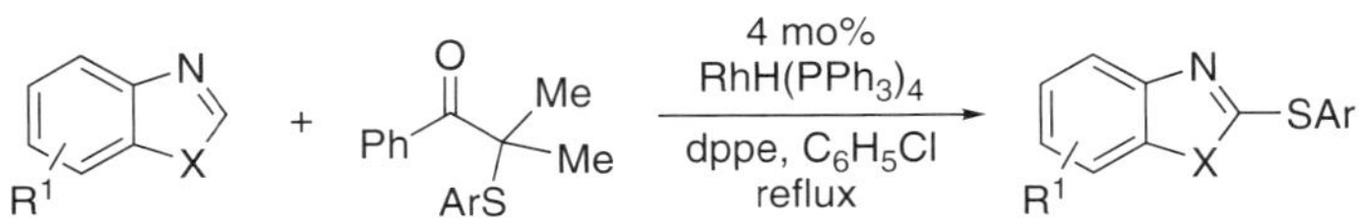

116, X = S, O **117** **118** (4.55)

2-(phenylthio)isobutyrophenone (**117**) via sp^3 C–S bond cleavage and sp^2 C–S bond formation, affording 2-phenylthio-heteroarenes **118** [Eq. (4.55)] [109]. Monocyclic heteroarenes, namely, 1-methyl-1,2,3,4-tetrazole and 2-cyanothiophene, were thus transformed to their corresponding 5-phenylthio derivatives. During the reactions, compounds **117** efficiently acted as the arylthio transfer reagents. Lewis acid $ZnCl_2$-mediated tandem Mukaiyama aldol lactonization of *O,S*-ketene acetals **119** with aldehydes provided a versatile and highly diastereoselective approach to *trans*-1,2-disubstituted β-lactones **120** [Eq. (4.56)] [110].

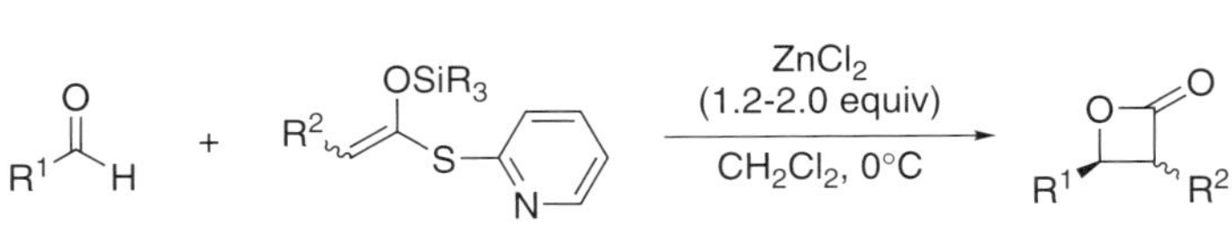

119 **120** (4.56)

It should be noted that noncatalytic transformations of organosulfur compounds also play an important role in modern synthetic chemistry. Diverse processes have been developed to activate and transform C–S bonds in organic and biosynthesis [111–116].

4.6 SUMMARY AND OUTLOOK

Transition-metal-catalyzed or -mediated reactions of organosulfur compounds play an important role in organic and biomimetic synthesis, and the relevant mechanistic study provides fundamental understanding of the C–S activation/cleavage in HDS processes. Using an appropriate co-catalyst or coupling partner can avoid poisoning of a C–S bond to the transition metal catalyst, establishing an efficient C–C cross-coupling process.

REFERENCES

1. Tan, R. Y., Song, D. T. (2011). C–H and C–S activations of quinoline-functionalized thiophenes by platinum complexes. *Organometallics* **30**, 1637–1645.
2. Kundu, S., Brennessel, W. W., Jones, W. D. (2011). C–S bond activation of thioesters using platinum(0). *Organometallics* **30**, 5147–5154.
3. Grochowski, M. R., Li, T., Brennessel, W. W., Jones, W. D. (2010). Competitive carbon–sulfur vs carbon–carbon bond activation of 2-cyanothiophene with $[Ni(dippe)H]_2$. *Journal of the American Chemical Society* **132**, 12412–12421.
4. Topsoe, H., Clausen, B. S., Massoth, F. E. (1996). *Hydrotreating Catalysis*, Springer., Berlin.
5. Dhar, G. M., Srinivas, B. N., Rana, M. S., Kumar, M., Maity, S. K. (2003). Mixed oxide supported hydrodesulfurization catalysts-a review. *Catalysis Today* **86**, 45–60.
6. Bej, S. K., Maity, S. K., Turaga, U. T. (2004). Search for an efficient 4,6-DMDBT hydrodesulfurization catalyst: A review of recent studies. *Energy & Fuels* **18**, 1227–1237.
7. Brunet, S., Mey, D., Pérot, G., Bouchy, C., Diehl, F. (2005). On the hydrodesulfurization of FCC gasoline: a review. *Applied Catalysis A: General* **278**, 143–172.
8. Wang, H. M., Iglesia, E. (2010). Thiophene hydrodesulfurization catalysis on supported Ru clusters: Mechanism and site requirements for hydrogenation and desulfurization pathways. *Journal of Catalysis* **273**, 245–256.
9. Ojo, W.-S., Pétillon, F. Y., Schollhammer, P., Talarmin, J. (2010). Carbon–sulfur and carbon–halogen bond cleavage of acyclic or cyclic thioethers, thiophenes, and dihaloalkanes with the trithiolato-bridged cation $[Mo_2Cp_2(\mu\text{-}SMe)_3(MeCN)_2]^+$. *Organometallics* **29**, 448–462.
10. Ornellas, S. De, Storr, T. E., Williams, T. J., Baumann, C. G., Fairlamb, I. J. S. (2011). Direct C–H/C–X coupling methodologies mediated by Pd/Cu or Cu: An examination of the synthetic applications and mechanistic findings. *Current Organic Synthesis* **8**, 79–101.

11. Nakata, N., Furukawa, N., Toda, T., Ishii, A. (2010). Cleavage of C–S and O–H bonds by platinum(0) complexes to give five-membered 1,2-oxaplatinacycles. *Angewandte Chemie International Edition* **49**, 5784–5787.

12. Atesin, T. A., Kundu, S., Skugrud, K., Lai, K. A., Swartz, B. D., Li, T., Brennessel, W. W., Jones, W. D. (2011). Predicting selectivity in oxidative addition of C–S bonds of substituted thiophenes to a platinum(0) fragment: An experimental and theoretical study. *Organometallics* **30**, 4578–4588.

13. Oster, S. S., Grochowski, M. R., Lachicotte, R. J., Brennessel, W. W., Jones, W. D. (2010). Carbon–sulfur bond activation of dibenzothiophenes and phenoxythiin by $[Rh(dippe)(\mu\text{-}H)]_2$ and $[Rh_2(dippe)_2(\mu\text{-}Cl)(\mu\text{-}H)]$. *Organometallics* **29**, 4923–4931.

14. Shibue, M., Hirotsu, M., Nishioka, T., Kinoshita, I. (2008). Ruthenium and rhodium complexes with thiolate-containing pincer ligands produced by C–S bond cleavage of pyridyl-substituted dibenzothiophenes. *Organometallics* **27**, 4475–4483.

15. Mathur, P., Boodida, S., Ji, R. S., Mobin, S. M. (2009). $Fe(CO)_5$ promoted C–S bond activation and formation of an unusual C_2S_3 ligand in $[\{Fe_2(CO)_6\}_2(\mu\text{-}C_2S_3)]$. *Journal of Organometallic Chemistry* **694**, 3043–3045.

16. Das, B., Sarkar, S., Zangrando, E., Chattopadhyay, P. (2009). C–S bond cleavage in pyridylmethylthioether systems promoted by oxorhenium(V) ion. *Inorganic Chemistry Communications* **12**, 1112–1115.

17. Singh, A. K., Mukherjee, R. (2008). Cobalt(II) and cobalt(III) complexes of thioether-containing hexadentate pyrazine amide ligands: C–S bond cleavage and cyclometallation reaction. *Dalton Transactions*, 260–270.

18. Schaub, T., Backes, M., Radius, U. (2007). Unusual nickel-mediated C–S cleavage of alkyl and aryl sulfoxides. *Chemical Communications*, 2037–2039.

19. Mills, R. C., Abboud, K. A., Boncella, J. M. (2001). Carbon–sulfur bond activation of thiophenes by $[W(NPh)\{o\text{-}(Me_3SiN)_2C_6H_4\}(pyridine)_2]$. *Chemical Communications*, 1506–1507.

20. Ballmann, J., Yeo, A., Mackay, B. A., Rijt, S. van, Patrick, B. O., Fryzuk, M. D. (2010). Complete disassembly of carbon disulfide by a ditantalum complex. *Chemical Communications* **46**, 8794–8796.

21. Iwasa, K., Seino, H., Mizobe, Y. (2008). C–S bond cleavage of 2-methoxythiophene by Ir-TMEDA complex (TMEDA = *N,N,N′,N′*-tetramethylethylenediamine). Formation of novel dinuclear iridathiacyclohexenyl complex. *Journal of Organometallic Chemistry* **693**, 3197–3200.

22. Huang, C. H., Gou, S. H., Zhu, H. B., Huang, W. (2007). Cleavage of C–S bonds with the formation of a tetranuclear Cu(I) cluster. *Inorganic Chemistry* **46**, 5537–5543.

23. Bradley, C. A., Veiros, L. F., Chirik, P. J. (2007). C–O and C–S bond cleavage in chelating diethers and thioethers promoted by η^9,η^5-bis(indenyl)zirconium sandwich complexes: A combined experimental and computational study. *Organometallics* **26**, 3191–3200.

24. Li, H. Z., Yu, K. Q., Watson, E. J., Virkaitis, K. L., Acchioli, J. S. D., Carpenter, G. B., Sweigart, D. A., Czech, P. T., Overly, K. R., Coughlin, F. (2002). Models for deep hydrodesulfurization of alkylated benzothiophenes. Reductive cleavage of C–S bonds mediated by precoordination of manganese tricarbonyl to the carbocyclic ring. *Organometallics* **21**, 1262–1270.

25. Goh, L. Y., Weng, Z. Q., Leong, W. K., Leung, P. H. (2001). C–S bond cleavage and C–C coupling in cyclopentadienylchromium complexes to give the first dithiooxamide-bridged and doubly dithiocarbamate-bridged double cubanes: $[Cp_6Cr_8S_8\{(C(S)NEt_2)_2\}]$ and $[Cp_6Cr_8S_8(S_2CNEt_2)_2]$. *Angewandte Chemie International Edition* **40**, 3236–3239.

26. Arliguie, T., Lescop, C., Ventelon, L., Leverd, P. C., Thuéry, P., Nierlich, M., Ephritikhine, M. (2001). C–H and C–S bond cleavage in uranium(III) thiolato complexes. *Organometallics* **20**, 3698–3703.

27. Beller, M., Blom, C., *Transition Metals for Organic Synthesis*, 2nd ed., VCH, Weinheim, 2004.

28. Dubbaka, S. R., Vogel, P. (2005). Organosulfur compounds: Electrophilic reagents in transition-metal-catalyzed carbon–carbon bond-forming reactions. *Angewandte Chemie International Edition* **44**, 7674–7684.

29. Prokopcová, H., Kappe, C. O. (2009). The Liebeskind–Srogl C–C cross-coupling reaction. *Angewandte Chemie International Edition* **48**, 2276–2286.

30. Liebeskind, L. S., Srogl, J., Savarin, C., Polanco, C. (2002). Bioinspired organometallic chemistry, *Pure and Applied Chemistry* **74**, 115–122.

31. Srogl, J., Allred, G. D., Liebeskind, L. S. (1997). Sulfonium salts. Participants *par excellence* in metal-catalyzed carbon–carbon bond-forming reactions. *Journal of the American Chemical Society* **119**, 12376–12377.

32. Liebeskind, L. S., Srogl, J. (2000). Thiol ester–boronic acid coupling. A mechanistically unprecedented and general ketone synthesis. *Journal of the American Chemical Society* **122**, 11260–11261.

33. Zhang, S. J., Zhang, D. W., Liebeskind, L. S. (1997). Ambient temperature, Ullmann-like reductive coupling of aryl, heteroaryl, and alkenyl halides. The *Journal of Organic Chemistry* **62**, 2312–2313.

34. Srogl, J., Liu, W. S., Marshall, D., Liebeskind, L. S. (1999). Bio-organometallic organosulfur chemistry. Transition metal-catalyzed cross-coupling using coenzyme M or thioglycolic acid as the leaving group. *Journal of the American Chemical Society* **121**, 9449–9450.

35. Savarin, C., Srogl, J., Liebeskind, L. S. (2000). Thiol ester–boronic acid cross-coupling. Catalysis using alkylative activation of the palladium thiolate intermediate. *Organic Letters* **2**, 3229–3231.

36. Savarin, C., Srogl, J., Liebeskind, L. S. (2001). Substituted alkyne synthesis under non-basic conditions: Copper carboxylate-mediated, palladium-catalyzed thioalkyne-boronic acid cross-coupling. *Organic Letters* **3**, 91–93.

37. Liebeskind, L. S., Srogl, J. (2002). Heteroaromatic thioether–boronic acid cross-coupling under neutral reaction conditions. *Organic Letters* **4**, 979–981.

38. Kusturin, C., Liebeskind, L. S., Rahman, H., Sample, K., Schweitzer, B., Srogl, J., Neumann, W. L. (2003). Switchable catalysis: Modular synthesis of functionalized pyrimidinones via selective sulfide and halide cross-coupling chemistry. *Organic Letters* **5**, 4349–4352.

39. Yu, Y., Liebeskind, L. S. (2004). Copper-mediated, palladium-catalyzed coupling of thiol esters with aliphatic organoboron reagents. The *Journal of Organic Chemistry* **69**, 3554–3557.

40. Kusturin, C. L., Liebeskind, L. S., Neumann, W. L. (2002). A new catalytic cross-coupling approach for the synthesis of protected aryl and heteroaryl amidines. *Organic Letters* **4**, 983–985.

41. Fausett, B. W., Liebeskind, L. S. (2005). Palladium-catalyzed coupling of thiol esters with aryl and primary and secondary alkyl organoindium reagents. The *Journal of Organic Chemistry* **70**, 4851–4853.

42. Yoshida, S., Yorimitsu, H., Oshima, K. (2009). 2-(2,2,2-Trifluoroethylidene)-1,3-dithiane monoxide as a trifluoromethylketene equivalent. *Organic Letters* **11**, 2185–2188.

43. Zhang, Z. H., Liebeskind, L. S. (2006). Palladium-catalyzed, copper(I)-mediated coupling of boronic acids and benzylthiocyanate. A cyanide-free cyanation of boronic acids. *Organic Letters* **8**, 4331–4333.

44. Yang, H., Li, H., Wittenberg, R., Egi, M., Huang, W. W., Liebeskind, L. S. (2007). Ambient temperature synthesis of high enantiopurity *N*-protected peptidyl ketones by peptidyl thiol ester-boronic acid cross-coupling. *Journal of the American Chemical Society* **129**, 1132–1140.

45. Villalobos, J. M., Srogl, J., Liebeskind, L. S. (2007). A new paradigm for carbon–carbon bond formation: Aerobic, copper-templated cross-coupling. *Journal of the American Chemical Society* **129**, 15734–15735.

46. Liebeskind, L. S., Yang, H.. Li, H. (2009). A copper-catalyzed, pH-neutral construction of high-enantiopurity peptidyl ketones from peptidic *S*-acylthiosalicylamides in air at room temperature. *Angewandte Chemie International Edition* **48**, 1417–1421.

47. Dandepally, S. R., Williams, A. L. (2010). Liebeskind–Srogl cross coupling mediated synthesis of verbenachalcone. *Tetrahedron Letters* **51**, 5753–5756.

48. Zhang, Z. H., Lindale, M. G., Liebeskind, L. S. (2011). Mobilizing Cu(I) for carbon-carbon bond forming catalysis in the presence of thiolate. Chemical mimicking of metallothioneins. *Journal of the American Chemical Society* **133**, 6403–6410.

49. Pan, L., Liu, Q. (2011). [5 + 1]-Annulation strategy based on alkenoyl ketene dithioacetals and analogues. *Synlett* **8**, 1073–1080.

50. Wang, H., Zhao, Y.-L., Ren, C.-Q., Diallo, A., Liu, Q. (2011). Tandem [5 + 1] annulation-isocyanide cyclization: Efficient synthesis of hydroindolones. *Chemical Communications* **47**, 12316–12318.

51. Wang, M., Sun, S. G., Liang, D. Q., Liu, B. Y., Dong, Y., Liu, Q. (2011). Sulfuric acid-catalyzed regioselective alkylation of indoles and β-naphthols with ketene dithioacetal-based allylic alcohols. *European Journal of Organic Chemistry* **2011**, 2466–2473.

52. Gualandi, A., Emer, E., Capdevila, M. G., Cozzi, P. G. (2011). Highly enantioselective α-alkylation of aldehydes with 1,3-benzodithiolylium tetrafluoroborate: A formal organocatalytic α alkylation of aldehydes by the carbenium ion. *Angewandte Chemie International Edition* **50**, 7842–7846.

53. Perrotta, R. R., Winter, A. H., Coldren, W. H., Falvey, D. E. (2011). 2-(3,5-Dinitrophenyl)-1,3-dithiane carbanion: A benzylic anion with a low energy triplet state. *Journal of the American Chemical Society* **133**, 15553–15558.

54. Smith, A. B., III, Han, H., Kim, W.-S. (2011). Diversity-oriented synthesis of 2,4,6-trisubstituted piperidines via type II anion relay chemistry. *Organic Letters* **13**, 3328–3331.

55. Xu, H.-C., Moeller, K. D. (2010). Intramolecular anodic olefin coupling reactions: Use of the reaction rate to control substrate/product selectivity. *Angewandte Chemie International Edition* **49**, 8004–8007.

56. Kobatake, T., Yoshida, S., Yorimitsu, H., Oshima, K. (2010). Reaction of 2-(2,2,2-trifluoroethylidene)-1,3-dithiane 1-oxide with ketones under pummerer conditions and its application to the synthesis of 3-trifluoromethyl-substituted five-membered heteroarenes. *Angewandte Chemie International Edition* **49**, 2340–2343.

57. Welker, M., Woodward, S., Veiros, L. F., Calhorda, M. J. (2010). Chemoselectivity as a delineator of cuprate structure in catalytic 1,4-addition of diorganozinc reagents to Michael acceptors. *Chemistry—A European Journal* **16**, 5620–5629.

58. Zhang, L. J., Xu, X. X., Tan, J., Pan, L., Xia, W. M., Liu, Q. (2010). Tandem Michael addition/intramolecular isocyanide [3 + 2] cycloaddition: Highly diastereoselective one pot synthesis of fused oxazolines. *Chemical Communications* **46**, 3357–3359.

59. Wang, M., Han, F., Yuan, H. J., Liu, Q. (2010). Tandem Nazarov cyclization–halovinylation of divinyl ketones under Vilsmeier conditions: synthesis of highly substituted cyclopentadienes. *Chemical Communications* **46**, 2247–2249.

60. Li, Y. F., Xu, X. X., Tan, J., Liao, P. Q., Zhang, J. P., Liu, Q. (2010). Polarity-reversible conjugate addition tuned by remote electronic effects. *Organic Letters* **12**, 244–247.

61. Tan, J., Xu, X. X., Zhang, L. J., Li, Y. F., Liu, Q. (2009). Tandem double-Michael-addition/cyclization/acyl migration of 1,4-dien-3-ones and ethyl isocyanoacetate: stereoselective synthesis of pyrrolizidines. *Angewandte Chemie International Edition* **48**, 2868–2872.

62. Kim, H., Park, Y., Hong, J. Y. (2010). Stereoselective synthesis of 2,6-*cis*-tetrahydropyrans through a tandem allylic oxidation/oxa-Michael reaction promoted by the *gem*-disubstituent effect: synthesis of (+)-neopeltolide macrolactone. *Angewandte Chemie International Edition* **48**, 7577–7581.

63. Ulshöfer, R., Podlech, J. (2009). Stereoelectronic effects in vinyl sulfoxides. *Journal of the American Chemical Society* **131**, 16618–16619.

64. Yu, H. F., Jin, W. W., Sun, C. L., Chen, J. P., Du, W. M., He, S. B., Yu, Z. K. (2010). Palladium-catalyzed cross-coupling of internal alkenes with terminal alkenes to functionalized 1,3-butadienes using C–H bond activation: Efficient synthesis of bicyclic pyridones. *Angewandte Chemie International Edition* **49**, 5792–5797.

65. Frontier, A. J., Collison, C. (2005). The Nazarov cyclization in organic synthesis. Recent advances. *Tetrahedron* **61**, 7577–7606.

66. Li, Y. F., Xu, X. X., Tan, J., Xia, C. Y., Zhang, D. W., Liu, Q. (2011). Double isocyanide cyclization: A synthetic strategy for two-carbon-tethered pyrrole/oxazole pairs. *Journal of the American Chemical Society* **133**, 1775–1777.

67. Misra, N. C., Panda, K., Ila, H., Junjappa, H. (2007). An efficient highly regioselective synthesis of 2,3,4-trisubstituted pyrroles by cycloaddition of polarized ketene *S,S*- and *N,S*-acetals with activated methylene isocyanides. *The Journal of Organic Chemistry* **72**, 1246–1251.

68. Zhao, Y.-L., Yang, S.-C., Di, C.-H., Han, X.-D., Liu, Q. (2010). Highly efficient synthesis of 3-amino-/alkylthio-cyclobut-2-en-1-ones based on the cyclization of acyl ketene dithioacetals. *Chemical Communications* **46**, 7614–7616.

69. Wang, M., Fu, Z. Q., Feng, H., Dong, Y., Liu, J., Liu, Q. (2010). Tandem [4 + 1 + 1] annulation and metal-free aerobic oxidative aromatization: Straightforward synthesis of

highly substituted phenols from one aldehyde and two ketones. *Chemical Communications* **46**, 9061–9063.

70. Fu, Z. Q., Wang, M., Dong, Y., Liu, J., Liu, Q. (2009). Direct synthesis of highly substituted 2-cyclohexenones and sterically hindered benzophenones based on a [5C + 1C] annulation. *The Journal of Organic Chemistry* **74**, 6105–6110.
71. Zhang, L., Liang, F. S., Cheng, X., Liu, Q. (2009). A new route to multifunctionalized *p*-terphenyls and heteroaryl analogues via [5C + 1C(N)] annulation strategy. *The Journal of Organic Chemistry* **74**, 899–902.
72. Bi, X. H., Dong, D. W., Liu, Q., Pan, W., Zhao, L., Li, B. (2005). [5 + 1] Annulation: A synthetic strategy for highly substituted phenols and cyclohexenones. *Journal of the American Chemical Society* **127**, 4578–4579.
73. Hu, J. L., Zhang, Q., Yuan, H. J., Liu, Q. (2008). Temperature-controlled synthesis of substituted pyridine derivatives via the [5C + 1N] annulation of 1,1-bisalkylthio-1,4-pentanedienes and ammonium acetate. *The Journal of Organic Chemistry* **73**, 2442–2445.
74. Liu, J. L., Xu, X. X., Li, D. W., Zhang, L., Zhang, K., Liu, Q. (2010). Efficient synthesis of highly functionalized vinylogous thiol esters. *Tetrahedron Letters* **51**, 6918–6920.
75. Yu, H. F., Yu, Z. K. (2009). Direct alkenylation of indoles with α-oxo ketene dithioacetals: Efficient synthesis of indole alkaloids meridianin derivatives. *Angewandte Chemie International Edition* **48**, 2929–2933.
76. Yuan, H. J., Wang, M., Liu, Y. J., Wang, L. L., Liu, J., Liu, Q. (2010). Unexpected hydrobromic acid-catalyzed C–C bond-forming reactions and facile synthesis of coumarins and benzofurans based on ketene dithioacetals. *Chemistry—A European Journal* **16**, 13450–13457.
77. Yadav, A. K., Peruncheralathan, S., Ila, H., Junjappa, H. (2007). Domino carbocationic rearrangement of α-[bis(methylthio)methylene]alkyl-2-(3/2-indolyl) cyclopropyl ketones. *The Journal of Organic Chemistry* **72**, 1388–1394.
78. Yuan, H.-J., Wang, M., Liu, Y.-J., Liu, Q. (2009). Copper(II)-catalyzed C–C bond-forming reactions of α-electron-withdrawing group-substituted ketene *S,S*-acetals with carbonyl compounds and a facile synthesis of coumarins. *Advanced Synthesis & Catalysis* **351**, 112–116.
79. Liu, Y. J., Wang, M., Yuan, H. J., Liu, Q. (2010). Copper(II) bromide/boron trifluoride etherate-cocatalyzed cyclization of ketene dithioacetals and *p*-quinones: a mild and general approach to polyfunctionalized benzofurans. *Advanced Synthesis & Catalysis* **352**, 884–892.
80. Liang, D. Q., Wang, M., Bekturhan, B., Xiong, B. B., Liu, Q. (2010). One-pot synthesis of polyfunctionalized 4*H*-chromenes and dihydrocoumarins based on copper(II) bromide-catalyzed C–C coupling of benzylic alcohols with ketene dithioacetals. *Advanced Synthesis & Catalysis* **352**, 1593–1599.
81. Jin, W. W., Du, W. M., Yang, Q., Yu, H. F., Chen, J. P., Yu, Z. K. (2011). Regio- and stereoselective synthesis of multisubstituted olefins and conjugate dienes by using α-oxo ketene dithioacetals as the building blocks. *Organic Letters* **13**, 4272–4275.
82. Dong, Y., Wang, M., Liu, J., Ma, W., Liu, Q. (2011). Aerobic, Cu-catalyzed desulfitative C–C bond-forming reaction of ketenedithioacetals/vinylogous thioesters and arylboronic acids. *Chemical Communications* **47**, 7380–7382.

83. Wang, Y. M., Bi, X. H., Li, W.-Q., Li, D. H., Zhang, Q., Liu, Q., Ondon, B. S. (2011). Iron-catalyzed aminolysis of β-carbonyl 1,3-dithianes: Synthesis of stereodefined β-enaminones and 3,4-disubstituted pyrazoles. *Organic Letters* **13**, 1722–1725.
84. Jin, W. W., Yu, H. F., Yu, Z. K. (2011). Regioselective synthesis of multisubstituted pyrazoles via cyclocondensation of β-thioalkyl-α,β-unsaturated ketones with hydrazines. *Tetrahedron Letters* **52**, 5884–5887.
85. Kobatake, T., Fujino, D., Yoshida, S., Yorimitsu, H., Oshima, K. (2010). Synthesis of 3-trifluoromethylbenzo[*b*]furans from phenols via direct *ortho* functionalization by extended pummerer reaction. *Journal of the American Chemical Society* **132**, 11838–11840.
86. Bi, X. H., Dong, D. W., Li, Y., Liu, Q., Zhang, Q. (2005). [5C + 1S] Annulation: A facile and efficient synthetic route toward functionalized 2,3-dihydrothiopyran-4-ones. *The Journal of Organic Chemistry* **70**, 10886–10889.
87. Singh, P. P., Yadav, A. K., Ila, H., Junjappa, H. (2009). Novel route to 2,3-substituted benzo[*b*]thiophenes via intramolecular radical cyclization. *The Journal of Organic Chemistry* **74**, 5496–5501.
88. Hegedus, L. L., McCabe, R. W., (1984) *Catalyst Poisoning*, Marcel Dekker, New York.
89. Hua, R. M., Takeda, H., Onozawa, S.-y., Abe, Y., Tanaka, M. (2001). Palladium-catalyzed thioesterification of alkynes with *O*-methyl *S*-phenyl thiocarbonate. *Journal of the American Chemical Society* **123**, 2899–2900.
90. Sugoh, K., Kuniyasu, H., Sugae, T., Ohtaka, A., Takai, Y., Tanaka, A., Machino, C., Kambe, N., Kurosawa, H. (2001). A prototype of transition-metal-catalyzed carbothiolation of alkynes. *Journal of the American Chemical Society* **123**, 5108–5109.
91. Yamashita, F., Kuniyasu, H., Terao, J., Kambe, N. (2008). Platinum-catalyzed regio- and stereoselective arylthiolation of internal alkynes. *Organic Letters* **10**, 101–104.
92. Umakoshi, K., Yamasaki, T., Fukuoka, A., Kawano, H., Ichikawa, M., Onishi, M. (2002). C–S bond cleavage of pyridine-2-thiol: A new functionality of quadruply bridged dinuclear platinum and palladium complexes. *Inorganic Chemistry* **41**, 4093–4095.
93. Hooper, J. F., Chaplin, A. B., González-Rodríguez, C., Thompson, A. L., Weller, A. S., Willis, M. C. (2012). Aryl methyl sulfides as substrates for rhodium-catalyzed alkyne carbothiolation: arene functionalization with activating group recycling. *Journal of the American Chemical Society* **134**, 2906–2909.
94. Itami, K., Mineno, M., Muraoka, N., Yoshida, J.-i. (2004). Sequential assembly strategy for tetrasubstituted olefin synthesis using vinyl 2-pyrimidyl sulfide as a platform. *Journal of the American Chemical Society* **126**, 11778–11779.
95. Muraoka, N., Mineno, M., Itami, K., Yoshida, J.-i. (2005). Rapid synthesis of CDP840 with 2-pyrimidyl vinyl sulfide as a platform. *The Journal of Organic Chemistry* **70**, 6933–6936.
96. Chung, M.-K., Schlaf, M. (2004). A catalytic synthesis of thiosilanes and silthianes: Palladium nanoparticle-mediated cross-coupling of silanes with thio phenyl and thio vinyl ethers through selective carbon–sulfur bond activation. *Journal of the American Chemical Society* **126**, 7386–7392.
97. Aguilar-Aguilar, A., Liebeskind, L. S., Peña-Cabrera, E. (2007). Pd-catalyzed, Cu(I)-mediated cross-couplings of bisarylthiocyclobutenediones with boronic acids and organostannanes. *The Journal of Organic Chemistry* **72**, 8539–8542.
98. Lee, K., Counceller, C. M., Stambuli, J. P. (2009). Nickel-catalyzed synthesis of oxazoles via C–S activation. *Organic Letters* **11**, 1457–1459.

99. Han, J. Y., Gonzalez, O., Aguilar-Aguilar, A., Peña-Cabrera, E., Burgess, K. (2009). 3- and 5-Functionalized BODIPYs via the Liebeskind–Srogl reaction. *Organic & Biomolecular Chemistry* **7**, 34–36.
100. Arshad, N., Hashim, J., Kappe, C. O. (2009). Palladium(0)-catalyzed, copper(I)-mediated coupling of cyclic thioamides with alkenylboronic acids, organostannanes, and siloxanes. *The Journal of Organic Chemistry* **74**, 5118–5121.
101. Yadav, A. K., Ila, H., Junjappa, H. (2010). Synthesis of novel substituted phenanthrenes and polycyclic heteroarenes by Pd-catalyzed, direct, intramolecular arylation/heteroarylation. *European Journal of Organic Chemistry* **2010**, 338–344.
102. Huang, H. N., Li, J., Lescop, C., Duan, Z. (2011). Palladium-catalyzed regioselective C–S bond cleavage of thiophenes. *Organic Letters* **13**, 5252–5255.
103. Itami, K., Yamazaki, D., Yoshida, J.-i. (2004). Pyrimidine-core extended π-systems: general synthesis and interesting fluorescent properties. *Journal of the American Chemical Society* **126**, 15396–15397.
104. Panda, K., Venkatesh, C., Ila, H., Junjappa, H. (2005). Efficient routes to acenaphthylene-fused polycyclic arenes/heteroarenes and heterocyclic fluoranthene analogues. *European Journal of Organic Chemistry* **2005**, 2045–2055.
105. Braun, M.-G., Quiclet-Sire, B., Zard, S. Z. (2011). A highly stereoselective, modular route to (*E*)-vinylsulfones and to (*Z*)- and (*E*)-alkenes. *Journal of the American Chemical Society* **133**, 15954–15957.
106. Ishizuka, K., Seike, H., Hatakeyama, T., Nakamura, M. (2010). Nickel-catalyzed alkenylative cross-coupling reaction of alkyl sulfides. *Journal of the American Chemical Society* **132**, 13117–13119.
107. Lauer, A. M., Mahmud, F., Wu, J. (2011). Cu(I)-catalyzed, α-selective, allylic alkylation reactions between phosphorothioate esters and organomagnesium reagents. *Journal of the American Chemical Society* **133**, 9119–9123.
108. Ramana, T., Saha, P., Das, M., Punniyamurthy, T. (2010). Copper-catalyzed domino intra- and intermolecular C–S cross-coupling reactions: synthesis of 2-(arylthio)arylcyanamides. *Organic Letters* **12**, 84–87.
109. Arisawa, M., Toriyama, F., Yamaguchi, M. (2011). Rhodium-catalyzed phenylthiolation reaction of heteroaromatic compounds using α-(phenylthio)isobutyrophenone. *Tetrahedron Letters* **52**, 2344–2347.
110. Zhao, C. X., Mitchell, T. A., Vallakati, R., Pérez, L. M., Romo, D. (2012). Mechanistic investigations of the $ZnCl_2$-mediated tandem Mukaiyama aldol lactonization: Evidence for asynchronous, concerted transition states and discovery of 2-oxopyridyl ketene acetal variants. *Journal of the American Chemical Society* **134**, 3084–3094.
111. Appel, R., Hartmann, N., Mayr, H. (2010). Scope and limitations of cyclopropanations with sulfur ylides. *Journal of the American Chemical Society* **132**, 17894–17900.
112. Creech, G. S., Kwon, O. (2010). Tandem 6π-electrocyclization and cycloaddition of nitrodienes to yield multicyclic nitroso acetals. *Journal of the American Chemical Society* **132**, 8876–8877.
113. Baciocchi, E., Bettoni, M., Giacco, T. D., Lanzalunga, O., Mazzonna, M., Mencarelli, P. (2011). Structure and C–S bond cleavage in aryl 1-methyl-1-arylethyl sulfide radical cations. *The Journal of Organic Chemistry* **76**, 573–582.
114. Minozzi, M., Nanni, D., Walton, J. C. (2004). Alkanethioimidoyl radicals: Evaluation of β-scission rates and of cyclization onto *S*-alkenyl substituents. *The Journal of Organic Chemistry* **69**, 2056–2069.

115. Ling, J. Y., Horsman, G. P., Huang, S.-X., Luo, Y. G., Lin, S. J., Shen, B. (2010). Enediyne antitumor antibiotic maduropeptin biosynthesis featuring a *C*-methyltransferase that acts on a CoA-tethered aromatic substrate. *Journal of the American Chemical Society* **132**, 12534–12536.

116. Hojo, H., Ozawa, C., Katayama, H., Ueki, A., Nakahara, Y., Nakahara, Y. (2010). The mercaptomethyl group facilitates an efficient one-pot ligation at xaa-ser/thr for (glyco)peptide synthesis. *Angewandte Chemie International Edition* **49**, 5318–5321.

CHAPTER 5

HOMOGENEOUS TRANSITION-METAL-CATALYZED C–O BOND ACTIVATION

DA-GANG YU, SHUANG LUO, FEI ZHAO, and ZHANG-JIE SHI
College of Chemistry and Molecular Engineering, Peking University, Beijing, People's Republic of China

Homogeneous Catalysis for Unreactive Bond Activation, First Edition. Edited by Zhang-Jie Shi.

5.1 INTRODUCTION

The C–O bond is one of the most important and common chemical bonds and exists widely in the chemicals in nature and the industry, including both small molecules (such as methanol, acetic acid, and acetone) and large molecules (such as protein, nucleic acid, lignin, cellulose, and other biomass). With less and less nonrenewable fossil resource, the use of renewable biomass to replace fosil energy has attractted more and more attention. However, there are still many challenges in efficient conversion of cellulose to industry alcohols. As the important resource of arenes, lignin is mostly used as the fuel and additive in buildings, if not directly poured to pollute rivers, due to the super inert and complicated polymeric structure. Moreover, most of the oxygen-containg compounds, including phenols, alcohols, ethers, and esters, are inexpensive and readily available. However, unlike the commonly used organohalides, these substrates are frequently used as the solvent and hardly applied in organic reactions to construct more important and useful chemicals because the C–O bonds are so inert that we have not yet thoroughly studied their reactivity and it is very difficult to achieve high selectivity. Therefore, it is quite important for chemists to investigate the reactivity and selectivity of C–O bonds transformation with novel methods.

Transition metal catalysis is one of the most powerful and important tools for chemists to use and synthesize useful chemicals with high selectivity and efficiency [1]. Most importantly, transition metal catalysis is especially powerful for activation and transformation of inert chemical bonds. If the biomass could be transformed to useful chemicals highly efficiently and selectively and the inexpensive oxygen-containg compounds from industry could be widely used in organic synthesis via transition-metal-catalyzed C–O bond activation, we can not only reduce the cost and pollution of production, but also provide a new way to solve the issues of energy, ecology, and the environment. Therefore, there have been lots of research groups and projects focusing on the transition-metal-catalyzed C–O bond activation, and much important progress has been achieved.

Due to huge reports on C–O bond cleavage and many excellent reviews in this field [2], in this chapter we would like to focus on "inert" C–O bond activation via a transition-metal-catalyzed oxidative addition pathway. The reactions via Lewis/Brønsted acids catalysis [3] or the "hydrogen-transfer" pathway [4], as well as those reactions of quite "active" C–O bonds, such as those in allylic esters [5] and orgnao-triflates [6], have been well reviewed and will only be mentioned here without extensive discussion.

To make the content clear, we wish to describe the reactions of inert C–O bond according to the different transition metals. In every subsection of one specific transition metal, different kinds of reactions were introduced in a sequence of formation of C–C and C–X (X = N, B, H, etc.) bonds. Each specific reaction will be introduced according to the reactivity of the C–O bonds, along which we could see how the catalytic system, especially the ligand, plays the important role in the transformations. Between different subsections, we could also contrast the different behaviors of different transition metals. Take the Pd and Ni complex, for example. In most cases of Pd-catalyzed C–O bond activation, low catalyst loading, wide substrate scope, good functional group, and moisture tolerance could be frequently observed. However, the palladium catalysts have rarely been reported to activate the highly inert C–O bonds, such as those in aryl ethers, carboxylates, and evern phenol themselves. For these substrates, more reactive and nucleophilic nickel complexes became excellent catalysts, especially in the presence of strongly electron-donating ligand such as PCy_3. However, high reactivity always leads to low selectivity. In Ni-catalyzed C–O bond activation, many functional groups, such as nitro, alkyne, C–S, C–F, and C–CN bonds, which could be well tolerated in Pd catalysis, undergo competitive side reaction. Moreover, Ni catalysis is more sensitive than Pd to the moisture and air, which makes the process much more strict and complicated. Furthermore, at the initial stage of Ni-catalyzed inert C–O bond activation, the reaction types are more limited and the catalyst loading is higher in contrast with Pd catalysis for most cases.

Based on the intensive studies on C–O bond activation, the main mechanisms of C–O bond cleavage could be divided into the following types: (1) oxidative addition of the C–O bond to low-valent transition metals; (2) transition-metal-catalyzed β–O elimination; (3) C–O bond cleavage promoted by Lewis/Brønsted acids; (4) C–O bond cleavage promoted by single electron transfer of early transition metals. In this chapter, the mechanism of oxidative addition will be focused on. Based on a detailed pathway, it could be mainly divided into the following pathways:

1. *Direct oxidative addition of a C–O bond to low-valent transition metals via three- or five-center transition state.* Similar to Pd-catalyzed oxidative addition of aryl carbon–halide bonds, the π-system of phenol and enol derivatives, such as carboxylates, carbomates, and sulfomates, were proposed to undergo similar η_2 coordination to low-valent transition metals, which further cleave the C–O bonds and generate C–M bond via three or five center transition state. Compared to the enol and naphthol, it is much more difficult for common phenols, which have a more stable aromatic character and a higher energy barrier to be destroyed, to form the η_2 coordination and undergo C–O bond cleavage. In addition, it is also not easy for some electron-rich and low-valent transition metals, such as a Ni(0), to dissociate from the product and coordinate to much more electron-rich aromatic ring of phenols.
2. *Nucleophilic attack of a low-valent transition metal to the C–O bond.* The cleavage of the acyl C–O bond in esters is proposed to arise from the nucleophilic attack of a low-valent transition metal to a carbonyl group and elimination. Moreover, it is also favored that S_N2-type nucleophilic attack of a low-valent transition metal to a C–O bond leads to the sp^3 C–O bond activation in alcohols, ethers, and esters.
3. *Double- or single-electron transfer.* Besides the η_2 coordination to a π-system, it is also possible for a low-valent transition metal to undergo nucleophilic addition to a double bond to generate a carbon anion intermediate, which undergoes β–O elimination to form the *ipso*-carbon–metal bond. Moreover, double single-electron transfer between a π-system and transition–metals, such as Ni(0), could also not be excluded.

To realize the C–O bond activation via oxidative addition, there are mainly three strategies:

1. *Lowering the electron density of C–O bond to increase its reactivity.* This is the intrinsic effect to improve the reactivity of a C–O bond and facilitate the oxidative addition. With phenol as an example, the reactivity of the C–O bond could be improved through reducing the electron density of aromatic ring or oxygen atom. The former strategy includes introduction of electron-withdrawing groups on the aromatic ring or coordination with the electron–deficient metal complexes (such as a three-carbonyl chromium). The latter could be realized through transferring the hydroxyl group to a good leaving group (such as sulfonates and phosphates), or coordinating with Lewis/Brønsted acids to form onium salts. For example, with the oxygen bonding with a strong electron-withdrawing trifluoromethylsufonyl (Tf) group, aryl triflates have been successfully used as electrophiles for a long time due to their relatively high reactivity, which is similar to that of aryl bromides. In contrast, other sulfonyl and phosphonyl groups are less electron-withdrawing so that the corresponding aryl phosphates and sulfonates, such as phosphates, tosylates, mesylates, sulfamates, and biaryl sulfonates, are less reactive and therefore similar to aryl chlorides. Furthermore, the C–O bonds of aryl carboxylates, carbamates, and ethers are even less reactive and more challenging. Kinetically, the lower electron density of an aromatic ring would also help the η_2 coordination or the electron transfer of electron-rich transition metals with a π-system.
2. *Introduction of a directing group.* The activation of C–O bond could be promoted by introducing directing groups both kinetically and thermodynamically. Kinetically, the coordination of directing group with transitionl metal could shorten the distance, thereby increasing the possibility and rate of reaction. Thermodynamically, more stable cyclic organometallic intermediates could be generated to promote the C–O bond cleavage in the presence of a directing group.
3. *Selecting a proper transition-metal complex.* In order to realize the C–O bond activation reaction with wide substrate scope, a proper transition metal complex should be chosen. Generally speaking, increasing the electron density of the transition metal center will be beneficial to the oxidative addition, and the combination of a low-valent transition metal with a strongly electron-donating ligand can promote C–O bond activation. However, it is necessary to choose the suitable transition metal and ligand with proper binging angle and electron donating ability in order to obtain a high catalytic efficiency. Moreover, the use of active organometallic reagents (such as Grignard reagents) as a reaction partner will be helpful to stabilize low-valent transition metal species and generate an adduct of a metalate with higher electron density and reactivity.

Using these strategies, many stochiometric transformations through C–O activation have been realized. For example, in 1999 Sweigart and co-workers found that the alkenyl C–O bond of benzofuran could undergo selective oxidative addtion to a Pt(0) complex when the arene was coordinated to a $Mn(CO)_3^+$ complex (Scheme 5.1) [7].

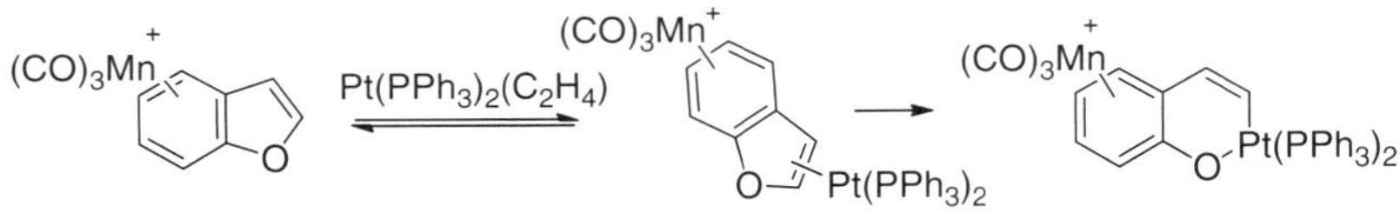

Scheme 5.1. Pt-mediated selective alkenyl C–O activation in the assistance of a $Mn(CO)_3^+$ complex.

Due to the realatively high reactivity, aryl triflates were studied by many organists in the Pd-catalyzed oxidative addition. However, because the products are unstable, addition of halides salts was used to promote this transformation and generate oxidative addition product with the halide as the couterion [8]. In 2003 Roy and Hartwig [9] reported the first oxidative addition of an unactivated aryl tosylate to a Pd(0) complex with powerful Josiphos ligands (Scheme 5.2). The transformation occurred efficiently at room temperature and showed a higher rate in a more polar solvent. The kinetic study indicated that this reaction was first order in aryl tosylate and inverse order in $P(o\text{-tolyl})_3$.

Besides improving the reactivity of a C–O bond, the introduction of a directing group could promote stoichiometric C–O activation and achieve a high selectivity. For example, during the mechanistic studies of directing-group-assisted ruthenium-catalyzed cross-coupling of a C–O bond in anisoles [10], Kakiuchi and co-workers [11] found that the aryl C–O bond could undergo oxidative addition to $RuH_2(CO)(PPh_3)_3$ to generate a cycloruthenium intermediate, which showed good catalytic activity in this cross-coupling (Scheme 5.3).

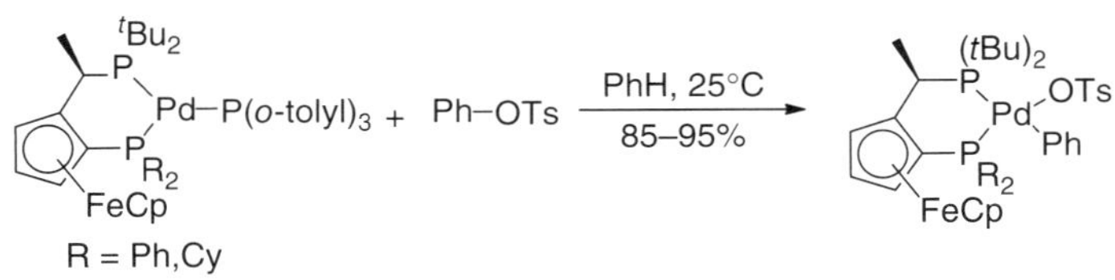

Scheme 5.2. Pd-mediated aryl C–O activation of phenyl tosylate.

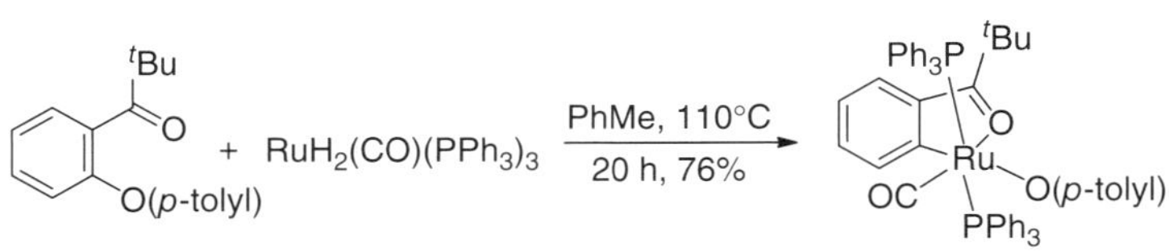

Scheme 5.3. Ru-mediated aryl C–O activation of aryl ether with the assistance of a carbonyl group.

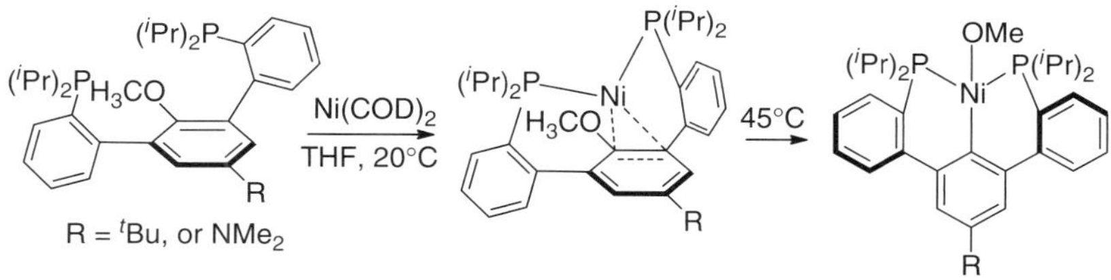

Scheme 5.4. Ni-mediated aryl C–O activation of an aryl ether with the assistance of a diphosphine ligand.

Similarly, during the mechanistic studies of the Ni-catalyzed hydrogenolysis of an aryl C–O bond, Agapie and co-workers [12] investigated the reaction of (diphosphine)aryl methyl ethers and isolated a Ni(0) complex containing Ni–arene interactions adjacent to the aryl–O bond, which was cleaved upon heating to generate a nickel–aryl–methoxide complex (Scheme 5.4).

Furthermore, selectivity in C–O bond activation also could be achieved by using different catalysts under various reaction conditions. For example, in 1976 Yamamoto and co-workers [13] found that selective activation of alkenyl and acyl C–O bonds in enol carboxylates could be realized with low-valent Ni species in different conditions. It was believed that the cleavage of an acyl C–O bond was reversible and probably arose from the nucleophilic attack on the carbonyl group by the low-valent Ni(0). In 1979, Tolman and Ittel discovered that the intramolecular methyl C–O bond activation in ether and ester was promoted by iron–hydride species to generate a Me–Fe complex (Scheme 5.5) [14].

In 1997 Milstein and co-workers [15] found that different selective C–O bond cleavage could be achieved by using various transition metals in the presence of bidentate direacting phosphine ligand (Scheme 5.6). When nucleophilic Rh(I) complex was used, the aryl C–OMe bond was activated to generate an Ar–Rh pincer complex, which further underwent β-H elimination to afford a Rh–H complex. On the other hand, if electrophilic and Lewis acidic Pd(II) or Ni(II) complex was used, the alkyl C–OAr was selectively cleaved [15].

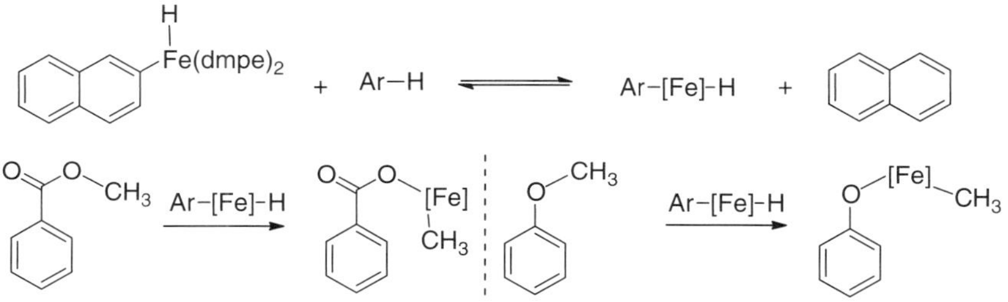

Scheme 5.5. Synthesis of aryl iron hydrides and further transformation via C–O cleavage.

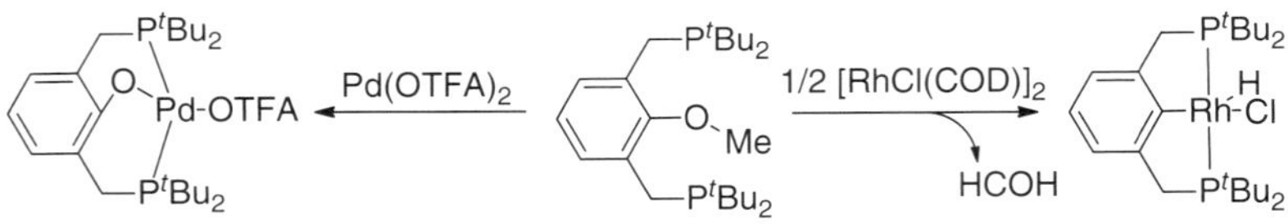

Scheme 5.6. Selective C–O cleavage with different transition metals.

5.2 PALLADIUM-CATALYZED C–O BOND ACTIVATION

5.2.1 Kumada–Tamao–Corriu Coupling

The cross-coupling reaction of organohalides with Grignard reagents catalyzed by a Ni or Pd catalyst was defined as the Kumada–Tamao–Corriu coupling, which was developed at the earliest stage and has been applied widely in organic synthesis [16]. Significant progress has been made to optimize such a transformation, which made it applicable with better group compatibility and higher efficiency [16c]. Similar to organohalides, phosphates and sulfonates also could undergo the Pd-catalyzed cross-coupling with Grignard reagents.

In 1986, Hayashi et al. [17] reported one example of $PdCl_2$[(R)-(S)-PPFA]-catalyzed enantioselective coupling of alkenyl phosphate with secondary alkyl Grignard reagents. Later, the Miller group developed a one-pot protocol to synthesize arylated olefins through cross-coupling of *in situ*-generated enol phosphates with Grignard reagents in the presence of $PdCl_2(PPh_3)_2$ as the catalyst [18]. In 2009, Skrydstrup and co-workers [19] found that this coupling could proceed well in the absence of any phosphine ligand (Scheme 5.7).

Besides alkenyl phosphates, heteroaromatic phosphates were also applied in this coupling with alkyl and aryl Grignard reagents by Gevorgyan and co-workers [20]. In this system, CyPF-^{t}Bu, a strongly electron-donating and chelating diphosphine ligand, was used to achieve high efficiency (Scheme 5.8).

Similar to phosphates, the C–O bonds in aryl and alkenyl sufonates also showed good reactivity in the Pd-catalyzed cross-coupling with Grignard reagents. Although triflates have never been reported to couple with Grignard reagents due to the instability, aryl tosylates, which are much more stable, have been successfully applied in the coupling with Grignard reagents. Based on the stoichiometric reaction and mechanistic study, Roy and Hartwig [9, 21] reported the first palladium-catalyzed cross-coupling of unactivated aryl tosylates with aryl Grignard reagents under mild conditions (Scheme 5.9). With PPF-^{t}Bu (a strongly electron-donating and sterically hindered bisphosphine) as ligand, the catalyst loading could be as low as 0.1 mol%

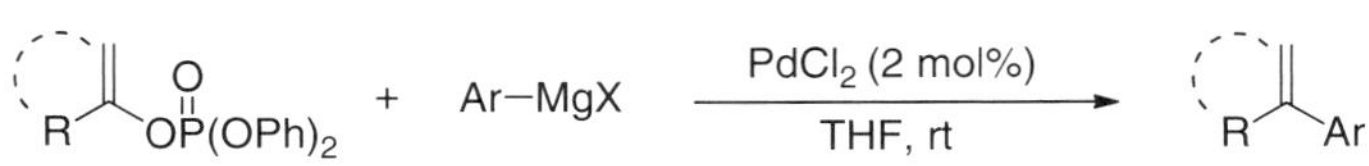

Scheme 5.7. Pd-catalyzed ligand-free Kumada–Tamao–Corriu coupling of alkenyl phosphates.

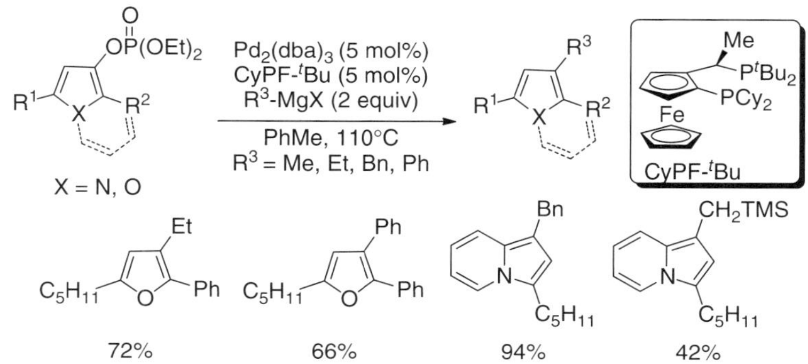

Scheme 5.8. Pd-catalyzed Kumada–Tamao–Corriu coupling of aryl phosphates.

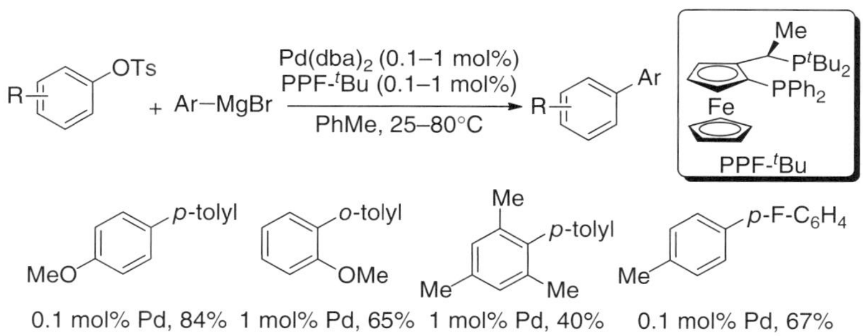

Scheme 5.9. Pd-catalyzed Kumada–Tamao–Corriu coupling of aryl tosylates.

and the substrate scope was significantly expanded to various aryl and alkenyl tosylates subsequently [22]. Using air-stable and easily available PinP(O)H as the ligand, Ackermann and Althammer [23] also realized a similar coupling with wide substrate scope and high efficiency (Scheme 5.10).

Using 1,3-diene as the additive, in 2003 Kambe and co-workers [24] successfully realized the palladium-catalyzed cross-coupling reaction of alkyl tosylates with alkyl and aryl Grignard reagents at room temperature (Scheme 5.11). A bis-allylpalladium complex was proposed to play an important role as active catalytic species in this transformation. Later, they found that a bis-allylpalladium indeed could catalyze this transformation efficiently [25].

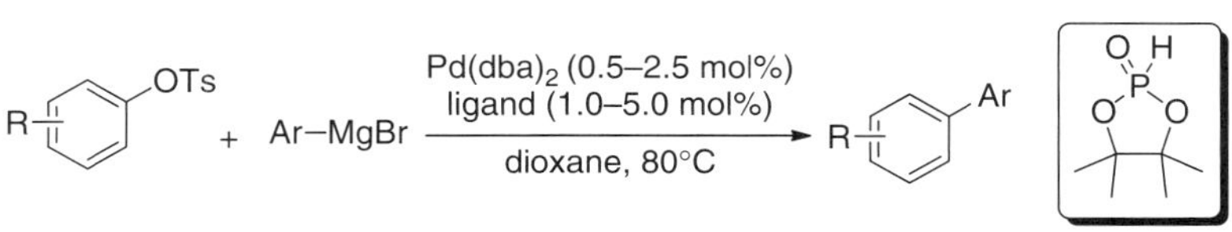

Scheme 5.10. Pd-PinP(O)H-catalyzed Kumada–Tamao–Corriu coupling of aryl tosylates.

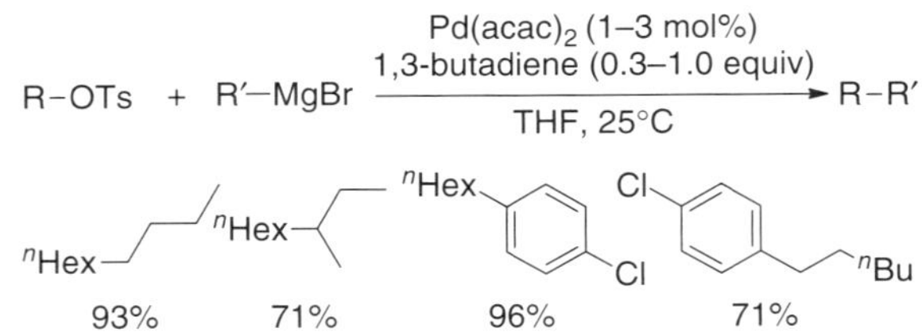

Scheme 5.11. Pd-catalyzed Kumada–Tamao–Corriu coupling of alkyl tosylates in the presence of 1,3-diene.

5.2.2 Negishi Coupling (Including Zinc, Aluminum, and Manganese Reagents)

The Ni- or Pd-catalyzed various cross-couplings of organo halides with organozinc, organoalanes, or organozirconium were referred to as Negishi coupling [26]. Because these organometallic reagents are more compatible with various electrophiles, Negishi coupling is well known for both the high efficiency and good functional groups tolerance.

In the early 1980s, Oshima and co-workers reported a series of work in this field. They found that $Pd(PPh_3)_4$ could catalyze the coupling of unactivated alkenyl phosphates with various kinds of aluminum reagents under mild conditions (Scheme 5.12A) [27]. Later, they also realized a similar coupling of alkenyl phosphates with organomanganese reagents [27d]. In 2001, Wu and Yang [28] studied the Pd-catalyzed coupling of 4-tosylcoumarin derivatives with aryl and alkyl zinc

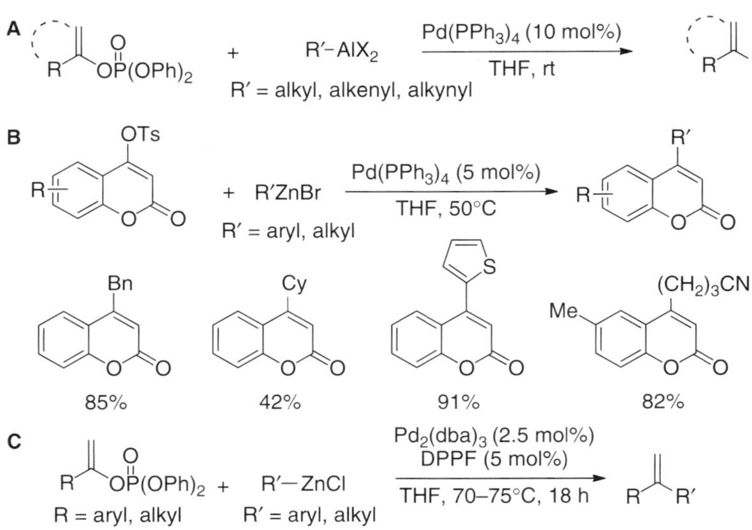

Scheme 5.12. Pd-catalyzed Negishi coupling of alkenyl phosphates and tosylates.

reagents (Scheme 5.12B). Although simple alkyl zinc reagents led to moderate yield due to the competitive β-H elimination, they found that those with a remote cyano group reacted quite well, in which the triple bond could act as an additional ligand of organopalladium intermediates to promote the productive reductive elimination. For these cases, moderately electron-donating PPh_3 was sufficient and was used for these transformations. Latter, Skrydstrup and co-workers [29] also successfully realized a similar coupling of alkenyl phosphates with organozinc reagents in the presence of DPPF as the ligand (Scheme 5.12C).

In 2003, Zhou and Fu [30] developed the first Pd-catalyzed cross coupling of primary alkyl tosylates with alkyl and alkenyl zinc reagents in the presence of $PCyp_3$ as the ligand (Scheme 5.13). Moreover, alkyl halides, including iodides, bromides, and chlorides, could also react efficiently in this condtion.

5.2.3 Stille Coupling

Due to the high selectivity, tolerance of most functional groups, and mild reaction condition, the cross-coupling of organostannanes with organic halides, referred to as the Migita–Kosugi–Stille reaction or the Stille reaction, is one of the most popular couplings in late-stage total synthesis [31].

In 1984 Stille and co-workers [32] reported the first $Pd(PPh_3)_4$-catalyzed cross-coupling reaction of alkenyl triflates with a variety of organostannanes under mild conditions. Many kinds of salts, such as LiCl, CsF, and Bu_4NI, were found to significantly promote this reaction. Mechanistic study indicated that an active Pd catalyst was generated in the presence of the salts. With LiCl as additive, Roth and Fuller [33] found that aryl fluorosulfonates had simiar reactivity with triflates in the coupling with aryl and alkenyl tin reagents. However, both triflates and fluorosulfonates are expensive to prepare and unstable. Thus, application of stable and less expensive arenesulfonates are much more desirable. For example, aryl *p*-fluorobenzenesulfonates were successfully applied in a cross-coupling reaction with alkyl-, vinyl-, allyl-, and aryl-stannanes by Badone et al. [34] in 1992 (Scheme 5.14). DPPP and Ph_2PMe showed good activity in this coupling. The substrates with electron-withdrawing groups on an aryl ring reacted faster than those with electron-donating groups because of better reactivity the aryl C–O bond.

In contrast, much more general tosylates showed lower reactivity. For the activated enol or heteroaryl tosylates, good results could be obtained with a Pd–PPh_3 system [35]. For example, coumarin, 1,2,3-triazolo[1,5-*a*]-quinazoline, and guanosine derivatives could be efficiently constructed via C–O bond activation of tosylates. For alkenyl mesylates, there was also a report of Pd-PPh_3-catalyzed cross-coupling

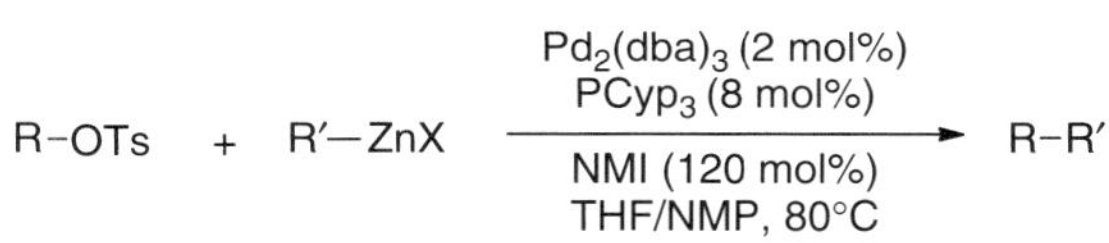

Scheme 5.13. Pd-catalyzed cross-coupling of alkyl tosylates with organozinc reagents.

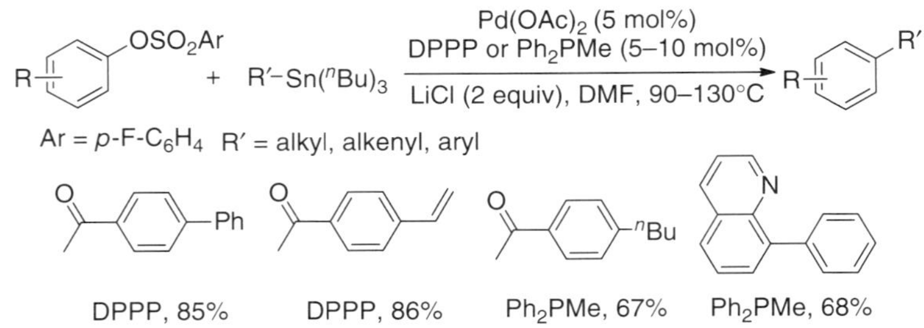

Scheme 5.14. Pd-catalyzed Stille coupling of aryl sufonates.

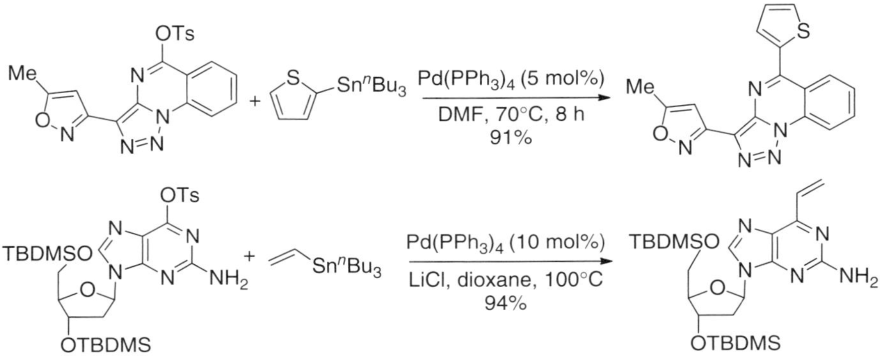

Scheme 5.15. Pd-catalyzed Stille coupling of imino and heteroaryl tosylates.

of β-mesyloxy enones with vinyl- and allylstannanes by Scott and co-workers [36] in 1991 (Scheme 5.15). However, it is not efficient for activation and transformation of the C–O bond in general aryl tosylates.

In 2010, Buchwald and co-workers [37] successfully realized the coupling of various aryl and heteroaryl sulfonates with aryl organostannanes using bulky and electron-rich X-Phos as ligand (Scheme 5.16). Nonactivated aryl tosylates worked well in this system. Moreover, more atom-economic and less reactive aryl mesylates are also good substrates. A variety of activated and nonactivated aryl mesylates could react with sterically hindered aryl- and heteroaryl stannanes in good yields.

Similar to alkenyl tosylates, alkenyl phosphates are also easy to prepare form corresponding carbonyl compounds and show good reactivity in Pd-catalyzed cross coupling with organostannanes. In 1997, Nicolaou et al. [38] applied the strategy of enol phosphanation-cross coupling sequence to construct medium and large cyclic ethers form lactones (Scheme 5.17). Although the catalytic system with Pd-PPh_3 as the catalyst and LiCl as additive is quite similar to those of alkenyl sulfonates, phosphates show much better reactivity for this kind of substrates. With this methodology, Brevitoxin A was successfully synthesized through double vinlyation from two directions of bis(lactone). Following this elegant work, many other groups used similar

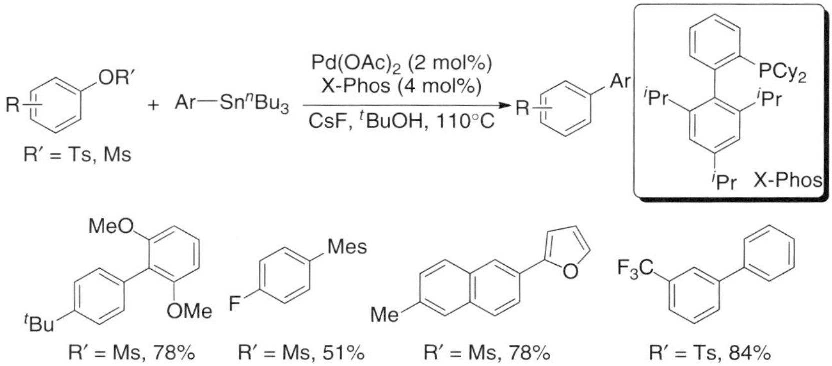

Scheme 5.16. Pd-catalyzed Stille coupling of imino and heteroaryl tosylates and mesylates.

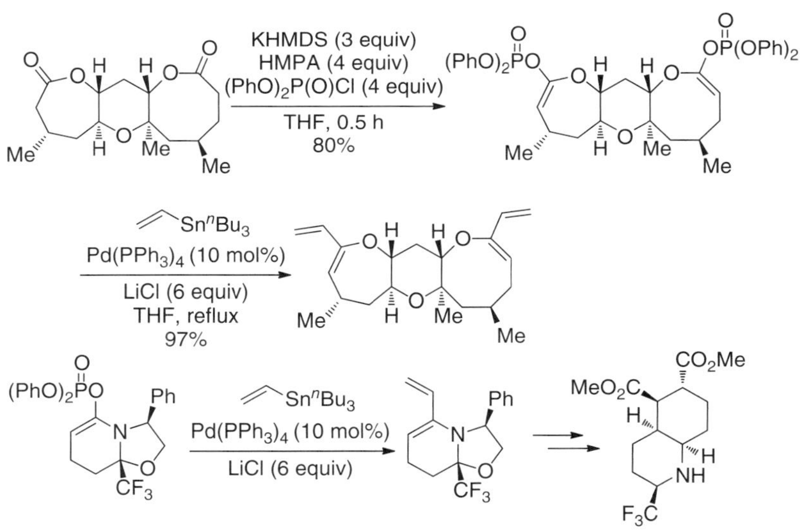

Scheme 5.17. Pd-catalyzed Stille coupling of alkenyl phosphates.

protocol to synthesize heterocycles from enol phosphinates and phosphonates [39]. For example, Jiang et al. [39e] realized the palladium-catalyzed coupling of cyclic enamine phosphate with vinyl stannane (Scheme 5.17).

5.2.4 Suzuki–Miyaura Coupling

5.2.4.1 Suzuki–Miyaura Coupling of Phosphates. The Pd-catalyzed cross-coupling of organoborane or boronic acid derivatives with electrophiles was referred to as Suzuki–Miyaura coupling, which is extremely useful to construct C–C bonds and particularly attractive among various cross-coupling reactions because of its mild reaction condition, high functional group tolerance, and easy availability of nontoxic and stable boronic acid derivatives [40].

The substrate scope of organophosphates in Pd-catalyzed Suzuki–Miyaura coupling is mainly limited to the relatively activated enol phosphates, phosphinate, phosphonates, and heteroaromatic phosphates. For these substrates, Pd–PPh_3 catalytic system is always efficient. For example, in 1999 Coudert and co-workers [41] reported the Suzuki–Miyaura coupling of cyclic α-phosphono enamides to provide arylated heterocycles efficiently (Scheme 5.18).

Independently, Sasaki and other groups extensively investigated the reactivity of the C–O bond in enol phosphates and successfully realized the Pd-catalyzed Suzuki–Miyaura coupling of cyclic α-phosphono enol ethers (Scheme 5.19) and acyclic α-phosphoryloxyenecarbamates [42].

For the nonactivated cyclic alkenyl phosphates, Pd catalysts showed lower reactivity in this coupling. Although using a powerful Buchwald-type biaryls phosphine ligand and harsh reaction condition, Begtrup and co-workers [43] realized a Pd-catalyzed Suzuki–Miyaura coupling of nonactivated cyclic alkenyl phosphates with in moderate efficiency (Scheme 5.20).

The C–O bonds in heterocyclic aryl phosphates show higher reactivity than those in normal aryl phosphates and could be activated by palladium catalysts smoothly. In 2008 Kang et al. [44] reported a novel Suzuki–Miyaura coupling

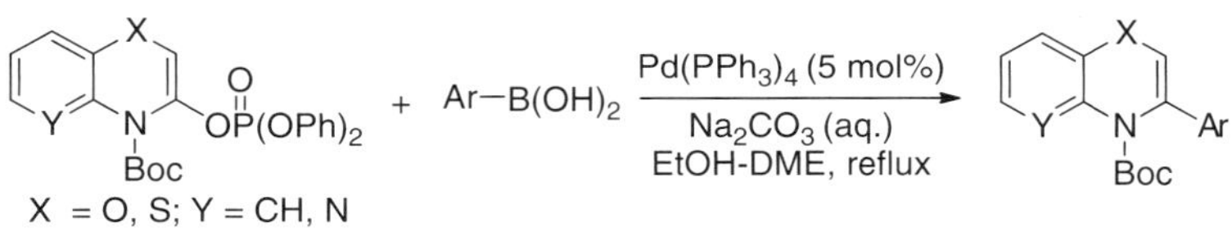

Scheme 5.18. Pd-catalyzed Suzuki–Miyaura coupling of cyclic α-phosphono enamides.

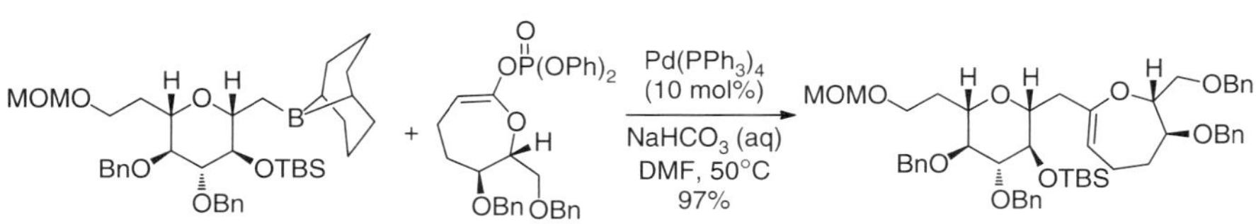

Scheme 5.19. Pd-catalyzed Suzuki–Miyaura coupling of cyclic α-phosphono enol ethers.

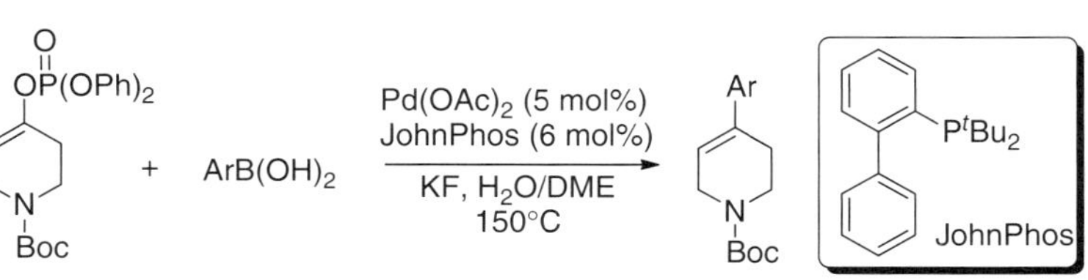

Scheme 5.20. Pd-catalyzed Suzuki–Miyaura coupling of alkenyl phosphates.

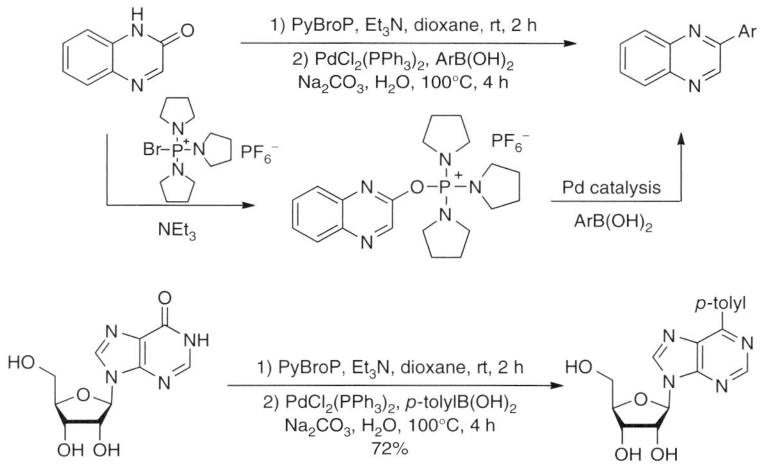

Scheme 5.21. Pd-catalyzed Suzuki–Miyaura coupling of tautomerizable heterocyles.

of tautomerizable heterocyles with aryl boronic acids via *in situ* generation of heterocycle–phosphonium salts (Scheme 5.21) [44]. It is important to note that biologically important 6-arylpurine ribonucleosides could be efficiently produced via such a cross-coupling of unprotected inosine and aryl boronic acid. Later, Han and co-workers [45] used this strategy and realized the Suzuki–Miyaura coupling of 2-hydroxypyridines and 2-hydroxyquinolines.

5.2.4.2 Suzuki–Miyaura Coupling of Tosylates, Sulfamates, and Mesylates. Similar to activated alkenyl phosphates, activated alkenyl tosylates and aryl imidazolylsulfonates also show good reactivity in Pd-catalyzed Suzuki–Miyaura coupling without the need for specific ligands [46]. For example, Yang and co-workers [46a] reported an efficient $PdCl_2(PPh_3)_2$-catalyzed cross-coupling of 4-tosyl-2(5H)-furanone with aryl and alkenyl boronic acids in 2003 (Scheme 5.22). Later, Wu and Peng and their co-workers also realized the Pd-catalyzed cross-coupling of 4-hydroxycoumarins, 4-hydroxy-pyrones, and 4-hydroxy-2(5H)-furanones with arylboronic acids through *in situ* generation of alkenyl tosylates (Scheme 5.23). The Suzuki–Miyaura coupling of more reactive imidazolylsulfonates was realized by the Albaneze–Walker group with Pd–DPPF catalytic system under mild conditions (Scheme 5.24) [48].

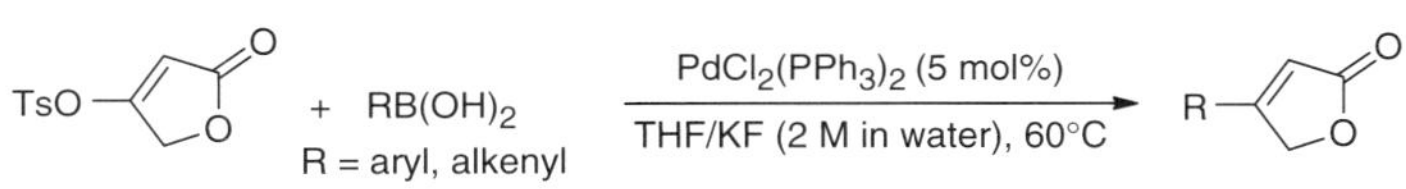

Scheme 5.22. Pd-catalyzed Suzuki–Miyaura coupling of alkenyl tosylates.

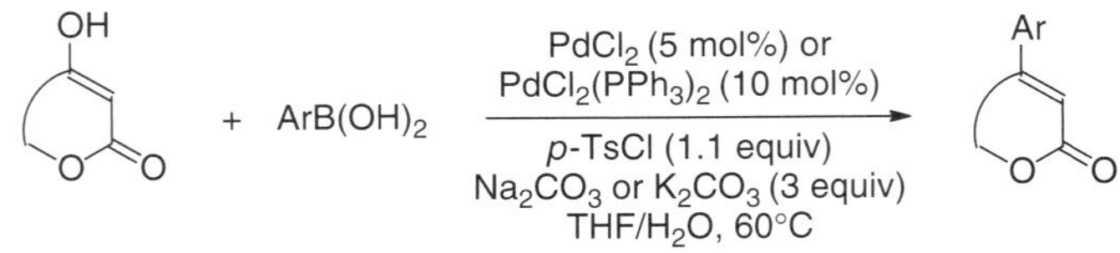

Scheme 5.23. Pd-catalyzed Suzuki–Miyaura coupling of *in situ* generated alkenyl tosylates.

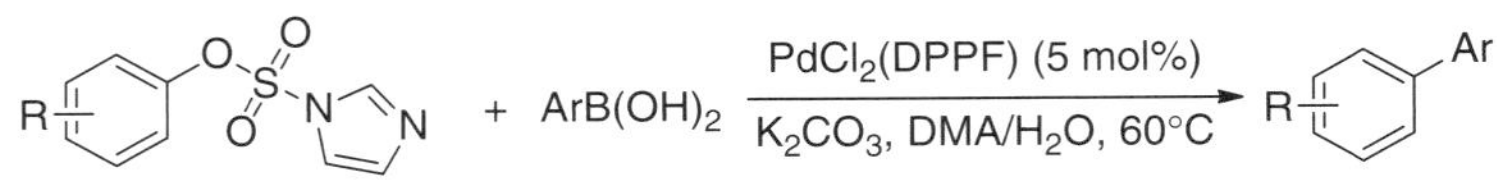

Scheme 5.24. Pd-catalyzed Suzuki–Miyaura coupling of aryl imidazolylsulfonates.

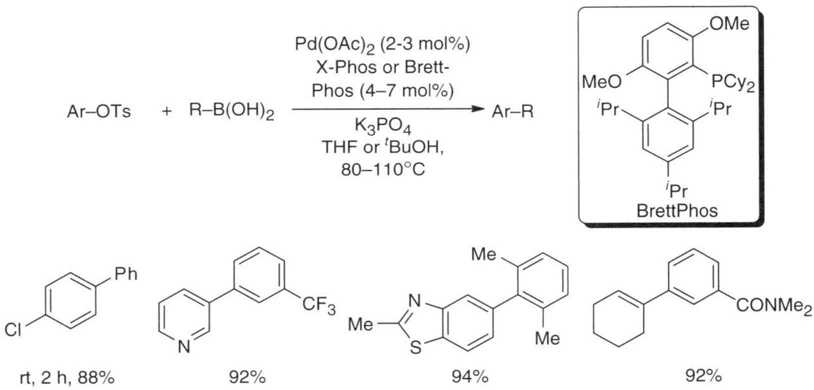

Scheme 5.25. Pd-catalyzed general Suzuki–Miyaura coupling of aryl tosylates.

However, powerful biaryl phosphine ligands are always required to promote the reactivy of Pd catalyst to activate the more inert C–O bonds in nonactivated alkenyl tosylates, normal aryl tosylates and mesylates. For example, Buchwald and co-workers [49] found that the combination of palladium catalyst with X-phos or BrettPhos was very effective for the Suzuki–Miyaura coupling of a wide range of aryl arenesulfonates, mesylates, and nonactivated alkenyl tosylates (Scheme 5.25). Most of the common functional groups were well tolerated. Notably, challenging highly steric hindered biaryls could also be constructed smoothly with this method.

In 2008, Kwong and co-workers [50] also realized similar couplings of aryl tosylates and mesylates with good functional group tolerance using $Pd(OAc)_2$ and indoyl phosphine ligand (CM-Phos), which was developed in the Kwong's group (Scheme 5.26).

In 2002 Netherton and Fu [51] developed a Pd-P^{t}Bu$_2$Me-catalyzed cross-coupling reaction of alkyl tosylates with primary alkyl and aryl boron reagents (Scheme 5.27).

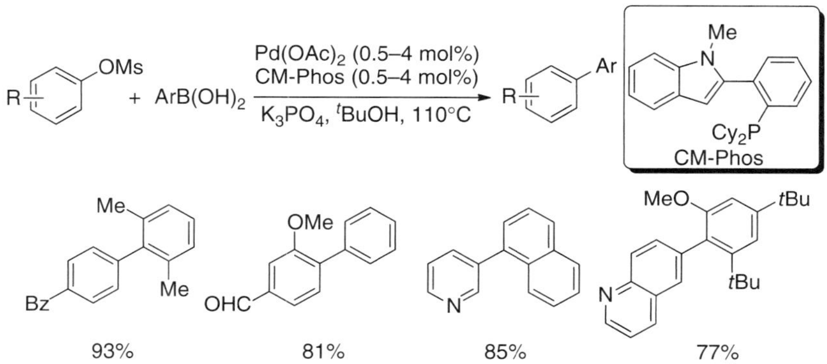

Scheme 5.26. Pd-catalyzed general Suzuki–Miyaura coupling of aryl mesylates.

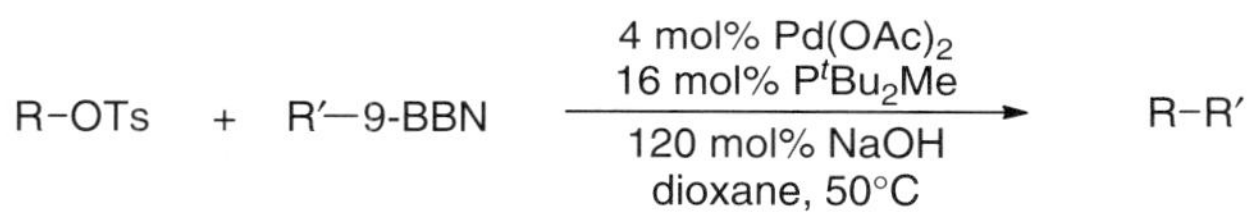

Scheme 5.27. Pd-catalyzed Suzuki–Miyaura coupling of alkyl tosylates.

Many functional groups, such as ester, nitrile, olefin, alcohol, ether, amide, and silicon, can be tolerated well in this transformation.

Besides alcohol derivatives, activated alcohols themselves also could take part in Pd-catalyzed cross coupling with boronic reagents. For example, In 2004 Tsukamoto et al. [52] reported the first $Pd(PPh_3)_4$-catalyzed cross coupling of allylic alcohols with aryl boronic acids in the absence of any additives. It was believed that the coordination of oxygen in alcohols to Lewis acidic boron center activated the C–O bonds (Scheme 5.28). It is interesting to note that when Tsukamoto et al. [53] modified the catalytic system, they found that higher yield was obtained if lower cat loading was employed, which was explained by formation of allylic Pd complex through direct attack by PPh_3. In 2005 Kobayashi and co-workers [54] developed a $[Pd(allyl)Cl]_2$-catalyzed cross-coupling of allylic alcohols and arylboronic acids in water at reflux with PPh_3 as ligand. Addition of a catalytic amount of a base allowed the reaction to proceed well at 40°C. Similarly, the first cross-coupling of allenic and proparglic alcohols with aryl boronic acids were also realized by Yoshida and Ihara's group in the same year (Scheme 5.29) [55].

5.2.5 Hiyama Coupling

The cross-coupling of electrophiles with various kinds of organosilanes, chloro- and fluorosilanes, alkoxysilanes, and silanols, referred to as Hiyama coupling, is an important method for C–C bond construction due to the unique advantages of

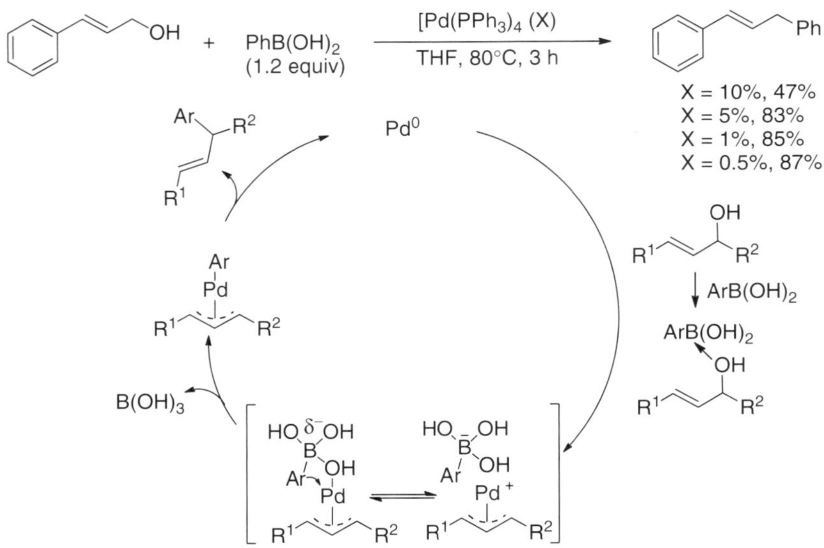

Scheme 5.28. Pd-catalyzed Suzuki–Miyaura coupling of allylic alcohols.

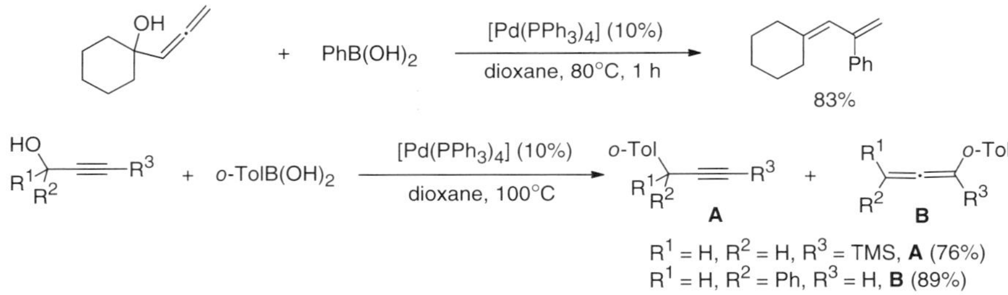

Scheme 5.29. Pd-catalyzed Suzuki–Miyaura coupling of allenic and proparglic alcohols.

organosilane reagents, including their high stability, low toxicity, good functional group tolerance, and ease of handling and purification [56].

Due to the low reactivity of organosilicon reagents, there are only few examples of Pd-catalyzed Hiyama coupling of phenol and enol derivatives. For example, in 2008 Zhang and Wu [57] realized the first Pd-catalyzed Hiyama coupling of aryl and alkenyl tosylates by using Buchwald ligand X-phos as the ligand (Scheme 5.30). The catalytic system was later successfully applied by the same group to Hiyama coupling of aryl mesylates, which is also quite efficient for a wide range of mesylates and silanes [58]. Independently, Kwong and co-workers [59] also realized a similarly efficient Hiyama coupling of aryl mesylates using their CM-phos ligand (Scheme 5.31).

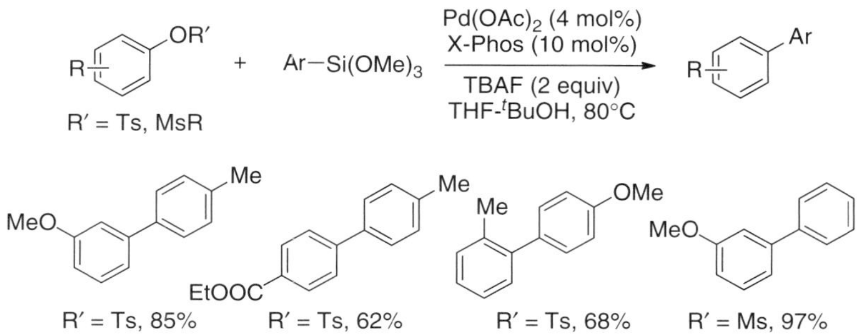

Scheme 5.30. Pd-catalyzed Hiyama coupling of aryl and alkenyl tosylates and mesylates.

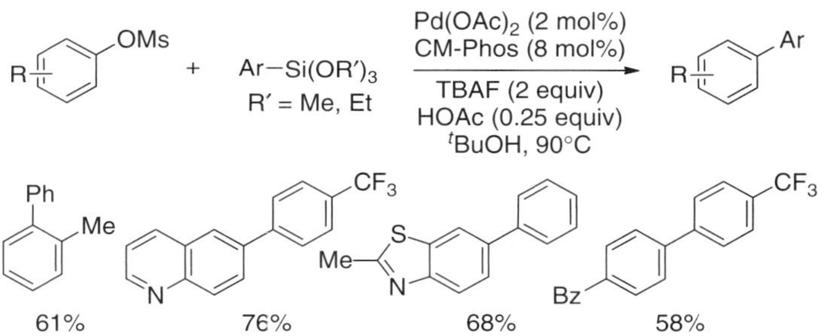

Scheme 5.31. Pd-catalyzed Hiyama coupling of aryl mesylates with CM-Phos.

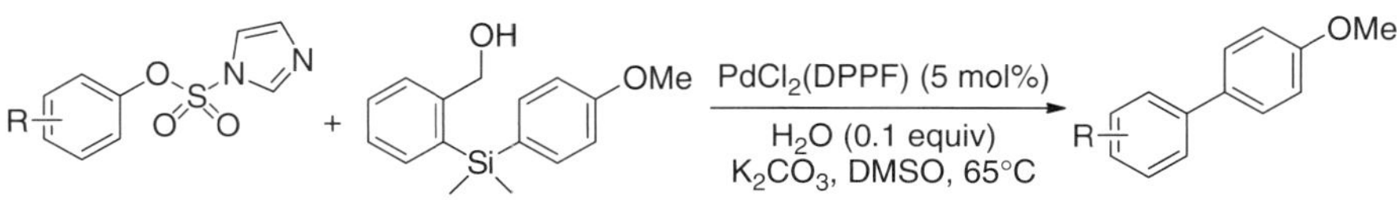

Scheme 5.32. Pd-catalyzed Hiyama coupling of aryl imidazolylsulfonates.

Similar to Suzuki–Miyaura coupling, the C–O bonds in aryl imidazolylsulfonates show better reactivity than tosylates in Hiyama coupling. For example, Williams and co-workers [60] reported an Pd-DPPF-catalzyed cross-coupling of aryl imidazolylsulfonates with Hiyama's silicon reagents in the absence of fluoride salts (Scheme 5.32).

5.2.6 Heck Reaction

The Heck reaction (or Mizoroki–Heck reaction) is one of the earliest cross-coupling reactions and a powerful method to synthesize highly functionalized alkenes from electrophiles with simple alkenes [61].

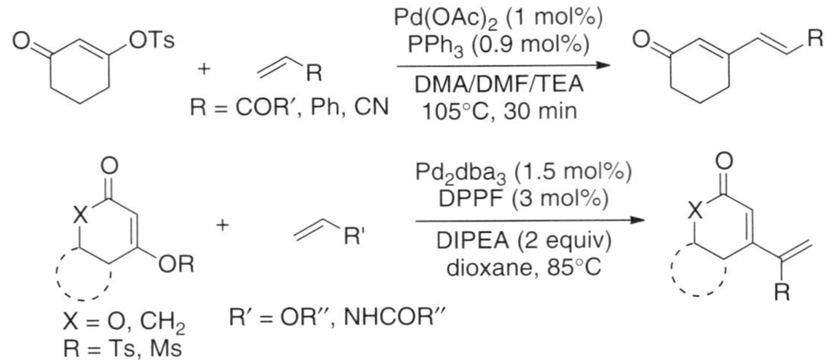

Scheme 5.33. Pd-catalyzed Heck reaction of activated alkenyl tosylates and mesylates.

Although the Pd-catalyzed Heck reaction of organohalides has been well investigated, the examples of phenol and enol derivatives are quite limited. Activated alkenyl phosphates and tosylates could undergo intramolecular and intermolecular Heck reaction smoothly. For example, Fu et al. [62] reported the first inermolecular Heck reaction of α,β-unsaturated tosylates with electron-poor olefins in the presence of a Pd-PPh_3 catalytic system (Scheme 5.33). Skrydstrup's group found that Pd-DPPF could catalyze the highly selective Heck reaction of α,β-unsaturated tosylates and mesylates with electron-rich alkenes (Scheme 5.33) [63].

One year later, Skrydstrup and co-workers [64] used P^tBu_3, a sterically hindered and electron-rich alkyl phosphine, to successfully realize the coupling of nonactivated alkenyl tosylates and phosphates with electron-poor alkenes (Scheme 5.34). Most of the reactions gave an *ipso*-alkenylative product. However, the highly steric hindered substrates, such as 1-*tert*-butylalkenyl tosylate or phosphate, would lead to *cine*-alkenylation through 1,2-migration of alkenyl–palladium intermediate.

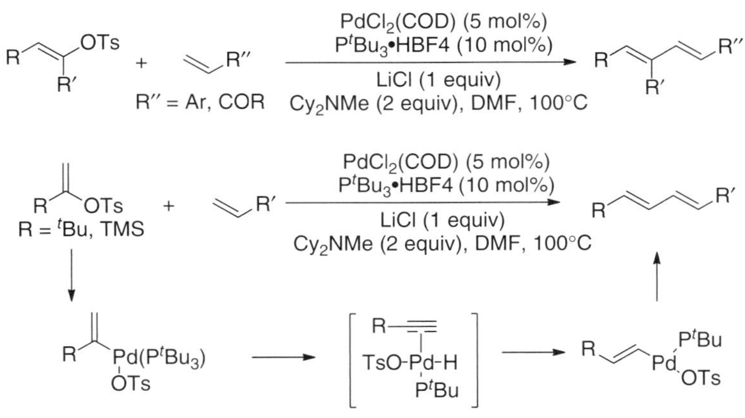

Scheme 5.34. Pd-catalyzed Heck reaction of nonactivated alkenyl tosylates.

Later, Skrydstrup and co-workers [65] also developed the Heck reaction of heteroaromatic tosylates and electron-rich alkenes using a Pd-DPPF catalytic system with complete α-regioselectivity. There is also one example of a Heck reaction of aryl mesylate, which was reported by Djakovitch and co-workers [66]. However, the conversion is only 30%.

5.2.7 Sonogashira Reaction

The Pd/Cu co-catalyzed reaction of electrophiles with alkynes in the presence of base was named as the Sonogashira reaction, which is highly efficient to synthesize important substituted alkynes [67].

For the activated substrates, such as alkenyl phosphates and tosylates, the catalytic system of Pd-PPh_3 could efficiently promote their Sonogashira rections to afford enynes [68]. For example, in 1998 Nicolaou et al. [68a] realized the first example of Pd-catalzyed Sonogashira reaction of cyclic enamine phosphate (Scheme 5.35). In 2001 Yang and co-workers reported a Pd-catalyzed coupling of 4-tosylcoumarins and terminal acetylenes under mild reaction conditions (Scheme 5.35).

Moreover, the Sonogashira reactions of tautomerizable heterocyles were also realized by Kang et al. [69a] (Scheme 5.36) and Shi and Aldrich [69b] independently through *in situ* protection of C-OH with a phosphonium salt. In the latter case, no copper catalyst was needed.

Using X-Phos as the liagand, Buchwald and co-workers [70a] realized the Sonogashira rection of electron-poor aryl tosylates and aryl imidazolylsulfonates

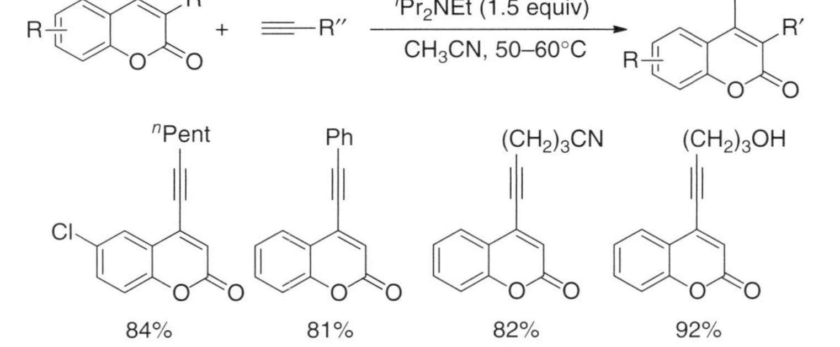

Scheme 5.35. Pd-catalyzed Sonogashira reaction of activated alkenyl phosphates and tosylates.

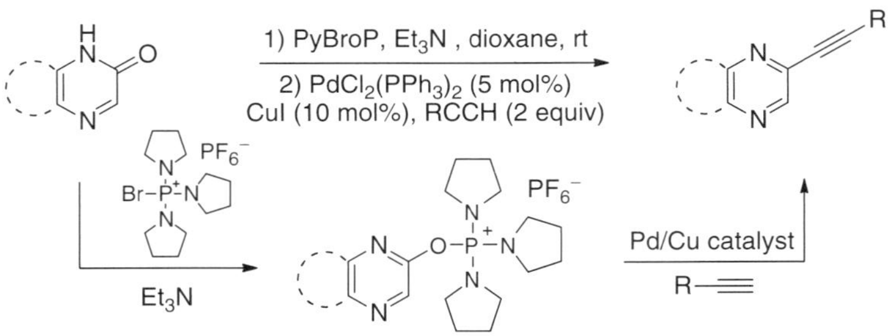

Scheme 5.36. Pd-catalyzed Sonogashira reaction of tautomerizable heterocyles.

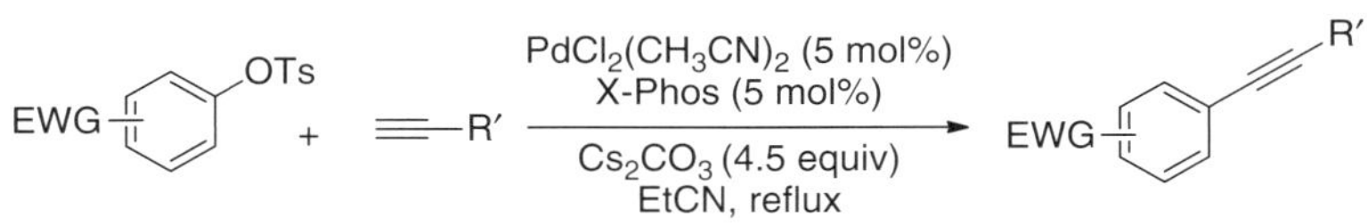

Scheme 5.37. Pd-catalyzed Sonogashira reaction of activated aryl tosylates.

(Scheme 5.37). Later, a similar work was also reported by Williams and co-workers [70b] independently.

The first general Sonogashira reaction of nonactivated aryl tosylates was reported simultaneously by two groups, namely, of Kwong and co-workers [71a] (Scheme 5.38) and Lindenschmidt and co-workers [71a] (Scheme 5.39) in 2010. The catalytic system of Kwong, who combined $Pd(OAc)_2$ with CM-phos, was also efficient for similar reactions of aryl and heteroaryl mesylates. Lindenschmidt and co-workers found that a catalytic system of $Pd(OTFA)_2$ and CyPF-*t*Bu ligand was efficient for the nonactivated, *ortho*-substituted aryl and heteroaryl tosylate. Both of these reactions showed good functional group tolerance and a wide substrate scope for both aryl tosylates and terminal alkynes.

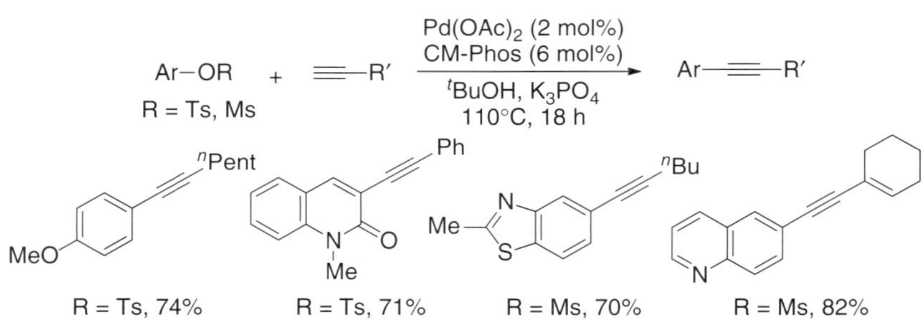

Scheme 5.38. Pd-catalyzed general Sonogashira reaction of aryl tosylates and mesylates.

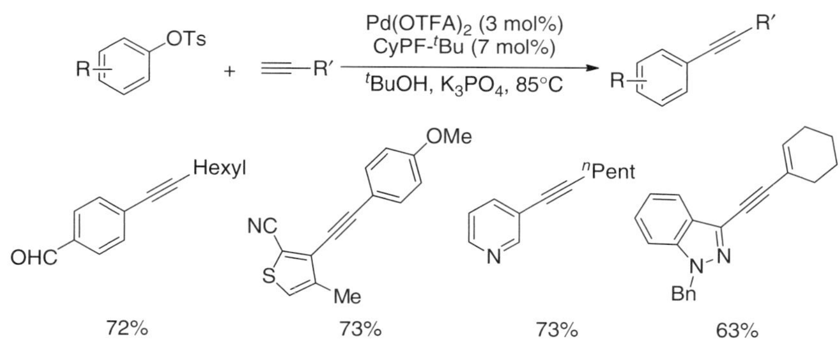

Scheme 5.39. Pd-catalyzed general Sonogashira reaction of aryl tosylates.

5.2.8 Cross-Coupling with Other C–H Bonds and Carboxylic Acids

5.2.8.1 Cross-Coupling with Aryl and Heteroaryl C–H Bonds. Although the cross-coupling with different kinds of organometallic reagents have been well developed for various electrophiles, the high price and unstability of these reagent hampers their wide application in industry. In contrast, direct reaction with C–H bonds in arenes [72] and carbonyl compounds [73] or carboxylative acids [74] shows great advantage due to their low price, low toxicity, easy handling, and higher step economy and atom economy.

In 2009 Ackermann et al. [75] reported a Pd-Xphos-catalyzed efficient direct arylation of heteroarenes, such as oxazole, caffeine, and 1,2,3-triazoles, with aryl tosylates (Scheme 5.40). Moreover, two examples of electron-poor aryl mesylates also underwent the reaction in moderate yields with this catalytic system

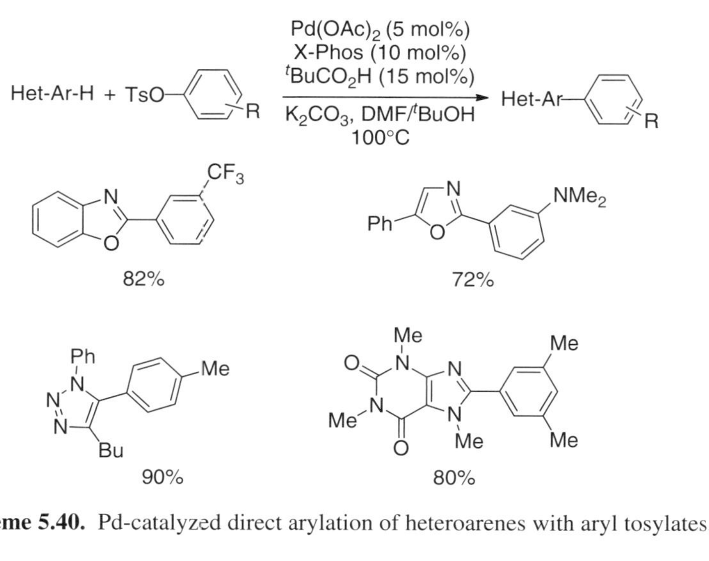

Scheme 5.40. Pd-catalyzed direct arylation of heteroarenes with aryl tosylates.

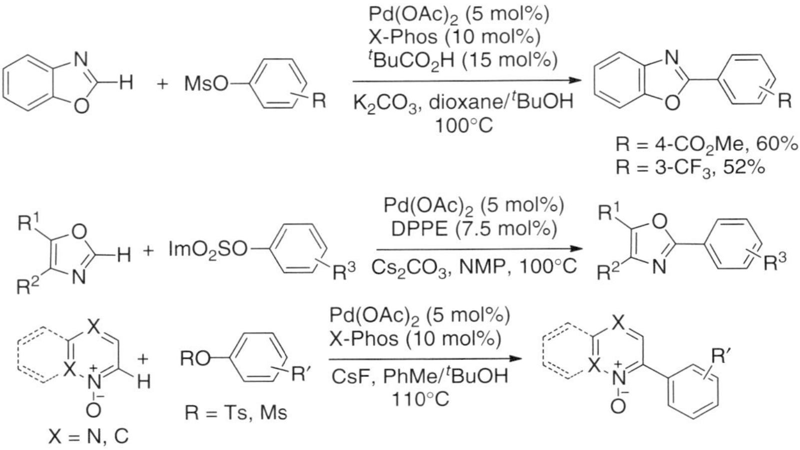

Scheme 5.41. Pd-catalyzed direct arylation of heteroarenes with other phenol derivatives.

(Scheme 5.41). Later, this group also successfully applied aryl imidazolylsulfonates, benzyl phosphates, alkenyl phosphates, and tosylates in the reaction with heteroarenes, electron-deficient *N*-oxides, and polyfluoroarenes (Scheme 5.41) [76].

Independently, the groups of Zhang [77] and Seayad [78] also reported the direct arylation of fluorinated aromatics with aryl sufonates including tosylates, triflates and mesylates. Both of these systems also employed eletron-rich Buchwald biaryl phosphine ligand (Scheme 5.42).

Besides intermolecular arylation, intramolecular C–H arylation of arene with tosylates and mesylates was reported by Kalyani recently (Scheme 5.43) [79]. Various heterocycles, such as furans, carbazoles, indoles, and lactams, underwent the reaction smoothly. Moreover, one-pot sequential tosylation–arylation of phenol derivatives was also realized for a variety of 2-aroxylphenols.

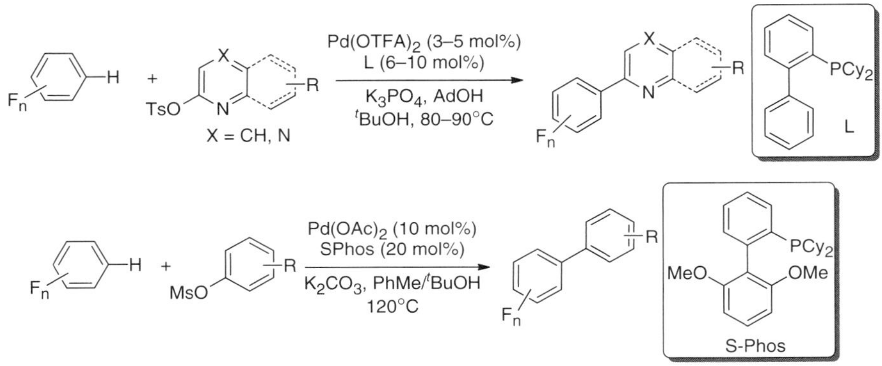

Scheme 5.42. Pd-catalyzed direct arylation of fluorinated aromatics with aryl sufonates.

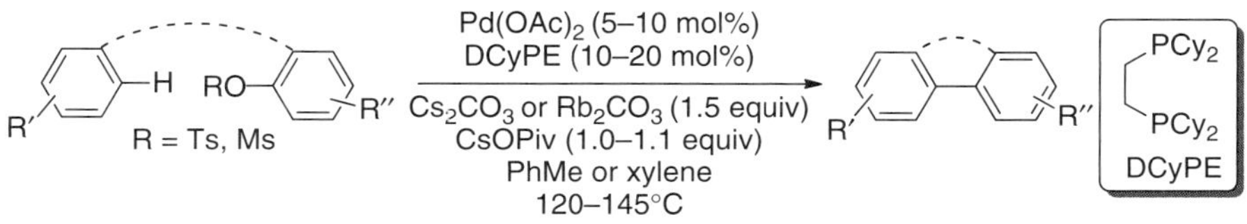

Scheme 5.43. Pd-catalyzed direct intramolecular arylation of aryl sufonates.

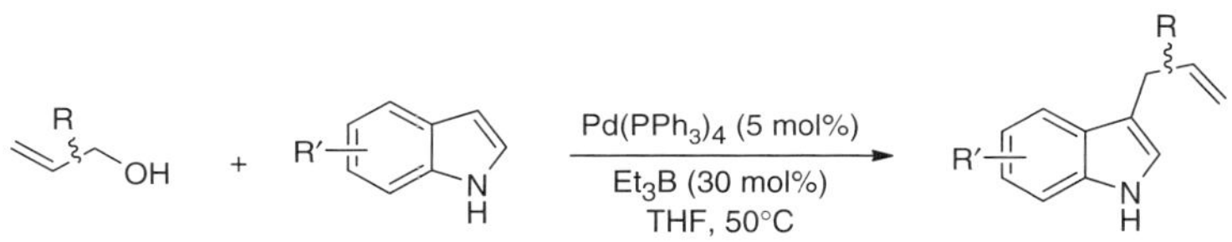

Scheme 5.44. Pd-catalyzed direct C3-allylation of indoles with allylic alcohols.

In 2005 Tamaru and co-workers [80] reported a Pd-catalyzed C3-selective allylation of indoles with allylic alcohols in the presence of Lewis acidic Et_3B (Scheme 5.44). The reaction shows high efficiency and a wide scope for both allyl alcohols and indoles. Later, Breit and co-workers [81] realized a similar reaction using a self-assembling ligand, which was based on complementary hydrogen bonding.

5.2.8.2 α-Arylation of Carbonyl Compounds. Buchwald–Hartwig α-arylation is also powerful method to construct important α-arylated carbonyl compounds. Besides aryl boromides and chlorides, one example of aryl tosylate was also reported by Kawatsura and Hartwig [82] in 1999 to undergo Pd-catalyzed α-arylation in the presence of an electron-rich and sterically hindered diphosphine ligand (Scheme 5.45). However, the substrate scope of aryl tosylate was not investigated.

In 2003 Buchwald and co-workers [83] employed electron-rich and sterically hindered X-Phos as the ligand and developed a general Pd-catalyzed α-arylation of various kinds of carbonyl compounds with aryl benzenesulonates as the electrophiles (Scheme 5.46). In this reaction, high selectivity for mono-arylation was observed.

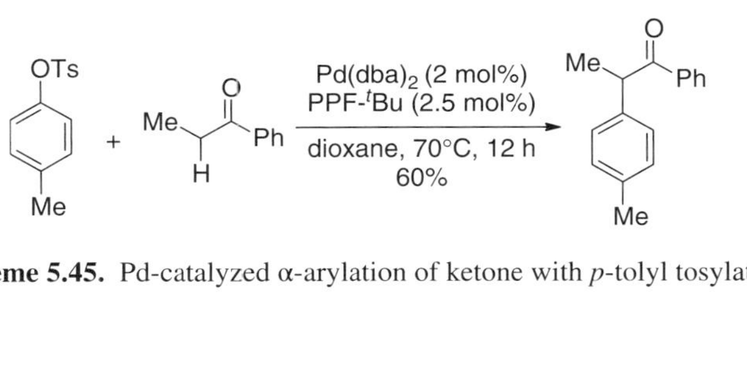

Scheme 5.45. Pd-catalyzed α-arylation of ketone with *p*-tolyl tosylate.

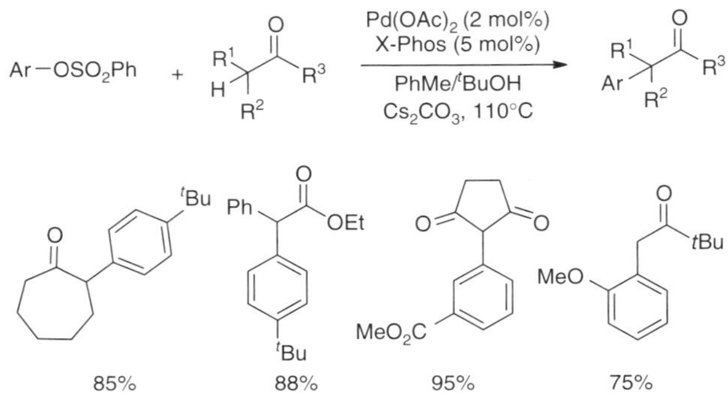

Scheme 5.46. Pd-catalyzed general α-arylation carbonyl compounds with aryl benzenesulfonates.

In 1989 Bergbreiter and Weatherford [84] reported the first allylation of active methylene compounds with allylic alcohols using $Pd(PPh_3)_4$ as a catalyst without prior activation of the allylic hydroxy group (Scheme 5.47). Later, the groups of Ozawa, Kobayashi, and Oshima independently realized allylic substitution of 1,3-dicarbonyl compounds with allylic alcohols with different catalytic systems [85].

In 2001 Tamaru and co-workers [86] reported the first direct α-allylic alkylation of aldehydes with allylic alcohols via palladium catalysis under mild reaction conditions (Scheme 5.48). Both Et_3N and BEt_3 were indispensable, the latter of which was proposed to activate the C–O bond of allylic alcohol and promote generation of enols from aldehydes in this system.

5.2.8.3 Cross-Coupling with Carboxylic Acids. Following their previous work of decarboxylative cross-couplings, Gooßen et al. [87] realized that Pd/Cu co-catalyzed decarboxylative cross-coupling of aryl tosylates and aromatic carboxylate salts in 2010 (Scheme 5.49) [87]. It is important to note that the tosylate anion, which was formed in the reaction, showed weak coordinating ability to the copper center and did not hinder the decarboxylation. A variety of aromatic carboxylate salts, regardless of their substitution pattern, have been coupled with aryl tosylates to give the corresponding biaryls in good yields.

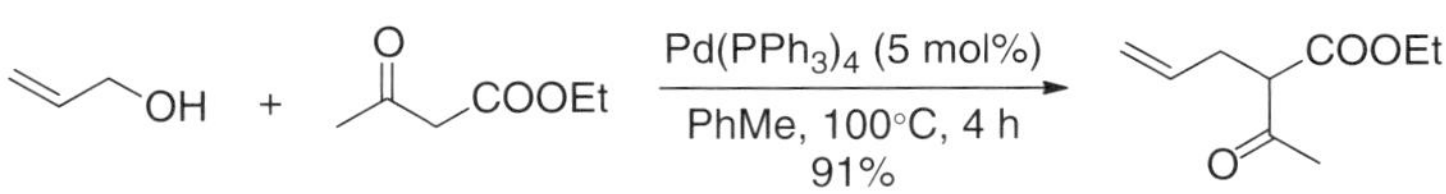

Scheme 5.47. Pd-catalyzed direct α-allylation of active methylene compounds with allylic alcohol.

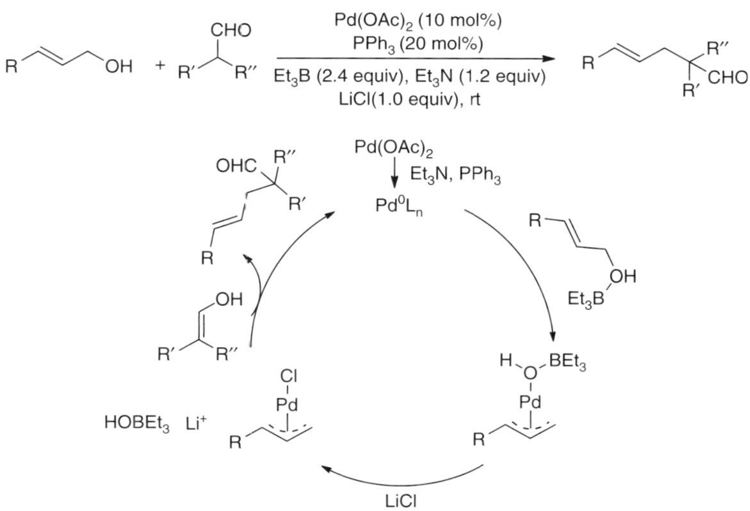

Scheme 5.48. Pd-catalyzed direct α-allylation of aldehydes with allylic alcohol.

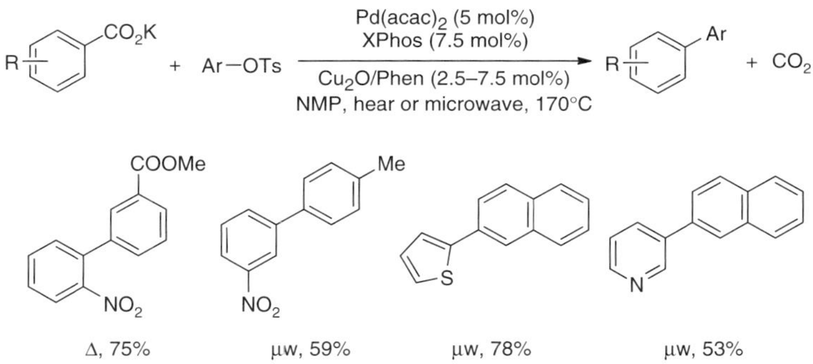

Scheme 5.49. Pd/Cu-co-catalyzed decarboxylative cross-coupling reaction of aryl tosylates with aromatic carboxylate salts.

5.2.9 Carbonylation Reaction

Carbon monoxide is one of the most important C_1 building blocks in organic synthesis. Pd-catalyzed carbonylation of various kinds of electrophiles and nuelophiles is an important and useful method to synthesize carbonyl compounds [88]. Similar to alkenyl and aryl halides, C–O bonds also could be employed in such a reaction. As realized by Ortar and co-workers [89] in the first Pd-catalyzed carbonylation of aryl triflates in 1986, alkoxycarbonylation and aminocarbonylation are two of the most classic modes of carbonylation, although carbonylative cross-couplings are also useful methods to synthesize functionalized ketones. Following this research, there are many other reports on Pd-catalyzed carbonylation of aryl triflates to synthesize aryl and alkenyl carboxylic acid derivatives [90]. Later, alkenyl phosphates

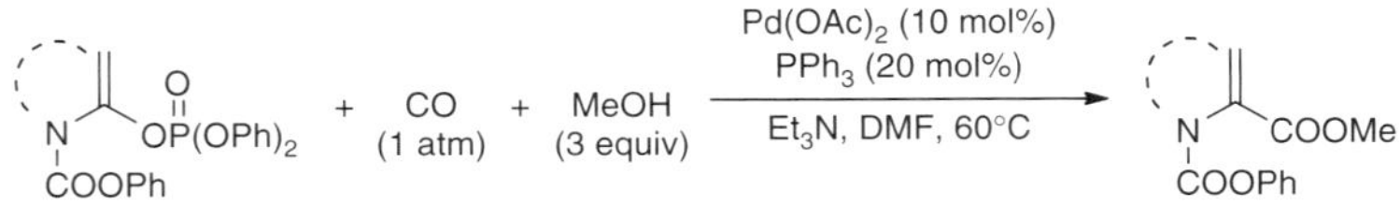

Scheme 5.50. Pd-catalyzed alkoxycarbonylation of alkenyl phosphates.

were also applied in similar carbonylation by Nicolaou et al. and other groups [91] (Scheme 5.50).

However, the development of less reactive aryl tosylates and phosphates was slower. In 1998 Sugi and co-workers [92] reported the first Pd-catalyzed carbonylation of electron-poor aryl tosylates in the presence of DPPP as the ligand (Scheme 5.51) [92]. However, unactivated substrates showed low reactivity in this transformation.

In 2006, Cai et al. [93] realized alkoxycarbonylation of aryl *p*-fluorobenzenesulfonate using the bidentate electron-rich Josiphos ligand (Scheme 5.52). Due to the presence of the electron-withdrawing fluoro atom, this kind of aryl sulfonate showed higher reactivity and the unactived phenyl tosylate also underwent the carbonylation with good efficiency.

In 2008 Buchwald and co-workers [94] reported a general catalytic system for the Pd-catalyzed alkoxycarbonylation of aryl tosylates and mesylates in the presence of a bidentate electron-rich alkyl phosphine ligand (Scheme 5.53). Both the electron-rich and electron-poor aryl tosylates showed good reactivity in 1 atmosphere of CO.

5.2.10 Buchwald–Hartwig Amination

Different from previous cross-coupling methods that construct C–C bonds, Pd-catalyzed amination of electrophiles to generate C–N bonds, referred to as Buchwald–Hartwig amination, is a unique and powerful method to afford important amines, amides, and other nitrogen-containing compounds [95]. However, similar to other cross-coupling, the significant development of Buchwald–Hartwig amination

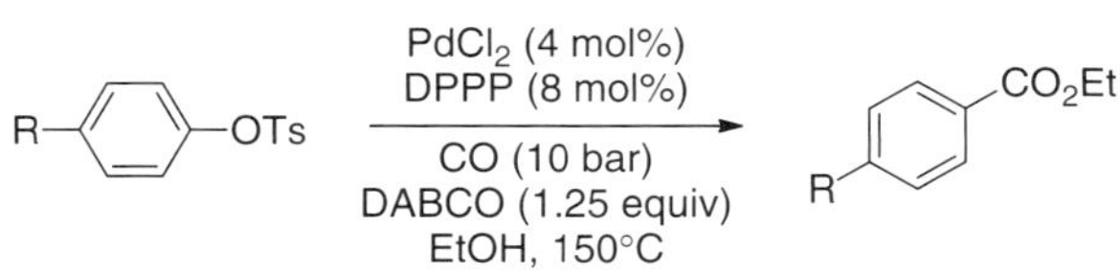

Scheme 5.51. Pd-catalyzed alkoxycarbonylation of aryl tosylates.

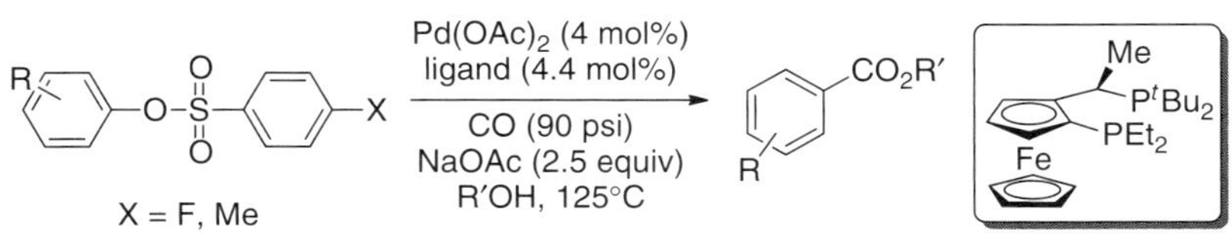

Scheme 5.52. Pd-catalyzed alkoxycarbonylation of aryl sufonates.

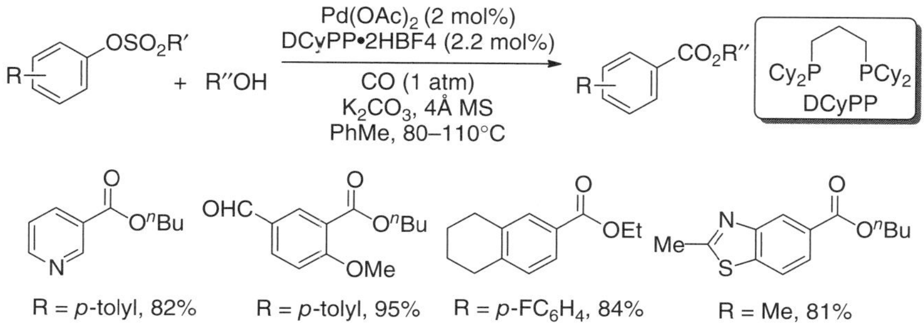

Scheme 5.53. Pd-catalyzed alkoxycarbonylation of aryl tosylates and mesylates.

relies on the improvement of the catalytic system, especially ligands design. The progress not only made unreactive aryl chloride possible but also realized efficient aminations of unreactive C–O bonds in tosylates and mesylates. Due to the higher energy barrier for the reductive elimination to form a C–N bond than to form a C–C bond, the important role of liagand is more obvious.

In 1998 Hamann and Hartwig [96] reported the first example amination of aryl tosylates with aniline and primary alkyl amine by using an electron-rich and chelating diphosphine ligand, which played an important role in promoting both the oxidative addition of a C–O bond and the reductive elimination to form a C–N bond (Scheme 5.54). Ten years later, Ogata and Hartwig [97] found that the catalytic system of $Pd(P(o\text{-tol})_3)_2$ and JoisPhos was highly efficient for the coupling of a variety of aryl and heteroaryl tosylates with primary amines at room temperature (Scheme 5.54). The sole selective amination of aryl C–OTs other than aryl C–Cl was observed. Moreover, the good selectivity for mono-arylation was contributed by the sluggish reaction of secondary amines. Later, using the same catalytic system, Vo and Hartwig

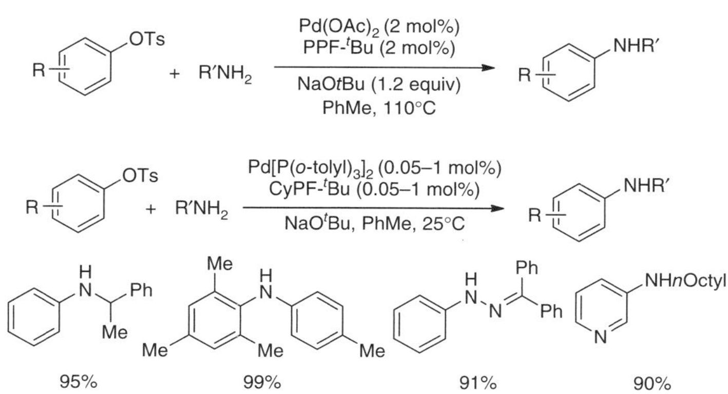

Scheme 5.54. Pd-catalyzed amination of aryl tosylates.

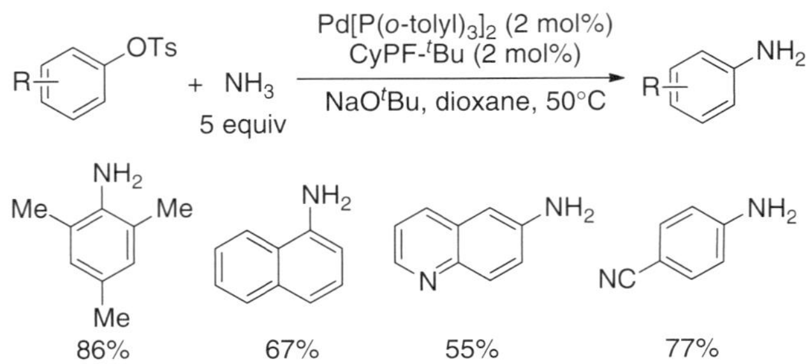

Scheme 5.55. Pd-catalyzed coupling of aryl tosylates with ammonia.

[98] further realized the first highly selective amination of aryl and heteroaryl tosylates with ammonia to generate primary amines (Scheme 5.55).

Beside the Hartwig group, many other groups also utilized chelating diphosphine ligand and P,N-ligand in amination of alkenyl and aryl tosylates. For example, in 2005 Klapars et al. [99] realized the Pd-catalyzed amidation of various enol tosylates in the presence of bidentate DIPF ligand (Scheme 5.56). In 2010 Skrydstrup and co-workers [100] reported the Pd-catalyzed amidation of a variety of heteroaromatic tosylates with amides, anilines, and indoles using DPPF as ligand. In the same year, bidentate P,N-ligand was also applied by Stradiotto and co-workers [101] in amination of aryl tosylates with ammonia at room temperature (Scheme 5.57). In 2011, Ackermann et al. [102] also realized a novel Pd-BINAP-catalyzed amination of imidazolylsulfonates with good functional group tolerance.

Buchwald biaryls phosphine ligand, which is significantly efficient to promote reductive elimination, was also highly powerful in amination of aryl tosylates. In 2003, Buchwald and co-workers [103] reported the first Pd-catalyzed amidation of arenesulfonates in the presence of XPhos as the ligand (Scheme 5.58). With the same

Scheme 5.56. Pd-catalyzed amidation of alkenyl tosylates.

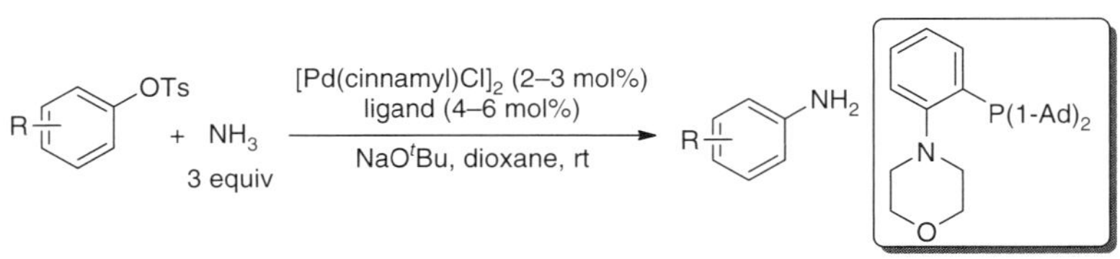

Scheme 5.57. Pd-catalyzed amination of aryl tosylates with ammonia.

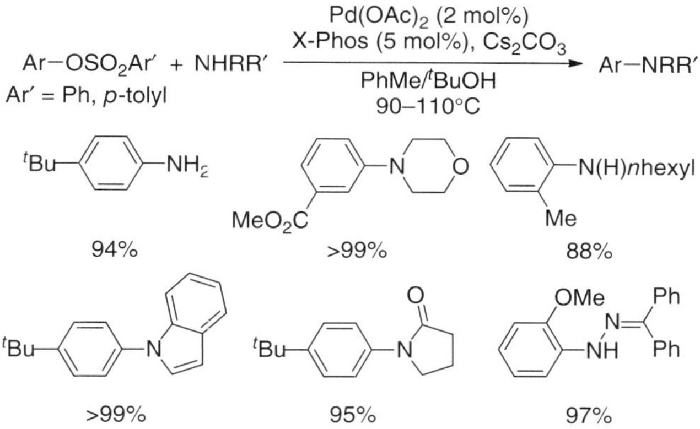

Scheme 5.58. Pd-catalyzed amination of aryl arenesulfonates.

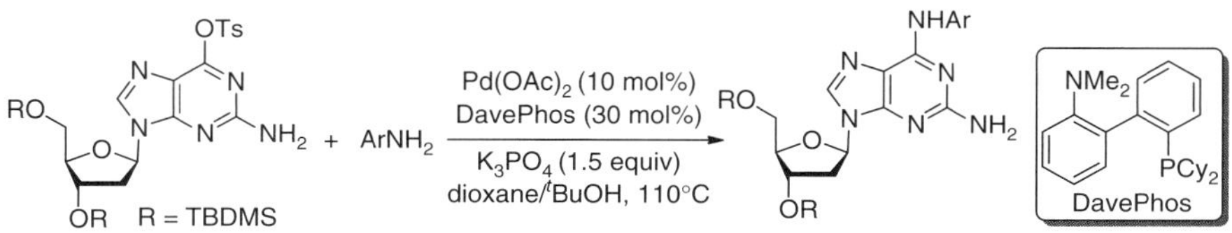

Scheme 5.59. Pd-catalyzed amination of heteroaryl tosylates.

catalytic system, various kinds of amines, indoles, benzophenone imines, and benzophenone hydrazones could also undergo efficient coupling with arenesulfonates.

Later, Lakshman and co-workers [104] also utilized the another biaryls phosphine ligand, DavePhos, to successfully realize the first efficient amination of nucleoside arenesulfonates with aryl amines to afford N^6-aryl-2,6-diaminopurine nucleoside analogues (Scheme 5.59).

Beside arenesulfonates, less reactive mesylates also could undergo amination in a similar reaction condition. For example, in 2008 Kwong and co-workers [105] reported the first Pd-catalyzed amination of aryl mesylates with various nitrogen heterocycles, such as indole, cabazole, and pyrrole, in the presence of anindolylphosphine ligand (Scheme 5.60). Independently, Buchwald and co-workers [106] also found that the BrettPhos ligand was highly efficient for Pd-catalyzed amination of aryl mesylates and *t*-BuBrettPhos for amidation (Scheme 5.61).

Besides the sp^2 C–O bond, activation of sp^3 C–O bond and sequent amination has also been investigated. For example, there have been many reports on Pd-catalyzed direct amination of allylic alcohols [107]. As early as the 1980s, Lu et al. [108] reported the Pd-catalyzed reaction of allylic alcohols with various nucleophiles through *in situ* generation of allyl arsenious ester or borate by using a catalytic amount of arsenic trioxide or equivalent boronic anhydride (Scheme 5.62).

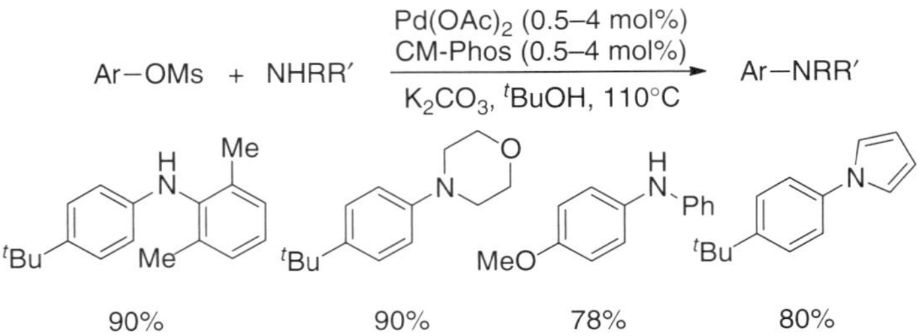

Scheme 5.60. Pd-catalyzed amination of aryl mesylates with CM-Phos as ligand.

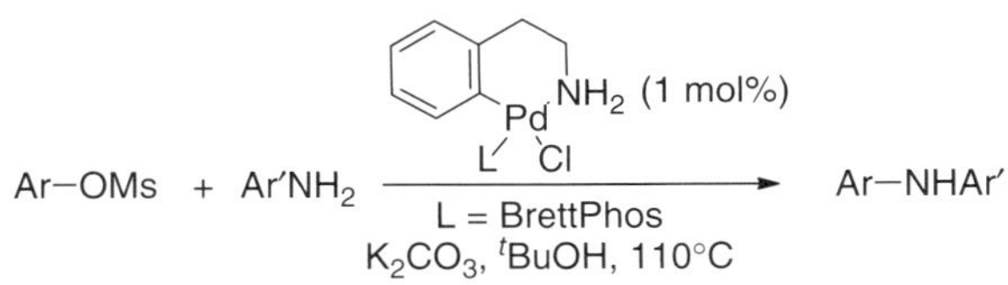

Scheme 5.61. Pd-catalyzed amination of aryl mesylates with BrettPhos as ligand.

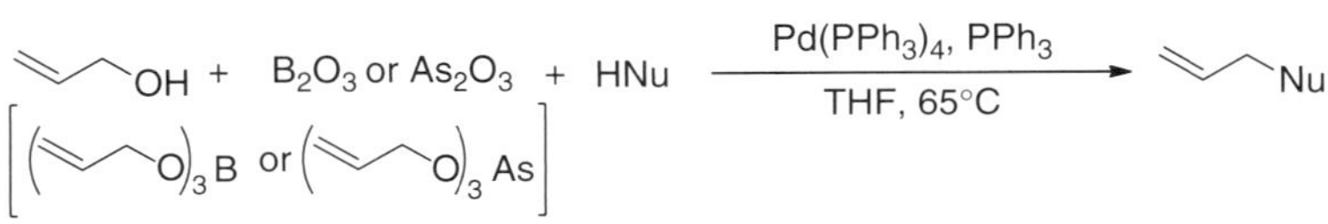

Scheme 5.62. Pd-catalyzed reactions of allylic alcohols with various nucleophiles via *in situ* generation of allylic esters.

Later, Yamamoto and co-workers [109] also realized a similar transformation through *in situ*-generated allyl carbonate in a CO_2 atmosphere.

Even without formation of ester *in situ*, stochiometric [110] and/or catalytic acids [111] were always used to increase the reactivity of allylic alcohols. Later, with the development of a catalytic system, such as an aqueous solution, no additive was neccessary (Scheme 5.63) [112].

In 2011, Hikawa and Yokoyama [113] reported a Pd-catalyzed di benzylation of 2-aminobenzoic acid with benzylic alcohols in water (Scheme 5.64). The H-bond between hydroxyl and water was proposed to activate the C–O bond in alcohol and promote oxidative addition to generate benzylic palladium complex, which was nucleophilicly attacked by aromatic amine to produce a monobenzylation product. Further dehydrogenation generated one molecule of benzyl imine, which was nucleophilicly attacked by a benzylic palladium complex to afford the dibenzylative product.

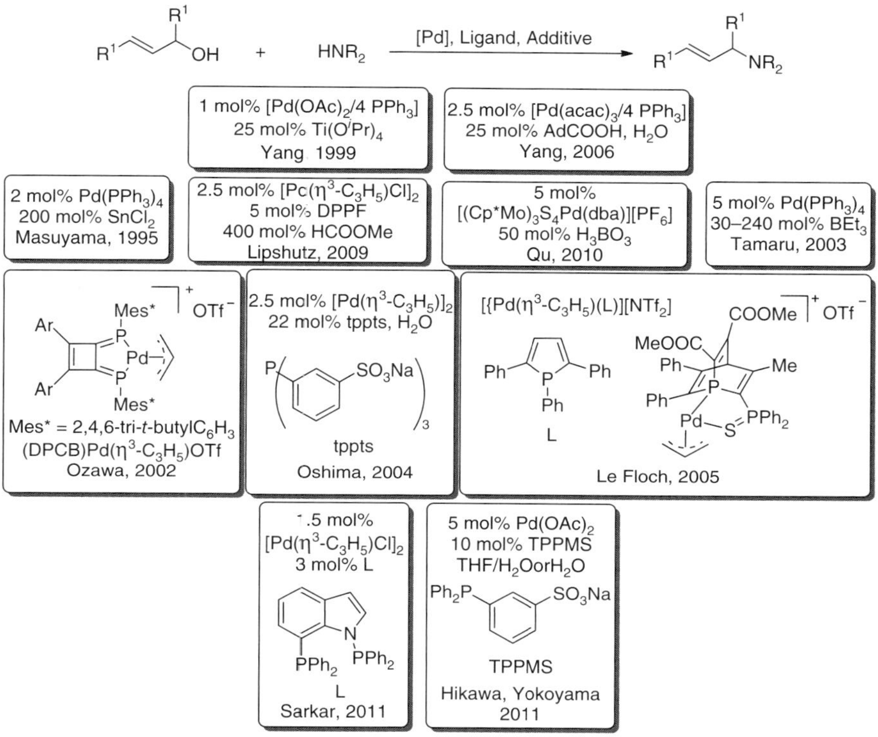

Scheme 5.63. Pd-catalyzed direct aminations of allylic alcohols.

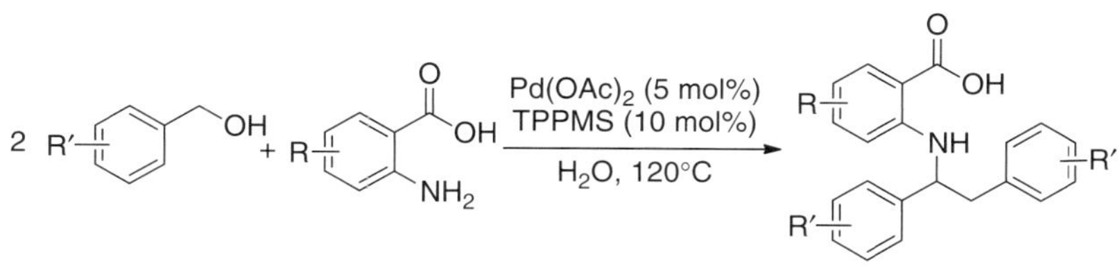

Scheme 5.64. Pd-catalyzed di-benzylation of 2-aminobenzoic acid with benzylic alcohols.

5.2.11 Other C–X Bond Formation Reactions

Besides C–C and C–N bond formation, C–O bonds also could be transformed to other C–X bonds (such as C–B, C–Si, C–P, C–S, C–H) and even halides, via Pd catalysis.

Organoboronic acids and derivatives are very useful organometallic reagents, which have been widely applied in various organic reactions including the famous Suzuki–Miyaura coupling. Besides organohalides, oxygen-containing compounds, such as sulfonates and phosphates, could also undergo borylation in the presence

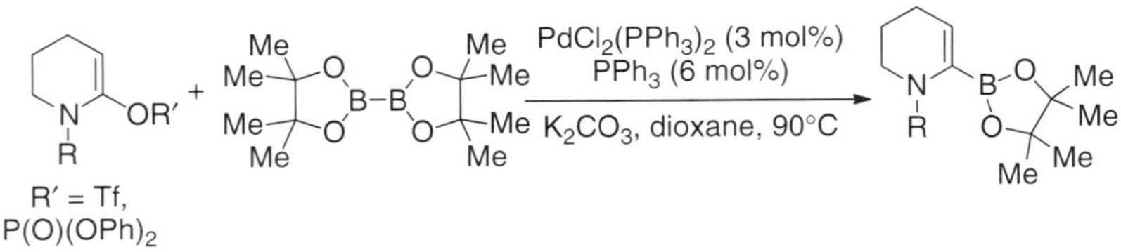

Scheme 5.65. Pd-catalyzed borylation of lactam-derived vinyl triflates and phosphates.

of Pd catalyst. For example, in 2005 Occhiato et al. [114] reported a Pd-catalyzed borylation of lactam-derived vinyl triflates and phosphates to generate corresponding alkenyl boronates (Scheme 5.65), which could be further efficiently coupled with various aryl and heteroaryl bromides and triflates, vinyl iodides and bromides, and acyl chlorides, to afford the corresponding piperidine derivatives.

Recently, Szabó and co-workers [115] have realized a Pd-catalyzed direct borylation and silylation of allylic alcohols (Scheme 5.66). The tetrafluoroborate counterion of the palladium catalyst is proposed to play an important role in both catalyst activation and as the rate-determining transmetalation step, in which BF_3 is generated to activate the hydroxyl group. Both reactions proceed with high regio- and stereoselectivity to generate the linear allylic products with *trans* configuration.

In 2009 Luo and Wu [116] realized a Pd-DPPP-catalyzed efficient aryl C–P bond-forming reaction to generate arylphosphonates from various aryl imidazolylsulfonates with good functional group tolerance (Scheme 5.67).

In 2003 Milne and Kocienski [117] reported a Pd-catalyzed C–S formation of cyclic enol phosphate with moderate yield and limited substrate scope (Scheme 5.68).

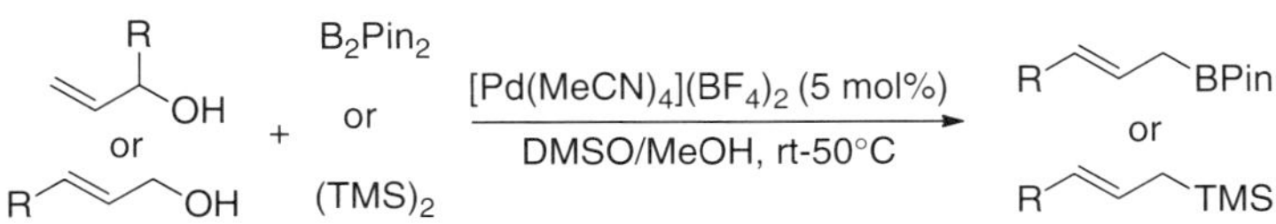

Scheme 5.66. Pd-catalyzed borylation and silylation of allylic alcohols.

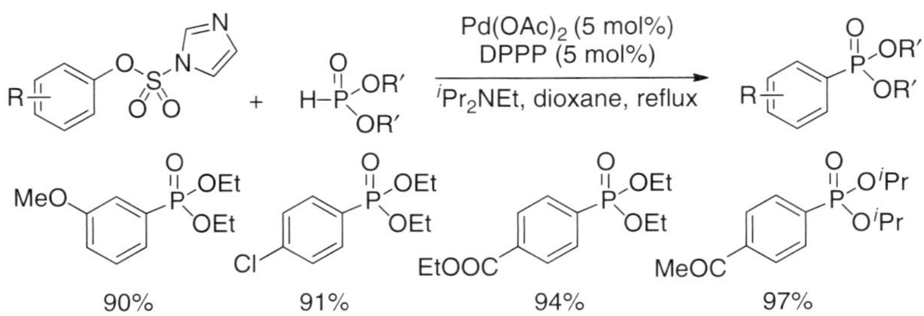

Scheme 5.67. Pd-catalyzed C–P formation of aryl imidazolylsulfonates.

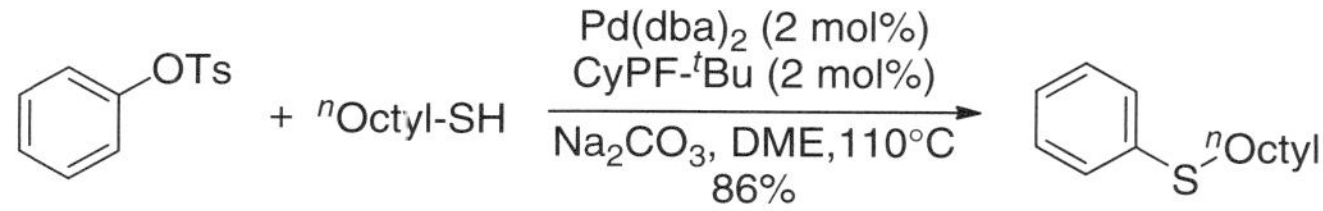

Scheme 5.68. Pd-catalyzed C–S formation of aryl tosylates.

Later, besides a highly efficient C–S bond formation reaction with aryl halides, two examples of triflate and tosylate were also reported to couple with alkylthiol by Hartwig and co-workers [118] in the catalyst system of $Pd(dba)_2$ and Josiphos ligand.

Moreover, the C–O bond of alkenyl phosphates could also be reduced to a C–H bond by a palladium catalyst in the presence of a reductant such as $AlEt_3$ [119] or $HSiEt_3$ (Scheme 5.69) [120].

Based on the stoichiometric study of reductive elimination of Ar–Pd–F complexes, in 2009 Buchwald and co-workers [121] realized the first conversion of aryl triflates to important aryl fluorides with CsF as the fluoride source and the combination of Pd and a highly sterically demanding and electron-donating biaryl ligand ^{t}BuBrettPhos as the catalytic system (Scheme 5.70). This reaction showed wide substrate scope and good functional group tolerace. In some cases, regioisomers could be obtained in relationship to the solvent polarity.

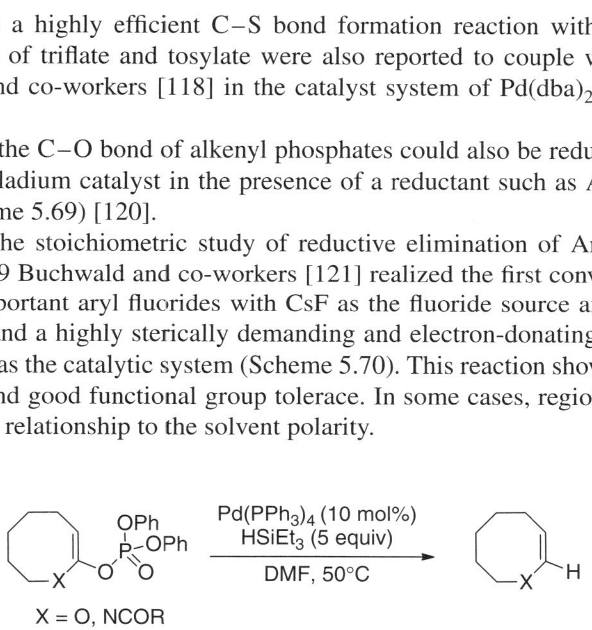

Scheme 5.69. Pd-catalyzed reduction of alkenyl phosphates.

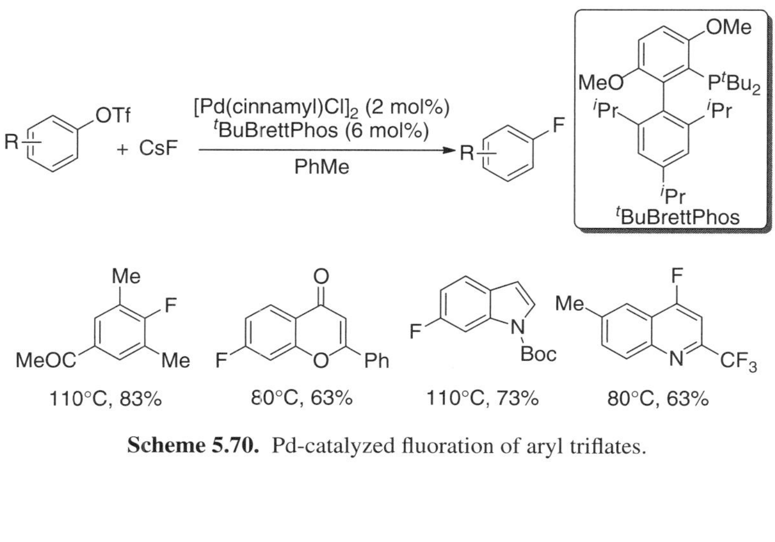

Scheme 5.70. Pd-catalyzed fluoration of aryl triflates.

5.3 NICKEL-CATALYZED C–O BOND ACTIVATION

5.3.1 Kumada–Tamao–Corriu Reaction

5.3.1.1 Kumada–Tamao–Corriu Coupling of Phosphates. Due to the high reactivity of Grignard reagents, the coupling of alkenyl and aryl phosphates also showed good reactivity and could proceed smoothly under mild conditions via Ni catalysis. For example, in 1981 Kumada and co-workers realized the Ni-catalyzed cross-coupling of alkenyl phosphates with TMSC(R)HMgX (Scheme 5.71) [122]. Since alkenyl phosphates could be easily prepared from carbonyl compounds and the cross-coupling products could be easily oxidized by H_2O_2 to afford important primary alcohols, this method was very useful and inspired many related works [123].

In the same year, Kumada and co-workers [124] also reported the first Ni-catalyzed cross-coupling of aryl phosphate with alkyl and aryl Grignard reagents under room temperature (Scheme 5.72). This method was later adopted by other groups in organic synthesis [125]. For example, in 2005 Cai and co-workers [125b] applied this method as a key step in synthesis of (±)-cryptotanshinone and its analogues.

In 2009, Nakamura and co-workers [126] developed a general Kumada coupling of aryl phosphates to synthesize important biaryls. With a well-defined bifunctional hydroxyphosphine ligand, the reaction could proceed smoothly with low catalyst loading under mild conditions. Nonactivated, deactivated, and *ortho*-substituted aryl phosphates also underwent this transformation in high yields (Scheme 5.73). The mechanistic study indicated that the high catalytic activity of this catalytic system was attributed to a nickel phosphine/magnesium alkoxide bimetallic species, which was formed *in situ* in the reaction system.

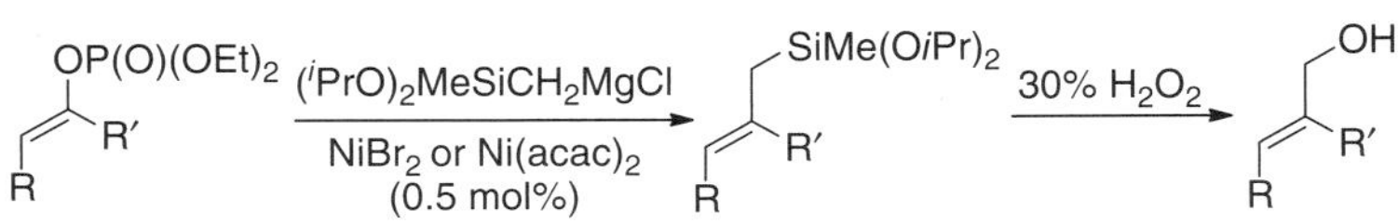

Scheme 5.71. Ni-catalyzed Kumada coupling of vinyl phosphates.

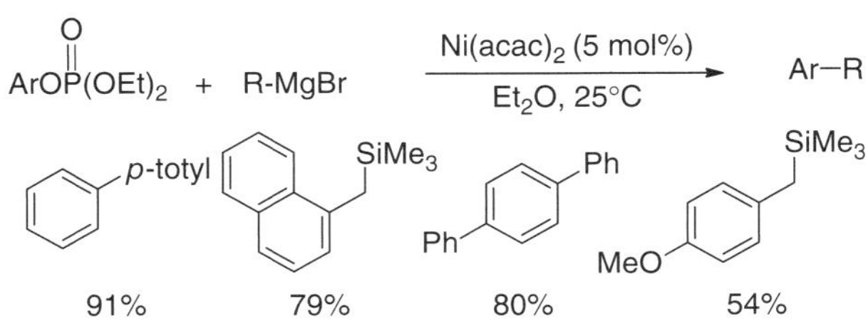

Scheme 5.72. Ni-catalyzed Kumada coupling of aryl phosphates.

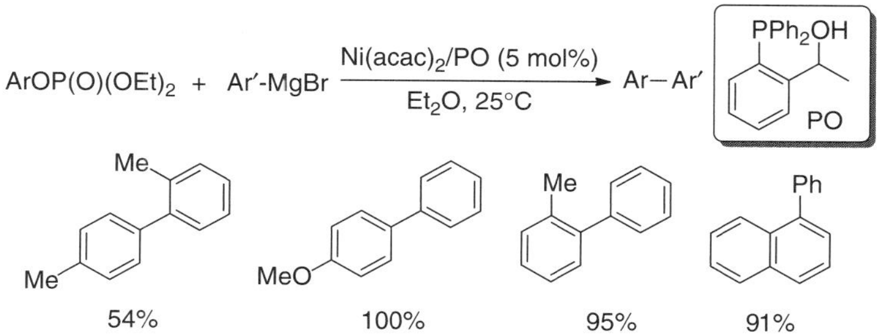

Scheme 5.73. Ni-catalyzed Kumada coupling of aryl phosphates.

5.3.1.2 *Kumada–Tamao–Corriu Coupling of Sulfonates.*

Similar to phosphates, aryl sulfonates also showed good reactivity in Kumada–Tamao–Corriu coupling. In 2005 Macklin and Snieckus [127] reported that aryl *O*-sulfamate could be used as an efficient electrophile in nickel-catalyzed cross-coupling with aryl Grignard reagents (Scheme 5.74). Direct *ortho* metalation (D*o*M) methodology was followed by the coupling of the sulfamate group. By using this traceless directing group strategy, various functionalized biaryls were synthesized. In a related work, When and Du Bois [128] reported that cyclic sulfamates, which were generated from Rh-catalzyed intramolecular C–H amination, could further undergo a Ni-catalyzed coupling with alkyl and aryl Grignard reagents in good efficiency.

In 2010, Shi and co-workers [129] reported the Ni-catalyzed Kumada coupling of diaryl sulfates with inorganic salts as the sole byproduct. (Scheme 5.75). Notably, both aryl groups could be used in this reaction and transformed into the desired products with excellent efficiency.

In contrast, reports about the Kumada coupling using aryl mesylates as substrate were rare. The sole example was reported by Percec et al. [130], who described the Ni-catalyzed cross-coupling of aryl mesylates with aryl Grignard reagents (Scheme 5.76). Although both aryl and alkyl Grignard reagents could be used, the relatively low yield and low selectivity limited the utility of this chemistry.

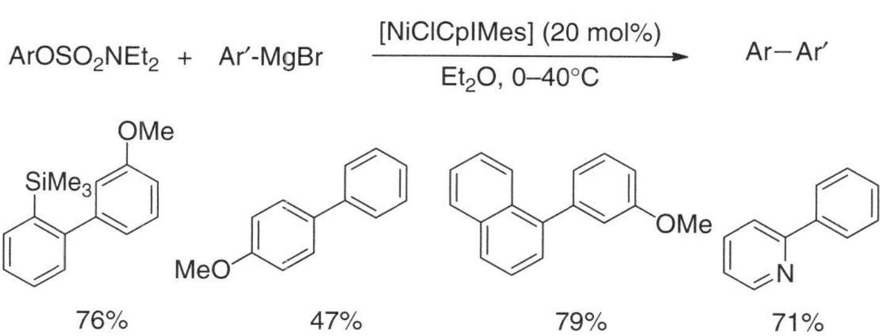

Scheme 5.74. Ni-catalyzed Kumada coupling of aryl sulfamates.

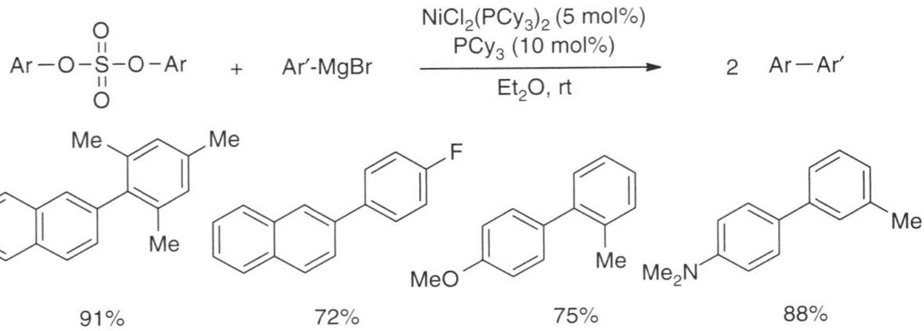

Scheme 5.75. Ni-catalyzed Kumada coupling of biaryl sulfates.

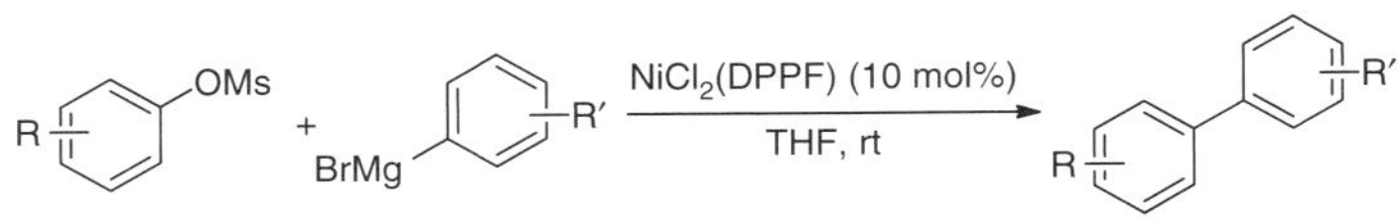

Scheme 5.76. Ni-catalyzed Kumada coupling of aryl mesylates.

5.3.1.3 Kumada–Tamao–Corriu Coupling of Carboxylates and Carbamates. As important derivatives of phenols, aryl carboxylates can be easily prepared from phenols and carboxylic anhydrides or acyl chlorides. Aryl esters and their derivatives show good directing ability in functionalization of arenes at the *ortho* or *para* positions via either *ortho* metallation [131] or electrophilic aromatic substitution [132], respectively, which makes the synthesis of diverse functionalized aromatic rings possible. However, for a long time, esters have only been used as protecting groups or employed as leaving groups in Tsuji–Trost reaction of allylic acetates [5] and benzylic derivatives [133]. To achieve the successful cross-coupling of aryl esters, the most challenging issue is how to selectively cleave the unreactive aryl C–OAc bond instead of the more reactive acyl C–O bond.

In 1989, Kocienski and Dixon [134] reported the first successful example of Ni-catalyzed cross-coupling reaction of enol carbamates with Grignard reagents, although the substrate scope was relatively limited. Later, their couplings with alkenyl, aryl, and acetylenyl Grignard reagents were realized by the groups headed by Betzer and Férézou [135]. In 1992, Snieckus and co-workers reported the Kumada–Tamao–Corriu coupling of aryl *O*-carbamates with $Ni(acac)_2$ as catalyst (Scheme 5.77) [136]. In this study the substrate scope was greatly expended. Carbamates bearing both electron-donating groups and electron-withdrawing groups reacted smoothly to give the desired products in moderate to good yields. However, the scope of the Grignard reagents was limited to PhMgX and $TMSCH_2MgCl$.

The bifunctional P–O ligand, developed by Nakamura and co-workers [126], was also highly effective to promote the Ni-catalyzed cross-coupling of aryl carbamates with aryl Grignard reagents. Various functionalized aryl carbamates and those with *ortho* substitutes could undergo this reaction efficiently (Scheme 5.78).

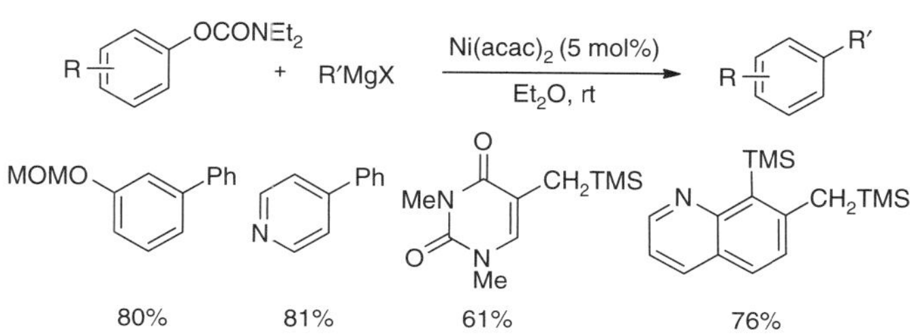

Scheme 5.77. Ni-catalyzed cross-coupling of aryl carbamate with Grignard reagent.

$$\mathrm{ArOCONEt_2} \quad + \quad \mathrm{Ar'\text{-}MgBr} \xrightarrow[\mathrm{Et_2O,\ 25^{\circ}C}]{\mathrm{Ni(acac)_2/PO\ (5\ mol\%)}} \mathrm{Ar-Ar'}$$

Scheme 5.78. Ni-catalyzed cross-coupling of aryl carbamate with aryl Grignard reagent.

5.3.1.4 Kumada–Tamao–Corriu Coupling of Ethers

KUMADA–TAMAO–CORRIU COUPLING OF ARYL/ALKENYL ETHERS. The C–C bond formation through the coupling of enol ethers with Grignard reagents or organocuprates has been discovered for decades [137]. Nucleophilic aromatic substitution (S_NAr) of aryl methyl ether with Grignard reagent is also known, but an ortho chelating group is required to enhance the reactivity [138]. In 1979, Wenkert reported the first $NiCl_2(PPh_3)_2$-catalyzed cross–coupling of enol ethers and aryl ethers with methyl and aryl Grignard reagents (Scheme 5.79) [139]. Although the substrate scope was limited, this transformation has afforded a conventional method to synthesize useful molecules [140]. When a secondary alkyl Grignard reagent such as ^{i}PrMgCl was used, reduction of enol ethers was observed [141]. In some cases, isomerization of the double–bond geometry occurred (Scheme 5.80). The alkenyl C–O bond in furan showed better reactivity than aryl ones and could be cleaved selectively, although *Z*/*E* isomerization also happened under this reaction condition [142].

In 1988, Johnstone and McLean [143] reported the coupling of aryl tetrazolyl ethers with alkyl and aryl Grignard reagents via nickel catalysis (Scheme 5.81). In this

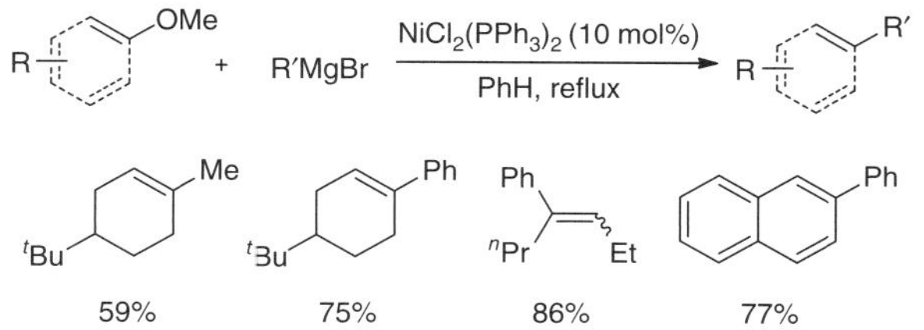

Scheme 5.79. Ni-catalyzed Kumada coupling of alkenyl/aryl methyl ethers.

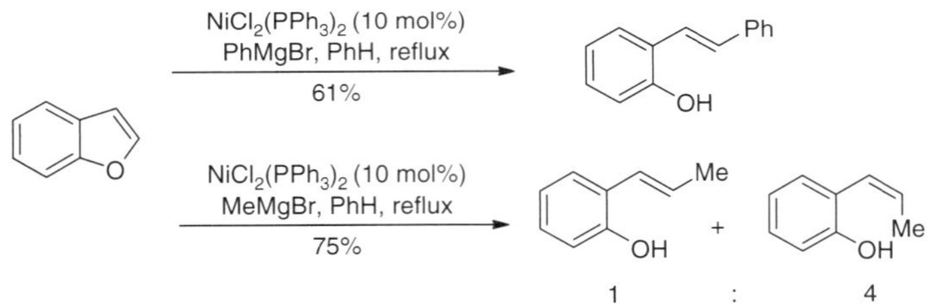

Scheme 5.80. Ni-catalyzed cross coupling of benzofuran with Grignard reagents.

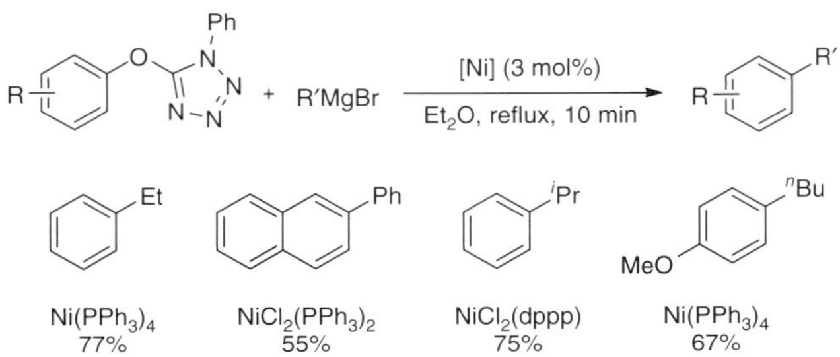

Scheme 5.81. Ni-catalyzed cross coupling of aryl tetrazolyl ethers with Grignard reagents.

case, due to the bond with a strong electron-withdrawing group, the aryl C–O bond is highly activated and even unactivated phenol derivate also could efficiently undergo this transformation. Importantly, methoxy group could be tolerated in this transformation. Later, it was found that the couplings of these substrates with organozinc and stannane reagents were also feasible [144].

Inspired by the pioneering work of Wenkert and others, Dankwardt and co-workers made a great contribution to this field. A wide range of aryl alkyl ethers were successfully applied in the Kumada coupling with $NiCl_2(PCy_3)_2$ or $NiCl_2(PhPCy_2)_2$ as catalyst (Scheme 5.82) [145a]. In this reaction, nonpolar solvents were essential to achieve high efficiency. Various functional groups (such as a free hydroxyl group),

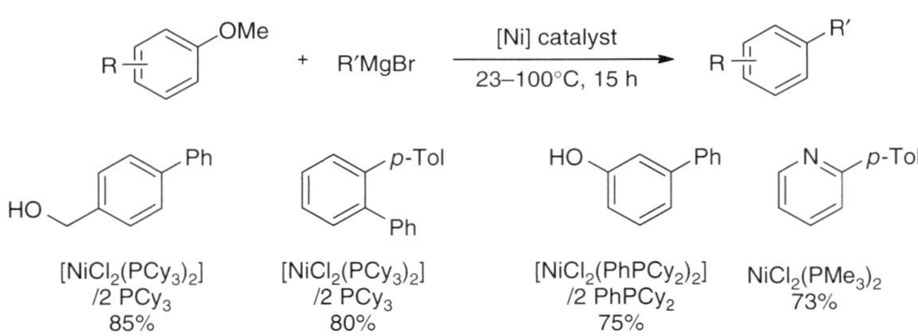

Scheme 5.82. Ni-catalyzed cross-coupling of anisoles with Grignard reagent.

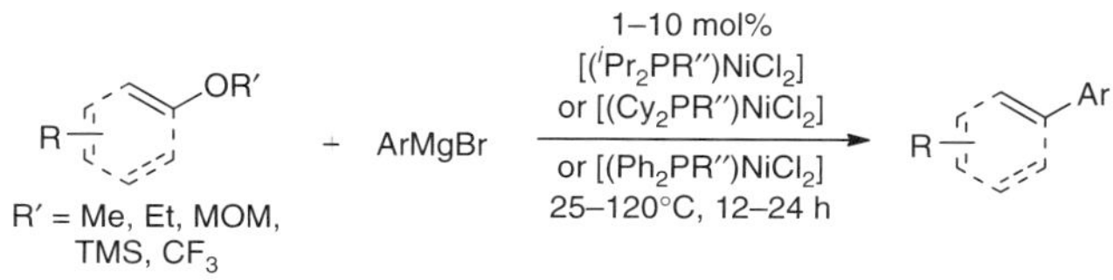

Scheme 5.83. Ni-catalyzed cross-coupling of aryl/alkenyl ethers with Grignard reagent.

enamine, and heteroaryls (including indole, pyridine and imidazole), were well tolerated. This methodology provided an important and useful method to synthesize the biaryl compounds from phenol ether derivatives. In 2011, Xie and Wang [145b] reported a general method for coupling of aryl, heteroaryl, and alkenyl with aryl Grignard reagent by using well-developed nickel complexes (Scheme 5.83). Aryltrifluoromethyl ether, which failed in Dankwardt's system, showed good reactivity in this reaction.

In 2008, Shi and co-workers [146] reported the methylation of aryl methyl ethers by nickel catalysis (Scheme 5.84). Besides naphthyl alkyl ethers, phenyl alkyl ethers could also give the desired product in high yields. Alkenyl group and free phenol can be well tolerated. Interestingly, the naphthyl methyl ether and phenyl ether could be differentiated by controlling the amount of the Grignard reagent.

Enol silyl ethers were also successfully used as a substrate in coupling reaction with a Grignard reagent. In 1980, Kumada and co-workers [147a] reported the first general method for cross-coupling of silyl enol ethers with alkyl and aryl Grignard reagents to form new carbon–carbon bonds (Scheme 5.85). Long-chain alkyl Grignard reagent could undergo this transformation smoothly, despite its tendency to undergo β-H elimination. The reaction proceeded with high stereoselectivity with respect to the alkene geometry. In 2011, Shi and co-workers [147b] reported the coupling of aryl and enol silyl ethers with aryl and methyl Grignard reagents under room temperature (Scheme 5.86). Vairous silyl ethers could react well in this reaction, but the effiency of phenyl silyl ether was somewhat low.

Kumada–Tamao–Corriu Coupling of Alkyl Ethers. In contrast with aryl and alkenyl ethers, cross-coupling of dialkyl ethers has been less investigated due to the difficulty of differentiating two similar sp^3 C–O bonds. In 2008, Shi and co-workers

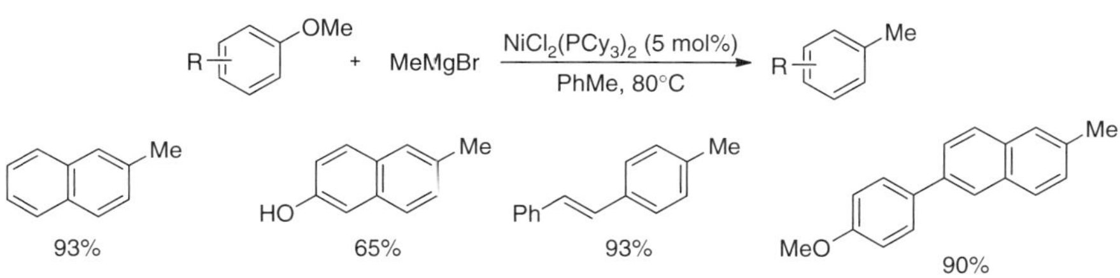

Scheme 5.84. Nickel-catalyzed cross-coupling of anisoles with methyl Grignard reagent.

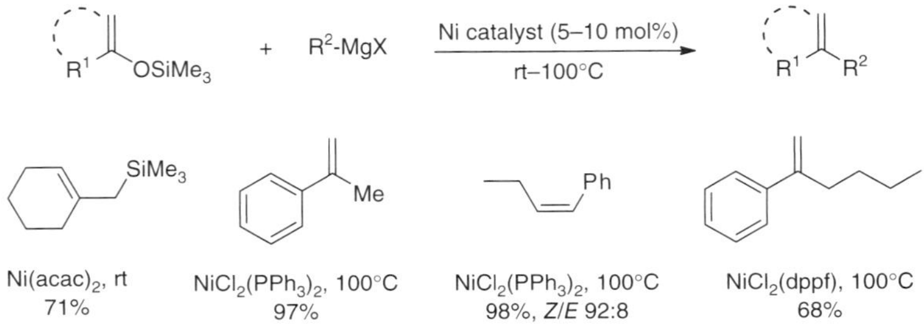

Scheme 5.85. Ni-catalyzed cross-coupling of silyl enol ethers with methyl Grignard reagent.

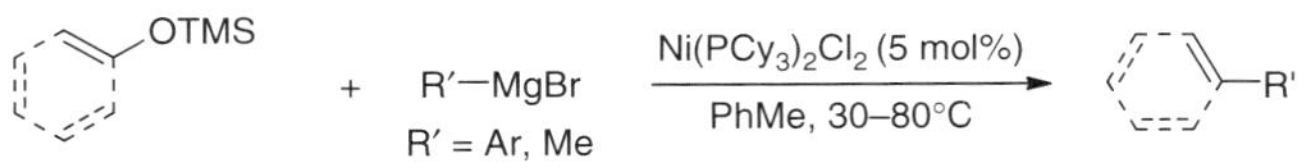

Scheme 5.86. Ni-catalyzed cross-coupling of aryl/enol silyl ethers with Grignard reagent.

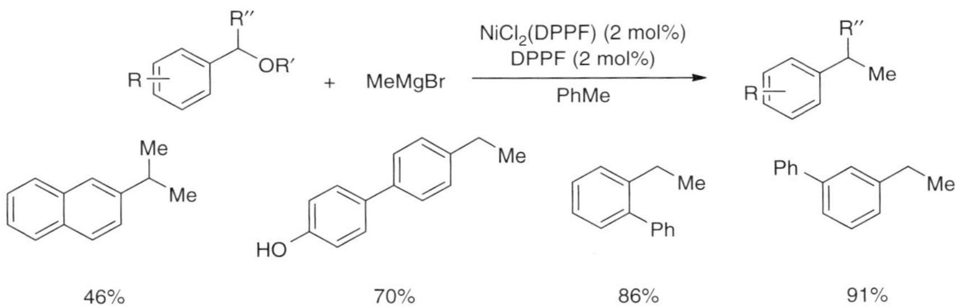

Scheme 5.87. Ni-catalyzed cross-coupling of benzyl methyl ethers with methyl Grignard reagent.

[148] reported a Ni-catalyzed methylation of benzyl methyl ethers with methyl Grignard reagent at room temperature (Scheme 5.87). When benzyl phenyl ether was subjected to the reaction, selective sp^3 C–O bond activation was achieved without cleavage of the sp^2 C–O bond. Various functional groups including free phenol were tolerated.

Later in 2011, Jarvo and co-workers [149] reported stereospecific nickel-catalyzed cross-coupling reactions of secondary benzylic ethers with methyl Grignard reagents (Scheme 5.88). The reaction proceeded with inversion of configuration and high stereochemical fidelity. Later, the synthesis of enantioenriched triarylmethanes by a nickel-catalyzed cross-coupling reaction of secondary benzylic ethers with various aryl Grignard reagents was developed by the same group (Scheme 5.89) [150].

5.3.1.5 *Kumada–Tamao–Corriu Coupling of Phenols and Phenolates.*

Obviously, the direct use of phenol or its inorganic salt as cross-coupling partners,

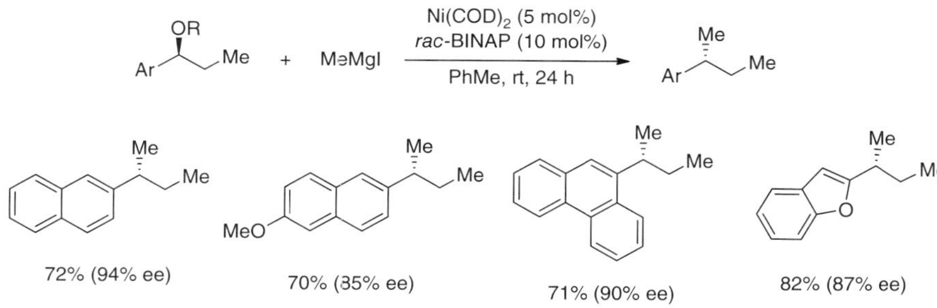

Scheme 5.88. Ni-catalyzed cross-coupling of secondary benzylic ethers with MeMgI.

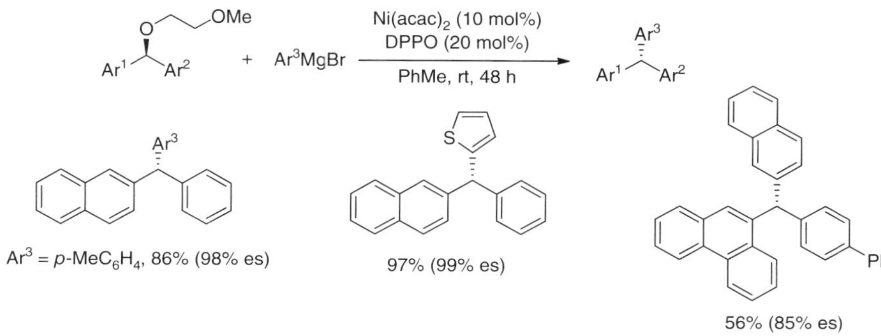

Scheme 5.89. Ni-catalyzed cross-coupling of secondary benzylic ethers with aryl Grignard reagents.

which avoided the extra procedure of group transfer and the generation of organic wastes, would be more atom- and step-economic. However, the direct cleavage of the aryl C–O bond in phenol was very difficult with respect to the high bond dissociation energy (BDE). The formation of the phenolic salt enhances the BDE, thus making the cleavage of C–O bonds more difficult. On the other hand, as a good σ-donor ligand, the phenolic anion can bind to the metal center and impede the transition-metal-induced cleavage of the C–O bond.

In 2010, Shi and co-workers [151] developed the unprecedented example of nickel-catalyzed cross-coupling reactions of naphtholates with aryl Grignard reagents (Scheme 5.90). In this reaction, 2-naphthol derivatives were treated sequentially with methyl Grignard reagent and an aryl Grignard reagent in the presence of NiF_2 and PCy_3 at 120°C. Various functional groups such as alkene, TMS, and a TBS-protected hydroxyl group could be tolerated. Unfortunately, the reaction was only successful with fused-ring substrates, and simple phenyl substrates were unable to participate in this reaction.

5.3.1.6 Kumada–Tamao–Corriu Coupling of Allylic and Benzylic Alcohols. The direct use of alcohols in coupling reactions is much more

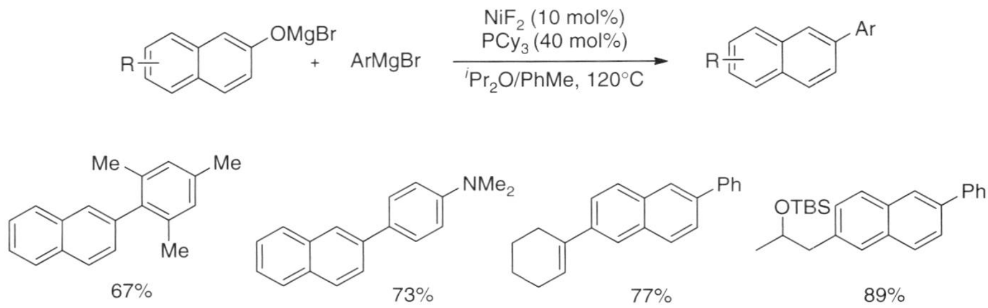

Scheme 5.90. Ni-catalyzed cross-coupling of aryl phenolates with aryl Grignard reagent.

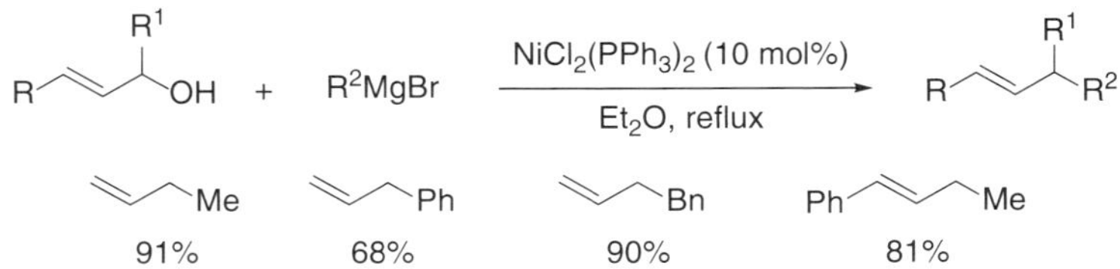

Scheme 5.91. Ni-catalyzed cross coupling of allyic alcohol with Grignard reagent.

step-economic, atom-economic, and environmentally benign than using allylic esters. For example, early in 1968, Felkin and co-workers [152] reported the nickel-catalyzed coupling of allylic alcohol with methyl, benzyl, and benzylic Grignard reagents (Scheme 5.91).

However, direct cross-couplings of benzyl alcohols have never been achieved until 2012. Shi and co-workers [153] reported a beautiful example of Ni-catalyzed cross-coupling of benzylic alcohol with Grignard reagent (Scheme 5.92). This transformation was accomplished with $NiCl_2(PCy_3)_2$ as a catalyst, PCy_3 as a ligand in $^{n}Bu_2O$, and toluene in 60°C for 24 h. Various benzylic alcohols involving fused rings afforded the desired product in moderate to good yields. Beside aryl Grignard

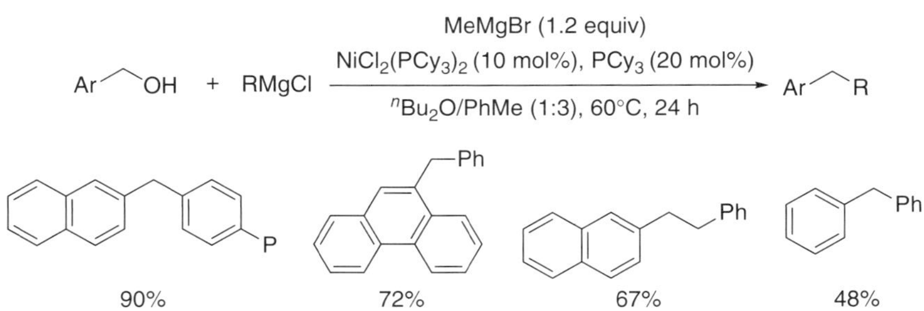

Scheme 5.92. Ni-catalyzed cross-coupling of benzylic alcohol with alkyl Grignard reagent.

reagents, benzyl and methyl Grignard reagent also showed good reactivity. However, other alkyl and alkenyl Grignard reagents were not suitable for this transformation.

5.3.2 Negishi Coupling (Including Zinc, Aluminum, Manganese, Copper, and Indium Reagents)

5.3.2.1 Negishi Coupling of Phosphates/Tosylates/Mesylates. Early in 1981, Kumada and co-workers [154] reported the coupling of aryl phosphates with aluminum reagents by nickel catalyst (Scheme 5.93).

Then the cross-couplings of various organometallic reagents were further developed. For example, nickel-catalyzed coupling of alkenyl phosphates/tosylates with organomanganese [155], zinc [156], and indium reagents [157] were developed by Oshima, Skrydstrup, Yang, and Wu, respectively. In 2010, a nickel-catalyzed coupling of aryl tosylates and mesylates with aryl indium reagents was also reported (Scheme 5.94) [158].

5.3.2.2 Negishi Coupling of Carboxylates. In 2008, Shi and co-workers [159] reported the first cross-coupling of aryl and alkenyl pivalates with organozinc reagents (Scheme 5.95). In order to avoid the direct nucleophilic attack of arylzinc

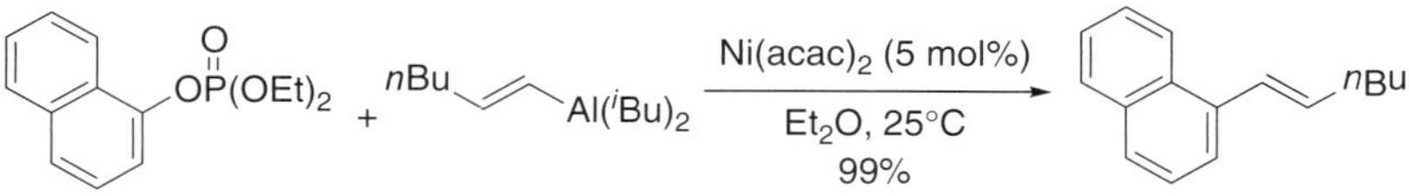

Scheme 5.93. Ni-catalyzed coupling of aryl phosphates with aluminum reagents.

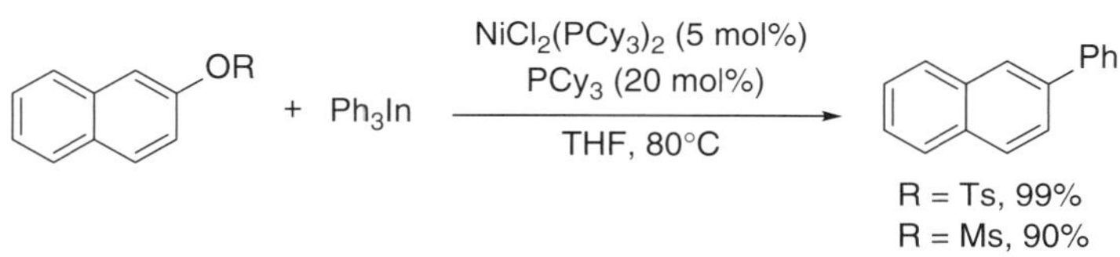

Scheme 5.94. Ni-catalyzed cross coupling of aryl tosylate/mesylate with aryl indium reagent.

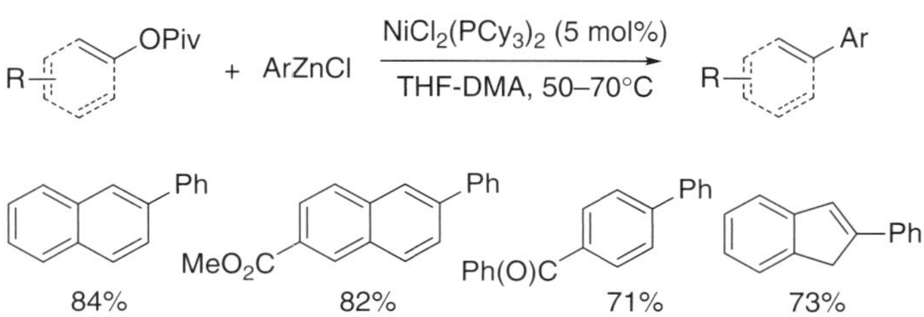

Scheme 5.95. Ni-catalyzed cross-coupling of aryl pivalates with organozinc reagents.

reagent to the carbonyl group, the more bulky aryl pivalates were used. The reaction could proceed under mild conditions due to the high reactivity of zinc reagents. However, the reaction was only suitable for pivalates bearing electron-withdrawing groups, and low yields were obtained with nonactivated aryl pivalates. Beside aryl and alkenyl pivalates, alkenyl acetates [160] also could be utilized in such a transformation under similar reaction conditions.

5.3.2.3 Negishi Coupling of Ethers. In 2011, Uchiyama and co-workers [161] reported the first Ni-catalyzed cross-coupling of aromatic ether and organozinc species through cleavage of the etheric C–O bond (Scheme 5.96). The usage of dianion-type zincates, which have high nucleophilicity and low basicity, allows the use of mild reaction conditions, especially room temperature. The reaction is compatible with a broad range of functional groups.

5.3.3 Suzuki–Miyaura Coupling

5.3.3.1 Suzuki–Miyaura Coupling of Phosphates. As a powerful method for C–C bond formation, Suzuki–Miyaura coupling is extremely useful in organic synthesis. In early reports of using C–O containing substrates as electrophiles, activated alkenyl phosphates were used as coupling partners, which are relatively easy to handle and synthetically useful.

Nonactivated cyclic alkenyl phosphates were also developed as coupling partners (Scheme 5.97). For example, the cross–coupling of cyclohexenyl phosphate with arylboronic acids was reported by using an *in situ* reduction nickel(0) catalyst [162].

In 2006, Skrydstrup and co-workers [163] reported the Suzuki–Miyaura coupling of acyclic alkenyl phosphates in the presence of $Ni(COD)_2$ and PCy_3 (Scheme 5.98). A double cross-coupling was also successful, providing 1,4-dialkenyl-substituted benzene in high efficiency.

In 2010, Han and co-workers [164] reported the nickel-catalyzed Suzuki–Miyaura coupling of aryl phosphoramides. By using a strongly electron-withdrawing phosphoramide, the C–O bond in the substrates were susceptible to cleave, thus enabling the reaction to proceed in high efficiency. Furthermore, since the advent of a Ni(II) pre-catalyst and air-stale materials, the reaction was operationally simple

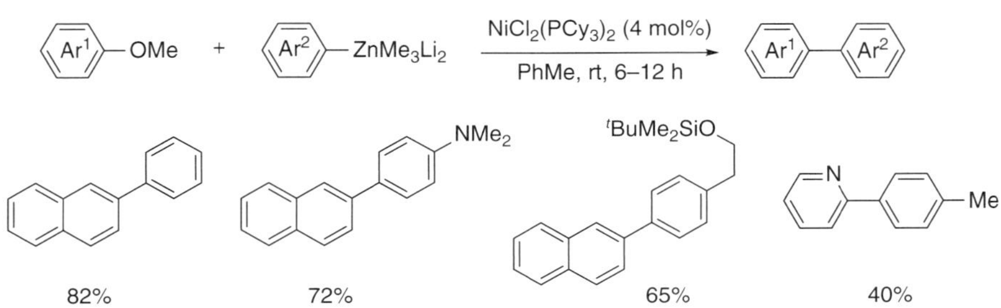

Scheme 5.96. Ni-catalyzed cross coupling of aryl methyl ethers with organozinc reagents.

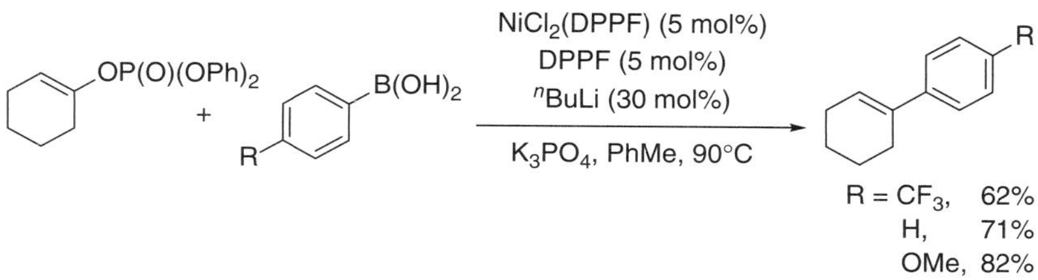

Scheme 5.97. Ni-catalyzed Suzuki–Miyaura coupling of nonactivated alkenyl phosphates.

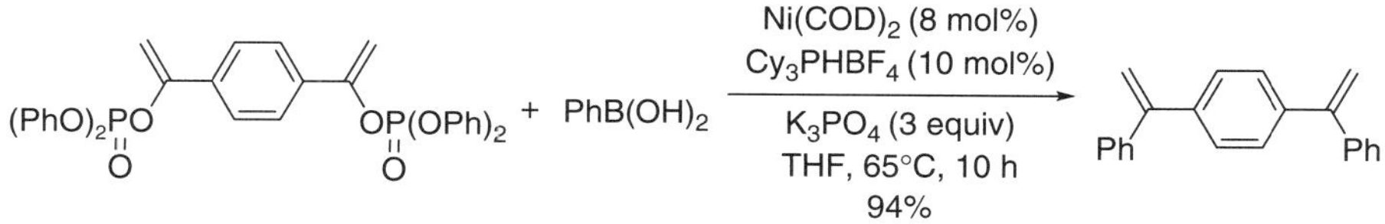

Scheme 5.98. Ni-catalyzed Suzuki–Miyaura coupling of alkenyl phosphates.

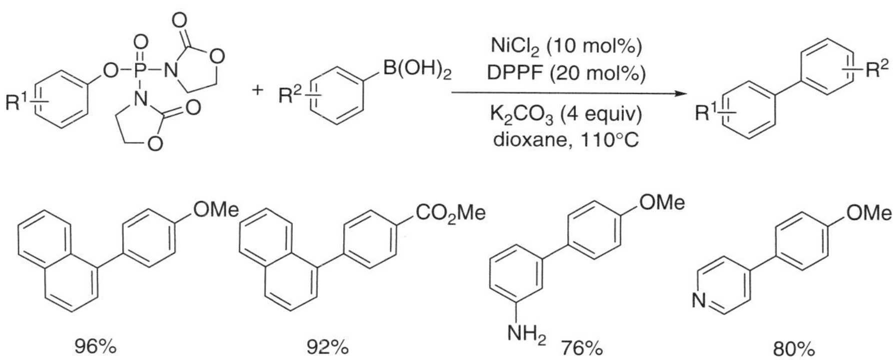

Scheme 5.99. Ni-catalyzed Suzuki–Miyaura coupling of aryl phosphates.

(Scheme 5.99). In 2011, Cheng and co-workers [165] reported the nickel-catalyzed Suzuki coupling of aryl phosphates with arylboronic acids.

5.3.3.2 Suzuki–Miyaura Coupling of Tosylates and Sulfamates. By using a reliable catalyst system $NiCl_2(PCy_3)_2/PCy_3$, the Suzuki coupling of aryl tosylates was developed (Scheme 5.100) [166]. Interestingly, no additional activating reagents were required to generate the active nickel(0) species. The active nickel(0) species might be formed through an aryl boronic acid transmetalation and reductive elimination sequence. Several experiments suggested that transmetalation between the catalyst and boronic acid was the rate-determining step.

Other catalyst systems for the cross-coupling of aryl tosylates were also developed (Scheme 5.101). A mixed ligand system for the coupling of various aryl tosylates was

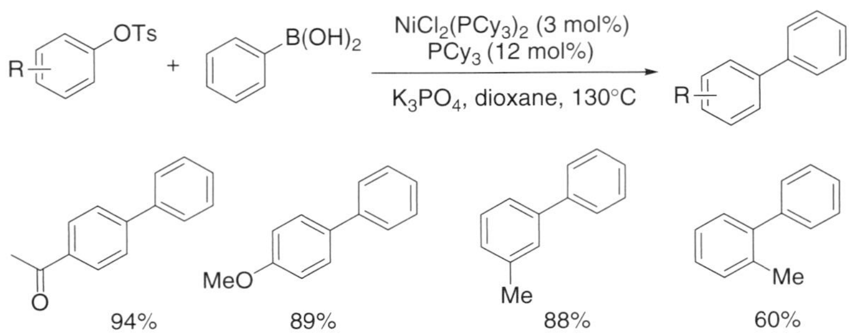

Scheme 5.100. Ni-catalyzed Suzuki–Miyaura coupling of aryl tosylates.

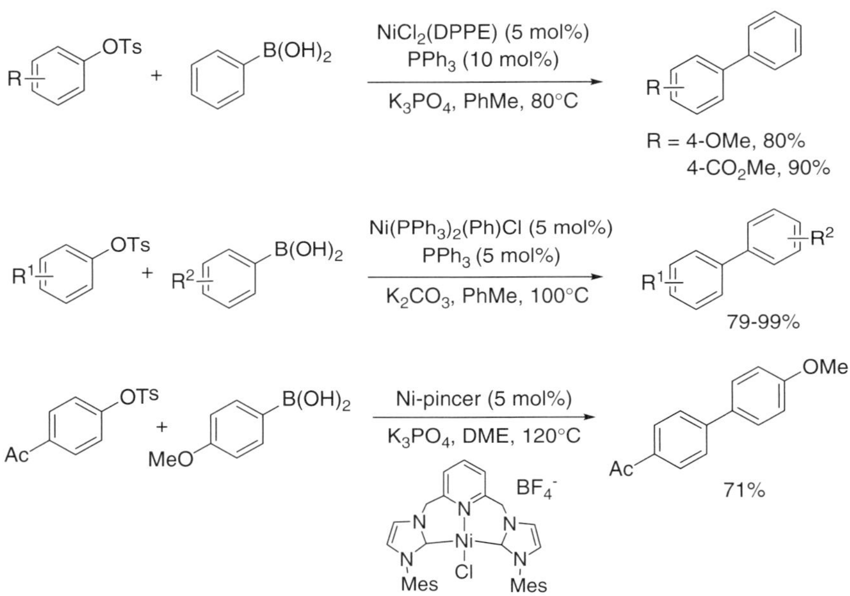

Scheme 5.101. Ni-catalyzed Suzuki–Miyaura coupling of aryl tosylates.

reported by Percec et al. [167]. The simultaneous presence of two kinds of ligand, the bidentate ligand dppe and the monodentate ligand PPh_3, was essential. Fan and Yang [168] also reported that a Ni^{II}–(σ-aryl) complex in combination with PPh_3 catalyzed the cross-coupling of various aryl tosylates with arylboronic acids. In addition, coupling of various activated aryl and alkenyl tosylates were realized in high efficiency by using a nickel–pincer complex derived from *N*-heterocyclic carbene [169].

Tang and Hu [170] reported the nickel-catalyzed cross-coupling of aryl arenesulfonates with aryl boronic acids at room temperature (Scheme 5.102). The reaction proceeded in the presence of $Ni(COD)_2$ and PCy_3. Various aryl arenesulfonates with

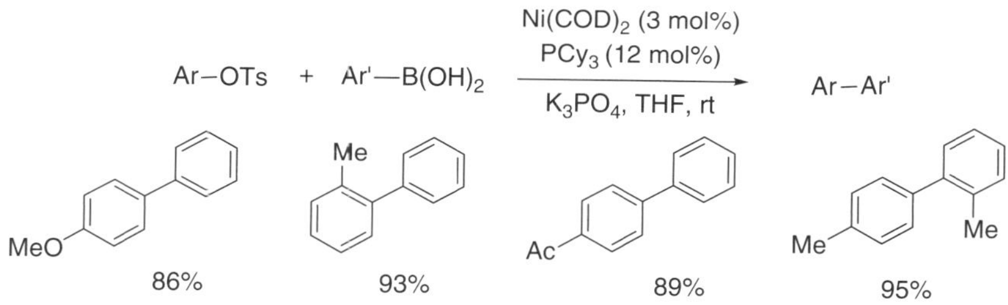

Scheme 5.102. Ni-catalyzed Suzuki–Miyaura coupling of aryl tosylates.

different electronic and steric properties underwent the coupling reaction smoothly. Later, the coupling of aryl tosylates with arylboronic acids was also reported by Hu and co-workers [171]. The reaction was conducted in the presence of ferrocenylmethylphosphines ligands. This method could be applied to functionalization activated alkenyl tosylates [172].

The Suzuki–Miyaura coupling of aryl tosylate could also be realized by heterogeneous nickel catalyst (Scheme 5.103). In 2008, Lipshutz et al. [173] reported the cross-coupling of activated aryl tosylates with various arylboronic acids by using nickel(II) mounted on graphite as the catalyst. The catalyst could be recycled without loss of activity.

Aryl sulfamates were also efficient electrophiles in the nickel-catalyzed Suzuki reactions. Garg and co-workers [174] reported the coupling of aryl sulfamates and boronic acids by using an easily available catalyst $NiCl_2(PCy_3)_2$. Various aryl sulfamates with electron-donating or -withdrawing groups reacted well. By using a sequence of D*o*M and cross-coupling reaction, *ortho* functionalized biaryls were obtained in good yields (Scheme 5.104). In 2012, Chen and Han [175] reported a similar work, using $NiCl_2$(dppf) as the catalyst.

5.3.3.3 *Suzuki–Miyaura Coupling of Mesylates.* Early in 1996, Percec et al. [176] reported the nickel-catalyzed reductive homocoupling of aryl mesylates, which suggested that the oxidative addition of the C–O bond of aryl mesylates to the nickel was feasible. Based on this result, they further developed the cross-coupling of aryl mesylates with aryl boronic acids. Zinc was used as a reducing reagent to generate the active low-valent nickel species (Scheme 5.105).

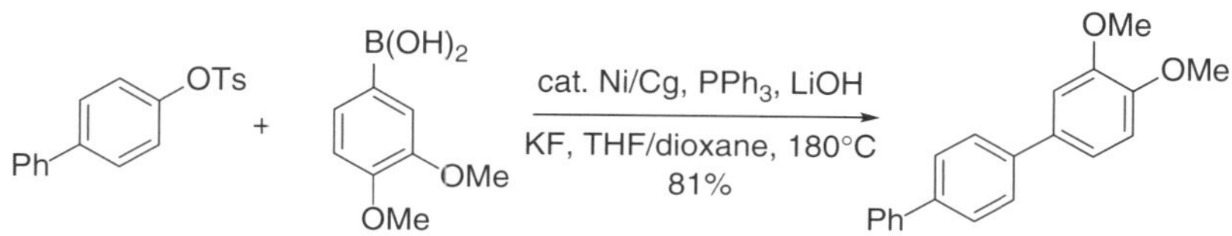

Scheme 5.103. Ni-catalyzed heterogeneous Suzuki–Miyaura coupling of aryl tosylates.

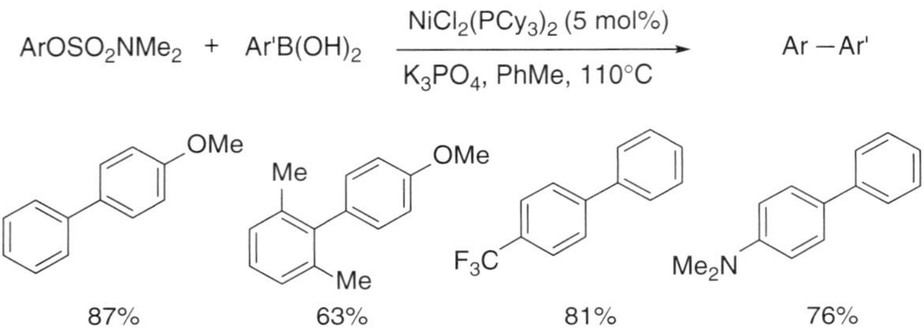

Scheme 5.104. Ni-catalyzed Suzuki–Miyaura coupling of aryl sulfamates.

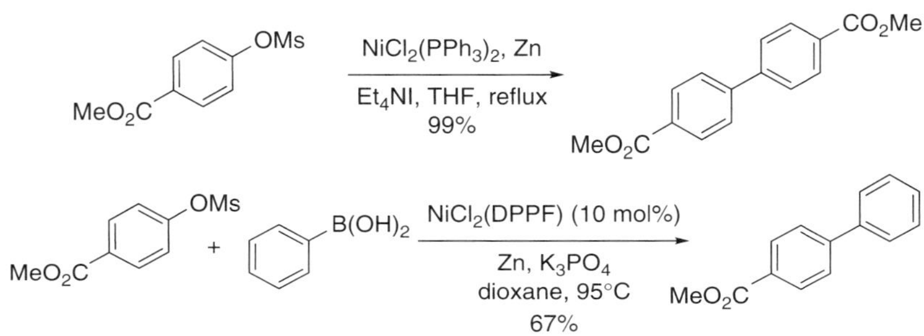

Scheme 5.105. Ni-catalyzed homocoupling and Suzuki–Miyaura coupling of aryl mesylates.

Kobayashi and Mizojiri [177] reported the coupling of lithium organoborate and aryl mesylates (Scheme 5.106). This reaction proceeded at room temperature without the addition of zinc due to the high reactivity of borate. However, this protocol was mainly limited to aryl mesylates with electron-withdrawing groups.

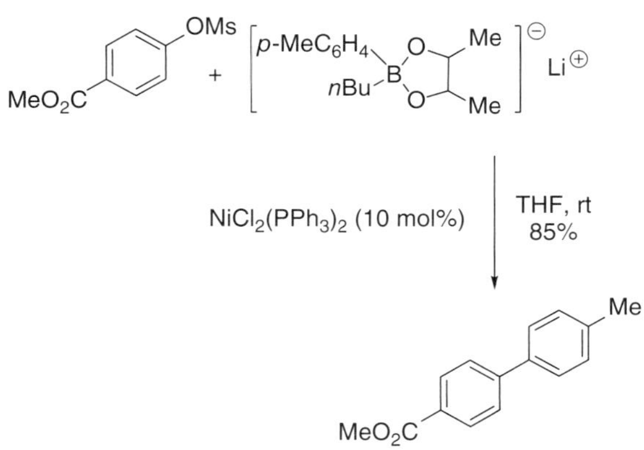

Scheme 5.106. Ni-catalyzed Suzuki–Miyaura coupling of aryl mesylates.

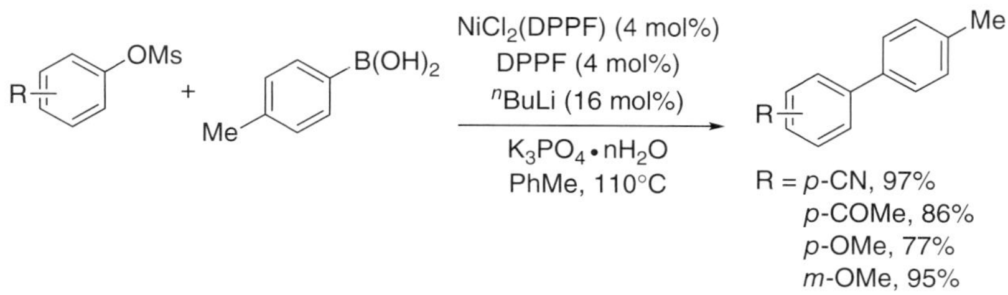

Scheme 5.107. Ni-catalyzed Suzuki–Miyaura coupling of aryl mesylates.

Subsequently, Miyaura and co-workers [178] found that aryl mesylates could couple with aryl boronic acids with high activity in the presence of a nickel(0) species generated by *in situ* reduction (Scheme 5.107). Both electron-donating- and electron-withdrawing-group-substituted aryl mesylates underwent smooth coupling reactions, providing various biaryls in high yields.

Percec et al. [179] developed a mixed ligand system for the coupling of aryl mesylates with phenylboronic acids. Using PCy_3 as a ligand, they reported the nickel-catalyzed coupling of activated aryl mesylates with arylboronic ester [180]. Furthermore, the coupling of nonactivated aryl mesylates was realized by a nickel–NHC complex [169b]. In 2010, Molander and Beaumard [181] investigated the nickel-catalyzed cross-coupling of aryl mesylates with aryltrifluoroborates and heteroaryltrifluoroborates (Scheme 5.108).

5.3.3.4 Suzuki–Miyaura Coupling of Alkenyl/aryl Carboxylates and Carbamates. In 2008, the cross-coupling of aryl carboxylates with boroxine and boronic acids was reported by the groups headed by Shi (Scheme 5.109) [182] and Garg (Scheme 5.110) [183], respectively. Commercially available $NiCl_2(PCy_3)_2$ was used as the pre-catalyst and K_3PO_4 was used as the base, and both of the coupling reactions proceeded with good yields. Shi and co-workers [182] also applied the method to modify natural products such as estrone and flavor derivatives (Scheme 5.111). In 2010, Molander and Beaumard [181] reported the cross-coupling

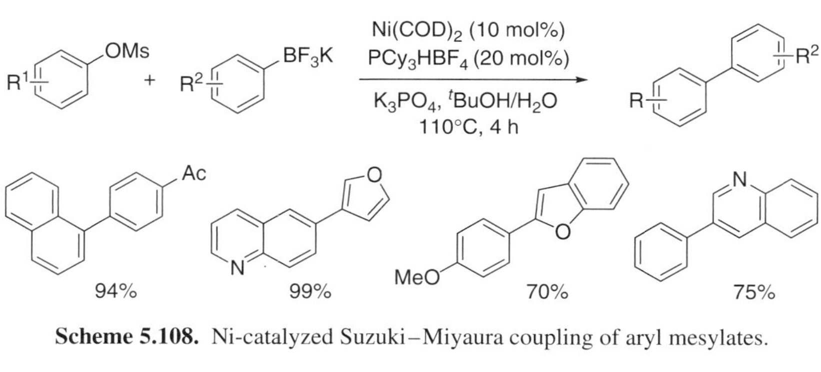

Scheme 5.108. Ni-catalyzed Suzuki–Miyaura coupling of aryl mesylates.

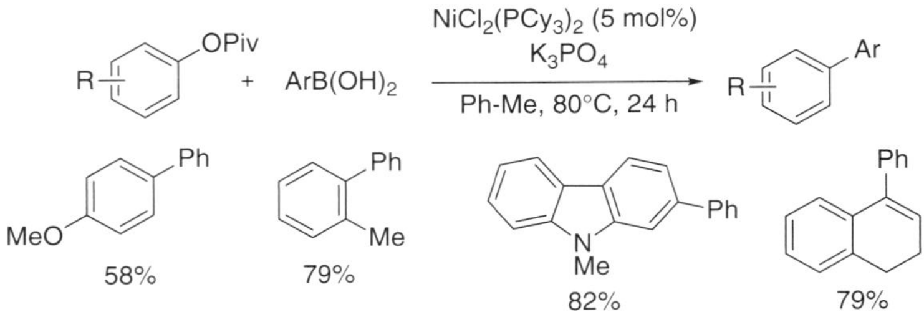

Scheme 5.109. Ni-catalyzed cross-coupling of aryl pivalates with organoboron reagents by Garg and co-workers [183].

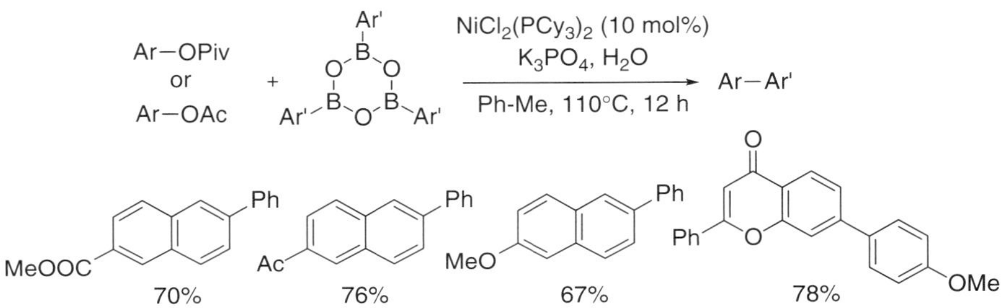

Scheme 5.110. Ni-catalyzed cross-coupling of aryl acetates/pivalates with aryl boroxines by Shi and co-workers [182].

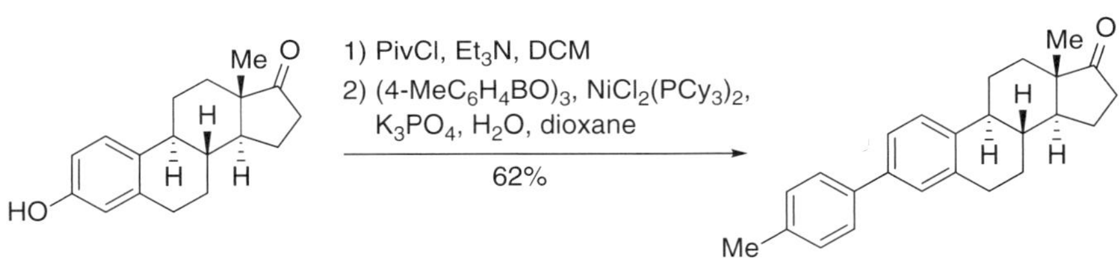

Scheme 5.111. Ni-catalyzed cross-coupling of estrone pivalate with *p*-tolyl boroxine by Shi and co-workers [182].

of aryl pivalates with heteroarylboron reagent. The use of aryltrifluoroborate as the coupling partner greatly reduced the amount of the boron reagent (Scheme 5.112).

A classical cross-coupling mechanism was proposed in these reactions. Computational studies of the mechanism were performed by Liu and co-workers [184]. They investigated selectivities of the C–O bond activation reaction in these reactions. The oxidative addition of the carbonyl C–OAr bond to Ni(0) is in equilibrium with the reverse reductive elimination process. However, transmetalation of the generated acyl Ni(II) intermediate was prohibited. Although the oxidative addition of the aryl C–OAc bond to Ni(0) is less favorable, this process is irreversible and the

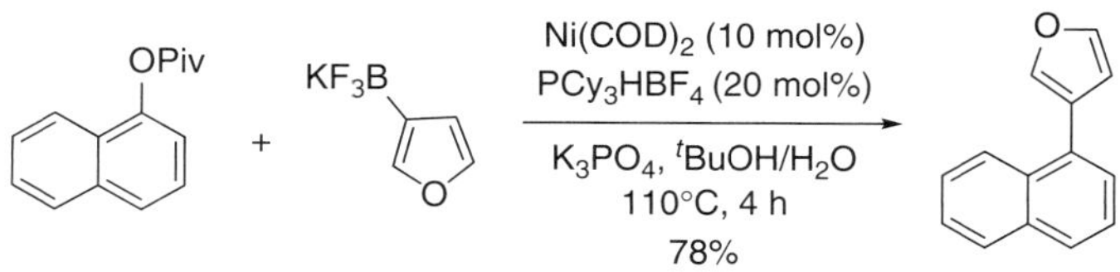

Scheme 5.112. Ni-catalyzed cross-coupling of aryl pivalate with trifluoroborate by Molander and Beaumard [181].

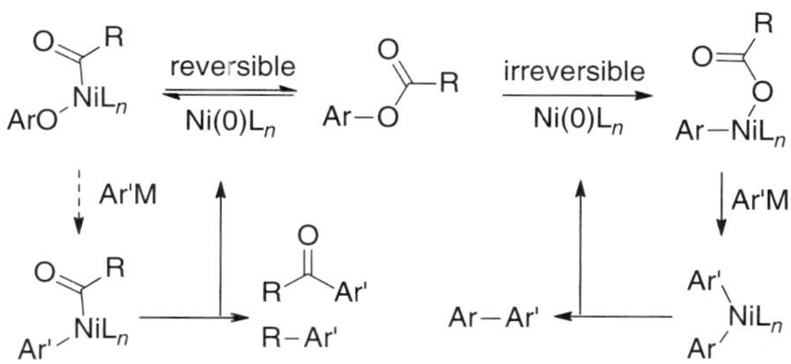

Scheme 5.113. Mechanistic consideration of the Ni-catalyzed cross-coupling of aryl esters.

generated aryl Ni(II) intermediate could undergo transmetalation with the activated boronic reagents quite smoothly (Scheme 5.113). The transmetalation was considered as the rate-determinining step of the catalytic cycle based on the computational analysis. Further support of this mechanism was provided by Yamamoto and co-workers [13], who investigated the stoichiometric reactions of nickel complexes with different esters. Together, these studies provided insight in understanding the mechanism and guideline for development of other coupling reactions of the aryl carboxylates.

Subsequently, the cross-coupling of aryl carbamates, which had superiorities in stability and good behavior in D*o*M [185], was further developed by the groups headed by Garg (Scheme 5.114) [186], Snieckus (Scheme 5.115) [187], and Shi

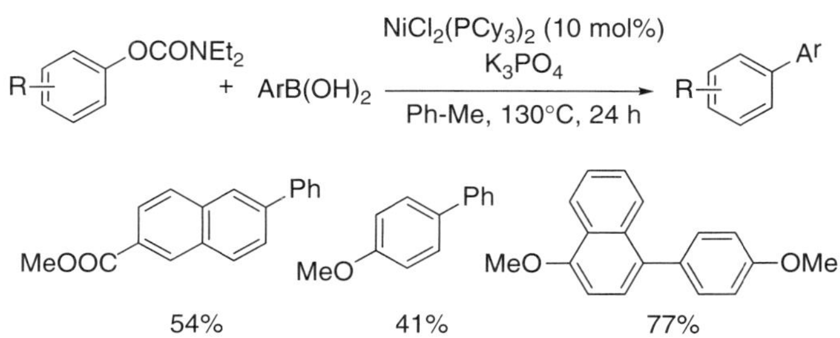

Scheme 5.114. Ni-catalyzed cross-coupling of aryl carbamates with organoboron reagents by Garg.

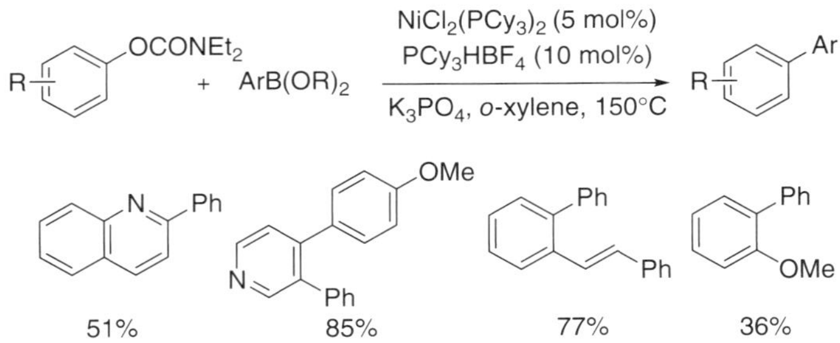

Scheme 5.115. Ni-catalyzed cross-coupling of aryl carbamates with organoboron reagent by Snieckus.

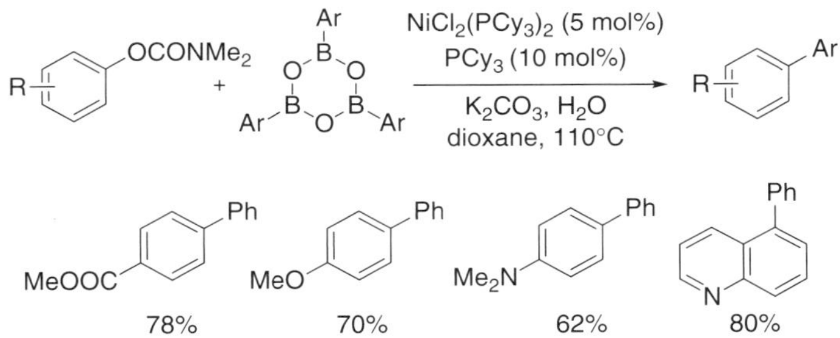

Scheme 5.116. Ni-catalyzed cross-coupling of aryl carbamates with organoboron reagent by Shi.

(Scheme 5.116) [188]. By combining directed *ortho* metalation and sequential Suzuki–Miyaura coupling, Snieckus and co-workers illustrated the superiorities of using carbamates as coupling partners. In Shi's work, deactivated electron-rich aryl carbamates could also undergo this transformation efficiently. In addition, Garg and co-workers showed that the coupling of fused aryl carbonates also worked well (Scheme 5.117).

In 2010, Shi and co-workers developed a general Suzuki–Miyaura reaction of alkenyl acetates [160] and carbamates [188] (Schemes 5.118 and 5.119). This method provides a versatile strategy to synthesize polysubstituted olefins from carbonyl compounds. Various aryl boroxines and alkenyl acetates/carbamates could be applied in this reaction.

5.3.3.5 *Suzuki–Miyaura Coupling of Ethers.*

The coupling of aryl methyl ethers with boronic esters was reported by Chatani and co-workers [189] in 2008. This reaction proceeded in the presence of Ni(COD)$_2$/PCy$_3$ without directing group assistance (Scheme 5.120). Both electron-deficient and electron-rich boronic esters afforded the coupling products in good yields. Although simple anisoles were unreactive, those bearing an electron-withdrawing group, such as 4-acetylanisole, gave

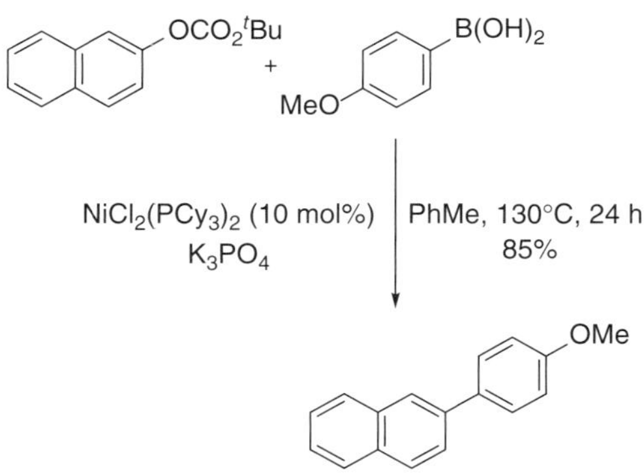

Scheme 5.117. Ni-catalyzed cross-coupling of aryl carbonates with organoboron reagent by Garg.

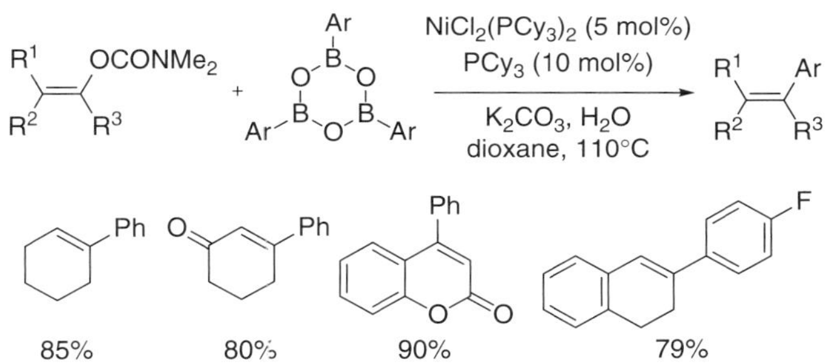

Scheme 5.118. Ni-catalyzed cross-coupling of alkenyl carbamates with organoboron reagents.

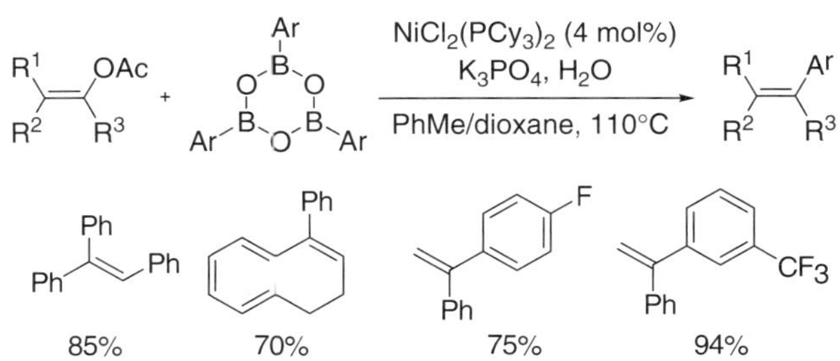

Scheme 5.119. Ni-catalyzed cross-coupling of alkenyl acetates with organoboron reagents.

the product in a moderate yield. The competitive experiment revealed that oxidative addition of the Ar–OMe bond was the rate-limiting step.

The catalytic system was further extended to alkenyl methyl ethers by the same group (Scheme 5.121) [190]. Different types of alkenyl methyl ethers could undergo

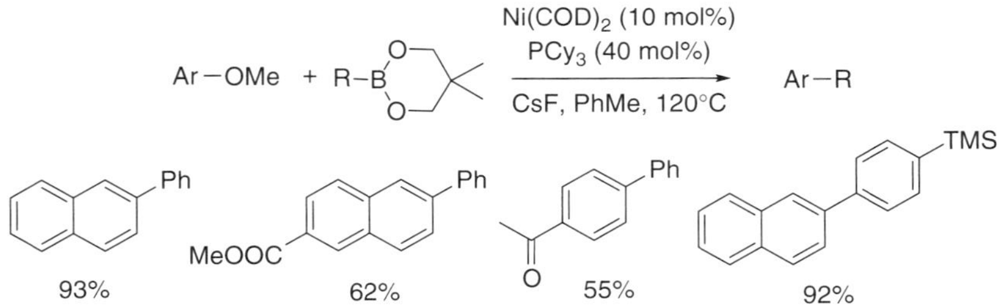

Scheme 5.120. Ni-catalyzed cross-coupling of anisoles with organoboron reagents.

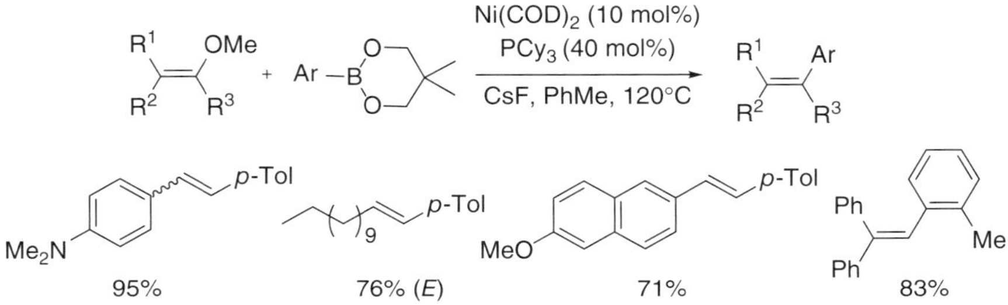

Scheme 5.121. Ni-catalyzed cross-coupling of alkenyl methyl ethers with organoboron reagents.

a cross-coupling reaction with a wide range of boronic esters to give the stilbene derivatives in good yields.

5.3.3.6 Suzuki–Miyaura Coupling of Alcohols and Phenols. Highly active alcohols, such as allylic alcohols, have been employed as coupling partners in Suzuki–Miyaura coupling. For example, one example of asymmetric coupling of allylic alcohol derivatives with aryl boronic acids was developed by Uemura and co-workers in 2000 (Scheme 5.122) [191].

In 2011, Shi and co-workers reported the first direct Suzuki–Miyaura cross-coupling of naphtholates with aryl boroxines to produce biaryl compounds (Scheme 5.123) [192]. Various naphthols derivatives could be applied in this reaction to afford moderate to good yield. The key point was the formation of the

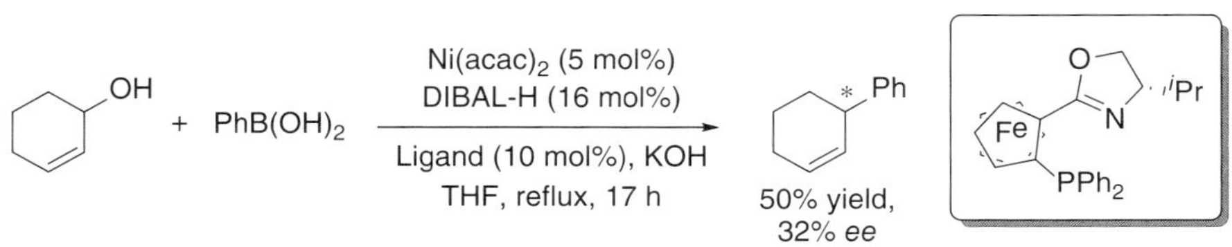

Scheme 5.122. Ni-catalyzed asymmetric coupling of allyic alcohol with aryl boroic acid.

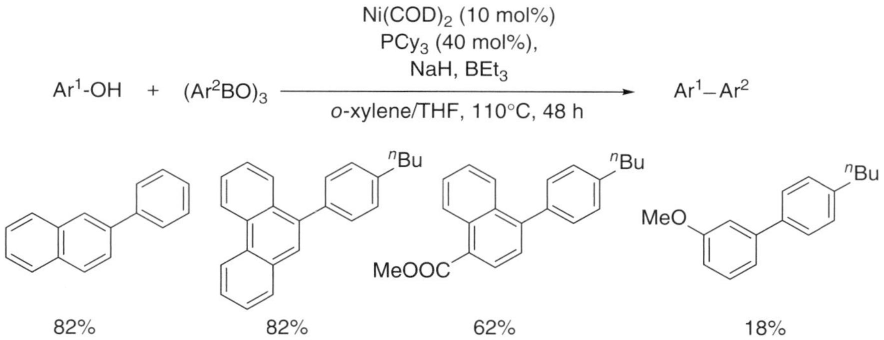

Scheme 5.123. Ni-catalyzed cross-coupling of phenols/phenolates with aryl boroxines.

borate, which offered the mutual activation of the two coupling partners. Additional pre-activation of phenols and the addition of strong bases was not required and thus made such a cross-coupling both step- and atom-economical. BEt_3 was used to promote the efficiency of the reaction, which implies a new pathway to activate the "inert" C–O bond by the double activation through the LA/LA interaction.

5.3.4 Heck Reaction

With similar strategy of reactions of allylic alcohols with nucleophiles, Matsubara and Jamison [193] developed the nickel-catalyzed coupling of allylic alcohol with ethylene to afford 1,4-diene in 2010 (Scheme 5.124).

In 2012, Watson and co-workers [194] developed the first C–C cross-coupling of Heck-type reaction of aryl pivalates with olefins by using a nickel catalyst (Scheme 5.125). This Heck reaction allowed facile preparation of 1,2-disubstituted olefins without use of an expensive triflate group or formation of a halogenated byproduct. The use of a bidentate phosphine ligand, DPPF (1,1′-bis(diphenylphosphino)ferrocene), can effectively promote the reaction, and a wide range of substrate scope could be compatible in this transformation with moderate to good yields.

5.3.5 Buchwald–Hartwig Amination

For the C–N bond formation, Buchwald–Hartwig amination is an extremely useful method, and nickel-catalyzed amination was realized by using aryl tosylates as

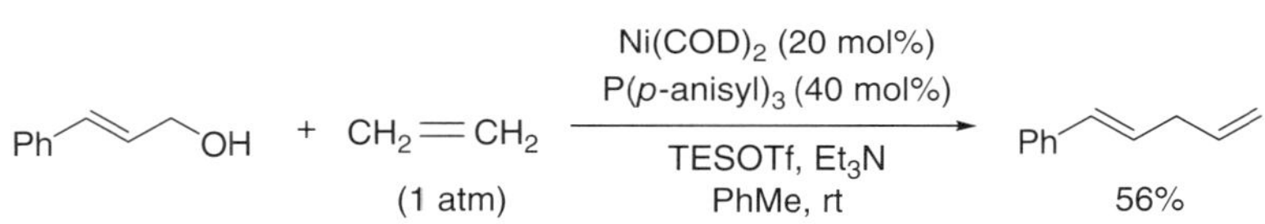

Scheme 5.124. Ni-catalyzed coupling of allyic alcohol with olefin.

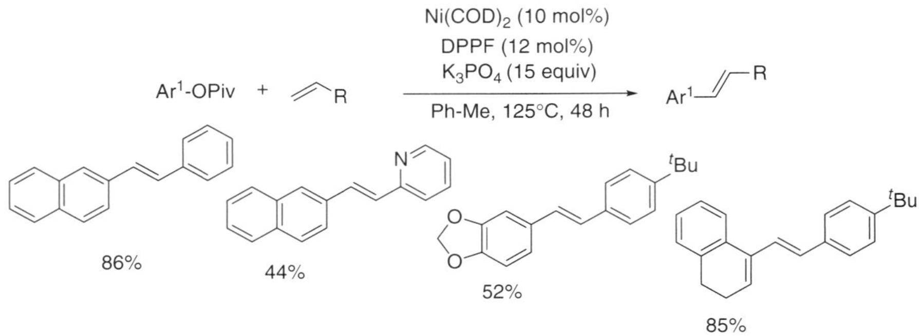

Scheme 5.125. Ni-catalyzed cross-coupling of aryl pivalates with olefins.

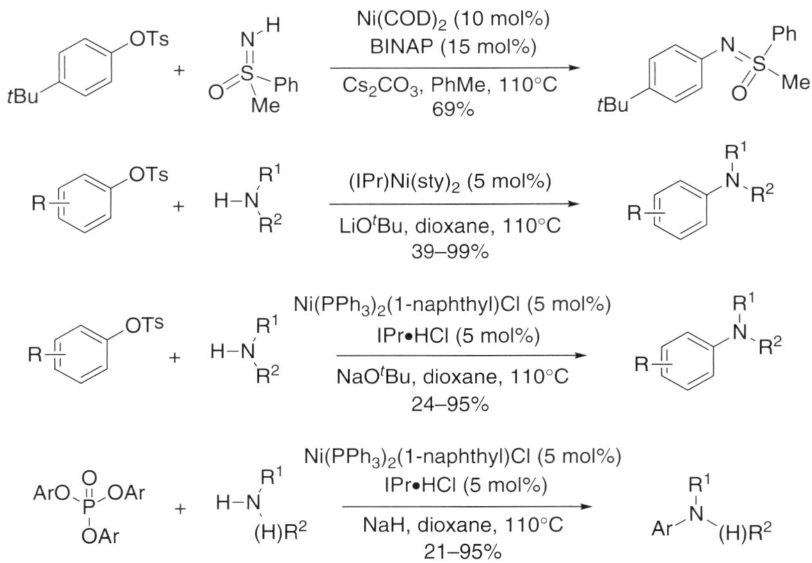

Scheme 5.126. Ni-catalyzed amination of aryl tosylates and phosphates.

coupling partners (Scheme 5.126). For example, Bolm et al. [195] reported the cross-coupling of sp^2-hybridized nitrogen nucleophiles such as sulfoximine, by using a Ni/BINAP complex. Recently, Nicasio and co-workers [196] reported a bis-styrene IPrNi(0) complex-catalyzed amination of aryl tosylates with secondary amine and anilines. In addition, Gao and Yang [197] developed the nickel-catalyzed amination of aryl tosylates with cyclic secondary amine and anilines using a combination of Ni(II)-(σ-aryl) complex/NHC ligand. In 2011, Huang and Yang group reported the amination of aryl phosphates by using the similar catalytic system.

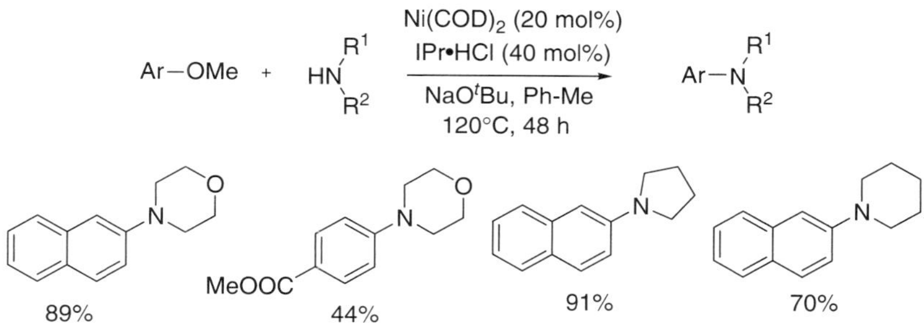

Scheme 5.127. Ni-catalyzed amination of anisoles.

In 2009, Chatani and co-workers [199] developed the Ni-catalyzed amination of anisoles with amines using a NHC ligand (Scheme 5.127). Although high temperature and high catalyst loading were required, this study offered an alternative route to synthesize aryl amines directly from readily available anisoles.

Later, Chatani and co-workers [200] further applied this method to aryl carboxylates (Scheme 5.128). Under similar conditions in a previous report, amination of aryl pivalates could react quite well. Both electron-rich and electron-poor aryl pivalates could produce the aminated products in good yields.

5.3.6 Borylation

Aryl boronic acids and their derivatives are useful starting material in various organic reactions. Using their previous mixed ligand system, which was highly effective for the borylation of aryl chlorides [201], the Percec group further realized the Ni-catalyzed borylation of aryl tosylates and mesylates with neopentylglycolborane (Scheme 5.129) [202]. It is important to note that the addition of zinc as a reductant in this reaction significantly improved the efficiency. However, it also induced a large amount of side reaction for the functional groups including aldehydes, ketones, and pyridines. Aryl mesylates worked quite successfully in this catalytic system, faring equally well with either electron-withdrawing or electron-donating groups.

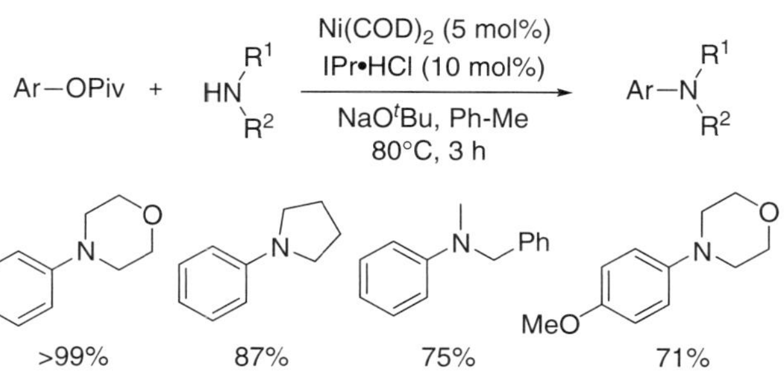

Scheme 5.128. Ni-catalyzed animation of aryl pivalates.

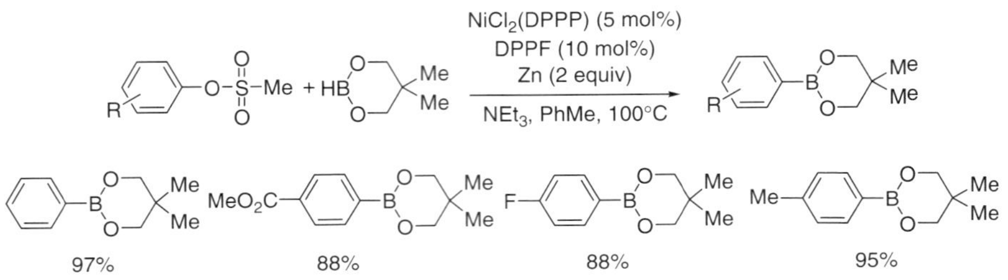

Scheme 5.129. Ni-catalyzed borylation of aryl tosylates and mesylates.

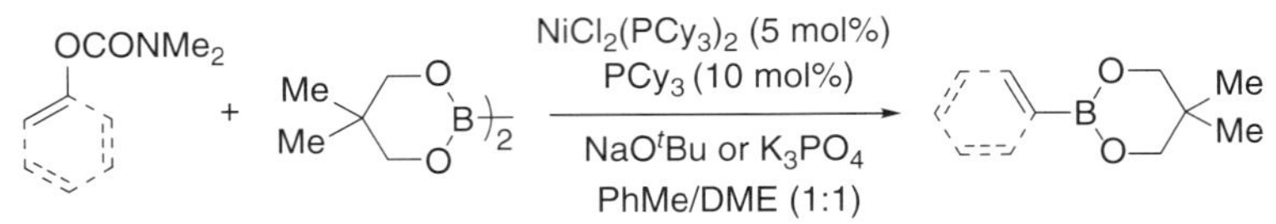

Scheme 5.130. Ni-catalyzed borylation of aryl and alkenyl carbamates.

In 2011, Shi and co-workers [203] developed a useful method of borylation starting from readily available and stable aryl/alkenyl carbamates (Scheme 5.130). With two sets of catalytic systems, the substrate scope was very broad, including aryl, alkenyl, and heterocyclic derivatives.

5.3.7 Direct Arylation

Recently, Itami and co-workers [204] reported the first nickel-catalyzed C–H bond arylation of azoles with phenol derivatives (Scheme 5.131). The reaction was accomplished by using $Ni(COD)_2$ as the catalyst, DCyPE (1,2-bis-[dicyclohexylphosphino] ethane) as the ligand, and Cs_2CO_3 as the base in 1,4-dioxane at 120°C for 12 h. Through the investigation of the reaction conditions, a dramatic ligand effect of dcype is notable, and the other ligand was ineffective. Various phenol derivatives, such as

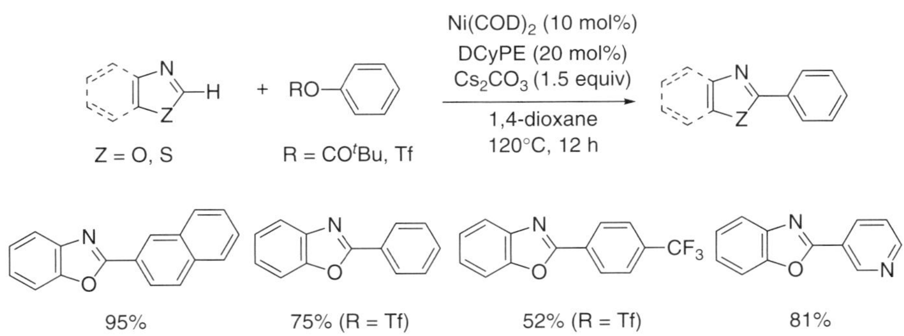

Scheme 5.131. Ni-catalyzed C–H bond arylation of azoles with phenol derivatives.

carbamates, carbonates, sulfamates, triflates, tosylates, and mesylates, were suitable substrates for this transformation. Notably, when aryl esters were used as substrates, the unprecedented decarbonylative reaction occurred and the coupling of azoles took place to afford biaryl compounds (Scheme 5.131) [205].

5.3.8 Reduction

Nickel-catalyzed reduction of various aryl tosylates was developed by using boronhydride as reductant [206]. Interestingly, the reduction of aryl tosylates and aryl mesylates could be achieved by a heterogeneous nickel catalyst (Scheme 5.132) [207].

Early in 2004, Milburn and Snieckus. [208] reported that tertiary aryl sulfonamides underwent Ni(0)-catalyzed reductive cleavage with β-hydride donors under especially mild conditions (Scheme 5.133). This transformation was accomplished by the usage of $Ni(acac)_2$ as a catalyst in the presence of $^{i}Pr_2Mg$ in diethyl ether at room temperature. Various functional groups were tolerated well in this mild condition. Notably, when conducting this reaction in refluxing toluene, the cross-coupling of aryl sulfonamides with RMgX occurred and the alkylation/arylation products were obtained.

Later in 2010, Martin and co-workers reported a more challenging example of Ni-catalyzed reduction of inert C–O bonds. This transformation proceeded well by

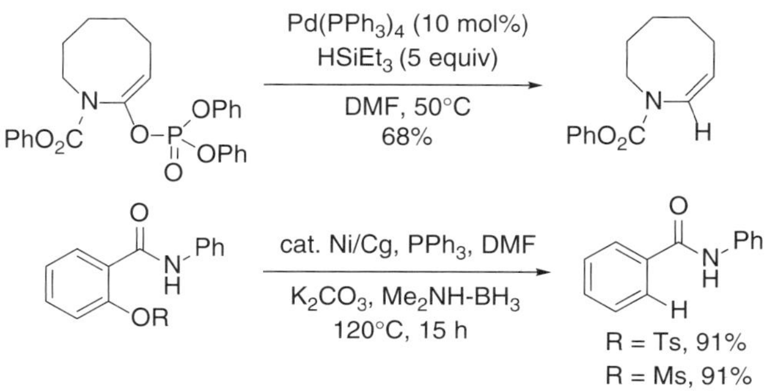

Scheme 5.132. Ni-catalyzed reduction of alkenyl phosphates and aryl sulfonates.

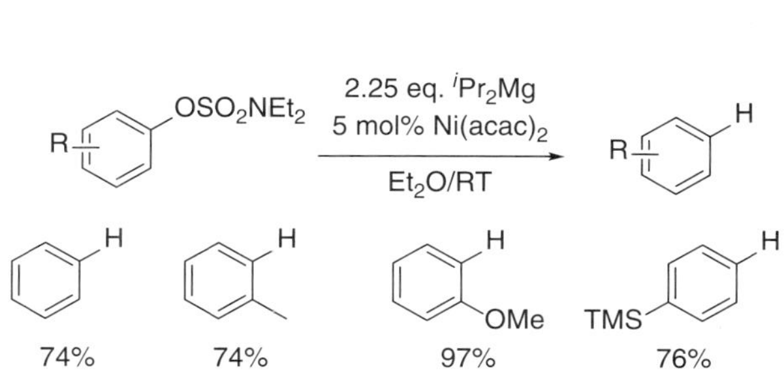

Scheme 5.133. Ni-catalyzed reduction of aryl sulfonamides with Grignard reagents.

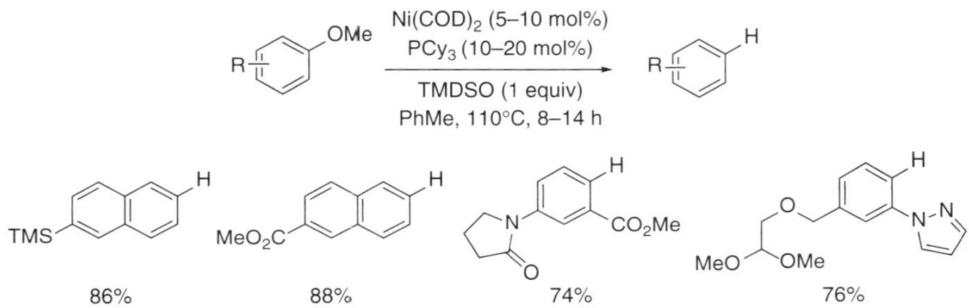

Scheme 5.134. Ni-catalyzed reduction of aryl methyl ethers.

the use of $Ni(COD)_2$ as the catalyst, PCy_3 as the ligand, and tetramethyldisiloxane (TMDSO) as the hydride source (Scheme 5.134) [209]. Various functional groups, such as silyl groups, esters, amides, acetals, tertiary amines, and nitrogen-containing heterocycles, were tolerated well. The ready availability of the substrates and the remarkable substrate scope observed make this method attractive to synthetic chemists.

Chatani and co-workers [210] reported a similar work in 2011. A nickel-catalyzed reductive deoxygenation of aryl alkyl ethers and aryl pivalates has been developed. Hydrosilanes served as a mild reducing agent. This protocol allowed the use of a pivalate group as a robust and traceless directing group in arene functionalization reactions.

Selective hydrogenolysis of the aromatic C–O bonds in aryl ethers is an unsolved synthetic problem important for the generation of fuels and chemical feedstocks from biomass and for the liquefaction of coal. In 2011, Sergeev and Hartwig [211] reported hydrogenolyses of aromatic C-O bonds in alkyl aryl and diaryl ethers that form exclusively arenes and alcohols (Scheme 5.135). This process was catalyzed by a soluble nickel carbene complex under 1 atm of hydrogen at 80–120°C. The reaction effectively and selectively cleave aromatic C–O bonds in the presence of other C–O bonds, without reduction of the arene units.

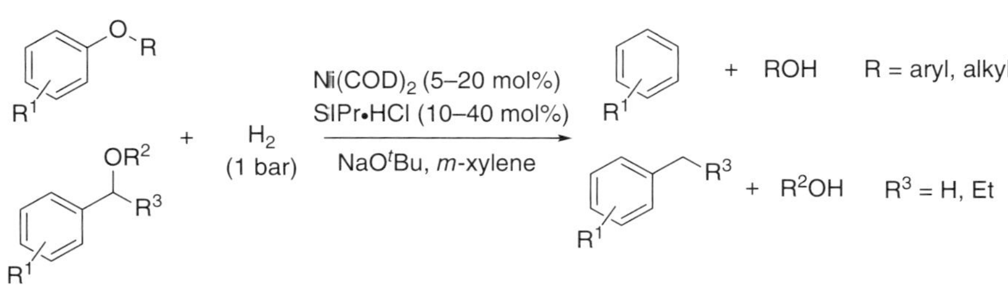

Scheme 5.135. Ni-catalyzed hydrogenolysis of aryl and benzyl ethers.

5.3.9 Other Reactions

Cyanation of aryl halides is a useful method for the synthesis of aryl nitriles. Percec et al. [212] reported nickel-catalyzed cyanation of aryl mesylate. Under the optimized conditions, the coupling reaction occurred in high yields. In general, electron-withdrawing groups on the aromatic ring increased the yields, while electron-donating groups decreased the yields (Scheme 5.136).

5.4 OTHER-TRANSITION-METAL-CATALYZED C–O BOND ACTIVATION

5.4.1 Fe-Catalyzed C–O Bond Activation

Iron catalysis has been playing important role in organic synthesis [213]. Due to their low price, low toxicity, and easy availability, Fe catalysts show great advantages over other frequently used noble metals. Moreover, iron catalysts exhibit significantly different reactivity in various transformations, which could not be easily performed with other transition metals. Therefore, Fe catalysis is attracting more and more attention.

Iron-mediated C–O bond cleavage has been observed for more than 30 years. However, there had been no report of Fe-catalyzed C–O activation until Cahiez and Avedissian [214] reported an example of iron-catalyzed coupling of alkenyl phosphate with butyl magnesium chloride in 1998 (Scheme 5.137). This catalytic system is also applicable for a variety of alkenyl chlorides, bromides, and iodides as well as various kinds of Grignard reagents.

Later, the groups headed by Cahiez, Begtrup, and Hayashi further expanded the substrate scope and apply this method in synthesis of various bioactive molecules [215]. For example, in 2008 Cahiez and co-workers further developed the coupling of dienol phosphates with primary Grignard reagents to generate terminal conjugated dienes under mild conditions. A variety of conjugated dienyl phosphates could be applied in this reaction, and the conformation of diene moiety in the products is

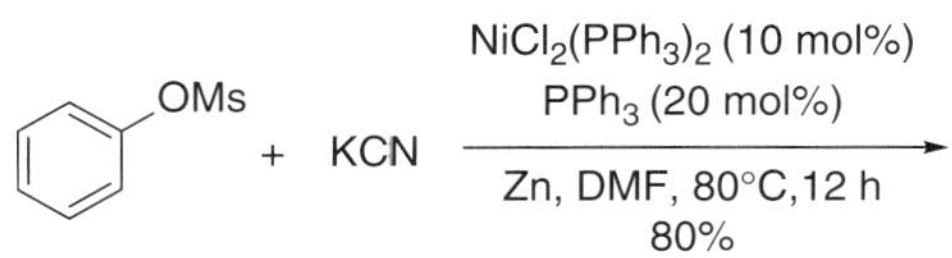

Scheme 5.136. Ni-catalyzed cyanation of aryl mesylates.

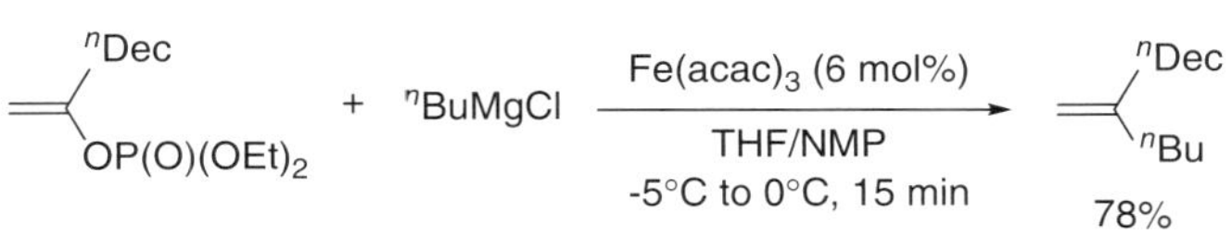

Scheme 5.137. Fe-catalyzed coupling of alkenyl phosphate with Grignard reagents.

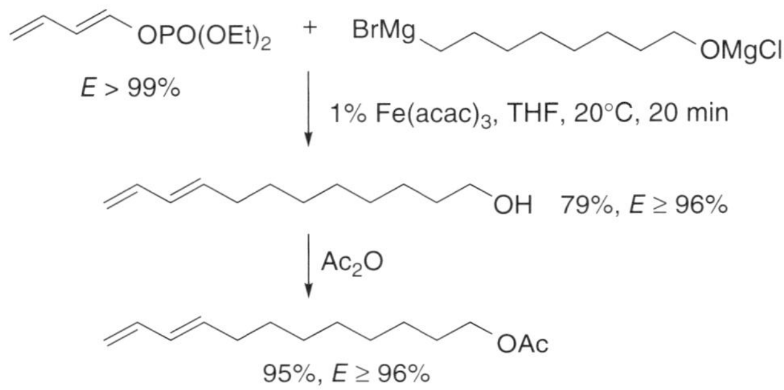

Scheme 5.138. Fe-catalyzed C–O activation of conjugated dienyl phosphates.

highly retained. With this method, an insect pheromone in the red bollworm moth, *Diparopsiscastanea*, was efficiently synthesized by this group (Scheme 5.138).

In the process of total synthesis of Latrunculin B, in 2003 Fürstner et al. [216a] realized the first Fe-catalyzed cross-coupling reactions of acetoacetate-derived alkenyl triflates with homopropargylic Grignard reagents. Later, the similar couplings of ketone- and acetoacetone-derived alkenyl triflates have also been developed by the same group and applied in various total synthesis (Scheme 5.139) [216].

In 2006 Dunet and Knochel [217] realized a $Fe(acac)_3$-catalyzed cross-coupling of alkenyl and dienyl sulfonates with functionalized arylcopper reagents

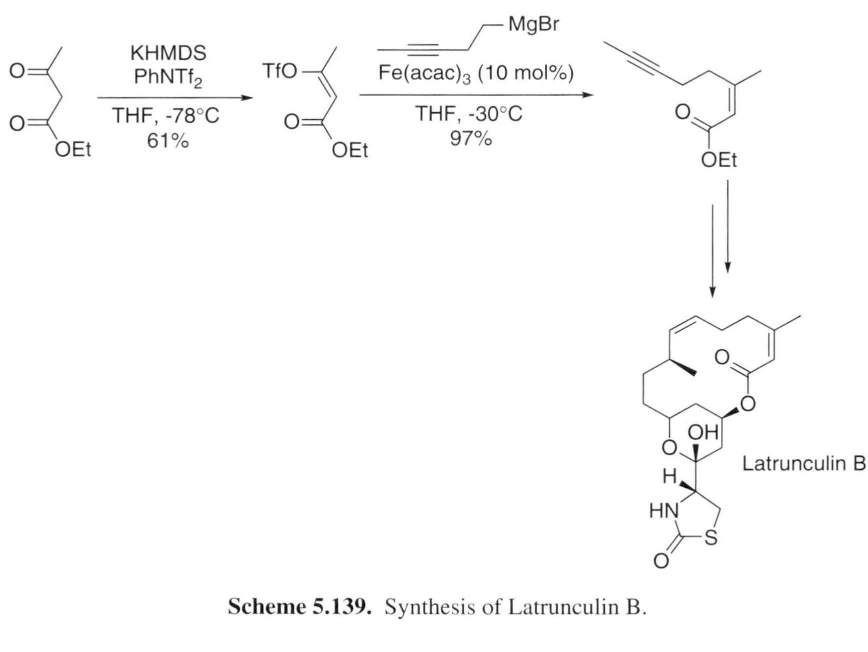

Scheme 5.139. Synthesis of Latrunculin B.

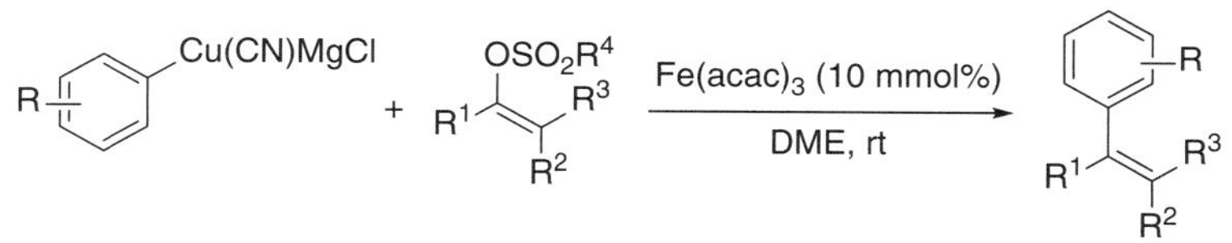

Scheme 5.140. Fe-catalyzed coupling of arylcoppers and alkenyl sulfonates.

(Scheme 5.140). This reaction took place under mild conditions and many functional groups, such as ester and cyano groups, were well tolerated.

In 2009, Shi and co-workers [218] reported an iron-catalyzed cross-coupling reaction of alkenyl carboxylates with Grignard reagents via C–O bond activation. In this report, they have accomplished an efficient method to break the inert alkenyl C–O bond rather than active acyl C–O bond by adding $FeCl_2$ as the catalyst. It was found that the combination of $FeCl_2$ and SIMes ligand was the most efficient. The reaction took place highly efficiently, especially when electron-deficient alkenyl carboxylates were applied. The authors suggested that a radical mechanism was likely involved (Scheme 5.141).

Beside alkenyl C–O bond activation, phenol derivatives have also been investigated via Fe catalysis. In 2002 Fürstner et al. [219] realized the first Fe-catalyzed cross-coupling reaction of aryl chlorides, triflates, and activated tosylates with alkyl Grignard reagents bearing β-H (Scheme 5.142). With the mixture of THF and NMP as the solvent, $Fe(acac)_3$ alone showed high activity in this reaction in the absence of any additional ligand. However, the use of [Fe(salen)Cl] is preferred to achieve high efficiency when secondary alkyl Grignard reagents were used. The reactions performed smoothly and many functional groups could be tolerated. Based on the mechanistic study, an inorganic Grignard reagent, $[Fe(MgX)_2]$, was proposed as the active catalyst (Scheme 5.142). This strategy was successfully applied by Fürstner and Leitner themselves in the total synthesis of odoriferous alkaloid (*R*)-(+)-muscopyridine (Scheme 5.143) [220] and adopted by Nagano and Hayashi [220d] in a relay cross-coupling.

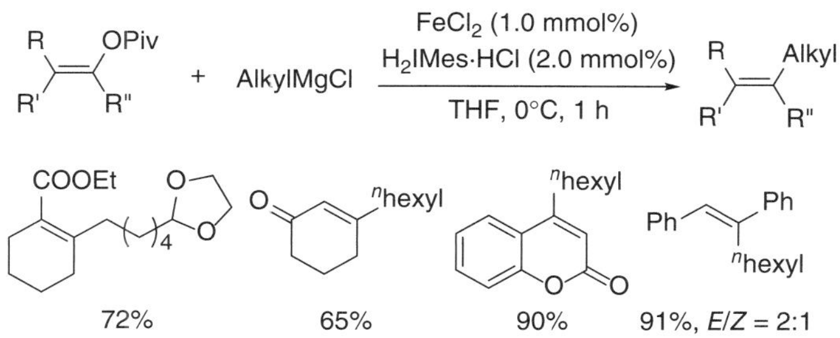

Scheme 5.141. Fe-catalyzed C–O bond activation of alkenyl/aryl carboxyzlates.

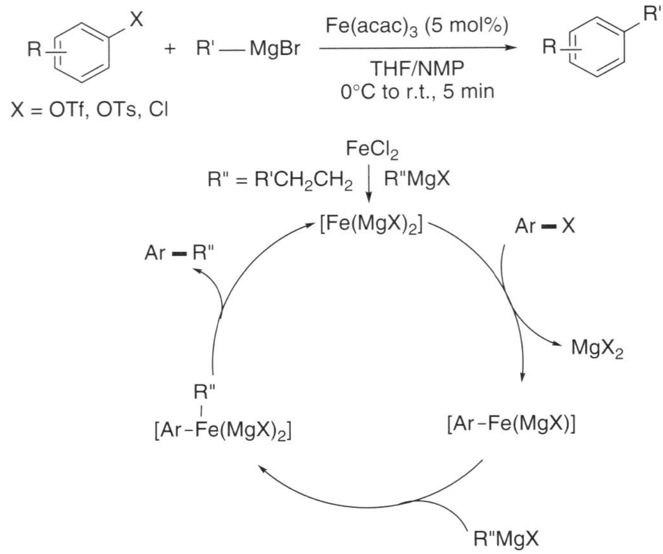

Scheme 5.142. Fe-catalyzed cross-coupling reactions of alkyl Grignard reagents with aryl tosylates, triflates, and chlorides.

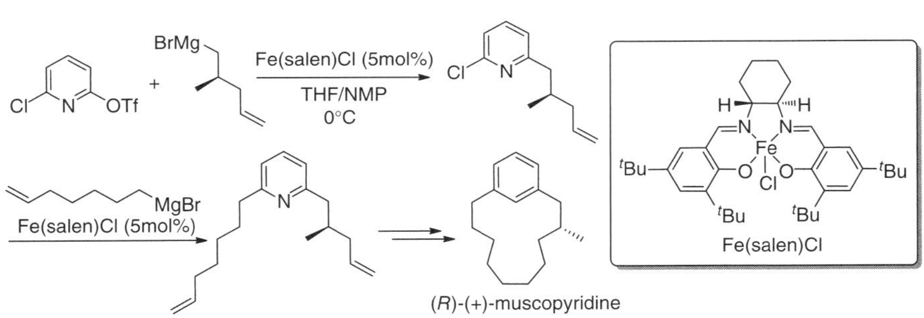

Scheme 5.143. Synthesis of (*R*)-(+)-muscopyridine.

With the same catalytic system of $FeCl_2$ and SIMes, Shi also realized the first example of coupling of 2-naphthyl pivalate and carbamate with alkyl Grignard reagent. The reaction was later investigated by Garg and co-workers [221] in 2012 and the substrate scope was expanded to a range of aryl carbamates and sulfamates, both of which showed good D*o*M ability to generate diverse arenes. Independently, Cook and co-workers also realized a similar transformation of aryl tosylates and sulfamates with primary and secondary alkyl Grignards in the presence of catalytic $FeF_3{\cdot}3H_2O$. For the secondary alkyl Grignards, the iron counterions had a significant influence on branched-to-linear ratios (Scheme 5.144) [222].

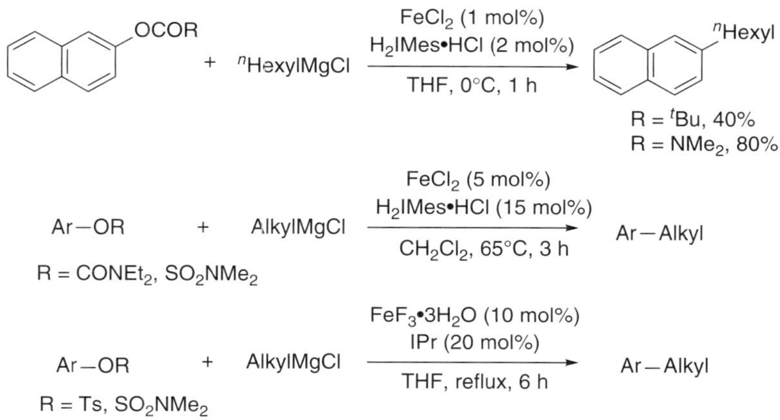

Scheme 5.144. Fe-catalyzed coupling of aryl pivalates, carbamates, and sulfamates with alkyl Grignard reagents.

Heteroaromatic sulfonates, sulfamates, and phosphates were also applied as electrophiles in an iron-catalyzed cross-coupling reaction via C–O bond activation by Skrydstrup and co-workers [223] in 2009. $FeCl_3$ was chosen as the catalyst, and no ligand was added in the co-solvent of THF and NMP. The reaction performed in low temperature with high efficiency and good functional group tolerance (Scheme 5.145).

Besides *sp*^{2} C–O activation, *sp*^{3} C–O activation of alkyl sulfate and even activated alcohols has also been realized with an iron catalyst. For example, in 2004 Hayashi and co-workers reported one example of $Fe(acac)_3$-catalyzed cross-coupling of *n*-octyltosylate with *p*-tolyl magnesium bromide in a moderate yield [224]. This system is more suitable for alkyl halides possessing β-H (Scheme 5.146). Later, Nakamura and co-workers [225] realized an iron-catalyzed cross-coupling of primary and secondary alkyl sulfonates with heteroaryl and electron-poor aryl zinc reagents in the presence of TMEDA and magnesium salts, in which more reactive alkyl iodides were generated to undergo cross-coupling with high selectivity (Scheme 5.147).

Recently, Shi and co-workers [153] demonstrated an iron-catalyzed direct magnesiation of benzyl alcohols (or their magnesium salts) with *n*-hexylMgCl via C–O activation (Scheme 5.148). Besides $FeBr_2$, other catalysts, such as $CoCl_2$ and $Ni(acac)_2$,

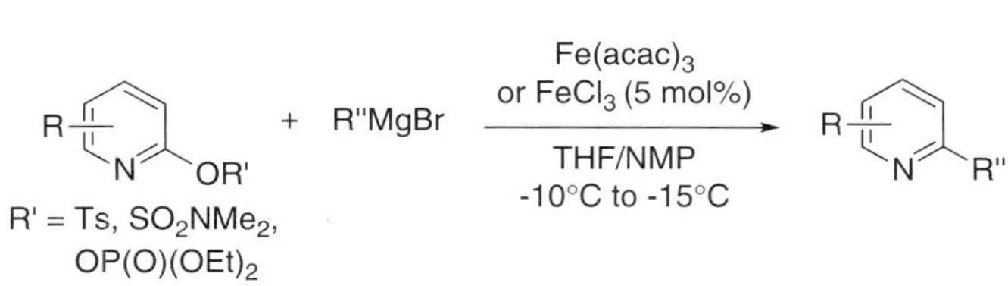

Scheme 5.145. Heteroaromatic sulfonates and phosphates as electrophiles in Fe-catalyzed cross-couplings.

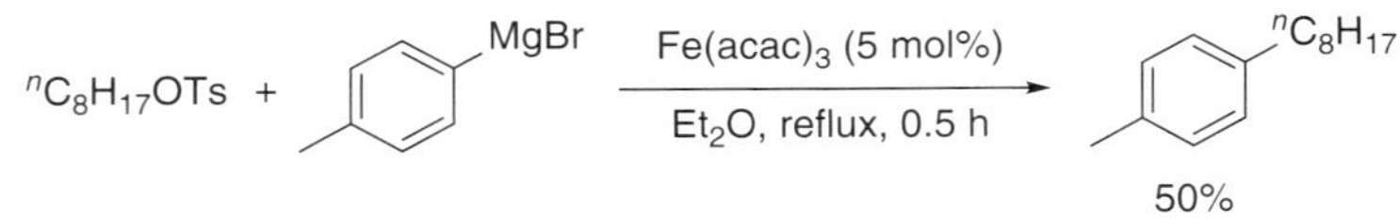

Scheme 5.146. Fe-catalyzed cross-coupling reaction of aryl Grignard reagents with alkyl tosylates and halides.

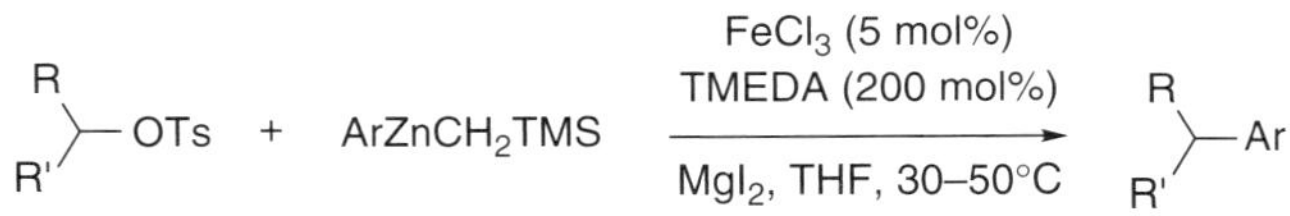

Scheme 5.147. Fe-catalyzed cross-coupling reaction of aryl zinc with alkyl tosylates.

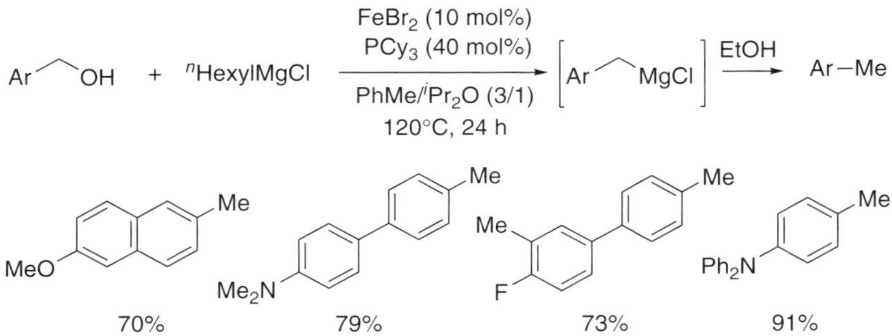

Scheme 5.148. Fe-catalyzed C–O bond activation/reduction of benzyl alcohols.

also showed good reactivity in this transformation. This is also the first example of transition metal-catalyzed magnesiation of a C–O bond in alcohols. The reversible transmetalation of benzyl–iron complex with *n*-hexylMgCl to generate a more stable benzyl Grignard reagent was proposed to be involved in this reaction.

5.4.2 Co-Catalyzed C–O Bond Activation

Similar to iron catalysts, cobalt catalysts are also inexpensive and play an important role in organic synthesis. There are also many examples of Co-catalyzed cross-coupling reactions [226], including those via C–O bond activation.

In as early as the 1970s, Co catalysts have been applied in the reaction of THF with silanes and CO. For example, in 1970 Chalk [227] found that the cationic polymerization of THF could take place in the presence of catalytic $Co_2(CO)_8$ and silanes. In 1977, Murai and co-workers [228] realized a $Co_2(CO)_8$-catalyzed ring-opening of THF to generate aldehydes or vinyl silyl ethers in the presence of excess $HSiEt_2Me$ and CO. In this reaction, $Et_2MeSiCo(CO)_4$ was proposed to be the active catalyst to generate alkyl–Co complex $RCo(CO)_4$ from THF. Further insertion of CO and

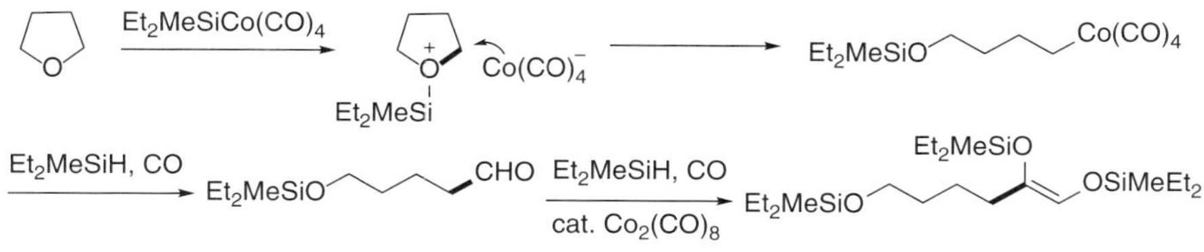

Scheme 5.149. Co-catalyzed C–O bond cleavage of THF.

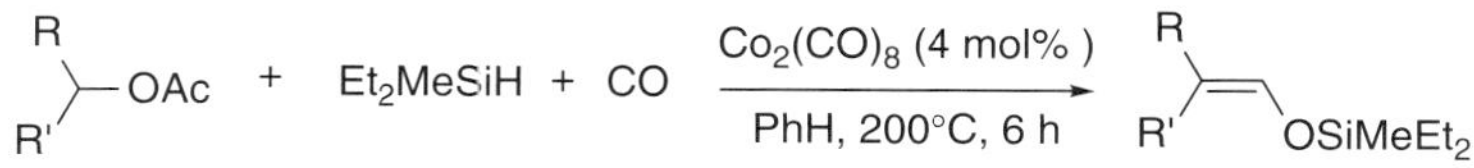

Scheme 5.150. Co-catalyzed C–O bond cleavage of secondary alkyl acetates.

reduction by $HSiEt_2Me$ would lead to aldehydes, which could further react with $Et_2MeSiCo(CO)_4$ to generate vinyl silyl ethers in the presence of excess $HSiEt_2Me$ (Scheme 5.149). Moreover, lactones and secondary alkyl acetates also could undergo such a transformation (Scheme 5.150) [229].

In 1991, Fuchikami and co-workers [230] reported a Co-catalyzed carbonylation reaction of unactivated alkyl sulfonates with carbon monoxide and alcohol in the presence of a catalytic amount of sodium iodide in tetramethylurea (TMU) solution. Low yield (35%) of desired ester could also be obtained in the absence of NaI. This is the first example of transition-metal-catalyzed carbonylation of alkyl sulfonates to generate esters as the products (Scheme 5.151) [230].

In 2006, Oshima and co-workers [231] realized a cobalt-catalyzed cross-coupling of cinnamyl methyl ethers with aryl magnesium bromide via allylic C–OMe bond activation. In this work the regioselectivity was controlled by the combination of the substrate, ligand, and Grignard reagent (Scheme 5.152). In the reaction of cinnamyl methyl ether, the formation of the linear coupling products was predominated. However, addition of a diphosphine ligand can change the regioselectivity to give the corresponding branched products in the cobalt-catalyzed allylation of allylic ethers.

Besides *sp*^{3} C–O bond activation, Co catalysis has also been applied in *sp*^{2} C–O bond activation. For example, in 2003, Gosmini and co-workers [232] described a cobalt-catalyzed electrochemical cross-coupling of vinyl acetates and aryl halides [232]. Later they developed this system to accommodate a purely chemical reaction (Scheme 5.153) [233]. In this work the author proposed that the catalytic cycle

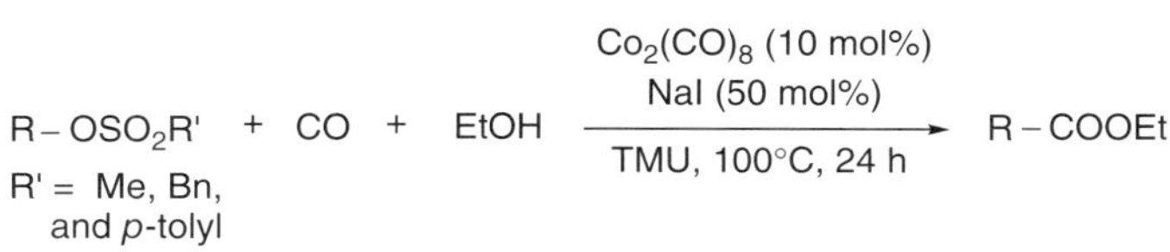

Scheme 5.151. Co-catalyzed carbonylation of unactivated alkyl sulfonates.

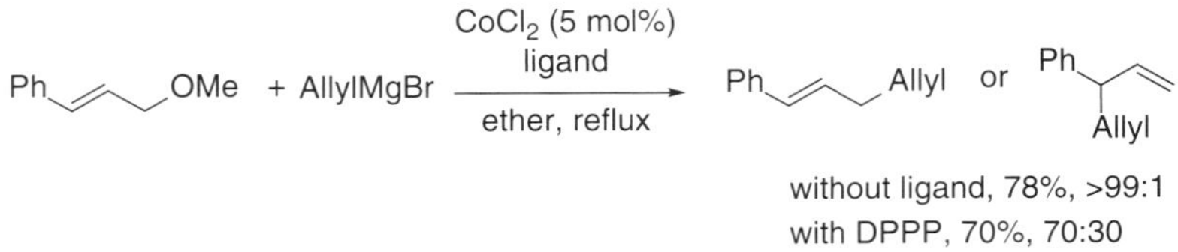

Scheme 5.152. Co-catalyzed cross-coupling of cinnamyl methyl ethers with aryl magnesium bromide.

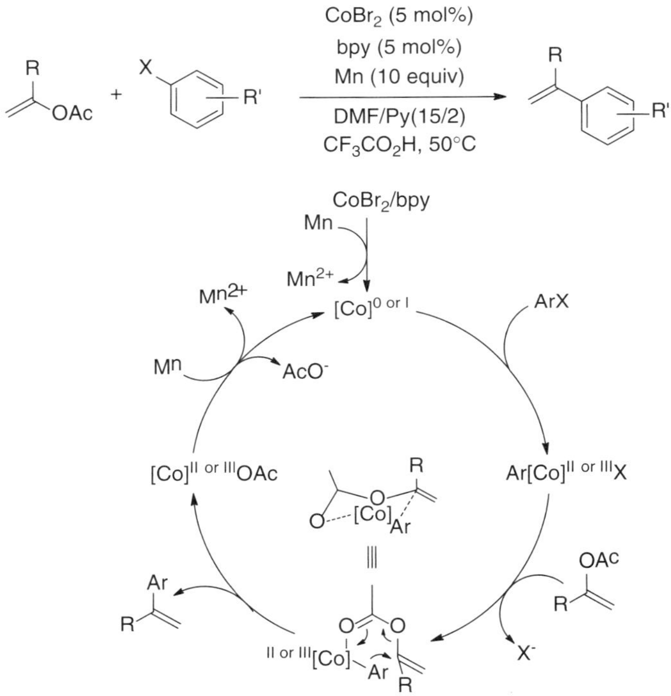

Scheme 5.153. Proposed mechanism for the vinylation of aryl halides.

started with a low valent cobalt intermediate, and the C–C bond formation proceeded through a six-membered-ring transition state (Scheme 5.153).

In 2006, Knochel and co-workers [234] developed a cobalt-catalyzed arylation of aryl fluorides and tosylates bearing an ortho carbonyl group with arylcopper reagents. $Co(acac)_2$ was used as the catalyst, and the promoters 4-fluorostyrene and Bu_4NI played important role in this reaction (Scheme 5.154).

In 2010, Knochel further developed a cobalt-catalyzed sulfonate/copper exchange reaction. Based on this method, organocopper reagents could be synthesized from aryl sulfonates via C–O bond cleavage. This reaction tolerated various kinds of sensitive

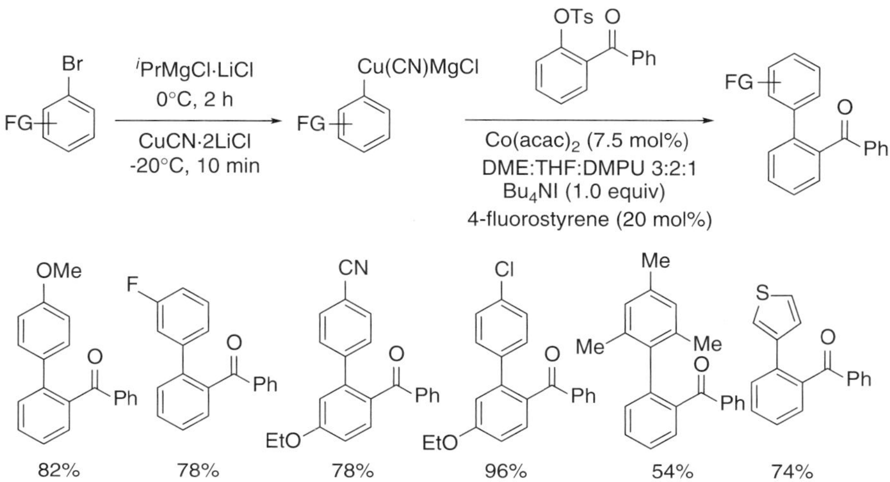

Scheme 5.154. Co(acac)$_2$-catalyzed arylation of aryl halides and aryl tosylates with arylcopper reagents.

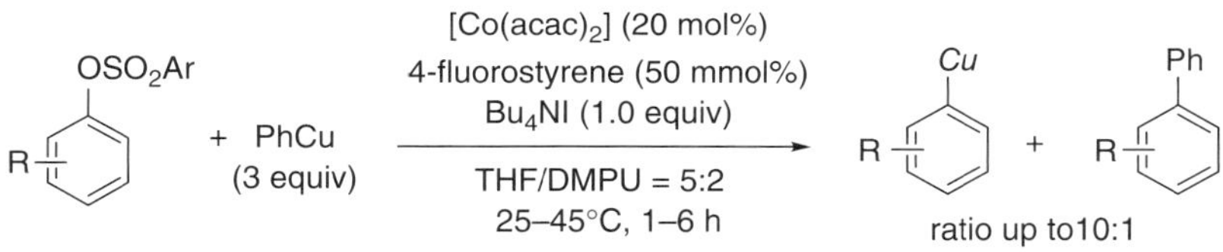

Scheme 5.155. Co-catalyzed sulfonate/copper exchange.

functional groups, including aldehyde, nitrile, and ester. It was proposed that a Co(IV) intermediate was involved in the transformation (Scheme 5.155) [235].

5.4.3 Cu-Catalyzed C–O Bond Activation

Although copper-catalyzed carbon–halogen bond activation has been widely applied in cross-coupling reactions, copper-catalyzed C–O bond activation, however, is very limited. Besides reports on reactions of organocopper reagent as nucleophiles to break the C–O bond in *O*-containing cyclic compounds [236], in 1983 Ishihara et al. [237] reported an organocuprate induced C–O cleavage of alkenyl phosphates to generate alkylated gem-difluoroolefins (Scheme 5.156).

In 2000, Erdik and Eroğlu [238] reported a copper-catalyzed cross-coupling reaction of aryl arenesulfonates with aryl Grignard reagents. In this reaction, both C–C and C–S formation could occur depending on different structure of the Grignard reagent (Scheme 5.157).

In 2010, a Cu-catalyzed alkyl C–O bond activation reaction was described by Ito et al. [239]. This work provides a useful method to construct borylated cyclobutanes

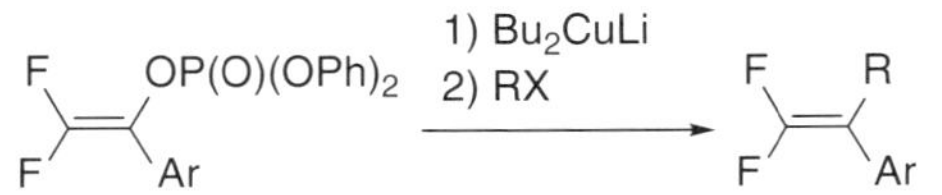

Scheme 5.156. Organocuprate-induced C–O activation of alkenyl phosphates.

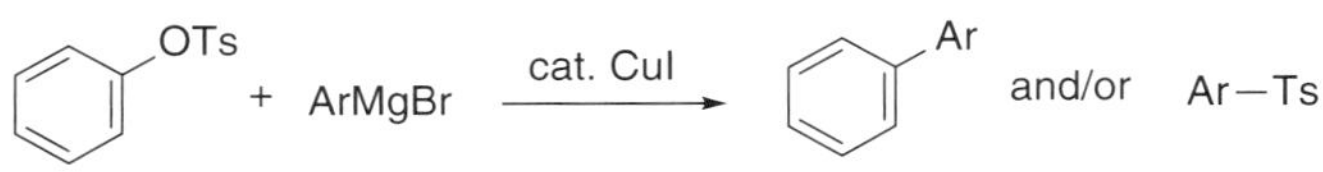

Scheme 5.157. Cu-catalyzed reaction of aryl Grignard reagents with aryl arenesulfonates.

and cyclopentanes, which are useful building blocks. In the presence of KO^{t}Bu and diborane, an active copper borate was generated and inserted into the double bond. Due to the significant lower yield in formation of a six-member ring ($n = 3$), the ring-closing step was proposed to proceed through Cu(III) metallacycle. The reaction is highly stereospecific with (*Z*)- and (*E*)-homoallylic mesylates being converted to the *trans* and *cis* products, respectively (Scheme 5.158).

Recently, there have been two elegant works of Cu-catalyzed C–OTs activation by Hu and Liu independently [240]. Both reports involve copper-catalyzed cross-coupling reaction of alkyl halides and tosylates with alkyl Grignard reagents, especially the challenging secondary and tertiary alkyl Grignard reagents. Although their system is slightly different, both reactions performed quite efficiently under mild conditions (Scheme 5.159). In Hu's work, the substrate scope was mainly focused on alkyl halides and only three examples of homobenzylic tosylates were presented. In Liu's report, a variety of challenging secondary alkyl halides and tosylates reacted with secondary and tertiary alkyl Grignard reagents in good yields. Many functional groups, including primary C–Cl and C–OBn bonds, aryl C–Br bond, alkene, and CF_3 groups were well tolerated. The use of TMEDA and LiOMe as additives is crucial for the success. Notably, it was demonstrated that such

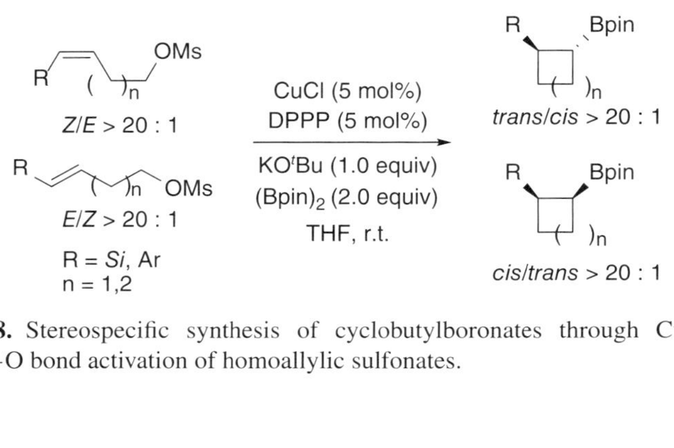

Scheme 5.158. Stereospecific synthesis of cyclobutylboronates through Cu(I)-catalyzed borylation/C–O bond activation of homoallylic sulfonates.

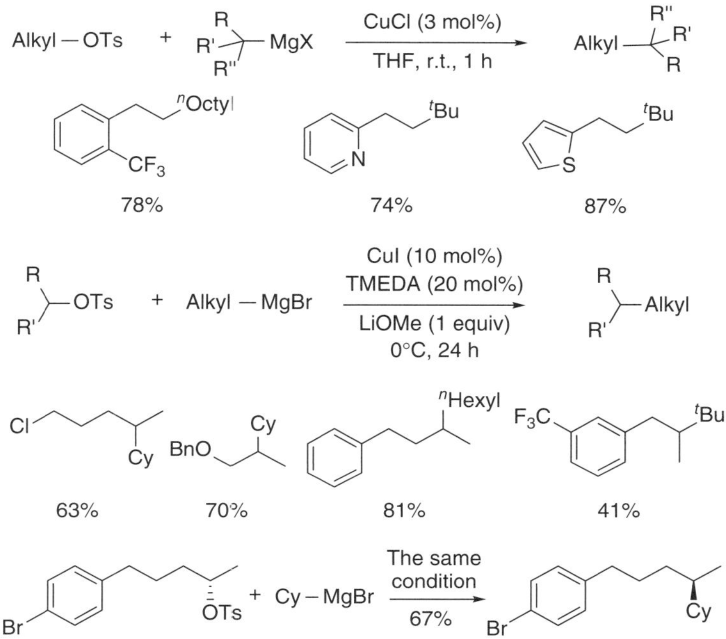

Scheme 5.159. Cu-catalyzed cross-coupling reaction of alkyl tosylates with alkyl Grignard reagents.

reaction underwent an S_N2 mechanism with inversion of configuration by a variety of stereo-controlled formation of C–C bonds from chiral secondary alcohols.

5.4.4 Rh-Catalyzed C–O Bond Activation

Besides one stoichiometric example of C–OMe bond cleavage with a rhodium–pincer complex by Milstein in 1997, recently there has been another example of Rh-catalyzed stoichiometric C–O bond cleavages by Ozerov and co-workers [241]. It was found that the ortho C–H bond of aryl carboxylates could undergo oxidative addition to rhodium–PNP complex and further migrate to give the product of C–O bond cleavage under thermolysis (Scheme 5.160) [241]. On the other hand, PhO_2CCF_3 chiefly underwent acyl–oxygen oxidative addition.

Similar to Co catalysis, Rh-catalyzed carbonylation of THF to generate lactone was also realized by Jenner et al. in 1984 [242]. With a similar mechanism, in 2008 Brookhart also reported an Ir-catalyzed reduction of ether and other dialkyl ether to corresponding alkanes and silyl ethers in the presence of Et_3SiH [243].

There are also a few examples of rhodium-catalyzed C–O bond activation of aryl and alkenyl tosylates. For example, in 2006, Wu et al. [244] realized the first $Rh(PPh_3)Cl$-catalyzed cross-coupling reaction of activated alkenyl tosylates with arylboronic acids or potassium aryltrifluoroborates under mild conditions in the

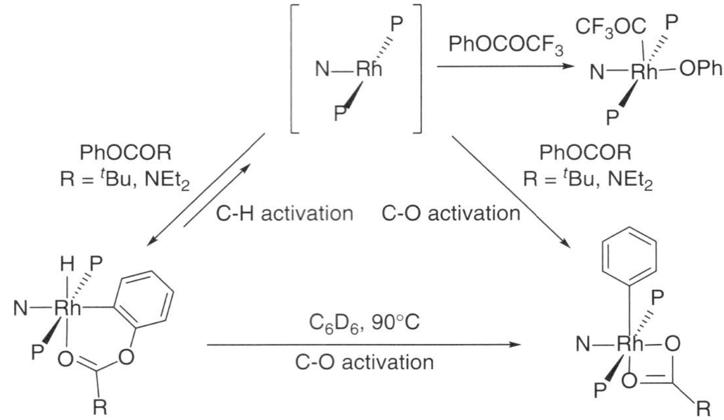

Scheme 5.160. C–H and C–O oxidative addition to a Rh–Pincer complex.

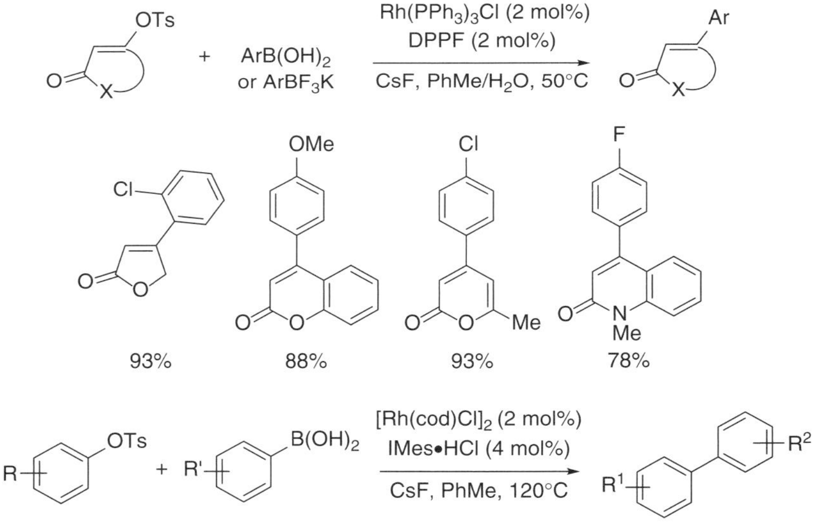

Scheme 5.161. Rh-catalyzed cross-coupling reactions of alkenyl and aryl tosylates with arylboron reagents.

presence of DPPF as the ligand. Although this system failed for the reaction of aryl tosylates, Zhang and Wu [245] further reported the cross-coupling of aryl tosylates with arylboronic acids with a rhodium–NHC system (Scheme 5.161).

In 2009, Kuwano and Kwong independently reported a rhodium-catalyzed of cross coupling of alkenyl acetateswith aryl boronic acids via C–O bond activation [246]. Both of them applied $[Rh(cod)Cl]_2$ and DPPB as the catalytic system. In Kuwano's reaction, a variety of aryl and vinyl boronates were tested with

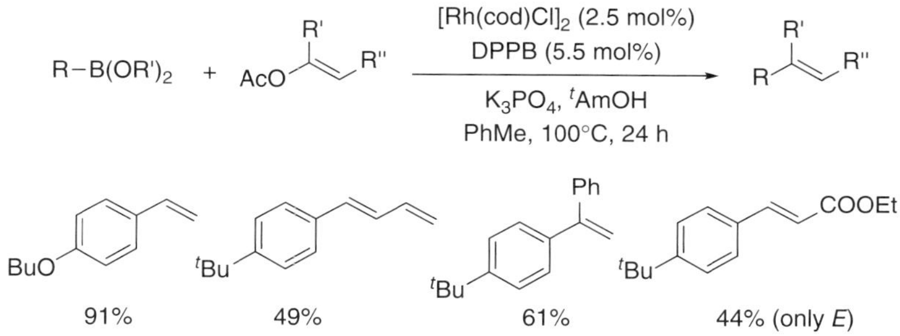

Scheme 5.162. Rh-catalyzed cross-coupling of organoboron compounds with vinyl acetates.

vinyl acetate to give corresponding products in moderate to good yields. However, the substrate scope of alkenyl acetates was quite limited (Scheme 5.162). Applying the catalyst system above to 1-phenylethenyl acetate would generate *ipso*-substitution products. In 2010, Kuwano and co-workers [247] found that *cine*-substitution products would be obtained if the catalyst system was changed to $[Rh(cod)(OAc)]_2$ as the catalyst and cod as the ligand (Scheme 5.163). The authors also found that the electron-withdrawing group on the arylboronates would favor the formation of *cine* substitution. A (hydrido)rhodium intermediate was proposed to be the key intermediate of the β-oxygen elimination mechanism.

In 2011, Rao and Li [248] reported a rhodium-catalyzed rearrangement of 2-aryloxybenzaldebydes involving C–O and C–H bond activation. $[RhCl(CO)_2]_2$ was the best catalyst and TBP was applied as the oxidant in this system (Scheme 5.164). A variety of functional groups could be tolerated in this reaction.

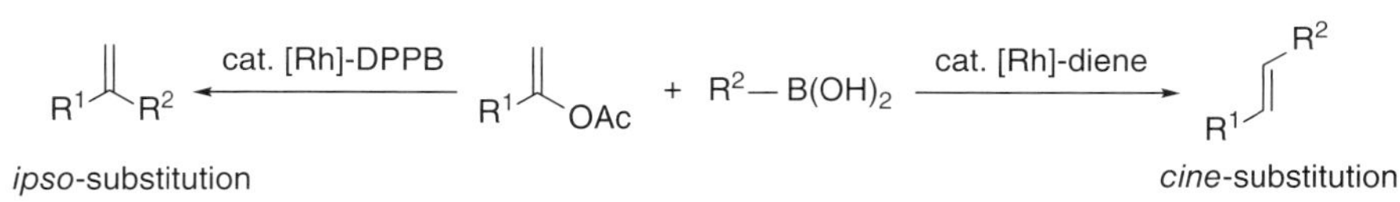

Scheme 5.163. *Cine*-substitution selectivity of 1-arylethenyl acetates-catalyzed by a diene–Rh system.

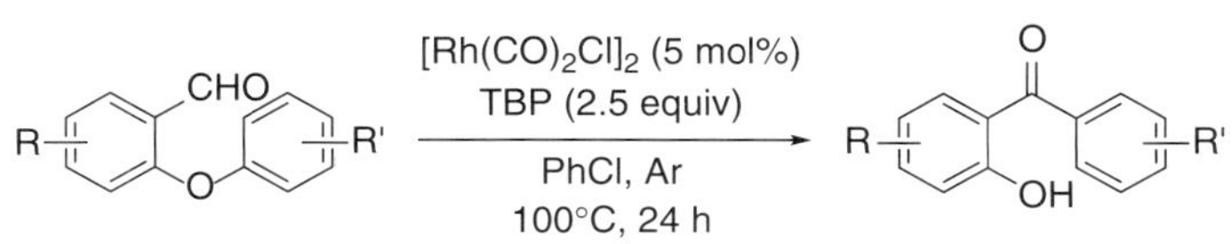

Scheme 5.164. Rh-catalyzed rearrangement of 2-aryloxybenzaldehydes.

5.4.5 Ru-Catalyzed C–O Bond Activation

In as early as 1987, Komiya et al. [249] demonstrated that vinyl acetate could undergo oxidative addition to Ru(COD)(COE) in the presence of PEt_3 to give mer-vinyl (acetato)tris(triethylphosphine)ruthnium(II) at 50°C in hexane.

For Ru-catalyzed acyl C–O bond cleavage, Murai and co-workers [250] have published their results as early as in 2001. In the presence of catalytic $Ru_3(CO)_{12}$ and excess $HCOONH_4$, the acyl C–O bond of 2-pyridylmethyl carboxylate was cleaved and, following decarboxylation, generated the corresponding reductive product (Scheme 5.165) [250].

Besides reduction in the presence of reductants, Ru-catalyzed reductive C–O bond cleavage could also happen in the redox-neuter condition. In 2010 Bergman and co-workers [251] reported a Ru-catalyzed redox neutral C–O bond cleavage of 2-aryloxy-1-arylethanols, which underwent a tandem dehydrogenation and reductive ether cleavage. It is important to note that a catalytic depolymerization of a lignin-related polymer gave the corresponding monomer in quantitative yield without any additives, which would play an important model in lignin degradation. Recently, Klankermayer and co-workers [252] also reported a [Ru(cod)(methallyl)$_2$]-catalyzed hydrogen-free reductive cleavage of C–O bond in 2-aryloxy-1-arylethanols in the presence of tripodal phosphine ligands. (Scheme 5.166).

One of the most successful systems in catalytic C–O bond activation is directing group-assisted ruthenium-catalyzed C–O bond cleavage of anisole derivatives (Scheme 5.167) [10]. In 2004 Kakiuchi and co-workers [253] found that $RuH_2(CO)(PPh_3)_3$ is a highly efficient catalyst for the carbonyl-directed C–O bond activation and cross-coupling reaction with various organoboronates, including aryl, alkenyl, and alkylboronates. Further investigation of the reaction intermediate

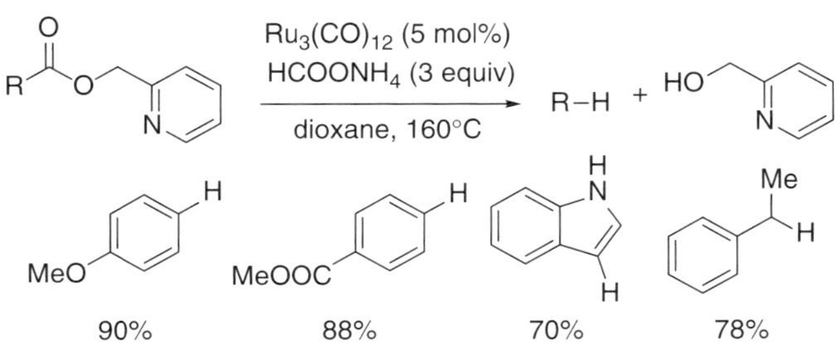

Scheme 5.165. Ru-catalyzed reductive decarboxylation of esters.

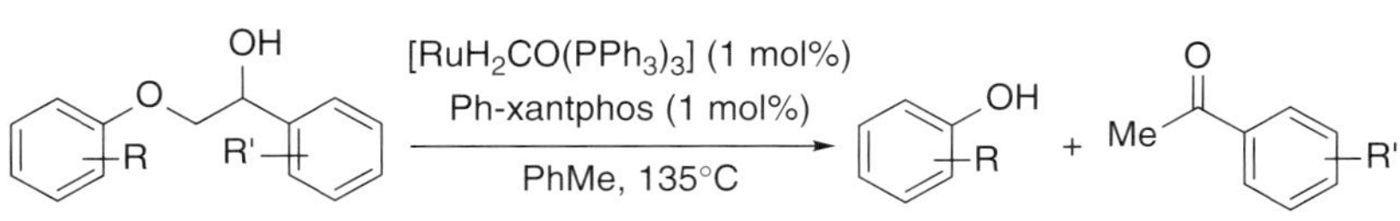

Scheme 5.166. Ru-catalyzed C–O bond cleavage of lignin model compounds.

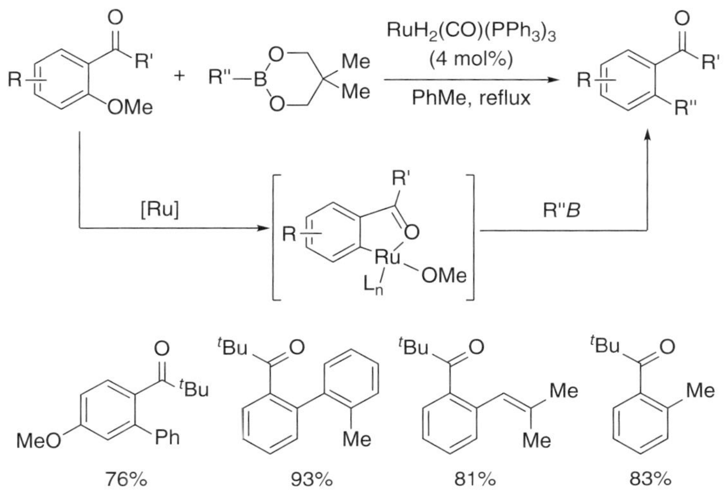

Scheme 5.167. Ru-catalyzed C–O bond activation by chelation assistance.

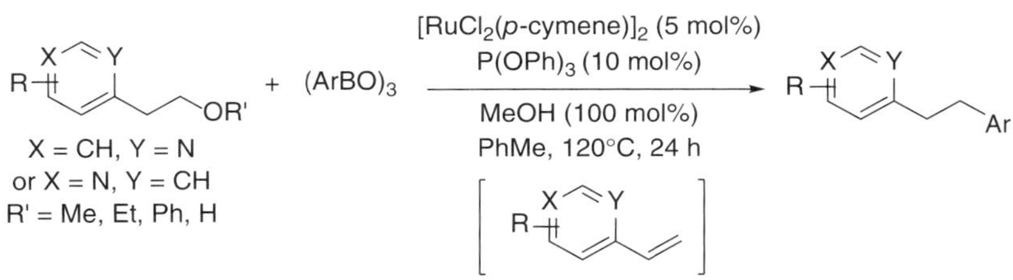

Scheme 5.168. Ru-catalyzed conversion of sp_3 C–O bonds to C–C bonds.

discovered the relative reactivity between aryl C–O bond and C–H bond [11]. A similar strategy was also applied on functionalization of acenes [253].

In 2011, Kochi and co-workers [254] reported a Ru-catalyzed formal coupling reaction of alkyl ethers bearing 2- or 4-pyridyl groups with aryl boroxines. In this reaction, C–O bond of 2-(2-methoxyethyl)pyridine was cleaved to release an olefin intermediate, which was then added with tirarylboroxines to generate the C–C bond (Scheme 5.168). Notably, the N-atom is indispensable for the transformation. Neither phenylethyl methyl ether nor 3-(2-methoxyethyl)-6-methylpyridine showed reactivity.

Besides the coupling with organometallic reagents, direct reactions with arenes were also realized through Ru-catalyzed C–O and C–H activation. For example, in 2006 Ackermann et al. [255] reported the first Ru-catalyzed cross-coupling of aryl tosylates and arenes in the presence of *N*-directing groups. A sterically hindered diamineophosphine oxide was applied as the ligand. A wide variety of functional groups could be well tolerated. Besides oxazoline, pyridine and pyrazole also showed

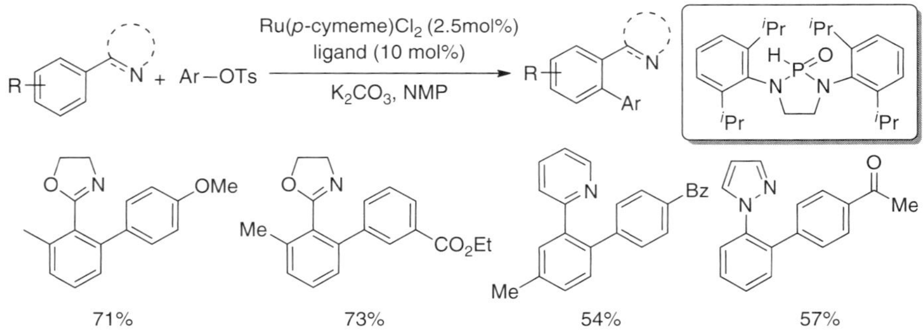

Scheme 5.169. Rh-catalyzed arylation of arenes with tosylates.

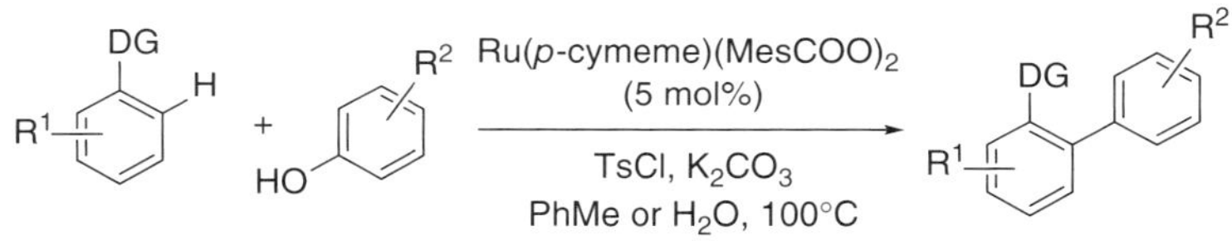

Scheme 5.170. Ru-catalyzed arylations of arenes with phenols via *in situ* generation of tosylates.

good directing ability in this reaction (Scheme 5.169). Using this strategy, Ackermann and Mulzer [256] further developed a dehydrative direct arylation of arenes with phenols assisted by *p*-TsCl. In this work, aryl tosylates were generated *in situ* (Scheme 5.170).

In 2007, Kakiuchi and co-workers [257] reported N-directing-group-assisted C–H bond activation and cross-coupling reaction with alkenyl acetates to generate functionalized styrene as products in the presence of catalytic ruthenium complex. In this reaction, *E*-isomers were isolated as the major products, regardless of the ratio of *E*/*Z* and different substitutes in the alkenyl acetates (Scheme 5.171).

Moreover, the significant progress also occurred in the area of direct activation of C–O bond in unactivated alcohols via Ru-catalysis. In 2011, Yi and co-workers [258] reported Ru-catalyzed selective C–H alkylation of alkenes with normal alcohols with good functional group tolerance (Scheme 5.172). A cationic ruthenium hydride complex $[(C_6H_6)(PCy_3)(CO)RuH]^+BF_4^-$ was applied as the catalyst in this system. It was

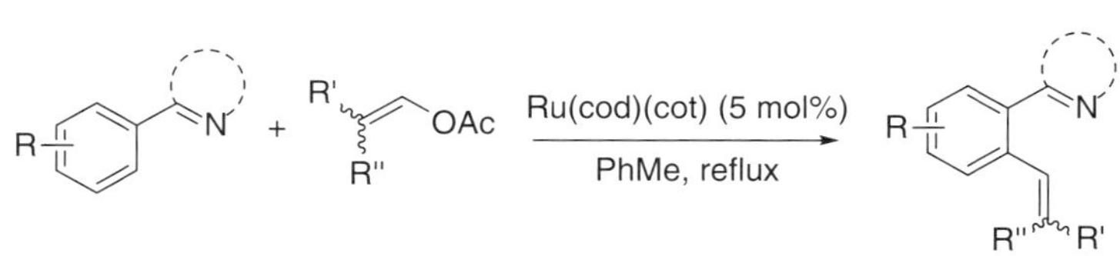

Scheme 5.171. Ru-catalyzed alkenylation of aromatic C–H bond with alkenyl acetates.

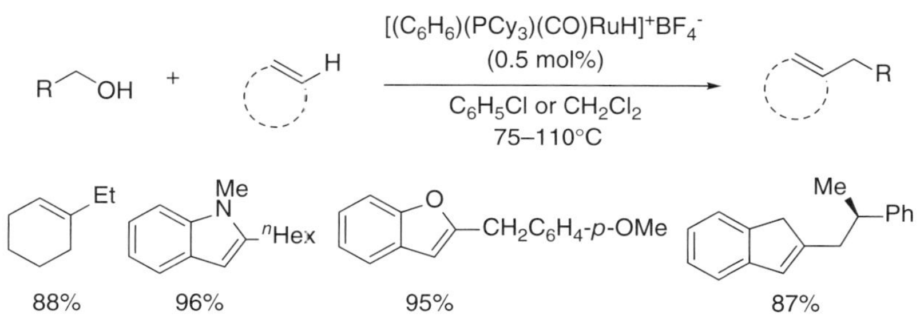

Scheme 5.172. Ru-catalyzed selective catalytic C–H alkylation of alkenes with alcohols.

proposed that this cationic ruthenium complex initially activated the alkenyl C–H bond to generate an alkenyl-Ru intermediate, which further underwent C–O bond cleavage and rate-determining reductive elimination to give the cross-coupling product. This reaction is highly atom- and step-economic and could be directly applied in late-stage functionalization of complex molecules.

Based on this cationic ruthenium catalyst, Yi and co-workers [259] further developed an alkylation and alkenylation of phenols with alcohols recently. In this report, the *ortho* C–H bond of phenols is activated. Primary and secondary alcohols, including cyclic secondary alcohols, were applied in this reaction efficiently. The ratio of alkylation and alkenylation is well controlled by a different amount of cyclopentene. Using 1,2-diols instead of monoalcohols in alkenylation conditions would generate benzofurans as products (Scheme 5.173).

Moreover, Ru-catalyzed C–O activation also could be used to generate C–X bonds, including C–N and carbon–halide bonds. In 2011 Dong and co-workers [260] realized a novel migratory rearrangement of alkyl groups in 2-alkoxyl heterocycles, such as pyridines, quinolines, thiazoles, and benzimidazoles, from oxygen to nitrogen atom to generate the corresponding *N*-alkyl heterocycles (Scheme 5.174). In addition to the activated benzyl alcohols, alkyl groups of general alcohols could also migrate efficiently. The mechanistic study indicated that the reaction underwent

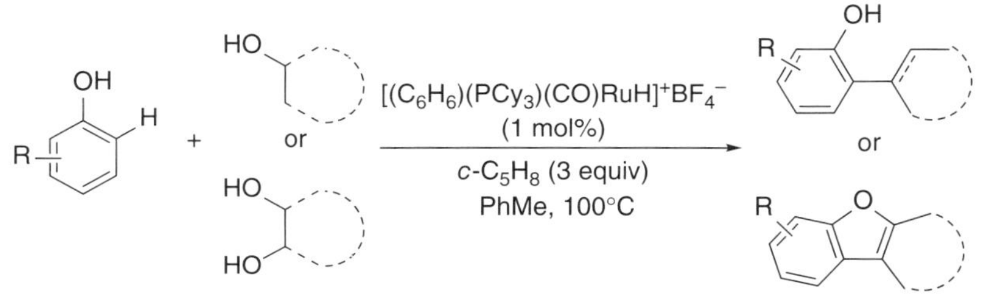

Scheme 5.173. Ru-catalyzed dehydrative C–H alkylation and alkenylation of phenols with alcohols.

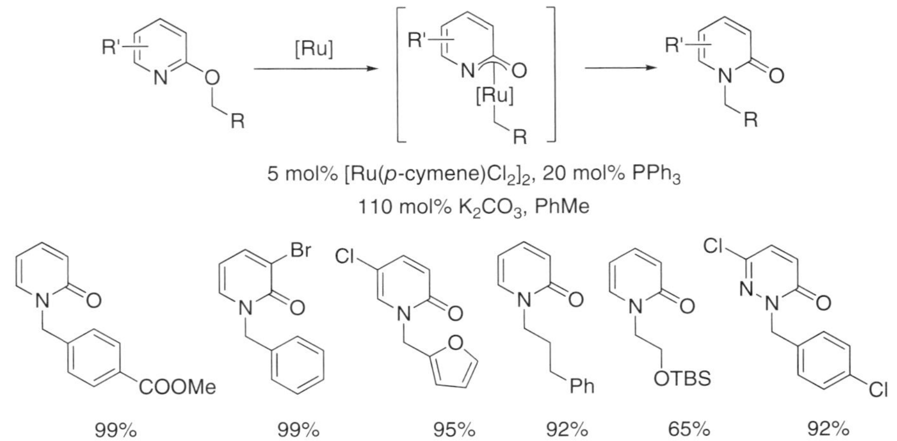

Scheme 5.174. Ru-catalyzed *O*- to *N*-alkyl migration in 2-alkoxypyridine and 3-alkoxy pyridazines.

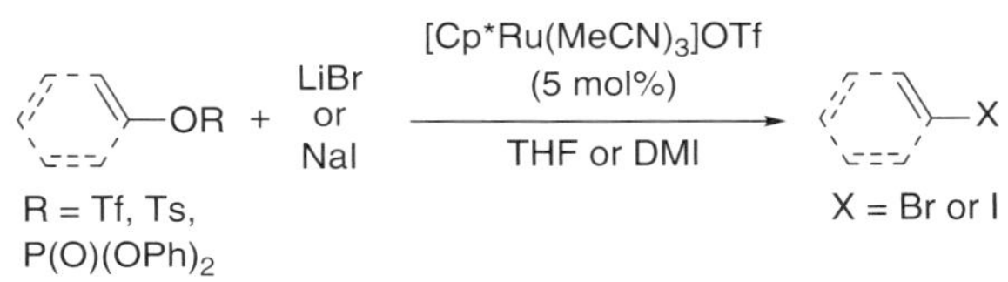

Scheme 5.175. Ru-catalyzed transformation of aryl and alkenyl triflates to halides.

oxidative addition of the C–H at the α position of oxygen, following steps of α-O elimination and hydride transfer process [260].

In 2012 Shirakawa and co-workers [261] reported a Ru-catalyzed efficient transformation of aryl and alkenyl sulfonates and phosphates to corresponding bromides and iodides under mild conditions in the presence of LiBr or NaI (Scheme 5.175). The [Cp*Ru(MeCN)$_3$]OTf was demonstrated to be highly active catalyst, which arose from the strong electron-donating ability of the Cp* ligand in support of the oxidative addition of C–O bond to Ru(II).

5.5 SUMMARY AND OUTLOOK

Although we have tried our best to cover most of the important progress in the area of inert C–O bond activation, it is extremely difficult to cover all of the development in this rapidly developing and hot field. There is no doubt that many contributions in this area were omitted in this chapter; however, there are many excellent reviews covering the details in different aspects, which have been cited in the chapter.

To our great delight, the development in C–O activation has made chemists easy to think of using oxygen-containing compounds instead of organohalides as the starting materials in organic synthesis. Although it is still far away from replacing organohalides comprehensively, it is obvious that more and more reactions via C–O activation will be applied in organic synthesis, including lignin degradation and late-stage drug modification.

To make the C–O activation more powerful and useful, there are still many limitations in this field. For example, the catalyst loading as high as 10 mol%, ligand loading up to 40%, harsh reaction conditions with high temperature and strong Lewis acids, and long reaction time are always required in the activation of inert C–O bonds. Moreover, there are still many kinds of reactions that could not employ oxygen-containing compounds but organohalides. Furthermore, due to the low reactivity of C–O bonds in phenols, enols, and alcohols, protection and pre-activation are the main strategy nowadays. Developing an efficient catalytic system, especially the highly reactive transition-metal catalyst and powerful ligands, will be the key to success. More attention also should be paid to the detailed mechanistic study. It is expected that, with the great efforts from chemists with different background, the chemistry of C–O activation will bloom and show their great power in the future synthetic world.

REFERENCES

1. (a) Diederich, F., Stang, P. J., eds. (1998). *Metal-Catalyzed Cross-Coupling Reactions, Wiley-VCH*, Weinheim, Germany. (b) Miyaura, N., ed. (2002). *Cross-Coupling Reactions. A Practical Guide*, Springer-Verlag, Berlin. (c) de Meijere, A., Diederich, F., eds. (2004) *Metal-Catalyzed Cross-Coupling Reactions*, Wiley-VCH, Weinheim, Germany. (d) Nolan, S. P., Navarro, O. (2006). *C–C Bond Formation by Cross-Coupling* in *Comprehensive Organometallic Chemistry III*, R. Crabtree, M. Mingos, eds., Elsevier B.V, Amsterdam. (e) Hegedus, L., Soderberg, B. (2009). *Transition Metals in the Synthesis of Complex Organic Molecules*, 3rd ed., University Science Books, Herndon, VA. (f) Hartwig, J. (2010). *Organotransition Metal Chemistry*, University Science Books, Sausalito. (g) C. C. C. Johansson Seechurn, Kitching, M. O., Colacot, T. J., Snieckus, V. (2012). *Angew. Chem. Int. Ed.* **51**, 5062.
2. (a) Lin, Y., Yamamoto, A. (1999). In *Topics in Organometallic Chemistry: Activation of Unreactive Bonds and Organic Synthesis*, Murai, S., ed., Springer, Berlin. (b) Gooßen, L. J., Gooßen, K., Stanciu, C. (2009). *Angew. Chem. Int. Ed.* **48**, 3569. (c) Knappke, C. E. I., Jacobi von Wangelin, A. J., (2010). *Angew. Chem. Int. Ed.* **49**, 3568. (d) Yu, D.-G., Li, B.-J., Shi, Z.-J. (2010). *Acc. Chem. Res.* **43**, 1486. (e) Rosen, B. M., Quasdorf, K. W., Wilson, D. A., Zhang, N., Resmerita, A.-M., Garg, N. K., Percec, V. (2011). *Chem. Rev.* **111**, 1346. (f) Li, B.-J., Yu, D.-G., Sun, C.-L., Shi, Z.-J. (2011). *Chem. Eur. J.* **17**, 1728. (g) McGlacken, G. P., Clarke, S. L.(2011). *ChemCatChem* **3**, 1260. (h) Tobisu, M., Chatani, N. (2011). *Chem Cat Chem* **3**, 1410. (h) So, C. M., Kwong, F. Y. (2011). *Chem. Soc. Rev.* **40**, 4963. (i) Yamaguchi, J., Muto, K., Itami, K. (2013). *Eur. J. Org. Chem.*, 19. (j) Guan, B.-T. (2009). *Transition Metal Catalyzed Activation of C-O Bonds and Molecular Oxygen*. Ph.D. Thesis. Peking University, P.R. China. (k) Yu, D.-G. (2012). *Transition*

Metal-Catalyzed Inert C-O and C-C Bonds Activation. Ph.D. Thesis. Peking University, P.R. China.

3. For recent reviews, see: (a) Emer, E., Sinisi, R., Capdevila, M. G., Petruzziello, D., De Vincentiis, F., Cozzi, P.G. (2011). *Eur. J. Org. Chem.,* 647. (b) Debleds, O., Gayon, E., Vrancken, E., Campagne, J.-M. (2011). *Beilstein J. Org. Chem.* **7**, 866.
4. For recent reviews, see: (a) Dobereiner, G. E., Crabtree, R. H. (2010). *Chem. Rev.* **110**, 681. (b) Guillena, G., Ramón, D. J., Yus, M. (2010). *Chem. Rev.* **110**, 1611. (c) Watson, A., Williams, J. M. J. (2010). *Science* **329**, 635. (d) Gunanathan, C., Milstein, D. (2011). *Acc. Chem. Res.* **44**, 588.
5. (a) Tsuji, J. (1969). *Acc. Chem. Res.* **2**, 144. (b) Tsuji, J., Minami, I. (1987). *Acc. Chem. Res.* **20**, 140. (c) Trost, B. M. (1977). *Tetrahedron* **33**, 2615. (d) Trost, B. M., Van Vranken, D. L. (1996). *Chem. Rev.* **96**, 395. (e) Trost, B. M., Crawley, M. L. (2003). *Chem. Rev.* **103**, 2921. (f) Lu, Z., Ma, S. (2007). *Angew. Chem. Int. Ed.* **47**, 258. (g) Tsuji, J. (2000) *Transition Metal Reagents and Catalysts*, Wiley, New York.
6. (a) Scott, W. J., McMurry, J. E. (1988). *Acc. Chem. Res.* **21**, 47. (b) Baraznenok, I. L., Nenajdenko, V. G., Balenkova, E. S. (2000). *Tetrahedron*, **56**, 3077.
7. Zhang, X., Watson, E. J., Dullaghan, C. A., Gorun, S. M., Sweigart, D. A. (1999). *Angew. Chem. Int. Ed.* **38**, 2206.
8. (a) Jutand, A., Mosleh, A. (1995). *Organometallics* **14**, 1810. (b) Lugdwig, M., Stromberg, S., Svensson, M., Akermark, B. (1999). *Organometallics* **18**, 970. (c) Hii, K. K., Caridge, T. D. W., Brown, J. M. (1997). *Angew. Chem. Int. Ed.* **36**, 984.
9. Roy, A. H., Hartwig, J. F., (2003) *J. Am. Chem. Soc.* **125**, 8704.
10. Kakiuchi, F., Usui, M., Ueno, S., Chatani, N., Murai, S. (2004). *J. Am. Chem. Soc.* **126**, 2706.
11. Ueno, S., Mizushima, E., Chatani, N., Kakiuchi, F. (2006). *J. Am. Chem. Soc.* **128**, 16516.
12. Kelley, P., Lin, S., Edouard, G., Day, M. W., Agapie, T. (2012). *J. Am. Chem. Soc.* **134**, 5480.
13. (a) Ishizu, J., Yamamoto, T., Yamamoto, A. (1976). *Chem. Lett.,* 1091. (b) Yamamoto, T., Ishizu, J., Kohara, T., Komiya, S., Yamamoto, A. (1980) *J. Am. Chem. Soc.* **102**, 3758.
14. Ittel, S. D., Tolman, C. A., English, A. D., Jesson, J. P. (1979). *J. Am. Chem. Soc.* **101**, 1742.
15. van der Boom, M. E., Liou, S.-Y., Ben-David, Y., Vigalok, A., Milstein, D. (1997). *Angew. Chem. Int. Ed.* **36**, 625. (b) van der Boom, M. E., Liou, S., Ben-David, Y., Shimon, L. J. W., Milstein, D. (1998). *J. Am. Chem. Soc.* **120**, 6531.
16. For reviews, see: (a) Kumada, M. (1980). *Pure Appl. Chem.* **52**, 669. (b) Tamao, K. (2002). *J. Organomet. Chem.* **653**, 23. (c) Knochel, P., Dohle, W., Gommermann, N., Kneisel, F. F., Kopp, F., Korn, T., Sapountzis, I., Anh Vu, V. (2003). *Angew. Chem. Int. Ed.* **42**, 4302. For pioneering works, see: (d) Tamao, K., Sumitani, K., Kumada, M. (1972). *J. Am. Chem. Soc.* **94**, 4374. (e) Corriu, R. J. P., Masse, J. P. (1972). *J. Chem. Soc., Chem. Commun.,* 144a.
17. Hayashi, T., Konishi, M., Okamoto, Y., Kabeta, K., Kumada, M. (1986). *J. Org. Chem.* **51**, 3772.
18. Miller, J. A., (2002). *Tetrahedron Lett.* **43**, 7111.
19. Gauthier, D., Beckendorf, S., Gøgsig, T. M., Lindhardt, A. T., Skrydstrup, T. (2009). *J. Org. Chem.* **74**, 3536.

20. Schwier, T., Sromek, A. W., Yap, D. M. L., Chernyak, D., Gevorgyan, V. (2007). *J. Am. Chem. Soc.* **129**, 9868.

21. Roy, A. H., Hartwig, J. F. (2004). *Organometallics* **23**, 194.

22. Limmert, M. E., Roy, A. H., Hartwig, J. F. (2005). *J. Org. Chem.* **70**, 9364.

23. Ackermann, L., Althammer, A. (2006). *Org. Lett.* **8**, 3457.

24. Terao, J., Naitoh, Y., Kuniyasu, H., Kambe, N. (2003). *Chem. Lett.* **32**, 890.

25. Terao, J., Naitoh, Y., Kuniyasu, H., Kambe, N. (2007). *Chem. Commun.*, 825.

26. (a) Negishi, E., Baba, S., (1976). *J. Chem. Soc. Chem. Commun.*, 596. (b) Negishi, E., (1976). *J. Am. Chem. Soc.* **98**, 6729. (c) Negishi, E., (1982). *Acc. Chem. Res.* **15**, 340. (d) Phapale, V. B., Cárdenas, D. J. (2011). *Chem. Soc. Rev.* **38**, 1598. (e) Negishi, E. (2011). *Angew. Chem. Int. Ed.* **50**, 6738.

27. (a) Takai, K., Oshima, K., Nozaki, H. (1980). *Tetrahedron Lett.* **21**, 2531. (b) Sato, M., Takai, K., Oshima, K., Nozakai, H. (1981). *Tetrahedron Lett.* **22**, 1609. (c) Takai, K., Sato, M., Oshima, K., Nozaki, M. (1984). *Bull. Chem. Soc. Jpn.* **57**, 108. (d) Fugami, K., Oshima, K., Utimoto, K. (1987). *Chem. Lett.*, 2203.

28. Wu, J., Yang, Z. (2001). *J. Org. Chem.* **66**, 7875.

29. Hansen, A. L., Ebran, J.-P., Gøgsig, T. M., Skrydstrup, T. (2007). *J. Org. Chem.* **72**, 6464.

30. Zhou, J., Fu, G. C. (2003). *J. Am. Chem. Soc.* **125**, 12527.

31. For reviews, see: (a) Stille, J. K. (1985). *Pure Appl. Chem.* **57**, 1771. (b) Stille, J. K. (1986). *Angew. Chem., Int. Ed.* **25**, 508. (c) Pereyre, M., Quintard, J. P., Rahm, A. (1987). *Tin in Organic Synthesis*, Butterworth, London, 1987. (d) Farina, V. (1995). In *Comprehensive Organometallic Chemistry* II, Abel, E. W., Stone, F. G. A., Wilkinson, G., eds., Elsevier, Oxford. (e) Farina, V., Krishnamurthy, V., Scott, W. J. (1997). *Org. React.* **50**, 1. (f) Farina, V., Krishnamurthy, V., Scott, W. J. (1998). *The Stille Reaction*, Wiley, New York. (g) Kugami, K., Kosugi, M. (2002) *Top. Curr. Chem.* **219**, 87. (h) Kosugi, M., Fugami, K. (2002). *J. Organomet. Chem.* **653**, 50. (i) Espinet, P., Echavarren, A. M. (2004). *Angew. Chem., Int. Ed.* **43**, 4704.

32. (a) Scott, W. J., Crisp, G. T., Stille, J. K. (1984). *J. Am. Chem. Soc.* **106**, 4630. (b) Scott, W. J., Stille, J. K. (1986). *J. Am. Chem. Soc.* **108**, 3033.

33. Roth, G. P., Fuller, C. E. (1991). *J. Org. Chem.* **56**, 3493.

34. Badone, D., Cecchi, R., Guzzi, U. (1992). *J. Org. Chem.* **57**, 6321.

35. (a) Schio, L., Chatreaux, F., Klich, M. (2000). *Tetrahedron Lett.* **41**, 1543. (b) Steinhuebel, D., Baxter, J. M., Palucki, M., Davies, I. W. (2005). *J. Org. Chem.* **70**, 10124. (c) Jones, P., Chambers, M. (2002). *Tetrahedron* **58**, 9973. (d) Nagatsugi, F., Uemura, K., Nakashima, S., Maeda, M., Sasaki, S. (1995). *Tetrahedron Lett.* **36**, 421. (e) Nagatsugi, F., Uemura, K., Nakashima, S., Maeda, M., Sasaki, S. (1997). *Tetrahedron* **53**, 3035. (f) Mathieu, J., Marsura, A. (2003). *Synth. Commun.* **33**, 409.

36. Hettrick, C. M., Kling, J. K., Scott, W. J. (1991). *J. Org. Chem.* **54**, 1489.

37. Naber, J. R., Fors, B. P., Wu, X., Gunn, J. T., Buchwald, S. L. (2010). *Heterocycles* **80**, 1215.

38. (a) Nicolaou, K. C., Shi, G.-Q., Gunzner, J. L., Gärtner, P., Yang, Z. (1997). *J. Am. Chem. Soc.* **119**, 5467. (b) Nicolaou, K. C., Shi, G.-Q., Namoto, K., Bernal, F. (1998). *Chem. Commun.*, 1757. (c) Nicolaou, K. C., Wallace, P. A., Shi, S., Ouellette, M. A., Bunnage, M. E., Gunzner, J. L., Agrios, K. A., Shi, G.-Q., Gärtner, P., Yang, Z. (1999). *Chem. Eur. J.* **5**, 618.

39. (a) Buon, C., Bouyssou, P., Coudert, G. (1999). *Tetrahedron Lett.* **40**, 701. (b) Claveau, E., Gillaizeau, I., Blu, J., Bruel, A., Coudert, G. (2007). *J. Org. Chem.* **72**, 4832. (c) Simas, A. B. C., de Sales, D. L., Pais, K. C., (2009). *Tetrahedron Lett.* **50**, 6977. (d) Steel, P. G., Woods, T. M., (2009). *Synthesis,* 3897. (e) Jiang, J., DeVita, R. J., Doss, G. A., Goulet, M. T., Wyvratt, M. J. (1999). *J. Am. Chem. Soc.* **121**, 593

40. For reviews: (a) Suzuki, A. (1982). *Acc. Chem. Res.* **15**, 178. (b) Miyaura, N., Suzuki, A. (1995). *Chem. Rev.* **95**, 2457. (c) Miyaura, N. (2002). *Top. Curr. Chem.* **219**, 11. (d) Yin, L.-X., Liebscher, J. (2007). *Chem. Rev.* **107**, 133. (e) Torborg, C., Beller, M. (2009). *Adv. Synth. Catal.* **351**, 3027. (f) Cheng, Y.-J., Yang, S.-H., Hsu, C.-S. (2009). *Chem. Rev.* **109**, 5868. (g) Suzuki, A. (2011). *Angew. Chem. Int. Ed.* **50**, 6722.

41. Lepifre, F., Buon, C., Rabot, R., Bouyssou, P., Coudert, C. (1999). *Tetrahedron Lett.* **40**, 6373.

42. (a) Sasaki, M., Fuwa, H., Ishikawa, M., Tachibana, K. (1999) *Org. Lett.* **1**, 1075. (b) Fuwa, H., Ebine, M., Bourdelais, A. J., Baden, D. G., Sasaki, M. (2006). *J. Am. Chem. Soc.* **128**, 16989. (c) Pedzisa, L., Vaughn, I. W., Pongdee, R. (2006). *Tetrahedron Lett.* **49**, 4142. (d) Mousset, D., Gillaizeau, I., Sabatié, A., Bouyssou, P., Coudert, G. (2006). *J. Org. Chem.* **71**, 5993. (e) Campbell, I. B., Guo, J., Jones, E., Steel, P. G. (2004). *Org. Biomol. Chem.* **2**, 2725. (f) Fuwa, H., Sasaki, M. (2007). *Org. Lett.* **9**, 3347.

43. Larsen, U. S., Martinyc, L., Begtrup, M. (2005). *Tetrahedron Lett.* **46**, 4261.

44. Kang, F.-A., Sui, Z., Murray, W. V. (2008). *J. Am. Chem. Soc.* **130**, 11300.

45. Li, S.-M., Huang, J., Chen, G.-J., Han, F.-S. (2011). *Chem. Commun.* **47**, 12840.

46. (a) Wu, J., Zhu, Q., Wang, L., Fathi, R., Yang, Z. (2003). *J. Org. Chem.* **68**, 670. (b) Wang, Z., Fan, R., Wu, J. (2007). *Adv. Synth. Catal.* **349**, 1943. (c) Baxter, J. M., Steinhuebel, D., Palucki, M., Davies, I. W. (2005). *Org. Lett.* **7**, 215. (d) Gøgsig, T. M., Søbjerg, L. S., Lindhardt, A. T., Jensen, K. L., Skrydstrup, T. (2008). *J. Org. Chem.* **73**, 3404. (e) Nakatsuji, H., Nishikado, H., Ueno, K., Tanabe, Y. (2009). *Org. Lett.* **11**, 4258.

47. (a) Luo, Y., Wu, J. (2009). *Tetrahedron Lett.* **50**, 2103. (b) Hu, Y., Ding, Q., Ye, S., Peng, Y., Wu, J. (2011). *Tetrahedron* **67**, 7258.

48. Albaneze-Walker, J., Raju, R., Vance, J. A., Goodman, A. J., Reeder, M. R., Liao, J., Maust, M. T., Irish, P. A., Espino, P., Andrews, D. R. (2009). *Org. Lett.* **11**, 1463.

49. (a) Nguyen, H. N., Huang, X., Buchwald, S. L. (2003). *J. Am. Chem. Soc.* **125**, 11818. (b) Bhayana, B., Fors, B. P., Buchwald, S. L. (2009). *Org. Lett.* **11**, 3954.

50. (a) So, C. M., Lau, C. P., Kwong, F. Y. (2008). *Angew. Chem.* **120**, 8179. (b) So, C. M., Lau, C. P., Kwong, F. Y. (2008) *Angew. Chem. Int. Ed.* **47**, 8059. (c) So, C. M., Lau, C. P., Chan, A. S. C., Kwong, F. Y. (2008). *J. Org. Chem.* **73**, 7731.

51. Netherton, M. R., Fu, G. C. (2002). *Angew. Chem. Int. Ed.* **41**, 3910.

52. Tsukamoto, H., Sato, M., Kondo, Y. (2004). *Chem. Commun.*, 1200.

53. Tsukamoto, H., Uchiyama, T., Suzuki, T., Kondo, Y. (2008). *Org. Biomol. Chem.* **6**, 3005.

54. Manabe, K., Nakada, K., Aoyama, N., Kobayashi, S. (2005). *Adv. Synth. Catal.* **347**, 1499.

55. (a) Yoshida, M., Gotou, T., Ihara, M. (2004). *Chem. Commun.*, 1124. (b) Yoshida, M., Gotou, T., Ihara, M. (2004). *Tetrahedron Lett.* **45**, 5573.

56. (a) Hiyama, T., Hatanaka, Y. (1994). *Pure. Appl. Chem.* **66**, 1471. (b) Denmark, S. E., Liu, J. H.-C., (2010). *Angew. Chem. Int. Ed.* **49**, 2978.

57. Zhang, L., Wu, J. (2008). *J. Am. Chem. Soc.* **130**, 12250.

58. Zhang, L., Qing, J., Yang, P., Wu, J. (2008). *Org. Lett.* **10**, 4971.

59. So, C. M., Lee, H. W., Lau, C. P., Kwong, F. Y. (2009). *Org. Lett.* **11**, 317.

60. Shirbin, S. J., Boughton, B. A., Zammit, S. C., Zanatta, S. D., Marcuccio, S. M., Hutton, C. A., Williams, S. J. (2010). *Tetrahedron Lett.* **51**, 2971.

61. (a) Beletskaya, I. P., Cheprakov, A. V. (2000). *Chem. Rev.* **100**, 3009. (b) Dounay, A. B., Overman, L. E. (2003). *Chem. Rev.* **103**, 2945. (c) Le Bras, J., Muzart, J., (2011). *Chem. Rev.* **111**, 1170.

62. Fu, X., Zhang, S., Yin, J., McAllister, T. L., Jiang, S. A., Tann, C.-H., Thiruvengadam, T. K., Zhang, F. (2002). *Tetrahedron Lett.* **43**, 573.

63. Hansen, A. L., Skrydstrup, T. (2005). *Org. Lett.* **7**, 5585.

64. (a) Hansen, A. L., Ebran, J.-P., Ahlquist, M., Norrby, P.-O., Skrydstrup, T. (2006). *Angew. Chem.* **118**, 3427. (b) Hansen, A. L., Ebran, J.-P., Ahlquist, M., Norrby, P. O., Skrydstrup, T. (2006). *Angew. Chem. Int. Ed.* **45**, 3349.

65. Gøgsig, T. M., Lindhardt, A. T., Grouleff, J., Skrydstrup, T. (2009). *Chem. Eur. J.* **15**, 5950.

66. Noël, S., Pinel, C., Djakovitch, L. (2006). *Org. Biomol. Chem.* **4**, 3760.

67. (a) Sonogashira, K. (1998). In *Metal-Catalyzed Cross-Coupling Reactions*, Diederich, F., Stang, P. J., eds., Wiley-VCH, New York, Chapter 5. (b) Chinchilla, R., Nájera, C. (1998). *Chem. Rev.* **107**, 874.

68. (a) Nicolaou, K. C., Shi, G.-Q., Namoto, K., Bernal, F. (2001). *Chem. Commun.* 1757. (b) Wu, J., Liao, Y., Yang, Z. (2001). *J. Org. Chem.* **66**, 3642. (c) Luo, Y., Wu, J., (2009). *Tetrahedron* **65**, 6810. (d) Braga, A. L., Emmerich, D. J., Silveira, C. C., Martins, T. L. C., Rodrigues, O. E. D. (2001). *Synlett* 369. (e) Gotoh, K., Matsuura, Y., Hatanaka, F., Sato, S., Kagabu, S., Yoshimatsu, M. (2004). *Chem. Lett.* **33**, 1322. (f) Fu, X., Zhang, S., Yin, J., Schumacher, D. P. (2002). *Tetrahedron Lett.* **43**, 6673.

69. (a) Kang, F.-A., Lanter, J. C., Cai, C., Sui, Z., Murray, W. V. (2010). *Chem. Commun.* **46**, 1347. (b) Shi, C., Aldrich, C. C. (2010). *Org. Lett.* **12**, 2286.

70. (a) Gelman, D., Buchwald, S. L. (2003). *Angew. Chem.* **115**, 6175. Gelman, D., Buchwald S. L. (**2003**). *Angew. Chem. Int. Ed.* 2003, **42**, 5993. (b) Shirbin, S. J., Boughton, B. A., Zammit, S. C., Zanatta, S. D., Marcuccio, S. M., Hutton, C. A., Williams, S. J. (2010). *Tetrahedron Lett.* **51**, 2971.

71. (a) Choy, P. Y., Chow, W. K., So, C. M., Lau, C. P., Kwong, F. Y. (2010). *Chem. Eur. J.* **16**, 9982. (b) R'kyek, O., Halland, N., Lindenschmidt, A., Alonso, J., Lindemann, P., Urmann, M., Nazaré, M. (2010). *Chem. Eur. J.* **16**, 9986.

72. (a) Dyker, G. (2005). *Handbook of C–H Transformations. Applications in Organic Synthesis*, Wiley-VCH, Weinheim. (b) Ackermann, L., Vicente, R., Kapdi, A. R. (2009). *Angew. Chem. Int. Ed.* **48**, 9792. (c) Giri, R., Shi, B.-F., Engle, K. M., Maugel, N., Yu, J.-Q., (2009). *Chem. Soc. Rev.* **38**, 3242. (d) Colby, D. A., Bergman, R. G., Ellman, J. A. (2010). *Chem. Rev.* **110**, 624. (e) Lyons, T. W., Sanford, M. S., (2010). *Chem. Rev.* **110**, 1147. (f) Yeung, C. S., Dong, V. M. (2011). *Chem. Rev.* **111**, 1215. (g) Newhouse, T., Baran, P. S. (2011). *Angew. Chem. Int. Ed.* **50**, 3362. (h) McMurray, L., O'Hara, F., Gaunt, M. J. (2011). *Chem. Soc. Rev.* **40**, 1885. (i) Cho, S. H., Kim, J. Y., Kwak, J., Chang, S. (2011). *Chem. Soc. Rev.* **40**, 5068. (j) K. Hirano, Miura, M. (2011). *Synlett*, 294. (k) Zhao, D., You, J., Hu, C. D. (2011). *Chem. Eur. J.* **17**, 5466. (l) Kuhl, N., Hopkinson, M. N., Wencel-Delord, J., Glorius, F. *Angew. Chem. Int. Ed.* **51**, 10236. (m) Li, B.-J., Shi, Z.-J., (2012). *Chem. Soc. Rev.* **41**, 5588.

73. (a) Kulkin, D. A., Hartwig, J. F. (2003). *Acc. Chem. Res.* **36**, 234. (b) Bellina, F., Rossi, R., (2010). *Chem. Rev.* **110**, 1082. (c) Johansson, C. C. C., Colacot, T. J. (2010). *Angew. Chem., Int. Ed.* **49**, 676. (d) Novák, P., Martin, R. (2011). *Curr. Org. Chem.* **15**, 3233.

74. (a) Gooßen, L. J., Rodríguez, N., Gooßen, K., (2008). *Angew. Chem. Int. Ed.* **47**, 3100. (b) Gooßen, L. J., Gooßen, K., Rodríguez, N., Blanchot, M., Linder, C., Zimmermann, B. (2008). *Pure Appl. Chem.*, **80**, 1725. (c) Gooßen, L. J., Collet, F., Gooßen, K. (2010). *Isr. J. Chem.* **50**, 617.

75. (a) Ackermann, L., Althammer, A., Fenner, S. (2009). *Angew. Chem.* **121**, 207. (b) Ackermann, L., Althammer, A., Fenner, S. (2009). *Angew. Chem. Int. Ed.* **48**, 201.

76. (a) Ackermann, L., Barfüsser, S., Pospech, J. (2010). *Org. Lett.* **12**, 724. (b) Ackermann, L., Fenner, S. (2011). *Chem. Commun.* **47**, 430.

77. Fan, S., Yang, J., Zhang, X. (2011). *Org. Lett.* **13**, 4374.

78. Chang, J. W. W., Chia, E. Y., Chai, C. L. L., Seayad, J. (2012). *Org. Biomol. Chem.* **10**, 2289.

79. Nervig, C. S., Waller, P. J., Kalyani, D. (2012). *Org. Lett.* **14**, 4838.

80. Kimura, M., Futamata, M., Mukai, R., Tamaru, Y. (2005). *J. Am. Chem. Soc.* **127**, 4592.

81. Usui, I., Schmidt, S., Keller, M., Breit, B. (2008). *Org. Lett.* **10**, 1207.

82. Kawatsura, M., Hartwig, J. F. (1999). *J. Am. Chem. Soc.* **121**, 1473.

83. Nguyen, H. N., Huang, X., Buchwald, S. L. (2003). *J. Am. Chem. Soc.* **125**, 11818.

84. Bergbreiter, D. E., Weatherford, D. A. (1989). *Chem. Commun.*, 883.

85. (a) Ozawa, F., Okamoto, H., Kawagishi, S., Yamamoto, S., Minami, T., Yoshifuji, M. (2002). *J. Am. Chem. Soc.* **124**, 10968. (b) Manabe, K., Kobayashi, S. (2003). *Org. Lett.* **5**, 3241. (c) Kinoshita, H., Shinokubo, H., Oshima, K. (2004). *Org. Lett.* **6**, 4085.

86. Kimura, M., Horino, Y., Mukai, R., Tanaka, S., Tamaru, Y. (2001). *J. Am. Chem. Soc.* **123**, 10401.

87. (a) Gooßen, L. J., Rodríguez, N., Lange, P. P., Linder, C. (2010). *Angew. Chem.* **122**, 1129. (b) Gooßen, L. J., Rodríguez, N., Lange, P. P., Linder, C. (2010). *Angew. Chem. Int. Ed.* **49**, 1111.

88. (a) Brennführer, A., Neumann, H., Beller, M. (2009). *Angew. Chem.* **121**, 4176. (b) Brennführer, A., Neumann, H., Beller, M. (2009). *Angew. Chem. Int. Ed.* **48**, 4114.

89. Cacchi, S., Ciattini, P. G., Morera, E., Ortar, G. (1986). *Tetrahedron Lett.* **27**, 3931.

90. (a) Dolle, R. E., Schmidt, S. J., Kruse, L. I., (1987). *J. Chem. Soc. Chem. Commun.* 904. (b) Cacchi, S., Lupi, A. (1992). *Tetrahedron Lett.* **33**, 3939. (c) Gerlach, U., Wollmann, T. (1992). *Tetrahedron Lett.* **33**, 5499. (d) Rahman, O., Kihlberg, T., Langström, B. (2003). *J. Org. Chem.* **68**, 3558. (e) Lou, R., VanAlstine, M., Sun, X., Wentland, M. P. (2003). *Tetrahedron Lett.* **44**, 2477.

91. (a) Nicolaou, K. C., Shi, G.-Q., Namoto, K., Bernal, F. (1998). *Chem. Commun.*, 1757. (b) Occhiato, E. G., Scarpi, D., Prandi, C. (2010). *Heterocycles* **80**, 697. (c) Lagerlund, O., Mantel, M. L. H., Larhed, M. (2009). *Tetrahedron* **65**, 7646. (d) Kubota, Y., Nakada, S., Sugi, Y., (1998). *Synlett.* 183.

92. Kubota, Y., Nakada, S., Sugi, Y. (1998). *Synlett* 183.

93. Cai, C., Rivera, N. R., Balsells, J., Sidler, R. R., McWilliams, J. C., Shultz, C. S., Sun, Y. (2006). *Org. Lett.* **8**, 5161.

94. Munday, R. H., Martinelli, J. R., Buchwald, S. L. (2008). *J. Am. Chem. Soc.* **130**, 2754.

95. (a) Hartwig, J. (2010). *Organotransition Metal Chemistry*, University Science Books, Sausalito, CA, **2010**, p. 907. (b) Hartwig, J. F. (2008). *Acc. Chem. Res.* 2008, **41**, 1534. (c) Surry, D. S., Buchwald, S. L. (2008). *Angew. Chem. Int. Ed.* **47**, 6338. (d) Buchwald, S. L., Mauger, C., Mignani, G., Scholz, U. (2006). *Adv. Synth. Catal.* **348**, 23. (e) Muci, A. R., Buchwald, S. L. (2002). *Top. Curr. Chem.* **219**, 131. (f) Busacca, C. A., Fandrick, D. R., Song, J. J., Senanayake, C. H. (2011). *Adv. Synth. Catal.* **353**, 1825.

96. Hamann, B. C., Hartwig, J. F. (1998). *J. Am. Chem. Soc.* **120**, 7369.

97. Ogata, T., Hartwig, J. F. (2008). *J. Am. Chem. Soc.* **130**, 13848.

98. Vo, G. D., Hartwig, J. F. (2009). *J. Am. Chem. Soc.* **131**, 11049.

99. Klapars, A., Campos, K. R., Chen, C., Volante, R. P. (2005). *Org. Lett.* **7**, 1185.

100. Mantel, M. L. H., Lindhardt, A. T., Lupp, D., Skrydstrup, T. (2010). *Chem. Eur. J.* **16**, 5437.

101. (a) Lundgren, R. J., Peters, B. D., Alsabeh, P. G., Stradiotto, M. (2010). *Angew. Chem.* **122**, 4165. Lundgren, R. J., Peters, B. D., Alsabeh, P. G., Stradiotto, M. (2010). *Angew. Chem. Int. Ed.* **49**, 4071. (b) Alsabeh, P. G., Lundgren, R. J., McDonald, R., Johansson Seechurn, C. C. C., Colacot, T. J., Stradiotto, M. (2013) *Chem. Eur. J.* **19**, 2131.

102. Ackermann, L., Sandmann, R., Song, W. (2011). *Org. Lett.* **13**, 1784.

103. Huang, X., Anderson, K. W., Zim, D., Jiang, L., Klapars, A., Buchwald, S. L. (2003). *J. Am. Chem. Soc.* **125**, 6653.

104. (a) Gunda, P., Russon, L. M., Lakshman, M. K. (2004). *Angew. Chem.* **120**, 6532. (b) Gunda, P., Russon, L. M., Lakshman, M. K. (2004). *Angew. Chem. Int. Ed.* **43**, 6372.

105. So, C. M., Zhou, Z., Lau, C. P., Kwong, F. Y. (2008). *Angew. Chem.* **120**, 6502; *Angew. Chem. Int. Ed.* **47**, 6402.

106. (a) Fors, B. P., Watson, D. A., Biscoe, M. R., Buchwald, S. L. (2008). *J. Am. Chem. Soc.* **130**, 13552. (b) Dooleweerdt, K., Fors, B. P., Buchwald, S. L. (2010). *Org. Lett.* **12**, 2350.

107. For a review, see: Muzart, J. (2007). *Eur. J. Org. Chem.*, 3077.

108. (a) Lu, X., Lu, L., Sun, J. *J. Mol. Catal.* **41**, 245. (b) Lu, X., Jiang, X., Tao, X., (1988). *J. Organomet. Chem.* **344**, 109.

109. Sakamoto, M., Shimizu, I., Yamamoto, A. *Bull. Chem. Soc. Jpn.* **69**, 1065.

110. (a) Masuyama, Y., Kagawa, M., Kurusu, Y. (1995). *Chem. Lett.* 1121. (b) Nishikata, T., Lipshutz, B. H. *Org. Lett.* **11**, 2377.

111. (a) Yang, S.-C., Hung, C.-W., (1999). *J. Org. Chem.* **64**, 5000. (b) Kimura, M., Futamata, M., Shibata, K., Tamaru, Y., *Chem. Commun.* 234. (c) Yang, S.-C., Hsu, Y.-C., Gan, K.-H., *Tetrahedron* **62**, 3949. (d) Tao, Y., Zhou, Y., Qu, J., Hidai, M. (2010). *Tetrahedron Lett.* **51**, 1982.

112. (a) Ozawa, F., Okamoto, H., Kawagishi, S., Yamamoto, S., Minami, T., Yoshifuji, M. (2002). *J. Am. Chem. Soc.* **124**, 10968. (b) Ozawa, F., Ishiyama, T., Yamamoto, S., Kawagishi, S., Murakami, H., Yoshifuji, M., (2004). *Organometallics* **23**, 1698. (c) Kinoshita, H., Shinokubo, H., Oshima, K., (2004). *Org. Lett.* **6**, 4085. (d) Piechaczyk, O., Doux, M., Ricard, L., Le Floch, P., (2005). *Organometallics* **24**, 1204. (e) Thoumazet, C., Grützmacher, H., Deschamps, B., Ricard, L., Le Floch, P. (2006). *Eur. J. Inorg. Chem.* 3911. (f) Hikawa, H., Yokoyama, Y., (2011). *J. Org. Chem.* **76**, 8433. (g) Ghosh, R., Sarkar, A. J. *Org. Chem.* **76**, 8508. (h) Utsunomiya, M., Miyamoto, Y., Ipposhi, J., Ohshima, T., Mashima, K. (2007). *Org. Lett.* **9**, 3371.

113. Hikawa, H., Yokoyama, Y. (2011). *Org. Lett.* **13**, 6512.

114. Occhiato, E. G., Galbo, F. L., Guarna, A. (2005) *J. Org. Chem.* **70**, 7324.

115. (a) Olsson, V. J., Sebelius, S., Selander, N., Szabó, K.J. (2006). *J. Am. Chem. Soc.* **128**, 4588. (b) Dutheuil, G., Selander, N., Szabó, K. J., Aggarwal, V. K. (2008). *Synthesis*, 2293. (c) Selander, N., Paasch, J. R., Szabó, K. J. (2010). *J. Am. Chem. Soc.* **133**, 409. (d) Larsson, J. M., Szabó, K. J. (2013). *J. Am. Chem. Soc.* **135**, 443.
116. Luo, Y., Wu, J., (2009). *Organometallics* **28**, 6823.
117. Milne, J. E., Kocienski, P. J. *Synthesis,* 584.
118. Fernández-Rodríguez, M. A., Shen, Q., Hartwig, J. F. (2006). *J. Am. Chem. Soc.* **128**, 2180.
119. Charbonnier, F., Moyano, A., Greene, A. E. (1987). *J. Org. Chem.* **52**, 2303.
120. Fuwa, H., Sasaki, M. (2008). *Heterocycles* **76**, 521.
121. Watson, D. A., Su, M., Teverovskiy, G., Zhang, Y., García-Fortanet, J., Kinzel, T., Buchwald, S. L. (2009). *Science* **325**, 1661.
122. (a) Hayashi, T., Fujiwa, T., Okamoto, Y., Katauro, Y., Kumada, M. (1981). *Synthesis* 1001. (b) Sahlberg, C., Quader, A., Claesson, A. (1983). *Tetrahedron Lett.* **24**, 5137. (c) Tamao, K., Ishida, N., Kumada, M. (1983). *J. Org. Chem.,* **48**, 2120.
123. (a) Karlström, A. S. E., Itami, K., Bäckvall, J.-E. (1999). *J. Org. Chem.* **64**, 1745. (b) William, A. D., Kobayashi, Y. (2002). *J. Org. Chem.* **67**, 8771. (c) Kobayashi, Y., Takeuchi, A., Wang, Y.-G. (2006). *Org. Lett.* **8**, 2699. (d) Verboom, R. C., Persson, B. A., Baeckvall, J.-E. *J. Org. Chem.* **69**, 3102. (e) Huang, W., Jiang, Y., Li, Q., Li, J., Li, J.Y., Lu, W., Cai, J. (2005). *Tetrahedron* **61**, 1863.
124. Hayashi, T., Katsuro, Y., Okamoto, Y., Kumada, M. (1981). *Tetrahedron Lett.* **22**, 449.
125. (a) Cossy, J., Belotti, D. (1999). *Tetrahedron* **55**, 5145. (b) Huang, W. G., Jiang, Y. Y., Li, Q., Li, J., Li, J. Y., Lu, W., Cai, J. C. (2005). *Tetrahedron* **61**, 1863.
126. Yoshikai, N., Matsuda, H., Nakamura, E. (2009). *J. Am. Chem. Soc.* **131**, 9590.
127. Macklin, T. K., Snieckus, V., (2005). *Org. Lett.* **7**, 2519.
128. When, P. M., Du Bois, J., (2005). *Org. Lett.* **7**, 4685.
129. Guan, B.-T., Lu, X.-Y., Zheng, Y., Yu, D.-G., Wu, T., Li, K.-L., Li, B.-J., Shi, Z.-J. (2010). *Org. Lett.* **12**, 396.
130. Percec, V., Bae, J.-Y., Hill, D. H. (1995). *J. Org. Chem.* **60**, 6895.
131. Snieckus, V., (1990). *Chem. Rev.* **90**, 879.
132. Smith, M. B., March, J. (2007). *March's Advanced Organic Chemistry*, 6th ed., Wiley, Hoboken, NJ, p. 668.
133. Kuwano, R. (2011). *J. Syn. Org. Chem. Jpn.* **69**, 1263.
134. P. Kocienski, Dixon, N. J. (1989). *Synlett,* 52.
135. (a) Porée, F.-H., Clavel, A., Betzer, J.-F., Pancrazi, A., Ardisson, J. (2003). *Tetrahedron Lett.* **44**, 7553. (b) Madec, D., Pujol, S., Henryon, V., Férézou, J. P. (1995). *Synlett,* 435.
136. Sengupta, S., Leite, M., Raslan, D. S., Quesnelle, C., Snieckus, V. (1992). *J. Org. Chem.* **57**, 4066.
137. (a) Hill, C. M., Walker, R. A., Hill, M. E. (1951). *J. Am. Chem. Soc.* **73**, 1663. (b) Hill, C. M., Senter, G. W., Haynes, L., Hill, M. E. (1954). *J. Am. Chem. Soc.* **76**, 4538. (c) Fujisawa, T., Kurita, Y., Kawashima, M., Sato, T. (1982). *Chem. Lett.* 1641.
138. (a) Meyers, A. I., Gabel R., and Mihelich, E. D. (1978). *J. Org. Chem.* **43**, 1372. (b) Hattori, T., Suzuki, M., Tomita, N., Takeda, A., Miyano, S. (1997). *J. Chem. Soc., Perkin Trans.* **1**, 1117. (c) Liard, A., Marek, I. (2000). *J. Org. Chem.* **65**, 7218.

139. (a) Wenkert, E., Michelotti, E. L., Swindell, C. S. (1979). *J. Am. Chem. Soc.* **101**, 2246. (b) Wenkert, E., Michelotti, E. L., Swindell, C. S., Tingoli, M. (1979). *J. Org. Chem.* **49**, 4894.

140. (a) Wenkert, E., Ferreira, V. F., Michelotti, E. L.,Tingoli, M. (1985). *J. Org. Chem.* **50**, 719. (b) Davis, T. L., Carlson, D. A. (1989). *Synthesis,* 936. (c) Kocieński, P. J., Pritchard, M., Wadman, S. N., Whitby, R. J., Yeates, C. L. *J. Chem. Soc. Perkin Trans.* **1**, 3419.

141. Ducoux, J.-P., Ménez, P. L., Kunesch, N. (1993). *J. Org. Chem.* **58**, 1290.

142. Wenkert, E., Leftin, M. H., Michelotti, E. L. (1984). *J . Chem. Soc., Chem. Commun.* 617.

143. Johnston, R. A. W., McLean, W. N. (1988). *Tetrahedron Lett.* **29**, 5553.

144. Brigas, A. F., Johnstone, R. A. W. *J. Chem. Soc., Perkin Trans.* **1**, 1735.

145. (a) Dankwardt, J. W. (2004). *Angew. Chem. Int. Ed.* **43**, 2428. (b) Xie, L. G., Wang, Z. X. (2011). *Chem. Eur. J.* **17**, 4972.

146. Guan, B., Xiang, S., Wu, T., Sun, Z., Wang, B., Zhao, K., Shi, Z. (2008). *Chem. Commun.* **12**, 1437.

147. (a) Hayashi, T., Katsuro, Y., Kumada, M. (1980). *Tetrahedron Lett.* **21**, 3915. (b) Zhao, F., Yu, D.-G., Zhu, R.-Y., Xi, Z.-F., Shi, Z.-J. *Chem. Lett.* **40**, 1001.

148. Guan, B., Xiang, S., Wang, B., Sun, Z., Wang, Y., Zhao, K., Shi, Z. (2008). *J. Am. Chem. Soc.* **130**, 3268.

149. Taylor, B., Swift, E., Waetzig, J., Jarvo, E. (2011). *J. Am. Chem. Soc.* **133**, 389.

150. Taylor, B., Harris, M., Jarvo, E. (2012). *Angew. Chem. Int. Ed.* **51**, 7790.

151. (a) Yu, D.-G., Li, B.-J., Zheng, S.-F., Guan, B.-T., Wang, B.-Q., Shi, Z.-J. (2010). *Angew. Chem.* **122**, 4670. (b) Yu, D.-G., Li, B.-J., Zheng, S.-F., Guan, B.-T., Wang, B.-Q., Shi, Z.-J. (2010). *Angew. Chem. Int. Ed.* **49**, 4566.

152. Chuit, C., Felkin, H., Frajerman, C., Roussi, G., Swierczewski, G. (1968). *J. Chem. Soc., Chem. Commun.,* 1604.

153. Yu, D.-G., Wang, X., Zhu, R.-Y., Luo, S., Zhang, X.-B., Wang, B.-Q., Shi, Z.-J. (2012). *J. Am. Chem. Soc.* **134***,* 14638.

154. (a) Hayashi, T., Fujiwa, T., Okamoto, Y., Katauro, Y., Kumada, M. (1981). *Synthesis,* 1001. (b) Hayashi, T., Katsuro, Y., Okamoto, Y., Kumada, M. (1981). *Tetrahedron Lett.* **22**, 4449.

155. Fugami, K., Oshima, K., Utimoto, K. (1987). *Chem. Lett.,* 2203.

156. (a) Wu, J., Liao, Y., Yang, Z. (2001). *J. Org. Chem.* **66***,* 3642. (b) Wu, J., Yang, Z. (2001). *J. Org. Chem.* **66**, 7875. (c) Wu, J., Sun, X., Zhang, L. (2005). *Chem. Lett.* **34**, 796. (d) Hansen, A. L., Ebran, J.-P., Gøgsig, T. M., Skrydstrup, T. (2007). *J. Org. Chem.* **72**, 6464. (e) Lindhardt, A. T., Gøgsig, T. M., Skrydstrup, T. (2009). *J. Org. Chem.* **74**, 135.

157. Gao, W., Luo, Y., Ding, Q., Peng, Y., Wu, J. (2010). *Tetrahedron Lett.* **51**, 136.

158. Zhang, L., Luo, Y., Wu, J. *Synlett,* 1845.

159. (a) Li, B.-J., Li, Y.-Z., Lu, X.-Y., Liu, J., Guan, B.-T., Shi, Z.-J. (2008). *Angew. Chem.* **120**, 10278. (b) Li, B.-J., Li, Y.-Z., Lu, X.-Y., Liu, J., Guan, B.-T., Shi, Z.-J. (2008). *Angew. Chem. Int. Ed.* **47**, 10124.

160. Sun, C.-L., Wang, Y., Zhou, X., Wu, Z.-H., Li, B.-J., Guan, B.-T., Z.-J. Shi, (2010). *Chem. Eur. J.* **16**, 5844.

161. Wang, C., Ozaki, T., Takita, R., Uchiyama, M. (2012). *Chem. Eur. J.* **18**, 3482.

162. Nan, Y., Yang, Z. (1999). *Tetrahedron Lett.* **40**, 3321.

163. Hansen, A. L., Ebran, J.-P., Gøgsig, T. M., Skrydstrup, T. (2006). *Chem. Commun.,* 4137.
164. Zhao, Y.-L., Li, Y., Li, Y., Gao, L.-X., Han, F.-S. (2010). *Chem. Eur. J.* **16**, 4991.
165. Chen, H., Huang, Z., Hu, X., Tang, G., Xu, P., Zhao, Y., Cheng, C. (2011). *J. Org. Chem.* **76**, 2338.
166. Zim, D., Lando, V. R., Dupont, J., Monteiro, A. L. (2001). *Org. Lett.* **3**, 3049.
167. Percec, V., Golding, G. M., Smidrkal, J., Weichold, O. *J. Org. Chem.* **69**, 3447.
168. Fan, X.-H., Yang, L.-M. (2010). *Eur. J. Org. Chem.* 2457.
169. (a) Kuroda, J., Inamoto, K., Hiroya, K., Doi, T. (2009). *Eur. J. Org. Chem.* 2251. (b) Tu, T., Mao, H., Herbert, C., Xu, M., Dötz, K. H. (2010). *Chem. Commun.* **46**, 7796.
170. Tang, Z.-Y., Hu, Q.-S. (2004). *J. Am. Chem. Soc.* **126**, 3058.
171. Tang, Z.-Y., Spinella, S., Hu, Q.-S. (2006). *Tetrahedron Lett.* **47**, 2427.
172. Tang, Z. Y., Hu, Q.-S. (2004). *Adv. Synth. Catal.* **346**, 1635.
173. Lipshutz, B. H., Butler, T., Swift, E. (2008). *Org. Lett.* **10**, 697.
174. Quasdorf, K. W., Riener, M., Petrova, K. V., Garg, N. K. (2009). *J. Am. Chem. Soc.* **131**, 17748.
175. Chen, G., Han, F. (2012). *Eur. J. Org. Chem.,* 3575.
176. Percec, V., Bae, J.-Y., Hill, D. H. (1996). *J. Org. Chem.* **60**, 1060.
177. Kobayashi, Y., Mizojiri, R. (1996). *Tetrahedron Lett.* **37**, 8531.
178. Ueda, M., Saitoh, A., Oh-tani, S., Miyaura, N. (1998). *Tetrahedron* **54**, 13079.
179. Percec, V., Golding, G. M., Smidrkal, J., Weichold, O. *J. Org. Chem.* 2004, **69**, 3447.
180. Wilson, D. A., Wilson, C. J., Rosen, B. M., Percec, V. (2008). *Org. Lett.* **10**, 4879.
181. Molander, G. A., Beaumard, F. (2010). *Org. Lett.* **12**, 4022.
182. Guan, B.-T., Wang, Y., Li, B.-J., Yu, D.-G., Shi, Z.-J. (2008). *J. Am. Chem. Soc.* **130**, 14468.
183. Quasdorf, K. W., Tian, X., Garg, N. K. (2008). *J. Am. Chem. Soc.* **130**, 14422.
184. Li, Z., Zhang, S.-L., Fu, Y., Guo, Q.-X., Liu, L. (2009). *J. Am. Chem. Soc.* **131**, 8815.
185. Snieckus, V. (1990). *Chem. Rev.* **90**, 879.
186. Quasdorf, K. W., Riener, M., Petrova, K. V., Garg, N. K. (2009). *J. Am. Chem. Soc.* **131**, 17748.
187. Antoft-Finch, A., Blackburn, T., Snieckus, V. (2009). *J. Am. Chem. Soc.* **131**, 17750.
188. Xu, L., Li, B.-J., Wu, Z.-H., Lu, X.-Y., Guan, B.-T., Wang, B.-Q., Zhao, K.-Q., Shi, Z.-J. (2010). *Org. Lett.* **12**, 884.
189. (a) Tobisu, M., Shimasaki, T., Chatani, N. (2008). *Angew. Chem.* **120**, 4944. (b) Tobisu, M., Shimasaki, T., Chatani, N. (2008). *Angew. Chem. Int. Ed.* **47**, 4866.
190. Shimasaki, T., Konno, Y., Tobisu, M., Chatani, N. (2009). *Org. Lett.* **11**, 4890.
191. Chung, K.-G., Miyake, Y., Uemura, S. (2000). *J. Chem. Soc., Perkin Trans.* **1**, 15.
192. Yu, D.-G., Shi, Z.-J. (2011). *Angew. Chem. Int. Ed.* **50**, 7097.
193. Matsubara, R., Jamison, T. F. (2010). *J. Am. Chem. Soc.* **132**, 6880.
194. Ehle, A., Zhou, Q., Watson, M. (2012). *Org. Lett.* **14**, 1202.
195. Bolm, C., Hildebrand, J. P., Rudolph, J. (2000). *Synthesis,* 911.
196. Jglesias, M., Blandez, J., Fructos, M., Prieto, A., Alvarez, E., Belderrain, T., Nicasio, M. C. (2012). *Organometallics* **31**, 6312.
197. Gao, C.-Y., Yang, L.-M. (2008). *J. Org. Chem.* **73**, 1624.

198. Huang, J., Yang, L. (2011). *Org. Lett.* **13**, 3750.

199. Tobisu, M., Shimasaki, T., Chatani, N. (2009). *Chem. Lett.* **38**, 710.

200. (a) Shimasaki, T., Tobisu, M., Chatani, N. (2010). *Angew. Chem.* **122**, 2991; (b) *Angew. Chem. Int. Ed.* **49**, 2929.

201. Moldoveanu, C., Wilson, D. A., Wilson, C. J., Corcoran, P., Rosen, B. M., Percec, V. (2009). *Org. Lett.* **11**, 4974

202. Wilson, D. A., Wilson, C. J., Moldoveanu, C., Resmerita, A.-M., Corcoran, P., Hoang, L. M., Rosen, B. M., Percec, V. (2010). *J. Am. Chem. Soc.* **132**, 1800.

203. Huang, K., Yu, D.-G., Zheng, S.-F., Wu, Z.-H., Shi, Z.-J. (2011). *Chem. Eur. J.* **17**, 786.

204. Muto, K., Yamaguchi, J., Itami, K. (2012). *J. Am. Chem. Soc.* **134**, 169.

205. Amaike, K., Muto, K., Yamaguchi, J., Itami, K. (2012). *J. Am. Chem. Soc.* **134**, 13573.

206. Kogan, V. (2006). *Tetrahedron Lett.* **47**, 7515.

207. (a) Lipshutz, B. H., Frieman, B. A., Butler, T., Kogan, V. (2006). *Angew. Chem.* **118**, 814. (b) Lipshutz, B. H., Frieman, B. A., Butler, T., Kogan, V. (2006). *Angew. Chem. Int. Ed.* **45**, 800.

208. Milburn, R., Snieckus, V. (2004). *Angew. Chem. Int. Ed.* **43**, 888.

209. Alvarez-Bercedo, P., Martin, R. (2010). *J. Am. Chem. Soc.* **132**, 17352.

210. Tobisu, M., Yamakawa, K., Shimasaki, T., Chatani, N. (2011). *Chem. Commun.* **47**, 2946.

211. Sergeev, A., Hartwig, J. (2011). *Science* **332**, 439.

212. Percec, V., Bae, J.-Y., Hill, D. H. (1995). *J. Org. Chem.* **60**, 6895

213. (a) Bolm, C., Legros, J., Le Paih, J., Zani, L. (2004). *Chem. Rev.* **104**, 6217. (b) Fürstner, A., Martin, R. (2005). *Chem. Lett.* **34**, 624. (c) Plietker, B., ed. (2008). *Iron Catalysis in Organic Chemistry: Reactions and Applications*, Wiley-VCH, Weinheim. (d) Sherry, B. D., Fürstner, A. (2008). *Acc. Chem. Res.* **41**, 1500. (e) Enthaler, S., Junge, K., Beller, M. (2008). *Angew. Chem., Int. Ed.* **47**, 3317. (f) Czaplik, W. M., Mayer, M., Cvengroš, J., Jacobivon Wangelin, A. (2009). *ChemSusChem* **2**, 396. (g) Nakamura, E., Yoshikai, N. (2010). *J. Org. Chem.* **75**, 6061. (h) Jana, R., Pathak, T. P., Sigman, M. S. (2011). *Chem. Rev.* **111**, 1417. (i) Mancheño, O. G. (2011). *Angew. Chem. Int. Ed.* **50**, 2216. (j) Sun, C.-L., Li, B.-J., Shi, Z.-J. (2011). *Chem. Rev.* **111**, 1293.

214. Cahiez, G., Avedissian, H. (1998). *Synthesis,* 1199.

215. (a) Larsen, U. S., Martiny, L., Begtrup, M. (2005). *Tetrahedron Lett.* **46**, 4261. (b) Cahiez, G., Gager, O., Habiak, V. (2008). *Synthesis* 2636. (c) Cahiez, G., Habiak, V., Gager, O., *Org. Lett.* **10**, 2389. (d) Hayashi, N., Nakada, M. (2009). *Tetrahedron Lett.* **50**, 232.

216. (a) Fürstner, A., De Souza, D., Parra-Rapado, L., Jensen, J. T. (2003). *Angew. Chem. Int. Ed.* **42**, 5358. (b) Scheiper, B., Bonnekessel, M., Krause, H., Fürstner, A. (2004). *J. Org. Chem.* **69**, 3943. (c) Maulide, N., Vanherck, J.-C., Mark, I. E. (2004). *Eur. J. Org. Chem.*, 3962. (d) Fürstner, A., Hannen, P. (2006). *Chem. Eur. J.* **12**, 3006. (e) Berthon-Gelloz, G., Hayashi, T. (2006). *J. Org. Chem.* **71**, 8957. (f) Hamajima, A., Isobe, M. (2006). *Org. Lett.* **8**, 1205. (g) Fürstner, A., Schlecker, A. (2008). *Chem. Eur. J.* **14**, 9181. (h) Le Marquand, P., Tsui, G. C., Whitney, J. C. C., Tam, W. (2008). *J. Org. Chem.* **73**, 7829. (i) Camacho-Dávila, A. A. (2008). *Synth. Commun.* **38**, 3823.

217. Dunet, G., Knochel, P. (2006). *Synlett* 407.

218. Li, B.-J., Xu, L., Wu, Z.-H., Guan, B.-T., Sun, C.-L., Wang, B.-Q., Shi, Z.-J. (2009). *J. Am. Chem. Soc.* **131**, 14656.

219. (a) Fürstner, A., Leitner, A. (2002). *Angew. Chem. Int. Ed.* **41**, 609. (b) Fürstner, A. Leitner, Méndes, M., Krause, H. (2002). *J. Am. Chem. Soc.* **124**, 13856.

220. (a) Fürstner, A., Leitner, A. (2003). *Angew. Chem. Int. Ed.* **42**, 308. (b) Fürstner, A., De Souza, D., Parra-Rapado, L., Jensen, J. T. (2003). *Angew. Chem. Int. Ed.* **42**, 5358. (c) Seidel, G., Laurich, D., Fürstner, A. (2004). *J. Org. Chem.* **69**, 3950. (d) Nagano, T., Hayashi, T. (2004). *Org. Lett.* **6**, 1297.
221. Silberstein, A. L., Ramgren, S. D., Garg, N. K. (2012). *Org. Lett.* **14**, 3796.
222. Agrawal, T., Cook, S. P. *Org. Lett.* 10.1021/ol303130j.
223. Gøgsig, T. M., Lindhardt, A. T., Skrydstrup, T. (2009). *Org. Lett.* **11**, 4886.
224. Nagano, T., Hayashi, T. (2004). *Org. Lett.* **6**, 1297.
225. Ito, S., Fujiwara, Y.-i., Nakamura, E., Nakamura, M. (2009). *Org. Lett.* **11**, 4306.
226. Gosmini, C., Bégouin, J.-M., Moncomble, A. (2008). *Chem. Commun.* 3221.
227. Chalk, A. J. (1970). *Chem. Commun.*, 847.
228. (a) Seki, Y., Murai, S., Yamamoto, I., Sonoda. N. (1977). *Angew. Chem. Int. Ed.* **16**, 789. (b) Seki, Y., Murai, S., Sonoda, N. (1979). *Angew. Chem. Int. Ed.* **17**, 119. (c) Murai, S., Sonoda, N. (1979). *Angew. Chem. Int. Ed.* **18**, 837.
229. (a) Chatani, N., Murai, S., Sonoda, N. (1983). *J. Am. Chem. Soc.* **105**, 1370. (b) Chatani, N., Fujii, S., Yamasaki, Y., Murai, S., Sonoda, N. (1986). *J. Am. Chem. Soc.* **108**, 7361. (c) Chatani, N., Ikeda, T., Sano, T., Sonoda, N., Kurosawa, H., Kawasaki, Y., Murai, S. (1988). *J. Org. Chem.* **53**, 3387. (d) Chatani, N., Kajikawa, Y., Nishimura, H., Murai, S. (1991). *Organometallics* **10**, 21.
230. Urata, H., Goto, D., Fuchikami, T.(1991). *Tetrahedron Lett.* **32**, 3091.
231. Yasui, H., Mizutani, K., Yorimitsu, H., Oshima, K. (2006). *Tetrahedron* **62**, 1410.
232. Gomes, P., Gosmini, C., Périchon, J. (2003). *Tetrahedron* **59**, 2999.
233. Amatore, M., Gosmini, C., Périchon, J. (2005). *Eur. J. Org. Chem.* 989.
234. (a) Korn, T. J., Knochel, P. (2005). *Angew. Chem. Int. Ed.* **44**, 2947. (b) Korn, T. J., Schade, M. A., Wirth, S., Knochel, P. (2006). *Org. Lett.* **8**, 725. (c) Korn, T. J., Schade, M. A., Cheemala, M. N., Wirth, S., Guevara, S. A., Cahiez, G., Knochel, P., (2006). *Synthesis,* 3547.
235. Rohbogner, C. J., Diène, C. R., Korn, T. J., Knochel, P. (2010). *Angew. Chem. Int. Ed.* **49**, 1874.
236. Fujisawa, T., Kurita, Y., Kawashima, M., Sato, T. (1982). *Chem. Lett.,* 1641.
237. (a) Ishihara, T., Yamana, M., Ando, T. (1983). *Tetrahedron Lett.* **24**, 5657. (b) Ishihara, T., Maekawa, T., Yamasaki, Y., Ando, T. (1987). *J. Org. Chem.* **52**, 300.
238. Erdik, E., Eroğlu, F. (2000). *Synth. React. Inorg. Met. Org. Chem.* **30**, 955.
239. Ito, H., Toyoda, T., Sawamura, M. (2010). *J. Am. Chem. Soc.* **132**, 5990.
240. (a) Ren, P., Stern, L.-A., Hu, X. (2012). *Angew. Chem. Int. Ed.* **51**, 9110. (b) Yang, C.-T., Zhang, Z.-Q., Liang, J., Liu, J.-H., Lu, X.-Y., Chen, H.-H., Liu, L. (2012). *J. Am. Chem. Soc.* **134**, 11124.
241. Zhu, Y., Smith, D. A., Herbert, D. E., Gatard, S., Ozerov, O. V. (2012). *Chem. Commun.* **48**, 218.
242. Jenner, G., Kheradmand, H., Kiennemann, A. (1984). *J. Organomet. Chem.* **277**, 427.
243. Yang, J., White, P. S., Brookhart, M. (2008). *J. Am. Chem. Soc.* **130**, 17509.
244. (a) Wu, J., Zhang, L., Gao, K. (2006). *Eur. J. Org. Chem.* 5260. (b) Wu, J., Zhang, L., Luo, Y. (2006). *Tetrahedron Lett.* **47**, 6747.
245. Zhang, L., Wu, J. (2008). *Adv. Synth. Catal.* **350**, 2409.

246. (a) Yu, J.-Y., Kuwano, R. (2009). *Angew. Chem. Int. Ed.* **48**, 7217. (b) Lee, H. W., Kwong, F. Y. (2009). *Synlett,* 3151.
247. Yu, J.-Y., Shimizu, R., Kuwano, R. (2010). *Angew. Chem. Int. Ed.* **49**, 6396.
248. Rao, H., Li, C.-J., (2011). *Angew. Chem. Int. Ed.* **50**, 8936.
249. (a) Komiya, S., Suzuki, J.-i., Miki, K., Kasai, N. (1987). *Chem. Lett.* 1287. (b) Planas, J. G., Marumo, T., Ichikawa, Y., Hirano, M., Komiya, S. (2000). *J. Chem. Soc., Dalton Trans.,* 2613.
250. Chatani, N., Tatamidani, H., Ie, Y., Kakiuchi, F., Murai, S. (2001). *J. Am. Chem. Soc.* **123**, 4849.
251. Nichols, J. M., Bishop, L. M., Bergman, R. G., Ellman, J. A. (2010). *J. Am. Chem. Soc.* **132***,* 12554.
252. von Stein, T., Weigand, T., Merkens, C., Klankermayer, J., Leitner, W., *Chem Cat Chem,* **5**, 439.
253. Matsumura, K., Kitazawa, K., Terai, S., Kochi, T., Ie, Y., Nitani, M., Aso, Y., Kakiuchi, F. (2012). *Org. Lett.* **14**, 3882.
254. Ogiwara, Y., Kochi, T., Kakiuchi, F. (2011). *Org. Lett.* **13**, 3254.
255. Ackermann, L., Althammer, A., Born, R. (2006). *Angew. Chem. Int. Ed.* **45**, 2619.
256. Ackermann, L., Mulzer, M. (2008). *Org. Lett.* **10**, 5043.
257. Matsuura, Y., Tamura, M., Kochi, T., Sato, M., Chatani, N., Kakiuchi, F. (2007). *J. Am. Chem. Soc.* **129**, 9858.
258. Lee, D.-H., Kwon, K.-H., Yi, C. S. (2011). *Science* **333**, 1613.
259. Lee, D.-H., Kwon, K.-H., Yi, C. S. (2012). *J. Am. Chem. Soc.* **134**, 7325.
260. Yeung, C. S., Hsieh, T. H. H., Dong, V. M. (2011). *Chem. Sci.* **2**, 544.
261. Imazaki, Y., Shirakawa, E., Ueno, R., Hayashi, T. (2012). *J. Am. Chem. Soc.* **134**, 14760.

CHAPTER 6

HOMOGENEOUS TRANSITION-METAL-CATALYZED C–H BOND FUNCTIONALIZATION

BI-JIE LI and ZHANG-JIE SHI

College of Chemistry and Molecular Engineering, Peking University, Beijing, People's Republic of China

Homogeneous Catalysis for Unreactive Bond Activation, First Edition. Edited by Zhang-Jie Shi.

6.1 OVERVIEW

The capability of direct installation of a new functionality to a C–H bond via direct C–H bond transformation is highly attractive in organic synthesis. Considering the ubiquitous nature of C–H bonds in organic substrates including hydrocarbons, drug molecules, polymers, and biological proteins, direct functionalization of C–H bond will impact both small-molecule synthesis and industrial chemical production. Selective C–H bond functionalization has been pursued since the early twentieth century and represents a highly desirable goal in organometallic chemistry.

The introduction of transition metals to the area of C–H activation has great impact and provides unlimited opportunities in this field. The term "C–H activation" is initially used to describe the cleavage of alkane C–H bond by metal complexes, but with time this term has become inaccurate. Typically the term "C–H bond activation" refers to the direct interaction of a C–H bond with the metal reagent to form a new C–M intermediate. This term could be either (a) a stoichiometric reaction of a C–H bond with metal to form a new product containing a C–M bond or (b) a catalytic reaction that comprises the stoichiometric cleavage of C–H bond by the metal center as an elementary step. By this definition, C–H activation distinguishes itself from (a) traditional ionic substitution such as a Friedel–Crafts reaction and (b) radical substitution such as halogenation of alkane.

Selective C–H functionalization faces two critical challenges. One is to achieve reactivity and the other is to control the selectivity. Many C–H bonds are inert and have high bond dissociation energy. Thus, the kinetic barrier of the cleavage of a C–H bond by a metal complex is usually high and consequently leads to slow or no C–H cleavage. Thus, the development of a more reactive metal complex is necessary to achieve the cleavage of an inert C–H bond. For example, the cleavage of an alkane C–H bond is only possible when highly electron-rich metal complexes are used. In addition, due to the high bond dissociation energy of a C–H bond, the overall transformation of C–H bond to another C–X bond is usually a thermodynamically disfavored process. For example, carbonylation of arene and alkane are slightly endothermic. Dehydrogenation of alkane is also thermodynamically disfavored. Therefore, new reagents or new reactions must be designed to overcome the thermodynamic issue.

Another key issue of C–H functionalization is the control of regioselectivity. Commonly in a complex molecular setting, there are multiple C–H bonds with similar electronic or steric properpties. To make the C–H functionalization synthetically useful, the reaction has to be regioselective. In many cases, the most unreactive C–H site is the one that is desired to be functionalized. Therefore, new modes of C–H cleavage and further functionalization must be discovered. One can harness the reactivity of a metal center or the electronic/steric properties of a particular C–H bond to achieve regioselectivity. Another commonly used strategy is to use a directing group. By incorporation of a directing group into the substrate, it coordinates to the metal center and directs the metal center to cleave a specific C–H bond through formation of a metallocycle. Many other strategies have also been developed to guide the design and application of regioselective C–H functionalization.

Through the understanding of the intrinsic reactivity of C–H bonds and the underlying mechanism, many metal-catalyzed C–H functionalization reactions have been developed. Some of these reactions can be applied to complex molecular settings. Usually, the synthetic steps to the target molecules can be significantly shortened by using a C–H functionalization strategy that allows the use of easily available C–H substrates as the starting material. In a retrosynthetic prospect, C–H functionalization strategy could inspire novel disconnection of a target molecule as compared with conventional retrosynthetic analysis that relies on functional group manipulation.

Catalytic C–H functionalization is a fast growing and rapidly evolving research field. New reactions emerge on a regular basis, and mechanistic understanding continues to change. Therefore, it extremely difficult to cover every aspect of C–H functionalization reactions. Dozens of reviews and monographs are available on this topic [1]. In the following sections, first we will discuss the mechanism of C–H cleavage. Then we will describe several types of C–H functionalization reactions that proceed through an organometallic intermediate. Smilar classification has been adopted in an excellent book [1x]. A mechanistic discussion is also included where possible. Literature is covered up to early 2013.

6.2 MECHANISM OF C–H CLEAVAGE

The reaction of a metal complex with C–H bond to form an organometallic intermediate is a crucial step in the C–H functionalization reaction. This step not only cleaves the C–H bond, but also generates an organometallic intermediate that could undergo further reaction to produce the functionalized product. In addition, the selectivity of a catalytic C–H fuctionalization is ususally governed by the selectivity in the reaction of a metal complex with a C–H bond. Generally, the C–H cleavage event can be classified into five categories according to their overall stoichiometry.

6.2.1 Oxidative Addition

Oxidative addition reaction usually occurs at electron-rich, low-valent late transition metal complexes such as Re, Fe, Ru, Os, Rh, Ir, and Pt. In this process, the reactive species has to be coordinatively unsaturated, generated *in situ* by thermal or photochemical decomposition of a transition metal complex precursor.

In early examples, the oxidative additions of C–H bonds in the ligand coordinated to the metal center were described. The intramolecular reactions have been observed for C–H bonds in the phosphine ligand [Eqs. (6.1)–(6.3)]. Metalation of both aryl and alkyl phosphines are known [2–8]. Intramolecular oxidative addition of an alkyl C–H bond of a bis-neopentyl iridium complex was also observed upon extrusion of neopentane [Eqs. (6.4) and (6.5)] [9–11].

Chatt and Davidson [12] observed an intermolecular arene C–H oxidative addition to a Ru complex containing chelating alkyl phosphine [Eq. (6.6)]. Reduction of *trans*-$[RuCl_2(PP)_2]$ by arene negative ions gave a series of hydrido (aryl) complexes. Green and Knowles [13] also observed arene C–H oxidative addition to a dicyclopentadienyl tungsten complex generated by photolysis of tungsten dihydride [Eq. (6.7)].

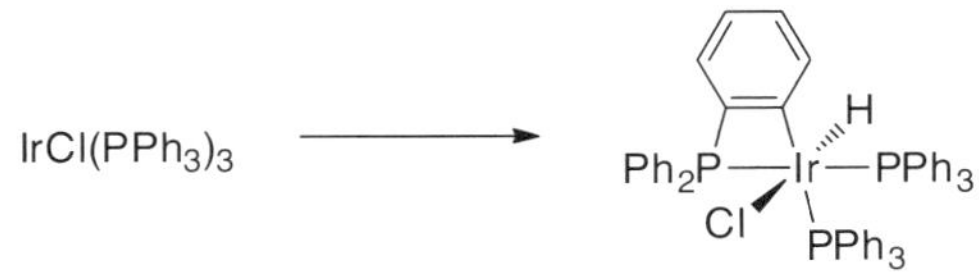

(6.1)

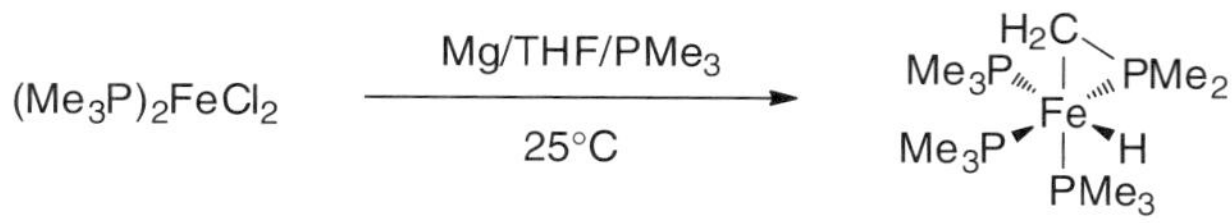

(6.2)

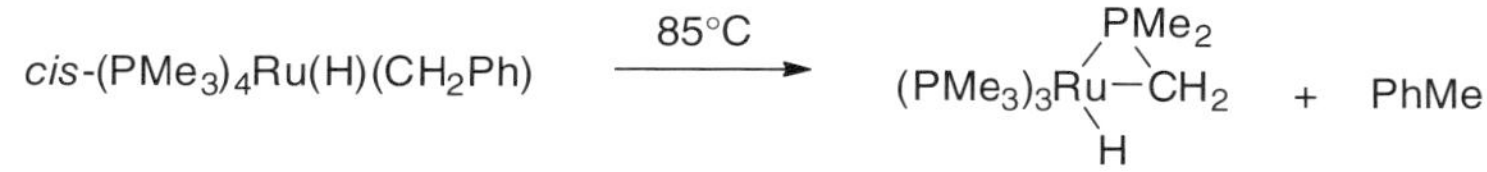

(6.3)

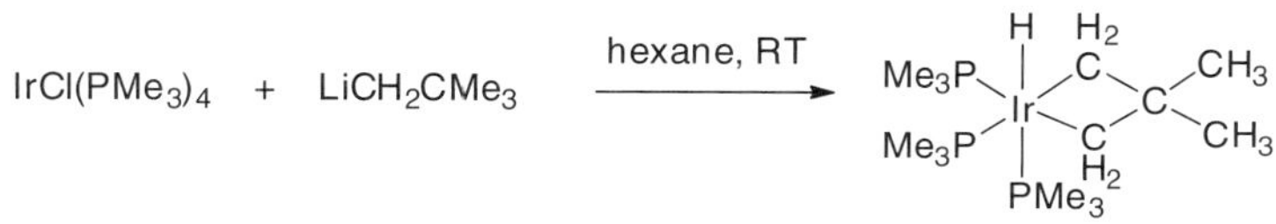

(6.4)

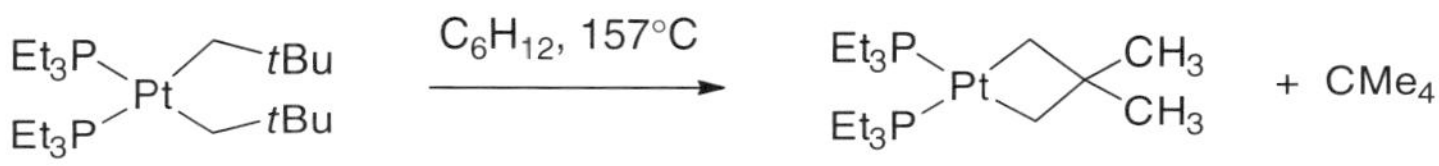

(6.5)

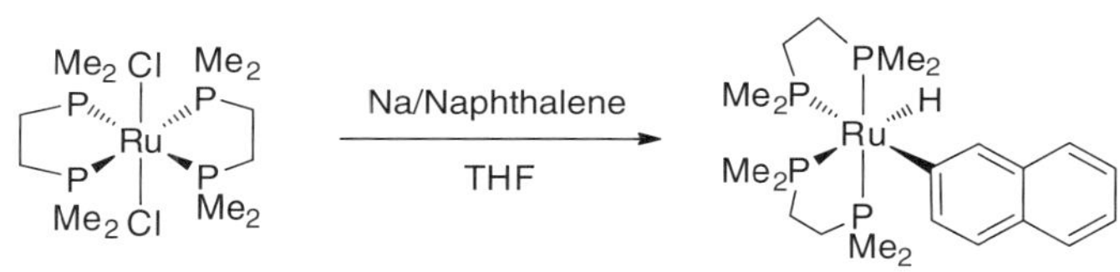

(6.6)

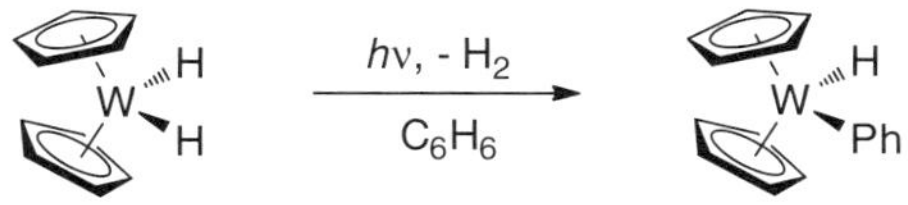

(6.7)

Although an aromatic C–H bond is stronger than an alkyl C–H bond, the oxidative addition of an aryl C–H bond is favored. The kinetic preference has been proposed to be a result of η^2-coordination of an arene to the metal center before C–H cleavage. Thermodynamically, the metal-aryl product of oxidative addition is much more stable than metal-alkyl product. This energy difference can compensate for the higher bond strength of aryl C–H bond than alkyl C–H bond.

Compared with the oxidative addition of an aromatic C–H bond, the intermolecular oxidative addition of an alkane C–H bond to a metal center is more difficult. This can be attributed to the lack of effective interaction between the alkane and the metal complex, along with the instability of the resulting alkyl metal species obtained by oxidative addition. Although intermolecular oxidative addition of alkane to a transition metal complex was not reported until the early 1980s, there is substantial evidence which suggests that this type of reaction could occur. Some metal complexes including a Pt complex were known to effect hydrogen–deuterium exchange and oxidation of the alkane C–H bond under relatively mild conditions [14]. Alkane dehydrogenation using soluble metal complexes has been observed [15, 16]. In addition, several examples of strong agostic interaction between a ligand C–H bond and the metal center were discovered [17]. These results strongly suggest the feasibility of the oxidative addition of alkane.

In 1982, Janowicz and Bergman [18] and Hoyano and Graham [19] independently reported the first examples of oxidative addition of an alkane C–H bond to a low-valent iridium center [Eq. (6.8)]. At nearly the same time, Jones and Feher [20] observed analogous oxidative addition of alkane to a Rh complex [Eq. (6.9)]. A significant feature of these reactions is the terminal selectivity. The strongest primary C–H bond underwent selective cleavage to afford terminal alkyl metal hydride complex. This type of regioselectivity clearly contrasts that of radical reactions and carbene insertion reactions, which favors tertiary and secondary C–H bonds over a primary C–H bond. The terminal alkyl metal complexes are typically more stable than the corresponding secondary and tertiary alkyl metal complexes [21].

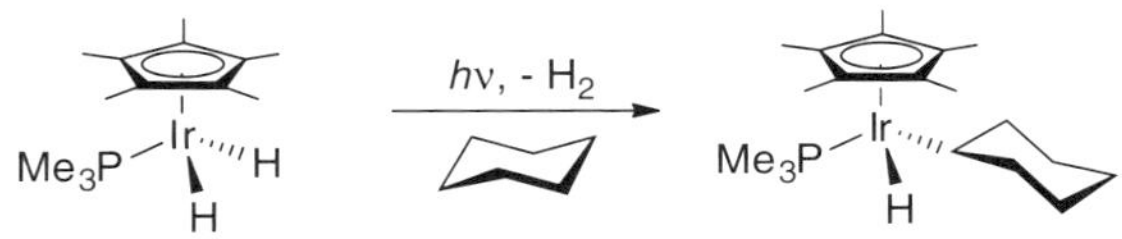

(6.8)

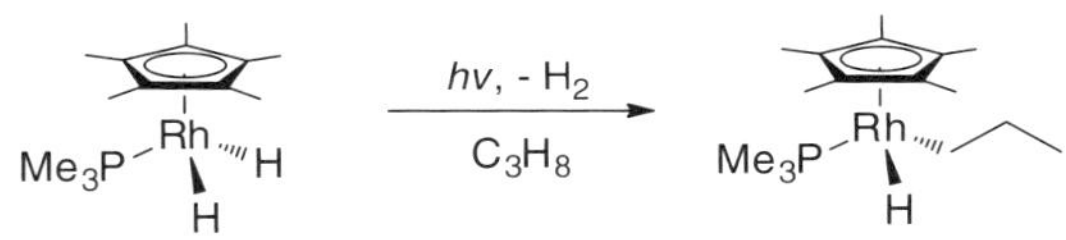

(6.9)

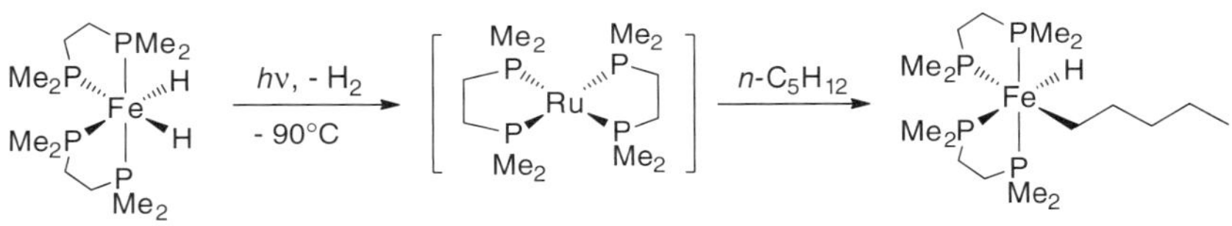

(6.10)

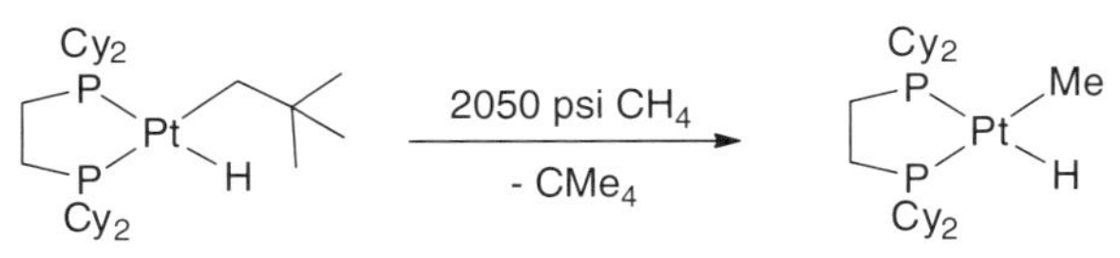

(6.11)

Subsequently, many other examples of alkane oxidative addition have been reported with iron [Eq. (6.10)] [22], rhodium [23, 24], rhenium [25, 26], iridium [27, 28], and platinum [Eq. (6.11)] [29] complexes. Most of these complexes require strong electron-donating ligands such as chelating alkylphosphine, Cp*, substituted tris-pyrazolylborate, and PCP-type ligand to make the metal center electron-rich enough to insert into alkane C–H bonds.

Substantial evidence suggests the existence of a sigma complex prior to the C–H cleavage [Eq. (6.12)] [30]. The formation of a sigma complex is a result of an electron donation from an alkane C–H bond to a metal center and back donation of electron density from a metal center to a sigma* orbital of a C–H bond. Although many complexes in which a C–H bond coordinates intramolecularly to the metal center have been observed by various types of spectroscopy in solution and even by X-ray in the solid state [31–33], no stable sigma complexes have been isolated. However, their existence as transient intermediates has been supported by isotope scrambling studies and an inverse kinetic isotope effect in the reductive elimination of alkyl metal hydride complexes [34]. The generation and spectroscopic detection of agnostic complexes between a hydrocarbon and a metal center have been carried out using an ultraviolet–visible spectrum and time-resolved infrared spectroscopy. Alkane complexes of CpReL(CO)(alkane) were generated by low-temperature photodissociation experiments and characterized by NMR spectroscopy [35]. A Rh(I)–methane sigma complex, obtained by protonation of a rhodium–methyl precursor at low temperature, was characterized by NMR spectroscopy as well [36].

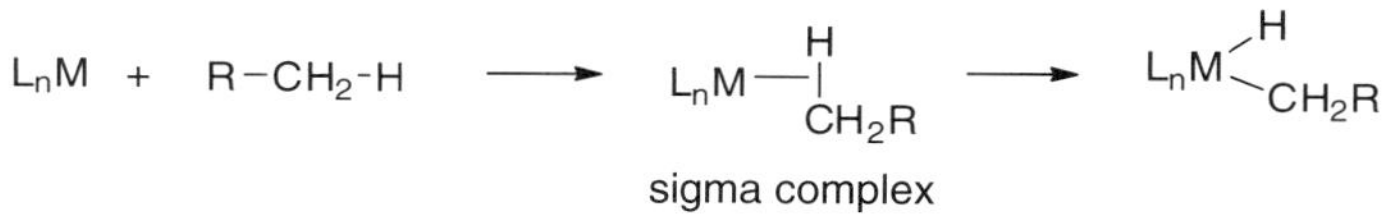

(6.12)

Once a sigma complex between alkane and the metal center has formed, the coordinated C–H bond is further cleaved to generate the organometallic product. Through the time-resolved infrared experiments, Frei and co-workers [37] provided a detailed picture of the C–H oxidative addition reaction [Eq. (6.13)]. The reaction of Tp*Rh(CO)$_2$ (Tp* = HB–Pz$_3$*, Pz* = 3,5-dimethylpyrazolyl) in an alkane solution began with dissociation of CO ligand to afford a coordination unsaturated intermediate which forms a sigma complex with alkane. After further dissociation of one nitrogen ligand on the Tp, the new intermediate cleaves the alkane C–H bond on a nanosecond timescale. Rapid recoordination of the nitrogen atom forms the final product (Scheme 6.1).

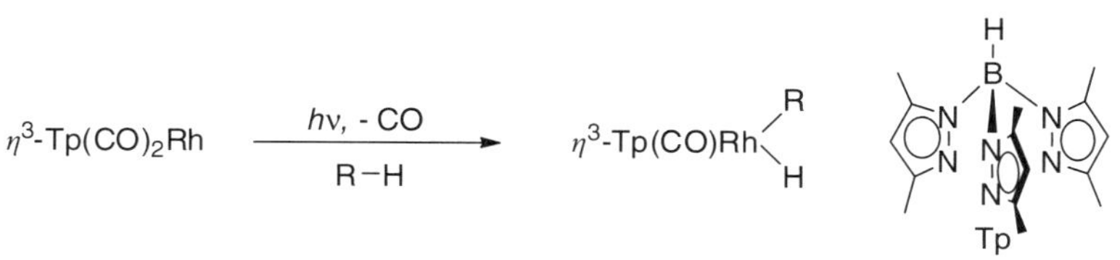

(6.13)

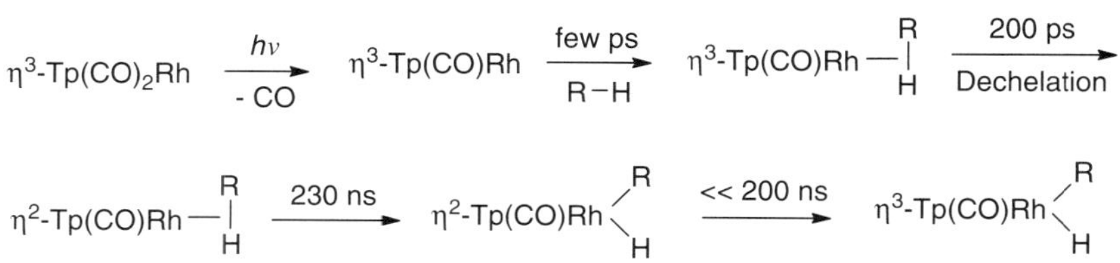

Scheme 6.1. C–H bond cleavage.

Similarly, the oxidative addition of arene C–H to a metal complex also begins with the initial interaction of the arene with the metal center [38, 39]. It is generally accepted that an η^2-arene π-complex is formed before the C–H bond is cleaved.

6.2.2 Electrophilic Substitution

Electrophilic substitution of a C–H bond usually occurs at an electron-deficient high valent metal center. A C–M bond is formed with the loss of a proton.

In as early as 1892, Volhard [40] showed the metalation of thiophene with mercury compounds. Later, this reaction was extended to various arenes [41]. In 1931, Kharasch and Isbell [42] reported that $AuCl_3$ can metalate aromatic hydrocarbons under very mild conditions [Eq. (6.14)]. These early works inspired numerous later studies of electrophilic metalation with metal complexes. A large variety of high-valent metal complexes of Tl^{3+}, Pb^{4+}, Pt^{4+}, Hg^{2+}, Ir^{3+}, Au^{3+}, and Pd^{2+} have been shown to metalate arenes [43]. However, direct electrophilic metalation of alkane to afford observable oganometallic intermediate is rare.

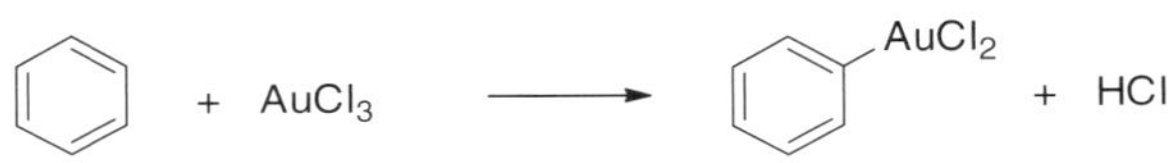

(6.14)

It is generally accepted that the electrophilic substitution starts with the formation of a π complex consisting of an arene and a metal center [Eq. (6.15)]. Subsequent electrophilic attack of the π electron by the electron-deficient metal forms a weireb intermediate. Finally, the loss of a proton and rearomatization provides the aryl–metal complex. NMR spectroscopy of the intermediate complexes in aromatic mercurations suggests that arene mercurinium ion complexes involve a rapid exchange of σ and π complexes [44].

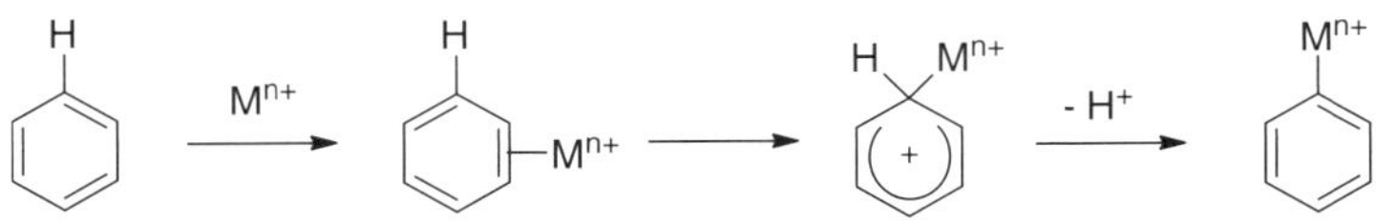

(6.15)

By attaching a basic functional group to the arene, it coordinates to the metal and directs the metalation to occur at the *ortho* position of the functional group [45, 46]. This process is commonly known as directing group-assisted *ortho* metalation and generates a metallocycle. Coordination with the directing group brings the metal to the proximity of C–H bond, facilitating the intramolecular metalation to occur [Eqs. (6.16) and (6.17)]. In addition, the presence of a directing group usually allows the metalation to occur regioselectively at the *ortho* position. This strategy has been used in many catalytic regioselective C–H functionalization reactions. A plethora of metallocycles have been synthesized through *ortho* metalation with high-valent metal complexes, with the largest number being reported with palladium [47]. Various directing groups such as pyridine, oxazoline, imine, amide, sulfide, and phosphine have been reported to facilitate the ortho metalation with palladium. In most cases, these reactions generate five- and six-membered metallocycles.

By attaching a directing group to an alkyl chain, metalation of alkyl C–H bond has also been achieved. Activated benzylic bonds could be cleaved in the presence of pyridine and phosphine as directing groups. An unactivated alkyl C–H bond in an oxime substrate also underwent cyclopalladation to give a palladacycle [Eq. (6.18)] [48]. Intermolecular electrophilic metalation of an alkane C–H bond by a metal complex to form an observable organometallic species has not been firmly established. However, this process is strongly indicated in catalytic alkane functionalization reactions using Pd^{2+}, Pt^{4+}, Hg^{2+}, Au^{3+} [49].

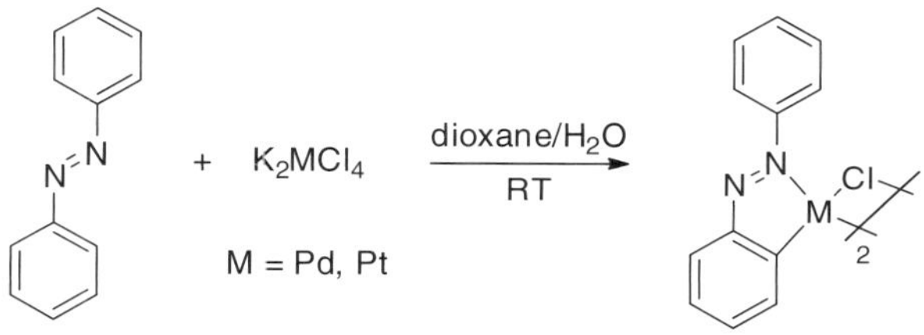

(6.16)

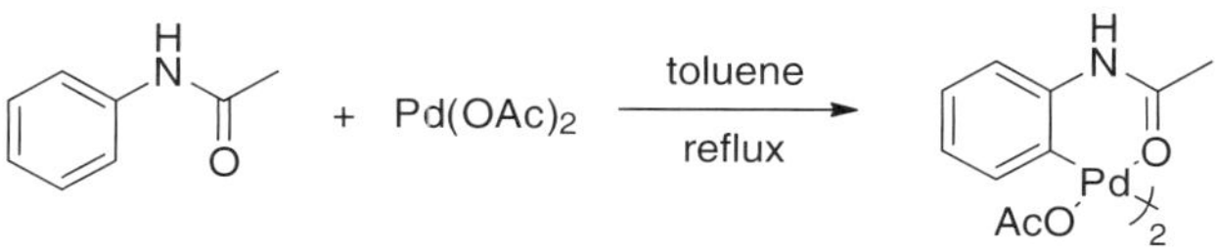

(6.17)

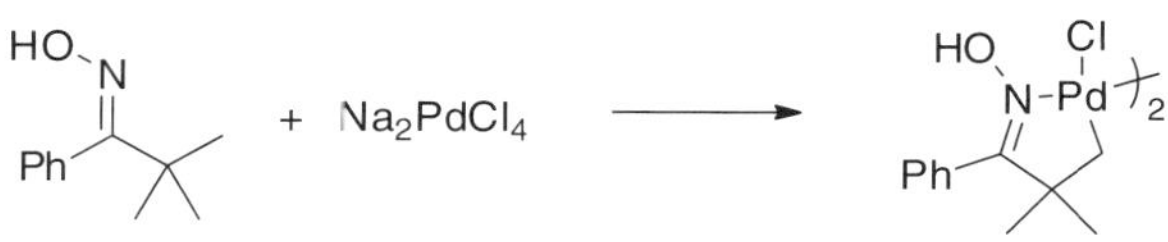

(6.18)

6.2.3 Sigma Bond Metathesis

A general scheme of sigma bond metathesis is shown below. Through a sigma bond metathesis between M–R and a C–H bond, a new M–C complex and a new C–H bond are formed [Eq. (6.19)]. These reactions usually occur with group 3 metals such as Sc, lanthanides and actinides [50], but some examples involving groups 4 and 5 metals are also known [51].

$$L_nM{-}R \quad + \quad R'{-}H \quad \longrightarrow \quad L_nM{-}R' \quad + \quad R{-}H \tag{6.19}$$

Early examples of sigma bond metathesis with an alkyl C–H bond are intramolecular reactions [Eq. (6.20)] [52]. Intermolecular sigma bond metathesis is now known for both arene and alkane. Even a methane C–H bond can be cleaved by lutetinum and yttrium–methyl complexes to form the product with the exchange of a methyl group [53]. A thoracyclobutane complex reacts with alkanes (including methane) through sigma bond metathesis to afford an alkyl thorium product [Eq. (6.21)] [54]. The reactivity of arene and alkane in a sigma bond metathesis pathway seems to correlate with those in oxidative addition reactions. For example, the reaction of arene is faster that that of alkane with scandium methyl complex, although an aryl C–H bond is stronger than an alkyl C–H bond [Eq. (6.22)] [50].

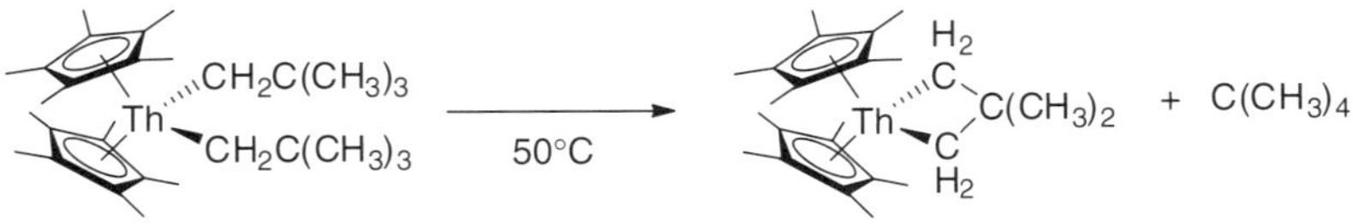

(6.20)

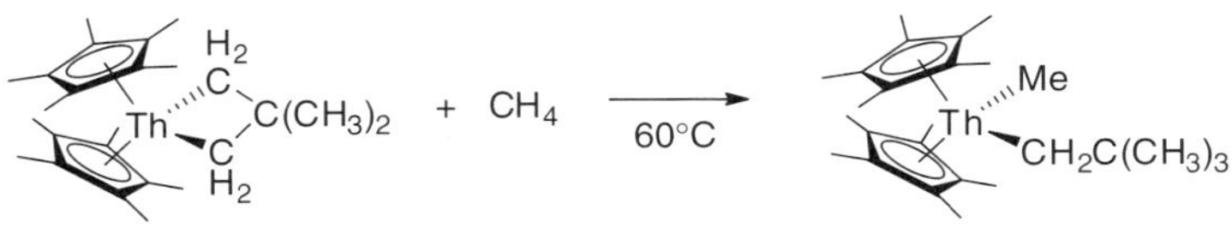

(6.21)

$$Cp^*_2Sc{-}Me \quad + \quad R{-}H \longrightarrow Cp^*_2Sc{-}R \quad + \quad CH_4 \tag{6.22}$$

R = aryl C-H faster than R = alkyl C-H

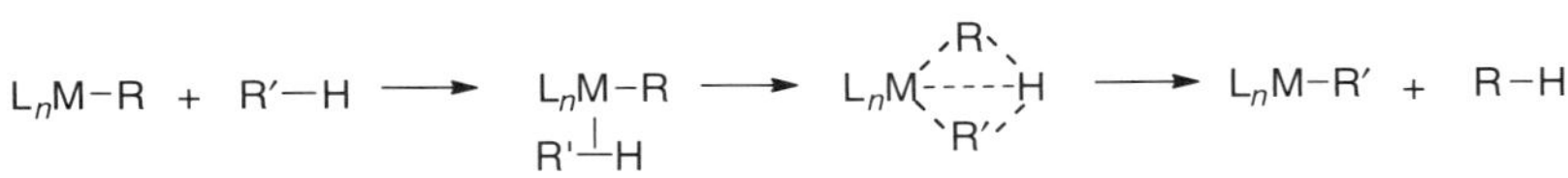

(6.23)

Computational studies indicated that a sigma bond metathesis occurs through multiple steps [55, 56]. It starts with the generation of a coordinatively unsaturated intermediate. Similar to that in oxidative addition, this intermediate coordinates with arene or alkane C–H bond to form a sigma complex. The sigma complex undergoes an intramolecular bond metathesis through a four-centered (skite) transition state to form a new sigma complex. Dissociation of the sigma complex generates the new M–C bond and a R–H bond [Eq. (6.23)].

6.2.4 1,2-Addition

A 1,2-addition reaction involves the addition of a C–H bond across a metal–ligand multiple bond to generate an organometallic compound containing a C–M bond. This type of reaction is most common for imido and carbene complexes of early and middle transition metal such as zirconium [57–59], titanium [60, 61], tantalum [62], tungsten [63], and molybdenum [63]. Late transition metal carbene and nitrene complexes react with hydrocarbon to produce a C–H insertion product which does not involve the formation of a new M–C bond [64, 65]. Thus, it is mechanistically distinct from 1,2-addition.

Early studies of 1,2-addition across a zirconium imodo complex were described by Bergman and Wolczanski. Bergman and co-workers [66] reported that a zirconocene imodo complex, generated *in situ* from a zirconocene amido complex, can undergo reaction with benzene to afford a 1,2-addition product [Eq. (6.24)]. Wolczanski and co-workers [67] showed that the *in situ*-generated silylamidozirconium–imido complex can cleave the C–H bonds in benzene and even methane to afford phenyl and methyl zirconium compounds [Eq. (6.25)]. With a related silylimido tungsten complex, Schaefer and Wolczanski [68] has shown that the formation of alkane complex precedes C–H bond activation, a similar process as observed in the oxidative addition of C–H bond. The selectivity toward different types of C–H bonds was probed in the reactions of a titanium–imido complex [Eq. (6.26)] [69]. A general reactivity trend follows sp^2 C–H bond > cyclopropane > primary sp^3 C–H bond > secondary sp^3 C–H bond, which is similar to the selectivity in oxidative addition and electrophilic substitution reactions. The stability of the resulting 1,2-addition products also follows the same trend.

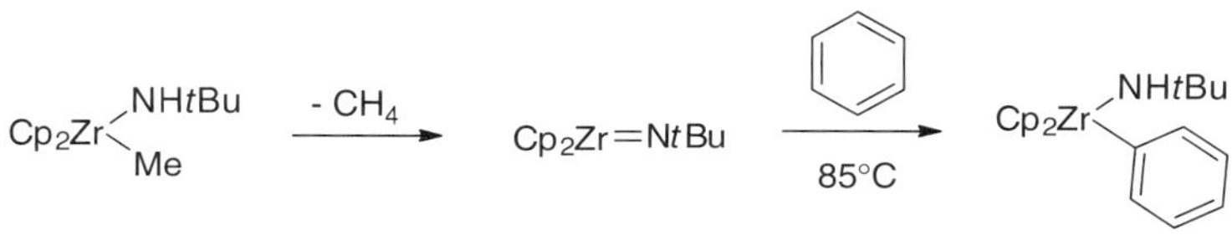

(6.24)

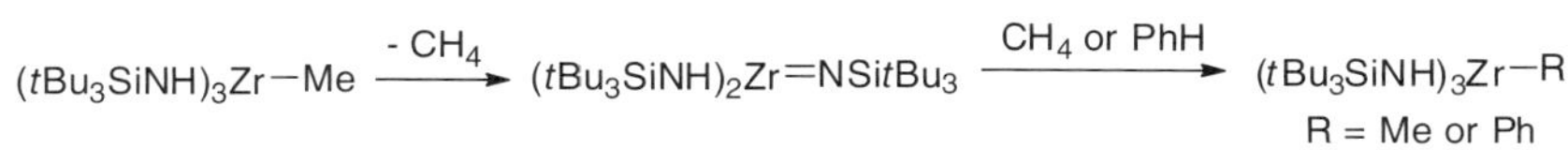

(6.25)

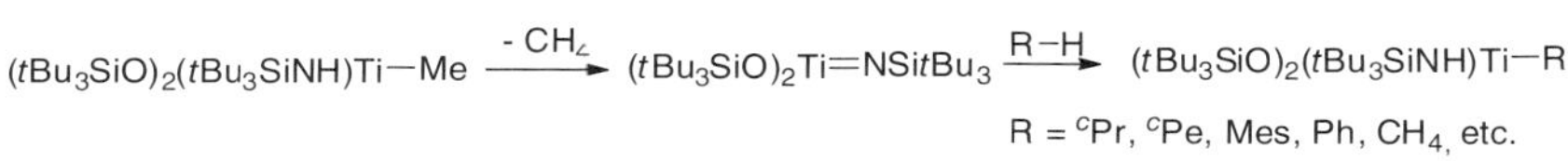

(6.26)

Metal–carbene complexes based on molybdenum and tungsten can also react with arene and alkane to form 1,2-addition products [70]. Thermolysis of Cp*M(NO)(hydrocarbyl)$_2$ precursors (M = Mo, W) results in loss of hydrocarbon and formation of 16-electron organometallic complexes such as Cp*M(NO)(alkylidene), Cp*M(NO)(η^2-benzyne), Cp*M(NO)(η^2-acetylene), and Cp*M(NO)(η^2-allene). These intermediates showed remarkable reactivity in C–H cleavage. For example, Cp*M(NO)(alkylidene) can react with benzene and tetramethylsilane to afford 1,2-addition products [Eq. (6.27)] [71]. The Cp*W(NO)(η^2-diene) complex cleaves an alkane C–H bond with exclusive terminal selectivity [Eq. (6.28)] [72]. In addition, some of these complexes can react further to effect double or triple C–H bond activations of the hydrocarbon substrates [Eq. (6.29)] [73]. Analogous 1,2-addition of a C–H bond across a M–C triple bond has been observed. The reaction of a PNP-ligand-supported titanium alkylidyne complex with benzene and ethane generates titanium–carbene complexes [Eq. (6.30)] [74, 75].

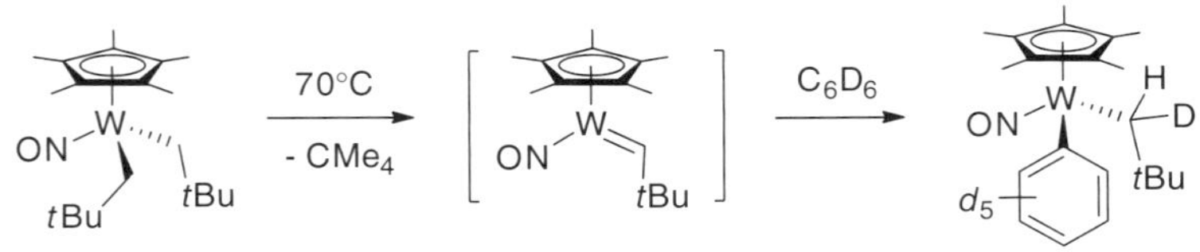

(6.27)

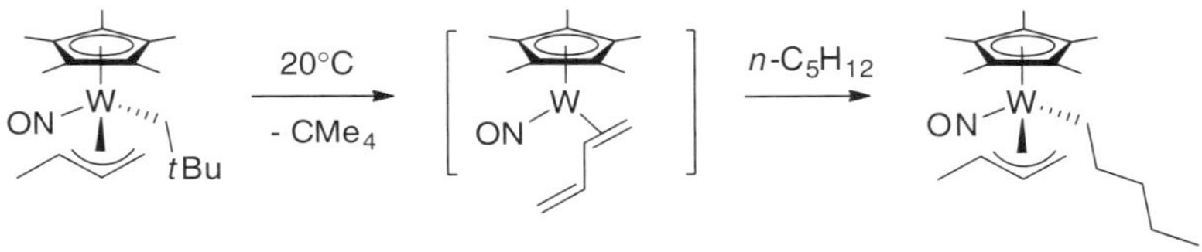

(6.28)

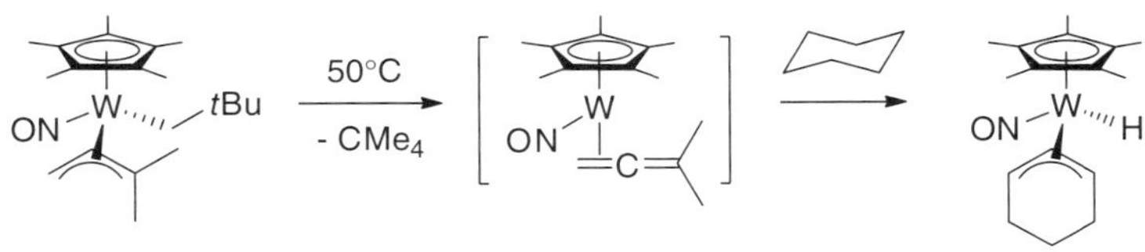

(6.29)

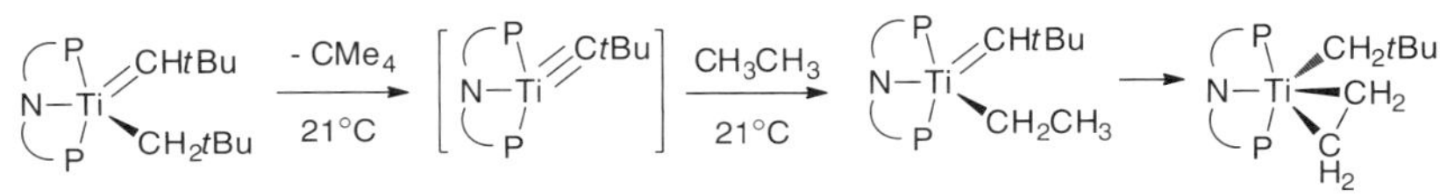

(6.30)

6.2.5 Metalloradical Activation

Rhodium(II)–porphyrin complexes, which either exist in monomer form or are generated by dissociation of a dimer, can cleave the C–H bond in methane to form a rhodium–hydride and a rhodium–methyl complex [Eq. (6.31)] [76, 77]. Kinetic studies suggest that a four-centered linear transition state has a relatively low activation enthalpy route for methane reacting with two metalloradicals. Although the benzylic C–H bond in toluene is weaker than that in methane, it reacted slower with metalloradicals because of the unfavorable steric interaction of the phenyl group with bulky metalloradicals in the transition state. An aromatic C–H bond is completely unreactive in this system as a result of the difficulty to align a linear Rh–C–H–Rh transition state for aromatic C–H bond.

$$2\,(\mathrm{TMP})\mathrm{Rh}\cdot + \mathrm{CH_4} \rightleftharpoons (\mathrm{TMP})\mathrm{Rh{-}H} + (\mathrm{TMP})\mathrm{Rh{-}CH_3}$$

TMP = Tetramesitylporphyrin

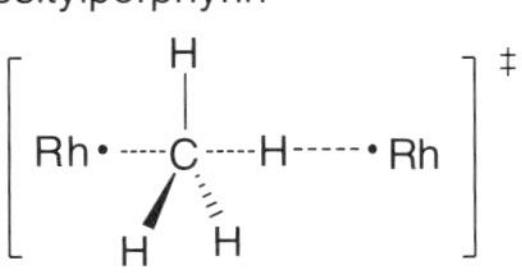

(6.31)

6.3 DIRECTED C–H OXIDATION

The cleavage of aromatic or aliphatic C–H bond in substrates attached to a functional group by a metal complex to form a metallocycle is known as cyclometalation. The formation of a specific ring size, usually a five- or six- membered ring, allows the metal to selectively cleave a C–H bond. But most of these processes are stoichiometric in metal complexes, and can hardly turn over. On the other hand, before 2000, there were several examples of metal-catalyzed C–H oxidation reactions to form a C–heteroatom bond, but they suffered from either low regioselectivity, low turn over numbers, or limited substrate scope [79, 80]. Now there are many examples of catalytic oxidative C–H functionalization reactions using directing-group-assisted cyclometalation strategy. The function of a directing group is severalfold. First, it allows regioselective cleavage and functionalization of a C–H bond *ortho* to the directing group through the formation of a metallocycle. Second, it enhances the reactivity of a target C–H bond by changing an intermolecular reaction to an intramolecular reaction through metal coordination. Third, the directing group usually can be converted to other functional groups, adding synthetic value to the C–H functionalization product.

One of the earliest examples of directing-group-assisted C–H oxidation is the chlorination of an *ortho* C–H bond in azobenzene by Cl_2 under Pd catalysis as reported by Fahey [81]. The reaction involves a putative palladacycle which reacts with Cl_2 to give the chlorination product [Eq. (6.32)]. Another elegant example is the Pt-catalyzed hydroxylation of amino acid developed by Sames and co-workers [82] [Eq. (6.33)]. Several amino acids can be cyclized under a modified Shilov catalyst system containing platinum(II) and copper oxidant. The C–H bond β to the carboxylic acid group was selectively cleaved to form a C–O bond.

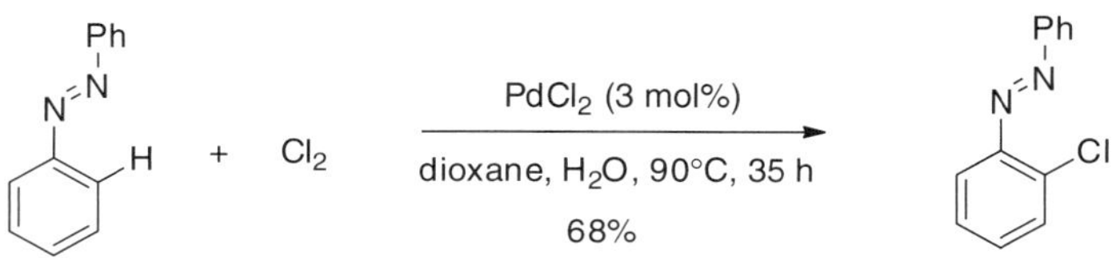

(6.32)

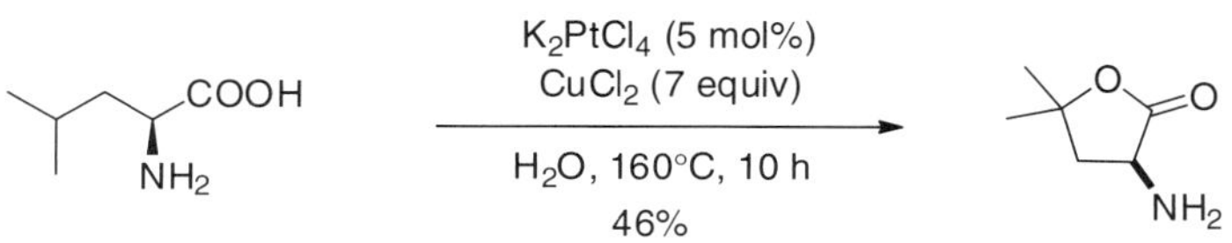

(6.33)

During the past decade, a number of groups revisited the directed oxidative C–H functionalization reactions and have developed a plethora of transition-metal (especially palladium)-catalyzed C–H oxygenation, amination, and halogenation reactions. Many transition-metal-catalyzed C–H oxidation reactions that convert both aryl and alkyl C–H bonds to C–O, C–N, and C–X bonds are developed. These studies exploit the use of various directing groups that enhance the synthetic utility of these processes.

6.3.1 Directed C–H Oxygenation

Sanford, Yu, and others reported Pd-catalyzed acetoxylation of aromatic and aliphatic C–H bonds directed by various nitrogen-based directing groups such as pyridine [Eq. (6.34)], oxazoline [Eq. (6.35)], and imine and oxygen-based directing groups such as amide, ester, and ketone. Strong oxidant such as $PhI(OAc)_2$ [83, 84], oxone [85–88] $K_2S_2O_8$ [89], and perester [90] were used in many cases. Related Pd-catalyzed directed alkoxylations of arene and alkyl C–H bonds were also developed using alcohol as the nucleophile in the presence of strong oxidants [Eq. (6.36)] [91–93]. Zhang and Yu [94] reported a Pd-catalyzed *ortho*-hydroxylation of benzoic acid using molecular oxygen as the oxidant [Eq. (6.37)]. The oxygen atom in dioxygen was directly incorporated into the substrate [94]. Ruthenium is also a competent catalyst for the *ortho* acetoxylation of arene containing amide, ester, and ketone groups [Eq. (6.38)] [95–97]. In addition, copper and rhodium complexes have also been used as catalysts in the directed *ortho*-acetoxylation of an aromatic C–H bond [Eq. (6.39)] [98, 99]. Pd-catalyzed intramolecular cyclization of phenethanol and 2-phenylacetic acid generate dihydrobenzofuran and benzofuranone derivatives, respectively, and the latter reaction can be carried out enantioselectively in the presence of a protected amino acid ligand [Eq. (6.40)] [100–102]. Pd-catalyzed directed C–S bond formation is achieved by using 2-phenylpyridine as the substrate and TsCl as the reagent [Eq. (6.41)] [103].

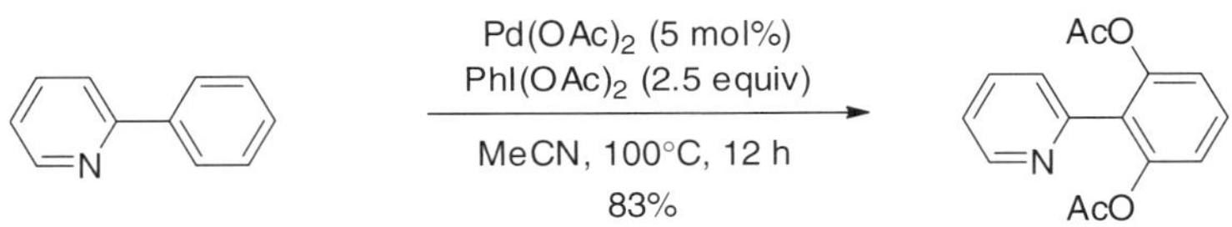

(6.34)

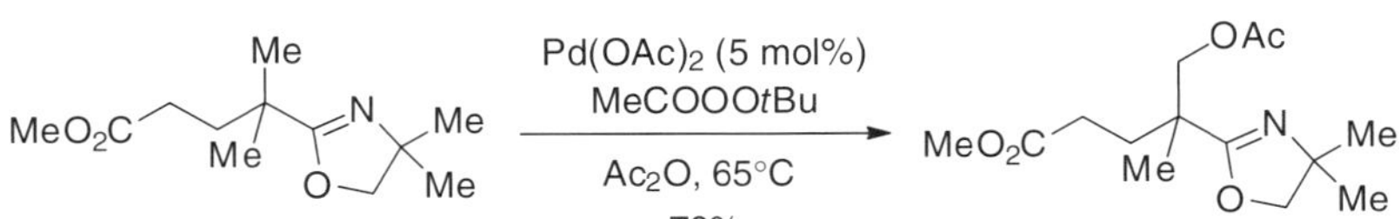

(6.35)

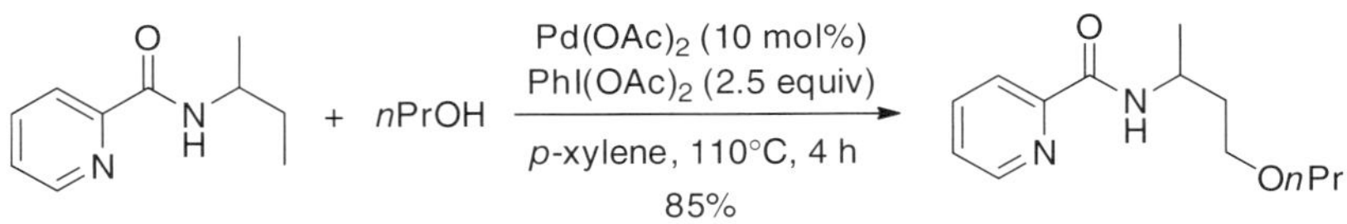

(6.36)

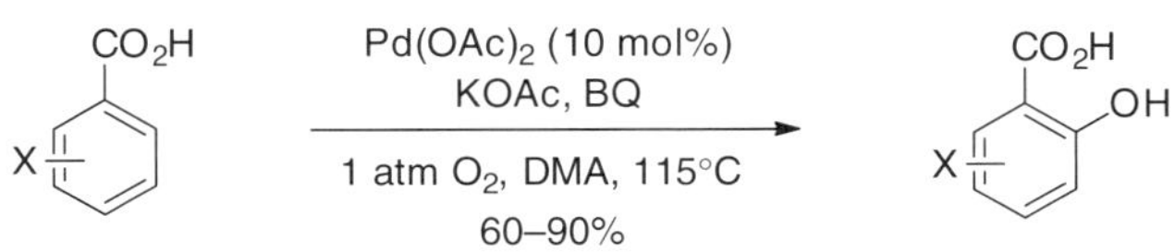

(6.37)

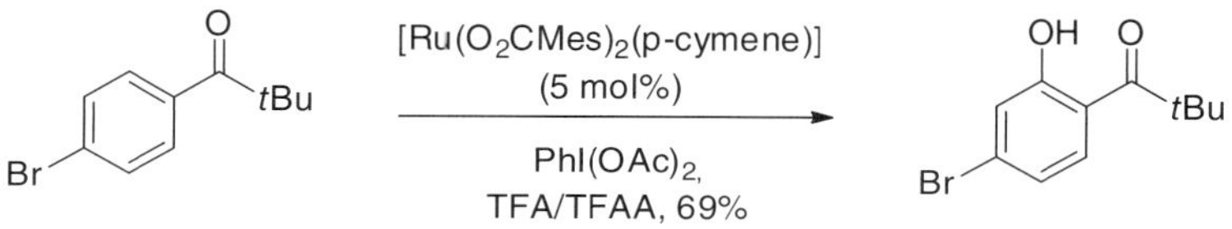

(6.38)

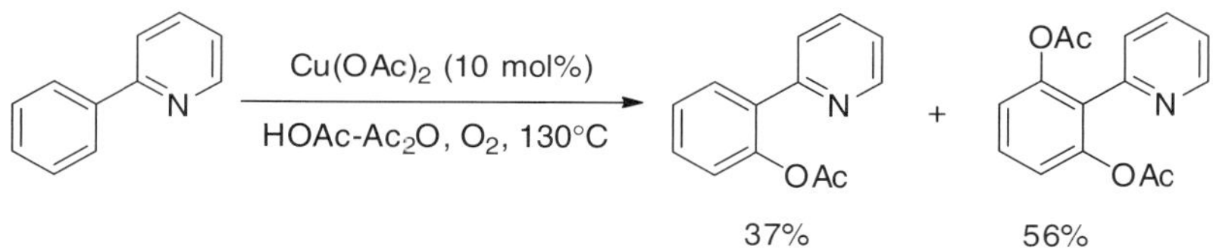

(6.39)

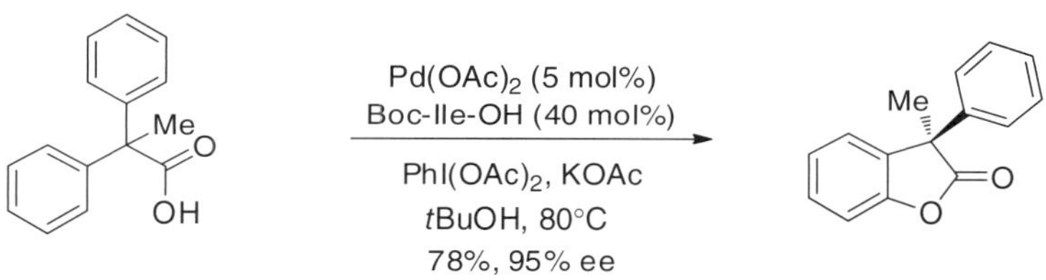

(6.40)

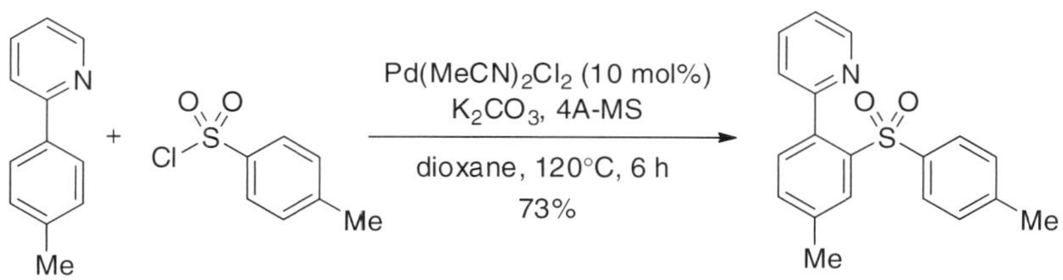

(6.41)

6.3.2 Directed C–H Amination

Similar strategy can also be applied to metal-catalyzed directed aryl and alkyl C–H amination reactions. In early studies, Buchwald and co-workers [104] developed a Pd-catalyzed intramolecular C–H amination reaction to form carbazole through amide-directed C–H cleavage [Eq. (6.42)]. Che reported a Pd-catalyzed intermolecular C–H amidation of both aromatic and aliphatic C–H bonds by using a combination of amide and persulfonate as the nitrogen source [Eq. (6.43)] [105]. Subsequently, intramolecular amination of aryl and alkyl C–H bond was developed to synthesize useful nitrogen-containing heterocycles such as carbazole [106, 107], benzimidazole [Eq. (6.44)] [108], benzoxazole [109], indazole [110], indoline [Eqs. (6.45) and (6.46)] [111, 112], and pyrroline [Eq. (6.47)] [113, 114] products. In most cases, palladium and copper were used as the catalysts. The intermolecular version has been realized with activated substrates such as azoles or polyfluorinated aromatics in many cases [Eq. (6.48)] [115–118], and directed amination of aromatic

C–H bonds with amide in combination with an oxidant [119] or with electrophilic nitrogen source [Eq. (6.49)] [120–123] are also known.

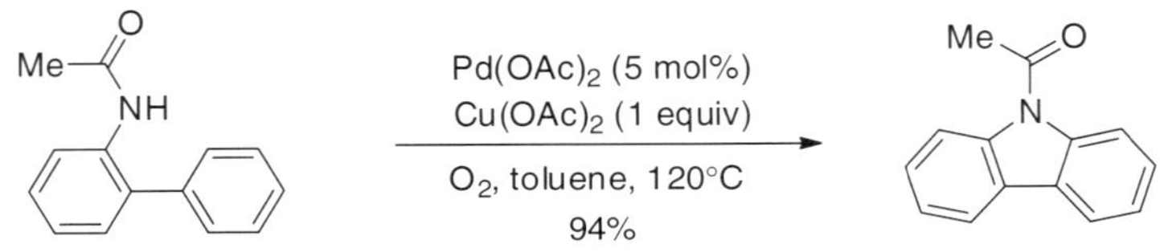

(6.42)

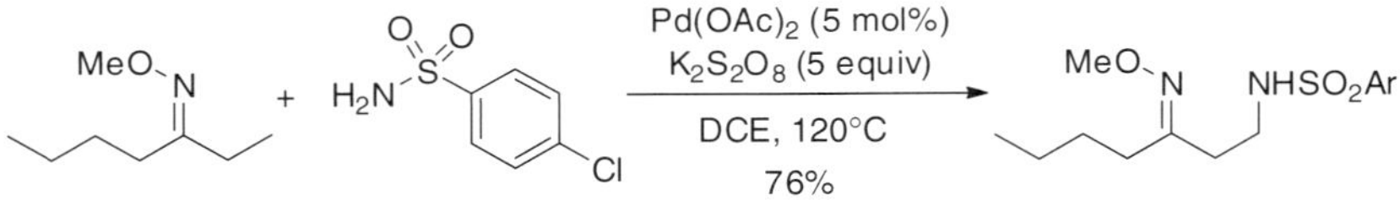

(6.43)

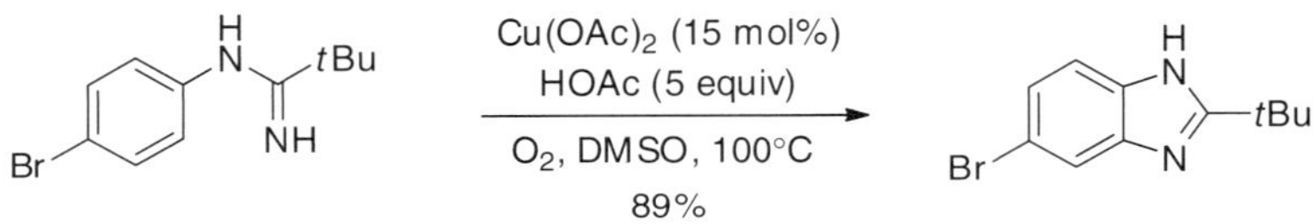

(6.44)

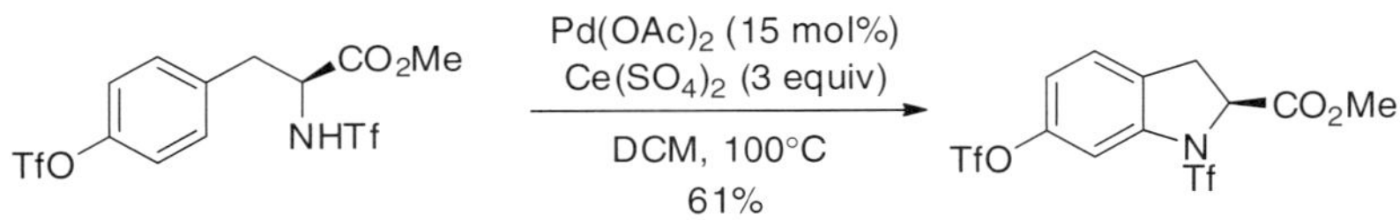

(6.45)

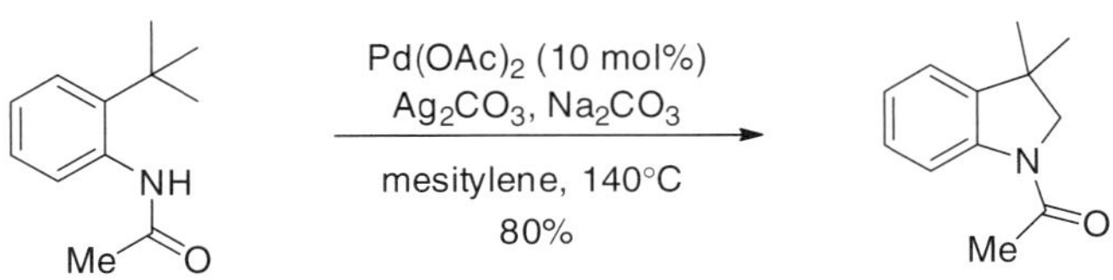

(6.46)

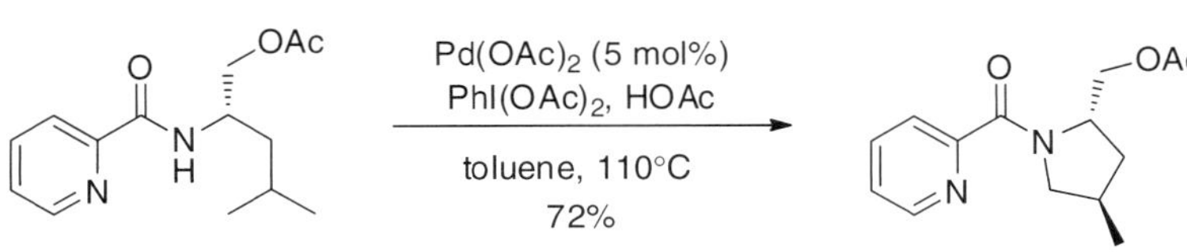

(6.47)

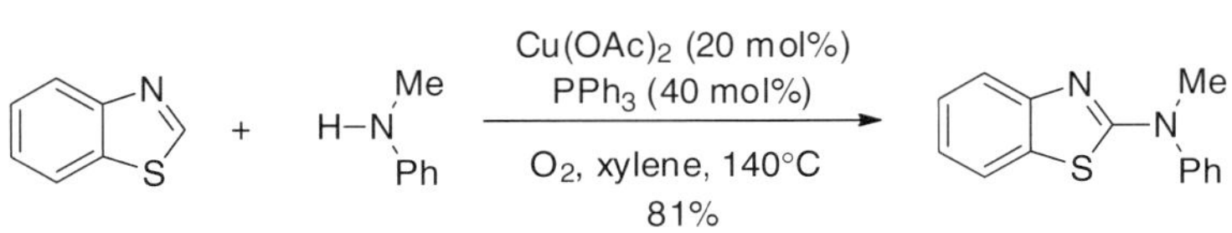

(6.48)

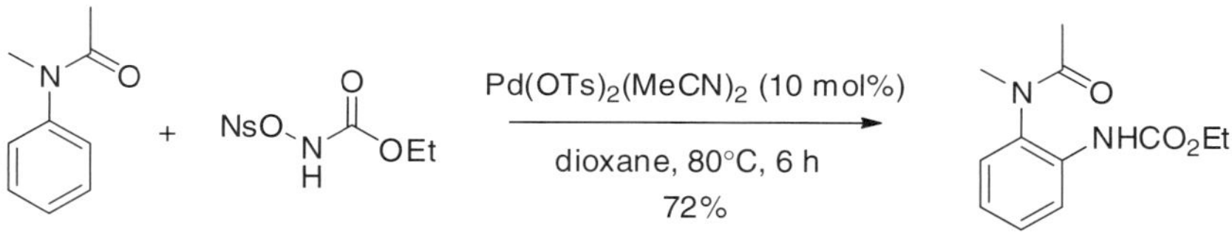

(6.49)

6.3.3 Directed C–H Halogenation

Regioselective halogenation of aryl and alkyl C–H bond has also been studied. A majority of these halogenation reactions were carried out with a Pd catalyst [Eqs. (6.50) and (6.51)], although Cu [124] and Rh [Eq. (6.52)] [125] catalysts were also reported. Different directing groups including pyridine [126–128], amide [129, 130], sulfamide [131], ketone, ester [125], and carboxylic acid [132, 133] have been investigated for aromatic C–H halogenation. Iodination of alkyl C–H bond was achieved using a similar strategy [Eq. (6.53)] [134, 135]. Either an electrophilic halogen reagent (such as NXS or IOAc) or a nucleophilic halogen anion together with an oxidant was employed to introduce the halogen atom. Similarly, use of electrophilic fluorination reagents allows regioselective fluorination of both aromatic and benzylic C–H bond in directing-group-containing substrates [Eq. (6.54)] [136–139].

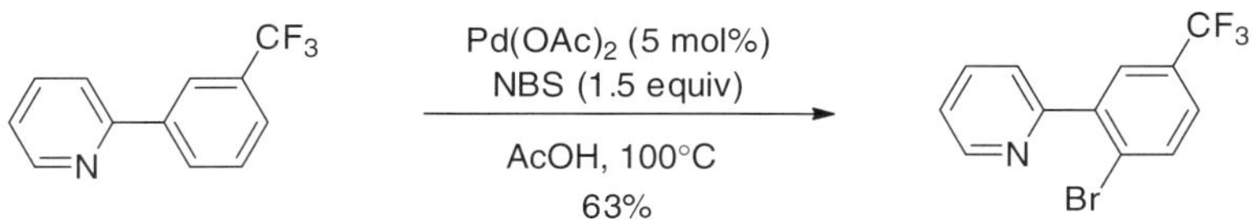

(6.50)

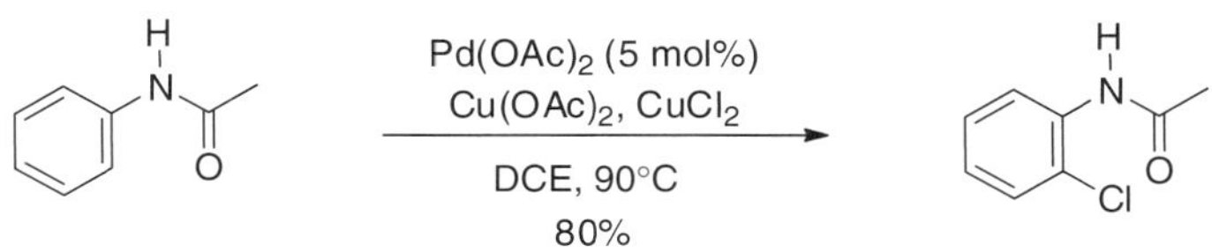

(6.51)

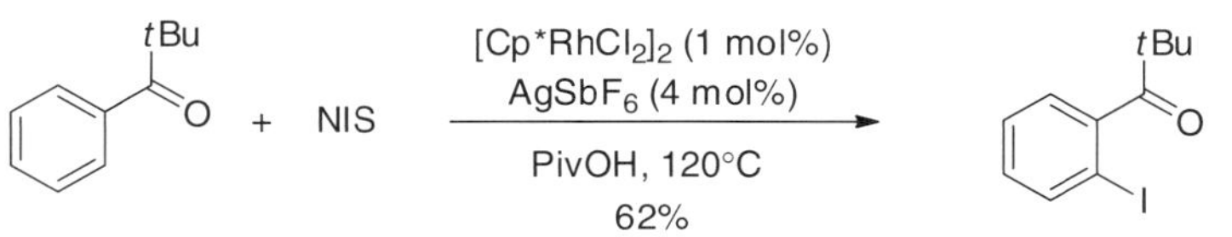

(6.52)

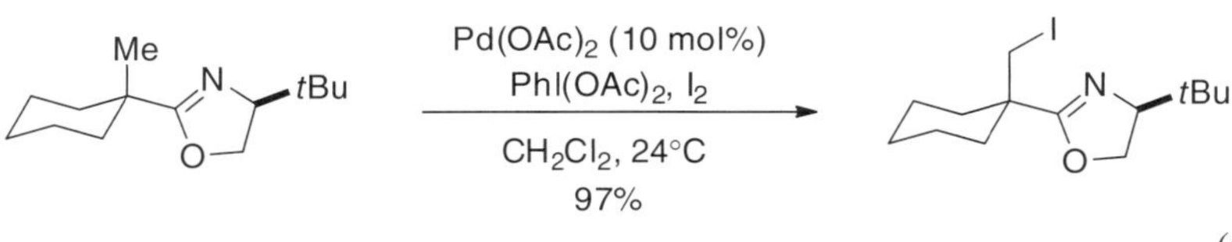

(6.53)

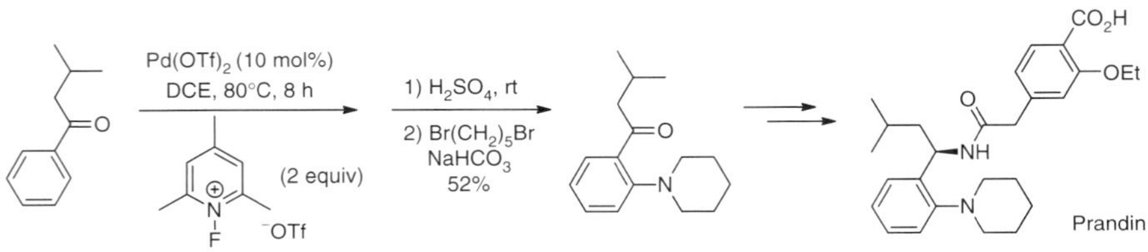

Scheme 6.2. Synthesis of Prandin.

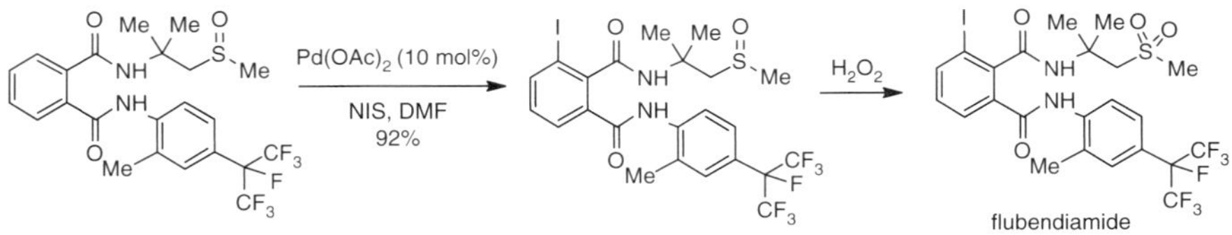

Scheme 6.3. Synthesis of flubendiamide.

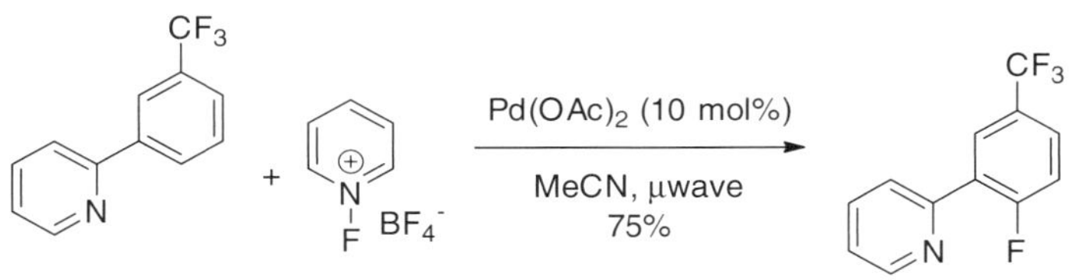

(6.54)

The palladium-catalyzed directed *ortho* C–H amidation of aromatic ketones was applied to the synthesis of a key intermediate for Prandin, a blood glucose-lowering drug (Scheme 6.2) [119]. 2-Aminophenyl ketone was easily obtained through C–H amidation and subsequent removal of the sulfonamide group. The synthesis of flubendiamide, a novel benzenedicarbamide insecticide, was greatly facilitated by a directed aromatic C–H iodination. The amide and sulfoxide possibly chelate to the Pd to achieve *ortho*-selective iodination with NIS (Scheme 6.3) [140].

6.3.4 Allylic C–H Oxidation

Allyllic C–H functionalization is also a useful strategy to control the regioselectivity because the allylic C–H bond is weakened through hyperconjugation and can be selectively cleaved after the metal coordination with alkene. In early studies, Pd-catalyzed allylic C–H oxidation was mainly limited to acetoxylation of internal olefin such as cyclohexene under acetic condition [Eq. (6.55)] [141–144]. White reported Pd-catalyzed allylic C–H acetoxylation of internal olefin to afford either linear or branched allylic acetates in the presence of a sulfone ligand [Eq. (6.56)] [145, 146]. Intramolecular allylic C–H oxidation can be used to synthesize macrolactone product [147]. By using a chiral Lewis acid additive, an asymmetric allylic C–H acetoxylation reaction was developed [Eq. (6.57)] [148]. Aerobic allylic C–H acetoxylation was achieved by a $PdCl_2$/DMA system [149] and a

4,5-diazafluorenone-ligated Pd catalyst [Eq. (6.58)] [150]. Similar strategy has been used to achieve both intramolecular [Eq. (6.59)] [151] and intermolecular [Eq. (6.60)] [152–155] allylic C–H amination in the presence of nitrogen nucleophile.

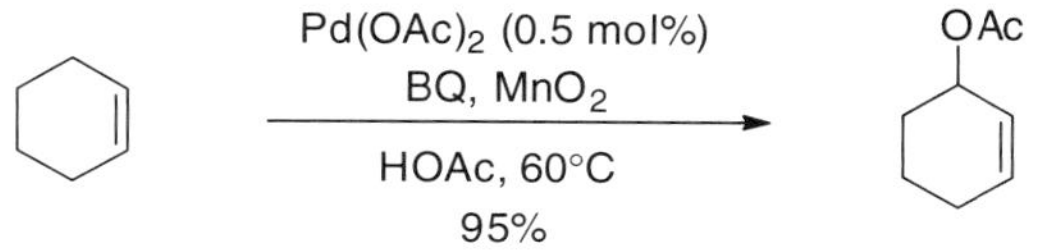

(6.55)

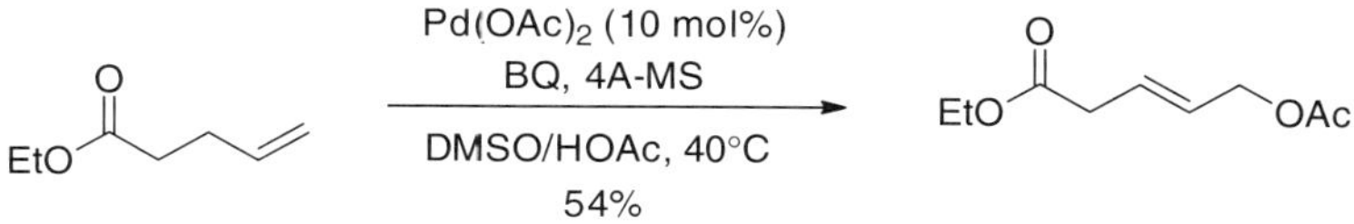

(6.56)

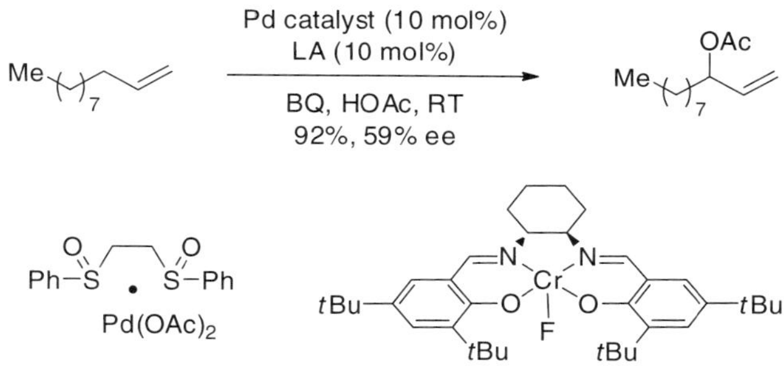

(6.57)

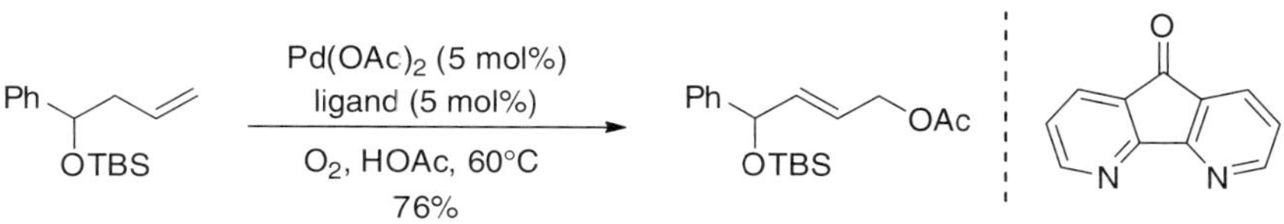

(6.58)

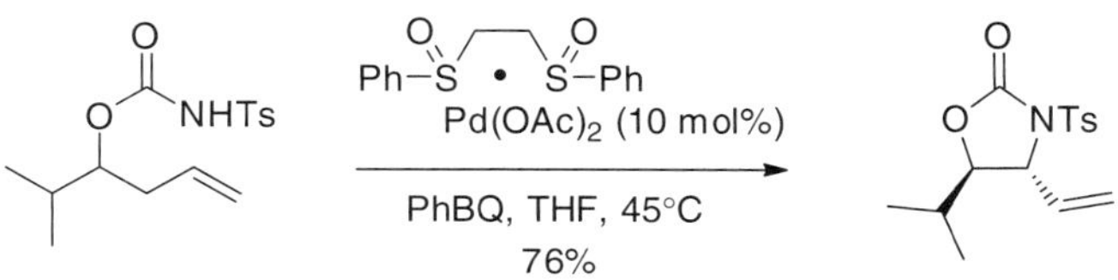

(6.59)

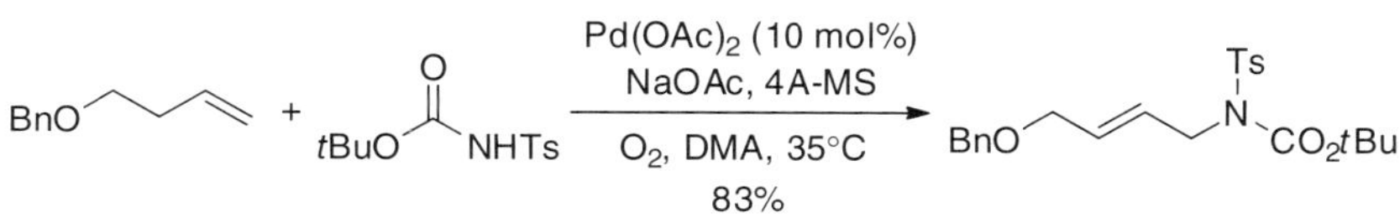

(6.60)

The mechanism of Pd-catalyzed allylic acetoxylation [156] and amination [157] has been studied. Stoichiometric reactions of palladium(II) with alkene substrates have been shown to generate allyl palladium complexes under mild conditions

[158, 159]. The catalytic reactions likely occur through an allyl-π Pd intermediate obtained by allylic C–H cleavage by palladium(II), which undergoes nucleophilic substitution by a heteroatom to give the allylic C–H oxidation product.

The synthetic utility of the Pd-catalyzed allylic C–H oxidation was demonstrated in the synthesis of a polyketide macrolide antibiotic 6-deoxyerythronolide B (Scheme 6.4) [160]. The macrocyclization through an intramolecular allylic C–H oxidation of a linear precursor proceeded well with excellent regio- and stereocontrol. Through this late-stage C–H oxidation strategy, synthesis of the natural product was accomplished in short steps.

6.3.5 Mechanism

In the oxidative C–H functionalization process involving a strong oxidant such as $PhI(OAc)_2$, peroxide, and other metal salts, it has been proposed that the reactions involve the formation of an aryl or alkyl Pd(II) species which is further oxidized to a Pd(IV) intermediate. In a model study, Sanford successfully obtained a diaryl Pd(IV) complex by the oxidation of bis(2-pyridylphenyl) palladium with $PdI(OAc)_2$, which is frequently used in various C–H functionalization reactions [Eq. (6.61)]. The Pd(IV) complex thus obtained is quite stable at room temperature, but undergoes reductive elimination at higher tempertaure to afford the aryl C–O formation product [161]. Detailed mechanistic studies suggested that the reaction likely proceeds through initial carboxylate dissociation and subsequent reductive elimination from the five-coordinate cationic intermediate [162]. C–Cl reductive elimination was observed from a related Pd(IV) complex obtained by oxidation of Pd(II) compound with NCS [163] or $PhICl_2$ [164]. Monoaryl Pd(IV) trifluoride [165] and pyridyl–sulfonamide Pd(IV) fluoride complexes [Eq. (6.62)] [166] have been isolated by Sanford and Ritter, respectively, and the later complex shows reductive elimination to form aryl fluoride [167]. Although oxidative addition to form Pd(IV) complexes and reductive elimination from these complexes are studied extensively, the relevance of these Pd(IV) model complex to catalytic reactions remains to be established because in most cases they contain specific supporting ligands.

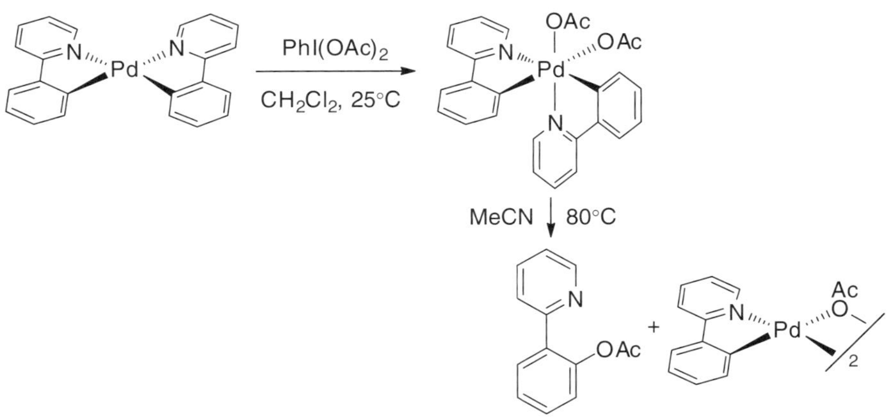

(6.61)

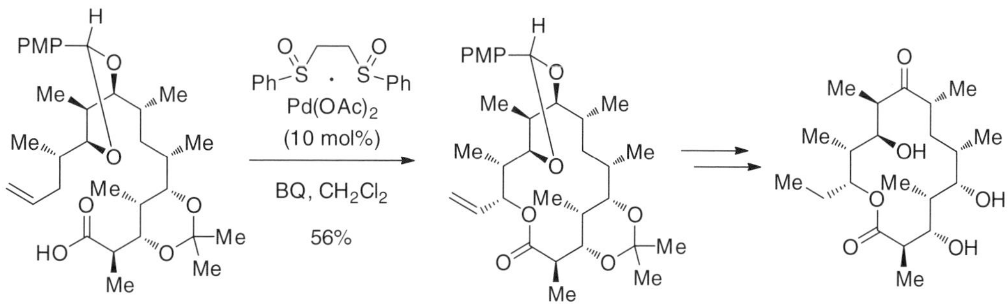

6-deoxyerythronolide B

Scheme 6.4. Synthesis of 6-deoxyerythronolide B.

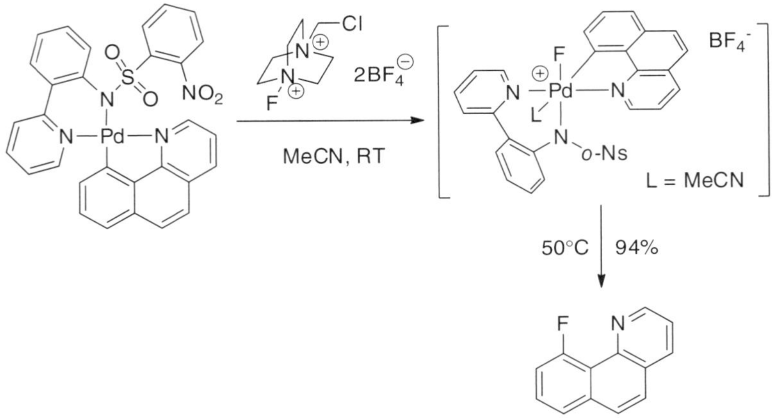

(6.62)

Ritter has proposed a dimeric Pd(III) intermediate as an alternative mechanistic scenario in directed C–H oxidation reactions [168, 169]. In this scenario, a discrete dimeric Pd(III) complex was obtained by the oxidation of palladacycle with different oxidants including $PhICl_2$, $PhI(OAc)_2$ and NCS, which are frequently used in oxidative C–H functionalization reactions [Eq. (6.63)]. These dimeric Pd(III) species have been observed under pseudo-catalytic conditions and have been shown to be kinetically competent catalyst for C–H oxidation reactions [170]. The dimeric Pd catalysis can lower the energy barrier for both (a) the oxidative addition of oxidant to the metal center and (b) reductive elimination from Pd(III) centers [171]. However, for certain dimeric Pd(III) complexes, it undergoes disruption of the metal center to form the monomeric Pd(IV) intermediate [Eq. (6.64)] [172]. The catalytic mechanism of these C–H oxidation reactions remains to be elucidated.

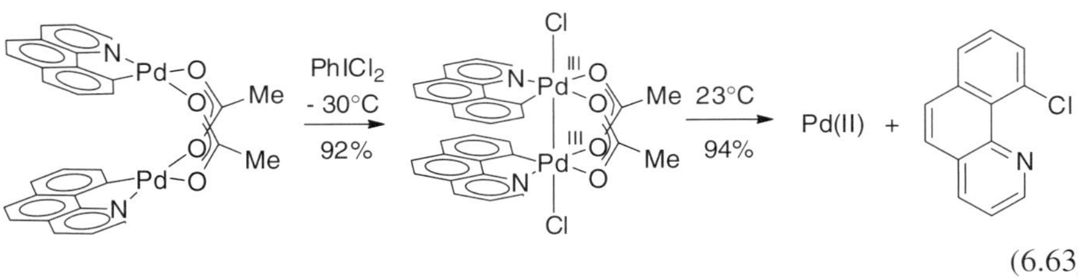

(6.63)

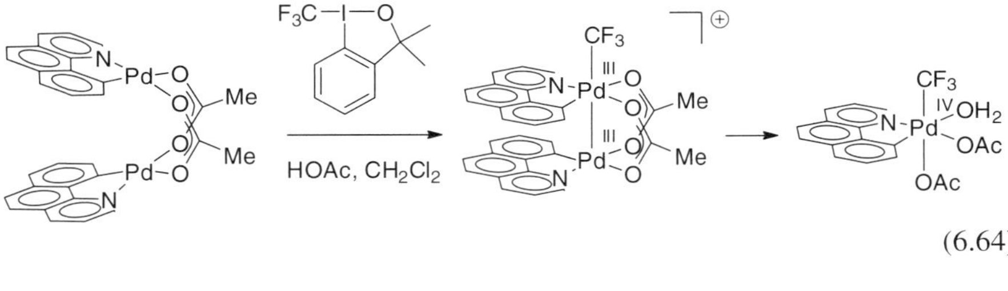

(6.64)

6.4 HYDROARYLATION

6.4.1 Directed Hydroarylation of Alkene

In 1993, Murai et al. [173] reported a ruthenium-catalyzed highly efficient addition of aryl C–H bond to ethylene and vinylsilane directed by a ketone group (Scheme 6.5). This is one of the earliest catalytic examples to control regioselectivity in C–H cleavage using a directing group. This study also led to a large body of subsequent works on hydroarylation reactions.

A variety of aromatic C–H substrates have been successfully employed in directed hydroarylation reactions using $RuH_2(CO)(PPh_3)_3$ and related catalysts. Ketone, imine, pyridine, ester, and nitrile groups can be used as the directing groups [Eq. (6.65)] [174–179]. The active Ru catalyst can be generated *in situ* by reduction of a stable Ru precursor $[RuCl_2(p\text{-cymene})]_2$ [Eq. (6.66)] [180]. Treatment of $RuH_2(CO)(PPh_3)_3$ with trimethylvinylsilane affords a cyclometalated Ru complex, which catalyzed the hydroarylation reaction at room temperature [181].

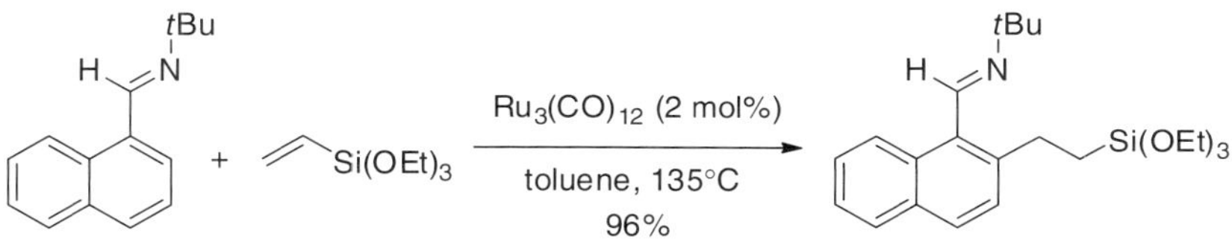

(6.65)

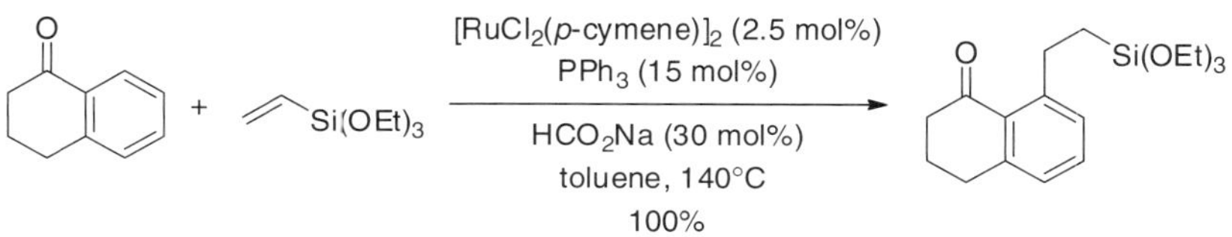

(6.66)

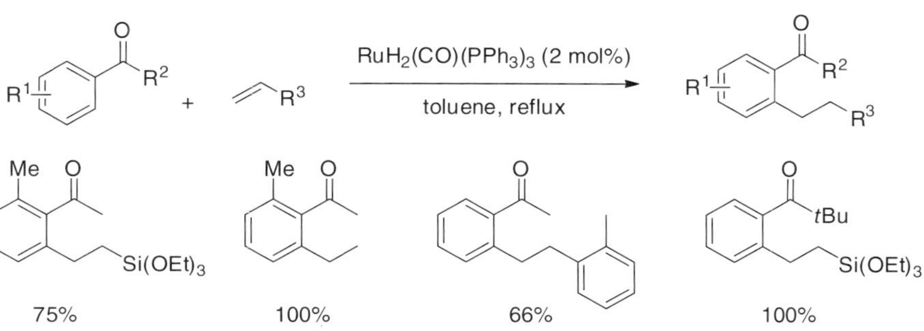

Scheme 6.5. Ruthenium-catalyzed hydroarylation.

Directed hydroarylation reactions can also be carried out with a Rh(I) catalyst and a phosphine ligand. Pyridine [Eq. (6.67)] [182, 183] and imine [Eq. (6.68)] [184, 185] can be used as directing groups. Alkylation with internal olefin afforded a linear product, probably due to the facile isomerization of the internal olefin to the terminal olefin. A related Rh/phosphine catalyst is also applied for the intramolecular hydroarylation with directing-group-containing arenes [Eq. (6.69)] [186, 187] and five-membered azoles [188] to synthesize various useful heterocycles. An enantioselective intramolecular hydroarylation reaction was achieved using Rh(I) and a chiral phosphoramidite ligand [Eq. (6.70)] [189]. The catalytic system was later extended to include the intermolecular hydroarylation with both five-membered azoles [Eq. (6.71)] [190, 191] and six-membered pyridine [192] and pyridine *N*-oxide [Eq. (6.72)] [193].

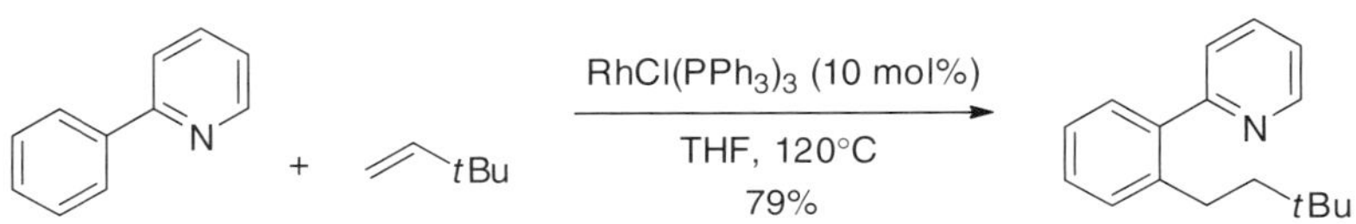

(6.67)

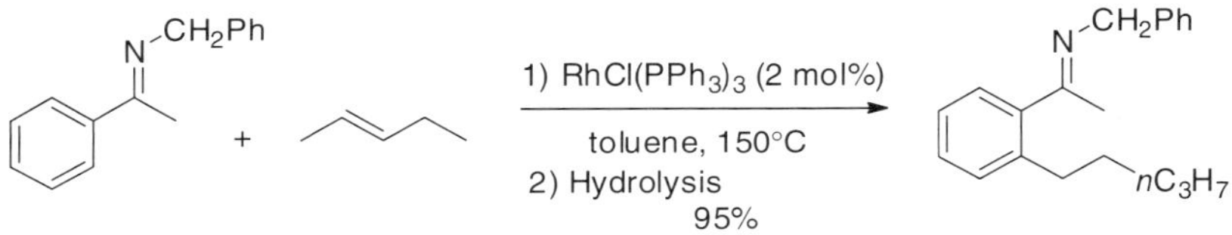

(6.68)

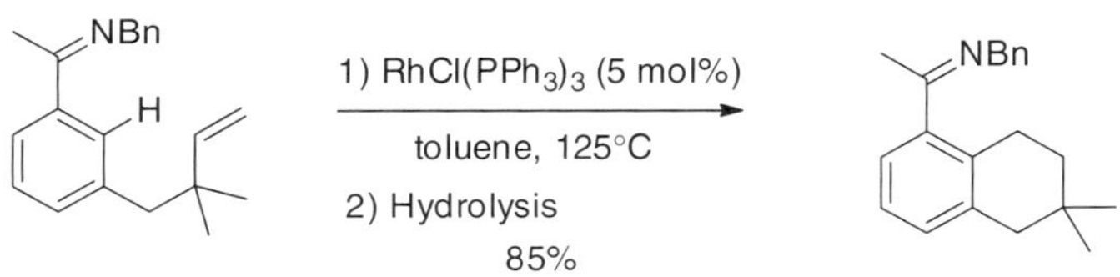

(6.69)

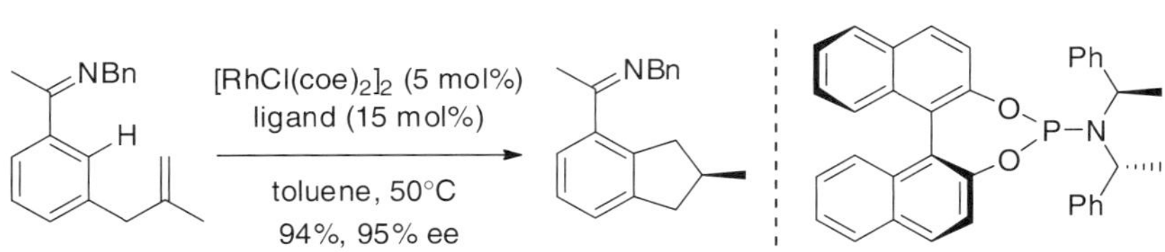

(6.70)

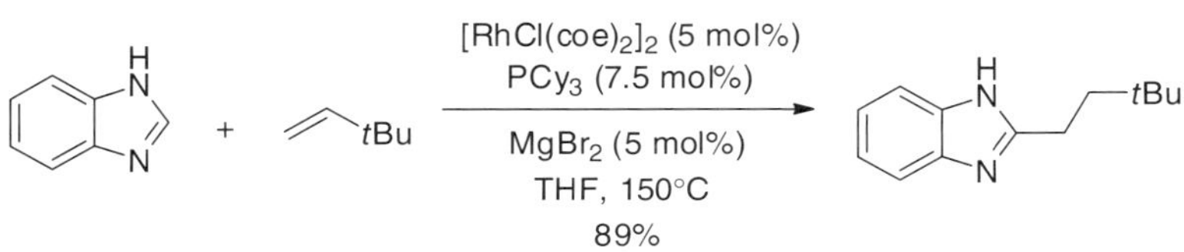

(6.71)

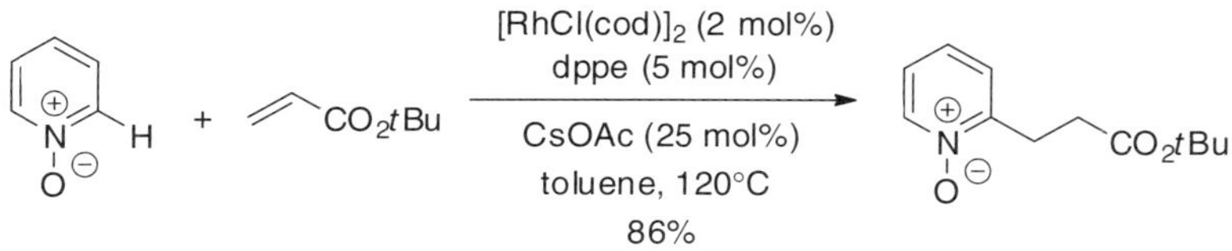

(6.72)

Regioselective C-2 alkylation of *N*-substituted indole derivatives with various alkenes has been developed using a cationic iridium catalyst, which selectively gives linear or branched 2-alkylindoles in high selectivity [Eq. (6.73)] [194]. A related iridium complex catalyzes the intermolecular asymmetric addition of the C–H bonds of indoles, thiophenes, pyrroles, and furans to bicycloalkenes in high yield and high enantiomeric excess with C-2 selectivity [Eq. (6.74)] [195].

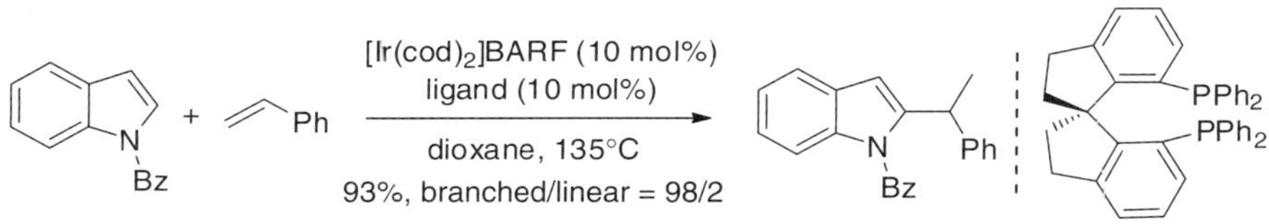

(6.73)

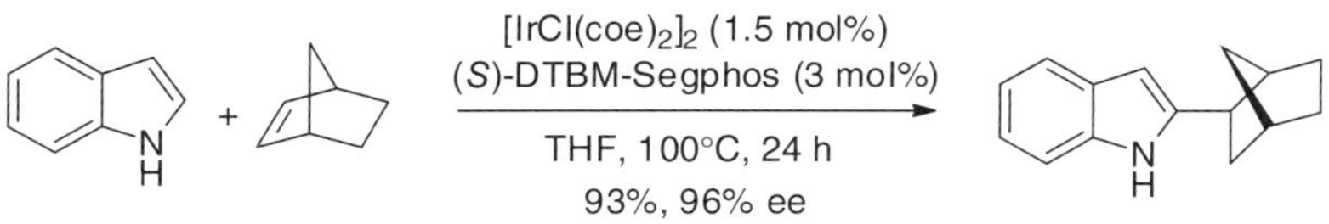

(6.74)

Besides low-valent late transition metal complexes such as Ru, Rh, and Ir, first-row transition metals also showed catalytic reactivity in hydroarylation of alkenes. A direct C-4-selective alkylation of pyridines is achieved by using a combination of nickel catalyst and Lewis acid [196]. A cobalt/phosphine catalyst can catalyze the directed hydroarylation of arene substrates containing pyridine and imine directing groups. The regioselectivity can be switched by changing different ligands [Eq. (6.75)] [197–199].

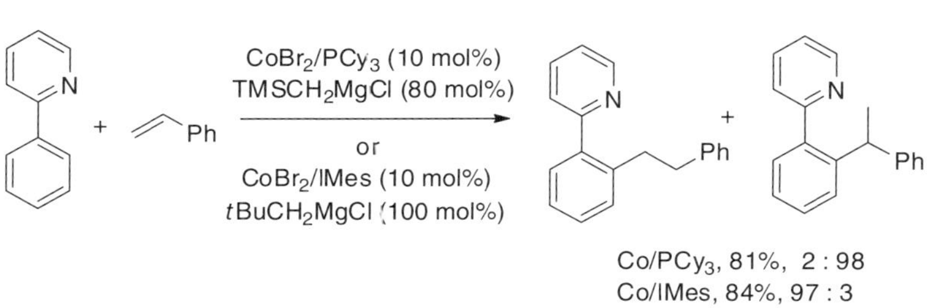

(6.75)

Rh-catalyzed intramolecular directed hydroarylation was used for the first total synthesis of (+)-lithospermic acid (Scheme 6.6) [200]. Chiral imine directed diastereoselective C–H alkylation followed by removal of the imine group provided an enantiopure hydroarylation product. Related intramolecular olefinic C–H alkylation has been applied to the concise total synthesis of (−)-incarvillateine [201]. Rh-catalyzed intramolecular hydroarylation of heterocycle was also applied in the synthesis of biologically active compounds such as PKC and JNK3 inhibitors (Scheme 6.7) [202, 203].

6.4.2 Mechanism of Alkene Hydroarylation

Directed hydroarylation reactions catalyzed by Ru and Rh catalyst have been studied in detail. The ruthenium-catalyzed directed hydroarylation starts with the oxidative addition of the ortho C–H bond to a low-valent Ru center. Monitoring of the catalytic reaction showed the presence of an *ortho*-ruthenated aromatic ketone complex with a carbonyl ligand during catalysis [181]. Then insertion of olefin to the ruthenium hydride bond generates an aryl alkyl ruthenium intermediate. Finally, reductive elimination would afford the hydroarylation product (Scheme 6.8). A deuterium labeling experiment suggests that the C–H cleavage and olefin insertion are relatively facile and reversible [174, 204]. The C–C bond forming reductive elimination is likely the rate-limiting step, which is supported by a carbon kinetic isotopic effect. Although generally an electron-withdrawing group retards the rate of reductive elimination, an electron-withdrawing group increases the rate of the reaction of aromatic ester with olefin. A σ–π coupling mechanism was invoked in which an alkyl group migration from the ruthenium center to the aromatic ring precedes the elimination of metal to retain aromaticity. Computational study supports such a mechanistic proposal [205, 206]. The energy barrier calculated for the σ–π coupling is lower than the direct reductive elimination from the aryl alkyl ruthenium species.

The hydroheteroarylation of olefin developed in Bergman and Ellman's lab was proposed to proceed through a different mechanism (Scheme 6.9) [207, 208]. The interaction of heterocycle such as benzimidazole with a Rh complex generates a Rh–carbene intermediate after an intramolecular hydrogen transfer. Computational study suggests that the Rh–carbene intermediate reacts with the pendant olefin through a [2 + 2] cycloaddition followed by hydrogen transfer to give an alkyl rhodium hydride complex. Upon reductive elimination, the final C–C formation product is obtained.

6.4.3 Undirected Hydroarylation of Alkene

Hydroarylation of olefin with simple arene without a directing group has also been studied. Although this reaction provides an alkylarene product that resembles a Friedel–Crafts reaction, the mechanism of hydroarylation reaction is completely different. In addition, a Friedel–Crafts reaction produces a Markovnikov addition product, while metal-catalyzed hydroarylation usually produces an anti-Markovnikov addition product.

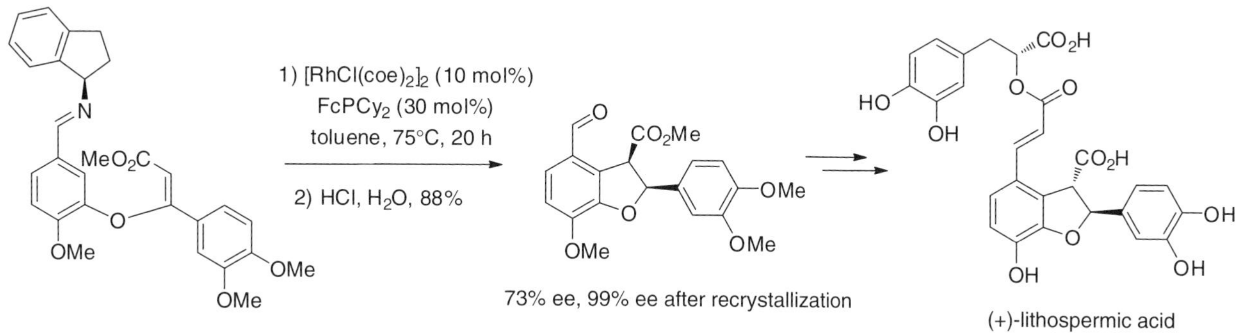

Scheme 6.6. Synthesis of (+)-lithospermic acid.

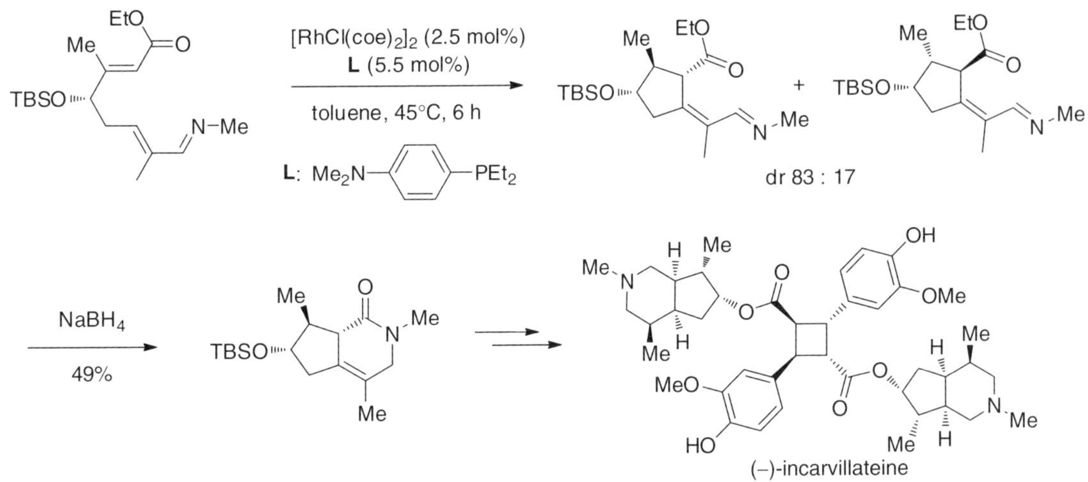

Scheme 6.7. Synthesis of (−)-incarvillateine.

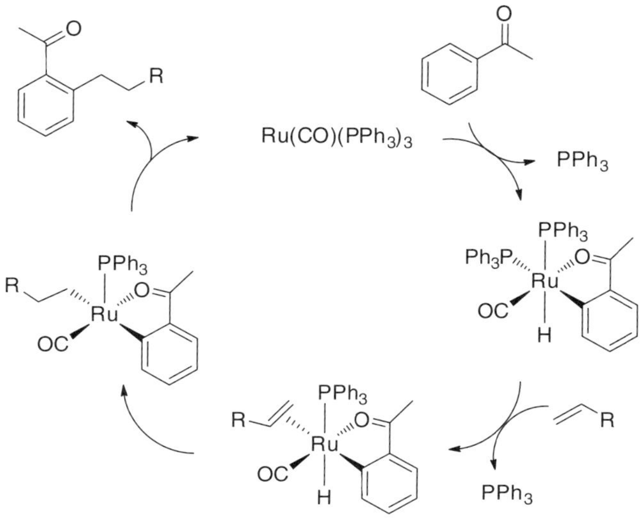

Scheme 6.8. Mechanism of ruthenium-catalyzed hydroarylation.

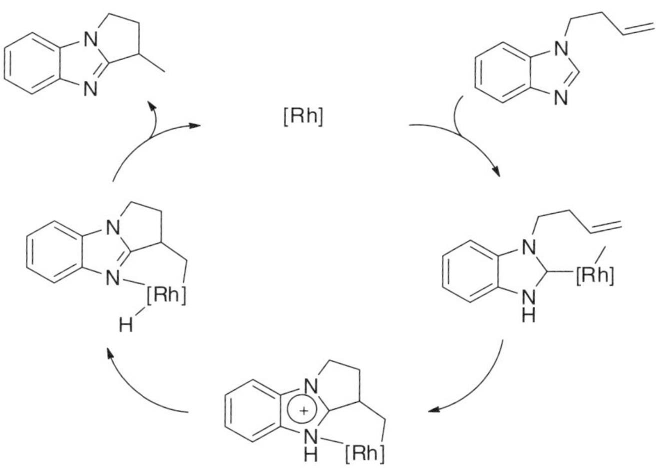

Scheme 6.9. Mechanism of rhodium-catalyzed hydroarylation.

Matsumoto initially reported a binuclear Ir complex for the hydroarylation of benzene with olefins. Although these reactions were carried out under relatively harsh conditions, anti-Markovnikov addition products were produced in good selectivity [Eq. (6.76)] [209, 210]. Later variation of this system by Periana et al. [211, 212] showed that a mononuclear Ir complex Ir(acac–O,O)$_2$ exhibited higher reactivity [Eq. (6.77)]. Gunnoe and co-workers [213, 214] reported a Ru catalyst for hydroarylation of arenes [Eq. (6.78)]. The Ru catalyst system is very reactive, and high turnover numbers can be achieved under mild conditions. When propene was used, an anti-Markovnikov addition product was slightly favored. Tilley and co-workers [215] reported a Pt–Ag catalyst system in which Pt complexes that were activated by Ag(I) salts have been shown to catalyze Markovnikov hydroarylations of unactivated olefins under mild conditions. Gunnoe and co-workers [216] reported a related hydroarylation of arene with olefin using a cationic Pt catalyst generated from the protonation of a diphenyl Pt complex.

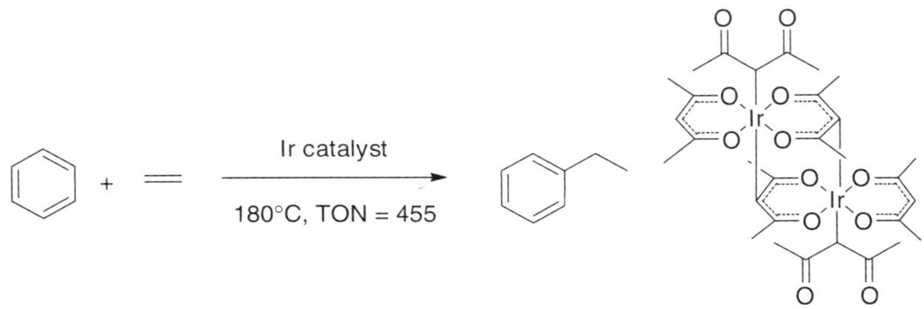

(6.76)

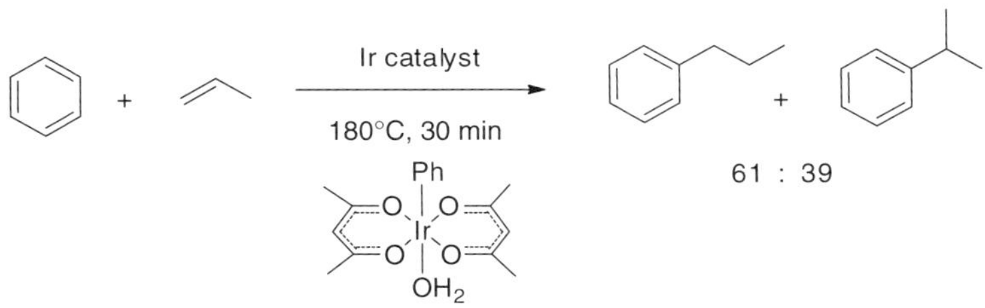

(6.77)

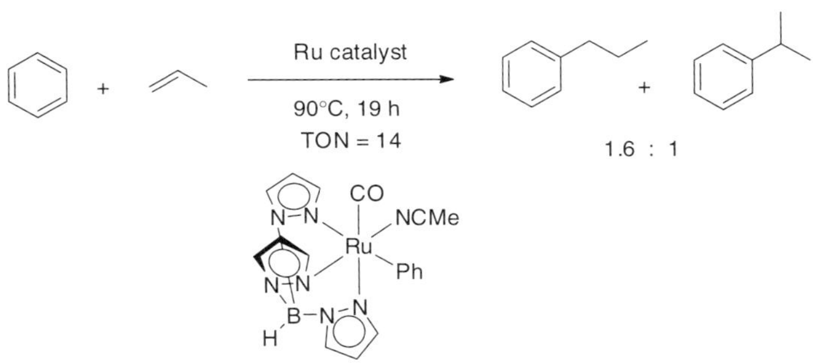

(6.78)

Experimental and computational studies have been carried out for these hydroarylation reactions [217–219]. Generally, these catalyst systems follow a similar catalytic pathway (Scheme 6.10). The reaction starts with insertion of olefin into an aryl metal complex to form an alkyl metal intermediate. This step has been studied experimentally in Ir, Ru and Pt [220] systems. The alkyl metal intermediate then reacts with an arene C–H bond to form the hydroarylation product and regenerates the aryl metal intermediate. The arene C–H cleavage by the alkyl metal complex has been

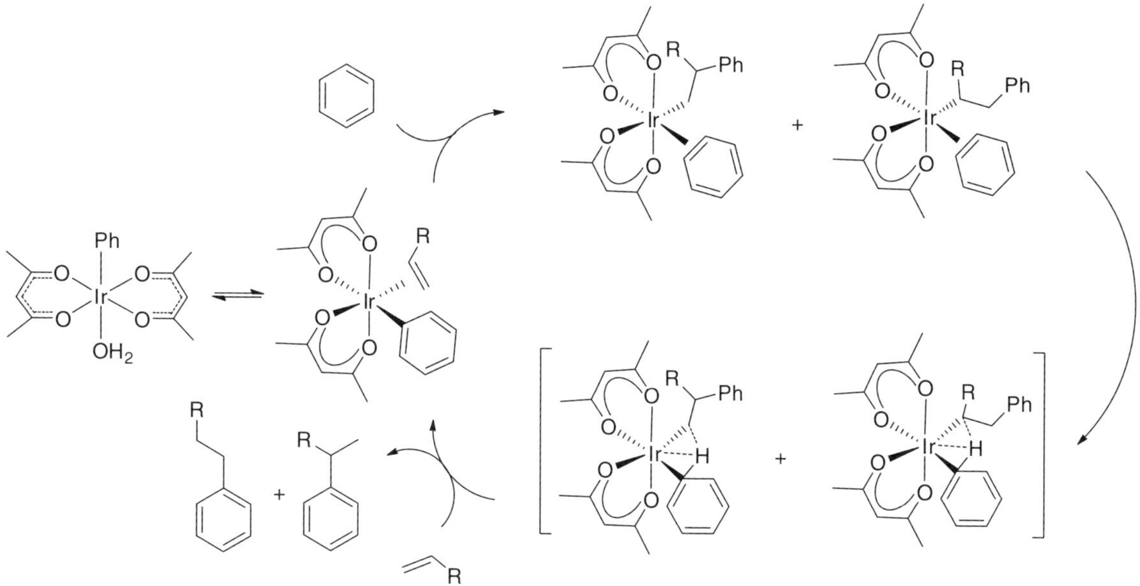

Scheme 6.10. Mechanism of iridium-catalyzed hydroarylation.

analyzed by computational studies. An oxidative hydrogen migration mechanism has been proposed for this step.

6.4.4 Undirected Hydroarylation of Alkyne

In a seminar study, Fujiwara and co-workers [221, 222] reported both intra- and intermolecular Pd-catalyzed hydroarylation of alkyne in trifluoroacetic acid [Eqs. (6.79) and (6.80)]. Although the *trans* addition selectivity is not well understood, it has been proposed that the reaction involves initial C–H cleavage by a cationic Pd(II) species. Many examples of both intra- and intermolecular hydroarylation using high-valent metal catalysts such as Pt [223, 224] and Au [225, 226] were developed afterwards [227]. Most of these reactions likely proceed through an electrophilic aromatic substitution pathway in which metal activates the alkyne toward nucleophilic attack by the arene [228]. There are also examples of high-valent metal-catalyzed hydroarylation of an alkyne with evidence supporting a C–H activation pathway. Gevorgyan and co-workers [229, 230] reported a Pd-catalyzed intramolecular hydroarylation of alkyne [Eq. (6.81)]. A large kinetic isotopic effect, non-Friedel–Crafts-type selectivity, and *cis*-hydroarylation product are all consistent with a pathway involving Pd-catalyzed C–H cleavage followed by alkyne insertion.

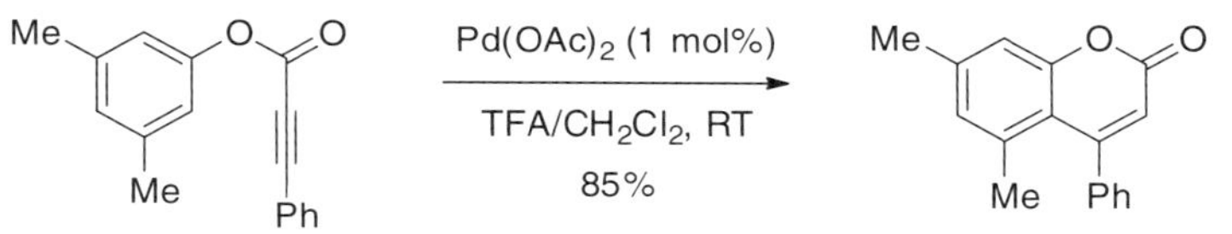

(6.79)

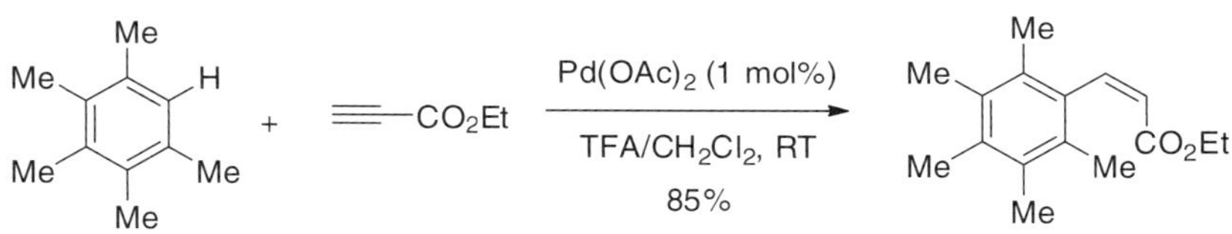

(6.80)

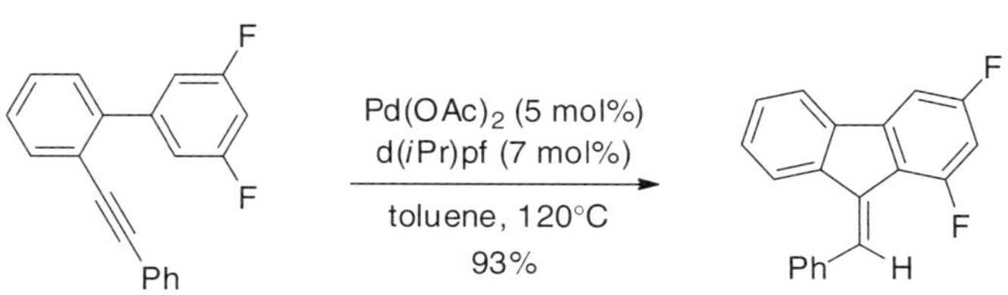

(6.81)

The hydroarylation of alkyne with simple benzene substrate without electronic bias has been realized using $Rh(PR_3)_2(CO)Cl$ under photolytic condition [Eq. (6.82)] [231, 232]. Some olefin product resulting from exclusively *syn* addition across the triple bond suggested a concerted addition of a hydridophenylrhodium species, which was obtained by insertion of Rh(I) to a benzene C–H bond. A Rh carbonyl complex was able to catalyze the hydroarylation of alkyne with simple arene under thermal conditions, but high temperature was required [233, 234]. The reaction can be carried out at lower temperature using a dinuclear palladium complex. The preference for electron-poor arenes as well as the unusual regioselectivity suggests a different mechanism other than electrophilic C–H cleavage [Eq. (6.83)] [235].

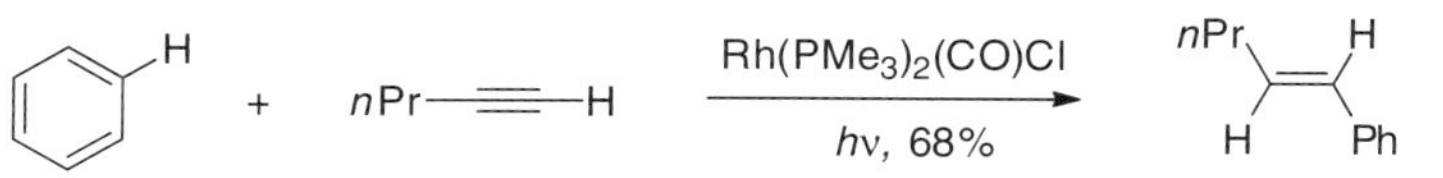

(6.82)

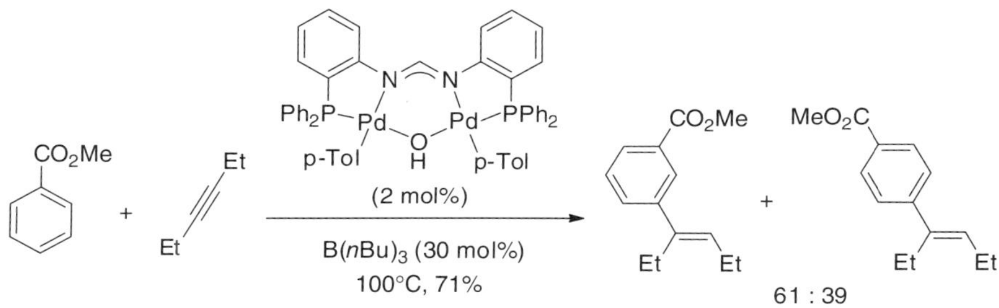

(6.83)

6.4.5 Directed Hydroarylation of Alkyne

The regioselectivity of alkyne hydroarylation can be controlled by a proper directing group. For example, the Ru and Rh catalysts developed for hydroarylation of alkenes can be extended to the hydroarylation of alkynes [236–239]. An aryl C–H bond *ortho* to ketone [Eq. (6.84)], imine, and pyridine [Eq. (6.85)] is selectively cleaved to give the *ortho*-alkenylated products. First-row transition metal complexes based on cobalt [Eq. (6.86)] [240–242], manganese [Eq. (6.87)] [243], and rhenium [244] can also catalyze the intermolecular hydroarylation of alkyne directed by azo, pyridine, and imine groups.

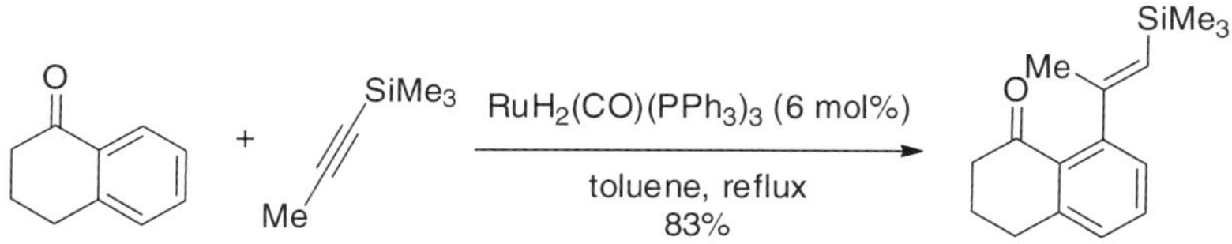

(6.84)

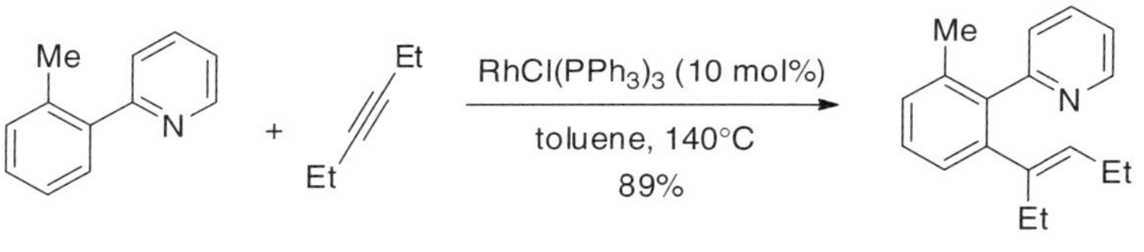

(6.85)

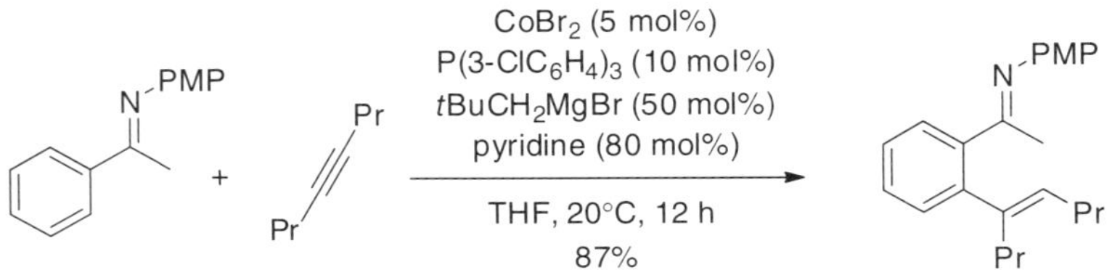

(6.86)

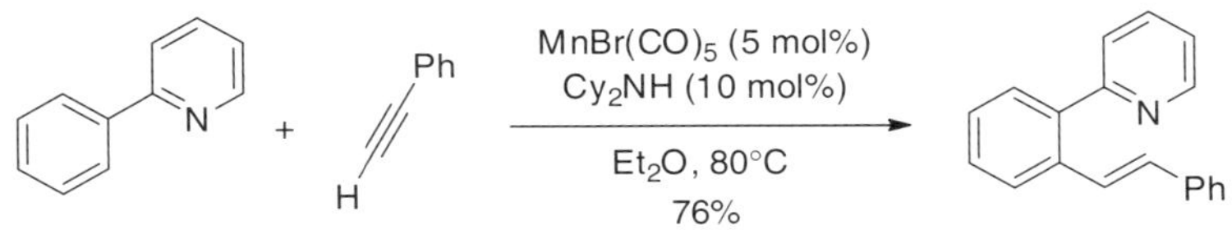

(6.87)

In addition to low-valent metal catalysts, a high-valent metal could also be used for directed alkyne hydroarylation. A C–H activation pathway was favored for an intermolecular hydroarylation catalyzed by a cationic Rh complex. The importance of a directing group, exclusive *syn* addition, and a deuterium labeling experiment all support initial C–H cleavage followed by alkyne insertion [Eq. (6.88)] [245]. A ruthenium-catalyzed hydroarylation of benzamides and phenylpyrazoles with alkynes in the presence of acetic acid proceeded regio- and stereoselectively to produce the corresponding ortho-alkenylated products [Eq. (6.89)] [246].

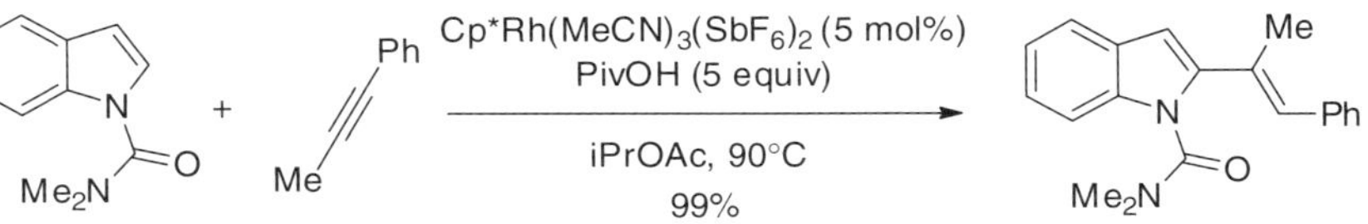

(6.88)

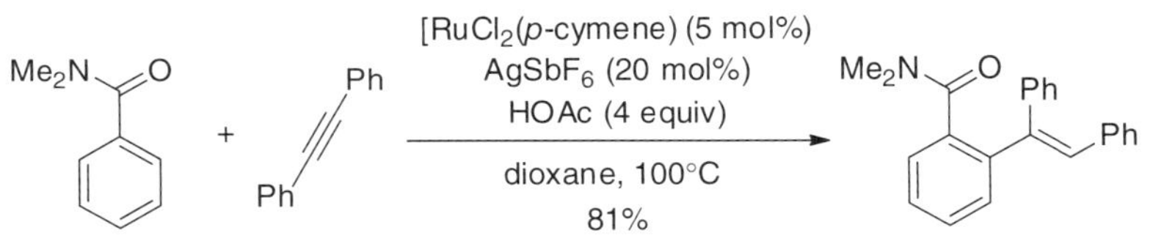

(6.89)

6.4.6 Oxidative Olefination

In the late 1960s, Fujiwara et al. [247–249] reported a Pd-catalyzed oxidative olefination of simple arenes with styrene in the presence of added oxidant [Eq. (6.90)]. Higher yields were obtained by the presence of electron-withdrawing groups on the olefins and electron-donating groups on the arene. The reaction generates mainly *beta*-aryl-*trans*-olefins. Although the catalytic reactivity of this system is not high and the regioselectivity for substituted arenes is not well controlled, this pioneering work provides the foundation for the subsequent studies on the coupling of aryl and alkyl C–H bonds with olefins.

Various oxidants including Cu^{2+}, Ag^{+}, and *t*-butyl perbenzoate [250], can be used for the Pd(II)-catalyzed olefination of simple arenes. High turnover numbers were obtained with BQ and *t*-butyl hydroperoxide as the oxidant [Eq. (6.91)] [251]. Olefination of simple arenes can be carried out using dioxygen as the sole oxidant in the presence of Pd [Eq. (6.92)] [252–254] or Ru [Eq. (6.93)] [255] catalyst, producing water as the only byproduct. Rh complexes were shown to catalyze the oxidative alkenylation of arenes in the late 1970s. Copper salts and O_2 were the terminal oxidants [256–258]. However, most of these reaction required harsh conditions or produced relatively low yields.

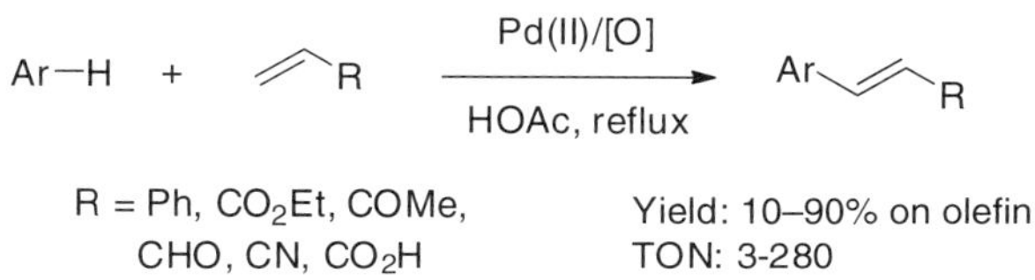

(6.90)

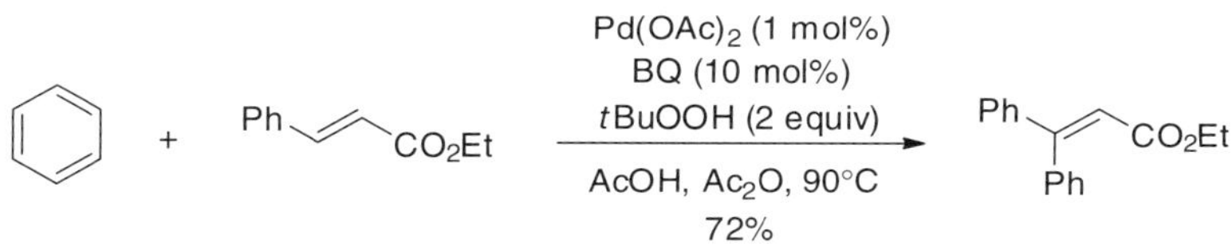

(6.91)

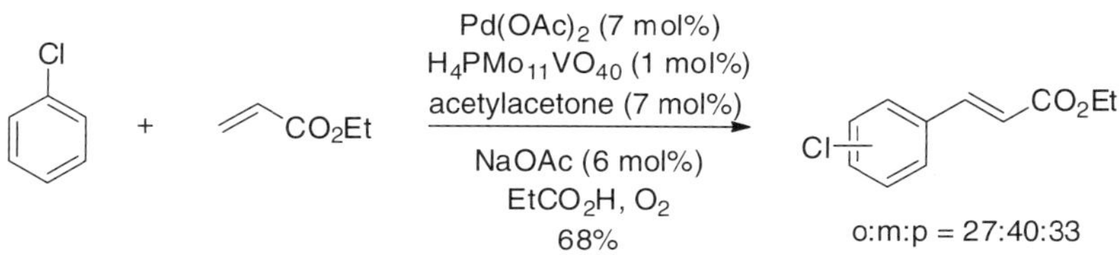

(6.92)

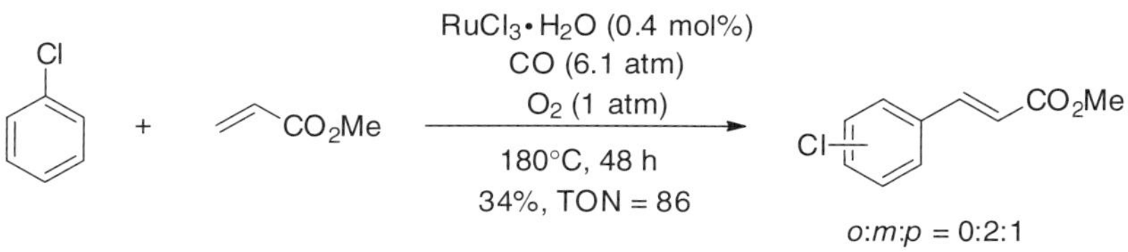

(6.93)

A major problem in the olefination of simple arenes is the low regioselectivity between different aromatic C–H bonds. One strategy to control the regioselectivity is to introduce a directing group to the substrate. In an early work, Miura et al. [259] reported a Pd-catalyzed oxidative coupling of 2-phenylphenol with alkene in the presence of copper and air as the oxidant [Eq. (6.94)]. Phenol acts as the directing group to enable regioselective alkenylation and subsequent Michael addition. Similar reactivity has been seen in the reactions of 2-phenylaniline sulfonamide and benzoic acid as the substrates [260]. de Vries and co-workers [261] reported a Pd-catalyzed *ortho*-selective olefination of acetanilide directed by an acetamino group [Eq. (6.95)]. Subsequent works have demonstrated the viability of Pd-catalyzed aryl C–H olefination using various directing groups including amide [262], amine [Eq. (6.96)] [263], sulfamide [Eq. (6.97)] [264], and carboxylic acid [Eq. (6.98)] [265, 266]. Yu and co-workers [267] achieved an enantioselective aryl C–H olefination through desymmetrization of diphenylacetic acid using protected amino acid as the chiral ligand. Besides palladium, other transition metal catalysts have also been reported for the directed oxidative olefination reaction. Satoh and Miura reported a Cp*Rh(III) catalyst system which is effective for the *ortho* olefination of various aromatic substrates including benzoic acid [Eq. (6.99)] [268], *N*-phenylpyrazole [269], benzamide [270], acetanilide [Eq. (6.100)] [271], benzoic ester [272], and acetophenone [Eq. (6.101)] [273]. Copper acetate and silver salts are usually used as the oxidant. Closely related oxidative olefination using different directing groups can also be carried out with less expensive ruthenium catalyst [Eqs. (6.102) and (6.103)] [274–277]. Directing-group-assisted arene olefination can be extended to olefination of alkyl C–H bond. Yu and co-workers [278] reported an amide-directed olefination of alkyl C–H bond in propionamides using copper and silver as the oxidants [Eq. (6.104)]. Sanford and co-workers [279] reported a Pd-catalyzed olefination of alkyl C–H bond using pyridine as the directing group and dioxygen as the oxidant [Eq. (6.105)].

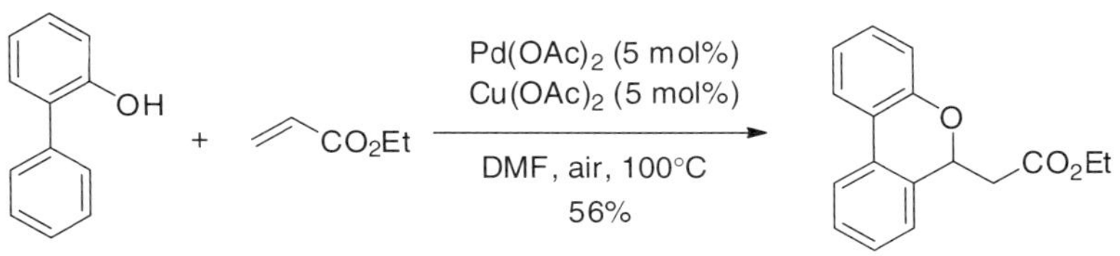

(6.94)

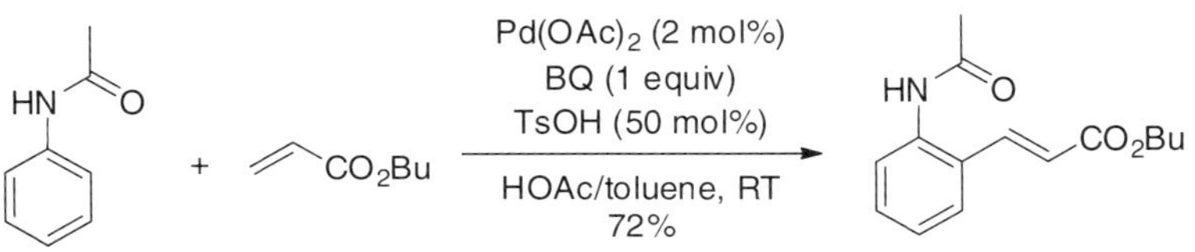

(6.95)

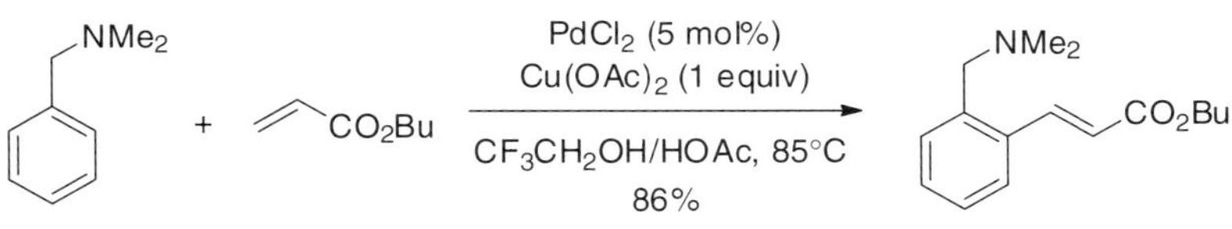

(6.96)

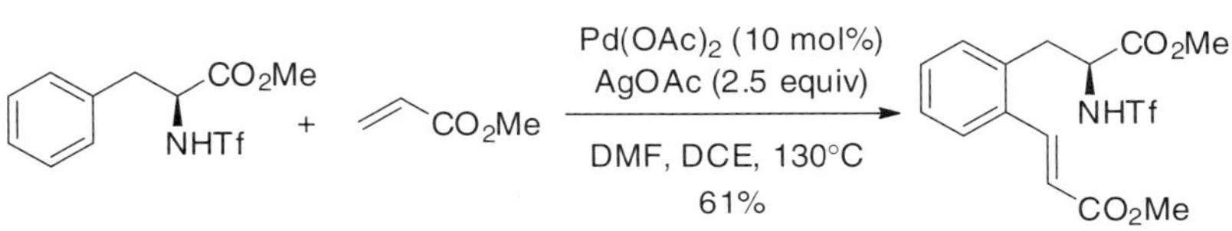

(6.97)

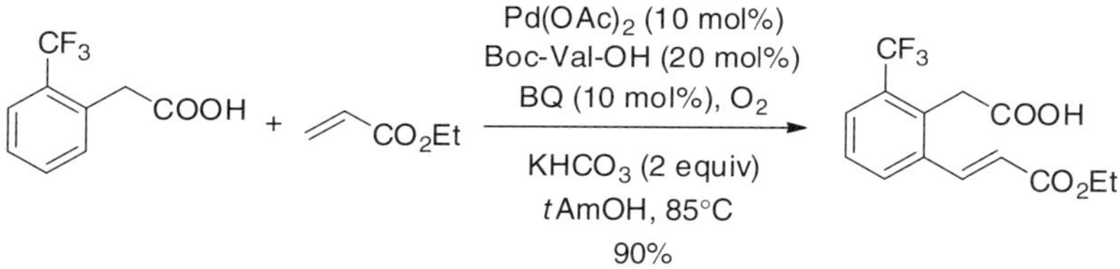

(6.98)

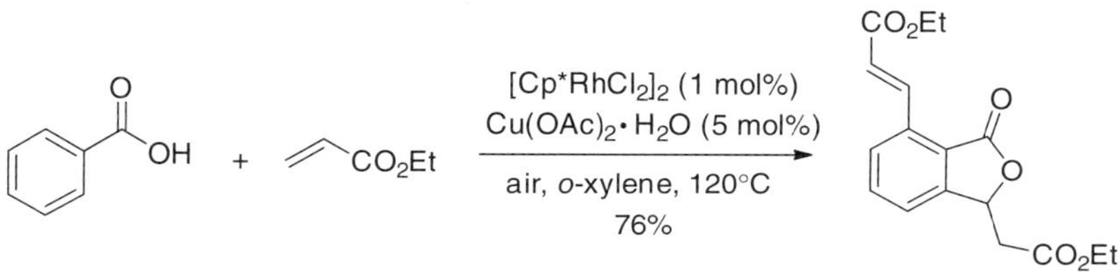

(6.99)

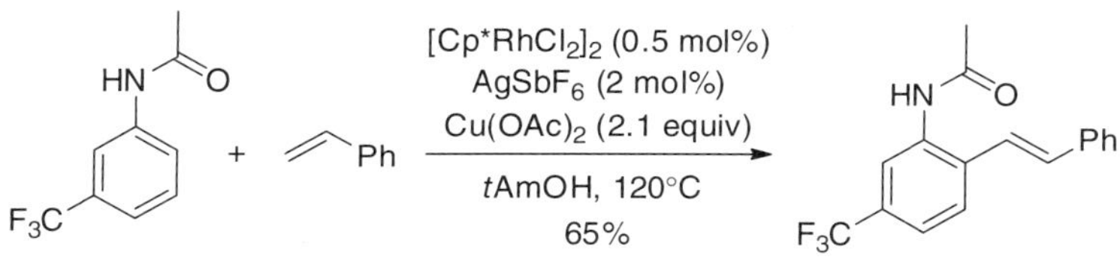

(6.100)

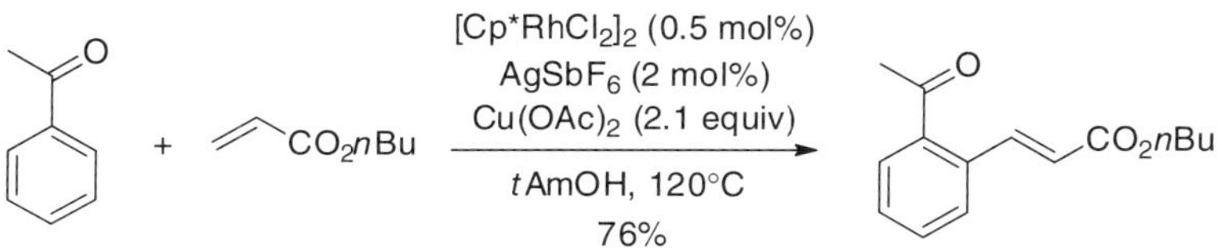

(6.101)

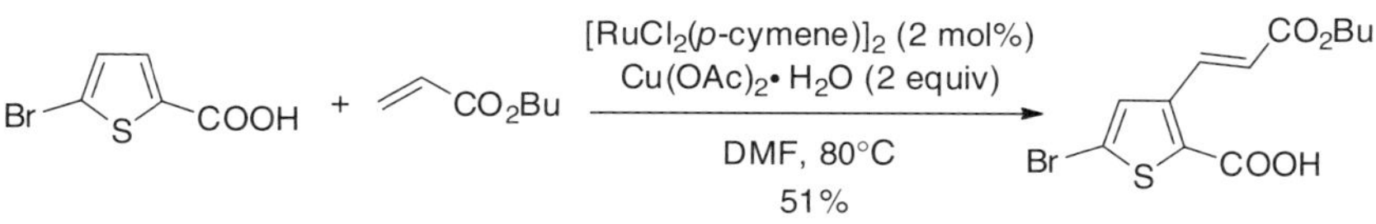

(6.102)

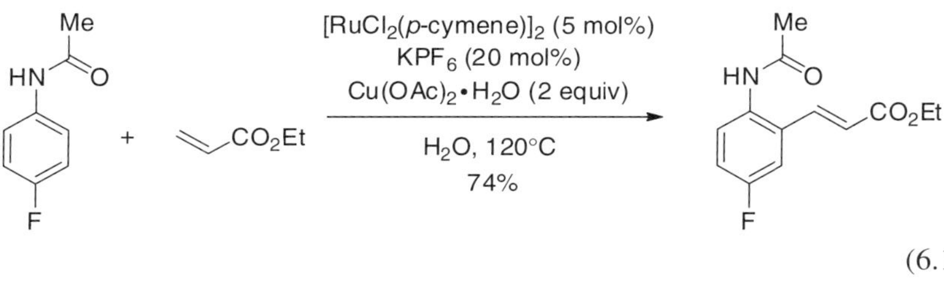

(6.103)

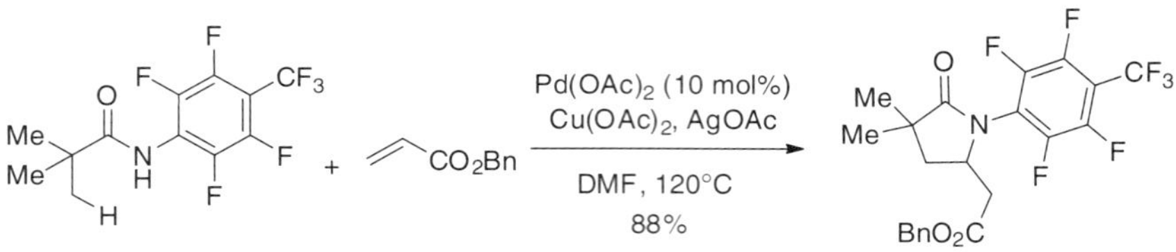

(6.104)

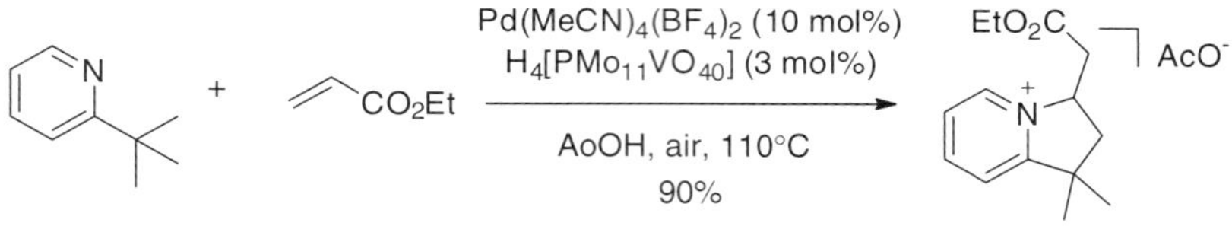

(6.105)

Another approach to control the regioselectivity takes advantage of the intrinsic electronic property of the arene. For example, electron-rich arene such as indole [Eq. (6.106)] [280, 281] pyrrole [Eq. (6.107)] [282], furan, thiophene [283–285], and azoles [Eq. (6.108)] [286, 287] can be regioselectively olefinated. Generally, the olefination takes place at the electron-rich position such as the C-2 position in furan, thiophene, and pyrrole and the C-3 position in indole, which can be explained by an electrophilic palladation mechanism during C–H cleavage. The regioselectivity can be tuned in some cases by the substrate structure or reaction conditions. Intramolecular aryl C–H olefinations are usually carried out with electron-rich substrates such as aniline [288], anisole [Eq. (6.109)] [289], and indole [290–292]. The regioselectivity is controlled by the position of the pendent olefin and the formation of a favorable cyclic structure. Pd-catalyzed oxidative olefination of eletron-deficient pyridine-*N*-oxide produces a C-2 olefination product regioselectively [Eq. (6.110)] [293, 294]. In the presence of palladium catalyst with a self-repulsive pyridine ligand, electron-deficient arenes containing trifluoromethyl, carbonyl, and nitro groups undergo *meta*-selective olefination [Eq. (6.111)]. The unique *meta*-selectivity could be attributed to an electrophilic substitution at the less electron-deficient *meta*-position or a proton abstraction process at the more acidic *meta* C–H bond [295]. Perfluorinated arenes undergo Pd-catalyzed olefination at the most acidic C–H bond [Eq. (6.112)] [296].

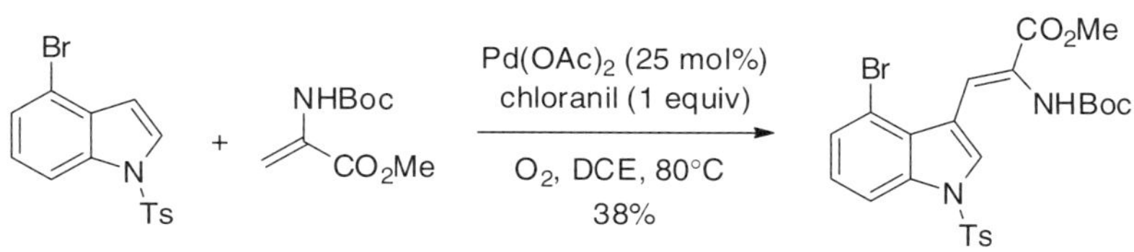

(6.106)

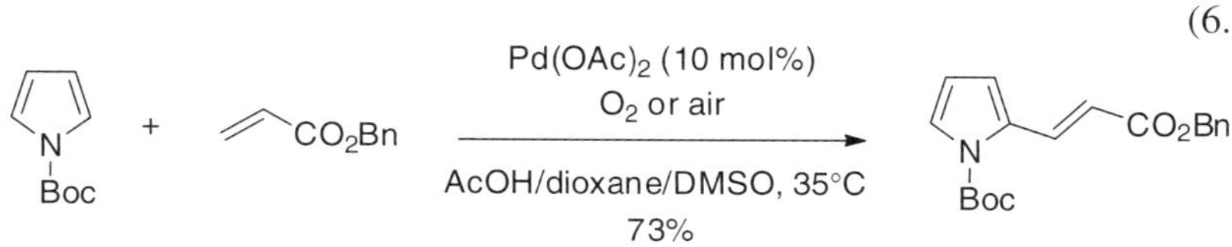

(6.107)

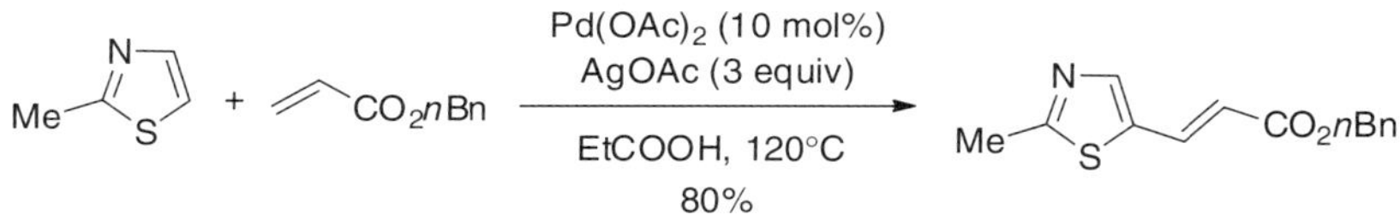

(6.108)

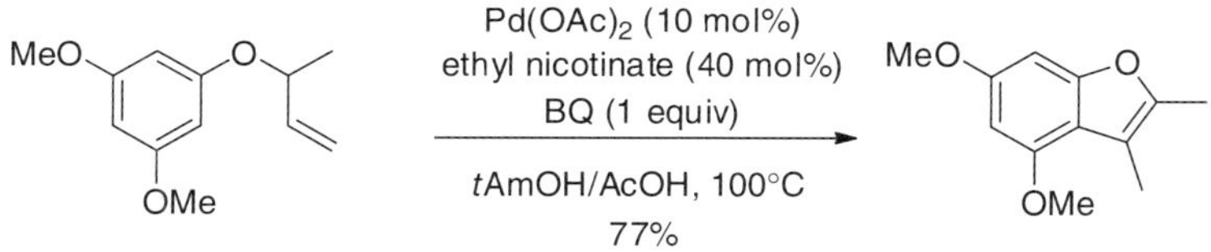

(6.109)

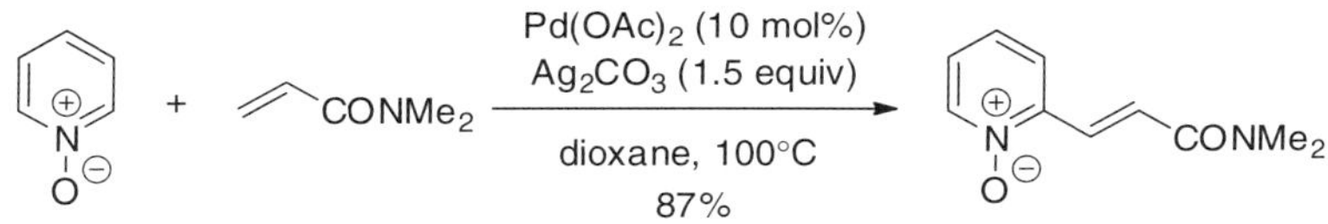

(6.110)

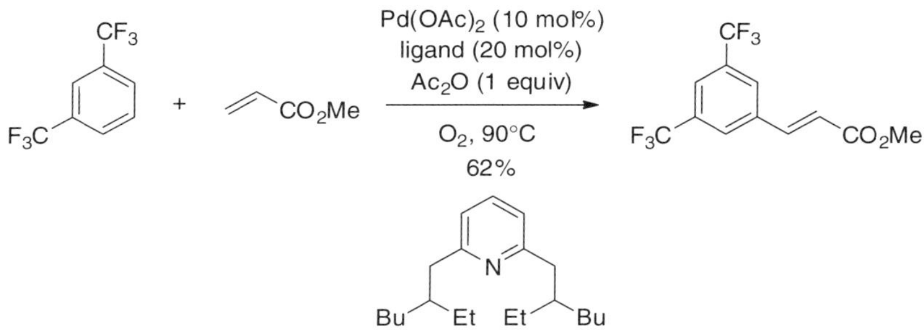

(6.111)

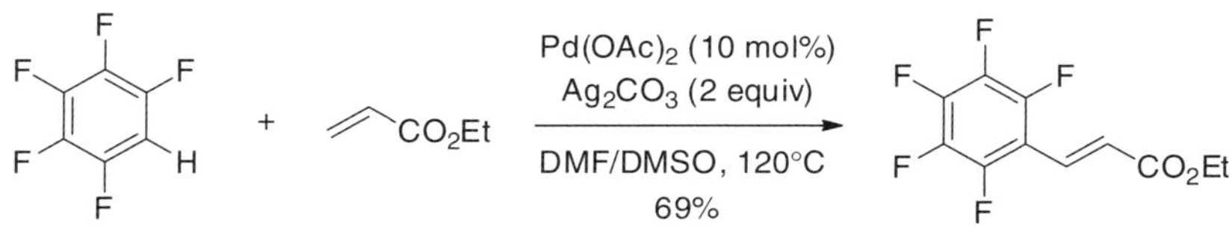

(6.112)

Transition-metal-catalyzed oxidative C–H olefination generally follows a similar mechanism (Scheme 6.11). It starts with the directed cleavage of aryl or alkyl C–H bond to form a metallacycle. Then olefin insertion and β-hydride elimination ensue. The metal catalyst is oxidized back to the initial oxidation state by different oxidants to close the catalytic cycle.

Pd-mediated oxidative intramolecular coupling of indole with a pendant internal olefin was used as the key step for the synthesis of okaramine N and its analogues (Scheme 6.12) [297, 298]. Although a stoichiometric amount of palladium was used, this strategy significantly shortened the synthetic steps. A similar Pd-mediated

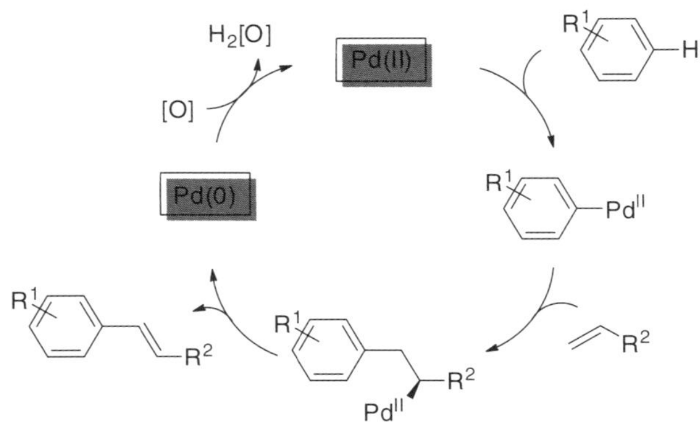

Scheme 6.11. Mechanism of Pd-catalyzed olefination.

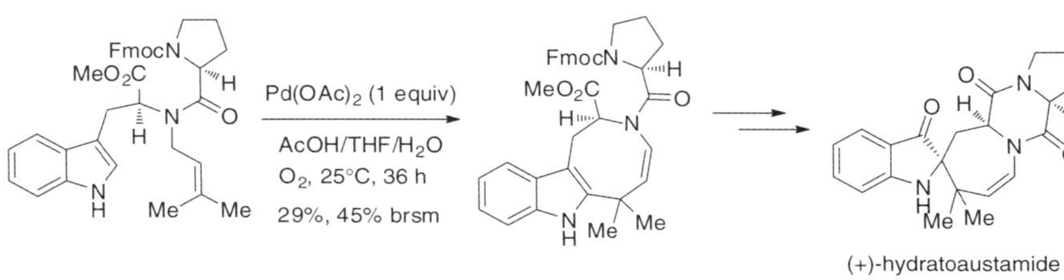

Scheme 6.12. Synthesis of (+)-hydratoaustamide.

oxidative coupling of pyrrole with internal olefin was used during the synthesis of (+)-dragmacidin F [299]. A catalytic oxidative coupling pyrrole with internal olefin was used during the synthesis of rhazinicine (Scheme 6.13) [300]. The total synthesis of (+)-lithospermic acid features a directed olefination of functionalized phenyl acetic acid in the presence of palladium catalyst with an amino acid ligand (Scheme 6.14) [301].

6.4.7 Annulation of C–H Bond with Alkene and Alkyne

In directed hydroarylation of alkene and alkyne, C–H bond is added across an alkene or alkyne, leaving the directing group intact after the reaction. If the directing group

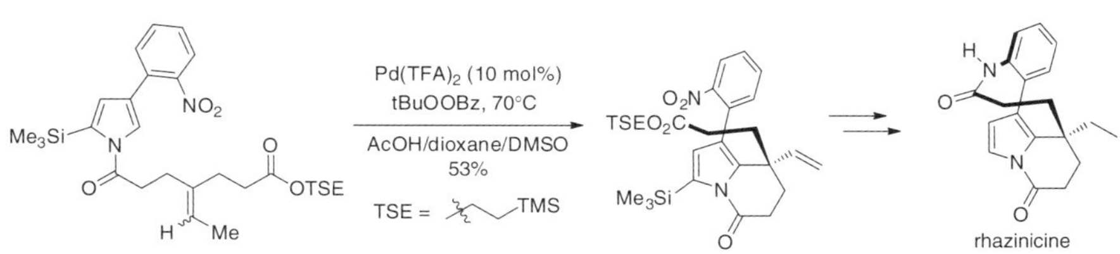

Scheme 6.13. Synthesis of rhazinicine.

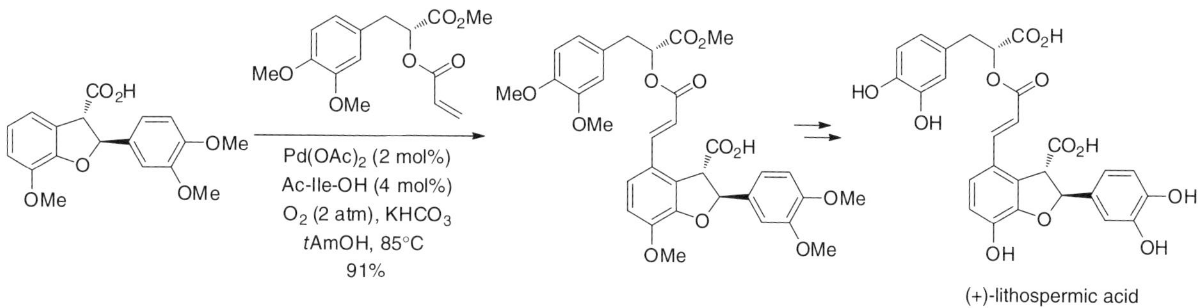

Scheme 6.14. Synthesis of (+)-lithospermic acid.

contains a functionality that can react with the organometallic intermediate during the hydroarylation process, the reaction would form the corresponding cyclization product with multiple bond constructions.

If the directing group of the substrate contains an electrophilic center, annulation with alkene or alkyne would proceed under redox-neutral condition. For example, Re [Eq. (6.113)] [302]- and Rh [303]-catalyzed annulation of aldimine with alkyne generates indanyl amine. An asymmetric variant of this annulation reaction was achieved by a combination of Rh(I) catalyst with chiral phosphine ligand [Eq. (6.114)] [304]. Related annulation of aryl ketone with alkyne catalyzed by cationic Rh provides an indanol product [Eq. (6.115)] [305, 306]. Cationic Rh-catalyzed annulation of aryl C–H substrate containing an ester or amide directing group with alkyne could give an indenone product [Eq. (6.116)] [307].

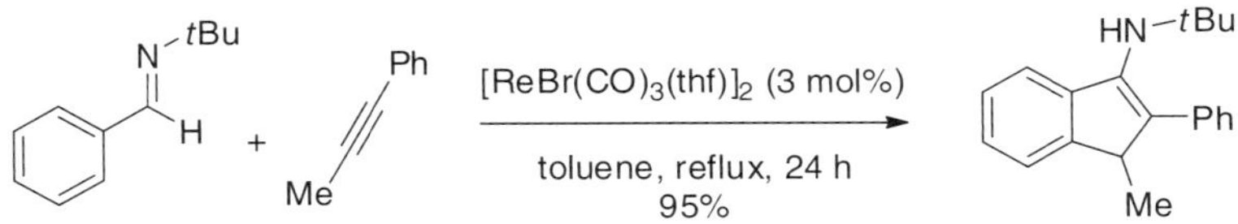

(6.113)

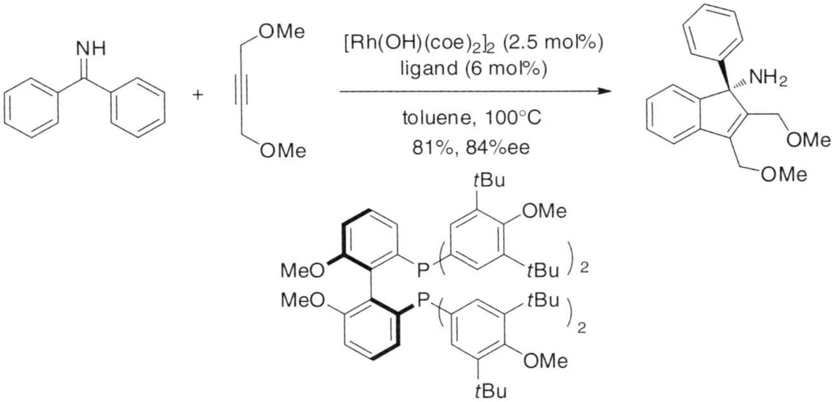

(6.114)

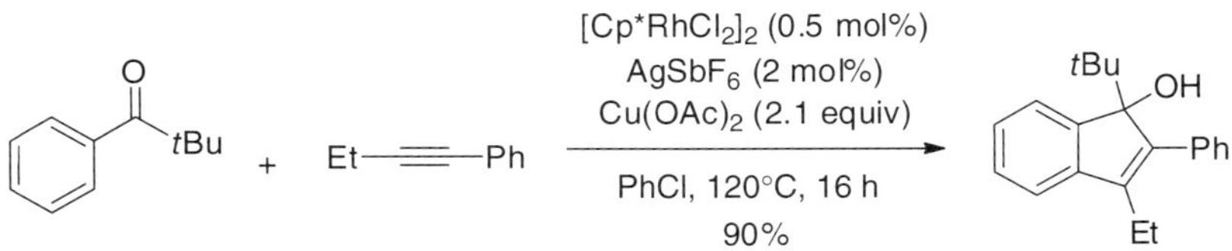

(6.115)

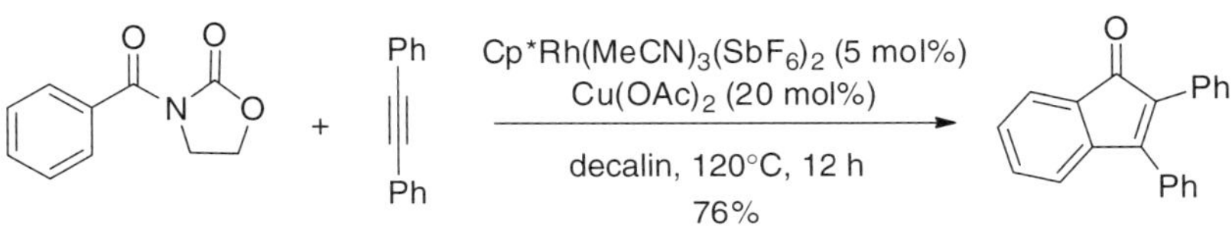

(6.116)

On the other hand, if the directing group of the substrate contains a nucleophilic center, annulation with an alkene or alkyne would require an oxidant to form an oxidative annulation product [308, 309]. In most cases, the nucleophilic centers are a N–H or an O–H bond. Pd(II), Cp*Rh(III), and [RuCl$_2$(*p*-cymene)]$_2$, among others, are particularly useful catalysts for this type of annulation reaction, affording a wide range of heterocycles. Various directing groups can be used for this purpose. Annulations of acetanilide [Eqs. (6.117) and (6.118)] [310–312], aryl imine [Eq. (6.119)] [313, 314], aryl carboxylic acid [Eq. (6.120)] [315–317], benzamide [Eq. (6.121)] [318–323], and sulfamide [Eq. (6.122)] [324] with alkene or alkyne produce indoline, indole, isoquinoline, isocoumarin, isoquinolinone, and benzosultam, respectively. When *N*-methoxy or *N*-pivolyl benzamide were used as the substrates, similar annulation with alkene and alkynes proceeded without addition of external oxidants [325–329]. Besides heterocycles, the annulation reaction can also provide carbocyles. Ruthenium [Eq. (6.124)] [330]- and rhodium [331]-catalyzed oxidative annulation between ketone compounds and alkyne proceed through enolate-directed C–H cleavage to give a carbocyclic product.

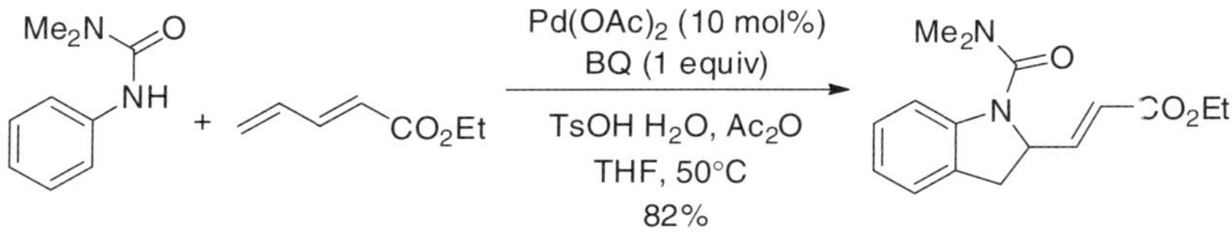

(6.117)

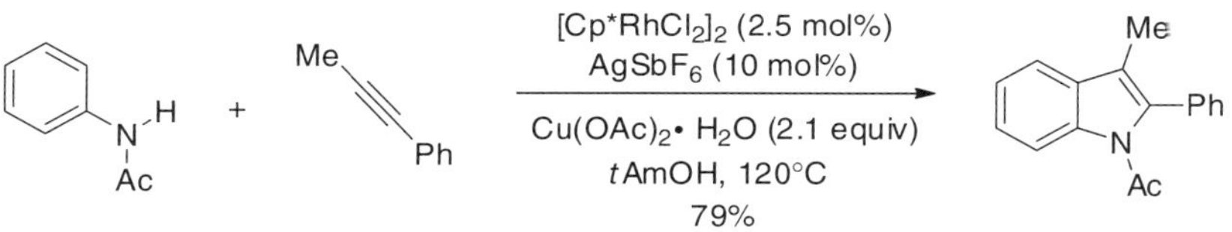

(6.118)

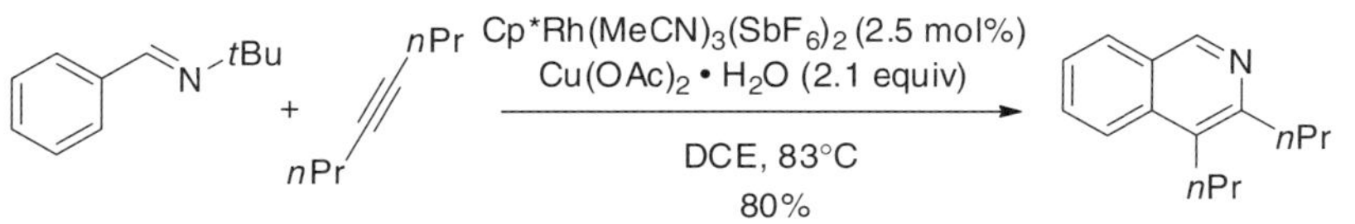

(6.119)

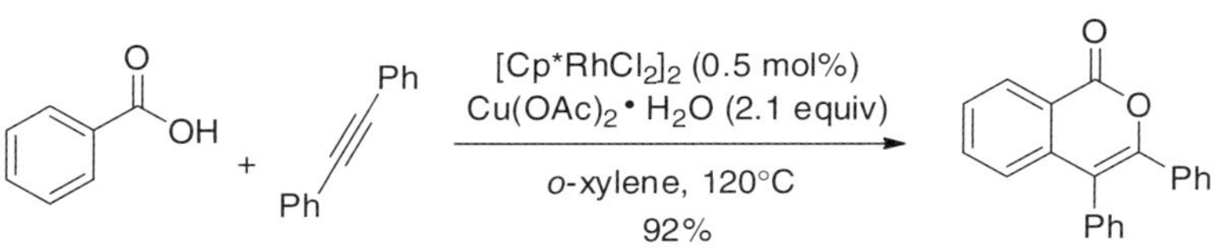

(6.120)

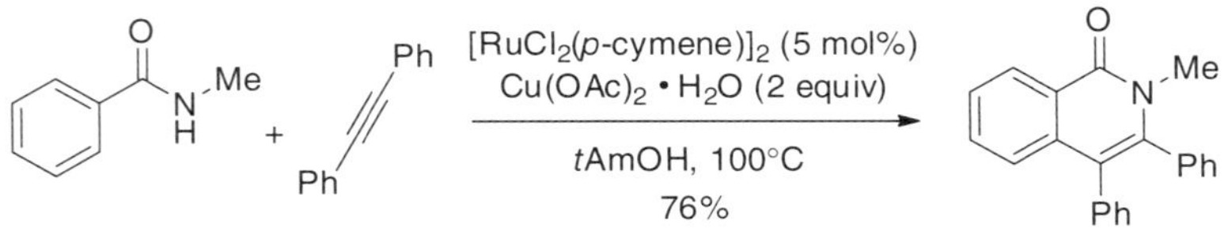

(6.121)

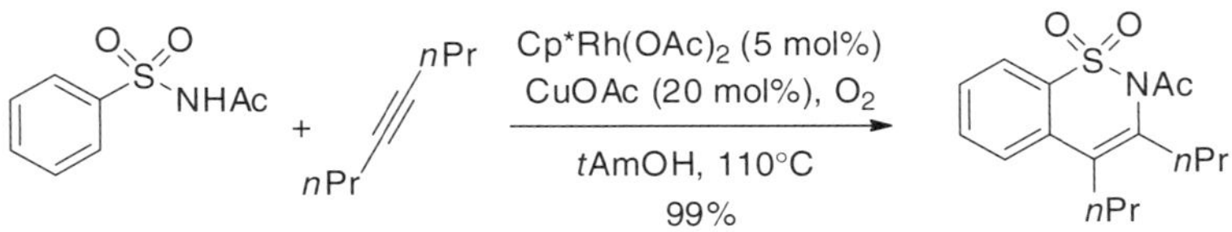

(6.122)

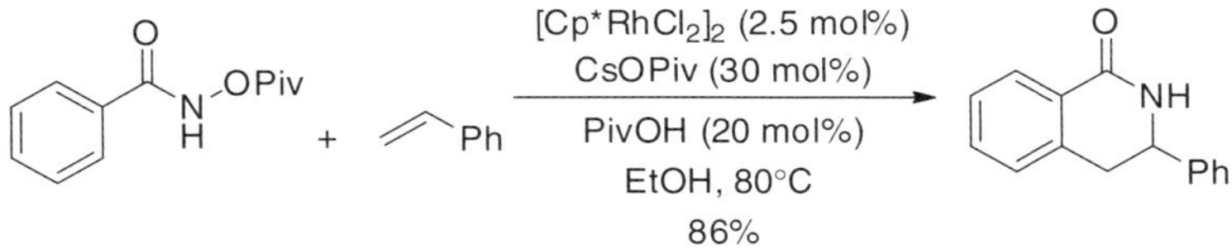

(6.123)

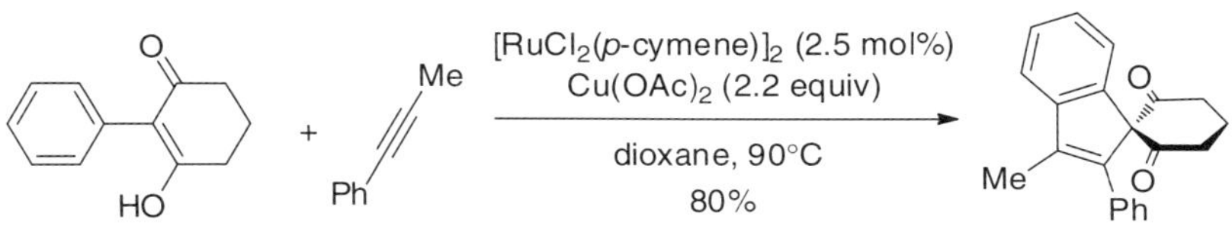

(6.124)

For the mechanism of these annulation reactions, it starts with *ortho* metalation of a C–H bond to form a metallocycle. The cyclometalation using Pd(II) and Cp*Rh(III) complexes are common. Then alkyne insertion forms an enlarged metallocycle which undergoes reductive elimination to form the cyclization product. Pd(0) or Rh(I) was oxidized to regenerate the catalyst.

6.5 DIRECT ARYLATION

Biaryls are valuable building blocks in different areas, such as biology, medicinal compound, and material sciences [332]. Their immense scientific and commercial value has stimulated vast research interest for making these compounds. Transition-metal-catalyzed cross-coupling between aryl halide and organometallic reagent is arguably the most powerful tool for the synthesis of bi(hetero)aryl structures [333–335]. However, the organohalide and organometallic reagent used in cross-coupling reactions are usually prepared from the corresponding arene substrates through a number of synthetic steps. In addition, some functionalized aryl halides and organometallic reagents are very expensive or not easily available. Therefore, direct use of arene substrates in cross-coupling reactions is advantageous from both environmental and economic points of view. The cross-coupling involving arene substrates, commonly described as direct arylation, has seen fast progress in the last decade [336, 337].

Catalytic direct arylations through C–H functionalization falls into three categories: the reaction between and arene C–H bond and an organometallic reagent, the reaction between two arene C–H bonds, and the reaction between an arene C–H bond and aryl halide. The first two reaction types are oxidative arylation and require a sacrificial oxidant, while the third reaction type is oxidative-neutral.

Since organic molecules usually contain different C–H bonds with comparable reactivity, the control of regioselectivity in direct arylation reactions is a significant challenge. Direct arylation of heteroarenes as well as electron-deficient arenes usually displays a high degree of regioselectivity since the regioselectivity is dominated by the intrinsic electronic character of their C–H bonds. Direct arylation of a simple arene without electronic bias is usually less selective. By attaching a directing group to the arene, the regioselectivity could be well controlled through cyclometalation of a specific C–H bond.

6.5.1 Direct Arylation with Organometallic Reagents

In an early example of direct arylation with organometallic reagents, Inoue and co-workers [338] reported the reaction of 2-arylpyridine with aryl stannane in the presence of Wilkinson catalyst to form arylation products [Eq. (6.125)]. The regioselectivity was controlled by the chelation of pyridine, and only an *ortho* arylation product was obtained. It is likely that the tetrachloroethane solvent acts as the oxidant for this arylation reaction.

Besides organostannane, less toxic organoboron reagent is also applicable in the direct arylation reaction with Rh catalyst. Miura reported a Rh-catalyzed arylation of imine substrate with tetraphenylborate, albeit in moderate yield [Eq. (6.126)] [339]. This type of Rh-catalyzed arylation reaction was later extended to include various substrates containing basic directing groups such as imine, imidazole, oxazoline, and pyridine [340, 341].

This type of arylation could start with the generation of aryl rhodium species from Rh catalyst and organoboron reagent. Then it undergoes oxidative addition to cleave the ortho C–H bond to form a metallocycle. After reductive elimination, the product is formed along with Rh–H, which is transformed to the catalytically active species by the oxidant. Alternatively, the ortho C–H bond could be cleaved by a Rh(III) species formed by oxidation of a Rh(I) precursor. Then transmetalation at the Rh(III) center followed by reductive elimination could also produce the product.

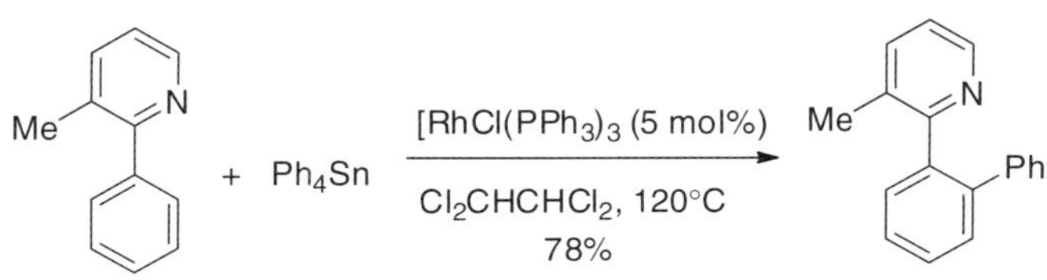

(6.125)

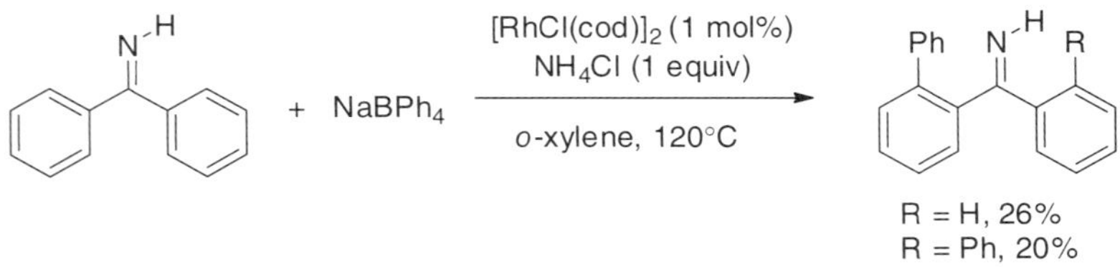

(6.126)

In an alternative approach, Kakiuchi et al. [342–344] developed a low-valent Ru-complex-catalyzed *ortho*-arylation of aromatic ketone with organoboron esters [Eq. (6.127)] [342–344]. Ketone acts as the hydride acceptor for this reaction, and a stoichiometric amount of alcohol derived from the reduction of the ketone was obtained. Thus, it was proposed that the C–H bond was cleaved through an oxidative addition to a ruthenium(0) species. The resulting aryl Ru–H reacted with a ketone to form a Ru–alkoxy species. Subsequent transmetalation with an organoboron reagent generated a biaryl ruthenium intermediate, which underwent reductive elimination to form the *ortho*-arylation product together with the Ru(0) catalyst (Scheme 6.15). By using imine as a directing group, Sames and co-workers [345] extended this protocol for the Ru-catalyzed arylation of sp^3 C–H bond alpha to a nitrogen atom with organoboronic ester [Eq. (6.128)].

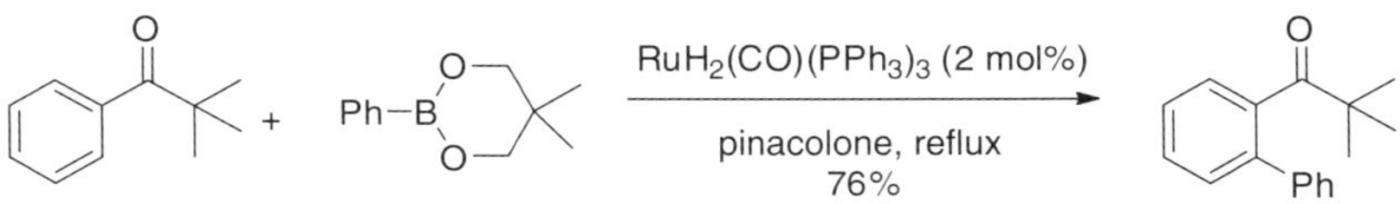

(6.127)

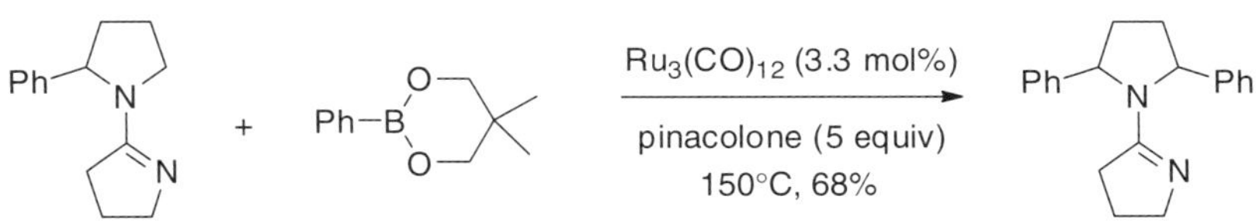

(6.128)

Palladium has been proved to be a useful catalyst for the reactions of both aromatic and aliphatic C–H bonds with organometallic reagents. In early studies, Yu discovered palladium catalyst systems for the alkylation of aryl C–H and alkyl C–H bonds with alkyl stannane or alkyl boron reagents [Eq. (6.129)] [346, 347]. Shi et al. [348, 349] reported Pd-catalyzed *ortho*-arylation of acetanilide derivatives with arylboronic acids [Eq. (6.130)] and aryl silanes [Eq. (6.131)] to form biaryls. Related arylations of both aryl and alkyl C–H bonds in carboxylic acid derivatives with arylboronic acids have been achieved by Yu [Eq. (6.132)] [350, 351]. In these reactions, a directing group is necessary to enable the reaction to occur regioselectively. In addition, an oxidant is required to achieve the oxidative coupling. In a general mechanistic scheme, the first step involves the cleavage of aryl or alkyl C–H bond with Pd(II) to form an aryl or alkyl palladacycle. Transmetalation of the palladacycle with

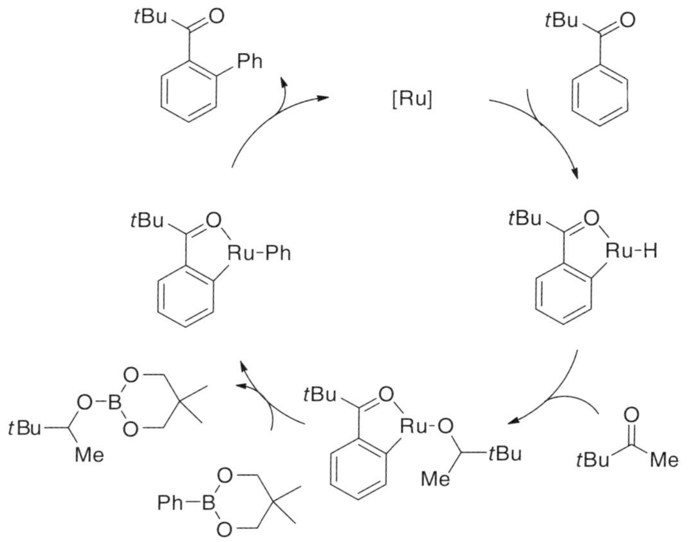

Scheme 6.15. Mechanism of Ru-catalyzed *ortho*-arylation.

orgaonoboronic acid followed by reductive elimination produces the arylation product together with Pd(0). Finally, oxidation of Pd(0) by the oxidant regenerates the Pd(II) catalyst (Scheme 6.16).

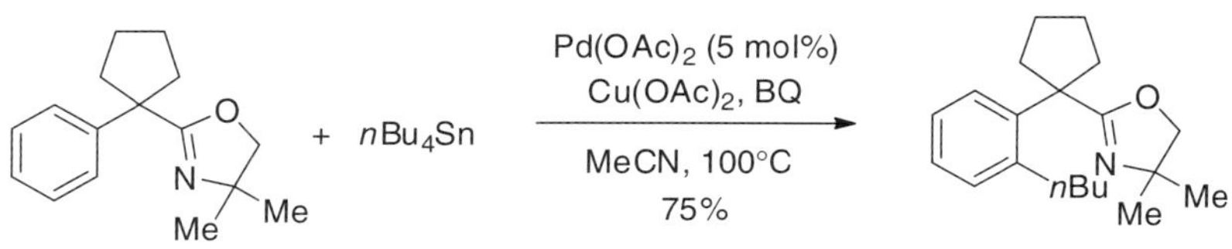

(6.129)

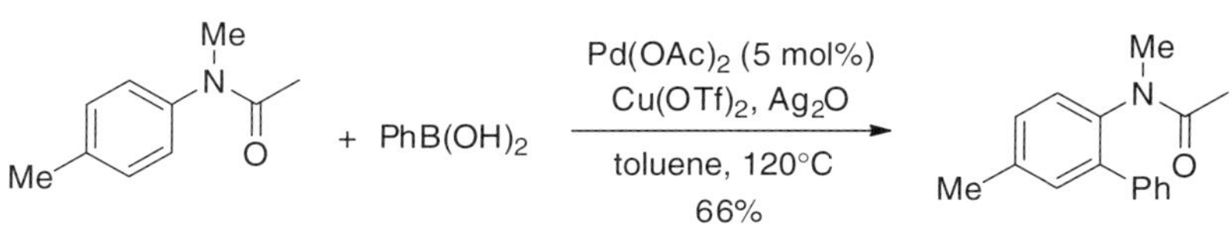

(6.130)

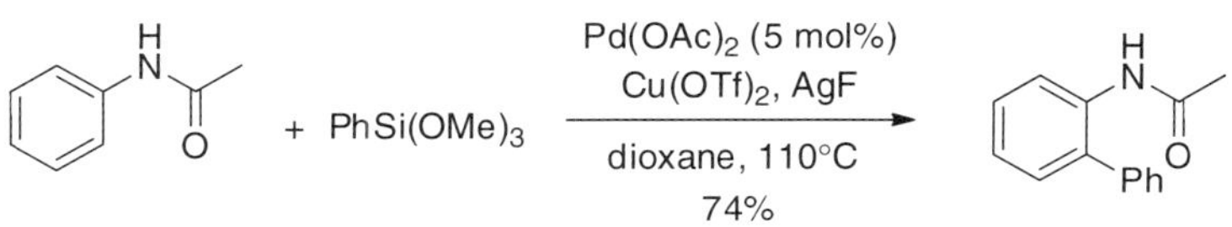

(6.131)

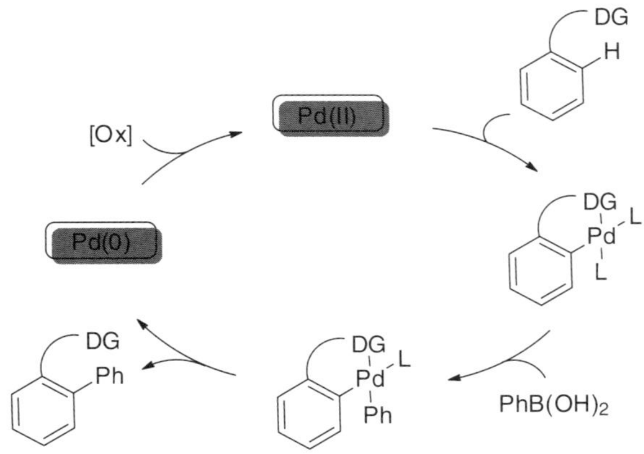

Scheme 6.16. Mechanism of Pd-catalyzed *ortho*-arylation.

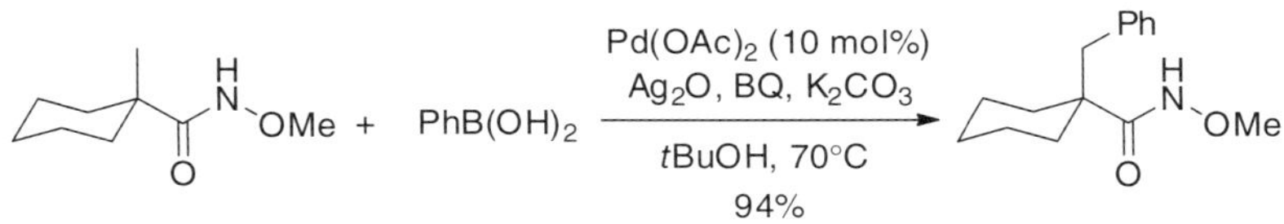

(6.132)

Various other directing groups such as pyridine [352], carboxylic acid [Eq. (6.133)] [353, 354], imine [Eq. (6.134)] [355, 356], and urea [Eq. (6.135)] [357] have also been shown to be effective for the Pd-catalyzed oxidative coupling of aryl C–H bonds with arylboronic acid derivatives for the synthesis of biaryl structures. Some reactions can occur at room temperature, low catalyst loading, making this reaction more efficient and practical. Enantioselective coupling of aryl and cyclopropyl C–H bonds with boronic acid through desymmetrization was realized by a combination of palladium catalyst with protected amino acid ligand [Eq. (6.136)] [358–360].

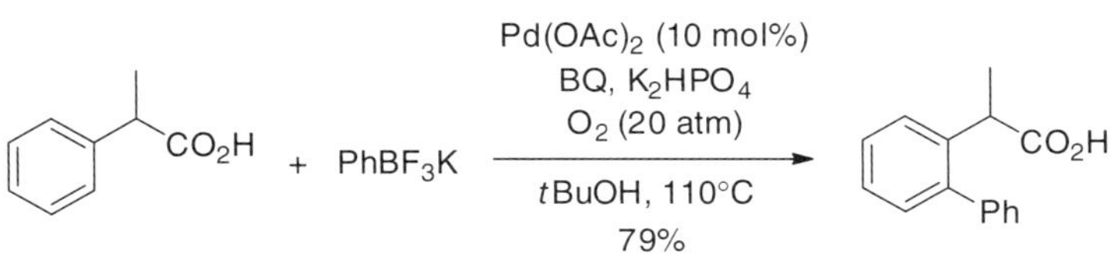

(6.133)

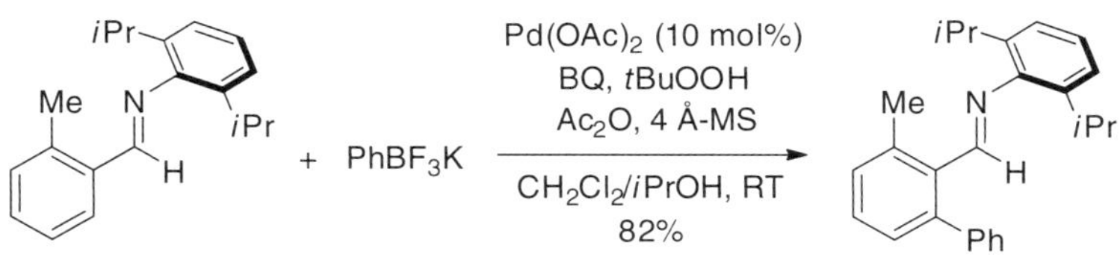

(6.134)

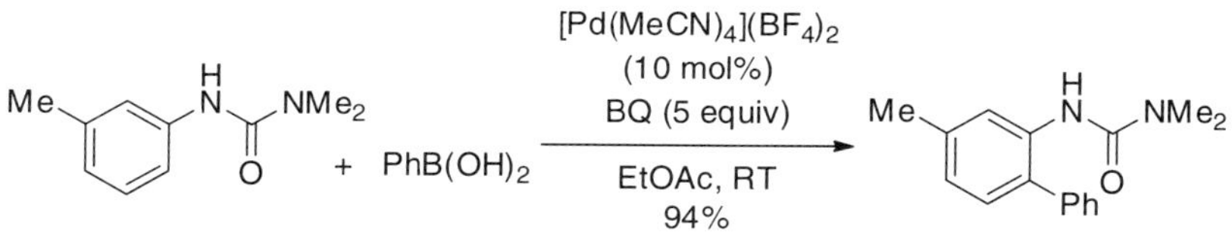

(6.135)

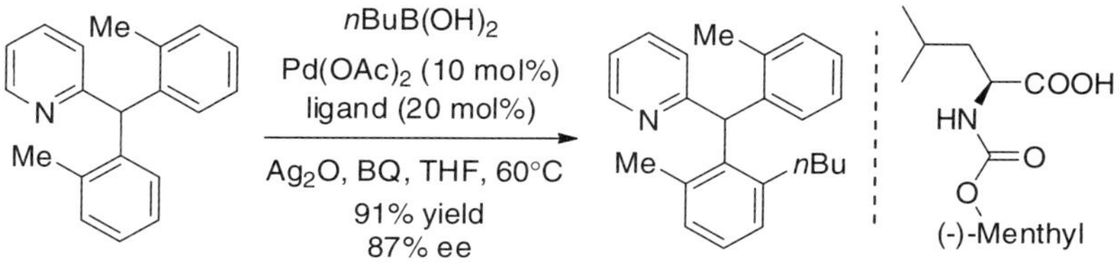

(6.136)

In the absence of a directing group, the regioselectivity of the C–H cleavage is mainly controlled by the intrinsic property of the arene such as electronic property. For example, the Pd-catalyzed arylation of a simple arene with organoboronic acids [Eq. (6.137)] [361], organostannes [362], and organosilanes [363] occurred preferentially at the more electron-rich site. A gold-catalyzed oxidative coupling of arenes with aryl silanes proceeded with high regioselectivity under mild conditions, favoring the electron-rich position of arenes [Eq. (6.138)] [364]. Itami and Scott [365, 366] have applied the Pd-catalyzed arylation of polyaromatics with boronic acids to form larger conjugated polyaromatics [Eq. (6.139)]. Pd-catalyzed arylation of perfluoroarene with organoboronic acids occurs at the most acidic C–H bond [Eq. (6.140)] [367].

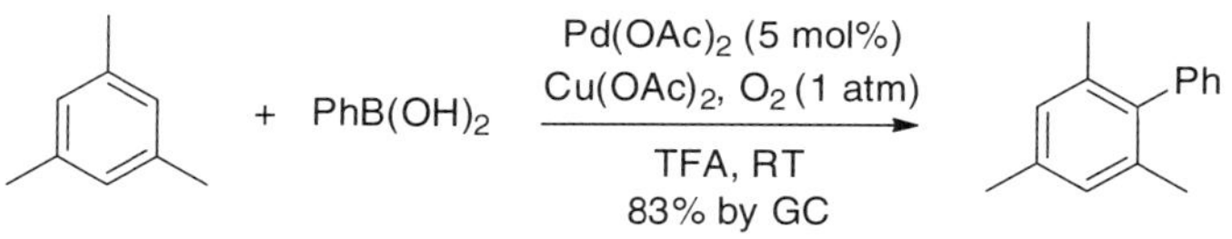

(6.137)

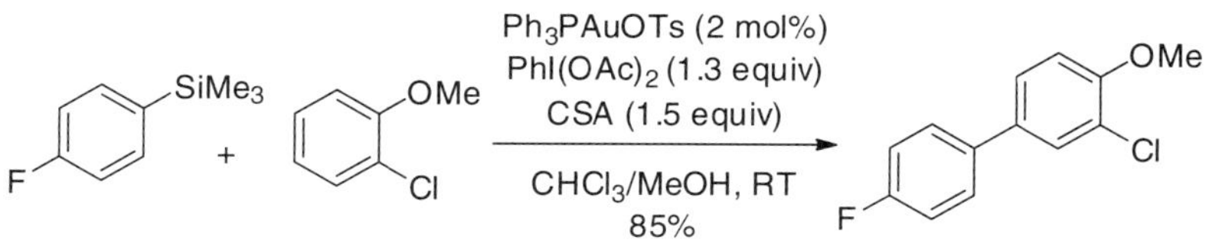

(6.138)

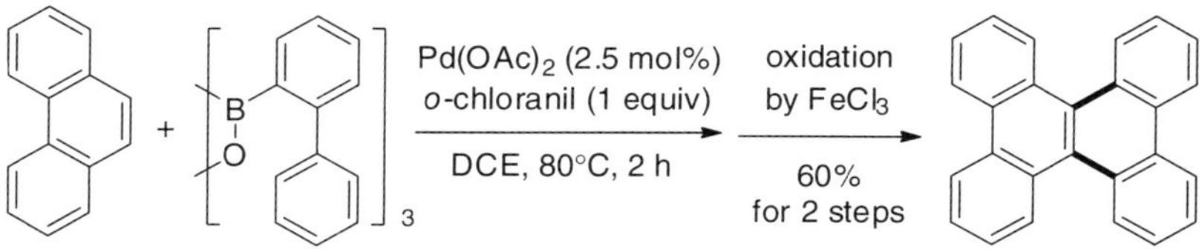

(6.139)

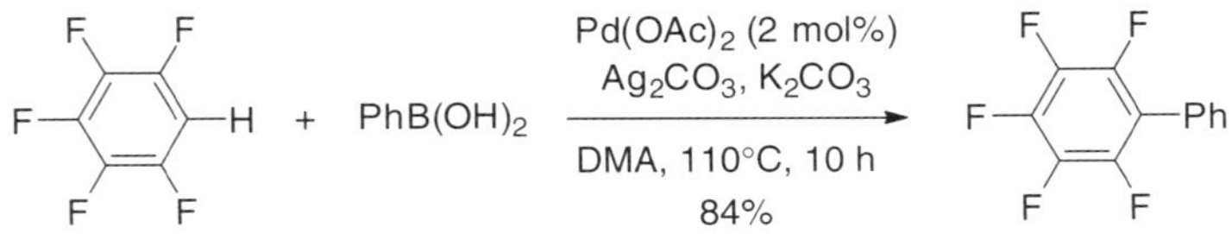
(6.140)

In heteroaromatic substrates, the regioselectivity can be well controlled. For example, oxidative arylation of indole with arylboronic acid or arylsiloxane selectively occurs at the C-2 position rather than at the more electron-rich C-3 position [Eq. (6.141)] [361, 368]. The mechanism of the C-2 selectivity is not well understood. In constrast, arylation of thiophene, benzothiophene, furan, and benzofuran occurs selectively at the more electron-rich C-2 position [369]. The regioselectivity of the Pd/bipyridine-catalyzed arylation of thiophene and thiazole can be switched to the C-4 position [Eq. (6.142)]. The presence of excess boronic acid is the key in achieving the unusual C-4 regioselectivity [370, 371]. By using bisoxazoline as a ligand, synthesis of sterically hindered biaryls was achieved in the C-4 arylation of thiophene with arylboronic acid. Enantioselective C–H arylation was also made possible by a chiral bisoxazoline ligand [Eq. (6.143)] [372].

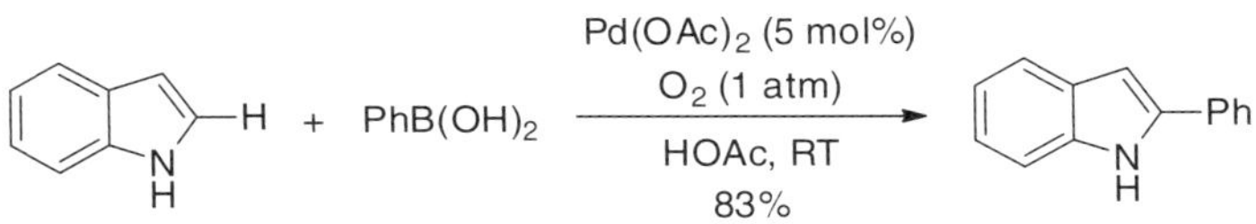
(6.141)

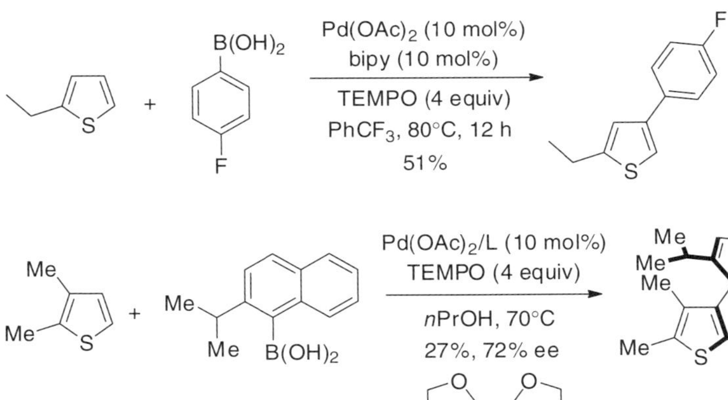
(6.142)

(6.143)

Compared with late transition metals, the use of abundant and less expensive first-row transition metals such as iron, cobalt, and nickel as catalysts in direct arylation is highly desirable. Nakamura developed an iron-catalyzed arylation of 2-arylpyridine derivatives with an arylzinc reagent under mild conditions

[Eq. (6.144)] [373]. The catalytic system can also be applied to the *ortho*-arylation of aryl imine [374]. In these reactions, 1,2-dichloroalkane was required as the oxidant to turn over the iron catalyst. The mechanism of these reactions, however, is not disclosed. Shi and Nakamura [375–377] reported *ortho*-arylation of 2-arylpyridines with Grignard reagent in the presence of cobalt and iron catalyst, respectively [Eq. (6.145)]. Arylation of C–H bonds in heterocycles such as azole and pyridine with orgaonsilane [378], organozinc [Eq. (6.146)] [379] and Grignard reagent [380] occurs with nickel catalyst.

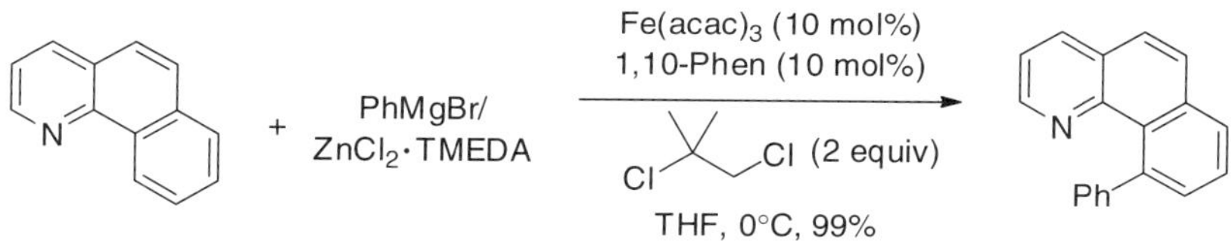

(6.144)

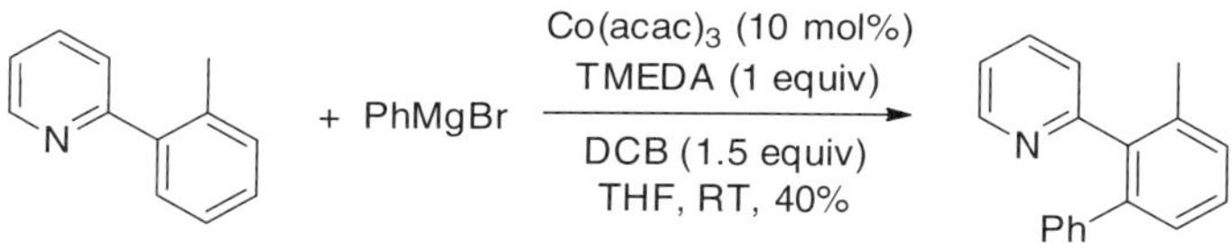

(6.145)

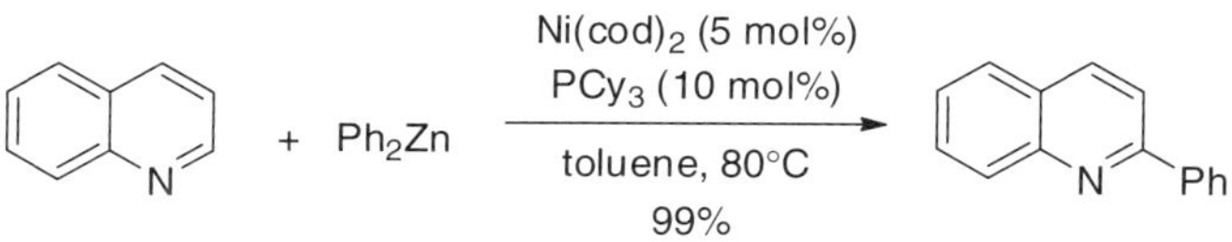

(6.146)

6.5.2 Oxidative Homocoupling and Cross-Coupling of Arenes

6.5.2.1 Intramolecular Oxidative Cyclization. Early studies on the cyclization of diarylamine, diaryl ether, benzophenone, and benzanilide to form carbazole, dibenzofuran, fluorenone, and phenanthridone were conducted with a stoichiometric amount of palladium [Eq. (6.147)] [381, 382]. Catalytic reactions were achieved using a proper oxidant. For example, Pd-catalyzed dehydrogenative cyclization of diphenylamine and diphenyl ether can be carried out using copper acetate or air as the oxidant [Eq. (6.148)] [383–385]. Related intramolecular C–H dehydrogenation was used to synthesize six-membered lactams from anilides [Eq. (6.149)] [386]. Besides formation of heterocycles, intramolecular arene coupling can also be used to synthesize carbocycles. Rh-catalyzed cyclization of diarylmethylamine and diarylacetic acid generates fluorene products [Eq. (6.150)] [387], while a Pd-catalyzed cyclization of diarylketone produces fluorenone derivatives [Eq. (6.151)] [388].

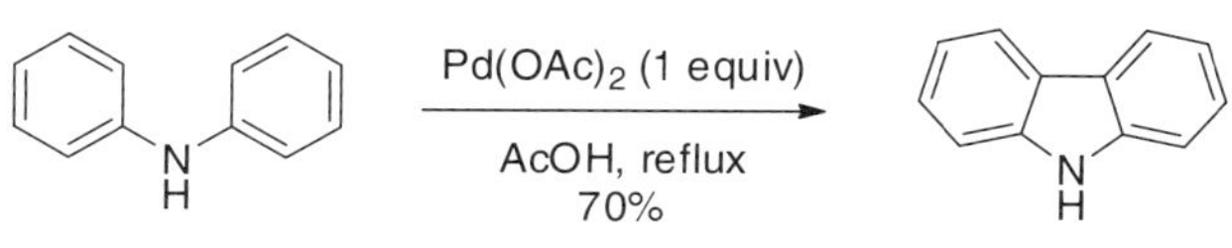

(6.147)

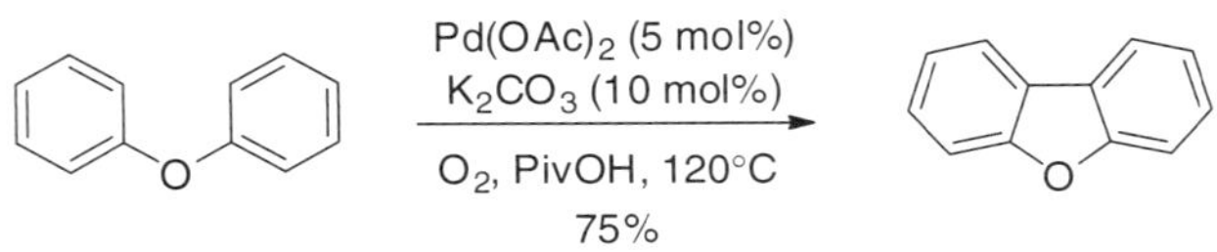

(6.148)

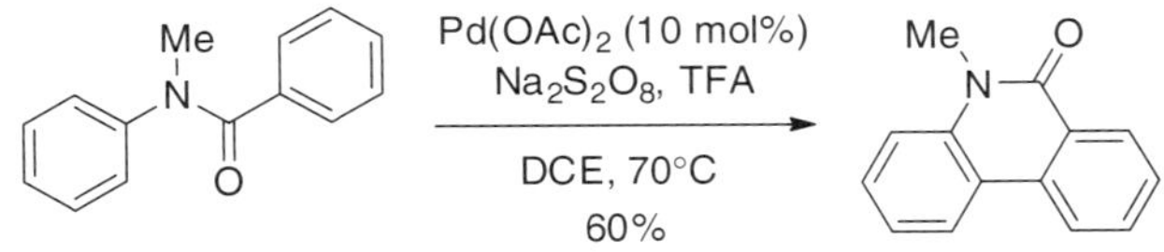

(6.149)

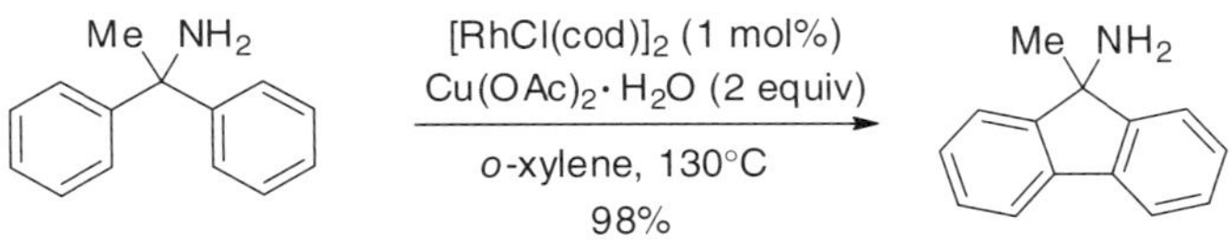

(6.150)

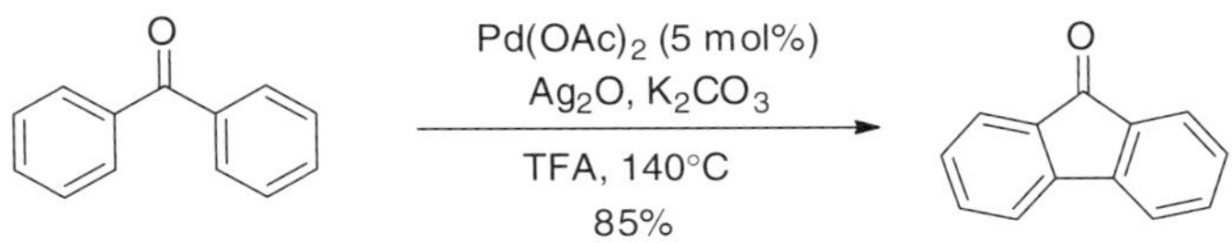

(6.151)

The intramolecular oxidative cyclization of arene substrates has been applied to the synthesis of natural products. For example, the bis-pyrrole structure in prodigiosin was constructed by a palladium-mediated oxidative coupling of 1,1′-carbonyldipyrrole (Scheme 6.17) [389]. Similarly, the bis-indole structure in rebeccamycin aglycone and its analogues can be accessed by intramolecular C–H/C–H coupling using palladium catalyst [390] or rhodium catalyst (Scheme 6.18) [391]. In addition, intramolecular oxidative cyclization of diarylamine has been used to synthesize various carbazole-based natural products [392, 393].

6.5.2.2 *Intermolecular Homocoupling.* Early works have shown that homocoupling of benzene and alkylbenzenes occurs in the presence of catalytic amount of palladium and oxygen [394]. However, a mixture of regioisomers were obtained

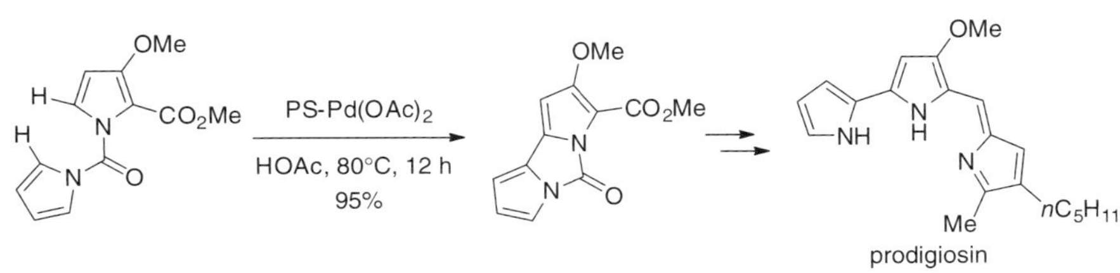

Scheme 6.17. Synthesis of prodigiosin.

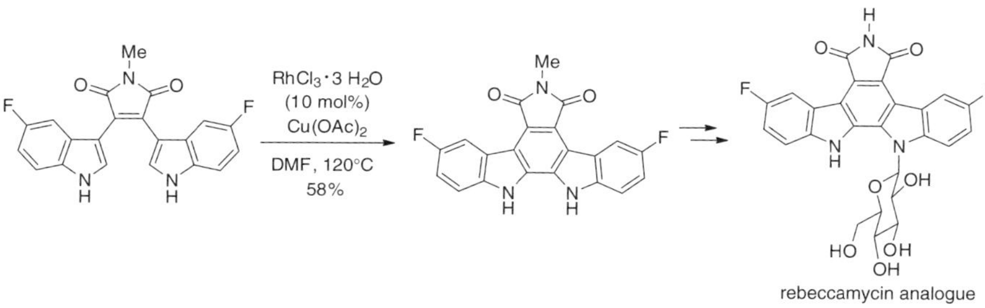

Scheme 6.18. Synthesis of rebeccamycin analogue.

when substituted benzenes were used, thus limiting the synthetic utility of this reaction. The regioselectivity can be well controlled by using heteroarenes or substrates containing a directing group. For example, Pd-catalyzed homocoupling of thiophene occurs regioselectively to afford oligothiophene [Eq. (6.152)] [395]. The regioselectivity of Pd-catalyzed indole homocoupling can be tuned by different oxidants to produce either a 3,3-coupling product or a 2,3-coupling product [Eq. (6.153)] [396, 397]. When arene substrates containing a directing group are used, homocoupling regioselectively occurs at the *ortho* position of the directing group. 2-Aryl pyridines could undergo palladium-catalyzed dehydrogenative homocoupling using oxone as the oxidant at ambient temperature [Eq. (6.154)] [398]. Mechanistic studies suggest that the first C–H activation occurs at the Pd(II) center while the second C–H activation occurs at a Pd(IV) center. Similar homocoupling of 2-arylpyridine can also be achieved in the presence of a stoichiometric amount of copper and iodine [399]. Using oxazoline as a directing group, a Ru-catalyzed homocoupling of arene also occurs regioselectively in the presence of allyl acetate as the oxidant [Eq. (6.155)] [400].

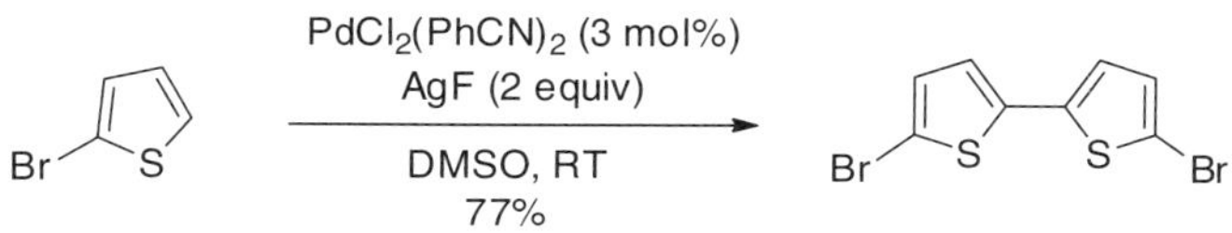

(6.152)

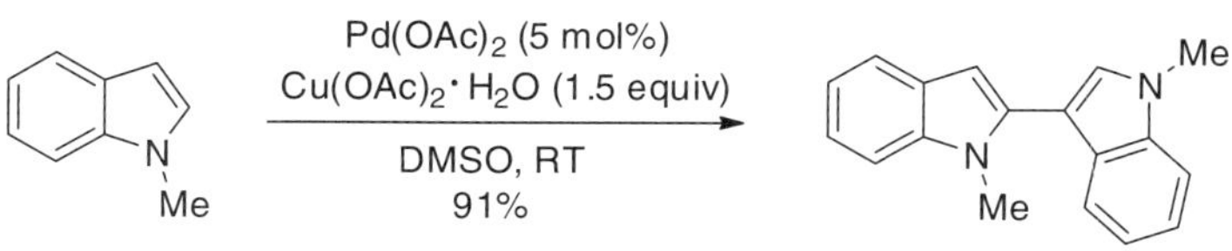

(6.153)

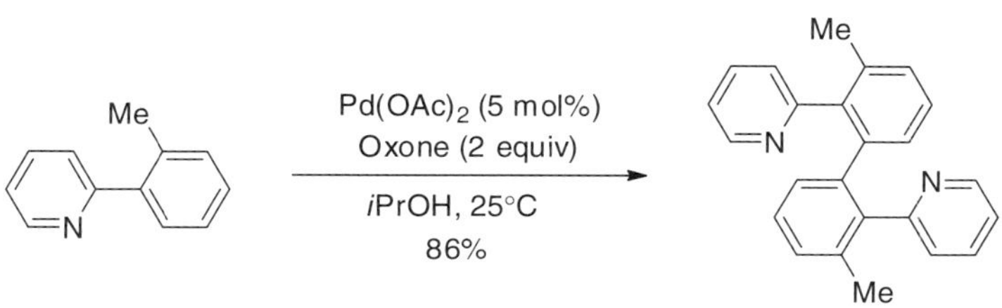

(6.154)

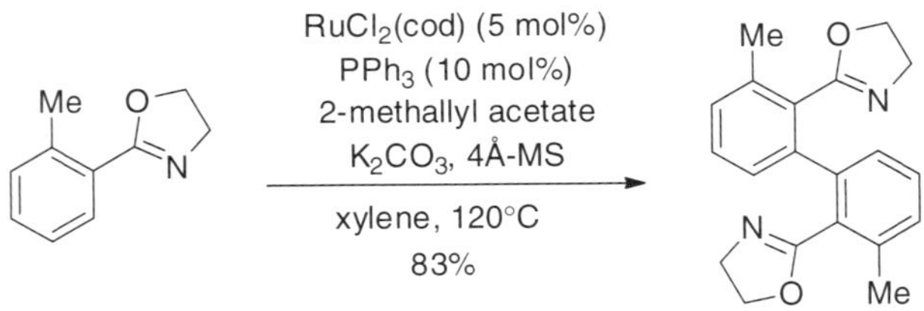

(6.155)

6.5.2.3 *Intermolecular Cross-Coupling.*

An ideal method to synthesize biaryls would be the catalytic intermolecular cross-coupling of two different arenes. However, this approach is challenging because homocoupling of each arene must be avoided and the regioselectivity of each arene must be controlled. An early example of this approach is illustrated by the stoichiometric amount of $Pd(OAc)_2$-mediated reaction of heteroarenes such as indole, pyrrole, furan, and thiophene with benzene to form arylated heterocycles [Eq. (6.156)] [401]. An early example of catalytic cross-coupling between naphthalene and arenes was achieved by Lu and co-workers [402] [Eq. (6.157)]. $K_2S_2O_8$ was used as the oxidant, and several turnover numbers were obtained. The reaction selectively occurred at the C-1 position of naphthalene, but produced several isomers with regard to the other arene. Stuart and Fagnou [403] developed an efficient Pd-catalyzed cross-coupling of *N*-acyl indole with simple arene [Eq. (6.158)]. The reaction selectively occurred at the electron-rich C-3 position of indole. Later it was found that the regioselectivity of indole can be tuned by different oxidants [404]. Similar cross-coupling of benzofuran or indole with arene was reported by DeBoef and co-workers [Eq. (6.159)] [383, 405].

The regioselectivity of catalytic arene cross-coupling can also be controlled by the presence of a directing group. Various directing groups including oxazoline [Eq. (6.160)] [406], pyridine [Eq. (6.161)] [407], acetamino [Eq. (6.162)] [408, 409], amide [Eqs. (6.163) and (6.164)] [410–413], and carbamate [414] can be used for this purpose. The regioselectivity of the other arene is usually controlled by steric effect or electronic effect. The cross-couplings of electron-deficient heteroarene [415] and perfluoroarene [416–418] with simple arenes have also been developed. Using a palladium acetate, copper acetate, or Cp*Rh as catalyst, cross-coupling of two heteroarenes selectively occurs to afford useful heteroarene products. This type of coupling encompasses the combination of electron-rich and electron-deficient heteroarenes [Eqs. (6.165)–(6.167)] [419–423] (e.g., between electron-rich five-membered heterocycles and azoles or pyridine *N*-oxides, etc.) and the combination of two structurally similar heteroarenes (e.g., between two electron-deficient azoles [419, 424–427] or between two electron-rich heteroarenes [428]) [Eqs. (6.168) and (6.169)].

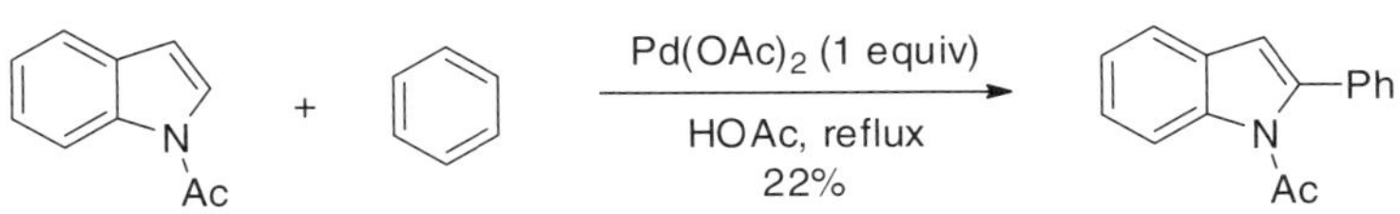

(6.156)

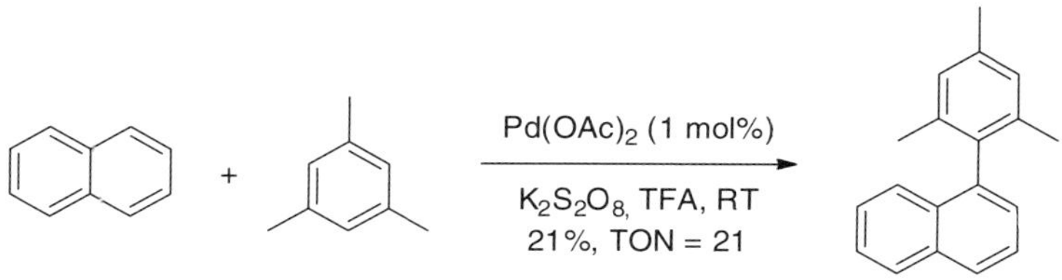

(6.157)

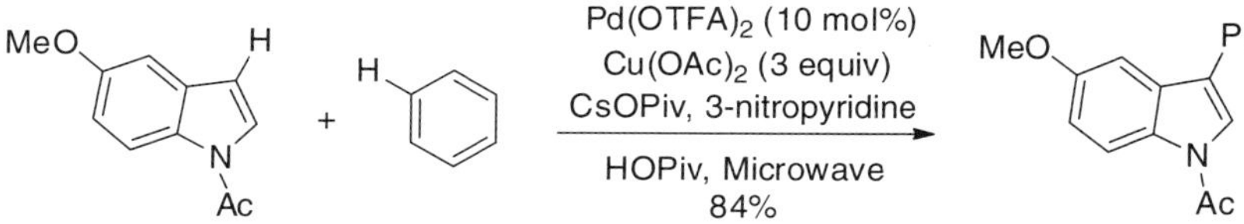

(6.158)

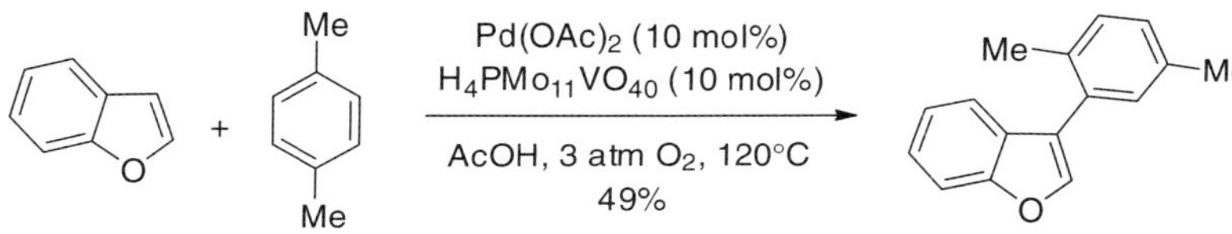

(6.159)

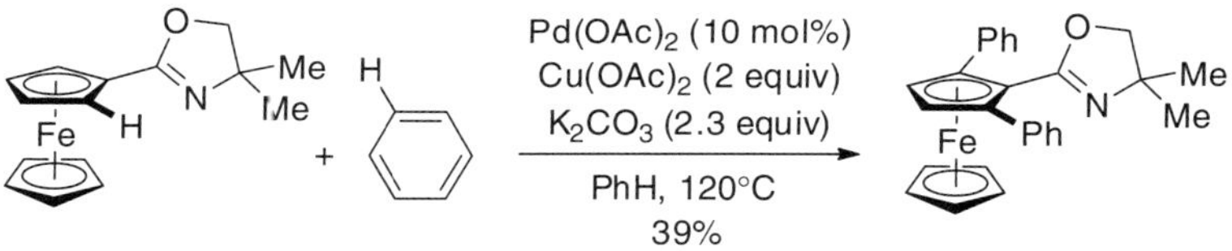

(6.160)

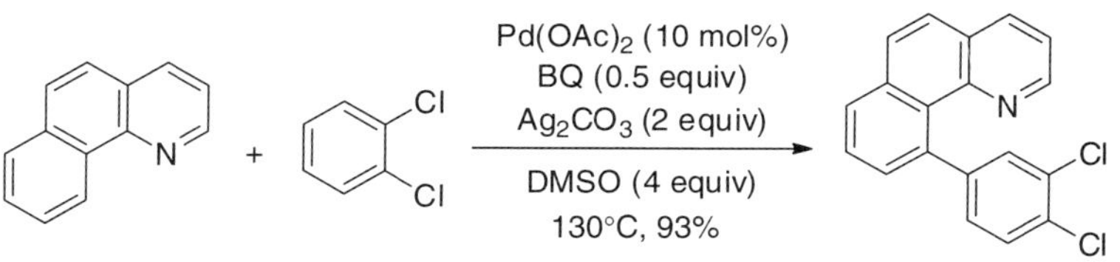

(6.161)

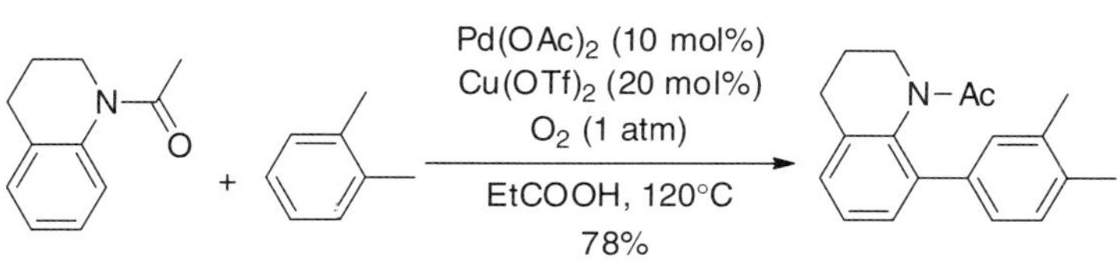

(6.162)

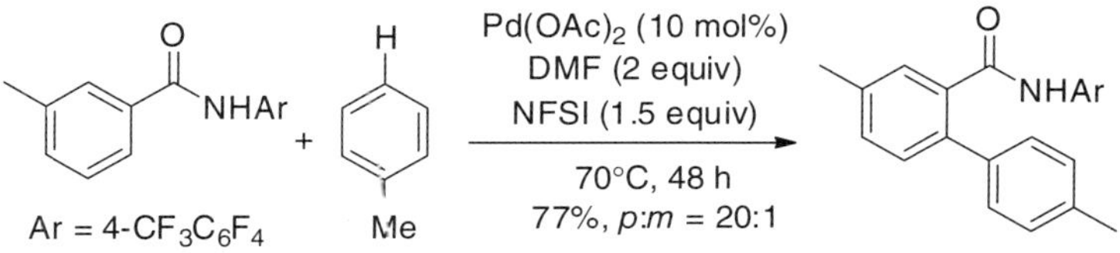

(6.163)

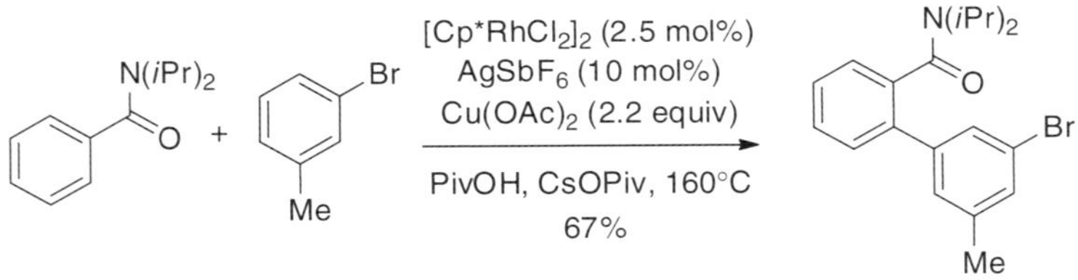

(6.164)

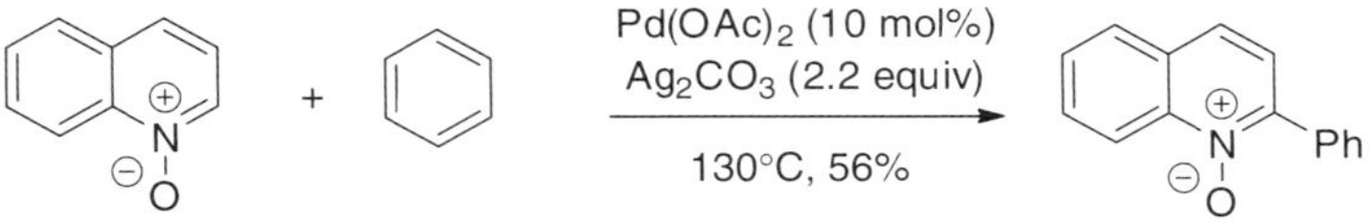

(6.165)

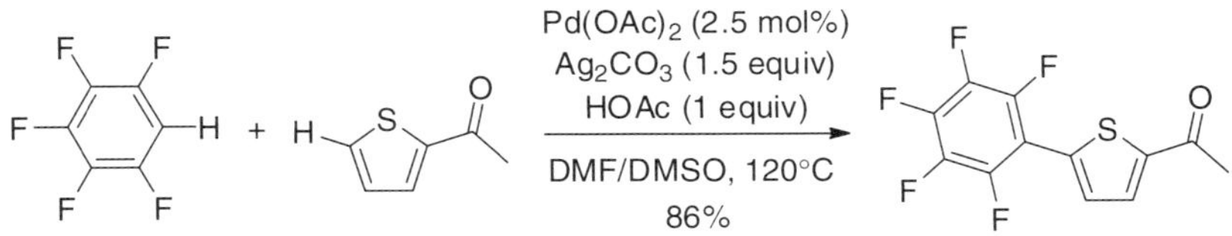

(6.166)

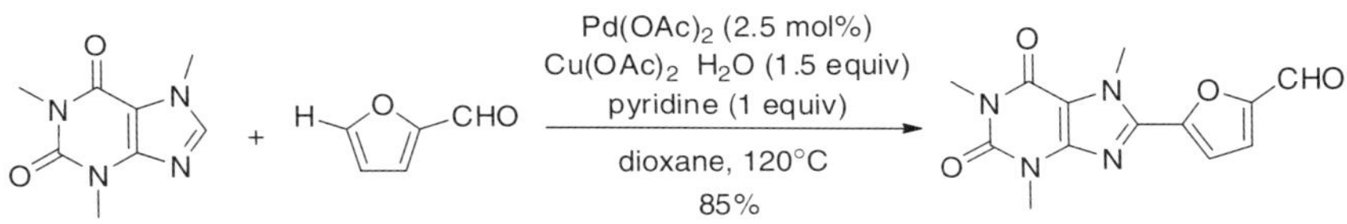

(6.167)

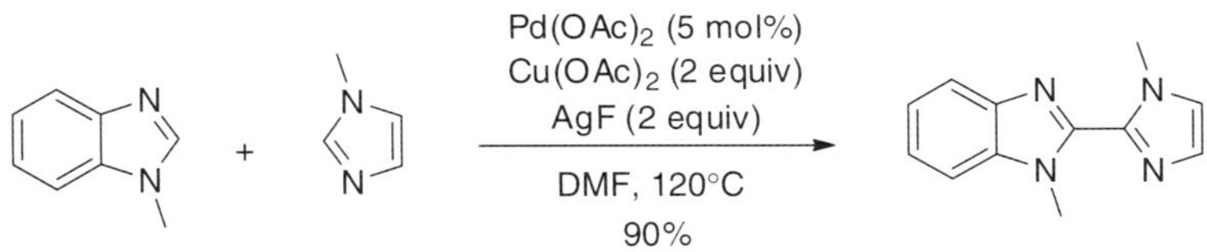

(6.168)

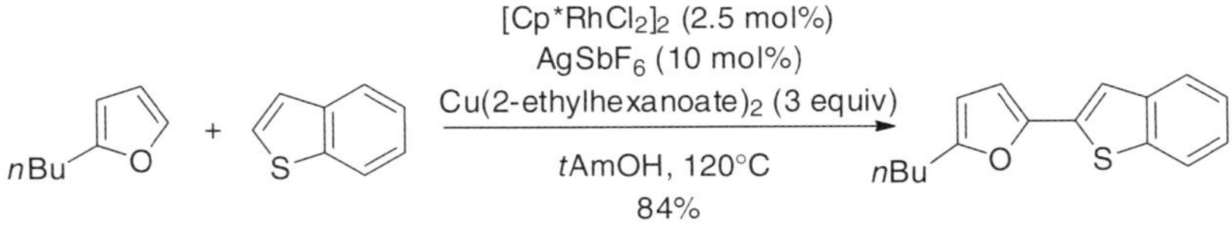

(6.169)

The synthesis of a complex marine natural product dragmacidin D was achieved by using three different C–H coupling steps (Scheme 6.19) [429]. One of them is a palladium-catalyzed oxidative cross-coupling of indole with pyrazine *N*-oxide, which occurred selectively at the C-3 position of functionalized indole and the C-2 position of pyrazine *N*-oxide. The functionalized indole was obtained through a C-4 selective direct arylation of thiophene and subsequent hydrogenation of the thiophene ring [430].

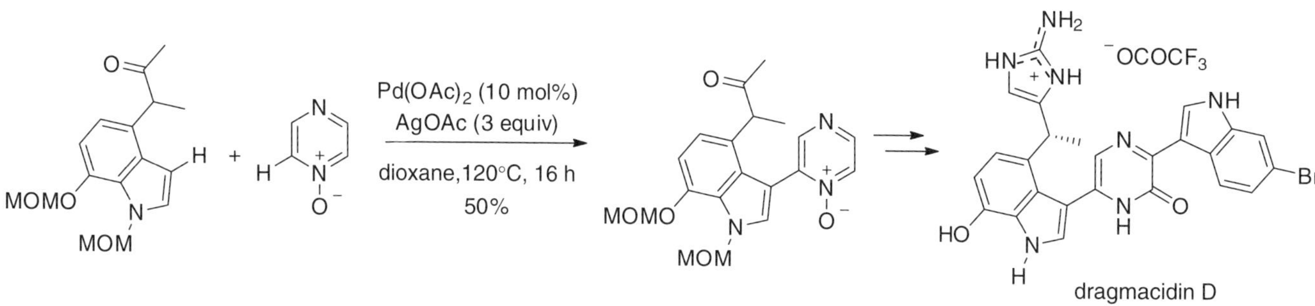

Scheme 6.19. Synthesis of dragmacidin D.

6.5.3 Direct Arylation with Aryl Halides and Pseudohalides

6.5.3.1 Early works. In an early example, Ames and Bull [431] discovered a Pd-catalyzed intramolecular arylation of heteroaryl bromide to form a cyclization product during investigations of palladium-catalyzed intermolecular Mizoroki–Heck reactions [Eq. (6.170)] [431]. This method was later extended to synthesize other five- and six-membered heterocycles, although under rather harsh conditions [432, 433].

Tajima and co-workers [434] reported the catalytic intermolecular direct arylation of isoxazoles with aryl iodides using a heterogeneous palladium catalyst [Eq. (6.171)]. Ohta and co-workers [435, 436] investigated the direct arylation of indole substrates with heteroaryl chloride [Eq. (6.172)]. C-2 arylation occurred with unsubstituted indole, while C-3 arylation occurred with *N*-tosyl indole.

Miura and co-workdeɾs [437] investigated the intermolecular arylation of 2-phenylphenol with aryl iodides [Eq. (6.173)]. The phenol group directed the palladium catalyst to regioselectively cleave the C–H bond at the *ortho* position.

These early discoveries laid the foundation for numerous studies on direct arylation reaction, especially in the last decades. The reaction types include intramolecular arylation, intermolecular arylation of heterocycles, directed intermolecular arylation, and arylation of simple arenes. The use of less reactive electrophiles such as aryl chloride, aryl tosylate, and mesylate is achieved by the development of novel phosphine and NHC ligands. Different metals including both late transiton metal and early transition metal Ru, Rh, Pd, Ir, Fe, Co, Ni, and Cu complexes have been employed as catalysts in direct arylation with aryl halides.

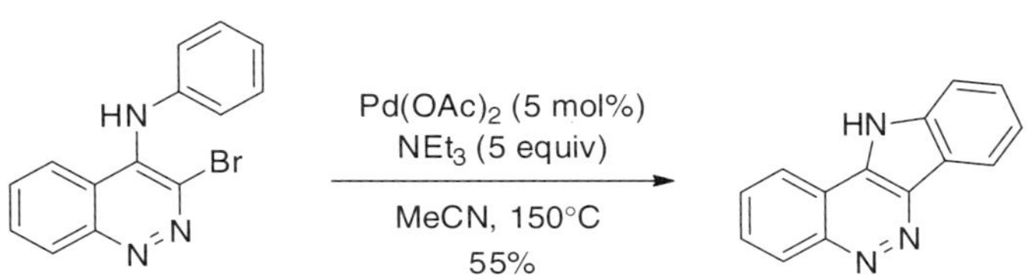

(6.170)

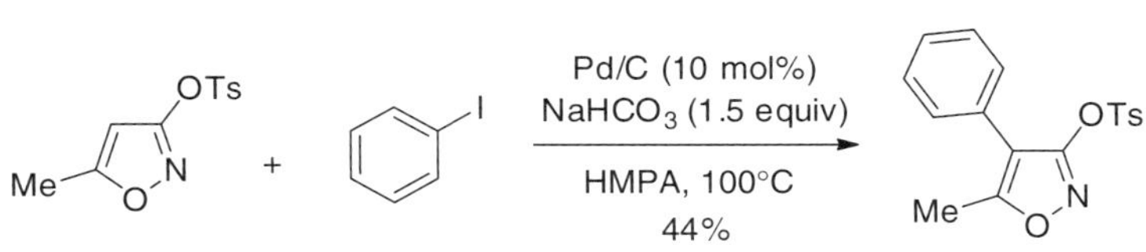

(6.171)

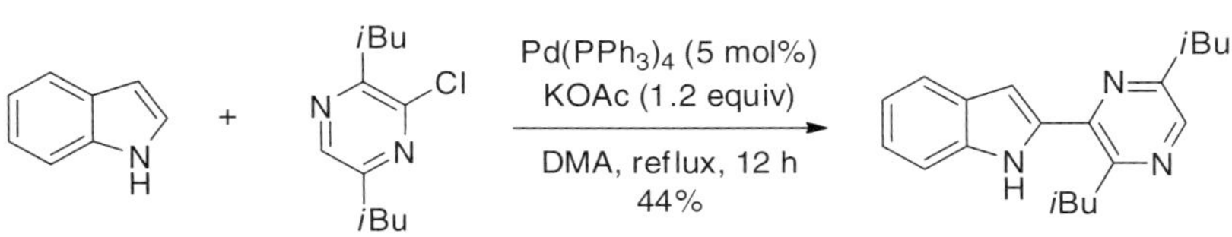

(6.172)

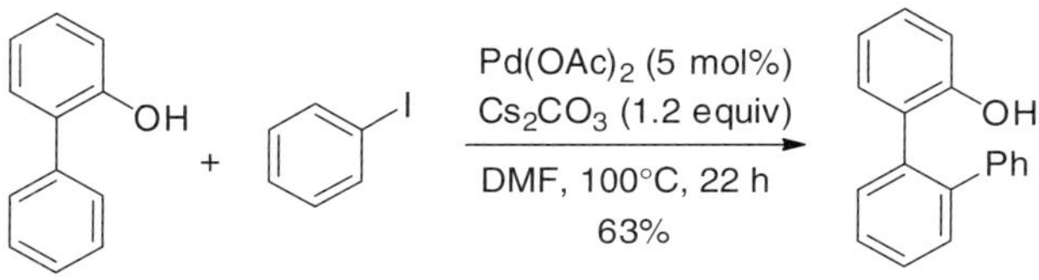

(6.173)

6.5.3.2 Intramolecular reaction. Intramolecular direct arylation of arene C–H bond can provide facile access to carbocycle or heterocycles [438]. Using nitrogen, oxygen, and amide as linkages, useful heterocycles containing biaryl strucutres can be synthesized [Eq. (6.174)–(6.176)] [439–442]. Arylation of heterocyclic C–H bond occurs to form an aryl–heteroaryl structure [Eq. (6.177)] [443–446]. The intramolecular direct arylation can also be used to synthesize a medium-sized ring system [Eq. (6.178)] [447–449]. Substantial efforts have been made to use less reactive aryl electrophiles such as aryl chloride [Eqs. (6.179) and (6.180)] [450–454]. The use of electron-rich phosphine and NHC ligands is the key to achieve reactivity with less reactive electrophiles. In the presence of palladium catalyst and a taddol-based phosphoramidite ligand, intramolecular direct arylation of vinyl triflates proceeds with high enantioselectivities [Eq. (6.181)] [455].

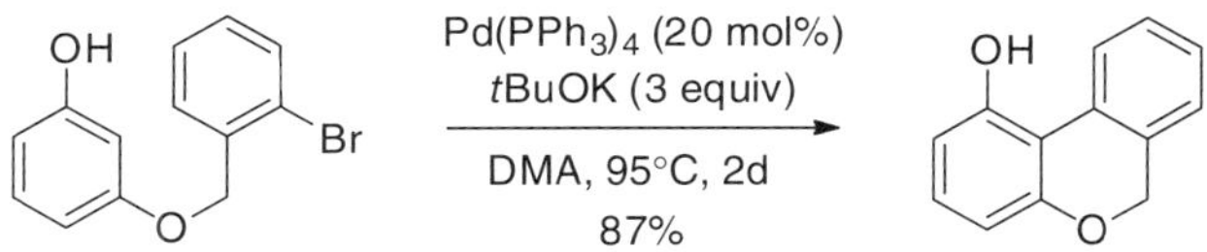

(6.174)

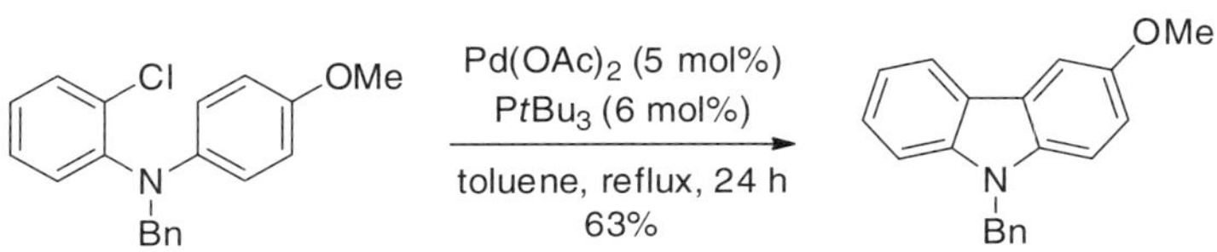

(6.175)

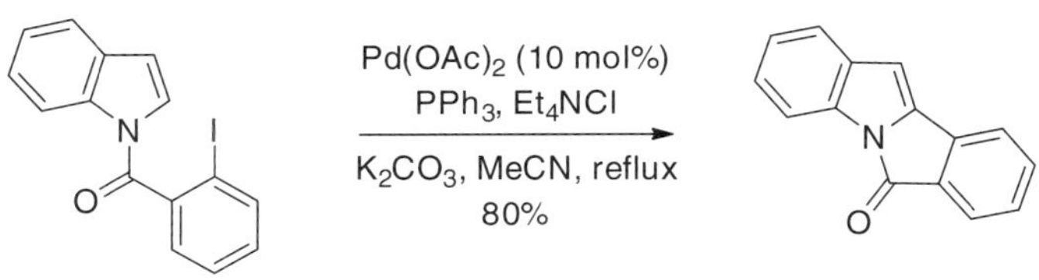

(6.176)

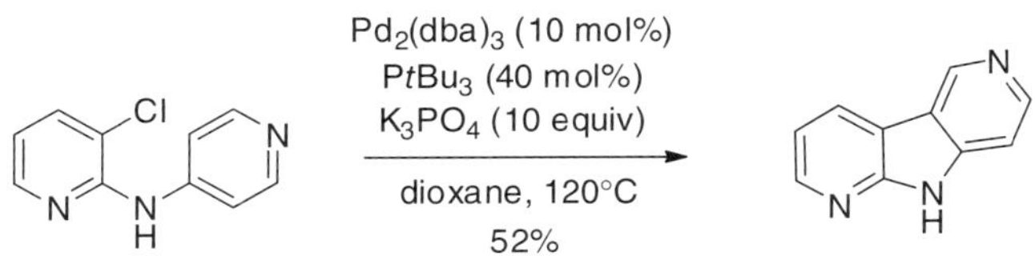

(6.177)

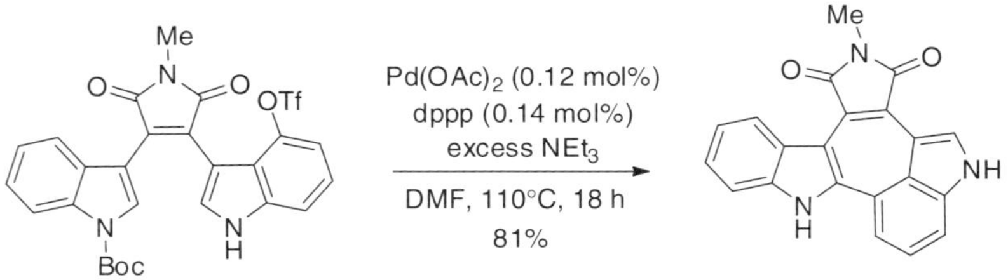

(6.178)

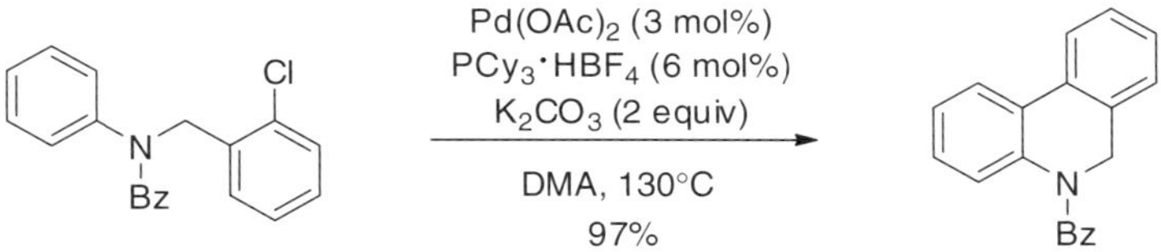

(6.179)

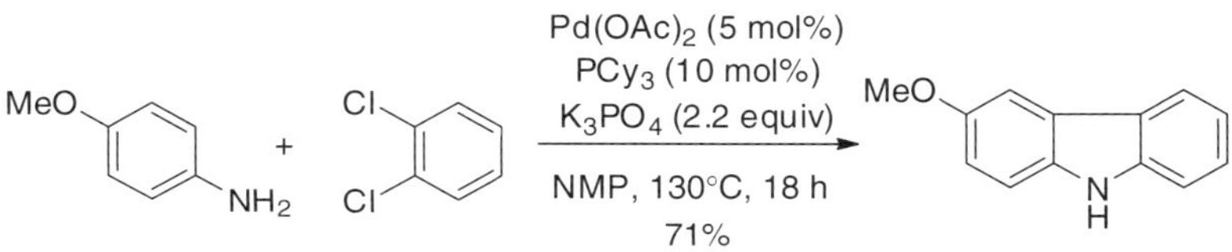

(6.180)

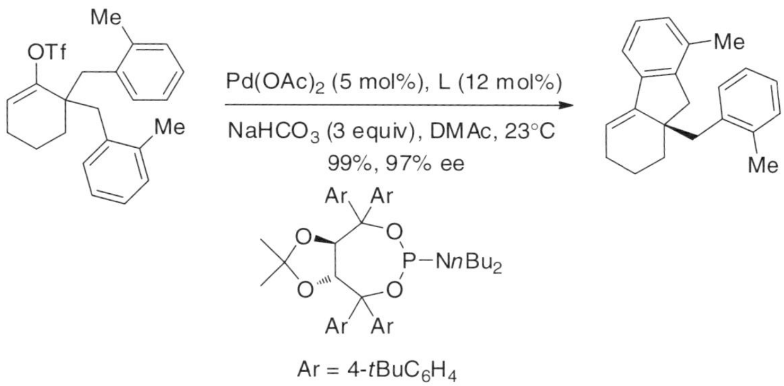

(6.181)

An early application of intramolecular C–H arylation in total synthesis was demonstrated by Bringmann et al. [456] in 1986. During the synthesis of (−)-ancistrocladine, a palladium-catalyzed intramolecular direct arylation was used to construct the biaryl linkage (Scheme 6.20). In another early example, Suzuki and co-workers [457] applied a similar palladium-catalyzed intramolecular direct arylation to construct the biaryl structure of (+)-gilvocarcin M (Scheme 6.21). Subsequently, many natural products have been synthesized based on intramolecular direct arylation of 2-haloarenes containing ester [458–462], amide [463], ketone [464], and alkyl [465] linkages. Synthesis of medium sized ring system through

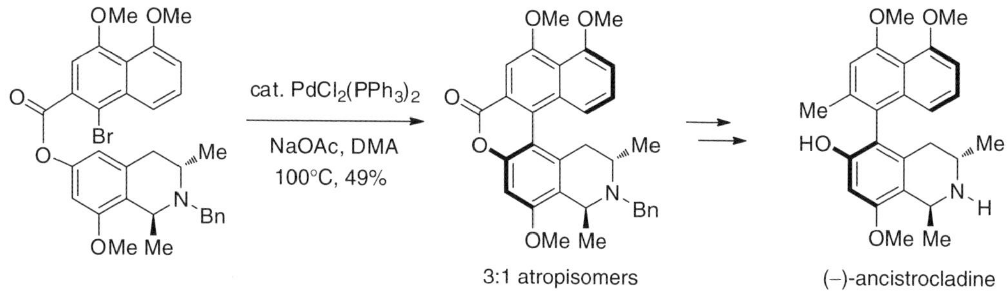

Scheme 6.20. Synthesis of (−)-ancistrocladine.

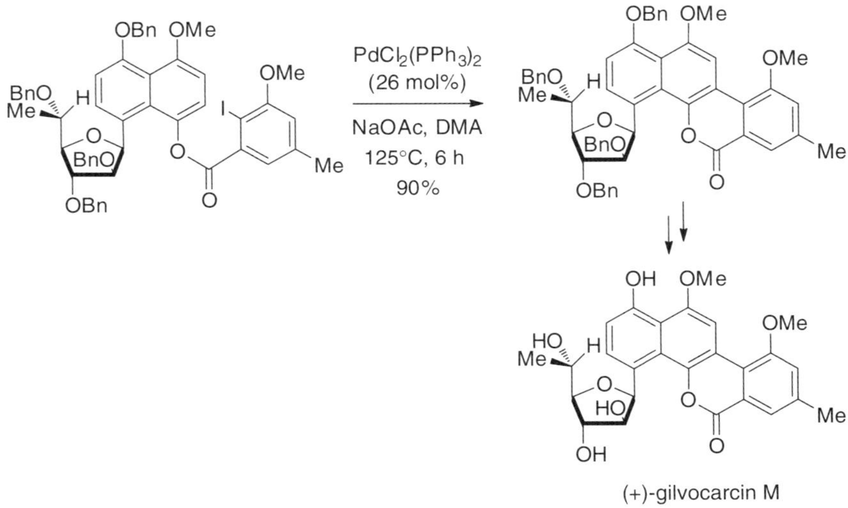

Scheme 6.21. Synthesis of (+)-gilvocarcin M.

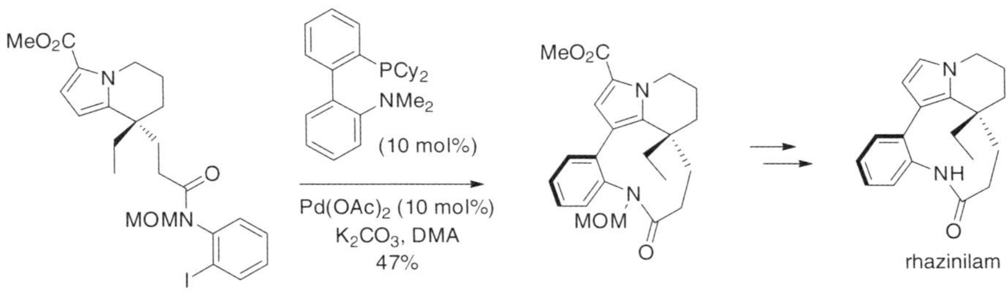

Scheme 6.22. Synthesis of rhazinilam.

direct arylation has been applied to the total synthesis of natural product such as (−)-allocochicine [452] and rhazinilam (Scheme 6.22) [466].

Intramolecular arylation of benzylic and aliphatic C–H bond has also been developed. In the presence of palladium catalyst and electron-rich phosphine ligand, intramolecular arylation of benzylic C–H bond occurs to give cyclic products [Eq. (6.182)] [467, 468]. Arylation of a sp^3 C–H bond adjacent to amide and sulfonamide was achieved using a palladium catalyst with a PCy_3 ligand [Eq. (6.183)] [469]. *Gem*-dialkyl groups on bromo- and iodobenzenes undergo C–H cleavage to give either olefins or benzocyclobutenes in the presence of a palladium catalyst [Eq. (6.184)] [470, 471]. Palladium/phosphine-catalyzed intramolecular arylation of an aryl halide containing an alkyl C–H group with O, N, and C linkages produced dihydrobenzofuran, indoline, cyclobutarene, and indane products [Eq. (6.185)]

[472–474]. Enantioselective intramolecular arylation of alkyl C–H bond was realized using palladium catalyst with chiral NHC ligand [Eq. (6.186)] [475, 476], mono dentate phosphine ligand [477–479], or bidentate phosphine ligand [480].

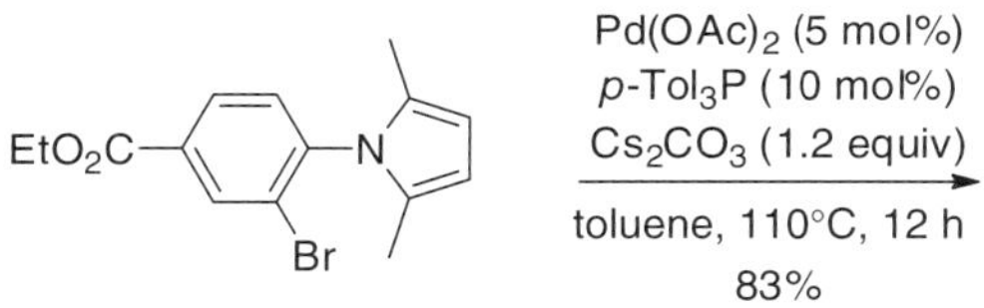

(6.182)

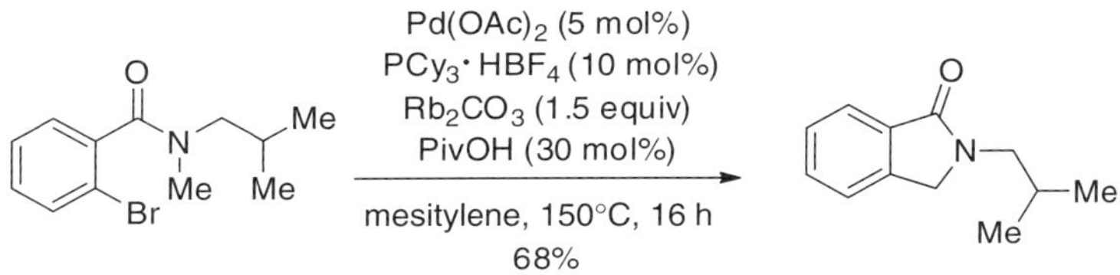

(6.183)

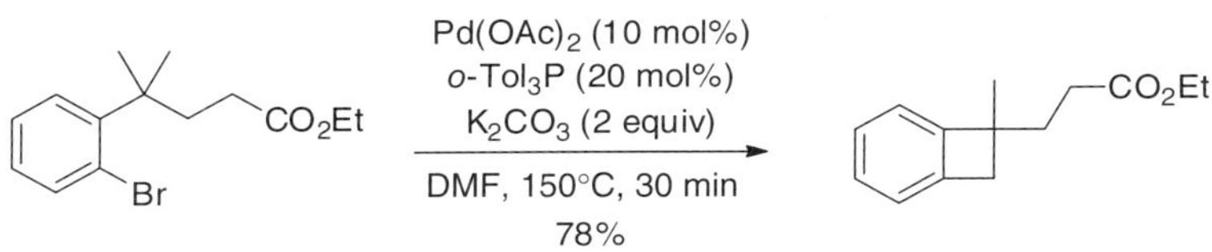

(6.184)

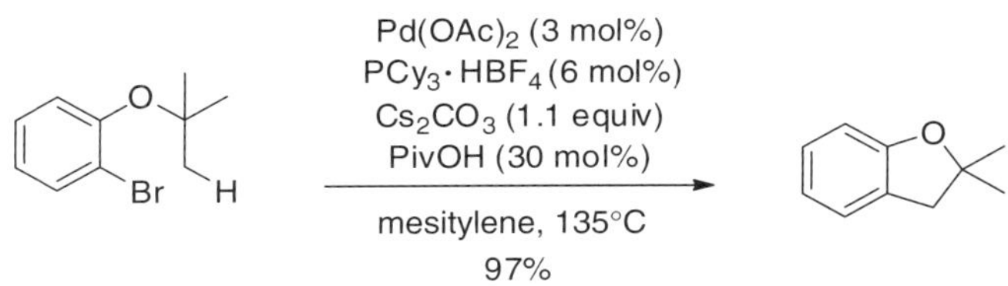

(6.185)

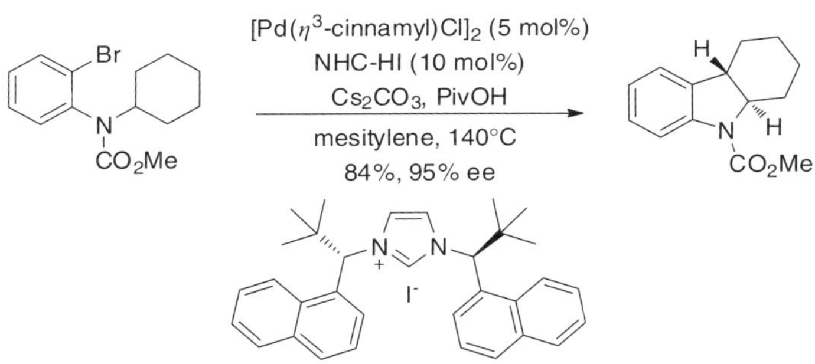

(6.186)

The synthesis of the tetrahydroisoquinoline alkaloid coralydine was achieved based on a palladium-catalyzed intramolecular aliphatic C–H arylation. The iminobenzocyclobutene, obtained by C–H arylation and subsequent functional group conversion, underwent electrocyclization to generate the core dihydroisoquinoline structure (Scheme 6.23) [481].

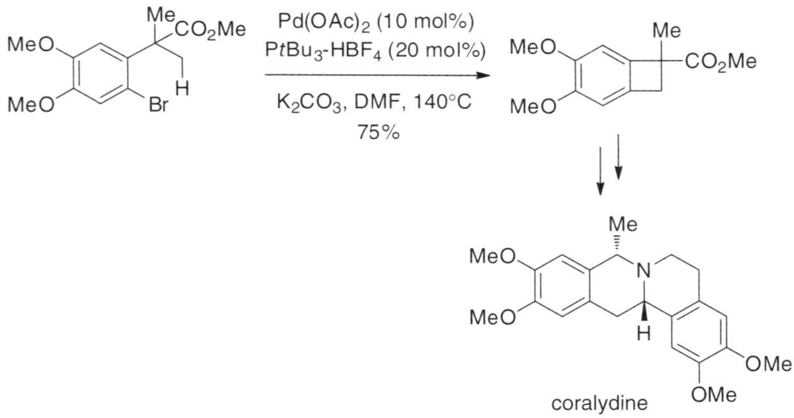

Scheme 6.23. Synthesis of coralydine.

6.5.3.3 Intermolecular Arylation of Heterocycles. The arylation of a wide range of heterocycles has been developed. These heterocycles include (a) five-membered rings such as pyrrole, thiophene, furan, azoles, and their related fused heterocycles and (b) six-membered ring systems such as pyridine, pyridine-*N*-oxide, azines, and fused heterocycles. These arylation reactions provide the corresponding useful arylated heterocycles. The selectivity of these arylation reactions is highly dependent on the types of the heterocycle, but it can also be influenced by the catalyst, the ligand, and the reaction conditions. Different selectivities have been seen for a single heterocycle, and sometimes the selectivity could be tuned by the dedicated choice of the reaction conditions.

Early works from Ohta and co-workers [435] has demonstrated the viability of intermolecular direct arylation of indole with activated chloroarene. The nitrogen-protecting group of indole had a significant influence on the regioselectivity. The use of *N*-alkyl indole and unprotected indole gave C-2 arylated products, while *N*-tosylindole favors the C-3 arylation. In addition, the *ortho* substituent on the heteroaryl chloride also influences the regioselectivity of the arylation for *N*-tosylindole. Sterically more hindered chloropyrazine favors the C-3 arylated product. Sames and co-workers [482, 483] developed a Pd-catalyzed arylation of *N*-alkyl indole using aryl iodide with good C-2 selectivity [Eq. (6.187)]. Similarly, when sterically hindered aryl halides were used, the reactions provide a mixture of C-2 and C-3 arylation products. The authors suggested that the reaction proceeds via initial electrophilic metalation at the C-3 position, followed by 1,2-palladium migration, deprotonation, and reductive elimination. The use of bulky iodobenzene led to slower 1,2-palladium migration, thus providing more C-3 arylation product by direct reductive elimination of the C-3 indolyl palladium aryl intermediate. Sames and co-workers further showed that C-3 arylation of *N*-unsubstituted indole can be achieved by using *in situ*-protected *N*-magnesium indole. Room temperature direct arylation was accomplished through the use of aryl iodides as arylating reagents along with *p*-nitrobenzoic acid and Ag_2O as additives [Eq. (6.188)] [484]. C-3-selective direct arylation of indole has been achieved using a palladium catalyst with a secondary phosphine oxide as the ligand [485, 486] or under a ligandless condition [Eq. (6.189)] [487].

Hypervalent iodinium reagents are analogues of aryl halides. Typically they are more reactive, thus allowing the arylation of an indole to occur under mild condition. For example, Sanford reported a Pd-catalyzed arylation of indole with diaryl iodonium salt which occurred at room temperature. Exclusive C-2 arylation was obtained [Eq. (6.190)] [488]. Copper also catalyzed the arylation of indole with diaryl iodonium salt. The regioselectivity is controlled by the protecting group on indole. Arylation occurs at the more electron-rich C-3 position for unsubstituted indole, while C-2-selective arylation was obtained for *N*-acylindole through the interaction of the acyl group with copper and brings it to the proximity to C-2 position [Eq. (6.191)] [489].

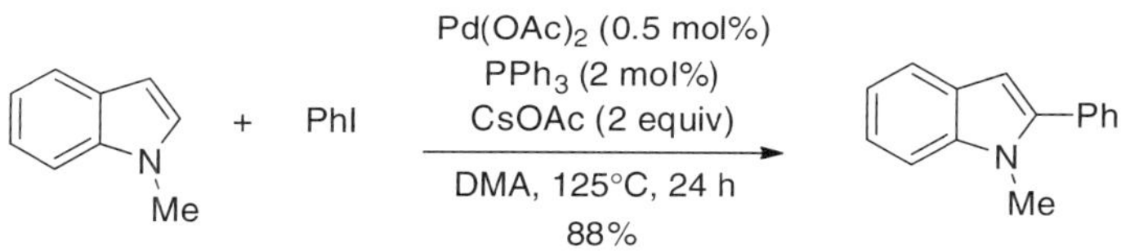

(6.187)

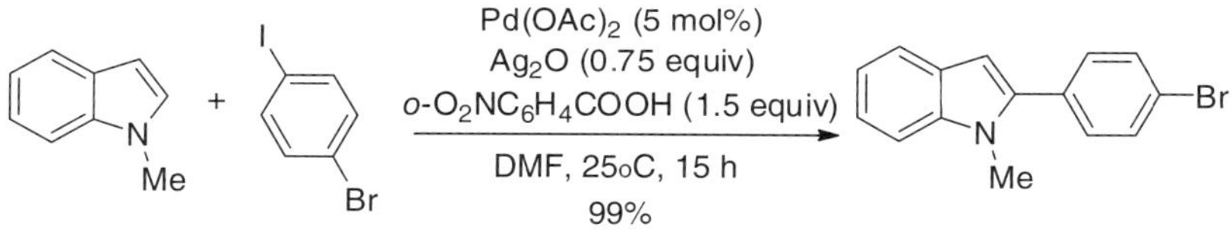

(6.188)

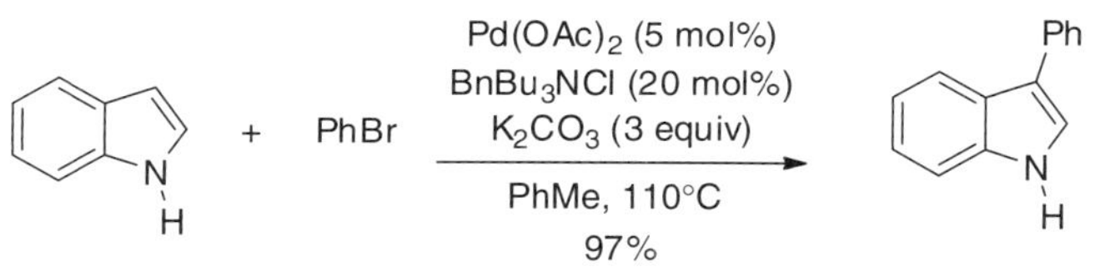

(6.189)

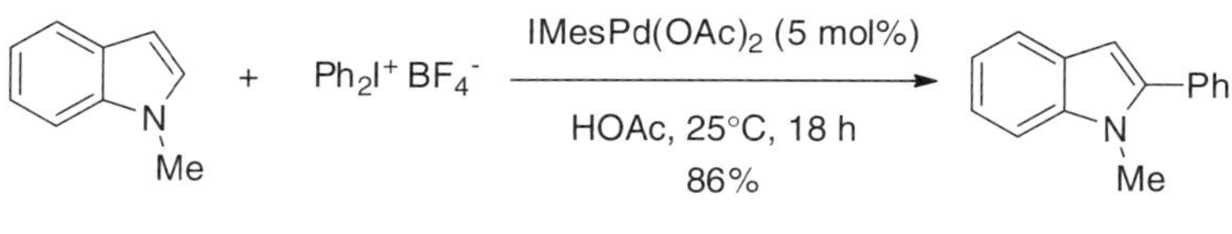

(6.190)

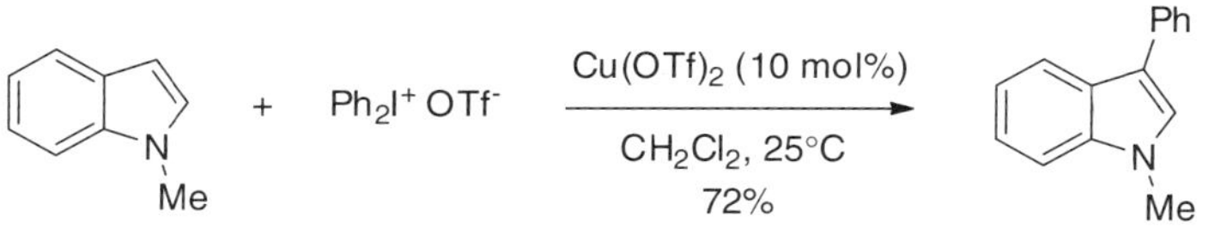

(6.191)

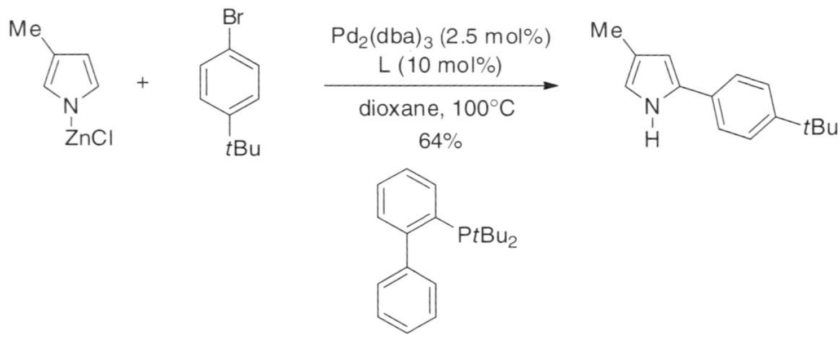

(6.192)

Direct arylation of pyrrole [490–492], furan, benzofuran, thiophene, and benzothiophene [493–496] occurs primarily at the more electron-rich C-2 position [Eq. (6.192)]. Palladium is the most frequently used catalyst for these arylation reactions. The C-2 selectivity can be explained based on an electrophilic substitution mechanism for the C–H cleavage. Recently, Fagnou and co-workers [497] suggested that the concerted-metallation-deprotonation (CMD) mechanism could also be used to explain the regioselectivity of the C–H cleavage in these electron-rich heterocycles. Computational studies showed that the regioselectivity matches well with the activation energy calculated from a CMD mechanism.

Direct arylation of (benzo)azoles were usually carried out using a metal catalyst with phosphine or NHC ligand in the presence of a base. Benzoazoles and C-5-substituted azoles usually undergo arylation at the C-2 position. Catalysts based on Pd [Eq. (6.193)] [495, 498, 499], Ni [Eq. (6.194)] [500, 501], and Cu [Eq. (6.195)] [495, 502] are known to catalyze the arylation of these substrates. Pd and Ni with an electron-rich phosphine ligand can catalyze the arylation using less reactive aryl chloride [Eq. (6.196)] [503, 504], tosylate and mesylate [505]. A Ni/diphosphine catalyst has been developed for the arylation of aryl tosylate, mesylate [Eq. (6.197)], and carboxylate [Eq. (6.198)] [506]. Arylation of azoles usually occurs preferentially at the C-5 position [507–509], because the C–H bond at the C-5 position is more electron-rich [Eq. (6.199)]. However, the regioselectivity is highly dependent on the catalyst and reaction conditions. C-2-selective arylation has also been observed in many arylation reactions, especially when CuI is used as an additive [495, 498, 510]. Through sequential C–H cleavage of azoles, multiple arylated azole product can be obtained.

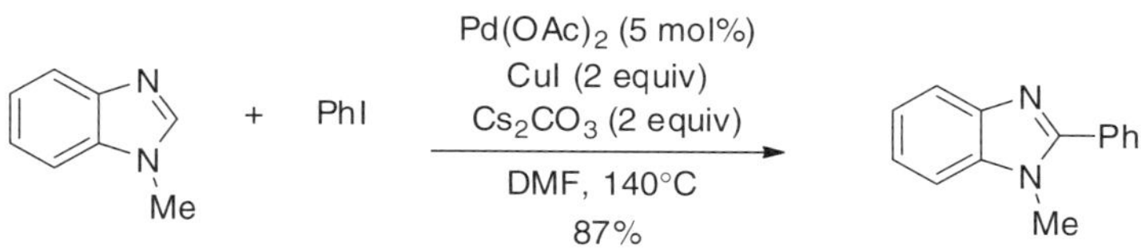

(6.193)

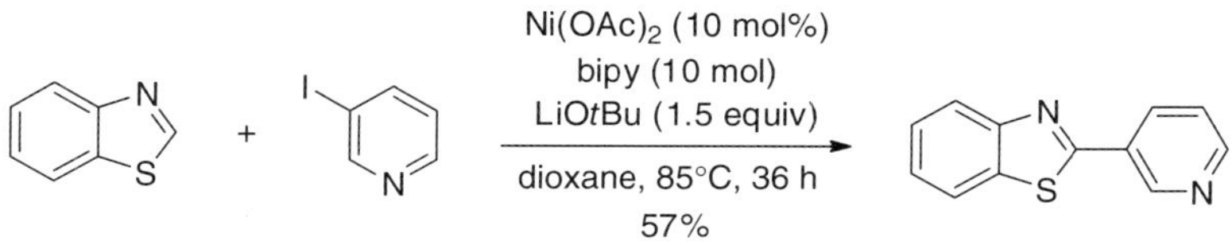

(6.194)

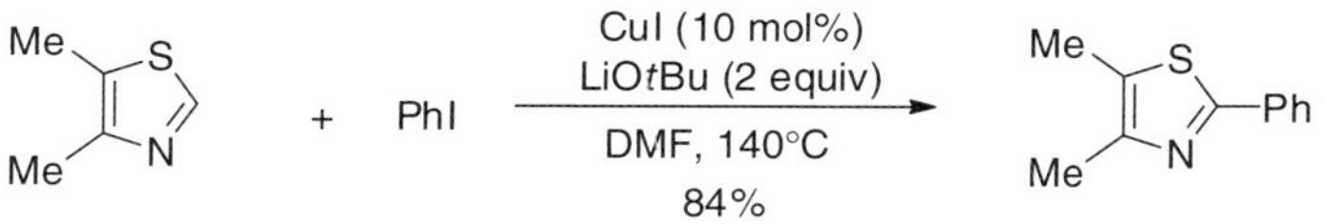

(6.195)

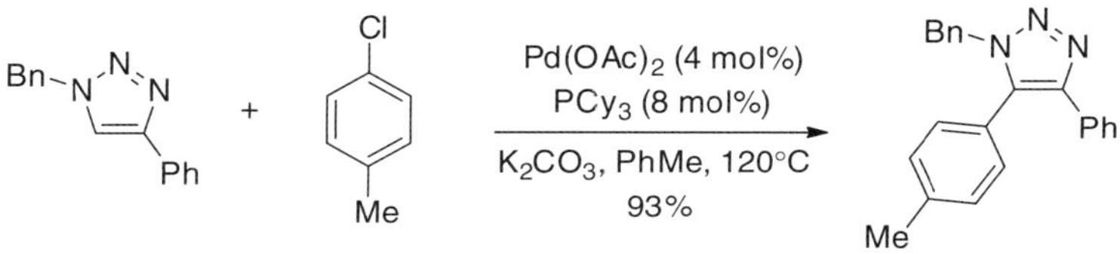

(6.196)

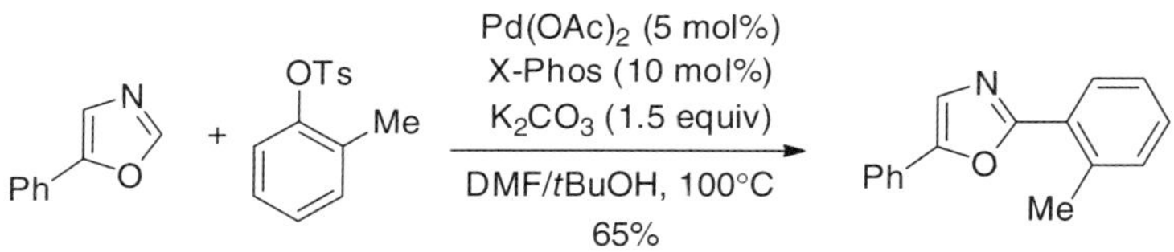

(6.197)

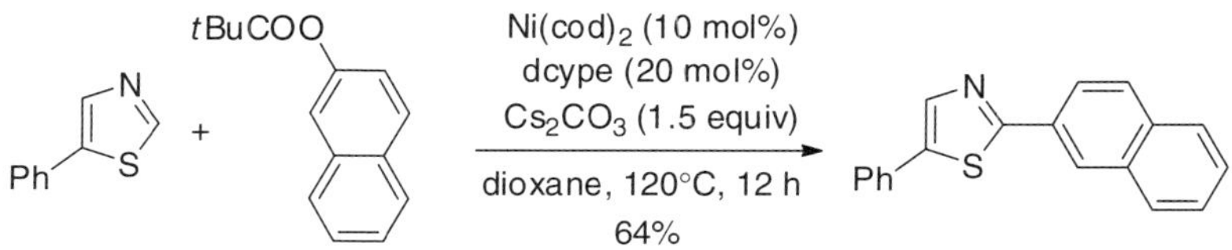

(6.198)

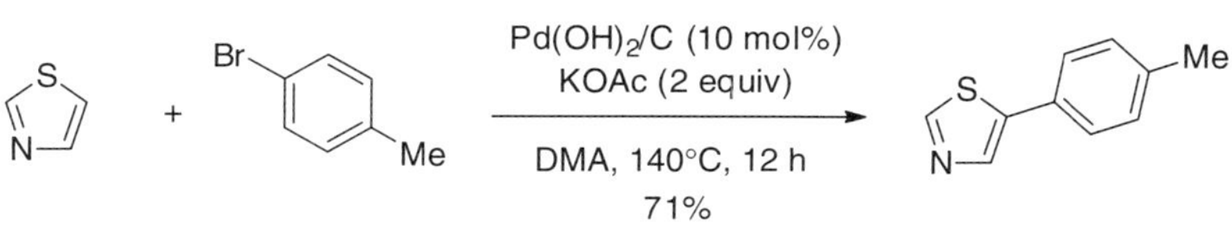

(6.199)

Ellmann and [511, 512] Bergman reported a Rh-catalyzed arylation of azoles through a different mechanism [Eq. (6.200)]. This reaction is based on his previous studies of hydroarylation of heterocycle which proceeds through a Rh-carbene intermediate. In the arylation reaction, it is likely that a similar carbene intermediate was involved.

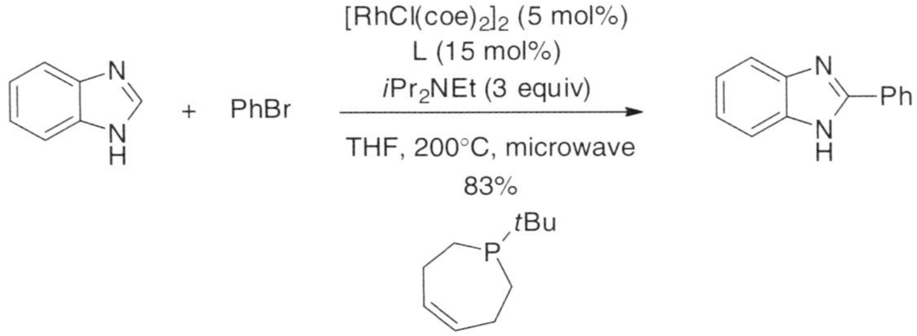

(6.200)

Early examples of arylation of pyridine were intramolecular reactions [513]. Intermolecular arylation of pyridine is more difficult because of the electron-deficient nature and coordinating ability of pyridine. An intermolecular arylation of pyridine has been achieved using palladium on charcoal in the presence of zinc metal in water [Eq. (6.201)] [514]. It was proposed that a radical mechanism was involved. Ellman and co-workers [515] reported a Rh-catalyzed arylation of pyridine and quinolone with aryl bromide under relatively harsh conditions [Eq. (6.202)]. The authors proposed that during the reaction, the interaction of pyridine C–H at the C-2 position with a Rh complex forms a carbene intermediate. Pd-catalyzed C-3- and C-4-selective arylation of pyridine with aryl bromide and iodide has been realized using phenanthroline or phosphine ligands [Eqs. (6.203) and (6.204)] [516, 517]. The selectivity is controlled by either the phenanthroline ligand or the electronic character of the substrate. Fagnou and co-workers [518] developed an alternative protocol for pyridine arylation by using pyridine-*N*-oxide as the substrate [Eq. (6.205)]. After deoxygenation, arylated pyridine could be obtained.

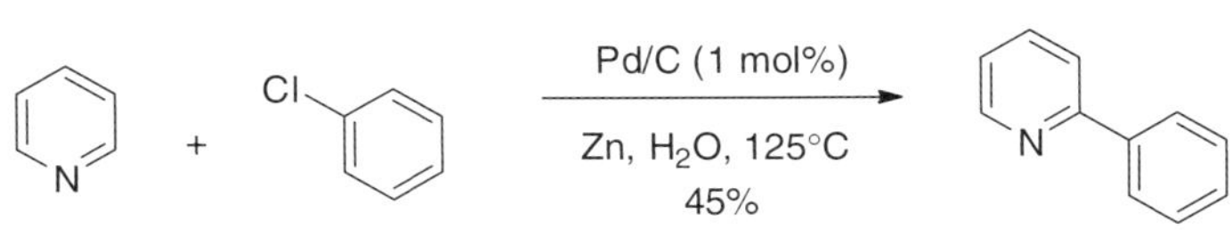

(6.201)

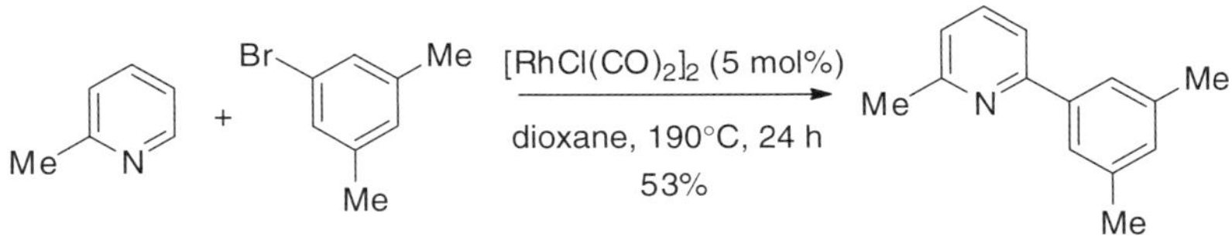

(6.202)

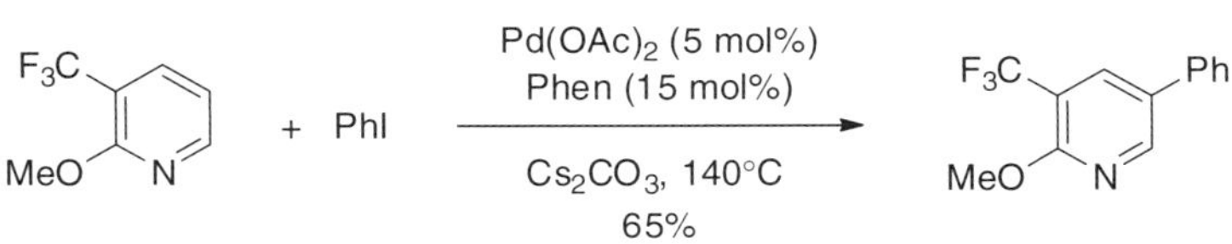

(6.203)

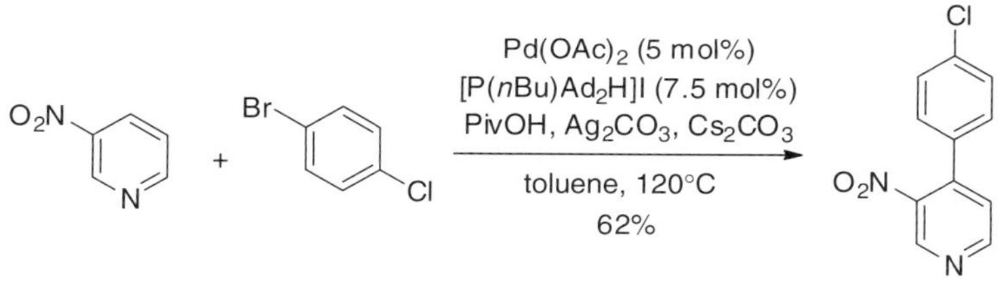

(6.204)

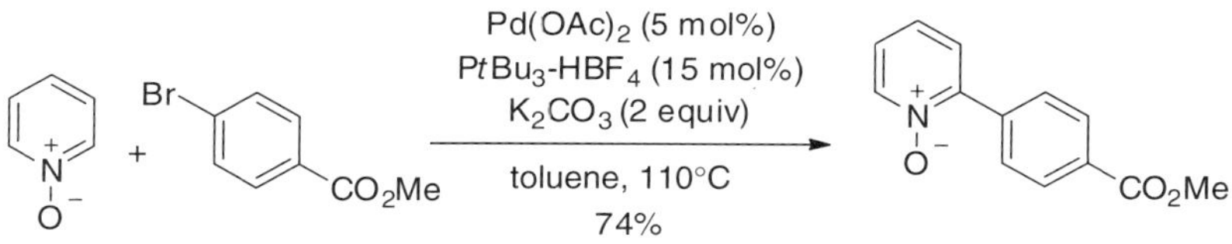

(6.205)

As many natural products and bioactive compounds contain arylated heterocycles, direct arylation of heterocycle can be an extremely useful method for the expedient synthesis of these compounds. In 2005, researchers at Merck demonstrated the utility of palladium-catalyzed intermolecular C–H arylation of heteroarene during the synthesis of GABA (γ-aminobutyric acid) agonists (Scheme 6.24) [519, 520]. The most nucleophilic C-3 position of imidazo[1,2-a]pyrimidine was regioselectively arylated with functionalized aryl bromide. Synthesis of diptoindonesin G [521] and Celecoxib

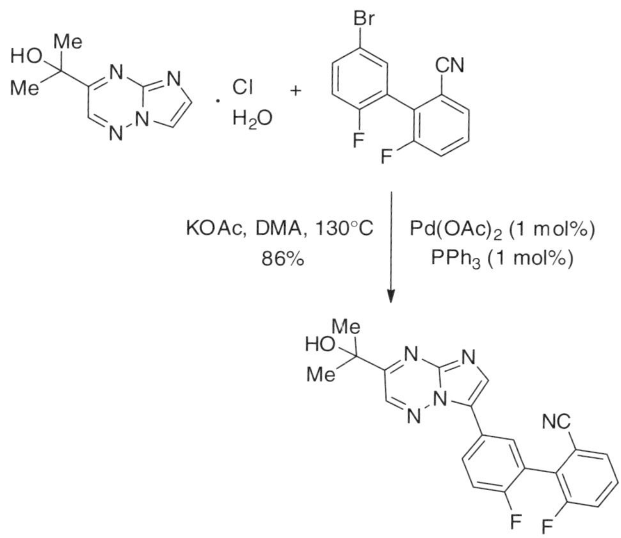

GABA agonist

Scheme 6.24. Synthesis of GABA (γ-aminobutyric acid) agonists.

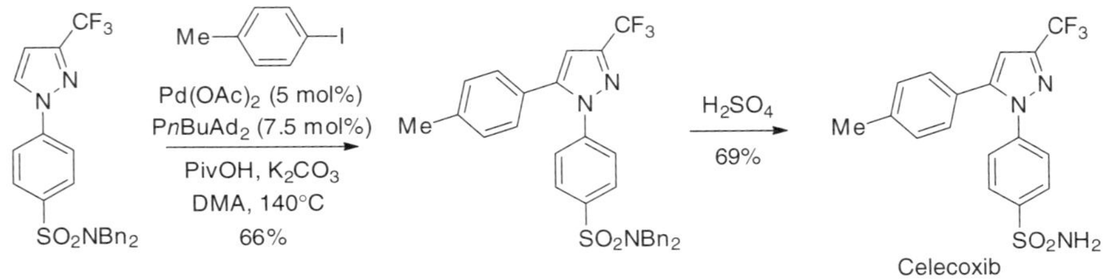

Scheme 6.25. Synthesis of Celecoxib.

[522] have been achieved by palladium-catalyzed regioselective arylation of benzofuran and pyrazole derivatives, respectively (Scheme 6.25). The Tie2 tyrosine kinase inhibitor was synthesized using multiple C–H arylations of imidazole [523] or imidazole *N*-oxide (Scheme 6.26) [524]. The first arylation occurred selectively at the C-2 position, while the second arylation occurred at the C-5 position.

6.5.3.4 *Arylation of Simple Arenes.*

An issue associated with the arylation of simple arene is the control of regioselectivity. In an early work, Dyker et al. [525] have shown that the electron-rich cyclopentadiene ring in azulene is regioselectively arylated in the presence of palladium catalyst [Eq. (6.206)]. Pentaarylated cyclopentadienes can be obtained by arylation of five-membered aromatic–containing metallocenes [526, 527]. Fujita et al. [528] reported an iridium-catalyzed direct arylation of simple arene C–H bonds in the absence of any directing or activating group [Eq. (6.207)]. The regioselectivities observed for the arylation of anisole and toluene are very similar to those for the radical aromatic substitution reactions, suggesting that a radical intermediate might be involved in the arylation reactions. Later it was found that many transition metals can catalyze the direct arylation of simple arenes in the presence of a base [529–532]. In most of these reactions, a radical mechanism seems to be likely. Lafrance and Fagnou [533] developed a palladium/phosphine catalyst system for the arylation of simple arenes [Eq. (6.208)]. The key feature of the catalyst system is the combination of K_2CO_3 and catalytic amount of weak acid which functions as the proton shuttle for the C–H cleavage. The regioselectivity of the arylation of anisole and toluene is different from a radical reaction. In addition, a large kinetic isotopic effect was obtained for benzene arylation. These data are inconsistent with electrophilic aromatic substitution and radical processes but correlate well with a proton-transfer pathway.

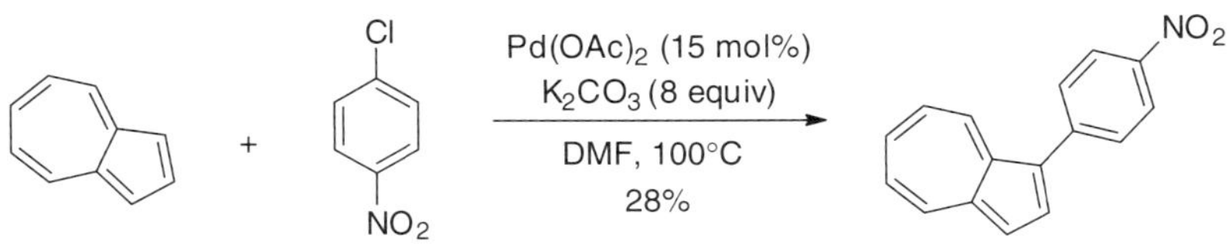

(6.206)

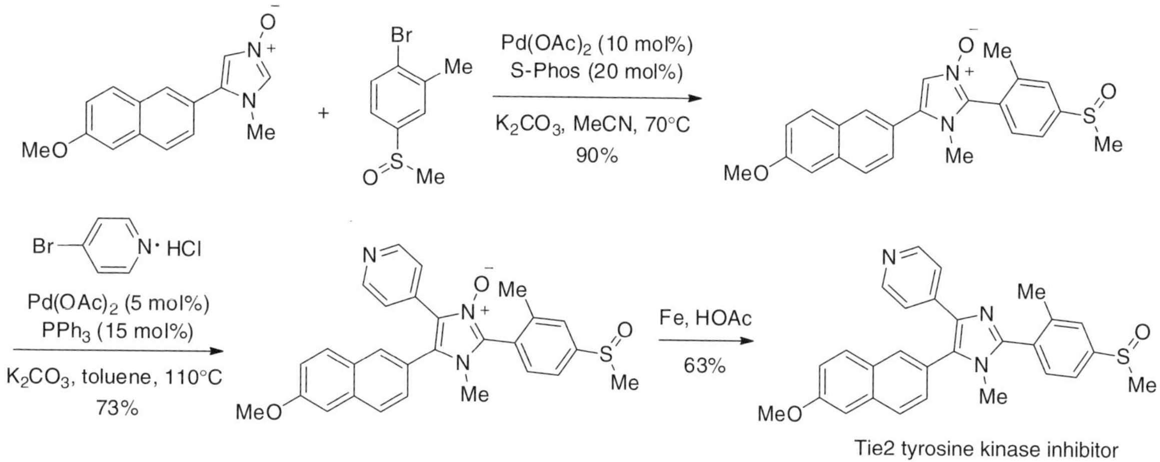

Scheme 6.26. Synthesis of Tie2 tyrosine kinase inhibitor.

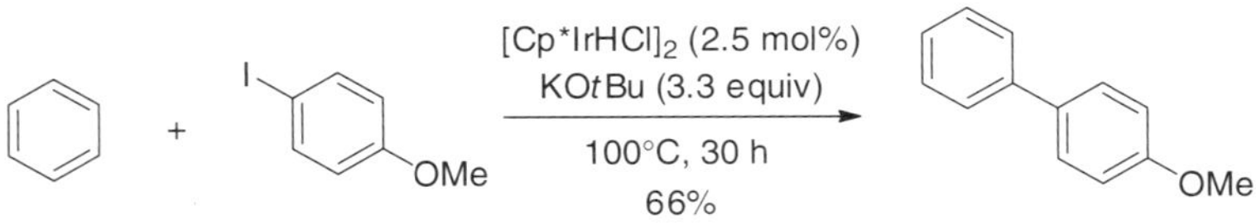

(6.207)

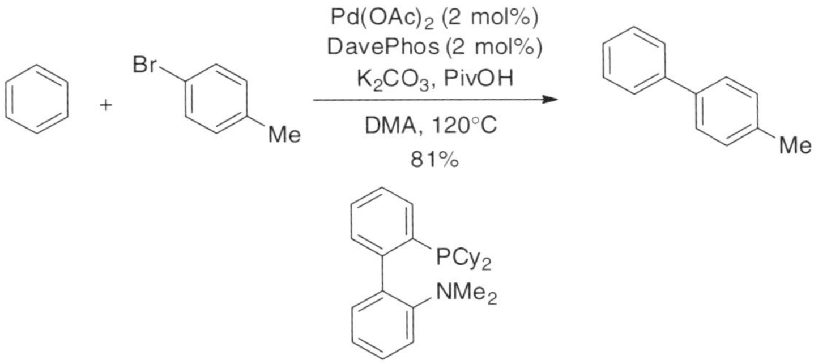

(6.208)

The CMD mechanism is very effective for C–H cleavage of electron-deficient arenes. Direct arylation of electron-deficient perfluoroarenes with aryl halides using $Pd(OAc)_2$ and electron-rich phosphine ligand was reported by Fagnou and co-workers [534–536] [Eq. (6.209)]. The reactions occur regioselectively at the more acidic C–H bond. The key C–H bond functionalization step occurs via a concerted arene metalation and proton abstraction. Similar reaction can be achieved using copper catalyst with a phenanthroline ligand [537].

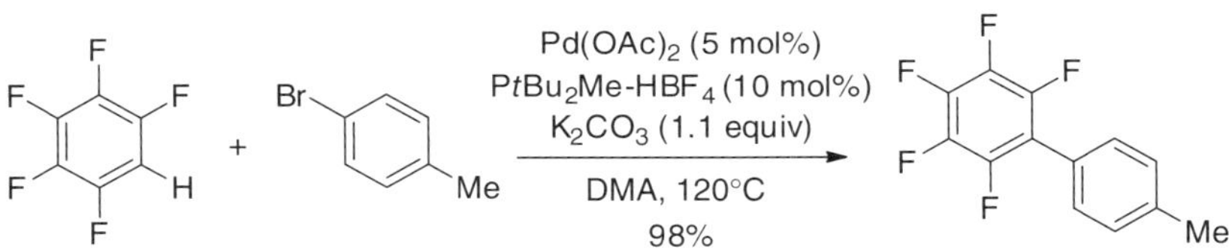

(6.209)

6.5.3.5 *Directed Arylation.* A directing group can be used to control the regioselectivity of direct arylation of both aryl and alkyl C–H bonds. In addition, usually a directing group can be transformed to other functionalities, thus adding synthetic value to the directed arylation reactions.

Miura and co-workers [538] discovered a Pd-catalyzed regioselective arylation of 2-phenylphenol with aryl iodide. The ratio of mono-arylated and di-arylated products can be tuned by changing the amount of aryl halides. It is proposed that the reaction started with oxidative addition of aryl halide to Pd(0). The resulting Pd(II) reacted with phenoxide to form a Pd(II) phenoxide, which underwent intramolecular C–H cleavage to produce a diaryl Pd(II) intermediate. Reductive elimination would afford the arylation product and Pd(0) catalyst.

The catalyst system comprising a Pd(0) catalyst, a ligand, and a base can be applied to the directed arylation of arene with other directing groups including arylmethanol

[Eq. (6.210)] [539, 540], benzaldehyde [541], phenone [542, 543], and benzamide [Eq. (6.211)] [544]. Many of these directed arylation reactions can be catalyzed by Ru(II) catalysts as well [Eq. (6.212)] [545–567]. Usually aryl iodide and aryl bromide are used as the arylation reagent. Arylation with less reactive aryl chloride and tosylate has been achieved with Pd(0) [Eq. (6.213)] [548] or Ru(II) [Eq. (6.214)] [549–551] catalyst ligated by electron-rich phosphine or NHC carbene ligand.

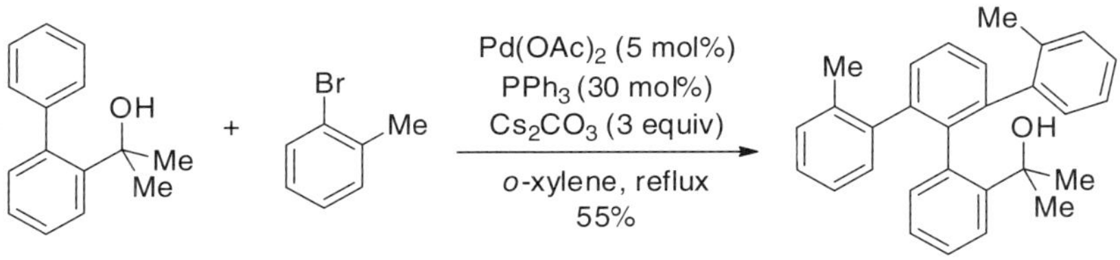

(6.210)

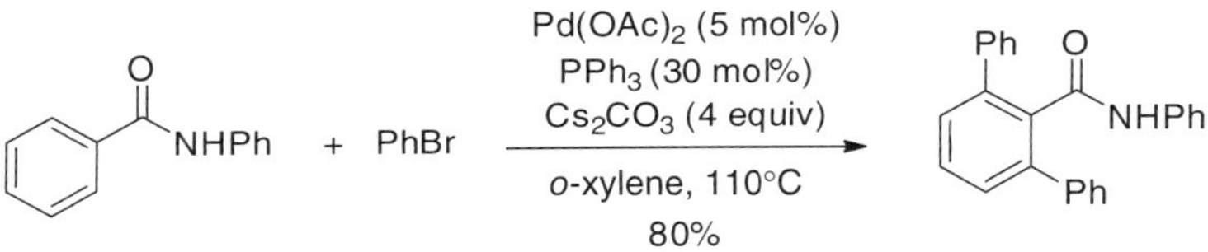

(6.211)

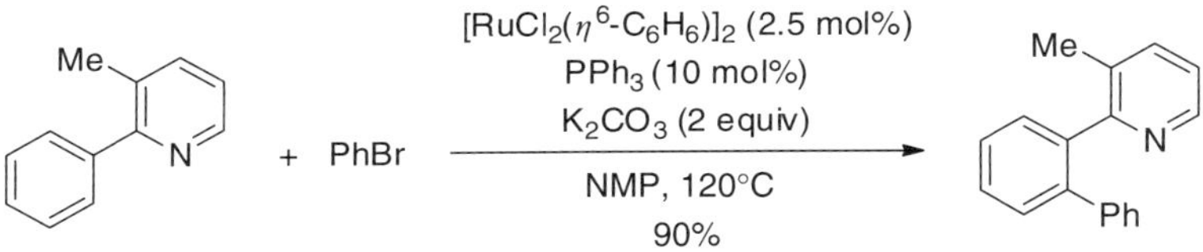

(6.212)

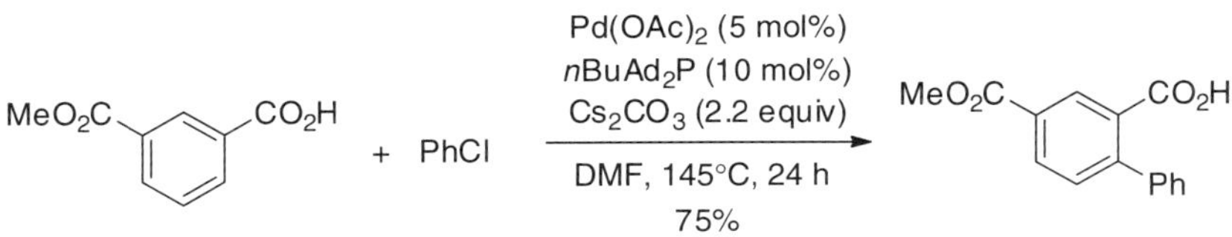

(6.213)

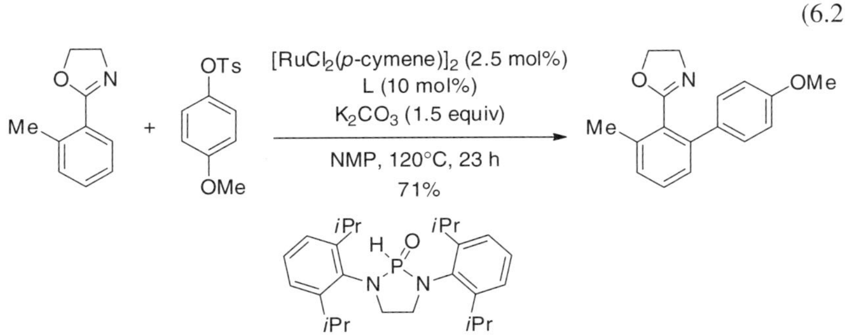

(6.214)

Sanford and co-workers [552] found that hypervalent iodine compounds can be used as the arylation reagents in palladium-catalyzed direct arylation reactions

[Eq. (6.215)]. Arenes bearing pyridine and amide groups underwent regioselective *ortho*-arylation. A Pd(II)/Pd(IV) catalytic cycle was invoked because hypervalent iodine compounds are known to oxidize Pd(II) to Pd(IV). Pyridine-directed benzylic C–H arylation was also achieved.

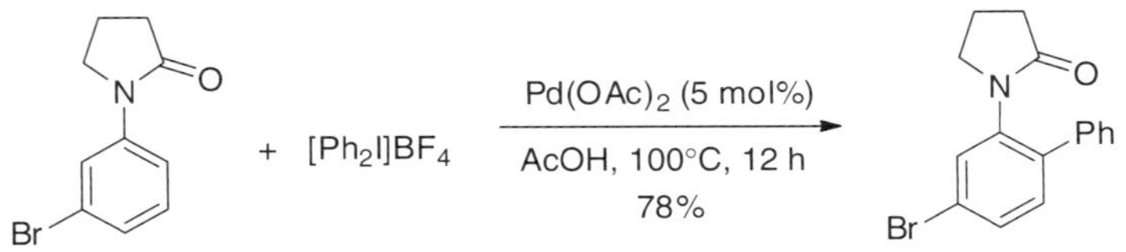

(6.215)

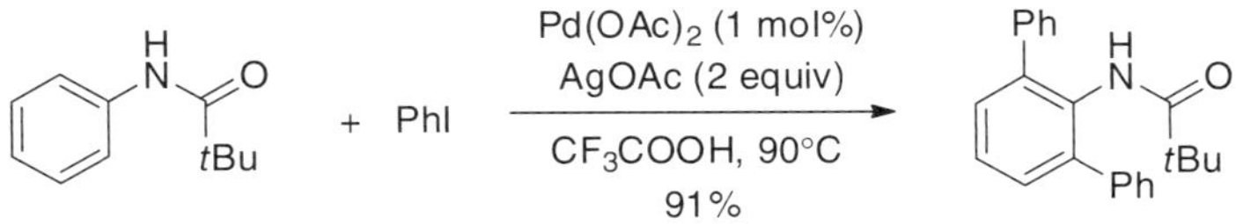

(6.216)

Daugulis and Zaitsev [553] developed a Pd-catalyzed arylation of anilides with aryl iodides. Silver salt such as AgOAc is of crucial importance for the reaction [Eq. (6.216)]. It acts as both a base and a halide scavenger. 2-Phenylpyridine was regioselectively arylated as well [554]. Using 8-aminopyridine as an auxiliary, the bidentate directing group allows *ortho*-arylation of benzamide to proceed with low catalyst loading [555]. A similar catalyst system has been applied to the *ortho*-arylation of benzylamine and benzamide [556, 557]. A mechanism of Pd(II)/Pd(IV) cycle was proposed. Until recently, examples of the oxidative addition of aryl halide to a cyclometalated palladacycle to afford a Pd(IV) complex are still rare [558]. The mechanism of this type of directed arylation reactions is still controversial.

Norbornene-mediated multiple C–H functionalization reactions was elegantly developed by Catellani et al. [559]. In this reaction, the two *ortho* C–H bonds can be alkylated by using alkyl halide [560, 561]. *Ortho*-arylation can be carried out using another aryl halide [Eq. (6.217)] [562, 563]. By controlling the reactivity of two aryl halides, *ortho* arylation of aryl iodide with aryl bromide followed by olefination, amination, or cyanation of the resulting aryl iodide can be achieved [Eq. (6.218)] [564–566]. In the proposed mechanism, the reaction starts with oxidative addition of aryl iodide to Pd(0). Insertion of norbornene to the aryl palladium followed by cyclometalation provides a pallacycle. Oxidative addition of aryl bromide to the palladacycle generates a Pd(IV) intermediate which undergoes reductive elimination and de-insertion of norbornene to give an aryl Pd(II) intermediate. Further Heck reaction or amination reaction occurs at the Pd(II) to close the catalytic cycle (Scheme 6.27).

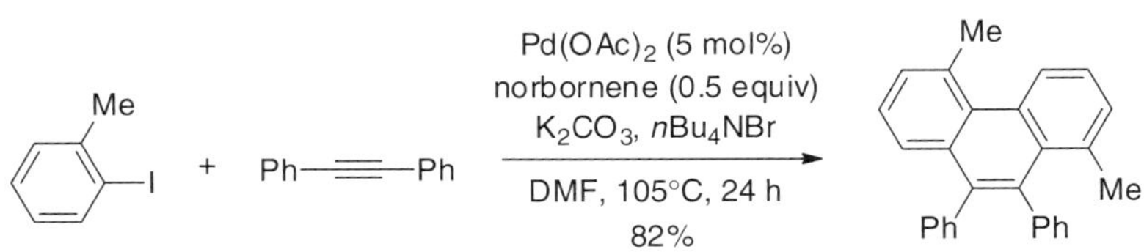

(6.217)

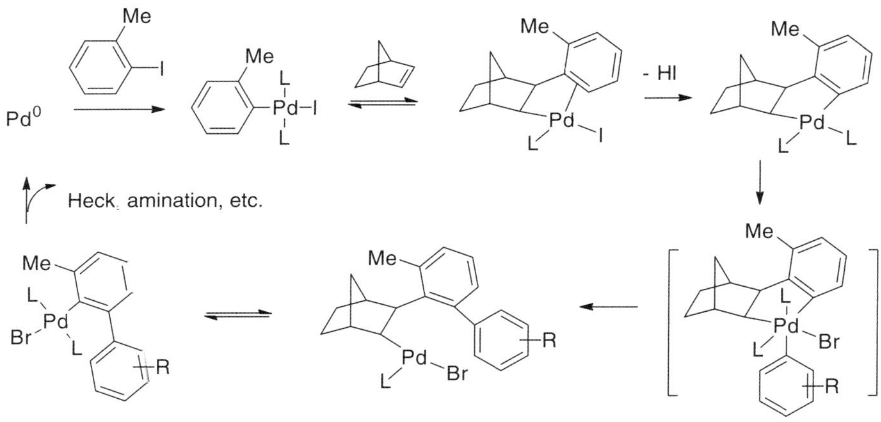

Scheme 6.27. Mechanism of norbornene-mediated reaction.

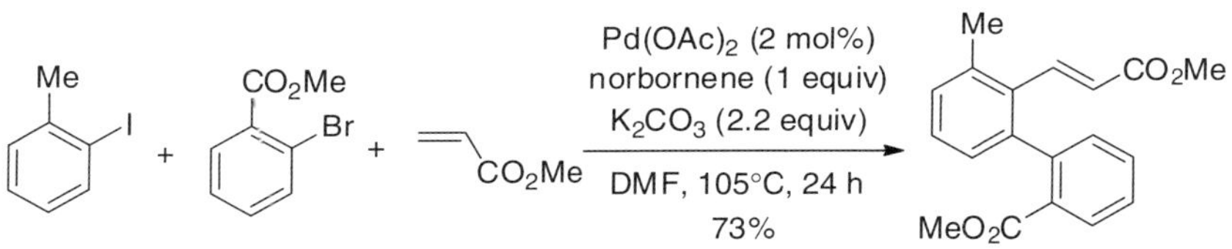

(6.218)

A Pd-catalyzed directed arylation reaction can be applied to an unactivated alkyl C–H bond as well. For example, using a bidentate pyridine-containing ligand, regioselective β-arylation of carboxylic acid derivatives and γ-arylation of amine derivatives were achieved [Eq. (6.219)] [555]. This protocol was later extended for the synthesis of nonnatural amino acids [Eq. (6.220)] [567]. Similar strategy has been adopted by He and Chen [568] for the structural diversification of various aliphatic scaffolds. Using carboxylic acid and amide as the directing groups, β-arylation of a sp^3 C–H bond with aryl iodide was achieved [Eq. (6.221)] [569]. Recently, Yu and co-workers [570] achieved the arylation of methylene C–H bond β to an electron-deficient amide directing group, enabled by the use of a 2-alkoxyquinoline ligand [Eq. (6.222)] [570].

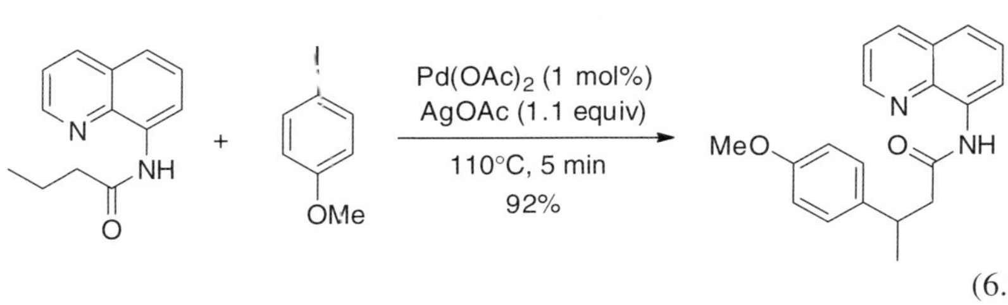

(6.219)

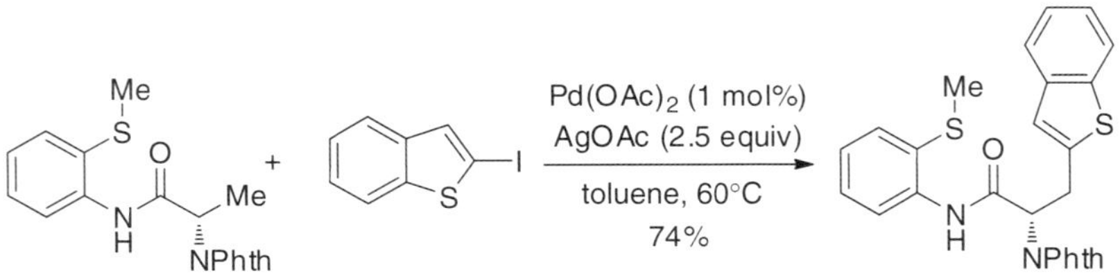

(6.220)

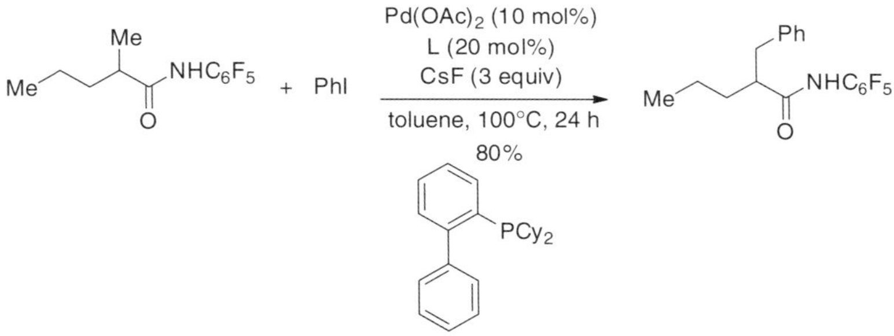

(6.221)

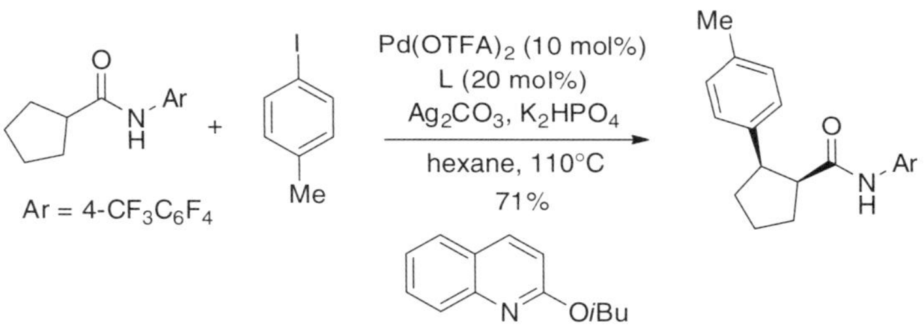

(6.222)

Transition-metal-catalyzed directed arylation assisted by directing groups has been applied to the synthesis of bioactive compounds. The core structure of anacetrapib, a potent and selective inhibitor of CETP (cholesterylester transfer protein), was constructed by a ruthenium-catalyzed directed arylation (Scheme 6.28) [571]. Oxazoline was used as the directing group to control the regioselectivity and was subsequently converted to other functional groups. A similar catalyst system can be applied to the synthesis of vaisartan (Diovan) [572] and losartan (Cozaar) [573] by using triazole as the directing group. A norbornene-mediated tandem direct arylation/*N*-arylation sequence was applied to the formal synthesis of nitidine and NK 109 (Scheme 6.29) [574].

The total synthesis of (−)-celogentin C features an intermolecular C(sp^3)-H arylation. By using 8-aminoquinoline as the auxiliary, β-arylation of amino acid with functionalized aryl iodide was carried out. After further elaboration of the resulting coupling product, celogentin C was accomplished in a total of 23 steps (Scheme 6.30) [575]. Piperarborenine B, a dimeric cyclobutane containing natural product, was synthesized by sequential C(sp^3)-H arylation reactions (Scheme 6.31) [576]. The use of

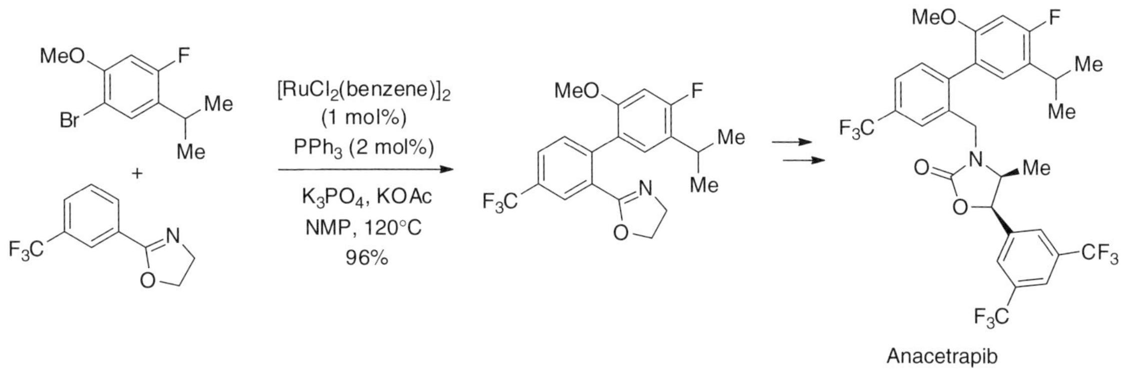

Scheme 6.28. Synthesis of anacetrapib.

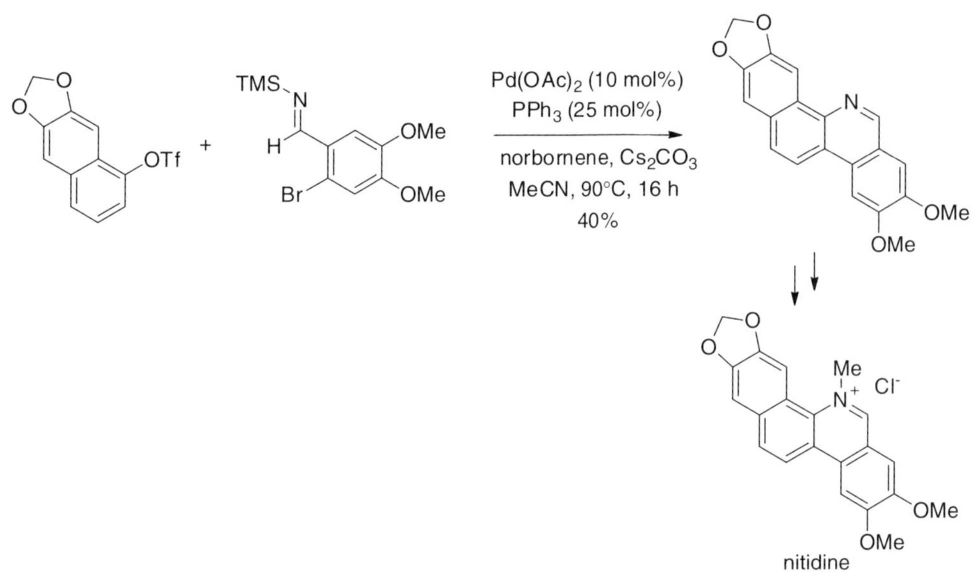

Scheme 6.29. Synthesis of nitidine.

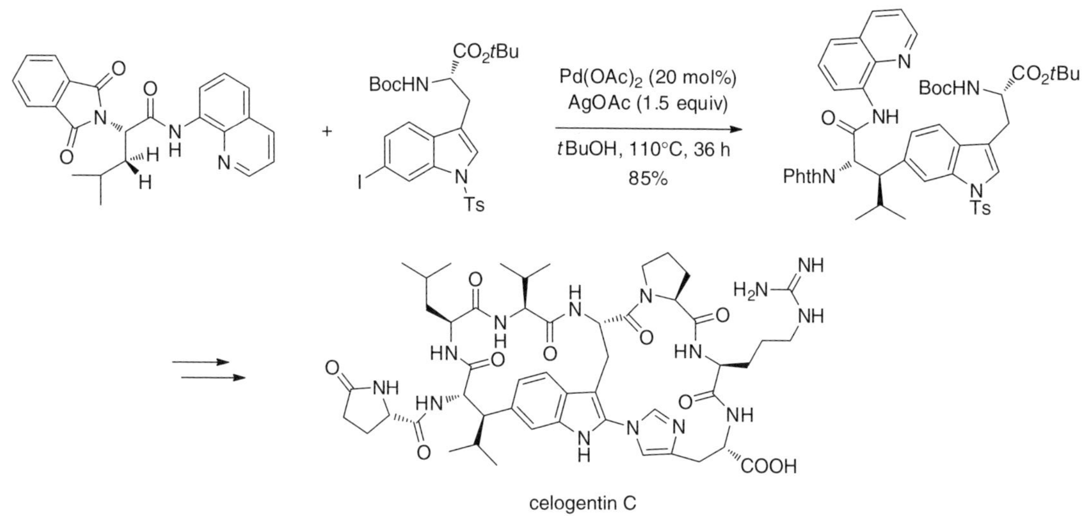

Scheme 6.30. Synthesis of (−)-celogentin C.

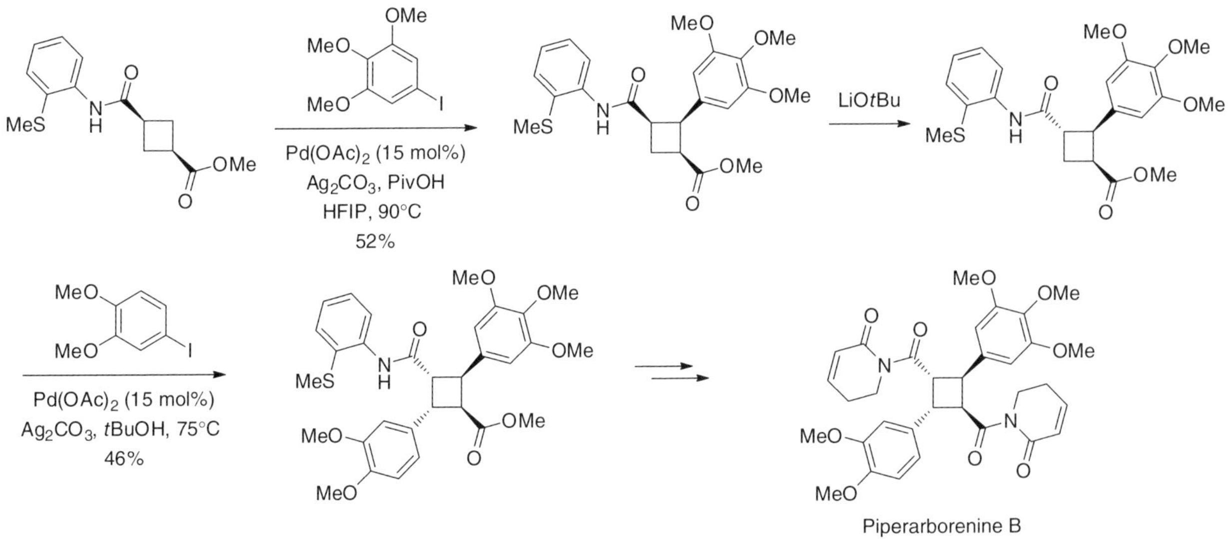

Scheme 6.31. Synthesis of Piperarborenine B.

2-methylthio anilide as the chelation group facilitates the first C–H arylation event in the presence of Pd catalyst and silver carbonate. After epimerization, the same directing group allows the second regio- and stereoselective C–H arylation. The resulting arylation product can be converted to the natural product in a few steps.

6.6 CARBONYLATION

6.6.1 Carbonylation to Form Aldehyde

The carbonylation of arene and alkane to form aldehyde is an endothermic reaction. To make the reaction thermodynamically favorable, it needs to be conducted under photochemical conditions. Photochemical carbonylation of benzene to benzaldehyde is achieved by $Rh(PPh_3)_2(CO)Cl$ catalyst and several related complexes [Eq. (6.223)] [577–580]. Related insertion of isocyanide to aryl C–H bond is more thermodynamically favored because the C–N triple bond is weaker than a C–O triple bond. A low-valent iron complex has been reported to catalyze the insertion of isocyanide to a benzene C–H bond [Eq. (6.224)] [581].

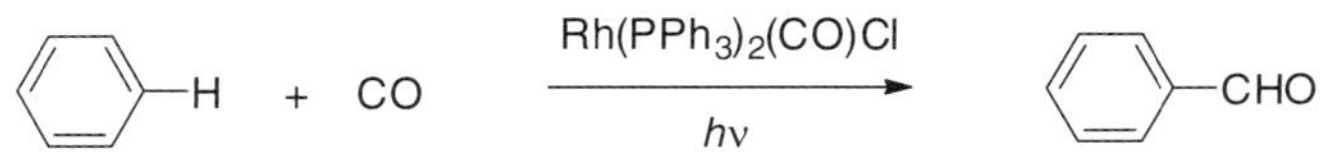

(6.223)

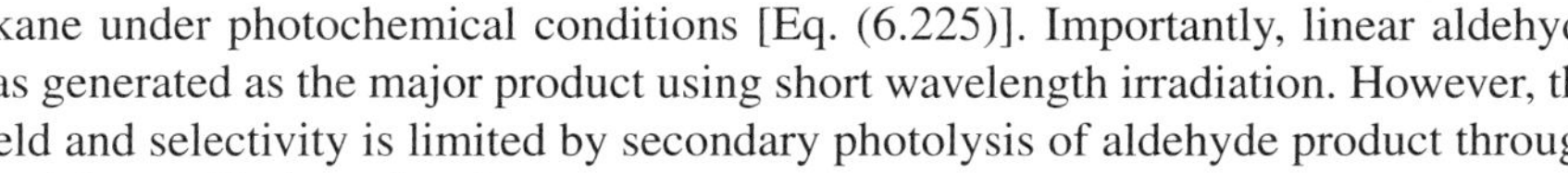

(6.224)

Using $Rh(PMe_3)_2(CO)Cl$, a more electron-rich analogue of the catalyst employed in arene carbonylation, Tanaka and co-workers [582–585] realized carbonylation of alkane under photochemical conditions [Eq. (6.225)]. Importantly, linear aldehyde was generated as the major product using short wavelength irradiation. However, the yield and selectivity is limited by secondary photolysis of aldehyde product through Norrish type II photochemistry.

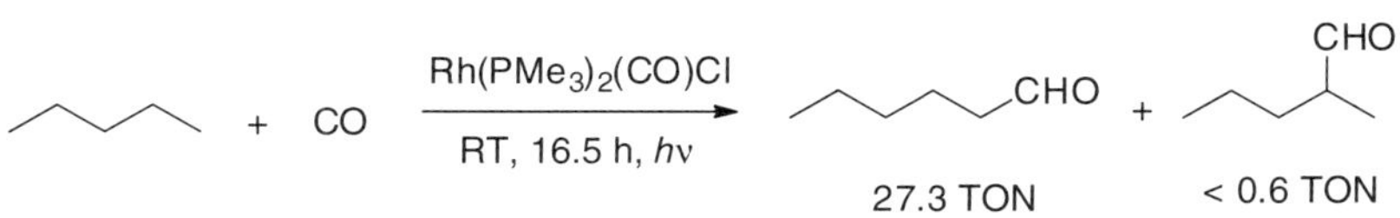

(6.225)

The mechanisms of both arene and alkane carbonylation have been studied. In Rh-catalyzed arene carbonylation [586], mechanistic study reveals that the shorter wavelength irradiation is required for two separate steps in the catalytic cycle (Scheme 6.32). The primary photo process does not involve loss of CO ligand, and the oxidative addition of the aryl C–H bond likely occurs at an intact photo excited state of $RhCl(CO)(PMe_3)_2$. Kinetics of the catalytic reaction is in agreement with the

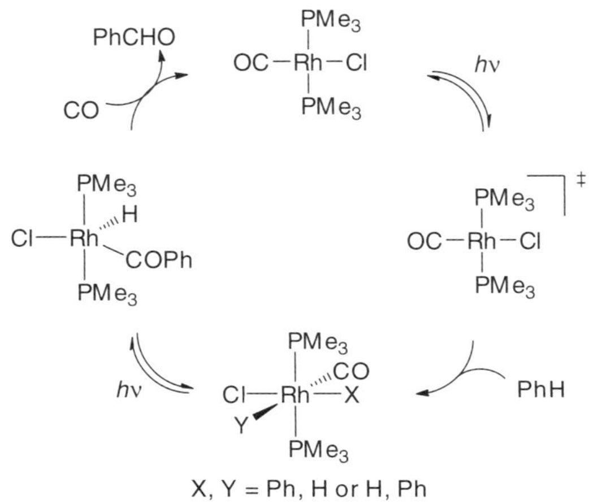

Scheme 6.32. Mechanism of Rh-catalyzed arene carbonylation.

involvement of six-coordinate phenyl hydrido complexes, $Rh(PMe_3)_2(Ph)(CO)HCl$, which has been studied by Field and co-workers [587] under low-temperature conditions. The carbonylation efficiency increases with CO pressure, which is attributed to the reaction of CO with the unsaturated benzoyl complex, $Rh(PMe_3)_2(COPh)HCl$.

In Rh-catalyzed photochemical carbonylation of alkanes, at least two different pathways are operative [588]. One is a one-photon radical process which produces a branched aldehyde. The other one involves a two-photon nonradical pathway to give primarily linear aldehyde product.

6.6.2 Carbonylation to Form Ketone

In contrast to the direct carbonylation of C–H bond to form aldehyde, three-component reaction between C–H substrate, CO, and alkene is thermodynamically favored. In 1979, Hong and Yamazaki [589] reported a $Rh_4(CO)_{12}$-catalyzed reaction of benzene with ethylene under CO pressure to give a small amount of propiophenone. Moore et al. [590] found that $Ru_3(CO)_{12}$-catalyzed carbonylation of pyridine in the presence of olefin occurred selectively at the α position of nitrogen [Eq. (6.226)]. Carbonylation of a five-membered heterocycle such as imidazole also occurred at the position *ortho* to the nitrogen [Eq. (6.227)] [591]. The regioselectivity of arene carbonylation in the presence of olefin can be controlled by the directing group such as pyridine [592, 593] imine [Eq. (6.228)] [594, 595] and pyrazole [596]. Murai and co-workers [597] found that carbonylation of some aza-heterocycle proceeded with unusual regioselectivity which occurred at the β position of the nitrogen atom [Eq. (6.229)] [597]. Besides arene carbonylation, similar reaction at alkyl C–H bond was also realized. By using pyridine as the directing group, Murai and co-workers [598] developed a Rh-catalyzed carbonylation of a sp^3 C–H bond α to nitrogen in alkylamine in the presence of olefin [Eq. (6.230)].

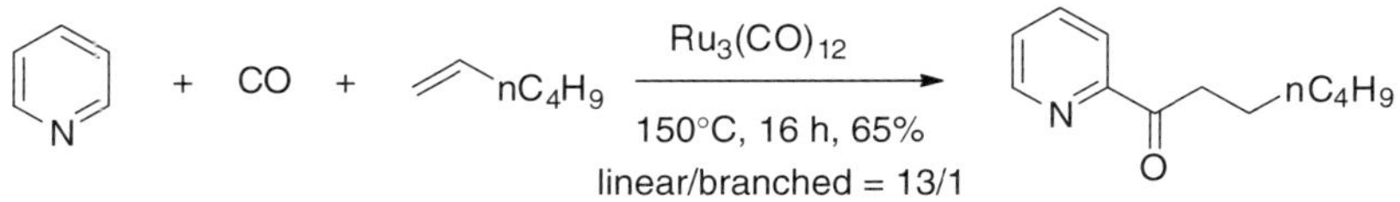

(6.226)

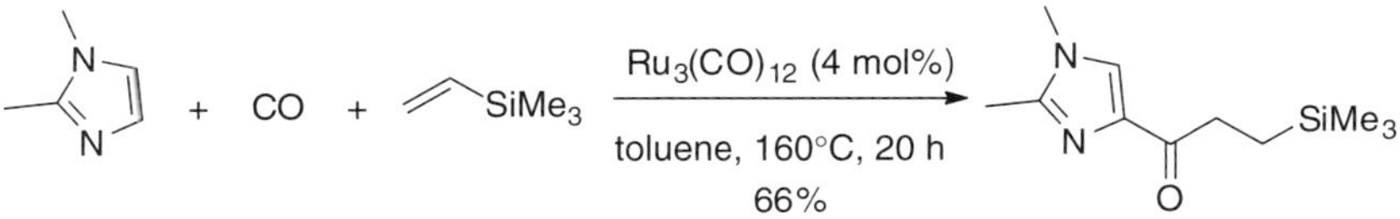

(6.227)

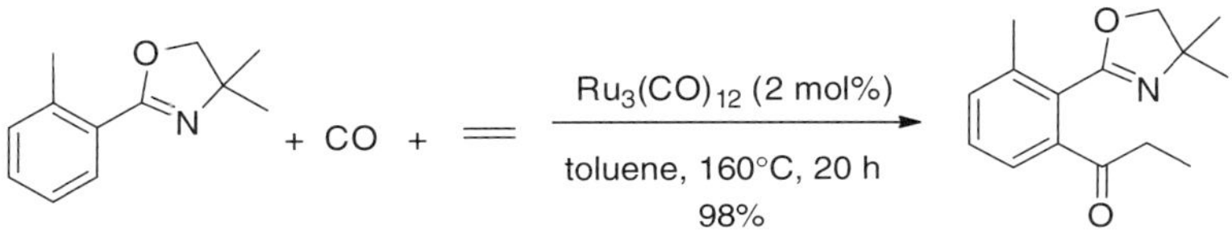

(6.228)

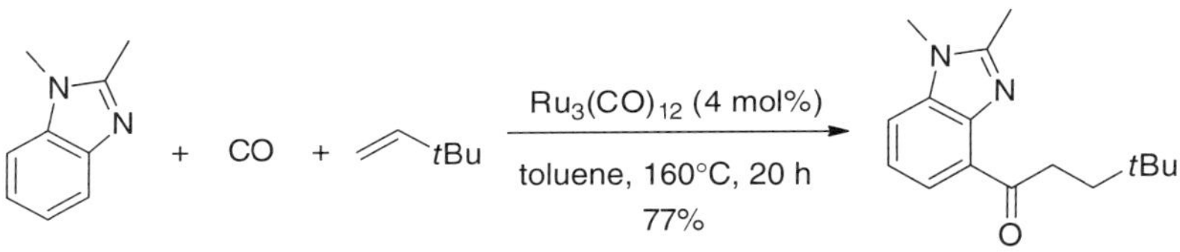

(6.229)

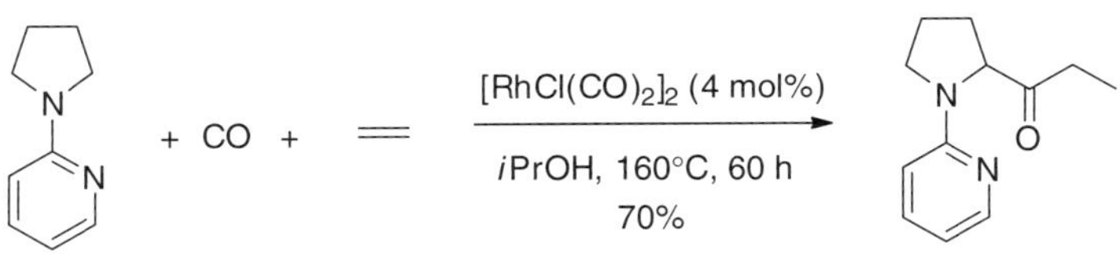

(6.230)

The carbonylation with C–H bond/CO/olefin likely occurs through initial oxidative addition of C–H bond to the low-valent metal center with or without directing group assistance to form the metal hydride. Then insertion of ethylene into the H-Rh bond followed by CO insertion gives the acyl Rh. Finally, reductive elimination gives the final product and the starting metal complex (Scheme 6.33).

6.6.3 Oxidative Carbonylation

Direct oxidative carbonylation of a C–H bond is a useful method to synthesize carboxylic acid derivatives. In the early 1980, Fujiwara et al. [599] reported the first direct carbonylation of arene to form carboxylic acid with a stoichiometric amount of palladium. The reaction can be made catalytic by adding an oxidant such as O_2, *t*BuOOH, or $K_2S_2O_8$ [600]. It was found that the reaction can be performed under mild conditions by using $K_2S_2O_8$ as the oxidant and TFA as the solvent [Eq. (6.231)]

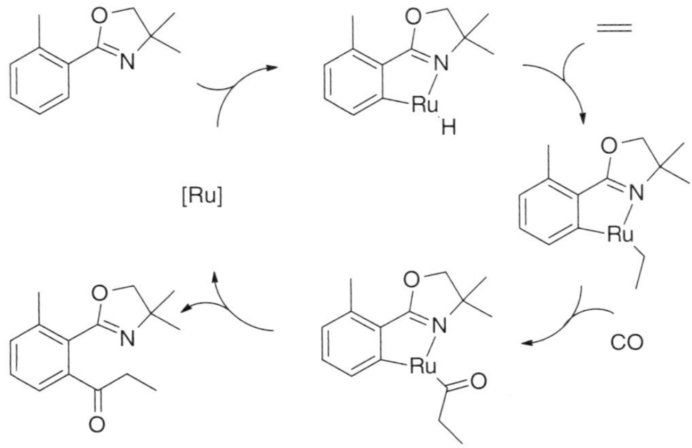

Scheme 6.33. Mechanism of Ru-catalyzed carbonylation.

[601, 602]. The regioselectivity is controlled by electrophilic C–H palladation as furans, thiophene reacted selectively at the C-2 position, and carbonylation of substituted arenes generates a mixture of regioisomers [603].

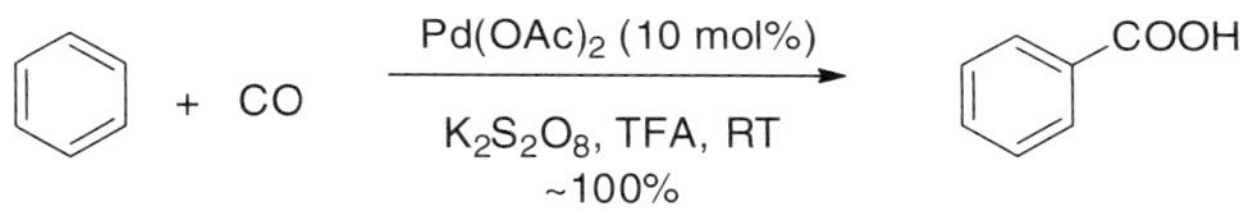

(6.231)

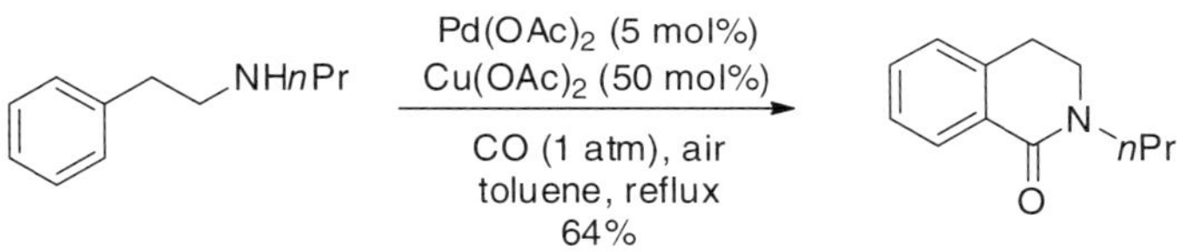

(6.232)

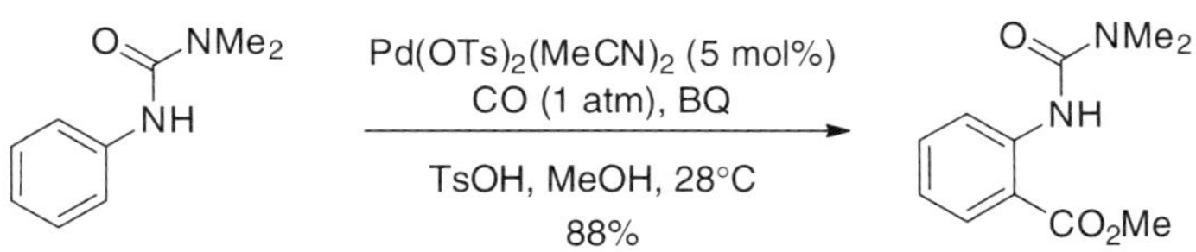

(6.233)

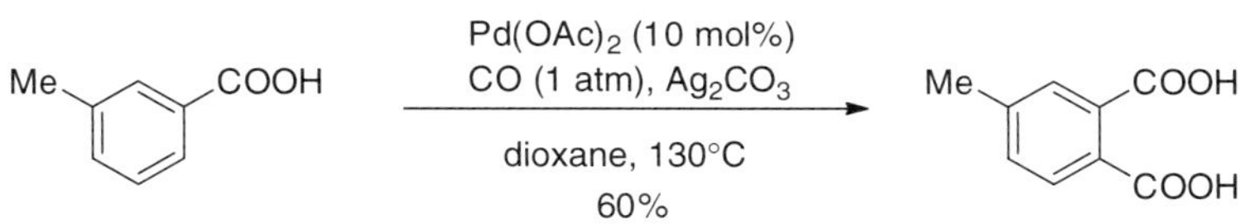

(6.234)

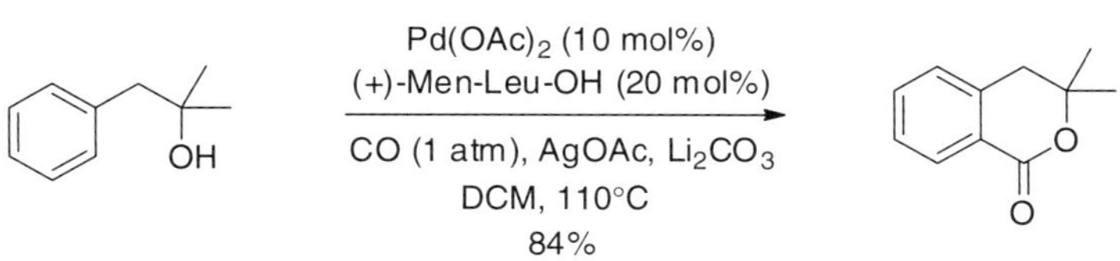

(6.235)

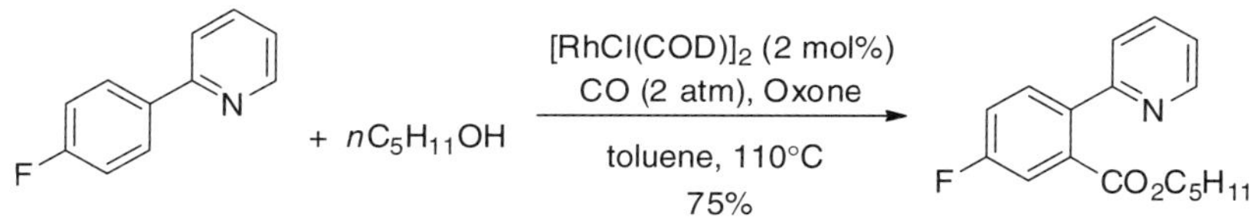

(6.236)

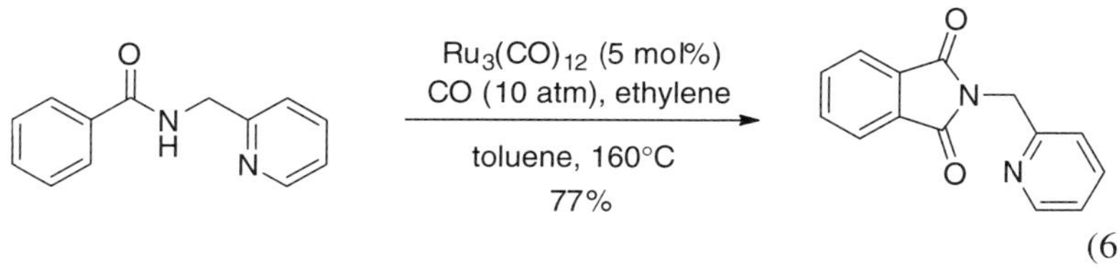

(6.237)

The regioselectivity of oxidative arene carbonylation can be controlled by a proper directing group. For example, Orito et al. [604] developed a Pd(II)-catalyzed carbonylation of arylalkylamine in which the amine acts as the directing group [Eq. (6.232)]. The carbonyl palladium intermediate was trapped by the amine to form a lactam product. Aromatic substrates containing other directing groups such as amide [Eq. (6.233)] [605–607], carboxylic acid [Eq. (6.234)] [608], alcohol [Eq. (6.235)] [609], and alkylamine [610–613] have also been successfully carbonylated to form carboxylic acid derivatives. Besides the palladium catalyst, Rh and Ru catalysts can also catalyze the carbonylation of 2-aryl pyridine [Eq. (6.236)] [614] and benzamide [Eq. (6.237)] [615, 616] substrates.

Carbonylation of alkyl C–H bond is also viable by using suitable directing groups. Yu and co-workers [617] reported a Pd-catalyzed carbonylation using a weakly coordinating amide as the directing group to form a lactam product [Eq. (6.238)]. A low-valent Ru complex was shown to catalyze the carbonylation of alkyl C–H bond using a bidentate pyridinyl amide directing group [Eq. (6.239)] [618]. In addition, the carbonylation of benzylic [Eq. (6.240)] [619] and allylic [620] C–H bond was achieved in the presence of Pd catalyst and a suitable oxidant.

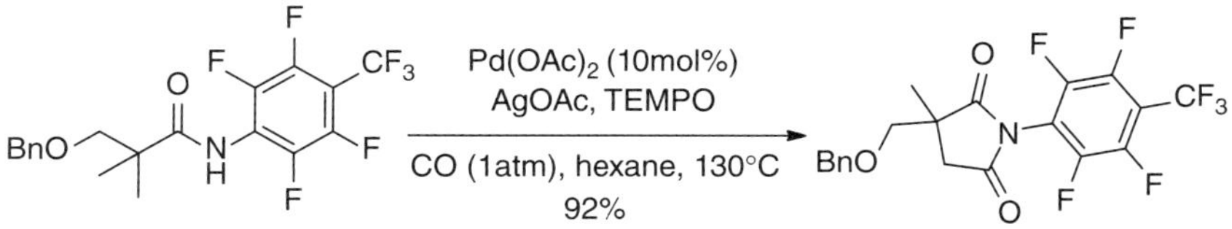

(6.238)

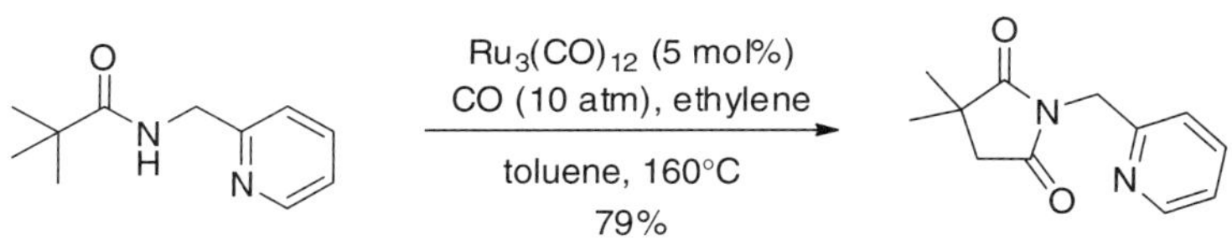

(6.239)

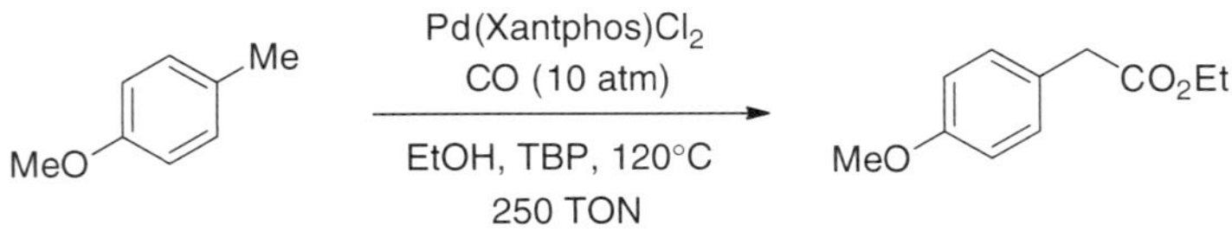

(6.240)

The reaction conditions reported by Fujiwara for the arene carbonylation were also applicable for alkane carbonylation. The reaction of cyclohexane with CO (20–40 atm) catalyzed by a Pd(II)/Cu(II) system in TFA at 80°C afforded cyclohexanecarboxylic acid in 20 turnover numbers and 4.3% yield based on cyclohexane [Eq. (6.241)] [621, 622]. The carbonylation of methane under similar conditions was also possible, but the yield was lower [623, 624]. The yield can be improved by using $VO(acac)_2$ and ligated vanadium complex as the catalyst [Eq. (6.242)] [625–627]. In addition, calcium chloride also showed high catalytic activity in the carbonylation of methane [Eq. (6.243)] [628]. Lin and Sen [629] reported a $RhCl_3$ catalyzed carbonylation of methane in aqueous media at 100°C using oxygen as the oxidant [Eq. (6.244)]. KI was a useful additive, and high yield of acetic acid was obtained. Periana et al. [630] developed a palladium-catalyzed direct synthesis of acetic acid from CH_4 in one step in the absence of CO [Eq. (6.245)]. In this system, CO seems to come from the oxidation of methane, which is supported by isotope labeling experiment.

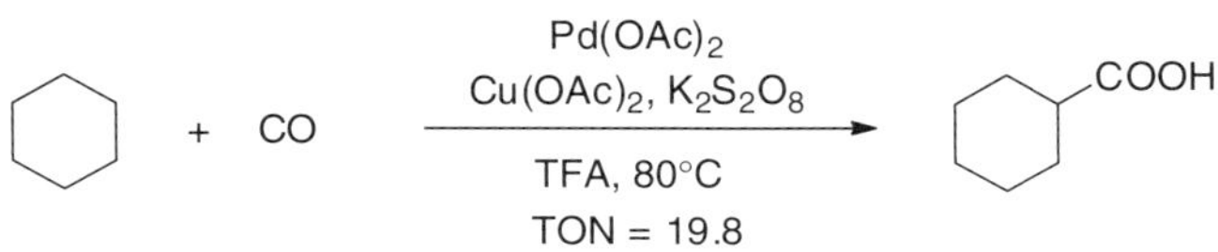

(6.241)

$$CH_4 + CO \xrightarrow[\text{H}_2\text{O, 100°C; TON = 49}]{\text{NaVO}_3\text{, air}} CH_3COOH \qquad (6.242)$$

$$CH_4 + CO \xrightarrow[\text{TFA/TFAA, 85°C; 93.8\% yield based on CH}_4]{\text{CaCl}_2\text{, K}_2\text{S}_2\text{O}_8} CH_3COOH \qquad (6.243)$$

$$CH_4 + CO \xrightarrow[\text{H}_2\text{O, 100°C}]{\text{RhCl}_3\text{, KI, O}_2} CH_3COOH \qquad (6.244)$$

$$CH_4 \xrightarrow[\text{180°C, 7 h; TON = 18}]{\text{PdSO}_4\text{/H}_2\text{SO}_4} CH_3COOH \qquad (6.245)$$

6.7 HYDROACYLATION

Hydroacylation formally involves the addition of a formyl C–H bond across a C–C multiple bond. Both intra- and intermolecular variants of the reactions are known. Intramolecular reactions generate cyclic ketone product, while intermolecular reactions generate linear ketone. The major problem associated with the development of an efficient hydroacylation reaction is the decarbonylation of the aldehyde group. Different strategies have been developed to suppress this undesired decarbonylation pathway. Intramolecular alkene hydroacylation can now be performed using low catalyst loadings under mild reaction conditions and has been applied in enantioselective synthesis. Cyclic ketone products with a variety of ring sizes can be prepared. Intermolecular alkene hydroacylation reactions are less developed, but a highly efficient catalyst system and an enantioselective version begin to appear. Compared with alkene hydroacylation, both intra- and intermolecular alkyne hydroacylations are less studied. Recent studies have demonstrated the possibility of efficient and enantioselective intramolecular reactions.

6.7.1 Intramolecular Hydroacylation of Alkene

Since Sakai et al.'s first report of a Rh-mediated hydroacylation reaction during the study on prostanoid synthesis [631], many Rh catalyst systems have been developed for intramolecular hydroacylation [Eq. (6.246)] [632–634]. The development of cationic rhodium complexes as catalysts by Bosnich represented a significant breakthrough in alkene hydroacylation [Eq. (6.247)] [635, 636]. A combination of cationic catalyst and chelating diphosphine ligand suppresses the undesired decarbonylation which poisons the catalyst and allows the use of low catalyst loading.

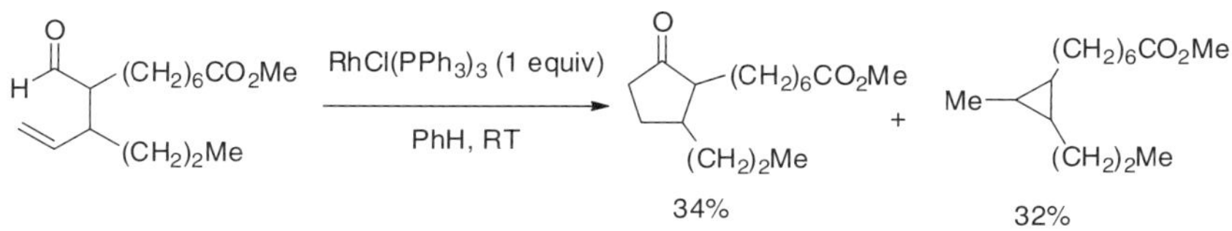

(6.246)

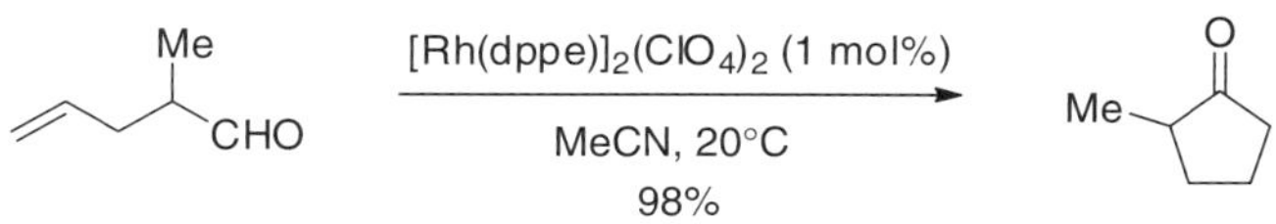

(6.247)

Generally, intramolecular hydroacylation favors the formation of five-membered ring system. Synthesis of a larger ring system has also been achieved using different strategies. Six-membered ring formation has been achieved using a rigid carbohydrate-derived scaffold [637]. By incorporating an additional alkene or cyclopropane group to the substrate, cycloheptenone, and cyclooctenone products were obtained after intramolecular hydroacylation [Eq. (6.248)] [638–640]. Sulfur [Eq. (6.249)] [641] and oxygen [642] linkage in the substrate facilitate the formation of chelation-stabilized intermediate, which provides medium-sized ring heterocycles after hydroacylation.

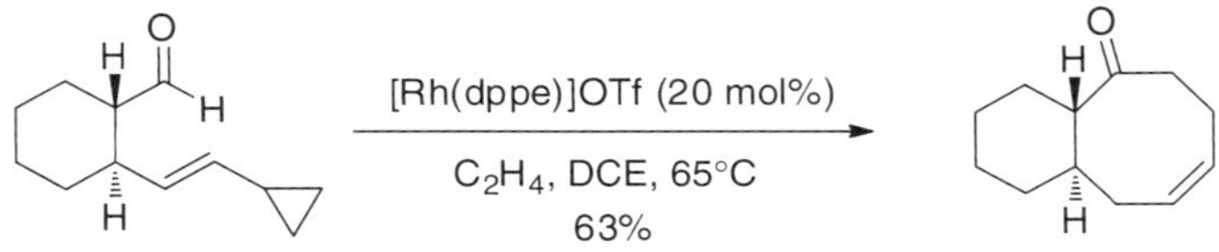

(6.248)

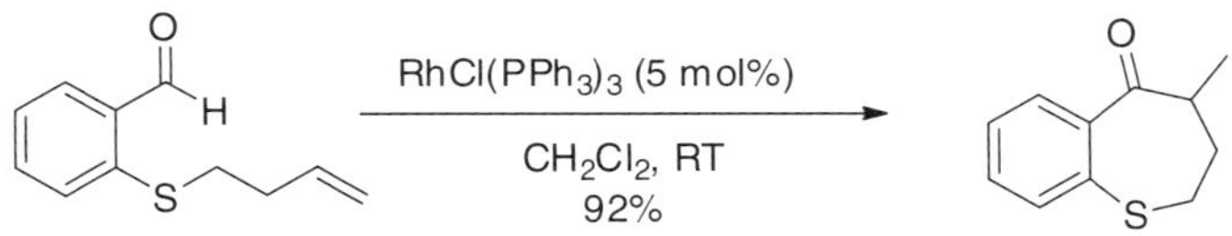

(6.249)

By using chiral diphosphines, enantioselective variants of intramolecular alkene hydroacylation have been achieved. James and Young [643, 644] reported enantioselective intramolecular hydroacylation via kinetic resolution using a chiral diphosphine. Sakai and co-workers [645, 646] reported the enantioselective cyclization of achiral 4-pentenals using a Rh catalyst and *trans*-1,2-bis-[(diphenylphosphino)methyl]cyclohexane (DIPMC) as the ligand with modest ee [Eq. (6.250)]. Both Sakai and co-workers [647] and Bosnich and co-workers [648–650] have obtained high enantioselectivity by using a cationic Rh catalyst and a BINAP ligand [Eqs. (6.251) and (6.252)] [651]. In addition, high ee's have been obtained by enantioselective desymmetrization [Eq. (6.253)] [652, 653] and kinetic resolution [654] using cationic Rh catalyst with chiral phosphine ligands. Sulfur, oxygen, and sulfur tethered substrates underwent enantioselective hydroacylation to provide medium-sized ring products with high enantioselectivity [Eq. (6.254)] [642].

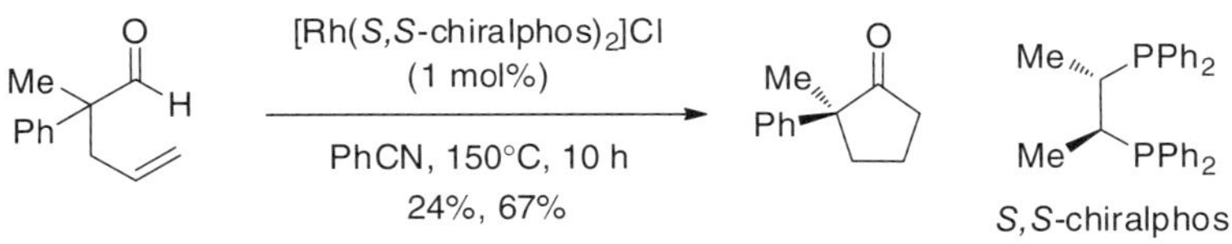

(6.250)

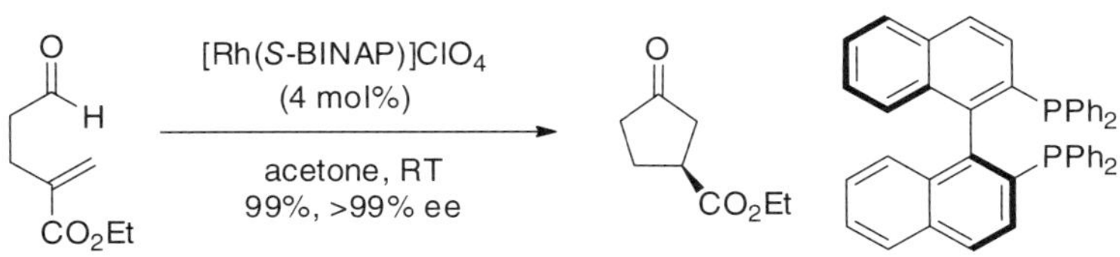

(6.251)

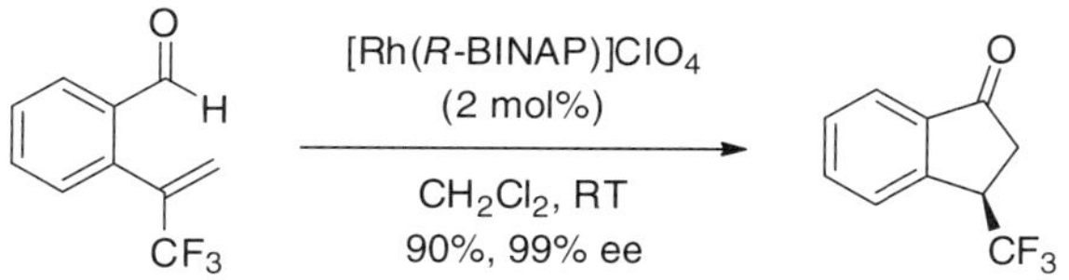

(6.252)

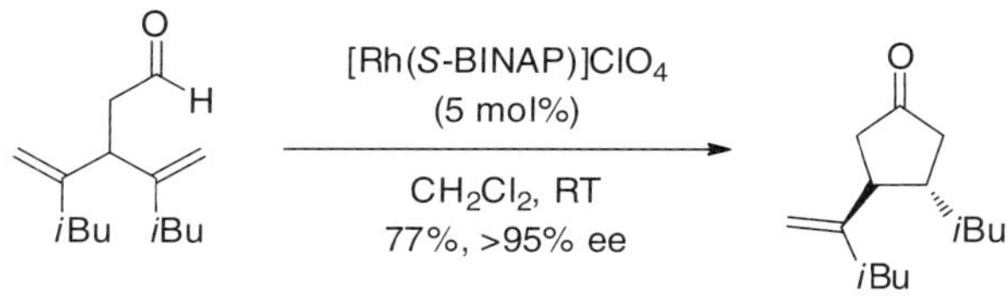

(6.253)

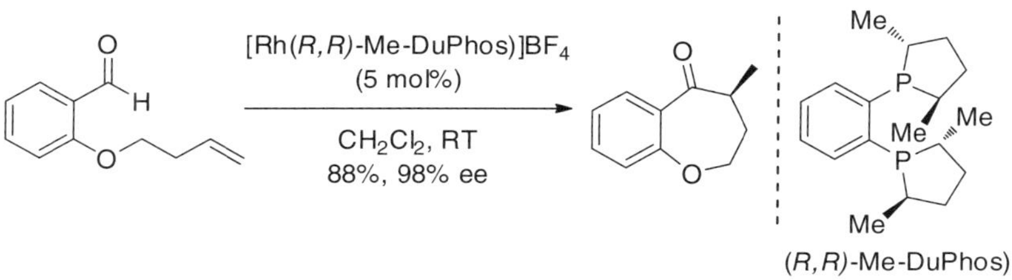

(6.254)

6.7.2 Intermolecular Hydroacylation of Alkene

Intermolecular hydroacylation is more challenging than the intramolecular version. During the study of cyclization of 4-pentenal substrate in ethylene-saturated solvent, Miller and co-workers [632, 655] observed the intermolecular hydroacylation with ethylene [Eq. (6.255)]. The alkene group in the substrate is important for the intermolecular reaction, presumably by coordination to the Rh center to prevent decarbonylation. Development of intermolecular hydroacylation using internal coordination strategy has been extended to other substrates. Miura and co-workers [656, 657] used salicylaldehyde as a chelating substrate for alkene hydroacylation with a Rh complex ligated by dppf ligand [Eq. (6.256)]. In a related Rh-catalyzed hydroacylation of salicylaldehyde with 1,5-diene, the generation of either linear or branched isomer can be controlled by changing different ligands [658]. The scope and efficiency of this hydroacylation reaction was greatly improved by using a phosphoramidite ligand, which selectively produced a linear hydroacylation product [Eq. (6.257)] [659]. Hydroacylation of allylic and homoallylic alcohols with salicylaldehydes in the presence of phosphinite selectively produced branched products. The regioselectivity was promoted by a catalytic amount of phosphinite which forms a dynamic covalent bond with the alcohol [660, 661]. Willis and co-workers [662–665] have exploited the chelation of a sulfur atom to develop intermolecular hydroacylation of S-containing aldehydes with both electron-deficient and electron-neutral alkenes [Eq. (6.258)]. Jun et al. [666] developed an intermolecular hydroacylation of both aryl and alkyl aldehyde with olefins in the presence of catalytic amount of Rh and picolyl amine [Eq. (6.259)]. *In situ* condensation of aldehyde and picolyl amine generates picolyl imine, which undergoes hydroacylation with pyridine coordination. Tanaka et al. [667] adopted an alternative strategy to stabilize the acyl rhodium species by using bidentate binding of the alkene substrate [Eq. (6.260)]. Using *N,N*-dialkyl acrylamides as the alkene component, effective intermolecular hydroacylation reactions were achieved with alkyl and aryl aldehydes. In the absence of a coordinating group on either the aldehyde or alkene substrate, hydroacylation of aryl and alkyl aldehyde with simple alkenes has been achieved using Cp*Co(I) [668, 669], Cp*Rh(I) [Eq. (6.261)] [670], or $Ru_3(CO)_{12}$ [671] as catalyst, although with limited scope.

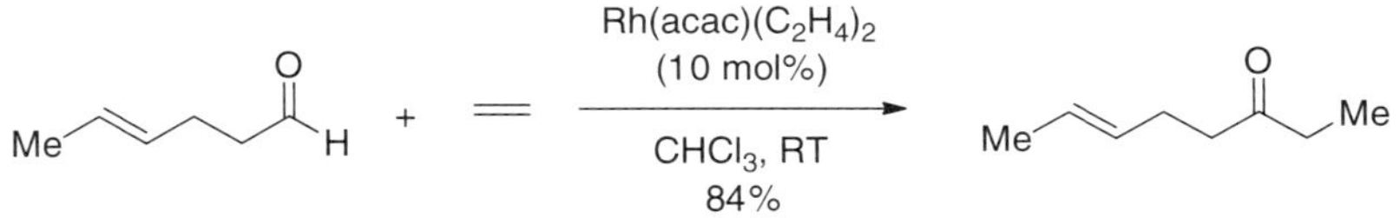

(6.255)

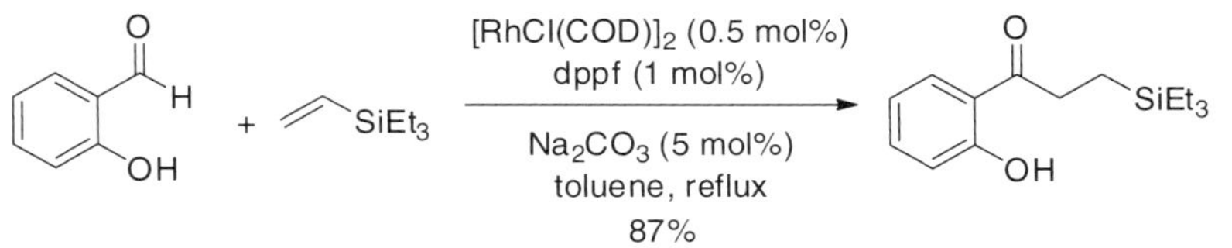

(6.256)

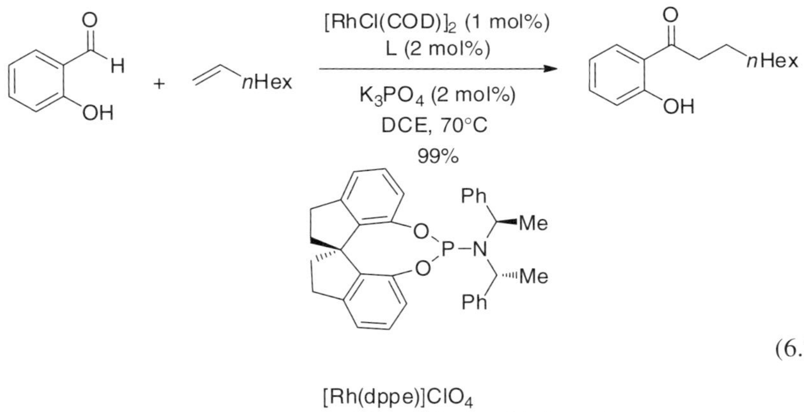

(6.257)

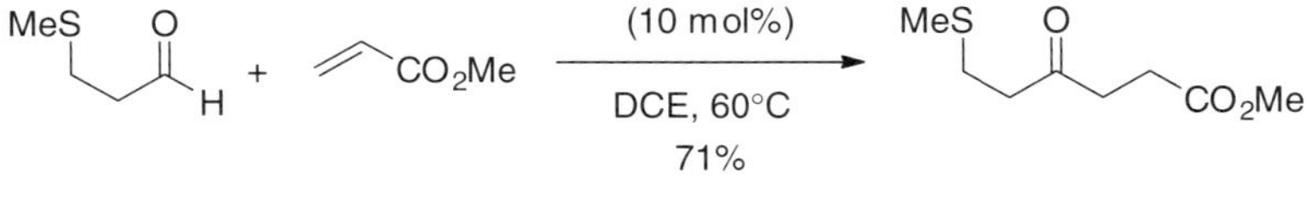

(6.258)

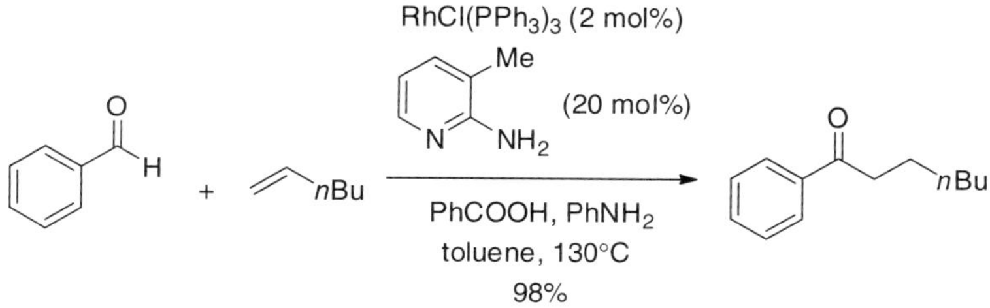

(6.259)

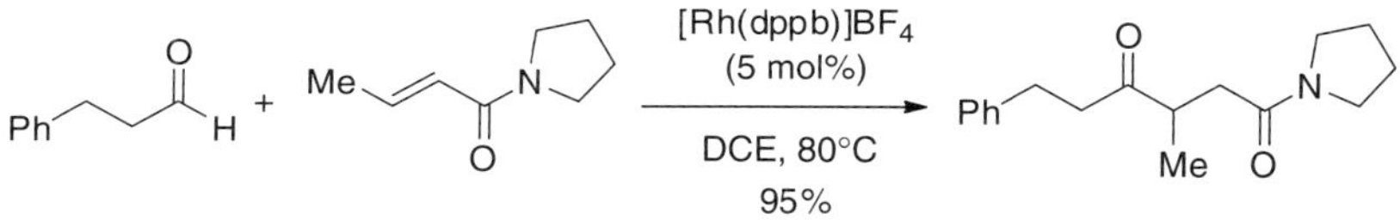

(6.260)

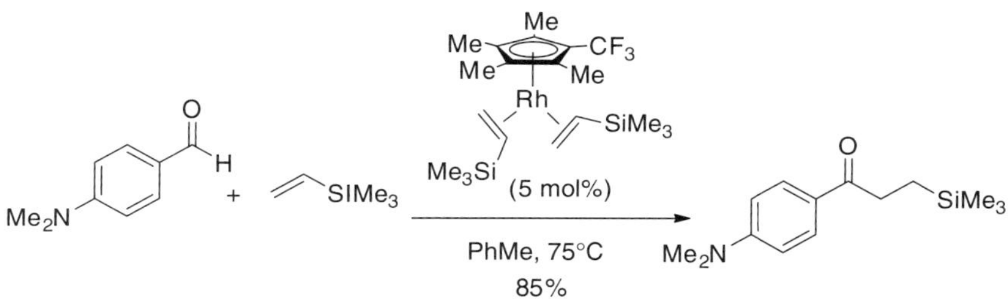

(6.261)

Comparing with many highly enantioselective intramolecular hydroacylations, asymmetric intermolecular versions are much less developed. Rh-catalyzed asymmetric intermolecular hydroacylation of salicylaldehyde with strained alkenes such as norbornadienes and cyclopropenes has been achieved by Stemmler and Bolm [672] [Eq. (6.262)] and Dong and co-workers [673] [Eq. (6.263)], respectively. Willis and co-workers [674] reported an enantioselective version of their intermolecular hydroacylation system employing β-S-substituted aldehydes. Shibata and Tanaka [675] reported that a cationic rhodium(I)/QuinoxP* complex catalyzes the highly enantioselective intermolecular hydroacylation of 1,1-disubstituted alkenes with unfunctionalized aldehydes [Eq. (6.264)]. 1,1-Disubstituted acrylamides presumably provide chelation sites to stabilize acylrhodium intermediates. A related intermolecular hydroacylation of olefin that uses substrate chelation was developed by Dong and co-workers [676], and it proceeded with high levels of both enantio- and regiocontrol [Eq. (6.265)].

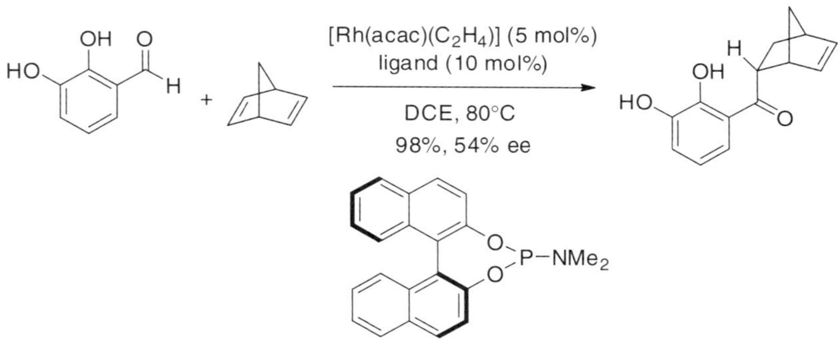

(6.262)

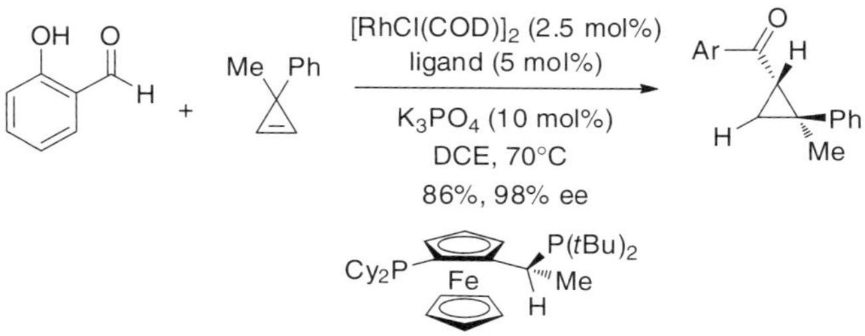

(6.263)

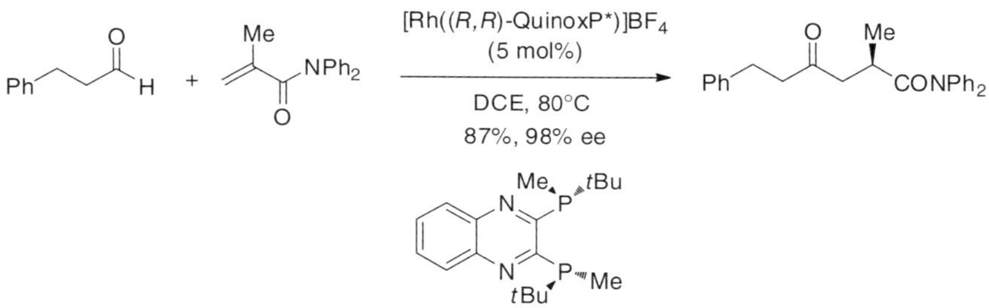

(*R,R*)-QuinoxP*

(6.264)

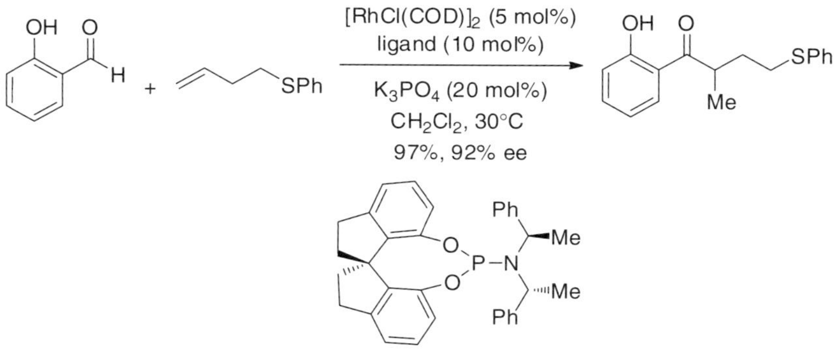

(6.265)

6.7.3 Intramolecular Hydroacylation of Alkyne

Tanaka and Fu [677] reported the cyclization of 4-alkynals using cationic rhodium complexes, which involves an unusual *trans* addition of the rhodium hydride across the alkyne [Eq. (6.266)]. In contrast, cyclization of 5- and 6-alkynals proceeds with *cis* addition across the triple bond [Eq. (6.267)] [678]. In the presence of Rh(I)/(*i*-Pr-DUPHOS) catalyst, racemic 4-alkynals bearing substituents in the 3-position underwent kinetic resolution to provide enantiomerically enriched cyclopentenones with good selectivities [Eq. (6.268)] [679]. Switch to a Rh(I)/(*i*-Pr-DUPHOS) catalyst resulted in parallel kinetic resolution, producing a cyclopentenone and a cyclobutenone with high ee [Eq. (6.269)] [680].

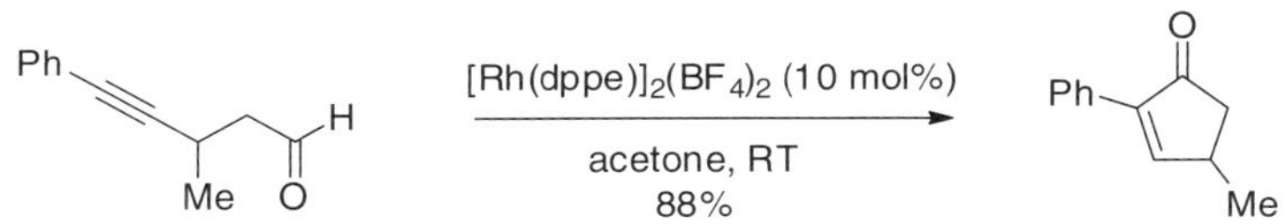

(6.266)

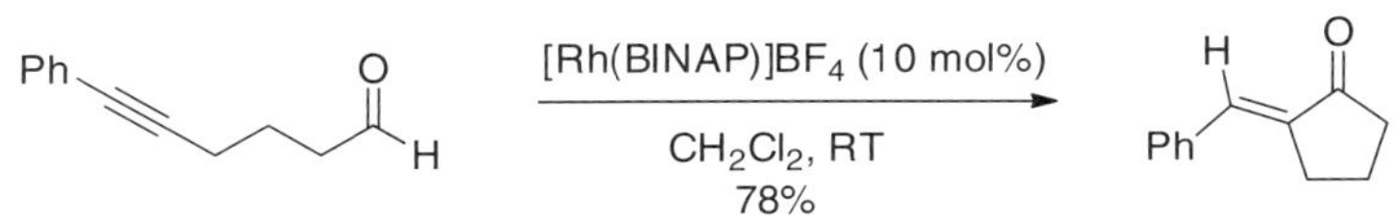

(6.267)

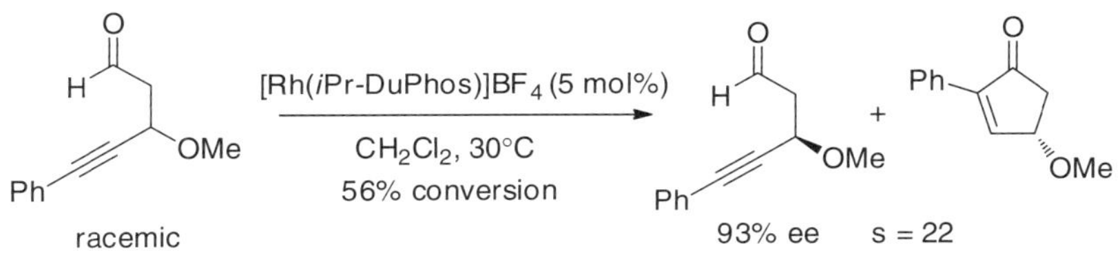

(6.268)

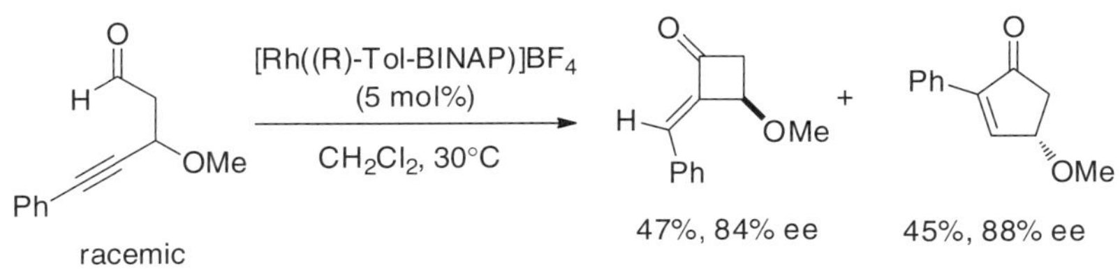

(6.269)

6.7.4 Intermolecular Hydroacylation of Alkyne

Tsuda et al. [681] demonstrated a Ni(0) complex-catalyzed hydroacylation of symmetrical internal alkynes with simple aryl and alkyl aldehydes [Eq. (6.270)]. The enone products were obtained in high *E*-selectivity. Similar to internal chelation-assisted hydroacylation of alkenes, intermolecular hydroacylation of alkynes has been conducted using substrates containing a coordination site. Hydroacylation of picolyl imine [682], salicylaldehyde [Eq. (6.271)] [683], and β-*S*-aldehydes [Eq. (6.272)] [662–684] with internal and terminal alkynes have been developed using N, O, S chelation, respectively. In some cases, the linear and branched selectivity can be tuned by different ligands [685]. An enantioselective intermolecular alkyne hydroacylation was achieved by kinetic resolution in the Rh-catalyzed reaction of β-S substituted aldehyde [686].

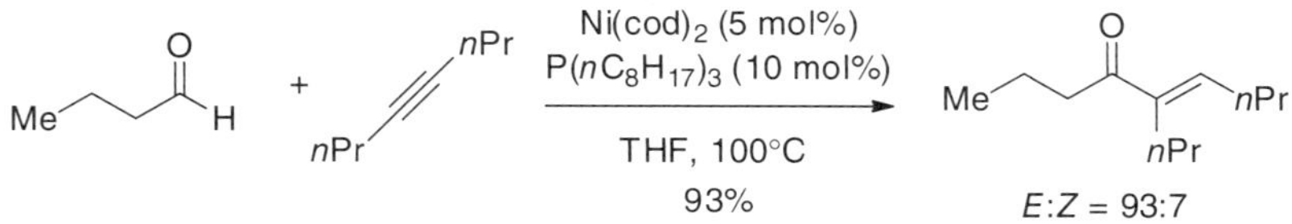

(6.270)

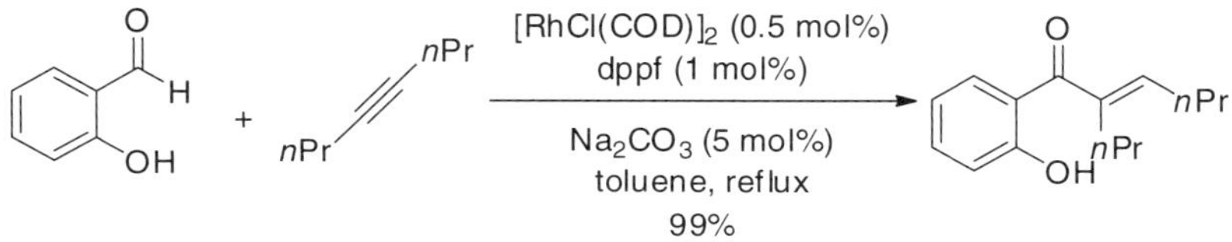

(6.271)

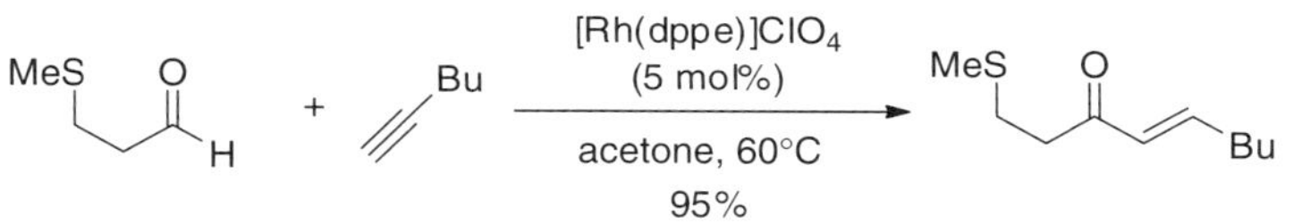

(6.272)

6.7.5 Mechanism

In a simplified general mechanism, hydroacylation of alkene occurs through initial oxidative addition of acyl C–H bond to a Rh center with or without chelation assistance to generate an acyl Rh hydride. Subsequent insertion of olefin or alkyne to the Rh hydride bond generates an acyl alkyl or acyl vinyl rhodium intermediate, which undergoes reductive elimination to afford the hydroacylation product (Scheme 6.34). The mechanism of Rh catalyzed intramolecular hydroacylation has been studied by Miller and co-workers [687, 688], Bosnich [689], and others [690]. Cyclization of 1-D-labeled 4-pentenal resulted in extensive deuterium scrambling. It was proposed that every step except the final reductive eliminations is reversible and that reductive elimination of the ketone product is rate-determining.

6.8 ALKANE DEHYDROGENATION

6.8.1 Alkane Dehydrogenation to Form Alkene

Alkanes are naturally abundant feedstocks in the chemical industry. Valuable alkene products can be directly produced from alkanes through dehydrogenation. Heterogeneous catalysts can effect dehydrogenation at high temperatures, but the

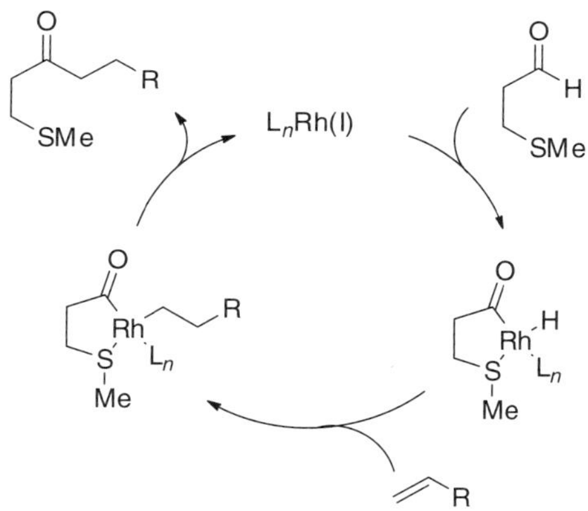

Scheme 6.34. Mechanism of Rh-catalyzed hydroacylation.

substrates are limited to ethane or ethylbenzene and the selectivity is low for higher alkanes [691].

In 1979, Crabtree et al. [692] reported the first stoichiometric dehydrogenation of alkanes by a soluble homogeneous metal complex [Eq. (6.273)]. The reaction of cationic complex [IrH_2(acetone)$_2$(PPh_3)$_2$][BF_4] with cyclopentane or cyclooctane produced the cyclopentadiene orcyclooctadiene iridium complexes. Felkin and co-workers [693] also reported similar dehydrogenation of cycloalkanes with (PPh_3)2ReH_7. Subsequently, Felkin and co-workers [694, 695] and Crabtree and co-workers [696, 697] reported catalytic alkane dehydrogenation by bis(trialkylphosphine) complexes of iridium in the presence of hydrogen acceptor such as *tert*-butylethylene. Various other metal catalysts have been developed for the transfer dehydrogenation of cylcooctane to cyclooctene as well [Eq. (6.274)] [698–705]. Acceptorless dehydrogenation can be carried out under reflux conditions [Eq. (6.275)] [700] or photolysis [706]. Under reflux conditions, hydrogen is extruded from the system, thus driving the equilibrium to the product formation. In photolytic examples, the reaction is driven by the photon energy through photodissociation of H_2 from the metal dihydride to regenerate the active catalyst.

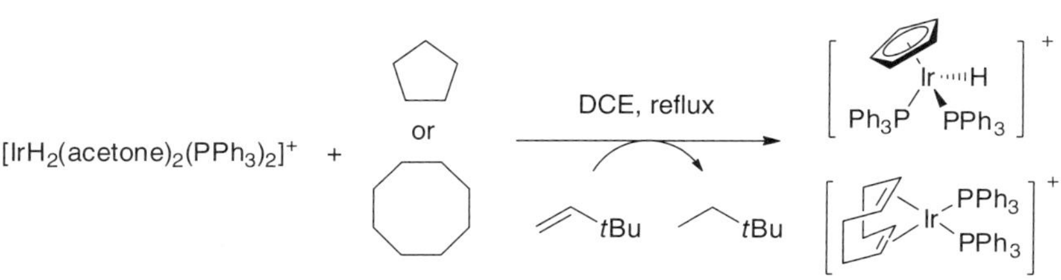

(6.273)

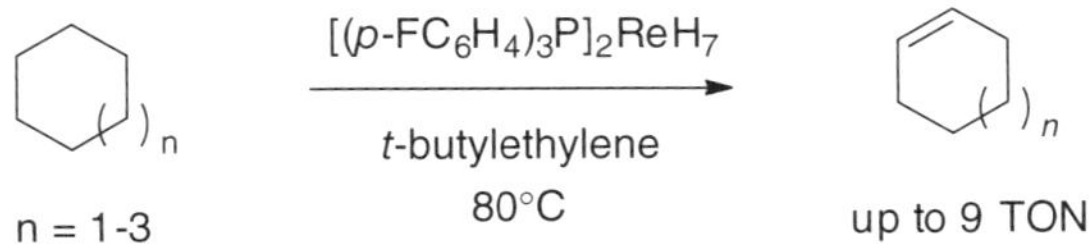

(6.274)

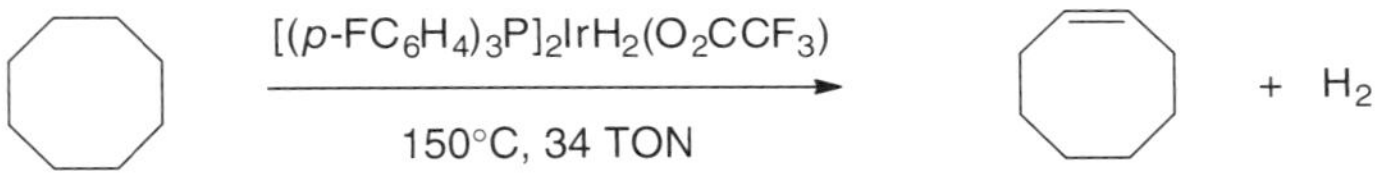

(6.275)

In previous systems, turnover numbers were limited by catalyst decomposition due to the high temperatures required for dehydrogenation and the tendency of highly reactive metal centers to attack their own ligands either intra- or intermolecularly. The application of a thermally stable pincer–metal complex has a profound impact on dehydrogenation reactions and has resulted in significantly improved turnover numbers.

The application of pincer–ligated iridium complexes to alkane dehydrogenation was first reported by Kaska and co-workers [707] in 1996 [Eq. (6.276)]. High turnover numbers were obtained in the dehydrogenation in the presence of TBE (tert-butylethylene) using iridium catalyst [708]. Acceptorless dehydrogenation can also be carried out in an open system using a pincer–iridium catalyst under reflux condition [709]. These catalysts show high kinetic selectivity for dehydrogenation of the terminal position of *n*-alkanes to give α-olefins. However, the selectivity is diminished by subsequent isomerization to form thermodynamically more stable internal olefins (Scheme 6.35) [710].

The thermal stability and catalytic reactivity of these pincer complexes can be further improved by using a rigid backbone or introducing bulky substituents to the phosphorus atom (Scheme 6.36) [711–714]. Another important modification of the PCP ligand is the bisphosphinite iridium–pincer complexes, which are discovered independently by Brookhart and co-workers [715, 716] and Jensen and co-workers [717]. These catalysts showed higher catalytic activity in the transfer dehydrogenation of COA (cyclooctane) and are among the most active catalysts for this reaction.

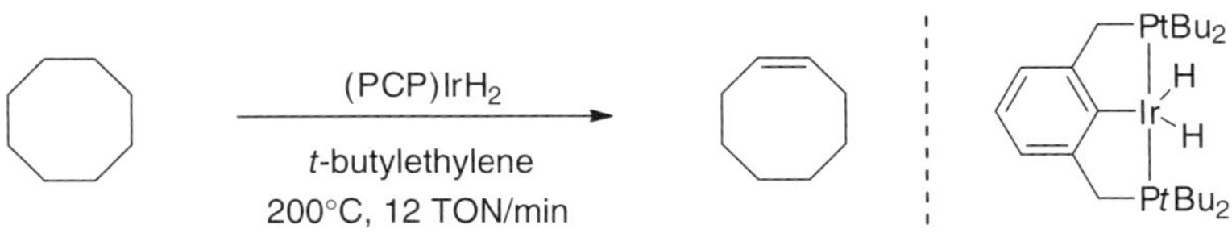

(6.276)

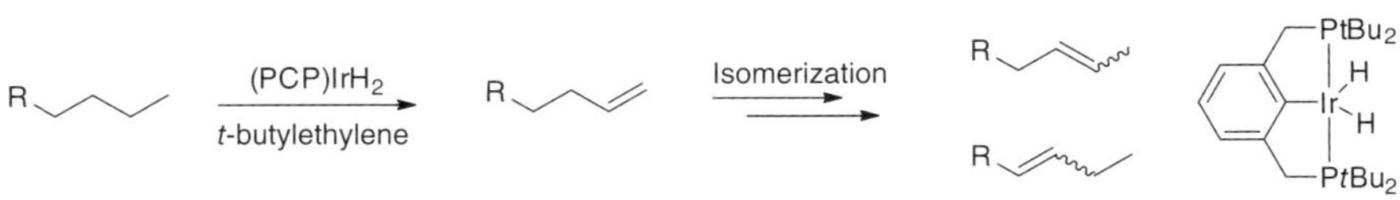

Scheme 6.35. Dehydrogenation of alkane and subsequent olefin isomerization.

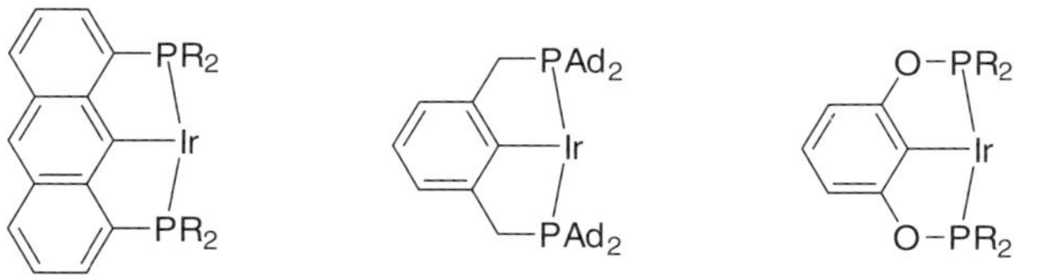

R = *i*Pr or *t*Bu R = *i*Pr or *t*Bu

Scheme 6.36. Pincer-Ir complexes.

6.8.2 Dehydroaromatization

During the investigation of *n*-alkane dehydrogenation using (MeO-*i*Pr_4PCP)Ir as a catalyst and norbornene (NBE) as a hydrogen acceptor, the formation of benzene from *n*-hexane was observed, although in low yield (2.3%) [711]. More recently, dehydroaromatization of *n*-alkanes to form substituted arene products has been studied in detail [Eq. (6.277)] [718]. Among various pincer–iridium complexes tested, a hybrid phosphine/phosphinite pincer–iridium complex is most effective, producing good yield of benzene from *n*-hexane after 120 h at 165°C in the presence of the TBE acceptor. *n*-Octane was converted to a mixture of *o*-xylene and ethylbenzene in high yield under the catalytic system.

The proposed mechanism for the dehydroaromatizations is shown for *n*-octane. Multiple dehydrogenation of *n*-alkane and isomerization produces a conjugated linear triene, which undergoes electrocyclic ring closure to yield the cyclic diene intermediate. Further dehydrogenation generates the final aromatic products (Scheme 6.37).

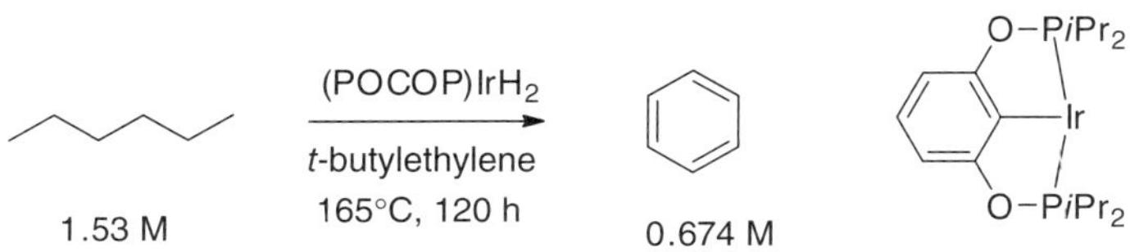

(6.277)

6.8.3 Alkane Metathesis

Goldman, Brookhart, and co-workers [719] developed the alkane metathesis by tandem alkane dehydrogenation and olefin metathesis using a combination of an

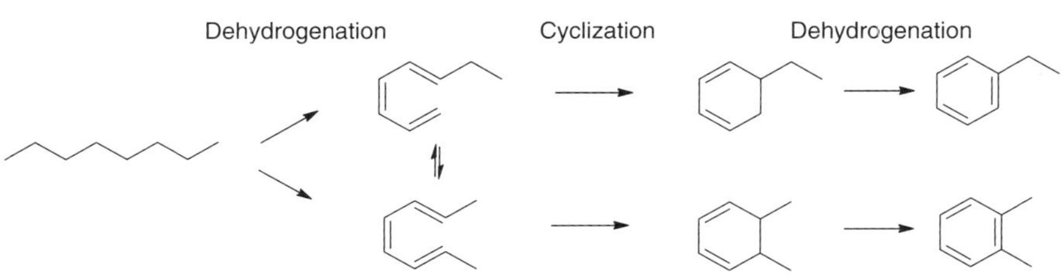

Scheme 6.37. Dehydroaromatization of alkane.

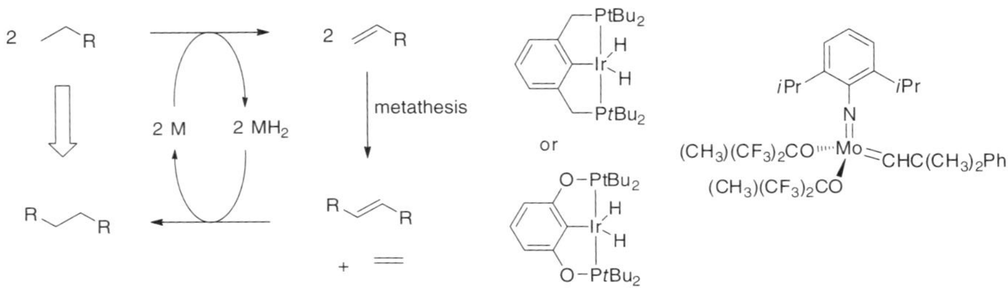

Scheme 6.38. Alkane metathesis.

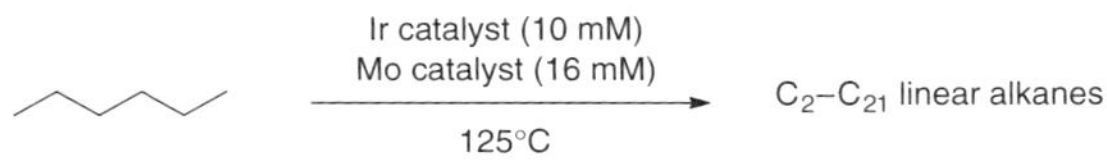

Ir catalyst	product concentration (mM)				total product (M)	$C_{10}/(C_7–C_{10})$ (%)
	$C_2–C_5$	$C_7–C_9$	C_{10}	$>C_{11}$		
(PCP)IrH_n	740	241	232	38	1.25	49.0
(POCOP)IrH_2	1350	548	95	53	2.05	14.8

Scheme 6.39. Product ratio of alkane metathesis.

iridium–pincer complex and the Schrock catalyst (Scheme 6.38). The alkene product obtained by dehydrogenation with the pincer–iridium complex undergoes metathesis to form a higher alkene, which is further hydrogenated by the hydrogen generated in the first dehydrogenation step. Ideally, this tandem alkane metathesis strategy would provide ethane and C_{2n-2} alkane from C_n alkane. In practice, alkanes with other molecular weights are also produced as a result of isomerization of terminal olefin to internal olefin and further metathesis of internal olefin (Scheme 6.39). This dual catalysts system is highly reactive and achieves a few hundred turnover numbers for the production of higher alkanes. To avoid secondary metathesis that decreases the overall selectivity, a two-pot system was designed. Excellent turnover numbers and moderate recyclability of the catalysts were achieved [720].

6.8.4 Mechanism

The mechanism of the catalytic dehydrogenation reactions has been studied in detail for both the (PCP)Ir [721] and (POCOP)Ir catalysts (Scheme 6.40) [715, 716]. The catalytic cycle starts with insertion of TBE into an Ir–H bond followed by reductive elimination of *tert*-butylethane (TBA) to generate the reactive 14-electron iridium species. The coordinative unsaturated iridium catalyst then undergoes oxidative addition of an alkane C–H bond. Subsequent β-hydride elimination produces the olefin product and regenerates the iridium hydride. The 14-electron species of (PCP)Ir can also undergo oxidative addition of the TBE vinylic C–H bond to generate an

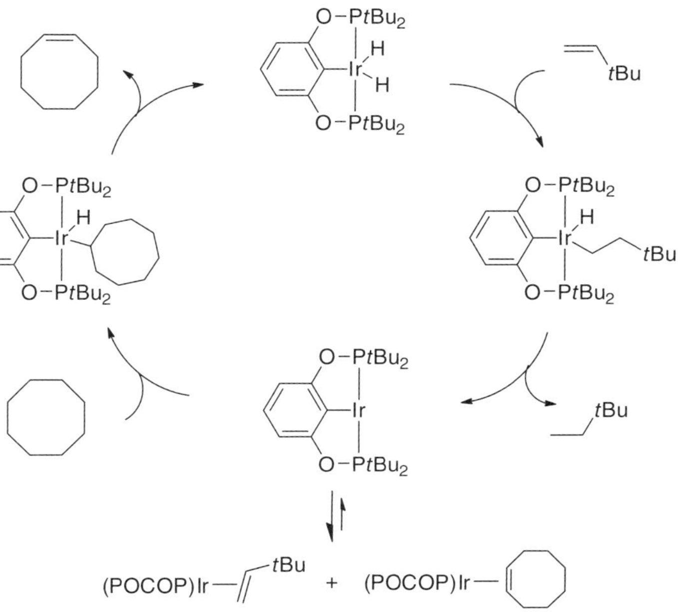

Scheme 6.40. Mechanism of alkane dehydrogenation.

alkenyl iridium hydride which lies off the catalytic cycle and is the resting state of the catalyst. In contrast, the 14-electron species of (POCOP)Ir reacts with olefin to form an η^2-olefin complex that is the resting state. The change of the resting state as well as the higher reactivity of (POCOP)Ir in dehydrogenation reaction can be attributed to the electronic and steric differences between these two catalysts. Although the oxygen atom reduces donating ability of the phosphorus atom, DFT calculations show that it donates substantial π-electron density to the aryl ring in the POCOP ligand [711]. Therefore, the iridium center with POCOP ligand is calculated to be slightly more electron-rich than that with PCP ligand. In addition, crystallographic characterization shows that (POCOP)Ir is less sterically hindered than (PCP)Ir [722]. The more open coordination site in (POCOP)Ir allows it bind strongly to olefin to form the resting state of the catalyst rather a vinyl hydride species. Moreover, steric effect probably also accounts for the faster rates of dehydrogenation of COA by (POCOP)Ir catalyst. However, (POCOP)Ir catalyst provides slower rates for linear alkanes dehydrogenation in the presence of TBE acceptors, presumably due to the formation of out-of-cycle olefin complex that lower the concentration of active iridium catalyst.

6.9 BORYLATION AND SILYLATION

6.9.1 Borylation of Alkyl C−H Bond

During the investigation of the reactivity of metal boryl complex, Hartwig and co-workers [723] observed the borylation of C−H substrates with various isolated metal boryl compounds under photochemical conditions. In particular, irradiation of tungsten–boryl complex in alkane led to alkane borylation in high yield with exclusive terminal selectivity [Eq. (6.278)] [724, 725]. The selective cleavage of a stronger primary C–H bond is noteworthy. Photocatalytic alkane borylation can be achieved using a rhenium catalyst under a CO atmosphere [726]. Later, they found that rhodium [Eq. (6.279)] [727, 728] and ruthenium [Eq. (6.280)] [729] catalysts are effective to catalyze alkane borylation under thermal conditions. In these catalytic reactions, borylation occurs selectively at the less hindered terminal position of an alkane chain. The Rh catalyst has also been applied to the borylation of polyolefins [730, 731]. Recently, borylation of a secondary C–H bond in THF was reported by the same group using an iridium catalyst with a substituted phenanthroline ligand [Eq. (6.281)] [732].

Selective borylation of a benzylic C–H bond in alkylbenzenes can be achieved by using either a Rh complex [Eq. (6.282)] [733] or a Pd/C [734] catalyst. Recently, Sawamura developed a supported phosphine ligand for amide directed borylation of sp^3 C–H bond adjacent to nitrogen atom [735]. Suginome reported an iridium-catalyzed borylation of methyl group in methylchlorosilanes to give (borylmethyl)chlorosilanes selectively [Eq. (6.283)] [736].

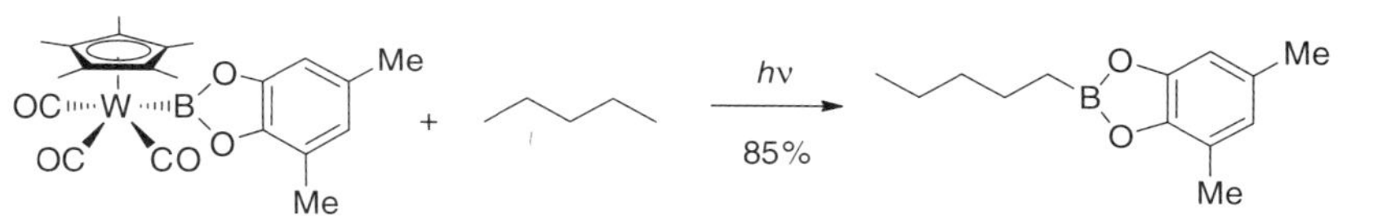

(6.278)

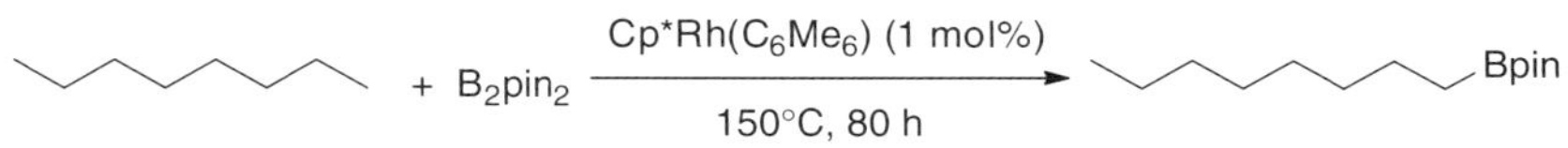

(6.279)

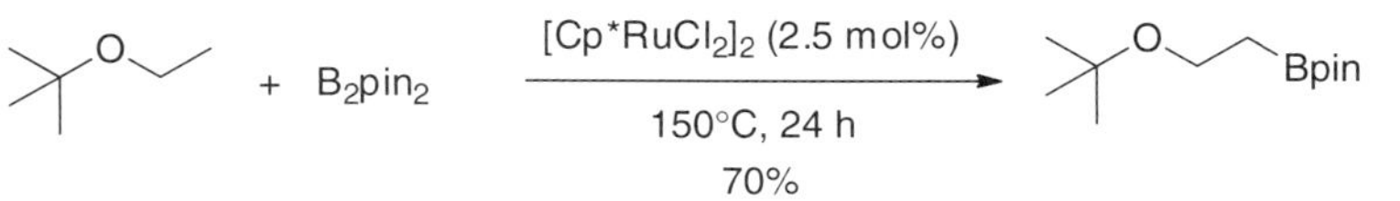

(6.280)

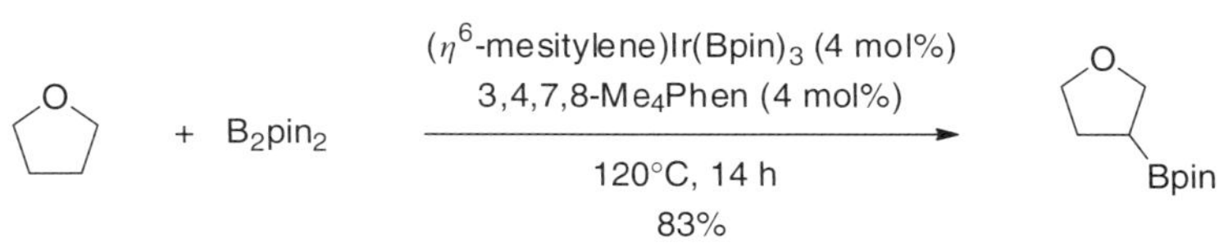

(6.281)

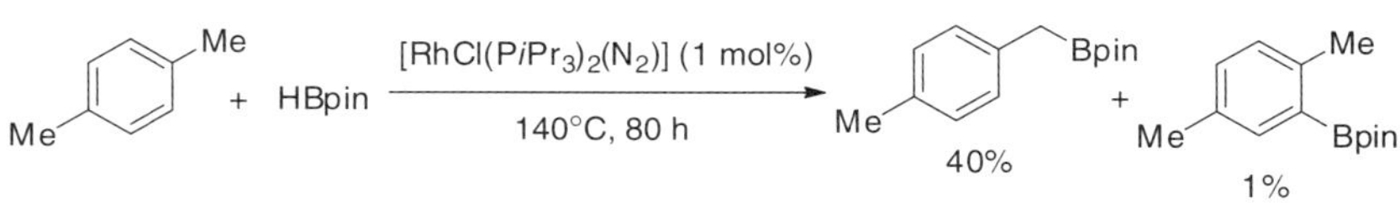

(6.282)

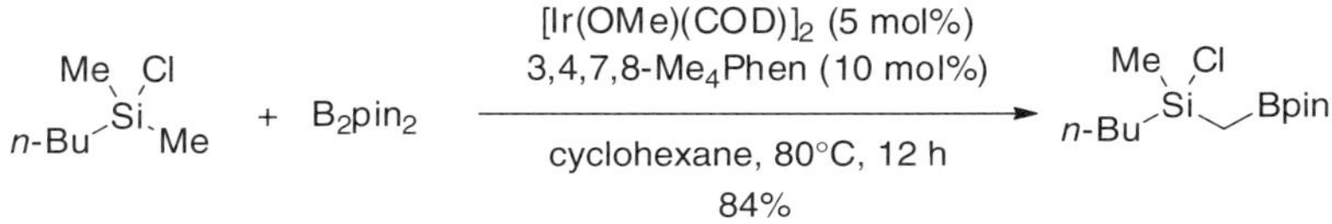

(6.283)

6.9.2 Borylation of Aryl C–H Bond

Hartwig and co-workers [723] observed stoichiometric borylation of benzene with an iron–boryl complex under photolysis. The first example of catalytic C–H borylation was reported by Iverson and Smith [737], although with only a few turnover numbers at high temperature [Eq. (6.284)]. Later, a catalyst system comprising trisboryl iridium and diphosphine was developed, which was very effective for borylation of various arenes at elevated temperature [Eq. (6.285)] [738]. The regioselectivity of the reaction is governed by steric effect. Borylation occurred selectively at the less hindered position. Hartwig and co-workers [739] found that a combination of $[Ir(COD)Cl]_2$ and bipyridine generates a highly active catalyst for arene boylation [Eq. (6.286)]. The reaction occurs at room temperature to 80°C with high turnover numbers. The use of $[Ir(COD)(OMe)]_2$ as the catalyst precursor allows the reaction to be carried out with a 1:1 ratio of arene to diboron at even room temperature [Eq. (6.287)] [740]. Again the regioselectivity was controlled by steric effect of arene substituents. Five-membered heterocycles such as indole, benzofuran, and benzothiophene also worked well, affording products with borylation at C-2 position [Eq. (6.288)] [741, 742]. The catalyst system can be applied to the borylation of pyridine and quinoline as well [743].

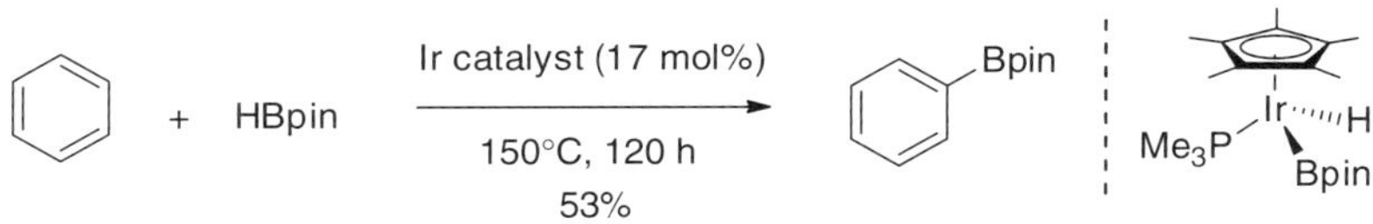

(6.284)

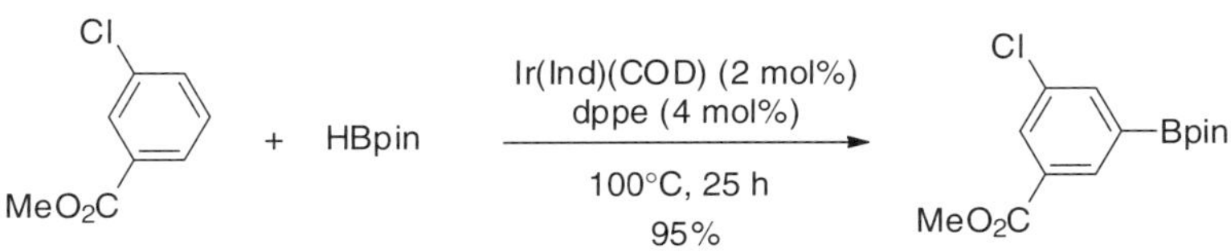

(6.285)

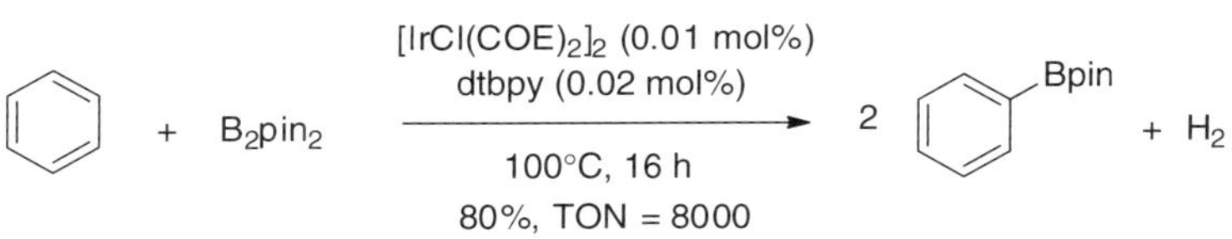

(6.286)

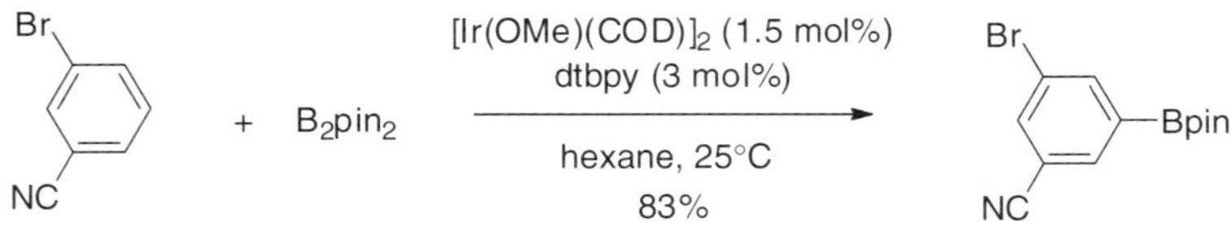

(6.287)

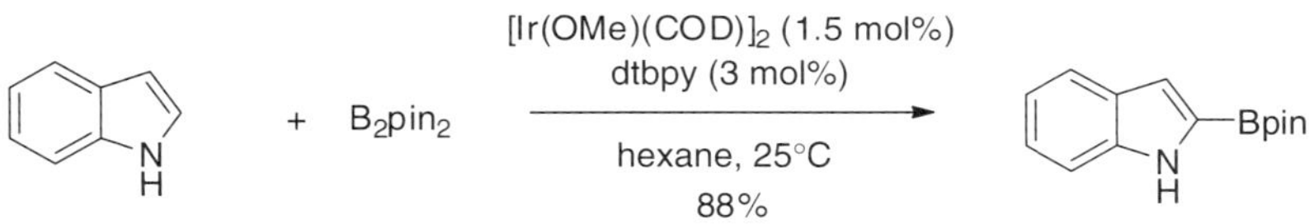

(6.288)

The steric controlled regioselectivity of arene borylation can be overridden by the presence of a directing group. For example, Hartwig and co-workers used a hydrosilane as the directing group for the *ortho*-borylation of benzyl silane, phenol, and aniline [Eq. (6.289)] [744] and for the C-9-selective borylation of indole [745]. Miyaura and co-workers [746] reported an Ir–phosphine catalyst for the *ortho*-borylation of benzoic ester [Eq. (6.290)]. Sawamura and co-workers [747] developed a silica-supported Ir catalyst for *ortho*-borylation of aryl C–H substrates containing ester, amide, ether, and halide directing groups [Eq. (6.291)]. Pyridine and imine can also be used as directing groups for aryl C–H borylation using a related Rh catalyst [748]. The use of hemilabile pyridine–hydrazone ligand allows mono borylation of 2-arylpyridine [Eq. (6.292)] [749] and diborylation of *N,N*-dimethylhydrazones [750]. A NHBoc group could direct the Ir-catalyzed *ortho*-selective C–H borylation of arenes [Eq. (6.293)] [751]. Based on experimental and computational studies, an outer sphere mechanism where the N–H proton bonds to a boryl ligand oxygen was proposed. Yu and co-workers [252] and Fu and co-workers [753] reported Pd-catalyzed borylation of an arene C–H bond with B_2Pin_2 in the presence of an oxidant [Eq. (6.294)]. The regioselectivity is controlled by the amide directing group attached to the arene.

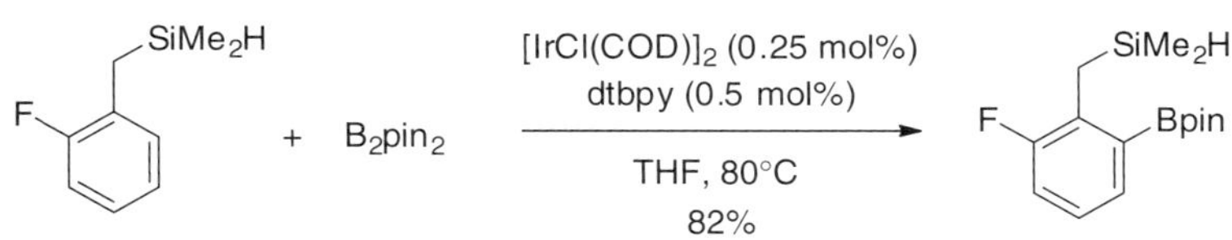

(6.289)

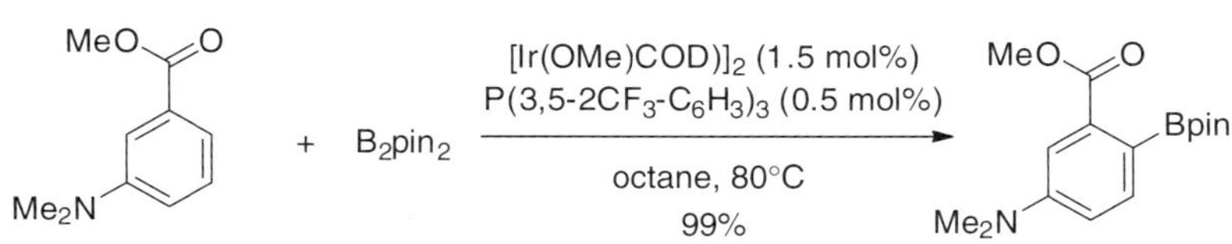

(6.290)

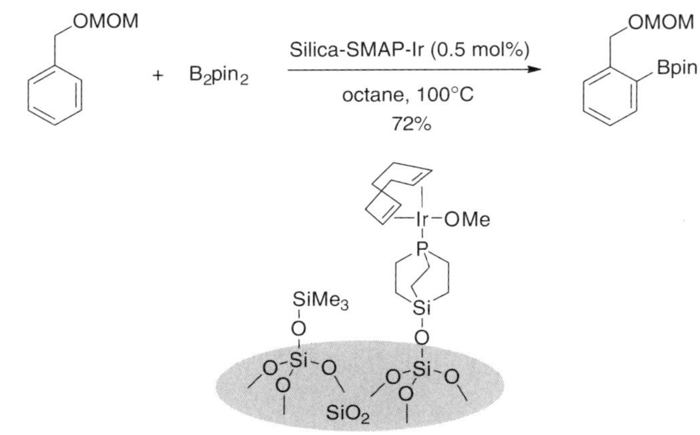

(6.291)

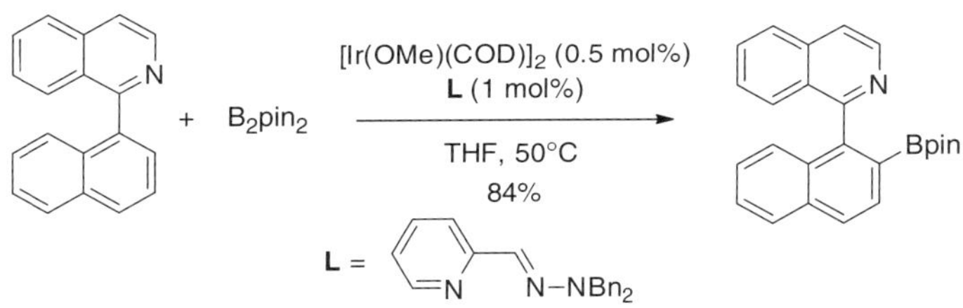

(6.292)

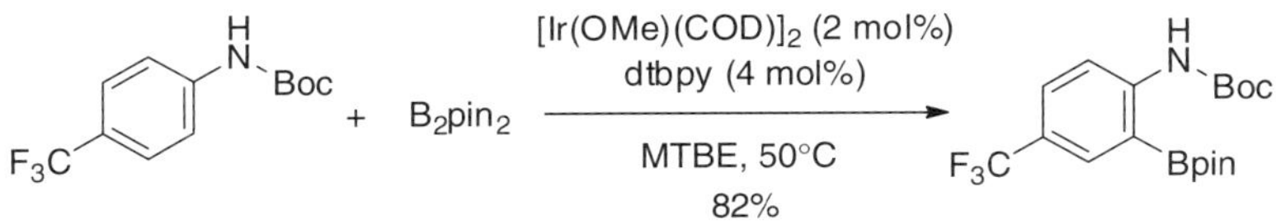

(6.293)

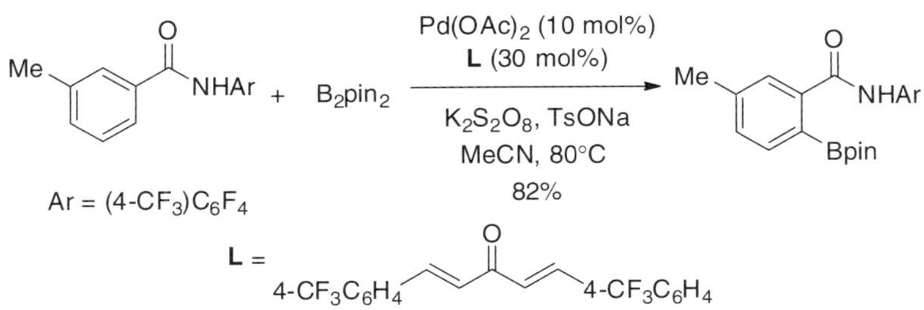

(6.294)

The synthetic utility of the Ir-catalyzed arene borylation was elegantly demonstrated in the synthesis of (+)-complanadine A (Scheme 6.41) [754]. Borylation of a functionalized pyridine substrate occurred regioselectively at the C-3 position. The resulting pyridyl boronate underwent cross-coupling with 2-pyridyl triflate to set the core structure, which can be easily converted to the target molecule. The first enantioselective synthesis of (−)-taiwaniaquinol B and (−)-taiwaniaquinone

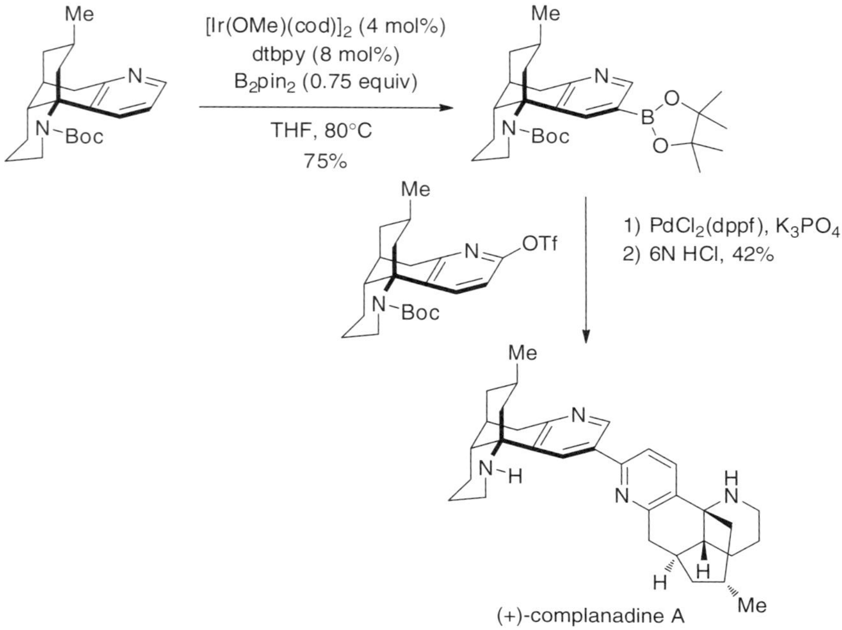

Scheme 6.41. Synthesis of (+)-complanadine A.

H also featured the use of Ir-catalyzed arene borylation (Scheme 6.42) [755]. An arene borylation/bromination sequence selectively generated the *meta*-bromination product, which underwent further enantioselective α-arylation.

6.9.3 Mechanism of Borylation

The mechanism of alkane borylation catalyzed by a Cp*Rh complex has been studied by Hartwig et al. [756]. Two bis-boryl Rh and tri-boryl Rh complexes have been independently synthesized from the reaction of catalyst precursor and hydroborane. Both rhodium boryl complexes react directly with arene and alkane to form the aryl boron ester and alkyl boron ester upon heating [Eq. (6.295)]. Detailed study of this step suggested that these rhodium boryl complexes undergo initial loss of borane to form a 16e rhodium boryl compound, which further reacts with alkane (Scheme 6.43). The *p*-orbital of the boron atom has been calculated to assist the C–H cleavage step. The origins of the selectivity for borylation of primary over secondary C–H bonds catalyzed by Cp*–Rh complexes were studied by both experiment and computation. The formation of the rhodium complex by cleavage of a secondary C–H bond is calculated to have a higher energy barrier than that by cleavage of a primary C–H bond. In addition, the barrier for B–C bond formation from the secondary alkyl Rh intermediate is higher than that from the primary alkyl Rh intermediate due to steric effects. The combined effects of selective C–H bond cleavage and selective C–B

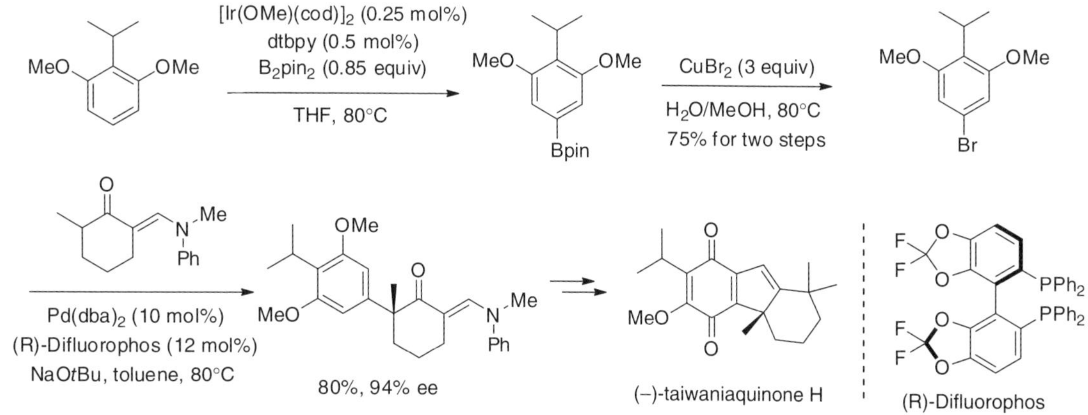

Scheme 6.42. Synthesis of (−)-taiwaniaquinone H.

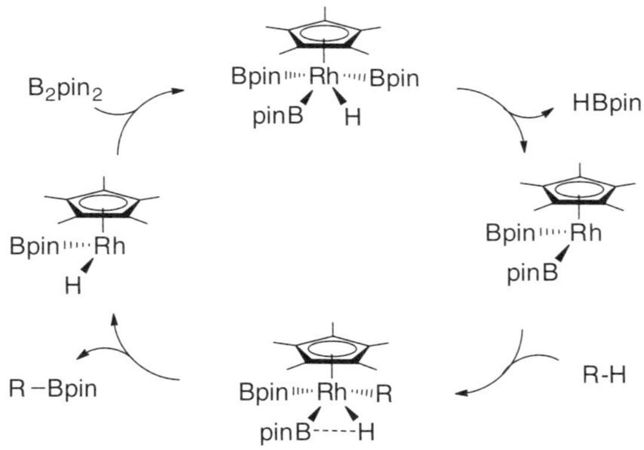

Scheme 6.43. Mechanism of Rh-catalyzed alkane borylation.

bond formation accounts for the primary selective C–H borylation observed in the reaction system. H/D exchange experiments are consistent with the computational analysis [757].

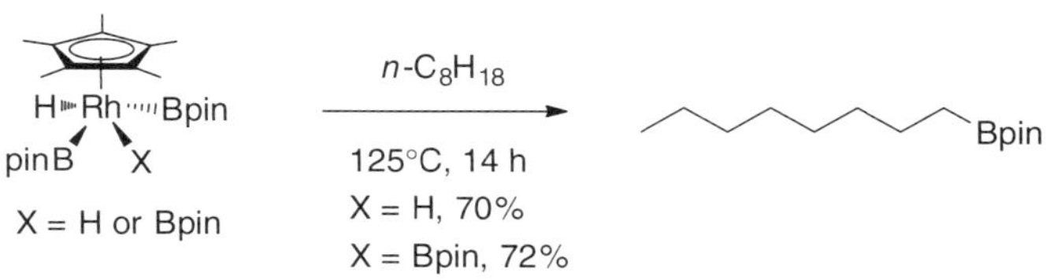

(6.295)

The mechanism of arene borylation catalyzed by an iridium complex and a dtbpy ligand has also been studied in detail by Hartwig and co-workers [758] (Scheme 6.44). A dtbpy-ligated iridium triboryl complex was synthesized by the reaction of an iridium precursor with borane. This iridium triboryl complex has also been observed in the catalytic reaction. The iridium triboryl complex reacts with benzene under mild condition to afford aryl boron ester in high yield. Mechanistic study suggests that this C–H cleavage event directly occurs with this iridium triboryl complex without prior dissociation of pinacol diboron. In a catalytic system, the C–H cleavage is the rate-limiting step, which could occur through either an oxidative addition–reductive elimination sequence or a sigma bond metathesis mechanism.

6.9.4 Silylation

Intermolecular C–H silylation was reported as early as 1982. Curtis and co-workers [759] reported a Vaska-type iridium complex-catalyzed silylation of benzene with hydridosiloxanes to give phenylsiloxanes, albeit in low yield. Tanaka and co-workers [760] have also reported a related Vaska-type rhodium complex-catalyzed dehydrocoupling of benzene and triethylsilane under photolytic conditions, which

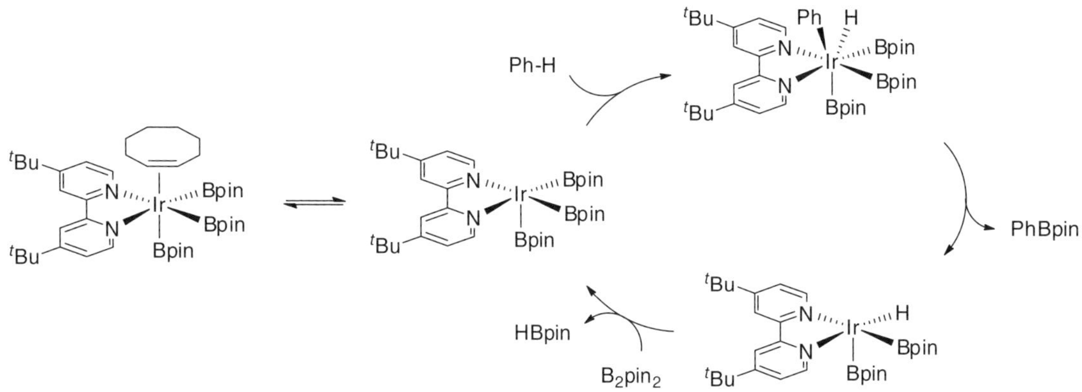

Scheme 6.44. Mechanism of Ir-catalyzed arene borylation.

also occurred in low yield. The yield was improved by using benzodisilacyclobutene [761] and *o*-bis(dimethylsilyl)benzene [762] as reagents for nickel- and platinum-catalyzed arene silylation [Eq. (6.296)]. Tsukada and Hartwig [763] reported a platinum-catalyzed reaction of aryl and alkyl C–H bonds with silane to form a new C–Si product [Eq. (6.297)]. To make the dehydrocoupling between arene and hydrosilane more thermodynamically favorable, a hydrogen acceptor such as *tert*-butylethylene was added to the Cp*Rh-catalyzed arene silylation reaction and significantly increased the yield [Eq. (6.298)] [764]. Silylation of arene with a disilane reagent also provides favorable thermodynamics. Miyaura and co-workers developed an iridium-catalyzed arene [Eq. (6.299)] [765, 766] and heteroarene [767] silylation using fluorodisilane. In most cases of arene C–H silylation, steric effects dominate the regioselectivity, as observed in arene C–H borylation reactions. The regioselectivity can be overrided by a proper directing group. Pyridine, pyrrazole, imine, and alkylamine have been employed as directing groups in Ru- or Rh-catalyzed arene silylation with silane or disilane [Eq. (6.300)] [768–772]. Pyridine-directed benzylic C–H silylation was also achieved using a Ru catalyst [Eq. (6.301)] [773]. Methane can be converted to methylsilane in the presence of a scandocene complex with diphenylsilane as the silylation reagent [Eq. (6.302)] [774].

Recently, Simmons and Hartwig [775] reported an Ir-catalyzed intramolecular aryl C–H silylation of (hydrido)silyl ether which was generated *in situ* by the reaction of aldehyde, ketone or alcohol with diethylsilane [Eq. (6.303)]. In this reaction, a hydroxyl group formally serves as the directing group for subsequent dehydrogenative cyclization. The silylation products easily underwent oxidation or cross-coupling reactions. The strategy was further extended to intramolecular alkyl C–H silylation directed by an alcohol or ketone [Eq. (6.304)]. The new C–Si bond obtained in these reactions can be oxidized to afford a useful alcohol group [776].

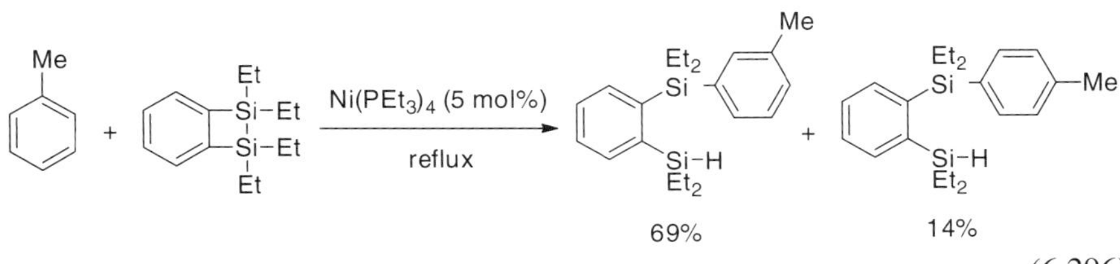

(6.296)

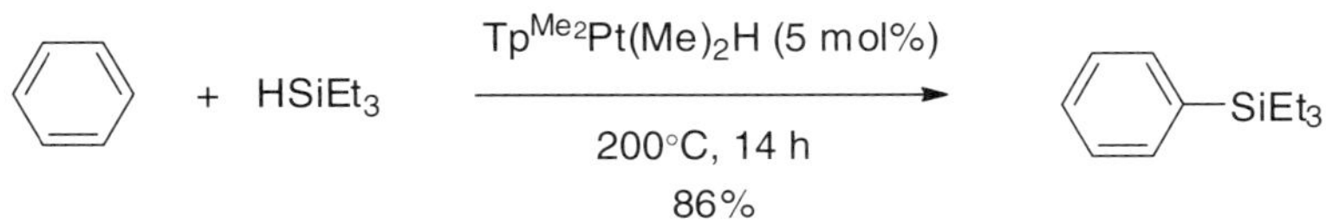

(6.297)

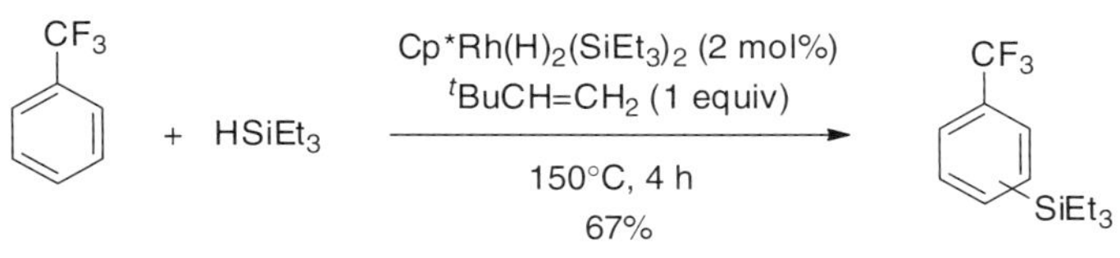

(6.298)

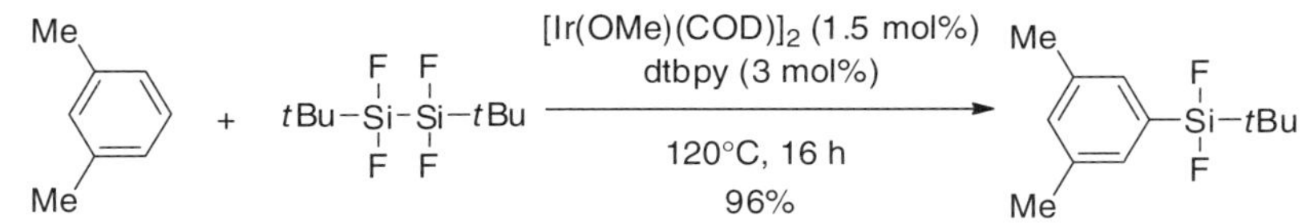

(6.299)

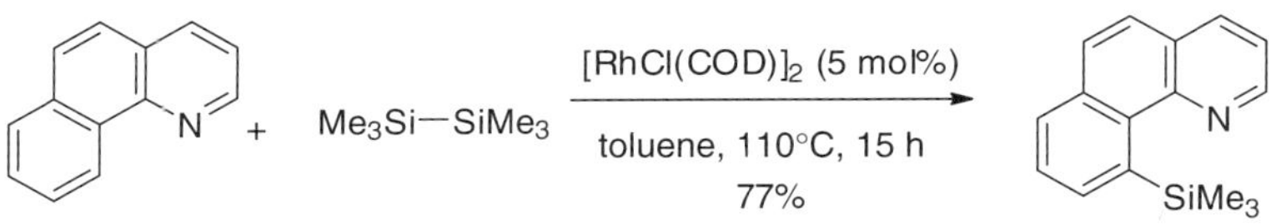

(6.300)

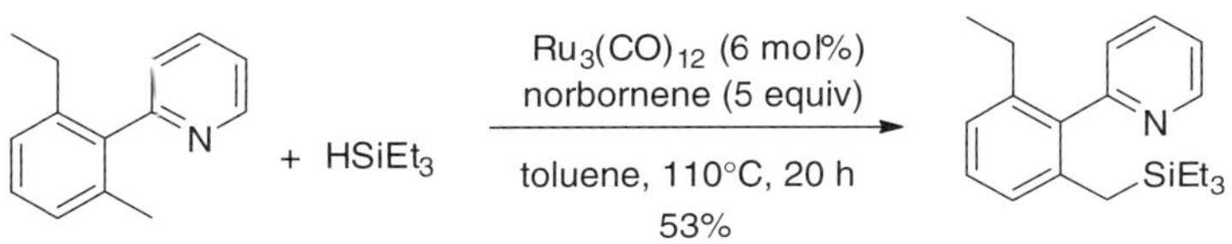

(6.301)

$$\mathrm{Ph_2SiH_2 + CH_4\ (150\ atm) \xrightarrow[\substack{C_6H_{12},\ 80^\circ C,\ 7\ d \\ TON\,=\,4}]{Cp^*ScMe\ (10\ mol\%)} Ph_2Si(Me)H + H_2} \qquad (6.302)$$

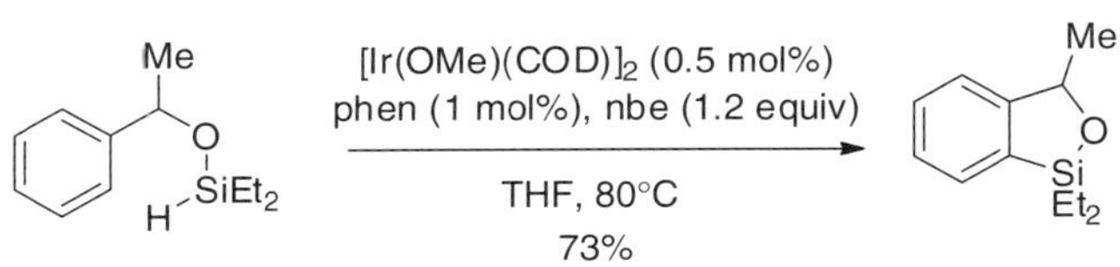

(6.303)

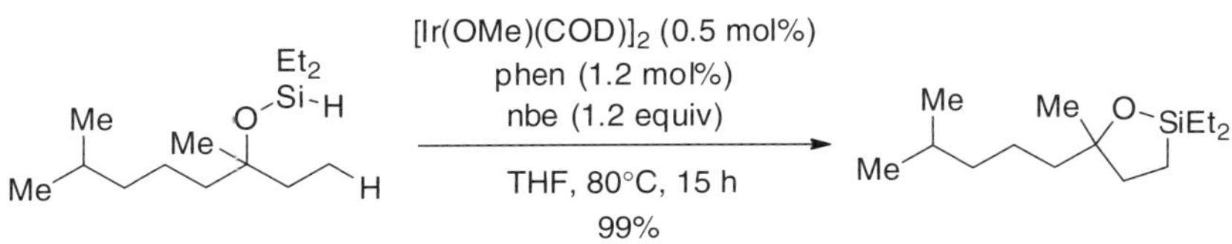

(6.304)

6.10 SUMMARY AND OUTLOOK

Transition-metal-catalyzed C–H functionalization is a rapidly evolving research field. Since C–H bonds are ubiquitous in nature and organic molecules, direct bond construction formation via C–H functionalization will minimize the pre-activation of substrates, alleviate the waste production, and shorten the synthetic steps. In a strategic level, it could provide novel and more efficient synthetic planning of a target molecule. During the past decades, impressive progress has been made in this area. In this chapter, we mainly focus on C–H functionalization through the intermediacy of C–M bond formation. Several types of reactions are discussed here, but certainly many more topics are also of great significance. One could anticipate that transition-metal-catalyzed C–H functionalization reactions will continue to be a hot research topic. The discovery of new reactions, study of the underlying mechanism, and application in organic synthesis will continue to impact the field of C–H functionalization. It will provide the chemists with powerful tools to solve the challenges in synthetic organic chemistry.

REFERENCES

1. (a) Jia, C., Kitamura, T., Fujiwara, Y. (2001). *Acc. Chem. Res.* **34**, 633. (b) Miura, M., Nomura, M. (2002). *Top. Curr. Chem.* **219**, 211. (c) Ritleng, V., Sirlin, C., Pfeffer, M. (2002). *Chem. Rev.* **102**, 1731. (d) Kakiuchi, F., Chatani, N. (2003). *Adv. Synth. Catal.* **345**, 1077. (e) Godula, K., Sames, D. (2006). *Science* **312**, 67. (f) Seregin, I. V., Gevorgyan, V. (2007). *Chem. Soc. Rev.* **36**, 1173. (g) Scott, M. E., Lautens, M. (2007). *Chem. Rev.* **107**, 174. (h) Chen, X., Engle, K.M., Wang, D.H., Yu, J.-Q. (2009). *Angew. Chem. Int. Ed.* **48**, 5094. (i) Daugulis, O., Do, H.-Q., Shabashov, D. (2009). *Acc. Chem. Res.* **42**, 1074. (j) Ackermann, L., Vicente, R., Kapdi, A.R. (2009). *Angew. Chem. Int. Ed.* **48**, 9792. (k) Li, C.-J. (2009). *Acc. Chem. Res.* **42**, 335. (l) Colby, D. A., Bergman, R. G., Ellman, J. A. (2010). *Chem. Rev.* **110**, 624. (m) Lyons, T. W., Sanford, M. S. (2010). *Chem. Rev.* **110**, 1147. (n) Sun, C.-L., Li, B.-J., Shi, Z.-J. (2010). *Chem. Commun.* **46**, 677. (o) Jazzar, R., Julien, H., Renaudat, A., Sofack-Kreutzer, J., Baudoin, O. (2010). *Chem. Eur. J.* **16**, 2654. (p) Mkhalid, I. A. I., Barnard, J. H., Marder, T. B., Murphy, J. M., Hartwig, J. F. (2010). *Chem. Rev.* **110**, 890. (q) Gutekunst, W. R., Baran, P. S. (2011). *Chem. Soc. Rev.* **40**, 1976. (r) Wencel-Delord, J., Dröge, T., Liu, F., Glorius, F. (2011). *Chem. Soc. Rev.* **40**, 4740. (s) Yeung, C. S., Dong, V. M. (2011). *Chem. Rev.* **111**, 1215. (t) Cho, S. H., Kim, J. Y., Kwak, J., Chang, S., (2011). *Chem. Soc. Rev.* **40**, 5068. (u) Yamaguchi, J., Yamaguchi, A. D., Itami, K. (2012). *Angew. Che. Int. Ed.* **51**, 8960. (v) *Handbook of C–H Transformations* Dyker G., ed. (2005). Wiley-VCH, Weinheim. (w) Goldman, A. S., Goldberg, K. I., eds. (2004). *Activation and Functionalization of C–H Bonds*, ACS Symposium Series 885, American Chemical Society, Washington, DC. (x) Hartwig, J. F. (2009). Catalytic C–H Functionalization, *in Organotransition Metal Chemistry: From Bonding to Catalysis*, University Science Books, South Orange, NJ.
2. Bennett, M. A., Milner, D. L. (1967). *Chem. Commun.* 581.
3. Karsch, H. H., Klein, H.-F., Schmidbaur, H. (1975). *Angew. Chem. Int. Ed.* **14**, 637.
4. Rathke, J. W., Muetterties, E. L. (1975). *J. Am. Chem. Soc.* **97**, 3272.
5. Harris, T. V., Rathke , J. W., Muetterties, E. L. (1978). *J. Am. Chem. Soc.* **100**, 6966.
6. Werner, H., Werner, R., (1981). *J. Organomet. Chem.* **209**, C060.
7. Desrosiers, P. J., Shinomoto, R. S., Flood, T. C. (1986). *J. Am. Chem. Soc.* **108**, 1346.
8. Hartwig, J. F., Andersen, R. A., Bergman, R. G. (1991). *J. Am. Chem. Soc.* **113**, 6492.
9. Tulip, T. H., Thorn, D. L. (1981). *J. Am. Chem. Soc.* **103**, 2448.
10. Foley, P., DiCosimo, R., Whitesides, G. M. (1980). *J. Am. Chem. Soc.* **102**, 6713.
11. Ibers, J. A., DiCosimo, R., Whitesides, G. M. (1982). *Organometallics* **1**, 13.
12. Chatt, J., Davidson, J. M. (1965). *J. Chem. Soc.*, 843.
13. Green, M. L., Knowles, P. J., (1970). *J. Chem. Soc. D*, 1677.

14. Shilov, A. E., Shteinman, A. A., (1977). Coord. *Chem. Rev.* **24**, 97.
15. Crabtree, R. H., Mihelcic, J. M., Quirk, J. M. (1979). *J. Am. Chem. Soc.* **101**, 7738.
16. Baudry, D., Ephritikhine, M., Felkin, H. (1980). *J. Chem. Soc., Chem. Commun,* 1243.
17. Brookhart, M., Green, M. L. H. (1983). *J. Organomet. Chem.* **250**, 395.
18. Janowicz, A. H., Bergman, R. G. (1982). *J. Am. Chem. Soc.* **104**, 352.
19. Hoyano, J. K., Graham, W. A. G. (1982). *J. Am. Chem. Soc.* **104**, 3723.
20. Jones , W. D., Feher, F. J. (1983). *Organometallics* **2**, 562.
21. Bergman, R. G. (1984). *Science* **223**, 902.
22. Baker, M. V., Field, L. D. (1987). *J. Am. Chem. Soc.* **109**, 2825.
23. Jones, W. D., Feher, F. J. (1984). *J. Am. Chem. Soc.* **106**, 1650.
24. Flood, T. C., Janak, K. E., Iimura, M., Zhen, H. (2000). *J. Am. Chem. Soc.* **122**, 6783.
25. Bergman, R. G., Seidler, P. F., Wenzel, T. T. (1985). *J. Am. Chem. Soc.* **107**, 4358.
26. Wenzel, T. T., Bergman, R. G. (1986). *J. Am. Chem. Soc.* **108**, 4856.
27. Janowicz, A. H., Bergman, R. G. (1983). *J. Am. Chem. Soc.* **105**, 3929.
28. Buchanan, J. M., Stryker, J. M., Bergman, R. G. (1986). *J. Am. Chem. Soc.* **108**, 1537.
29. Hackett, M., Whitesides, G. M. (1988). *J. Am. Chem. Soc.* **110**, 1449.
30. Hall, C., Perutz, R. (1996). *Chem. Rev.* **96**, 3125.
31. Crabtree, R. H., Hamilton, D. G. (1988). *Adv. Organomet. Chem.* **28**, 299.
32. Brookhart, M., Green, M. L. H. (1983). *J. Organomet. Chem.* **250**, 395.
33. Braga, D., Grepioni, F., Biradha, K., Desiraju, G. R. (1996). *J. Chem. Soc., Dalton Trans*, 3925.
34. Gross, C. L., Girolami, G. S. (1998). *J. Am. Chem. Soc.* **120**, 6605.
35. Geftakis, S., Ball, G. E. (1998). *J. Am. Chem. Soc.* **120**, 9953.
36. Bernskoetter, W. H., Schauer, C. K., Goldberg, K. I., Brookhart, M. (2009). *Science* **326**, 553.
37. Bromberg, S. E., Yang, H., Asplund, M. C., Lian, T., McNamara, B. K., Kotz, K. T., Yeston, J. S., Wilkens, M., Frei, H., Bergman, R. G., Harris, C. B. (1997). *Science* **278**, 260.
38. Chin, R. M., Dong, L., Duckett, S. B., Partridge, M. G., Jones, W. D., Perutz, R. N. (1993). *J. Am. Chem. Soc.* **115**, 7685.
39. Reinartz, S., White, P. S., Brookhart, M., Templeton, J. L. (2001). *J. Am. Chem. Soc.* **123**, 12724.
40. Volhard, J., (1892). *Liebigs Ann. Chem.* **267**, 172.
41. Kimroth, O., (1902). *Ber. Dtsch. Chem. Ges.* **35**, 2853.
42. Kharasch, M. S., Isbell, H. S., (1931). *J. Am. Chem. Soc.* **53**, 3053.
43. Shilov, A. E., Shul'pin, G. B. (2000). *Activation and Catalytic Reactions of Saturated Hydrocarbons in the Presence of Metal Complexes*, Kluwer Academic Publishers, Dordrecht.
44. Olah, G. A., Yu, S. H., Parker, D. G. (1976). *J. Org. Chem.* **41**, 1983.
45. Kleiman, J. P., Dubeck, M., (1963). *J. Am. Chem. Soc.* **85**, 1544.
46. Cope, A. C., Siekman, R. W. (1965). *J. Am. Chem. Soc.* **87**, 3272.
47. Ryabov, A. D. (1985). *Synthesis* 233.

48. Constable, A. G., McDonald, W. S., Sawkins, L. C., Shaw, B. L. (1978). *J. Chem. Soc., Chem. Commun.* 1061.
49. Stahl, S. S., Labinger, J. A., Bercaw, J. E. (1998). *Angew. Chem. Int. Ed.* **37**, 2180.
50. Thompson, M. E., Baxter, S. M., Bulls, A. R., Burger, B. J., Nolan, M. C., Santarsiero, B. D., Schaefer, W. P., Bercaw, J. E. (1987). *J. Am. Chem. Soc.* **109**, 203.
51. Rothwell, I. P., (1988). *Acc. Chem. Res.* **21**, 153.
52. Bruno, J. W., Marks, T. J., Day, V. W., (1982). *J. Am. Chem. Soc.* **104**, 7357.
53. Watson, P. L., (1983). *J. Am. Chem. Soc.* **105**, 6491.
54. Fendrick, C. M., Marks, T. J. (1984). *J. Am. Chem. Soc.* **106**, 2214.
55. Ziegler, T., Folga, E., Berces, A. (1993). *J. Am. Chem. Soc.* **115**, 636.
56. Rabaa , H., Saillard, J. Y., Hoffmann, R. (1986). *J. Am. Chem. Soc.* **108**, 4327.
57. Schaller, C. P., Bonanno , J. B., Wolczanski, P. T. (1994). *J. Am. Chem. Soc.* **116**, 4133.
58. Schaller, C, P., Cummins, C. C., Wolczanski, P. T. (1996). *J. Am. Chem. Soc.* **118**, 591.
59. Hoyt, H. M., Michael, F. E., Bergman, R. G. (2004). *J. Am. Chem. Soc.* **126**, 1018.
60. Cummins, C. C., Schaller, C. P., Van Duyne, G. D., Wolczanski, P. T., Chan , A. W. E., Hoffmann, R. (1991). *J. Am. Chem. Soc.* **113**, 2985.
61. Cundari, T. R., Klinckman, T. R., Wolczanski, P. T. (2002). *J. Am. Chem. Soc.* **124**, 1481.
62. Schaller, C. P., Wolczanski, P. T. (1993). *Inorg. Chem.* **32**, 131.
63. Pamplin, C. B., Legzdins, P. (2003). *Acc. Chem. Res.* **36**, 223.
64. Davies, H. M. L., Beckwith, R. E. J. (2003). *Chem. Rev.* **103**, 2861.
65. Doyle, M. P., Duffy, R., Ratnikov, M., Zhou, L. (2010). *Chem. Rev.* **110**, 704.
66. Walsh, P. J., Hollander, F. J., Bergman, R. G., (1988). *J. Am. Chem. Soc.* **110**, 8729.
67. Cummins, C. C., Baxter, S. M., Wolczanski, P. T. (1988). *J. Am. Chem. Soc.* **110**, 8731.
68. Schafer, D. F. II, Wolczanski, P. T. (1998). *J. Am. Chem. Soc.* **120**, 4881.
69. Bennett, J. L., Wolczanski, P. T. (1997). *J. Am. Chem. Soc.* **119**, 10696.
70. Pamplin, C. B., Legzdins, P. (2003). *Acc. Chem. Res.* **36**, 223.
71. Adams, C. S., Legzdins, P., Tran, E. (2001). *J. Am. Chem. Soc.* **123**, 612.
72. Tsang, J. Y. K., Buschhaus, M. S. A., Graham, P. M., Semiao, C. J., Semproni, S. P., Kim, S. J., Legzdins, P. (2008). *J. Am. Chem. Soc.* **130**, 3652.
73. Ng, S. H. K., Adams, C. S., Hayton, T. W., Legzdins, P., Patrick, B. O. (2003). *J. Am. Chem. Soc.* **125**, 15210.
74. Bailey, B. C., Fan, H., Huffman, J. C., Baik, M.-H., Mindiola, D. J. (2007). *J. Am. Chem. Soc.* **129**, 8781.
75. Cavaliere, V. N., Crestani, M. G., Pinter, B., Pink, M., Chen, C.-H., Baik, M.-H., Mindiola, D. J. (2011). *J. Am. Chem. Soc.* **133**, 10700.
76. Sherry, A. E., Wayland, B. B. (1990). *J. Am. Chem. Soc.* **112**, 1259.
77. Wayland, B. B., Ba, S., Sherry, A. E. (1991). *J. Am. Chem. Soc.* **113**, 5305.
78. Davidson, J. M., Triggs, C. (1966). *Chem. Ind.* 457.
79. Tisue, T., Downs, W. J. (1969). *J. Chem. Soc. D* 410a.
80. Yoneyama, T., Crabtree, R. H. (1996). *J. Mol. Catal. A: Chem.* **108**, 35.
81. Fahey, D. R. (1971). *J. Organomet. Chem.* **27**, 283.
82. Dangel, B. D., Johnson, J. A., Sames D. (2001). *J. Am. Chem. Soc.* **123**, 8149.
83. Dick, A. R., Hull, K. L., Sanford, M. S. (2004). *J. Am. Chem. Soc.* **126**, 2300.

84. Kalyani, D., Sanford, M. S. (2005). *Org. Lett.* **7**, 4149.
85. Desai, L. V., Malik, H. A., Sanford, M. S. (2006). *Org. Lett.* **8**, 1141.
86. Reddy, B. V. S. L., Reddy, R., Corey, E. J. (2006). *Org. Lett.* **8**, 3391.
87. Mo, F., Trzepkowski, L. J., Dong, G. (2012). *Angew. Chem. Int. Ed.* **51**, 13075.
88. Shan, G., Yang, X., Ma, L., Rao, Y. (2012). *Angew. Chem. Int. Ed.* **51**, 13070
89. Wang, G., Yuan, T., Wu, X.-L. (2008). *J. Org. Chem.* **73**, 4717.
90. Giri, R., Liang, J., Lei, J.-G., Li, J.-J., Wang, D.-H., Chen, X., Naggar, I. C., Guo, C., Foxman, B. M., Yu, J.-Q. (2005). *Angew. Chem. Int. Ed.* **44**, 7420.
91. Desai, L. V., Malik, H. A., Sanford, M. S. (2006). *Org. Lett.* **8**, 1141.
92. Wang, G.-W., Yuan, T.-T., (2010). *J. Org. Chem.* **75**, 476.
93. Zhang, S.-Y., He, G., Zhao, Y., Wright, K., Nack, W. A., Chen, G. (2012). *J. Am. Chem. Soc.* **134**, 7313.
94. Zhang, Y.-H., Yu, J.-Q. (2009). *J. Am. Chem. Soc.* **131**, 14654.
95. Yang, Y., Lin, Y., Rao, Y. (2012). *Org. Lett.* **14**, 2874.
96. Thirunavukkarasu, V. S., Hubrich, J., Ackermann, L. (2012). *Org. Lett.* **14**, 4210.
97. Thirunavukkarasu, V. S., Ackermann, L. (2012). *Org. Lett.* **14**, 6206.
98. Chen, X., Hao, X., Goodhue, C. E., Yu, J.-Q. (2006). *J. Am. Chem. Soc.* **128**, 6790.
99. Ye, Z., Wang, W., Luo, F., Zhang, S., Cheng, J. (2009). *Org. Lett.* **11**, 3974.
100. Wang, X., Lu, Y., Dai, H.-X., Yu, J.-Q. (2010). *J. Am. Chem. Soc.* **132**, 12203.
101. Cheng, X.-F., Li, Y., Su, Y.-M., Yin, F., Wang, J.-Y., Sheng, J., Vora, H. U., Wang, X.-S., Yu, J.-Q., (2013). *J. Am. Chem. Soc.* **135**, 1236.
102. Yang, M., Jiang, X., Shi, W.-J., Zhu, Q.-L., Shi, Z.-J. (2013). *Org. Lett.* **15**, 690.
103. Zhao, X., Dimitrijević, E., Dong, V. M. (2009). *J. Am. Chem. Soc.* **131**, 3466.
104. Tsang, W. C. P., Zheng, N., Buchwald, S. L. (2005). *J. Am. Chem. Soc.* **127**, 14560.
105. Thu, H.-Y., Yu, W.-Y., Che, C.-M. (2006). *J. Am. Chem. Soc.* **128**, 9048.
106. Jordan-Hore, J. A., Johansson, C. C. C., Gulias, M., Beck, E. M., Gaunt, M. J. (2008). *J. Am. Chem. Soc.* **130**, 16184.
107. Cho, S. H., Yoon, J., Chang, S. (2011). *J. Am. Chem. Soc.* **133**, 5996.
108. Brasche, G., Buchwald, S. L. (2008). *Angew. Chem. Int. Ed.* **47**, 1932.
109. Ueda, S., Nagasawa, H. (2008). *Angew. Chem. Int. Ed.* **47**, 6411.
110. Inamoto, K., Saito, T., Katsuno, M., Sakamoto, T., Hiroya, K. (2007). *Org. Lett.* **9**, 2931.
111. Mei, T.-S., Wang, X., Yu, J.-Q. (2009). *J. Am. Chem. Soc.* **131**, 10806.
112. Neumann, J., Rakshit, S., Droge, T., Glorius, F. (2009) *Angew. Chem., Int. Ed.* **48**, 6892.
113. Nadres, E. T., Daugulis, O. (2012). *J. Am. Chem. Soc.* **134**, 7.
114. He, G., Zhao, Y., Zhang, S., Lu, C., Chen, G. (2012). *J. Am. Chem. Soc.* **134**, 3.
115. Monguchi, D., Fujiwara, T., Furukawa, H., Mori, A. (2009). *Org. Lett.* **11**, 1607.
116. Cho, S. H., Kim, J. Y., Lee, S. Y., Chang, S. (2009). *Angew. Chem., Int. Ed.* **48**, 9127.
117. Kawano, T., Hirano, K., Satoh, T., Miura, M. (2010). *J. Am. Chem. Soc.* **132**, 6900.
118. Liu, X.-Y., Gao, P., Shen, Y.-W., Liang, Y.-M. (2011). *Org. Lett.* **13**, 4196.
119. Xiao, B., Gong, T.-J., Xu, J., Liu, Z.-J., Liu, L. (2011). J. Am.Chem. Soc. **133**, 1466.
120. Ng, K.-H., Chan, A. S. C., Yu, W.-Y., (2010). *J. Am. Chem. Soc.* **132**, 12862.
121. Sun, K., Li, Y., Xiong, T., Zhang, J., Zhang, Q. (2011). *J. Am. Chem. Soc.* **133**, 1694.

122. Yoo, E. J., Ma, S., Mei, T.-S., Chan, K. S. L., Yu, J.-Q. (2011). *J. Am. Chem. Soc.* **133**, 7652.
123. Grohmann, C., Wang, H., Glorius, F. (2012). *Org. Lett.* **14**, 656.
124. Wang, W., Pan, C., Chen, F., Cheng, J. (2011). *Chem. Commun.* **47**, 3978.
125. Schröder, N., Wencel-Delord, J., Glorius, F. (2012). *J. Am. Chem. Soc.* **134**, 8298.
126. Dick, A. R., Hull, K. L., Sanford, M. S. (2004). *J. Am. Chem. Soc.* **126**, 2300.
127. Kalyani, D., Dick, A. R., Anani, W. Q., Sanford, M. S. (2006). *Org. Lett.* **8**, 2523.
128. Kakiuchi, F., Kochi, T., Mutsutani, H., Kobayashi, N., Urano, S., Sato, M., Nishiyama, S., Tanabe, T. (2009). *J. Am. Chem. Soc.* **131**, 11310.
129. Wan, X., Ma, Z., Li, B., Zhang, K., Cao, S., Zhang, S., Shi, Z. (2006). *J. Am. Chem. Soc.* **128**, 7416.
130. Bedford, R. B., Haddow, M. F., Mitchell, C. J., Webster, R. L. (2011). *Angew. Chem. Int. Ed.* **50**, 5524.
131. Li, J.-J., Mei, T.-S., Yu, J.-Q. (2008). *Angew. Chem. Int. Ed.* **47**, 6452.
132. Mei, T.-S., Giri, R., Maugel, N., Yu, J.-Q. (2008). *Angew. Chem. Int. Ed.* **47**, 5215.
133. Mei, T.-S., Wang, D.-H., Yu, J.-Q., (2010). *Org. Lett.* **12**, 3140.
134. Giri, R., Chen, X., Yu, J.-Q. (2005). *Angew. Chem. Int. Ed.* **44**, 2112.
135. Wang, D.-H., Hao, X.-S., Wu, D.-F., Yu, J.-Q. (2006). *Org. Lett.* **8**, 3387.
136. Hull, K. L., Anani, W. Q., Sanford, M. S. (2006). *J. Am. Chem. Soc.* **128**, 7134.
137. McMurtrey , K. B., Racowski , J. M., Sanford, M. S. (2012). *Org. Lett.* **14**, 4094.
138. Wang, X., Mei, T.-S., Yu, J.-Q. (2009). *J. Am. Chem. Soc.* **131**, 7520.
139. Chan, K. S. L., Wasa, M., Wang, X., Yu, J.-Q. (2011) *Angew. Chem. Int. Ed.* **50**, 9081.
140. Kodama, H., Katsuhira, T., Nishida, T., Hino, T., Tsubata, K. Patent WO2001083421A1, 2001.
141. Heumann, A., Åkermark, B. (1984). *Angew. Chem. Int. Ed. Engl.* **23**, 453.
142. McMurry, J. E., Kočotovsky, P. (1984). *Tetrahedron Lett.* **25**, 4187.
143. Jia, C., Müller, P., Mimoun, H. (1995). *J. Mol. Catal. A: Chem.* **101**, 127.
144. El Firdoussi, L., Baqqa, A., Allaoud, S., Allal, B. A., Karim, A., Castanet, Y., Mortreux, A. (1998). *J. Mol. Catal. A: Chem.* **135**, 11.
145. Chen, M. S., White, M. C. (2004). *J. Am. Chem. Soc.* **126**, 1346.
146. Chen, M. S., Prabagaran, N., Labenz, N. A., White, M. C. (2005). *J. Am. Chem. Soc.* **127**, 6970.
147. Fraunhoffer, K. J., Prabagaran, N., Sirois, L. E., White, M. C. (2006). *J. Am. Chem. Soc.* **128**, 9032.
148. Covell, D. J., White, M. C. (2008). *Angew. Chem. Int. Ed.* **47**, 6448.
149. Mitsudome, T., Umetani, T., Nosaka, N., Mori, K., Mizugaki, T., Ebitani, K., Kaneda, K. (2006). *Angew. Chem. Int. Ed.* **45**, 481.
150. Campbell, A. N., White, P. B., Guzei, I. A., Stahl, S. S. (2010). *J. Am. Chem. Soc.* **132**, 15116.
151. Fraunhoffer, K. J., White, M. C. (2007). *J. Am. Chem. Soc.* **129**, 7274.
152. Liu, G., Yin, G., Wu, L. (2008). *Angew. Chem. Int. Ed.* **47**, 4733.
153. Reed, S. A., White, M. C. (2008). *J. Am. Chem. Soc.* **130**, 3316.
154. Rice, G. T., White, M. C. (2009). *J. Am. Chem. Soc.* **131**, 11707.

155. Reed, S. A., Mazzotti, A. R., White, M. C. (2009). *J. Am. Chem. Soc.* **131**, 11701.
156. Grennberg, H., Bäckvall, J.-E. (1998). *Chem.-Eur. J.* **4**, 108.
157. Yin, G., Wu, Y., Liu, G. (2010). *J. Am. Chem. Soc.* **132**, 11978.
158. Trost, B. M., Fullerton, T. J. (1973). *J. Am. Chem. Soc.* **95**, 292.
159. Trost, B. M., Metzner, P. J. (1980). *J. Am. Chem. Soc.* **102**, 3572.
160. Stang, E. M., White, M. C. (2009). *Nat. Chem.* **1**, 547.
161. Dick, A. R., Kampf, J., Sanford, M. S. (2005). *J. Am. Chem. Soc.* **127**, 12790.
162. Racowski, J. M., Dick, A. R., Sanford, M. S. (2009). *J. Am. Chem. Soc.* **131**, 10974.
163. Whitfield, S. R., Sanford, M. S. (2007). *J. Am. Chem. Soc.* **129**, 15142.
164. Arnold, P. L., Sanford, M. S., Pearson, S. M. (2009). *J. Am. Chem. Soc.* **131**, 13912.
165. Ball, N. D., Sanford, M. S. (2009). *J. Am. Chem. Soc.* **131**, 3796.
166. Furuya, T., Ritter, T. (2008). *J. Am. Chem. Soc.* **130**, 10060.
167. Furuya, T., Benitez, D., Tkatchouk, E., Strom, A. E., Tang, P., Goddard, W. A., III, Ritter, T., (2010). *J. Am. Chem. Soc.* **132**, 3793.
168. Powers, D. C., Ritter, T. (2009). *Nat. Chem.* **1**, 302.
169. Powers, D. C., Xiao, D. Y., Geibel, M. A. L., Ritter, T. (2010). *J. Am. Chem. Soc.* **132**, 14530.
170. Powers, D. C., Geibel, M. A. L., Klein, J. E. M. N., Ritter, T. (2009). *J. Am. Chem. Soc.* **131**, 17050.
171. Powers, D. C., Benitez, D., Tkatchouk, E., Goddard, W. A., III, Ritter, T. (2010). *J. Am. Chem. Soc.* **132**, 14092.
172. Powers, D. C., Lee, E., Ariafard, A., Sanford, M. S., Yates, B. F., Canty, A. J., Ritter, T. (2012). *J. Am. Chem. Soc.* **134**, 12002.
173. Murai, S., Kakiuchi, F., Sekine, S., Tanaka, Y., Kamatani, A., Sonoda, M., Chatani, N. (1993). *Nature* **366**, 529.
174. Kakiuchi, F., Sekine, S., Tanaka, Y., Kamatani, A., Sonoda, M., Chatani, N., Murai, S. (1995). *Bull. Chem. Soc. Jpn.* **68**, 62.
175. Sonoda, M., Kakiuchi, F., Chatani, N., Murai, S. (1997). *Bull. Chem. Soc. Jpn.* **70**, 3117.
176. Murai, S., Chatani, N., Kakiuchi, F. (1997). *Pure Appl. Chem.* **69**, 589.
177. Sonoda, M., Kakiuchi, F., Kamatani, A., Chatani, N., Murai, S. (1996). *Chem. Lett.* 109.
178. Kakiuchi, F., Yamauchi, M., Chatani, N., Murai, S. (1996). *Chem. Lett.* 111.
179. Kakiuchi, F., Sonoda, M., Tsujimoto, T., Chatani, N., Murai, S. (1999). *Chem. Lett.* 1083.
180. Martinez, R., Simon, M.-O., Chevalier, R., Pautigny, C., Genet, J.-P., Darses, S. (2009). *J. Am. Chem. Soc.* **131**, 7887.
181. Kakiuchi, F., Kochi, T., Mizushima, E., Murai, S. (2010). *J. Am. Chem. Soc.* **132**, 17741.
182. Lim, Y.-G., Kim, Y. H., Kang, J.-B. (1994). *J. Chem. Soc., Chem. Commun.* 2267.
183. Lim, Y.-G., Kang, J.-B., Kim, Y. H. (1996). *J. Chem. Soc., Perkin Trans.* **1**, 2201.
184. Jun, C.-H., Hong, J.-B., Kim, Y.-H., Chung, K.-Y., (2000). *Angew. Chem. Int. Ed.* **39**, 3440.
185. Jun, C.-H., Moon, C. W., Hong, J.-B., Lim, S.-G., Chung, K.-Y., Kim, Y.-H. (2002). *Chem. Eur. J.* 485.
186. Thalji, R. K., Ahrendt, K. A., Bergman, R. G., Ellman, J. A. (2001). *J. Am. Chem. Soc.* **123**, 9692.

187. Thalji, R. K., Ahrendt, K. A., Bergman, R. G., Ellman, J. A. (2005). *J. Org. Chem.* **70**, 6775.
188. Tan, K. L., Bergman, R. G., Ellman, J. A. (2001). *J. Am. Chem. Soc.* **123**, 2685.
189. Thalji, R. K., Ellman, J. A., Bergman, R. G. (2004). *J. Am. Chem. Soc.* **126**, 7192.
190. Tan, K. L., Bergman, R. G., Ellman, J. A. (2002). *J. Am. Chem. Soc.* **124**, 13964.
191. Tan, K. L., Park, S., Ellman, J. A., Bergman, R. G. (2004). *J. Org. Chem.* **69**, 7329.
192. Lewis, J. C., Bergman, R. G., Ellman, J. A. (2007). *J. Am. Chem. Soc.* **129**, 5332.
193. Ryu, J., Cho, S. H., Chang, S. (2012). *Angew. Chem. Int. Ed.* **51**, 3677.
194. Pan, S., Ryu, N., Shibata, T. (2012). *J. Am. Chem. Soc.* **134**, 17474.
195. Sevov, C. S., Hartwig, J. F. (2013). *J. Am. Chem. Soc.* **135**, 2116.
196. Nakao, Y., Yamada, Y., Kashihara, N., Hiyama, T. (2010). *J. Am. Chem. Soc.* **132**, 13666.
197. Gao, K., Yoshikai, N. (2011). *J. Am. Chem. Soc.* **133**, 400.
198. Gao, K., Yoshikai, N. (2011). *Angew. Chem. Int. Ed.* **50**, 6888.
199. Lee, P.-S., Yoshikai, N. (2013). *Angew. Chem. Int. Ed.* **52**, 1240.
200. Malley, S. J. O, Tan, K. L., Watzke, A., Bergman, R. G., Ellman, J. A. (2005). *J. Am. Chem. Soc.* **127**, 13496.
201. Tsai, A. S., Bergman, R. G., Ellman, J. A. (2008). *J. Am. Chem. Soc.* **130**, 6316.
202. Wiedemann, S. H., Ellman, J. A., Bergman, R. G. (2006). *J. Org. Chem.* **71**, 1969.
203. Rech, J. C., Yato, M., Duckett, D., Ember, B., LoGrasso, P. V., Bergman, R. G., Ellman, J. A. (2007). *J. Am. Chem. Soc.* **129**, 490.
204. Kakiuchi, F., Ohtaki, H., Sonoda, M., Chatani, N., Murai, S. (2001). *Chem. Lett.* 918.
205. Matsubara, T., Koga, N., Musaev, D. G., Morokuma, K. (1998). *J. Am. Chem. Soc.* **120**, 12692.
206. Matsubara, T., Koga, N., Musaev, D. G., Morokuma, K. (2000). *Organometallics* **19**, 2318.
207. Tan, K. L., Bergman, R. G., Ellman, J. A. (2002). *J. Am. Chem. Soc.* **124**, 3202.
208. Wiedemann, S. H., Lewis, J. C., Ellman, J. A., Bergman, R. G. (2006). *J. Am. Chem. Soc.* **128**, 2452.
209. Matsumoto, T., Taube, D. J., Periana, R. A., Taube, H., Yoshida, H. (2000). *J. Am. Chem. Soc.* **122**, 7414.
210. Matsumoto, T., Periana, R. A., Taube, D. J., Yoshida, H. (2002). *J. Mol. Catal. A.: Chem.* **180**, 1.
211. Periana, R. A., Liu, Y. X., Bhalla, G. (2002). *Chem. Commun.*, 3000.
212. Oxgaard, J., Muller, R. P., Periana, R. A., Goddard, W. A., III. (2004). *J. Am. Chem. Soc.* **126**, 352.
213. Lail, M., Arrowood, B. N., Gunnoe, T. B. (2003). *J. Am. Chem. Soc.* **125**, 7506.
214. Oxgaard, J., Goddard, W. A., III. (2004). *J. Am. Chem. Soc.* **126**, 442.
215. Karshtedt, D., Bell, A. T., Tilley, T. D. (2004). *Organometallics* **23**, 4169.
216. McKeown, B. A., Foley, N. A., Lee, J. P., Gunnoe, T. B. (2008). *Organometallics* **27**, 4031.
217. Foley, N. A., Lee, J. P., Ke, Z., Gunnoe, T. B., Cundari, T. R. (2009). *Acc. Chem. Res.* **42**, 585.
218. Oxgaard, J., Muller, R. P., Periana, R. A., Goddard, W. A., III. (2004). *J. Am. Chem. Soc.* **126**, 352.

219. Oxgaard, J., Goddard, W. A., III. (2004). *J. Am. Chem. Soc.* **126**, 442.
220. McKeown, B. A., Gonzalez, H. E., Friedfeld, M. R., Gunnoe, T. B., Cundari, T. R., Sabat, M. (2011). *J. Am. Chem. Soc.* **133**, 19131.
221. Jia, C., Piao, D., Oyamada, J., Lu, W., Kitamura, T., Fujiwara, Y. (2000). *Science* **287**, 1992.
222. Jia, C., Lu, W., Oyamada, J., Kitamura, T., Matsuda, K., Irie, M., Fujiwara, Y. (2000). *J. Am. Chem. Soc.* **122**, 7252.
223. Nevado, C., Echavarren, A. M. (2005). *Chem.-Eur. J.* **11**, 3155.
224. Pastine, S. J., Youn, S. W., Sames, D. (2003). *Org. Lett.* **5**, 1055.
225. Reetz, M. T., Sommer, K. (2003). *Eur. J. Org. Chem.* 3485.
226. Shi, Z., He, C. (2004). *J. Org. Chem.* **69**, 3669.
227. Nevado, C., Echavarren, A. M. (2005). *Synthesis* 167.
228. Tunge, J. A., Foresee, L. N. (2005). *Organometallics* **24**, 6440.
229. Chernyak, N., Gevorgyan, V. (2008). *J. Am. Chem. Soc.* **130**, 5636.
230. Jiang, T.-S., Tang, R.-Y., Zhang, X.-G., Li, X.-H., Li, J.-H. (2009). *J. Org. Chem.* **74**, 8834.
231. Tokunaga, Y., Sakakura, T., Tanaka, M. (1989). *J. Mol. Catal.* **56**, 305.
232. Boese, W. T., Goldman, A. S. (1991). *Organometallics* **10**, 782.
233. Hong, P., Cho, B.-R., Yamazaki, H. (1979). *Chem. Lett.* 339.
234. Hong, P., Cho, B.-R., Yamazaki, H., (1980). *Chem. Lett.* 507.
235. Tsukada, N., Mitsuboshi, T., Setoguchi, H., Inoue, Y. (2003). *J. Am. Chem. Soc.* **125**, 12102
236. Kakiuchi, F., Yamamoto, Y., Chatani, N., Murai, S. (1995). *Chem. Lett.*, 681.
237. Harris, P. W. R., Richard, C. E. F., Woodgate, P. D. (1999). *J. Organomet. Chem.* **589**, 168.
238. Aulwurm, U. R., Melchinger, J. U., Kisch, H. (1995). *Organometallics* **14**, 3385.
239. Lim, Y.-G., Lee, K.-H., Koo, B. T., Kang, J.-B. (2001). *Tetrahedron Lett.* **42**, 7609.
240. Halbritter, G., Knoch, F., Wolski, A., Kisch, H. (1994). *Angew. Chem. Int. Ed. Engl.* **33**, 1603.
241. Lee, P.-S., Fujita, T., Yoshikai, N. (2011). *J. Am. Chem. Soc.* **133**, 17283.
242. Ding, Z., Yoshikai, N. (2012). *Angew. Chem. Int. Ed.* **51**, 4698.
243. Zhou, B., Chen, H., Wang, C. (2013). *J. Am. Chem. Soc.* **135**, 1264.
244. Kuninobu, Y., Tokunaga, Y., Kawata, A., Takai, K. (2006). *J. Am. Chem. Soc.* **128**, 202.
245. Schipper, D. J., Hutchinson, M., Fagnou, K. (2010). *J. Am. Chem. Soc.* **132**, 6910.
246. Hashimoto, Y., Hirano, K., Satoh, T., Kakiuchi, F., Miura, M. (2013). *J. Org. Chem.* **78**, 638.
247. Moritani, I., Fujiwara, Y. (1967). *Tetrahedron Lett.* **8**, 1119.
248. Fujiwara, Y., Moritani, I., Matsuda, M., Teranishi, S. (1968). *Tetrahedron Lett.* **9**, 3863.
249. Fujiwara, Y., Moritani, I., Danno, S., Asano, R., Teranishi, S. (1969). *J. Am. Chem. Soc.* **91**, 7166.
250. Tsuji, J., Nagashima, H. (1984). *Tetrahedron* **40**, 2699.
251. Jia, C., Lu, W., Kitamura, T., Fujiwara, Y. (1999). *Org. Lett.* **1**, 2097.
252. Shue, R.-S. (1971). *J. Chem. Soc., Chem. Commun.* 1510.

253. Yokota, T., Tani, M., Sakaguchi, S., Ishii, Y. (2003). *J. Am. Chem. Soc.* **125**, 1476.
254. Dams, M., De Vos, D. E., Celen, S., Jacobs, P. A. (2003). *Angew. Chem. Int. Ed.* **42**, 3512.
255. Weissman, H., Song, X., Milstein, D. (2001). *J. Am. Chem. Soc.* **123**, 337.
256. Hong, P., Yamazaki, H. (1979). *Chem. Lett.* 1335.
257. Matsumoto, T., Yoshida, H. (2000). *Chem. Lett.* 1064.
258. Matsumoto, T., Periana, R. A., Taube, D. J., Yoshida, H. (2002). *J. Catal.* **206**, 272.
259. Miura, M., Tsuda, T., Satoh, T., Nomura, M. (1997). *Chem. Lett.* 1103.
260. Miura, M., Tsuda, T., Satoh, T., Pivsa-Art, S., Nomura, M. (1998). *J. Org. Chem.* **63**, 5211.
261. Boele, M. D. K., van Strijdonck, G. P. F., de Vries, A. H. M., Kamer, P. C. J., de Vries, J. G., van Leeuwen, P. W. N. M. (2002). *J. Am. Chem. Soc.* **124**, 1586.
262. Nishikata, T., Lipshutz, B. H. (2010). *Org. Lett.* **12**, 1972.
263. Cai, G., Fu, Y., Li, Y., Wan, X., Shi, Z. (2007). *J. Am. Chem. Soc.* **129**, 7666.
264. Li, J.-J., Mei, T.-S., Yu, J.-Q. (2008). *Angew. Chem. Int. Ed.* **47**, 6452.
265. Wang, D.-H., Engle, K. M., Shi, B.-F., Yu, J.-Q. (2010). *Science* **327**, 315.
266. Engle, K. M., Wang, D.-H., Yu, J.-Q. (2010). *Angew. Chem. Int. Ed.* **49**, 6169.
267. Shi, B.-F., Zhang, Y.-H., Lam, J. K., Wang, D.-H., Yu, J.-Q. (2010). *J. Am. Chem. Soc.* **132**, 460.
268. Ueura, K., Satoh, T., Miura, M. (2007). *Org. Lett.* **9**, 1407.
269. Umeda, N., Hirano, K., Satoh, T., Miura, M. (2009). *J. Org. Chem.* **74**, 7094.
270. Wang, F., Song, G., Li, X. (2010). *Org. Lett.* **12**, 5430.
271. Patureau, F. W., Glorius, F. (2010). *J. Am. Chem. Soc.* **132**, 9982.
272. Park, S. H., Kim, J. Y., Chang, S. (2011). *Org. Lett.* **13**, 2372.
273. Patureau, F. W., Besset, T., Glorius, F. (2011). *Angew. Chem. Int. Ed.* **50**, 1064.
274. Ackermann, L., Pospech, J. (2011). *Org. Lett.* **13**, 4153.
275. Ackermann, L., Wang, L., Wolfram, R., Lygin, A. V. (2012). *Org. Lett.* **14**, 728.
276. Ueyama, T., Mochida, S., Fukutani, T., Hirano, K., Satoh, T., Miura, M. (2011). *Org. Lett.* **13**, 706.
277. Li, B., Ma, J., Wang, N., Feng, H., Xu, S., Wang, B. (2012). *Org. Lett.* **14**, 736.
278. Wasa, M., Engle, K. M., Yu, J.-Q. (2010). *J. Am. Chem. Soc.* **132**, 3680.
279. Stowers, K. J., Fortner, K. C., Sanford, M. S. (2011). *J. Am. Chem. Soc.* **133**, 6541.
280. Yokoyama, Y., Matsumoto, T., Murakami, Y. (1995). *J. Org. Chem.* **60**, 1486.
281. Grimster, N. P., Godfrey, C. D., Gaunt, M. J. (2005). *Angew. Chem. Int. Ed.* **44**, 3125.
282. Beck, E. M., Grimster, N. P., Hatley, R., Gaunt, M. J. (2006). *J. Am. Chem. Soc.* **128**, 2528.
283. Fujiwara, Y., Maruyama, O., Yoshidomi, M., Taniguchi, H. (1981). *J. Org. Chem.* **46**, 851.
284. Maehara, A., Satoh, T., Miura, M. (2008). *Tetrahedron* **64**, 5982.
285. Aouf, C., Thiery, E., Bras, J. L., Muzart, J. (2009). *Org. Lett.* **11**, 4096.
286. Miyasaka, M., Hirano, K., Satoh, T., Miura, M. (2010). *J. Org. Chem.* **75**, 5421.
287. Cui, S., Wojtas, L., Antilla, J. C. (2011). *Org. Lett.* **13**, 5040.
288. Iida, H., Yuasa, Y., Kibayashi, C. (1980). *J. Org. Chem.* **45**, 2938.
289. Zhang, H., Ferreira, E. M., Stoltz, B. M. (2004). *Angew. Chem. Int. Ed.* **43**, 6144.

290. Ferreira, E. M., Stoltz, B. M. (2003). *J. Am. Chem. Soc.* **125**, 9578.
291. Kong, A., Han, X., Lu, X. (2006). *Org. Lett.* **8**, 1339.
292. Schiffner, J. A., Machotta, A. B., Oestreich, M. (2008). *Synlett*, 2271.
293. Cho, S. H., Hwang, S. J., Chang, S. (2008). *J. Am. Chem. Soc.* **130**, 9254.
294. Wu, J., Cui, X., Chen, L., Jiang, G., Wu, Y. (2009). *J. Am. Chem. Soc.* **131**, 13888.
295. Zhang, Y.-H., Shi, B.-F., Yu, J.-Q. (2009). *J. Am. Chem. Soc.* **131**, 5072.
296. Zhang, X., Fan, S., He, C.-Y., Wan, X., Min, Q.-Q., Yang, J., Jiang, Z.-X. (2010). *J. Am. Chem. Soc.* **132**, 4506.
297. Baran, P. S., Corey, E. J. (2002). *J. Am. Chem. Soc.* **124**, 7904.
298. Baran, P. S., Guerrero, C. A., Corey, E. J. (2003). *J. Am. Chem. Soc.* **125**, 5628.
299. Garg, N. K., Caspi, D. D., Stoltz, B. M. (2004). *J. Am. Chem. Soc.* **126**, 9552.
300. Beck, E. M., Hatley, R., Gaunt, M. J. (2008). *Angew. Chem. Int. Ed.* **47**, 3004.
301. Wang, D.-H., Yu, J.-Q. (2011). *J. Am. Chem. Soc.* **133**, 5767.
302. Kuninobu, Y., Tokunaga, Y., Kawata, A., Takai K. (2006). *J. Am. Chem. Soc.* **128**, 202.
303. Sun, Z.-M., Chen, S.-P., Zhao, P. (2010). *Chem.-Eur. J.* **16**, 2619.
304. Tran, D. N., Cramer, Nb. (2011). *Angew. Chem. Int. Ed.* **50**, 11098.
305. Muralirajan, K., Parthasarathy, K., Cheng, C.-H. (2011). *Angew. Chem. Int. Ed.* **50**, 4169.
306. Patureau, F. W., Besset, T., Kuhl, N., Glorius, F. (2011). *J. Am. Chem. Soc.* **133**, 2154.
307. Li, B.-J., Wang, H.-Y., Q.-L. Zhu; Shi, Z.-J. (2012). *Angew. Chem. Int. Ed.* **51**, 3948.
308. Satoh, T., Miura, M. (2010). *Chem. Eur. J.* **16**, 11212.
309. Song, G., Wang, F., Li, X., (2012). *Chem. Soc. Rev.* **41**, 3651.
310. Houlden, C. E., Bailey, C. D., Ford, J. G., Gagné, M. R., Lloyd-Jones, G. C., Booker-Milburn, K. I. (2008). *J. Am. Chem. Soc.* **130**, 10066.
311. Stuart, D. R., Bertrand-Laperle, M., Burgess, K. M. N., Fagnou, K. (2008). *J. Am. Chem. Soc.* **130**, 16474.
312. Stuart, D. R., Alsabeh, P., Kuhn, M., Fagnou, K. (2010). *J. Am. Chem. Soc.*, **132**, 18326.
313. Li, L., Brennessel, W. W., Jones, W. D. (2008). *J. Am. Chem. Soc.* **130**, 12414.
314. Guimond, N., Fagnou, K. (2009). *J. Am. Chem. Soc.* **131**, 12050.
315. Ueura, K., Satoh, T., Miura, M. (2007). *J. Org. Chem.* **72**, 5362.
316. Ueura, K., Satoh, T., Miura, M. (2007). *Org. Lett.* **9**, 1407.
317. Ackermann, L., Pospech, J., Graczyk, K., Rauch, K. (2012). *Org. Lett.* **14**, 930.
318. Mochida, S., Umeda, N., Hirano, K., Satoh, T., Miura, M. (2010). *Chem. Lett.* **39**, 744.
319. Hyster, T. K., Rovis, T. (2010). *J. Am. Chem. Soc.* **132**, 10565.
320. Song, G., Chen, D., Pan, C.-L., Crabtree, R. H., Li, X. (2010). *J. Org. Chem.* **75**, 7487.
321. Ackermann, L., Lygin, A. V., Hofmann, N. (2011). *Angew. Chem. Int. Ed.* **50**, 6379.
322. Zhong, H., Yang, D., Wang, S., Huang, J. (2012). *Chem. Commun.* **48**, 3236.
323. Shiota, H., Ano, Y., Aihara, Y., Fukumoto, Y., Chatani, N., (2011). *J. Am. Chem. Soc.* **133**, 14952.
324. Pham, M. V., Ye, B., Cramer, N. (2012). *Angew. Chem. Int. Ed.* **51**, 10610.
325. Guimond, N., Gouliaras, C., Fagnou, K. (2010). *J. Am. Chem. Soc.* **132**, 6908.
326. Guimond, N., Gorelsky, S. I., Fagnou, K. (2011). *J. Am. Chem. Soc.* **133**, 6449.
327. Rakshit, S., Grohmann, C., Besset, T., Glorius, F. (2011). *J. Am. Chem. Soc.* **133**, 2350.

328. Wang, H., Grohmann, C., Nimphius, C., Glorius, F. (2012). *J. Am. Chem. Soc.* **134**, 19592.
329. Ackermann, L., Fenner, S. (2011). *Org. Lett.* **13**, 6548.
330. Chidipudi, S. R., Khan, I., Lam, H. W. (2012). *Angew. Chem. Int. Ed.* **51**, 12115.
331. Tan, X., Liu, B., Li, X., Li, B., Xu, S., Song, H., Wang, B. (2012). *J. Am. Chem. Soc.* **134**, 16163.
332. Hassan, J., Sévignon, M., Gozzi, C., Schulz, E., Lemaire, M. (2002). *Chem. Rev.* **102**, 1359.
333. Diederich, F., Stang, P. J., eds. (1998). *Metal-Catalyzed Cross- Coupling Reactions*, Wiley-VCH, New York.
334. de Meijere, A., Diederich, F. (2004). *Metal-Catalyzed Cross-Coupling Reactions*, Wiley-VCH, Weinheim.
335. Ackermann, L. (2009). *Modern Arylation Methods*, Wiley-VCH, Weinheim.
336. Alberico, D., Scott, M. E., Lautens, M. (2007). *Chem. Rev.* **107**, 174.
337. Ackermann, L., Vicente, R., Kapdi, A. R. (2009). *Angew. Chem. Int. Ed.* **48**, 9792.
338. Oi, S., Fukita, S., Inoue, Y. (1998). *Chem. Commun.* 2439.
339. Ueura, K., Satoh, T., Miura, M. (2005). *Org. Lett.* **7**, 2229.
340. Miyamura, S., Tsurugi, H., Satoh, T., Miura, M. (2008). *J. Organomet. Chem.* **693**, 2438.
341. Vogler, T., Studer, A. (2008). *Org. Lett.* **10**, 129.
342. Kakiuchi, F., Kan, S., Igi, K., Chatani, N., Murai, S. (2003). *J. Am. Chem. Soc.* **125**, 1698.
343. Kakiuchi, F., Matsuura, Y., Kan, S., Chatani, N. (2005). *J. Am. Chem. Soc.* **127**, 5936.
344. Kitazawa, K., Kochi, T., Sato, M., Kakiuchi, F. (2009). *Org. Lett.* **11**, 1951.
345. Pastine, S. J., Gribkov, D. V., Sames, D. (2006). *J. Am. Chem. Soc.* **128**, 14220.
346. Chen, X., Li, J. J., Hao, X. S., Goodhue, C. E., Yu, J.-Q. (2006). *J. Am. Chem. Soc.* **128**, 78.
347. Chen, X., Goodhue, C. E., Yu, J.-Q. (2006). *J. Am. Chem. Soc.* **128**, 12634.
348. Shi, Z., Li, B., Wan, X., Cheng, J., Fang, Z., Cao, B., Qin, C., Wang, Y. (2007). *Angew. Chem. Int. Ed.*, **46**, 5554.
349. Yang, S., Li, B., Wan, X., Shi, Z. (2007). *J. Am. Chem. Soc.* **129**, 6066.
350. Giri, R., Maugel, N., Li, J.-J., Wang, D.-H., Breazzano, S. P., Saunders, L. B., Yu, J.-Q. (2007). *J. Am. Chem. Soc.* **129**, 3510.
351. Wang, D.-H., Wasa, M., Giri, R., Yu, J.-Q. (2008). *J. Am. Chem. Soc.* **130**, 7190.
352. Chu, J.-H., Lin, P.-S., Wu, M.-J. (2010). *Organometallics* **29**, 4058.
353. Wang, D.-H., Mei, T.-S., Yu, J.-Q. (2008). *J. Am. Chem. Soc.* **130**, 17676.
354. Engle, K. M., Thuy-Boun, P. S., Dang, M., Yu, J.-Q. (2011). *J. Am. Chem. Soc.* **133**, 18183.
355. Sun, C.-L., Liu, N., Li, B.-J., Yu, D.-G., Wang, Y., Shi, Z.-J. (2010). *Org. Lett.* **12**, 184.
356. Tredwell, M. J., Gulias, M., Bremeyer, N. G., Johansson, C. C. C., Collins, B. S. L., Gaunt, M. J. (2011). *Angew. Chem., Int. Ed.* **50**, 1076.
357. Nishikata, T., Abela, A. R., Huang, S., Lipshutz, B. H. (2010). *J. Am. Chem. Soc.* **132**, 4978.
358. Shi, B.-F., Maugel, N., Zhang, Y.-H., Yu, J.-Q. (2008). *Angew. Chem., Int. Ed.* **47**, 4882.
359. Gao, D.-W., Shi, Y.-C., Gu, Q., Zhao, Z.-L., You, S.-L. (2013). *J. Am. Chem. Soc.* **135**, 86.

360. Wasa, M., Engle, K. M., Lin, D. W., Yoo, E. J., Yu, J.-Q. (2011). *J. Am. Chem. Soc.* **133**, 19598.
361. Yang, S. D., Sun, C. L., Fang, Z., Li, B. J., Li, Y.-Z., Shi, Z.-J. (2008). *Angew. Chem., Int. Ed.* **47**, 1473.
362. Kawai, H., Kobayashi, Y., Oi, S., Inoue, Y. (2008). *Chem. Commun.*, 1464.
363. Funaki, K., Kawai, H., Sato, T., Oi, S. (2011). *Chem. Lett.* **40**, 1050.
364. Ball, L. T., Lloyd-Jones, G. C., Russell, C. A. (2012) *Science.* **337**, 1644.
365. Mochida, K., Kawasumi, K., Segawa, Y., Itami, K. (2011). *J. Am. Chem. Soc.* **133**, 10716.
366. Zhang, Q., Kawasumi, K., Segawa, Y., Itami, K., Scott, L. T. (2012). *J. Am. Chem. Soc.* **134**, 15664.
367. Wei, Y., Kan, J., Wang, M., Su, W., Hong, M. (2009). *Org. Lett.* **11**, 3346.
368. Liang, Z., Yao, B., Zhang, Y. (2010). *Org. Lett.* **12**, 3185.
369. Liu, B., Qin, X., Li, K., Li, X., Guo, Q., Lan, J., You, J. (2010). *Chem. Eur. J.* **16**, 11836.
370. Kirchberg, S., Tani, S., Ueda, K., Yamaguchi, J., Studer, A., Itami, K. (2011). *Angew. Chem. Int. Ed.* **50**, 2387.
371. Schnapperelle, I., Breitenlechner, S., Bach, T. (2011). *Org. Lett.* **13**, 3640.
372. Yamaguchi, K., Yamaguchi, J., Studer, A., Itami, K. (2012). *Chem. Sci.* **3**, 2165.
373. Norinder, J., Matsumoto, A., Yoshikai, N., Nakamura, E. (2008). *J. Am. Chem. Soc.* **130**, 5858.
374. Yoshikai, N., Matsumoto, A., Norinder, J., Nakamura, E. (2009). *Angew. Chem. Int. Ed.*, **48**, 2925.
375. Li, B., Wu, Z.-H., Gu, Y.-F., Sun, C.-L., Wang, B.-Q., Shi, Z.-J. (2011). *Angew. Chem., Int. Ed.* **50**, 1109.
376. Yoshikai, N., Asako, S., Yamakawa, T., Ilies, L., Nakamura, E. (2011). *Chem. Asian J.* **6**, 3059.
377. Ilies, L., Asako, S., Nakamura, E. (2011). *J. Am. Chem. Soc.* **133**, 7672.
378. Hachiya, H., Hirano, K., Satoh, T., Miura, M. (2010). *Angew. Chem., Int. Ed.* **49**, 2202.
379. Tobisu, M., Hyodo, I., Chatani, N. (2009). *J. Am. Chem. Soc.* **131**, 12070.
380. Qu, G.-R., Xin, P.-Y., Niu, H.-Y., Wang, D.-C., Ding, R.-F., Guo, H.-M. (2011). *Chem. Commun.* **47**, 1140.
381. Åkermark, B., Eberson, L., Jonsson, E., Pettersson, E. (1975). *J. Org. Chem.* **40**, 1365.
382. Shiotani, A., Itatani, H. (1978). *Angew. Chem., Int. Ed. Engl.* **13**, 471.
383. Dwight, T. A., Rue, N. R., Charyk, D., Josselyn, R., DeBoef, B. (2007). *Org. Lett.* **9**, 3137.
384. Liegault, B., Lee, D., Huestis, M. P., Stuart, D. R., Fagnou, K. (2008). *J. Org. Chem.* **73**, 5022.
385. Watanabe, T., Oishi, S., Fujii, N., Ohno, H. (2009). *J. Org. Chem.* **74**, 4720.
386. Yeung, C. S., Borduas, N., Zhao, X., Dong, V. M. (2010). *Chem. Sci.* **1**, 331.
387. Morimoto, K., Itoh, M., Hirano, K., Satoh, T., Shibata, Y., Tanaka, K., Miura, M. (2012). *Angew. Chem. Int. Ed.* **51**, 5359.
388. Li, H., Zhu, R.-Y., Shi, W.-J., He, K.-H., Shi, Z.-J. (2012). *Org. Lett.* **14**, 4850.
389. Boger, D. L., Patel, M. (1988). *J. Org. Chem.* **53**, 1405.
390. Wang, J., Rosingana, M., Watson, D. J., Dowdy, E. D., Discordia, R. P., Soundarajan, N., Li, W.-S., (2001). *Tetrahedron Lett.* **42**, 8935.

391. Witulski, B., Schweikert, T., (2005). *Synthesis* 1959.

392. Miller, R. B., Moock, T. (1980). *Tetrahedron Lett.* **21**, 3319.

393. Forke, R., Jöger, A., Knölker, H.-J. (2008). *Org. Biomol. Chem.* **6**, 2481

394. Yoshimoto, H., Itatani, H., (1973). *Bull. Chem. Soc. Jpn.* **46**, 2490.

395. Takahashi, M., Masui, K., Sekiguchi, H., Kobayashi, N., Mori, A., Funahashi, M., Tamaoki, N. (2006). *J. Am. Chem. Soc.* **128**, 10930.

396. Li, Y., Wang, W.-H., Yang, S.-D., Li, B.-J., Feng, C., Shi, Z.-J. (2010). *Chem. Commun.* **46**, 4553.

397. Liang, Z., Zhao, J., Zhang, Y. (2010). *J. Org. Chem.* **75**, 170.

398. Hull, K. L., Lanni, E. L., Sanford, M. S. (2006). *J. Am. Chem. Soc.* **128**, 14047.

399. Chen, X., Dobereiner, G., Hao, X.-S., Giri, R., Maugel, N., Yu, J.-Q. (2009). *Tetrahedron* **65**, 3085.

400. Oi, S., Sato, H., Sugawara, S., Inoue, Y. (2008). *Org. Lett.* **10**, 1823.

401. Itahara, T., (1985). *J. Org. Chem.* **50**, 5272.

402. Li, R., Liang, L., Lu, W. (2006). *Organometallics* **25**, 5973.

403. Stuart, D. R., Fagnou, K. (2007). *Science* **316**, 1172.

404. Stuart, D. R., Villemure, E., Fagnou, K. (2007). *J. Am. Chem. Soc.* **129**, 12072.

405. Potavathri, S., Pereira, K. C., Gorelsky, S. I., Pike, A., LeBris, A. P., DeBoef, B. (2010). *J. Am. Chem. Soc.* **132**, 14676.

406. Xia, J.-B., You, S.-L. (2007). *Organometallics* **26**, 4869.

407. Hull, K. L., Sanford, M. S. (2007). *J. Am. Chem. Soc.* **129**, 11904.

408. Li, B.-J., Tian, S.-L., Fang, Z., Shi, Z.-J. (2008). *Angew. Chem., Int. Ed.* **47**, 1115.

409. Brasche, G., García-Fortanet, J., Buchwald, S. L. (2008). *Org. Lett.* **10**, 2207.

410. Yeung, C. S., Borduas, N., Zhao, X., Dong, V. M. (2010). *Chem. Sci.* **1**, 331.

411. Wang, X., Leow, D., Yu, J.-Q. (2011). *J. Am. Chem. Soc.* **133**, 13864.

412. Wencel-Delord, J., Nimphius, C., Patureau, F. W., Glorius, F. (2012). *Angew. Chem. Int. Ed.* **51**, 2247.

413. Wencel-Delord, J., Nimphius, C., Wang, H., Glorius, F. (2012). *Angew. Chem. Int. Ed.* **51**, 13001.

414. Zhao, X., Yeung, C. S., Dong, V. M. (2010). *J. Am. Chem. Soc.* **132**, 5837.

415. Cho, S. H., Hwang, S. J., Chang, S. (2008). *J. Am. Chem. Soc.* **130**, 9254.

416. He, C.-Y., Fan, S., Zhang, X. (2010). *J. Am. Chem. Soc.* **132**, 12850.

417. Wei, Y., Su, W. (2010). *J. Am. Chem. Soc.* **132**, 16377.

418. Li, H., Liu, J., Sun, C.-L., Li, B.-J., Shi, Z.-J. (2011). *Org. Lett.* **13**, 276.

419. Xi, P., Yang, F., Qin, S., Zhao, D., Lan, J., Gao, G., Hu, C., You, J. (2010). *J. Am. Chem. Soc.* **132**, 1822.

420. Wang, Z., Li, K., Zhao, D., Lan, J., You, J. (2011). *Angew. Chem., Int. Ed.* **50**, 5365.

421. Mandal, D., Yamaguchi, A. D., Yamaguchi, J., Itami, K. (2011). *J. Am. Chem. Soc.* **133**, 19660.

422. Gong, X., Song, G., Zhang, H., Li, X. (2011). *Org. Lett.* **13**, 1766.

423. Nishino, M., Hirano, K., Satoh, T., Miura, M. (2012). *Angew. Chem., Int. Ed.* **51**, 6993.

424. Han, W., Mayer, P., Ofial, A. R. (2011). *Angew. Chem., Int. Ed.* **50**, 2178.

425. Dong, J., Huang, Y., Qin, X., Cheng, Y., Hao, J., Wan, D., Li, W., Liu, X., You, J. (2012). *Chem. Eur. J.* **18**, 6158.
426. Mao, Z., Wang, Z., Xu, Z., Huang, F., Yu, Z., Wang, R. (2012). *Org. Lett.* **14**, 3854.
427. Fan, S., Chen, Z., Zhang, X. (2012). *Org. Lett.*, **14**, 4950.
428. Kuhl, N., Hopkinson, M. N., Glorius, F. (2012). *Angew. Chem., Int. Ed.* **51**, 8230.
429. Mandal, D., Yamaguchi, A. D., Yamaguchi, J., Itami, K. (2011). *J. Am. Chem. Soc.* **133**, 19660.
430. Ueda, K., Yanagisawa, S., Yamaguchi, J., Itami, K. (2010). *Angew. Chem. Int. Ed.* **49**, 8946.
431. Ames, D. E., Bull, D. (1982). *Tetrahedron* **38**, 383.
432. Ames, D. E., Opalko, A. (1983). *Synthesis* 234.
433. Ames, D. E., Opalko, A. (1984). *Tetrahedron* **40**, 1919.
434. Nakamura, N., Tajima, Y., Sakai, K. (1982). *Heterocycles* **19**, 235.
435. Akita, Y., Inoue, A., Yamamoto, K., Ohta, A., Kurihara, T., Shimizu, M. (1985). *Heterocycles* **23**, 2327.
436. Akita, Y., Itagaki, Y., Takizawa, S., Ohta, A. (1989). *Chem. Pharm. Bull.* **37**, 1477.
437. Satoh, T., Kawamura, Y., Miura, M., Nomura, M. (1997). *Angew. Chem. Int. Ed. Engl.* **36**, 1740.
438. Hennings, D. D., Iwasa, S., Rawal, V. H. (1997). *J. Org. Chem.* **62**, 2.
439. Iwaki, T., Yasuhara, A., Sakamoto, T. (1999). *J. Chem. Soc., Perkin Trans.* **1**, 1505.
440. Bedford, R. B., Cazin, C. S. J. (2002). *Chem. Commun.* 2310.
441. Huang, Q., Campo, M. A., Yao, T., Tian, Q., Larock, R. C. (2004). *J. Org. Chem.* **69**, 8251.
442. Nishioka, H., Shoujiguchi, Y., Abe, H., Takeuchi, Y., Harayama, T. (2004). *Heterocycles* **64**, 463.
443. Grigg, R., Sridharan, V., Stevenson, P., Sukirthalingam, S., Worakun, T. (1990). *Tetrahedron* **46**, 4003.
444. Desarbre, E., Mérour, J.-Y. (1995). *Heterocycles* **41**, 1987.
445. Jonckers, T. H. M., Maes, B. U. W., Lemière, G. L. F., Rombouts, G., Pieters, L., Haemers, A., Dommisse, R. A. (2003). *Synlett* **5**, 615.
446. Hodgetts, K. J., Kershaw, M. T. (2003). *Org. Lett.* **5**, 2911.
447. Brenner, M., Mayer, G., Terpin, A., Steglich, W. (1997). *Chem. Eur. J.* **3**, 70.
448. Gómez-Lor, B., Echavarren, A. M. (2004). *Org. Lett.* **6**, 2993.
449. Beccalli, E. M., Broggini, G., Martinelli, M., Paladino, G., Rossi, E. (2006). *Synthesis* 2404.
450. Campeau, L.-C., Parisien, M., Jean, A., Fagnou, K., (2006). *J. Am. Chem. Soc.* **128**, 581.
451. Campeau, L.-C., Thansandote, P., Fagnou, K. (2005). *Org. Lett.* **7**, 1857.
452. Leblanc, M., Fagnou, K. (2005). *Org. Lett.* **7**, 2849.
453. Ackermann, L., Althammer, A. (2007). *Angew. Chem. Int. Ed.* **46**, 1627.
454. Quimby, J. M., Scott, L. T. (2009). *Adv. Synth. Catal.* **351**, 1009.
455. Albicker, M. R., Cramer, N. (2009). *Angew. Chem. Int. Ed.* **48**, 9139.
456. Bringmann, G., Jansen, J. R., Rink, H.-P. (1986). *Angew. Chem. Int. Ed. Engl.* **25**, 913
457. Matsumoto, T., Hosoya, T., Suzuki, K. (1992). *J. Am. Chem. Soc.* **114**, 3568.

458. Rao, A. V. R., Chakraborty, T. K., Joshi, S. P. (1992). *Tetrahedron Lett.* **33**, 4045.
459. Molander, G. A., George, K. M., Monovich, L. G. (2003). *J. Org. Chem.* **68**, 9533.
460. Ohmori, K., Tamiya, M., Kitamura, M., Kato, H., Oorui, M., Suzuki, K. (2005). *Angew. Chem. Int. Ed.* **44**, 3871.
461. Abe, H., Fukumoto, T., Horino, Y., Harayama, T. (2010). *Heterocycles* **82**, 851.
462. Butler, J. R., Wang, C., Bian, J., Ready, J. M. (2011). *J. Am. Chem. Soc.* **133**, 9956.
463. Harayama, T., Akamatsu, H., Okamura, K., Miyagoe, T., Akiyama, T., Abe, H., Takeuchi, Y., (2001). *J. Chem. Soc. Perkin Trans.* **1**, 523.
464. Qabaja, G., Jones, G. B. (2000). *J. Org. Chem.* **65**, 7187.
465. Lafrance, M., Blaquière, N., Fagnou, K. (2004). *Chem. Commun.* 2874.
466. Bowie, A. L., Jr., Hughes, C. C., Trauner, D. (2005). *Org. Lett.* **7**, 5207.
467. Ren, H., Knochel, P. (2006). *Angew. Chem., Int. Ed.* **45**, 3462.
468. Dong, C.-G., Hu, Q.-G. (2006). *Angew. Chem., Int. Ed.* **45**, 2289.
469. Rousseaux, S., Gorelsky, S. I., Chung, B. K. W., Fagnou, K. (2010). *J. Am. Chem. Soc.* **132**, 10692.
470. Baudoin, O., Herrbach, A., Guéritte, F. (2003). *Angew. Chem., Int. Ed.* **42**, 5736.
471. Chaumontet, M., Piccardi, R., Audic, N., Hitce, J., Peglion, J.-L., Clot, E., Baudoin, O. (2008). *J. Am. Chem. Soc.* **130**, 15157.
472. Lafrance, M., Gorelsky, S. I., Fagnou, K. (2007). *J. Am. Chem. Soc.* **129**, 14570.
473. Watanabe, T., Oishi, S., Fujii, N., Ohno, H. (2008). *Org. Lett.* **10**, 1759.
474. Rousseaux, S., Davi, M., Sofack-Kreutzer, J., Pierre, C., Kefalidis, C. E., Clot, E., Fagnou, K., Baudoin, O. (2010). *J. Am. Chem. Soc.* **132**, 10706.
475. Nakanishi, M., Katayev, D., Besnard, C., Kündig, E. P. (2011). *Angew. Chem. Int. Ed.* **50**, 7438.
476. Katayev, D., Nakanishi, M., Bürgi, T., Kündig, E. P. (2012). *Chem. Sci.* **3**, 1422.
477. Saget, T., Lemouzy, S., Cramer, N. (2012). *Angew. Chem. Int. Ed.* **51**, 2238.
478. Saget, T., Cramer, N. (2012). *Angew. Chem. Int. Ed.* **51**, 12842.
479. Martin, N., Pierre, C., Davi, M., Jazzar, R., Baudoin, O. (2012). *Chem. Eur. J.* **18**, 4480.
480. Anas, S., Cordi, A., Kagan, H. B. (2011). *Chem. Commun.* **47**, 11483.
481. Chaumontet, M., Piccardi, R., Baudoin, O. (2009). *Angew. Chem. Int. Ed.* **48**, 179.
482. Lane, B. S., Brown, M. A., Sames, D. (2005). *J. Am. Chem. Soc.* **127**, 8050.
483. Lane, B. S., Sames, D. (2004). *Org. Lett.* **6**, 2897.
484. Lebrasseur, N., Larrosa, I. (2008). *J. Am. Chem. Soc.* **130**, 2926.
485. Zhang, Z., Hu, Z., Yu, Z., Lei, P., Chi, H., Wang, Y., He, R. (2007). *Tetrahedron Lett.* **48**, 2415.
486. Ackermann, L., Barfüsser, S. (2009). *Synlett*, 808.
487. Bellina, F., Benelli, F., Rossi, R. (2008). *J. Org. Chem.* **73**, 5529.
488. Deprez, N. R., Kalyani, D., Krause, A., Sanford, M. S., (2006). *J. Am. Chem. Soc.* **128**, 4972.
489. Phipps, R. J., Grimster, N. P., Gaunt, M. J. (2008). *J. Am. Chem. Soc.* **130**, 8172.
490. Rieth, R. D., Mankad, N. P., Calimano, E., Sadighi, J. P. (2004). *Org. Lett.* **6**, 3981.
491. Aoyagi, Y., Inoue, A., Koizumi, I., Hashimoto, R., Tokunaga, K., Gohma, K., Komatsu, J., Sekine, K., Miyafuji, A., Kunoh, J., Honma, R., Akita, Y., Ohta, A. (1992). *Heterocycles* **33**, 257.

492. Touré, B. B., Lane, B. S., Sames, D. (2006). *Org. Lett.* **8**, 1979.
493. Ohta, A., Akita, Y., Ohkuwa, T., Chiba, M., Fukunaga, R., Miyafuji, A., Nakata, T., Tani, N., Aoyagi, Y. (1990). *Heterocycles* **31**, 1951.
494. Basnak, I., Takatori, S., Walker, R. T. (1997). *Tetrahedron Lett.* **38**, 4869.
495. Pivsa-Art, S., Satoh, T., Kawamura, Y., Miura, M., Nomura, M. (1998). *Bull. Chem. Soc. Jpn.* **71**, 467.
496. McClure, M. S., Glover, B., McSorley, E., Millar, A., Osterhout, M. H., Roschangar, F. (2001). *Org. Lett.* **3**, 1677.
497. Gorelsky, S. I., Lapointe, D., Fagnou, K. (2008). *J. Am. Chem. Soc.* **130**, 10848.
498. Bellina, F., Cauteruccio, S., Rossi, R. (2006). *Eur. J. Org. Chem.* 1379.
499. Gallagher, W. P., Maleczka, R. E., Jr. (2003). *J. Org. Chem.* **68**, 6775.
500. Canivet, J., Yamaguchi, J., Ban, I., Itami, K. (2009). *Org. Lett.* **11**, 1733.
501. Hachiya, H., Hirano, K., Satoh, T., Miura, M. (2009). *Org. Lett.* **11**, 1737.
502. Do, H.-Q., Daugulis, O. (2007). *J. Am. Chem. Soc.* **129**, 12404.
503. Ackermann, L., Vicente, R., Born, R. (2008). *Adv. Synth. Catal.* **350**, 741.
504. Yamamoto, T., Muto, K., Komiyama, M., Canivet, J., Yamaguchi, J., Itami, K. (2011). *Chem. Eur. J.* **17**, 10113.
505. Ackermann, L., Althammer, A., Fenner, S. (2009). *Angew. Chem. Int. Ed.* **48**, 201.
506. Muto, K., Yamaguchi, J., Itami, K. (2012). *J. Am. Chem. Soc.* **134**, 169.
507. Parisien, M., Valette, D., Fagnou, K. (2005). *J. Org. Chem.* **70**, 7578.
508. Yokooji, A., Okazawa, T., Satoh, T., Miura, M., Nomura, M. (2003). *Tetrahedron* **59**, 5685.
509. Bellina, F., Cauteruccio, S., Mannina, L., Rossi, R., Viel, S. (2005). *J. Org. Chem.* **70**, 3997.
510. Bellina, F., Cauteruccio, S., Mannina, L., Rossi, R., Viel, S. (2006). *Eur. J. Org. Chem.* 693.
511. Lewis, J. C., Wiedemann, S. H., Bergman, R. G., Ellman, J. A. (2004). *Org. Lett.* **6**, 35.
512. Lewis, J. C., Berman, A. M., Bergman, R. G., Ellman, J. A. (2008). *J. Am. Chem. Soc.* **130**, 2493.
513. Ames, D. E., Opalko, A. (1984). *Tetrahedron* **40**, 1919.
514. Mukhopadhyay, S., Rothenberg, G., Gitis, D., Baidossi, M., Ponde, D. E., Sasson, Y. (2000). *J. Chem. Soc., Perkin Trans.* **2**, 1809.
515. Berman, A. M., Lewis, J. C., Bergman, R. G., Ellman, J. A. (2008). *J. Am. Chem. Soc.* **130**, 14926.
516. Ye, M., Gao, G.-L., Edmunds, A. J. F., Worthington, P. A., Morris, J. A., Yu, J.-Q. (2011). *J. Am. Chem. Soc.* **133**, 19090.
517. Guo, P., Joo, J. M., Rakshit, S., Sames, D. (2011). *J. Am. Chem. Soc.* **133**, 16338.
518. Campeau, L.-C., Rousseaux, S., Fagnou, K. (2005). *J. Am. Chem. Soc.* **127**, 18020.
519. Gauthier, D. R., Jr., Limanto, J., Devine, P. N., Desmond, R. A., Szumigala, R. H., Jr., Foster, B. S., Volante, R. P. (2005). *J. Org. Chem.* **70**, 5938.
520. Cameron, M., Foster, B. S., Lynch, J. E., Shi, Y.-J., Dolling, U.-H. (2006). *Org. Process Res. Dev.* **10**, 398
521. Kim, K., Kim, I. (2010). *Org. Lett.* **12**, 5314.
522. Gaulier, S. M., Mckay, R., Swain, N. A. (2011). *Tetrahedron Lett.* **52**, 6000.

523. Campeau, L., Stuart, D., Leclerc, J., Bertrand-Laperle, M., Villemure, E., Sun, H., Lasserre, S., Guimond, N., Lecavallier, M., Fagnou, K. (2009). *J. Am. Chem. Soc.* **131**, 3291.

524. Shibahara, F., Yamaguchi, E., Murai, T. (2011). *J. Org. Chem.* **76**, 2680.

525. Dyker, G., Borowski, S., Heiermann, J., Körning, J., Opwis, K., Henkel, G., Köckerling, M. (2000). *J. Organomet. Chem.* **606**, 108.

526. Miura, M., Pivsa-Art, S., Dyker, G., Heiermann, J., Satoh, T., Masakatsu, N. (1998). *Chem. Commun.* 1889.

527. Dyker, G., Heiermann, J., Miura, M. (2003). *Adv. Synth. Catal.* **345**, 1127.

528. Fujita, K.-I., Nonogawa, M., Yamaguchi, R. (2004). *Chem. Commun.* 1926.

529. Liu, W., Cao, H., Lei, A. (2010). *Angew. Chem. Int. Ed.* **49**, 2004.

530. Vallee, F., Mousseau, J. J., Charette, A. B. (2010). *J. Am. Chem. Soc.* **132**, 1514.

531. Li, H., Sun, C.-L., Yu, M., Yu, D.-G., Li, B.-J., Shi, Z.-J. (2011). *Chem. Eur. J.* **16**, 3593.

532. Liu, W., Cao, H., Xin, J., Jin, L., Lei, A. (2011). *Chem. Eur. J.* **17**, 3588.

533. Lafrance, M., Fagnou, K. (2006). *J. Am. Chem. Soc.* **128**, 16496.

534. Lafrance, M., Rowley, C. N., Woo, T. K., Fagnou, K. (2006). *J. Am. Chem. Soc.* **128**, 8754.

535. Lafrance, M., Shore, D., Fagnou, K. (2006). *Org. Lett.* **8**, 5097.

536. René, O., Fagnou, K. (2010). *Org. Lett.* **12**, 2116.

537. Do, H.-Q., Daugulis, O. (2008). *J. Am. Chem. Soc.* **130**, 1128.

538. Satoh, T., Inoh, J.-I., Kawamura, Y., Miura, M., Nomura, M. (1998). *Bull. Chem. Soc. Jpn.* **71**, 2239.

539. Terao, Y., Wakui, H., Satoh, T., Miura, M., Nomura, M. (2001). *J. Am. Chem. Soc.* **123**, 10407.

540. Terao, Y., Wakui, H., Nomoto, M., Satoh, T., Miura, M., Nomura, M. (2003). *J. Org. Chem.* **68**, 5236.

541. Gürbüz, N., Özdemir, I., Cetinkaya, B. (2005). *Tetrahedron Lett.* **46**, 2273.

542. Satoh, T., Kametani, Y., Terao, Y., Miura, M., Nomura, M. (1999). *Tetrahedron Lett.* **40**, 5345.

543. Satoh, T., Miura, M., Nomura, M. (2002). *J. Organomet. Chem.* **653**, 161.

544. Kametani, Y., Satoh, T., Miura, M., Nomura, M. (2000). *Tetrahedron Lett.* **41**, 2655.

545. Oi, S., Fukita, S., Hirata, N., Watanuki, N., Miyano, S., Inoue, Y. (2001). *Org. Lett.* **3**, 2579.

546. Oi, S., Ogino, Y., Fukita, S., Inoue, Y. (2002). *Org. Lett.* **4**, 1783.

547. Oi, S., Aizawa, E., Ogino, Y., Inoue, Y. (2005). *J. Org. Chem.* **70**, 3113.

548. Chiong, H. A., Pham, Q.-N., Daugulis, O. (2007). *J. Am. Chem. Soc.* **129**, 9879.

549. Ackermann, L. (2005). *Org. Lett.* **7**, 3123.

550. Ackermann, L., Vicente, R., Althammer, A. (2008). *Org. Lett.* **10**, 2299.

551. Ackermann, L., Althammer, A., Born, R. (2006). *Angew. Chem. Int. Ed.* **45**, 2619.

552. Kalyani, D., Deprez, N. R., Desai, L. V., Sanford, M. S. (2005). *J. Am. Chem. Soc.* **127**, 7330.

553. Daugulis, O., Zaitsev, V. G. (2005). *Angew. Chem., Int. Ed.* **44**, 4046.

554. Shabashov, D., Daugulis, O. (2005). *Org. Lett.* **7**, 3657.

555. Zaitsev, V. G., Shabashov, D., Daugulis, O. (2005). *J. Am. Chem. Soc.* **127**, 13154.
556. Lazareva, A., Daugulis, O. (2006). *Org. Lett.* **8**, 5211.
557. Shabashov, D., Daugulis, O. (2006). *Org. Lett.* **8**, 4947.
558. Vicente, J., Arcas, A., Juliá-Hernández, F., Bautista, D. (2011). *Angew. Chem. Int. Ed.* **50**, 6896.
559. Catellani, M., Motti, E., Della Ca, N., (2008). *Acc. Chem. Res.* **41**, 1512.
560. Catellani, M., Frignani, F., Rangoni, A. (1997). *Angew. Chem., Int. Ed. Engl.* **36**, 119.
561. Lautens, M., Piguel, S. (2000). *Angew. Chem., Int. Ed.* **39**, 1045.
562. Catellani, M., Motti, E., Baratta, S. (2001). *Org. Lett.* **3**, 3611.
563. Motti, E., Mignozzi, A., Catellani, M. (2003). *J. Mol. Catal. A: Chem.* **204**, 115.
564. Faccini, F., Motti, E., Catellani, M. (2004). *J. Am. Chem. Soc.* **126**, 78.
565. Ferraccioli, R., Carenzi, D., Rombola, O., Catellani, M. (2004). *Org. Lett.* **6**, 4759.
566. Mariampillai, B., Alliot, J., Li, M., Lautens, M. (2007). *J. Am. Chem. Soc.* **129**, 15372.
567. Tran, L. D., Daugulis, O. (2012). *Angew. Chem. Int. Ed.* **51**, 5188.
568. He, G., Chen, G. (2011). *Angew. Chem. Int. Ed.* **50**, 5192.
569. Wasa, M., Engle, K. M., Yu, J.-Q. (2009). *J. Am. Chem. Soc.* **131**, 9886.
570. Wasa, M., Chan, K. S. L., Zhang, X.-G., He, J., Miura, M., Yu, J.-Q. (2012). *J. Am. Chem. Soc.* **134**, 18570.
571. Ouellet, S. G., Roy, A., Molinaro, C., Angelaud, R., Marcoux, J.-F., Shea, P. D. O., Davies, I. W. (2011). *J. Org. Chem.* **76**, 1436.
572. Seki, M. (2011). *ACS Catal.* **1**, 607.
573. Seki, M., Nagahama, M. (2011). *J. Org. Chem.* **76**, 10198.
574. Blanchot, M., Candito, D. A., Larnaud, F., Lautens, M. (2011). *Org. Lett.* **13**, 1486.
575. Feng, Y., Chen, G. (2010). *Angew. Chem. Int. Ed.* **49**, 958.
576. Gutekunst, W. R., Baran, P. S. (2011). *J. Am. Chem. Soc.* **133**, 19076.
577. Fisher, B. J., Eisenberg, R. (1983). *Organometallics* **2**, 764.
578. Kunin, A. J., Eisenberg, R. (1986). *J. Am. Chem. Soc.* **108**, 535.
579. Kunin, A. J., Eisenberg, R. (1988). *Organometallics* **7**, 2124.
580. Boese, W. T., Goldman, A. S. (1992). *J. Am. Chem. Soc.* **114**, 350.
581. Jones, W. D., Foster, G. P., Putinas, J. M. (1987). *J. Am. Chem. Soc.* **109**, 5047.
582. Sakakura, T., Tanaka, M. J. (1987). *Chem. Soc. Chem. Commun.* 758.
583. Sakakura, T., Tanaka, M. (1987). *Chem. Lett.* 249.
584. Sakakura, T., Sasaki, K., Tokunaga, M., Wada, K., Tanaka, M. (1988). *Chem. Lett.* 155.
585. Sakakura, T., Sodeyama, T., Sasaki, K., Wada, K., Tanaka, M. (1990). *J. Am. Chem. Soc.* **112**, 7221.
586. Rosini, G. P., Boese, W. T., Goldman, A. S. (1994). *J. Am. Chem. Soc.* **116**, 9498.
587. Boyd, S. E., Field, L. D., Partridge, M. G. (1994). *J. Am. Chem. Soc.* **116**, 9492.
588. Rosini, G. P., Zhu, K., Goldman, A. S. (1995). *J. Organomet. Chem.* **504**, 115.
589. Hong, P., Yamazaki, H. (1979). *Chem. Lett.* 1335.
590. Moore, E. J., Pretzer, W. R., O'Connell, T. J., Harris, J., LaBounty, L., Chou, L., Grimmer, S. S. (1992). *J. Am. Chem. Soc.* **114**, 5888.
591. Chatani, N., Fukuyama, T., Kakiuchi, F., Murai, S. (1996). *J. Am. Chem. Soc.* **118**, 493.

592. Chatani, N., Ie, Y., Kakiuchi, F., Murai, S., (1997). *J. Org. Chem.* **62**, 2604,
593. Chatani, N., Yorimitsu, S., Asaumi, T., Kakiuchi, F., Murai, S. (2002). *J. Org. Chem.* **67**, 7557.
594. Fukuyama, T., Chatani, N., Kakiuchi, F., Murai, S. (1997). *J. Org. Chem.* **62**, 5647.
595. Ie, Y., Chatani, N., Ogo, T., Marshall, D. R., Fukuyama, T., Kakiuchi, F., Murai, S. (2000). *J. Org. Chem.* **65**, 1475.
596. Asaumi, T., Chatani, N., Matsuo, T., Kakiuchi, F., Murai, S. (2003). *J. Org. Chem.* **68**, 7538.
597. Fukuyama, T., Chatani, N., Tatsumi, J., Kakiuchi, F., Murai, S. (1998). *J. Am. Chem. Soc.* **120**, 11522.
598. Chatani, N., Asaumi, T., Ikeda, T., Yorimitsu, S., Ishii, Y., Kakiuchi, F., Murai, S. (2000). *J. Am. Chem. Soc.* **122**, 12882.
599. Fujiwara, Y., Kawanchi, T., Taniguchi, H. (1980). *J. Chem. Soc., Chem. Commun.*, 220.
600. Jintoku, T., Taniguchi, H., Fujiwara, Y. (1987). *Chem. Lett.* 1159.
601. Taniguchi, Y., Yamaoka, Y., Nakata, K., Takaki, K., Fujiwara, Y. (1995). *Chem. Lett.* 345.
602. Lu, W., Yamaoka, Y., Taniguchi, Y., Kitamura, T., Fujiwara, Y. (1999). *J. Organomet. Chem.* **580**, 290.
603. Jintoku, T., Fujiwara, Y., Kawata, I., Taniguchi, H. (1990). *J. Organomet. Chem.* **385**, 297.
604. Orito, K., Horibata, A., Nakamura, T., Ushito, H., Nagasaki, H., Yuguchi, M., Yamashita, S., Tokuda, M. (2004). *J. Am. Chem. Soc.* **126**, 14342.
605. Houlden, C. E., Hutchby, M., Bailey, C. D., Ford, J. G., Tyler, S. N. G., Gagne, M. R., Lloyd-Jones, G. C., Booker-Milburn, K. I. (2009). *Angew. Chem., Int. Ed.* **48**, 1830.
606. Wrigglesworth, J. W., Cox, B., Lloyd-Jones, G. C., Booker-Milburn, K. I. (2011). *Org. Lett.* **13**, 5326.
607. Giri, R., Yu, J. K., Lam, J.-Q. (2010). *J. Am. Chem. Soc.* **132**, 686.
608. Giri, R., Yu, J.-Q. (2008). *J. Am. Chem. Soc.* **130**, 14082.
609. Lu, Y., Leow, D., Wang, X., Engle, K. M., Yu, J.-Q. (2011). *Chem. Sci.* **2**, 967.
610. Li, H., Cai, G.-X., Shi, Z.-J. (2010). *Dalton Trans.* **39**, 10442.
611. López, B., Rodriguez, A., Santos, D., Albert, J., Ariza, X., Garcia, J., Granell, J. (2011). *Chem. Commun.* **47**, 1054.
612. Haffemayer, B., Gulias, M., Gaunt, M. J. (2011). *Chem. Sci.* **2**, 312.
613. Guan, Z.-H., Chen, M., Ren, Z.-H. (2012). *J. Am. Chem. Soc.* **134**, 17490.
614. Guan, Z.-H., Ren, Z.-H., Spinella, S. M., Yu, S., Liang, Y.-M., Zhang, X. (2009). *J. Am. Chem. Soc.* **131**, 729.
615. Du, Y., Hyster, T. K., Rovis, T. (2011). *Chem. Commun.* **47**, 12074.
616. Inoue, S., Shiota, H., Fukumoto, Y., Chatani, N. (2009). *J. Am. Chem. Soc.* **131**, 6898.
617. Yo, E. J., Wasa, M., Yu, J.-Q. (2010). *J. Am. Chem. Soc.* **132**, 17378.
618. Hasegawa, N., Charra, V., Inoue, S., Fukumoto, Y., Chatani, N. (2011). *J. Am. Chem. Soc.* **133**, 8070.
619. Xie, P., Xie, Y., Qian, B., Zhou, H., Xia, C., Huang, H. (2012). *J. Am. Chem. Soc.* **134**, 9902.
620. Chen, H., Cai, C., Liu, X., Li, X., Jiang, H. (2011). *Chem. Commun.* **47**, 12224.
621. Fujiwara, Y., Jintoku, T., Uchida, Y. (1989). *New J. Chem.* **13**, 649.

622. Fujiwara, Y., Takaki, K., Uchida, Y., Taniguchi, H. (1989). *Chem. Lett.* 1687.

623. Nishiguchi, T., Nakata, K., Takaki, K., Fujiwara, Y. (1992). *Chem. Lett.* 1141.

624. Nakata, K., Yamaoka, Y., Miyata, T., Taniguchi, Y., Takaki, K., Fujiwara, Y. (1994). *J. Organomet. Chem.* **473**, 329.

625. Nizova, G. V., Süss-Fink, G., Stanislas, S., Shul'pin, G. B. (1998). *Chem. Commun.* 1885.

626. Taniguchi, Y., Hayashida, T., Shibasaki, H., Piao, D., Kitamura, T., Yamaji, T., Fujiwara, Y. (1999). *Org. Lett.* **1**, 557.

627. Reis, P. M., Silva, J. A. L., Palavra, A. F., Fraústo da Silva, J. J. R., Kitamura, T., Fujiwara, Y., Pombeiro, A. J. L. (2003).*Angew. Chem., Int. Ed.* **42**, 821.

628. Asadullah, M., Kitamura, T., Fujiwara, Y. (2000). *Angew. Chem., Int. Ed.* **39**, 2475.

629. Lin, M., Sen, A. (1994). *Nature* **368**, 613.

630. Periana, R. A., Mironov, O., Taube, D., Bhalla, G., Jones, C. J. (2003). *Science* **301**, 814.

631. Sakai, K., Ide, J., Oda, O., Nakamura, N. (1972). *Tetrahedron Lett.*, 1287.

632. Lochow, C. F., Miller, R. G. (1976). *J. Am. Chem. Soc.* **98**, 1281.

633. Larock, R. C., Oertle, K., Potter, G. F. (1980). *J. Am. Chem. Soc.* **102**, 190.

634. Morgan, J. P., Kundu, K., Doyle, M. P. (2005). *Chem. Commun.*, 3307.

635. Fairlie, D. P., Bosnich, B. (1988). *Organometallics* **7**, 936.

636. Fairlie, D. P., Bosnich, B. (1988). *Organometallics* **7**, 946.

637. Gable, K., Benz, G. A. (1991). *Tetrahedron Lett.* **32**, 3473.

638. Sato, Y., Oonishi, T., Mori, M. (2002). *Angew. Chem., Int. Ed.* **41**, 1218.

639. Oonishi, Y., Mori, M., Sato, Y. (2007). *Synthesis* 2323.

640. Aloise, A. D., Layton, M. E., Shair, M. D. (2000). *J. Am. Chem. Soc.* **122**, 12610.

641. Bendorf, H. D., Colella, C. M., Dixon, E. C., Marchetti, M., Matukonis, A. N., Musselman, J. D., Tiley, T. A. (2002). *Tetrahedron Lett.* **43**, 7031.

642. Coulter, M. M., Dornan, P. K., Dong, V. M. (2009). *J. Am. Chem. Soc.* **131**, 6932.

643. James, B. R., Young, C. G. (1983). *J. Chem. Soc., Chem. Commun.* 1215.

644. James, B. R., Young, C. G. (1985). *J. Organomet. Chem.* **285**, 321.

645. Taura, Y., Tanaka, M., Funakoshi, K., Sakai, K. (1989). *Tetrahedron Lett.* **30**, 6349.

646. Taura, T., Masakazu, T., Wu, X.-M., Funakoshi, K., Sakai, K. (1991). *Tetrahedron* **47**, 4879.

647. Wu, X.-M., Funakoshi, K., Sakai, K. (1992). *Tetrahedron Lett.* **33**, 6331.

648. Barnhart, R. W., Wang, X., Noheda, P., Bergens, S. H., Whelan, J., Bosnich, B. (1994). *J. Am. Chem. Soc.* **116**, 1821.

649. Barnhart, R. W., McMorran, D. A., Bosnich, B. (1997). *Chem. Commun.*, 589.

650. Barnhart, R. W., McMorran, D. A., Bosnich, B. (1997). *Inorg. Chem. Acta* **263**, 1.

651. Kundu, K., McCullagh, J. V., Morehead, A. T., Jr. (2005). *J. Am. Chem. Soc.* **127**, 16042

652. Wu, X.-M., Funakoshi, K., Sakai, K. (1993). *Tetrahedron Lett.* **34**, 5927.

653. Tanaka, M., Imai, M., Fujio, M., Sakamoto, E., Takahashi, M., Eto-Kato, Y., Wu, X. M., Funakoshi, K., Sakai, K., Suemune, H. (2000). *J. Org. Chem.* **65**, 5806.

654. Imai, M., Tanaka, M., Suemune, H. (2001). *Tetrahedron* **57**, 1205.

655. Vora, K. P., Lochow, C. F., Miller, R. G. (1980). *J. Organomet. Chem.* **192**, 257.

656. Kokobu, K., Matsumasa, K., Miura, M., Nomura, M. (1997). *J. Org. Chem.* **62**, 4564.

657. Kokobu, K., Matsumasa, K., Nishinaka, Y., Miura, M., Nomura, M. (1999). *Bull. Chem. Soc. Jpn.* **72**, 303.
658. Imai, M., Tanaka, M., Tanaka, K., Yamamoto, Y., Imai-Ogata, N., Shimowatari, M., Nagumo, S., Kawahara, N., Suemune, H. (2004). *J. Org. Chem.* **69**, 1144.
659. von Delius, M., Le, C. M., Dong, V. M. (2012). *J. Am. Chem. Soc.* **134**, 15022.
660. Murphy, S. K., Coulter, M. M., Dong, V. M. (2012). *Chem. Sci.* **3**, 355.
661. Murphy, S. K., Petrone, D. A., Coulter, M. M., Dong, V. M. (2011). *Org. Lett.* **13**, 6216.
662. Willis, M. C., McNally, S. J., Beswick, P. J. (2004). *Angew. Chem., Int. Ed.* **43**, 340.
663. Willis, M. C., Randell-Sly, H. E., Woodward, R. L., Currie, G. S. (2005). *Org. Lett.* **7**, 2249.
664. Moxham, G. L., Randell-Sly, H. E., Brayshaw, S. K., Woodward, R. L., Weller, A. S., Willis, M. C. (2006). *Angew. Chem., Int. Ed.* **45**, 7618.
665. Chaplin, A. B., Hooper, J. F., Weller, A. S., Willis, M. C. (2012). *J. Am. Chem. Soc.* **134**, 4885.
666. Jun, C.-H., Lee, D.-Y., Lee, H., Hong, J.-B. (2000). *Angew. Chem., Int. Ed.* **39**, 3070.
667. Tanaka, K., Shibata, Y., Suda, T., Hagiwara, Y., Hirano, M. (2007). *Org. Lett.* **9**, 1215.
668. Lenges, C. P., Brookhart, M. (1997). *J. Am. Chem. Soc.* **119**, 3165.
669. Lenges, C. P., White, P. S., Brookhart, M. (1998). *J. Am. Chem. Soc.* **120**, 6965.
670. Roy, A. H., Lenges, C. P., Brookhart, M. (2007). *J. Am. Chem. Soc.* **129**, 2082.
671. Kondo, T., Akazome, M., Tsuji, Y., Watanabe, Y. (1990). *J. Org. Chem.* **55**, 1286.
672. Stemmler, R. T., Bolm, C. (2007). *Adv. Synth. Catal.* **349**, 1185.
673. Phan, D. H. T., Kou, K. G. M., Dong, V. M. (2010). *J. Am. Chem. Soc.* **132**, 16354.
674. Osborne, J. D., Randell-Sly, H. E., Currie, G. S., Cowley, A. R., Willis, M. C. (2008). *J. Am. Chem. Soc.* **130**, 17232.
675. Shibata, Y., Tanaka, K. (2009). *J. Am. Chem. Soc.* **131**, 12552.
676. Coulter, M. M., Kou, K. G. M., Galligan, B., Dong, V. M. (2010). *J. Am. Chem. Soc.* **132**, 16330.
677. Tanaka, K., Fu, G. C. (2001). *J. Am. Chem. Soc.* **123**, 11492.
678. Takeishi, K., Sugishama, K., Sasaki, K., Tanaka, K. (2004). *Chem. Eur. J.* **10**, 5681.
679. Tanaka, K., Fu, G. C. (2002). *J. Am. Chem. Soc.* **124**, 10296.
680. Tanaka, K., Fu, G. C. (2003). *J. Am. Chem. Soc.* **125**, 8078.
681. Tsuda, T., Kiyoi, T., Saegusa, T. (1990). *J. Org. Chem.* **55**, 2554.
682. Jun, C.-H., Lee, H., Hong, J.-B., Kwon, B.-I. (2002). *Angew. Chem., Int. Ed.* **41**, 2146.
683. Kokobu, K., Matsumasa, K., Miura, M., Nomura, M. (1997). *J. Org. Chem.* **62**, 4564.
684. Lenden, P., Entwistle, D. A., Willis, M. C. (2011). *Angew. Chem. Int. Ed.* **50**, 10657.
685. Gonzalez-Rodriguez, C., Pawley, R. J., Chaplin, A. B., Thompson, A. L., Weller, A. S., Willis, M. C. (2011). *Angew. Chem. Int. Ed.* **50**, 5134.
686. González-Rodríguez, C., Parsons, S. R., Thompson, A. L., Willis, M. C. (2010). *Chem. Eur. J.* **16**, 10950.
687. Campbell, R. E., Jr. Lochow, C. F., Vora, K. P., Miller, R. G. (1980). *J. Am. Chem. Soc.* **102**, 5824.
688. Campbell, R. E., Jr. Miller, R. G. (1980). *J. Organomet. Chem.* **186**, C27.
689. Bosnich, B. (1998). *Acc. Chem. Res.* **31**, 667.

690. Hyatt, I. F. D., Anderson, H. K., Morehead, A. T., Jr. Sargent, A. L. (2008). *Organometallics* **27**, 135.
691. Grabowski, R. (2006). *Catal. Rev.-Sci. Eng.* **48**, 199.
692. Crabtree, R. H., Mihelcic, J. M., Quirk, J. M. (1979). *J. Am. Chem. Soc.* **101**, 7738.
693. Baudry, D., Ephritikhine, M., Felkin, H. (1980). *J. Chem. Soc., Chem. Commun.* 1243.
694. Baudry, D., Ephritikhine, M., Felkin, H., Holmes-Smith, R. (1983). *J. Chem. Soc., Chem. Commun.* 788.
695. Felkin, H., Fillebeen-Khan, T., Gault, Y., Holmes-Smith, R., Zakrzewski, J. (1984). *Tetrahedron Lett.* **25**, 1279.
696. Burk, M. J., Crabtree, R. H., Parnell, C. P., Uriarte, R. J. (1984). *Organometallics* **3**, 816.
697. Burk, M. J., Crabtree, R. H. (1987). *J. Am. Chem. Soc.* **109**, 8025.
698. Belli, J., Jensen, C. M. (1996). *Organometallics* **15**, 1532.
699. Nomura, K., Saito, Y. (1988). *J. Chem. Soc., Chem. Commun.* 161.
700. Fujii, T., Saito, Y. (1990). *J. Chem. Soc., Chem. Commun.* 757.
701. Fujii, T., Higashino, Y., Saito, Y. (1993). *J. Chem. Soc., Dalton Trans.* 517.
702. Nomura, K., Saito, Y. (1988). *J. Chem. Soc., Chem. Commun.* 161.
703. Sakakura, T., Sodeyama, T., Tokunaga, M., Tanaka, M. (1988). *Chem. Lett.* 263.
704. Maguire, J. A., Boese, W. T., Goldman, A. S. (1989). *J. Am. Chem. Soc.* **111**, 7088.
705. Maguire, J. A., Petrillo, A., Goldman, A. S. (1992). *J. Am. Chem. Soc.* **114**, 9492.
706. Burk, M. W., Crabtree, R. H., McGrath, D. V. (1985). *J. Chem. Soc., Chem. Commun.* 1829.
707. Gupta, M., Hagen, C., Flesher, R. J., Kaska, W. C., Jensen, C. M. (1996). *Chem. Commun.* 2083.
708. Gupta, M., Hagen, C., Kaska, W. C., Cramer, R. E., Jensen, C. M. (1997). *J. Am. Chem. Soc.* **119**, 840.
709. Xu, W.-W., Rosini, G. P., Gupta, M., Jensen, C. M., Kaska, W. C., Krogh-Jespersen, K., Goldman, A. S. (1997). *Chem. Commun.* 2273.
710. Liu, F., Pak, E. B., Singh, B., Jensen, C. M., Goldman, A. S. (1999). *J. Am. Chem. Soc.* **121**, 4086.
711. Zhu, K., Achord, P. D., Zhang, X., Krogh-Jespersen, K., Goldman, A. S. (2004). *J. Am. Chem. Soc.* **126**, 13044.
712. Haenel, M. W., Oevers, S., Angermund, K., Kaska, W. C., Fan, H.-J., Hall, M. B. (2001). *Angew. Chem., Int. Ed.* **40**, 3596.
713. Kundu, S., Choliy, Y., Zhuo, G., Ahuja, R., Emge, T. J., Warmuth, R., Brookhart, M., Krogh-Jespersen, K., Goldman, A. S. (2009). *Organometallics* **28**, 5432.
714. Punji, B., Emge, T. J., Goldman, A. S. (2010). *Organometallics* **29**, 2702.
715. Göttker-Schnetmann, I., White, P., Brookhart, M. (2004). *J. Am. Chem. Soc.* **126**, 1804.
716. Göttker-Schnetmann, I., Brookhart, M. (2004). *J. Am. Chem. Soc.* **126**, 9330.
717. Morales-Morales, D., Redon, R., Yung, C., Jensen, C. M. (2004). *Inorg. Chim. Acta* **357**, 2953.
718. Ahuja, R., Punji, B., Findlater, M., Supplee, C., Schinski, W., Brookhart, M., Goldman, A. S., (2011). *Nature Chem.* **2**, 167.
719. Goldman, A. S., Roy, A. H., Huang, Z., Ahuja, R., Schinski, W., Brookhart, M. (2006). *Science* **312**, 257.

720. Huang, Z., Rolfe, E., Carson, E. C., Brookhart, M., Goldman, A. S., El-Khalafy, S. H., MacArthur, A. H. R. (2010). *Adv. Synth. Catal.* **352**, 125.
721. Renkema, K. B., Kissin, Y. V., Goldman, A. S. (2003). *J. Am. Chem. Soc.* **125**, 7770.
722. Kuklin, S. A., Sheloumov, A. M., Dolgushin, F. M., Ezernitskaya, M. G., Peregudov, A. S., Petrovskii, P. V., Koridze, A. A. (2006). *Organometallics* **25**, 5466.
723. Waltz, K. M., He, X., Muhoro, C. N., Hartwig, J. F. (1995). *J. Am. Chem. Soc.* **117**, 11357.
724. Waltz, K. M., Hartwig, J. F. (1997). *Science* **277**, 211.
725. Waltz, K. M., Hartwig, J. F. (2000). *J. Am. Chem. Soc.* **122**, 11358.
726. Chen, H., Hartwig, J. F. (1999). *Angew. Chem., Int. Ed.* **38**, 3391.
727. Chen, H., Schlecht, S., Semple, T. C., Hartwig, J. F. (2000). *Science* **287**, 1995.
728. Lawrence, J. D., Takahashi, M., C. Bae, Hartwig, J. F. (2004). *J. Am. Chem. Soc.* **126**, 15334.
729. Murphy, J. M., Lawrence, J. D., Kawamura, K., Incarvito, C., Hartwig, J. F. (2006). *J. Am. Chem. Soc.* **128**, 13684.
730. Bae, C., Hartwig, J. F., Boaen, N. K., Long, R. O., Anderson, K. S., Hillmyer, M. A. (2005). *J. Am. Chem. Soc.* **127**, 767.
731. Bae, C., Hartwig, J. F., Chung, H., Harris, N. K., Switek, K. A., Hillmyer, M. A. (2005). *Angew. Chem., Int. Ed.* **44**, 6410.
732. Liskey, C. W., Hartwig, J. F. (2012). *J. Am. Chem. Soc.* **134**, 12422.
733. Shimada, S., Batsanov, A. S., Howard, J. A. K., Marder, T. B. (2001). *Angew. Chem., Int. Ed.* **40**, 2168.
734. Ishiyama, T., Ishida, K., Takagi, J., Miyaura, N. (2001). *Chem. Lett.* **30**, 1082.
735. Kawamorita, S., Miyazaki, T., Iwai, T., Ohmiya, H., Sawamura, M. (2012). *J. Am. Chem. Soc.* **134**, 12924.
736. Ohmura, T., Torigoe, T., Suginome, M. (2012). *J. Am. Chem. Soc.* **134**, 17416.
737. Iverson, C. N., Smith, M. R. III. (1999). *J. Am. Chem. Soc.* **121**, 7696.
738. Cho, J.-Y., Tse, M. K., Holmes, D., Maleczka, R. E., Jr., Smith, M. R. III. (2002). *Science* **295**, 305.
739. Ishiyama, T., Takagi, J., Ishida, K., Miyaura, N., Anastasi, N. R., Hartwig, J. F. (2002). *J. Am. Chem. Soc.* **124**, 390.
740. Ishiyama, T., Takagi, J., Hartwig, J. F., Miyaura, N. (2002). *Angew. Chem., Int. Ed.* **41**, 3056.
741. Takagi, J., Sato, K., Hartwig, J. F., Ishiyama, T., Miyaura, N. (2002). *Tetrahedron Lett.* **43**, 5649.
742. Ishiyama, T., Takagi, J., Yonekawa, Y., Hartwig, J. F., Miyaura, N. Adv (2003). *Synth. Catal.* **345**, 1103.
743. Mkhalid, I. A. I., Coventry, D. N., Albesa-Jove, D., Batsanov, A. S., Howard, J. A. K., Marder, T. B., Perutz, R. N. (2006). *Angew. Chem., Int. Ed.* **45**, 489.
744. Boebel, T. A., Hartwig, J. F. (2008). *J. Am. Chem. Soc.* **130**, 7534.
745. Robbins, D. W., Boebel, T. A., Hartwig, J. F. (2010). *J. Am. Chem. Soc.* **132**, 4068.
746. T. Ishiyama, H. Isou, T. Kikuchi, N. Miyaura, (2010). *Chem. Commun.* **46**, 159.
747. Kawamorita, S., Ohmiya, H., Hara, K., Fukuoka, A., Sawamura, M. (2009). *J. Am. Chem. Soc.* **131**, 5058.

748. Kawamorita, S., Miyazaki, T., Ohmiya, H., Iwai, T., Sawamura, M. (2011). *J. Am. Chem. Soc.* **133**, 19310.

749. Ros, A., Estepa, B., López-Rodríguez, R., Álvarez, E., Fernández, R., Lassaletta, J. M. (2011). *Angew. Chem., Int. Ed.* **50**, 11724.

750. Ros, A., López-Rodríguez, R., Estepa, B., Álvarez, E., Fernández, R., Lassaletta, J. M. (2012). *J. Am. Chem. Soc.* **134**, 4573.

751. Roosen, P. C., Kallepalli, V. A., Chattopadhyay, V. B., Singleton, D. A., Maleczka, R. E., Jr., Smith M. R., III. (2012). *J. Am. Chem. Soc.* **134**, 11350.

752. Dai, H.-X., Yu, J.-Q. (2012). *J. Am. Chem. Soc.* **134**, 134.

753. Xiao, B., Li, Y.-M., Liu, Z.-J., Yang, H.-Y., Fu, Y. (2012). *Chem. Commun.* **48**, 4854.

754. Fischer, D. F., Sarpong, R. (2010). *J. Am. Chem. Soc.* **132**, 5926.

755. Liao, X., Stanley, L. M., Hartwig, J. F. (2011). *J. Am. Chem. Soc.* **133**, 2088.

756. Hartwig, J. F., Cook, K. S., Hapke, M., Incarvito, C. D., Fan, Y., Webster, E. C., Hall, M. B. (2005). *J. Am. Chem. Soc.* **127**, 2538.

757. Wei, C. S., Jiménez-Hoyos, C. A., Videa, M. F., Hartwig, J. F., Hall, M. B. (2010). *J. Am. Chem. Soc.* **132**, 3078.

758. Boller, T. M., Murphy, J. M., Hapke, M., Ishiyama, T., Miyaura, N., Hartwig, J. F. (2005). *J. Am. Chem. Soc.* **127**, 14263.

759. Gustavson, W. A., Epstein, P. S., Curtis, M. D. (1982). *Organometallics* **1**, 884.

760. Sakakura, T., Tokunaga, Y., Sodeyama, T., Tanaka, M. (1987). *Chem. Lett.* 2375.

761. Ishikawa, M., Okazaki, S., Naka, A., Sakamoto, H. (1992). *Organometallics* **11**, 4135.

762. Uchimaru, Y., El Sayed, A. M. M., Tanaka, M. (1993). *Organometallics* **12**, 2065.

763. Tsukada, N., Hartwig, J. F. (2005). *J. Am. Chem. Soc.* **127**, 5022.

764. Ezbiansky, K., Djurovich, P. I., LaForest, M., Sinning, D. J., Zayes, R., Berry, D. H. (1998). *Organometallics* **17**, 1455.

765. Ishiyama, T., Sato, K., Nishio, Y., Miyaura, N. (2003). *Angew. Chem., Int. Ed.* **42**, 5346.

766. Saiki, T., Nishio, Y., Ishiyama, T., Miyaura, N. (2006). *Organometallics* **25**, 6068.

767. Ishiyama, T., Sato, K., Nishio, Y., Saiki, T., Miyaura, N. (2005). *Chem. Commun.*, 5065.

768. Kakiuchi, F., Igi, K., Matsumoto, M., Chatani, N., Murai, S. (2001). *Chem. Lett.*, 422.

769. Kakiuchi, F., Igi, K., Matsumoto, M., Hayamizu, T., Chatani, N., Murai, S. (2002). *Chem. Lett.* 396.

770. Kakiuchi, F., Matsumoto, M., Tsuchiya, K., Igi, K., Hayamizu, T., Chatani, N., Murai, S. (2003). *J. Organomet. Chem.* **686**, 134.

771. Tobisu, M., Ano, Y., Chatani, N. (2008). *Chem. Asian J.* **3**, 1585.

772. Ihara, H., Suginome, M. (2009). *J. Am. Chem. Soc.* **131**, 7502.

773. Kakiuchi, F., Tsuchiya, K., Matsumoto, M., Mizushima, E., Chatani, N. (2004). *J. Am. Chem. Soc.* **126**, 12792.

774. Sadow, A. D., Don Tilley, T. (2003). *Angew. Chem., Int. Ed.* **42**, 803.

775. Simmons, E.M., Hartwig, J. F. (2010). *J. Am. Chem. Soc.* **132**, 17092.

776. Simmons, E.M., Hartwig, J. F. (2012). *Nature* **483**, 70.

CHAPTER 7

CATALYSIS IN C–C ACTIVATION

HU LI and ZHANG-JIE SHI
College of Chemistry and Molecular Engineering, Peking University, Beijing, People's Republic of China

7.1 INTRODUCTION: IMPORTANCE AND CHALLENGES IN C–C ACTIVATION

Selective C–C bond activation by transition metal complexes has attracted much attention, due not only to its fundamental scientific interest but also to its potential utility in organic synthesis. The C–C bond constructs the framework of organic compounds, thus C–C activation provides a powerful tool to approach organic

Homogeneous Catalysis for Unreactive Bond Activation, First Edition. Edited by Zhang-Jie Shi.

synthesis employing readily available and inexpensive raw materials from nature [1]. For example, C–C activation is involved in two important processes in petroleum chemistry—catalytic cracking and catalytic reforming. On the other hand, considering the demands of diminishing the global "white pollution" problem, C–C activation offers the potential to cleaving and degrading synthetic polymers and biomass to produce valuable chemicals [2, 3].

C–C bond activation by a transition metal (e.g., oxidative addition) is the reverse process of C–C bond formation (e.g., reductive elimination); however, the former has rarely been reported compared with the latter in organic synthesis. The formation of a C–C bond (90 kcal mol^{-1}) is more thermodynamically favored than two weak metal–carbon bonds (20–30 kcal mol^{-1} per bond) [4]. In order to facilitate C–C bond activation, two basic strategies can be applied. One is to increase the energy state of starting materials, and the other is to lower the energy state of C–C bond cleaved intermediates or products.

The first strategy is to use high-energy starting materials such as four- or four-membered ring compounds. Relief of the ring strain in these molecules can facilitate the C–C bond activation by forming more thermodynamically stable ring-expanded metallacyclic complexes.

On the other hand, for C–C bond activation of unstrained molecules, special driving forces are required to generate stable organometallic complexes, which is a strategy for lowering the energy state of the C–C bond cleaved intermediates or products. In particular, employing chelating assistance to undergo cyclometalation is widely used, because a five-membered ring metallacyclic complex has a special thermodynamically stability. Besides, several specially designed molecules (e.g., pincer-type model compounds) may undergo C–C bond activation in the presence of a variety of transition metal complexes under very mild conditions.

There are two kinds of elementary steps available for generation of reactive organometallic intermediates through cleavage of C–C bonds with transition metals. The first one is oxidative addition of a C–C linkage to a low-valent transition metal, which generates a bis(organyl)metal complex with an increase of the formal oxidation state of the metal by two (Scheme 7.1, A). The other elementary step is β-carbon elimination occurring with an intermediate of an M–X–C–C structure (M = metal, X = C, N, O, etc.), in which the carbon γ to the metal migrates onto the

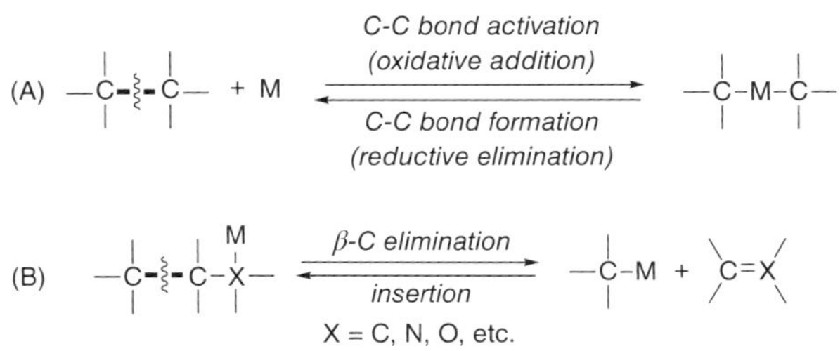

Scheme 7.1.

metal with the extrusion of an X=C bond (Scheme 7.1, B). The formal oxidation state of the metal remains unchanged during this step. Both elementary steps generate an organotransition metal species that is amenable to further transformations including C–C bond formation.

Despite significant developments in C–C bond activation, most of these reactions focus on the stoichiometric ones. Only a few catalytic examples have been reported, and suitable substrates are strictly limited. Therefore, an energy-releasing step is required at the C–C bond cleaved organometallic intermediates, producing stable organic compounds in order to be a catalytic reaction. In general, the final products should be more stable than the starting materials. In this chapter, the basic principle and mechanism of homogeneous transition-metal-catalyzed C–C bond cleavage reactions, in which carbon–metal bonds are generated, will be discussed. Alkene and alkyne metatheses are beyond the scope of this chapter in spite of their abundant applications and impact on organic synthesis.

7.2 C–C ACTIVATION OF STRAINED MOLECULES

The first activation mode for C–C bond cleavage is using high-energy starting materials such as strained molecules. Relief of the ring strain can facilitate C–C bond activation by forming more thermodynamically stable ring-expanded metallacyclic complexes through metal insertion into the strained C–C bond in cyclopropane, cyclobutane, cubane, quadricyclene, and so on. For this reason, coordinatively unsaturated 14e or 16e transition metal complexes could also be used as good reacting partners.

By this strategy, most of the examples were demonstrated by relieving the ring-strain energy of three- or four-membered rings to generate more stable four- or five-membered metallacyclic intermediates through the direct insertion of a metal into the C–C bond of cyclopropyl or cyclobutyl skeletons. Throughout the catalytic reaction, ring-opened or five- to six-membered ring organic compounds are produced.

7.2.1 C–C Activation of Cyclopropane Substrates

For an early example of stoichiometric activation of cyclopropane, ultraviolet irradiation of rhodium(III) dihydride complex **7.2.1.1** generated coordinatively unsaturated 16-electron species **7.2.1.2** with the releasing of H_2, which further reacted with cyclopropane to afford the hydrido cyclopropyl rhodium(III) complex **7.2.1.3** through C–H bond activation. Upon raising the temperature, rearrangement of **7.2.1.3** led to a thermodynamically more stable C–C bond-cleaved four-membered ring metallacyclic complex **7.2.1.4** (Scheme 7.2) [5]. From this result, the C–H insertion product is kinetically favored, and the C–C insertion product is thermodynamically favored. The kinetic preference for C–H insertion clearly demonstrates the greater steric accessibility of the C–H bond compared with the C–C bond.

A catalytic example of C–C bond activation of a cyclopropane derivative **7.2.1.5** was developed using Rh(I) species (Scheme 7.3) [6]. Initially, a non-sterically

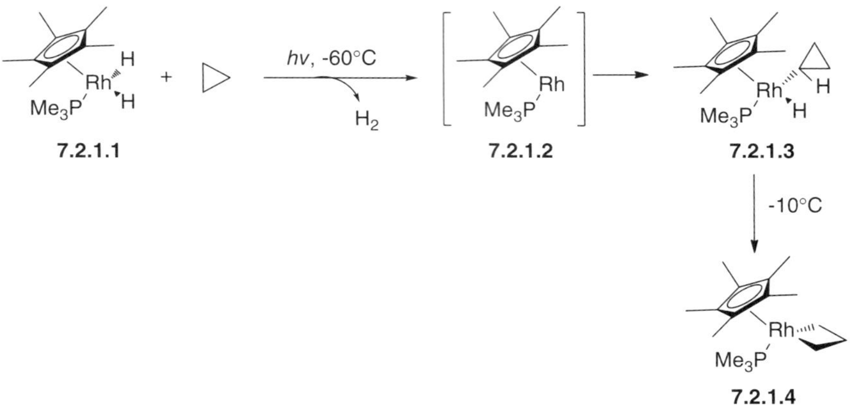

Scheme 7.2.

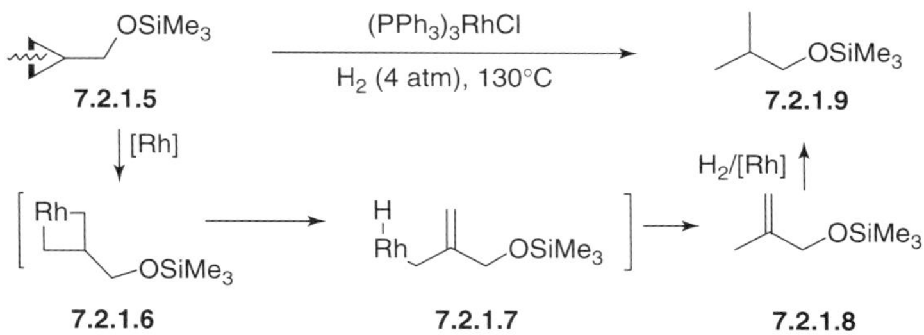

Scheme 7.3.

hindered C–C bond in **7.2.1.5** was cleaved by $(PPh_3)_3RhCl$ to generate the rhodacyclobutane complex **7.2.1.6**, and sequential β-hydride elimination and reductive elimination gave a branched alkene derivative **7.2.1.8**, which was further reduced to a branched alkane derivative **7.2.1.9** in the presence of H_2.

The Rh(I) catalytic system could be extended to cascade cleavage of two C–C bonds in spiropentane derivative **7.2.1.10** (Scheme 7.4) [7]. After cleavage of a non-sterically hindered C–C bond, carbon monoxide underwent insertion to generate a five-membered rhodacyclic carbonyl intermediate **7.2.1.12**, which further underwent β-carbon elimination to afford another C–C bond-cleaved six-membered rhodacyclic carbonyl intermediate **7.2.1.13**. Sequential reductive elimination and isomerization gave a cyclopentenone **7.2.1.15**.

Other than ordinary cyclopropanes, a number of investigations have appeared documenting the metal-mediated cleavage of a variety of cyclopropane derivatives, particularly those which are more reactive due to additional structural strain, unsaturation, and/or substituents. In a pioneering study, 2-methylcyclopropenone **7.2.1.16** was shown to react with $Pt(CH_2{=}CH_2)(PPh_3)_2$ to furnish platinacyclobutenone complex **7.2.1.18** (Scheme 7.5) [8].

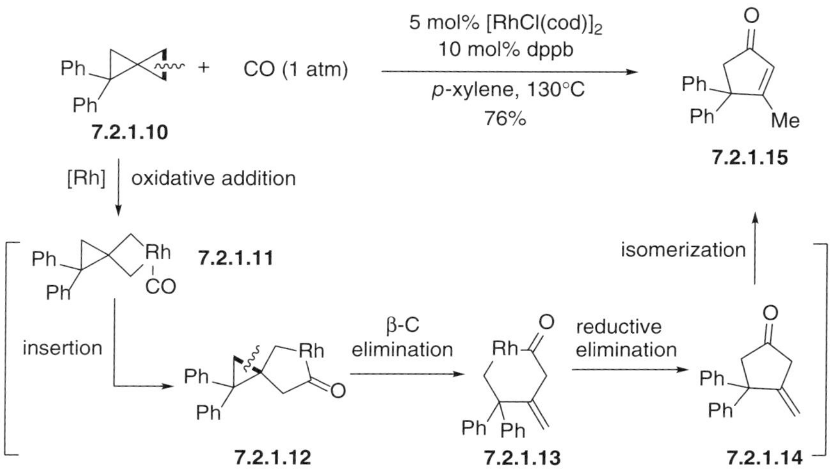

Scheme 7.4.

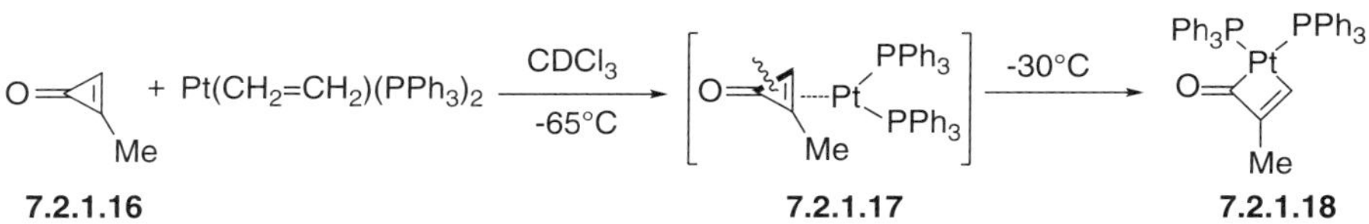

Scheme 7.5.

The reaction of cyclopropenone **7.2.1.19** with ketene could take place in the presence of catalytic amount of $Ni(CO)_4$, affording cyclopentenedione as a mixture of regioisomers **7.2.1.21** and **7.2.1.22** (Scheme 7.6) [9].

Highly strained methylenecyclopropanes are also favorite substrates for Rh-promoted C–C bond activation. A typical reaction is isomerization of methylenecyclopropane **7.2.1.23** to 1,3-diene **7.2.1.27** in the presence of rhodium hydride as catalyst (Scheme 7.7) [10]. The rearrangement was initiated by addition of a Rh–H bond to a C=C bond to form (cyclopropylmethyl)Rh(I) complex **7.2.1.24**, which was converted into a thermodynamically more stable 3-butenylrhodium(I) complex **7.2.1.25** through the C–C bond cleavage by β-alkyl elimination accompanied by cleavage of the less sterically hindered C–C bond of the three-membered ring. Although this step was reversible, the 3-butenylrhodium complex **7.2.1.25** could not undergo β-hydride elimination, and the alternative cleavage of the more sterically hindered C–C bond of **7.2.1.24** resulted in the formation of 1,3-diene **7.2.1.27** and the rhodium hydride.

C–C bond activation of vinylcyclopropanes constitutes one of the most important processes in organic synthesis, because it is possible to obtain various products depending on the reaction conditions (Scheme 7.8).

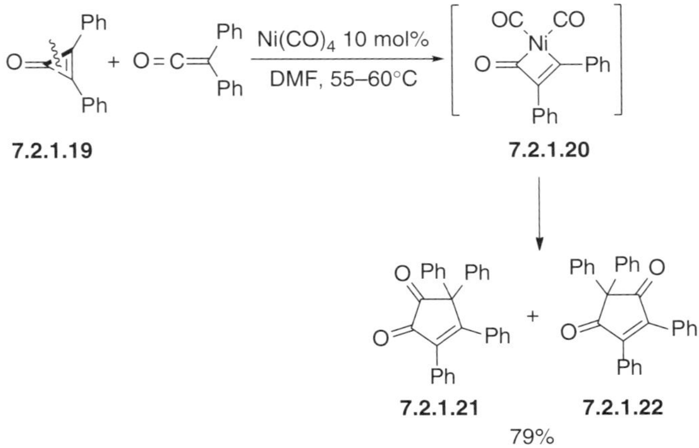

Scheme 7.6.

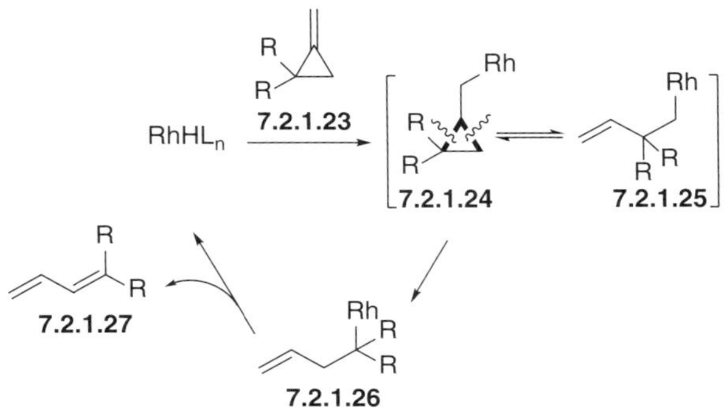

Scheme 7.7.

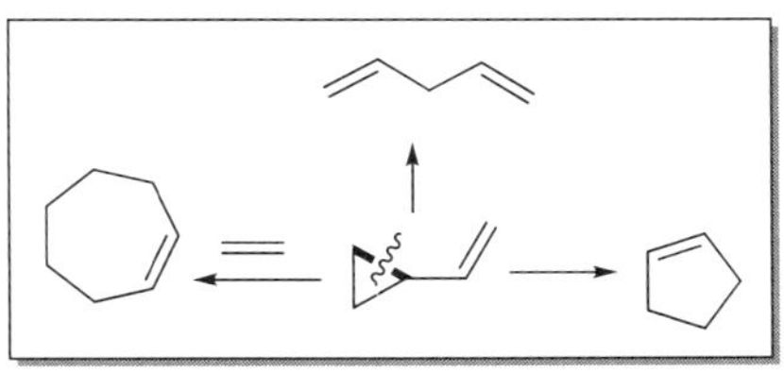

Scheme 7.8.

The initially oxidative addition of a strained C–C bond of the three-membered ring to a Rh(I) complex generated an allylrhodium(III)–alkyl intermediate **7.2.1.30** reversibly. The η^1-allyl form of this intermediate, a rhodacyclohexene **7.2.1.33**, must have sufficient lifetime to allow conformational reorganization, which gave the epimerized vinylcyclopropane **7.2.1.32** through reductive elimination of Rh(I)

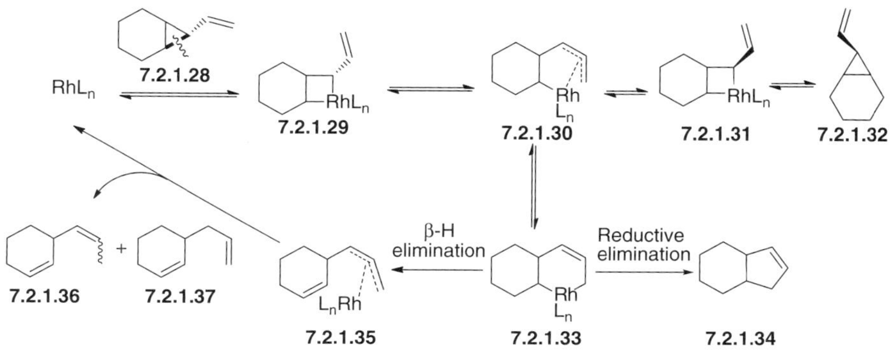

Scheme 7.9.

complex **7.2.1.31**. The rhodacyclohexene **7.2.1.33** might alternatively undergo reductive elimination to afford cyclopentene **7.2.1.34**, or stereospecific *cis*-β-hydride elimination, leading to ring-cleaved dienes **7.2.1.36** and **7.2.1.37** (Scheme 7.9) [11].

One of the early examples was the $[Rh(CO)_2Cl]_2$-catalyzed rearrangement of [7.1.0]bicyclononatriene **7.2.1.38**, which cleanly afforded *cis*-8,9-dihydroindene **7.2.1.39** (Scheme 7.10) [12].

Interestingly, an attempt to rearrange a vinylcyclopropane derivative **7.2.1.41** resulted in just the regioselective and exclusive less substituted C–C bond cleaved ring-opening product **7.2.1.42** (Scheme 7.11) [13].

One of the most important processes in synthetic application of the vinylcyclopropane rearrangement is the intra- or intermolecular [5 + 2]-cycloaddition with alkenes or alkynes in the presence of a Rh(I) catalyst, where vinylcyclopropane can serve as five-carbon unit building blocks. Two plausible mechanisms have been proposed as shown in Scheme 7.12 [14]. In path A it is assumed that the cleavage of the

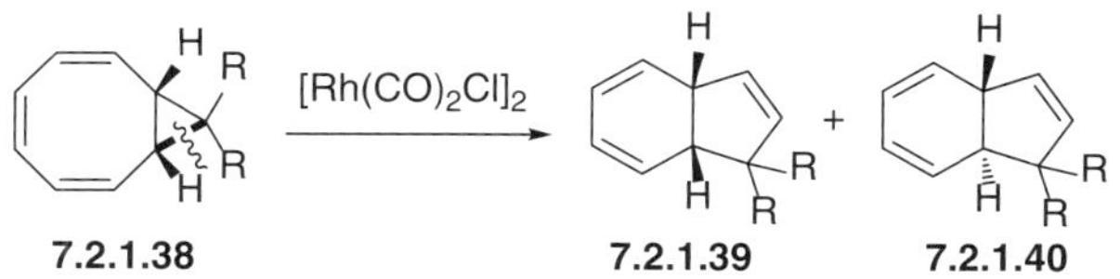

	7.2.1.39	7.2.1.40
R = H, 35°C	99%	8%
R = Me, 140°C	80%	

Scheme 7.10.

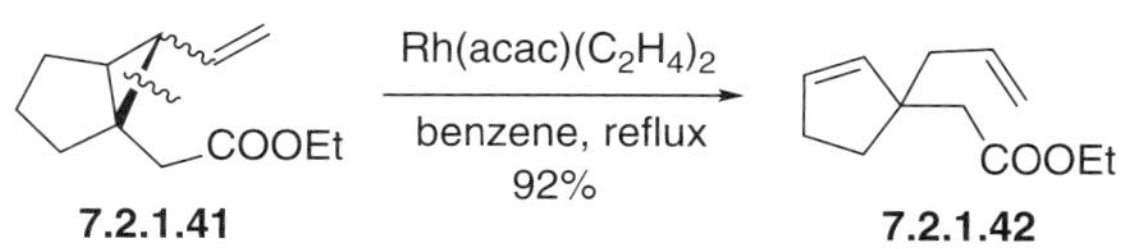

Scheme 7.11.

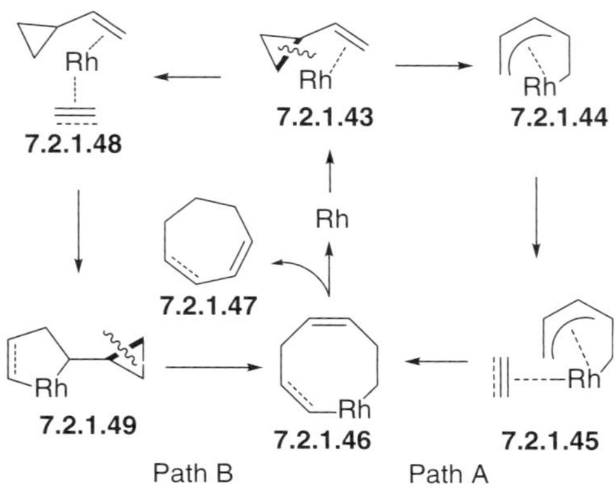

Scheme 7.12.

cyclopropyl group of the Rh(I)–alkene complex **7.2.1.43** yields a Rh(III)-η^3-complex **7.2.1.44**, followed by the insertion of an unsaturated substrate (alkyne or alkene) to form an eight-membered ring rhodacycle **7.2.1.46**. Finally reductive elimination generates the cycloheptene derivative **7.2.1.47** and Rh(I) species. And DFT calculations favor path A for an intermolecular reaction.

The [5 + 2]-cycloaddition is of great synthetic value for approaching seven-membered ring structural. An early synthetic application was demonstrated in the total synthesis of dictamol, a trinorguaiane sesquiterpene isolated from the roots of *Dictamnus dascarpus* TURCZ. The key synthetic step was based on $Rh(PPh_3)_3Cl$-catalyzed intramolecular [5 + 2]-cycloaddition of an allenyl-vinylcyclopropane **7.2.1.50** (Scheme 7.13) [15].

7.2.2 C–C Activation of Cyclobutane Substrates

The C–C activation of cyclobutane and its derivatives proceed less readily than that of cyclopropanes. The main reason is the decreased ring strain. Nevertheless, this area has attracted considerable attention, especially since it has been found that presence of other functionalities on the cyclobutane ring, such as a carbonyl group, enhances activation of the C–C bond.

The cleavage of biphenylene could be achieved using a Ni(0) complex. The ease of the insertion of Ni species into a C–C bond depended largely on the ligands attached

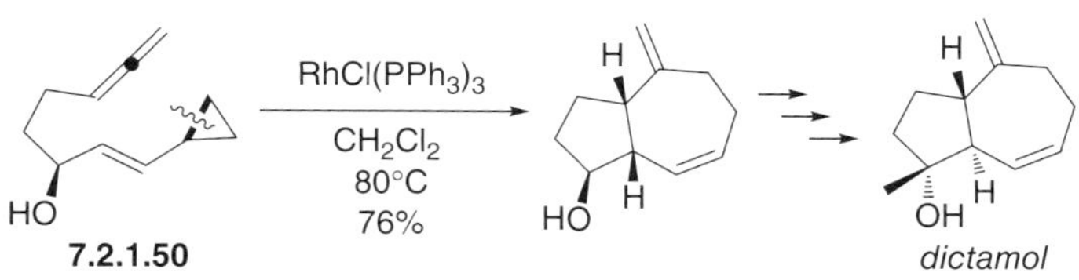

Scheme 7.13.

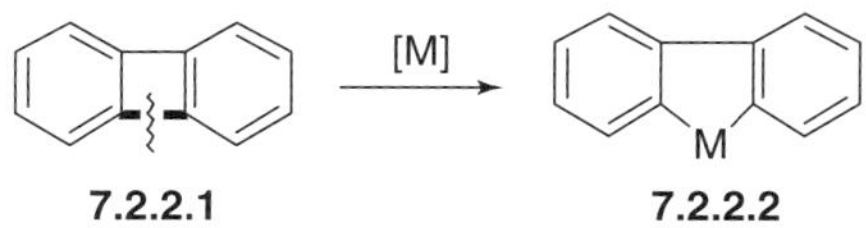

M = Ni, Cr, Ir, Co, Pt, and Pd

Scheme 7.14.

to Ni, with more basic ligands being preferred [16]. Besides, Cr [17], Ir [18], Co [19], Pt, and Pd [20] complexes also had the ability to cleave the aryl–aryl bond of biphenylene to afford a stable metallacyclic complex (Scheme 7.14).

A coordinatively unsaturated Rh(I) complex generated from **7.2.2.3** reacted with biphenylene **7.2.2.1** to give a C–H bond inserted complex **7.2.2.4** as the kinetic product, which was thermally converted to the C–C bond inserted complex **7.2.2.5** [21]. This result may be relevant to the reaction of cyclopropane shown in Scheme 7.15 in that a C–H bond insertion product is kinetically favored, and a C–C bond insertion product is thermodynamically favored.

As mentioned, a variety of late transition metal complexes have been found to cleave the C–C bond of biphenylene **7.2.2.1**, giving insertion complexes **7.2.2.2** to relieve ring-strain energy. These metal complexes participate in various insertion reactions with small unsaturated molecules, such as H_2, CO, olefin, and alkyne, to give functionalized products (Scheme 7.16) [22]. All these products are more stable than biphenylene itself through conjugation, aromatization, ring-expansion, and so on.

Cyclobutanones have also been targets for C–C bond cleaving reactions. Interestingly, for diketones, different transition metals exhibit various regioselectivity. For example, treatment of benzocyclobutane-1,2-dione **7.2.2.6** with $Pt(PPh_3)_4$ caused unsymmetrical C–C bond cleaved product **7.2.2.7** [23]. Unlike platinum(0), rhodium,

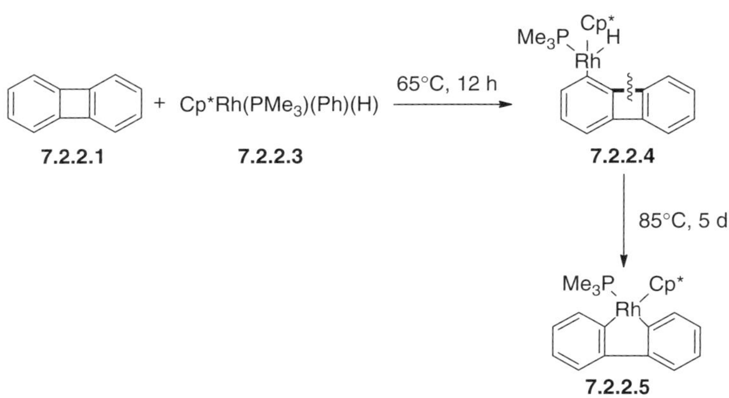

Scheme 7.15.

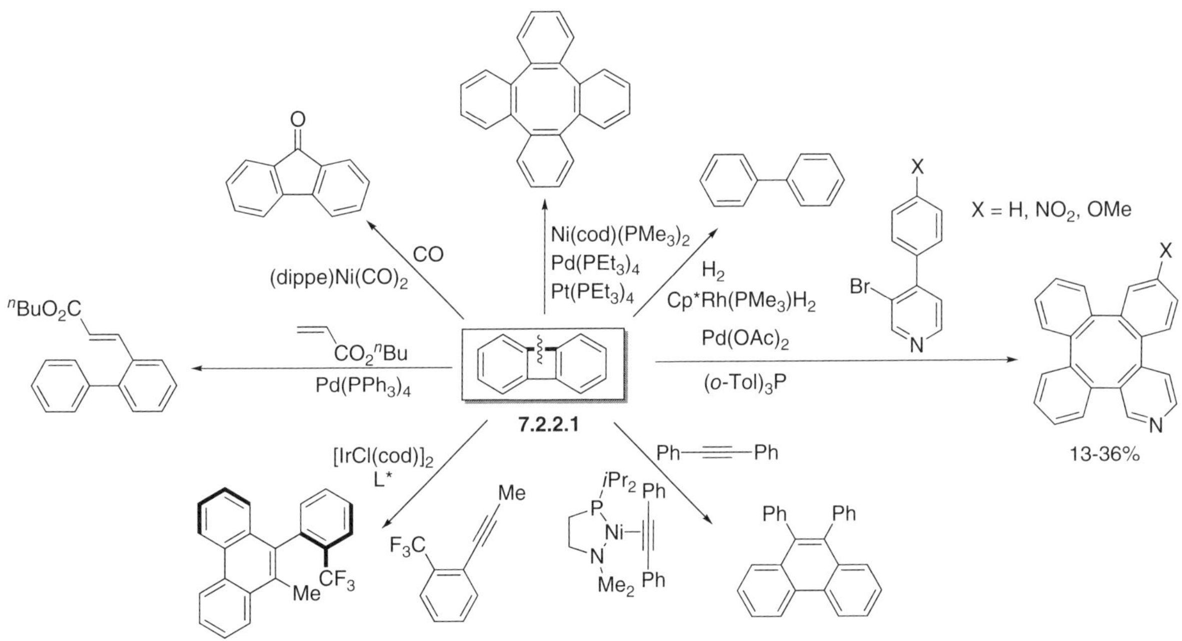

CO
(dippe)Ni(CO)2
Ni(cod)(PMe3)2
Pd(PEt3)4
Pt(PEt3)4
H2
Cp*Rh(PMe3)H2
Pd(OAc)2
(o-Tol)3P
X = H, NO2, OMe
Br
X
N
13-36%
ⁿBuO2C
CO2ⁿBu
Pd(PPh3)4
7.2.2.1
[IrCl(cod)]2
L*
Me
F3C
CF3
iPr2
P
Ni
N
Me2
Ph
Ph—≡—Ph

Scheme 7.16.

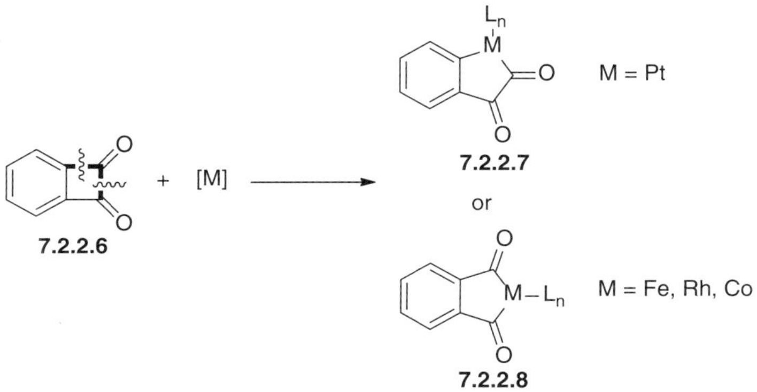

Scheme 7.17.

cobalt, and iron afforded symmetrical complexes **7.2.2.8** resulting from insertion between two carbonyl groups of **7.2.2.6** (Scheme 7.17) [24].

Rh-mediated or -catalyzed C–C bond activation of cyclobutanone **7.2.2.9** usually proceeds by oxidative addition of the C–C bond adjacent to the carbonyl group to afford a rhodacyclopentanone intermediate **7.2.2.10**. Applications of the Rh-catalyzed processes (reported below) in target-oriented synthetic chemistry are still few in number; however, owing to rich reaction patterns of cyclobutanone chemistry the C–C bond activation is of considerable synthetic value (Scheme 7.18).

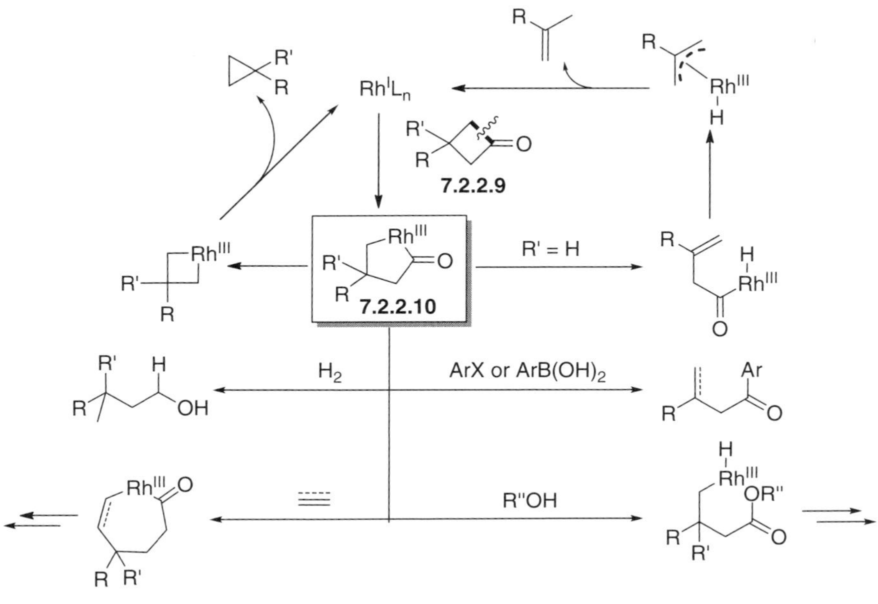

Scheme 7.18.

One of the first examples of Rh-mediated or -catalyzed decarbonylation of cyclobutanones resulted in the formation of cyclopropanes or alkenes. The decarbonylative ring contraction is a sensitive reaction depending on the ligand environment around the central Rh atom as well as on the substitution pattern on the cyclobutane ring. The fine ligand tuning revealed that the course of the reaction for the cyclobutane **7.2.2.11** possessing a hydrogen atom in position 3 (i.e., decarbonylative ring contraction and decarbonylative ring opening) enabled the development of pathways to either the cyclopropane (derivatives) or the alkene derivatives. The use of triphenylarsine as the ligand led to the formation of the cyclopropane derivative **7.2.2.12**, whereas dppp afforded alkene derivative **7.2.2.13** (Scheme 7.19) [25].

The following two examples are synthetic applications of the ring cleavage of cyclobutanone derivatives, **7.2.2.14** and **7.2.2.18**, bearing *o*-styryl and *o*-phenol groups (Schemes 7.20 and 7.21).

Initially formed five-membered acylrhodium intermediate **7.2.2.15** was subjected to an intramolecular alkene insertion reaction, reacting with the vinyl group to give a bicyclo[3.2.1]octanone skeleton **7.2.2.16**. Further reductive elimination gave rise to **7.2.2.17** [26]. This type of insertion of olefin into a C–C single bond is very rare in organometallic reactions. On the other hand, five-membered acylrhodium intermediate **7.2.2.19** reacted with the phenolic hydroxyl group to give a six-membered ring

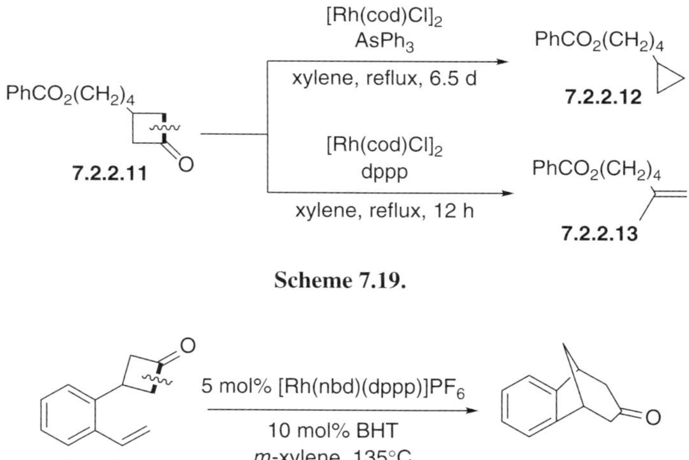

Scheme 7.19.

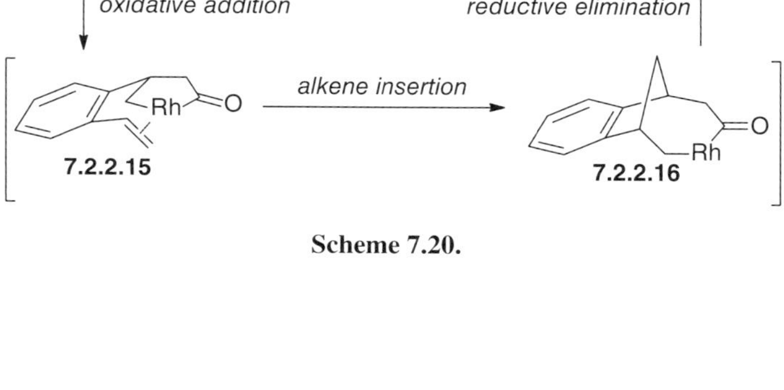

Scheme 7.20.

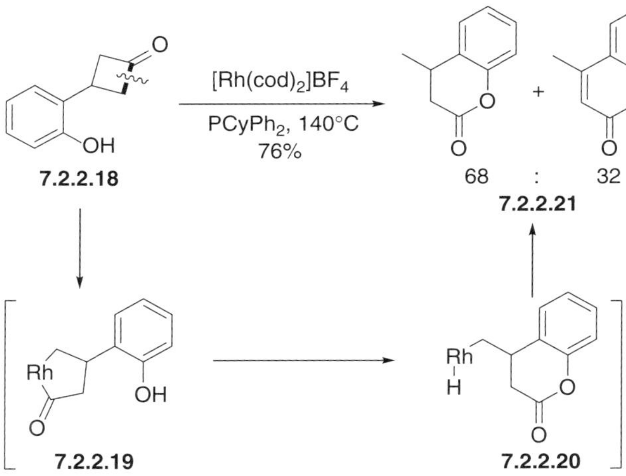

Scheme 7.21.

lactone **7.2.2.20**. Direct reductive elimination or β-hydride elimination followed by isomerization gave lactones **7.2.2.21** [27].

Besides a directly oxidative addition of a C–C bond to the metal center, when a strained molecule is bonded to a metal through a carbon, nitrogen, or oxygen atom, β-alkyl elimination commonly occurs to give ring-opened alkyl–metal intermediates. This is another way of relieving the ring-strain energy of a strained molecule; thus the β-carbon elimination that occurred with cyclobutane derivatives is also an important pathway. The Rh-catalyzed reaction of cyclobutanone **7.2.2.22** with arylboronic acids gave butyrophenone derivatives **7.2.2.26** via rhodium(I) cyclobutanolates **7.2.2.23** (Scheme 7.22) [28]. 1,2-Addition of arylrhodium species to the carbonyl group was followed by β-carbon elimination to open the four-membered carbocycle.

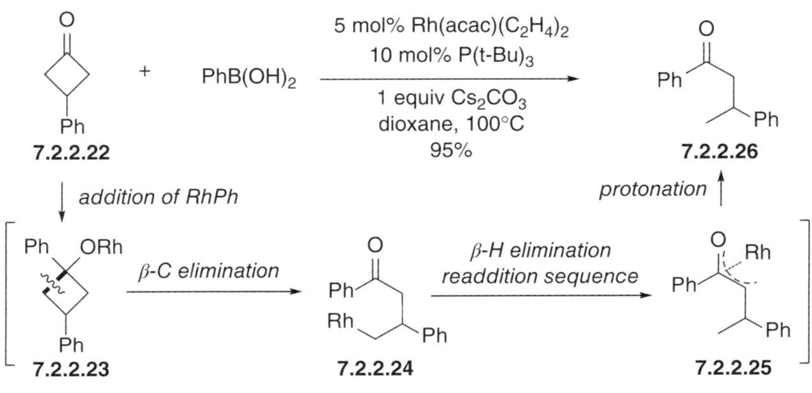

Scheme 7.22.

A related reaction with *o*-hydroxyphenyl-substituted cyclobutanones **7.2.2.27** catalyzed by a chiral rhodium complex provided access to dihydrocoumarins (Scheme 7.23) [29]. The phenolic OH of **7.2.2.27** formed a rhodium alcoholate, promoting the generation of bicyclic species **7.2.2.28** via hemiketalization, followed by sequential β-carbon elimination and 1,4-rhodium shift generating aryl-rhodium species **7.2.2.30**, which either underwent protodemetalation to form dihydrocoumarins with quaternary stereogenic center **7.2.2.31** or reacted in a Michael-type fashion to give rise to substituted dihydrocoumarins **7.2.2.32**. The different pathways of this transformation highlight the power and versatility of such an activation to obtain useful compounds that are difficult to access by common synthetic methods.

Formal insertion of alkynes and alkenes into the C–C bond of cyclobutanone could be achieved by using Ni(0) catalysts (Scheme 7.24) [30]. Bicyclic eight-membered ring ketone **7.2.2.37** was obtained by [4 + 2 + 2] annulation of cyclobutanone **7.2.2.34** with diyne **7.2.2.33**, which incorporated a process of β-carbon elimination of a nickel cyclobutanolate **7.2.2.35** for ring expansion.

Apart from cyclobutanones, cyclobutanols are also found to be good substrates for C–C bond activation. Pioneering work and also the first enantioselective example reported by Pd-catalyzed arylation of *tert*-cyclobutanols **7.2.2.38** with arylating reagents involving the enantioselective C–C bond cleavage afforded optically active γ-arylated ketones **7.2.2.41** in excellent yields with high enantioselectivity (Scheme 7.25) [31]. In this reaction, the key step is the β-carbon elimination of palladium(II) alcoholate **7.2.2.39** formed *in situ*. The chiral bidentate ligand plays a dominant role in controlling the direction of the enantioselective C–C bond cleavage.

Rearrangement of allenyl-*tert*-cyclobutyl alcohols **7.2.2.42** into substituted cyclohexenones **7.2.2.46** having a quaternary stereogenic center in the β-position could be achieved by Rh(I) catalysis (Scheme 7.26) [32]. The key step was the enantioselective insertion of the chiral rhodium center into one of the two prostereogenic C–C bonds of **7.2.2.43**, leading to the metallacycloheptanone **7.2.2.44**. A subsequent reductive elimination resulted in the formation of methylene cyclohexanone **7.2.2.45** as the primary product. An isomerization of the exocyclic double bond converted **7.2.2.45** into more stable conjugated enone **7.2.2.46**.

Alkoxyallenyl substituted *tert*-cyclobutanols **7.2.2.47** can be rearranged with a catalytic amount of a chiral palladium complex into cyclopentanones with α-chiral *O*-tertiary centers **7.2.2.50** (Scheme 7.27) [33]. Although the substrate class is closely related to the previously described reaction, its mechanistic picture is completely different. In this case, the suggested mechanism started with a hydropalladation of the starting alkoxy allene moiety **7.2.2.48** to form π-allyl palladium intermediate **7.2.2.49**. Upon deprotonation, this intermediate participated in a facile Wagner–Meerwein shift to give the ring-expanded cyclopentanone **7.2.2.50** in excellent enantioselectivities.

In the presence of CO, allenylcyclobutanols **7.2.2.51** underwent intramolecular rearrangement into seven-membered ring compounds **7.2.2.52** (Scheme 7.28) [34]. From the mechanistic point of view, it can be considered to be a [6 + 1]-cycloaddition process. It was necessary to carry out the reaction under positive pressure of CO to

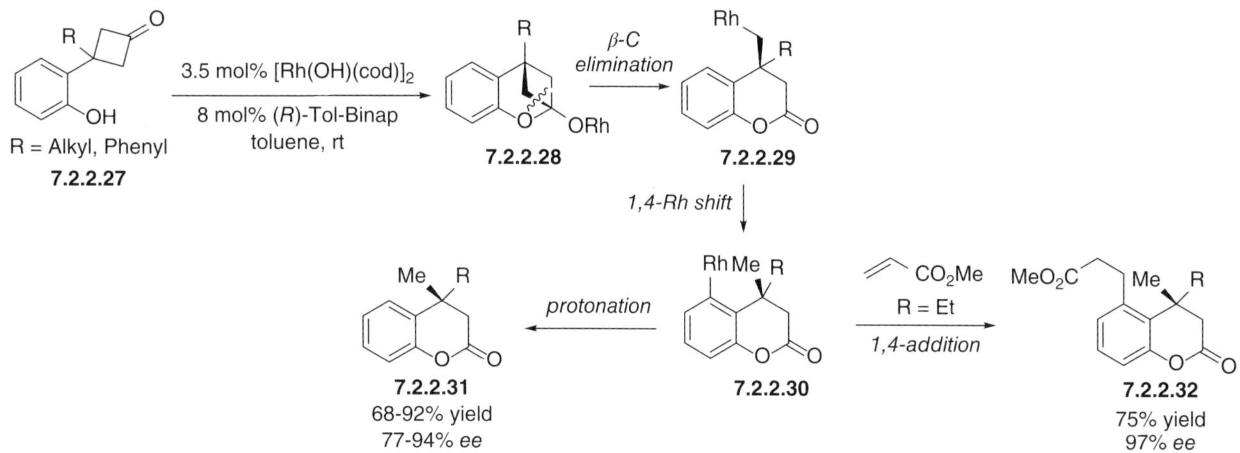

R = Alkyl, Phenyl
7.2.2.27
3.5 mol% [Rh(OH)(cod)]2
8 mol% (R)-Tol-Binap
toluene, rt
7.2.2.28
β-C
elimination
7.2.2.29
1,4-Rh shift
7.2.2.30
protonation
7.2.2.31
68-92% yield
77-94% ee
CO2Me
R = Et
1,4-addition
7.2.2.32
75% yield
97% ee

Scheme 7.23.

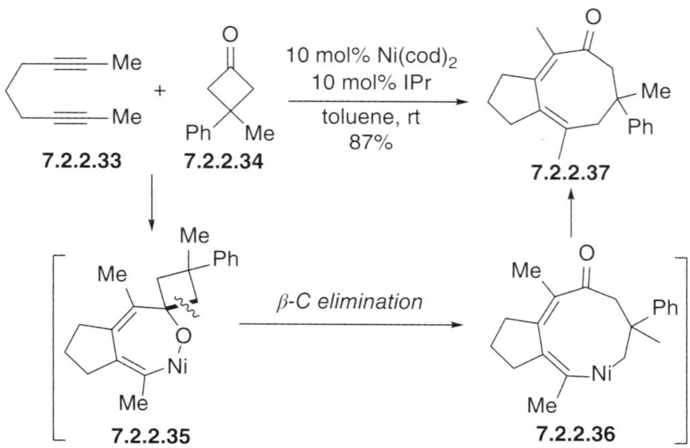

Scheme 7.24.

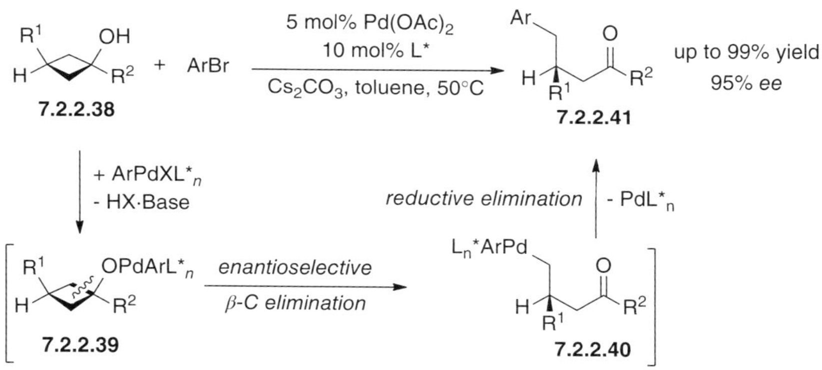

Scheme 7.25.

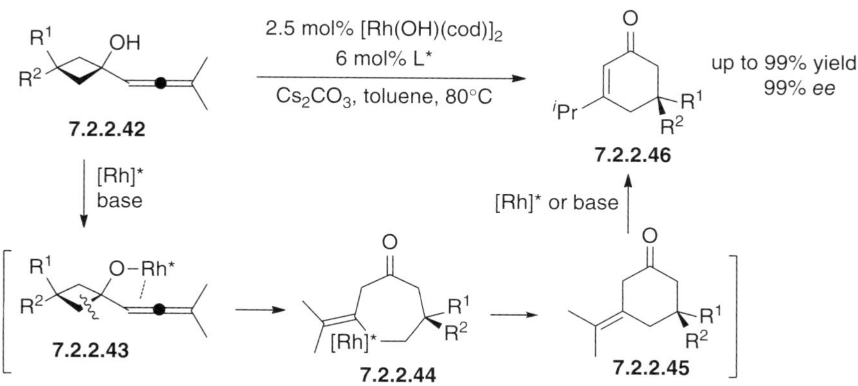

Scheme 7.26.

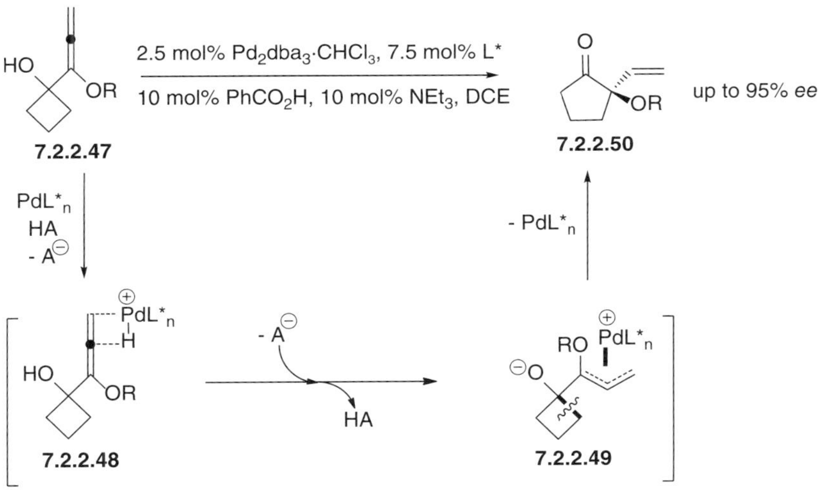

Scheme 7.27.

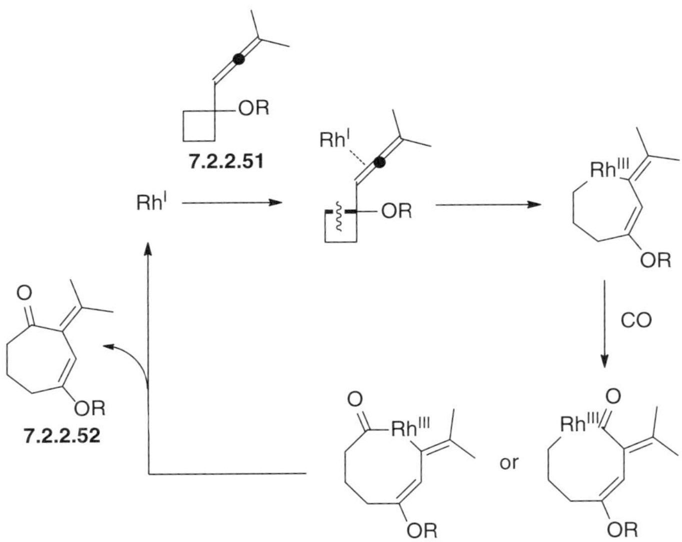

Scheme 7.28.

avoid side reactions. This transformation has enormous synthetic potential, because it could open a short pathway to a number of compounds bearing seven-membered rings.

One of the note worthy processes of a synthetic potential is intramolecular [6 + 2]-cycloaddition of vinylcyclobutanes **7.2.2.53** with a double bond, which affords

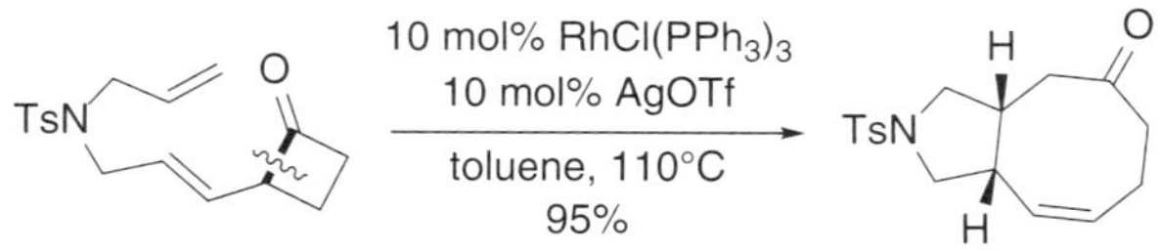

Scheme 7.29.

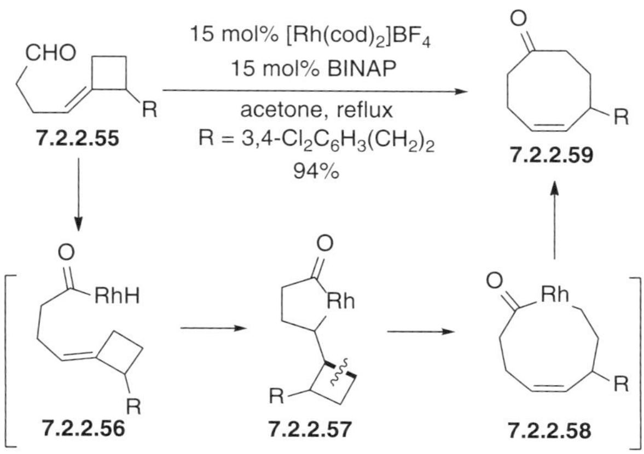

Scheme 7.30.

bicyclic compounds with eight-membered rings **7.2.2.54** (Scheme 7.29) [35]. The process is similar to the previously mentioned [5 + 2]-cycloaddition.

An example of Rh-catalyzed hydroacylation of alkylidenecyclobutanes **7.2.2.55** achieved by β-carbon elimination occurred with an intermediary 5-cyclobutylrhodacyclopentan-2-one **7.2.2.57** to open the four-membered ring (Scheme 7.30) [36]. Reductive elimination of the resulting nine-membered rhodacycle **7.2.2.58** led to the formation of an eight-membered ring ketone **7.2.2.59**.

Another interesting example is the C–C bond activation of cyclobutanone *O*-acyloxime **7.2.2.60** into nitriles (Scheme 7.31) [37]. The cyclobutaniminopalladium(II) complexes **7.2.2.61**, which were generated from an oxidative addition of Pd(0) species to the N–O bond of **7.2.2.60**, underwent β-alkyl elimination and successive β-hydride elimination to give nitriles **7.2.2.63** and **7.2.2.64**.

7.3 C–C ACTIVATION OF UNSTRAINED MOLECULES

As mentioned, for C–C bond activation of unstrained molecules, special driving forces are required to generate stable organometallic complexes, which is a strategy for lowering the energy state of the C–C bond cleaved intermediates or products. In general, the final products should be more stable than the starting materials.

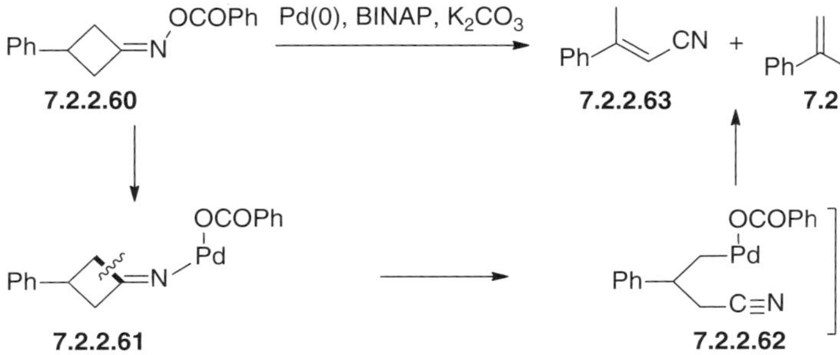

Scheme 7.31.

Different type of substrates, such as nitriles, ketones, imines, alcohols, allylic compounds, and pincer-type, substrates, will be discussed herein.

7.3.1 C–C Activation of C–CN Bonds

The C–CN bond in aromatic and vinylic nitriles can be cleaved in the presence of transition metals. Generally there are two pathways to undergo C–CN bond activation (Scheme 7.32): One pathway is direct oxidative addition of a C–CN bond to the metal center, like organic halides, followed by insertion of unsaturated bonds, or transmetalation with another organometallic reagent to undergo a cross-coupling reaction; The other pathway proceeds in the presence of disilanes as necessary coreactants, affording the corresponding silylated derivatives of arenes or alkenes and silylcyanide.

As mentioned, a series of insertion reactions of unsaturated organic molecules into C–CN bonds could be achieved. For example, acyl cyanides **7.3.1.1** underwent oxidative addition to transition metals like rhodium and palladium, generating acylmetal species **7.3.1.2**, which led to catalytic acylcyanation of terminal alkynes to afford **7.3.1.3** (Scheme 7.33) [38].

Moreover, aryl and allyl cyanides underwent 1,2-addition onto alkynes in the presence of Ni(0) catalysts with high regio- and stereoselectivities (Scheme 7.34) [39]. Even acetonitrile underwent the carbocyanation with the aid of Lewis acids $AlMe_3$ as co-catalyst.

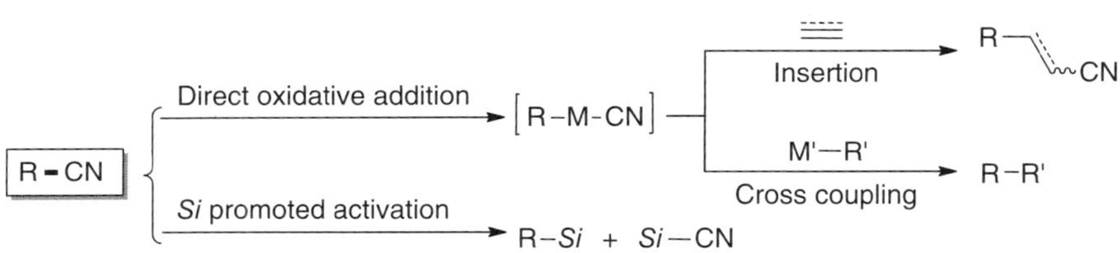

Scheme 7.32.

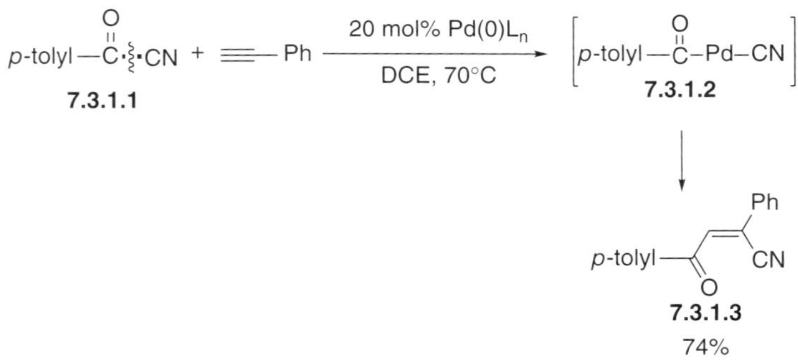

Scheme 7.33.

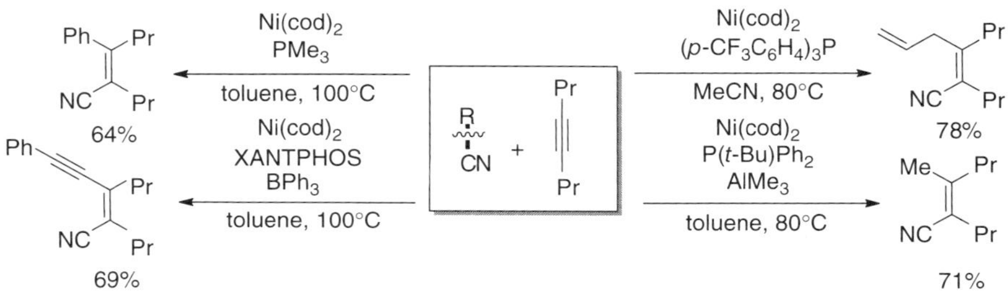

Scheme 7.34.

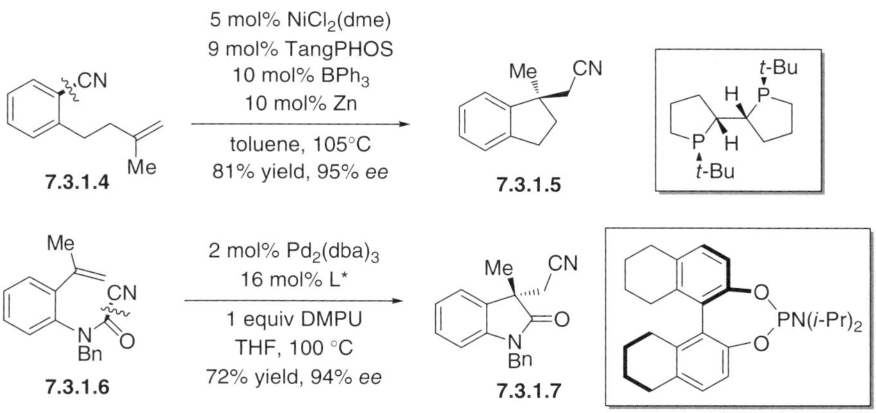

Scheme 7.35.

Arylcyanation across alkenes could occur in an intramolecular version (Scheme 7.35). Asymmetric arylcyanation of alkene-tethered benzonitriles **7.3.1.4** was promoted by a Ni(0)–TangPHOS–BPh_3 catalyst system to afford 1-indanylacetonitrile derivatives **7.3.1.5** in excellent enantiomeric excesses [40]. Besides, activation of a C–C bond between a carbonyl sp^2 carbon and a cyano sp carbon is also possible.

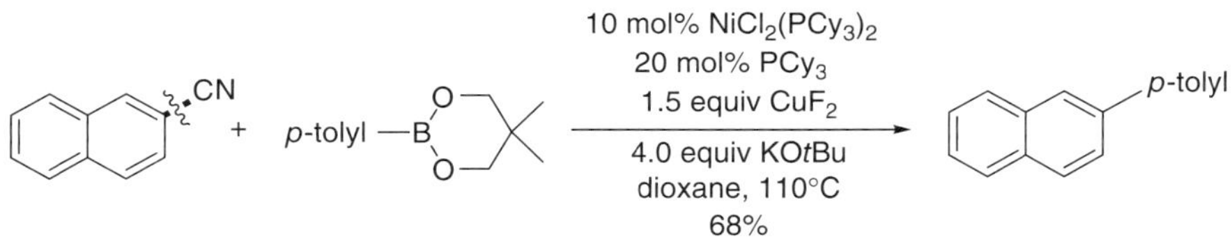

Scheme 7.36.

Carbamoyl cyanides **7.3.1.6** bearing an alkene moiety cyclized in an enantioselective manner to give chiral 3,3-disubstituted oxindole derivatives **7.3.1.7** [41].

One interesting example of C–CN bond cleavage of aryl nitriles is the Ni-catalyzed Suzuki–Miyaura coupling with aryl/alkenyl boronic esters (Scheme 7.36) [42]. With this report, the cyano group might be applied as a leaving group to construct biaryl and polyarene scaffolds.

Besides the direct oxidative addition of a C–CN bond to a transition metal center, the other pathway proceeds in the presence of disilanes as co-reactants. For example, aryl cyanides **7.3.1.8** and hexamethyldisilane **7.3.1.9** underwent a substitution reaction to afford the corresponding arylsilanes **7.3.1.13** and silyl isocyanide (Scheme 7.37) [43]. The first step was substitution of an anion of the Rh(I) complex with silyl group to give a silyl–Rh(I) complex **7.3.1.10**. Then C–C bond activation proceeded by silylmetalation of the cyano group to form η^2-iminoacyl complex **7.3.1.11**. The subsequent deinsertion of silyl isocyanide gave rise to an arylrhodium(I) complex **7.3.1.12**, which reacted with the disilane **7.3.1.9** to regenerate the silyl–rhodium complex **7.3.1.10** with the concomitant formation of an arylsilane **7.3.1.13** and silyl isocyanide, which should isomerize to thermodynamically more stable silyl cyanide.

7.3.2 C–C Activation of C–C=X Bonds (X = O, N)

As mentioned, for C–C bond activation of unstrained molecules, employing chelating assistance to undergo cyclometallation is facile, because a five-membered ring metallacyclic complex has a special thermodynamically stability for lowering the energy

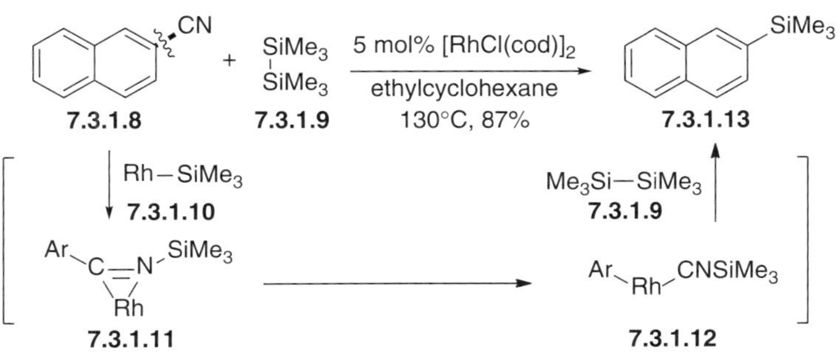

Scheme 7.37.

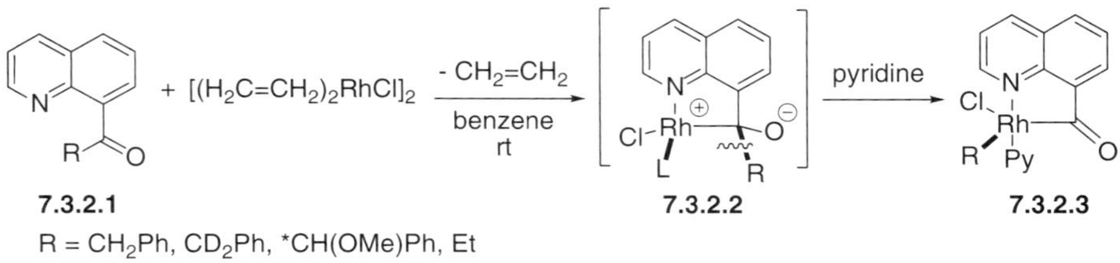

Scheme 7.38.

state of the C–C bond cleaved intermediates or products. This strategy is used in a variety of substrates, such as ketones, imines, and alcohols.

An early example is 8-quinolyl alkyl ketones **7.3.2.1** as substrates for C–C bond cleavage, as shown in Scheme 7.38 [44]. The formation of complexes **7.3.2.3** by insertion between the carbonyl carbon and the α-carbon is favored due to the general preference for five-membered chelate rings observed for cyclometalated complexes. Interestingly, the chirality of the α-carbon is retained during the insertion process as exemplified. It was proposed that a tetrahedral intermediate **7.3.2.2** arising from direct attack of rhodium on the carbonyl carbon occurs en route to C–C bond cleavage without the intervention of a C–H activation process.

Carbonyl groups are also utilized in catalytic C–C bond cleaving reactions. Under catalytic conditions, 8-quinolyl phenyl ketone **7.3.2.4** reacted with ethylene to give 8-quinolyl ethyl ketone **7.3.2.7** and styrene in quantitative yield (Scheme 7.39) [45]. Styrene was formed by cleavage of the phenyl–carbonyl bond to afford **7.3.2.5**, followed by ethylene insertion into the resultant phenyl–rhodium bond and then β-hydride elimination. The accompanying formation of a rhodium–hydride complex **7.3.2.6** was followed by incorporation of ethylene to furnish the ethyl ketone **7.3.2.7**.

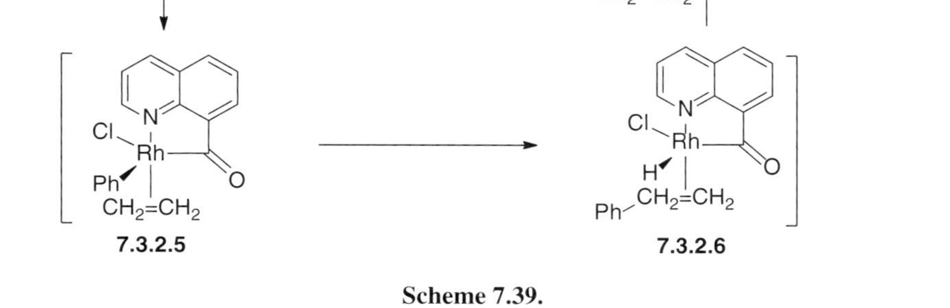

Scheme 7.39.

Interestingly, C–C bond activation of ketones can also be achieved bearing non coordinating functionalities by conducting the temporary installation of an appropriate chelation auxiliary—for example, 2-amino-3-picoline, which can be easily removed from the product by a simple manipulation such as hydrolysis [46]. In this case, benzylacetone **7.3.2.8** reacted with excess olefin, *tert*-butylethylene, under co-catalysts of Wilkinson's complex and 2-amino-3-picoline to give an alkyl group-exchanged ketone **7.3.2.9** and a trace of styrene (Scheme 7.40).

The first step might be the formation of ketimine **7.3.2.10** by the condensation of ketone **7.3.2.8** and 2-amino-3-picoline (Scheme 7.41). Then, the C–C bond of ketimine **7.3.2.10** was cleaved by the Rh(I) complex to generate an (iminoacyl)rhodium(III) phenethyl **7.3.2.11**, which underwent β-hydride elimination giving an (iminoacyl)rhodium(III) hydride **7.3.2.12** and styrene. The hydrometalation of **7.3.2.12** into *tert*-butylethylene and the subsequent reductive elimination afforded ketamine **7.3.2.14**. Hydrolysis of ketimine **7.3.2.14** led to the formation of ketone **7.3.2.9** with regeneration of 2-amino-3-picoline.

One interesting example of this chelation-assisted C–C bond activation is the skeletal rearrangement of cyclic ketone **7.3.2.15** to afford a mixture of the ring-contracted cycloalkanones, **7.3.2.16** and **7.3.2.17** (Scheme 7.42) [47]. The C–C bond cleavage by the Rh(I) complex followed by β-hydride elimination led to an (iminoacyl)rhodium(III) hydride **7.3.2.18**. The intramolecular hydride insertion in **7.3.2.18** gave an (iminoacyl)-rhodium(III) alkyl **7.3.2.19**, which was transformed into **7.3.2.20** through reductive elimination. Likewise, the ketimine **7.3.2.22** was

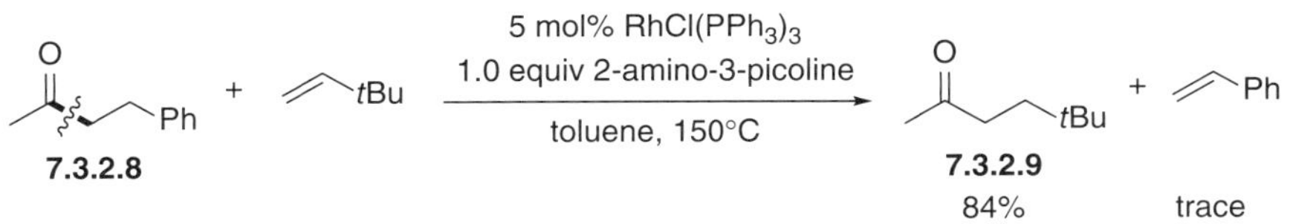

Scheme 7.40.

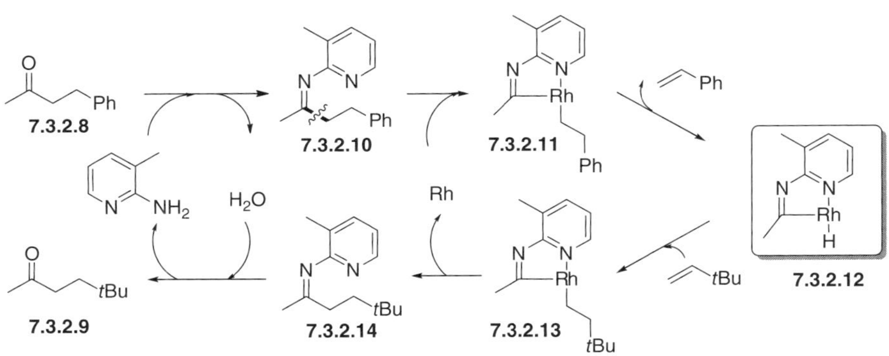

Scheme 7.41.

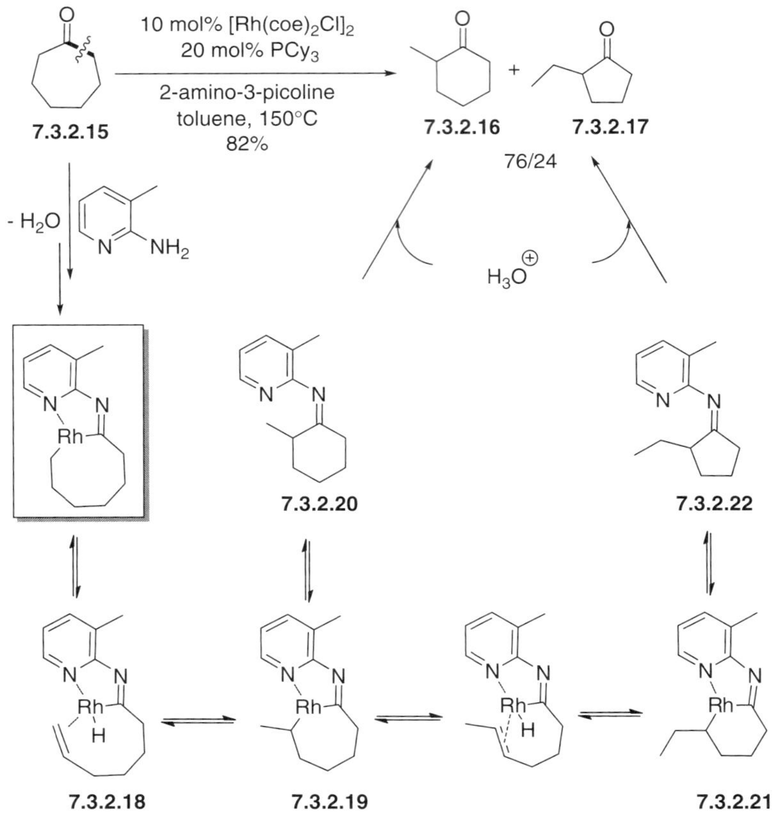

Scheme 7.42.

also formed from **7.3.2.19** via **7.3.2.21**. Since all these intermediates were in thermodynamic equilibrium, a hydrometalation by Markovnikov's rule (i.e., **7.3.2.18** to **7.3.2.19**) was an unfavorable process. However, additional stabilities of five- and six-membered rings of ketimines (**7.3.2.20** and **7.3.2.22**) compared with that of the seven-membered ring of cycloheptanoketimine seemed to drive a forward reaction.

Although the C–C activation of C–C=X bond is the main topic in this section, it is noteworthy that the chelation-assisted strategy can also be applied to secondary alcohol **7.3.2.23** under a similar catalytic system (Scheme 7.43) [48]. The reaction consisted of two consecutive reactions: a hydrogen transfer reaction of secondary alcohol **7.3.2.23** and a chelation-assisted C–C bond activation to afford the ketone **7.3.2.9**.

Allylamine **7.3.2.24** can be regarded as a masked form of formaldehyde since it undergoes C–C bond activation as well as C–H bond activation by the Rh(I) catalyst. The reaction of an allylamine **7.3.2.24** and *tert*-butylethylene yielded symmetric dialkyl ketone **7.3.2.30** as a major product after hydrolysis of the resulting ketimine

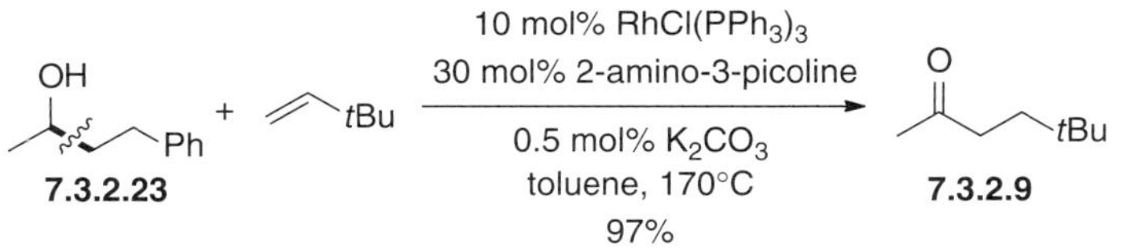

Scheme 7.43.

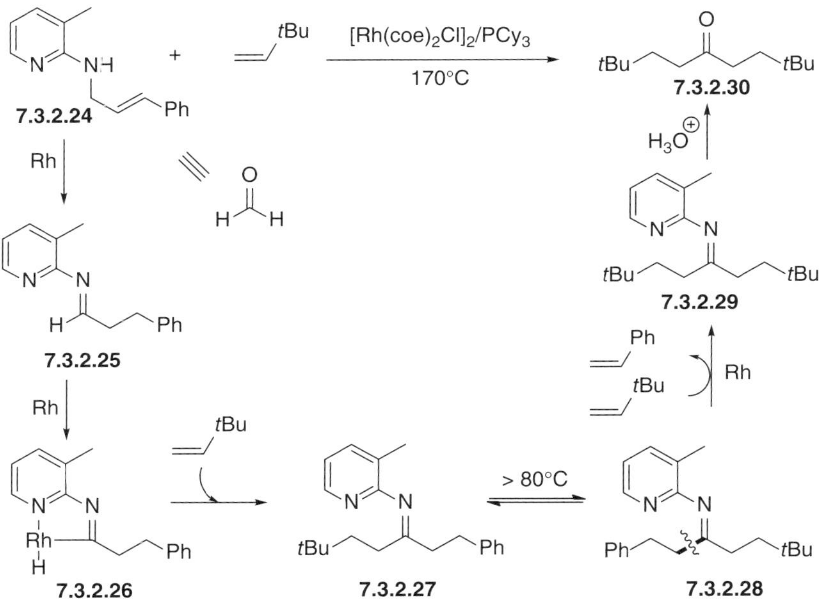

Scheme 7.44.

7.3.2.29 (Scheme 7.44) [49]. The first step was an isomerization of **7.3.2.24** to an aldimine **7.3.2.25**, followed by chelation-assisted C–H bond activation. A hydride insertion of the resulting (iminoacyl)-rhodium(III) hydride **7.3.2.26** into olefin and the subsequent reductive elimination of (iminoacyl)rhodium(III) alkyl led to the formation of ketimine **7.3.2.27**. It is known that above 80°C an imine like **7.3.2.27** underwent *syn–anti* isomerization allowing the rhodium catalyst to cleave the C–C bond in **7.3.2.28**. The catalytic C–C bond activation of **7.3.2.28** in the reaction with *tert*-butylethylene resulted in symmetric dialkyl ketimine **7.3.2.29**.

A further application of allylamine **7.3.2.24** is a facile synthesis of cycloalkanone derivative **7.3.2.32** from diene **7.3.2.31** (Scheme 7.45) [50]. With this synthetic strategy, from five- to seven-membered ring cycloalkanones could be prepared from 1,3- to 1,5-dienes.

One interesting example is Rh(I)-catalyzed decarbonylation process through dual C–C cleavage of biaryl ketones **7.3.2.33** with the assistance of a pyridinyl group

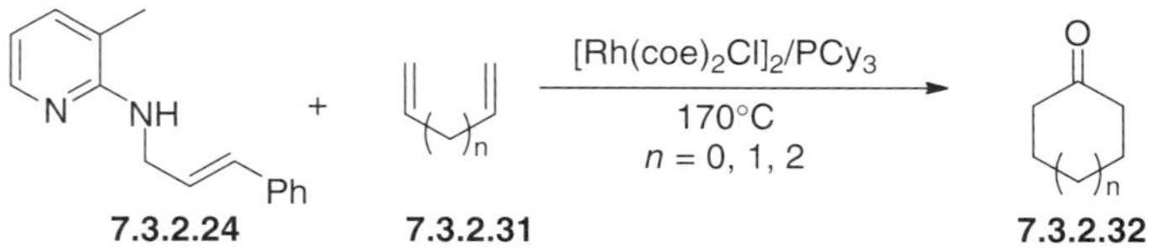

Scheme 7.45.

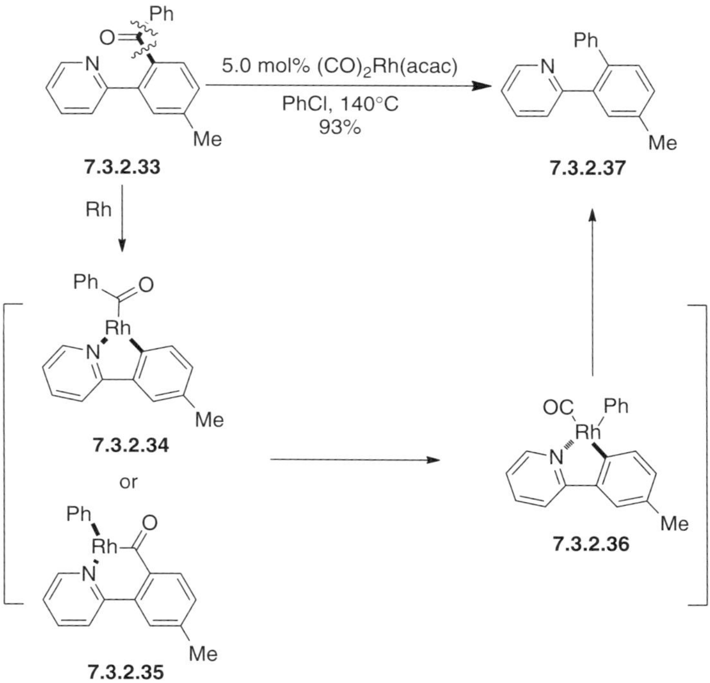

Scheme 7.46.

(Scheme 7.46) [51]. The decarbonylation was initiated from the oxidative cleavage of the C–C bond in the square-planar Rh(I) complex with the assistance of a pyridinyl group to form the octahedral acylrhodium(III) species **7.3.2.34** or **7.3.2.35**. After the reverse migratory insertion to afford **7.3.2.36**, reductive elimination produced the final product **7.3.2.37**. This method offers an alternative way to synthesize biaryl structures.

In the presence of Cu(I) catalyst, an efficient arylation/C–C activation process of β-diketones **7.3.2.38** with aryl halides could occur to afford α-aryl ketones **7.3.2.43** (Scheme 7.47) [52]. Aryl halide could oxidatively add to the Cu(I) complex **7.3.2.39** to generate the Cu(III) intermediate **7.3.2.40**. In the presence of H_2O, the C–C bond activation/cleavage occurred, which led to the formation of intermediate **7.3.2.41**

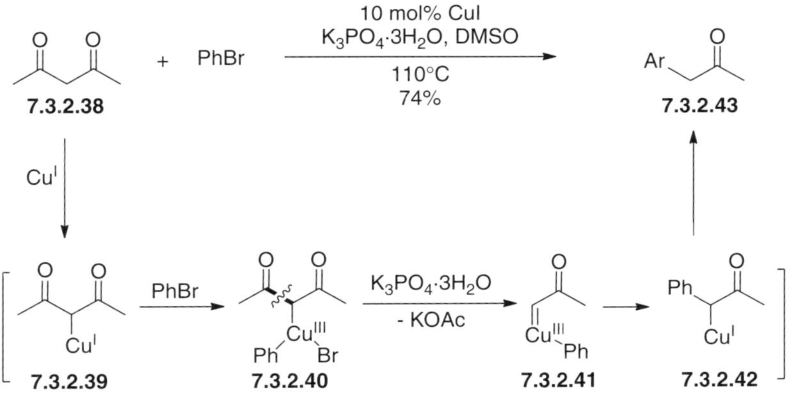

Scheme 7.47.

and the release of KOAc. Reductive elimination of intermediate **7.3.2.41** could produce the other Cu(I) intermediate **7.3.2.42**, which finally produced the α-aryl ketones **7.3.2.43**. This method is simple, general, and practical, which complemented the classic method for the rapid construction of C–C bonds to a carbonyl moiety.

7.3.3 C–C Activation of C–C–OH Bonds in Tertiary Alcohol Substrates

In the previous section, β-carbon elimination is a good way to relieve the ring-strain energies in strain molecules. Nevertheless, even with unstrained molecules it is possible to cleave the C–C bond through β-carbon elimination. Tertiary alcohols are one of the successful substrates for this selective C–C bond cleavage.

An example of stoichiometric C–C activation of tertiary alcohols was achieved by the β-aryl eliminations of rhodium(I) tertiary alkoxide complexes **7.3.3.1** under mild conditions, forming rhodium aryl complexes **7.3.3.3** and free ketones **7.3.3.4** (Scheme 7.48) [53]. After dissociation of a phosphine from the trisphosphine complexes **7.3.3.1**, the resulting bisphosphine-ligated rhodium alkoxide complex **7.3.3.2**, stabilized by Rh-C(phenyl) interactions, underwent irreversible β-phenyl elimination, which was consistent with kinetic results. The structure of the bisphosphine β-aryl alkoxide complex **7.3.3.2** illustrated the pathway for C–C bond cleavage.

An analogous C–C bond activation occurred with tertiary benzylic alcohols **7.3.3.5** catalyzed by palladium species (Scheme 7.49) [54]. The β-aryl elimination

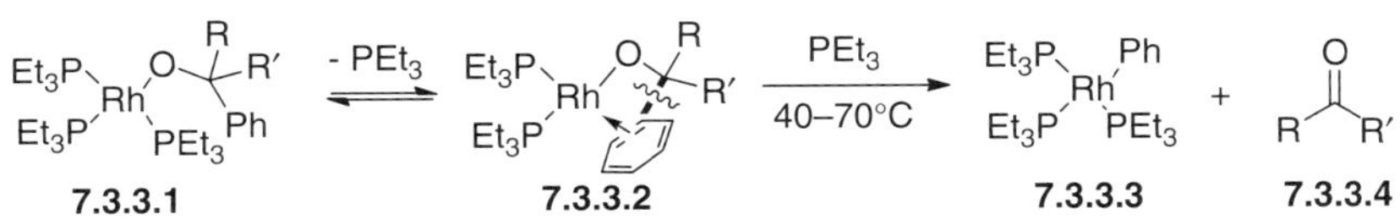

Scheme 7.48.

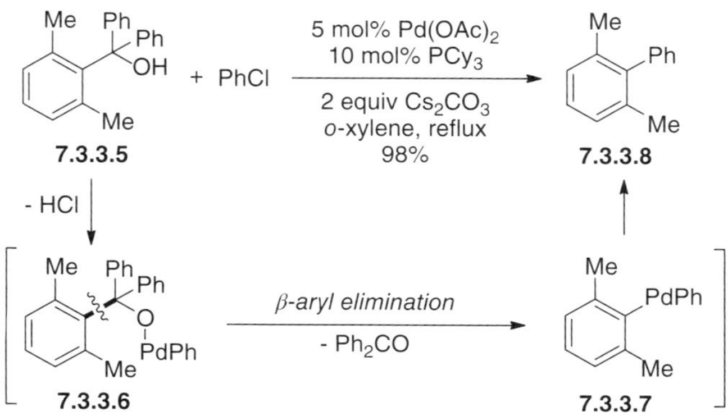

Scheme 7.49.

of palladium(II) alcoholate **7.3.3.6** formed *in situ* was an important step in which a strong aryl sp^2 carbon–palladium bond in **7.3.3.7** was generated, followed by reductive elimination to furnish biaryls **7.3.3.8**.

Chelation-assisted strategy can also be employed in the cleavage of a $C(sp^3)$–$C(sp^3)$ bond of tertiary alcohols. For example, treatment of 2-(2-pyridyl) ethanol derivatives **7.3.3.9** with aryl chlorides in the presence of a palladium catalyst resulted in the transfer of the pyridylmethyl moiety of the alcohol to yield the corresponding (2-pyridylmethyl)arene **7.3.3.13** (Scheme 7.50) [55]. With the aid of the coordination, the aryl palladium alkoxide intermediate **7.3.3.10** was likely to undergo cleavage of the $C(sp^3)$–$C(sp^3)$ bond to yield palladium amide **7.3.3.11**, which would immediately isomerize to aryl(2-pyridylmethyl)palladium **7.3.3.12** to recover aromaticity. Further reductive elimination generated the product **7.3.3.13**.

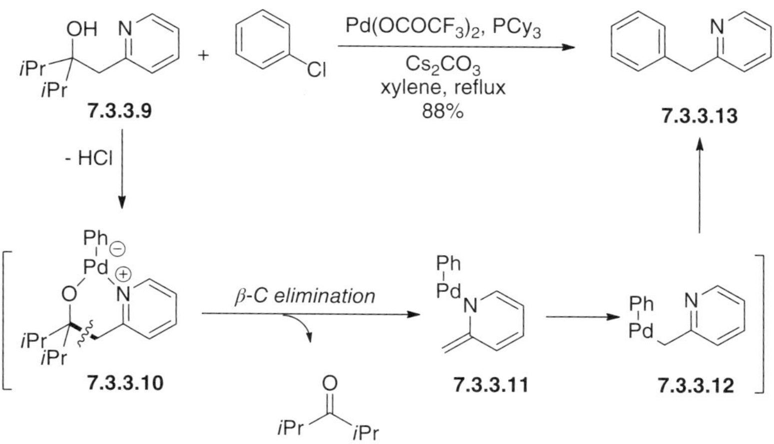

Scheme 7.50.

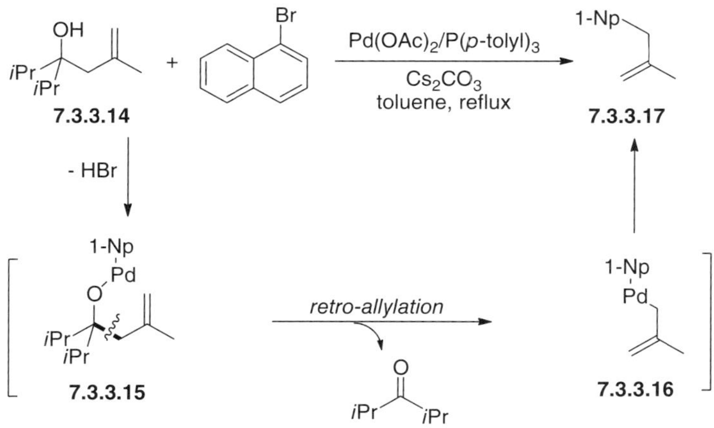

Scheme 7.51.

A similar example is the treatment of tertiary homoallyl alcohol **7.3.3.14** with aryl halide under palladium catalysis that resulted in the transfer of the allyl moiety of the homoallyl alcohol to aryl halide and yielded the corresponding cross-coupling product **7.3.3.17** stereo- and regiospecifically (Scheme 7.51) [56]. The transfer process included retro-allylation via a conformationally regulated six-membered transition state from aryl palladium alkoxide intermediate **7.3.3.15** to the σ-allylpalladium **7.3.3.16**.

Another example of C–C activation for tertiary homoallyl alcohols is Ru-catalyzed deallylation of **7.3.3.18**, which was achieved via the oxidative addition of a hydroxyl group and subsequent β-allyl elimination of **7.3.3.19** to afford acetophenone and allylruthenium(II) species **7.3.3.20**. Reductive elimination afforded propene as the final product (Scheme 7.52) [57].

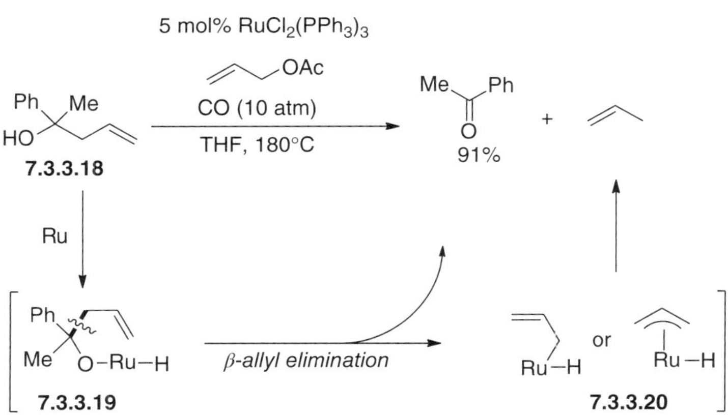

Scheme 7.52.

A 2-hydroxy-2-methylbutynyl group serves as a masked anionic ethynyl group because acetone is easily expelled on treatment with a base such as NaOH. Similarly, tertiary propargylic alcohols can be utilized as the precursor for a transition metal acetylide *via* β-alkynyl elimination with the release of a ketone. One interesting example, **7.3.3.21** reacted with ethyl acrylate to give enyne compounds **7.3.3.24** under the palladium catalyst (Scheme 7.53) [58]. During the reaction, a very strong alkynyl *sp* C–Pd bond in **7.3.3.23** was generated by β-alkynyl elimination of **7.3.3.22**.

Another example, a Rh-catalyzed asymmetric rearrangement of tertiary alkenyl(alkynyl)carbinols **7.3.3.25** proceeded via β-alkynyl elimination and the following 1,4-addition to the resulting α,β-carbonyl compounds **7.3.3.28** (Scheme 7.54) [59]. As a result, the alkynyl group underwent 1,3-migration from intermediate **7.3.3.26** to product **7.3.3.28**.

7.3.4 C–C Activation of C–C–OH Bonds in Secondary and Primary Alcohol Substrates

Unlike tertiary alcohols, catalytic C–C bond cleavage of secondary alcohols has rarely been reported. The major challenge remaining in cleaving the C–C bond

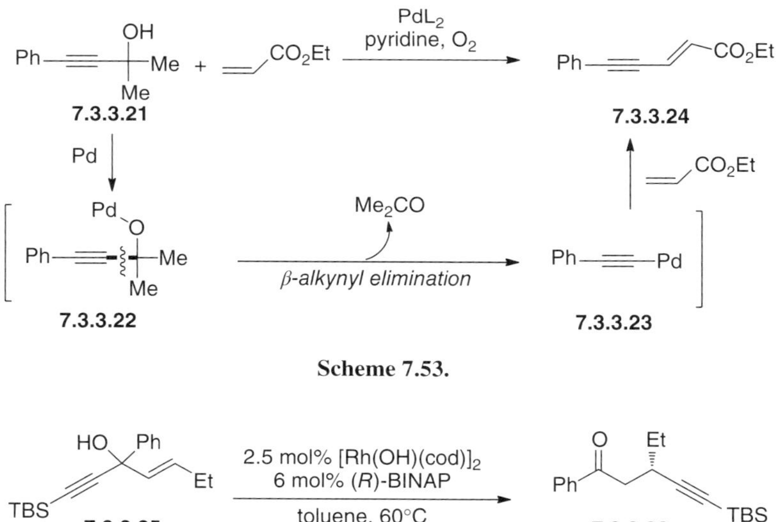

Scheme 7.53.

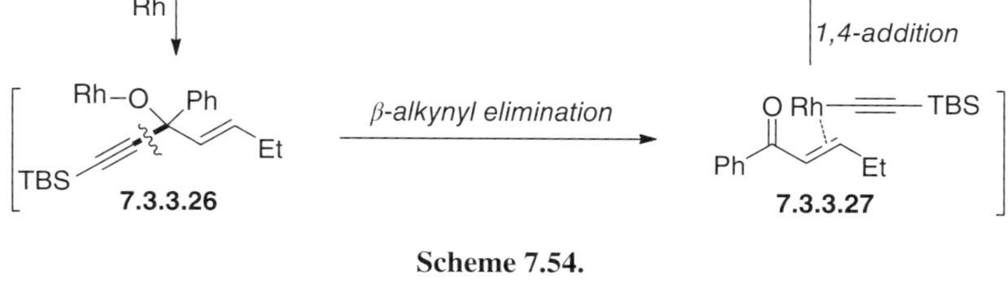

Scheme 7.54.

adjacent to the hydroxyl group of secondary alcohols is to avoid hydrogen transfer, which is the typical transformation of secondary alcohols in the presence/absence of external oxidants. For example, chelation-assisted C–C bond activation can be applied after the hydrogen transfer reaction of a secondary alcohol (Scheme 7.43).

When 5-azido-cyclopent-2-enol **7.3.4.1** was treated with a palladium(II) catalyst, the resulting palladium(II) alkoxide **7.3.4.2** underwent β-carbon elimination to cleave an unstrained C–C bond. The following extrusion of dinitrogen from an azide afforded alkylideneaminopalladium(II) species **7.3.4.3**, which then underwent intramolecular nucleophilic addition to generate dihydropyridine **7.3.4.4**. Protonation followed by dehydration gave rise to a pyridine derivative **7.3.4.6** (Scheme 7.55) [60]. In this reaction, coordination of the nitrogen might facilitate the β-carbon elimination to generate intermediate **7.3.4.3**.

As mentioned, a proper substrate for C–C bond cleavage should satisfy geometrical requirements by generating a thermodynamically favorable metallacyclic complex as the key intermediate. Recently, a series of research reports on C–C bond cleavage of secondary alcohols **7.3.4.7** through Rh(III)-catalyzed β-carbon elimination directed by a pyridinyl group have been published. In this reaction, a 5-membered rhodacycle **7.3.4.9** generated from β-carbon elimination of rhodium(III) alkoxide **7.3.4.8**, was proposed as a key intermediate, which underwent further alkenylation to afford C–C bond reforming product **7.3.4.10** (Scheme 7.56) [61].

Interestingly, because the five-membered rhodacycle intermediate **7.3.4.9** has also been demonstrated in C–H activation, "dual activation" of secondary alcohol **7.3.4.11** was achieved by sequential C–C bond cleavage/C–H activation by stepwise addition of styrene and 4-methylstyrene in one pot, affording di-alkenylated product **7.3.4.12** with good efficiency and selectivity (Scheme 7.57). This unique activity highlights the potential utility of this C–C cleavage in postsynthetic elaboration of complex natural products. Thus, rationally designed alcohol substrates and their alkenylated products can serve as versatile synthetic intermediates.

Furthermore, in the presence of H_2, reductive cleavage of the C(sp^2)–C(sp^3) bond of secondary alcohols **7.3.4.7** took place smoothly (Scheme 7.58) [62].

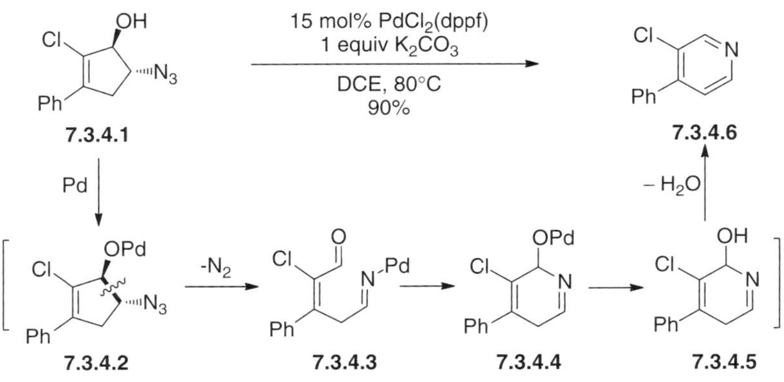

Scheme 7.55.

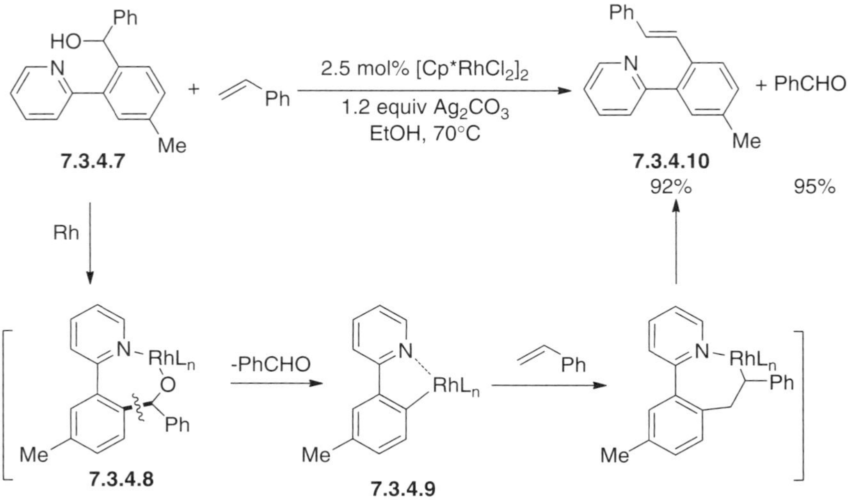

Scheme 7.56.

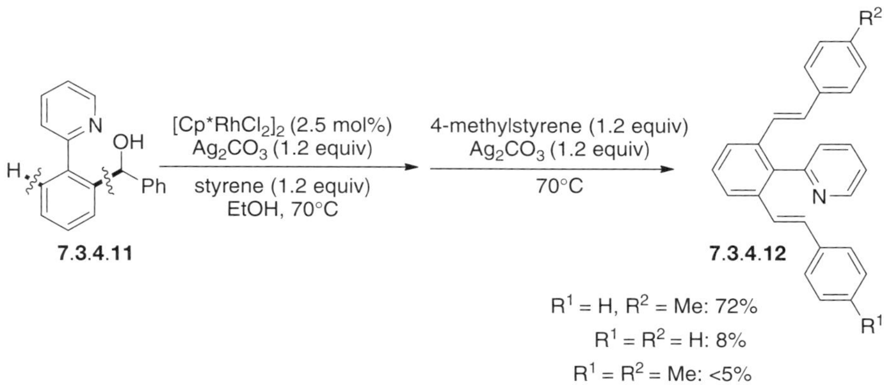

Scheme 7.57.

Preliminary studies indicated that the five-membered rhodacycle intermediate **7.3.4.9** converted into a rhodium(III) hydride species **7.3.4.13** for the reduction. Both pyridinyl and pyrazolyl groups could serve as directing groups.

Moreover, the five-membered rhodacycle intermediate **7.3.4.9** could undergo oxidative arylation with arylsilanes. It is noteworthy that C–C bond cleavage of primary alcohol **7.3.4.15** was achieved, albeit with low efficiency (Scheme 7.59) [63].

Interestingly, based on the concept of nucleophilic addition of five-membered rhodacycle intermediate to imines or aldehydes, direct transformation of secondary

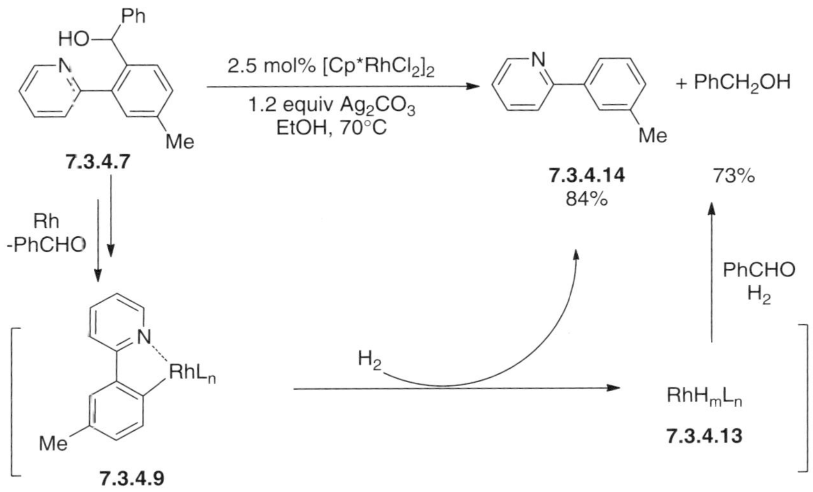

Scheme 7.58.

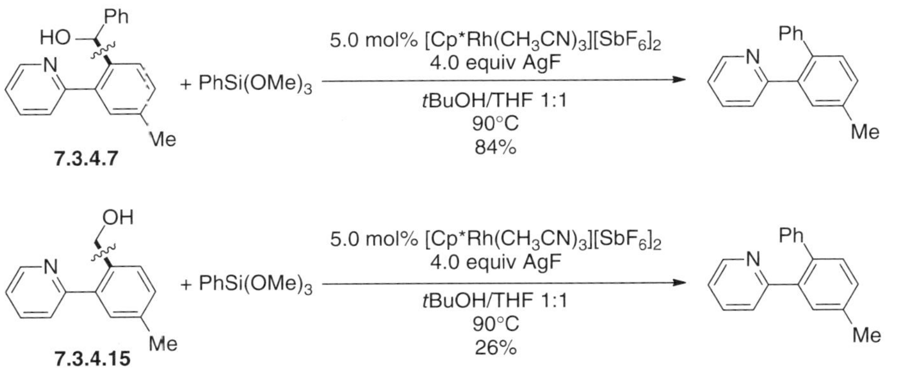

Scheme 7.59.

alcohols to tosylamines or other alcohols by group substitution could be achieved by the C–C bond cleavage of initial substrates (Scheme 7.60) [64]. The reaction conditions were redox-economic and no additive was required. Like the "transplant," this method offers a new concept that can be used to synthesize the desired products through direct reorganization of the skeleton.

Another important example for C–C bond activation of secondary alcohols was the generation of ω-dienyl aldehydes **7.3.4.18** from 1,3-diols **7.3.4.16**, based on Pd(0)-catalyzed dehydration fragmentation (Scheme 7.61) [65]. The allylic secondary alcohol **7.3.4.16** could be catalytically transformed into a π-allylpalladium species **7.3.4.17** in the presence of 9-PhBBN. The palladium(II) of the thus formed

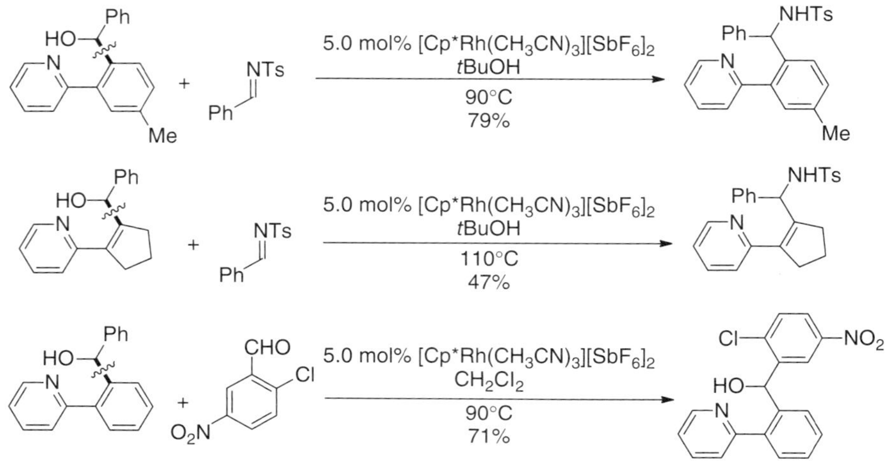

Scheme 7.60.

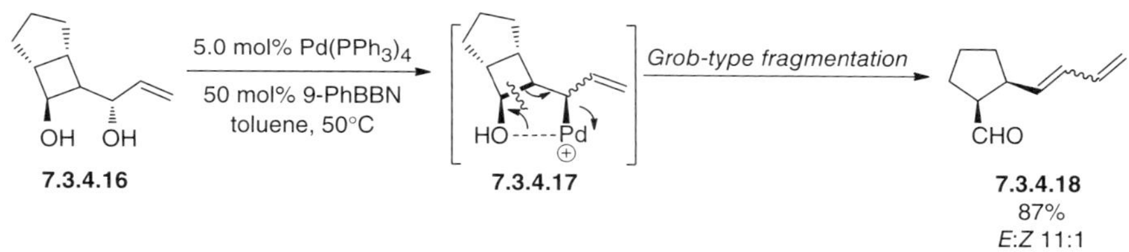

Scheme 7.61.

π-allylpalladium intermediate **7.3.4.17** might serve as a leaving group to facilitate the Grob-type fragmentation, thereby regenerating the ω-dienyl aldehyde **7.3.4.18**. This reaction demonstrates the ring-opening of an array of cycloalkanols, ranging from cyclobutanol to cyclodecanol under essentially neutral conditions.

Secondary (*E*)-allyl *vic*-diols **7.3.4.19** could undergo C–C bond cleavage in the presence of the second-generation Grubbs catalyst, affording the corresponding α,β-unsaturated aldehydes **7.3.4.21** (Scheme 7.62) [66]. The C=C bond adjacent

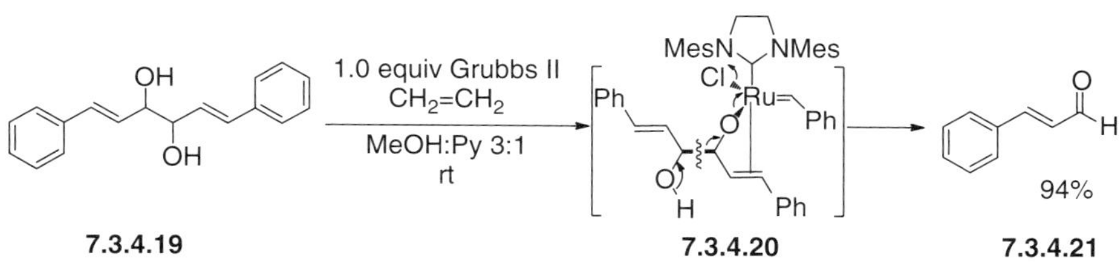

Scheme 7.62.

to a *vic*-diol plays a crucial role in the cleavage reaction as the coordination to the ruthenium center in intermediate **7.3.4.20**. This strategy is very useful to stereostructural analysis in natural products chemistry.

7.3.5 C–C Activation of C–Allyl Bonds

The bond between the carbon atoms α and β to a C=C bond can be broken by a transition metal with formation of a π-allyl intermediate providing the driving force. For example, isomerization of allylic malonate **7.3.5.1** to the thermodynamically favored linear isomer **7.3.5.3** took place on treatment with a palladium catalyst (Scheme 7.63) [67]. Formation of a stabilized carbanion and π-(allyl)palladium species **7.3.5.2** facilitated the C–C bond cleavage. This result demonstrates that the transition-metal-catalyzed nucleophilic substitution of an allylic substrate with a carbon nucleophile is reversible, if the cleaved nucleophile is sufficiently stabilized.

Another important example, the Rh-catalyzed protocol, was suitable for deallylation of various monoallyl- and diallyl-substituted malonates bearing the unsubstituted double bond at the allylic moiety, in the presence of excess of organoaluminums (Scheme 7.64) [68]. First, a rhodium hydride, generated from the chloride complex and triethylaluminum, hydrometalated the C=C bond of an allylmalonate **7.3.5.4** to form the *sec*-alkylrhodium species **7.3.5.5** with coordinated triethylaluminum to the

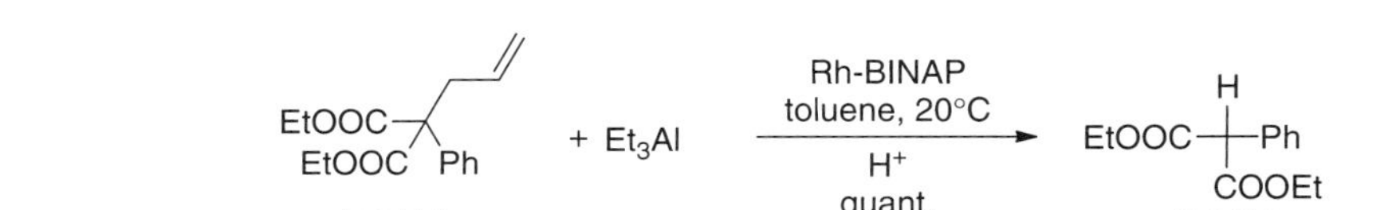

Scheme 7.63.

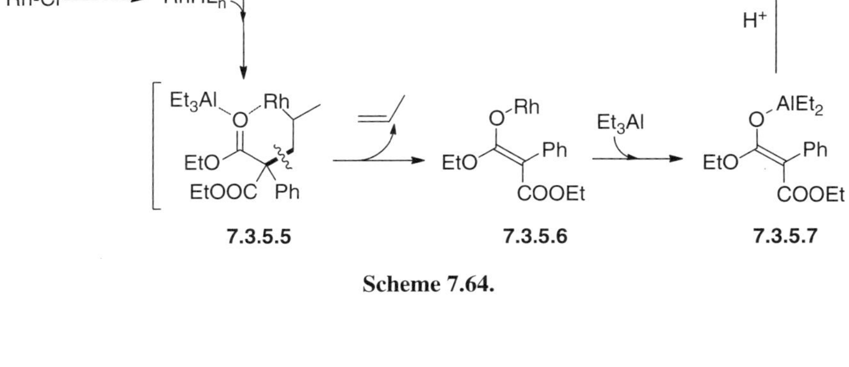

Scheme 7.64.

carbonyl group. Then a sequence of C–C, C–O, C–Rh, and Rh–O bond formations and cleavages through a six-membered transition state promoted by coordination of triethylaluminum to the carbonyl group followed, resulting in the formation of a rhodium enolate **7.3.5.6**. During this process the allyl moiety was released as an alkene. Finally, the transmetalation of the rhodium enolate with triethylaluminum furnished an aluminum enolate **7.3.5.7**, which underwent protonation to afford deallylation product **7.3.5.8**.

7.3.6 C–C Activation of Pincer-Type Substrates

The presence of coordinating functionalities in a target molecule induces "precoordination," whereby a transition metal is brought into the close proximity of a specific C–C bond, thus facilitating insertion of metal. The diphosphine pincer-type chelating ligands can undergo selective cleavage of an alkyl group attached to an aromatic ring.

The first example of metal insertion into an unstrained, unactivated C–C bond of pincer-type substrates was demonstrated with [HRh(PPh$_3$)$_4$]. The use of a metal hydride complex was intended to provide a substantial driving force to the overall C–C activation process by the expected methane elimination following insertion into the C–C bond. Reaction of [HRh(PPh$_3$)$_4$] with **7.3.6.1** at room temperature resulted in elimination of H_2 and formation of the kinetic C–H activation product **7.3.6.2** (Scheme 7.65) [69]. Significantly, the C–H activation process can be reversed by heating of this product under a mild hydrogen pressure, resulting in quantitative C–C cleavage product **7.3.6.3** and methane elimination.

As the complex with the activated C–C bond does not react with methane, the overall C–C activation process is irreversible and thermodynamically more favorable than the overall C–H activation sequence. The use of hydrogen to drive the reaction toward C–C activation could potentially be utilized as a general strategy to overcome the problem of C–H activation which is expected to accompany C–C activation in hydrocarbons.

Employment of hydrogen in the above-mentioned process masks the relative thermodynamic stability of the C–H and C–C activation products, leaving open the questions of the thermodynamic feasibility of the oxidative addition of a strong C–C bond in the absence of added reagents and the thermodynamic interplay between C–C and C–H oxidative addition. Direct C–C insertion in the absence of hydrogen proved to be thermodynamically feasible with the pincer-type substrates. Heating **7.3.6.4** with [Rh(PEt$_3$)$_3$Cl] led to the product **7.3.6.5** of direct C–C activation quantitatively (Scheme 7.66) [70].

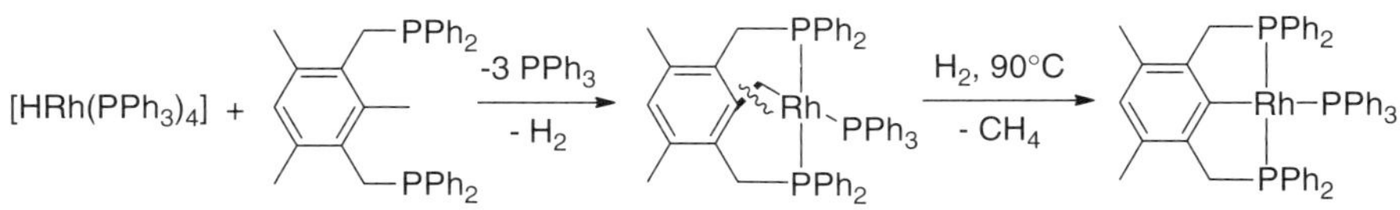

7.3.6.1 **7.3.6.2** **7.3.6.3**

Scheme 7.65.

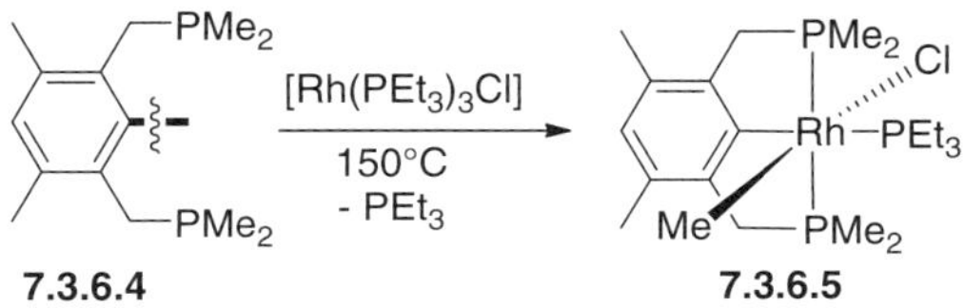

Scheme 7.66.

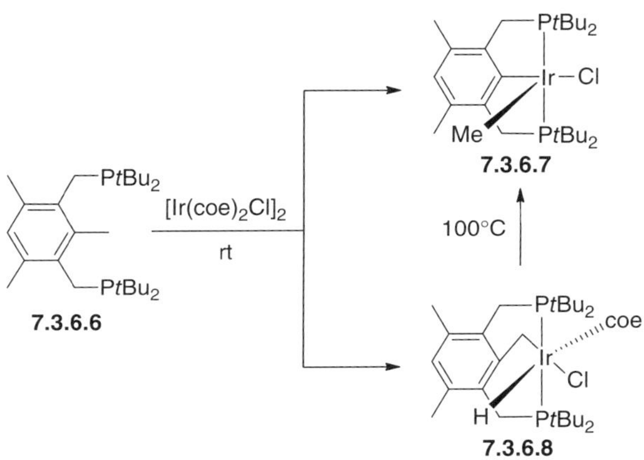

Scheme 7.67.

Concurrent C–H and C–C activation took place in the reaction of $[Ir(coe)_2Cl]_2$ with the pincer type substrate **7.3.6.6** at room temperature. The product of C–H activation, complex **7.3.6.8**, was quantitatively converted into the C–C activation product, complex **7.3.6.7**, upon moderate heating (Scheme 7.67) [71].

The above-mentioned reactivity unequivocally proves that the insertion of iridium and rhodium into the C–C bond is thermodynamically more favorable than insertion into the C–H bond. The C–H and C–C insertion products were formed irreversibly within the temperature range of 20–60°C, indicating that the C–C and C–H activation processes are kinetically controlled. The C–H activation product **7.3.6.8** is not an intermediate in the C–C activation process, as verified by the observation that it does not convert into the C–C activation product under the reaction conditions. Both the C–C and C–H activation processes were shown to proceed through a common intermediate with the two phosphane arms coordinated to the metal center, thus taking into account that the complexes are formed in two independent, concurrent processes.

The C–C bond in pincer-type substrate **7.3.6.9** was also cleaved by platinum species (Scheme 7.68) [72]. The kinetic C–H activation product **7.3.6.10** was quantitatively converted into the C–C activation product **7.3.6.11** by addition of HCl.

For the substrate **7.3.6.9**, the C–C bond can be catalytically cleaved by $[Rh(coe)_2Cl]_2$ under a mild H_2 pressure or with an excess of $HSi(OEt)_3$, resulting in

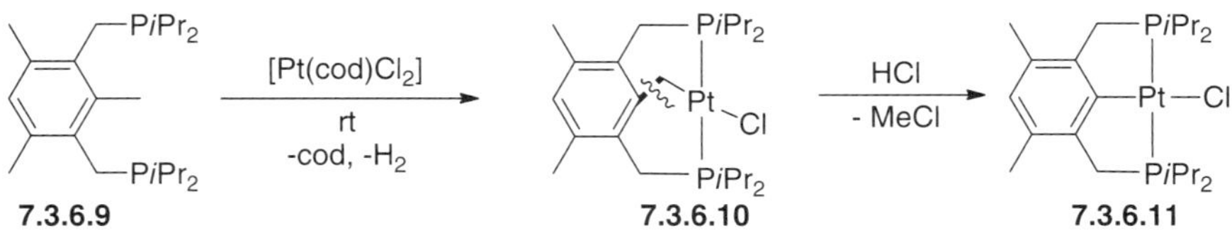

Scheme 7.68.

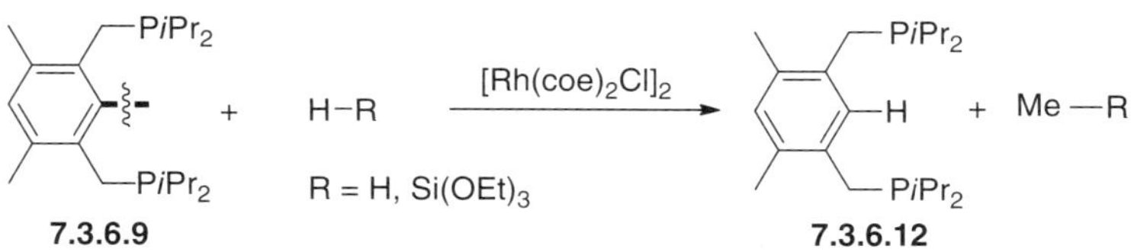

Scheme 7.69.

the product **7.3.6.12** (Scheme 7.69) [73]. Although the catalytic reactions were not optimized, more than 100 turnovers were observed in the case of H_2.

7.3.7 C–C Activation of Miscellaneous Substrates

Besides the above-mentioned, there are also other pathways and substrates that can affect C–C bond cleavage. Substitution reactions of propargylic amines proceeded smoothly in the presence of Cu(I) catalysts (Scheme 7.70) [74]. Mechanistic studies showed that C(*sp*)–C(sp^3) bond cleavage of propargylic amines **7.3.7.1** assisted by nitrogen lone-pair electrons was essential for generating copper acetylides **7.3.7.2** and iminium intermediates **7.3.7.3**, which underwent fragment exchange with additional amines, aldehydes, and alkynes to give rise to substitution products **7.3.7.4** (Scheme 7.71). Because iminium intermediates are key to aldehyde–alkyne–amine coupling reactions, this transformation is effective not only for reconstruction of propargylic amines but also for chiral induction of racemic compounds in the presence of chiral catalysts.

The ability of transition metal species to bring about the loss of carbon dioxide/carbon monoxide from various carboxyl/carbonyl compounds provides a

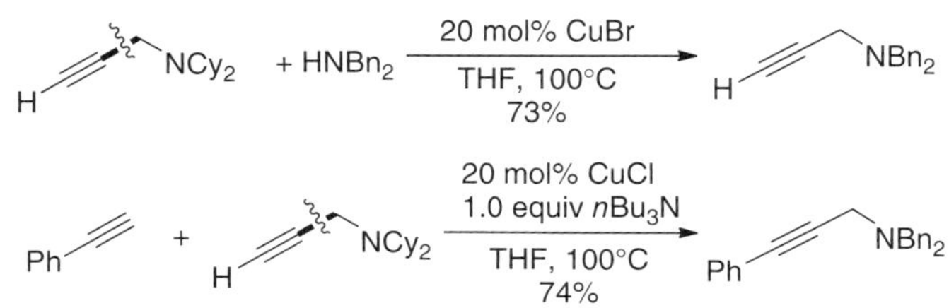

Scheme 7.70.

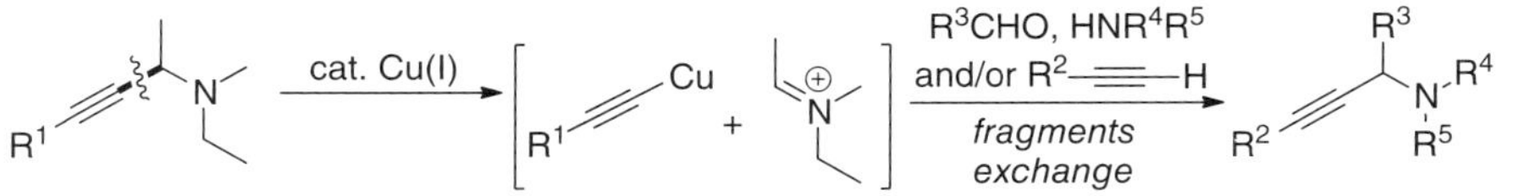

7.3.7.1 7.3.7.2 7.3.7.3 7.3.7.4

Scheme 7.71.

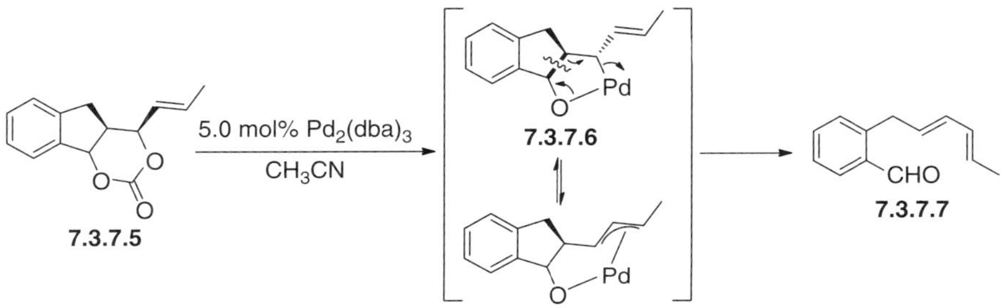

7.3.7.5 7.3.7.6 7.3.7.7

Scheme 7.72.

unique opportunity to carry out C–C bond activation. Herein only selected examples will be discussed. A six-vinyl cyclic carbonate **7.3.7.5** underwent Pd(0)-catalyzed decarboxylative C–C bond cleavage to furnish ω-diene aldehyde **7.3.7.7** (Scheme 7.72) [75]. This reaction was believed to proceed via a β-decarbopalladation, not via β-alkyl elimination. Decarboxylation of the allylic carbonate moiety provided the driving force for production of the intermediate five-membered hetero-palladacycle **7.3.7.6**, which was stabilized by π-allyl coordination to the metal catalyst.

A bimetallic catalytic decarboxylation reaction of aryl carboxylic acids **7.3.7.8** with aryl halides could give rise to a convenient cross-coupling strategy for the large-scale synthesis of biaryls **7.3.7.13**, where a copper complex capable of mediating the strongly endothermic extrusion of carbon dioxide from arenecarboxylates combines with a palladium catalyst capable of catalyzing the cross-coupling with an aryl halide (Scheme 7.73) [76].

The copper derivative would thus initially coordinate to the arylcarboxylic acid salts as **7.3.7.9** and then would shift to the aryl π-system and insert into the C–C bond under extrusion of CO_2 to form a stable aryl–copper intermediate **7.3.7.10**. Then the transmetalation with aryl palladium species **7.3.7.11** afforded diarylpalladium **7.3.7.12**, followed by reductive elimination to form the desired biaryl **7.3.7.13**.

Interestingly, it was shown that acyl halides **7.3.7.14** were decarbonylated in the presence of catalytic amounts of $[Rh(cod)Cl]_2$ and PPh_3 to react with terminal alkynes to give the corresponding chloroarylation products **7.3.7.19** (Scheme 7.74) [77]. The reaction was initiated by oxidative addition of aroyl chloride **7.3.7.14** to a Rh(I) species followed by coordination of a terminal alkyne to form an acylrhodium(III) complex **7.3.7.15**. The subsequent decarbonylation gave an intermediate arylrhodium(III) carbonyl complex **7.3.7.16**. Then, two possible pathways

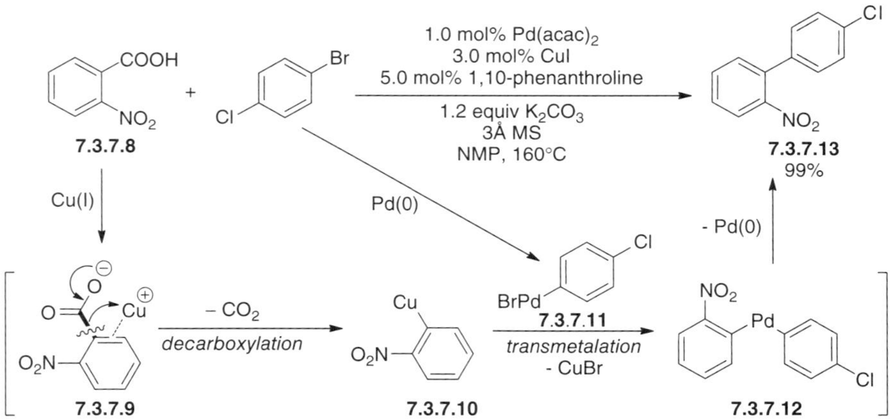

Scheme 7.73.

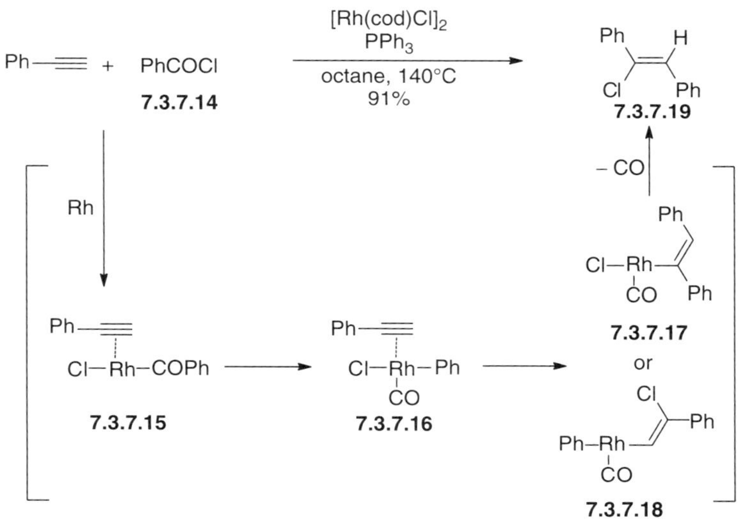

Scheme 7.74.

could take place. One could be arylrhodation in the course of which the aryl moiety migrated to the coordinated alkyne to give an alkenylrhodium(III) chloride complex **7.3.7.17**. The other is presumed to be chlororhodation where the chlorine migrated to the alkyne to give an (aryl)alkenylrhodium(III) complex **7.3.7.18**. Regardless of the arylrhodation or chlororhodation pathway, reductive elimination afforded the product **7.3.7.19** along with carbon monoxide. On the other hand, the reaction with internal alkynes afforded 2,3-disubstituted-1-indenones, involving an elimination–insertion sequence of carbon monoxide in the course of the reaction.

The acyl–rhodium complexes generated from carboxylic acid derivatives yielded decarbonylation to give rise to alkyl- and arylrhodium complexes, which could then take part in various C–C bond-forming reactions—for example, the Mizoroki–Heck reaction [78] and the Suzuki–Miyaura reaction [79].

The driving force of aromaticity can be exploited for C–C bond cleavage in prearomatic systems. For example, the reaction of 1,1-dimethylcyclopentane **7.3.7.20** with cationic iridium complex **7.3.7.21** in the presence of an olefin as a hydrogen scavenger afforded the dehydrogenated cationic 1,1-dialkyl-substituted cyclopentadienyliridium complex **7.3.7.22**, which subsequently underwent C–C bond cleavage of the ligand to give complex **7.3.7.23** (Scheme 7.75) [80]. Interestingly, no C–C cleavage was observed with the analogous saturated 18e complex, containing a dialkylcyclohexadienyl ligand, under various reaction conditions. Thus, coordinative unsaturation was suggested to play a key role in the process of C–C activation.

The energy gain from aromatization in the systems described above renders the processes thermodynamically feasible. The orientation of the alkyl group toward the metal center in the rigidly coordinated cyclopentadiene is favorable for C–C cleavage, making the process also kinetically feasible. The *endo*-alkyl group which is closer to the metal center is always the one transferred to the metal. A number of variations of this reaction have been reported with Mo [81], Fe [82], Mn [83], and Re [84].

7.4 SUMMARY AND PERSPECTIVE

In this chapter, the recent developments of C–C bond activation by homogenous transition metal catalysis are discussed. In spite of the inertness of C–C bonds in organic molecules, there has been growing interest in catalytic C–C bond activation. In order to achieve catalytic C–C bond activation several requirements need to be fulfilled. In terms of the thermodynamic stabilities of the organic substrate, the product should be more stable than the starting organic molecules. The intermediate generated from the starting organic molecules and the transition metal catalyst should be moderately stabilized, for example, by generating stable five-membered metallacyclic complexes, stable π-allyl transition metal complexes, or very strong metal–carbon bonds. The stabilization of intermediate complexes might lower the activation enthalpy in the reaction profile for transformation of the starting molecules to the intermediates. In general, there are two kinds of elementary steps available for generation of reactive

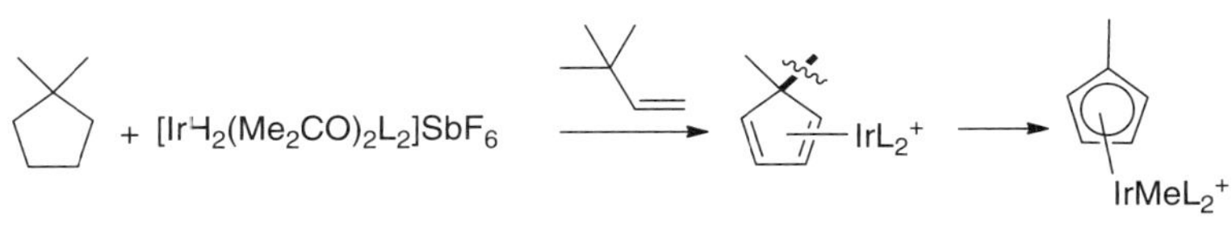

7.3.7.20 **7.3.7.21** **7.3.7.22** **7.3.7.23**

Scheme 7.75.

organometallic intermediates through cleavage of C–C bonds with transition metals: oxidative addition and β-carbon elimination.

Since C–C oxidative addition can be thermodynamically favorable, an important question is whether it is possible to bring unstrained, unactivated C–C bonds of substrates that are not bound to a metal through a chelation group in close proximity of the metal center in order to achieve metal insertion, the entire process being thermodynamically driven. Such a process (inter- or intramolecular) would be of primary importance in the light of possible catalytic applications to simple organic molecules. On the other hand, the insertion of a transition metal into an unstrained bond between two sp^3 carbon atoms in a selective fashion offers the most difficult and ultimate challenge. The rational design of ligands and thermodynamically downhill reaction systems will represent a true adventure in this field.

In spite of the many limitations to achieving catalytic C–C bond activation, this area of research is on the threshold of opening up new strategies for organic synthesis. It is evident that transition-metal-catalyzed C–C bond cleavage offers an attractive opportunity for restructuring carbon frameworks. The unique potential of the C–C bond cleavage has made a wonderful addition to the synthetic toolbox. The direct goal of chemistry of C–C bond cleavage will be satisfying the practicality required for the synthesis of general targets and also for the application in polymer degradation.

REFERENCES

1. For recent reviews, see: (a) Rybtchinski, B., Milstein, D. (1999). *Angew. Chem. Int. Ed.* **38**, 870. (b) Murakami, M., Ito, Y. (1999). *Top. Organomet. Chem.* **3**, 97. (c) van der Boom, M. E., Milstein, D. (2003). *Chem. Rev.* **103**, 1759. (d) Jun, C.-H. (2004). *Chem. Soc. Rev.* **33**, 610. (e) Satoh, T., Miura, M. (2005). *Top. Organomet. Chem.* **14**, 1. (f) Jun, C.-H., Park, J.-W. (2007). *Top. Organomet. Chem.* **24**, 117. (g) Nečas, D., Kotora, M. (2007). *Curr. Org. Chem.* **11**, 1566. (h) Seiser, T., Cramer, N. (2009). *Org. Biomol. Chem.* **7**, 2835. (i) Murakami, M., Matsuda, T. (2011). *Chem. Commun.* **47**, 1100.
2. For degrading synthetic polymers, see: (a) Caulfield, M. J., Qiao, G. G., Solomon, D. H. (2002). *Chem. Rev.* **102**, 3067. (b) Gritsenko, K. P., Krasovsky, A. M. (2003). *Chem. Rev.* **103**, 3607. (c) Celina, M. C., Wiggins, J. F., Billingham, N. C., eds. ACS Symposium Series: *Polymer Degradation and Performance*, American Chemical Society, New York.
3. For degrading biomass, see: (a) Huber, C. W., Iborra, S., Corma, A. (2006). *Chem. Rev.* **106**, 4044. (b) Corma, A., Iborra, S., Velty, A. (2007). *Chem. Rev.* **107**, 2411. (c) Navarro, R. M., Pena, M. A., Fierro, J. L. G. (2007). *Chem. Rev.* **107**, 3952. (d) Klemm, D., Heublein, B., Fink, H., Bohn, A. (2005). *Angew. Chem. Int. Ed.* **44**, 3358. (e) de Lasa, H., Salaices, E., Mazumder, J., Lucky, R. (2011). *Chem. Rev.* **111**, 5404.
4. Halpern, J. (1982). *Acc. Chem. Res.* **15**, 238.
5. (a) Periana, R. A., Bergman, R. G. (1984). *J. Am. Chem. Soc.* **106**, 7272. (b) Periana, R. A., Bergman, R. G. (1986). *J. Am. Chem. Soc.* **108**, 7346.
6. Bart, S. C., Chirik, P. J. (2003). *J. Am. Chem. Soc.* **125**, 886.
7. Matsuda, T., Tsuboi, T. Murakami, M. (2007). *J. Am. Chem. Soc.* **129**, 12596.

8. (a) Wong, W., Singer, S. J., Pitts, W. D., Watkins, S. F., Baddley, W. H. (1972). *J. Chem. Soc. Chem. Commun.*, 672. (b) Visser, J. P., Ramakers-Blom, J. E. (1972). *J. Organomet. Chem.* **44**, C63.
9. Baba, A., Ohshiro, Y., Agawa, T. (1976). *J. Organomet. Chem.* **110**, 121.
10. Itazaki, M., Nishihara, Y., Takimoto, H., Yoda, M., Osakada, K. (2005). *J. Mol. Catal. A* **241**, 65.
11. Salomon, R. G., Salomon, M. F., Kachinski, J. L. C. (1977). *J. Am. Chem. Soc.* **99**, 1043.
12. Grigg, R., Hayes, R., Sweeney, A. (1971). *J. Chem. Soc. Chem. Commun.*, 1248.
13. Gassman, P. G., Lee, C.-J. (1994). *Synth. Commun.* **24**, 1457.
14. (a) Wender, P. A., Takahashi, H., Witulski, B. (1995). *J. Am. Chem. Soc.* **117**, 4720. (b) Wender, P. A., Rieck, H., Fuji, M. (1998). *J. Am. Chem. Soc.* **120**, 10976.
15. Wender, P. A., Fuji, M., Husfeld, C. O., Love, J. L. (1999). *Org. Lett.* **1**, 137.
16. Eisch, J. J., Piotrowski, A. M., Han, K. I., Krüger, C., Tsay, Y. H. *Organometallics* **4**, 224.
17. Atkinson, E. R., Levins, P. L., Dickelman, T. E. (1964). *Chem. Ind.* 934.
18. Lu, Z., Jun, C.-H., de Gala, S. R., Sigalas, M. P., Eisenstein, O., Crabtree, R. H. (1995). *Organometallics* **14**, 1168.
19. Perthuisot, C., Edelbach, B. L., Zubris, D. L., Jones, W. D. (1997). *Organometallics* **16**, 2016.
20. Edelbach, B. L., Lachicotte, R. J., Jones, W. D. (1998). *J. Am. Chem. Soc.* **120**, 2843.
21. Perthuisot, C., Jones, W. D. (1994). *J. Am. Chem. Soc.* **116**, 3647.
22. (a) Perthuisot, C., Edelbach, B. L., Zubris, D. L., Simhai, N., Iverson, C. N., Müller, C., Satoh, T., Jones, W. D. (2002). *J. Mol. Catal. A: Chem.* **189**, 157 and references therein. (b) Masselot, D., Charmant, J. P. H., Gallagher, T. (2006). *J. Am. Chem. Soc.* **128**, 694. (c) Shibata, T., Nishizawa, G., Endo, K. (2008). *Synlett* 765.
23. Evans, J. A., Everitt, G. F., Kemmitt, R. D. W., Russell, D. R. (1973). *J .Chem. Soc. Chem. Commun.*, 158.
24. (a) Liebeskind, L. S., Baysdon, S. L., South, M. S. (1980). *J. Organomet. Chem.* **202**, C73. (b) Liebeskind, L. S., Baysdon, S. L., South, M. S., Iyer, S., Leeds, J. P. (1985). *Tetrahedron* **41**, 5839.
25. (a) Murakami, M., Amii, H., Ito, Y. (1994). *Nature* **370**, 540. (b) M. Murakami, H. Amii, K. Shigeto, Ito, Y. (1996). *J. Am. Chem. Soc.* **118**, 8285.
26. Murakami, M., Itahashi, T., Ito, Y. (2002). *J. Am. Chem. Soc.* **124**, 13976.
27. Murakami, M., Tsuruta, T., Ito, Y. (2000). *Angew. Chem. Int. Ed.* **39**, 2484.
28. Matsuda, T., Makino, M., Murakami, M. (2004). *Org. Lett.* **6**, 1257.
29. Matsuda, T., Shigeno, M., Murakami, M. (2007). *J. Am. Chem. Soc.* **129**, 12086.
30. M. Murakami, S. Ashida, T. Matsuda, (2006). *J. Am. Chem. Soc.* **128**, 2166.
31. Matsumura, S., Maeda, Y., Nishimura, T., Uemura, S. (2003). *J. Am. Chem. Soc.* **125**, 8862.
32. Seiser, T., Cramer, N. (2008). *Angew. Chem. Int. Ed.* **47**, 9294.
33. (a) Trost, B. M., Xie, J. (2006). *J. Am. Chem. Soc.* **128**, 6044. (b) Trost, B. M., Xie, J. (2008). *J. Am. Chem. Soc.* **130**, 6231.
34. Wender, P. A., Deschamps, N. M., Sun, R. (2006). *Angew. Chem. Int. Ed.* **45**, 3957.
35. Wender, P. A., Correa, A. G., Sato, Y., Sun, R. (2000). *J. Am. Chem. Soc.* **122**, 7815.
36. Crépin, D., Dawick, J., Aissa, C. (2010). *Angew. Chem. Int. Ed.* **49**, 620.

37. Nishimura, T., Uemura, S. (2000). *J. Am. Chem. Soc.* **122**, 12049.
38. Nozaki, K., Sato, N., Takaya, H. (1996). *Bull. Chem. Soc. Jpn.* **69**, 1629.
39. (a) Nakao, Y., Oda, S., Yada, A., Hiyama, T. (2006). *Tetrahedron* **62**, 7567 (b) Hirata, Y., Yukawa, T., Kashihara, N., Nakao, Y., Hiyama, T. (2009). *J. Am. Chem. Soc.* **131**, 10964 (c) Nakao, Y., Hirata, Y., Tanaka, M., Hiyama, T. (2008). *Angew. Chem. Int. Ed.* **47**, 385 (d) Nakao, Y., Yada, A., Ebata, S., Hiyama, T. (2007). *J. Am. Chem. Soc.* **129**, 2428.
40. Watson, M. P., Jacobsen, E. N. (2008). *J. Am. Chem. Soc.* **130**, 12594.
41. Reddy, V. J., Douglas, C. J. (2010). *Org. Lett.* **12**, 952.
42. Yu, D-G., Yu, M., Guan, B.-T., Li, B.-J., Zheng, Y., Wu, Z.-H., Shi, Z.-J. (2009). *Org. Lett.* **11**, 3374.
43. (a) Tobisu, M., Kita, Y., Chatani, N. (2006). *J. Am. Chem. Soc.* **128**, 8152. (b) Tobisu, M., Kita, Y., Ano, Y., Chatani, N. (2008). *J. Am. Chem. Soc.* **130**, 15982.
44. (a) Suggs, J. W., Jun, C.-H. (1984). *J. Am. Chem. Soc.* **106**, 3054. (b) Suggs, J. W., Jun, C.-H., (1986). *J. Am. Chem. Soc.* **108**, 4679.
45. Suggs, J. W., Jun, C.-H., (1985). *J. Chem. Soc. Chem. Commun.*, 92.
46. (a) Jun, C.-H., Lee, H. (1999). *J. Am. Chem. Soc.* **121**, 880. (b) Jun, C.-H., Lee, D.-Y., Lee, H., Hong, J.-B. (2000). *Angew. Chem. Int. Ed.* **39**, 3070. (c) Jun, C.-H., Moon, C. W., Lee, D.-Y. (2002). *Chem. Eur. J.* **8**, 2423.
47. Jun, C.-H., Lee, H., Lim, S.-G. (2001). *J. Am. Chem. Soc.* **123**, 751.
48. Jun, C.-H., Lee, D.-Y., Kim, Y.-H., Lee, H., (2001). *Organometallics* **20**, 2928.
49. (a) Jun, C.-H., Hong, J.-B. (1999). *Org. Lett.* **1**, 887 (b) Jun, C.-H., Lee, H., Park, J.-B., Lee, D.-Y. (1999). *Org. Lett.* **1**, 2161.
50. Lee, D.-Y., Kim, I.-J., Jun, C.-H. (2002). *Angew. Chem. Int. Ed.* **41**, 3031.
51. Lei, Z.-Q., Li, H., Li, Y., Zhang, X.-S., Chen, K., Wang, X., Sun, J., Shi, Z.-J. (2012). *Angew. Chem. Int. Ed.* **51**, 2690.
52. He, C., Guo, S., Huang, L., Lei, A. (2010). *J. Am. Chem. Soc.* **132**, 8273.
53. Zhao, P., Incarvito, C. D., Hartwig, J. F. (2006). *J. Am. Chem. Soc.* **128**, 3124.
54. (a) Terao, Y., Wakui, H., Satoh, T., Miura, M., Nomura, M. (2001). *J. Am. Chem. Soc.* **123**, 10407. (b) Terao, Y., Wakui, H., Nomoto, M., Satoh, T., Miura, M., Nomura, M. (2003). *J. Org. Chem.* **68**, 5236.
55. Niwa, T., Yorimitsu, H., Oshima, K. (2007). *Angew. Chem. Int. Ed.* **46**, 2643.
56. Hayashi, S., Hirano, K., Yorimitsu, H., Oshima, K. (2006). *J. Am. Chem. Soc.* **128**, 2210.
57. Kondo, T., Kodoi, K., Nishinaga, E., Okada, T., Morisaki, Y., Watanabe, Y., Mitsudo, T.-a. (1998). *J. Am. Chem. Soc.* **120**, 5587.
58. Nishimura, T., Araki, H., Maeda, Y., Uemura, S., (2003). *Org. Lett.* **5**, 2997.
59. Nishimura, T., Katoh, T., Takatsu, K., Shintani, R., Hayashi, T. (2007). *J. Am. Chem. Soc.* **129**, 14158.
60. Chiba, S., Xu, Y.-J., Wang, Y.-F. (2009). *J. Am. Chem. Soc.* **131**, 12886.
61. Li, H., Li, Y., Zhang, X.-S., Chen, K., Wang, X., Shi, Z.-J. (2011). *J. Am. Chem. Soc.* **133**, 15244.
62. Chen, K., Li, H., Lei, Z.-Q., Li, Y., Ye, W.-H., Zhang, L.-S., Sun, J., Shi, Z.-J. (2012). *Angew. Chem. Int. Ed.* **51**, 9851.
63. Chen, K., Li, H., Li, Y., Zhang, X.-S., Lei, Z.-Q., Shi, Z.-J. (2012). *Chem. Sci.* **3**, 1645.
64. Zhang, X.-S., Li, Y., Li, H., Chen, K., Lei, Z.-Q., Shi, Z.-J. (2012). *Chem. Eur. J.* **18**, 16214.

65. Kimura, M., Mori, M., Tamaru, Y. (2007). *Chem. Commun.*, 4504.
66. Han, C., Uemura, D. (2008). *Tetrahedron Lett.* **49**, 6988.
67. Nilsson, Y. I. M., Andersson, P. G., Bäckvall, J. E. (1993). *J. Am. Chem. Soc.* **115**, 6609.
68. (a) Nečas, D., Turský, M., Kotora, M. (2004). *J. Am. Chem. Soc.* **126**, 10222 (b) Turský, M., Nečas, D., Drabina, P., Sedlák, M., Kotora, M. (2006). *Organometallics* **25**, 901.
69. Gozin, M., Weisman, A., Ben-David, Y., Milstein, D. (1993). *Nature* **364**, 699.
70. Liou, S.-Y., Gozin, M., Milstein, D. (1995). *J. Am. Chem. Soc.* **117**, 9774.
71. Rybtchinski, B., Vigalok, A., Ben-David, Y., Milstein, D. (1996). *J. Am. Chem. Soc.* **118**, 12406.
72. van der Boom, M. E., Kraatz, H.-B., Ben-David, Y., Milstein, D. (1996). *Chem. Commun.*, 2167.
73. Liou, S.-Y., van der Boom, M. E., Milstein, D. (1998). *Chem. Commun.*, 687.
74. Sugiishi, T., Kimura, A., Nakamura, H. (2010). *J. Am. Chem. Soc.* **132**, 5332.
75. (a) Harayama, H., Kuroki, T., Kimura, M., Tanaka, S., Tamaru, Y. (1999). *Angew. Chem. Int. Ed.* **36**, 2352. (b) Tamaru, Y. (1999). *J. Organomet. Chem.* **576**, 215.
76. Gooßen, L. J., Deng, G., Levy, L. M. (2006). *Science* **313**, 662.
77. Kokubo, K., Matsumasa, K., Miura, M., Nomura, M. (1996). *J. Org. Chem.* **61**, 6941.
78. Sugihara, T., Satoh, T., Miura, M., Nomura, M. (2003). *Angew. Chem. Int. Ed.* **42**, 4672.
79. Gooßen, L. J., Paetzold, L. (2004). *Adv. Synth. Catal.* **346**, 1665.
80. (a) Crabtree, R. H., Dion, R. P. (1984). *J. Chem. Soc. Chem. Commun.* 1260. (b) Crabtree, R. H., Dion, R. P., Gibboni, D. J., McGrath, D. V., Holt, E. M. (1986). *J. Am. Chem. Soc.* **108**, 7222.
81. Benfield, F W. S., Green, M. L. H. (1974). *J. Chem. Soc. Dalton Trans.*, 1324.
82. Eilbracht, P., Dahler, P. (1980). *Chem. Ber.* **113**, 542.
83. Hemond, R. C., Hughes, R. P., Locker, H. B. (1986). *Organometallics* **5**, 2391.
84. Jones, W. D., Maguire, J. A. (1987). *Organometallics* **6**, 1301.

INDEX

Homogeneous Catalysis for Unreactive Bond Activation, First Edition. Edited by Zhang-Jie Shi.

Chemistry